1.1 *ICONS throughout Discover Biology indicate topics that are also covered as activities or tutorials on the Student Website/CD. A "T" on the list below indicates an Animated Tutorial.*

Boxes

Media and Supplements Overview

Student Supplements

Student Website/CD

Discover Biology, Second Edition offers an enhanced, comprehensive student media resource that is available as a website at **www.discoverbiology.com,** and as a CD included with the book. This thoroughly updated website/CD provides students with a variety of interactive tools for learning the important concepts introduced to them in the textbook. The contents of the website and CD are identical, with the exception of the *Biology in the News* feature, which is only available on the website (a link is provided from the CD). Features include:

Intuitive Interface: The interface has been designed to present the review material and animated activities in a logical, intuitive manner.

Essential Ideas: The Main Message and all the major concepts from each chapter are summarized and presented in a bulleted list for quick review.

Activities: Over 200 animated and interactive activities present key concepts, processes, and techniques in an accessible format. Icons in *Discover Biology* direct students to activities on the Website/CD.

Animated Tutorials: There are more than 40 in-depth Tutorials, comprised of three parts: an Introduction that sets up the concept to be presented; an Animation that presents the concept in a clear, easy-to-follow narrative using animations based on textbook art; and a Quiz that tests the student's understanding of the material presented.

Other **Activities** include interactive exercises, simulations of models and experiments, and drag-and-drop labeling exercises.

What's News? Activities are thought exercises based on developments in the news. These activities connect the important concepts in the textbook to real-world news, providing the student with a relevant context for their study of biology

Biology in the News: The *Discover Biology* website provides a constantly updated database of news articles in the biological sciences, cross-referenced to the key concepts within each textbook chapter.

Key Terms: All of the key terms from the textbook are included for each chapter. Each term is linked to its Glossary entry for quick access to definitions, and page references are provided for further review.

Quizzes: Self-Quizzes for each chapter test the student's knowledge *and* provide the feedback he or she needs to master the material. **Practice Quizzes** for each chapter are self-administered multiple-choice tests that offer the student the option to submit the results to their instructor via email.

Glossary: A comprehensive glossary of all the important terms introduced in the textbook. Audio pronunciations are provided for over 100 difficult-to-pronounce words.

Study Guide

Jerry Waldvogel and V. Christine Minor (both of Clemson University), David Demers (University of Hartford), Ed Dzialowski (University of North Texas), and Betty McGuire (Smith College)
This study aid for students includes chapter-by-chapter essential ideas keyed to the *Discover Biology*, Second Edition textbook, plus a wealth of factual and conceptual review questions. The Study Guide also includes Related Activities that help the student think more deeply about the concepts they are learning and how they relate to each other and the world around them.

Art Notebook

The Art Notebook is an invaluable tool for taking effective notes during lecture. All of the line art figures from the textbook are reproduced in black-and-white, with ample space for taking notes. This allows the student to focus on the lecture, and not worry about trying to copy drawings from the blackboard or projector.

Instructor Supplements

Instructor's Resource CD

This set of CDs includes a wealth of electronic resources to enhance the lecture and facilitate course planning and assessment. Contents include:

Browser Interface: A convenient way to preview all of the images and animations on the CD.

Images: All of the line art figures and tables from the book in high- and low-resolution JPEG format.

Photo Collection: An excellent collection of over 500 photos selected to illustrate the full range of topics presented in the textbook.

Animations: All the animations and activities from the Second Edition Student Website/CD (over 200) are included.

Videos: A collection of video segments showing key biological processes and phenomena.

Cross-Platform Electronic Test Bank: Provided both as Microsoft Word documents and in test-creation software.

Chapter Outlines: Provided as Microsoft Word documents.

Lecture Notes: Provided as Microsoft Word documents for easy customization and lecture development.

Instructor's Manual

Erica Bergquist (Holyoke Community College), Asa Oudes (Washington State University), and Eric Strate
The Instructor's Manual provides instructors with resources for planning the course and developing the lecture. Contents include the following for each chapter:

- **Overview & Outline:** A thorough summary and outline of the important points in the chapter.
- **Lecture Notes:** Detailed notes that form a complete lecture for each chapter. All important concepts are included, as well as references to textbook figures and tables.
- **Teaching Suggestions:** Useful tips on how to present the material and activities and exercises to help engage students in the lecture.
- **Discussion Questions:** Thought-provoking questions to stimulate discussion and provide context.

Test File/Computerized Test File

Susan Weinstein, Marshall University
The Test File contains a complete set of test questions for each chapter of the textbook. Each chapter has a set of multiple-choice, fill-in-the-blank, and true-false ques-

tions, including factual recall and conceptual questions. Each chapter includes additional in-depth, conceptual questions added for the Second Edition. Using the electronic test file (on the Instructor's Resource CD), instructors can create customized tests using any combination of these questions and their own questions.

Overhead Transparencies

The *Discover Biology* transparency set includes all the four-color line art figures from the textbook in a relabeled and resized format that is optimal for projection and produces excellent image quality.

Laboratory Manual

*Ralph W. Preszler and Laura Lowell Haas
(both of New Mexico State University)*
Designed specifically to accompany *Discover Biology*, Second Edition, this lab manual reinforces textbook concepts by asking students to generate and test predictions derived from hypotheses. In keeping with the text's focus on applications, the Lab Manual incorporates hypothesis evaluation into discussion of societal issues. Also included are unique "Outside the Laboratory" activities, as well as introductions, methods sections, and discussion questions. The Lab Manual is also available in an **Instructor's Version**, which provides conceptual background for the labs, a guide to each lab exercise, and time and material guidelines.

Labs on Demand—Customized Laboratory Manuals

Alison Morrison-Shetlar and Virginia Bennett (both of Georgia Southern University), Robert Shetlar, Ralph Preszler and Laura Haas (both of New Mexico State University)
Available to qualified adopters of *Discover Biology*, Second Edition, Labs on Demand allows the instructor to choose from a database of hundreds of proven lab experiments and high-quality illustrations and photos, and then create his or her own custom lab manual with help from our content specialists and layout and design professionals.

Biology Video Library

This extensive videotape library, available to qualified adopters of *Discover Biology*, Second Edition, includes material on cells and cellular function, development, embryology, predation, and food chains.

DISCOVER BIOLOGY
Second Edition

DISCOVER
BIOLOGY

SECOND EDITION

MICHAEL L. CAIN
Rose-Hulman Institute of Technology

HANS DAMMAN
Chilliwack, British Columbia

ROBERT A. LUE
Harvard University

CAROL KAESUK YOON
Bellingham, Washington

Sinauer Associates, Inc.
Sunderland, Massachusetts

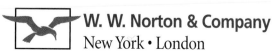
W. W. Norton & Company
New York • London

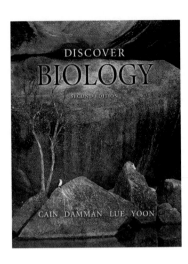

About the cover

A Pied Cormorant (*Phalacrocorax melanoleucos*) in the Australian outback. Photograph © Art Wolfe, Photo Researchers, Inc.

About the book

Editor: Andrew D. Sinauer
Project Editor: Kerry L. Falvey
Copy Editor: Norma Roche
Editorial Assistant: Sydney Carroll
Production Manager: Christopher Small
Book Layout and Production: Jefferson Johnson, Janice Holabird, and Joan Gemme
Illustration Program: Precision Graphics
Pen and Ink Illustrations: Nancy Haver
Book and Cover Design: Jefferson Johnson
Photo Research: David McIntyre
Color Separations: Burt Russell Litho
Book and Cover Manufacture: Courier Companies, Inc.

Discover Biology, Second Edition

Copyright © 2002 by Sinauer Associates, Inc. All rights reserved.
This book may not be reproduced in whole or in part without permission.

Address editorial correspondence to:
Sinauer Associates, Inc., P. O. Box 407
23 Plumtree Road, Sunderland, MA, 01375 U.S.A.
Fax: 413-549-1118
Internet: www.sinauer.com; Email: publish@sinauer.com

Address orders to:
W. W. Norton & Company, 500 Fifth Avenue
New York, NY, 10110-0017 U.S.A.
Phone: 1-800-233-4830, Fax: 1-800-458-6515
Internet: www.wwnorton.com

Library of Congress control number: 2002281076

10 9 8 7 6 5 4 3 2

About the Authors

Michael L. Cain is Associate Professor of Applied Biology at Rose-Hulman Institute of Technology, where he teaches introductory biology and a broad range of other biology courses. He received his Ph.D. in Ecology and Evolutionary Biology from Cornell University in 1989 and did postdoctoral research at Washington University in molecular genetics. Dr. Cain's interest in biology began early, stimulated by walks along the coast of Maine and by the expert guidance of his high school biology teacher, Clayton Farraday. Dr. Cain has published dozens of scientific articles on such topics as genetic variation in plants, insect foraging behavior, long-distance seed dispersal, and factors that promote speciation in crickets. Dr. Cain is the recipient of numerous fellowships, grants, and awards, including the Pew Charitable Trust Teacher-Scholar Fellowship and research grants from the National Science Foundation.

Hans Damman received a Ph.D. in Entomology in 1986 from Cornell University. He was until 1999 an Associate Professor of Biology at Carleton University in Ottawa. At Carleton University he twice received the Faculty of Science Award for Excellence in Teaching. His teaching included courses in Animal Form and Function, Plant and Animal Interactions, and Ecology, and he supervised a large number of undergraduate Honors research projects. His research work began with a focus on the interactions between plants and the insects that eat them, and has recently expanded to include the population biology of plants. Dr. Damman currently lives in Chilliwack, British Columbia.

Robert A. Lue is Senior Lecturer on Molecular and Cellular Biology and the Director of Undergraduate Studies in the Biological Sciences at Harvard University, where he has received several teaching awards from the Faculty of Arts and Sciences and the Division of Continuing Education. He received his Ph.D. from Harvard and has taught undergraduate courses there since 1988. Dr. Lue has published award-winning educational multimedia on HIV and AIDS, and has chaired educator conferences on college biology, most recently for the National Science Foundation. Dr. Lue is a molecular cell biologist, and his own research interests include the role of cytoskeletal proteins in human cancer. His research has been funded by grants from the National Institutes of Health.

Carol Kaesuk Yoon received her Ph.D. from Cornell University and has been writing about biology for *The New York Times* since 1992. Her articles have also appeared in *The Washington Post*, *The Los Angeles Times*, and *Science* magazine. Dr. Yoon has taught writing as a Visiting Scholar with Cornell University's John S. Knight Writing Program, working with professors to help teach critical thinking in biology classes. She has also served as science consultant to Microsoft for their children's CD-ROM on tropical rainforests, part of the Magic School Bus series. Recent articles include stories on worldwide declines in penguins, a new date for the evolution of land plants, and the discovery of a new species of deep sea squid.

Preface

It was a joy to write this book, from the first planning stages to the final touches still being made as the book went off to press. This joy stems from the fact that all four authors find biology to be a gripping subject, one that is simultaneously intensely interesting and critically important. Because scientific understanding of many fundamental principles is growing by leaps and bounds, this an exciting time to write, teach, and learn about all areas of the biological sciences.

For example, researchers are beginning to unlock the secrets of the human genome, allowing us to explore in unprecedented detail both the causes of and potential treatments for a host of human genetic disorders. And ecologists have begun to make real progress in understanding and predicting events that affect all of the world's ecosystems, such as the occurrence of El Niño events and the contribution of human activities to global warming.

Discoveries such as these, along with a host of others, mean that biology is a subject of news stories on an almost daily basis. In recent weeks, for example, the news has focused on the dangers of bioterrorism; on a remarkably simple and inexpensive procedure—bubbling nitrogen into ballast water to deplete the water of oxygen—that may reduce the billions of dollars lost each year when pest species are inadvertently spread from port to port in the ballast water released by cargo and cruise ships; and on biomedical developments that may allow doctors to transplant the organs of other species, such as pigs, into humans without danger of immune system rejection. These and other aspects of biology in the news open up a world of applied issues with important ramifications not only for individuals but for all human societies. Biology touches our lives every day, and its rapid progress ensures that we will hear more, not less, about biology in the years ahead.

The Philosophy of this Book Remains "Less Is More"

Unfortunately, the very things that make biology so interesting—the rapid pace of new discoveries and the critical uses and applications of these discoveries by human societies—can make it a difficult subject to teach and to learn. When we set out to write the Second Edition of *Discover Biology*, we asked ourselves, How can we convey biology's excitement, breadth, and relevance to students without burying them in facts and definitions? We have achieved this goal by writing clear, streamlined chapters that focus on essential concepts.

Over years of teaching, we have found that students learn best from short, focused chapters. We think this book's audience—introductory students who are not science majors—does not need to be overwhelmed with a mass of detail that is hard to sort through and place in context, nor do these students need to temporarily memorize exhaustive lists of new terms and vocabulary. We firmly believe that *less is more*: Students learn more when presented with less material (a belief well supported by the education literature).

As described in detail below, to put our "less is more" view into practice, we begin all our chapters with a single main message and a small set of key concepts. We develop the main message to a depth that only slightly exceeds what can be covered in one class period. We make heavy use of applied topics to illustrate key concepts while maintaining a focus on the overarching theoretical nature of biology, without which practical applications would not emerge.

Changes in the Second Edition: A Labor of Love

As we began to revise the successful First Edition of *Discover Biology*, we listened carefully to comments from students and instructors who were using the book. They provided invaluable help in both describing the book's strengths and pointing out places where there was room for improvement. Students were particularly helpful in identifying sentences and examples that could use a bit more elaboration to make the meaning crystal clear. Instructors helped us tackle the challenge of selecting the best possible examples with which to illustrate a wide range of concepts. During the long hours spent carefully revising each chapter, the guidance and encouragement we received from these comments served us extremely well. We also benefited greatly from the thoughtful responses to survey questions that dozens of instructors provided us. These instructors took the time to examine the book closely and provided us with extensive feedback on all of its features.

The message from all these comments was loud and clear: stay with the existing model of short, focused chapters that emphasize key concepts, applied examples, and breaking news stories, while adding more human examples and more of the sidebars on "Biology in Our Lives" and "The Scientific Process."

In response to this input, we wrote two new, highly relevant chapters: "Harnessing the Human Genome" (Chapter 18) and "From Birth to Death" (Chapter 38). In conjunction with revised material on genome biology in Chapter 16, the new chapter on the human genome exposes students to the revolutionary changes brewing as scientists continue to obtain and analyze the DNA sequence data of entire genomes. "From Birth to Death" offers a fresh look at what it means to be human, as it describes the human life cycle and seeks to explain why we exist in the form we do.

We applied the same principles that guided the writing of new chapters to our revision of the existing chapters. We added new boxed essays on topics such as cyanide poisoning of coral reef organisms (p. 699) and the recent discovery of a gene that could be used to genetically engineer coffee beans to produce a drink with all of the taste but none of the "jolt" of caffeinated coffee (p. 262). As we rewrote each chapter, we constantly sought new and interesting applied examples to illustrate key concepts. For instance, in revising Chapter 45, "Global Change," we incorporated the most recent examples of how people affect global ecosystems and added a new section on one of the greatest challenges faced by human societies: How to change our behaviors, laws, and economies so that the impact of *Homo sapiens* on our planet becomes sustainable (see p. 749).

Our obsession with incorporating applied examples was matched only by our obsession that key concepts be conveyed in a clear and accessible manner. We revised and polished many aspects of the writing, including the order in which material is presented within and between chapters, the figures, and the text prose. To give just one of many possible examples, in Unit I, "The Diversity of Life," we rearranged the material and order of Chapters 2 and 3, including the discussion of the Linnaean classification system, so that now students learn first how organisms are classified into groups and how the groups relate to each other in the new Chapter 2, "Understanding and Organizing the Diversity of Life." The vast diversity of life forms is then described in the revised

Chapter 3, "Major Groups of Living Organisms." To help students keep the context provided by Chapter 2 in perspective, we added small, iconic evolutionary trees to Chapter 3, to remind students where each group fits in on the "tree of life."

Features of the Second Edition: Approaches and Tools to Improve Learning

A primary goal of the Second Edition is to improve the biological literacy of our readers

Topics such as genetic engineering, new cancer treatments, the disappearance of rain forests, and global warming are increasingly in the news. We think it is critical that students understand the biology behind the news; only in this way can they make informed choices when the issues involve biology—as more and more issues do.

As part of our effort to demonstrate the application of biology to day-to-day living, each chapter in *Discover Biology* includes a unique feature called **Evaluating "The News."** These sections offer a brief original article, editorial, or "letter to the editor," often based on actual news stories, and highlight how applications of biology create issues that confront all citizens. Each chapter's Evaluating "The News" includes questions that encourage students to think about the social and ethical impli-

Each chapter ends with an Evaluating "The News" feature, accompanied by thoughtful questions

cations of many biological issues, providing a forum for students to develop critical thinking skills and apply what they learn in the textbook to very real problems.

Each of the book's six units ends with a chapter devoted to a significant issue confronting society: biodiversity, cancer, genomics, human evolution, the human lifespan, and global change. These **capstone chapters** draw on the biological concepts introduced earlier in the unit and give an in-depth consideration of how biology is connected to the rest of our lives.

The illustration program: An opportunity and a challenge

Biology is certainly the most visual of all academic disciplines. In creating the First Edition of *Discover Biology*, we were overwhelmed by the vast number of gorgeous photographic images available to illustrate biological phenomena. The Second Edition provides the opportunity to improve on an already stunning array of photographs.

The line art in an introductory biology book does much of the "heavy lifting" of making new and challenging concepts clear. For the First Edition, we worked closely with a team of talented biological illustrators. Employing "balloon" captions that place explanatory copy directly on the art, we feel we succeeded in creating illustrations that are simple and effective. For the Second Edition, we used the advice from adopters and reviewers to hone and refine a large number of figures, making them even clearer.

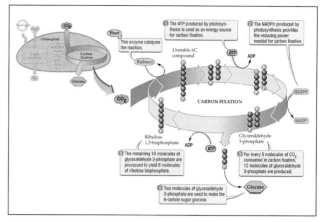

"Balloon" captions bring explanations of difficult concepts right into the art

A unique feature of *Discover Biology* is its use of fine art to introduce each chapter. By this device, we intend to help students connect biology to the rest of the liberal arts curriculum. It was so much fun selecting the paintings and sculptures that adorn the chapters that we chose many new ones for the Second Edition.

Chapters remain short and focused on a few key concepts

Along with the fine art, each chapter's **main message** appears on the chapter-opening spread. The chapter text also begins here with a **hook**—a vignette designed to

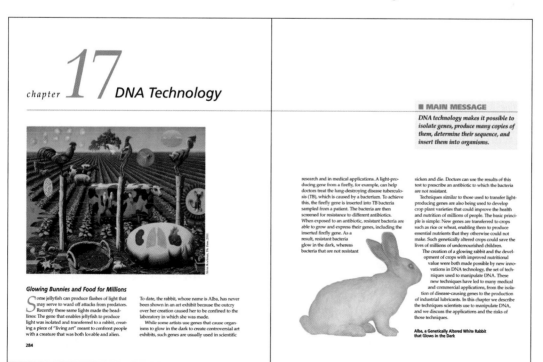

chapter **17** *DNA Technology*

■ MAIN MESSAGE

DNA technology makes it possible to isolate genes, produce many copies of them, determine their sequence, and insert them into organisms.

Glowing Bunnies and Food for Millions

Some jellyfish can produce flashes of light that may serve to ward off attacks from predators. Recently these same lights made the headlines: The gene that enables jellyfish to produce light was isolated and transferred to a rabbit, creating a piece of "living art" meant to confront people with a creature that was both lovable and alien.

To date, the rabbit, whose name is Alba, has never been shown in an art exhibit because the outcry over her creation caused her to be confined to the laboratory in which she was made.

While some artists use genes that cause organisms to glow in the dark to create controversial art exhibits, such genes are usually used in scientific

research and in medical applications. A light-producing gene from a firefly, for example, can help doctors treat the lung-destroying disease tuberculosis (TB), which is caused by a bacterium. To achieve this, the firefly gene is inserted into TB bacteria sampled from a patient. The bacteria are then screened for resistance to different antibiotics. When exposed to an antibiotic, resistant bacteria are able to grow and express their genes, including the inserted firefly gene. As a result, resistant bacteria glow in the dark, whereas bacteria that are not resistant

sicken and die. Doctors can use the results of this test to prescribe an antibiotic to which the bacteria are not resistant.

Techniques similar to those used to transfer light-producing genes are also used to develop crop plant varieties that could improve the health and nutrition of millions of people. The basic principle is simple: New genes are transferred to crops such as rice or wheat, enabling them to produce essential nutrients that they otherwise could not make. Such genetically altered crops could save the lives of millions of undernourished children.

The creation of a glowing rabbit and the development of crops with improved nutritional value were both made possible by new innovations in DNA technology, the set of techniques used to manipulate DNA. These new techniques have led to many medical and commercial applications, from the isolation of disease-causing genes to the production of industrial lubricants. In this chapter we describe the techniques scientists use to manipulate DNA, and we discuss the applications and the risks of those techniques.

Alba, a Genetically Altered White Rabbit that Glows in the Dark

284

Each chapter begins with a two-page "hook" and includes a work of fine art

■ THE SCIENTIFIC PROCESS

For the Extinct Elephant Bird, Was Incubation a Crushing Disaster?

The eggs of birds, when not serving as part of a nutritious breakfast, provide a wonderful example of conflicting selection pressures at work. The shell of an egg, for example, acts both as an external skeleton and as a filter through which materials enter and leave the egg. The shell must be weak enough that a tiny chick can break it open when ready to hatch, yet strong enough that it can support the weight of an incubating adult bird. The shell must be thin and porous enough to allow in the oxygen that the developing chick needs, yet it must resist the loss of water so that the developing chick does not dry out.

The largest bird in the world, the 450-kilogram elephant bird of Madagascar, laid the biggest eggs in the world, weighing 9 kilograms. Sadly, the elephant bird went extinct in the seventeenth century, leaving almost no records of its natural history in its wake. We can use what we know about the eggs of other bird species, however, to reconstruct some of the selection pressures that shaped the nesting biology of this remarkable bird.

Among living birds, large species have eggshells that are thin relative to egg weight. The relatively thinner eggshells of large eggs appear to compensate for their smaller surface area-to-volume ratio as compared with small eggs. In other words, a thick shell on a large egg would keep the developing chick from getting enough oxygen.

Shell thickness, however, directly determines how much force an egg can tolerate before it breaks. We know that the eggs of small birds can support a weight many times that of the parent bird. The eggs of large birds, however, are relatively fragile. An egg of the ostrich, the largest living bird species, cannot quite support the weight of a single adult. Ostriches deal with this problem by laying 10 or more eggs in a nest, so that each egg supports just a fraction of the parent's weight.

Birds as big as the elephant bird, however, must have faced tremendous problems during incubation. A single elephant bird egg would break under a weight 1/200 that of an adult bird. In other words, an elephant bird nest would have to contain more than 200 eggs before a parent could safely settle in. Because the elephant bird is extinct, we may never know how this species managed to incubate its eggs.

The extinct elephant bird of Madagascar laid eggs weighing about 9 kilograms.

Boxes throughout the book focus on either the process of science or how biology affects our lives

capture students' interest and lead them into the material they will be learning about in the chapter. Hook topics range from life in extreme environments to the global spread of AIDS to the use of forensic genomics to determine whether the Russian Grand Duchess Anastasia escaped execution in 1918 (she did not).

Following the hook is an **overview** of the topics to be covered and a list of the chapter's **key concepts**. Two levels of heading organize the text in a clear and straightforward fashion; second-level headings are usually in the form of a declarative sentence that conveys a key point or summarizes an important message. Each chapter closes with a **highlight** that either explores the "hook" in more depth or discusses a new topic chosen to put the chapter's material in a memorable context.

With the exception of the six capstones, every chapter includes a boxed essay that illuminates an important biological phenomenon, with either a human-interest focus ("Biology in Our Lives") or a glimpse of how scientists work ("The Scientific Process").

Chapters end with a summary of important points, a list of key terms, a self-quiz, and a set of review questions, along with the Evaluating "The News" feature mentioned earlier. Answers to the self-quizzes and review questions are in the back of the book, along with "Suggested Readings" for further exploration, a glossary, and the index.

Student Website/CD

No textbook about a subject as broad as biology can stand alone. Some topics require additional information

---

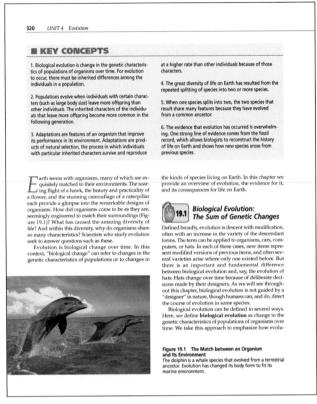

■ KEY CONCEPTS

1. Biological evolution is change in the genetic characteristics of populations of organisms over time. For evolution to occur, there must be inherited differences among the individuals in a population.

2. Populations evolve when individuals with certain characters (such as large body size) leave more offspring than other individuals. The inherited characters of the individuals that leave more offspring become more common in the following generation.

3. Adaptations are features of an organism that improve its performance in its environment. Adaptations are products of natural selection, the process in which individuals with particular inherited characters survive and reproduce

at a higher rate than other individuals because of those characters.

4. The great diversity of life on Earth has resulted from the repeated splitting of species into two or more species.

5. When one species splits into two, the two species that result share many features because they have evolved from a common ancestor.

6. The evidence that evolution has occurred is overwhelming. One strong line of evidence comes from the fossil record, which allows biologists to reconstruct the history of life on Earth and shows how new species arose from previous species.

Earth teems with organisms, many of which are exquisitely matched to their environments. The soaring flight of a hawk, the beauty and practicality of a flower, and the stunning camouflage of a caterpillar each provide a glimpse into the remarkable designs of organisms. How did organisms come to be as they are, seemingly engineered to match their surroundings (Figure 19.1)? What has caused the amazing diversity of life? And within this diversity, why do organisms share so many characteristics? Scientists who study evolution seek to answer questions such as these.

Evolution is biological change over time. In this context, "biological change" can refer to changes in the genetic characteristics of populations or to changes in

the kinds of species living on Earth. In this chapter we provide an overview of evolution, the evidence for it, and its consequences for life on Earth.

19.1 *Biological Evolution: The Sum of Genetic Changes*

Defined broadly, evolution is descent with modification, often with an increase in the variety of the descendant forms. The term can be applied to organisms, cars, computers, or hats. In each of these cases, new items represent modified versions of previous items, and often several varieties arise where only one existed before. But there is an important and fundamental difference between biological evolution and, say, the evolution of hats: Hats change over time because of deliberate decisions made by their designers. As we will see throughout this chapter, biological evolution is not guided by a "designer" in nature, though humans can, and do, direct the course of evolution in some species.

Biological evolution can be defined in several ways. Here, we define **biological evolution** as change in the genetic characteristics of populations of organisms over time. We take this approach to emphasize how evolu-

Figure 19.1　The Match between an Organism and Its Environment
The dolphin is a whale species that evolved from a terrestrial ancestor. Evolution has changed its body form to fit its marine environment.

More than 200 media icons identify activities and tutorials on the Student Website/CD

or hands-on practice. In these cases, alternative media such as animation and video can be used in activities designed to help students understand difficult material. For the Second Edition, *Discover Biology* offers an improved and comprehensive **student media resource** that is available online at www.discoverbiology.com and as a CD.

More than 200 concepts that are covered on the Student Website/CD are referenced in the book with an **icon** that directs the reader to a specific media activity. A description of the Student Website/CD and a list of these activities is provided at the front of the book. Of special interest is the addition of more than 40 new **animated tutorials**, including "Anthrax Bacteria Release Deadly Toxins," "Sequencing the Human Genome," "*T* Pressure Flow in Plants," and "The Carbon Cycle."

This Book is a Work in Progress

Like the field of biology itself, any textbook aiming to describe it is a work in progress. We hope you will contact us with your comments, questions, and ideas as you use the book. You can reach any or all of the authors with an email message to discoverbiology@sinauer.com, or by writing to us at Discover Biology, Sinauer Associates, 23 Plumtree Road, Sunderland, MA 01375 U.S.A.

We Thank the Many People Who Helped Us with this Revision

As we revised *Discover Biology*, each of us contacted scientists from around the world to ask about their research or about breaking stories in their field of expertise. We are grateful for the time they spent answering our many questions. We are also grateful for the scientific papers, illustrations, and photos they provided. We have communicated with many more scientists than we can thank individually, but collectively we thank them here. We are deeply in their debt.

As discussed above, a great many individuals provided critiques of the First Edition, answered a daunting questionnaire between editions, and reviewed the Second Edition manuscript. These colleagues are listed on the next page, and we thank them all. Thanks also to Susan McGlew for coordinating all of this invaluable feedback.

Our copublishers again made writing this book a stimulating experience. At Sinauer Associates, Andy Sinauer put together the team and coordinated all the major decisions. Kerry Falvey, our production editor, handled that complex task magnificently. Norma Roche is a gifted manuscript editor and we were most fortunate to have her work on our words. David McIntyre again served as photo editor and again provided an abundance of superb images for us to choose from. Christopher Small coordinated art and production with authority, and Jeff Johnson again led the team that produced alluring page spreads. Jason Dirks, with lots of help from Gayle Sullivan and Mara Silver, put together the excellent media and supplements package for the Second Edition.

At W.W. Norton, John Byram and Margaret Barrett provided the eyes and ears to bring us significant feedback from the marketplace. They both participated in the myriad decisions that go into shaping the publishing plan for a new edition. Dan Bartell, National Sales Manager, and Roby Harrington, Director of the College Department, provided continuous support and enthusiasm.

MICHAEL L. CAIN

HANS DAMMAN

ROBERT A. LUE

CAROL KAESUK YOON

Reviewers

Reviewers for the Second Edition

Gregory Beaulieu, University of Victoria

Robert Bernatzky, University of Massachusetts/Amherst

Nancy Berner, University of the South

Sarah Bruce, Towson University

Neil Buckley, SUNY/Plattsburgh

David Byres, Florida Community College/Jacksonville—South Campus

Van Christman, Ricks College

Jerry L. Cook, Sam Houston State University

Paul da Silva, College of Marin

Judith D'Aleo, Plymouth State College

Sandra Davis, University of Louisiana/Monroe

Pablo Delis, Hillsborough Community College

Deborah Fahey, Wheaton College

Wendy Garrison, University of Mississippi

Aiah A. Gbakima, Morgan State Univeristy

Kajal Ghoshroy, Museum of Natural History/Las Cruces

Jack Goldberg, University of California/Davis

Andrew Goliszek, North Carolina Agricultural and Technological State University

Glenn A. Gorelick, Citrus College

Bill Grant, North Carolina State University

Harry W. Greene, Cornell University

Barbara Hager, Cazenovia College

Robert Harms, Saint Louis Community College/Meramec

Chris Haynes, Shelton State Community College

Thomas E. Hemmerly, Middle Tennessee State University

Tasneem F. Khaleel, Montana State University/Billings

Hans Landel, North Seattle Community College

Katherine C. Larson, University of Central Arkansas

Ann S. Lumsden, Florida State University

Joyce Maxwell, California State University/Northridge

Bob McMaster, Holyoke Community College

Ali Mohamed, Virginia State University

Brenda Moore, Truman State University

Ruth S. Moseley, S.D. Bishop Community College

Benjamin Normark, University of Massachusetts/Amherst

Douglas Oba, Brigham Young University

Mary O'Connell, New Mexico State University

Marcy Osgood, University of Michigan

Donald Padgett, Bridgewater State College

Anthony Palombella, Longwood College

Snehlata Pandey, Hampton University

Robert P. Patterson, North Carolina State University

Jeffrey Podos, University of Massachusetts/Amherst

Richard A. Ring, University of Victoria

Kurt Schwenk, University of Connecticut

Erik P. Scully, Towson University

Cara Shillington, Eastern Michigan University

Barbara Shipes, Hampton University

Shaukat Siddiqi, Virginia State Univeristy

Julie Snyder, Hudson High School

Neal Stewart, University of North Carolina/Greensboro

Tim Stewart, Longwood College

Marshall D. Sundberg, Emporia State University

Cheryl Vaughan, Harvard University

John Vaughan, St. Petersburg College

Alain Viel, Harvard Medical School

William A. Velhagen, Jr., Longwood College

Mary Vetter, Luther College at the University of Regina

Carol H. Weaver, Union University

Jerry Waldvogel, Clemson University

Reviewers for the First Edition

Brief Table of Contents

Contents

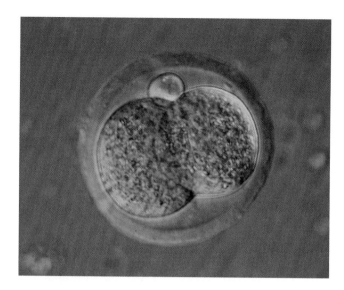

Unit 2: Cells: The Basic Units of Life

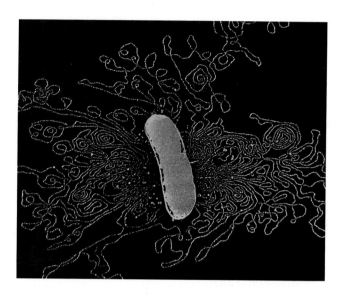

Unit 3: Genetics

Unit 4: Evolution

Unit 5: Form and Function

25 Size and Complexity 418

26 Support and Shape 432

27 Movement 448

Unit 6: Interactions with the Environment

40 Why Organisms Live Where They Do 666

41 Growth of Populations 680

42 Interactions among Organisms 694

43 Communities of Organisms 708

Alexandre-Isidore Leroy de Barde, *Still Life with Exotic Birds,* 1804.

UNIT *1*

Diversity of Life

1 The Nature of Science and the Characteristics of Life

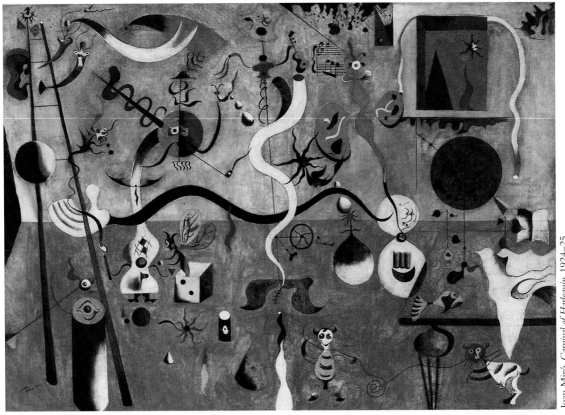

Joan Miró, *Carnival of Harlequin*, 1924–25.

It's Alive! Or Is It?

Off the coast of Western Australia, in ancient rock buried 3 miles below the ocean floor, lie what some scientists claim are the smallest living organisms ever discovered. These minuscule life forms are known as nanobes because they are so tiny that they are measured in billionths of meters, or nanometers. But while some scientists are hailing this finding as an important discovery,

others argue that nanobes are not living organisms and could not possibly be.

The Australian researchers who recently discovered the nanobes say that the tiny organisms carry hereditary material, known as DNA, just as other living organisms do. These scientists also report that, like other forms of life, nanobes grow. In fact, nanobes grow so quickly that within weeks they go from

There is an underlying unity to the diversity of life.

being visible only with the world's most powerful microscopes to being visible to the naked eye as expanding colonies of threadlike mats.

Skeptics, however, contend that nanobes are much too small to be alive, too small to contain the materials and machinery basic to life.

Any schoolchild can distinguish between an inanimate stone and the living being who skips it across a pond. How can it be that scientists cannot agree on whether something is alive or not?

What is life? Deceptively simple, this question is in many ways one of the most profound. It underlies medical controversies ranging from abortion and when life begins to the right to die and when life ends. It reaches deep into the sciences as we seek to understand when life on our planet first took hold and whether life has ever existed on other planets.

Scientists and philosophers alike have found that life defies easy, airtight definition. What really does distinguish life from nonlife?

Tempest in a Teapot
Is this nanobe the smallest living organism ever discovered? Or is it not alive at all?

■ KEY CONCEPTS

1. Biology is the scientific study of life. To understand living organisms, biologists develop possible explanations, known as hypotheses, about phenomena they observe in nature. These hypotheses are then subjected to rigorous tests, and changed or rejected as appropriate.

2. Because all living organisms, diverse though they are, descended from a single common ancestor, there is a great unity in the characteristics of living organisms.

3. Shared characteristics define life: Living organisms are built of cells, reproduce, grow and develop, capture energy from their environment, sense stimuli in their environment and respond to them, show a high level of organization, and evolve.

4. A biological hierarchy encompasses all components of living organisms and their environments, from molecules to cells, tissues, organs and organ systems, to individual organisms, populations, species, communities, ecosystems, and biomes, and finally the biosphere.

5. More than 3.5 billion years ago, life arose from nonlife. How this happened remains a matter of intense interest and research.

Biology is the scientific study of life. The main goal of biology is to improve our understanding of living organisms, from microscopic bacteria to giant redwood trees to human beings. In this chapter we begin with an exploration of science and how scientists ask and answer questions about living organisms. Then we move on to the most fundamental of biological questions: What is life? We will see that all living things, diverse though they are, share characteristics that unify them, and that all living organisms are part of a greater biological hierarchy of life.

1.1 Asking Questions, Testing Answers: The Work of Science

Science is a method of inquiry, a rational way of discovering truths about the natural world. This powerful way of understanding nature holds a central place in modern society. For scientists and nonscientists alike, understanding how nature works can be exciting and fulfilling. In addition, applications of scientific knowledge influence all aspects of modern life. Every time we turn on a light, fly in an airplane, enjoy a vase of hothouse flowers, take medicine, or work at a computer, we are enjoying the benefits of science.

Yet few of us have a good picture of how science works, how it generates knowledge, and what its powers and limits are. This lack of understanding is unfortunate for several reasons. First, an understanding of science can be personally rewarding. It can add to our appreciation of day-to-day events, leading to a sense of awe about how nature works.

A second reason is that science plays an increasingly important role in decisions made by society as a whole, as well as in personal decisions made by individuals. As a society, we must evaluate the discoveries made by scientists when making decisions about issues such as global warming, the control of pollution, and even whether teachers in our public schools can provide their students with the most current forms of scientific knowledge. As individuals, we must also evaluate what scientists tell us in order to make decisions about such things as whether we will eat genetically engineered foods or treat our bodies using new drugs or medical procedures. To make good decisions on these and many other issues, everyone—not just scientists—benefits by understanding how the scientific process works.

In the sections that follow we first describe, in a general way, the methods scientists use to answer questions and learn about the natural world. Then we give an example of how one scientist answered a question.

Scientists follow well-defined steps to search for answers

The natural world is extremely complex. To deal with this complexity, scientists attempt to explain the natural world by developing a simplified concept, or "model," of how some part of nature works. No scientist attempts to study every conceivable thing that could influence the aspects of the natural world that interest him or her. Such a task would be impossible. Instead, scientists must master the "art" of simplifying nature in ways that can reveal exciting and important information.

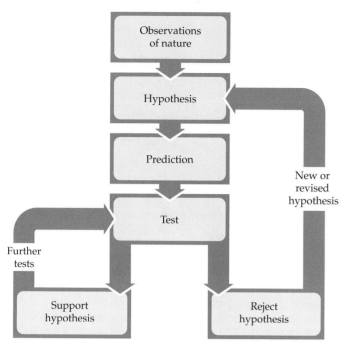

Figure 1.1 The Scientific Method

tions are not supported by the results of the tests. Together these steps are called the **scientific method**.

An experiment with honeysuckle demonstrates the scientific method

Dr. Katherine Larson, a botanist at the University of Central Arkansas, was interested in two species of vines: a species native to the United States, called coral honeysuckle, and a species from Asia, the Japanese honeysuckle (Figure 1.2). The Japanese honeysuckle has gorgeous flowers and was originally brought to the southeastern United States as a cultivated plant. But now it is considered a pest species. It has escaped from cultivation and is spreading so quickly that scientists are concerned it will drive native plant species to extinction.

Most vines must find supports to grow on (such as trees, shrubs, or fence posts) in order to survive. Dr. Larson wanted to understand why the Japanese honeysuckle was able to grow so well in the wild, even in areas with few supports. As she observed the 2 species in the wild, Dr. Larson noticed that Japanese honeysuckle

To construct such a simplified model of nature, scientists typically begin by observing, describing, and asking questions about the natural world (Figure 1.1). In many ways, this beginning can be the most important step of a scientific study. How closely a scientist's model of nature matches reality can hinge on the quality of his or her initial observations.

The next step in the scientific process is to come up with a possible explanation—known as a **hypothesis**—of the phenomenon being studied. The hypothesis must have logical consequences that can be proved true or false. That is, the hypothesis must lead to a prediction that can be tested rigorously. If the results of the test match the prediction, the hypothesis is supported, but not proved true. Proving a hypothesis true is not possible, because it always might fail if subjected to a different test. If the test does not uphold the prediction, the hypothesis must be rejected or changed. This last point is central to understanding how progress is made in science: Hypotheses are constantly being tested; the good ones are kept, the bad ones rejected. In this way, science can correct itself. Scientists develop a hypothesis, test its predictions by performing experiments, then change or discard the hypothesis if the predic-

Figure 1.2 A Scientist and Her Experimental Subjects
Dr. Katherine Larson's experiments with honeysuckles are shedding light on why one species of honeysuckle, introduced from Asia, is able to spread more quickly than the honeysuckle species that is native to its environment. The stem with red flowers is the native species; the white-flowered stem is the pest species.

spread rapidly across the ground, whereas the native honeysuckle tended to spread less quickly. Based on her observation, Dr. Larson came up with a possible explanation, or hypothesis. She hypothesized that the Japanese honeysuckle was more successful than the native species because it was better at locating new supports, especially supports that were far from where the vine was growing.

Dr. Larson's hypothesis had logical predictions that could be tested: (1) that Japanese honeysuckle would grow across areas without supports more rapidly than the native coral honeysuckle would, and (2) that Japanese honeysuckle would find distant supports more often than coral honeysuckle would. In order to test her hypothesis and its predictions, Dr. Larson designed an **experiment**—a controlled, repeated manipulation of nature. In her experiment, she grew plants from each of the two honeysuckle species in a garden in which supports were placed both close to and far away from the plants.

Though still in progress, this experiment has already shown that Japanese honeysuckle does spread more rapidly across areas without supports than the native species does. Results from this experiment also suggest that Japanese honeysuckle is more effective at locating distant supports than the native species. Thus, the results support both of Dr. Larson's predictions, and her hypothesis.

In addition, Dr. Larson's research has led to some new and unexpected findings. While conducting the honeysuckle experiment, she noticed differences in the details of how the plants grew. She knew that most plants rotate slowly about a central axis as they grow (Figure 1.3), and that vines show a pronounced version of this behavior. Dr. Larson observed that Japanese honeysuckle rotated differently than the native species. The way Japanese honeysuckle rotated as it grew along the ground allowed it to form roots more often and to spread more rapidly than the native species. Plants obtain nutrients and water through their roots. Thus, having the ability to form roots more often may allow the Japanese honeysuckle to survive in areas where the native species cannot.

To summarize, Dr. Larson conducted an experiment to test her hypothesis that the Japanese honeysuckle is outcompeting the native species because it spreads more rapidly and finds distant supports more easily. So far, the results of the experiment support her hypothesis. In addition, Dr. Larson made new observations about the details of plant growth that helped explain the results of the experiment. In this way Dr. Larson's work illustrates an important point about how the scientific method works in practice: Chance observations and new discoveries often take the scientist in unex-

Figure 1.3   Doing the Twist
This time-lapse photograph shows the position of a single growing tip of a honeysuckle vine as it rotates on a central axis. The way Japanese honeysuckle rotates may explain why it is more successful than native honeysuckle.

pected directions, and that is part of what makes science so exciting.

Like all scientific studies, Dr. Larson's work raises as many questions as it answers. What else might be contributing to the success of Japanese honeysuckle? Are there insects in its native environment in Asia that attack the honeysuckle to keep it from growing so heartily and spreading so quickly? Is an absence of such enemies in the United States helping the Japanese honeysuckle thrive in its new environment? Clearly, Dr. Larson's studies are just the beginning in helping scientists to understand why these plants have been able to invade so successfully.

> ■ Scientists observe the natural world and then form hypotheses that make predictions about how that world might operate. Scientists then design experiments to test these predictions. When the experiment upholds the predictions, a hypothesis gains strength and support. When the predictions are not upheld, the hypothesis must be rejected or modified.

We have discussed the methods biologists use to study the living world. But what exactly is the living world? How do biologists define life?

1.3 The Characteristics that All Living Organisms Share

From the frigid Antarctic to the burning Sahara Desert, from the highest mountaintop to the deepest regions of the sea, life is everywhere. How can all the world's living creatures, from whales to bacteria, fit into a single definition of life? In fact, the great diversity of body forms, habits, and sizes of organisms makes a simple, single-sentence definition of life impossible. But despite this diversity, all living organisms are thought to be descendants of a single common ancestor that arose billions of years ago (see the box on page 8). For this reason, certain characteristics unify all forms of life. Biologists define life by this set of shared characteristics (Table 1.1), which we describe in the sections that follow.

Living organisms are built of cells

The first organisms were single cells that existed billions of years ago, and the simplest of organisms still are made up of just a single cell. The cell remains the smallest and most basic unit of life. Enclosed by a membrane, **cells** are tiny, self-contained units that make up every living organism.

Cells can be viewed as building blocks. A bacterium is made up of one cell, a single, self-sustaining building block. **Multicellular** organisms such as toads or oak trees are made up of many different kinds of specialized cells. A toad, for example, has within it skin cells, muscle cells, brain cells, and so on. Whatever the beast, plant, or bug, it is always at its most basic level a collection of cells (Figure 1.4).

Bacterium

Living organisms reproduce themselves using the hereditary material DNA

One of the key characteristics of living organisms is that they can replicate, or reproduce, themselves. Many single-celled organisms, such as bacteria, reproduce by dividing into two new genetically identical copies of themselves, like a pair of identical twins. (Genetics will be discussed in detail in Unit 3.) In contrast, multicellular organisms reproduce in a variety of ways. For example, sunflowers reproduce when their flowers generate seeds that can germinate to become new sunflowers. And humans reproduce by producing infants who eventually grow up to be adult humans, like those that produced them.

1.1 The Shared Characteristics of Life

All living organisms
- are built of cells
- reproduce themselves using the hereditary material DNA
- grow and develop
- capture energy from their environment
- sense their environment and respond to it
- show a high level of organization
- evolve

Whether organisms produce seeds, lay eggs, or just split in two, they all reproduce using a molecule known as **DNA** (**deoxyribonucleic acid**). DNA is the hereditary, or genetic, material that carries information from parents to offspring. We will discuss DNA in detail in Unit 3. Briefly, for our purposes here, the DNA molecule functions as a blueprint for building an organism. It is shaped like a double helix—imagine a ladder that is twisted along its length.

The intestinal cells shown here are just one of many different kinds of cells in this monkey.

Figure 1.4   The Basic Building Block of Life: The Cell
Like all organisms, this Sykes' monkey is composed of cells.

■ THE SCIENTIFIC PROCESS

Cooking Up a Theory for the Origin of Life: One of Biology's Classic Experiments

At the time Earth formed, 4.6 billion years ago, the planet was lifeless. Three and a half billion years ago, the first cellular life had appeared. But what happened in the meantime—how life arose from non-life—remains one of the most puzzling and hotly debated issues in science.

Scientists have found different amounts of evidence to support different hypotheses about the origin of life. Some researchers have hypothesized that life originated in hot springs deep on the ocean floor. Others have suggested that life's birthplace lies near underwater volcanoes or in hot-water geysers on land, such as Yellowstone's Old Faithful. Still others have suggested that life, or its building blocks, did not arise on Earth, but that living creatures or key molecules for constructing life reached Earth by traveling through space from Mars, or some other planet, on an asteroid or meteorite.

All of these competing hypotheses have their strengths and weaknesses. The best-known and longest-standing hypothesis, however, is what is sometimes referred to as the "soup theory" of life.

In 1953, in a now famous experiment considered a classic in modern biology, Stanley L. Miller, then a young graduate student, set up an ingenious test. Miller attempted to re-create the conditions of early Earth—hot seas and lightning-filled skies—to see if he could re-create the beginnings of life.

To make his "primordial soup," Miller began with water, which he kept boiling. The water vapor rose into a simulated atmosphere, which contained the gases methane, ammonia, and hydrogen—some of the gases

that Earth's early atmosphere could well have harbored, and the very sort that could have belched forth from ancient volcanoes. Miller added an electric spark to simulate lightning. He then cooled the sparking vapors until they turned to liquid, and the resulting liquid, like rain falling back to the sea, was returned to the boiling water. With the apparatus set up to let the water travel continuously from boiling sea to sparking sky and back to the sea again, Miller let the whole thing cook for a week.

Surprisingly, when Miller examined his primordial soup after a week of cooking, he discovered that from nothing but water and simple gases, he had created an array of unexpected molecules critical to the origin of life, including two amino acids, the building blocks of proteins. Since then, other such "soup" experiments attempting to recreate Earth's early conditions have produced other key biological molecules, including all the common amino acids, sugars, and lipids, as well as the basic building blocks of DNA and RNA.

Perhaps most importantly, Miller's simple experiment illustrates that even though life originated billions of years ago, scientists can find ways to study those early processes in the laboratory today.

Many questions remain: How closely do the conditions set up in the primordial soup experiments mimic early

Earth conditions? Once there is a soup of critical molecules, how do those molecules get organized into larger molecules and into the first cell able to gather energy and reproduce? Did the first cell float freely in a soup, or was it organized first on a surface, such as the ocean bottom?

Researchers are continuing to try to answer these and other questions in many different ways. Some study ancient fossils, searching for the first hints of cellular life. Others study the family tree of all living organisms in an attempt to understand better what kind of organism was most likely to have been the ancestor of us all. Still others study meteorites and other extraterrestrial objects for hints of what might once have landed on Earth. How Earth made the crucial leap from planet of barren stone to cradle of life remains one of science's most difficult and most fascinating questions.

Stanley Miller with the apparatus he used to cook up a primordial soup nearly half a century ago.

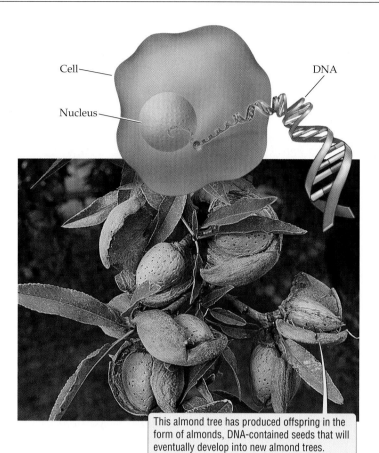

Cell

Nucleus

DNA

This almond tree has produced offspring in the form of almonds, DNA-contained seeds that will eventually develop into new almond trees.

Figure 1.5   The DNA Molecule: A Blueprint for Life
DNA is a hereditary blueprint found in the cells of every living organism. DNA provides a set of instructions that an individual organism can use to grow and develop, and which it passes on to its offspring so that they, too, can grow and develop.

The DNA molecule (Figure 1.5) contains a wealth of information—all the information necessary for an organism to create more cells like itself or to grow from a fertilized egg into a complex multicellular organism that will eventually produce offspring like itself. DNA is found in every cell in every living thing. Life, no mat-

ter how simple or how complex, uses this inherited blueprint. Since all living organisms today reproduce using DNA, we can infer that our original ancestor also reproduced using DNA.

Living organisms grow and develop

Using DNA as a blueprint, organisms grow and build themselves anew every generation. A human begins life as a simple, single cell that eventually grows and develops inside its mother, emerging after 9 months as a living, breathing baby. This miraculous transformation is part of the process known as **development** (Figure 1.6). All organisms go through a process of development, whether they complete their growth as a single cell or continue to develop into something as complicated as a cactus or an octopus.

Living organisms capture energy from their environment

To carry out their growth and development, and simply to persist, organisms need energy, which they must capture from their environment. Organisms use a great variety of methods to capture energy.

Plants are among the organisms that can capture the energy of sunlight through a chemical process known as photosynthesis, by which they produce sugars and starches. (We will discuss photosynthesis in detail in Chapter 8.) Some bacteria can also harness energy from chemical sources such as iron or ammonia through an entirely different chemical reaction. However, many organisms, such as animals, can capture energy only by consuming other organisms.

Figure 1.6   Growing Up: The Process of Development
All organisms develop. A monarch butterfly develops from egg to caterpillar to chrysalis to flying adult.

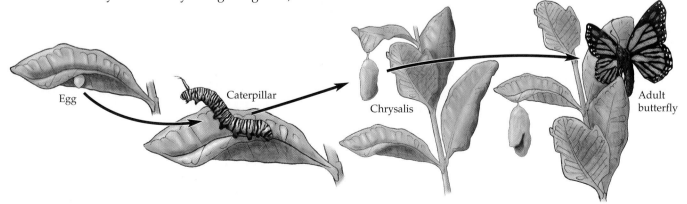

Egg

Caterpillar

Chrysalis

Adult butterfly

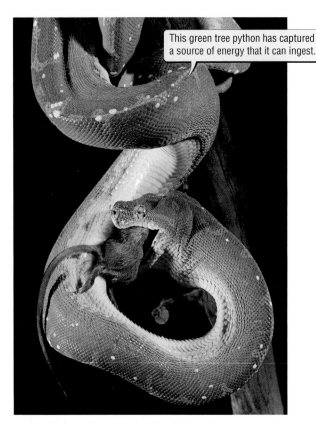

This green tree python has captured a source of energy that it can ingest.

Figure 1.7   Capturing Energy
While plants can capture energy through photosynthesis, animals, such as this green tree python, must get their energy by eating other organisms.

Animals have many different ways of capturing energy efficiently (Figure 1.7). Some insects, for example, have mouthparts that they use to suck the nutritive juices from plants. Cheetahs can run so quickly that they can chase and capture a fast-moving source of energy such as a gazelle.

Living organisms sense their environment and respond to it

In order to survive, different organisms must sense a wide range of different phenomena in their environment, from danger to potential mates. Like humans, many organisms can smell, hear, taste, touch, and see their environments. But many organisms can see things humans cannot see, such as ultraviolet or infrared light. Others can hear sounds that humans cannot hear. Still others have senses that are entirely different from any human senses. Some bacteria, for example, can sense which direction is north and which direction is up or down using magnetic particles within their cells.

Organisms must not only gather information about their environment by sensing it, but must also respond appropriately to that information (Figure 1.8). Some responses don't need to be learned; for example, a dog automatically pants when it is hot. But learning itself is an excellent example of sensing an environmental stimulus and responding to it. After touching a stinging nettle plant once, many organisms learn never to touch one again. Humans are particularly good at this kind of response because we have large brains relative to our body size. We can even learn very abstract lessons from our environment, such as how to read or sing a song.

Stinging nettle

Living organisms show a high level of organization

Living organisms are complex, functioning beings composed of numerous essential components that are organized in a very specific way. Human bodies, for example, have highly organized internal organs and tissues that allow the body to function properly. Organs or tissues in disarray, whether as a result of disease or accident, can lead to illness or death. In the same way, the structure of a flower or a bacterial cell is highly organized. Such organization is required for all organisms to function properly (Figure 1.9).

Figure 1.8   Here Comes the Sun
All organisms must be able to sense and respond to stimuli in their environment. These Maryland sunflowers have all detected rays of sunshine and have turned toward their light and warmth.

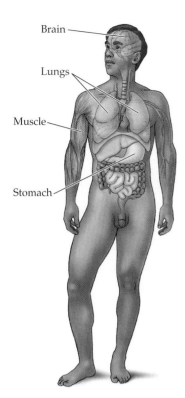

Figure 1.9   Staying Organized
Spatial organization—having each part in its proper place—is crucial to an organism's functioning. The internal organs of a human being, such as the stomach, lungs, and brain, must be highly organized to function properly.

Living organisms evolve

In the developmental process, individual organisms change over short spans of time, developing from seeds into mature trees or from eggs into adult butterflies. Over longer time spans, whole groups of organisms change. A **species** is a group of organisms that can interbreed to produce fertile offspring (that is, offspring that themselves can reproduce) and that do not breed with other organisms outside that group. Mountain lions, monarch butterflies, and Douglas fir trees are all distinct species of organisms.

The characteristics of a species can change over time—a process known as **evolution**. The pronghorn, for example, is the fastest-running creature in North America. Over time, these antelope became more fleet of foot because only the ones that could outrace their predators survived to reproduce. The offspring of these speedy survivors tended to be speedy themselves because they shared much of their DNA with their parents.

In the struggle to survive and the contest to reproduce, characteristics of species—such as the average speed at which pronghorn can run—tend to change over time (Figure 1.10). Advantageous features that help an organism survive or reproduce, such as the ability of a pronghorn to run quickly or the protective prickly hairs of a stinging nettle plant, are known as **adaptations**.

In evolution, not only can existing species change, but new species can come into being. For example, one species can split into two different species. Unit 4 will focus on all these aspects of evolution in detail.

> ■ Because all living organisms descended from a common ancestor, they share common characteristics. For example, all organisms are built of cells. They all reproduce using DNA, grow and develop, capture energy from their environment, sense and respond to their environment, show a high level of organization, and evolve.

Levels of Biological Organization

Biologists organize the great array of living organisms, the many components that make up living organisms, and the environments they live in into a biological hierarchy (Figure 1.11).

At its lowest level, the biological hierarchy begins with molecules, such as the molecules of DNA that carry the blueprint for building an organism. Many such specialized molecules are organized into a cell, the basic unit of life. Some organisms, such as bacteria, consist of only a single cell. Multicellular organisms can contain specialized and coordinated collections of cells known as **tissues**, such as muscle tissues or nerve tissues, that per-

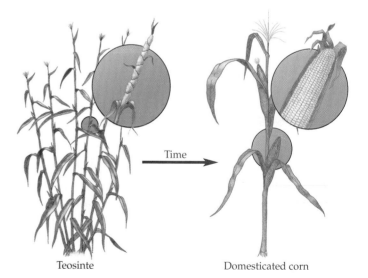

Teosinte   Domesticated corn

Figure 1.10   Living Organisms Evolve
Over time, species of living organisms evolve. Domesticated corn plants, for example, have many large kernels, or seeds, on large cobs. These familiar corn plants evolved from the wild species known as teosinte, which has fewer, smaller seeds.

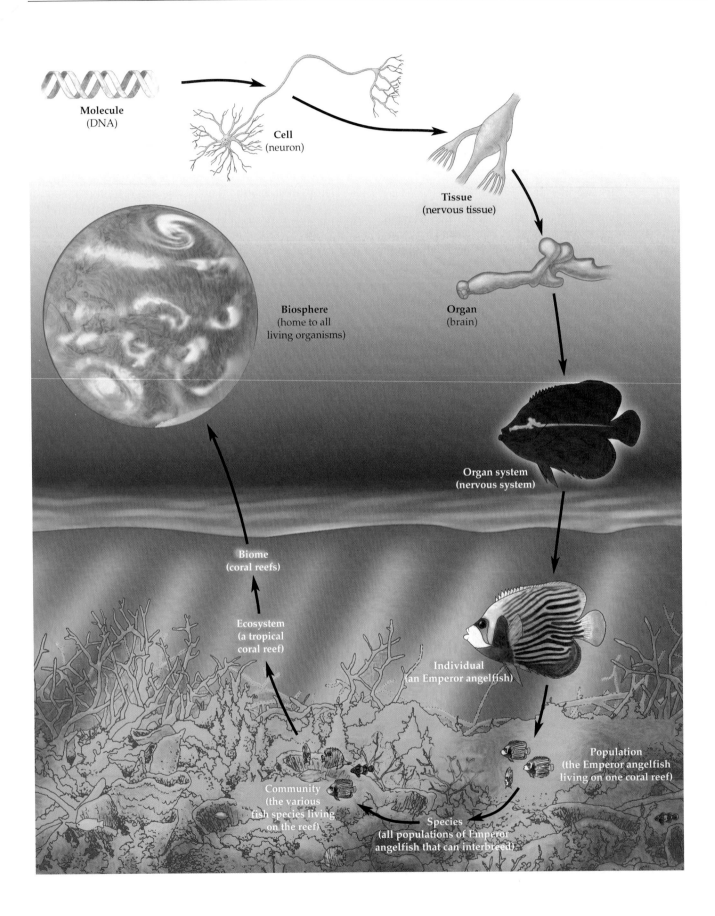

Molecule
(DNA)

Cell
(neuron)

Tissue
(nervous tissue)

Biosphere
(home to all
living organisms)

Organ
(brain)

Organ system
(nervous system)

Biome
(coral reefs)

Ecosystem
(a tropical
coral reef)

Individual
(an Emperor angelfish)

Population
(the Emperor angelfish
living on one coral reef)

Community
(the various
fish species living
on the reef)

Species
(all populations of Emperor
angelfish that can interbreed)

Figure 1.11   The Biological Hierarchy
Life in this tropical coral reef can be organized into a number of levels, from molecules to cells, to tissues, to organs, to organ systems, to individuals, to populations, to species, to communities, to ecosystems, to biomes, and finally to the biosphere, which includes all living organisms and the places where they live.

form particular, specialized functions in the body. Sometimes these tissues are organized into **organs**, such as hearts or brains. Groups of organs can work together in **organ systems**, the way the stomach, liver, and intestines are all part of the digestive system. Groups of organ systems work together in the functioning of a single organism—an **individual**.

Each individual organism is a member of a larger group of similar organisms, known as a **population**—for example, a population of field mice in one field or a population of blueberry plants on a mountaintop. As we have already learned, all the populations of all similar organisms in the world that can successfully breed with one another form a species. For example, all the humans in the world form one species, known as *Homo sapiens*. A group of different species that live and interact in a particular area is known as a **community**—for example, the community of insect species living in a forest.

Communities plus the physical environments in which they are situated are known as **ecosystems**; for example, a river ecosystem includes the river itself as well as the community of organisms living in it. At an even larger scale are the large regions of the world known as **biomes**. On land, biomes are defined by vegetation type, and in water, they are defined by the physical characteristics of the environment. The coral reef and the arctic tundra are two different kinds of biomes. Finally, each biome is part of the **biosphere**, which is defined as all the world's living organisms and the places where they live.

> ■ Living organisms are just one part of a biological hierarchy. At the lowest level are molecules, then come cells, tissues, organs, organ systems, individuals, populations, species, communities, ecosystems, biomes, and finally the biosphere.

Energy Flow through Biological Systems

Energy flows continually through living systems. Plants and other photosynthesizing organisms are called **producers** because they take energy from sunlight and produce chemical energy in the form of sugars and starches.

That energy is then harvested by **consumers**, such as animals and fungi. Consumers eat either plants or other organisms whose energy ultimately derives from plants. **Decomposers** obtain energy by decomposing, or breaking down, dead organisms. As a result, energy flows almost entirely in one direction through the biosphere: from the sun to producers, which form the basis of the energy flow, and then to consumers and decomposers, which give off energy as heat (Figure 1.12). A diagram of energy flow through an ecosystem showing which species are eating which other species is known as a **food web**.

> ■ Energy flows from the sun to producers and then to consumers. Food webs depict the complex interrelationships of these organisms.

■ HIGHLIGHT

Life on the Edge: Viruses

Every winter millions of people are laid low by the influenza virus. If you come down with the flu, the symptoms are all too familiar: You are achy. You cough and sneeze. The reason you're suffering is that an influenza virus is infecting your body's cells, reproducing throughout your nose, throat, and lungs.

The specialized cells of your immune system (see Chapter 32) fight back, attacking and destroying cells already infected by the virus and raising your temperature (hence your high fever) to prevent the virus from reproducing. But the microscopic flu virus can knock you out for days or weeks, replicating itself again and again at your body's expense while evading your body's defenses. The influenza virus also evolves rapidly over time, changing so quickly that defending against it is extremely difficult. A virus is thus a formidable enemy.

In fact, the flu virus has been a deadly foe throughout human history. In 1918 the Spanish flu epidemic killed more than half a million people in the United States, 10 times the toll of U.S. soldiers killed in World War I. Worldwide, the flu epidemic of 1918 killed between 20 million and 40 million people in just half a year. But there is much more to viruses than those that cause influenza in humans. There are many different kinds of viruses that infect all the different forms of life.

These powerful foes certainly seem alive, just like the many organisms they attack. Like living organisms, viruses reproduce, show a high level of organization, and evolve. Yet all viruses, including the influenza virus,

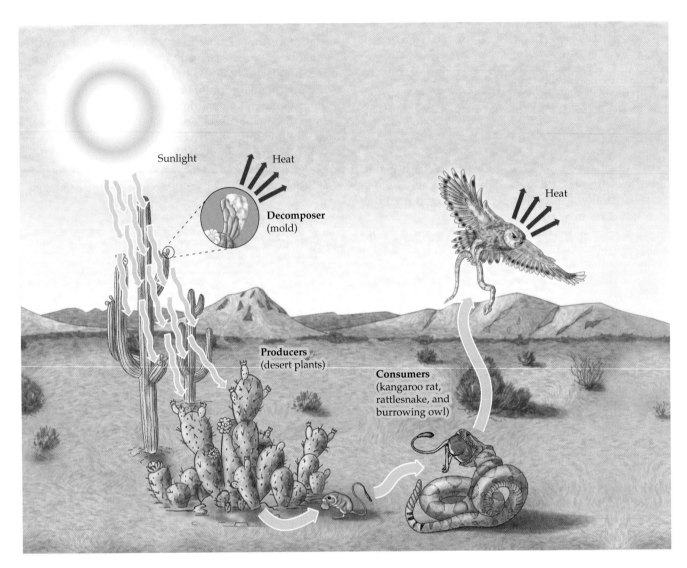

Figure 1.12 Energy Flow through an Ecosystem
Sunlight is captured by producers—in this case, desert plants. Energy then flows to consumers, such as the fruit-eating kangaroo rat, that eat those producers. Those consumers are then eaten by other consumers, such as the rattlesnake, which can then be eaten by still other consumers. Decomposers, such as the mold on the fruit, likewise get their energy either from plants or from organisms whose energy ultimately derives from plants. Decomposers and consumers give off energy in the form of heat. Energy flows from the sun to producers and then to consumers and decomposers throughout this food web.

lack some of the basic characteristics of life. For example, unlike living organisms, viruses are not made up of cells. A virus is simply a hereditary blueprint, a piece of genetic material wrapped in a coat of protein. Unlike cells or organisms made from cells, viruses lack the structures necessary, for example, to gather energy, reproduce, and do nearly all the things living organisms do. In order to reproduce, viruses must take over the cellular machinery of the hosts they invade.

Another unusual feature of viruses is that, unlike living organisms, the genetic material they pass from one generation to the next is not always DNA. Some viruses employ another molecule, known as RNA, or ribonucleic acid. (We will learn more about RNA in Unit 3.)

So, are viruses inanimate or alive?

In fact, viruses raise many of the same questions as nanobes, the incredibly tiny organisms described at the start of this chapter. Like viruses, nanobes show some characteristics of living organisms. Nanobes appear to carry genetic material and to grow. But, like viruses, nanobes do not appear to share all the characteristics of living organisms. Some scientists say, for example, that nanobes are too small to be made up of cells.

Much more remains to be learned about nanobes, but like the well-known viruses, they defy easy definition. Many scientists consider viruses to be nonliving. Other biologists place viruses in a gray zone between life and nonlife. However we choose to define such fascinating entities, viruses and nanobes both force us to stretch the limits of our definitions while making it clear just how diverse life on Earth can be.

> ■ Viruses test the limits of our definition of life, as they have some characteristics of living organisms, but lack others.

SUMMARY

Asking Questions, Testing Answers: The Work of Science

- To answer questions about the natural world, scientists begin with observations of nature, formulate hypotheses from those observations, test those hypotheses, and then reject or modify them as necessary.

- Scientists test their hypotheses by performing controlled, repeated manipulations of nature known as experiments.

The Characteristics that All Living Organisms Share

- The great diversity of life on Earth is unified by a set of shared characteristics.

- All living organisms are built of cells, reproduce using DNA, grow and develop, capture energy from their environment, sense and respond to their environment, show a high level of organization, and evolve.

Levels of Biological Organization

- Living organisms are part of a biological hierarchy with levels from molecules through cells, tissues, organs, organ systems, individuals, populations, species, communities, ecosystems, and biomes to the biosphere.

Energy Flow through Biological Systems

- Energy flows through biological systems from producers, such as plants, that create chemical energy in the form of sugars and starches from sunlight, to consumers, such as animals, that eat plants or other organisms whose energy ultimately derives from plants.

Highlight: Life on the Edge: Viruses

- Viruses, which have some, but not all, of the characteristics of living organisms, illustrate how difficult it can be to define life precisely.

KEY TERMS

adaptation p. 11	food web p. 13
biome p. 13	hypothesis p. 5
biosphere p. 13	individual p. 13
cell p. 7	multicellular p. 7
community p. 13	organ p. 13
consumer p. 13	organ system p. 13
decomposer p. 13	population p. 13
development p. 9	producer p. 13
DNA (deoxyribonucleic acid) p. 7	science p. 4
ecosystem p. 13	scientific method p. 5
evolution p. 11	species p. 11
experiment p. 6	tissue p. 11

CHAPTER REVIEW

Self-Quiz

1. Which of the following is *not* an essential element of the scientific method?
   a. observations
   b. conjecture
   c. experiments
   d. hypotheses

2. After reading about Dr. Larson's research, another scientist said, "Japanese honeysuckle might also be doing well in the United States because it isn't being attacked by any of its natural enemies, such as insects, from Asia." This statement is an example of
   a. an experiment.
   b. a hypothesis.
   c. a test.
   d. a prediction.

3. Which of the following are both universal characteristics of life?
   a. the ability to grow and the ability to reproduce using DNA
   b. the ability to reproduce using DNA and the ability to capture energy directly from the sun
   c. the ability to move and the ability to sense the environment
   d. the ability to sense the environment and the ability to grow indefinitely

4. Which of the following is the basic unit of life?
   a. virus
   b. DNA
   c. cell
   d. organism

5. Which of the following can reproduce without its own DNA?
   a. human being
   b. virus
   c. single-celled organism
   d. none of the above

6. Which of the following is a multicellular organism?
   a. beetle
   b. brain
   c. bacterium
   d. forest ecosystem

Review Questions

1. What are the observations, hypotheses, and experiments in Dr. Larson's honeysuckle work?

2. What are the elements of the biological hierarchy, and how are they arranged in their proper relationship with respect to one another, from smallest to largest?

3. How does energy flow through biological systems?

1   𝔗𝔥𝔢 𝔇𝔞𝔦𝔩𝔶 𝔊𝔩𝔬𝔟𝔢

Creationism Fights for a Place alongside Evolution in the Classroom

WEST HARTSVILLE, TN—A group of concerned citizens crowded the local elementary school gymnasium yesterday to hear arguments about whether "creation science" should be taught alongside evolutionary biology in this small town's classrooms. In contrast to biologists, creation scientists purport that species do not change over time and that all organisms are unchangeable and designed by a divine creator.

"Creation science is not a science at all, but a set of religious beliefs," said Dr. Naomi Latte, evolutionary biologist at Tennessee State College, speaking before the school board. "It should not be taught alongside evolutionary biology." Dr. Latte argued that faith "has its place in the home, the church, and in private schools, but not in public schools," where the law requires a separation of church and state.

Dr. Tim Garter, of the Creation Science Foundation, countered that creation science is a real science. "Though people like Miss Latte will tell you otherwise, creation science is a legitimate field of study, and we should not be censored by university biologists, so many of whom are the worst kind of atheists. Students have the right to know the full range of scientific ideas, however noxious those ideas might be to some people."

Evaluating "The News"

1. The discovery that Japanese honeysuckle can find distant supports more quickly than native honeysuckle is an example of scientific knowledge. The statement that angels are awaiting us in heaven is a religious belief. What distinguishes science from religion?

2. Imagine you are a citizen of West Hartsville and you want the school board to make a fair decision about whether creation science should be taught in the local schools. Everyone agrees that if creation science is religion it should not be taught in public school, but if it's a science, it should. How could you decide whether creation science is a science or not?

3. What kinds of questions can't science answer?

2 Understanding and Organizing the Diversity of Life

Alexis Rockman, *The Bounty*, 1991.

The Iceman Cometh

On a September day in 1991, hikers high in the Alps near the border between Austria and Italy came across a remarkable sight: the ancient, mummified body of a man in the melting glacial ice. Instantly dubbed the Iceman, the well-preserved body appeared to be a prehistoric hiker dating to the Stone Age, possibly a shepherd or traveler. When he died on the mountainside, his body was captured by the ice, which preserved it for some 5000 years.

Extremely well preserved—right down to his underwear—this one-and-a-half-meter-tall visitor from another time promised a tantalizing peek into the past. Sporting tattoos on his body, he wore a loincloth, a leather belt and pouch, leggings and a jacket made of animal skin, a cape of woven grasses, and

All living things have a place in the tree of life and in the classification system known as the Linnaean hierarchy.

calfskin shoes lined with grass. He carried a bow and arrows, an axe, knives, and two pieces of birch fungus, perhaps a kind of prehistoric penicillin.

Perfectly suited for scientific examination, this first-ever corpse from the Stone Age found right in the path of hikers seemed too good to be true. Researchers began to fear that the Iceman might be an elaborate hoax. Speculation was rising that the body was a transplanted Egyptian or pre-Columbian American mummy. How could biologists determine the true identity of this long-lost wanderer?

By studying the DNA still preserved in his tissues, researchers were able to show that the Iceman was not a transplant. In fact, he turned out to be a close relative of contemporary northern European peoples from the same area, much more closely related to them than to any other human group. The authenticity of the Iceman was established by comparing his DNA to that of other groups of people. Thus he was placed in the human family tree and was proclaimed a northern European.

In the same way, whenever biologists find any new organism and want to know what it is, they must first ask themselves what the organism's closest relatives are. Is it most closely related to plants, animals, fungi, or bacteria? If it is a plant, is it more closely related to a cycad or a bluebell? If it is an animal, is it more closely related to a guppy or an eagle? Is it a species already known, or something entirely new? Once scientists know where an organism belongs in the family tree of life, they can place it in the grand classification scheme of life and define what it is. So, whether scientists are collecting new species from the rainforest or sleuthing out the identity of a disease bacterium, their first question is always, Where does this organism fit in the family tree of life?

The prehistoric mummy known as the Iceman as he was discovered in melting ice by hikers in the Alps in 1991, some 5000 years after he died.

■ KEY CONCEPTS

1. Systematics is the science of naming and classifying groups of organisms and determining the relationships among them.

2. Because living things have evolved from a common ancestor, their relationships can be depicted on a kind of family tree called an evolutionary tree.

3. Biologists use all sorts of characteristics of organisms, including the shape of body structures, behaviors, and DNA, to help them decipher the evolutionary relationships among species.

4. Cladistics is now widely accepted as the method of choice for building evolutionary trees. In cladistic analyses, scientists use shared, novel features of organisms to determine evolutionary relationships.

5. Once researchers know the closest relatives of an organism and where that organism belongs on an evolutionary tree, they can begin to make predictions about the biology of the organism, with expectations that much of its biology will be similar to that of its close relatives.

Genealogists show how different people are related by using a family tree. Such a diagram shows which people in a family are most closely related to which others and how they are related (see, for example, Figure 2.1*a*). In the same way, biologists called **systematists** study relationships between groups of different organisms, figuring out which groups are most closely related to which others. Using the information they uncover about evolutionary relationships, they create what are known as **evolutionary trees** (Figure 2.1*b*). Systematists also give organisms their scientific names and place them in a classification system that encompasses all living organisms.

In this chapter we examine how systematists build evolutionary trees. We look at what kinds of features of organisms are useful for determining their relationships and what kinds are not. Next we examine the Linnaean hierarchy, the classification scheme used by biologists today to organize and name the organisms in the great tree of life. Finally we look at some of the more interesting and important branches on the evolutionary tree of life that scientists are working to decipher.

Let's begin by imagining a systematist studying a diverse array of plants or animals. Her goal is to figure out their evolutionary tree. Where should she begin?

2.1 Looking for Clues to Evolutionary Relationships

As anyone who has been to a family reunion knows, the more closely related two people are, the more similar they tend to be to one another in the way they look and sometimes even in the way they act—showing the same

smile or sneezing in just the same way. Close relatives even tend to be similar in how their bodies work, often exhibiting the same physical strengths or being prone to the same kinds of illnesses. In fact, their similarities extend right down to the level of a person's genetic material, or DNA, and for good reason.

Recall from Chapter 1 that DNA is the genetic blueprint for the development of an individual that is passed down from one generation to the next. We inherited these blueprints for body structures and behaviors from our parents, who inherited theirs from their parents, and so on. As a result, we exhibit many of the same characteristics—a particular smile or a tendency to hay fever, for example—that our relatives exhibit. In the same way, closely related groups of organisms tend to resemble one another.

Unit 4 provides a detailed discussion of evolution and the origin of species. For now, suffice it to say that evolution is the process by which species can change, including giving rise to new species. Recall from Chapter 1 that all living organisms descended from a common ancestor. Over time, a species can evolve, splitting into two new species. In this way, an ancestor species gives rise to new species that are its descendants. On an evolutionary tree, these descendants are depicted as the tips of branches. Each group of close relatives branching off from the tree is considered to be a single **lineage** (like a lineage within a human family). The branches trace back and meet at a point where the descendants shared their last common ancestor (see Figure 2.1*b*).

Descendants often share key features because they share a common ancestor. In human families, for example, a father's distinctive nose may be seen on the faces of all his children. Similarly, all the vertebrate animals—includ-

(*a*) Family tree of Britain's royalty

(*b*) Swallowtail butterfly evolutionary tree

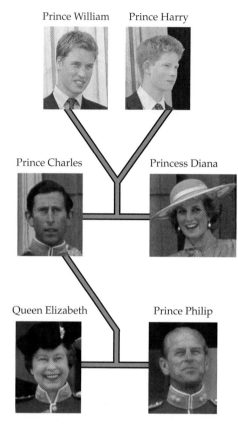

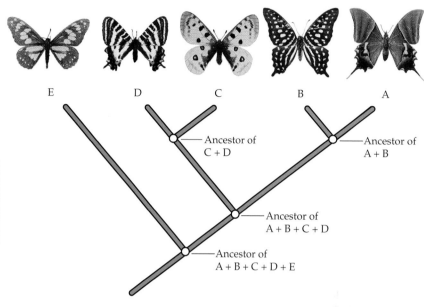

Figure 2.1   Family Trees versus Evolutionary Trees
(*a*) A family tree depicts relationships between ancestors and descendants. This tree depicts the relationships between some of the familiar members of Britain's royal family, including Queen Elizabeth and Prince Philip, their son Charles, Princess Diana, and Charles and Diana's two sons. (*b*) An evolutionary tree is read quite differently, but it, too, depicts relationships, in this case the evolutionary relationships of swallowtail butterflies. Each group of butterflies is represented by a photograph at the tip of a branch of the tree. Trace any branch back to where it meets another group's branch. That meeting point represents the common ancestor from which those groups descended. When we trace a branch backward, we are really tracing back through time. The farther down the tree we go, the farther into history we are delving. This tree indicates that butterfly groups A and B are each other's closest relatives. These two butterfly groups descended from a common ancestor depicted on the tree at the point where their branches meet. Thus, from the common ancestor, groups A and B branched off and evolved into 2 distinct lineages, as the photographs show. Groups A, B, C, and D all descended from the common ancestor depicted at the next branching point down the tree, and so on.

ing humans, birds, snakes, frogs, and fish—can be grouped together on an evolutionary tree because they all have the backbone that was a feature of their **most recent common ancestor**, the animal from which all these descendant animals sprang.

Systematists can find such shared features in various aspects of an organism's biology. Traditionally, systematists have compared species by looking at inherited structural characteristics of the body: numbers of legs, the structure of a flower, the anatomy of the heart, and so on. In recent years, however, more researchers have begun searching for similarities in other features, including the behaviors of organisms. The most powerful new feature available to systematists is an organism's genetic material, or DNA. Recent advances in techniques for studying DNA have revolutionized all of biology, including the study of evolutionary relationships.

Using these new techniques, systematists have been able to study the relationships of many different groups that they were unable to study before. For example, by looking at DNA, researchers have been able to make stunning progress in understanding the relationships among major groups such as bacteria, plants, and animals—groups whose body parts are so different that they are very difficult to compare. In addition, researchers are learning much more about groups, such as parasitic worms, that were difficult to study previously because they are so simple structurally that they do not exhibit many distinctive features that can be compared.

Parasitic worm

■ Closely related groups of organisms that descended from a common ancestor tend to share features that they inherited from that common ancestor. To determine which groups of organisms are most closely related, systematists examine all aspects of an organism's biology—its body structure, its behavior, its DNA—to look for such inherited similarities.

Just looking for similarities in various features, however, does not provide the answer to the systematist's question. What a systematist often finds is that while species A shares certain features with species B, it also shares certain other features with species C. Which, then, is A's closest relative? Should A and B be placed closest together on the evolutionary tree (as in Figure 2.2*a*), or should A and C (as in Figure 2.2*b*)? Because there is only one true evolutionary history, there is only one correct evolutionary tree for any given group of organisms. Which shared features are most likely to show us the true tree of life?

Building Evolutionary Trees

Over the years there have been several different schools of thought on which features of organisms best reveal their evolutionary relationships. In this section we dis-

cuss two of them. We first discuss the oldest approach to this subject, known as evolutionary taxonomy. Next we tackle cladistics, the school of thought that has come to prominence today.

Evolutionary taxonomy is the oldest approach to systematics

In the traditional approach to systematics, known as **evolutionary taxonomy**, scientists used their expert knowledge to decide whether groups of organisms were closely or distantly related. Without using any specific formula or rationale, these scientists determined the relatedness of two groups by attempting to assess their general, overall similarity.

Figure 2.2   Showing Relationships on an Evolutionary Tree
(*a*) In this tree, A and B are shown as more closely related than either is to C, sharing their most recent common ancestor at the point depicted. The two also share a common ancestor with C, but farther down the tree, meaning deeper in the evolutionary past. Note that switching the positions of A and B does not change how the tree is read. (*b*) This tree indicates that species A and C are more closely related to one another than either is to B, because they share a common ancestor more recently than either one does with B. Again, switching the positions of A and C here does not change how the tree is read.

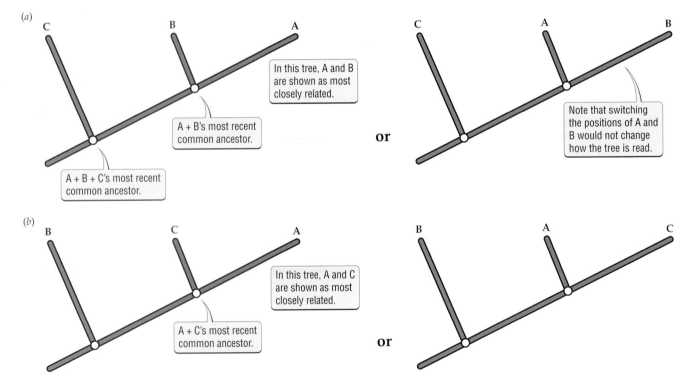

In addition, traditional systematists sometimes singled out particular characteristics of groups as the most "important" and gave those characteristics greater weight. An evolutionary taxonomist might decide that two species were each other's closest relatives because they shared one special feature that the scientist considered either extremely important to the biology of the group or so unusual that it was a very reliable indicator of their relatedness. For example, two species might share a certain feather structure, or an unusual arrangement of internal organs.

Choosing the most important features without explicit criteria can become a matter of personal opinion, leading to controversies that are difficult to resolve. More importantly, such decisions, like judgments about overall similarity, do not necessarily lead to the correct evolutionary tree. Since the 1950s, evolutionary taxonomy has slowly been replaced by a more explicit, rigorous method, known as cladistics.

Cladistics uses shared derived features

In **cladistics**, systematists rely on one particular class of features to determine relationships: those unique features that evolved for the first time in an ancestor species and were then passed down to all of its descendants. The presence of such novel, distinctive traits clearly identifies an ancestor organism's descendants as a closely related group. Systematists recognize that organisms that share many such novel, distinctive features are more closely related to one another than they are to other organisms that lack those features. Less closely related organisms, not descended from the same common ancestor, do not display those features. Instead, they display the original ancestral traits and different novel features that are unique to themselves and their own close relatives.

Figure 2.3 depicts the relationships of some familiar groups of animals, identifying some of the shared dis-

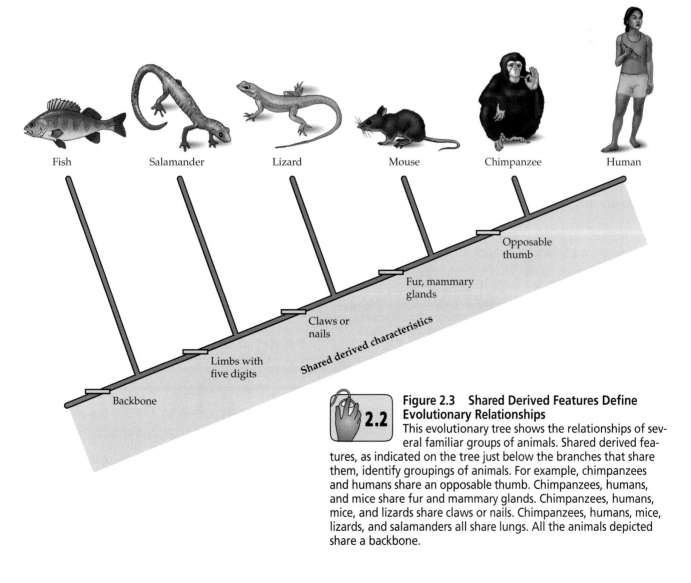

Fish   Salamander   Lizard   Mouse   Chimpanzee   Human

Opposable thumb

Fur, mammary glands

Claws or nails

Shared derived characteristics

Limbs with five digits

Backbone

Figure 2.3   Shared Derived Features Define Evolutionary Relationships
This evolutionary tree shows the relationships of several familiar groups of animals. Shared derived features, as indicated on the tree just below the branches that share them, identify groupings of animals. For example, chimpanzees and humans share an opposable thumb. Chimpanzees, humans, and mice share fur and mammary glands. Chimpanzees, humans, mice, and lizards share claws or nails. Chimpanzees, humans, mice, lizards, and salamanders all share lungs. All the animals depicted share a backbone.

tinctive features that define these groups. Starting at the upper right-hand corner of the tree, the first shared feature is an opposable thumb (a thumb that is capable of being placed against one or more of the other fingers on the hand), one of the many features that set chimpanzees and humans apart from the other animals on the tree. No other organisms on the tree share this feature. Most likely we share it with chimpanzees because our common ancestor had an opposable thumb. Thus, this shared feature is one of the pieces of evidence that humans and chimpanzees share a more recent common ancestor than do the other animals on the tree.

Working down the tree, the next closest relatives of chimpanzees and humans are mice. Distinctive features shared by chimpanzees, humans, and mice are hair and the mammary glands that females use to produce milk for their young. These are features shared by the most recent common ancestor of these three groups, the original mammal. Other features define larger groups as we move down the tree. At each point where lineages trace back to a common ancestor, a shared feature derived from that ancestor is indicated. These features—unique to that common ancestor and then passed down to all of its descendants, clearly defining the descendants as a group—are called **shared derived features**.

Shared ancestral features are not useful

In addition to sharing derived traits, the organisms of any particular group share many traits that are *not* novel traits derived from their most recent common ancestor—**shared ancestral features**. Shared ancestral features are not useful for identifying close evolutionary relationships. For example, if a systematist were comparing mice, chimpanzees, and humans, she might consider the fact that neither mice nor chimpanzees have the ability to use language. Without distinguishing between ancestral and derived traits, she could easily decide that mice and chimpanzees are more closely related to each other than either one is to humans and that they shared a more recent common ancestor—simply because they do not possess language. An evolutionary tree based on this information would be wrong.

The reason is that the lack of language is not a novel trait that evolved in the most recent common ancestor of mice and chimpanzees. Instead, the lack of language is an ancestral trait that any number of organisms—from bacteria to plants to mice and chimpanzees—share, and thus it does not distinguish any pair of them as being more closely related to one another than they are to humans.

Shared ancestral features do not help systematists understand the correct relationships between groups of organisms and cannot be used to build evolutionary trees. There is one more class of shared features that is not useful for building trees: convergent features.

Convergent features do not indicate relatedness

Sometimes organisms share features not because they share a common ancestor, but because they all evolved the same feature independently. Such features are known as **convergent features**. For example, many desert plants (such as cacti and some spurges) resemble one another because they have all evolved characteristics that help them live under the same parching sun. Desert-dwelling plants typically have very small leaves, or no leaves at all, and a shape that reduces water loss. These features evolved independently in each separate lineage of plants. A systematist might be tempted to conclude that cacti and spurges are close relatives because they share so many features. However, they share these traits only because these plants "converged" on a similar strategy for surviving in the desert. Convergent features can mislead systematists rather than guiding them to the correct evolutionary tree.

Spurge

While convergent features might seem obvious and easy to avoid, in practice they can be difficult to identify, whether one is looking at leaf size or at DNA. This difficulty in identifying convergent features is one of the major obstacles that systematists face in trying to decipher the true tree of life from the data at hand.

> ■ Cladistics is the most prominent school of thought in systematics today. In cladistics, systematists use shared derived features to build evolutionary trees. Two types of shared features that are not good indicators of relatedness are shared ancestral features and convergent features.

Using Evolutionary Trees to Predict the Biology of Organisms

Once systematists discover the correct evolutionary relationships among the organisms in a group, the resulting evolutionary tree is useful for more than just describing and organizing knowledge already gained about the organisms. The tree also has predictive power, because researchers can expect that close relatives will share

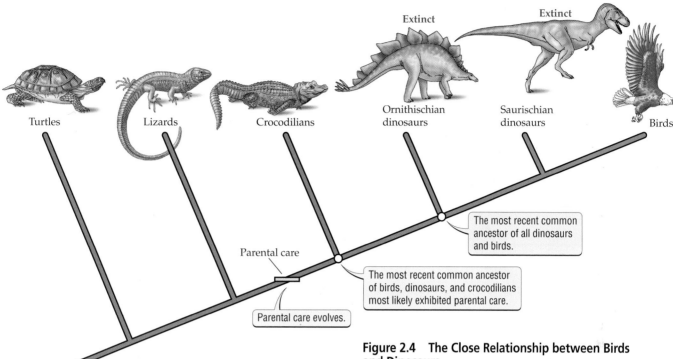

Extinct

Extinct

Turtles

Lizards

Crocodilians

Ornithischian dinosaurs

Saurischian dinosaurs

Birds

Parental care

The most recent common ancestor of all dinosaurs and birds.

The most recent common ancestor of birds, dinosaurs, and crocodilians most likely exhibited parental care.

Parental care evolves.

Figure 2.4   The Close Relationship between Birds and Dinosaurs
As this evolutionary tree shows, birds and the two lineages of dinosaurs—ornithischians (plant eaters, such as *Stegosaurus*, characterized by horny beaks and a lack of teeth) and saurischians (animals such as *Tyrannosaurus*, characterized by a long, mobile S-shaped neck)—are closely related. The next most closely related group depicted on the tree is the crocodilians, which includes crocodiles and alligators.

many of the same novel features passed down by their common ancestor.

As surprising as it might seem, there is now overwhelming evidence from living and fossil animals that the closest relatives of birds are the extinct creatures that we know as dinosaurs (Figure 2.4). Of the animals shown in Figure 2.4, the next closest relatives of birds, after dinosaurs, are crocodiles and alligators, a group known as crocodilians. Knowing the relationships among these groups has made it possible for biologists to make predictions about the behavior of long-extinct dinosaurs.

Both crocodilians and birds are known to be dutiful parents. They build nests and defend their eggs and young. Scientists reasoned that if crocodilians and birds both provide parental care to their young, then most likely their common ancestor exhibited this behavior as well. Because dinosaurs share this same common ancestor, scientists were able to predict that these long-extinct and unobservable creatures probably tended their eggs and hatchlings as well—a shocking notion for creatures with a reputation for being big, vicious, and pea-brained.

In 1994, Mark Norell, a paleontologist at the American Museum of Natural History in New York City, and an international team of colleagues confirmed that

dinosaurs exhibited parental care. They discovered a fossilized dinosaur that died 80 million years ago—while sitting on a nest of its own eggs. This dinosaur, which had actually been unearthed from the Gobi Desert in 1923, was originally thought to have been eating the eggs, and hence was given the name *Oviraptor*, which means "egg seizer." Biologists at the time did not know that birds and dinosaurs were close relatives, so they did not expect the two groups to show similar behaviors such as parental care. As a result, the idea that a dinosaur could brood a nest of eggs was inconceivable. In fact, however, *Oviraptor* appears to have died in a sandstorm protecting its nest, its limbs encircling its unhatched young in a posture as protective as that of any bird (Figure 2.5).

■ Evolutionary trees not only depict the relationships of groups of organisms, but can also be used to predict behaviors and other attributes of organisms.

(a)

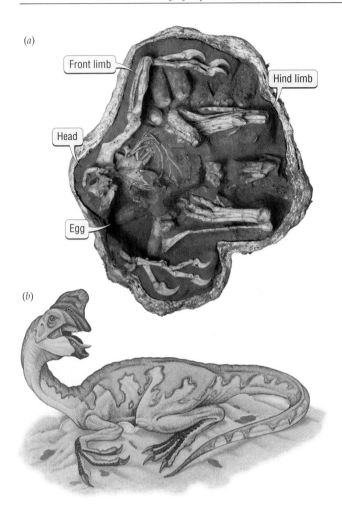

Front limb

Hind limb

Head

Egg

(b)

Figure 2.5 Parental Care by Dinosaurs
(a) This fossil shows the remains of an *Oviraptor* dinosaur, which died in a sandstorm, protecting its nest of eggs. (b) An artist's rendition of *Oviraptor* brooding its eggs shows how the dinosaur might have looked shortly before the sandstorm began. Compare this drawing with (c), an ostrich as it would look on a nest of eggs today.

(c)

A Classification System for Organizing the Abundance of Life: The Linnaean Hierarchy and Beyond

In addition to building evolutionary trees, biologists have organized the world's species into a classification system known as the **Linnaean hierarchy**. In this section we will explore how the Linnaean hierarchy is organized.

The Linnaean hierarchy was developed in the 1700s by a Swedish biologist named Carolus Linnaeus. In this system, the species is the smallest unit of classification. Closely related species are grouped together to form a genus (plural genera) (Figure 2.6). Closely related genera are grouped together into a family. Closely related families are grouped into an order. Closely related orders are grouped into a class. Closely related classes are grouped into a phylum (plural phyla). Finally, closely related phyla are grouped together into the largest category in the hierarchy, a **kingdom**. An easy way to remember the hierarchy is to memorize the following sentence, in which the first letter of each word stands for each descending level in the hierarchy: *K*ing *P*hillip *C*leaned *O*ur *F*ilthy *G*ym *S*horts.

Every species is given a unique, two-word Latin name, also known as a scientific name. The first word tells what genus the organism belongs to, and the second defines its unique species. For example, the scientific name for humans is *Homo sapiens*—*Homo* meaning "man" and *sapiens* meaning "wise." *Homo* indicates our genus, and *sapiens* is our particular species name within the genus. We are the only living species in our genus. Other species in the genus are *Homo erectus* ("upright man") and *H. habilis* ("handy man"), both of which are extinct. (The genus name is often abbreviated to its first letter, as in *H. habilis*, when it is repeated in a discussion.)

Homo habilis

Kingdoms are the highest level of the Linnaean hierarchy. Originally Linnaeus described just two kingdoms, plants and animals, but the kingdoms have been revised over time as biologists have learned more about relationships among living organisms. Today's classification schemes recognize anywhere from five to six or more kingdoms, depending on how finely they divide up the world's organisms (Figure 2.7). In this book we employ the widely used six-kingdom scheme. (We will explore the nature of the six kingdoms and the distinctions between them in Chapter 3.)

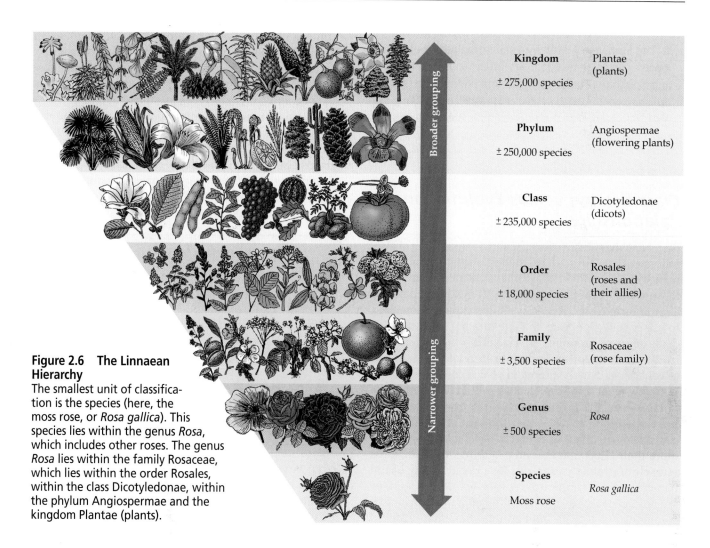

	Kingdom ± 275,000 species	Plantae (plants)
	Phylum ± 250,000 species	Angiospermae (flowering plants)
	Class ± 235,000 species	Dicotyledonae (dicots)
	Order ± 18,000 species	Rosales (roses and their allies)
	Family ± 3,500 species	Rosaceae (rose family)
	Genus ± 500 species	*Rosa*
	Species Moss rose	*Rosa gallica*

Figure 2.6 The Linnaean Hierarchy

The smallest unit of classification is the species (here, the moss rose, or *Rosa gallica*). This species lies within the genus *Rosa*, which includes other roses. The genus *Rosa* lies within the family Rosaceae, which lies within the order Rosales, within the class Dicotyledonae, within the phylum Angiospermae and the kingdom Plantae (plants).

In addition to kingdoms, most biologists have also begun using an even higher level of organization, called **domains** (Figure 2.7). The three domains are the Bacteria (which include familiar disease-causing bacteria), the Archaea (bacteria-like organisms best known for living in extremely harsh environments), and the Eucarya (which includes all the rest of the living organisms from amoebas to plants to fungi to animals). Although they were highly controversial at first, the three domains are now widely accepted. They are considered to describe the most basic and ancient divisions among living organisms.

Three-domain system: Bacteria | Archaea | Eucarya

Six-kingdom system: Bacteria | Archaea | Protista | Plantae | Fungi | Animalia

Figure 2.7 Domains and Kingdoms

In this book we employ the three-domain system and the widely used six-kingdom classification scheme. The domain Bacteria is equivalent to the kingdom Bacteria, and the domain Archaea is equivalent to the kingdom Archaea. The domain Eucarya encompasses four kingdoms in this scheme: Protista (or protists, including organisms such as amoebas and algae), Plantae (or plants), Fungi (including mushroom-producing species), and Animalia (or animals). For more on these groups, see Chapter 3.

■ The Linnaean hierarchy goes from species at the lowest level to genera, families, orders, classes, phyla, and kingdoms. The kingdom system has evolved over time and continues to evolve as we learn more. Biologists also recognize the three-domain system as the most basic division of living organisms.

Classification versus Evolutionary Relationships

Based on what we have learned about the building of evolutionary trees, it might seem logical for systematists to name genera, families, kingdoms, and so forth only when those sets of organisms constitute all the descendants of a common ancestor. That is, we might expect them never to use partial groupings that are missing some descendants or mixed groupings that include some descendants from one ancestor and some from another. With the advent of cladistics, naming only such complete groups is exactly what some systematists would like to do.

However, the Linnaean classification system was developed long before either the idea of evolution or cladistics came to prominence. In fact, biologists still use more than 11,000 names and groupings that Linnaeus himself originally proposed. As a result, many of the named groups of organisms with which we are familiar are not the complete groups of descendants of a single, common ancestor, the "**real groups**" that many systematists would like to use.

To see how systematists define real groups, imagine an evolutionary tree as an actual tree. If we were to make any single cut in the tree, all the branches that came off together from that single cut would constitute a real group. A growing number of scientists want only such real groups to be recognized and named, and want all other names to be abandoned. For many others, abandoning familiar names and familiar groups and using only real groups as we now understand them would be too radical a move.

Dinosaurs, for example, are a well-known group of creatures that most elementary school students can identify. However, they are not a real group consisting of all the descendants of their most recent common ancestor. As Figure 2.4 shows, the most recent common ancestor of both lineages of dinosaurs is also the most recent common ancestor of the birds. If you try to make a single "cut" to the tree to get both dinosaur groups, you get the birds as well. Thus dinosaurs are not a "real group" unless they include the living birds. If groups are to correspond only to complete groups of descendants, either the term

"dinosaurs" must be eliminated from our language—a practical impossibility—or we must call the robin we see in the backyard a dinosaur. In fact, many scientists now consider the modern birds, along with such beasts as *Tyrannosaurus rex*, to be types of saurischian dinosaurs.

Systematists are encountering an increasing number of such problems as new information about evolutionary relationships and the push for using nothing but real groups clashes with traditional classifications. The idea of doing away with familiar groups and names remains controversial, and scientists continue to debate the merits and problems of different systems.

■ There is a controversy in biology today about the naming and classification of groups of organisms. Some biologists believe that only real groups should be named. Others believe that doing away with long-standing and familiar groupings will cause more problems than it will solve.

2.3 Branches on the Tree of Life: Interpreting Relatedness

What have scientists learned about the relationships among different groups of organisms? The study of evolutionary relationships is one of the fastest-moving areas of biology, with new evolutionary trees being deciphered all the time. In this section we examine two of the trees of greatest interest to scientists: the tree that includes all living organisms, and the tree that depicts the relationships of humans and their closest relatives.

The tree of all living things reveals some surprising relationships

The more distantly related two organisms are—for example, a bacterium and a human—the more difficult it is to compare them. What part of a bacterium would show any similarity to any part of a person? One of the few features that can easily be compared across such wide divides is DNA, which all organisms use as their hereditary blueprint. By using DNA to make their comparisons, systematists have revolutionized the study of evolutionary relationships. They can now compare even extremely distantly related groups, such as domains and kingdoms, to reconstruct the arrangement of the major branches on the tree of life.

The two trees depicted in Figure 2.8 and 2.9 are the result of these ongoing studies. In Figure 2.8, the pro-

Three-domain system

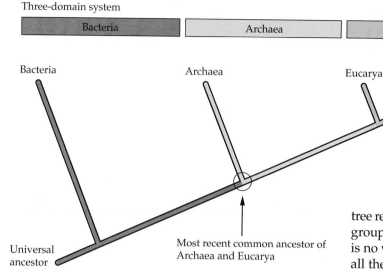

Figure 2.8   The Three Domains
This evolutionary tree shows the relationships of the three domains. At the root of the tree is the universal ancestor of all living things. The first evolutionary split came between the Bacteria and the lineage that would give rise to the Archaea and Eucarya. The next split was between the Archaea and the Eucarya, making these two domains more closely related to each other than they are to the Bacteria.

posed evolutionary relationships of the three domains are depicted. Note that in the more modern system of domains, each of the domains is a real group, made up of a single ancestor and all its descendants. This tree shows the most basic divisions among living organisms and traces some of the early events in the evolution of life. The Bacteria were the first group to evolve, splitting apart from the lineage that would later give rise to the Archaea and the Eucarya. The Archaea and Eucarya evolved more recently than the Bacteria. (There has been some controversy about the simplicity of this tree, however; see the box on page 32.)

The tree in Figure 2.9 provides the next level of detail, showing the hypothesized relationships of the six kingdoms. In general, the relationships among the kingdoms are not surprising. The most distant relatives of plants, animals, and fungi are the two kingdoms Bacteria and Archaea, single-celled creatures that seem like very distant cousins of ours, at best.

Note, however, that while we recognize six kingdoms, there are more than six groups on the tree. The kingdom known as Protista (which includes such things as amoebas and algae) includes three separate lineages: diplomonads, ciliates, and diatoms. An examination of the

tree reveals, however, that the protists are not a complete group of all the descendants of a single ancestor. There is no way to make a single cut on the tree that includes all the protists—and only the protists. The most recent common ancestor of the diplomonads, ciliates, and diatoms is the ancestor to the Eucarya. This means that the smallest real group that the protists are a part of is the domain Eucarya. Like the dinosaurs (see Figure 2.4), the protists are an example of the conflicts that can arise between classification and systematics as we gain new information about the evolutionary relationships of well-established groups.

Finally, notice the relationships of plants, fungi, and animals: The two most closely related groups among the three are fungi and animals. For years, fungi were thought to be more closely related to plants. Unable to move and very unlike animals, these faceless organisms, most familiar to us as mushrooms, seemed more akin to trees, shrubs, and mosses. As a result, it came as a huge surprise when recent DNA studies showed that fungi are actually more closely related to animals—including humans—than they are to plants. That is, fungi share a more recent common ancestor with animals than they do with plants (see Figure 2.9). Put another way, the mushrooms on your pizza are more closely related to you than they are to the green peppers sitting next to them.

Plants and fungi were lumped together as closest relatives not because they shared unique derived features, but because biologists mistakenly based their grouping on shared ancestral features. Plants and fungi are similar simply because they have not evolved some of the unique characteristics that animals have evolved. Animals and fungi, on the other hand, belong together because they share certain novel features.

Could the slime in your bathroom shower really be more closely related to you than it is to a plant? How much do we really have in common with the likes of bread mold or yeast or a mushroom at the grocery store? A lot, it turns out. The finding that fungi and animals are more closely related to each other than either

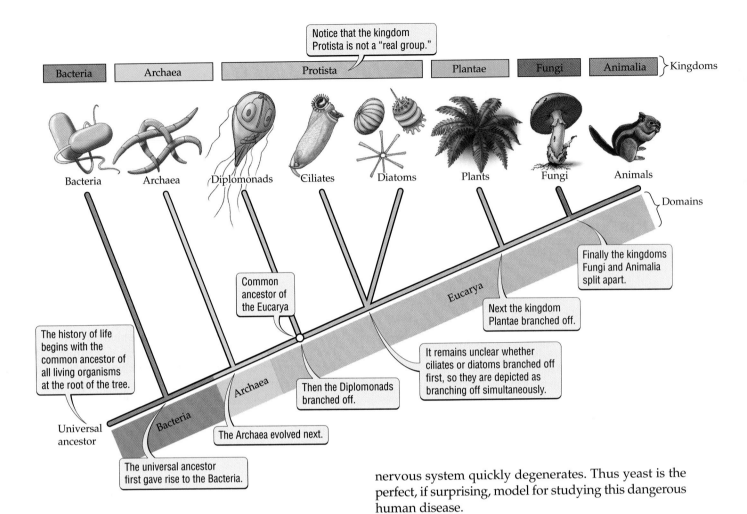

Figure 2.9   The Tree of All Life
This evolutionary tree shows the relationships of the six kingdoms as well as the three domains. Each of the groups branching off the tree can be thought of as a cluster of close relatives—a lineage, just like a lineage in a human family.

is to plants solved the long-standing mystery of why doctors often have such a difficult time treating fungal infections, particularly internal infections, in which a fungus has begun living inside the human body. Because fungi and animals are such close relatives, human cells and fungal cells work similarly. Thus anything a doctor might use to kill off a fungus could harm or kill its human host as well.

So similar are humans and fungi that there is even a disorder in yeasts (which are fungi) that is similar to Lou Gehrig's disease, a fatal disease of humans in which the nervous system quickly degenerates. Thus yeast is the perfect, if surprising, model for studying this dangerous human disease.

The primate family tree reveals the closest relative of humans

When we visit the ape house at a zoo, the striking similarities between the beings standing outside the cage looking in and those inside the cage looking out become obvious. But which of our primate relatives is our closest relation? Over the years this question has generated intense interest and heated controversy. Researchers have studied everything from bone structure to behavior to DNA in attempts to determine which primate is humankind's closest kin.

While controversy still remains, something of a consensus has been reached on the basis of DNA evidence. The emerging consensus suggests that our closest relative is the chimpanzee, a fellow tool user with whom we share a remarkable degree of similarity in our DNA (Figure 2.10). More distantly related to us are gorillas, and beyond that, orangutans, gibbons, Old World and New World monkeys (such as the spider monkey), and, most distantly of all, lemurs.

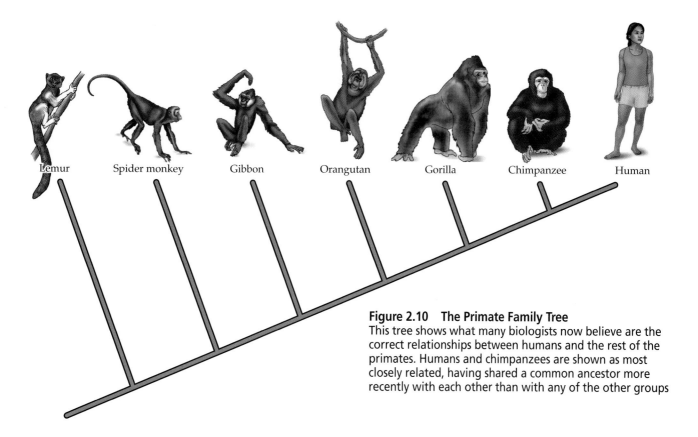

Lemur   Spider monkey   Gibbon   Orangutan   Gorilla   Chimpanzee   Human

Figure 2.10   The Primate Family Tree
This tree shows what many biologists now believe are the correct relationships between humans and the rest of the primates. Humans and chimpanzees are shown as most closely related, having shared a common ancestor more recently with each other than with any of the other groups

The evolutionary tree of life is a work in progress

Although many of the major branching patterns on the evolutionary tree of life are well established and unlikely to change, scientists view evolutionary trees as working hypotheses, the best approximations given what we know today. Biologists continue to study and reevaluate the relationships of all organisms using new information as it is found. In fact, the vast majority of relationships among the world's millions of different species of plants, animals, fungi, bacteria, and protists remain to be worked out in detail.

■ Systematists are making dramatic progress in deciphering the evolutionary relationships of the world's many organisms, in part due to the increasing use of DNA. As biologists study the tree of all living organisms, they are finding surprising relationships, such as the unexpectedly close relationship between animals and fungi. Biologists studying the most controversial of evolutionary trees are beginning to come to a consensus that humanity's closest relative is the chimpanzee.

■ HIGHLIGHT

Everything in Its Proper Place: The Contribution of Systematics

Though little appreciated for many years, systematics has become one of biology's most vital sciences, crucial for the study of everything from agriculture to human diseases. The reason that systematics is so important is that before researchers can begin to ask serious questions about an organism, they must know what that organism is, which means identifying its close relatives, determining where it belongs on the tree of life, and providing it with a name.

Like the Iceman described at the beginning of this chapter, new, unidentified organisms often find their way into our lives. Gypsy moths, which had long been ravaging the trees of the northeastern United States, appeared in the Pacific Northwest for the first time in 1991. Researchers needed to know which kind of gypsy moth was invading—the European gypsy moth, which has long been a problem in the eastern United States, or the Asian type. Unlike the European gypsy moth female, the Asian gypsy moth female can fly, making the species much more

Gypsy moth

■ THE SCIENTIFIC PROCESS

A Tangled Web at the Root of the Tree of Life?

Biologists today tend to agree about the arrangement of the major branches of the tree of life. But in recent years, a number of curious findings have caused biologists to question whether this tidy arrangement might not look a lot less like a tree than like a highly interconnected web, particularly at its base.

As biologists study the DNA of more and more organisms, they are finding unexpected DNA in unexpected places. For example, scientists have uncovered distinctly bacterial DNA in members of the Archaea and Eucarya. How can this be, if the Archaea and Eucarya diverged from the Bacteria long ago and became entirely separate from that lineage? If that is indeed what happened, each lineage should have evolved to be quite distinct. How can scientists explain the fact that the DNA of some organisms looks like a grab-bag of DNA from across the tree of life?

One possible explanation comes from Dr. W. Ford Doolittle, a biochemist at Dalhousie University in Canada. He has made the revolutionary suggestion that throughout the early history of life, before and after the three major lineages (Bacteria, Archaea, and Eucarya) evolved, those three lineages were freely exchanging DNA. So, in addition to DNA being passed down through a lineage from one generation to the next, or vertically, it also appears to have been moving from one entirely distinct lineage to another—or horizontally.

Dr. Doolittle and some colleagues believe that these kinds of direct exchanges—known as horizontal gene transfer—were so common that it may not make sense to think of life as having descended from some single, universal ancestor into three neat lineages. Instead, he and others have begun to suggest that the base of the tree of life may instead have been a loosely knit community of very primitive cells that were freely exchanging DNA, creating curious mixtures of genes that persist in organisms across the tree of life even today.

Biologists are rapidly learning more about the genes of organisms across the three domains, promising to shed light on just how important horizontal gene transfer has been in the history of life. In the meantime, scientists are continuing to debate this startling new hypothesis.

Arrows represent multiple events in which genes apparently moved between the lineages Bacteria, Archaea, and Eucarya, even after the three lineages were well established as separate.

---

mobile and quick to spread. Asian gypsy moths can also feed on and damage a wider variety of tree species.

By studying their DNA, Richard Harrison and Steven Bogdanowicz of Cornell University were able to show that the newly invading moths were close relatives of moths from Asia. By determining which pests they were up against, and thus knowing how quickly the moths could spread and which plants they might attack, researchers were able to give forest managers a better chance of controlling the moths effectively.

It was also systematists who solved the mystery of whether a dentist infected with HIV (the human immunodeficiency virus, which causes AIDS) had transmitted the virus to his patients. Against a backdrop of contro-

versy about whether health care workers with HIV are a risk to their patients, 10 patients of one infected Florida dentist tested positive for HIV. Four of the patients, however, had lifestyles or habits that put them at risk for HIV infection by other means. Was the dentist to blame or not?

Researchers created an evolutionary tree of the viruses taken from each of the patients, from the dentist, and from other infected people by comparing each person's DNA that was derived from the virus. If the dentist had infected his patients, the viruses from his patients should be mostly closely related to his virus and less closely related to viruses sampled from other people.

The results showed that all six patients who were HIV positive but were not at risk of contracting HIV for other reasons were carrying a virus very closely related to the virus from the dentist. Those whose lifestyles or habits had exposed them to HIV in other ways had viruses that were not closely related to the dentist's virus. Scientists concluded that, at least in the case of six of the patients, the dentist was the source of their infection.

Like the scientists who identified the Iceman, many systematists are having a wider and wider reach, as these examples illustrate. When it comes to solving biological problems in the real world, systematics—knowing where an organism fits in the tree of life—is the first order of business.

> ■ Systematics is an active and vital part of biology that is helping answer a vast array of questions in areas as wide-ranging as evolutionary biology, insect pest management, and the spread of disease.

SUMMARY

Looking for Clues to Evolutionary Relationships

- Species give rise to descendant species with which they share characteristics.

- Descendant species often share key features, such as distinctive physical structures or behaviors, because they share DNA inherited from a common ancestor.

- Evolutionary trees depict the evolutionary relationships of groups of organisms.

- Systematists look for features that groups of organisms share to determine their relationships. DNA has proved to be a particularly useful new feature for such studies.

Building Evolutionary Trees

- Evolutionary taxonomy, the oldest approach to systematics, lacks a well-defined method.

- Cladistics uses shared derived features to identify close relatives and to determine evolutionary relationships.

- Shared ancestral features are not useful for determining evolutionary relationships.

- Convergent features can mislead systematists into thinking that distantly related groups are closely related, and vice versa.

Using Evolutionary Trees to Predict the Biology of Organisms

- Evolutionary trees can predict and give insight into the biology of organisms and lead to surprising discoveries such as the behavior of long-extinct dinosaurs.

A Classification System for Organizing the Abundance of Life: The Linnaean Hierarchy and Beyond

- The Linnaean hierarchy provides a classification scheme that organizes all the species on Earth.

- The hierarchy begins with species, then moves up through genera, families, orders, classes, phyla, and kingdoms.

- Biologists now also use a level of classification above kingdoms, known as domains.

Classification versus Evolutionary Relationships

- Some systematists want to name only "real groups" and do away with other familiar names and groupings. Such a move remains controversial.

Branches on the Tree of Life: Interpreting Relatedness

- Studies using DNA have allowed scientists to make great strides in deciphering the tree of life, which has provided some interesting surprises.

- Scientists have greatly improved our understanding of the primate family tree. There is growing agreement that chimpanzees are humans' closest relatives.

- Evolutionary trees can best be thought of as working hypotheses or works in progress.

Highlight: Everything in Its Proper Place: The Contribution of Systematics

- Systematics puts organisms in their proper place in the tree of life, answering what are sometimes otherwise unanswerable biological questions and providing a starting point for asking other biological questions.

KEY TERMS

cladistics p. 23	**Linnaean hierarchy p .26**
convergent feature p. 24	**most recent common**
domain p. 27	**ancestor p. 21**
evolutionary taxonomy p. 22	**real group p. 28**
evolutionary tree p. 20	**shared ancestral feature p. 24**
kingdom p. 26	**shared derived feature p. 24**
lineage p. 20	**systematist p. 20**

CHAPTER REVIEW

Self-Quiz

1. In Figure 2.2, which evolutionary trees depict A and B as most closely related?
   a. a
   b. b
   c. a and b
   d. none of the above

2. The most powerful new feature being studied by systematists today is
   a. behavior.
   b. the cell.
   c. DNA.
   d. organs.

3. Evolutionary taxonomy
   a. is the newest school of systematics.
   b. is more rigorous than cladistics.
   c. requires the study of DNA.
   d. uses no specific formula or rationale to define its methodology.

4. The most useful features for cladistic analysis are
   a. convergent features.
   b. shared ancestral features.
   c. shared derived features.
   d. shared features of any kind.

5. In Figure 2.10, the closest relative of humans is the
   a. chimpanzee.
   b. gorilla.
   c. orangutan.
   d. lemur.

6. Dinosaurs and crocodilians most likely exhibit similar parental behaviors because
   a. they are both scaly.
   b. they both lay eggs.
   c. they are both closely related to birds.
   d. they share a common ancestor that exhibited parental behaviors.

Review Questions

1. How do systematists identify an unknown organism and place it on the tree of life?

2. What is a "real group," and why do some systematists think real groups are important?

3. How is an evolutionary tree like a hypothesis?

2 𝕿𝖍𝖊 𝕯𝖆𝖎𝖑𝖞 𝕲𝖑𝖔𝖇𝖊

New Species of Deer Discovered in Da Nang, Vietnam

DA NANG, VIETNAM—A team of scientists from the University of Eastern Texas emerged from the Annamite Mountains yesterday, saying they had discovered the world's newest species of large mammal.

"It is an incredible thrill to see a huge beast that you had no idea even existed," said Dr. Sean Vanderveld, mammalogist at the University of Eastern Texas. "She was clearly a new kind of deer. She stood just 10 feet away and munched grass while we photographed her." The researchers then shot the animal, took blood samples, and carried the en-

tire carcass back to the city of Da Nang. They plan to do systematic studies by examining DNA from the animal's blood as well as by studying the carcass itself.

The region in which the animal was found, known as the Annamite Mountains, is remote and rugged, forming a forbidding natural border between Vietnam and Laos. Dr. Vanderveld and colleagues note that while new to Western science, the animal is well known to locals, who call it a honinh. The researchers said they went in search of this apparently rare mammal

after seeing its antlers displayed in a number of homes in mountain villages.

Amid the excitement, Dr. Morris Berger, a mammalogist at the University of Los Angeles, said that the new deer, whose photographs he has now seen, is no different from a Vietnamese subspecies of deer described in 1945.

"People shouldn't be going and getting all excited just yet," said Dr. Berger, who has already requested a chance to examine the blood samples and the carcass.

Evaluating "The News"

1. How can scientists determine if the newly discovered deer is really a different species from the one known since 1945?

2. What does systematics have to offer society—and the researchers studying the new deer—other than the names of organisms?

3. There is a growing controversy among scientists as to whether newly discovered and apparently rare animals should be killed for study. Some say that without the organism in hand, there is no way to definitively identify it. Others ask, What good is a name for an

animal if you kill the last one, or perhaps the last male or female? Should these researchers have photographed their discovery and then set the deer free?

3 Major Groups of Living Organisms

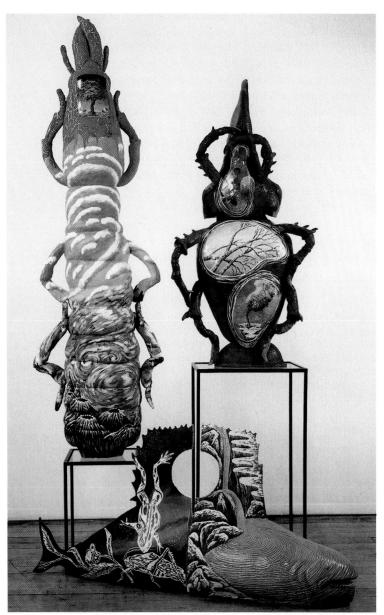

Michael Lucero, *Termite (Soldier), Hercules Beetle, Unidentified Fish*, 1986.

Here, There, and Everywhere

In 1994 scientists drilling deep into the earth made a shocking finding. Two miles beneath the surface, in what was thought to be nothing but barren rock, the researchers discovered life—abundant life. What they found was a world of microscopic organisms known as bacteria that had been isolated from the rest of the world's organisms for millions of years. Among the species discovered was a bacterium for which scientists immediately proposed the name *Bacillus infernus*, which means literally "bacterium from the inferno" or "bacterium from hell."

Such dramatic finds are becoming more common as biologists probe the last dark corners where life is hiding, as they peer not only into unexplored habitats, such as the deep sea, but into internal worlds, such as a giraffe's gut. Everywhere scientists are finding new organisms of all sorts. As the world's species become better known, the lesson biologists continue to learn is that life is hardy and resourceful, thriving almost everywhere, even in the most unlikely places.

■ MAIN MESSAGE

The world is home to a huge diversity of organisms that exhibit an astounding variety of structures, lifestyles, and behaviors.

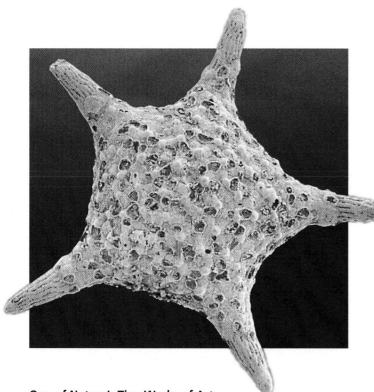

One of Nature's Tiny Works of Art
A microscopic organism called a foraminiferan.

In the heat of the Sonoran Desert, cacti point their scorched green limbs to the sky, sustained by the gallons of water that are stored safely inside them. High in the Himalayas, a woolly flying squirrel the size of a woodchuck glides through the skies. Wildflowers in the snow-covered Swiss Alps turn their blossoms to follow the movement of the sun and capture its precious warming rays. In the steaming waters of Yellowstone National Park's most famous geyser, Old Faithful, heat-loving bacteria comfortably persist, well suited to such extremes of temperature. And everywhere insects, the world's most abundant animals, fly and crawl, making meals out of everything from the nectar of flowers to poison-filled plant leaves to human blood to cow dung.

Life abounds in countless forms on Earth, and biologists are trying to study it all. In this chapter we take a look at the diversity of life, the riot of living organisms that inhabit our planet.

■ KEY CONCEPTS

1. All of life can be assembled into an evolutionary tree. Life can also be divided into three domains: Bacteria, Archaea, and Eucarya. In the Linnaean hierarchy, those domains can be divided into kingdoms. We recognize six kingdoms: Bacteria, Archaea, Protista, Plantae, Fungi, and Animalia.

2. The prokaryotes, Bacteria and Archaea, are microscopic, single-celled organisms. Prokaryotes are more diverse in the ways they obtain their nutrition than are the members of the Eucarya.

3. Eukaryotic cells contain tiny structures with specialized functions known as organelles. Prokaryotes do not have organelles in their cells.

4. The Protista includes the most ancient groups within the Eucarya. Protists represent early stages in the evolution of the eukaryotic cell and in the evolution of multicellularity.

5. The Plantae pioneered life on land. Plants are multicellular, and they photosynthesize, using the energy of sunlight and carbon dioxide to make sugars. As producers, plants form the base of all food webs on land.

6. The Fungi include mushroom-producing species as well as molds and yeasts. Fungi are among the world's key decomposers, many using dead and dying organisms as their food.

7. The Animalia include a wide range of multicellular organisms. Animals have evolved specialized tissues, organs and organ systems, body plans, and behaviors.

As described in Chapter 1, all living organisms share a basic set of characteristics. Biologists believe that life shares this set of common properties because all living organisms descended from a single common ancestor. But life first appeared on this planet more than 3.5 billion years ago. Since that time, life has evolved in many directions. From a living world of nothing more than individual cells, a planet full of wildly different organisms has evolved. We are part of a great diversity of life that is still far from being completely known, counted, or named (see the box on page 40). Most biologists estimate that there are between 3 million and 30 million species on Earth.

In this chapter we are not aiming for a comprehensive examination of the world's many species, which would be an impossible goal. Instead, we attempt to familiarize you with the diversity of life by introducing you to the major groups of organisms.

The Major Groups in Context

The major groups of organisms are the Bacteria (which include familiar disease-causing bacteria), the Archaea (bacteria-like organisms that are best known for living in extreme environments), the Protista (a diverse group that includes amoebas and algae), the Plantae (plants), the Fungi (mushrooms, molds, and yeasts), and the Animalia (animals).

In order to provide a framework for thinking about the world's many organisms, we have placed the major groups in context in Figure 3.1 in three different ways: (1) on the evolutionary tree of life, (2) in the Linnaean hierarchy, and (3) in the domain system. We saw how the evolutionary tree of life is constructed, and were introduced to the three domains and the six kingdoms of the Linnaean hierarchy, in Chapter 2.

As we also saw in Chapter 2, some systematists have argued against the use of the Linnaean hierarchy for organizing and naming living organisms, in part because it contains many groups that are not "real groups." Despite this controversy over how to organize and name organisms, all three systems shown in Figure 3.1 are widely used, so it is important to learn all three. To help you keep track of where you are in each of these systems, at the beginning of each section of text that describes a major group we provide an illustration that highlights that group's position on the tree of life and in the kingdom and domain systems.

In this chapter, we describe the key features that characterize each major group. In particular, we stress the features evolved by each group that allow its members to live and reproduce successfully, features known as **evolutionary innovations**. In addition, we describe some of the more important and interesting members of each group.

A photo gallery accompanies the description of each major group. It provides a broad overview, an evolu-

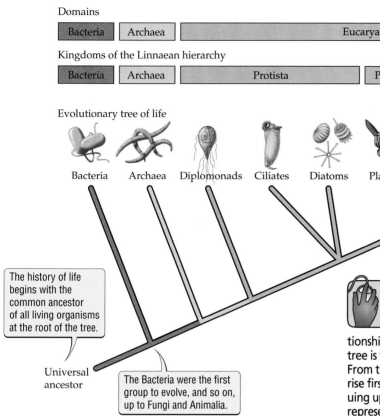

Domains

| Bacteria | Archaea | Eucarya |

Kingdoms of the Linnaean hierarchy

| Bacteria | Archaea | Protista | Plantae | Fungi | Animalia |

Evolutionary tree of life

Bacteria   Archaea   Diplomonads   Ciliates   Diatoms   Plantae   Fungi   Animalia

The history of life begins with the common ancestor of all living organisms at the root of the tree.

Universal ancestor

The Bacteria were the first group to evolve, and so on, up to Fungi and Animalia.

3.1

Figure 3.1 The Major Groups on the Tree of Life, in the Linnaean Hierarchy, and in the Domain System
Depicted on this tree are the evolutionary relationships of the major groups of organisms. At the root of the tree is the universal ancestor, the ancestor of all living things. From there tracing up the tree, the universal ancestor gives rise first to Bacteria, the oldest of the groups depicted. Continuing up the tree, Archaea arose next. Next shown are three representatives of the many members of the Protista. First the diplomonads branched off, then ciliates and diatoms. It remains unclear whether ciliates or diatoms branched off first, so they are depicted as branching off simultaneously. Next the Plantae branched off, and lastly the Fungi and Animalia split apart. Each of the groups branching off the tree can be thought of as a cluster of close relatives—a lineage, just like a lineage in a human family (see Chapter 2 for more on reading evolutionary trees). Above the evolutionary tree, the kingdom to which each group belongs is shown. Each of the kingdoms comprises a single group on the evolutionary tree, except the Protista, which includes three of the groups shown. The domain to which each group belongs is also shown.

tionary tree of the subgroups within that major group, and a description of some of the prominent subgroups within that group. This gallery provides a framework for the discussion of the group's evolutionary innovations. It should also serve as a reference as you read later chapters and want to take a second look at groups you are learning about.

3.2 *The Bacteria and the Archaea: Tiny, Successful, and Abundant*

| BACTERIA | ARCHAEA | EUCARYA |
| BACTERIA | ARCHAEA | PROTISTA | PLANTAE | FUNGI | ANIMALIA |

When life arose on Earth more than 3.5 billion years ago, the first living organisms were tiny, simple, single cells. The first modern lineage to arise—that is, the first to branch off the tree of life—was the **Bacteria**. The Bacteria are probably most familiar to you as single-celled organisms that cause diseases in humans. Then the **Archaea** arose. These single-celled bacteria-like organisms are best known as so-called extremophiles, meaning that many of them can live in

extreme environments such as boiling hot geysers, highly acidic waters, or the freezing cold waters off Antarctica. When the Archaea arose, they split from the lineage **Eucarya**, which would give rise to all the rest of the world's organisms (Figure 3.2).

The Bacteria and Archaea are each recognized as both a kingdom and a domain. The Eucarya, in contrast, is a domain that encompasses four kingdoms.

The Bacteria and the Archaea are distinguished from the Eucarya—in other words, from all the rest of living organisms—by the structure of their cells. We will discuss these structural differences in greater detail below and in later chapters, but for now, suffice it to say that the cells of the Eucarya contain compartments, each of which has a specialized function. One compartment,

■ THE SCIENTIFIC PROCESS

Biological Exploration: When Getting There Is Half the Fun

Biologists seeking to catalog the great diversity of life on Earth have their work cut out for them. Many species have been able to elude scientists for years because they live in inaccessible habitats.

To catalog the entire biosphere, biologists are discovering that they must become daring explorers, venturing into extreme habitats, a feat that often requires not only a sense of adventure but more than the usual laboratory equipment.

Among the most challenging of biological explorations are the expeditions that researchers are taking into the forest canopy, an aerial world of leaves and branches. Biologists estimate that this complex habitat has the same total surface area as the entire Earth. Those who have managed to get there have found abundant life, including new species from all the major groups. In order to reach the canopy, some biologists use dirigibles to float through the air and work on inflated rafts that are gently lowered onto the treetops. Others use hooks attached to their ankles to work their way up a tree like a clawed animal. Still others use huge cranes that move them around their treetop study sites as if they were doing construction work.

The ocean bottom, most of which is still unexplored, is another entirely new world for biologists. To venture into these crushing depths beneath miles of water, biologists descend in submersibles. They also dredge what they can from boats. One researcher described pulling up net after net full of creatures, with each dredge pulling up an entirely different group of organisms—lampshells, peanut worms, moss animals, ribbon worms, beard worms, and many more creatures without any such common names—many of which he had never seen before.

Peanut worm

Yet despite the increased effort to reach these remote places, many of the amazing creatures that call them home—such as a squid the length of a city bus—have still never been seen alive in their natural habitats. This beast, known as the giant squid, or *Architeuthis*, is a monstrous representative of the molluscs. It has become known to scientists only as a titan that occasionally washes ashore, dead. Still elusive in its natural surroundings, the giant squid is a symbol of the plentiful mysteries of the biosphere that still remain.

A

B

Some scientists use dirigibles (A) to lower themselves onto the forest canopy.
(B) Dr. Nalini Nadkarni of The Evergreen State College uses ropes and pulleys to hoist herself into the treetops.

Figure 3.2 The Prokaryotes: Bacteria and Archaea

The microscopic, single-celled Bacteria and Archaea are the most ancient forms of life. The Bacteria branched off first, then the Archaea, from the lineage that would lead to the rest of the living organisms, or the Eucarya.

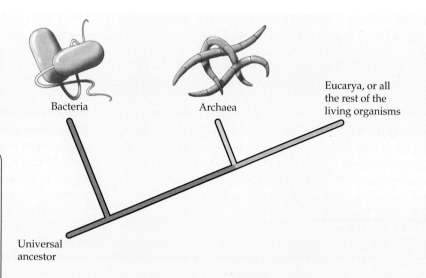

Bacteria

Archaea

Eucarya, or all the rest of the living organisms

Universal ancestor

Number of species discovered to date: ~4,800

Functions within ecosystems: Producers, consumers, decomposers

Economic uses: Many, including producing antibiotics, cleaning up oil spills, treating sewage

Funky factoid: The number of bacteria in your digestive tract outnumber all the humans that have lived on Earth since the beginning of time.

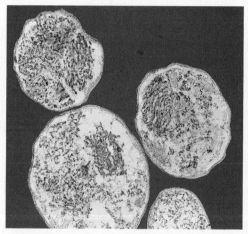

The bacterium *Chlamydia trachomatis* causes the most common sexually transmitted disease, chlamydia.

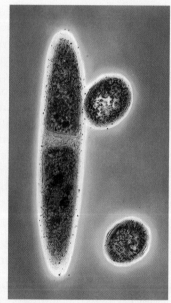

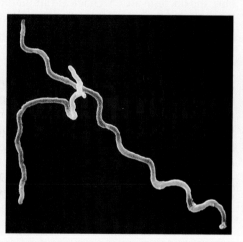

These bacteria, *Borrelia burgdorferi*, known as spirochetes because of their spiral-shaped cells, cause Lyme disease, which is transmitted to humans through a tick bite.

These archaeans, known as *Methanospirillum hungatii*, are shown in cross section (the two circular shapes) and as an elongated cell that is about to fission into two cells.

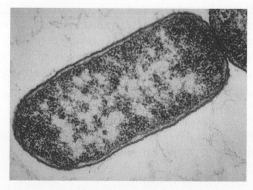

This bacterium, known as *Escherichia coli*, is usually a harmless inhabitant of the human gut. However, toxic strains can contaminate and multiply on foods, such as raw hamburger, and cause illness or death in humans who eat them.

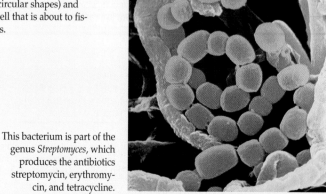

This bacterium is part of the genus *Streptomyces*, which produces the antibiotics streptomycin, erythromycin, and tetracycline.

called the nucleus (plural nuclei), holds the genetic material, or DNA. In fact, the name of the Eucarya, or **eukaryotes**, comes from the Greek *eu*, meaning "true," and *karyote*, meaning "kernel," referring to the nucleus. In contrast, the Bacteria and the Archaea do not have compartments in their cells. As a result, the Bacteria and Archaea are often referred to collectively as the **prokaryotes** (*pro*, "before"; *karyote*, "kernel") because they evolved before the evolution of the nucleus and other compartments. (Note that the two terms "eukaryote" and "prokaryote" refer to the structure of an organism's cells, and are not part of the domain, kingdom, or evolutionary tree system.)

While the members of both groups are small and single-celled, the Bacteria and Archaea show some key differences. For example, archaeans are distinguished from bacteria by having larger components of DNA and by structural differences in their cells. However, as prokaryotes, the Bacteria and Archaea are similar in many ways. For that reason, we begin our introduction to the major groups of life by describing the world of the prokaryotes, the Bacteria and the Archaea.

The Bacteria and Archaea are arguably both the simplest living organisms and the most successful at colonizing Earth. The prokaryotic cells that make up these two groups can be shaped like spheres, rods, or corkscrews, but they all share a basic structural plan (Figure 3.3). The picture of efficiency, these stripped-down organisms are nearly always single-celled and microscopic. They typically have much less DNA than the cells of eukaryotes have. Eukaryotic genomes are often full of what appears to be extra DNA that serves no function. In contrast, prokaryotic genomes contain only DNA that is actively in use for the survival and reproduction of the cell. Prokaryotic reproduction is similarly uncomplicated; bacteria typically reproduce by splitting in two.

Simplicity translated into success

While most people think of biodiversity in terms of butterflies, redwood trees, and other complex organisms built of many cells, the vast majority of life on Earth is single-celled and prokaryotic. Scientists estimate that the number of individual bacteria on Earth is about 5,000,000,000,000,000,000,000,000,000,000, or 5×10^{30}. That success is due, in part, to the speed with which they reproduce. Overnight, a single *Escherichia coli*, a bacterium that normally lives harmlessly in the human gut, can divide to produce a population of 10 million bacteria.

Prokaryotes are also the most widespread of organisms, able to live nearly anywhere. They can persist in places where most organisms would perish, such as the lightless ocean depths, the insides of boiling hot geysers, and miles below Earth's surface. Because of their small size, prokaryotes also live on and in other organisms. Scientists estimate that 1 square centimeter of healthy human skin is home to between 1000 and 10,000 bacteria.

In addition, while many prokaryotes are aerobes (*aer*, "air") and need oxygen to survive, many others are anaerobes (*an*, "without") and can survive without oxygen. This ability to exist in both oxygen-rich and oxygen-free environments also increases the number of habitats in which prokaryotes can persist. But the real key to the success of these groups is the great diversity of ways in which they obtain and use nutrients.

Prokaryotes exhibit unmatched diversity in their methods of obtaining nutrition

Every organism needs two things to grow and survive: a source of energy and a source of carbon. Carbon is the chemical building block used to make critical molecules for living, such as proteins and DNA. The Bacteria and Archaea are distinguished by having the most diverse methods of obtaining energy and carbon of any groups of organisms on Earth.

When humans and other animals eat, we consume other organisms, from which we get both energy, in the form of chemical bonds, and carbon, in the form of carbon-containing molecules. In fact, many organisms,

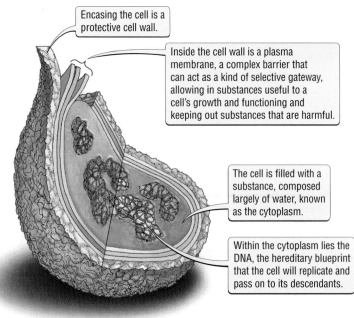

Encasing the cell is a protective cell wall.

Inside the cell wall is a plasma membrane, a complex barrier that can act as a kind of selective gateway, allowing in substances useful to a cell's growth and functioning and keeping out substances that are harmful.

The cell is filled with a substance, composed largely of water, known as the cytoplasm.

Within the cytoplasm lies the DNA, the hereditary blueprint that the cell will replicate and pass on to its descendants.

Figure 3.3 The Basic Structure of the Prokaryotic Cell Prokaryotic cells tend to be small, about 10 times smaller than the cells of organisms in the Eucarya, and have much less DNA.

Figure 3.4 Pond Scum: Bacteria that Photosynthesize
These photosynthetic cyanobacteria can be found growing as slimy mats on freshwater ponds.

ents by using light energy and carbon dioxide in the process of photosynthesis to produce sugars. (Photosynthesis will be described in detail in Chapter 8.) Prokaryotes can carry out photosynthesis as well. Cyanobacteria, more commonly known as pond scum, are a familiar example of prokaryotes that live on sunlight and carbon dioxide (Figure 3.4).

But while the Eucarya are restricted to either eating or carrying out photosynthesis, prokaryotes are also able to survive in two other ways. Some prokaryotes use light as an energy source, the way plants and cyanobacteria do, but do not get carbon from carbon dioxide. Instead, these prokaryotes get their carbon from organic compounds. Finally, there are also prokaryotes that use carbon dioxide as a carbon source, as do plants and cyanobacteria, but get their energy from such unlikely materials as iron and ammonia (Figure 3.5).

Prokaryotes can thrive in extreme environments

Prokaryotes are well known for their ability to live in nearly any kind of environment. While some bacteria thrive in unusual environments, the Archaea is the group best known for the extreme lifestyles of some of its members. Some are extreme thermophiles (*thermo,*

including all animals, all fungi, and some protists, get their energy and carbon by consuming other organisms. Prokaryotes can do the same. A familiar example is *Clostridium botulinum*, a bacterium that can cause food poisoning. It lives in and consumes food that humans have stored and produces a toxin that can make humans that eat the bacteria-laden food sick. But while the rest of the world's consumers are restricted to consuming other organisms, or parts or products of other organisms, some prokaryotes can live by consuming carbon-containing compounds that are not part of other organisms, such as pesticides.

While animals, fungi, and many protists must eat to live, plants produce their own nutrition. Plants get their energy from the light of the sun and their carbon from carbon dioxide, the gas that humans and other animals exhale. Plants and some protists make their own nutri-

Figure 3.5 An Unusual Eater
The archaean *Sulfolobus* needs neither other organisms nor light to make its living. It gets its carbon from carbon dioxide, and it gets its energy by chemically processing inorganic materials such as iron. This extremophile is living in a volcanic vent on the island of Kyushu in Japan.

Figure 3.6 Better than a Bag of Chips
For those who love salt, such as archaeans that are extreme halophiles, nothing beats a salt farm. Here in Thailand, seawater is being evaporated, making a more and more concentrated salt solution and creating an environment that only an archaean could love.

"heat"; *phile*, "lover") that live in geysers and hot springs. Others are salt lovers, or extreme halophiles (*halo*, "salt"). These prokaryotes grow where nothing else can live—for example, in the Dead Sea and on fish and meat that have been heavily salted to keep most bacteria away (Figure 3.6).

Not all archaeans, however, are so remote from our daily experience. Members of one group, the methanogens (*methano*, "methane"; *gen*, "producer"), inhabit animal guts and produce the methane gas in such things as human flatulence (intestinal gas) and cow burps.

Prokaryotes are important in the biosphere and for human society

Because of their ability to use such a variety of substances as food and energy sources and to live under such a variety of conditions, prokaryotes play numerous and important roles in ecosystems and in human society. For example, plants require nitrogen, but they cannot take up and use nitrogen gas from the air. For this reason, they depend on bacteria that can convert nitrogen gas to a chemical form called nitrate, which the plants can use. Without these bacteria there would be no plant life, and without plant life there would be no life on land.

Like plants, some bacteria, such as cyanobacteria, can photosynthesize. This ability makes them producers, the organisms at the base of food webs. Other bacteria are important decomposers, breaking down dead organisms by using them as food. Oil-eating bacteria can be used to clean up ocean oil spills. Bacteria that can live on sewage are used to help decompose human waste so that it can be safely and usefully returned to the environment. Bacteria also live harmlessly in animal guts (including our own), helping animals digest their food.

Of course, not all prokaryotes are helpful. Many bacteria cause diseases. Some are the stuff of nightmares, such as the flesh-eating bacteria that can destroy human flesh at frightening rates. With their ability to use almost anything as food, bacteria can also attack crops, stored foods, domesticated animals, and nearly every other living organism.

Bacteria and Archaea exhibit key differences

While they are similar in many ways, the Bacteria and the Archaea are distinct lineages. In recent years, biologists have learned more about archaeans, the more newly recognized of the two groups, and described their differences in more detail. One key feature of archaeans is their DNA. Biologists have found that much of archaean DNA is unlike the DNA found in any other organisms.

In addition, there are structural differences between the cells of the two groups. All cells, as we will see in Chapter 6, are surrounded by a plasma membrane, which is made up of chemical building blocks called lipids. All prokaryotic cells have a cell wall surrounding their plasma membrane, but Bacteria and Archaea use different molecules to make up those walls. In addition, the two groups employ different types of lipids in their plasma membranes.

■ While distinct, the two prokaryotic groups , Bacteria and Archaea, are similar in many ways. They both are considered a domain and a kingdom. They both consist of simple, microscopic, single-celled organisms. Of all organisms on Earth, their members are the most numerous (in terms of numbers of individuals), the most widespread, and the most diverse in their methods of obtaining nutrition. Prokaryotes can act as decomposers, producers, and consumers.

The Evolution of More Complex Cells: The Eukaryotes Arise from Prokaryotes

The prokaryotes have played a critical role in the evolution of the Eucarya, the domain that includes all the other living organisms (the Protista, Plantae, Fungi, and Animalia). The prokaryotes (the Bacteria and the Archaea), as we have seen, have cells that do not contain internal compartments. The cells of the Eucarya do contain such compartments, called organelles. Each organelle is a tiny structure that performs a specialized function. For example, organelles known as chloroplasts are the sites where photosynthesis takes place in plant cells. The nucleus is an organelle that contains a eukaryotic cell's DNA.

While these organelles function as specialized structures within eukaryotic cells today, they appear to have descended from what were once free-living prokaryotes. Biologists have hypothesized that sometime deep in the evolutionary past, ancient prokaryotic cells engulfed other prokaryotic cells. Plant cells, for example, are thought to have acquired their ability to photosynthesize by engulfing cyanobacterial cells, which have evolved to become chloroplasts. The captured cells eventually evolved to function as parts of the cells that had engulfed them; that is, they became organelles (see Figure 6.13). Over time, the organelles became critical to the survival of the cell that housed them, and became unable to function on their own outside that cell. Thus a combination of prokaryotes appears to have created the eukaryotes, the group from which multicellular, complex organisms evolved.

■ The cells of the Eucarya contain organelles. Eukaryotic cells arose when some prokaryotic cells engulfed others. The engulfed prokaryotic cells eventually evolved into organelles.

The Protista: A Window into the Early Evolution of the Eucarya

The **Protista** is a kingdom within the domain Eucarya. Its diverse members are the most ancient of the eukaryotic groups, making the Protista the oldest kingdom within the Eucarya. The protists consist largely of single-celled eukaryotes, but include some multicellular eukaryotes as well. They include some familiar groups, such as single-celled amoebas and multicellular kelps, as well as many groups with which most people are unfamiliar. One of the few generalizations that can be made about this hard-to-define group is that its members are very diverse in size, shape, and lifestyle.

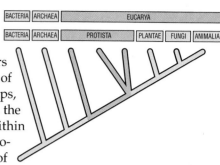

One of the reasons for their great diversity is the fact that the protists are not all members of a single lineage (see Figure 3.1), as plants, animals, and fungi are. Instead, protists are composed of different groups of organisms that branched off at separate times from the rest of the tree of life. Therefore, they are not a complete set of the descendants of a single common ancestor. In other words, they are not a "real group," as described in Chapter 2.

In fact, there is much that remains unknown about the evolutionary relationships of the protists. Figure 3.7 presents a hypothetical evolutionary tree for some of the major groups of protists. The diplomonads are shown branching off first. Then three major groups (one including dinoflagellates, apicomplexans, and ciliates; another including diatoms and water molds; and the last comprising the green algae) are shown splitting off at the same time. That is because systematists still do not know which of these groups branched off first. Other groups, such as the slime molds, are not shown on the tree at all because their placement is even more poorly understood.

Protists show great variety in their lifestyles. There are plantlike protists (such as green algae) that can photosynthesize. There are also animal-like protists (such as ciliates) that move and hunt for food. Still others (such as the slime molds) are more like fungi. Most protists are single-celled (as are paramecia), but there are also many multicellular protists (such as the multicellular algae).

Protists represent early stages in the evolution of the eukaryotic cell

As already described, the eukaryotic cell evolved by a process in which some prokaryotic cells engulfed oth-

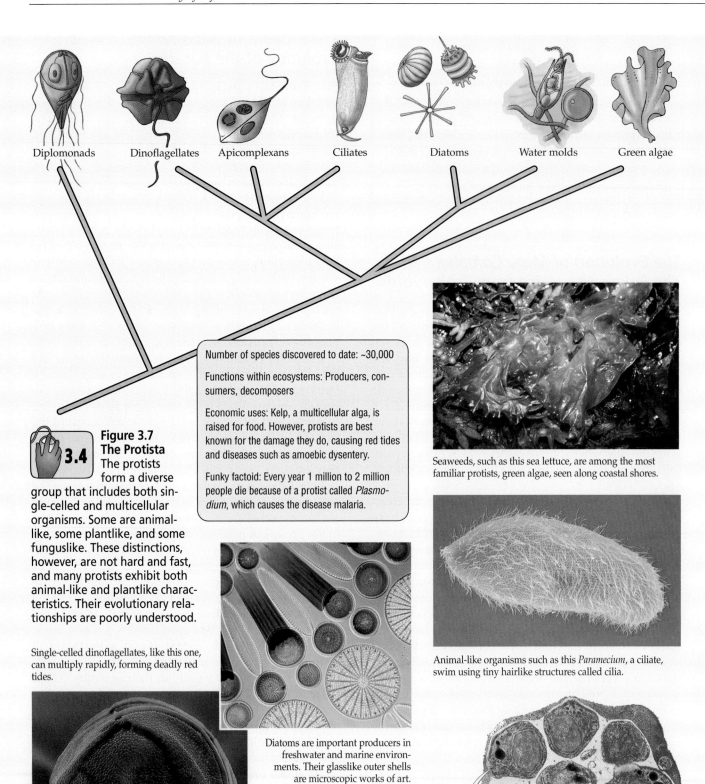

Diplomonads   Dinoflagellates   Apicomplexans   Ciliates   Diatoms   Water molds   Green algae

3.4

**Figure 3.7
The Protista**
The protists form a diverse group that includes both single-celled and multicellular organisms. Some are animal-like, some plantlike, and some funguslike. These distinctions, however, are not hard and fast, and many protists exhibit both animal-like and plantlike characteristics. Their evolutionary relationships are poorly understood.

Number of species discovered to date: ~30,000

Functions within ecosystems: Producers, consumers, decomposers

Economic uses: Kelp, a multicellular alga, is raised for food. However, protists are best known for the damage they do, causing red tides and diseases such as amoebic dysentery.

Funky factoid: Every year 1 million to 2 million people die because of a protist called *Plasmodium*, which causes the disease malaria.

Seaweeds, such as this sea lettuce, are among the most familiar protists, green algae, seen along coastal shores.

Single-celled dinoflagellates, like this one, can multiply rapidly, forming deadly red tides.

Diatoms are important producers in freshwater and marine environments. Their glasslike outer shells are microscopic works of art.

Animal-like organisms such as this *Paramecium*, a ciliate, swim using tiny hairlike structures called cilia.

This is one *Plasmodium* individual.

A gathering of apicomplexans, a group of protist parasites. Shown here is *Plasmodium*, which causes malaria.

ers. Over time, these captured cells evolved to become less like free-living life forms and more like tiny organs within the cells that had captured them. For example, in addition to nuclei, eukaryotic cells typically contain mitochondria (singular mitochondrion), organelles that produce energy. Eukaryotic cells and their organelles now depend entirely on each other (just as a human body and its organs do), and cannot survive apart from each other.

Scientists now believe that the engulfing of one cell by another was not a one-time event. Increasing evidence shows that prokaryotes were engulfed by other prokaryotes many times in the history of life. Among the protists, different species illustrate the variety of experimentation that has gone on over time.

Within the protist group known as Diplomonads, for example, is the species *Giardia lamblia*. This protist lives in streams and other water sources. It can cause a painful ailment of the digestive tract when it is consumed by humans, causing diarrhea and flatulence that has a rotten-egg odor. Most eukaryotic cells contain a nucleus and mitochondria, and if they photosynthesize, chloroplasts. *Giardia*, however, appears to be a curious experiment in putting together a single-celled organism: it has two nuclei, no chloroplasts, and has lost the mitochondria it once had.

Giardia

Protists provide insight into the early evolution of multicellularity

Some protists are of interest to biologists because they have evolved from being single-celled creatures, like bacteria, to forming multicellular groupings that function to greater or lesser degrees like more complex multicellular individuals. Among the more interesting of these experiments in the evolution of multicellularity are the slime molds, protists that were once thought to be fungi (hence their name, since molds belong to another major group, the Fungi). Commonly found on rotting vegetation, slime molds make their living by digesting bacteria. But these curious organisms can live their lives in two phases: as independent, single-celled creatures and as members of a multicellular body. Like other protists that do not live either strictly as single-celled organisms or as typical multicellular organisms, slime molds are studied by biologists who hope to gain insight into the transition from single-celled to multicellular living.

Protists had sex first

Like bacteria, many protists reproduce simply by splitting in two—a form of asexual (nonsexual) reproduction. Under some circumstances, however, bacteria can receive DNA (genetic material) from other bacteria. Some scientists consider such combining of genetic material to be a form of prokaryotic sexual reproduction. But sex is more often defined as a process in which two individual organisms produce specialized reproductive cells known as gametes (for example, human eggs and sperm). These gametes fuse together, combining the DNA contributions from both parents into one individual offspring. One of the most noteworthy achievements of the protists was the invention of this kind of eukaryotic sex. Although many protists reproduce asexually, it was among the protists that sexual reproduction, using gametes from different individuals that fuse to form a new individual, first appeared.

Protists are best known for their disease-causing abilities

Although most protists are harmless to other organisms, many of the best-known protists cause diseases. Dinoflagellates are microscopic plantlike protists that live in the ocean and sometimes rapidly increase in numbers, a phenomenon known as blooms. Occasional blooms of toxic dinoflagellates cause dangerous "red tides." During red tides, shellfish eat toxic dinoflagellates, and humans can be poisoned by eating the shellfish. An animal-like protist, *Plasmodium*, is the organism that causes malaria. Finally, the protists left their mark on human history forever when one, a water mold, attacked potato crops in Ireland in the 1800s, causing the disease known as potato blight. The resulting widespread loss of potato crops caused a devastating famine.

> ■ The Protista is recognized as a kingdom in the six kingdom system. It is the oldest of the major groups in the domain Eucarya. The protists are a diverse group of single-celled and multicellular organisms, some animal-like, some plantlike, and some funguslike. This group includes species that represent early stages in the evolution of the eukaryotic cell and of multicellularity.

The Plantae: Pioneers of Life on Land

To some people, there is nothing more mundane than a plant, and greenery represents little more than garnishes at meals and decorative houseplants. But plants are among evolution's great pioneers. Life on Earth began in the seas, and there it stayed for 3 billion years. It was only when the **Plantae**, or plants, evolved that life colonized the land. These first colonists turned barren ground into a green paradise in which a whole new world of land-dwelling organisms evolved.

The Plantae is a kingdom of multicellular organisms within the domain Eucarya. Figure 3.8 shows the basic structure of a plant. Life on land required special structures for obtaining and conserving water. One of the features of plants that allows them to live on land is a root system, a collection of fingerlike growths that absorbs water and nutrients from the soil. Another such feature is a waxy covering over the stem and leaves, known as the cuticle. The cuticle prevents the plant's tissues from drying out, even when they are exposed to sun and air. Plant cells also have a rigid cell wall that is composed of an organic compound known as cellulose. These cell walls provide support for a stem growing out of the ground.

The key feature of plants, however, is their ability to use light (energy from the sun) and carbon dioxide (a gas in the air) to produce food. As described in Chapter 1, plants are producers, and as such, these organisms form the base of essentially all terrestrial (land-based) food webs. In addition, one of the useful by-products of photosynthesis is oxygen, a gas that is critical for the survival of many other organisms. Most of a plant's photosynthesizing takes place in its leaves, which typically grow in ways that maximize their ability to capture sunlight.

The diversity of plants today ranges from the most ancient lineages, the mosses and their close relatives, to ferns, which evolved next, to gymnosperms and angiosperms (Figure 3.9). In the sections that follow we examine three major innovations in the evolution of plants: vascular systems, seeds, and flowers. These innovations were critical to what has been a highly successful colonization of land by plants.

Vascular systems allowed plants to grow to great heights

Early in their evolution, plants grew close to the ground. Mosses and their close relatives, the most ancient of plant lineages, represent those early days in the history of plants. These plants rely on the absorption of water directly by each of their cells. Thus, the innermost cells

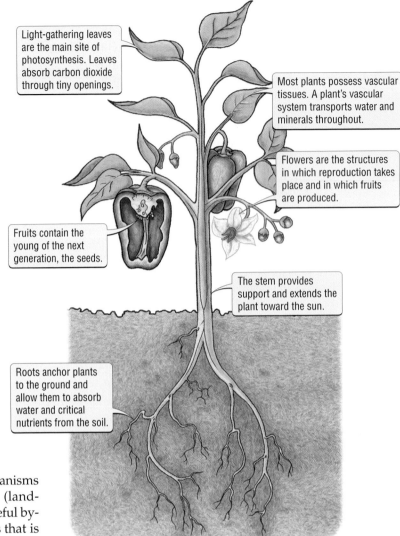

Light-gathering leaves are the main site of photosynthesis. Leaves absorb carbon dioxide through tiny openings.

Most plants possess vascular tissues. A plant's vascular system transports water and minerals throughout.

Flowers are the structures in which reproduction takes place and in which fruits are produced.

Fruits contain the young of the next generation, the seeds.

The stem provides support and extends the plant toward the sun.

Roots anchor plants to the ground and allow them to absorb water and critical nutrients from the soil.

Figure 3.8 The Basic Form of a Plant

of their bodies receive water only after it has managed to pass through every cell between them and the outermost layer of the plant. Because such movement of water from cell to cell, like the movement of water through a kitchen sponge, is relatively inefficient, these plants cannot grow to great heights or sizes.

Ferns and their close relatives, which arose later, can grow taller because they have evolved vascular systems. Vascular systems are networks of specialized hollow cells that extend from the roots throughout the body of a plant. Such collections of specialized cells, as you will recall from Chapter 1, are known as tissues. The vascular tissues of a plant can efficiently transport fluids and nutrients, much as the human circulatory system of veins and arteries transports blood. In addition, this network of water-filled vessels can make a plant firmer. By providing both structural support and efficient circulation

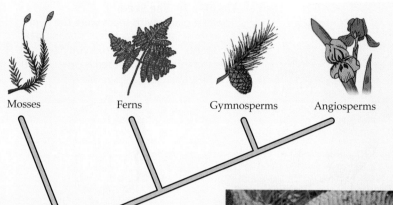

Mosses   Ferns   Gymnosperms   Angiosperms

Number of species discovered to date: ~250,000

Functions within ecosystems: Producers

Economic uses: Flowering plants provide all our crops: corn, tomatoes, rice, and so on. Fir trees and other conifers provide wood and paper. Plants also produce important chemicals, such as morphine, caffeine, and menthol.

Funky factoid: Of the 250,000 species of plants, at least 30,000 have edible parts. In spite of this abundance of potential foods, just three species—corn, wheat, and rice—provide most of the food the world's human populations eat.

Figure 3.9 The Plantae
Plants are multicellular organisms that make their living by photosynthesis. They are a diverse group that pioneered life on land.

Ferns and their close relatives evolved vascular systems that allowed them to grow to greater heights. This Ama'uma'u fern grows only in Hawaii.

The most familiar gymnosperms are the conifers. Many conifers, such as this giant sequoia, are important wood and paper producers.

Mosses and their close relatives, the most ancient group of plants, do not have vascular systems and do not grow more than a few inches high.

The orchids, the most species-rich family of angiosperms in the world, also produce some of the world's most beautiful flowers.

On Mt. Sago in Sumatra, the angiosperm *Rafflesia arnoldii* produces the world's largest blossoms, measuring as much as 1 meter across.

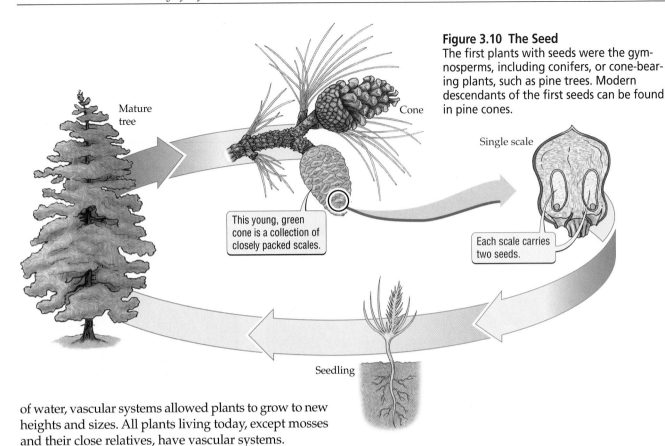

Mature tree

Cone

Figure 3.10 The Seed
The first plants with seeds were the gymnosperms, including conifers, or cone-bearing plants, such as pine trees. Modern descendants of the first seeds can be found in pine cones.

Single scale

This young, green cone is a collection of closely packed scales.

Each scale carries two seeds.

Seedling

of water, vascular systems allowed plants to grow to new heights and sizes. All plants living today, except mosses and their close relatives, have vascular systems.

Gymnosperms evolved seeds to protect their young

Following mosses and ferns, the next group of plants to evolve was the **gymnosperms**. This group includes pine trees and other cone-bearing plants, known as conifers, as well as cycads and ginkgos.

Gymnosperms were the first plants to evolve **seeds**, structures that provide young plants (embryos) with a protective covering and a stored supply of nutrients (Figure 3.10). These plants get their name (*gymno*, "naked"; *sperm*, "seed") from the fact that their seeds are less well protected than those of angiosperms, which we describe in the next section. Gymnosperms were the dominant plants 250 million years ago, and biologists believe that the evolution of seeds was an important part of their success. Seeds provide nutrients for plant embryos until they have grown and developed enough to produce their own food via photosynthesis. Seeds also afford protection from drying out or rotting and from attack by predators.

Angiosperms produced the world's first flowers

Although typically we think of flowers when we think of plants, flowering plants are a relatively recent development in the history of life, having evolved only about

200 million years ago. Today the flowering plants, known as **angiosperms**, are the most dominant and diverse group of plants on Earth, including organisms as different as orchids, grasses, corn plants, and apple and maple trees. As already mentioned, angiosperms produce seeds that have more protective tissues than those of gymnosperms (*angio* means "vessel," referring to the protective tissues that encase the plant's embryo).

Highly diverse in size and shape, angiosperms live in a wide range of habitats—from mountaintops to deserts to salty marshes and fresh water. Almost any plant we can think of that is not a moss, a fern, or a cone-producing tree is an angiosperm. The defining feature of angiosperms is the **flower**, a specialized structure for sexual reproduction in which the male and the female gametes meet (Figure 3.11).

Angiosperms display a wide variety of flowers, which aid the reproduction of the plants in many ways. Many flowers provide foods, such as the sugary liquid known as nectar, to attract animals, which visit the flowers and in the process transport pollen—containing male gametes—from one flower to another. The transported pollen can fertilize a flower's female gametes. Thus mobile animals can provide a means of sexual reproduction between immobile plants, even very distant ones. Flower evolution has reached its extreme in the elaborate blooms of orchids, many of which are designed to deceive animal pollinators. One orchid has

Figure 3.11 The Flower
A flower is really an attractive meeting place for male and female gametes, a very publicly displayed reproductive structure.

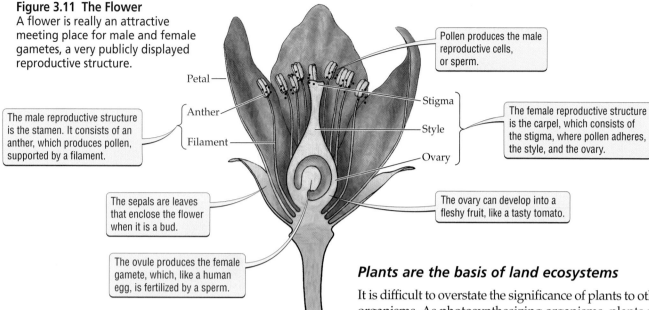

Pollen produces the male reproductive cells, or sperm.

The male reproductive structure is the stamen. It consists of an anther, which produces pollen, supported by a filament.

The female reproductive structure is the carpel, which consists of the stigma, where pollen adheres, the style, and the ovary.

Petal

Anther

Filament

Stigma

Style

Ovary

The sepals are leaves that enclose the flower when it is a bud.

The ovary can develop into a fleshy fruit, like a tasty tomato.

The ovule produces the female gamete, which, like a human egg, is fertilized by a sperm.

evolved flowers that look so much like female bees that male bees repeatedly attempt to mate with them, unintentionally pollinate them in the process.

Angiosperms have also evolved fleshy, tasty fruits that attract animals. While the seeds of angiosperms are developing, the surrounding ovary (see Figure 3.11) may develop into a ripening fruit. Animals eat the fruits and later excrete the seeds in heaps of feces. These nutrient-rich wastes provide a good place for the seeds to sprout and start life, often far away from their parent plant, where they will not compete with it for water, nutrients, or light. But immobile plants don't always need the help of animals to get their young off to a good start; plants have evolved many other ways to move their seeds to distant locations (Figure 3.12).

Plants are the basis of land ecosystems

It is difficult to overstate the significance of plants to other organisms. As photosynthesizing organisms, plants use sunlight and air to make food that they and the organisms that eat them can use. All other organisms on land ultimately depend on plants, eating plants or eating other organisms (such as animals) that eat plants, and so on. Many organisms live on or in plants, or on or in soils made up of decomposed plants.

Flowering plants provide humans with materials such as cotton for clothing and with pharmaceuticals such as morphine. Essentially all agricultural crops are flowering plants, and the entire floral industry rests on the work of angiosperms. Gymnosperms such as pines, spruces, and firs are the basis of forestry industries, providing wood and paper.

As valuable as plants are to us when harvested, they are also valuable when left in nature. By soaking up rainwater in their roots and other tissues, for example, plants prevent runoff and erosion of soils that can con-

(a)

(b)

Figure 3.12 Getting Around
Plants have evolved many ways of spreading to new areas. (*a*) Consider the magnificent fruit known as the coconut. A palm tree seed in a coconut can float for hundreds of miles until it reaches a new beach where it can take root and grow. (*b*) Seeds can have wings or other structures, such as the feathery growths on these milkweed seeds, which help them to be carried great distances by the wind.

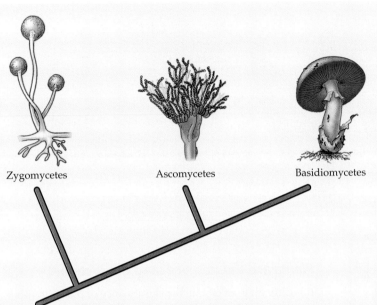

Zygomycetes   Ascomycetes   Basidiomycetes

Number of species discovered to date: ~70,000

Function within ecosystems: Decomposers and consumers

Economic uses: Mushrooms are used for food, yeasts for producing alcoholic beverages and bread. Some fungi also produce antibiotics, drugs that help fight bacterial infections.

Funky factoid: Highly sought-after mushrooms known as truffles can sell for $600 a pound.

Figure 3.13 The Fungi
The fungi are most familiar to us as mushrooms, but the main bodies of such fungi typically are hidden underground or in another organism's tissues. Some fungi are decomposers, breaking down dead and dying organisms. Others are parasites, living on or in other organisms and harming them, or mutualists, living with other organisms to their mutual benefit. The relationships of the three major groups of fungi are shown in the evolutionary tree.

These foul-smelling basidiomycetes, known as stinkhorn mushrooms, attract flies, which get covered with their sticky spores and then scatter the spores as they fly to other locations.

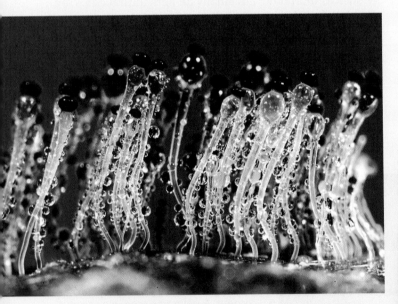

Pilobolus, a zygomycete that lives on dung, can shoot its spores out at an initial speed of 50 kilometers per hour.

This ascomycete, *Penicillium*, is a relative of the original species that produced the antibiotic penicillin, a drug that fights bacterial infections and has saved the lives of countless people.

taminate streams and harm fish populations. Plants also produce the crucial gas oxygen.

> ■ Plants, a kingdom within the domain Eucarya, were the first of the major groups to colonize the land. Plants evolved diverse shapes and sizes after they evolved vascular systems. Two other evolutionary innovations for plants were seeds and flowers. Plants are essential components of land ecosystems. As producers, they provide the food that all other organisms eventually use.

3.6 The Fungi: A World of Decomposers

The **Fungi** is a kingdom within the domain Eucarya. Most people are familiar with fungi as the mushrooms on their pizza or on their lawns, but this group includes not only mushroom-producing organisms, but also molds and yeasts. In fact, the familiar mushroom is only the very visible reproductive structure sprouting from what is usually the much larger main body of a fungus. Most of the tissues of a fungus typically are woven through the tissues of whatever other organism—whether dead or alive—the fungus is digesting and making its meal. Because most of the tissues of a fungus are usually hidden from view, the fungi are among the most enigmatic and poorly understood of the major groups of organisms.

As Figure 3.13 shows, the fungi can be divided into three distinct groups: the zygomycetes, which evolved and branched off first, the ascomycetes, and the basidiomycetes. These three groups differ in, and are named for, specialized reproductive structures that are specific to each group. In the discussion that follows, we examine the structure of the main body of a fungus and see how that structure makes fungi good decomposers.

Fungi have evolved a structure that makes them highly efficient decomposers

The main body of a fungus is known as the **mycelium** (plural mycelia). It is made up of a mat of threadlike projections known as **hyphae** (singular hypha), which typically grow hidden, either within the soil or through the tissues of the organism the fungus is digesting (Figure 3.14).

Hyphae are composed of cell-like compartments that are encased in a cell wall. Unlike the plant cell wall, the

fungal cell wall is composed of chitin, the same material that makes up the hard outer skeleton of insects. Furthermore, unlike the cells of other multicellular organisms the cells of fungi are usually only partially separated from one another. In fact, these cell-like compartments typically are separated only by a partial divider known as a septum (plural septa), which allows organelles to pass from one compartment to another. Even nuclei can move around from one compartment to another.

Like animals, fungi rely on other organisms for both energy and nutrients. But unlike animals, which ingest food through the mouth and digest it internally, fungi digest their food externally and then absorb the nutrients. In much the same way as our stomachs release digestive juices and proteins, the fungal hyphae growing through the tissues of a plant or animal release special digestive proteins that break down those tissues. The hyphae then absorb the released nutrients for the fungus to use.

The ability of fungi to grow through things makes them well suited to the role of **decomposer**—that is, an organ-

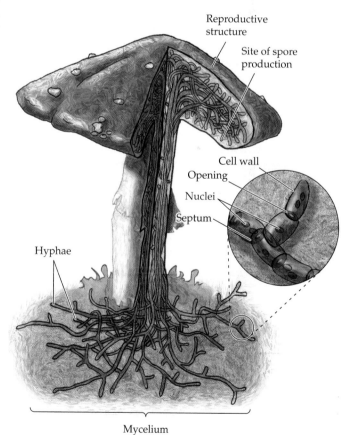

Figure 3.14 Fungi Grow with Hyphae
Hyphae are threadlike projections that form dense mats. Mats of hyphae form the main feeding body of a fungus, which is known as the mycelium. The hyphae are composed of cell-like compartments that are encased in a cell wall.

Figure 3.15 Fungi Spread via Spores
The powdery dust on this orange is a coating of fungal spores.

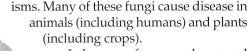

ism that attacks and breaks down dead organisms. In fact, fungi are among the most important groups of decomposers, recycling a large proportion of the world's dead and dying organisms. For example, shelf fungi, which earn their name from their stacked, shelflike shape, are key decomposers of dead and dying trees.

The reproduction of fungi also sets them apart from other organisms. Characterized by complex mating systems, fungi come not in male and female sexes, but in a variety of mating types. Each mating type can mate successfully only with a different mating type.

Another, more familiar aspect of fungal reproduction is **spores**, reproductive cells that typically are encased in a protective coating that keeps them from drying out or rotting. Known to most of us as the powdery dust on moldy food (Figure 3.15), spores, like plant seeds, are scattered into the environment by wind, water, and animals. Once carried to new locales, spores can begin growing as new, separate individuals.

The same characteristics that make fungi good decomposers make them dangerous parasites

Although most fungi are decomposers, some are **parasites**, organisms that live on or in another organism at that organism's expense. Parasitic fungi grow their hyphae through the tissues of other living organ-

isms. Many of these fungi cause disease in animals (including humans) and plants (including crops).

In humans, for example, several different species of fungi can cause athlete's foot, and *Pneumocystis carinii* causes a deadly fungal pneumonia that is a leading killer of people suffering from AIDS. Fungi also attack plants. *Ceratocystis ulmi* causes Dutch elm disease, which has nearly eliminated the elm trees of the United States. Rusts and smuts are fungi that attack crops. Still other fungi are specialized for eating insects, and biologists are looking for ways to use these fungi to kill off insects that are pests of crops (Figure 3.16).

Some fungi live in beneficial associations with other species

Some fungi are **mutualists**; that is, they live with other organisms in a close association that is helpful to both organisms. Morels, for example, are highly prized mushrooms that are the products of fungi living in mutually beneficial associations with plant roots. Such fungi are known as mycorrhizal fungi, and they can be found in all three groups of fungi: zygomycetes, ascomycetes, and basidiomycetes. In these associations, called mycorrhizae, the fungi live in or on the roots of the plants, forming thick mycelial mats. The fungi receive sugars and amino acids—the building blocks of proteins—from the plants. The roots of the plants, infested with the spongelike hyphae, can absorb more water and nutrients than they could without the fungi.

Another familiar fungal association is the lichen. These often lacy, gray-green growths seen on tree trunks or rocks are actually associations of algae (which are photosynthetic protists, as we learned earlier) and fungi (Figure 3.17). Both ascomycetes and basidiomycetes are known to form lichens. The algae and fungi in lichens grow with their tissues intimately entwined. The fungi receive sugars and other carbon

Morel

Figure 3.16 An Insect-Eating Fungus
Some fungi are parasites, making their living by attacking the tissues of other living organisms. This beetle, a weevil in Ecuador, has been killed by a *Cordyceps* fungus, the stalks of which are growing out of its back.

Figure 3.17 Lichens
A lichen consists of an alga and a fungus living intimately entwined in a mutually beneficial association. These lichens, known as British soldiers, are growing on an old log.

compounds from the algae; in turn, the fungi produce what are known as lichen acids, mixtures of chemicals that scientists believe may function to protect both the fungi and the algae from being eaten by predators.

Fungi have important roles in the biosphere

As decomposers of dead and dying organisms, fungi are crucial components of terrestrial ecosystems. They are also critical to plants; more than 95 percent of ferns and their relatives, gymnosperms, and angiosperms have mycorrhizal fungi living in association with their roots.

Fungi can be costly to human society. Some cause deadly diseases, contaminate crops, spoil food, and force us to clean our bathrooms more often than we might like. Other fungi add to the quality of human life. They provide us with pharmaceuticals, including antibiotics such as penicillin. Yeasts, which are single-celled fungi that feed on sugars, produce two important products: alcohol and the gas carbon dioxide. What would life be like without yeasts such as *Saccharomyces cerevisiae*, which produces the carbon dioxide that makes bread rise and the alcohol in beer? Fungi also provide highly sought-after delicacies, such as the morels mentioned above and a mushroom known as the truffle, whose underground growing locations can be sniffed out only by specially trained dogs and pigs.

■ The main body of a fungus is a mycelial mat composed of hyphae. Members of the Fungi, a kingdom within the domain Eucarya, digest food externally and then absorb it. Fungi are critical components of many ecosystems as decomposers and as mycorrhizal mutualists. Other fungi can attack humans and other animals, crops, and stored food.

 ### 3.8 The Animalia: Complex, Diverse, and Mobile

The **Animalia**, or the animals, is the most familiar major group and the one to which we humans belong. The Animalia is a kingdom within the domain Eucarya. All of the Animalia are multicellular creatures,

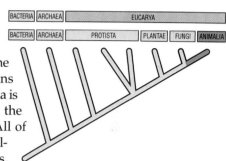

many of them quite complex. They include flashy creatures such as Bengal tigers and yourself. They also include worms, sea stars, snails, insects, and other creatures that are less obviously animal-like, such as sponges and corals. Figure 3.18 illustrates the major groups of animals.

The evolutionary tree in Figure 3.18 shows the sponges, the most ancient of the animal lineages, branching off first. Next, cnidarians (such as jellyfish and corals) evolved, and then flatworms. The next set of groups to evolve are known collectively as the protostomes. The protostomes encompass more than 20 separate groups, including molluscs (such as snails and clams), annelids (the segmented worms), and arthropods (such as crustaceans and insects). Of these three, arthropods branched off first, then molluscs and annelids split apart. Next to evolve were the deuterostomes, including the echinoderms (sea stars and the like) and the vertebrates (animals with backbones, such as fish, frogs, birds, and humans).

Like all fungi and some bacteria and protists, animals make their living off the tissues of other organisms, getting their carbon and energy from the organisms they eat. Animals differ from fungi and plants in that their cells do not have cell walls surrounding their plasma membranes.

Typically mobile and often in search of either food or mates, animals have evolved a huge diversity of ways of getting what they need. As we'll see, over the past 500 million years, different lineages of animals have evolved bodies that have collections of specialized and coordinated cells organized into tissues (such as human skin),

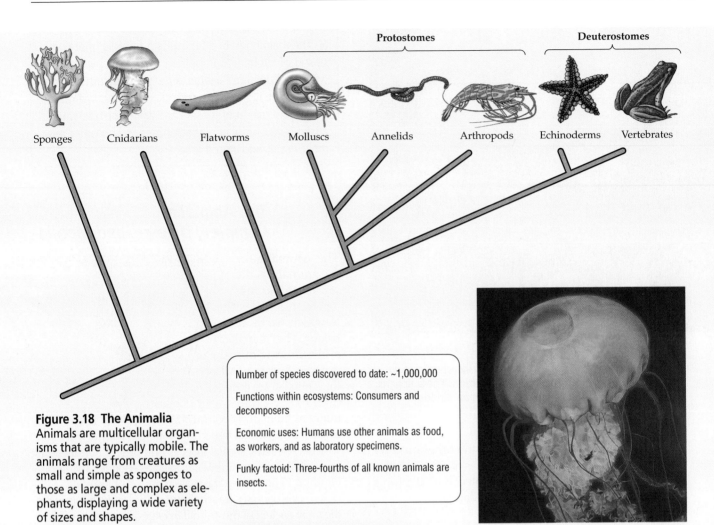

Protostomes

Deuterostomes

Sponges Cnidarians Flatworms Molluscs Annelids Arthropods Echinoderms Vertebrates

Figure 3.18 The Animalia
Animals are multicellular organisms that are typically mobile. The animals range from creatures as small and simple as sponges to those as large and complex as elephants, displaying a wide variety of sizes and shapes.

Number of species discovered to date: ~1,000,000

Functions within ecosystems: Consumers and decomposers

Economic uses: Humans use other animals as food, as workers, and as laboratory specimens.

Funky factoid: Three-fourths of all known animals are insects.

Cnidarians include jellyfish, like the one seen swimming here, as well as anemones and corals. Members of this group, the first organisms to evolve true tissues, are named for their stinging cells, which they use for protection and to disable prey.

Sponges are ancient aquatic animals. They have evolved some specialized cells, but no true tissues.

Molluscs include snails, slugs, and octopi, as well as this giant clam from a tropical reef. As is typical of many molluscs, this clam's tender flesh is protected by a hard outer shell.

Flatworms, like this oceangoing flatworm from the U.S. West Coast, were among the earliest animals to evolve true organs and organ systems.

A key feature of annelids, also known as segmented worms, is segmentation. This body plan of repeating units can be seen as the series of distinct segments in this fire worm. The segmented body plan, which is also seen in arthropods and vertebrates, facilitated the evolution of many different body forms.

Echinoderms include sea stars, like the one from Indonesia shown here, and sea urchins. They are closely related to the vertebrates.

Arthropods include crustaceans, like lobsters and crabs, as well as millipedes, spiders, and the most species-rich of all groups, the insects. This *Morpho* butterfly, an inhabitant of the tropical rainforest, is one of the most spectacularly beautiful insects on Earth.

Amphibians, slimy creatures that include frogs, like this poison arrow frog from Costa Rica, and salamanders, typically spend part of their lives in water and part on land.

Vertebrates are the animals that have backbones, including fish, reptiles, amphibians, birds, and mammals. Shown here is a coral reef fish from Thailand. Fish were the earliest vertebrate animals.

Primates include monkeys, apes, and humans. In this group we find our closest relative, the chimpanzee, shown here, and the gorilla.

Mammals are characterized by milk-producing mammary glands in females, as well as young that are born live (rather than being born in an egg that later hatches). These kangaroos are mammals, as are bears, dogs, lions, and humans.

and collections of specialized tissues organized into organs and organ systems (such as the human digestive system). Like plants, which also have specialized tissues (such as vascular tissues), animals have evolved features that have helped them live successfully in a variety of environments. Animals have also evolved a diversity of sizes and shapes, many of them variations on a few themes. In addition, animals display an astounding array of behaviors that help them survive and reproduce.

Animals evolved tissues

Sponges are among the simplest of animals. They represent a time in the evolution of animals before specialized tissues had evolved. Sponges are little more than loose collections of cells. In fact, if a sponge is put through a sieve, breaking it apart into individual cells, it will slowly reassemble itself. These animals live off amoebas and other tiny organisms that they filter out of the water that surrounds them (Figure 3.19). While they are widespread and highly successful, they are not very efficient; they must filter a ton of water just to grow an ounce.

One of the earliest animal groups to evolve true tissues was the group that includes jellyfish, corals, and anemones. The name of this group, Cnidaria, comes from the Greek word for "nettle," a stinging plant. Cnidarians are so named because they are characterized by stinging cells that they use for protection and for immobilizing prey. Jellyfish, like other cnidarians, have specialized nervous tissues (Figure 3.20), musclelike tissues, and digestive tissues. This innovation allowed them to do things that require the coordination of many cells, such as gracefully and rapidly swimming away from predators.

Animals evolved organs and organ systems

After tissues, the next level of complexity to evolve was organs and organ systems. Organs are body parts composed of different tissues that are organized to carry out

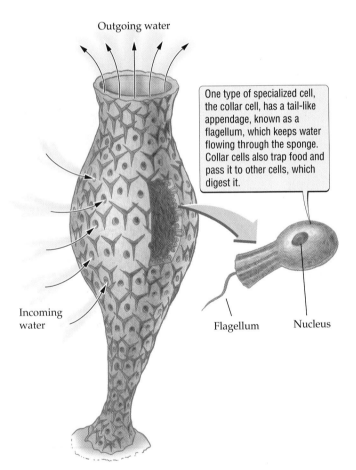

Outgoing water

One type of specialized cell, the collar cell, has a tail-like appendage, known as a flagellum, which keeps water flowing through the sponge. Collar cells also trap food and pass it to other cells, which digest it.

Incoming water

Flagellum   Nucleus

Figure 3.19 Sponges Have Specialized Cells but Lack Tissues
Unlike most animals, sponges are loose associations of cells. Although some of these cells are specialized, none are organized into tissues.

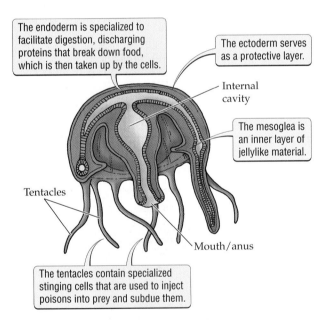

The endoderm is specialized to facilitate digestion, discharging proteins that break down food, which is then taken up by the cells.

The ectoderm serves as a protective layer.

Internal cavity

The mesoglea is an inner layer of jellylike material.

Tentacles

Mouth/anus

The tentacles contain specialized stinging cells that are used to inject poisons into prey and subdue them.

Figure 3.20 The Tissue Layers of a Jellyfish
Cnidarians were one of the earliest groups to evolve specialized tissues. These animals have an outer layer of tissue known as the ectoderm (ecto, "outer"; derm, "skin") and an inner layer known as the endoderm (endo, "inner"). Sandwiched between the ectoderm and the endoderm is an inner layer of secreted material known as the mesoglea (meso, "middle"; glea, "jelly"). The endoderm and ectoderm are also organized to contract like muscle tissue. The tentacles bring food into the internal cavity through a single opening, which serves as both a mouth and an anus.

specialized functions. Usually organs have a defined boundary and a characteristic size and shape. An example is the human stomach.

An organ system is a collection of organs working together to perform a specialized task. The human digestive system, for example, is an organ system that includes the stomach as well as other digestive organs, such as the pancreas, liver, and intestines, all of which work together to digest food. Flatworms, a group of fairly simple worm-like animals, were one of the earliest groups of animals to evolve true organs and organ systems (Figure 3.21).

Animals evolved complete body cavities

Still later in their history, animals evolved a complete body cavity—an interior space with a mouth at one end and an anal opening at the other. The two distinct evolutionary lineages that exhibit such cavities are known as the protostomes and the deuterostomes (see the tree in Figure 3.18). The protostomes include animals such as insects, worms, and snails. The deuterostomes include animals such as sea stars and all the animals with backbones (vertebrates), such as humans, fish, and birds.

Protostomes and deuterostomes are named for the way the body cavity forms as they develop from a fertilized egg into an embryo and then into a juvenile animal. In **protostomes** (*proto*, "first"; *stome*, "opening"), the first opening to form in the embryo becomes the mouth, and the anus forms elsewhere later. In **deuterostomes** (*deutero*, "second"), the second opening becomes the mouth, and the first becomes the anus. This develop-

mental difference has led to very different organizations of tissues in these two groups. Among these two groups, there has been a great flowering of animal sizes and shapes. Interestingly, many of these can be seen as variations on a few basic themes, as described in the next section.

Animal body forms are variations on a few themes

Animals exhibit a great variety of shapes and sizes, many of which are variations on a few basic body plans.

The **arthropods** (*arthro*, "jointed"; *pod*, "foot") are a group that includes millipedes, crustaceans (such as lobsters and crabs), insects, and spiders. With their hard outer skeleton (which is made of chitin, the same material that is found in the cell walls of fungi), arthropods are a wonderful illustration of how evolution can take a basic body plan and run with it (Figure 3.22). Among the arthropods are the **insects**, six-legged organisms, such as

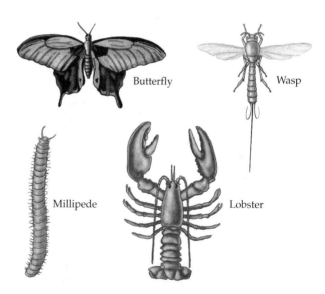

Butterfly

Wasp

Millipede

Lobster

Figure 3.22 Variations on a Theme
From a simple segmented body plan, arthropods have evolved a huge diversity of body forms and sizes. The millipede can be viewed as the simplest form of these segmented animals, with all its segments similar. As segments have evolved and diversified, a variety of organisms have arisen, from lobster to swallow-tailed butterfly to parasitoid wasp. Looking just at the evolution of the last abdominal segments (the rear ends of these creatures), one can see that evolutionary changes in this one set of segments have resulted in a huge variety of shapes and lifestyles. The last abdominal segments have evolved into the delicate abdomen of the butterfly, the abdomen of the wasp, which has a huge structure for inserting its eggs deep into another animal, and the delicious tail of the lobster.

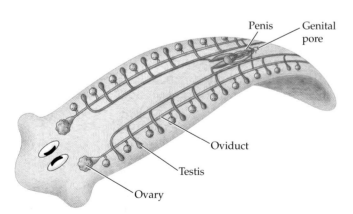

Penis Genital pore

Oviduct

Testis

Ovary

Figure 3.21 Animals Have Evolved Organs and Organ Systems
The reproductive system, one of the flatworm's organ systems, contains both male (penis) and female (genital pore) organs. All of these organs work together to help the flatworm reproduce. Every flatworm can function as both a male and a female.

grasshoppers, beetles, and ants, that live on land. Whereas prokaryotes dominate Earth in terms of sheer numbers of individuals, insects dominate in terms of numbers of species, having many more species than any other group of organisms.

One of the features that has facilitated the evolution of arthropod bodies is the fact that they are divided into segments. The different segments have evolved differently over time, resulting in a huge number of different types of animals. Such body segmentation can also be seen in the segmented worms, or annelids, such as earthworms, whose bodies are made up of a repeated series of segments (see Figure 3.18). The **vertebrates**—animals that have a backbone—are also built on a segmented body plan (Figure 3.23).

(a)

(b)

Figure 3.23 Many Animals Are Segmented
Segmentation, a type of body plan in which segments are repeated and often can evolve independently of one another, is shown here in (*a*) a giant African millipede and (*b*) a vertebrate.

The vertebrates encompass several subgroups, including fish, amphibians (such as frogs and salamanders), reptiles (such as snakes and lizards), birds, and mammals (such as humans and kangaroos). This group illustrates other ways in which a variety of very different forms can evolve from one basic body plan. The front appendage of vertebrates, for example, has evolved in humans as an arm, in birds as a wing, in whales as a flipper, in snakes as an almost nonexistent nub, and in salamanders as a front leg.

Animals exhibit an astounding variety of behaviors

Another fascinating characteristic of animals is their ability to move and behave. Animals have evolved many behaviors to capture and eat prey, avoid being captured, attract mates, and move to new environments. Animals can be quite useful to immobile organisms, such as plants, which, as we have seen, have evolved ways to get animals to carry their pollen and seeds. Humans also find the mobility of other animals convenient, as, for example, when we travel by horse or camel or send messages via carrier pigeon. We will discuss animal behavior in detail in Chapter 35.

Animals play key roles in ecosystems and provide useful products for humans

Because they live by eating other organisms, and because most of them are mobile, animals play many roles in ecosystems. Most serve as consumers, preying on plants and other animals. Some animals, such as carrion beetles, serve as decomposers of dead organisms. Animals help spread plant seeds and fungal spores. Animals can also spread diseases; for example, ticks spread the protist parasite that causes Lyme disease. Animals, especially insects, can be crop pests, such as the tomato hornworm, a caterpillar that attacks tomatoes. Domesticated animals, such as cattle, provide humans with food and material for clothing as well as work and transportation. We are even beginning to use animals as sources of organs for transplantation into humans.

Carrion beetle

Homo sapiens is the animal species that has the greatest impact on life on Earth. As members of a species with rapidly growing populations and the ability to drastically and rapidly modify Earth with our cities, agriculture, and industries, we are having dramatic effects on other species and their environments, as we will see in the next chapter.

■ Animals, a kingdom within the domain Eucarya, are multicellular organisms that are typically mobile. Animals include a wide variety of organisms from simple sponges to complex vertebrates. Animals have evolved specialized tissues, organs and organ systems, body cavities, and a wide variety of shapes and sizes. Animals also exhibit an astounding variety of behaviors. Insects are the most species-rich group of organisms on Earth. Most animals are consumers; some are decomposers. Animals have many effects on human society, which is itself composed of animals.

The Difficulty of Viruses

Viruses are not placed on the evolutionary tree of life or in any kingdom or domain. As we saw in Chapter 1, viruses are simply protein wrapped around a fragment of DNA or RNA. They lack many of the characteristics of living organisms, occupying a gray zone between life and nonlife. Viruses may have arisen as pieces of DNA from many different organisms. That is, rather than arising from any single point on the evolutionary tree, new viruses may be appearing all the time, evolving from different kinds of organisms across the tree of life. Difficult to classify and lacking a clear evolutionary relationship to any one group, viruses are not easily placed into any of the existing classification schemes.

■ Viruses are not placed on the tree of life or classified into any kingdom or domain, since they lack clear evolutionary relationships with other organisms and since it is difficult to define whether they are living or not.

■ HIGHLIGHT

Where the Major Groups Meet

Like the "bacterium from hell" found 2 miles beneath the surface of the earth, which we introduced in the beginning of this chapter, more and more organisms are being discovered in unexpected places and situations. And while biologists attempt to neatly define and separate major groups, such as kingdoms and domains, in order to organize this explosion of life, organisms in nature often ignore these boundaries to live together, sometimes in surprisingly intimate associations.

Lichens, for example, are actually complex mixtures of organisms from two different major groups (Fungi and Protista) living intricately entwined (see Figure 3.17). Other examples of ways in which the major groups meet include photosynthetic corals. Corals are cnidarians (Animalia) that have incorporated photosynthesizing algae (Protista) into their tissues. While most animals have to hunt down their food or seek out plants to eat, corals have set up shop with photosynthesizers living in their bodies. These algae provide both color and sugar, the key product of photosynthesis, to the corals. When the corals are stressed—for example, when they experience unusually high water temperatures—these mutualist algae can abandon their hosts, taking the coloration of the corals with them, leading to the increasingly common phenomenon known as coral bleaching.

Even humans can be considered a cross-kingdom, even a cross-domain grab bag. Every human body is host to millions of bacteria, both externally on the skin and internally in the digestive system. In addition, animals abound on humans in the form of hundreds of microscopic arthropods. Some of these associates are beneficial, while some cause disease or are parasitic. The human body is also being continually invaded by viruses, making each person a densely populated community of living organisms.

■ Some apparent individuals are composed of organisms from entirely separate major groups.

SUMMARY

The Major Groups in Context

- The six major groups of living organisms are the Bacteria, Archaea, Protista, Plantae, Fungi, and Animalia.

- All living organisms can be placed in the context of the evolutionary tree of life, the three-domain system, and the Linnaean hierarchy. All six major groups are kingdoms within the Linnaean hierarchy.

The Bacteria and the Archaea: Tiny, Successful, and Abundant

- The Bacteria and the Archaea are simple, microscopic single-celled organisms.

- The Bacteria and the Archaea are both prokaryotes; that is, organisms that do not have compartments in their cells.

- Prokaryotes are the most abundant and widespread form of life on Earth.

- Some prokaryotes, including many archaeans, thrive in extreme (for example, very hot or very salty) environments.

- Prokaryotes exhibit unmatched diversity in their methods of getting and using nutrients.

- Prokaryotes perform key roles in ecosystems. They are useful to humanity in many ways, but they also cause deadly diseases.
- The Bacteria and Archaea, while similar, are distinct in key ways. They differ in their DNA and in key components of their cell walls and plasma membranes.

The Evolution of More Complex Cells: The Eukaryotes Arise from Prokaryotes

- In the past, some prokaryotic cells engulfed others. These engulfed cells eventually evolved into organelles.
- Cells containing organelles evolved into the Eucarya, or eukaryotes, and gave rise to the first multicellular organisms.

The Protista: A Window into the Early Evolution of the Eucarya

- The Protista are highly diverse, in part because they are a collection of separate evolutionary lineages that do not make up a "real group."
- The Protista include plantlike, animal-like, and funguslike organisms.
- The evolutionary relationships among members of the Protista remain poorly understood.
- Protists represent early stages in the evolution of the eukaryotic cell.
- Protists provide insight into the early evolution of multicellularity.
- The sexual reproduction that is characteristic of eukaryotes evolved in protists.
- Although protists include many harmless organisms, such as kelps, they also include many disease-causing organisms, such as *Plasmodium*, which causes malaria.

The Plantae: Pioneers of Life on Land

- The Plantae are multicellular, photosynthesizing organisms.
- Several important evolutionary innovations, including vascular systems, seeds, and flowers, permitted plants to colonize land successfully.
- Gymnosperms evolved the first seeds, and angiosperms evolved the first flowers.
- As producers, plants are the ultimate food source for other organisms and thus are critical components of land-based food webs.
- Plants produce food, pharmaceuticals, shelter, oxygen, and many other important benefits for humans.

The Fungi: A World of Decomposers

- Fungi grow using hyphae to penetrate, digest, and absorb food.

- Most Fungi are decomposers, but fungi are also important parasites and mutualists.
- Fungi cause diseases of humans, animals, and crops, but they also produce valuable products such as foods and pharmaceuticals.

The Animalia: Complex, Diverse, and Mobile

- The Animalia make a living by eating other organisms and are typically mobile.
- Important evolutionary innovations in animals are tissues, organs and organ systems, body cavities, and behaviors.
- Animals exhibit a great variety of forms and sizes, many of them variations on a few themes; they also exhibit a great variety of behaviors.
- Animals play a variety of roles in ecosystems. Most are consumers; some are decomposers. Some spread disease; others are pests of crops. Animals also provide food, clothing, and other products to human society.

The Difficulty of Viruses

- Viruses are hard to categorize. They are not placed on the tree of life or in any kingdom or domain, since they lack clear evolutionary relationships and are difficult to define as living or nonliving.

Highlight: Where the Major Groups Meet

- Sometimes what appears to be a single organism, such as a lichen or coral or person, is actually a combination of organisms from more than one major group.

KEY TERMS

angiosperm p. 50	hypha p. 53
Animalia p. 55	insect p. 59
Archaea p. 39	mutualist p. 54
arthropod p. 59	mycelium p. 53
Bacteria p. 39	parasite p. 54
decomposer p. 53	Plantae p. 48
deuterostome p. 59	prokaryote p. 42
Eucarya p. 39	Protista p. 45
eukaryote p. 42	protostome p. 59
evolutionary innovation p. 38	seed p. 50
flower p. 50	spores p. 54
Fungi p. 53	vertebrate p. 60
gymnosperm p. 50	virus p. 61

CHAPTER REVIEW

Self-Quiz

1. Which group is the most abundant in terms of numbers of individuals?
   a. Animalia
   b. Eucarya
   c. Protista
   d. Prokaryotes

2. Eukaryotes differ from prokaryotes in which of the following ways?
   a. They do not have organelles in their cells, as prokaryotes do.
   b. They exhibit a much greater diversity of nutritional modes than prokaryotes do.
   c. They have organelles in their cells and prokaryotes do not.
   d. They are more widespread than prokaryotes.

3. Which of the following groups contains organisms that represent early stages in the evolution of the eukaryotic cell?
   a. Archaea
   b. Protista
   c. Fungi
   d. Animalia

4. Which of the following groups was the first to colonize the land?
   a. Plantae
   b. Animalia
   c. Bacteria
   d. Fungi

5. What were the key evolutionary innovations of the Plantae?
   a. seeds, organelles, flowers
   b. roots, cuticle, seeds, flowers
   c. roots, hyphae, flowers
   d. hyphae, cuticle, organelles

6. Fungi grow using
   a. hyphae.
   b. chloroplasts.
   c. angiosperms.
   d. prokaryotes.

Review Questions

1. Name three factors that probably contributed to the success of prokaryotes.

2. Why are *Giardia* and slime molds of particular interest to biologists interested in the early evolution of eukaryotes?

3. Describe the evolution of specialized cells, tissues, and organs in the Animalia, including the name of the animal group that first showed each evolutionary innovation.

4. To what kingdom and domain do viruses belong and why?

5. Name the three domains and kingdoms within them.

6. Draw the tree of life showing a representative from each of the six kingdoms.

3 **The Daily Globe**

A New Method for Treating Malaria?

GLASGOW, MONTANA—Researchers reported today that they may have discovered the key to fighting a huge group of human parasites, including the organism that causes the often deadly disease malaria.

Deep inside the cells of these parasites, which the researchers describe as animal-like protists, they have discovered something that does not seem very animal-like at all. Inside these parasites' cells are organelles known as plastids, ancient chloroplasts similar to those found in plants.

The researchers say this finding raises the possibility that malaria and other diseases caused by this group of parasites could be treated with herbicides, chemicals normally used to kill plants.

"It was like a bolt from the blue," said Dr. Bob D. Wartheim, a biologist at Central Montana University, who was one of the authors of the new report and who described the finding as a complete shock. Researchers say they still don't know what the function of the plastid is.

So far it appears to be nonfunctional and is clearly unable to carry out photosynthesis. "We were about to give up studying these organelles because we had been unable to get funding from any source for the past 5 years. Everyone rejected the research as uninteresting and without important applications or ramifications. You just never know what you're going to find."

Evaluating "The News"

1. Why would herbicides be a potentially powerful drug for treating malaria in humans?

2. Some elected officials have complained bitterly about the use of taxes for basic research—for example, the study of odd little organelles inside of little-understood organisms—proclaiming them a waste of hard-earned tax dollars. Should taxpayers be funding basic research, or should all research be aimed at solving a particular societal woe, such as human disease or environmental problems?

3. Scientists point to cases such as this one as proof that basic research is valuable to society. But the vast majority of basic scientific research does not result in such applied findings. Is it still useful to fund basic research? Why or why not?

chapter 4 *Biodiversity*

Martin Johnson Heade, *Study of an Orchid*, 1872.

Where Have All the Frogs Gone?

*I*n 1987, in a tropical forest high in the mountains of Costa Rica, the golden toad, a spectacularly beautiful creature, could be found in abundance. That year hundreds of the brightly colored animals were seen in the Monteverde Cloud Forest Preserve, the only spot in the world from which the toads were known. The next year just a few toads were found. Within a few years the golden toad had disappeared entirely, never to be seen again.

While it is always a concern when a species plummets into extinction, most extinctions are easier to understand than the loss of the golden toad. When a forest-dwelling bird goes extinct because its forest is cut down, there is no lingering mystery. But the golden toads were living high in a montane forest preserve, a pristine area far from deforestation or development. These toads, it seemed, should not have gone extinct.

■ MAIN MESSAGE

The diversity of life has risen and fallen naturally and drastically in the past, but is now declining rapidly as a result of human activity.

Since the time when the golden toad was last seen, biologists around the world have documented declines in populations of other amphibians (the group that includes frogs, toads, and salamanders), many in preserved natural areas. In the United States, for example, both in and around Yosemite National Park, where frogs and toads were once abundant, numerous species have declined or disappeared. For frogs living at high altitudes, increases in their exposure to damaging ultraviolet light may be a problem. In many parts of the world, fungal diseases are killing huge numbers of frogs. In other cases, researchers believe that pollution is the cause of the frogs' decline. Whatever the specific problem, amphibians are being lost around the world. Many scientists believe that amphibians are more sensitive than other animals to environmental damage, and that their deaths are signs of a deteriorating environment.

Meanwhile, as amphibians mysteriously disappear, biologists are finding that many other species are rapidly going extinct. Everywhere we hear warnings about the loss of species around the globe. How serious are these species losses, and, in the end, do they really matter?

Once Abundant on a Mountaintop in Costa Rica, This Amphibian Mysteriously Went Extinct in the Late 1980s
Here the orange-colored male mates with the very differently colored and larger female.

■ KEY CONCEPTS

1. Estimating the number of species on Earth is difficult, but not impossible. Although the total number remains a question, biologists generally agree that the vast majority of species have yet to be discovered.

2. Biologists generally agree that we are on our way toward a mass extinction, one of a few times in the history of Earth during which huge numbers of species have been lost. If species losses continue at their current rate, the result could be the most rapid mass extinction in the history of the planet.

3. Deforestation and other habitat losses, invasions of non-native species, climate change, and the resulting decrease in species numbers are problems not just in the tropical rainforest, but in our own backyards.

4. Biodiversity matters to the health of forests, grasslands, rivers, oceans, and other ecosystems on which human society depends. Therefore, biodiversity matters to the health of humans.

We hear constantly that the world's species are rapidly becoming extinct, and that the world's ecosystems—Earth's many habitats and the organisms that live in them—are under threat. But how many species are there, and what exactly is happening to them? In this chapter we examine how species numbers are changing now and how they have changed during Earth's history. We also explore the question of what value, if any, the world's many species—which constitute its **biodiversity**—have to humanity.

How Many Species Are There on Earth?

Despite intense worldwide interest, scientists do not know the exact number of species on Earth. Estimates range widely, from 3 million to 100 million species. Most estimates, however, fall into the range of 3 million to 30 million species.

Scientists use indirect methods to estimate species numbers

Up to now, the total number of species that have been collected, identified, named, and placed in the Linnaean hierarchy is around 1.5 million. But despite this great cataloging of organisms, which has been going on for more than two centuries, many biologists believe that they have barely scratched the surface of Earth's species. Some estimates suggest that 90 percent or more of all living organisms remain to be discovered, identified, and named by biologists.

Biologists use indirect methods to estimate how many species remain unknown. For example, in 1952 a researcher at the U.S. Department of Agriculture estimated, on the basis of the rate at which unknown

insects were pouring into museums, that there were 10 million insect species in the world. Terry Erwin, an insect biologist at the Smithsonian Institution, shocked the world in 1982 with his estimate that the arthropods alone (the group that includes insects, spiders, and their relatives) numbered more than 30 million species, 20 times the number of all the previously named species of all kinds on Earth. Most of these arthropods, Erwin said, were living in the nearly inaccessible tops of tropical rainforest trees, a region of the forest known as the **canopy**.

Erwin based his estimate on actual counts of arthropods that he obtained by a method known as fogging (Figure 4.1). He blew a biodegradable insecticide high into the top of a single rainforest tree, then collected and counted the dead and dying insects that rained down to the ground. Erwin found more than 1100 species of beetles alone living in the top of one particular species of tropical tree. He estimated that 160 of these beetle species were likely to be specialists, making their living only in that particular tree species. There are an estimated 50,000 or so tropical tree species in the world. Thus, if the tree species Erwin was studying were typical, the beetle species living in tropical trees should number 8,000,000 ($50,000 \times 160$).

Beetle species are thought to make up about 40 percent of all arthropod species. If that is the case, then the total number of arthropod species in the tropical rainforest canopy should be 20 million. Many scientists believe that the total number of arthropod species in the canopy is double the number found in other parts of tropical forests, suggesting that there are another 10 million arthropod species in noncanopy tropical forest environments. Assuming these numbers, the total number of arthropod species in tropical forests should be 30 million. That means, of course, that there should be

are based, but one thing is certain: The 1.5 million species discovered and named to date are far from the total number living on Earth.

Some groups of organisms are well known and others are poorly studied

About half of the 1.5 million known species (750,000) are insects. All the remaining animals make up a mere 280,000 species or so. The next largest group is the plants, with about 250,000 known species. There are also approximately 69,000 named fungi, 57,700 protists, and some 4800 prokaryotes (Bacteria and Archaea) (Figure 4.2).

Among these groups of organisms, some have been very well studied because they are large, easy to capture, or popular with biologists. Others have been studied very poorly, often because they are microscopic or otherwise hard to collect and identify. In addition, particularly with poorly studied organisms, it can be difficult to determine whether two organisms are both members

Figure 4.1 Fogging to Count Species
Tropical biologist Terry Erwin fogs the canopy of a tree to collect its many insects. On the basis of his studies of insects from tropical treetops, Erwin estimated that there could be as many as 30 million species of tropical arthropods alone.

even more species of all kinds in the rest of the Tropics and in the rest of the world.

Like all such estimates, Erwin's is based on numerous assumptions. Changes in these assumptions could drastically alter the final numbers. While Erwin's is among the most famous of these estimates, such indirect measures are typical of how the numbers of species yet to be discovered are estimated, since it is impossible to count them directly. Scientists continue to argue over the exact figures and the assumptions on which they

Figure 4.2 A Piece of the Pie
This pie chart shows a breakdown of all the known species on Earth. Animals (particularly insects) and plants make up the vast majority of the known species, but many more remain to be discovered, particularly in other groups.

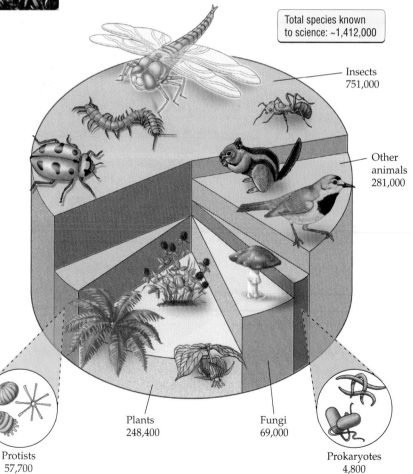

Total species known to science: ~1,412,000

Insects
751,000

Other animals
281,000

Protists
57,700

Plants
248,400

Fungi
69,000

Prokaryotes
4,800

of the same species or if they are two different species, making the tallying of total numbers of species even more complex. The birds, for example, total 9000 species and are among the best-studied organisms on Earth, with relatively few new species remaining to be discovered. Insects, on the other hand, remain poorly known; the majority, possibly the vast majority, are still undiscovered and unidentified.

Other groups, including whole kingdoms and domains, such as the Fungi, Bacteria, and Archaea, are also poorly known (Figure 4.3). Scientists have estimated that there may be as many as 10,000 species of bacteria in a single gram of Maine soil. That's about 5000 more species than have so far been named by biologists worldwide. The studies that form the basis of this estimate counted the different types of bacterial DNA found in the soil, each of which represents a different bacterial species.

Biologists continue to discover new species even in relatively well-known groups. For example, about 100 new fish species are discovered each year. In another case, although scientists had thought that all the large land mammals were accounted for, in 1992 a large deer species was found in Vietnam. Just 2 years later another deer species, of a type known as a barking deer, was discovered there as well.

The barking deer

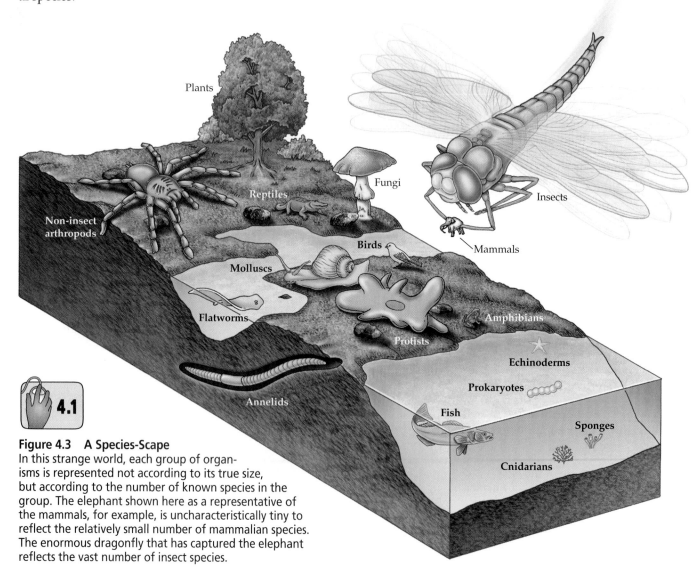

Figure 4.3   A Species-Scape
In this strange world, each group of organisms is represented not according to its true size, but according to the number of known species in the group. The elephant shown here as a representative of the mammals, for example, is uncharacteristically tiny to reflect the relatively small number of mammalian species. The enormous dragonfly that has captured the elephant reflects the vast number of insect species.

■ The exact number of species on Earth remains unknown. Scientists use indirect methods to estimate the total number of species, with most estimates falling between 3 million and 30 million. Half of the 1.5 million known species are insects. There is great variation in how well documented different groups of organisms are, and many, including fungi and bacteria, remain poorly known.

The Beginnings of a Present-Day Mass Extinction

The history of life on Earth includes a handful of drastic events during which huge numbers of species have gone extinct, events known as **mass extinctions**. Today, even as biologists struggle to get a total species count, many biologists believe that we are on our way toward losing many of those species. In fact, many biologists believe that Earth is now at the beginning of the most rapid mass extinction in its history. As with the total number of species on Earth, extinction rates are estimates. But even by conservative estimates, species are being lost at a staggering rate. The cause of this mass extinction is abundantly clear: the activity of the ever-increasing number of human beings living on Earth.

Numerous studies have documented recent species declines caused by humans. One large-scale study found that 20 percent of the bird species that existed 2000 years ago are no longer alive. Of the remaining bird species, 10 percent are estimated to be endangered—that is, in danger of extinction (Figure 4.4). Twenty percent of the freshwater fish species known to be alive in recent history have either gone extinct or are nearly extinct. And in 1998, scientists announced that one in every eight plant species in the world is under threat of extinction.

The most devastating and obvious losses of species are occurring in wet, lush forests in the tropical regions, known as **rainforests** (Figure 4.5). Known to be cradles of huge numbers of unique species found nowhere else in the world, tropical rainforests are quickly being burned or cut. By 1989, less than half of the rainforest that existed in prehistoric times remained. And what remained was disappearing at the rate of one football field's worth of rainforest every second, or an area the size of Florida each year. How do such large-scale losses of natural areas translate into species lost? According to Edward O. Wilson, a biologist at Harvard University, even conservative estimates put the yearly

Bachman's warbler

The former range of this once widespread species.

Figure 4.4   Going, Going, Gone?
Bachman's warbler, if it has not already gone extinct, is the most endangered of North America's songbirds.

Figure 4.5   In the Rainforest
Tropical rainforests, like this one in Hawaii, are typically home to numerous species found nowhere else.

species losses from rainforests at 27,000—an average of 74 per day, or 3 species every hour.

Although it's easy to imagine that disappearing species are an issue only in faraway places, studies indicate problems in our own backyard as well. In the United States, in addition to the disappearing frogs of Yosemite that were mentioned at the beginning of this chapter, 200 plant species known to have existed in recent history have already gone extinct, and at least 600 more are predicted to disappear within the next few years. In North America overall, 29 percent of freshwater fish and 20 percent of freshwater mussels are endangered or extinct.

In fact, evidence suggests that humans have been driving species to extinction for a very long time. The fossil record shows that around the times humans arrived in North America, Australia, New Zealand, and Madagascar, large animal species began to disappear. Although some scientists suggest that these species extinctions were due to climate changes, the coincidence of species losses with the arrival of humans in three different parts of the world is striking and consistent.

Of the genera of large mammals that roamed Earth 10,000 years ago, 73 percent are now extinct. (Remember that genera are the next category above species in the Linnaean hierarchy, as we learned in Chapter 2.) Many of these large mammals, such as mammoths, ground sloths, camels, and horses, would have made a hearty meal for prehistoric hunters. Other creatures, such as saber-toothed cats, may have suffered from having to compete with humans for prey. A similar number of birds, particularly flightless species, which would have been the easiest prey for humans, disappeared. One species lost, for example, was a flightless duck—literally a sitting duck. Numerous other animals that depended on the larger animals for survival, such as vulture species that fed on the carcasses of the dead beasts, also went extinct.

> ■ Many studies indicate that species are currently going extinct at an extremely rapid rate and that we are at the beginning of a mass extinction—possibly the most rapid mass extinction in the history of Earth.

Like total species number, rates of species losses are impossible to determine with certainty. However, biologists have amassed a wealth of data on species already gone or on their way out, and even conservative analyses suggest that huge numbers of species have been and continue to be lost. Why are these species disappearing?

4.2 The Many Threats to Biodiversity

The remaining species of the world face continuing challenges to their survival. In this section we examine some of the forces that are threatening and destroying biodiversity around the globe, including in our own backyards.

Habitat loss and deterioration are the biggest threats to biodiversity

Foremost among the direct threats to biodiversity is the destruction or deterioration of places where species can live—that is, their **habitats**. Habitats for nonhuman species continue to disappear or be radically changed as human homes, farms, and industries spring up where natural areas once existed.

The term "habitat loss" quickly conjures up images of burning rainforest in the Amazon, but the problem is much more widespread. Every time a suburban housing development goes up where once there was a forest or field, habitat is destroyed. So widespread is the impact of growing human populations in urban and suburban areas that species are disappearing even from parks and preserves in these areas. For example, Richard Primack and Brian Drayton, ecologists at Boston University, found that in a large preserve in the midst of increasing suburban development, 150 of the park's native plant species had disappeared. The species were most likely lost as a result of trampling and other disturbances as more and more people used the park (Figure 4.6a). In addition, pollution, erosion, and other effects of human activity and human population growth are altering natural habitats to the point where many species can no longer inhabit them (Figure 4.6b).

Introduced nonnative species can wipe out native species

Another threat to biodiversity is the introduction of foreign, or **nonnative**, **species**—that is, species that do not naturally live in an area but are brought there accidentally or on purpose by humans. Researchers estimate that 50,000 such nonnative species have been introduced into the United States. Often these invaders are able to sweep through a landscape, eating, outcompeting, or otherwise wiping out native species as they go.

In Africa's Lake Victoria, more than 300 species of native cichlid fish evolved over some 10,000 years, making this lake a treasure trove of fish diversity. Fewer than half of those species now remain, and many of the sur-

(a)

(b)

Figure 4.6 The Threats of Habitat Loss and Deterioration
(a) Researchers discovered that 150 plant species had disappeared from the Middlesex Fells Reservation in Massachusetts, a 100-year-old preserve, despite the ban on development within the park. Many of the plants (which still exist in other locations) are suspected to have been killed as a result of increasing use of the park by people from the growing suburban area surrounding it. (b) Usually habitat loss or deterioration is much more obvious.

viving species are close to extinction. The Nile perch, a voracious predator, which was introduced into the lake as a food fish, is largely responsible for the extinction of the cichlid species.

In Guam, the brown tree snake, another introduced species, has drastically reduced the numbers of most of Guam's forest birds, leaving an eerie quiet where once the forest was noisy with tropical birdsong. The tree snake is thought to have been introduced accidentally, brought by U.S. military planes from New Guinea, where it occurs naturally.

In Hawaii, introduced domesticated pigs that have escaped and are living in the wild (Figure 4.7a) are devouring the native plant species. Domesticated cats and mongooses, also introduced species, have killed many of Hawaii's native birds, especially the ground-dwelling species whose nests are easy targets. Purple loosestrife (Figure 4.7b), eucalyptus trees, and Scotch broom are introduced plant species that are choking out native plants in various parts of the United States.

Climate changes threaten species

Changes in climate, which many scientists believe are caused by human activity, constitute another threat that seems to be affecting many species. For example, in Austria, biologists have found whole communities of plants moving slowly up the Alps, apparently in response to global warming, as they are able to survive only at ever higher, cooler elevations. These plants are moving upward at an average rate of about a meter per decade. If the climate continues to warm, these plants, which exist nowhere but on these mountaintops, will eventually run out of mountain and become extinct (Figure 4.8).

Some threats are difficult to identify or define

Biologists now agree that many frogs, salamanders, and other amphibians—as well as many other kinds of organisms—are disappearing around the world. In some cases, as with the amphibians, the reason for these disappearances remains unclear. For some amphibians, as we have seen, pollution appears to be the culprit. In other cases, increased exposure to ultraviolet radiation seems to be killing the amphibians. In still other instances diseases are wiping them out. According to one study, climate change may be at the root of the problem, making amphibians more susceptible to both ultraviolet light and disease. Scientists studying toads in the northwestern United States found evidence that changing climate patterns were resulting in less rain falling in the mountains

(a)

(b)

Figure 4.7   The Threat of Nonnative Species
Introduced nonnative species threaten native ecosystems everywhere. (*a*) Feral pigs brought by humans from the mainland to Hawaii are devouring the native plants. (*b*) Purple loosestrife brought from Europe to the United States threatens native plants across the country. Furthermore, this plant quickly takes over mudflats, destroying critical feeding grounds for migrating birds.

of that region. Frog and toad eggs developing in what were now sometimes very shallow pools of water were subjected to stronger ultraviolet light. These weakened eggs became vulnerable to disease, including infection by a deadly fungus.

Despite what scientists have learned about many of the specific cases of amphibian demise, it remains

unclear why so many amphibians all around the world are dying off at the same time. Is one force weakening them all? Lacking an obvious "smoking gun," biologists continue to confront a variety of potential threats without knowing which ones cause the most serious problems for the populations they study.

4.3   Human population growth underlies threats to biodiversity

Perhaps the biggest threat to nonhuman species is the growth of human populations, since many of the problems that we have mentioned here are the direct result of human activities. For example, the growth of human populations is what spurs continuing habitat deterioration as natural areas are converted to the housing developments, farms, roads, and factories that sup-

Figure 4.8   Plants with Nowhere to Go
The Alpine androsace is one in a community of plant species that has been moving up the mountainsides of the Alps. For example, on Mount Hohe Wilde in the Austrian Alps, this mountain wildflower has been moving upward at a rate of about 2 meters per decade. Researchers believe that if global warming continues, this species will soon run out of mountaintop to climb and go extinct.

port human life. As more people drive more cars, burn more gas and oil, and use more paper, plastic, and other products, more pollution finds its way into the water and air, and more waste crowds the landfills, all of which hastens the demise of other species. In the search for solutions to these problems, we need to look not only at deforestation itself, for example, but also at the human population growth that drives it.

The factors we have discussed in this section are causing extinctions around the globe. Many more species have dwindled to dangerously low levels because of these same problems. While these species may continue to hang on for a while, their low numbers make them much more susceptible to future extinction.

> ■ The world's species are threatened by numerous forces, including habitat loss and deterioration, the introduction of nonnative species, climate change, and other forces still unidentified. The ever-increasing number of humans on Earth underlies most, if not all, of these challenges to the persistence of biodiversity.

The current mass extinction is not, however, the first in the history of life on Earth. Next we examine the other times in Earth's history when biodiversity has been devastated.

Mass Extinctions of the Past

Since the time when life first took hold on this planet, more than 3.5 billion years ago, there have been five well-documented mass extinctions. These mass extinctions took place about 440, 350, 250, 206, and 65 million years ago.

The most devastating mass extinction occurred 250 million years ago, at the end of the Permian geologic time period. At that time, some 80 to 90 percent of all marine species went extinct. The most famous mass extinction took place 65 million years ago, at the end of the Cretaceous and the beginning of the Tertiary periods. At that time, the last of the dinosaurs (not including the birds!—see Chapter 2) disappeared.

Unlike the current mass extinction, the previous mass extinctions were not caused by humans (which had not yet evolved). Scientists have hypothesized other causes for the different mass extinctions, including climate change, increased volcanic activity, reductions in sea level, and in the case of the dinosaurs' demise, the after-effects of an asteroid impact that filled Earth's atmosphere with a thick cloud of dust.

During these mass extinctions, certain groups of organisms disappeared, while others survived appar-

ently unscathed. And after each mass extinction, species numbers eventually rebounded as whole new groups of organisms evolved and colonized the planet. For example, when dinosaurs were the dominant creatures on Earth, there were only a few kinds of small mammals around. But when the dinosaurs went extinct, mammals evolved into many new species. These species began living in new kinds of habitat and exhibiting new behaviors. In fact, this great diversification of the mammals eventually resulted in the evolution of our own species.

If Earth has recovered from mass extinctions in the past, why should we be worried about the current mass extinction? Scientists suggest a variety of reasons for concern. First, it may be much more difficult for life to recover from the rapid and large-scale changes, such as the cutting of forests or the building of cities, that we humans are creating than it was to recover from the natural changes of past mass extinctions. Second, although new species might eventually evolve to increase species numbers again, those that are lost will be gone forever.

In addition, recovery from past mass extinctions required many millions of years. Ocean reefs, for example, have been destroyed and have recovered from mass extinctions multiple times in the past 500 million years, but their recovery time was on the order of 5 to 10 million years. Are we willing to wait that long? And are we willing to take the chance that one particular species—humans—might not survive the ongoing mass extinction?

> ■ There have been five previous mass extinctions in the history of life on Earth. Unlike the current mass extinction, the previous ones were not caused by humans. Species numbers recovered from each mass extinction over many millions of years.

Does the current mass extinction really threaten the survival of our own species? Are there other reasons why conserving biodiversity is important? In the next section we examine some of the arguments that are made for the value of biodiversity.

The Importance of Biodiversity

Many people wonder whether the loss of one mouse species here or one beetle species there really makes a difference to humanity. One question that has particularly interested biologists is how, if at all, biodiversity affects the forests, fields, wetlands, oceans, rivers, and other ecosystems whose plants produce the oxygen we breathe, clean the water we drink, produce much of the food we eat, and so on. During the 1990s, in various experimental studies, researchers showed that biodi-

Figure 4.9 Testing the Importance of Biodiversity
The ecotron is a tiny experimental ecosystem created in England to test the importance of biodiversity under controlled laboratory conditions. This constructed ecosystem includes grasses, wildflowers, snails, flies, and other organisms.

versity can indeed contribute to the health and stability of ecosystems.

Biodiversity can make ecosystems more productive, stable, and resilient

From tiny experimental ecosystems in a laboratory in England (Figure 4.9) to experimental prairies in the midwestern United States, researchers have found that, in many ways, the more species that are present in an ecosystem, the healthier the ecosystem appears to be. One measure of the health or vigor of an ecosystem is its **productivity**, the actual mass of plant matter—leaves, stems, fruits—that a plot can produce from the available nutrients and sunlight. Researchers looking at a variety of ecosystems have found that the more biodiversity there is, the more productive the ecosystem tends to be.

Why might this be? One explanation is that different species are good at using the different resources in a habitat. Some plant species grow best in bright sunlight. Others do best in the shade, and still others grow best in dappled light. The more different kinds of species there are, the more productively all the resources, from shade to bright sun, can be used. The same goes for other resources, such as water, the availability of which varies considerably within a habitat, from wet, muddy areas to dry, parched areas to areas that are occasionally very dry and at other times soaked. The more species there are in a habitat, the more likely it is that at least one of those species can use each of its different resources productively.

Researchers have also found evidence that the more species a prairie contains, the better able it is to recover and return to a healthy state following a drought. Researchers are continuing to study the effects of biodiversity, looking at its possible effects on an ecosystem's ability to resist disease and invasion by nonnative species.

The rivet hypothesis suggests that ecosystems can collapse quickly and unexpectedly

Although the presence of more species has enabled experimental ecosystems to function better and to be more stable, the question remains: How many species are required to keep an ecosystem from collapsing? That is, do we really need them all?

The **rivet hypothesis** provides some insight into this question. First put forward by Paul Ehrlich, an ecologist at Stanford University, this notion suggests that as species are lost from an ecosystem, at first there may be no noticeable effect, just as a few rivets lost from an airplane might not be noticed. After losing enough rivets, though, the plane would abruptly become a disintegrating bucket of metal crashing down out of the sky. Loss of species from an ecosystem might follow the same pattern as the loss of rivets from an airplane. Many species might be lost without any visible effect, but then, as a few more are lost, the ecosystem could suddenly, unexpectedly collapse.

Researchers already know that there are many different interactions between different groups of organisms in an ecosystem, creating a broad interdependence among them, just as there is among the many parts and pieces of an airplane. Put another way, just as rivets and other parts perform specific functions in the plane, species perform particular functions in ecosystems. For example, as we saw in Chapter 3, bacteria provide nitrogen in a usable form to plants, and plants form the basis of most terrestrial food webs (see Chapter 1 for a discussion of food webs). Many animals, fungi, and bacteria are important decomposers, returning nutrients from dead organisms to ecosystems in usable form. Plants provide food for many animals, which in turn harbor numerous forms of bacterial, protist, arthropod, and other life.

Like the airplane that loses rivet after rivet, after losing enough species, an ecosystem that has withstood the previous insults of species extinctions with no sign of harm or drastic change could suddenly, seemingly inexplicably, collapse under the cumulative loss of its members.

In addition, many scientists argue that while researchers can identify some of the ways in which biodiversity is useful to ecosystems, the infinite number of possible benefits of species cannot possibly be tested or even conceived of. For that reason, some argue, biodiversity should be preserved as a kind of insurance policy, in case we should ever need it.

Given the many known and potential interconnections among the living organisms in the ecosystems that provide us with oxygen, clean water, and so many other benefits, conserving species clearly will help maintain the health of these important ecosystems and, consequently, the continued existence of humanity.

> ■ Biodiversity increases the productivity of ecosystems, and some evidence suggests that it makes ecosystems more stable and resilient as well. The rivet hypothesis suggests that while the loss of individual species may not appear to matter, cumulative losses of species may lead to unexpected and sudden ecosystem collapse.

■ HIGHLIGHT

Harvesting the Fruits of Nature

When we lose a species, such as the golden toad pictured at the start of this chapter, what are we really losing? In addition to providing critical products such as oxygen and clean water, many of the world's nonhuman species have much to offer humanity, from drugs to foods to spiritual sustenance. Let's look at some of these important contributions to our existence.

The biosphere directly provides many products used by humans. One-fourth of all prescription drugs dispensed by pharmacies are extracted from plants. Nearly as many come from animals, fungi, or microscopic organisms such as bacteria. Quinine, used as an antimalarial drug, comes from a plant called yellow cinchona. Taxol, an important drug for treating cancer, comes from the Pacific yew tree. Bromelain, a substance that controls tissue inflammation, comes from pineapples. Wild species also provide chemicals that are useful as glues, fragrances, pesticides, and flavorings.

Wood for constructing homes and furniture is provided by many different tree species. Even the most basic requirements of human life are provided by other organisms. Plants produce the oxygen that we breathe. Every bit of food we eat is provided by other species.

A wide variety of crops and livestock, such as tomatoes and cattle, have been domesticated from wild species. In addition, these wild relatives can be of great importance to plant and animal breeders. Researchers looking to increase the sweetness of tomatoes, for example, might look for a very sweet wild relative of tomatoes to breed with domesticated plants. And, in many societies, species that remain wild provide important foodstuffs. Insects are an important source of protein for many peoples around the world. In Central America, some people dine on the green iguana, a huge lizard that likes to sunbathe in treetops. Known as the chicken of the trees, this lizard has been a food source for 7000 years. In some parts of South America, the capybara, the world's largest rodent, is a prized source of meat.

Capybara

Biodiversity also provides the world with aesthetic gifts. Scarlet macaws, parrot fish, sea anemones, tulip trees, and shooting-star wildflowers are among the species that make it clear that if there is a value to beauty, then biodiversity is worth a lot. For many people, the benefits of such a rich living world go beyond mere beauty to providing spiritual refreshment and rejuvenation. Consider also the argument that biodiversity does not have to fulfill any human need in order to have value—that each species has the right to exist unperturbed and not to be destroyed by other species, such as our own.

As great as the bounty of nature already is, with so many species yet undiscovered, the vast majority of nature's wealth remains untapped. If the sheer numbers of species yet to be discovered are any indication, many more treasures await us—food, shelter, medicine, beauty—if we can find them before they disappear along with the golden toad and so many other species.

■ Biodiversity provides many products to human society: food, shelter, clothing, medicines, and, for some, spiritual rejuvenation.

SUMMARY

How Many Species Are There on Earth?

■ Scientists do not know the exact number of species on Earth.

■ Using indirect methods, scientists have generated estimates of the total number of Earth's species, with most estimates falling between 3 million and 30 million.

■ A large proportion of the world's species remain unknown.

■ Of the 1.5 million species already discovered and named, about half are insects.

The Beginnings of a Present-Day Mass Extinction

■ Although it is difficult to measure numbers of extinct species and extinction rates precisely, abundant evidence suggests that Earth is in the early stages of a mass extinction, which could end up being the most rapid in the history of life.

■ Humans are the cause of the current mass extinction.

The Many Threats to Biodiversity

■ Among the threats to biodiversity today are habitat loss and deterioration, invasion of nonnative species, and climate change.

■ Many of the threats to biodiversity today are a result of the continuing growth of human populations.

Mass Extinctions of the Past

■ Each of the five previous mass extinctions on Earth wiped out millions of species, devastating the planet's biodiversity.

■ The previous mass extinctions were not caused by humans.

■ After each mass extinction of the past, the number of species on Earth rebounded slowly, over millions of years.

The Importance of Biodiversity

■ Humanity depends on healthy ecosystems for oxygen and clean water, among many other things.

■ Biodiversity can make ecosystems more productive, stable, and resilient.

■ The rivet hypothesis suggests that although the loss of one species here or there may have no noticeable effect, the cumulative effect of continuing species losses could result in the sudden, seemingly inexplicable collapse of ecosystems.

Highlight: Harvesting the Fruits of Nature

■ The many species on Earth provide humanity with a multitude of products, including foods, medicines, material for shelter and clothing, and other useful and beautiful items.

KEY TERMS

biodiversity p. 68	nonnative species p. 72
canopy p. 68	productivity p. 76
habitat p. 72	rainforest p. 71
mass extinction p. 71	rivet hypothesis p. 76

CHAPTER REVIEW

Self-Quiz

1. Estimates of the total number of species on Earth range between
   a. 1 billion and 3 billion.
   b. 50 million and 500 million.
   c. 3 million and 100 million.
   d. 1000 and 3000.

2. The number of species that have been collected, identified, and named is about
   a. 15,000.
   b. 150,000.
   c. 1,500,000.
   d. 15,000,000.

3. The known species of insects, plants, and fungi each number in the
   a. hundreds.
   b. thousands.
   c. millions.
   d. billions.

4. Which of the following has been hypothesized to be a cause of past mass extinctions?
   a. human population growth
   b. water pollution
   c. an asteroid impact
   d. disappearance of rainforests

5. Which of the following is *not* considered a threat to species today?
   a. rainforest destruction
   b. human population growth
   c. climate change
   d. scientific research

6. Which of the following have scientists shown to be a benefit that biodiversity offers ecosystems?
   a. increased resistance to invasion by nonnative species
   b. increased productivity
   c. increased resistance to pollution
   d. increased resistance to disease

Review Questions

1. How did Terry Erwin estimate the number of arthropod species in the world, and why are such estimates so difficult to do?

2. How has biodiversity risen and fallen over time—when and in what way?

3. How has human population growth helped spur other threats to biodiversity?

The Daily Globe

Protect the Rights of Humans, Not Animals

To the Editor:
The most useful thing Congress could do this year is throw out the Endangered Species Act. This law, which supposedly protects endangered species from being destroyed by human activities, has everything completely upside down. I live in the Pacific Northwest, where many of us have long and proud family traditions in the timber industry. Now all of a sudden we can't cut trees because conservationists are worried that a bird, like the spotted owl or the marbled murrelet, might be killed off in the process. People who are living in houses built of wood that my grandfather risked his life to harvest are telling us we can't go into these forests to make a living anymore!

The same kind of thing is happening everywhere, every day, with insects, worms, and fish getting in the way of progress. What's more important—a person making a living, or a bug that no one's ever heard of? I am sick of reading in your paper about how the world's species are going extinct and how terrible that is. No one would care if these species went extinct, but I care about providing for my family. When you write about the destruction of tropical rainforests, what you never say is that most of the species in rainforests are insects that we all hate anyway! Biologists don't even know exactly how many species there are on Earth, let alone how many are going extinct. I have no doubt that most species —except humans—are doing just fine. What's really endangered are the rights of people.

Steven S. Gotling

Evaluating "The News"

1. Do nonhuman species have a right to exist? If so, where do the rights of human beings end and the rights of other species begin? Should logging or other industries or development ever be curtailed to protect, for example, a "bug that no one's ever heard of"?

2. In the past, when endangered species were threatened by a new building or other development, that development was often forbidden by the courts and the Endangered Species Act. Now, more often, when development conflicts with the survival of an endangered species, developers and conservationists compromise. Some land is used for building, and some is set aside for the endangered species. Why might such a compromise be a good idea? Why not?

3. The author of the letter to the editor is correct in stating that biologists don't know either the exact number of species on Earth or the exact rate at which species are going extinct. Given their lack of precise information, should we take seriously biologists' concerns about worldwide extinction of species? Why or why not?

4. Write a short letter to the editor, arguing either for or against the importance of preserving biodiversity and backing up your statements with evidence wherever possible. This issue is alive and kicking. Every community has its conflicts between biodiversity and development, and every voice counts. Send your letter to the editor of the local newspaper.

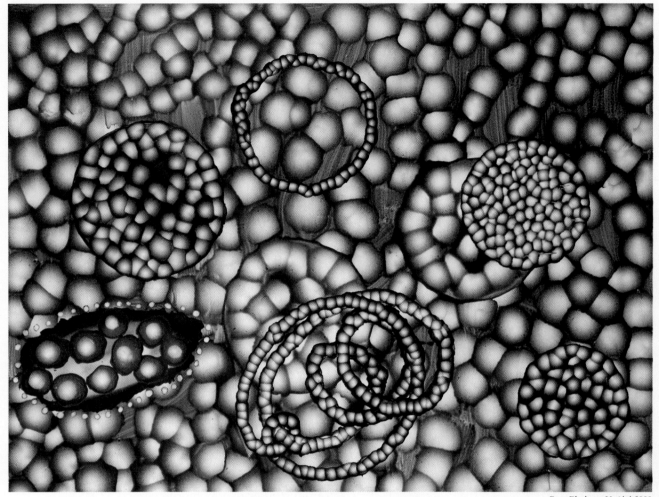

Ross Bleckner, *Untitled*, 2000.

Cells: The Basic Units of Life

5 Chemical Building Blocks

chapter

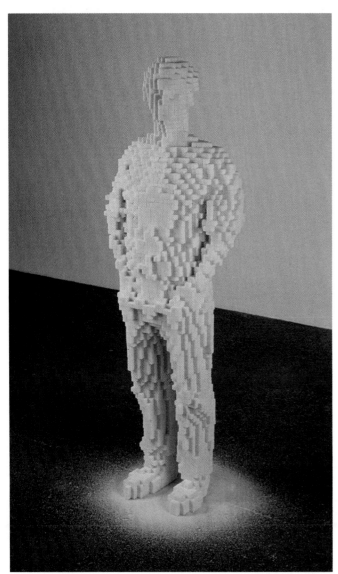

Tom Friedman, *Untitled*, 1999.

Looking to the Stars

Throughout recorded history, human beings have gazed up at the stars and wondered what secrets those distant points of light might hold. The idea that the stars might tell us something about our origins has gained support in recent years, thanks to several unlikely discoveries in the ice of Antarctica and Canada. These discoveries, in the form of meteorite fragments embedded in ice, have yielded insights into the chemical composition of the universe and even the possibility of life on other planets.

In 1996, people who believe that extraterrestrial life may exist received a tremendous boost from studies of a meteorite named ALH84001. It was ejected from the planet Mars more than 100 million years ago and landed in Antarctica about 13,000 years ago. There it lay buried in glacial ice until its discovery in the twentieth century. More recent studies of ALH84001 have thrust it into the scientific and public spotlight.

How could a simple rock provide evidence of life on another planet? Perhaps the most dramatic discovery thus far has been shapes embedded in the rock that resemble microscopic fossils. These shapes look remarkably like bacteria, and one in particular appears almost wormlike. The shock of seeing what looks like the remains of living things from another planet jolted the scientific world and injected new life into the public's belief in extraterrestrial life.

The idea that the observed forms were once alive has been met with skepticism. Some sci-

82

entists point to the need for other kinds of evidence, such as the chemical compounds found in ALH84001. If certain compounds that contain carbon are found, they might be ancient traces of biological activity on Mars—the chemical remains of Martian life. With this in mind, scientists are applying what they have learned about the chemical compounds that make up life on Earth to search for similar compounds in ALH84001.

Another important meteorite discovery was made in a frozen lake in northwestern Canada. On a cold January night in 2000, an ancient meteorite entered Earth's atmosphere, creating a shower of fireballs over Tagish Lake. The spectacle was witnessed by dozens of people on the ground and even

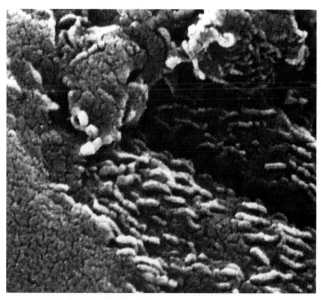

Have the Martians Landed?
Tiny lifelike shapes can be seen in the ALH84001 meteorite.

by American military satellites. Given that roughly a thousand meteorites have been witnessed entering Earth's atmosphere over the past 200 years, what made this event so significant? The answer lies in the fact that this meteorite was an ancient C1 chondrite, which is not only a very rare type of meteorite, but also contains interstellar matter dating from the birth of our solar system.

One of the witnesses of the Tagish Lake event, Jim Brook, carefully gathered frozen samples a few days later and stored them in his freezer. By maintaining the fragments in a frozen state, Brook prevented the loss of chemical compounds that would have turned to gas on thawing and avoided contamination with terrestrial compounds. These pristine meteorite samples are remarkable for the insights they give us into the organic compounds present at the birth of our solar system. The compounds found in the samples, made up of carbon, nitrogen, and sulfur, may even be similar to those that fell on the early Earth and contributed to the origin of life. Thus, studying such meteorite fragments amounts to studying the chemical building blocks that may have gone into the first living systems on Earth.

Both of these exciting studies of meteorite fragments assume that all life consists of a limited number of chemical compounds that interact with one another in well-established ways—a reasonable assumption, as we'll see in this chapter.

■ KEY CONCEPTS

1. Living organisms are composed of atoms linked together by chemical bonds. Arrangements of linked atoms (molecules) are essential for the processes of life.

2. Covalent bonds are the strongest chemical linkages that can form between two atoms. Most molecules found in living organisms are arrangements of atoms such as carbon, nitrogen, hydrogen, oxygen, phosphorus, and sulfur that are held together by covalent bonds.

3. Noncovalent bonds, which are weaker and more easily broken, may form between two or more separate molecules and between parts of a single large molecule. These bonds help give living systems their dynamic properties.

4. The chemical characteristics of water are essential to the chemistry of life. Water is both the primary medium for and a key participant in life-supporting chemical reactions.

5. Chemical reactions change the arrangement of atoms in molecules. These changes are responsible for the many different processes observed in living organisms.

6. The four major classes of chemical building blocks found in living organisms are sugars, nucleotides, amino acids, and fatty acids. Each class has several functions in living systems, ranging from information storage to energy transfer to structural support.

For all its remarkable diversity, life as we know it is based on a rather limited number of chemical elements, which are found throughout the universe. The fact that such complexity can come from such simple components shows how we humans could have arisen from the chemical soup of a primitive Earth, using the same chemical elements that make up the very Earth we stand on. More than anything else, our origin in Earth's chemical elements reminds us of the fundamental unity between our bodies and our surroundings.

In this chapter we begin to explore the tremendous complexity of living organisms by first identifying their simplest chemical components. Then we examine how these simple components are linked together to form the many levels of organization in both the physical structures and chemical processes of life.

How Atoms Make Up the Physical World

All the components of the physical world are made up of 92 natural chemical **elements**. These elements, such as oxygen and calcium, are the simplest building blocks of the physical world. Each element is identified by a one- or two-letter symbol: for example, oxygen is identified as O, calcium as Ca. Each element exists in units so small that more than a trillion of them could easily fit on the head of a pin. These tiny units are called **atoms**, and there are 92 different kinds, one for each natural element. An atom is therefore the smallest unit of an element that still has the characteristic chemical properties of that element.

It stands to reason that the properties of an element such as oxygen must depend on the properties of oxygen atoms. So what makes the atoms of one element different from those of another? The answer lies in the different combinations of what are called subatomic particles, bits of matter that make up the structure of individual atoms. Two kinds of subatomic particles of particular interest are **protons** and **electrons**. Both are electrically charged: Protons have a positive charge (+), and electrons have a negative charge (−). These opposite charges determine how atoms behave in the physical world and interact with one another: Like charges repel each other; opposite charges attract each other.

A single atom has a central core that contains one or more protons and is thus positively charged. That core is surrounded by one or more negatively charged electrons (Figure 5.1). The positive core is called the nucleus of the atom, and electrons move around it like moons orbiting a planet. However, the troupe of electrons in a given atom is not fixed; the atom can lose or gain electrons, or even share them with another atom. When an atom loses one or more of its negative electrons, it becomes positively charged. Likewise, when an atom gains one or more electrons, it becomes negatively charged. Atoms that become charged due to the loss or gain of electrons are called **ions**.

The number of electrons associated with an atom determines the chemical properties of the element. These properties allow the atoms of one element to form linkages either with one another or with atoms of other elements. Orderly associations of atoms of different elements form what are known as **chemical compounds**. The smallest unit of such a compound with the required arrangement of atoms is called a **molecule**.

Chemists have developed a simple system of molecular formulas to represent compounds made up of mol-

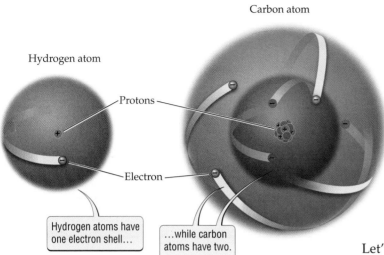

Carbon atom

Hydrogen atom

Protons

Electron

Hydrogen atoms have
one electron shell...

...while carbon
atoms have two.

5.1

Figure 5.1   Atoms
The subatomic particles that make up these carbon and hydrogen atoms are shown greatly enlarged in relation to the size of the whole atom. Negatively charged electrons move rapidly around the positively charged nucleus in defined orbits that lie in layers, or shells, around the nucleus. Each shell can contain a certain maximum number of electrons. The differences in the numbers of subatomic particles in these two atoms give them different chemical properties.

ecules ranging in size from two atoms to many thousands of atoms. Table salt, for example, is a chemical compound that has equal numbers of sodium (Na) and chlorine (Cl) atoms. The molecular formula for salt is therefore NaCl. When the ratio of the different atoms is not one to one, a subscript number is listed after any atoms that are present more than once. For example, each molecule of water is made up of two hydrogen (H) atoms and one oxygen (O) atom, so the molecular formula for water is H_2O. The same notation is used for more complex compounds, such as table sugar (sucrose), which has 12 carbons, 22 hydrogens, and 11 oxygens per molecule ($C_{12}H_{22}O_{11}$).

■ All components of the physical world, including living organisms, are made up of atoms, which are linked together in specific associations called molecules.

Covalent Bonds: The Strongest Linkages in Nature

All chemical linkages that hold atoms together to form molecules are called bonds. The strongest of these linkages are **covalent bonds**, which link the atoms within a single molecule. In covalent bonds, atoms share electrons (Figure 5.2a). As Figure 5.1 shows, the electrons around the nucleus of every atom move in defined orbits that lie in layers, or shells, around the nucleus. Each shell can contain a certain maximum number of electrons, and the atom is in its most stable state when all the shells are filled to capacity. One way for an atom to fill its outermost shell is to share electrons with a neighboring atom. This sharing of electrons links the two atoms, forming a strong, stable covalent bond.

Let's look at the makeup of some chemical compounds to see how these bonds work. Water and the natural gas called methane have the molecular formulas H_2O and CH_4, respectively. These formulas reveal the atomic components of each compound, but they say nothing about how the various atoms are bonded together and arranged in space. Another type of notation, known as a structural formula, is used to indicate both the atomic components and the bonding arrangement of compounds. As Figure 5.2b shows, a water molecule is held together by covalent bonds between its single oxygen atom and each of its two hydrogen atoms. Likewise, methane has four covalent bonds that link its lone carbon atom to each of its four hydrogens.

The bonding properties of each element are determined by its electrons, so the bonding arrangement of atoms in a particular chemical compound is never accidental or random. Thus, there will never be a water molecule in which one hydrogen is bonded to another hydrogen, which is then bonded to oxygen. When two hydrogen atoms share electrons, they form hydrogen gas (H_2) and have no electrons left to share with an oxygen atom. Oxygen, on the other hand, has room in its outer shell of electrons to form two covalent bonds with hydrogen atoms (H_2O). As these examples show, covalent bonds occur in well-defined ways that establish the spatial arrangement of the atoms and hence the three-dimensional shape of the molecule.

Carbon, nitrogen, hydrogen, oxygen, phosphorus, and sulfur are the most common elements found in living organisms, and their atoms can all form covalent bonds. Consequently, combinations of all or some of these atoms are found in the molecules that make up living organisms.

■ Covalent bonds, formed by the sharing of electrons between atoms, connect the atoms within a molecule. They are the strongest chemical bonds in nature.

(a)

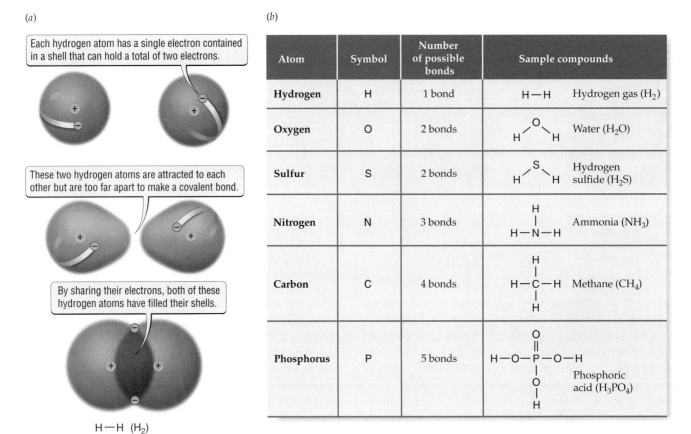

Each hydrogen atom has a single electron contained in a shell that can hold a total of two electrons.

These two hydrogen atoms are attracted to each other but are too far apart to make a covalent bond.

By sharing their electrons, both of these hydrogen atoms have filled their shells.

H—H (H$_2$)

(b)

Atom	Symbol	Number of possible bonds	Sample compounds	
Hydrogen	H	1 bond	H—H	Hydrogen gas (H$_2$)
Oxygen	O	2 bonds		Water (H$_2$O)
Sulfur	S	2 bonds		Hydrogen sulfide (H$_2$S)
Nitrogen	N	3 bonds		Ammonia (NH$_3$)
Carbon	C	4 bonds		Methane (CH$_4$)
Phosphorus	P	5 bonds		Phosphoric acid (H$_3$PO$_4$)

Figure 5.2   Covalent Bonds
(a) Two hydrogen atoms can share electrons, forming a covalent bond. The length of a covalent bond is such that the electrons can be shared, while the two positively charged nuclei are still kept far enough apart that they don't repel each other. (b) The number of covalent bonds that an atom can form depends on the number of extra electrons needed to fill its outermost shell. An oxygen atom, for example, has six electrons in an outer shell that can hold eight of them; therefore, it requires two more electrons to fill its shell and thus can form two covalent bonds. A carbon atom requires four more electrons and thus can form four covalent bonds.

5.2 Noncovalent Bonds: Dynamic Linkages between Molecules

Whereas covalent bonds link atoms to form molecules, **noncovalent bonds** are the most common linkages between separate molecules and between different parts of a single large molecule. By bringing molecules together in specific configurations, noncovalent bonds promote the complex organization that we perceive as the diverse physical structures and activities characteristic of living organisms. Despite this important role, noncovalent bonds are far weaker than covalent bonds and do not involve the

direct sharing of electrons between atoms. Instead, the various kinds of noncovalent bonds are based on other chemical properties of two or more atoms.

Because of the relative weakness of noncovalent bonds, many such bonds are often required to hold two molecules together, and even to link two parts of the same large molecule. In this case, however, weakness is a virtue, because it allows molecules to come together and pull apart easily. The resulting changeability of molecular arrangements is necessary for many living processes. When your skin is pinched and then released, for example, its ability to stretch and then spring back depends on the breaking and re-forming of noncovalent bonds between many different classes of molecules.

Hydrogen bonds are important temporary bonds

The **hydrogen bond** is one of the most important kinds of noncovalent bonds in nature. Each hydrogen bond is about 20 times weaker than a covalent bond. During rapidly changing life processes such as muscle contraction, hydrogen bonds are broken and re-formed moment by moment.

The simplest example of a hydrogen bond can be found between water molecules (Figure 5.3). Hydro-

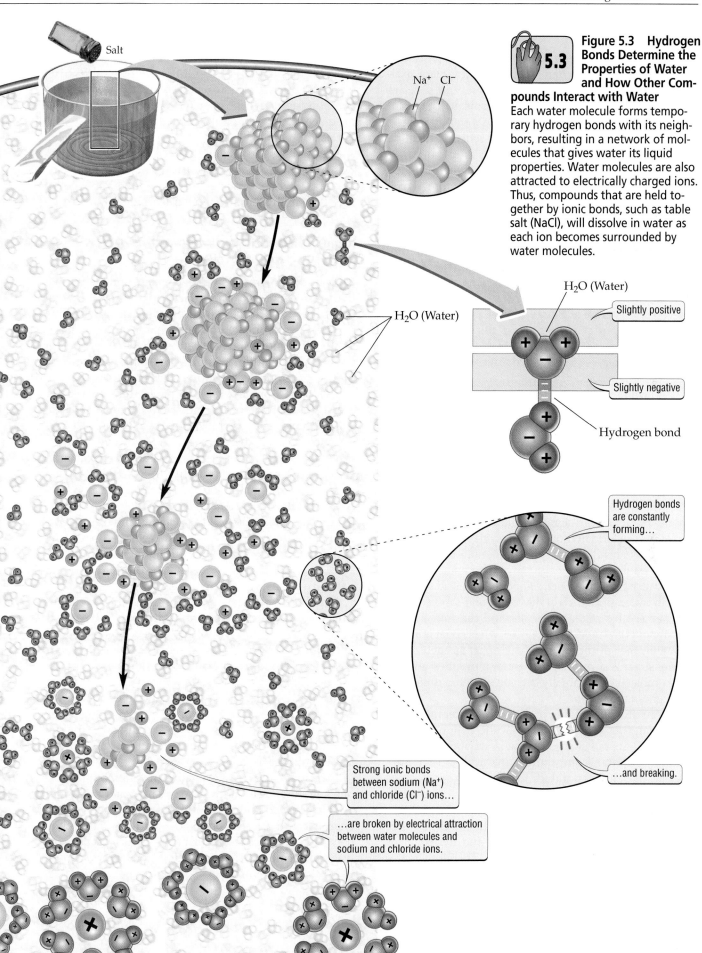

5.3

Figure 5.3   Hydrogen Bonds Determine the Properties of Water and How Other Compounds Interact with Water
Each water molecule forms temporary hydrogen bonds with its neighbors, resulting in a network of molecules that gives water its liquid properties. Water molecules are also attracted to electrically charged ions. Thus, compounds that are held together by ionic bonds, such as table salt (NaCl), will dissolve in water as each ion becomes surrounded by water molecules.

Salt

Na^+ Cl^-

H_2O (Water)

H_2O (Water)

Slightly positive

Slightly negative

Hydrogen bond

Hydrogen bonds are constantly forming…

…and breaking.

Strong ionic bonds between sodium (Na^+) and chloride (Cl^-) ions…

…are broken by electrical attraction between water molecules and sodium and chloride ions.

gen bonding is an important attribute of water and is based on a particular chemical property of water molecules. The properties of water, as we will see, are extremely important to living organisms. Most organisms consist of more than 70 percent of water by weight, and nearly every chemical process associated with life occurs in water.

As discussed earlier, each molecule of water is made up of two hydrogen atoms and one oxygen atom, which are held together by covalent bonds. The positively charged nucleus of the oxygen atom tends to attract the negative electrons of the two hydrogen atoms, causing each water molecule to have an uneven distribution of electrical charge. That is, the oxygen atom carries a slightly negative charge, while the hydrogen atoms in turn carry a slightly positive charge. Molecules with such an uneven distribution of charge are described as **polar**.

Because opposite charges attract, the slightly positive hydrogen atoms of one water molecule are attracted to the slightly negative oxygen atom of a neighboring water molecule. This attraction forms a hydrogen bond between the two water molecules. Those molecules, in turn, can form hydrogen bonds with other neighboring water molecules (see Figure 5.3). The resulting network of water molecules, in which hydrogen bonds are constantly forming and breaking, is responsible for the liquid properties of water.

Any polar compound that contains hydrogen atoms can form hydrogen bonds with a neighboring polar compound that contains a slightly negative atom. Thus water molecules can form hydrogen bonds with other polar compounds. As a result, when such compounds are placed in water, their molecules break apart and become surrounded by water molecules. Such compounds are said to be **soluble** in water, since each of the compound's molecules is surrounded by water molecules and can move freely throughout the liquid. For example, when sugar is added to water, the solid crystals break apart as each molecule of sugar is surrounded by water molecules, and the sugar molecules eventually become scattered uniformly throughout the water.

The hydrogen-bonding properties of water molecules also mean that they will not associate with other molecules that are not charged—that is, that are **nonpolar**. When nonpolar compounds are added to water, instead of being separated, like the sugar molecules above, they are pushed into clusters. This is exactly what happens when olive oil is added to a salad dressing. In the molecules that make up the oil, the distribution of electrons between carbon and hydrogen is nearly equal, making them nonpolar. Therefore, water

molecules do not mix with the oil molecules. Instead, the oil molecules tend to cluster into tiny floating droplets when the dressing is shaken. Waxes are also nonpolar, and automobile enthusiasts wax their cars not just to make them look shiny, but also to repel water and reduce the risk of rusting.

Molecules such as sugar that interact freely with water are said to be **hydrophilic** (*hydro*, "water"; *philic*, "loving"), while those such as oil that are repelled by water are **hydrophobic** (*phobic*, "fearing").

Ionic bonds form between atoms of opposite charge

Ionic bonds are important noncovalent bonds that, like hydrogen bonds, rely on the fact that opposite electrical charges attract. In this case, however, the attraction is between two atoms of opposite charge, and it has different consequences. The electrons in the outer shell of one atom are completely transferred to the outer shell of the second atom. As mentioned earlier, the loss of electrons by one atom and the gain of electrons by another means that both atoms become charged ions.

Ionic bonds between molecules dissolved in water are relatively weak. Like hydrogen bonds, however, they are essential for many temporary associations between molecules. For example, our ability to taste what we eat depends on the ionic bonds that form between food molecules dissolved in water and other specialized molecules in our taste buds. The rapid association and dissociation of multiple food molecules also enable us to discern several different tastes in a short amount of time. If these associations were not brief, every meal would be dominated by the taste of the first bite.

Ionic bonds between molecules under dry conditions can be very strong. For example, dry table salt consists of countless sodium ions (Na^+) linked by ionic bonds to chloride ions (Cl^-). In the absence of water, these ions pack tightly to form the hard, three-dimensional structures we know as salt crystals. When salt is added to water, however, the polar water molecules are attracted to the charges surrounding both types of ions. This interaction with water breaks up and dissolves the salt crystals, scattering both positive and negative ions throughout the liquid (see Figure 5.3).

Like sugar, then, salt is soluble in water. But there is one key difference: In the case of salt, water molecules are attracted to and surround each ion because of its charge, but hydrogen bonds do not form. Hydrogen bonds form only between polar compounds, not between water molecules and ions. Therefore, ions and polar molecules form two different classes of hydrophilic compounds that dissolve in water.

> ■ Noncovalent bonds, such as hydrogen bonds and ionic bonds, are temporary linkages that are responsible for the dynamic character of biological molecules, which is necessary for living processes.

How Chemical Reactions Rearrange Atoms in Molecules

The arrangements of atoms found in molecules are the simplest examples of physical structures found in living organisms. Molecules, as you will recall from Chapter 1, are the first level of the biological organizational hierarchy. We have identified the covalent bonds that link atoms into molecules and the noncovalent bonds that exist between molecules. However, molecules are not static arrangements of atoms. Living processes require that atoms break existing connections and form new ones with other atoms. Consider the additional molecules that must be made to provide a growing plant with components for new leaves and stems. The plant does not produce these new molecules from raw atoms; rather, it acquires and rearranges preexisting molecules. The processes by which bonds between atoms are formed and broken are known as **chemical reactions**.

The standard notation for chemical reactions describes changes in the arrangement of atoms. Nitrogen and hydrogen, for example, can combine to produce ammonia gas (NH_3), which gives laundry bleach its sharp odor. This chemical reaction is written as follows:

$$3 H_2 + N_2 \rightarrow 2 NH_3$$

The numbers before the molecular symbols show how many molecules participate in the reaction. In this case, three molecules of hydrogen gas (H_2) combine with one molecule of nitrogen gas (N_2) to produce two molecules of ammonia. The arrow indicates the direction of the reaction; that is, it shows that hydrogen and nitrogen are converted to ammonia, not the reverse.

Notice that the reaction begins and ends with the same number of each type of atom. The ratio of atoms that make up the molecules at the start of the reaction (known as the reactants) must be the same as the ratio of atoms that make up the molecules at the end of the reaction (known as the products). That is, the total number of each kind of atom on the left side of the reaction must match the number of that atom on the right side. In the example here, there are six hydrogen atoms and two nitrogen

atoms on each side of the reaction. All chemical reactions rearrange atoms in this way; chemical reactions can neither create nor destroy atoms.

Later in this chapter we'll see how chemical reactions promote complex molecular organization. First, however, we focus on the chemical reactions that determine how ions and molecules interact with water.

Acids, bases, and pH

5.4 Nearly all chemical reactions that support life occur in water. Of particular importance are reactions that involve two classes of compounds: acids and bases. An **acid** is a polar compound that gives up one or more hydrogen ions (H^+) when it dissolves in water. These hydrogen ions are lost because they are attracted by the surrounding water molecules, with which they form positively charged hydronium ions (H_3O^+). Because the formation of hydronium ions is easily reversed, however, hydrogen ions are constantly being exchanged between water molecules and other molecules dissolved in water. Thus, some free hydrogen ions will be present in the water at any time. Hydroxyl ions (OH^-) are left when water molecules lose hydrogen ions.

Bases, which are also polar, are compounds that accept hydrogen ions from water molecules, reducing the number of free hydrogen ions that are present. Thus acids and bases interact with water molecules in different ways and have opposite effects on the amount of free hydrogen ions in water.

The concentration of free hydrogen ions in water influences the chemical reactions of many other molecules. This hydrogen ion concentration is expressed on a scale of numbers from 0 to 14, where 0 represents the highest concentration of free hydrogen ions and 14 represents the lowest. This scale is called the **pH** scale, and each unit represents a 10-fold increase or decrease in the concentration of hydrogen ions.

When water contains no acids or bases, the concentrations of free hydrogen ions and hydroxyl ions are equal, and the pH is said to be neutral, or in the middle of the scale, at pH 7. Below pH 7, the solution is said to be acidic because the concentration of free hydrogen ions is higher as a result of donations from an acid. Above pH 7, the solution is said to be basic because the concentration of free hydrogen ions is lower as a result of the acceptance of hydrogen ions by a base.

We have all experienced acidic and basic substances. The tartness of lemon juice in a good homemade lemonade is due to the acidity of the juice (about pH 3). Our stomach juices are able to break down food because they are very acidic (about pH 2). At this low pH, many bonds between molecules are disrupted by the high concentration of free hydrogen ions, which associate with atoms that would otherwise be bonded to other atoms. At the other extreme is a substance such as oven cleaner, which is very basic (about pH 13).

Buffers prevent large changes in pH

Most living systems function best at an internal pH that is close to 7. Any change in pH to a value significantly below or above 7 adversely affects many life processes. Because hydrogen ions can move so freely from one molecule to another, organisms need to control the pH levels of their internal environments. This need is met by compounds called **buffers**, which maintain the concentration of hydrogen ions within narrow limits. Most buffers are weak acids or bases or both, and can therefore release or accept free hydrogen ions to maintain a relatively constant concentration in the internal environment.

> ■ Chemical reactions are processes by which chemical bonds between atoms are formed and broken, altering the arrangement of atoms in molecules and generating new chemical properties. The chemical characteristics of water dictate the properties of acids and bases and the pH of the environment.

The Chemical Building Blocks of Living Systems

If we removed the water in any living organism, we would be left with four major classes of chemical compounds: sugars, nucleotides, amino acids, and fatty acids. These chemical compounds consist of different arrangements of carbon atoms, though hydrogen, oxygen, nitrogen, phosphorus, and sulfur atoms may also be present.

Carbon is the predominant element in living systems, partly because it can form large molecules that contain thousands of atoms. A single carbon atom, as we have seen, can form strong covalent bonds with up to four other atoms. Even more importantly, carbon can bond to carbon, forming long chains, branched trees, or even rings (Figure 5.4*a*). Living organisms can therefore make a wide variety of large and complex structures using carbon-containing mol-

ecules. All carbon-containing compounds found in living organisms are described as **organic** compounds.

Small carbon-containing molecules can either remain as individual molecules or bond with other small molecules to form larger structures called **macromolecules** (*macro*, "large"). Living organisms often follow this basic principle of building very large and complex structures from small units. Individual small molecules, containing about 20 atoms or fewer, are called **monomers** (*mono*, "one"; *mer*, "part"). Macromolecules, which can contain hundreds of monomers bonded together (Figure 5.4*b*), are called **polymers** (*poly*, "many"). Polymers account for most of an organism's dry weight and are essential for every structure and chemical process that we associate with life.

In living organisms, fewer than 70 different organic monomers are combined in a nearly endless variety of ways to produce polymers with many different properties. Polymers are therefore a step up from monomers in organizational complexity, and they have chemical properties that are not possible for a monomer. Furthermore, many organic polymers acquire chemical properties from attached functional groups (Table 5.1). As their

5.1 Some Important Functional Groups Found in Organic Molecules

Functional group	Formula	Ball-and-stick model
Amino	NH_2	
Carboxyl	COOH	
Hydroxyl	OH	
Phosphate	PO_4	

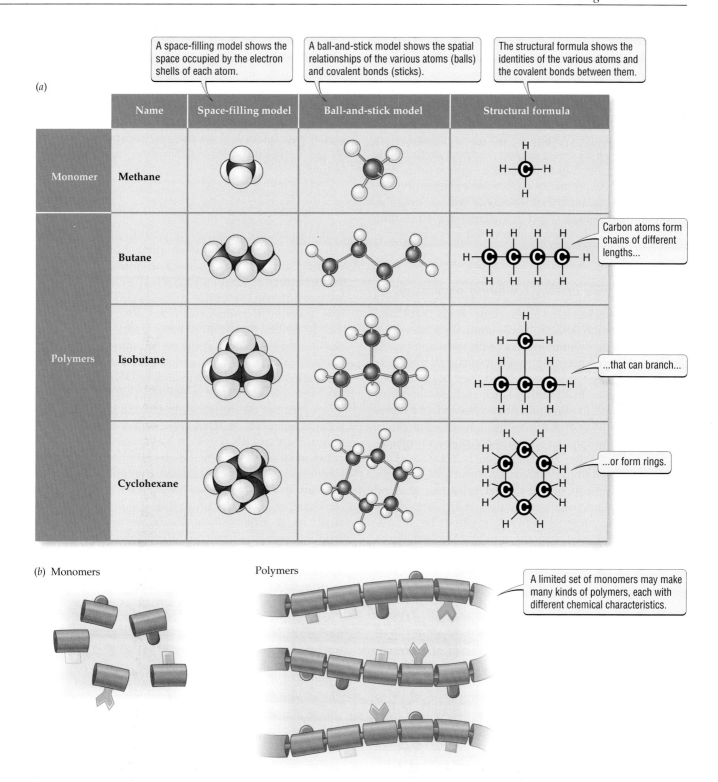

Figure 5.4 Assembling Complex Structures from Smaller Components
Living organisms build large and complex structures from small components. (*a*) A single carbon atom can form four covalent bonds with other atoms. When carbon atoms form bonds with other carbon atoms, a variety of chains, branches, and rings can be formed. (*b*) Small molecules can form covalent bonds with one another, giving rise to larger molecules with specific chemical properties. The small constituent compounds are called monomers, while the resulting large assemblage is called a polymer.

name implies, **functional groups** are groups of covalently linked atoms that have specific chemical properties wherever they are found. Some functional groups help establish covalent linkages between monomers; others have more general effects on the chemical characteristics of a polymer.

The importance of these properties is illustrated in each of the four classes of organic compounds found in all living systems: sugars, nucleotides, amino acids, and fatty acids. As we consider each of these classes of compounds in the sections that follow, we will see how the tremendously complex and changeable structures that make up living organisms can be based on such a limited range of organic compounds.

Sugars provide energy for living organisms

Sugars are familiar to us as the compounds that make foods taste sweet. Although only some sugars are perceived as sweet by human taste buds, most sugars are important food sources and a major means of storing energy in living organisms. Sugars and their polymers are referred to as **carbohydrates**, since each carbon atom (*carbo-*) is linked to two hydrogen atoms and an oxygen atom—the equivalent of a molecule of water (*-hydrate*).

The simplest sugar molecules are called **monosaccharides** (*mono*, "one"; *sacchar*, "sugar"). They are made up of units containing carbon, hydrogen, and oxygen atoms in the ratio of 1:2:1. This ratio can also be expressed as the molecular formula $(CH_2O)_n$, with n representing the number of units ranging from 3 to 7. Sugars are often referred to by the number of carbon atoms they contain; for example, a sugar with the chemical formula $(CH_2O)_5$ is called a five-carbon sugar.

The best-known example of a monosaccharide is **glucose**. Glucose has a key role in short-term energy storage, and nearly all of the chemical reactions that produce energy for living organisms involve the synthesis and breakdown of glucose. Glucose is such a major player, particularly in the metabolism of animal cells, that elaborate control mechanisms regulate the concentration of glucose in the body in response to changing energy needs. The role of glucose in providing energy for life will be discussed further in Chapter 8.

Monosaccharides can combine to form larger, more complex molecules. For example, two covalently bonded molecules of glucose form maltose (Figure 5.5*a*). Up to thousands of monosaccharides can be linked together in a similar way to form a polymer called a **polysaccharide**.

Carbohydrates perform several different functions in living organisms. A polysaccharide called cellulose forms strong parallel fibers that help support the leaves and stems of plants (Figure 5.5*b*). Other carbohydrates, such as starch, store energy, and are the basis for "carbo loading" before a strenuous activity such as running a marathon (Figure 5.5*c*). Just as nucleic acids have more than one essential function in the life of an organism, sugars play multiple roles, ranging from providing energy to forming important structural components of the body.

Nucleotides store information and energy

Nucleotides are organic compounds that consist of linked rings of atoms. In each nucleotide (Figure 5.6), one ring structure, known as a **nucleotide base**, is covalently bonded to another ring structure, a sugar. There are five different nucleotide bases, named adenine, cytosine, guanine, thymine, and uracil. Two different sugars can be attached to those bases. Deoxyribose sugars can bond to adenine, cytosine, guanine, or thymine bases, forming the building blocks for a stringlike polymer called deoxyribonucleic acid, or **DNA**. Ribose sugars can bond to adenine, guanine, cytosine, and uracil bases, forming the building blocks for a polymer called ribonucleic acid, or **RNA**. (We will learn a lot more about DNA and RNA in Unit 3, on genetics.) The third component of each nucleotide is the phosphate group, a functional group that consists of a phosphate atom and four oxygen atoms (see Table 5.1). A nucleotide can have one, two, or three phosphate groups bound to the sugar. Nucleotides in these three states are described as monophosphate, diphosphate, or triphosphate nucleotides, respectively.

Nucleotides have two essential functions in the cell: information storage and energy transfer. Both of these functions highlight key characteristics of living systems. Every living organism, as we saw in Chapter 1, is defined by a DNA "blueprint" that dictates what chemical building blocks are produced to make up the organism and how they are assembled. The order in which different nucleotides are hooked together in DNA or RNA polymers determines the physical attributes of, and the chemical reactions that occur in, living organisms. The structures and roles of these polymers will be discussed in Chapter 14.

In addition to storing genetic information, nucleotides store and transfer energy. A key player in energy transfer is adenosine triphosphate, or **ATP**, which is made up of an adenine base, a ribose sugar, and three phosphate groups. Although it is not directly eaten, ATP is the universal fuel for living organisms, and many chemical reactions require energy from ATP in order to proceed.

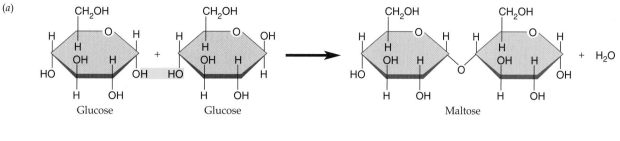

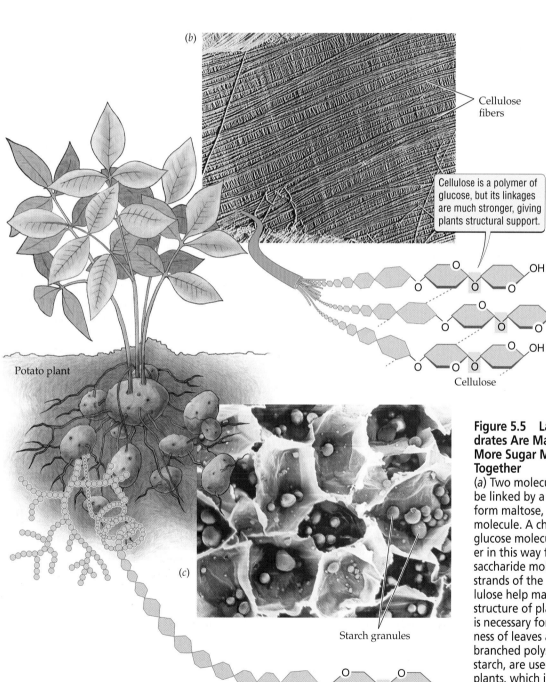

Figure 5.5 Large Carbohydrates Are Made Up of Two or More Sugar Molecules Joined Together

(*a*) Two molecules of glucose can be linked by a covalent bond to form maltose, releasing one water molecule. A chain of thousands of glucose molecules bonded together in this way forms a single polysaccharide molecule. (*b*) Parallel strands of the polysaccharide cellulose help maintain the rigid structure of plant cell walls, which is necessary for the physical soundness of leaves and stems. (*c*) Highly branched polysaccharides, such as starch, are used to store energy in plants, which is why starch-rich foods such as potatoes are good sources of energy.

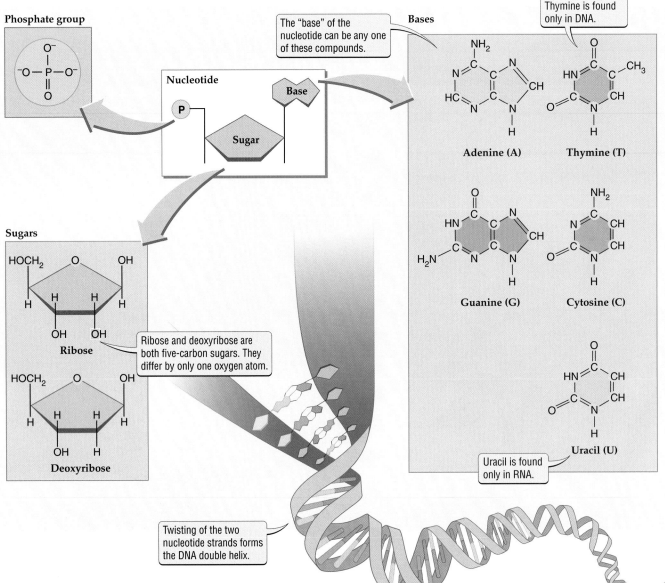

Phosphate group

The "base" of the nucleotide can be any one of these compounds.

Bases

Thymine is found only in DNA.

Nucleotide

Adenine (A)   **Thymine (T)**

Sugars

Ribose and deoxyribose are both five-carbon sugars. They differ by only one oxygen atom.

Ribose

Deoxyribose

Guanine (G)   **Cytosine (C)**

Uracil (U)

Uracil is found only in RNA.

Twisting of the two nucleotide strands forms the DNA double helix.

Figure 5.6   Nucleotide Components
Each nucleotide consists of a nitrogen-containing base linked to a sugar and one or more phosphate groups. The bases adenine, guanine, cytosine, and thymine, when linked to deoxyribose sugar, form the building blocks of DNA. The bases adenine, guanine, cytosine, and uracil, when linked to ribose sugar, form the building blocks of RNA.

The energy of ATP is stored in the covalent bonds that link its three phosphate groups together (Figure 5.7). The breaking of the bond between the second and third phosphates releases energy, which can be used to drive other chemical reactions. The production of ATP and how it is used as an energy carrier will be discussed in detail in Chapter 8.

Amino acids are the building blocks of proteins

Of the many different kinds of chemical compounds found in living organisms, **proteins** may be the most familiar to you. These polymers make up more than half the dry weight of living things. The steaks we throw on the grill in the summer consist mainly of proteins, and our ability to run to the finish line of a race depends on the coordinated actions of many proteins in our muscles. Our bodies are made up of thousands of different proteins. Some of those proteins form physical structures, such as hair; others, called enzymes, help regulate the chemical reactions that drive living processes. We will see how enzymes work in Chapter 7.

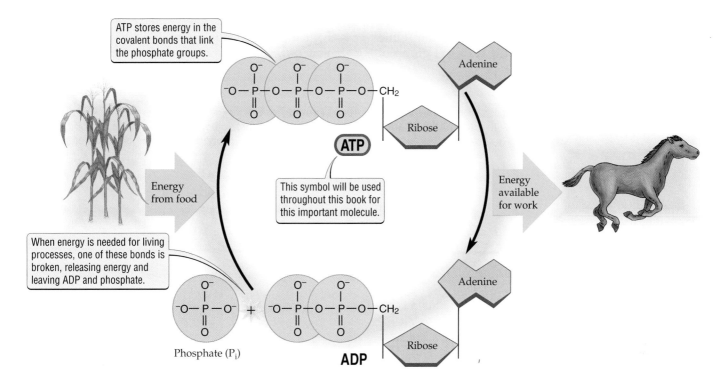

Figure 5.7   Production of ATP
ATP is the major means of short-term energy storage in living organisms. The energy obtained from food is used to form a covalent bond between ADP (adenosine diphosphate) and a phosphate group, making an ATP molecule. When that bond is broken, the energy stored in ATP is released.

Amino acids are the monomers used to build proteins. There are 20 different amino acids, all of which share some structural characteristics. All amino acids have a carbon atom, called the alpha carbon, that forms a central attachment site for several functional groups: an amino group (—NH_2), a carboxyl group (—COOH), and a chemical side chain called the reactive or R group (Figure 5.8).

The different R groups give the different amino acids their different chemical properties. R groups range from a single hydrogen atom to complex arrangements of carbon chains and ring structures (Figure 5.9). Thus organisms have a diverse pool of building blocks from which they can make proteins with many different properties.

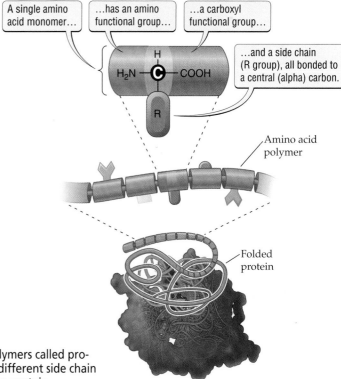

Figure 5.8   Amino Acids Make Up Proteins
Amino acids are the building blocks that form polymers called proteins. There are twenty amino acids, each with a different side chain that contributes specific chemical properties to the protein.

To form proteins, linear chains of amino acids are linked together by covalent bonds between the amino group of one amino acid and the carboxyl group of another. When amino acids are linked together in such a chain, the chain bends and folds so that the R groups can interact. These interactions define the overall three-

Figure 5.9   The Diversity of Amino Acids
Each of the 20 amino acids has a different R group bonded to the alpha carbon. The various R groups give each amino acid chemical properties that can be classified as either hydrophobic, hydrophilic, or special.

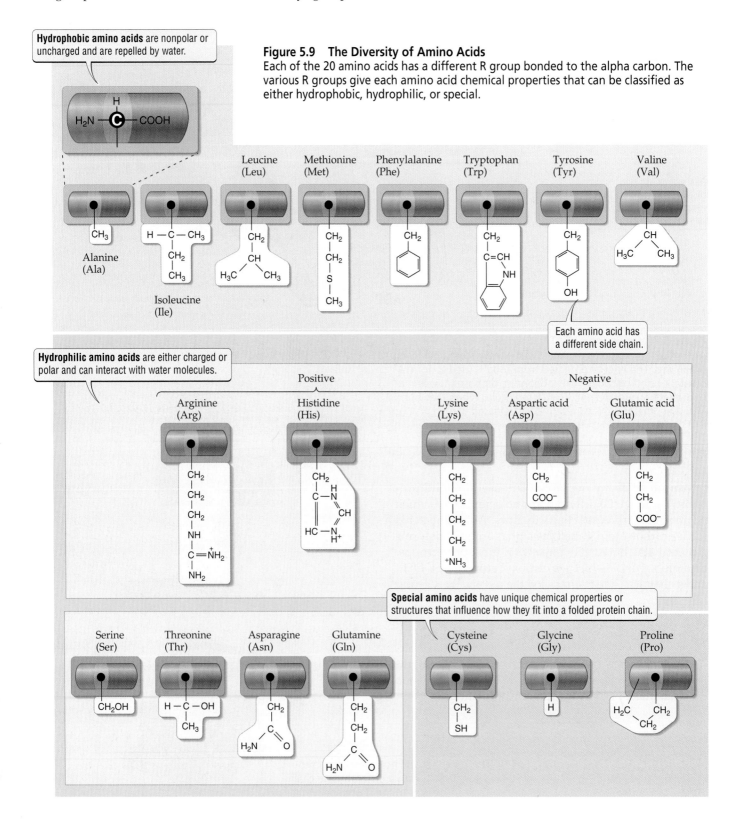

dimensional structure and chemical properties of the protein. Both noncovalent and covalent bonds between the R groups of the individual amino acids in the protein help keep the linear chain folded and maintain its compact three-dimensional structure. For most proteins, this specific folded structure is necessary for normal function, whether that function is a specific chemical activity or the structural support of an organism's shape (see Figure 5.8).

In recent years our understanding of protein structure has grown tremendously, thanks to sophisticated new methods of visualizing a protein's three-dimensional shape. These breakthroughs have helped unravel the mysterious chemical forces that shape proteins, paving the way for the design of improved synthetic proteins. We return to this important idea at the end of the chapter.

Fatty acids store energy and form membranes

Fatty acids are molecules composed primarily of long chains of carbon and hydrogen atoms, referred to as **hydrocarbons**, and ending with a carboxyl group. They are the key components of fats and lipids. **Fats** function in the long-term storage of energy for living organisms and are familiar to some of us as the prime targets of weight loss programs. The role of fats in energy storage will be discussed in Chapter 7. **Lipids**, on the other hand, help form essential physical boundaries both inside organisms and between organisms and their external environments. These lipid-based boundaries, called **membranes**, establish the surface structure of living cells and control the exchange of materials between cells and their environment.

The long hydrocarbon chains found in fatty acids usually contain 16 or 18 carbon atoms, which can vary in the way they are covalently bonded together. Fatty acids in which all the carbon atoms are linked together by single covalent bonds are said to be **saturated** because each carbon is also bonded to a full complement of hydrogens (Figure 5.10*a*). When one or more of the carbon atoms are linked together by double covalent bonds, the fatty acid is said to be **unsaturated** because a bond to hydrogen must be sacrificed to create each double covalent bond between two carbon atoms (Figure 5.10*b*).

The significance of the double bonds in unsaturated fatty acids goes beyond simple differences in the number of hydrogens linked to the carbon chain. Single bonds tend to adopt a straight configuration in space, while double bonds tend to introduce kinks into the hydrocarbon chain. Consequently, the straighter saturated fatty acids can pack very tightly together, forming solids such as fats and waxes at room temperature. In contrast, the double-bond kinks in unsaturated fatty acids prevent such tight packing, so these compounds tend to be liquid (oils) at room temperature.

The role of fatty acids in energy storage and cell structure depends on their covalent bonding to a simple three-carbon compound named glycerol. Glycerol has

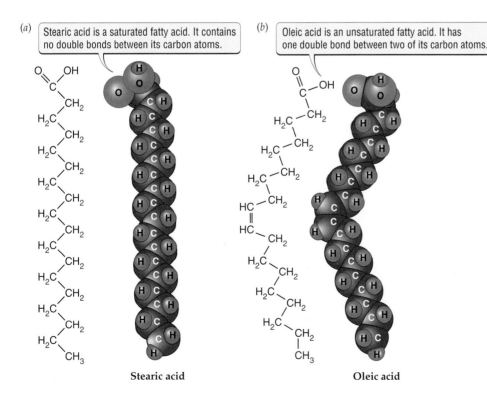

(*a*) Stearic acid is a saturated fatty acid. It contains no double bonds between its carbon atoms.

(*b*) Oleic acid is an unsaturated fatty acid. It has one double bond between two of its carbon atoms.

Stearic acid

Oleic acid

Figure 5.10 Saturated and Unsaturated Fatty Acids
(*a*) Stearic acid is a saturated fatty acid. The structural formula shows that this molecule has no double bonds between its carbon atoms. The space-filling model to the right of it shows that the molecule is straight, so that it can pack tightly to form a waxy solid at room temperature. (*b*) Oleic acid is an unsaturated fatty acid. The double covalent bond in this compound forms a kink, preventing its molecules from packing as tightly as they do in stearic acid. Oleic acid therefore tends to be more liquid at room temperature and is commonly found in the storage fat of humans.

three OH groups, each of which can form a covalent bond with the carboxyl group (—COOH) at the end of a fatty acid chain. When all three OH groups are bonded to a fatty acid, the resulting compound is the most common storage fat in animals and plants (Figure 5.11). Fats contain significantly more energy than does an equal amount of glucose.

The linkage of two fatty acid chains and a negatively charged phosphate group to glycerol produces another class of compounds, called **phospholipids**. Phospholipids are the major component of all membranes in living organisms. The negatively charged phosphate group on one end of a phospholipid allows this region of the molecule to interact with polar water molecules or with positively charged ions, and thus this end of the molecule is hydrophilic. In contrast, the fatty acid chains are hydrophobic. They are entirely nonpolar, and therefore are repelled by water molecules.

The resulting dual character of phospholipids allows them to form double-layered sheet in water, exposing their hydrophilic heads to the water while keeping their hydrophobic tails isolated in the middle. This double-layered sandwich of molecules is called a phospholipid bilayer, and is the basis of all biological membranes (Figure 5.12). Membranes are a clear demonstration of how the chemical properties of molecules in water define a physical structure that is essential for living organisms. Moreover, the ability of these molecules to organize themselves may hold important lessons for the development of new technologies (see the box on p. 100). The structure and roles of biological membranes will be discussed in Chapter 6.

■ Sugars, nucleotides, amino acids, and fatty acids are the molecular building blocks that make up all living organisms.

■ HIGHLIGHT

Proteins That Can Take the Heat

5.6 Chemical reactions are as common in the kitchen as they are in the laboratory. For example, a hard-boiled egg is different in texture from a raw egg because chemical changes during cooking cause the egg to change from a thick, sticky liquid rich in the nutrients needed by a developing chick embryo, to the firm white and yellow mass that tastes so good to humans. What compounds are responsible for these changes? The answer is proteins, which simply cannot withstand heat.

Proteins are very particular about the temperatures at which they will remain folded and function normally. Most properly folded proteins tend to have hydrophilic R groups exposed on their surfaces and hydrophobic R groups isolated on the inside. This careful arrangement allows the protein to form hydrogen bonds with surrounding water molecules and remain dissolved.

When proteins are heated beyond a certain temperature, weak non covalent bonds between the R groups break,

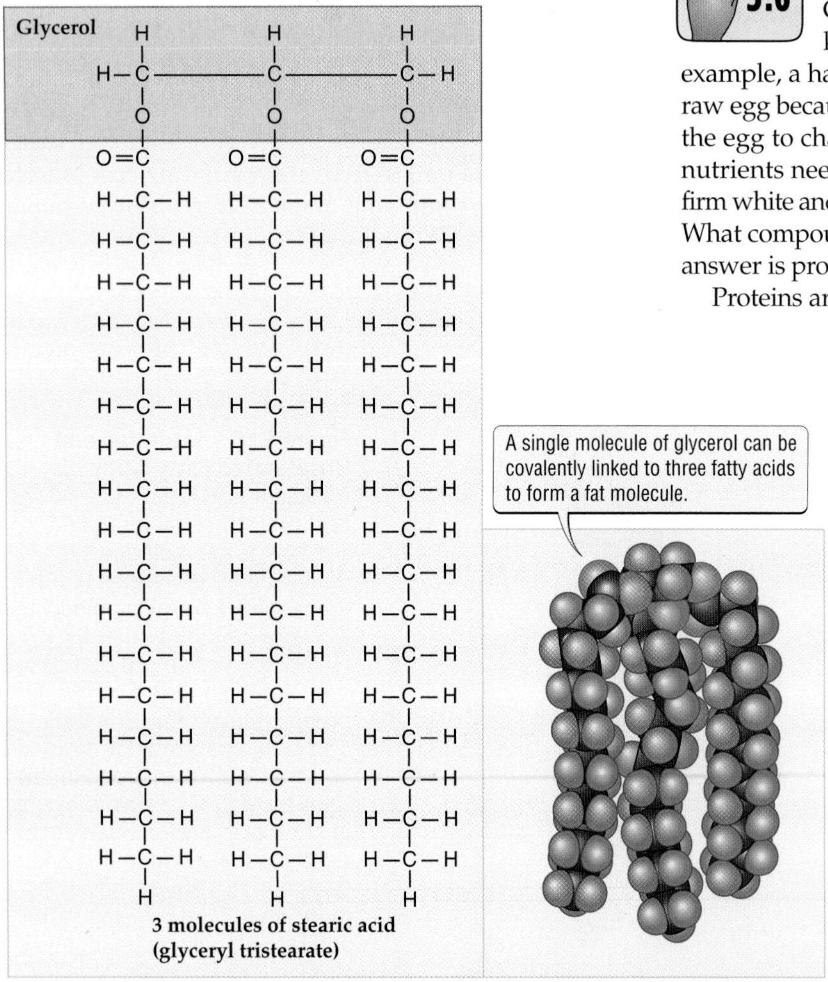

A single molecule of glycerol can be covalently linked to three fatty acids to form a fat molecule.

Glycerol

3 molecules of stearic acid
(glyceryl tristearate)

Figure 5.11   Fats
When the fatty acid molecules that bond to glycerol are stearic acid, the resulting fat is glyceryl tristearate, the most common storage form of fat in animals and plants.

(a)

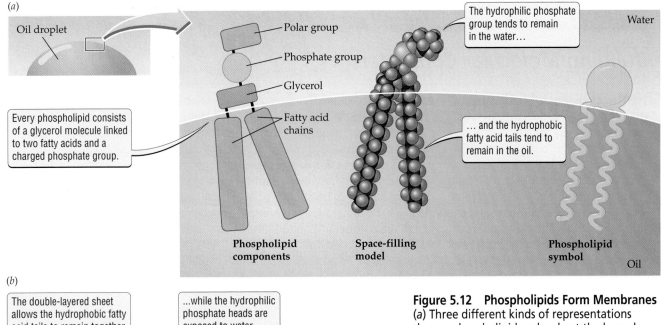

Oil droplet

Every phospholipid consists of a glycerol molecule linked to two fatty acids and a charged phosphate group.

Polar group

Phosphate group

Glycerol

Fatty acid chains

The hydrophilic phosphate group tends to remain in the water…

Water

… and the hydrophobic fatty acid tails tend to remain in the oil.

Phospholipid components

Space-filling model

Phospholipid symbol

Oil

(b)

The double-layered sheet allows the hydrophobic fatty acid tails to remain together and away from water…

…while the hydrophilic phosphate heads are exposed to water.

Water

Heads

Tails

Water

Phospholipid bilayer

Figure 5.12 Phospholipids Form Membranes (a) Three different kinds of representations show a phospholipid molecule at the boundary between an oil droplet and water. At an oil–water boundary, phospholipids are oriented in a specific way because of the different chemical properties of their component parts. (b) A bilayer is the preferred orientation of phospholipids surrounded by water. Phospholipid bilayers form all the membranes of living organisms.

and the proteins unfold, losing their regular three-dimensional structure. However, the covalent bonds that link one amino acid to another remain intact. In the ensuing chaos, hydrophobic R groups from several protein molecules randomly cluster together away from water, forming a featureless blob. In the case of an egg in boiling water, proteins such as albumin that were previously dissolved in the watery environment of the egg are unfolded by the heat and randomly cluster together into the firm white of the boiled egg.

Until the 1970s it was accepted that proteins could not be heated significantly above body temperature and be expected to function normally. Then came the discovery of bacteria that live and thrive in hot springs at temperatures just at the boiling point of water (100°C). To survive, these bacteria must have proteins that remain functional at 100°C. To improve our understanding of how proteins might be able to withstand such high temperatures, biologists set out to isolate and study the proteins from these heat-loving organisms. They discovered several proteins that promote chemical reactions in these bacteria and that are fully active at 100°C. One of these heat-stable proteins has become the basis of the forensic technique known as the polymerase chain reaction, which has figured so prominently in recent criminal trials in which matching of DNA samples is important evidence.

Biologists are now learning how these proteins are able to take the heat, and they are trying to alter proteins from other organisms that live at room temperature to be just as active at 100°C. This endeavor is not meant merely to satisfy scientific curiosity. Learning how to engineer a heat-resistant protein would mean discovering what kinds of chemical bonds form a folded structure that is exceptionally stable. Thus, proteins that were engineered to survive high temperatures would also be far more stable at room temperature and could last longer on a supermarket shelf.

How do biologists make a protein more heat-resistant, or, in other words, better folded? The answer lies in understanding how the chemical properties of amino

■ BIOLOGY IN OUR LIVES

Building Molecular Computers

The challenge of creating complex systems out of many small components applies to both living systems and computers. If the quest to create an artificial "brain" is ever to succeed, computer scientists must come to grips with the current limitations of solid-state semiconductor technology. The size and complexity of circuits that can be achieved with today's technology is limited. This may seem surprising, given that the computing speed of our integrated circuits has been doubling every 18 to 24 months. However, for a computer based on today's circuitry to even remotely approach the cognitive abilities of a human brain, it would have to contain so many transistors that it would run into problems of size, overheating, and fabrication costs. Thus, scientists are seeking ways of making circuits from smaller components, taking some lessons from the incredible density of circuitry in living systems.

The search for denser, smaller, and more complex computer circuits has spawned a new field called molecular electronics. In this field, the comparatively bulky transistors and diodes of traditional circuits are being replaced with single molecules. Such techniques could create processors the size of individual macromolecules. By way of comparison, if today's smallest transistor were the size of a football field, an equivalent molecular device would be the size of a blade of grass.

So how can a single molecule serve as a changeable memory device? The answer lies in a molecule that can switch between different states on demand. This binary principle is used in all modern computers, where the numbers "1" and "0" represent the "on" and "off" states of a current switch. As an early step in the development of molecular computers, researchers have synthesized switches in the form of organic molecules. These single molecules can alter their electrical conductivity on demand by trapping electrons. But one molecular switch does not a computer make, and the added challenge is to figure out how to combine millions of such switches in a well-ordered arrangement.

One possible solution to this problem has already been put to good use in living organisms. In a process known as self-assembly, many biological molecules, such as the lipids that form a phospholipid bilayer, spontaneously arrange themselves into regular patterns without external intervention. Understanding how the chemical properties of molecules affect this process has allowed researchers to modify synthetic compounds in a manner that promotes self-assembly. Such modified compounds could therefore be used to create complex arrays of molecular switches without having to physically arrange them. While science is just starting down the road to developing molecular computers, future decades are likely to see a revolution in chip design based on the insights gained from the way biological molecules assemble themselves and function as switches and memory devices in the cell.

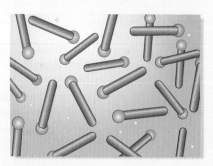
Molecules in solution

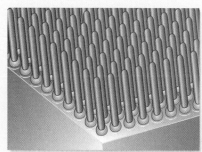
Self-assembled monolayer

Molecules with head groups possessing specific chemical properties can assemble themselves on a suitable surface that chemically interacts with those groups. Different side-to-side interactions between tail groups can also affect the overall orientation of these molecules on the surface.

acids form the bonds that fold proteins into their highly stable three-dimensional shapes. One recent test case that proves the possibility of such an approach involves a protein called thermolysin.

Found in bacteria that live at room temperature, thermolysin helps break down other proteins. Biologists have changed 8 of the total 319 amino acids in this protein to produce a reengineered thermolysin that is fully active at 100°C. By making these amino acid changes, the researchers caused the formation of a new covalent bond that locked the three-dimensional structure of the protein in place, making it heat-resistant. In comparison to the normal protein, the reengineered thermolysin was 340 times more stable at 100°C.

Although we may not want to engineer an egg that will not hard-boil, the ability to engineer proteins that remain active at high temperatures opens up exciting new possibilities. Not only could protein-rich food products be engineered to last far longer without chemical preservatives, but even medically important proteins such as vaccines could be modified so that they no longer required cold storage conditions, an important benefit for developing countries.

> ■ Understanding how the chemical properties of amino acids define protein shapes has allowed biologists to engineer proteins that are more stable.

SUMMARY

How Atoms Make Up the Physical World

■ The physical world is made up of atoms. There are 92 different kinds of atoms, one for each natural chemical element.

■ Individual atoms are made up of subatomic particles, which include positively charged protons in the nucleus and negatively charged electrons surrounding the nucleus. The particular combination of subatomic particles associated with an atom determines its chemical properties.

■ When atoms lose or gain electrons, they become positively or negatively charged ions.

■ Chemical compounds are formed by specific arrangements of atoms of different elements. These arrangements depend on the unique chemical characteristics of the atoms involved.

Covalent Bonds: The Strongest Linkages in Nature

■ Atoms are linked together by different kinds of chemical bonds.

■ Covalent bonds, formed by the sharing of electrons between atoms, connect atoms within a molecule. They are the strongest bonds in nature.

■ The proton core of every atom is surrounded by a specific arrangement of electrons, which move in defined orbits or shells. Atoms share electrons with other atoms to fill their outer electron shells to capacity. The bonding properties of an atom are determined by its particular arrangement of electrons.

■ Covalent bonds establish the spatial arrangement of atoms in a molecule and hence the molecule's three-dimensional shape.

Noncovalent Bonds: Dynamic Linkages between Molecules

■ Noncovalent bonds link separate molecules and serve as secondary links between atoms within a single large molecule. They are weaker than covalent bonds and do not involve the sharing of electrons between atoms.

■ Hydrogen bonds are noncovalent bonds that form between polar compounds, which have an unequal distribution of electrical charge. Slightly positive hydrogen atoms in one compound form hydrogen bonds with slightly negative atoms in another compound. Hydrogen bonding between water molecules accounts for the physical properties of water and is essential for life.

■ Compounds that interact freely with water are hydrophilic; nonpolar compounds, which are repelled by water, are hydrophobic. Nonpolar, hydrophobic compounds tend to group together in water.

■ Ionic bonds are noncovalent bonds that form between two atoms when electrons from one atom are transferred to another atom.

How Chemical Reactions Rearrange Atoms in Molecules

■ Chemical reactions break and form bonds between atoms. They neither create nor destroy atoms, but merely alter the arrangements of atoms in molecules.

■ Nearly all chemical reactions that support life occur in water. Specialized classes of compounds called acids and bases affect the concentration of free hydrogen ions in water. Acids give up hydrogen ions to water molecules, forming hydronium ions; bases accept hydrogen ions from water molecules, forming hydroxyl ions.

■ The concentration of free hydrogen ions in water is expressed as a number on the pH scale. The pH scale ranges from 0 (highest concentration of hydrogen ions—that is, most acidic) to 14 (lowest concentration—that is, most basic). Most living organisms require a pH of about 7.

■ Buffers are compounds that can both give up hydrogen ions to and accept them from water molecules. They help maintain the constant internal pH that is necessary for the chemical reactions of life.

The Chemical Building Blocks of Living Systems

■ Carbon compounds form the physical framework for all biological molecules. The ability of carbon atoms to form large and complex polymers has an important role in the generation of diverse biological structures.

■ Sugars provide both energy and physical support for living organisms. Carbohydrates consist of simple sugars (monosaccharides) and more complex polymers (polysaccharides).

■ Nucleotides are the building blocks of DNA and RNA. Specialized nucleotides such as ATP function in the storage and transfer of energy from one chemical reaction to another.

■ Amino acids are the building blocks of proteins. The chemical properties of the different amino acids are determined by their different R groups. The function and three-dimensional shape of a protein are defined by the chemical properties of the amino acids it contains.

- Fatty acids are the building blocks of fats and lipids. Fats are an important means of long-term energy storage; lipids are the basic components of biological membranes.

Highlight: Proteins That Can Take the Heat

- Most proteins lose their three-dimensional folded structure when heated.

- Changing a few amino acids in a protein may allow it to withstand high temperatures and to be more stable at room temperature.

- Reengineered heat-resistant proteins will have important uses in food products and medicine.

KEY TERMS

acid p. 89	lipid p. 97
amino acid p. 95	macromolecule p. 90
atom p. 84	membrane p. 97
ATP p. 92	molecule p. 84
base p. 89	monomer p. 90
buffer p. 90	monosaccharide p. 92
carbohydrate p. 92	noncovalent bond p. 86
chemical compound p. 84	nonpolar p. 88
chemical reaction p. 89	nucleotide p. 92
covalent bond p. 85	nucleotide base p. 92
DNA p. 92	organic p. 90
electron p. 84	pH p. 89
element p. 84	phospholipid p. 98
fat p. 97	polar p. 88
fatty acid p. 97	polymer p. 90
functional group p. 92	polysaccharide p. 92
glucose p. 92	protein p. 94
hydrocarbon p. 97	proton p. 84
hydrogen bond p. 86	RNA p. 92
hydrophilic p. 88	saturated p. 97
hydrophobic p. 88	soluble p. 88
ion p. 84	sugar p. 92
ionic bond p. 88	unsaturated p. 97

CHAPTER REVIEW

Self-Quiz

1. The atoms of a single element
   a. have the same number of electrons.
   b. can form linkages only with atoms of the same element.
   c. can have different numbers of electrons.
   d. can never be part of a chemical compound.

2. Two atoms can form a covalent bond
   a. by sharing protons.
   b. by swapping nuclei.
   c. by sharing electrons.
   d. by sticking together on the basis of opposite electrical charges.

3. Which of the following statements about molecules is true?
   a. A single molecule can contain atoms from only one element.
   b. The chemical bonds that link atoms into a molecule are arranged randomly.
   c. Molecules are found only in living organisms.
   d. Molecules can contain as few as two atoms.

4. Which of the following statements about ionic bonds is not true?
   a. They cannot exist without water molecules.
   b. They are not the same as hydrogen bonds.
   c. They require the loss of electrons.
   d. They are more temporary than covalent bonds.

5. Hydrogen bonds are especially important for living organisms because
   a. they occur only inside of organisms.
   b. they are very strong and maintain the physical stability of molecules.
   c. they allow biological molecules to dissolve in water, which is the universal medium for living processes.
   d. once formed, they never break.

6. Glucose is an important example of a
   a. protein.
   b. carbohydrate.
   c. fatty acid.
   d. nucleotide.

Review Questions

1. All the major chemical building blocks found in living organisms form polymers. Why are polymers especially useful in the organization of living systems?

2. A sample of pure water contains no acids or bases. Predict the pH of the water and explain your reasoning.

3. Describe the chemical properties of carbon atoms that make them especially suitable for forming biological polymers.

4. Polymers of amino acids have chemical characteristics that are important for life. How are these characteristics used by a living organism?

5

The Daily Globe

Hard Candy Could Save Your Life

RESEARCH TRIANGLE PARK, NC— How would you like to preserve yourself forever? Imagine a time when individuals with a fatal disease can be preserved in suspended animation until a cure is found. Today this idea may conjure up images from science fiction movies, but thanks to recent discoveries made in plants, the idea may soon be more than just fantasy.

Corn seeds can survive adverse conditions by entering a sort of suspended animation. If kept dry, the seed can remain in this state for decades and then sprout vigorously when it receives water. Biologists have discovered that the delicate proteins necessary for all the seed's living functions are carefully preserved in a protective coating of hard sugar. When the seed comes in contact with water, the sugar dissolves, the proteins become active, and the seed sprouts.

Certain insects also encase themselves in sugars to survive long winters in a frozen state of suspended animation. But will this ever work for humans? The answer from Ivana Livalot, president and CEO of Methuselah Technologies, is a resounding yes.

In a recent press conference, Livalot announced a joint venture with researchers at Yoakum University that would test the use of complex sugars to produce suspended animation in small mammals. When asked about the source of funding for this venture, Livalot admitted that most of it came from private investors with a "strong personal interest" in the technology.

Evaluating "The News"

1. If preserving proteins is so important for the success of suspended animation, what does this tell you about the role of proteins in living processes?

2. For Methuselah Technologies to develop a suspended animation method for humans, it must eventually be tested on human subjects. Which individuals should have the highest priority as test subjects for suspended animation?

3. Do you think that the greater complexity of our brains is likely to make waking up from suspended animation more difficult for humans than for insects? Would this affect your willingness to be a human volunteer in a clinical trial?

chapter 6 Cell Structure and Compartments

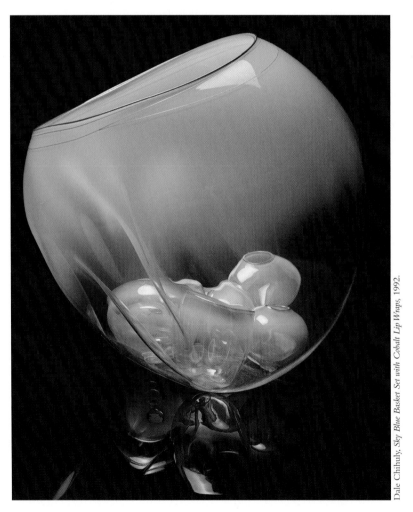

Dale Chihuly, *Sky Blue Basket Set with Cobalt Lip Wraps*, 1992.

Inner Space Travel

*I*n 1997, Planet Earth was treated to a spectacular astronomical event: the comet Hale-Bopp. You did not need to be an astronomer with a telescope to observe this event, since the comet was easily visible to the naked eye. In fact, this bright particle with its trailing luminous tail was a fixture in the night sky for weeks on end.

Many of the millions of people who gazed up at this magnificent galactic intruder would have been surprised to know that another invader with a remarkably similar appearance can sometimes be found in the human body. In those persons unfortunate enough to have been exposed to a bacterium named *Listeria monocytogenes*, comet-shaped invaders move around various organs of the body much as the comet Hale-Bopp wanders from one solar system to another.

Listeria infects humans and can cause severe gastroenteritis and damage to the nervous system. Nearly 2000 people in the United States become seriously ill from *Listeria* each year, and more than 400 die from the infection. The bacterium is found in contaminated foods, and recent outbreaks have been traced to sources as seemingly harmless as chocolate milk. What is most remarkable about *Listeria*, however, is that under a

104

Prokaryotes and eukaryotes differ in the way their cells form internal compartments to concentrate and facilitate chemical reactions.

microscope it can be seen traveling around in organ samples taken from infected individuals.

Since every organ in the body is made up of small, membrane-enclosed units called cells, the bacteria must cross membranes to move from cell to cell. Under the microscope, the moving *Listeria* look very much like comets with rod-shaped bodies and trailing tails. Biologists now know that the "comet tail" of *Listeria* is made up of proteins that are captured from the infected cell and made to form fibers that extend from the body of the bacterium. The lengthening of these fibers provides the propulsive force for the bacterium, literally pushing it through the cell. This force is so significant that when *Listeria* hits the inside of a membrane, it pushes the membrane out to form a spike, allowing it to leap to another cell and spread the infection.

Although the beautiful sight of a comet moving through space bears a coincidental resemblance to the movement of *Listeria* through the inner space of the cell, the relationship of the tail to the body is very different in the comet. The tail of a comet is nothing more than a trail of gases and ice crystals, which contributes nothing to its movement. In the case of *Listeria*, the tail drives the movement.

The evolution of *Listeria*'s ability to use the cell's own proteins against it underscores the rich possibilities that lie in the components of the cell. In this chapter we will explore some of these components and their normal functions in the life of a cell.

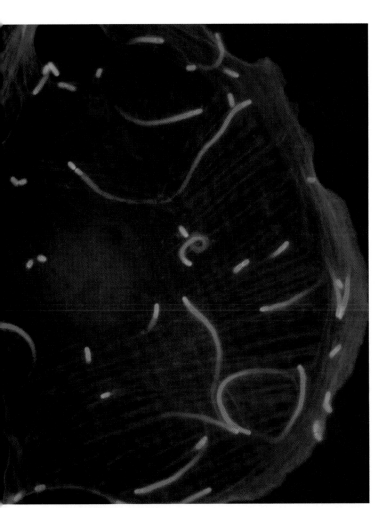

Invaders in Inner Space
The bacteria *Listeria monocytogenes* moving inside of an infected cell.

■ KEY CONCEPTS

1. All living organisms are made up of one or more basic units called cells.

2. A plasma membrane forms a boundary around every cell. It limits the movement of molecules both into and out of the cell, and it determines how the cell communicates with the external environment.

3. Prokaryotes are single-celled organisms that lack a nucleus and have very little internal organization. Eukaryotes are single-celled or multicellular organisms that have a nucleus and several other specialized internal compartments.

4. The specialized internal compartments of the eukaryotic cell are called organelles. They concentrate and transport the macromolecules necessary for the chemical processes of life.

5. The cytoskeleton is composed of two distinct systems of filaments and their associated proteins, which have important roles in supporting cell shape and movement.

6. Organelles such as mitochondria and chloroplasts are probably descendants of primitive prokaryotes that were engulfed by the ancestors of eukaryotes.

---

*E*very complex structure imaginable can be broken down into smaller and simpler parts. Even something as impressive as the Empire State Building in New York City can be reduced to simple components such as concrete blocks and steel girders. Based on this assumption, we can begin to understand a complex system by first identifying and examining its elementary components.

The basic principle of building complex structures out of simple components applies to living systems as much as it applies to skyscrapers. Chapter 5 discussed the atoms and molecules that make up the building blocks of life. These chemical components must be further arranged into living units before a complex organism such as a human being can exist. In other words, macromolecules such as proteins and DNA must be organized into more complex arrangements that promote the chemical reactions required for life. This chapter explores that next level of organization in living systems.

Cells: The Basic Units of Life

The **cell** is the basic unit of life. In the same way that molecules such as proteins and fatty acids are made up of atoms, every living organism known is made up of from one to trillions of membrane-enclosed units called cells. Bacteria are single cells, while a complex organism such as a human being contains about 10 trillion cells. Cells make up every organ in our bodies, and they determine how we look, move, and function as organisms. Even the proteins that make up your hair and the surface of your skin depend on cells to synthesize them.

Given the wide variety of organs and specialized parts in the human body, it is not surprising that more than 200 different kinds of specialized cells are required to make up all these different parts. Let's compare the cells that make up muscles with those that form the clear lens of the eye. Muscle cells have the important task of generating the movements we experience as muscle contractions and relaxations. They have specific protein components that allow them to change shape and generate physical force. In contrast, the cells that make up the lens of the eye do not need to generate physical force. Instead, their task is to help focus light into the eye. They therefore contain specific protein components that focus light as it passes through the cell in much the same way that the glass in a lens focuses light into a camera.

The diversity of cells is far greater than the range of cell types found in the human body. In fact, none of the cell types found in our bodies are found in any plant. With millions of different species on Earth, the variety of cell types is enormous. Yet even among so many different kinds of cells, certain basic components and structures are shared by all of them. Our ability to see these structures under the microscope led to the discovery of cells (see the box on p. 108).

In this chapter we tour the cell and examine its major structural components. We begin at the physical boundary of the cell and work our way inside to discover the structures and compartments that allow cells to support living processes.

> ■ Cells are the basic units that make up all living organisms.

The Plasma Membrane: Separating Cells from the Environment

One of the key characteristics of life is the existence of a boundary that separates the organism from its nonliving environment. If all the molecules and proteins that we discussed in Chapter 5 were allowed to diffuse freely through the environment, they would not encounter one another frequently enough for life-sustaining reactions to take place. Thus, an enclosed compartment—the cell—concentrates all the required compounds in a limited space.

The boundary structure that defines all cells is called the **plasma membrane**. As we saw in Chapter 5, biological membranes are composed mainly of a bilayer of phospholipid molecules oriented such that their hydrophilic heads are exposed to the watery environments both inside and outside of the cell and their hydrophobic fatty acid tails are grouped together in the interior of the membrane.

If the plasma membrane had no function other than to define the boundary of the cell and keep all its contents inside, a simple phospholipid bilayer would suffice. However, living cells need a plasma membrane that also allows them to communicate with the outside environment, capture essential molecules, and release unwanted waste products. The plasma membrane must selectively allow some molecules to pass through, prevent other molecules from entering or leaving the cell, and receive signals from the outside environment.

The selective permeability of plasma membranes depends on various proteins that are part of the membrane. As Figure 6.1 shows, some of these proteins form channels that allow the passage of selected ions and molecules into and out of the cell. Others are used by the cell to recognize changes in the outside environment, including signals from other cells. We will learn much more about how these proteins work in Unit 5.

Proteins in the plasma membrane may be inserted either all the way through or partially through the phospholipid bilayer. These proteins are kept in place by

Figure 6.1   Proteins Are Embedded in the Plasma Membrane
Proteins in the phospholipid bilayer allow the cell to communicate with the outside environment, capture essential molecules, and release unwanted waste products. They are inserted in the bilayer in a variety of ways.

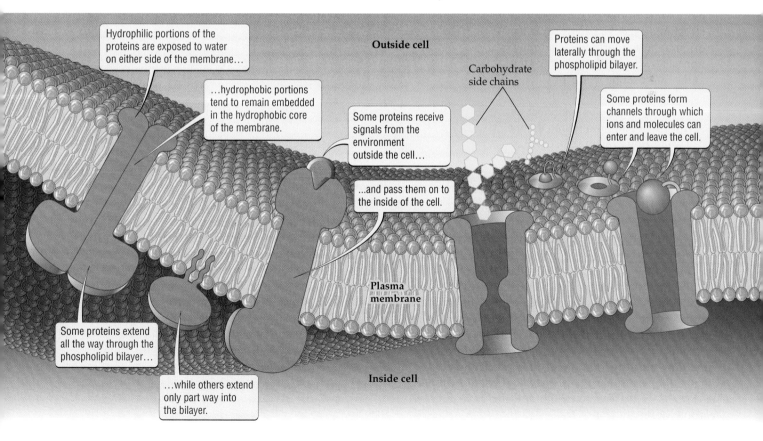

■ THE SCIENTIFIC PROCESS

Exploring Cells under the Microscope

To see something is to begin to know it. This simple statement is just as true of biology as it is of fine art. Our awareness of cells as the basic units of life is based largely on our ability to see them. The instrument that opened the eyes of the scientific world to the existence of cells was the light microscope, which was invented in the last quarter of the sixteenth century. The key components of early light microscopes were ground-glass lenses that bent incoming rays of light to produce magnified images of tiny specimens.

The study of magnified images began in the seventeenth century when Robert Hooke examined a piece of cork under a microscope and saw that it was made up of little compartments. Hooke described these structures as small rooms, or cells, originating the term we use today. Ironically, what Hooke saw under the early microscope were not living cells, since cork is dead plant tissue. Instead, the small chambers that he saw were nothing more than empty cell walls. However, the discovery of previously invisible living things pro-

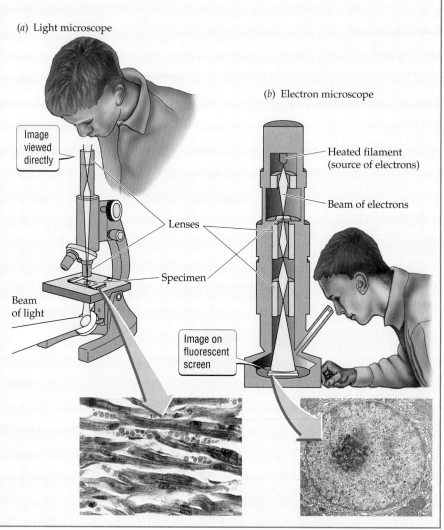

(a) Light microscope

Image viewed directly

Beam of light

(b) Electron microscope

Heated filament (source of electrons)

Beam of electrons

Lenses

Specimen

Image on fluorescent screen

their hydrophobic portions, which tend to remain embedded in the hydrophobic core of the bilayer. Unless these proteins are anchored to structures inside the cell, however, they are free to move sideways through the highly fluid phospholipid bilayer. This lateral freedom of movement supports what is referred to as the **fluid mosaic model**, which describes the plasma membrane as a highly mobile mixture of phospholipids and proteins. This lateral mobility is essential for many cellular functions, including whole-cell movement and the ability to detect external signals. Specific subsets of proteins and lipids in the plasma membrane also have attached carbohydrates.

Although the plasma membrane is a common feature of all cells, the set of proteins and carbohydrates found in the membrane varies from one cell type to another. The specific combination of proteins in the membrane determines how a cell interacts with its external environment and contributes to the unique properties of each cell type.

> ■ Every cell is surrounded by a plasma membrane that separates the chemical components of life from the nonliving environment. Proteins and carbohydrates in the plasma membrane allow the cell to exchange materials with and respond to its environment.

■ THE SCIENTIFIC PROCESS *(continued)*

ceeded rapidly, opening up a new world to scientific exploration.

While the light microscope has a place in the early history of biology, similar instruments are just as important in ongoing research today. The basic principles that enable light microscopes to magnify the image of a specimen remain the same, but the quality of current lenses has improved significantly. Thus, the 200- to 300-fold magnification achieved in the seventeenth century has been improved to the well over a thousand-fold magnification achieved by today's standard light microscopes. This degree of magnification allows us to distinguish structures as small as 1/2,000,000 of a meter, or 0.5 micrometer (μm). Light microscopes therefore reveal not just animal and plant cells (5–100 μm), but also organelles such as mitochondria and chloroplasts (2–10 μm) and tiny organisms such as bacteria (1 μm).

Since the 1930s, an even more dramatic increase in magnification has been achieved by the replacement of visible light with streams of electrons that are focused by powerful magnets instead of glass lenses. Called electron microscopes, these instruments can magnify a specimen by more than 100,000-fold, revealing the internal structure of cells and even individual molecules such as proteins and nucleic acids. Both types of microscopy give us insights into how cells are organized and how different types of cell are physically adapted to specific functions in the body of a multicellular organism.

The ability to distinguish the various parts of a specimen under a microscope depends on the existence of a way to create contrast. This has been a challenge for light microscopy because cells are generally transparent and their different structures do not contrast with one another, and as a result they tend to be indistinguishable.

The earliest solution to this problem was to stain the cells with dyes. Different parts of the cells take up different dyes, altering the light passing through them and producing different colors, thus distinguishing one structure from another. Today similar methods are still being used to visualize cellular structures and even the distribution of specific proteins within the cell.

Comparing Prokaryotes and Eukaryotes

All the living organisms that we know today can be classified into two groups based on the internal structure of their cells, as we saw in Chapter 3. Organisms whose cells lack internal membrane-enclosed compartments are known as **prokaryotes**. Those whose cells have such compartments are known as **eukaryotes**.

Prokaryotic cells have very little internal organization, and they were probably the first cells to arise during evolution. Today, all prokaryotes are single-celled bacteria or archaeans. They are usually spherical or rod-shaped. They have a tough cell wall that forms outside the plasma membrane and helps maintain the shape and structural integrity of the prokaryotic cell (Figure 6.2).

A typical bacterial cell is filled with a watery substance known as the **cytosol**. The cytosol contains a multitude of molecules, including DNA, RNA, proteins, and enzymes, all suspended in water. These components, together with a host of free ions, support the chemical reactions necessary for life. There are so many small and large molecules crowded into the cytosol that it behaves more like a thick jelly than like a free-flowing liquid. In fact, the cytosol is probably similar to the rich soup that first supported life billions of years ago.

The well-studied bacterium *Escherichia coli*, a common resident of the human intestine, is only 2 millionths of a meter (2 micrometers) long. The relatively small size of prokaryotes may account for their ability to get along without further internal organization. Since the chemical components are contained in such a small volume of cytosol, they do not need to be further concentrated to carry out their particular activities.

Eukaryotes can exist as single cells (such as yeasts) or large multicellular organisms (such as humans). All multicellular organisms are collections of eukaryotic cells that are specialized for different functions. The major distinction between prokaryotic and eukaryotic cells is the presence in eukaryotic cells of internal membrane-enclosed compartments. The most distinctive of these compartments is the **nucleus**. This structure houses most of the cell's DNA, effectively separating it from the remainder of the cell's components.

All eukaryotic cells are even further organized by the presence of several other membrane-enclosed compartments (see Figure 6.2). Like the nucleus, each compartment in the eukaryotic cell is formed by membranes and has a specific function. Since most eukaryotic cells are about a hundred times larger than prokaryotes, they cannot rely on chemical components being close enough

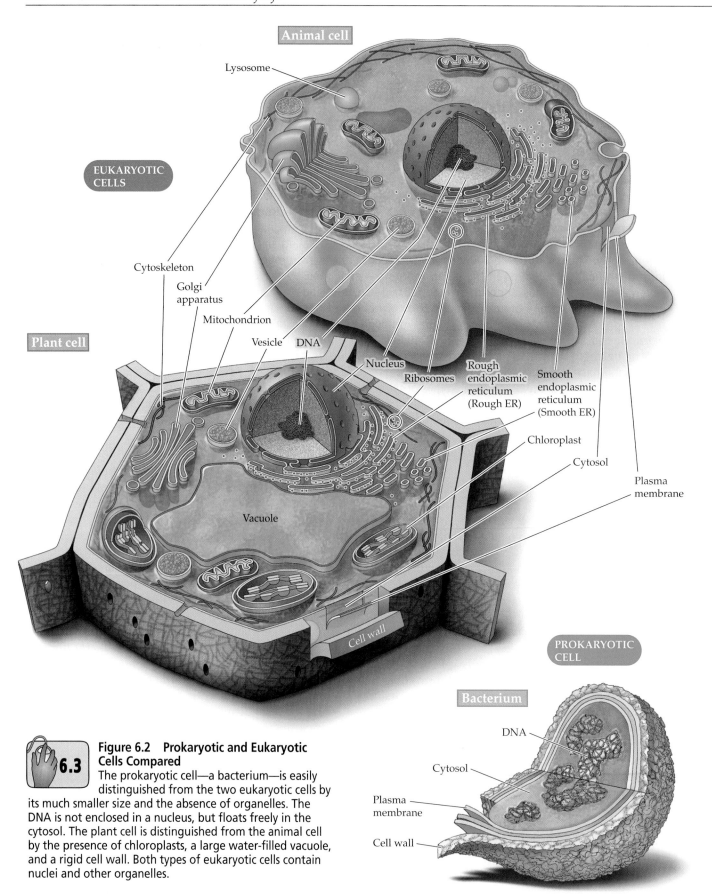

Figure 6.2   Prokaryotic and Eukaryotic Cells Compared
The prokaryotic cell—a bacterium—is easily distinguished from the two eukaryotic cells by its much smaller size and the absence of organelles. The DNA is not enclosed in a nucleus, but floats freely in the cytosol. The plant cell is distinguished from the animal cell by the presence of chloroplasts, a large water-filled vacuole, and a rigid cell wall. Both types of eukaryotic cells contain nuclei and other organelles.

6.3

together for the necessary reactions to occur. Their division into compartments concentrates these components where they are needed and promotes the chemical reactions of life.

> ■ Living organisms can be classified as either prokaryotes or eukaryotes. The cells of eukaryotes have internal membrane-enclosed compartments such as the nucleus; prokaryotic cells do not. The larger size and greater diversity of eukaryotic cells require a more complex internal organization than is found in prokaryotes.

6.7 The Specialized Internal Compartments of Eukaryotic Cells

The typical eukaryotic cell can be compared to a large manufacturing plant with many different departments. Each department must have a specific function and internal organization that contributes to the overall "life" of the corporation. Specific subdivisions represent an effective means of accomplishing particular tasks. For example, if all the members of the assembly line were scattered throughout the building, it would be difficult for the plant to assemble goods in a timely fashion. Thus, the workers and packers are all located in a centralized assembly department, effectively concentrating and coordinating their efforts.

The eukaryotic cell faces challenges similar to our hypothetical corporation, since specific life processes need to be carried out efficiently by specialized "departments." The goals of enhanced efficiency, including the faster manufacturing of products, are shared by the corporation and the cell. The specialized departments of the cell are the various membrane-enclosed compartments that divide its contents into smaller spaces. These smaller spaces contain, isolate, and concentrate the proteins and smaller molecules necessary for different processes. For example, processes such as ATP production and protein synthesis occur in different compartments.

The cell's membrane-enclosed compartments are called **organelles**, a name that is especially appropriate since they are the "little organs" of the cell. In the same way that organs such as the heart and lungs have different and unique functions in the human body, each organelle has specific duties in the life of the cell. Unlike the prokaryotic cell, in which all the contents inside the plasma membrane form the cytosol, the contents of a eukaryotic cell are divided between the cytosol and the organelles. In other words, the eukaryotic cytosol consists of all the cell contents inside the plasma membrane, excluding the organelles. **Cytoplasm** is another common term used to describe all the contents of the eukaryotic cell, excluding only the nucleus.

The nucleus is the storehouse for genetic information

The nucleus is the most distinctive organelle of eukaryotic cells. As we saw earlier, it is this membrane-enclosed compartment that distinguishes eukaryotes from prokaryotes, which do not have a nucleus. Returning to our comparison with a manufacturing plant, the nucleus of the cell is equivalent to the administrative offices of the plant. In other words, the nucleus is the specialized compartment that directs the activities and physical appearance of the cell. It fulfills this function by housing the cell's DNA, which carries the information necessary to build all the structures and carry out all the activities of the cell (Figure 6.3).

Inside the nucleus, long polymers of DNA are packed with proteins into a remarkably small space. Since eukaryotic cells can have more than a thousand times more DNA than prokaryotes, careful packing of DNA with proteins is necessary for it to fit inside the nucleus.

The arrangement of the membranes that surround the nucleus is different from that of the plasma membrane that surrounds the whole cell. The boundary of the nucleus, called the **nuclear envelope**, is a double membrane that contains small openings called **nuclear pores** (see Figure 6.3). These pores are the gateways that allow the movement of molecules into and out of the nucleus, enabling it to communicate with the rest of the cell. The pores are essential features of the nuclear envelope because the transfer of information encoded by DNA depends on the movement of molecules out of the nucleus. Likewise, how and when this DNA information is used depends on specialized proteins that move into the nucleus. How the DNA in the nucleus dictates the activities of the cell will be discussed in Chapter 15.

The endoplasmic reticulum manufactures proteins and lipids

If the nucleus functions as the administrative offices of the cell, the endoplasmic reticulum is the factory floor where many of the cell's chemical building blocks are manufactured. The **endoplasmic reticulum (ER)** is surrounded by a single membrane that is connected to the outer membrane of the nuclear envelope. Unlike the nucleus, which is usually an irregular spherical structure, the ER is an extensive and complex network of interconnected tubes and flattened sacs. You can visu-

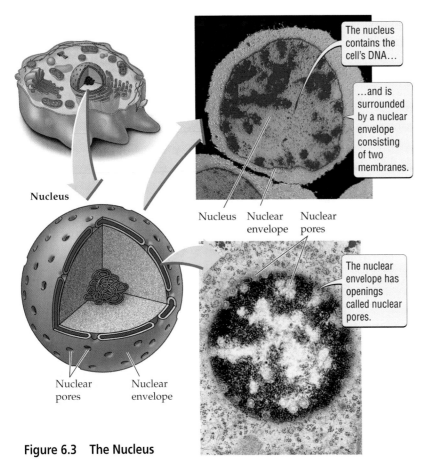

Nucleus

Nucleus

Nuclear pores

Nuclear envelope

The nucleus contains the cell's DNA...

...and is surrounded by a nuclear envelope consisting of two membranes.

Nucleus Nuclear envelope Nuclear pores

The nuclear envelope has openings called nuclear pores.

Figure 6.3   The Nucleus

When viewed under a microscope, ER membranes have two different appearances: rough and smooth (see Figure 6.4). In most cells, the majority of ER membranes appear to have small rounded particles associated with them that are exposed to the cytosol. Such ER is referred to as **rough ER**, and the particles are called **ribosomes**. Each ribosome can manufacture proteins from amino acid building blocks using instructions originating from the DNA in the nucleus. The ribosomes attached to the rough ER manufacture proteins that are destined for the ER lumen or for insertion into a membrane. Other ribosomes that float free in the cytosol manufacture proteins that remain in the cytosol. The process of protein synthesis by ribosomes will be discussed in Chapter 15.

In most cells, a small percentage of ER membrane lacks ribosomes and is called **smooth ER** (see Figure 6.4). The smooth ER is connected to the rough ER and marks sites where portions of the ER membrane actively bud off to produce small enclosed membranous bags called **vesicles**. Since each vesicle is formed from a patch of ER membrane that encloses a portion of the ER lumen, vesicles are an effective means of moving proteins that are either embedded in the ER mem-

alize it as a series of membranous shapes—some like tubes, others like hot water bottles—all stacked and connected to each other.

As Figure 6.4 shows, the ER has a multi-chambered appearance. Its chambers produce the various lipids and proteins destined for other cellular compartments or for export from the cell. The internal space enclosed by the ER membrane is called the **lumen**, and it contains free-floating proteins. In contrast, proteins and lipids that are destined to reside in membranes are inserted into the membrane of the ER.

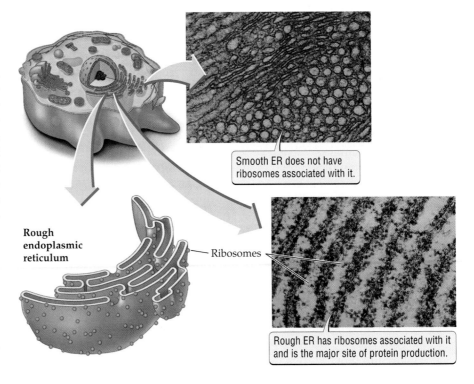

Smooth ER does not have ribosomes associated with it.

Rough endoplasmic reticulum

Ribosomes

Rough ER has ribosomes associated with it and is the major site of protein production.

Figure 6.4   The Endoplasmic Reticulum of a Eukaryotic Cell
The ER forms a series of flattened membrane sacs that are major sites for the synthesis of proteins and lipids.

brane or floating free in the lumen (Figure 6.5). You could think of vesicles as the carts that are used to move goods between different departments of a factory, since they are used to move proteins and lipids from one organelle to another, as well as to the plasma membrane and the outside of the cell.

The Golgi apparatus directs proteins and lipids to their final destinations in the cell

Another membranous organelle, the **Golgi apparatus**, directs proteins and lipids produced by the ER to their final destinations, either inside the cell or out to the external environment. The Golgi therefore functions as a sorting station, much as the shipping department in a factory does. In a shipping department, goods destined for different locations must have address labels that indicate where they should be sent. Something similar happens in the Golgi, where the addition of specific chemical groups to proteins and lipids helps target them to other destinations in the cell. These cellular address labels include carbohydrates and phosphate groups.

Under the electron microscope, the Golgi looks like a series of flattened membrane sacs stacked together and surrounded by many small vesicles (Figure 6.6). These vesicles transport proteins from the ER to the Golgi and between the various sacs of the Golgi. Vesicles are therefore the primary means by which proteins and lipids move through the Golgi apparatus and to their final destinations.

Lysosomes and vacuoles are specialized compartments

Not all the proteins produced and sorted by the ER and Golgi apparatus are destined for the cell surface. In animal cells, distinct subsets of proteins are sent to the lumens of other organelles, called lysosomes. **Lysosomes** are specialized vesicles that contain

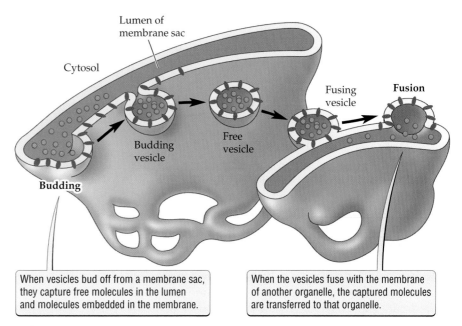

When vesicles bud off from a membrane sac, they capture free molecules in the lumen and molecules embedded in the membrane.

When the vesicles fuse with the membrane of another organelle, the captured molecules are transferred to that organelle.

Figure 6.5   How Vesicles Move Proteins and Lipids
Proteins and lipids move between the ER and the Golgi apparatus, and between the various membrane sacs of the Golgi, in smaller sacs of membrane called vesicles. The budding and fusion of vesicles moves proteins from the ER through the Golgi apparatus and to other compartments of the cell.

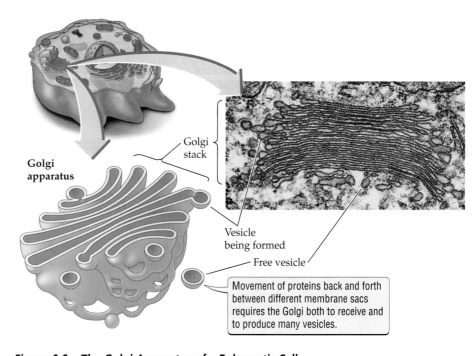

Movement of proteins back and forth between different membrane sacs requires the Golgi both to receive and to produce many vesicles.

Figure 6.6   The Golgi Apparatus of a Eukaryotic Cell
The Golgi apparatus consists of flattened membrane sacs, in which proteins are directed to their final destinations in the cell.

Figure 6.7   Lysosomes in an Animal Cell
Lysosomes are specialized vesicles filled with enzymes that break down macromolecules. The cell shown here is from the stomach lining; it uses its lysosomes to break down food materials.

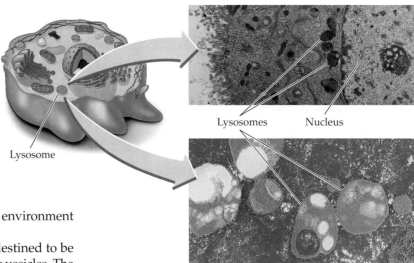

Lysosome

Lysosomes   Nucleus

enzymes used to break down macromolecules such as sugars, proteins, and fats. They can adopt a variety of irregular shapes (Figure 6.7), but all are characterized by an acidic lumen with a pH of about 5. This acidic pH is the optimum environment for the lysosomal enzymes.

The various macromolecules that are destined to be broken down are delivered to lysosomes by vesicles. The breakdown products, which include amino acids and lipids, are then transported across the lysosomal membrane into the cytosol for use by the cell.

Plants and fungi have a different class of organelles, called **vacuoles**, which are related to lysosomes. Vacuoles are significantly larger than lysosomes, usually occupying more than a third of a plant cell's total volume (Figure 6.8). Besides containing enzymes that break down macromolecules, some vacuoles can store nutrients for later use by the plant cell. In the case of seeds, vacuoles in specialized cells store nutrient proteins that are later broken down to provide amino acids for the growth of the plant embryo during germination. As shown by the fact that seeds from the burial tombs of Egyptian kings have been successfully germinated thousands of years later, storage vacuoles are very good at preserving their contents.

In many plant cells, large vacuoles filled with water contribute to the overall rigidity of the plant by applying pressure against the cell walls. Vacuoles therefore make many different contributions to the life of the plant cell, and a single cell can easily have several vacuoles with different functions.

Mitochondria are the power plants of the cell

We have now explored the administrative offices, factory floor, and shipping department of our cellular manufacturing plant. However, none of the specialized departments in a factory could function without a source of energy to run the machines that produce the goods. The eukaryotic cell is no different: All the cellular processes discussed so far require a source of energy.

The primary producers of this energy are organelles called **mitochondria** (singular: mitochondrion), which function as the cell's power plants. They use chemical reactions to transform the energy from many different molecules into ATP. As we saw in Chapter 5, ATP is the universal fuel of the cell. ATP contains stored energy in the form of covalent bonds. When this energy is released, it can be used to drive the many chemical reactions in the cell.

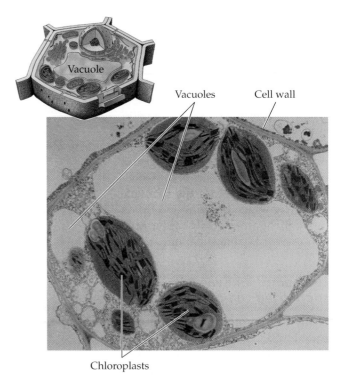

Vacuole

Vacuoles   Cell wall

Chloroplasts

Figure 6.8   Vacuoles in a Plant Cell
Vacuoles are large vesicles in plant cells that are used to store water, nutrients, or enzymes.

Figure 6.9 Energy-Producing Organelles: Mitochondria

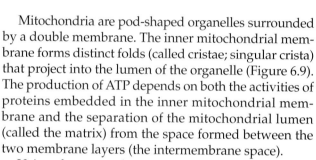

Mitochondria are the major energy-producing organelles in all eukaryotic cells—that is, in the cells of animals, plants, fungi, and protists. Each mitochondrion has a double membrane, and the inner membrane is highly folded. The infoldings (cristae) of the inner membrane contain enzymes that participate in energy production. The inner lumen of the mitochondrion is called the matrix.

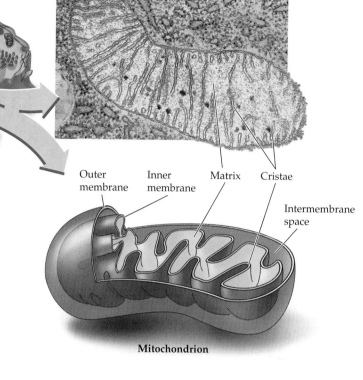

Outer membrane Inner membrane Matrix Cristae Intermembrane space

Mitochondrion

Mitochondria are pod-shaped organelles surrounded by a double membrane. The inner mitochondrial membrane forms distinct folds (called cristae; singular crista) that project into the lumen of the organelle (Figure 6.9). The production of ATP depends on both the activities of proteins embedded in the inner mitochondrial membrane and the separation of the mitochondrial lumen (called the matrix) from the space formed between the two membrane layers (the intermembrane space).

Using these membrane proteins and the physical structure of the organelle, mitochondria are able to harness the energy released by the chemical breakdown of sugar molecules to synthesize energy-rich ATP. In the process, oxygen is consumed and carbon dioxide is released, just as it is in our own breathing or respiration. The details of cellular respiration in mitochondria will be discussed in Chapter 8.

Chloroplasts capture energy from sunlight

Both animal and plant cells have mitochondria to provide them with life-sustaining ATP, but plant cells have additional energy-producing organelles, called **chloroplasts**. Unlike mitochondria, which break down sugars to produce ATP, chloroplasts capture the energy of sunlight and use it to synthesize sugar molecules from carbon dioxide and water. This process is called photosynthesis and results in the release of oxygen as a waste product. Many organisms depend on that oxygen, as we will see in Chapter 8.

Chloroplasts are enclosed by a double membrane, within which lies a separate internal membrane system that is arranged like stacks of pancakes (Figure 6.10). These stacked membrane discs, called thylakoids, contain specialized pigments, of which chlorophyll is the most common. Chlorophyll enables chloroplasts to cap-

ture energy from sunlight. The green color of chlorophyll also accounts for the green coloration of most plants. Enzymes present in the space surrounding the thylakoids use energy, water, and carbon dioxide to produce carbohydrates. The mechanism of energy production by photosynthesis in plants will be discussed in Chapter 8.

■ Each organelle inside a eukaryotic cell makes a specific contribution to the life of the cell. The nucleus stores the genetic blueprint of the cell. The endoplasmic reticulum and Golgi apparatus produce proteins and direct them to various destinations inside and outside the cell. Lysosomes and vacuoles break down macromolecules, and some vacuoles also provide physical support for plant cells. The production of energy for use by the cell depends on mitochondria in animal cells and on both mitochondria and chloroplasts in plant cells.

The Cytoskeleton: Providing Shape and Movement

If a eukaryotic cell consisted only of the plasma membrane and organelle compartments, it would be a limp bag of cytosol with no sustainable organization. Thanks

Figure 6.10   Energy-Transforming Organelles: Chloroplasts
Found only in plant cells, chloroplasts capture energy from sunlight. Each chloroplast has both a double outer membrane and a third inner membrane system, which consists of stacked discs called thylakoids. Each stack of thylakoids contains the proteins and pigments used to harness energy from light.

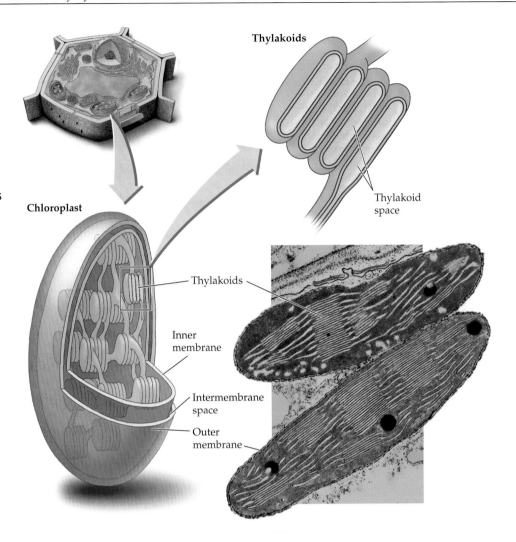

to a system of structural supports called the **cytoskeleton**, however, this is not the case. As its name implies, the cytoskeleton is an internal support system for the cell. It maintains both cell shape and the distribution of organelles in the cytosol. Furthermore, the cytoskeleton is ever changing and dynamic, allowing some cells to change shape and move around on their own. Unlike the bone skeleton of an adult human, which has fixed connections between bones, the cytoskeleton has many noncovalent associations between its proteins, which can break, re-form, and reshape the overall structure of the cell. The cytoskeleton is based on distinct systems of protein filaments, which include microtubules and actin filaments.

Microtubules support movement inside the cell

Microtubules are the thickest of the cytoskeleton filaments, with diameters of about 25 nanometers. Each microtubule is a helical polymer of the protein mono-

mer **tubulin** and has two distinct ends. Microtubules can grow or shrink in length by adding or losing tubulin monomers at either end. This ability allows microtubules to form dynamic structures capable of rapidly reorganizing the cell when necessary. The microtubules in most animal cells radiate out from the center of the cell and end at the inner face of the plasma membrane (Figure 6.11*a*). This radial pattern of microtubules serves as an internal scaffold that helps position organelles such as the ER and the Golgi apparatus.

Microtubules also define the paths along which vesicles move in their travels from one organelle to another or from organelles to the cell surface. The ability of microtubules to act as "railroad tracks" for vesicles depends on **motor proteins** that attach to both vesicles and microtubules. These specialized proteins convert the energy of ATP into mechanical movement, which allows them to move along a microtubule in a specific direction, carrying an attached vesicle like cargo (Figure 6.11*b*).

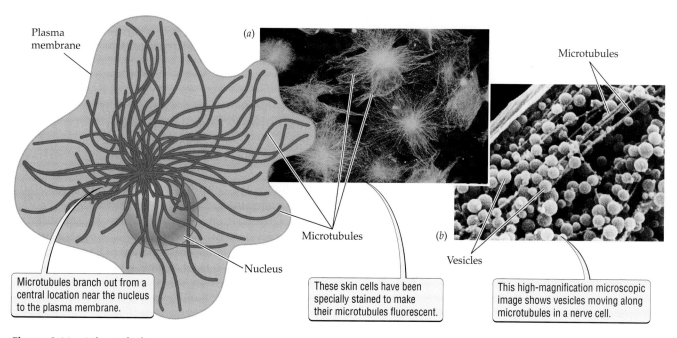

Figure 6.11   Microtubules
(*a*) Microtubules form radial patterns in most animal cells, stretching from the center of the cell to the plasma membrane. (*b*) Microtubules can function as tracks along which vesicles are shuttled around the cell.

Actin filaments allow cells to change shape and move

Actin filaments are the thinner of the cytoskeleton filaments, but they are the most important when it comes to supporting cell movements. Each actin filament is a polymer of actin monomers. Like microtubules, actin filaments can rapidly change length by adding or losing monomers.

Cell movements often depend on rapid rearrangements of complex networks of actin filaments. Perhaps the best example of their action can be found in a cell moving across a flat surface. Certain skin cells called fibroblasts have the ability to move independently. This ability is an important part of wound healing, since fibroblasts migrate into the area of a wound to assist in closing up the damaged edges. When observed under a microscope, fibroblasts can be seen crawling around on the surface of the microscope slide. They achieve this motion by extending flattened sheets of plasma membrane called **pseudopodia** (*pseudo*, "false"; *podia*, "feet"; singular pseudopodium) in the direction they are moving, while their rear ends retract behind them (Figure 6.12).

Rearrangements of actin filaments drive the movements of these cells. Actin filaments in a protruding pseudopod tend to point all in the same direction, toward the plasma membrane. When these well-organized filaments lengthen, they push on the plasma membrane and extend the pseudopod in the direction the cell is moving. At the same time, on the other side of the cell, the actin filaments tend to run in all directions and shorten. This process loosens and breaks down the actin filament network at that end of the cell, resulting in retraction of the plasma membrane and what appears like the cell pulling up its rear end behind it (see Figure 6.12).

Cells are accustomed to lengthening and shortening actin filaments as part of their daily activities, and this mechanism is what the invading *Listeria* bacteria that we talked about at the beginning of the chapter take over and use. *Listeria* is able to use the actin proteins and natural processes of the host cell for its own purposes. In fact, biologists have identified proteins on the surface of *Listeria* that capture actin monomers and start the process of polymerization to form filaments. The explosive propulsion created by this cometlike tail of growing actin filaments allows *Listeria* to move through the infected cell and to jump from one cell to another, thus spreading the infection. As remarkable as this hijacking of the cell's machinery by *Listeria* might seem, other disease-causing bacteria, such as *Shigella*, also use actin-based propulsion to infect host cells, causing dysentery.

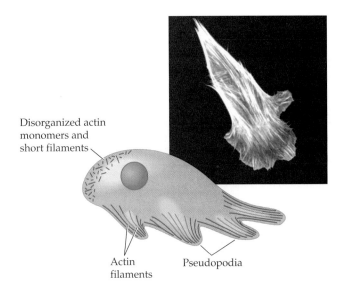

Disorganized actin monomers and short filaments

Actin filaments

Pseudopodia

Figure 6.12   Actin Filaments
Networks of actin filaments help cells crawl on surfaces by allowing them to extend flattened projections called pseudopodia.

■ The cytoskeleton provides eukaryotic cells with structural support and the ability to change shape and move. Protein filaments make up most cytoskeletal networks. Microtubules are essential for the movement of vesicles and other organelles inside the cell. Actin filaments allow cells to change shape and move.

■ HIGHLIGHT

The Evolution of Organelles

6.11   The use of the cell's own resources by an invading prokaryote such as *Listeria* to further its own agenda might seem terribly unfair. However, we eukaryotes may not have the right to pass judgment on *Listeria*'s behavior. In the distant evolutionary past, the ancestors of eukaryotic cells probably did a similar thing by capturing their prokaryotic neighbors and using them for their own purposes (Figure 6.13).

The mitochondria and chloroplasts in eukaryotic cells bear a striking physical resemblance to primitive prokaryotes. Both organelles have their own DNA and are able to make some proteins. They also reproduce independently of the eukaryotic cell, by simply dividing in two. These striking characteristics imply that mitochondria and chloroplasts were once free-living prokaryotes that were engulfed by other cells, eventually forming a mutually beneficial relationship.

How did early single-celled eukaryotes benefit from capturing primitive prokaryotes? The answer to this question may lie in the environment of Earth more than 3.5 billion years ago. At that time, when the first cells are thought to have arisen, there was virtually no oxygen gas (O_2) in the atmosphere, and primitive prokaryotes could break down sugars in the absence of oxygen. However, as time passed, some prokaryotes evolved the ability to use the energy of sunlight to make organic compounds from carbon dioxide and water. The production of oxygen as a by-product of these photosynthetic reactions slowly changed the atmosphere of Earth.

As oxygen gas accumulated in the early atmosphere, some prokaryotes evolved mechanisms for using this gas to break down sugars and release energy, giving them a significant advantage over earlier prokaryotes. At this turning point in evolution, about 2 billion years ago, a primitive single-celled eukaryote is thought to have captured a smaller prokaryote and gained the ability to use atmospheric oxygen in energy production. The descendants of these captured prokaryotes are the mitochondria that we now find in every eukaryotic cell. Likewise, the ancestors of chloroplasts were probably primitive cyanobacteria that had evolved the ability to photosynthesize.

The eukaryotic cells we know today all depend on mitochondria for their ability to use oxygen to produce

Figure 6.13   How Primitive Eukaryotes May Have Acquired Mitochondria

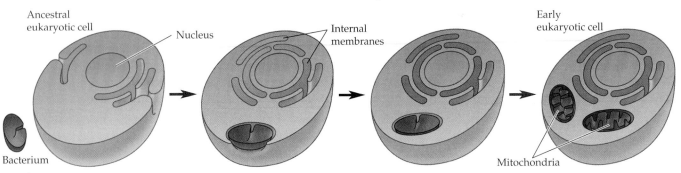

Ancestral eukaryotic cell

Nucleus

Internal membranes

Early eukaryotic cell

Bacterium

Mitochondria

energy. Likewise, plant cells depend on chloroplasts for photosynthesis. Thus we are reminded that eukaryotic cells are made up of very specialized compartments and chemical components that function together to support the processes of life. Since the division of labor seen in all eukaryotic cells may be based on the capture of useful prokaryotes in the evolutionary past, it is unsurprising that the same principle can in turn be used by current invaders such as *Listeria*. These phenomena confirm the old saying: What goes around comes around.

> ■ Mitochondria and chloroplasts are probably descendants of primitive prokaryotes that were engulfed by other cells billions of years ago.

SUMMARY

Cells: The Basic Units of Life

- Cells are the basic units that make up all living organisms.
- All multicellular organisms are made up of many different types of specialized cells.

The Plasma Membrane: Separating Cells from the Environment

- Every cell is surrounded by a plasma membrane that separates the chemical components of life from the nonliving environment.
- The basic component of the plasma membrane is a phospholipid bilayer. The membrane also contains proteins that can move sideways in the plane of the bilayer.
- The proteins in the plasma membrane allow the cell to exchange materials with and respond to its environment.

Comparing Prokaryotes and Eukaryotes

- Living organisms can be classified as either prokaryotes or eukaryotes.
- The cells of prokaryotes lack internal membrane-enclosed compartments; the cells of eukaryotes have internal compartments, such as the nucleus.
- Prokaryotes were the first living organisms to evolve. Today's prokaryotes are all single-celled bacteria and archaeans.
- Eukaryotic cells are approximately a hundred times larger than prokaryotic cells and require internal compartments to concentrate the chemical components of life.

The Specialized Internal Compartments of Eukaryotic Cells

- The specialized membrane-enclosed compartments inside a eukaryotic cell are known as organelles. Each organelle makes a specific contribution to the life of the cell.

- The nucleus is the most distinctive organelle in eukaryotic cells. It houses the DNA-encoded instructions that control every activity and structural feature of the cell.
- The endoplasmic reticulum manufactures proteins and lipids for use by the cell or for export to the environment. The Golgi apparatus receives proteins and lipids from the endoplasmic reticulum and directs them to their final destinations inside or outside of the cell.
- Vesicles transport proteins and lipids between organelles and between an organelle and the plasma membrane.
- Lysosomes break down organic macromolecules such as proteins into simpler compounds that can be used by the cell. Vacuoles are similar to lysosomes but can also lend physical support to plant cells.
- Mitochondria produce energy for use by all eukaryotic cells. Chloroplasts harness the energy of sunlight for use by plant cells.

The Cytoskeleton: Providing Shape and Movement

- Eukaryotic cells depend on the cytoskeleton for structural support and the ability to change shape and move.
- The cytoskeleton consists of different types of protein filaments, including microtubules and actin filaments.
- Microtubules and actin filaments can change length frequently and rapidly. They are essential for the movement of organelles inside the cell and for movement of the entire cell.

Highlight: The Evolution of Organelles

- Mitochondria and chloroplasts are similar to primitive prokaryotes.
- Eukaryotic cells probably acquired mitochondria and chloroplasts by engulfing prokaryotes at some time in their evolution.

KEY TERMS

actin filament p. 117	motor protein p. 116
cell p. 106	nuclear envelope p. 111
chloroplast p. 115	nuclear pore p. 111
cytoplasm p. 111	nucleus p. 109
cytoskeleton p. 116	organelle p. 111
cytosol p. 109	plasma membrane p. 107
endoplasmic reticulum (ER) p. 111	prokaryote p. 109
eukaryote p. 109	pseudopodium p. 117
fluid mosaic model p. 108	ribosome p. 112
Golgi apparatus p. 113	rough ER p. 112
lumen p. 112	smooth ER p. 112
lysosome p. 113	tubulin p. 116
microtubule p. 116	vacuole p. 114
mitochondrion p. 114	vesicle p. 112

CHAPTER REVIEW

Self-Quiz

1. Which of the following statements about the plasma membrane is true?
   a. It is a solid layer of protein that protects the contents of the cell.
   b. The plasma membrane of a bacterium has none of the same components as the plasma membrane of an animal cell.
   c. It is a rigid and unmoving layer of phospholipids and proteins.
   d. It allows selected molecules to pass into and out of the cell.

2. Which of the following cellular components can be used to distinguish a prokaryotic cell from a eukaryotic cell?
   a. nucleus
   b. plasma membrane
   c. DNA
   d. proteins

3. One key function of nuclear pores is to
   a. allow cells to communicate with one another.
   b. aid in the production of new nuclei.
   c. allow molecules such as proteins to move into and out of the nucleus.
   d. form connections between different organelles.

4. Vesicles are essential for the normal functioning of the Golgi apparatus because
   a. they provide energy for chemical reactions.
   b. they move proteins and lipids between different parts of the organelle.
   c. they contribute to the structural integrity of the organelle.
   d. they produce the sugars that are added to proteins.

5. Which of the following statements is *not* true?
   a. Both mitochondria and chloroplasts provide energy to cells in the same way.
   b. Both mitochondria and chloroplasts have more than one membrane.
   c. Only chloroplasts contain the pigment chlorophyll.
   d. Both animal and plant cells contain mitochondria.

6. Actin filaments contribute to cell movement by
   a. providing energy in the form of ATP.
   b. lengthening and pushing against the plasma membrane.
   c. forming a stable and unchanging network inside the cell.
   d. allowing organelles to change position inside the cell.

Review Questions

1. Proteins embedded in the plasma membrane have several important functions in the life of the cell. Describe two of these functions and explain why they are important to the cell.

2. Vacuoles can have a wider variety of functions in plant cells than do lysosomes in animal cells. Describe one function that vacuoles perform in plant cells that lysosomes do not in animal cells.

3. You have discovered an enzyme in the lumen of lysosomes that enables them to break down polysaccharides. Where in the cell do you think the ribosome that produced this enzyme is located—free in the cytosol or associated with the endoplasmic reticulum? Justify your answer.

4. Describe one common characteristic shared by microtubules and actin filaments. Relate this characteristic to the function of these filaments in the cell.

6   𝕿𝖍𝖊 𝕯𝖆𝖎𝖑𝖞 𝕲𝖑𝖔𝖇𝖊

New Protein Provides Hope for Muscular Dystrophy

WASHINGTON, DC—Researchers reported at a news conference today that genetically engineered mice with muscular dystrophy–like symptoms suffer less muscle cell damage when they produce more of a newly discovered protein known as utrophin. Researchers say the new study suggests that children with Duchenne's muscular dystrophy (DMD) might also benefit from drugs that increase their production of this protein.

DMD is an inherited disease that afflicts more than 20,000 male infants each year worldwide. These children usually die from heart failure before reaching adulthood. The muscle cells of these children degenerate due to the absence of a cytoskeletal protein called dystrophin.

Biologists have been investigating DMD by altering the genetic profile of mice so that they, too, lack dystrophin and experience DMD symptoms. These mice become models for the disease and can be studied and subjected to experimental treatments that are too risky for humans.

"Based on what we see in DMD mice, a drug therapy that increases the production of utrophin could slow muscle degeneration and prolong these children's lives," said Dr. Sarah Benning, DMD researcher at Massachusetts Charitable Hospital. "This is potentially a huge breakthrough."

Evaluating "The News"

1. The similarity of utrophin to dystrophin and its potential ability to act as a substitute in affected cells implies that the cytoskeleton has built-in backup systems. What do you think might be the benefit of having backup proteins in a cell?

2. The development of mice that lack dystrophin was an important breakthrough in understanding DMD in humans. Why might studying these so-called model organisms be so useful in medically relevant research?

3. The genetic modification of animals such as mice to experience human diseases is now a common method used to study such diseases. Do the benefits of learning more about human diseases justify breeding animals that are doomed to suffer from those same diseases? When it comes to improving human health, do animals used in research have any rights?

chapter 7 Energy and Enzymes

Ben Shahn, *Helix and Crystal*, 1957.

"Take Two Aspirin and Call Me in the Morning"

*I*f someone told you there was a wonder drug that reduced pain, fever, and inflammation and helped combat heart disease and cancer, would you believe them? As amazing as it may sound, such a drug does exist—in fact, it recently celebrated its hundredth birthday. It is the drug known to all of us as aspirin. Even more remarkable is that our understanding of how aspirin works and how it can have so many remarkable effects on the human body is still in its infancy.

The active ingredient in aspirin, called salicylic acid, is found naturally in willow bark and leaves, and has long been used by humans as an antiseptic and fever-reducing agent. Our chemical knowledge of aspirin started with the laboratory synthesis of salicylic acid in 1860. Unfortunately, salicylic acid also damages the stomach lining, so in 1899 a more palatable version was developed, which we know today as aspirin. In the years that followed, aspirin became an important means of lowering pain and reducing fevers and inflammation. For decades these therapeutic benefits were more than enough to make aspirin a staple in every doctor's bag and hospital around the world. It even spawned the well-known phrase "take two aspirin and call me in the morning," implying that this wonder drug could handle most medical problems, at least overnight.

Today we know that aspirin can have an even more amazing effect on human health than its well-established use as a pain reliever. Well-publicized studies show that low doses of aspirin taken daily can reduce the risk of heart disease. In 1996, the importance of aspirin was further bolstered by studies showing that people who take aspirin regularly have lower rates of colon cancer. However, taking aspirin for prolonged periods does have negative side effects, including damage to the stomach and kidneys. Given all the possible health benefits of taking aspirin, researchers had some obvious questions: How does this wonder drug work, and can we make it better by reducing the side effects?

The first ideas about how aspirin works emerged almost 75 years after its development. Today we know that the ability of aspirin to affect so many processes in the body is directly related to its blocking of the chemical synthesis of certain chemical messengers, or hormones. Ironically, this blocking of chemical reactions in our bodies also accounts for the negative side effects of aspirin. Armed with a growing knowledge of how aspirin affects these chemical reactions , researchers are now seeking to improve this old wonder drug by eliminating the negative aspects of its action. In this chapter, we will look at the fun-

damental principles that control every chemical reaction in our bodies, and we will see how our understanding of these processes could pave the way for the development of a new superaspirin.

An Apothecary's Kit from the Late Nineteenth Century, Which Would Have Contained Aspirin

■ KEY CONCEPTS

1. Living organisms must obey the universal laws of energy conversion and chemical change.

2. The sun is the primary source of energy for living organisms. Photosynthetic cells in organisms such as plants capture energy from the sun and use it to synthesize sugars from carbon dioxide and water. All cells break down sugars to release energy.

3. Metabolism consists of the chemical reactions that produce complex macromolecules such as sugars and proteins, and which break down those macromolecules to yield smaller molecules and usable energy.

4. Enzymes control the speed of chemical reactions in cells. Metabolic pathways are sequences of enzyme-controlled chemical reactions.

All living processes require energy, which living organisms must extract from their environment. They use this energy to manufacture and transform the various chemical compounds that make up living cells. Both the capture and the use of energy by living organisms involve thousands of chemical reactions, which together are known as **metabolism**. The metabolic reactions that create complex molecules out of smaller compounds are described as biosynthetic; those that break down complex molecules to produce usable energy are described as catabolic.

All the chemical reactions that occur in cells can be grouped into a surprisingly small number of metabolic pathways. In the same way that chemical building blocks fall into a limited number of categories, the reactions that allow the cell to manufacture and transform these building blocks are also limited in number. In this chapter, we examine the role played by energy in the chemical reactions that maintain living systems. We also discuss the role of specialized proteins that speed up chemical reactions that would otherwise take too long for us to survive.

The Role of Energy in Living Systems

The discussion of any chemical process in the cell is a discussion about energy. The idea that energy is behind every activity in the cell seems natural and unsurprising, since all of us are accustomed to thinking of energy as a form of fuel. However, energy is more than just fuel, since its properties dictate which chemical reactions can occur and how molecules can be organized into living systems.

The laws of thermodynamics apply to living systems

The relationship between energy and the cell's activities is governed by the same physical laws that apply to everything else in the universe. These laws of thermodynamics define the ways cells transform chemical compounds and interact with the environment. The **first law of thermodynamics** states that the total energy of a system always remains constant. In other words, energy cannot be either created or destroyed, only converted from one form to another (Figure 7.1a). For example, mitochondria convert energy from food molecules such as sugars into the energy of covalent bonds in ATP (ADP + phosphate + energy → ATP) (Chapter 5). Thus, mitochondria do not create energy from nothing. Instead, they convert energy from one form (sugars) to another (ATP), which can then be used by the cell. The way the energy stored in ATP is used to drive other chemical reactions is discussed later in this chapter.

The **second law of thermodynamics** describes how each cell relates to its environment and how it maintains a well-ordered internal organization. This law states that systems, such as a cell or even the whole universe, tend to become more disorderly. This statement may seem like a law written for an adolescent's room, which usually tends toward disorder and chaos, but it is true of all systems, including the internal organization of the cell (Figure 7.1b).

As we saw in Chapters 5 and 6, a cell is made up of many chemical compounds assembled into complex structures. Such a high level of organization flies in the face of disorder and must be compensated for. The tremendous structural complexity of the cell and its organelles exists in a constant struggle against chaos. Thus, to counteract the natural tendency toward disorder, the cell must transfer some of its disorder elsewhere.

Living systems pass off or transfer disorder by releasing heat into the environment. Heat is a form of energy that causes rapid and random movement of molecules, a condition that is highly disordered. Thus, when cells release heat, they increase the degree of disorder in the molecules of the environment, which compensates for

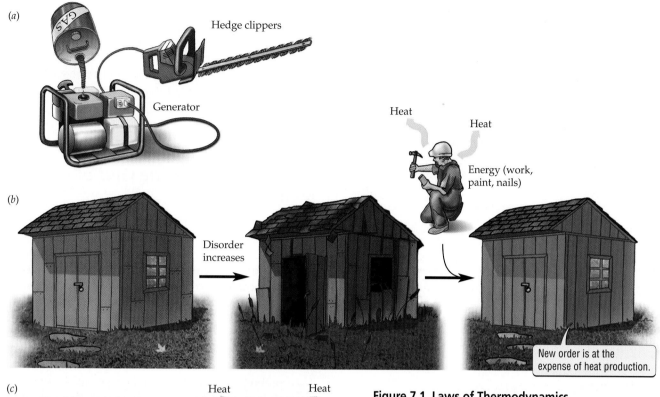

(a)

Hedge clippers

Generator

Heat

Heat

Energy (work, paint, nails)

(b)

Disorder increases

New order is at the expense of heat production.

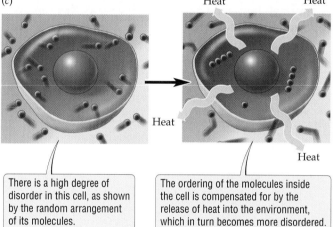

(c)

Heat

Heat

Heat

Heat

There is a high degree of disorder in this cell, as shown by the random arrangement of its molecules.

The ordering of the molecules inside the cell is compensated for by the release of heat into the environment, which in turn becomes more disordered.

Figure 7.1 Laws of Thermodynamics
(*a*) Energy is neither created nor destroyed. The chemical energy contained in the covalent bonds of the gasoline molecules is converted into electrical energy by the generator. The electrical energy, in turn, is converted into the mechanical energy of motion in the hedge clippers. Neither the generator nor the hedge clippers creates or destroys energy. (*b*) The disorder of a system always tends to increase. Left unattended, all structures, such as this wooden shed, tend to lose their order and become disarrayed. The input of energy, by way of human effort in this case, is needed to maintain the order of the structure. (*c*) Cells maintain their organization through a continuous input of energy from the environment. Thus, they obey the second law of thermodynamics, which states that the disorder of a system always increases unless there is an input of energy.

the increasing order inside the cell (Figure 7.1*c*). There is a direct connection between cell organization and the transfer of energy because the chemical processes used to build well-ordered structures are the same ones that produce the heat. Hence, the generation of order is directly coupled with the release of heat energy.

The flow of energy and the cycling of carbon connect living things with the environment

Where does the energy that creates order in the cell come from? We know from the first law of thermodynamics that the cell cannot create energy from nothing; thus, it must come from outside of the cell. In other words, ener-

gy must be transferred into the cell in some fashion. In the case of photosynthetic organisms, the energy comes from sunlight. By using that energy to synthesize sugar molecules from carbon dioxide and water, those organisms convert it into the chemical bonds of sugars. For nonphotosynthesizing organisms, energy comes from the chemical bonds in food molecules, such as sugars and fats.

The two statements in the preceding paragraph reveal the chemistry of the relationship between producers and consumers. As we saw in Chapter 1, photosynthetic producers, such as plants, capture energy from sunlight, and nonphotosynthetic consumers, such as animals, obtain this energy by consuming plants or other animals that have eaten plants. This means that, thanks to photosynthesis, the sun is the primary energy source for living organisms.

Now, don't feel sorry for plants. They are more than consumable energy sources for nonphotosynthetic organisms. First, plants also use the sugars they make by photosynthesis, especially at night, when there is no sunlight and no photosynthesis. Second, the carbon dioxide (CO_2) that most organisms produce as a by-product of the energy-harnessing process called respiration (which will be discussed in Chapter 8) is a source of carbon for photosynthesis. So, during photosynthesis, a producer uses carbon dioxide as a carbon source for its production of sugars, which in turn are used either by the producer itself or by consumers. Thus, carbon atoms are continually cycled from carbon dioxide in the atmosphere to sugars made by producers and back to carbon dioxide released by respiring producers and consumers (Figure 7.2). This recycling occurs not only for carbon, but also for other atomic building blocks of life, such as nitrogen and phosphorus.

■ Living organisms obey the same laws of energy that apply to the physical world. Organisms must obtain energy from their environment and must convert it into usable forms and structural components. Carbon atoms are recycled between living organisms and the environment.

Using Energy from the Controlled Burning of Food

Living systems obtain energy from food by burning organic molecules such as sugars to form carbon dioxide and water. If our cells were to convert food into carbon dioxide and water in a single chemical reaction, we would burst into flame like a lit match. Here's the chemical equation that describes what happens when the match burns:

$$Wood + O_2 \rightarrow CO_2 + H_2O + energy \text{ (heat and light)}$$

This combustion reaction is similar to what occurs when our cells burn food, but fortunately for us, there are some important differences.

Figure 7.2 Carbon Cycling
Carbon atoms cycle among producers, consumers, and the environment. Carbon atoms become parts of different kinds of molecules as they cycle between living organisms and the environment.

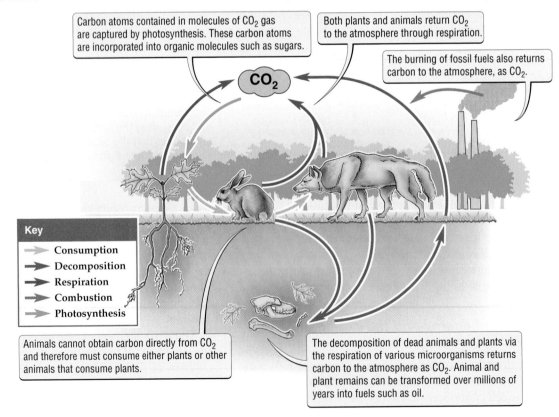

Carbon atoms contained in molecules of CO_2 gas are captured by photosynthesis. These carbon atoms are incorporated into organic molecules such as sugars.

Both plants and animals return CO_2 to the atmosphere through respiration.

The burning of fossil fuels also returns carbon to the atmosphere, as CO_2.

Key
→ Consumption
→ Decomposition
→ Respiration
→ Combustion
→ Photosynthesis

Animals cannot obtain carbon directly from CO_2 and therefore must consume either plants or other animals that consume plants.

The decomposition of dead animals and plants via the respiration of various microorganisms returns carbon to the atmosphere as CO_2. Animal and plant remains can be transformed over millions of years into fuels such as oil.

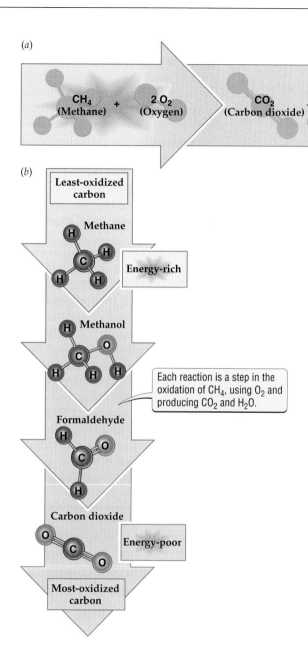

Figure 7.3 Oxidation of Methane
(*a*) When methane gas encounters a spark, it undergoes an explosive oxidation, or combustion, reaction. Energy is released to the environment as heat. (*b*) Alternatively, methane can be oxidized in a series of smaller steps. The single carbon atom in each compound becomes increasingly oxidized with each step. That is, the carbon atom is gradually surrounded by more oxygen atoms that tend to attract its electrons, leaving it more oxidized.

Capturing energy from foods requires the transfer of electrons

In the multiple chemical reactions that allow cells to capture energy from foods, electrons are transferred from one molecule or atom to another. **Oxidation** is the loss of electrons from one molecule or atom to another; **reduction** is just the opposite, the gain of electrons by one molecule or atom from another.

Let's consider the carbon atom contained in the gas methane (CH_4) (Figure 7.3*a*). This carbon atom is bonded to four hydrogen atoms and is in a reduced, or electron-rich, state. If we burn some of the methane gas, the products of the combustion reaction are carbon dioxide (CO_2) and water (H_2O). If we compare the electron-rich carbon atom in methane to the one in the CO_2 produced by combustion, we find that the oxygen atoms tend to attract electrons away from the carbon atom, leaving it in an oxidized, or electron-poor, state. Thus, the combustion of an organic compound such as methane is an oxidation reaction.

In contrast to the combustion of methane, biological oxidation takes place in a series of small steps, not all at once. Instead of jumping from being part of a complex sugar molecule to being part of a simple CO_2 molecule, a carbon atom passes through several chemical reactions and intermediate compounds. Each intermediate compound is a little more oxidized than the preceding one (Figure 7.3*b*). This stepwise and controlled combustion allows the cell to couple each small energy-releasing oxidation reaction with other reactions, which store some of the released energy in newly formed chemical bonds. This transfer of energy from one compound to another is the basis of the reactions in metabolism.

Energy in a living system is transferred via the universal energy carrier, ATP. As we saw in Chapter 5, when ATP is produced from ADP and a phosphate group (see Figure 5.7), energy is stored in its chemical bonds. The

The energy released from the burning match is dispersed into the environment as heat and cannot be regained by the match in any fashion. In contrast, living systems need to capture that released energy. They can do so only by controlling the combustion reaction and breaking it down into a series of much smaller chemical reactions. This breakdown not only saves us from bursting into flame, but also gives our cells the opportunity to capture the small amounts of energy released at each step. In this section we review the key characteristics of the chemical reactions that release energy from food molecules for use by the cell.

energy released from ATP when it is broken down to ADP and phosphate is used to carry out the activities of the cell. ATP production is an urgent priority for the human body, since each cell consumes its entire supply of ATP almost every minute. In fact, in nearly every chemical reaction in the cell, ATP is consumed or synthesized.

Metabolic reactions that create complex molecules out of smaller compounds are described as **biosynthetic** reactions; those that break down complex molecules to produce usable energy are described as **catabolic** reactions. The energy released from ATP when it is broken down to ADP and phosphate is used for such activities as moving molecules and ions between various cellular compartments, generating mechanical force in a crawling cell, and manufacturing complex macromolecules from simpler chemical compounds. This final activity comprises the biosynthetic reactions of metabolism. The catabolic reactions in the cell that release energy and harness it in the form of ATP are tightly coupled to the biosynthetic reactions, forming the two sides of metabolism: releasing energy by breaking things down and using energy to build things up. For an unexpected consequence of a high rate of metabolism, see the box on page 130.

Chemical reactions are governed by simple energy laws

How does the cell control such a powerful event as combustion and break it down into smaller, more manageable and useful steps? The answer to this question lies in the very nature of chemical reactions. Let's review some of the fundamental principles that govern chemical reactions, as represented by the following generic example:

$$A + B \rightarrow C + D$$

A and B are the starting materials, or reactants; C and D are the products that are formed by the reaction. As we saw in Chapter 5, a chemical reaction changes the arrangement of atoms in molecules. All chemical reactions tend to proceed in the direction that will result in products with greater stability and a lower energy state. Products in a lower energy state than their reactants have less energy stored in the form of well-ordered bonds and are therefore in a less ordered state, which is favored by the second law of thermodynamics. This tendency toward less order and a lower energy state favors a particular direction of change in chemical reactions. That is, the process of "going downhill" energetically from products to reactants is a good thing according to the laws of the physical world.

It is important to note that just because A and B are present together does not mean they will react. The reason is that all compounds tend to be in a semi-stable state. Thus, they need to be destabilized, or activated ("jump-started"), before a chemical reaction can begin and proceed. The jump start that is required is the input of a small amount of energy, called the **activation energy** of the reaction. The activation energy is required to alter the chemical configuration of the reactants so that the reaction can take place. Once the reactants have overcome this activation energy barrier, the reaction proceeds, and a product of lower energy is produced (Figure 7.4).

Where do chemical reactions get the activation energy required for them to proceed? Returning to the example of a burning match, the activation energy required to light the match can come from the friction generated by moving the head of the match across a rough surface. Chemical reactions in cells can acquire the activation energy they need from random collisions between molecules floating in the cytosol. These collisions become more frequent and energetic as the temperature increases and molecules in the cytosol move faster. But at the normal body temperatures of most organisms, these collisions do not release enough energy to drive all the reactions required for life. To compensate, cells contain a class of specialized proteins called enzymes, which directly promote chemical reactions in living organisms.

> ■ Living organisms capture and use energy released from foods by the stepwise oxidation of organic compounds. To proceed, each chemical reaction requires the input of a small amount of activation energy.

7.3   How Enzymes Speed Up Chemical Reactions

Chemical reactions are a crucial part of life. As we have seen, some chemical reactions are involved in the production of macromolecules that make up the structures of the cell; others allow the cell to obtain energy from food molecules. These are just a few of the different chemical reactions that support life. Yet, if we were to depend on these chemical reactions occurring unassisted, they would happen so slowly that life would not be possible.

To solve this problem, cells use a specialized class of proteins called **enzymes** to speed up chemical reactions.

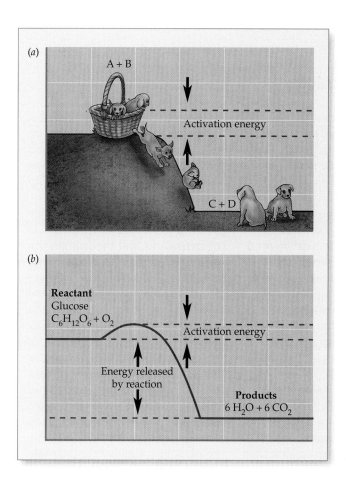

Figure 7.4 Getting Over the Activation Energy Barrier
(*a*) Imagine that the reactants of a chemical reaction (A + B) are a group of frisky puppies trying to get out of a basket sitting on a slope. In this analogy, the sides of the basket represent the activation energy barrier, and the end of the slope represents the lower energy state of the products. As each puppy tries to scramble over the edge of the basket, those that manage it are like the reactant molecules that receive enough energy from the environment to make it over the activation energy barrier. Once the successful puppies are at the bottom of the slope, it is more difficult for them to make it back up the slope and into the basket; getting back in requires more effort. Thus, the preferred direction is for the puppies to move out of the basket and not back in. Likewise, the products of the chemical reaction (C + D) are in a lower energy state than the reactants, so a much larger input of energy would be required to reverse the direction of the reaction. (*b*) The oxidation of glucose during respiration to produce water and carbon dioxide must overcome an activation energy barrier.

Nearly all chemical reactions that take place in living organisms require the assistance of enzymes. Our human cells contain several thousand different enzymes, each of which affects a specific chemical reaction.

To increase the rate at which a chemical reaction proceeds, each enzyme binds to specific reactants, called its **substrates**. By doing so, it lowers the amount of activation energy that the substrates require to react with each other. In fact, when an enzyme binds reactants, it can bring them together in an orientation that favors the making or breaking of bonds required to form the products. For a particular enzyme to bind the correct reactants and alter them appropriately for the chemical reaction, it must have a high degree of specificity for those reactants. In other words, each enzyme is specifically tailored to promote only one of the thousands of possible reactions that occur in the cell.

In the presence of many reactants, the presence of one enzyme rather than another can determine which chemical reaction takes place. However, an enzyme cannot make an impossible reaction happen. Nor can it promote a particular reaction by changing the amount of energy

associated with the reactants or the products. It can only affect the rate at which a reaction occurs by lowering the activation energy barrier. Since enzymes affect the rate at which reactions occur, but remain unchanged by the reactions themselves, they are called **catalysts**.

The control that enzymes exercise over the rate of chemical reactions is necessary for all living processes. The removal of carbon dioxide from cells in the human body is essential for life. This process depends on an enzyme in red blood cells. Before we can exhale carbon dioxide, the gas must first be transferred from our cells into the bloodstream and then into the lungs. For this transfer to occur, carbon dioxide must react with water so that it can be transported in the blood as bicarbonate ions (HCO_3^-):

$$H_2O + CO_2 \rightarrow HCO_3^- + H^+$$

This simple reaction, called the hydration of carbon dioxide (since it involves the addition of water), is necessary for normal respiration. The enzyme responsible is called carbonic anhydrase; it speeds up the hydration of carbon dioxide by a factor of nearly 10 million. In fact, a single carbonic anhydrase enzyme can hydrate more than 10,000 molecules of carbon dioxide in a single second. Needless to say, without carbonic anhydrase, the rate of carbon dioxide hydration would be so slow that we would not be able to rid our bodies of carbon dioxide fast enough to survive. When the circulating bicarbonate ions arrive at the lungs, they are converted back into carbon dioxide, which we then exhale as CO_2 gas.

■ BIOLOGY IN OUR LIVES

Take It Easy, You Might Live Longer!

The idea of immortality has fascinated human beings throughout recorded history. In the past, people interested in extending their life span turned to alchemists or wizards. Today, researchers are discovering which biological factors limit our life span and how those factors might be controlled to let us live longer. One key factor that has emerged is the overall metabolic rate of an organism.

In general, small animals have faster metabolic rates and shorter life spans than large animals. Furthermore, laboratory tests have shown that mice with a restricted diet, and thus a slower metabolism, live longer than mice that are allowed to eat as much as they like. The idea that a higher metabolic rate can shorten

one's life span may seem contradictory, since metabolism includes all of the chemical reactions that maintain living organisms. How might metabolism shorten the life span of an organism?

The answer lies not in the reactions of metabolism themselves, but rather in the toxic chemical by-products that are sometimes accidentally produced when these reactions occur. These chemical compounds react with and damage cellular components such as DNA. The gradual accumulation of this cellular damage is an important contributing factor to aging and, ultimately, death.

The link between a slower metabolism and a longer life holds true for a nematode worm named *Caenorhabditis elegans* as well as for mice. Worms that lack proteins responsible for maintaining a normal metabolic rate, and which have an abnormally low metabolic rate as a result, live up to five times longer than they should. These worms also take longer to

mature and display a slowing of specific behaviors. Similar phenomena have been observed in fruit flies and in yeasts, implying that a broad range of species are subject to life-limiting metabolic accidents.

Given the supporting evidence gathered from several species, it is not surprising that the human life span may also be affected by metabolic rates. Women have slower metabolic rates than men, and they have a higher life expectancy (in the United States, 79 years versus 72 years for men). Even more striking is the fact that 9 out of 10 individuals 100 years old or older are women. Thus, although genetic factors that affect overall health also have a role in determining life expectancy, the accumulated mistakes from a lifetime of metabolism clearly take their toll. Perhaps if we can limit the metabolic activities that run the highest risk of producing toxic by-products, we will finally achieve the goal of extending our life span.

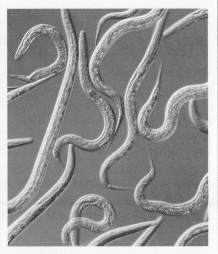

Studies of women, mice, and the nematode worm *C. elegans* show that a slower metabolism may lengthen an organism's life span.

The shape of an enzyme directly determines its activity

The specificity that enzymes have for their substrates depends on the three-dimensional shapes of both substrate and enzyme molecules. In the same way that a particular lock accepts only a key with just the right shape, each enzyme has an **active site** that fits only substrates with the correct three-dimensional shape and chemical characteristics (Figure 7.5*a*). The matching of an enzyme's active site to one or more substrates guarantees that a specific reaction will take place to yield the expected products.

Carbonic anhydrase, for example, is able to bind molecules of both carbon dioxide and water in its active site. By bringing the two substrates ($H_2O + CO_2$) together in just the right positions, the active site of carbonic anhydrase promotes the hydration reaction (Figure 7.5*b*). All enzymes have active sites that are specific for their substrates and will not bind other molecules.

The action of carbonic anhydrase demonstrates how specific binding of two substrates by an enzyme can push them together so that a chemical reaction takes place between them. Thus, we can think of enzymes as molecular matchmakers that bring the right substrates together. If there were no enzyme present, the two substrates would need to collide with each other in just the right way before the hydration reaction could take place. These sorts of molecular collisions do occur all the time, but not nearly as frequently as would be required for the continuous and rapid transfer of carbon dioxide from cells into the blood.

Enzyme chain reactions have energetic advantages

So far, we have discussed the activity of a single enzyme acting alone to promote a single chemical reaction, but this state of affairs is not so common in the cell. Instead, groups of enzymes usually catalyze multiple steps in a sequence of chemical reactions known as a **chemical pathway**. This arrangement presents both advantages and challenges that illustrate important aspects of how enzymes usually behave in the cell.

Let's begin with a particularly noteworthy advantage granted by multi-step metabolic pathways. A pathway permits the product formed by one enzyme to immediately become a reactant for the next reaction. In other words, the product of the first reaction is the immediate substrate for another enzyme and is rapidly consumed in a second catalyzed reaction. In general, a pathway of enzyme-catalyzed steps ensures a particular outcome from a sequence of chemical reactions. Such a sequence can be represented as follows:

$$A \xrightarrow{\text{E1}} B \xrightarrow{\text{E2}} C \xrightarrow{\text{E3}} D$$

where enzyme E1 catalyzes the conversion of A to B, enzyme E2 catalyzes the conversion of B to C, and so on, ensuring that D will be produced in the end. Metabolic

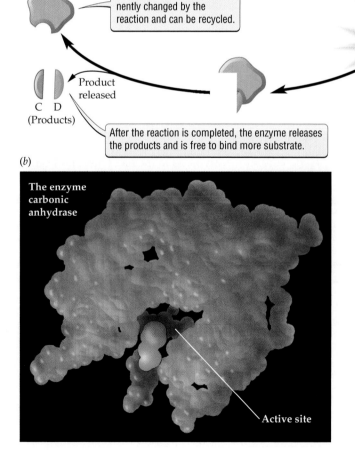

(*a*)

A B
(Substrates)

The active site of the enzyme has the shape and chemical properties to bind to the reactants (substrates) and facilitate the reaction.

Enzyme

Substrate binds

The enzyme is not permanently changed by the reaction and can be recycled.

Catalysis

Product released

C D
(Products)

After the reaction is completed, the enzyme releases the products and is free to bind more substrate.

(*b*)

The enzyme carbonic anhydrase

Active site

Figure 7.5 Enzymes as Molecular Matchmakers
(*a*) An enzyme brings together two reactants (A and B) such that a chemical reaction proceeds to form the products (C and D). (*b*) Carbonic anhydrase catalyzes the reaction of carbon dioxide and water to form bicarbonate.

pathways like this one produce most of the chemical building blocks of the cell, such as amino acids and nucleotides, and are necessary for the harnessing of energy from food or sunlight.

The challenge faced by all enzyme-catalyzed pathways is the need for the products of each reaction to find the enzyme that catalyzes the next reaction in a timely fashion. Enzymes and their substrates do not actively swim after each other like sharks looking for prey. Instead, they depend on random encounters or collisions inside the cell. Within the crowded environment of the cell, an enzyme collides with many molecules every second, but most of these molecules are not its substrates and do not fit into its active site, so nothing happens. Catalysis occurs only when the enzyme encounters appropriate substrates that fit its active site. Thus, although molecular collisions do happen frequently in the cell, only some of these collisions result in enzyme-catalyzed reactions.

One way to increase the efficiency of a chemical pathway is to increase the frequency of molecular collisions between enzymes and their substrates. The enzymes involved in many of the multi-step pathways that are so common in metabolism are located close together. This physically close arrangement of enzymes means that the products of one reaction are close to the next enzyme that uses them as substrates, increasing the likelihood of their collision with it and promoting the next chemical reaction.

At the level of cellular organization, the enzymes necessary for a particular chemical pathway can be contained inside a specific organelle (Figure 7.6). As discussed in Chapter 6, organelles concentrate the proteins and chemical compounds required for specific life processes. Mitochondria, for example, are the sites where the breakdown products of foods are oxidized, generating most of the cell's ATP. The efficient production of ATP requires that the necessary enzymes and substrates be concentrated inside a small compartment. Several enzymes floating in the mitochondrial matrix participate in a series of reactions called the citric acid cycle, which forms the first part of the pathway that produces ATP, as we will see in Chapter 8. Other enzymes involved in the production of ATP are associated with the inner mitochondrial membrane in such a way that they are arranged in sequence.

On the molecular level, several enzymes can be physically connected in a single giant multienzyme complex (see Figure 7.6). Many enzymes involved in the biosynthesis of cellular building blocks such as fatty acids and proteins function as large aggregates of multiple enzymes.

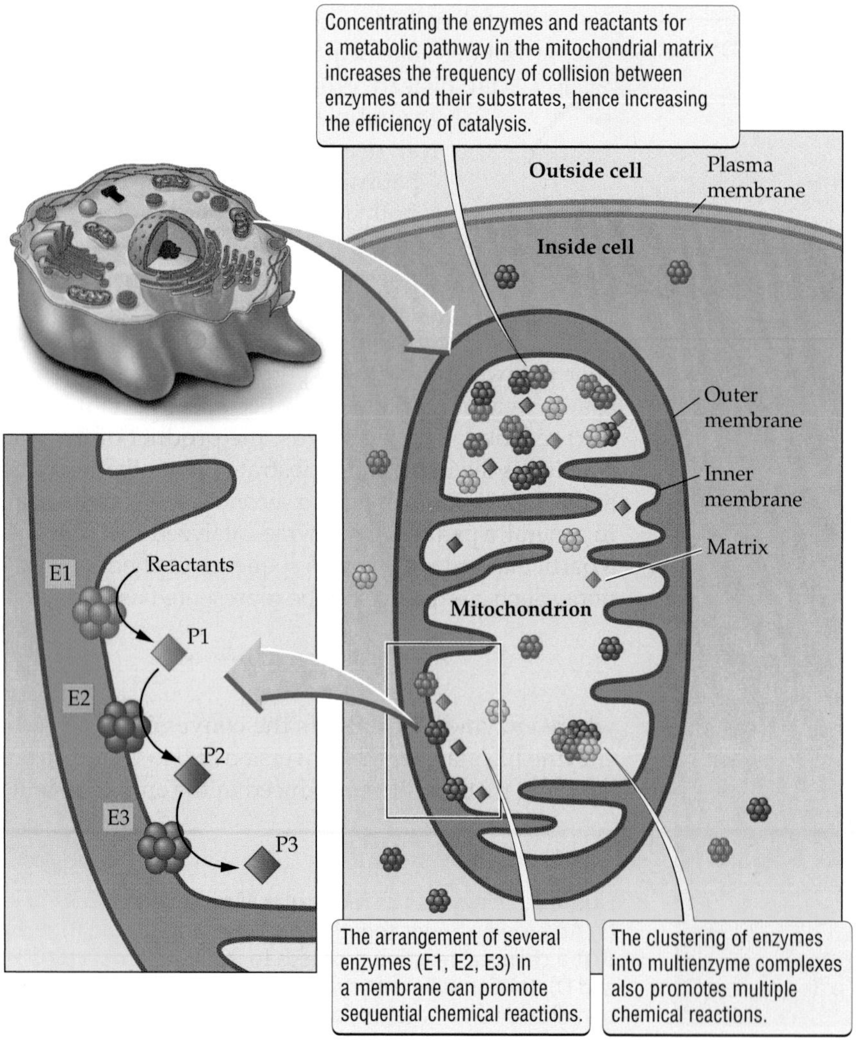

Concentrating the enzymes and reactants for a metabolic pathway in the mitochondrial matrix increases the frequency of collision between enzymes and their substrates, hence increasing the efficiency of catalysis.

Outside cell

Plasma membrane

Inside cell

Outer membrane

Inner membrane

Matrix

Mitochondrion

E1

Reactants

P1

E2

P2

E3

P3

The arrangement of several enzymes (E1, E2, E3) in a membrane can promote sequential chemical reactions.

The clustering of enzymes into multienzyme complexes also promotes multiple chemical reactions.

Figure 7.6 Grouping of Enzymes in the Cell
Enzymes are arranged in the cell so as to promote multiple chemical reactions. These arrangements include localization of enzymes in organelles, in membranes, and as parts of multienzyme complexes.

■ All chemical reactions in the cell require the assistance of enzymes to proceed rapidly enough to support life. Enzymes speed up chemical reactions by lowering the amount of activation energy required for the reaction. Metabolism consists of many multi-step pathways, each of which requires several different enzymes.

Enzymes and Energy in Use: The Building of DNA

Enzyme-catalyzed metabolic pathways are involved in the synthesis and breakdown of most complex molecules in the cell. The use of ATP and enzymes is a common feature of metabolic pathways, such as those involved in the biosynthesis of DNA. As we saw in Chapter 5, DNA is a polymer made up of nucleotides that are linked together by covalent bonds. Before cells can divide, their DNA must be duplicated so that each daughter cell can receive a complete set of the genetic material. This essential process also ensures that our

genetic blueprint is passed from one generation to the next. Duplicating a cell's DNA requires the synthesis of new DNA polymers, a process that involves ATP and several enzymes. In this case, ATP is used to convert individual nucleotides to an activated form that can be added to a growing DNA chain.

As we saw in Chapter 5, nucleotides can have one, two, or three phosphate groups bound to the sugar molecule, and are described as monophosphate, diphosphate, or triphosphate nucleotides, respectively (see Figure 5.6). Energy-rich triphosphate nucleotides are used for DNA synthesis. They are produced from monophosphate nucleotides with the help of two enzymes and ATP. Each of the enzymes catalyzes the transfer of a phosphate group from a molecule of ATP to the monophosphate nucleotide (Figure 7.7a). Since the high-energy phosphate bonds of two ATPs are consumed in these reactions, the triphosphate nucleotide is now energy-rich and can form a covalent bond with another nucleotide. The chemical reaction that adds a triphosphate nucleotide to the end of a DNA chain is catalyzed by a third enzyme and results in the release of the two terminal phosphate groups (Figure 7.7b).

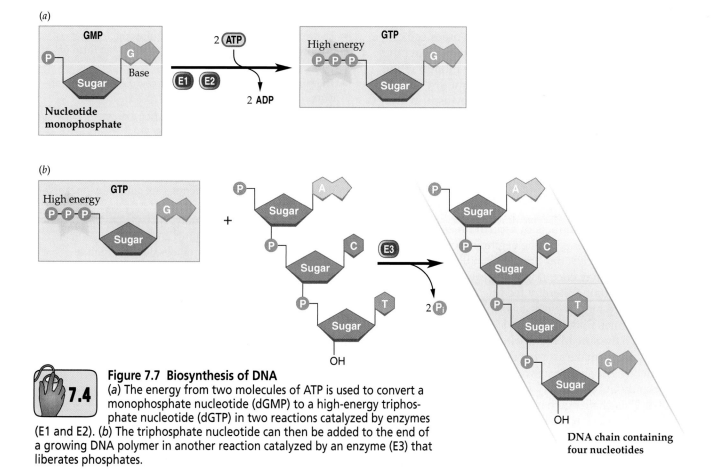

Figure 7.7 Biosynthesis of DNA
(a) The energy from two molecules of ATP is used to convert a monophosphate nucleotide (dGMP) to a high-energy triphosphate nucleotide (dGTP) in two reactions catalyzed by enzymes (E1 and E2). (b) The triphosphate nucleotide can then be added to the end of a growing DNA polymer in another reaction catalyzed by an enzyme (E3) that liberates phosphates.

DNA chain containing four nucleotides

In addition to DNA, other complex molecules, such as sugars and fatty acids are produced using ATP and specific enzymes. These and other biosynthetic pathways create and maintain the complex structures found in every living cell.

■ Enzymes are necessary for the synthesis and breakdown of complex molecules in the cell. The synthesis of DNA requires the action of at least three enzymes and energy from ATP.

■ HIGHLIGHT

Making a Wonder Drug Even Better

Enzymes control the rates of specific chemical reactions in the body, so altering enzyme activity sometimes causes illness and sometimes promotes healing. The effects of aspirin, it turns out, are due to its blocking of the activity of two important enzymes, COX-1 and COX-2. COX-1 is continuously produced in the body and catalyzes the biosynthesis of hormones that help maintain the lining of the stomach. In contrast, COX-2 is produced only when an injury occurs, and it catalyzes the biosynthesis of different hormones that promote inflammation, fever, and the sensation of pain throughout the body. Although both enzymes are inhibited by aspirin, they participate in two different biosynthetic pathways and play different biological roles. It is clear that the therapeutic benefits of aspirin (reduction of pain, inflammation, and fever) are due to its blocking of COX-2, while the negative side effects (damage to the stomach lining) are due to its blocking of COX-1.

The blocking of COX-2 activity is probably also responsible for the effects of aspirin on colon cancer, which were mentioned at the beginning of the chapter. Some cancerous cells have an abnormally high level of COX-2, which may encourage the growth of blood vessels into the tumor, thereby feeding it and allowing it to grow into a more serious cancer. By blocking COX-2 activity, aspirin may limit the blood supply to tumors and reduce the spread of the cancer.

How can we make aspirin better, with fewer negative side effects? The simple answer is to develop a drug that blocks only COX-2 activity and has little or no effect on COX-1. This challenge has been enthusiastically taken up by many research laboratories around the world, resulting in the recent development of a first generation of superaspirins.

The first successful step in developing superaspirins was based on an understanding of the three-dimensional shape of COX-2. As we have already seen, the shape of an enzyme defines its catalytic activity. Knowing the shape of the COX-2 enzyme allowed researchers to design inhibitors that bind only to COX-2 and not to COX-1. Over a dozen new compounds have been developed that bind to COX-2 and block its activity while having no significant effect on COX-1. This first generation of superaspirins is already being tested for anticancer properties. Although the results are pending, the development of ways to limit the metabolic activities of COX-2 is likely to yield significant health benefits.

■ Aspirin blocks the activities of both COX-1 and COX-2 enzymes. Superaspirins will have the beneficial effects of blocking the COX-2 enzyme without the negative effects of blocking the COX-1 enzyme.

SUMMARY

The Role of Energy in Living Systems

- Living organisms obey the same laws of thermodynamics that apply to the physical world.

- The creation of biological order requires the transfer of disorder to the environment, most often in the form of heat.

- The sun is the primary energy source for all living organisms.

- Atomic building blocks such as carbon are cycled between living organisms and the environment.

Using Energy from the Controlled Burning of Food

- In oxidation, electrons are lost from one molecule or atom to another. In reduction, electrons are gained by one molecule or atom from another.

- Catabolic reactions, which break down macromolecules and harness energy, are tightly coupled to biosynthetic reactions, which build macromolecules and require energy.

- All chemical reactions require the input of a small amount of activation energy to proceed.

How Enzymes Speed Up Chemical Reactions

- Enzymes greatly increase the rate at which chemical reactions proceed by lowering the amount of activation energy they require.

- Most chemical reactions that support life require the assistance of enzymes.

- The activity of enzymes is highly specific. Each enzyme binds to a specific set of substrates and catalyzes a specific chemical reaction.

- The specificity of an enzyme depends on its three-dimensional shape and the chemical characteristics of its active site.
- Several enzymes may catalyze multiple steps in a metabolic pathway.

Enzymes and Energy in Use: The Building of DNA

- DNA consists of nucleotides that are covalently bonded together to form a polymer.
- Before it can be added to a DNA polymer, a monophosphate nucleotide must be converted to an energy-rich triphosphate form by the enzyme-catalyzed transfer of two phosphate groups from two molecules of ATP.
- The addition of triphosphate nucleotides to the end of a growing DNA chain also requires the action of an enzyme.

Highlight: Making a Wonder Drug Even Better

- Aspirin blocks the activity of two enzymes, COX-1 and COX-2.
- The therapeutic benefits of aspirin (relief of pain, inflammation, and fever) arise from the blocking of COX-2, while the negative side effects (damage to the stomach) arise from the blocking of COX-1.
- New and improved superaspirins are chemical compounds that specifically block COX-2 without affecting COX-1.

KEY TERMS

activation energy p. 128	**first law of thermo-dynamics p. 124**
active site p. 131	
biosynthetic p. 128	**metabolism p. 124**
catabolic p. 128	**oxidation p. 127**
catalyst p. 129	**reduction p. 127**
chemical pathway p. 131	**second law of thermo-dynamics p.124**
enzyme p. 128	
	substrate p. 129

CHAPTER REVIEW

Self-Quiz

1. Which of the following statements is true?
   a. Cells are able to produce their own energy from nothing.
   b. Cells use energy only to generate heat and move molecules around.
   c. Cells obey the same physical laws of energy as the nonliving environment.
   d. Photosynthetic plants have no effect on the way animals obtain energy.

2. Living organisms use energy to
   a. organize chemical compounds into complex biological structures.
   b. decrease the disorder of the surrounding environment.
   c. cancel the laws of thermodynamics.
   d. cut themselves off from the nonliving environment.

3. The carbon atoms contained in organic molecules such as proteins
   a. are manufactured by cells for use in the organism.
   b. are recycled from the nonliving environment.
   c. differ from those found in CO_2 gas.
   d. cannot be oxidized under any circumstances.

4. Oxidation is
   a. the removal of oxygen atoms from a chemical compound.
   b. the gain of electrons by an atom.
   c. the loss of electrons by an atom.
   d. the synthesis of complex molecules.

5. The small input of energy required before a chemical reaction can proceed
   a. is called the activation energy.
   b. is independent of the laws of thermodynamics.
   c. is provided by an enzyme.
   d. always takes the form of heat.

6. The active site of an enzyme
   a. has the same shape for all known enzymes.
   b. can bind both its substrate and other kinds of molecules.
   c. does not play a direct role in catalyzing the reaction.
   d. can bring molecules together such that a chemical reaction takes place.

Review Questions

1. Explain why it is important for cells to oxidize food molecules gradually in multiple steps instead of doing it all at once in a single reaction.

2. Why is the release of heat by cells so important to their organization?

3. For a chemical reaction to occur, the reactants must collide with each other. Compare the way higher temperatures facilitate this process and speed up the reaction with the way an enzyme does.

4. Cells use several methods to increase the efficiency of enzyme catalysis in metabolic pathways. Describe two of these methods and how they apply to mitochondria.

5. Enzymes can be found in laundry detergents, where they assist in the removal of stains from clothing. How might this use of enzymes be similar to what they do in the cell?

The Daily Globe

Is Spinach Mightier than the Sword?

BETHESDA, MD—Everyone, from Popeye to vegetable-detesting children, knows about the benefits of eating spinach, a vegetable rich in vitamins and iron. Now, according to a new study in the journal *Advanced Military Science*, this humble vegetable might also have the power to disarm explosives.

Spinach, researchers have discovered, contains a powerful enzyme known as nitroreductase, a substance that can neutralize dangerous explosives such as TNT. The spinach enzyme reduces TNT to other compounds that can then be converted to carbon dioxide gas through additional chemical reactions.

Experts say this discovery could be great news for the United States military, which has been struggling to find a safe way to dispose of more than half a million tons of stockpiled explosives. Researchers note that the by-product, carbon dioxide gas, is less harmful to the environment than the usual chemical by-products of TNT degradation.

A spokesman at DeArm, a company that specializes in products that defuse explosives, warned that its research indicates that nitroreductase is not the environmentally friendly cure-all that the new research suggests. In fact, DeArm scientists claim the spinach enzyme is more harmful to the environment than traditional methods, since it releases carbon dioxide, a gas implicated in global warming.

But many were undeterred by this potential flaw. "This is a fantastic finding," said Dr. John J. Blowemup, a chemist at Southern Michigan University. "Nitroreductase may provide the most environmentally friendly method for getting rid of unwanted explosives. It's cheap, it's safe and it's absolutely silent. Who'd have guessed spinach could do all that?"

Evaluating "The News"

1. The possibility of using a spinach enzyme to break down explosives has many advantages. Describe an advantage from each of the following viewpoints: the environment, the taxpayer, and the U.S. military.

2. The DeArm company claims that the spinach enzyme is not an environmentally friendly solution, while this new study claims it is. What do you need to know to evaluate these two conflicting claims?

3. Why do you think a plant like spinach would have an enzyme with such reducing power?

chapter 8 *Photosynthesis and Respiration*

Lois Guarino, *The Prayer* (from The Fertility Series), 1992.

Food for Thought

The next time you feel hungry after skipping a meal, keep in mind that much of your body's demand for food is being made by your brain. A distinctive feature of the human brain is its size and need for energy. Although other animals, such as whales, certainly have larger brains by weight, the human brain is the largest when compared to the size of the human body. In other words, humans have the highest ratio of brain to body weight, which contributes to our status as the most intelligent of animals.

A daily challenge of having such a large brain is the urgent need to supply it with energy. Your brain consumes a large amount of energy while sending and receiving nerve impulses. Its energy demand is so high that more than half of the nourishment consumed by an infant is used up by the brain.

Given the tremendous energy demands made by your brain, how does your body manage to keep it satisfied? The answer ultimately lies in how energy is captured from sunlight by photosynthetic organisms and used to manufacture sugars. These sugars are used in turn by all organisms, including humans, to provide the energy that supports life processes.

In this chapter we discuss how energy is converted from sunlight into sugars, and how those sugars are broken down and used to supply energy to living cells. Perhaps by understanding these complex metabolic processes, we can answer the question of how we keep our brains running.

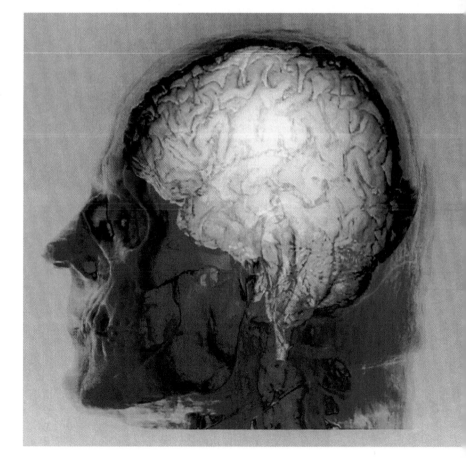

A 3-D Reconstruction of the Human Brain

■ KEY CONCEPTS

1. The temporary storage and transfer of usable energy in the cell requires the production of energy carrier molecules, including ATP, NADH, and NADPH.

2. Photosynthesis is a series of chemical reactions that use sunlight to manufacture sugars from atmospheric carbon

dioxide gas and water. The process also releases oxygen gas into the environment.

3. Catabolism is a series of chemical reactions that break down food molecules and produce ATP. Three stages of catabolism are glycolysis, the citric acid cycle, and oxidative phosphorylation.

Given all the complex structures and chemical reactions of every living system, obviously some kind of fuel must power it all. The capture of energy from the environment is one of the fundamental processes that support life. Ultimately, the sun is the primary source of energy for life on Earth. Solar energy is used by plants and other photosynthetic organisms to make sugars from carbon dioxide gas and water. Animals that consume plants acquire this chemical energy, which in turn is passed on to other animals that eat those animals (Figure 8.1).

As we saw in Chapter 7, the chemical reactions that transfer energy from one molecule to another and from one organism to another form the basis of metabolism. In this chapter, we first explore the biosynthetic

processes of metabolism by discussing how plants capture solar energy and use it to form the chemical bonds found in food molecules such as sugars. We then discuss the catabolic processes of metabolism, in which food molecules are oxidized and broken down to produce usable forms of energy. Before we can understand the chemical reactions of metabolism, however, we must first understand how energy is transferred from one molecule to another.

8.1 *Energy Carriers: Powering All Activities of the Cell*

One method of energy transfer commonly found in the physical world depends on heat. When water in a kettle is boiled on a stove, the energy from the flame is transferred in the form of heat to the metal of the kettle and then to the water molecules. Organisms generally cannot use such violent means of energy transfer. Instead, they use specialized molecules that transfer the chemical energy stored in covalent bonds.

The temporary storage and transfer of energy in the cell depends on several so-called **energy carrier** molecules. The most commonly used energy carrier is ATP, which was introduced in Chapter 5 (see Figure 5.7). ATP stores energy in the form of covalent bonds between phosphate groups. The addition of a phosphate group to an organic molecule, as when a phosphate group is added to ADP to make ATP, is called phosphorylation. ATP can then contribute its stored energy to another molecule by transferring one of its phosphate groups to that molecule. This trans-

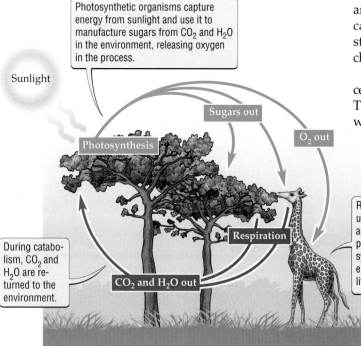

Photosynthetic organisms capture energy from sunlight and use it to manufacture sugars from CO_2 and H_2O in the environment, releasing oxygen in the process.

Sunlight

Sugars out

O_2 out

Photosynthesis

Respiration

Respiring organisms use both the sugars and the oxygen produced by photosynthesis to generate energy for living processes.

During catabolism, CO_2 and H_2O are returned to the environment.

CO_2 and H_2O out

Figure 8.1   The Flow of Energy from the Environment to Living Organisms

fer energizes the recipient molecule, enabling it to change shape or react chemically with other molecules. An example of how ATP is used to synthesize a complex macromolecule such as DNA is discussed in Chapter 7.

Chemical bond energy is not the only form of energy that is transferred by energy carrier molecules. Other important carriers—nicotinamide adenine dinucleotide phosphate ($NADP^+$) and nicotinamide adenine dinucleotide (NAD^+)—can pick up high-energy electrons and donate them to oxidation–reduction reactions. (Oxidation and reduction were introduced in Chapter 7.) Each of these carriers can pick up two high-energy electrons with a hydrogen ion (H^+), forming compounds known as NADPH and NADH, respectively. The ability of these compounds to donate these electrons to other molecules in turn means that NADPH and NADH can reduce other molecules. That is, they become oxidized by losing electrons, while the other compounds that accept the electrons are reduced.

Although NADPH and NADH have equal abilities as reducing agents, there is a difference of one phosphate group between them. That small difference determines which target molecules they bind to and react with and which metabolic pathways they affect. Later in this chapter, we will see that NADPH is used in the biosynthetic reactions that manufacture sugar from carbon dioxide, and NADH is used in the catabolic reactions that produce ATP from the breakdown of sugars.

Two of the organelles that were introduced in Chapter 6—chloroplasts and mitochondria—produce the energy carriers that power the activities of the cell. In this chapter we will see how their membrane arrangements facilitate the capture of usable energy. Chloroplasts cap-

ture energy from sunlight and use it to synthesize sugars; mitochondria capture energy from sugars and use it to synthesize energy carriers (Figure 8.2).

> ■ The chemical reactions that constitute life require energy. Energy is transferred within living organisms via specialized compounds called energy carriers. ATP, the most commonly used energy carrier, donates the energy of chemical bonds to chemical reactions. The energy carriers NADH and NADPH donate electrons to oxidation–reduction reactions.

Photosynthesis: Capturing Energy from Sunlight

The next time you walk outside, look at the plants around you and try to appreciate the critical role they play in supporting the web of life that includes human beings. Plants and other photosynthetic organisms, such as green algae and some bacteria, are able to capture energy from sunlight in the form of chemical bonds. The process of **photosynthesis** uses solar energy to synthesize energy carriers such as ATP, and complex, energy-rich molecules such as sugars.

The chemical reactions of photosynthesis also result in the chemical splitting of water and the release of oxygen gas (O_2) into the environment. The O_2 by-product of

8.2 **Figure 8.2 The Exchange of Molecules between Chloroplasts and Mitochondria Produces Energy Carriers**

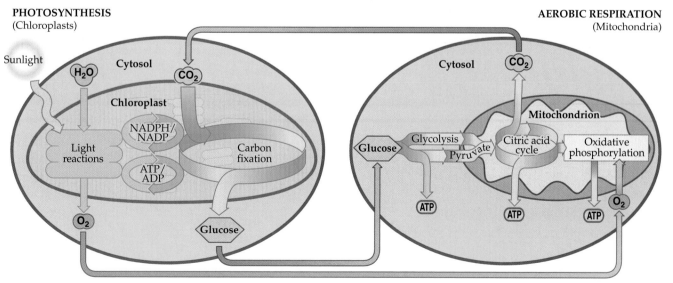

PHOTOSYNTHESIS
(Chloroplasts)

AEROBIC RESPIRATION
(Mitochondria)

photosynthesis is essential for all **aerobic**, or oxygen-dependent, life forms—another reason why plants are worthy of our respect. In other words, plants help support all animal life, including humans, since animals either indirectly or directly depend on plants for both food and oxygen.

Chloroplasts are the sites of photosynthesis

Photosynthesis takes place inside organelles called **chloroplasts**. These specialized membrane-enclosed compartments were introduced in Chapter 6; here we focus on how the membranes and associated proteins of chloroplasts contribute to the chemical reactions of photosynthesis.

Chloroplasts are found in the cytoplasm of plant cells. These important organelles have a distinctive arrange-

ment of membranes that is necessary for photosynthesis (Figure 8.3). Like mitochondria, chloroplasts are surrounded by a double membrane, which divides the organelle into compartments. The first compartment, between the two membranes, is appropriately called the **intermembrane space**; the second compartment, enclosed by the inner membrane, is called the **stroma**. Inside the stroma is yet another compartment in the form of flattened interconnected sacs formed by the **thylakoid membrane**. This third compartment, called the **thylakoid space**, is not found in mitochondria.

Each internal compartment of the chloroplast has a specific role in the overall photosynthetic process. Both the thylakoid membrane and the thylakoid space house enzymes and other molecules needed to transfer solar energy to energy carriers such as ATP and NADPH. The stroma contains enzymes that use these energy carriers

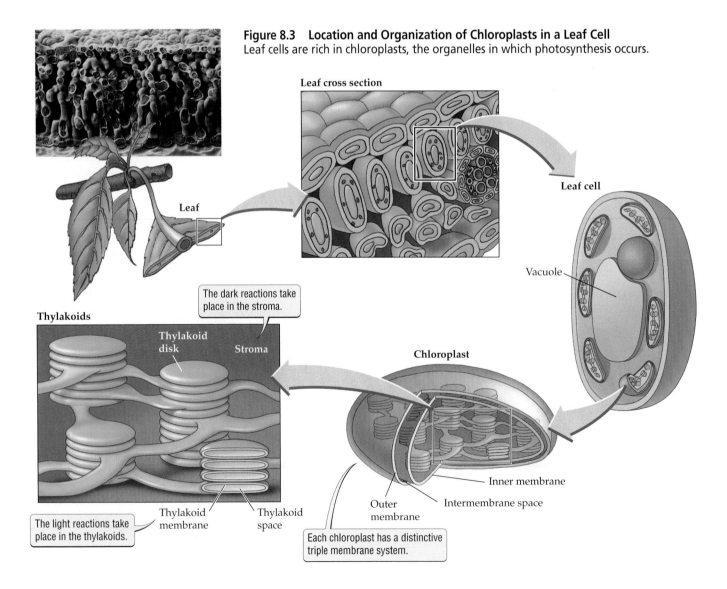

Figure 8.3   Location and Organization of Chloroplasts in a Leaf Cell
Leaf cells are rich in chloroplasts, the organelles in which photosynthesis occurs.

Leaf cross section

Leaf

Leaf cell

Vacuole

Thylakoids

The dark reactions take place in the stroma.

Thylakoid disk

Stroma

Chloroplast

The light reactions take place in the thylakoids.

Thylakoid membrane

Thylakoid space

Outer membrane

Inner membrane

Intermembrane space

Each chloroplast has a distinctive triple membrane system.

to manufacture sugars from carbon dioxide and water. Thus, the chloroplast is a well-organized power plant and factory that segregates its metabolic activities into specialized compartments.

The process of photosynthesis can be divided into two sets of reactions. The first series of reactions that we will consider takes place in the thylakoid membrane and directly captures energy from sunlight. Since these reactions require light, they are collectively known as the **light reactions**. The reactions that use this captured energy to synthesize sugars are known as the **dark reactions**, since they do not require light to proceed (although they actually occur in both the light and the dark). In the following sections we describe the light and dark reactions, emphasizing the key chemical events that characterize each series of reactions.

The light reactions capture energy from sunlight

The capture of energy from sunlight is essential to photosynthesis and supports life throughout the biosphere. In this process, solar energy is converted to the energy contained in the electrons of organic compounds. Some specialized pigments, of which **chlorophyll** is the most common, carry out this process. Chlorophyll accounts for the green color of most plant foliage.

Energy is captured by molecules of chlorophyll embedded in the thylakoid membrane. When exposed to light, the electrons associated with the covalent bonds of a chlorophyll molecule absorb energy from the light and become more energized. These energized electrons are often said to be "excited." To guarantee that the energy captured in this way is not lost to the environment and wasted, hundreds of chlorophyll molecules are arranged together in a formation called the **antenna complex**—an appropriate name given its role in capturing solar energy (Figure 8.4). Excited electrons pass energy from one chlorophyll molecule to another in the antenna complex until ultimately they are passed to a group of electron-accepting proteins located nearby in the thylakoid membrane. These groups of proteins are collectively called **electron transport chains**, or **ETCs**, and together with the antenna complexes, they make up more than half of the thylakoid membrane by weight.

Electron transport chains are series of proteins embedded in a membrane, and we will see them at work in other metabolic processes that involve the transfer of energy. As electrons are passed down the ETC from one protein component to another, small amounts of energy are released and used to generate the energy carriers ATP and NADPH. The combination of an antenna complex with a neighboring ETC is called a **photosystem**.

There are two distinct types of photosystems in the thylakoid membrane. They are called photosystem I and photosystem II, and each is associated with a different ETC (see Figure 8.4). Photosynthesis operates by inte-

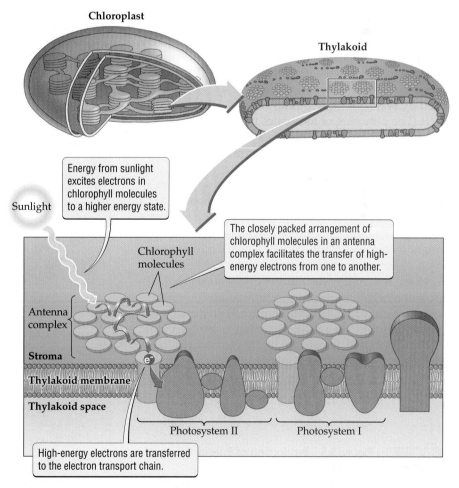

Energy from sunlight excites electrons in chlorophyll molecules to a higher energy state.

The closely packed arrangement of chlorophyll molecules in an antenna complex facilitates the transfer of high-energy electrons from one to another.

High-energy electrons are transferred to the electron transport chain.

Figure 8.4 Arrangement of Photosystems in the Thylakoid Membrane
The special arrangement of molecules in the thylakoid membrane facilitates the transfer of electrons (e⁻) from the antenna complex (a collection of chlorophyll molecules) to the electron transport chain (colored purple) of photosystems I and II.

grating the activities of both photosystems. Photosystem I is primarily responsible for the production of the powerful reducing agent NADPH; the reactions of photosystem II lead to the production of ATP. Let's consider how each photosystem contributes to photosynthesis as a single integrated process.

As Figure 8.5 shows, the journey of electrons along the ETC in photosystem II includes their transfer to a channel protein that spans the thylakoid membrane. The channel protein uses the energy of the electrons to pump protons (H^+) from the stroma into the thylakoid space. The protons accumulate inside the thylakoid space, causing an imbalance in the proton concentration across the thylakoid membrane. In other words, protons become abundant inside the thylakoid space and scarcer in the stroma.

This imbalance, called a **proton gradient**, is used to manufacture ATP. All molecules have a tendency to move from a region of higher concentration to one of lower concentration. Since the thylakoid membrane will not allow protons to pass through it, the only way for them to return to their region of lower concentration in the stroma is to move through another channel protein that spans the thylakoid membrane called ATP synthase. This movement of protons releases the energy that is used by ATP synthase to phosphorylate ADP to form ATP. (The pumping of molecules to create a concentration gradient is a common means of harnessing energy for cellular processes; we will learn more about this mechanism in Chapter 24.)

Photosystem I receives electrons from the last protein in the ETC of photosystem II. Therefore, photosystem I uses electrons transferred from photosystem II to replace the electrons lost from chlorophyll molecules in its own antenna complex. Photosystem I eventually transfers those electrons to an ETC protein, which in turn transfers them to $NADP^+$. The transfer of electrons to $NADP^+$ gives it a negative charge, and it takes up H^+ from the stroma to form NADPH (see Figure 8.5). In this manner, the two photosystems work together to produce both ATP and NADPH, which are used in the dark reactions. The release of oxygen gas by photosynthesis results from the necessary replacement of electrons lost by the antenna complex of photosystem II. These replacement elec-

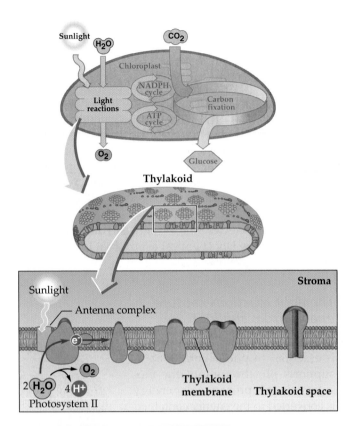

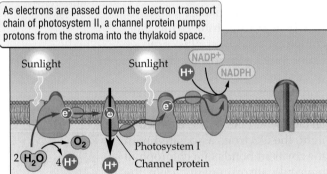

As electrons are passed down the electron transport chain of photosystem II, a channel protein pumps protons from the stroma into the thylakoid space.

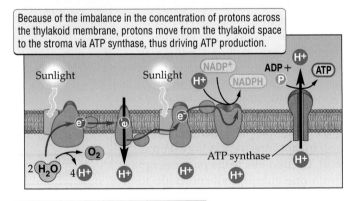

Because of the imbalance in the concentration of protons across the thylakoid membrane, protons move from the thylakoid space to the stroma via ATP synthase, thus driving ATP production.

When photosystems I and II are operating in series, the electrons ultimately are transferred to $NADP^+$, which is reduced to NADPH.

8.3

Figure 8.5   Production of Energy Carriers by the Light Reactions
The transfer of electrons between photosystems I and II results in the production of NADPH and ATP, along with the splitting of water molecules and the release of oxygen gas. Both the ATP and the NADPH produced are later used in the dark reactions.

trons are donated by water molecules, which split as a result to produce hydrogen ions and oxygen gas (O_2).

The dark reactions manufacture sugars

The energy carriers, ATP and NADPH, that are produced by the light reactions are used in the dark reactions to synthesize sugars from carbon dioxide. The dark reactions capture inorganic carbon atoms from CO_2 gas in the atmosphere and incorporate them into the organic compounds found in sugars—and thus in all living organisms. This process, known as **carbon fixation**, further emphasizes the interconnectedness of life with its environment.

As mentioned earlier, the dark reactions are catalyzed by enzymes that float freely in the stroma. The most important of these enzymes is known as **rubisco**. Rubisco catalyzes the first reaction of carbon fixation, in which a molecule of the one-carbon (1C) compound CO_2 combines with a five-carbon (5C) compound called ribulose 1,5-bisphosphate to produce two three-carbon compounds ($1C + 5C = 2 \times 3C$). This first reaction is followed by a multi-step cycle of many reactions designed both to manufacture sugars and to regenerate more ribulose

1,5-bisphosphate to keep the dark reactions going. The entire process requires the input of both energy from ATP and hydrogen ions from NADPH.

Three turns of the carbon fixation cycle produce one molecule of the three-carbon sugar glyceraldehyde 3-phosphate. One can follow this process by tracking the number of carbon atoms as they are rearranged into different compounds at each step of the cycle (Figure 8.6). For every three molecules of CO_2 ($3 \times 1C = 3C$) combined with three ribulose 1,5-bisphosphate molecules ($3 \times 5C = 15C$), six molecules of the three-carbon compound ($6 \times 3C = 18C$) are produced. These molecules eventually produce three ribulose 1,5-bisphosphates ($3 \times 5C = 15C$) and one molecule of glyceraldehyde 3-phosphate (3C). As the math indicates, it takes three turns of the cycle to produce one three-carbon sugar molecule, with the other carbon atoms being recycled to ribulose 1,5-bisphosphate. The formation of one molecule of glyceraldehyde 3-phosphate also requires the input of nine molecules of ATP and six molecules of NADPH.

Glyceraldehyde 3-phosphate is the chemical building block used to manufacture glucose, from which other carbohydrates needed by the cell can be made. Most of the glyceraldehyde 3-phosphate made in the chloro-

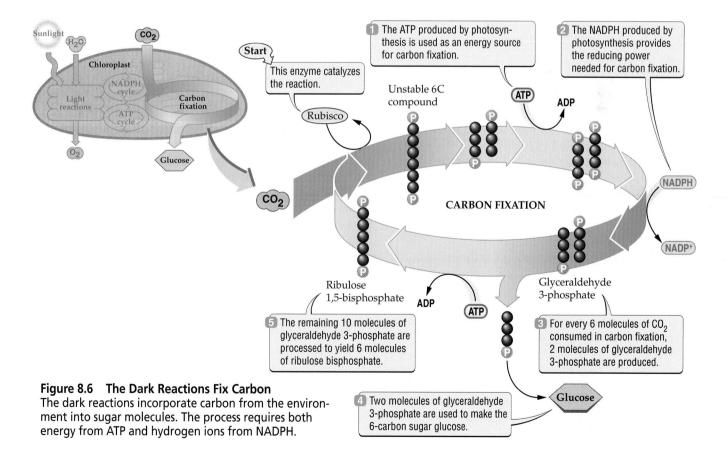

Figure 8.6   The Dark Reactions Fix Carbon
The dark reactions incorporate carbon from the environment into sugar molecules. The process requires both energy from ATP and hydrogen ions from NADPH.

plasts is exported from those organelles and eventually consumed in chemical reactions that produce ATP. Some molecules of glyceraldehyde 3-phosphate that are not immediately consumed in ATP synthesis are used to manufacture sucrose in the cytoplasm. Sucrose is the sugar most often used as a sweetener in desserts. It is an important food source for all the cells in a plant and is transported from the leaves, where photosynthesis takes place, to all other parts of the plant. Significant amounts of sucrose are stored in the cells of sugarcane and sugar beets, which is why these two plants are the major crops of the sugar industry worldwide.

Not all the glyceraldehyde 3-phosphate is exported from the chloroplasts, however. Some of it is converted into starch by enzymes in the stroma. Starch is a polymer of glucose and is an important form of stored energy in plants. It accumulates in chloroplasts during the day and is then broken down into simple sugars at night. This setting aside of food to generate energy at night compensates for the lack of sunlight, and hence of photosynthesis, during that time. Plant seeds, roots, and tubers are also rich in stored starch, which provides the energy required for germination and growth. Indeed, the energy-rich nature of these plant components explains why they are such an important food source for animals.

> ■ Photosynthesis captures energy from sunlight and uses it to synthesize sugars from CO_2 and H_2O. It occurs in chloroplasts and produces as a by-product the O_2 that is essential for all oxygen-breathing organisms on Earth. The photosynthetic reactions that capture the energy of sunlight and produce ATP and NADPH are called the light reactions. The ATP and NADPH produced by the light reactions are used in the dark reactions to synthesize sugars.

Catabolism: Breaking Down Molecules for Energy

The conversion of food into useful energy, as we saw in Chapter 7, requires the breakdown and gradual oxidation of food molecules. This process, called **catabolism**, is constantly occurring in our bodies.

The first stage of catabolism is the breakdown of large, complex food molecules such as carbohydrates, proteins, and fats into their simpler components. This is the digestive process that occurs in our stomach and intestines after every meal. The simpler compounds that

are released by digestion, such as amino acids and simple sugars, are absorbed by the intestine and passed on to other cells in the body via the bloodstream. Digestion is discussed in Chapter 28; here we are more concerned with how the simple sugars supplied by digestion are converted into fuel for use by the cell. This catabolic process consists of three major stages: glycolysis, the citric acid cycle, and oxidative phosphorylation.

Glycolysis is the first stage in the breakdown of sugars

The first stage of catabolism that occurs inside the cell is known as **glycolysis** (literally, "the splitting of sugar"). From an evolutionary standpoint, glycolysis was probably the earliest means of producing ATP from food molecules, and it is still found in prokaryotes (see the box on page 148). In most eukaryotic organisms today, however, glycolysis is only a necessary preparation for more efficient ways of producing ATP in the mitochondria.

Glycolysis is a series of chemical reactions that take place in the cytosol (Figure 8.7). These reactions break down the simple six-carbon sugar glucose to provide the mitochondria with the substrate molecules they need to generate ATP. A series of enzyme-catalyzed reactions partially oxidizes glucose into pyruvate, which is then transported into the mitochondria for further processing. Partway through the series of glycolytic reactions, a six-carbon sugar intermediate formed from glucose is split into two three-carbon sugars, which are then converted into two three-carbon molecules of pyruvate.

During glycolysis, for each molecule of glucose consumed, four molecules of ADP are phosphorylated to four molecules of ATP, and electrons are donated to two molecules of NAD^+, generating two molecules of NADH. Since the early steps of glycolysis consume two molecules of ATP per glucose molecule, a single glucose molecule produces a net yield of two ATP molecules and two NADH molecules (see Figure 8.7). Glycolysis does not require oxygen gas (O_2). For complete oxidation using O_2, the pyruvate molecules enter the mitochondria for use in the eukaryotic cell's most efficient ATP-producing system.

Fermentation sidesteps the need for oxygen

Since glycolysis does not require O_2 from the atmosphere, its reactions are described as **anaerobic**, and they were probably essential for early life forms in the oxygen-poor atmosphere of primitive Earth. The reactions that take place in the mitochondria are aerobic; that is,

Figure 8.7   Glycolysis

In glycolysis, a single six-carbon glucose molecule is converted into two three-carbon molecules of pyruvate. The process produces a relatively low net yield of two ATP molecules and two NADH molecules per glucose molecule. If oxygen is present, the pyruvate and the NADH move into the mitochondria for use in aerobic respiration. If oxygen is absent, they are used in fermentation reactions.

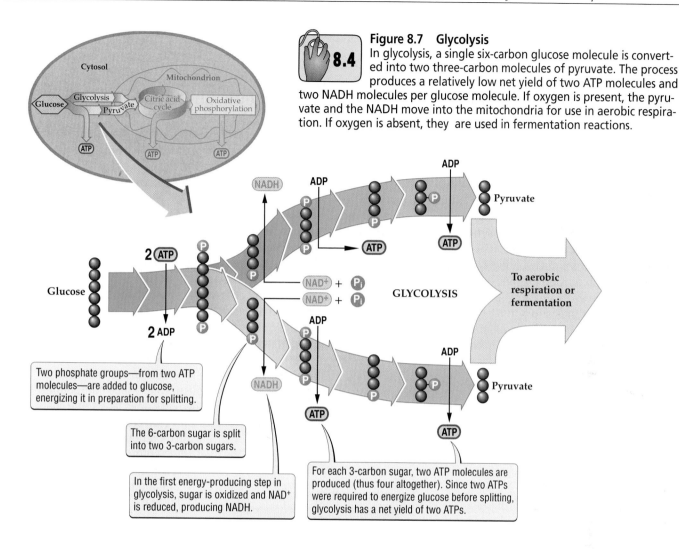

Two phosphate groups—from two ATP molecules—are added to glucose, energizing it in preparation for splitting.

The 6-carbon sugar is split into two 3-carbon sugars.

In the first energy-producing step in glycolysis, sugar is oxidized and NAD^+ is reduced, producing NADH.

For each 3-carbon sugar, two ATP molecules are produced (thus four altogether). Since two ATPs were required to energize glucose before splitting, glycolysis has a net yield of two ATPs.

---

they require O_2. These aerobic reactions dominate ATP production in most organisms living under Earth's present oxygen-rich atmosphere. The catabolic process in which aerobic reactions follow glycolysis is called **aerobic respiration**.

Some organisms, however, still use glycolysis as their only means of generating ATP. These organisms are anaerobic; that is, they can live without O_2. In these organisms, the pyruvate and the NADH produced by glycolysis remain in the cytosol. The pyruvate is converted into alcohol and CO_2, and the NADH is converted back to NAD^+, which can be used again in the reactions of glycolysis. The catabolic process in which these anaerobic reactions follow glycolysis is known as **fermentation**.

Yeast is a familiar anaerobic fungus that has an essential role in the production of beer. Fermentation in yeast converts pyruvate into ethanol and CO_2 gas, giving beer

its alcohol content and foamy effervescence (Figure 8.8). This production of CO_2 also explains the role of baker's yeast in making bread rise, since the released gas expands the bread dough.

Fermentation is not limited to anaerobic organisms, however. It also takes place in the human body. When we exercise hard and push our muscles to the point of exhaustion, the pain we feel is due largely to fermentation processes in the muscles. A rapid burst of strenuous exercise can exhaust the ATP stores in muscle in a matter of seconds. Both marathon runners and Olympic cyclists know from firsthand experience that to sustain strenuous physical activity, muscle cells must use both aerobic respiration and fermentation to generate more ATP. However, athletes' muscles do not produce alcohol and CO_2, as yeasts do; instead, pyruvate is converted into lactic acid, which causes the burning sensation in aching muscles.

■ THE SCIENTIFIC PROCESS

Harnessing Glycolysis for Commercial Gain

Making use of single-celled organisms for industrial purposes has a long history. Their applications range from using yeast to ferment alcoholic beverages to using bacteria to breakdown biological refuse. When it comes to synthesizing complex compounds, however, chemists, not microbiologists, have clearly dominated the industrial process for over 60 years. The rule of chemical synthesis in a test tube may soon end, however, with the clever re-engineering of bacteria as miniature chemical factories that we can control.

In order to turn bacteria into prime producers of chemical compounds, one must have a thorough understanding of their metabolic pathways. In the same way that organic chemists perform a sequence of chemical reactions to synthesize a final product, bacterial metabolic pathways utilize a sequence of reaction steps. If there was some way to alter what chemical steps are performed by bacterial metabolism, one could harness bacteria to execute multiple reaction steps in a single cell instead of the multiple reaction vessels used by traditional chemists. The small size of bacterial cells is not a problem, since one can grow huge numbers of bacteria in a nutrient medium, making up in numbers what they lack in size.

So how do you go about changing the chemical reaction steps that occur in bacteria? The answer lies in modifying their genetic profile such that new enzymes are produced, which in turn catalyze chemical reactions that normally would not occur in the cell. Ideally, these new steps should fit into a preexisting metabolic pathway, such as glycolysis, making use of chemical substrates that are available to the bacteria.

Recent efforts by the chemical company DuPont have focused on engineering bacteria to produce an enzyme that alters the reactions of glycolysis. Instead of using glucose to synthesize pyruvate, these bacteria break down glucose to produce 1,3-propanediol. Monomers of 1,3-propanediol can then be used to manufacture a form of polyester found in many commercial textiles. On another front, similarly engineered bacteria are transforming glucose into a chemical compound used in the synthesis of vitamin C. In this case, a one-step production in bacteria replaces a more costly seven-step chemical synthesis in the factory.

The re-engineering of bacterial metabolism to produce compounds for our own use does face some challenges. One such challenge is how to redirect the metabolic effort of bacteria without killing them in the process. That is, how can one alter the path of bacterial metabolism without starving them of life-supporting ATP? One possible solution is to test the different metabolic pathways present in bacteria and utilize those that are not essential. Furthermore, additional genetic engineering may be used to shut down other nonessential metabolic pathways to maximize the cell's ability to produce the desired chemical. Clearly a current understanding of bacterial metabolism, coupled with the power of genetic engineering, may allow microbiologists to compete with chemists in the arena of industrial chemical synthesis.

Synthesis in the lab

Synthetic bacteria

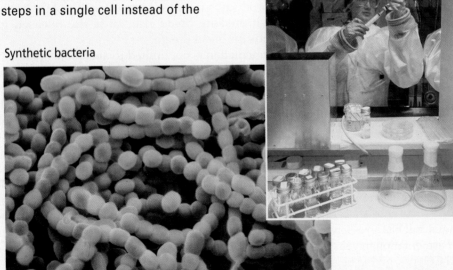

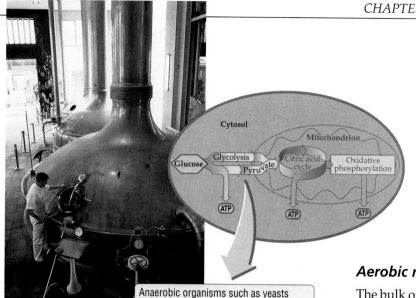

Figure 8.8   Fermentation Has a Variety of Uses

Fermentation produces ATP without using oxygen. The production of ethanol by alcoholic fermentation is the reason that yeasts are used to brew alcoholic beverages like beer (top photo). A similar process occurs in strenuously exercised muscles (bottom photo), but instead of alcohol, lactic acid is produced.

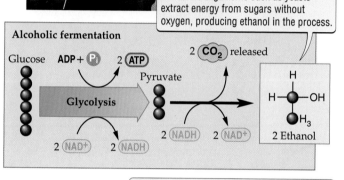

Anaerobic organisms such as yeasts extract energy from sugars without oxygen, producing ethanol in the process.

Alcoholic fermentation

Glucose — $ADP + P_i$ — 2 ATP — $2 CO_2$ released — Pyruvate — Glycolysis — $2 NAD^+$ — $2 NADH$ — $2 NADH$ — $2 NAD^+$ — 2 Ethanol

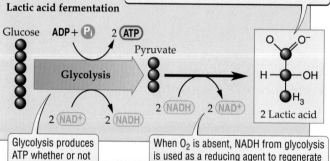

During strenuous activity, muscle cells resort to fermentation to produce extra ATP quickly. The process yields lactic acid instead of ethanol.

Lactic acid fermentation

Glucose — $ADP + P_i$ — 2 ATP — Pyruvate — Glycolysis — $2 NAD^+$ — $2 NADH$ — $2 NADH$ — $2 NAD^+$ — 2 Lactic acid

Glycolysis produces ATP whether or not O_2 is present, as long as the cell's supply of NAD^+ is replenished.

When O_2 is absent, NADH from glycolysis is used as a reducing agent to regenerate NAD^+, thus replenishing the cell's supply.

Aerobic respiration produces the most ATP

The bulk of ATP production in most eukaryotic organisms depends on those energy powerhouses, the mitochondria (see Figure 6.9). These organelles use both pyruvate and O_2 in a series of oxidation reactions that release energy, which is used to phosphorylate ADP to ATP. Mitochondria are therefore said to perform aerobic respiration, which produces many more molecules of ATP per molecule of glucose consumed than fermentation does.

The important connections between oxygen, mitochondria, and ATP production turn up repeatedly in different cells in all living organisms. The muscle cells of the human heart, for example, have an exceptionally large number of mitochondria to produce the enormous amounts of ATP needed to keep the heart beating. Blind mole-rats that live underground in oxygen-poor burrows and must dig and dig every day have also optimized their means of using oxygen to generate ATP in their muscles. In comparison to those of white laboratory rats, the muscles of blind mole-rats have 50 percent more mitochondria and 30 percent more blood capillaries. This means that their muscles are supplied with more blood to maximize the transfer of available oxygen, and have more mitochondria to maximize the resulting production of ATP.

Blind mole-rat

The citric acid cycle produces NADH and CO_2

In aerobic eukaryotes such as humans, all the energy-producing steps that follow the production of pyruvate by glycolysis take place inside mitochondria. Once pyruvate enters a mitochondrion, it is converted into an acetyl group and bound to a large carrier molecule called coenzyme A; in the process, one of its three carbon atoms is released in the form of CO_2. The high-energy compound that is formed, called acetyl CoA, is an

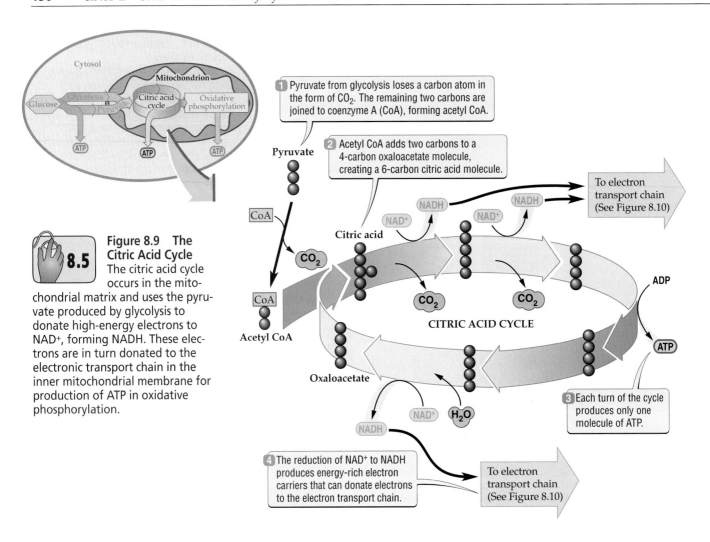

Pyruvate from glycolysis loses a carbon atom in the form of CO_2. The remaining two carbons are joined to coenzyme A (CoA), forming acetyl CoA.

Acetyl CoA adds two carbons to a 4-carbon oxaloacetate molecule, creating a 6-carbon citric acid molecule.

Each turn of the cycle produces only one molecule of ATP.

The reduction of NAD^+ to NADH produces energy-rich electron carriers that can donate electrons to the electron transport chain.

To electron transport chain (See Figure 8.10)

To electron transport chain (See Figure 8.10)

CITRIC ACID CYCLE

Pyruvate

CoA

CO_2

CoA

Acetyl CoA

Citric acid

NAD^+ NADH NAD^+ NADH

CO_2 CO_2

ADP

ATP

Oxaloacetate

NAD^+ H_2O

NADH

Figure 8.9 The Citric Acid Cycle
The citric acid cycle occurs in the mitochondrial matrix and uses the pyruvate produced by glycolysis to donate high-energy electrons to NAD^+, forming NADH. These electrons are in turn donated to the electronic transport chain in the inner mitochondrial membrane for production of ATP in oxidative phosphorylation.

important substrate for a series of eight oxidation reactions called the **citric acid cycle**, all of which take place in the mitochondrial matrix (Figure 8.9). Its name is based on the fact that citric acid is the product of the first reaction involving acetyl CoA. The major consequence of this oxidation cycle is the production of high-energy electrons stored in NADH; CO_2 is released as a by-product. The molecules of NADH are then used in the next stage of ATP production.

Most of the oxidation reactions that take place in the cell are part of the citric acid cycle, emphasizing the importance of this process in energy production. In addition to sugars from glycolysis, stored fats can enter the citric acid cycle. They are first broken down into fatty acids, which enter the mitochondria and are converted to acetyl CoA by a different set of reactions. Thus, the processes that produce energy from both sugars and fats come together at the beginning of the citric acid cycle.

Oxidative phosphorylation uses O_2 and NADH to produce ATP in quantity

The jackpot of ATP is generated in the third and last stage of aerobic respiration. This is also the stage at which the physical structure of mitochondria really comes into play. As we saw in Chapter 6, mitochondria have a double membrane that forms two separate compartments, the intermembrane space and the matrix. The NADH molecules produced in the matrix by the citric acid cycle donate their high-energy electrons to an electron transport chain (ETC) embedded in the inner membrane of the mitochondrion. A component of the ETC phosphorylates ADP to produce ATP. Since the electrons carried by the NADH molecules were gained by the oxidation of pyruvate in the citric acid cycle, and the components of the ETC are oxidized by the transfer of electrons, the whole process is appropriately called **oxidative phosphorylation** (Figure 8.10).

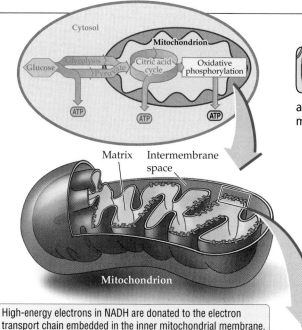

Figure 8.10 Oxidative Phosphorylation

Oxidative phosphorylation is the last stage of aerobic respiration and produces the most ATP of any process in the cell. Electrons from NADH produced in the citric acid cycle are donated to the electron transport chain in the inner mitochondrial membrane, where their energy is used in ATP production.

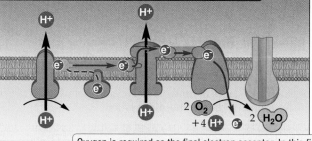

High-energy electrons in NADH are donated to the electron transport chain embedded in the inner mitochondrial membrane.

As electrons are passed down the electron transport chain, protons are pumped from the matrix into the intermembrane space, creating a proton imbalance across the inner mitochondrial membrane.

Oxygen is required as the final electron acceptor. In this final electron transfer, O_2 picks up protons and forms water.

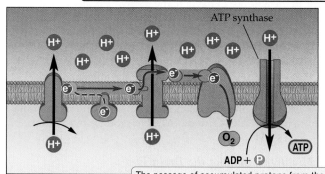

The passage of accumulated protons from the intermembrane space to the matrix through ATP synthase drives ATP synthase to produce ATP.

If this process seems familiar, the reason is that the ETC in the inner mitochondrial membrane has a function similar to that of the ETC found in the thylakoid membrane of chloroplasts. In mitochondria, the electrons donated by NADH are passed along a series of ETC components, releasing energy that is used to pump protons from the matrix into the intermembrane space. The resulting buildup of protons in the intermembrane space has the same effect as the buildup of protons in the thylakoid space: It forms a proton gradient across the membrane. As is the case in chloroplasts, the proton gradient causes protons to move through an ATP synthase in the inner mitochondrial membrane, catalyzing the phosphorylation of ADP to form ATP (see Figure 8.10). The similarities in how ATP is produced in chloroplasts and mitochondria demonstrates the evolutionary selection of a common biological mechanism.

In the final step of oxidative phosphorylation, electrons are donated to O_2 that has diffused into the mitochondrion. When O_2 accepts these electrons, it also combines with H^+ in the matrix to form water (H_2O). Thus, O_2 is the last electron acceptor in the series of electron transfers and is required for oxidative phosphorylation (see Figure 8.10).

Aerobic respiration (glycolysis, the citric acid cycle, and oxidative phosphorylation) has a net production of well over 30 ATP molecules per molecule of glucose. It is clearly more efficient than fermentation, which has a net production of only two ATP molecules per molecule of glucose consumed. This marked contrast explains the need for aerobic respiration in larger multicellular organisms with higher energy demands.

■ Catabolism is a series of oxidation reactions that break down food molecules and produce ATP. The three stages of catabolism are glycolysis, the citric acid cycle, and oxidative phosphorylation. Glycolysis occurs in the cytosol and splits sugar molecules, producing pyruvate and a small amount of ATP. The citric acid cycle and oxidative phosphorylation occur in mitochondria and produce most of the cell's ATP.

■ HIGHLIGHT

Solving the Brain Drain

Now that we know how energy is obtained from food, how do we meet the energy demands of our particularly large brains? One way might be to increase our rate of metabolism so that we consume more food and generate more ATP. However, humans do not generally have a higher rate of metabolism than do other mammals that have far smaller brains, such as sheep. So the answer does not lie in any special alteration of the catabolic processes described in this chapter. Instead, to meet the high energy demands of the human brain, we have evolved ways to redistribute energy to our various organ systems and to make higher energy demands on our mothers early in life.

Satisfying the energy needs of a living organism involves more than the mere production of ATP. The ATP that is available to a multicellular organism must be carefully distributed to satisfy the varying needs of different cell types. Approximately 70 percent of our resting metabolism (that is, that which is not being used for physical activity) supplies ATP to the heart, liver, kidneys, gastrointestinal tract, and brain. To supply our large brains, one or more of the other organs must either give up some of its precious energy resources or be smaller. In fact, when compared with those of other primates, the human gastrointestinal tract is nearly 40 percent smaller than expected. Thus, the larger energy demands of our brains are compensated for by our smaller gastrointestinal tracts. During evolution, early humans switched from a strictly vegetarian diet to a more varied diet that included easily digested foods such as meat. Thus, there was no longer an advantage to having the large gastrointestinal tract that is needed for slowly digesting vegetation.

The evolution of a smaller gastrointestinal tract was probably not the only change necessary to accommodate our large brains. The amount of energy that a mother contributes to her child during pregnancy and early infancy is equally important to the development of the large human brain. By the time a child reaches age 4, he or she has a brain that is 85 percent of its adult size. The human brain requires enormous amounts of energy during early development to support such rapid growth, and the only source for such metabolic resources is the mother. During pregnancy the fetus greedily pulls nutrients from its mother to supply its growing energy needs, and this transfer of energy continues after birth with the intake of breast milk. Thus, we probably owe the development of our impressive brains to the evolu-

tionary rearrangement of energy needs in the body and the generous investment of energy by our mothers. Thanks, Mom!

> ■ Our large human brains require a significant share of the body's ATP production, as well as a high investment of energy from our mothers early in life.

SUMMARY

Energy Carriers: Powering All Activities of the Cell

■ The chemical reactions that constitute life require energy.

■ Energy is transferred within living organisms via specialized compounds called energy carriers.

■ ATP, the most commonly used energy carrier, donates energy stored in chemical bonds to chemical reactions.

■ The energy carriers NADH and NADPH donate electrons to oxidation–reduction reactions.

Photosynthesis: Capturing Energy from Sunlight

■ Photosynthesis captures energy from sunlight and uses it to synthesize sugars from CO_2 and H_2O.

■ Photosynthesis occurs in chloroplasts and produces O_2 as a by-product.

■ The photosynthetic reactions that capture energy from sunlight are called the light reactions. They produce ATP and NADPH.

■ The ATP and NADPH produced by the light reactions are used in the dark reactions to synthesize sugars from CO_2.

Catabolism: Breaking Down Molecules for Energy

■ Catabolism is a series of oxidation reactions that break down food molecules and produce ATP.

■ The three stages of catabolism are glycolysis, the citric acid cycle, and oxidative phosphorylation.

■ Glycolysis occurs in the cytosol and splits sugar molecules to produce pyruvate and a small amount of ATP and NADH.

■ In the absence of oxygen, the pyruvate remains in the cytosol and is converted into alcohol and CO_2, or lactic acid, by fermentation. In the presence of oxygen, the pyruvate is used by mitochondria to generate many additional molecules of ATP.

■ The citric acid cycle occurs inside the mitochondrial matrix and uses pyruvate to produce NADH and CO_2.

■ Oxidative phosphorylation occurs in the intermembrane space and uses O_2 and NADH to produce most of the cell's ATP.

Highlight: Solving the Brain Drain

■ Satisfying the energy needs of a living organism involves more than the mere production of ATP.

■ Our large brains require a high investment of energy from our mothers early in life, as well as a significant share of the body's ATP production.

KEY TERMS

aerobic p. 142	fermentation p. 147
aerobic respiration p. 147	glycolysis p. 146
anaerobic p. 146	intermembrane space p. 142
antenna complex p. 143	light reactions p. 143
carbon fixation p. 145	oxidative phosphorylation p. 150
catabolism p. 146	photosynthesis p. 141
chlorophyll p. 143	photosystem p. 143
chloroplast p. 142	proton gradient p. 144
citric acid cycle p. 150	rubisco p. 145
dark reactions p. 143	stroma p. 142
electron transport chain (ETC) p. 143	thylakoid membrane p. 142
	thylakoid space p. 142
energy carrier p. 140	

CHAPTER REVIEW

Self-Quiz

1. Energy carriers such as ATP
   a. transfer energy from the environment into cells.
   b. transfer energy to molecules in the cell, allowing chemical reactions to occur.
   c. enable organisms to get rid of excess heat.
   d. are synthesized only by mitochondria.

2. Photosynthesis
   a. is found exclusively in plants.
   b. breaks down sugars with energy from sunlight.
   c. releases CO_2 into the atmosphere.
   d. releases O_2 into the atmosphere.

3. Which of the following compounds is produced by the light reactions and used by the dark reactions to synthesize sugars?
   a. ATP
   b. NaCl
   c. CO_2
   d. NADH

4. Which of the following statements is *not* true?
   a. Glycolysis produces most of the ATP required by aerobic organisms.
   b. Glycolysis produces pyruvate, which is consumed by the citric acid cycle.
   c. Glycolysis occurs in the cytosol of the cell.
   d. Glycolysis is the first stage of aerobic respiration.

5. Electron transport chains
   a. are found in both chloroplasts and mitochondria.
   b. are clumps of protein that float freely in the cytosol.
   c. pass protons from one ETC component to another.
   d. have no role in ATP production.

6. Which of the following is essential for oxidative phosphorylation?
   a. rubisco
   b. NADPH
   c. a proton gradient
   d. chlorophyll

Review Questions

1. The transfer of electrons down an ETC produces a similar chemical event in both chloroplasts and mitochondria. Describe this chemical event and how it contributes to the production of ATP by both organelles.

2. The cells in the root of a plant do not contain chloroplasts. How do these cells obtain energy without the benefit of photosynthesis?

3. Oxidative phosphorylation is the final stage of aerobic respiration. What important chemical reaction in oxidative phosphorylation directly consumes O_2?

4. Certain drugs allow protons to pass through the thylakoid membrane on their own, without the involvement of a channel protein. How do you think these drugs affect ATP synthesis?

8   𝕿𝖍𝖊 𝕯𝖆𝖎𝖑𝖞 𝕲𝖑𝖔𝖇𝖊

Tropical Rainforests May Be Source of Pollution

SAN JOSE, COSTA RICA—For years, the accepted scientific dogma has been that the greenhouse effect—global warming that is caused by increases in carbon dioxide in the atmosphere—is combated by tropical rainforests, which consume carbon dioxide. Now a series of new reports suggests that the rainforest of Costa Rica may have turned traitor, producing more carbon dioxide than it consumes, itself becoming a significant source of this greenhouse gas.

"This goes against everything we had thought," said Dr. Theodore Michois, ecologist at La Bruja Ecological Station in Costa Rica, who called the finding "alarming." "We must rethink the role of rainforests."

Rainforest plants photosynthesize during the day, consuming carbon dioxide and releasing oxygen. At night, however, there is no photosynthesis, and aerobic respiration dominates, consuming oxygen and releasing carbon dioxide. The findings at La Bruja suggest that rainforests may have become net carbon dioxide producers, rather than consumers, due to global warming that has already occurred. Higher-than-normal temperatures at night caused by global warming may stress the trees, increasing their aerobic respiration and causing them to release more carbon dioxide. Meanwhile, levels of photosynthesis remain the same, tipping the balance sheet toward rainforests producing more carbon dioxide than they consume.

"If higher levels of carbon dioxide are driving the production of even more carbon dioxide in rainforests," said Dr. Will Helmina, climatologist at Purdue College, "then we are really in trouble. It could be the beginning of a truly vicious cycle."

Evaluating "The News"

1. Faced with the apparent increase in CO_2 levels caused by the Costa Rican rainforest, some people have suggested deforestation as a solution. Is this an acceptable solution? Would it work?

2. Trees in cooler climates seem to grow faster when levels of CO_2 in the atmosphere are higher. How might this phenomenon affect the global climate?

3. Does faster growth of trees in these regions make higher levels of CO_2 a plus?

chapter

9

How Cells Communicate

Ross Bleckner, *Healthy Spot*, 1996.

Sleeping with the Enemy

Since the dawn of recorded history, humans have waged wars, both with one another and with deadly bacteria in the environment. Although the threat of nuclear war has decreased in the past decade with the end of the Cold War between the United States and the Soviet Union, there has been an upsurge in smaller ethnic conflicts around the world and in the threat of biological warfare. In an ironic twist of fate, our puzzling tendency to wage war with one another has now overlapped with our struggle against two types of bacteria.

Recent ethnic conflicts in East Africa have resulted in the creation of refugee camps to house fleeing civilians. As these camps swell with thousands of refugees, crowded conditions and poor sanitation have led to serious epidemics of disease. A particularly unfortunate consequence of these conditions has been the resurgence of cholera.

Cholera is caused by the bacterium *Vibrio cholerae*, which produces a deadly toxin responsible for severe diarrhea and eventual death by dehydration. As alarming reports of antibiotic-resistant strains of cholera continue to flow from refugee camps, it appears that human warfare has presented an opportunity for a deadly bacterium to spread.

Perhaps even more disturbing is evidence that some individuals might harness the deadly properties of bacteria for use as biological weapons of war. In recent years, there has been increasing fear that international terrorists or misguided leaders might use anthrax in this way. During the Gulf War in 1991, there were concerns that missile warheads armed with anthrax might be used on the battlefield. More recently, the tragic terrorist attacks on the United States of September 11, 2001, have been fol-

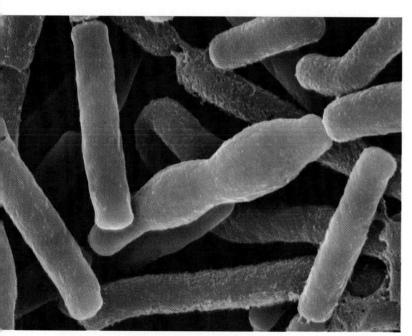

The Bacterium that Causes Anthrax
This electron micrograph shows *Bacillus anthracis*. The reddish individual is about to produce spores.

■ MAIN MESSAGE

Multicellular organisms must coordinate their specialized cells through cell-to-cell communication.

lowed at the time of this writing by a second wave of terrorism in the form of anthrax-contaminated mail sent to newspaper and government offices. But what exactly is this deadly biological weapon?

Anthrax is caused by the bacterium *Bacillus anthracis*, which, like cholera, produces a deadly toxin that kills infected animals. In the case of anthrax infections, the bacterium requires the death of its unfortunate host to release spores and spread from one individual to another. Fortunately, the passing of anthrax to humans is relatively rare under normal conditions, but the major concern today is how to counteract anthrax exposure caused by use of the spores as a biological weapon.

To develop effective countermeasures, we must understand how bacterial toxins damage and ultimately kill so many cells in the body of an infected individual. Thanks to intensive biological research, we now know that both the cholera and the anthrax toxins exercise their deadly effects by interfering with cell-to-cell communication. Multicellular eukaryotic organisms such as humans depend on proper communication between all the different specialized cells that make up our bodies. In this chapter, we see how cells receive and respond to signals sent by other cells. Then we see how the cholera and anthrax toxins can disrupt these essential processes, with deadly consequences.

■ KEY CONCEPTS

1. Multicellular organisms require specialized cells to perform the various processes of life efficiently. Specialization requires communication between cells.

2. Cells use a variety of different signaling molecules to communicate with one another, including proteins, fatty acid derivatives, and even gases.

3. To receive and interpret signals, cells have specific receptor proteins that bind signaling molecules and trigger certain chemical reactions inside the cell.

4. Some receptor proteins reside in the cytosol and bind to signaling molecules that pass through the plasma membrane; others reside in the plasma membrane and bind only to external signaling molecules.

5. One signaling molecule can induce many chemical reactions, amplifying the cell's response to the signal.

6. Different signaling molecules can affect some of the same chemical reactions in the cell, resulting in the combining of different signals.

In the preceding chapters we have seen how chemical compounds are organized, and how they function in living systems. We know how a limited selection of chemical components, such as nucleic acids, sugars, amino acids, and lipids, are arranged into the complex structures of the cell. We also have learned how energy is captured from the environment and used to power every living process. Armed with this understanding of the cell as a functioning unit, we can now discuss how cells communicate with one another and form large multicellular organisms like us.

Sometimes it is tempting, and even useful, to consider cells in isolation and to think of multicellular organisms as nothing more than collections of individual cells. However, this portrayal could not be further from the truth. Like any complex community, multicellular organisms benefit from the specialization of their cells, and must coordinate the activities of all the cells in the body. Over the course of evolution, this simple need for coordination presented a tremendous challenge to living systems. It took almost 3 billion years before primitive single-celled life forms evolved the systems of communication and cooperation that led to the first multicellular organisms. Clearly, cooperation is not an easy thing to achieve, especially when it involves billions upon billions of cells.

Specialization and Communication in the Community of Cells

Two organizational principles apply to every large multicellular organism: the specialization of cell function and the need for communication between cells. The principle of **cell specialization** means that the cells found in a multicellular organism are not all the same, nor do they

have the same function. Indeed, the larger the organism, the greater the need for different types of cells with different structures and functions. Having specialized types of cells that are especially well suited to specific duties ensures that all the processes necessary for the life of the organism are carried out quickly and efficiently.

Specialized cells fulfill a wide variety of needs in multicellular organisms. For example, just as every cell needs the structural support provided by the cytoskeleton, as described in Chapter 6, every multicellular organism requires specialized cells that support, maintain, and strengthen its physical form. In the case of humans, specialized cells deposit bone in the body to form a skeleton, which both supports the body and allows the movement of limbs. In plants, as we saw in Chapter 3, specialized hollow cells in the roots, stems, and leaves transport water between different parts of the plant as well as providing structural support.

Cell specialization does not mean that each type of cell functions in isolation. On the contrary, most cell types function in close cooperation with other cell types to fulfill particular functions for the body. A set of cooperating cell types that performs a particular, specialized function in the body, as we saw in Chapter 1, is called a tissue. For example, the cells that make up our lungs are collectively known as lung tissue, since they work together to facilitate the exchange of carbon dioxide and oxygen that we need to carry out aerobic respiration (see Chapter 8). However, not all the cells that make up lung tissue are the same. Some of the cells are very thin and thus are specialized to exchange the gases inside the lungs; other cells form the lining of the capillaries that bring blood to the lungs (Figure 9.1). The various cell types that make up different tissues and how they work together are discussed in Unit 5 (see also Figure 16.1).

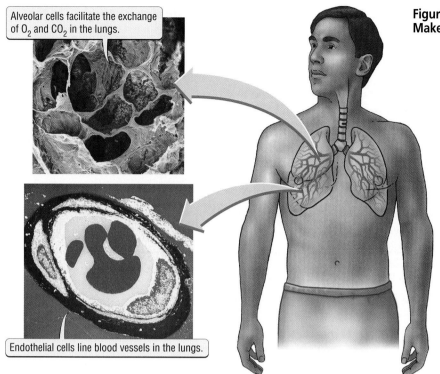

Alveolar cells facilitate the exchange of O_2 and CO_2 in the lungs.

Endothelial cells line blood vessels in the lungs.

Figure 9.1 Different Types of Specialized Cells Make up Lung Tissue

The reflex just described uses several different specialized cell types and means of communication. Specialized nerve cells in the skin of the hand sense the heat and transmit signals to other nerve cells in the spinal cord. The nerve cells in the spinal cord then transmit signals to other nerve cells, which instruct the muscles of the arm to contract, resulting in the quick jerk of the hand away from the heat. Remarkably, this entire series of cell-to-cell communications occurs without the brain or consciousness being involved in the decision. Such simple reflex responses are discussed further in Chapter 34.

The nerve cells involved in this reflex use small molecules called neurotransmitters to communicate with one another and with the muscles of the arm. These molecules are released by one cell

The second organizational principle that applies to all multicellular organisms is that of **cell communication**. Given the diversity of cells and tissues found in a multicellular organism, they must have some way to coordinate their activities, and hence must be able to communicate with one another. The comparison to a corporation that was applied to a single cell in Chapter 6 also holds true for a multicellular organism, which must establish communication networks among its various cells and tissues. For example, if you touch an open flame, the perception of heat must be transferred to your spinal cord, which then tells your hand to jerk back quickly. The simple reflex of jerking away from a painful stimulus that we have all experienced is based on a rapid series of cell-to-cell communications (Figure 9.2).

Figure 9.2 A Simple Reflex Requires Communication between Nerve Cells
The perception of heat results in the passing of signals from the sensory nerve to the spinal cord, which in turn signals muscle cells to contract via a motor nerve.

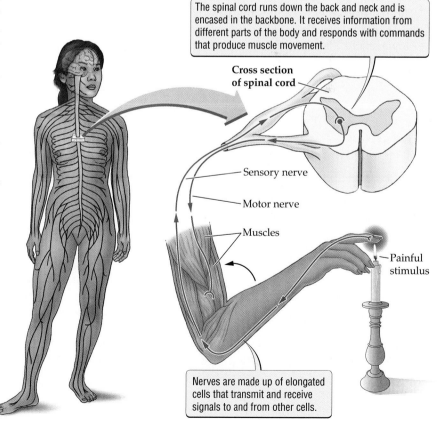

The spinal cord runs down the back and neck and is encased in the backbone. It receives information from different parts of the body and responds with commands that produce muscle movement.

Cross section of spinal cord

Sensory nerve

Motor nerve

Muscles

Painful stimulus

Nerves are made up of elongated cells that transmit and receive signals to and from other cells.

and attach to an adjacent cell, resulting in the transmission of a signal. (We'll learn more about neurotransmitters and communication among nerve cells in Chapters 33 and 34.) The example of a simple reflex involves a very specific set of nerve cells, but the principle of using a small molecule to transmit signals between cells is widely applied in multicellular organisms. In general, communication between cells uses small proteins or other molecules that are released by one cell and received by another cell, which is called the **target cell**. These so-called **signaling molecules** form the language of cellular communication, as we'll see in the next section.

> ■ Multicellular organisms require specialized cells to perform specific functions. The coordination of these functions depends on communication between different types of specialized cells. Cells communicate by releasing small signaling molecules, which are received by target cells.

9.1 The Roles of Signaling Molecules in Cell Communication

The passing of signaling molecules from one cell to another is an effective way for two cells to communicate. However, the identities of these signaling molecules, and their effects on target cells, are highly variable. The particular signaling molecule that is used often depends on the speed with which a particular signal must move and the distance that it must travel between cells.

Returning to our example of a corporation, a simple internal memo that has to travel only between persons within the same building can take the form of a single typewritten sheet of paper. In contrast, a letter that has to be mailed out to customers in different states must be placed in a protective envelope with a complete address label.

The same is true of signaling molecules that travel only between neighboring cells in a tissue versus those that travel through the bloodstream to reach cells in different parts of the body. In the former situation, the signaling molecule can be relatively fragile and short-lived, much like the internal memo; in the latter situation, the signaling molecule must have a longer life span and must be able to withstand environmental stresses. When a signaling molecule reaches a target cell, that cell must have a specific means of receiving it and acting on its message. These tasks are the responsibility of a class of proteins called **receptors**. Since there are many kinds of signaling molecules, each of which has a specific effect on the cell, there must be a matching diversity of receptors. Some receptors are located on the surface of the cell and encounter their matching signaling molecules there (Figure 9.3). Other receptors are located in the cytosol or inside the nucleus of the cell. To reach these receptors, signaling molecules must cross the plasma membrane and perhaps the nuclear envelope.

Signaling molecules are important coordinators of a broad range of cellular processes (Table 9.1). In the sections that follow, we identify and discuss some of the different kinds of signaling molecules and receptors that cells use to communicate with one another, and the different effects that each can have on the target cell. We'll start with a small short-range signaling molecule, the gas nitric oxide.

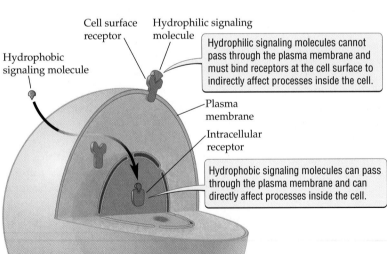

Cell surface receptor

Hydrophilic signaling molecule

Hydrophobic signaling molecule

Hydrophilic signaling molecules cannot pass through the plasma membrane and must bind receptors at the cell surface to indirectly affect processes inside the cell.

Plasma membrane

Intracellular receptor

Hydrophobic signaling molecules can pass through the plasma membrane and can directly affect processes inside the cell.

Figure 9.3 Two Classes of Receptors
Two major classes of receptors bind to signaling molecules. Cell surface receptors are embedded in the plasma membrane and bind to signaling molecules that cannot cross the membrane. Intracellular receptors reside in the cytosol and bind to signaling molecules that can cross the plasma membrane.

9.1 *Examples of Signaling Molecules*

Name of molecule	Type of molecule	Site(s) of synthesis	Function(s)
Nitric oxide	Gas	Endothelial cells in blood vessel walls; nerve cells	Promotes relaxation of blood vessel walls
Testosterone	Steroid	Testes	Promotes the development of secondary male sex characteristics
Progesterone	Steroid	Ovaries	Prepares the uterus for implantation; promotes mammary gland development in females
Insulin	Protein	Beta cells of the pancreas	Promotes the uptake of glucose by cells
Adrenaline	Amino acid derivative	Adrenal glands	Promotes release of stored fuels and increased heart rate
Thyroxine	Amino acid derivative	Thyroid gland	Promotes increased metabolic rate
Platelet-derived growth factor (PDGF)	Protein	Many cell types	Promotes cell multiplication
Nerve growth factor (NGF)	Protein	Tissues richly supplied with nerves	Promotes nerve growth and survival
Auxin	Acetic acid derivative	Most plant cells	Promotes root formation and stem elongation
Acetylcholine	Choline derivative	Nerve cells	Assists signal transmission from nerves to many muscles

Nitric oxide is a short-range signaling molecule

Signaling molecules often take the form of proteins or fatty acid derivatives. In the past decade, however, biologists have shown that even a gas can function as a signaling molecule inside the human body. This gas is **nitric oxide** (NO), and it has an important role in lowering blood pressure. How does a gas serve as a signal to lower blood pressure? The answer lies in how specific target cells respond to nitric oxide by changing their physical shape.

Blood pressure is directly related to the internal volume (capacity) of blood vessels. The larger the internal volume of a blood vessel, the lower will be the pressure of the blood inside it. Specific nerve signals induce the cells that form the lining of blood vessels, called endothelial cells, to produce nitric oxide. Small molecules of NO gas rapidly spread into the neighboring muscle cells, causing them to relax. Relaxation of the muscle cells allows the blood vessel to expand, increasing the volume of blood that it can carry and decreasing blood pressure (Figure 9.4). This process is known as **vasodilation** (*vaso*, "vessel"; *dilation*, "expansion").

Although we have only recently discovered the role of nitric oxide as a signaling molecule, we have long used it as a therapy without knowing it. Since the nineteenth century, nitroglycerin has been administered to patients suffering from angina (chronic chest pain), which is due in part to the narrowing of the blood vessels that deliver blood to the heart. Although nitrogly-cerin is often associated with dangerous explosives such as TNT, the human body converts administered nitroglycerin into nitric oxide, resulting in increased blood flow to the heart and a reduction in angina pain.

The direct inhalation of nitric oxide has even been used to treat newborns with pulmonary hypertension. This serious disorder is characterized by abnormally high blood pressure in the vessels that supply blood to the lungs, usually caused by an inherited heart defect. Before the identification of nitric oxide as a signaling molecule, customary therapies often lowered the infant's overall blood pressure to dangerous levels. In contrast, inhaling nitric oxide affects only the blood vessels of the lungs, causing them to expand and reducing the high blood pressure in that tissue alone. The effect of inhaled nitric oxide is highly localized to the lungs because the gas acts only over short distances. Once nitric oxide diffuses out of the cell in which it is produced, it is broken down within a matter of seconds. Thus, only those target cells in the immediate vicinity are close enough to receive and respond to the signal. The inhaled nitric oxide does not travel far enough to affect blood vessels outside of the lungs, and thus it does not have the undesirable effect of lowering blood pressure throughout the body.

9.2 Hormones are long-range signaling molecules

The short life span of nitric oxide makes it an effective signaling molecule over short distances. But

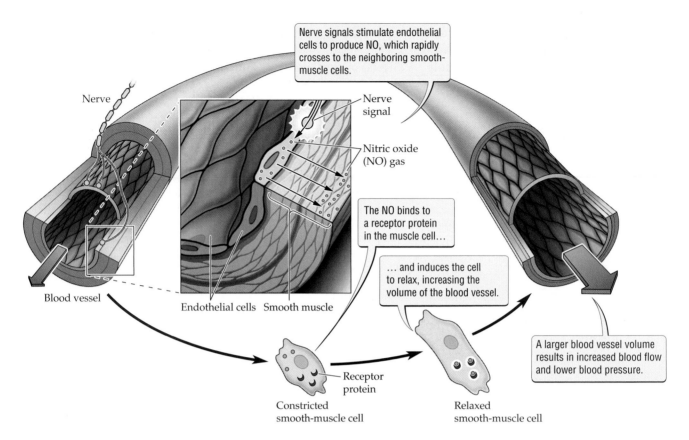

Nerve signals stimulate endothelial cells to produce NO, which rapidly crosses to the neighboring smooth-muscle cells.

Nerve

Nerve signal

Nitric oxide (NO) gas

The NO binds to a receptor protein in the muscle cell…

… and induces the cell to relax, increasing the volume of the blood vessel.

Blood vessel

Endothelial cells Smooth muscle

A larger blood vessel volume results in increased blood flow and lower blood pressure.

Receptor protein

Constricted smooth-muscle cell

Relaxed smooth-muscle cell

Figure 9.4   Nitric Oxide (NO) Gas Lowers Blood Pressure
The endothelial cells that form the lining of blood vessels are surrounded by smooth-muscle cells, which control the volume of the blood vessel by contracting and relaxing.

communication is not limited to cells that are near one another. Depending on the size of the organism, cells that are anywhere from a few centimeters to several meters apart must also be able to communicate with one another. Such long-distance communication requires a different class of signaling molecules, called **hormones**.

Hormones are used by all multicellular organisms to coordinate the activities of different cells and tissues. In contrast to nitric oxide, hormones are produced by cells in one part of the body and transported to target cells in another part of the body. We'll learn more about the many functions of hormones in Chapter 33; here, we focus on how they transmit their messages to cells.

The transportation of hormones from their site of production to their target cells often depends on the circulation of fluids inside the organism. In plants, some hormones are dissolved in the sap, which moves between the roots and the rest of the plant. In animals, hormones are dissolved in the blood, which circulates throughout the body, ensuring rapid distribution.

Steroid hormones can cross cell membranes

Steroid hormones are an important class of signaling molecules that are essential for many growth processes, including the normal development of reproductive tissues in mammals. The hormone **progesterone**, for example, is produced in the ovaries of female mammals, circulates through the bloodstream, and ensures that the cells of the uterus are ready to support an embryo.

All steroid hormones are derived from a lipid called cholesterol and are hydrophobic. Because they are hydrophobic, steroid hormones can pass easily through the hydrophobic core of the target cell's plasma membrane and enter the cytosol. But their hydrophobic character also means that they cannot move unaided through the bloodstream. Steroids must be packaged with proteins that help them dissolve in the watery environment of the body and extend their life span in the bloodstream.

The ability of steroids to survive and be transported in the bloodstream for up to several days improves the likelihood that they will reach distant target cells. Survival in the blood allows progesterone, for example, to affect different cells in parts of the body very distant from the site of its synthesis in the ovaries. In addition to preparing the cells of the uterus for pregnancy, progesterone can induce development of the more distant mammary glands to prepare them for milk production.

Once steroids reach their target cells, they cross the plasma membrane and alter the production of specific proteins inside the cell (Figure 9.5). In order to do this, they must bind to receptors in the cytosol. Together, the steroid and receptor form an active complex that can enter the nucleus and affect the target cell's DNA. As we saw in Chapter 6, the DNA in the nucleus carries the instructions for making all of the proteins needed by the cell. A gene is a segment of DNA that carries the instructions for making a particular protein, as we will see in Chapter 12. The steroid–receptor complex acts on specific genes, activating the production of the proteins they encode. The genes that are activated by progesterone in target cells in the uterus, for example, encode proteins that are necessary for the proper growth of the uterine lining. Thus, the action of steroid hormones in effect coordinates the production of proteins necessary for specific changes in the cell.

Some hormones require the help of cell surface receptors

Not all signaling molecules enter the cell as steroids do. Some hormones send their signals into the cell via cell surface receptors. Some of these hormones are small proteins; others are chemical derivatives of amino acids and fatty acids.

Adrenaline is one such hormone, whose effects we have all felt when startled or frightened. It is derived from the amino acid tyrosine. When an event or a situation frightens us, our adrenal glands release adrenaline into the bloodstream to enhance the body's ability to respond to a threat (Figure 9.6).

Adrenaline binds to receptors on the surfaces of cells in the liver and fatty tissues of the body. This binding results in changes in the metabolism of these cells, promoting the breakdown of polysaccharides in the liver and of fats in fatty tissue. The glucose and fatty acids that are produced are released into the bloodstream for use as fuel by all of the cells in the body during times of stress.

Adrenaline has a different effect on muscle cells in the heart. It increases the rate at which heart muscle contracts and relaxes, resulting in a faster heartbeat. When the heartbeat increases, blood circulates more rapidly through the body, delivering more oxygen and fuel to cells. Once again, the ability of the body to respond to the frightening stimulus is enhanced. Thus, adrenaline is a signaling molecule that has different but complementary effects on different types of cells.

Growth factors induce cell division

Growth factors are another class of signaling molecules that use receptors on the cell surface. The proliferation of cells in the human body is controlled largely by growth factors, which both initiate and maintain the processes needed for cell growth and cell division. Scores of proteins that function as growth factors are produced by cells; in most cases, their effects are confined to neighboring cells. In this respect, growth factors are like nitric oxide, but unlike NO, they do not enter the target cell. Instead, they depend on cell surface receptors to transmit a message into the cell and alter its function.

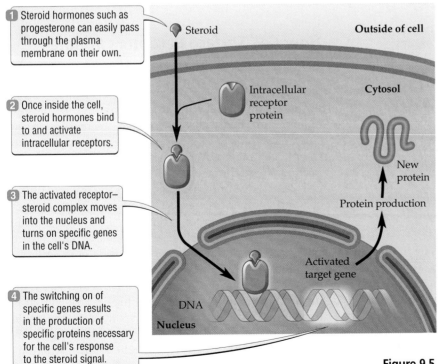

1 Steroid hormones such as progesterone can easily pass through the plasma membrane on their own.

2 Once inside the cell, steroid hormones bind to and activate intracellular receptors.

3 The activated receptor–steroid complex moves into the nucleus and turns on specific genes in the cell's DNA.

4 The switching on of specific genes results in the production of specific proteins necessary for the cell's response to the steroid signal.

Steroid

Outside of cell

Intracellular receptor protein

Cytosol

New protein

Protein production

Activated target gene

DNA

Nucleus

Figure 9.5   A Cell's Response to a Steroid Hormone

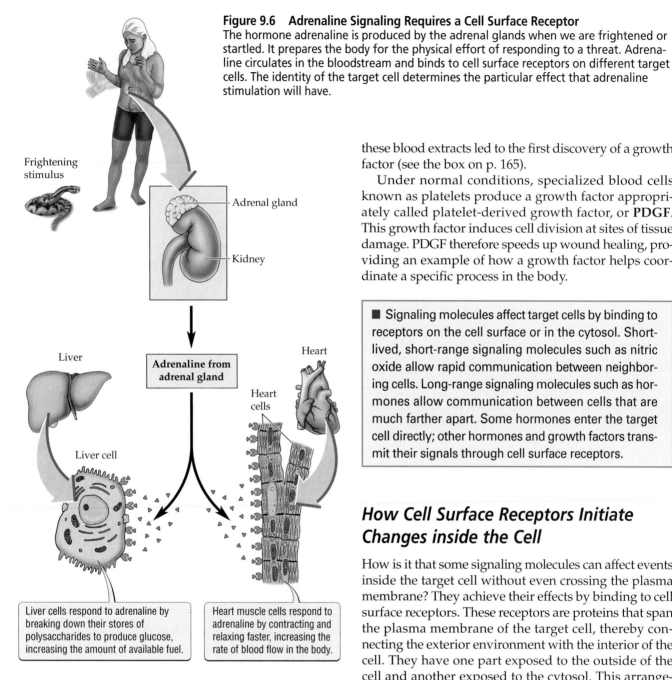

Figure 9.6   Adrenaline Signaling Requires a Cell Surface Receptor
The hormone adrenaline is produced by the adrenal glands when we are frightened or startled. It prepares the body for the physical effort of responding to a threat. Adrenaline circulates in the bloodstream and binds to cell surface receptors on different target cells. The identity of the target cell determines the particular effect that adrenaline stimulation will have.

Frightening stimulus

Adrenal gland

Kidney

Liver

Adrenaline from adrenal gland

Heart

Heart cells

Liver cell

Liver cells respond to adrenaline by breaking down their stores of polysaccharides to produce glucose, increasing the amount of available fuel.

Heart muscle cells respond to adrenaline by contracting and relaxing faster, increasing the rate of blood flow in the body.

these blood extracts led to the first discovery of a growth factor (see the box on p. 165).

Under normal conditions, specialized blood cells known as platelets produce a growth factor appropriately called platelet-derived growth factor, or **PDGF**. This growth factor induces cell division at sites of tissue damage. PDGF therefore speeds up wound healing, providing an example of how a growth factor helps coordinate a specific process in the body.

> ■ Signaling molecules affect target cells by binding to receptors on the cell surface or in the cytosol. Short-lived, short-range signaling molecules such as nitric oxide allow rapid communication between neighboring cells. Long-range signaling molecules such as hormones allow communication between cells that are much farther apart. Some hormones enter the target cell directly; other hormones and growth factors transmit their signals through cell surface receptors.

How Cell Surface Receptors Initiate Changes inside the Cell

How is it that some signaling molecules can affect events inside the target cell without even crossing the plasma membrane? They achieve their effects by binding to cell surface receptors. These receptors are proteins that span the plasma membrane of the target cell, thereby connecting the exterior environment with the interior of the cell. They have one part exposed to the outside of the cell and another exposed to the cytosol. This arrangement allows the receptor to bind the signaling molecule on the outside of the cell, change its shape, and induce a specific chemical reaction on the inside of the cell.

The triggering of a specific chemical reaction inside the cell in response to an external signal is known as **signal transduction**. As the name implies, the signal is transferred, or transduced, from the outside to the inside of the cell. When a signaling molecule binds to a cell surface receptor, it triggers a series of chemical reactions that activate proteins inside the cell. This simple but effective transduction process is referred to as a **signal cascade**. Since the collection of signaling molecules that

Growth factors were first discovered by biologists attempting to grow cells outside of the body. Cells can live outside of the body in suitable culture dishes if they are supplied with appropriate nutrients in a liquid medium, but they will not divide and multiply, and they will eventually die. This presented a problem for biologists interested in long-term studies of cells. Eventually it was discovered that the addition of blood extracts to the liquid medium induced cells to divide, permitting the proliferation of cells outside of the body. Further analysis of

■ THE SCIENTIFIC PROCESS

Rita Levi-Montalcini: How One Woman Pursued Biology against All Odds

In 1986, two remarkable scientists, Rita Levi-Montalcini and Stanley Cohen, shared the Nobel prize in physiology or medicine. The prize was awarded in recognition of their discovery of the first known growth factors, which revealed an essential means of communication used by cells. Prior to their work in the 1950s, scientists knew that cells could be successfully grown outside the body in liquid media supplemented with blood extracts. However, the substances in the extracts responsible for this effect remained mysterious.

Rita Levi-Montalcini was born in Italy in 1909, the daughter of Adamo Levi, an electrical engineer, and Adele Montalcini, a gifted painter. During her childhood, Rita's father prevented the three Levi-Montalcini sisters from pursuing a university education. He valued intellectual pursuits, but felt that women should not pursue careers that could distract them from their duties as wives and mothers. By age 20, Rita was dissatisfied with her father's view of a traditional feminine role, and she successfully persuaded him to let her attend medical school in Turin. In 1936, she graduated from medical school with highest honors—the very same year that Mussolini declared it illegal for non-Aryan Italian citizens to hold academic appointments. This did not prevent Rita from pursuing her interests in neurobiology. By 1940, she had set up a small laboratory in her bedroom and had begun her groundbreaking work on

the development of chick embryos. After the invasion of Italy by the Nazis, Rita and her family fled to Florence, where they lived in hiding until the end of World War II.

When the war ended, Levi-Montalcini moved to America and joined Viktor Hamburger's laboratory in St. Louis, where she continued her work on chick embryos. She observed that certain tumors from mice, when transplanted into chick embryos, could induce the growth of nerves, even when there was no direct contact between the tumor cells and the nerve cells. She concluded that the tumor cells must secrete some kind of substance that signaled specific nerve cells to grow. She named this substance nerve growth-promoting factor, or NGF, and went on to develop extracts that allowed her to measure its powerful growth-inducing properties. An American biochemist named Stanley Cohen joined the research effort in 1953. Armed with Levi-Montalcini's method for measuring growth-inducing activity, he successfully purified NGF and proved that it was a protein.

The collaborative effort that led to the discovery and characterization of NGF was a major milestone in cell biology. It marked the first time a signaling molecule was shown to control a complex biological process such as cell growth, fundamentally changing our view of how growth and development are regulated. The discovery of NGF and other growth

factors also opened up a new field of therapeutic medicine that is still blossoming today.

Rita Levi-Montalcini went on to serve as the director of the Institute of Cell Biology in Rome until her retirement in 1979. Even now she remains an outspoken advocate for research funding in Italy, as well as an enduring symbol of how a stubborn passion for biology can lead to great discoveries despite daunting obstacles.

Rita Levi-Montalcini

affect cells is diverse, the range of receptors that bind them and initiate specific events inside the cell is equally diverse. In other words, each type of signaling molecule binds to a specific type of receptor on the cell surface, causing a specific signal cascade that activates specific proteins.

In the case of a hormone such as adrenaline, binding of the signal to its receptor causes a change in the shape of the receptor protein, converting it into an active form. The portion of the receptor that is exposed to the cytosol is then able to bind and activate a G protein at the inner face of the plasma membrane (Figure 9.7).

G proteins are so named because they bind to and are regulated by nucleotides with a guanine base: GDP (guanosine diphosphate) and GTP (guanosine triphosphate). When a G protein is bound to GDP, it is inactive. When an inactive G protein binds to an activated receptor, however, it is able to release GDP and bind GTP from the cytosol. This process of exchanging one guanine nucleotide for another alters the shape of the G protein and activates it.

The G protein is just the first of several types of proteins that must be activated in order for the signal to be transduced inside the target cell. The active G protein activates an enzyme called adenylate cyclase, which catalyzes the conversion of ATP into a form called cAMP (cyclic adenosine monophosphate, or cyclic AMP). As the amount of cAMP in the cytosol increases, several enzymes and other proteins in the cell bind to cAMP and become activated themselves (see Figure 9.7). Among the enzymes that are activated by cAMP are those required for the adrenaline response. Thus, unlike steroid hormones, which directly activate genes and affect protein production when bound to their cytosolic receptors, adrenaline depends on a series of protein activations in the cytosol to transmit its message.

The proteins in other signal cascades can be activated in various ways, many of which do not depend on GTP or cAMP. In the case of a growth factor such as PDGF, the cell surface receptor that binds the signaling molecule is itself a type of enzyme, called a **kinase**. Binding of the signaling molecule by the kinase results in the enzyme-catalyzed transfer of phosphate groups from ATP to several proteins in the cytosol (Figure 9.8).

In the case of growth factor receptors, the part of the kinase that binds the growth factor is on the outside of the cell, and the part that catalyzes the phosphorylation reaction is on the inside of the cell. Therefore, binding of the growth factor to the exposed portion of the kinase on the cell surface activates the enzymatic activity of the receptor, which leads to multiple phosphorylation events on the inside of the cell.

The first protein to become phosphorylated in a growth factor signal cascade is the receptor kinase itself. This process of autophosphorylation (*auto*, "self") targets specific amino acids in the cytosolic portion of the receptor (see Figure 9.8). Once phosphorylated, these amino acids become suitable

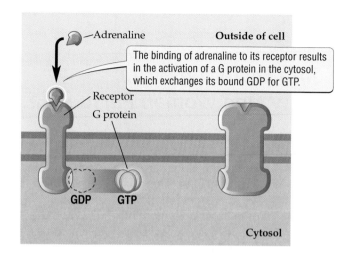

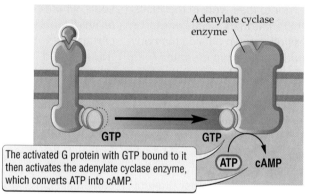

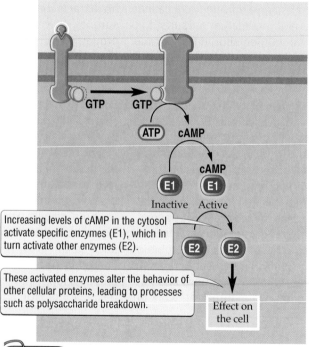

Figure 9.7 A G Protein Signal Cascade

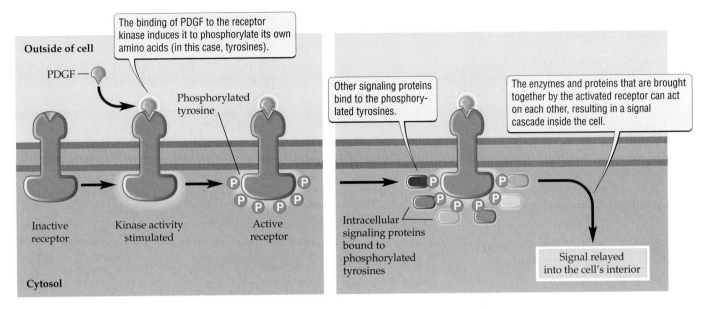

Figure 9.8 A Receptor Kinase Signal Cascade

binding sites for other proteins involved in the growth factor signal cascade, which themselves become phosphorylated. Thus, instead of merely activating a target protein, the kinase creates binding sites on itself for other proteins. This process of assembling or concentrating the proteins necessary for the next steps of the signal cascade is yet another way of transducing a signal.

How does phosphorylation result in the transduction of an external signal? Upon phosphorylation, some enzymes are activated and can catalyze reactions faster, while others are inhibited and catalyze reactions more slowly, or not at all. Phosphorylation is a process that is well suited to making such changes, since the addition of a highly charged phosphate group (PO_4^{3-}) to one or more amino acids in a protein changes the three-dimensional shape, and hence the activity, of the protein (Figure 9.9). As we saw in Chapter 5, the chemical characteristics of the amino acids that make up a given protein determine its folded three-dimensional shape. Thus, when a kinase adds a charged phosphate group to specific amino acids of a target protein, it dramatically changes their chemical characteristics, resulting in altered shape and activity of the protein.

■ In signal transduction, specific chemical reactions are triggered inside a cell by an external signal, causing a series of protein activations known as a signal cascade. Cell surface receptors trigger signal cascades by binding external signaling molecules, changing shape, and then activating or inactivating specific proteins inside the cell.

Figure 9.9 Phosphorylation Can Change a Protein's Activity
The addition of a highly charged phosphate group to the R group of an amino acid can change the overall chemical character of a protein. This addition, in turn, changes the three-dimensional shape of the protein.

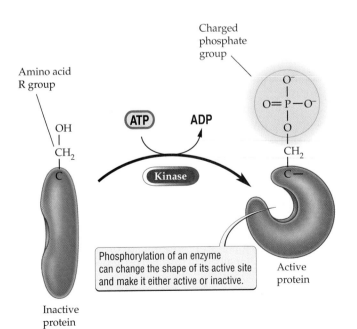

Amplifying and Combining Signals inside the Cell

Imagine a situation in which a single hormone or growth factor binds to a cell surface receptor. If this binding event resulted in the activation or phosphorylation of only one protein inside the cell, many binding events at the cell surface would be needed in order to bring about any significant change in the cell. To avoid such an inefficient arrangement, most signal cascades greatly amplify the initial signal. A signal cascade is like a huge avalanche on a snow-covered mountain that is started by one small icicle falling off a tree at the top of the slope.

Signal amplification depends on a cascade of enzyme-catalyzed reactions

For the signaling molecules described above, the binding of the signal to its cell surface receptor fulfills the role of the falling icicle. If we consider the case of a hormone such as adrenaline, each activated receptor can activate several G protein molecules, each of which in turn can act on several enzymes. Likewise, a single activated receptor kinase leads to the phosphorylation and activation of many target proteins. This simple principle is apparent in many different signal cascades inside the cell, and explains why the binding of one signaling molecule to a single receptor can ultimately activate thousands of target proteins. If signal amplification did not occur, the resulting inefficiency of cell-to-cell communication would not allow multicellular organisms to exist.

The fact that a multitude of cellular processes are regulated by different external signals means that both the timing of the signal cascade and the identity of the target molecules can vary a great deal. So far we have discussed

signal cascades that ultimately lead either to the activation of cellular enzymes or to the turning on of genes and the production of specific proteins. In the former case, which is well illustrated by the example of adrenaline, the enzymes necessary for the release of glucose are already present in the cytosol, and they are activated in a matter of seconds after the hormone binds to its receptor on the surface of a liver cell (Figure 9.10*a*). Since adrenaline production is associated with a perception of danger and the possible need to flee, it makes sense that the levels of available glucose in the bloodstream would need to increase rapidly to provide energy for an escape response.

In contrast, the promotion of uterine growth by progesterone depends on the turning on of genes and the

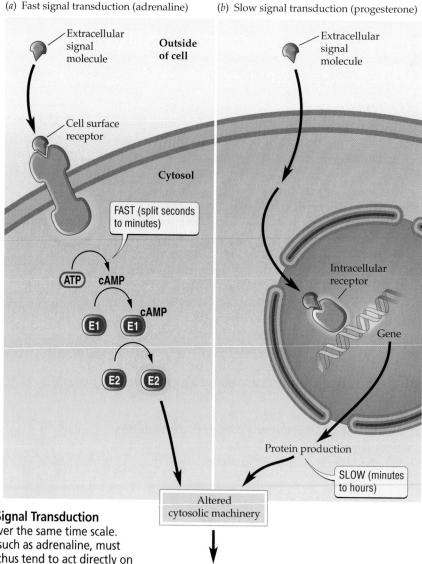

(*a*) Fast signal transduction (adrenaline)

(*b*) Slow signal transduction (progesterone)

Figure 9.10   The Timing of Signal Transduction
Not all signal cascades occur over the same time scale. (*a*) Some signaling molecules, such as adrenaline, must produce rapid responses, and thus tend to act directly on existing proteins in the cell. (*b*) Other signaling molecules, such as some steroids, act more slowly because they need to turn on genes, which in turn produce the necessary proteins.

production of proteins, which occur in a matter of hours rather than seconds (Figure 9.10*b*). This process does not have the immediate urgency of the adrenaline response, and given the importance of ensuring the appropriate timing and extent of uterine growth, it is not surprising that the body uses slower and more regulated signaling pathways to control the process.

Different signals can be combined inside the cell

In the previous section we addressed only a few examples of signal cascades and the processes they control. Almost every activity a cell undertakes involves some sort of regulatory input from external signals. Even the shapes that cells adopt are regulated by external signals, which lead to the phosphorylation and rearrangement of proteins in the cytoskeleton. Yet not all signal cascades that are started by particular signaling molecules stand alone. Indeed, several different signaling molecules can share some of the components of a single signal cascade in a manner that results in cooperation between different signals. This process of **signal integration** means that one signal can either enhance or inhibit the effects of another signal.

Several kinases, first discovered as part of the growth factor signal cascade, also participate in the transduction of signals from several other signaling molecules, including hormones. These kinases promote such diverse processes as cell growth in humans and sexual reproduction in yeasts. Therefore, the effect that activation of such kinases can have on the cell must depend in part on the differing combinations of proteins that are present in different cells. The combining of different signals gives cells an increased range of possible responses to their environment.

■ The activation of many proteins at each step of a signal cascade greatly amplifies the original signal. Signal cascades that modify existing proteins inside the cell occur in a matter of seconds; those that activate genes to produce new proteins can take several hours. Certain proteins can participate in multiple signal cascades, allowing the cell to combine different external signals.

■ HIGHLIGHT

 ### 9.5 *Defeating Deadly Bacterial Toxins*

The signal transduction events that we have discussed in this chapter are frequently the targets of harmful bacteria. We now know, for example, that the toxin pro-

duced by the cholera bacterium specifically affects a G protein found in the cells that line the intestine. Once the toxin enters an intestinal cell, it chemically modifies the G protein such that it becomes permanently bound to GTP. This means that the cholera toxin permanently activates the G protein, and the signal cascade in which it participates thus cannot be turned off.

This abnormal state of affairs leads to excess activity on the part of the enzymes normally regulated by the G protein. The activities of these enzymes include the opening of membrane channels that facilitate the movement of water out of the bloodstream, through the intestinal cell, and into the lumen of the intestine. The resulting excessive loss of water from the blood to the intestine causes the diarrhea and dehydration that are characteristic of cholera infections. As antibiotic-resistant strains of cholera bacteria appear, new drug therapies that limit the effect of the toxin on intestinal cells must be developed. Such drugs could save tens of thousands of lives each year.

Instead of overactivating a signal cascade, as the cholera toxin does, the anthrax toxin blocks the action of a kinase used by several signaling molecules. The anthrax toxin is an enzyme that cuts up other proteins, making them inactive; enzymes that do this are called proteases. One target of the anthrax protease happens to be a kinase that has important functions in many types of cells. Once the kinase is inactivated, the cell can no longer respond to multiple external signals; the result is the deregulation of many normal cellular processes and ultimately cell death. One of the most difficult problems encountered in the treatment of anthrax is that by the time the bacteria are killed with antibiotics, they have already released so much toxin into the body that the host animal still dies.

Now that scientists have uncovered the mechanism by which the anthrax toxin kills cells, it may be possible to develop a drug that directly blocks its activity. The idea of a drug inhibiting a protease is nothing new. One of the most effective therapies for HIV infection includes a drug that blocks a viral protease necessary for the HIV life cycle. In the case of anthrax exposure, such a drug could prevent the anthrax toxin from cutting apart the kinase. It could be administered together with antibiotics, providing a therapy that would both neutralize the effects of the toxin and clear the bacterial infection from the body.

■ The toxins produced by cholera and anthrax bacteria cause disease by affecting signal cascades inside the cell. The cholera toxin overactivates a G protein signal cascade, while the anthrax toxin inhibits an important kinase. New drug therapies are in development that will specifically block these effects.

SUMMARY

Specialization and Communication in the Community of Cells

- Multicellular organisms require specialized cells to perform specific functions in the body.

- The coordination of activities in a multicellular organism depends on communication between different types of specialized cells.

- Communication between cells depends on small signaling molecules that are produced by specific cells and received by other target cells.

The Roles of Signaling Molecules in Cell Communication

- Signaling molecules affect target cells by binding to receptor proteins either on the cell surface or in the cytoplasm.

- Short-lived signaling molecules such as nitric oxide allow rapid communication between neighboring cells.

- Hormones are long-range signaling molecules that allow communication between cells that are far apart in a large multicellular organism.

- Some hormones (such as steroids) enter cells directly; others (such as adrenaline) use cell surface receptors to send their signals into cells.

- Growth factors are signaling molecules that induce the target cell to divide, leading to cell multiplication.

How Cell Surface Receptors Initiate Changes inside the Cell

- In signal transduction, specific chemical reactions are triggered inside a cell by an external signal, causing a series of protein activations known as a signal cascade.

- Cell surface receptors trigger signal cascades by binding external signaling molecules, changing shape, and then activating or inactivating specific proteins inside the cell.

- The cytosolic proteins that participate in a signal cascade include G proteins and specific enzymes.

- Activation of proteins in a signal cascade often involves enzyme-catalyzed phosphorylation or the binding of specific nucleotides such as GTP.

Amplifying and Combining Signals inside the Cell

- The activation of many proteins at each step of a signal cascade greatly amplifies the original signal.

- Signal cascades that modify existing proteins inside the cell occur in a matter of seconds; those that activate genes to produce new proteins can take several hours.

- Certain proteins can participate in multiple signal cascades, allowing the cell to integrate different external signals.

Highlight: Defeating Deadly Bacterial Toxins

- Knowing how signal cascades are affected by deadly bacterial toxins will allow biologists to design drug therapies that specifically block their effects.

KEY TERMS

adrenaline p. 163	progesterone p. 162
cell communication p. 159	receptor p. 160
cell specialization p. 158	signal cascade p. 164
G protein p. 166	signal integration p. 169
growth factor p. 163	signal transduction p. 164
hormone p. 162	signaling molecule p. 160
kinase p. 166	steroid p. 162
nitric oxide p. 161	target cell p. 160
PDGF p. 164	vasodilation p. 161

CHAPTER REVIEW

Self-Quiz

1. Multicellular organisms have different specialized cell types because
   a. cells are unable to coordinate their activities with those of other cells.
   b. only some cells can produce ATP for use by the entire organism.
   c. the products of cell division are variable.
   d. specialized cells are better able to perform specific functions required for the life of the organism.

2. Signaling molecules
   a. are found exclusively in animals.
   b. allow cells to communicate only with the nonliving environment.
   c. are always proteins.
   d. are produced by specific cells and received by target cells.

3. Which of the following statements is true?
   a. Signaling molecules bind to specific receptors, which in turn affect chemical reactions inside the cell.
   b. All receptors are located on the cell surface.
   c. G proteins are important receptors for hormone signaling molecules.
   d. Steroid hormones require cell surface receptors to affect target cells.

4. Which of the following statements is *not* true?
   a. Hormones tend to be long-range signaling molecules.
   b. Hormones are not found in plants.
   c. Hormones move from one part of the body to another through the bloodstream.
   d. Hormones can affect the production of specific proteins in a target cell.

5. Growth factors are signaling molecules that
   a. stimulate target cells to grow larger and not divide.
   b. were first discovered by biologists attempting to grow cells outside of the body.
   c. freely pass through the plasma membrane of the target cell.
   d. affect only cells in distant parts of the body.

6. Nitric oxide lowers blood pressure by
   a. causing the heart to beat more slowly.
   b. increasing water loss via the intestine.
   c. reducing the volume of blood circulating in the body.
   d. relaxing the muscle cells that line blood vessels.

Review Questions

1. The cellular location of a receptor depends on the chemical characteristics of the signaling molecule to which it binds. Where would you expect the receptor for a hydrophobic molecule to be located?

2. Signal cascades greatly amplify the effect of a single signaling molecule binding to a receptor. Why is signal amplification so important for a cell's ability to receive communications from other cells?

3. Why is phosphorylation by a kinase such an effective way to alter the activity of an enzyme?

4. G proteins and adenylate cyclase both participate in the signal cascade induced by adrenaline. How do these two classes of proteins use nucleotides in promoting the signal cascade?

9

The Daily Globe

Equal Opportunity Viagra: Is What's Good for the Gander Good for the Goose?

PARAMUS, NJ—For decades, male erectile dysfunction has been the subject of veiled sneers and stand-up comedy. But that all changed when the U.S. Food and Drug Administration approved the sale of the drug Viagra in 1998. Today, erectile dysfunction is big business as well as the subject of vigorous public debate—and even television advertising.

Viagra relieves erectile dysfunction by increasing blood flow to the genital area. The drug achieves this effect by enhancing the signal cascade induced by nitric oxide. Since nitric oxide signals prompt blood vessels to expand, Viagra prolongs the resulting increase in blood flow.

Dr. Phyllis Hoffman of the New Jersey Medical Center firmly believes that Viagra can successfully treat female sexual dysfunction. "It is true that women are wired differently than men," Dr. Hoffman says. "But, increased blood circulation to the genital area is just as important in the female sexual response as it is in males."

While Viagra is still being tested in women, the attention brought to the biology of female sexuality has been applauded by many researchers. "Whether or not Viagra works in women," says Dr. Kris Carmichael of the Female Reproductive Biology Unit at Mt. Sinai Hospital, "these clinical studies will greatly improve our limited understanding of female sexual dysfunction."

Evaluating "The News"

1. Given what you know about nitric oxide as a signaling molecule, do you think that drugs like Viagra might have negative side effects?

2. The wide use of Viagra among men has exceeded even the most optimistic predictions of its manufacturer. The huge number of men using Viagra has led some physicians to fear that its use has become more recreational than therapeutic. Should the use of drugs like Viagra be regulated?

3. Why do you think human sexual dysfunction has received so little attention from the scientific community until now?

chapter *10* *Cell Division*

Hilma af Klint, *Group 4. The Ten Greatest no. 7 Manhood (Lnr 108)*, 1907.

An Army in Revolt

Your body is constantly under siege. With every breath you take, foreign invaders enter your body and must be repelled by its defenses. The continuous struggle against dangerous microorganisms in the environment is a normal part of your body's day-to-day existence. The complex organization and functioning of the body require a great deal of effort to maintain and protect. Given the many kinds of specialized cells that make up the body, it is not surprising that a specific group of cells is specialized to defend the body against foreign invaders. These cells are known collectively as the immune system. They serve as both the guards that protect exposed body surfaces and the soldiers that seek out and destroy any invaders that get inside.

Unfortunately, the immune system, like any army, is not perfect. In fact, many dangerous organisms have developed ways not only to evade the immune system's defenses, but also to use them to damage the body. One such organism is the bacterium *Borrelia burgdorferi*, which is the primary cause of Lyme disease. Tiny ticks that live on ani-

■ MAIN MESSAGE

Cell division is the means by which organisms grow and maintain their tissues and by which genetic information is passed on from one generation to the next.

mals such as deer and sheep carry these microscopic invaders. Each year more than 16,000 persons are bitten by infected ticks while camping or merely walking through tall grass. Just one bite from an infected tick can pass *Borrelia* to an unfortunate individual. The resulting Lyme disease begins with skin rashes and flulike symptoms; over time, it can lead to serious arthritic and neurological disorders.

Fortunately, *Borrelia* is easily killed by powerful antibiotics. However, some patients experience severe arthritic pain in their joints long after ridding the body of the bacteria. This mysterious pain is caused by an inappropriate attack on the joints by cells of the immune system. Infection by *Borrelia* bacteria allows abnormal immune cells that attack components of the body to divide rapidly, increase in number, and in effect turn your own immune system against you.

This alarming event emphasizes the importance of cell division in diverse biological processes and the need for ways of regulating it. In this chapter we explore the mechanics of cell division, and we see how cell division is involved in such diverse processes as reproduction, the replacement of worn-out tissues, and the defense of the body by the immune system.

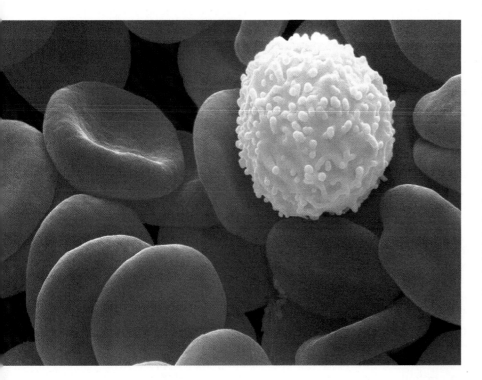

A Single Immune System Cell among Many Red Blood Cells

■ KEY CONCEPTS

1. Cells divide to produce two daughter cells. This process is necessary for growth and development, the ability to maintain and replace tissues, and the passing on of genetic information.

2. Cell division has several distinct stages collectively known as the cell division cycle. Each stage is marked by physical changes in the cell that either prepare it for division or directly participate in the process.

3. Most of the events that prepare the cell for division occur during the interphase stage of the cell division cycle. The physical events of cell division occur during the stages of mitosis or meiosis.

4. Mitosis ensures that both daughter cells receive identical and complete sets of chromosomes. The number and shape of the chromosomes is uniquely characteristic of each species. Cells that make up the body but are not involved in sexual reproduction undergo mitosis.

5. Meiosis consists of two cycles of cell division. Its end product is four daughter cells, each with half the number of chromosomes of the parent cell. Meiosis occurs exclusively in cells that produce sperm and eggs.

A s you sit and read this book, you are probably not aware of the millions of cells in your body that are responding to growth signals, dividing, and replacing old cells with new ones. Cells proliferate both when tissue is damaged or lost, as is the case when you fall and scrape your knee, and as a natural part of body repair and maintenance. The production of new cells is an essential activity for every multicellular organism.

In Chapter 9 we discussed how cells communicate with one another using signaling molecules. One of the most significant consequences of cell-to-cell communication is cell division. Why, and when, is cell division important? The most obvious answer is during the growth and development of a fertilized egg into a mature multicellular organism. The increase in the size and complexity of the body during development requires not just more cells, but different kinds of cells. Thus, as a multicellular organism develops, a tremendous amount of cell division and proliferation takes place to expand existing tissues and create new tissue types. We will learn more about how cells communicate to guide one another through the development process in Chapter 37.

Once an organism has achieved its mature size, cells must still divide to replace worn-out and damaged tissue. Skin, for example, must be continuously renewed as the dead cells on the surface are worn away. Simply put, the upper portion of the skin consists of multiple layers of cells that gradually move to the surface as old surface layers are lost to wear and tear. In the process, the cells undergo dramatic physical changes, such that by the time they reach the surface layer, they form dead, flattened scales of protein (Figure 10.1). As cells move up through the layers and are lost from the skin surface, they are replaced by new cells produced by the division of specialized cells in the deepest layer of the skin, called stem cells (see the box on p. 176).

Stages of the Cell Division Cycle

What exactly does cell division mean? How do we go from one cell to two cells? The simple answer is that a single cell divides to form two so-called daughter cells. But although the outcome of cell division is simple, the process requires a great deal of cellular preparation. In order for the daughter cells to have the complete set of proteins they need to live and function normally, both must receive the genetic material that contains the blueprint for all of those proteins. In other words, both daughter cells must receive the full complement of DNA in the form of chromosomes that is characteristic of that organism. In addition, the parent cell must be large enough to divide in two and still contribute sufficient cytosol and organelles to each daughter cell.

Both requirements mean that before cell division takes place, key cellular components must be duplicated, including DNA, proteins, and lipids. These preparations are normally accomplished in a series of steps that make up the life cycle of every eukaryotic cell.

The cell division cycle has two major stages

For a single eukaryotic cell, the life cycle both begins and ends with cell division and is therefore called the **cell division cycle**. In its simplest terms, the cell division cycle consists of two major stages that are very differ-

Figure 10.1 Cell Division Replenishes the Skin

Rapid cell division in the deepest layer of the skin is necessary for the replacement of dead cells lost at the surface of the skin. This loss can be due to normal wear or physical damage, such as the sunburn shown here.

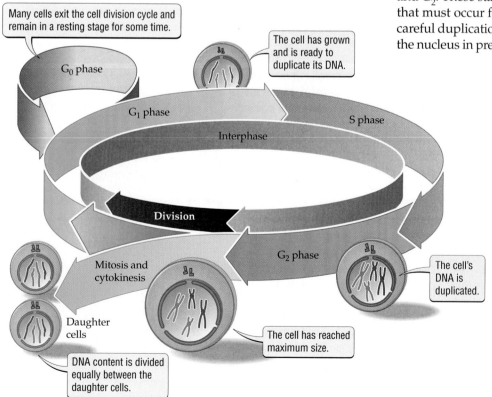

As new cells are produced by the rapid division of stem cells deep in the skin…

…older cells gradually move closer to the exposed surface…

…where the outermost layers of dead skin flake off.

Direction of cell movement

Many cells exit the cell division cycle and remain in a resting stage for some time.

The cell has grown and is ready to duplicate its DNA.

G_0 phase

G_1 phase

S phase

Interphase

Division

Mitosis and cytokinesis

G_2 phase

The cell's DNA is duplicated.

Daughter cells

The cell has reached maximum size.

DNA content is divided equally between the daughter cells.

ent from each other (Figure 10.2). The stage whose events are easily distinguished under the microscope is called **mitosis**. Mitosis ends with the physical division of the parent cell into two daughter cells, and lasts about an hour in most mammalian cells. The stage between two successive mitotic divisions is called **interphase**, and it lasts about 10 to 14 hours in most actively dividing mammalian cells.

It was once thought that the relatively long period of interphase was uneventful for the cell. Today we know that interphase is an active stage during which the cell prepares itself for division.

Interphase prepares the cell for division

Interphase consists of three major stages: S, G_1, and G_2. These stages are defined by a key event that must occur for cell division to proceed: the careful duplication of the entire DNA content of the nucleus in preparation for partitioning to the

Figure 10.2 The Cell Division Cycle

The cell division cycle consists of two major stages (shown by the central circle): division and interphase. The division stage, which is called mitosis, is the period during which the cell divides into two daughter cells. Interphase can be subdivided into three phases, as shown by the outer circle. The cell prepares itself for division by increasing in size during G_1 and G_2 phases and by duplicating its DNA during S phase.

■ THE SCIENTIFIC PROCESS

Stem Cells and the Regeneration of the Human Body

We have seen how the skin must regenerate itself as layer after layer of skin cells is worn away. This need for regeneration is not confined to the skin, as most tissues in the body go through a related process. For example, the food-absorbing cells that line the small intestine live for less than a week and must therefore be continuously replaced with new cells. In other words, cells have a limited life span and must be replaced as they wear out, become less effective at performing their duties, or die.

Depending on the tissue, cells are regenerated in one of two ways. Some tissues, such as the endothelial cells that form the inner walls of blood vessels, regenerate by simple cell division. Other tissues, such as the skin and the lining of the small intestine, depend on a special class of cells called **stem cells**.

The second type of regeneration involves two key features. First, stem cells have the ability to keep dividing for the lifetime of the organism, making them an ideal source of new cells. Second, stem cells are unspecialized, lacking the specific characteristics of any particular tissue. They divide to produce new cells that then become specialized in response to external signals from other cells. This process of becoming specialized is called **differentiation**.

Stem cells are responsible for the production of the specialized cells that circulate in the bloodstream. The bone marrow contains stem cells that divide to produce all the different cell types found in the blood, including the red blood cells that carry oxygen and the T cells that defend the body from foreign microorganisms. The discovery of bone marrow stem cells in 1991 was an important breakthrough for cancer patients undergoing radiation therapy. Bone marrow is easily destroyed by radiation therapy, resulting in fewer blood cells and severe anemia. Transplantation of donated stem cells into cancer patients enables their bodies to replace the lost blood cells.

Other kinds of stem cells have been discovered that can produce nearly any kind of tissue if given the right chemical signals. Transplantation of these stem cells could eventually let us regenerate any damaged or aged organ at will, creating our own fountain of youth. While most would agree that research on these stem cells could end up curing many human diseases, the ethics of such research has divided the scientific community. The reason for the fierce debate is that these stem cells are usually derived from human embryos.

In vitro fertilization clinics routinely discard frozen embryos left over after conception is successful; these embryos have been an essential source of stem cells for research. In 2000, the Clinton administration announced that research using such stem cells was permitted as long as federal monies were not used to harvest the cells from human embryos. This decision distressed anti-abortion advocates, who called for a complete ban on all funding for embryonic stem cell research. On the other hand, many patient groups argued that it would be unethical to deny the possibility of life-saving therapy to the thousands of individuals suffering from degenerative diseases such as Alzheimer's and Parkinson's.

More recently, President George W. Bush announced that federally funded research on embryonic stem cells would be allowed to continue as long as it was restricted to cell lines developed before August 9, 2001. Cell lines are populations of cells that grow and divide in culture. This compromise position has sparked further debate on whether there are sufficient preexisting cell lines to support future research in the field. To further complicate matters, questions remain as to the number and regenerative properties of these cell lines, as well as their intellectual ownership, given that several were developed by biotechnology companies. At the time of this writing, stem cell researchers were still waiting for the National Institutes of Health to release the definitive list of approved embryonic stem cell lines. Once this is done, the effort to determine whether future research in this critical area is truly hindered by these restrictions will begin.

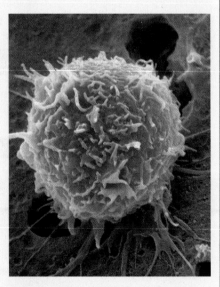

A Stem Cell

daughter cells during mitosis. Since duplication requires the synthesis of new DNA, this stage is called **S phase** (S for "synthesis").

In most cells, S phase does not come immediately before or after mitosis. Two other stages separate mitosis and S phase in the cell division cycle. The first, known as **G_1 phase** (G for "gap"), occurs after mitosis but before S phase begins. The second, **G_2 phase**, occurs between the end of S phase and the start of mitosis (see Figure 10.2). Although their names describe these stages as mere gaps between mitosis and S phase, many essential processes occur during both G_1 and G_2.

The G_1 and G_2 phases are important periods of growth, during which the size of the cell and its protein content increase. Furthermore, during the G_1 phase, particular proteins must be made and activated for S phase to occur. Once the cell is large enough, these proteins promote the production of enzymes that directly synthesize DNA. Thus, G_1 is a time of preparation for S phase. Similarly, G_2 prepares the cell for mitosis, as another set of proteins promotes the cellular events necessary for the physical dividing of the cell. These events include physical changes in the nucleus that are required before the parent cell's DNA can be split equally between the two daughter cells.

The time it takes to complete the cell division cycle depends on the type of cell and the life stage of the organism. Dividing cells in tissues that require frequent replenishing, such as the skin or the lining of the intestine, require about 12 hours to complete the cell division cycle. Most other actively dividing tissues in the human body require about 24 hours to complete the cycle. By contrast, a single-celled eukaryote such as yeast can complete the cycle in only 90 minutes.

Not all of the cells in your body go through the cell division cycle. If they did, it would be difficult to control the size of your body and its organs because many tissues do not require such rapid replenishing of cells. Instead, the cells that make up most tissues pause in the cell division cycle somewhere between mitosis and S phase for periods ranging from days to years. This resting stage is called the **G_0 phase** and is easily distinguished from G_1 by the absence of preparations for DNA synthesis (see Figure 10.2).

Cells in G_0 have exited the cell division cycle. Liver cells remain in this resting state for up to a year before undergoing cell division. Other cells, such as those that form the lens of the eye, remain in G_0 for life, thus forming a nondividing tissue. Most of the nerve cells that make up the brain also exist in this nondividing state, which explains why brain cells lost as a result of physical trauma or chemical damage are usually not replaced.

■ The cell division cycle has two distinct stages: interphase and mitosis. Interphase prepares the cell for division and consists of three stages: S, G_1, and G_2. The cell division cycle ends in mitosis, which culminates in the physical division of the parent cell into two identical daughter cells.

10.2 Mitosis and Cytokinesis: From One Cell to Two Identical Cells

Mitosis is the climax of the cell division cycle. The central event of mitosis is the equal distribution of the parent cell DNA to two daughter nuclei, each of which is destined for a daughter cell. This process, called **DNA segregation**, is a distinctly physical process that requires the coordinated actions of several structural proteins. But before discussing the details of DNA segregation during mitosis, we must consider the earlier preparation and packing of the DNA that occurs in the nucleus.

The DNA in the nucleus is not a random tangle of nucleotide polymers, but rather is highly packed and organized into distinct individual structures. This packing is essential because about 1 to 2 meters of DNA must fit into the average nucleus, which has a diameter of less than 5 micrometers. DNA and proteins are packed together to form thicker and more complex strands; this complex of DNA and proteins is called **chromatin**. Chromatin, in turn, is further looped and packed to form even more complex structures, called **chromosomes** (Figure 10.3). Each chromosome contains a single molecule of DNA that carries a defined set of genes. At the beginning of mitosis, the chromatin is packed and condensed even more densely than usual, and it is at this stage that chromosomes can be seen under the microscope. (We will look at DNA packing in greater detail in Chapter 16.)

Each species has a distinctive karyotype

Every species has a characteristic number of chromosomes in each cell. When chromosomes become visible during mitosis, they adopt characteristic shapes that allow them to be identified under the microscope (Figure 10.4a). The portrait formed by the number and shapes of chromosomes found in a species is known as that species' **karyotype**.

The cells of the human body, with the exception of eggs and sperm, contain 46 chromosomes each. In contrast, cells from horses have 64 chromosomes, and cells

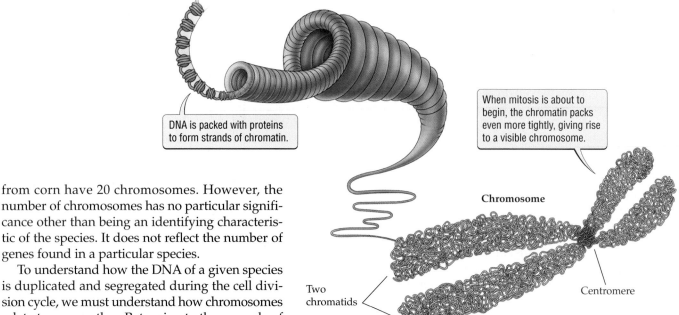

DNA is packed with proteins to form strands of chromatin.

When mitosis is about to begin, the chromatin packs even more tightly, giving rise to a visible chromosome.

Chromosome

Two chromatids

Centromere

Figure 10.3
The Packing of DNA into a Chromosome

from corn have 20 chromosomes. However, the number of chromosomes has no particular significance other than being an identifying characteristic of the species. It does not reflect the number of genes found in a particular species.

To understand how the DNA of a given species is duplicated and segregated during the cell division cycle, we must understand how chromosomes relate to one another. Returning to the example of the human karyotype, our 46 chromosomes can be arranged in 23 pairs, 22 of which are described as homologous pairs (numbered 1 to 22), plus one pair of sex chromosomes (individually lettered X or Y). Each pair of **homologous chromosomes** consists of two chromosomes, one inherited from the mother and the other from the father. The two homologues are alike in

(*a*)

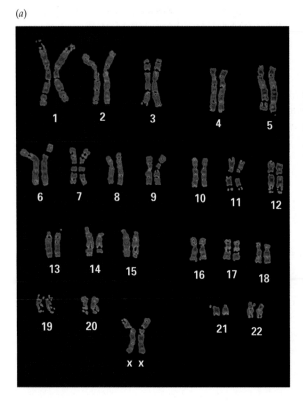

(*b*)

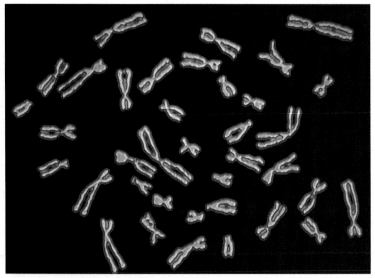

Figure 10.4   The Karyotype of a Human Female
(*a*) In this preparation, the 46 chromosomes have been arranged in homologous pairs and numbered (or lettered, in the case of the sex chromosomes). (*b*) These duplicated chromosomes are shown at the beginning of mitosis. Each pair consists of two sister chromatids attached to each other at a centromere.

terms of length, shape, and the set of genes they carry. On the other hand, the two different types of **sex chromosomes** (X or Y) can form a nonhomologous pair (XY) or a homologous pair (XX). They are called sex chromosomes because they determine the gender of the organism; for example, humans with two X chromosomes are female and those with one X and one Y chromosome are male. (The determination of gender by the sex chromosomes is described in Chapter 13.)

Before cell division can proceed, the DNA of the parent cell must be duplicated so that each daughter cell receives a complete set of chromosomes. This duplication occurs during S phase and produces chromosomes made up of two identical, side-by-side strands called **chromatids**. Thus, at the beginning of mitosis, the nucleus of a human cell contains twice the usual amount of DNA, since each of 46 chromosomes consists of a pair of identical sister chromatids, held together at a constriction called the **centromere** (Figure 10.4*b*). The important roles of chromatids and centromeres in mitosis will become obvious as we describe the stages of mitosis: prophase, prometaphase, metaphase, anaphase, and telophase. Mitosis is usually followed by cytokinesis, which is the process of physically dividing the cell into two daughter cells.

Chromosomes become visible during prophase

Mitosis is divided into five stages, each of which features easily identifiable events that are visible under the light microscope (Figure 10.5). The first stage of mitosis, called **prophase** (*pro*, "before"; *phase*, "appearance"), is characterized by the first appearance of visible chromosomes. In an interphase cell, the chromatin is well dispersed throughout the nucleus, and specific chromosomes cannot be distinguished. As the cell moves from G_2 phase into prophase, the chromatin condenses, and the chromosomes become visible in the nucleus, looking like a tangled ball of spaghetti.

Important changes occur in the cytosol during prophase as well. Two protein structures called **centrosomes** (*centro*, "center"; *some*, "body") begin to move around the nucleus, finally halting at opposite sides in the cell. As will be obvious later, this arrangement of centrosomes defines the opposite ends, or poles, of the cell between which it will eventually separate to form two daughter cells.

At the same time that the centrosomes are moving toward the poles of the cell, microtubules are growing outward from each centrosome. These filaments are the beginnings of a structure called the **mitotic spindle**, which will later guide the movement of the chromosomes.

Chromosomes are attached to the spindle during prometaphase

In the next stage of mitosis, **prometaphase**, the nuclear envelope breaks down (see Figure 10.5). In the process, the mitotic spindle radiating from the centrosomes, which have now reached the poles, extends and enters the region of the cell that was once within the nucleus. The spindle microtubules then attach to the chromosomes at their centromeres, effectively linking each chromosome to both centrosomes.

The physical structure of the centromere dictates how each chromosome will be attached to the spindle microtubules. Each centromere has two plaques of protein, called **kinetochores**, that are oriented on opposite sides of the centromere. Each kinetochore forms a site of attachment for single microtubules, so that the two chromatids that make up a chromosome end up being attached to opposite poles of the cell. This arrangement is essential for the later segregation of DNA.

Chromosomes line up in the middle of the cell during metaphase

Once each chromosome is attached to both poles of the spindle, the lengthening and shortening of its microtubule attachments moves it toward the middle of the cell. There the chromosomes eventually line up in a single plane that is equally distant from both spindle poles. This stage of mitosis is called **metaphase** (*meta*, "after"), and the plane in which chromosomes are arranged is called the metaphase plate (see Figure 10.5). Metaphase is so visually distinctive that its appearance is used as an indicator of actively dividing cells in the examination of tissue under a microscope.

Chromatids separate during anaphase

During the next stage of mitosis, called **anaphase** (*ana*, "up"), DNA segregation takes place. At the beginning of anaphase, the sister chromatids separate (see Figure 10.5). Once separated, each chromatid is now considered to be a chromosome. The gradual shortening of the microtubules pulls the newly separated chromosomes to opposite poles of the cell. This remarkable event results in the equal segregation of chromosomes between the two daughter cells.

In a human cell, each of the 46 chromosomes is duplicated during S phase, yielding 46 pairs of identical sister chromatids (see Figure 10.4*b*). Thus, when the chromatids separate at anaphase, identical sets of 46 chromosomes arrive at each spindle pole, ready to become a complete genetic blueprint for each daughter cell.

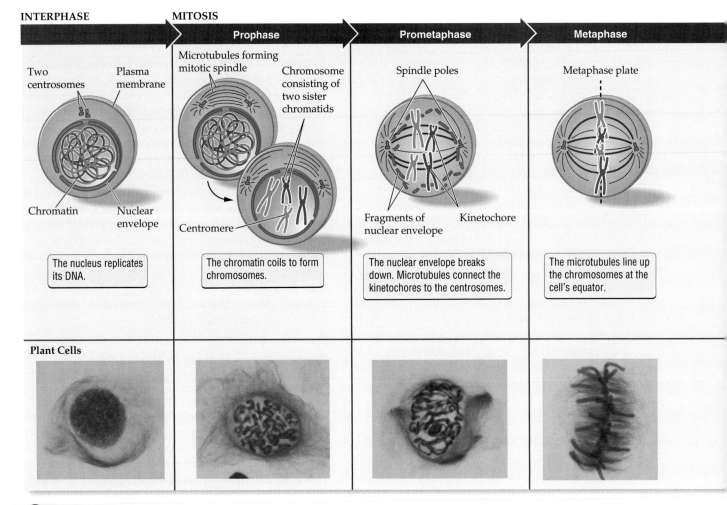

INTERPHASE

MITOSIS

Prophase

Prometaphase

Metaphase

Two centrosomes

Plasma membrane

Chromatin

Nuclear envelope

The nucleus replicates its DNA.

Microtubules forming mitotic spindle

Chromosome consisting of two sister chromatids

Centromere

The chromatin coils to form chromosomes.

Spindle poles

Fragments of nuclear envelope

Kinetochore

The nuclear envelope breaks down. Microtubules connect the kinetochores to the centrosomes.

Metaphase plate

The microtubules line up the chromosomes at the cell's equator.

Plant Cells

Figure 10.5   The Stages of Mitosis
Diagrams of a dividing animal cell (top) and microscopic images of a dividing plant cell (bottom). Note the presence of the cell wall in the plant cell that encloses the two daughter cells after cytokinesis (see also Figure 10.6).

New nuclei form during telophase

Telophase (*telo*, "end") begins when a complete set of chromosomes arrives at a spindle pole. Major cytosolic changes also occur in preparation for division into two new cells. Once the new sets of chromosomes arrive at the opposite poles of the parent cell, the spindle microtubules break down, and nuclear envelopes begin to form around each set of chromosomes (see Figure 10.5). As the two new nuclei become increasingly distinct in the cell, the chromosomes within them start to unfold, becoming less visible under the microscope. This is the last stage of mitosis, and the cell is now ready to physically divide into two daughter cells.

In most plants, additional changes occur during telophase that prepare the plant cell for the physical process of dividing in two (Figure 10.6). Vesicles containing cell wall components accumulate in the region previously occupied by the metaphase plate and begin to fuse with one another. As they fuse and share their contents, a new cell wall begins to form down the middle of the cell.

Cell division occurs during cytokinesis

Cytokinesis (*cyto*, "hollow vessel"; *kinesis*, "movement"), features the division of the parent cell into two daughter cells (see Figure 10.5). In animal cells, the physical act of separation is performed by a ring of actin filaments that lies against the inner face of the plasma membrane like a belt at the equator of the cell. When the actin ring contracts, it pinches the cytoplasm of the cell and divides it in two. Since the plane of constriction by the actin ring lies between the two newly formed nuclei, successful division results in two daughter cells, each with its own nucleus.

In plant cells, the new cell wall that began forming in telophase is completed, dividing the cell into two independent daughter cells (see Figure 10.6). Cytokinesis marks the end of the cell division cycle, and once it is completed, the daughter cells are free to enter G_1 phase and start the process anew.

> ■ DNA in the nucleus is packed with proteins into chromatin, which in turn is packed into specific structures called chromosomes. The number and shape of the chromosomes constitutes a unique karyotype for each species. During mitosis the parent cell DNA is separated such that each daughter cell receives a complete set of chromosomes.

10.4 Meiosis: Halving the Chromosome Number

The remarkable process of mitosis occurs throughout your body during your entire lifetime. However, certain specialized cells in your body undergo a related but significantly different cell division cycle. This cycle, called **meiosis** (*meio*, "less"), produces daughter cells with half the number of chromosomes found in the parent cell. Human sex cells, or **gametes** (sperm in males and eggs in females), are the only cells in the body produced by meiosis. What is it about these cells that requires them

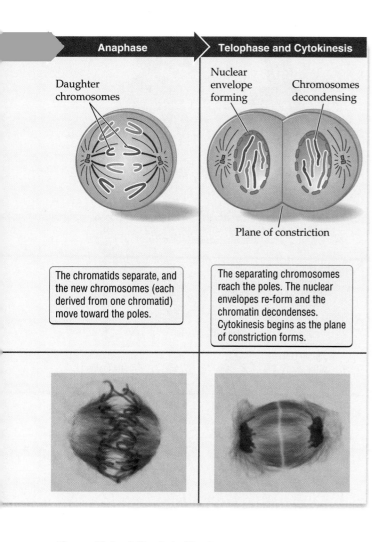

Anaphase	Telophase and Cytokinesis
Daughter chromosomes	Nuclear envelope forming Chromosomes decondensing
	Plane of constriction
The chromatids separate, and the new chromosomes (each derived from one chromatid) move toward the poles.	The separating chromosomes reach the poles. The nuclear envelopes re-form and the chromatin decondenses. Cytokinesis begins as the plane of constriction forms.

Figure 10.6   Mitosis in Plants
The tough cell wall that surrounds the cells of most plants requires a special kind of cytokinesis. Instead of pinching in two, as animal cells do, plant cells divide by building a new cell wall down the middle. Vesicles filled with cell wall components accumulate in the middle of the cell at the start of telophase. As the vesicles fuse, the new cell wall then forms from the vesicles' contents, dividing the original cell into two new daughter cells at cytokinesis.

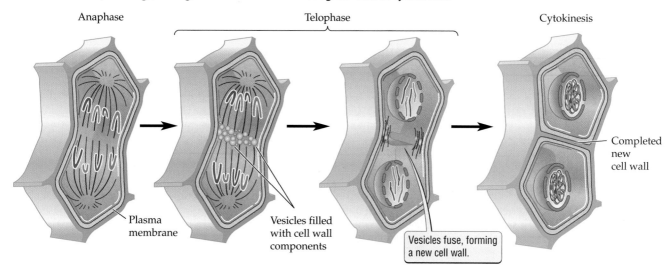

Anaphase   Telophase   Cytokinesis

Plasma membrane

Vesicles filled with cell wall components

Vesicles fuse, forming a new cell wall.

Completed new cell wall

to undergo a different kind of division cycle? The answer lies in the role they play in sexual reproduction.

Gametes contain half the number of chromosomes

The creation of a new organism via sexual reproduction requires the fusion of two gametes in a process known as **fertilization** (see Chapter 36). A successful union of two gametes forms a single cell called a **zygote** (Figure 10.7). The zygote then undergoes multiple rounds of mitosis to form the embryo that will develop into a new organism.

If both the sperm and the egg contained a complete set of chromosomes (46 for humans), the zygote would have double that number (92 chromosomes for humans), and this karyotype would be duplicated and passed on in all the later cell divisions. The resulting embryos would have double the normal number of chromosomes—an abnormality that would be lethal. Therefore, if offspring are to have the same karyotype as their parents, the fusion of gametes must produce the normal number of chromosomes in the zygote after fertilization.

The simple solution to this problem is for the gametes to contain half the number of chromosomes found in other cells (see Figure 10.7). This arrangement is possible thanks to the organization of the karyotype into pairs of homologous chromosomes. In the case of humans, the 22 homologous pairs of chromosomes are found in all body cells. Each gamete a person produces, however, contains only one chromosome from each homologous pair. In addition, depending on the gender of the person, the female gametes (eggs) all contain a single X chromosome, and the male gametes (sperm) contain either an X or a Y chromosome.

Because each gamete contains just one chromosome from each homologous pair plus one sex chromosome, the zygote formed by fertilization will contain a complete set of 22 homologous pairs of chromosomes and one pair of sex chromosomes—that is, a normal human karyotype. Furthermore, each pair of homologous chromosomes in the zygote will consist of one chromosome from the father and one from the mother. This equal con-

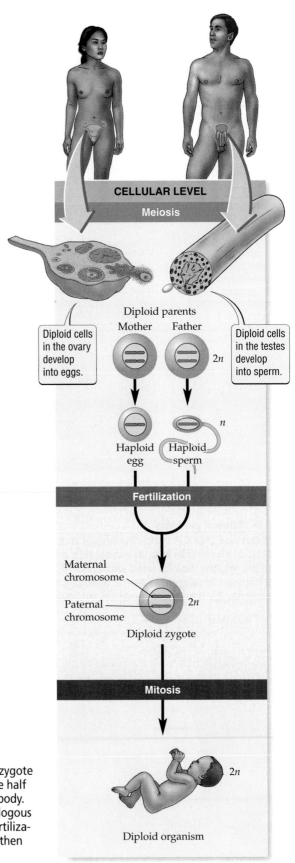

Figure 10.7 Sexual Reproduction Requires a Reduction in Chromosome Number
The fusion of a sperm and an egg at fertilization must produce a zygote with the normal karyotype. Thus the two gametes must each have half the number of chromosomes found in the cells of the rest of the body. Therefore, each gamete receives only one member of each homologous pair of chromosomes, and is said to be haploid (n). Thus, when fertilization occurs, a full homologous pair re-forms in the zygote, which then has a diploid karyotype ($2n$).

tribution of chromosomes by each parent is the basis for genetic inheritance, as we will see in Chapter 13.

Gametes are haploid, while other cells are diploid

Human gametes, which have a chromosome number of 23 (*n*), are said to be **haploid**. The cells that make up the rest of the human body, which have a chromosome number of 46 (2*n*), are said to be **diploid**. The symbol *n* designates the haploid number for a species; 2*n* designates the diploid number.

Since gametes are the only cells that go on to form a new organism, the continued propagation of a species depends on the haploid state. On the other hand, the day-to-day processes that maintain the life of an individual organism depend on the diploid state. So, for the sake of propagating the species, there must be a way of generating haploid gametes from diploid cells. This process is meiosis (Figure 10.8), the specialized cell division cycle in which a single diploid cell ultimately yields four haploid gametes.

The stages of meiosis are very similar to those of mitosis. However, unlike mitosis, in which a single nuclear division is sufficient, meiosis involves two nuclear divisions. These successive divisions are called meiosis I and meiosis II, and each has a distinct role in the important task of producing haploid cells from a diploid parent.

Meiosis I is the reduction division

Let's begin with the diploid cells that form the reproductive tissue responsible for the production of gametes. The first step in producing haploid gametes from these cells is the halving of the chromosome number. This reduction in chromosome number occurs during **meiosis I**. It is achieved by a pairing of homologous chromosomes that is not seen in mitosis. Otherwise, meiosis I has all the same stages as mitosis, and has a similar overall appearance (see Figure 10.8). In fact, the stages of meiosis I have the same names as those of mitosis, except that the Roman numeral I is added to distinguish them from similar stages in meiosis II.

How does chromosome pairing during meiosis I reduce the number of chromosomes? The answer lies in both the preparation for and the mechanics of chromosome segregation. During prophase I, each chromosome—now consisting of two chromatids—pairs with its homologue. In other words, both copies of chromosome 6 pair up, as do both copies of chromosome 3, and so on. These pairs are called **bivalents** and have a total of four chromatids each. The formation of bivalents provides an opportunity for the exchange of genetic sequences between homologous chromosomes (this important process is discussed in Chapter 13).

An important consequence of the formation of bivalents becomes obvious during the later stages of prophase I, when the spindle microtubules extend to meet the chromosomes. In contrast to mitotic prophase, microtubules from only one pole attach to each chromosome of a homologous pair. Thus, when the bivalents move into position at the metaphase plate during metaphase I, the two chromosomes of the bivalent are attached to opposite spindle poles. When the microtubules shorten at anaphase I, they therefore pull apart the homologous chromosomes of each bivalent to opposite sides of the cell. This process is very different from what happens at anaphase of mitosis, in which the individual chromatids of each chromosome are separated to form new chromosomes (see Figure 10.8).

After anaphase I of meiosis, the events of telophase I follow the same patterns seen in mitosis, and cytokinesis produces two daughter cells. Unlike those formed after mitosis, however, the daughter cells of meiosis I contain only half the number of chromosomes found in the parent cell. That is, each daughter cell has only one chromosome from each homologous pair, and is haploid. Meiosis I is therefore called a reduction division because it reduces the chromosome number by half: from diploid (2*n*) to haploid (*n*).

Meiosis II is similar to mitosis

The two haploid cells formed after meiosis I go through a second round of division, called **meiosis II**. This time the stages of the division cycle are more like those of mitosis. In particular, the chromatids separate at anaphase II, leading to an equal segregation of chromosomes into the daughter cells (see Figure 10.8). In this manner, the two haploid cells produced after meiosis I give rise to a total of four haploid cells. These haploid cells are gametes, and they contain the appropriate number of chromosomes such that fertilization will produce a diploid zygote.

The remarkable reduction in chromosome number observed after meiosis I offsets the combining of chromosomes during fertilization and is an elegant way of maintaining the constant chromosome number of a species during sexual reproduction.

> ■ Meiosis produces daughter cells with half the number of chromosomes found in the parent cell. The only cells in the body that are produced by meiosis are gametes. Each gamete contains only one chromosome from each pair of homologous chromosomes and is haploid (*n*). Other cells in the body contain a full karyotype and are diploid (2*n*). Meiosis has two stages: meiosis I (the reduction division) and meiosis II (which is similar to mitosis).

■ HIGHLIGHT

Immune Cell Proliferation and Lyme Disease

Cell division is essential for the replacement of worn-out cells and tissues. Less obvious is the role that it plays in defending the body from foreign invaders such as harmful bacteria, viruses, and fungi. These undesirable guests are called pathogens (*patho*, "suffering"; *gen*, "producing"), in contrast to microorganisms that are helpful to the body, such as beneficial intestinal bacteria.

Figure 10.8   Differences and Similarities between Meiosis and Mitosis
The major difference between meiosis and mitosis can be seen in meiosis I. The homologous chromosomes are paired during prophase I through metaphase I, resulting in a separation of the homologues at the end of meiosis I. Meiosis I is the reduction division. In contrast, meiosis II is more similar to mitosis. For simplicity, not all the stages are shown.

The specialized cells collectively known as the immune system have the job of patrolling the body and defending it against pathogens. (We will examine the immune system in greater detail in Chapter 32.) Whereas human armies react to an attack on their country by recruiting new soldiers, these defenders of the body must undergo cell division to proliferate when they are needed. Thus, cell division is not confined to growth and tissue replacement; it also determines how an organism responds to challenges from the external environment.

Cytotoxic T cells are among the crucial soldiers of the immune system, and they are prime candidates for proliferation when a pathogen attacks. They are called T cells because they originate from the thymus, and they are described as cytotoxic (*cyto*, "hollow vessel"; *toxic*, "poison") because their duty is to kill other cells that have been damaged by pathogens. Their ability to recognize cells that are infected by a virus, for example, is based on the virus proteins that always appear on the surface of infected cells. Once these virus proteins are recognized as foreign, the T cell binds to them and kills

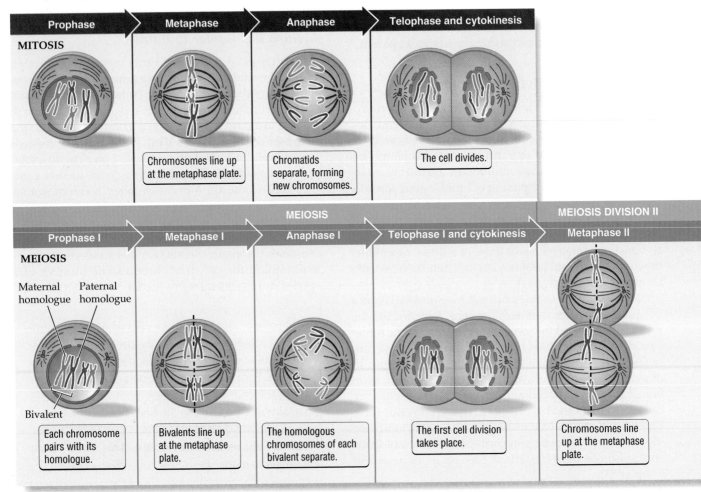

the infected cell. This is how the body rids itself of infected cells and limits the illness caused by the pathogen.

This immune response depends on the body's ability to produce specific cytotoxic T cells that can recognize and destroy cells infected by a particular pathogen. These T cells are tailored to detect certain pathogen components, usually during a previous exposure to the pathogen. After the first exposure, pathogen-specific T cells persist in the body as memory cells. If the pathogen reappears, the memory cells are stimulated to divide rapidly and create a large population of T cells that can both detect and destroy infected cells.

Normally, T cells are able to distinguish pathogen proteins from the body's own proteins, which they do not attack. This ability is an essential characteristic of the immune system, known as self-tolerance. In fact, the body has protective mechanisms for quickly eliminating any T cells that bind to self components, since every army must clearly distinguish friends from foes. However, certain pathogens, such as *Borrelia*, the bacteria that causes Lyme disease, are able to derail this normal process of self-tolerance.

Borrelia burgdorferi

As described at the start of this chapter, the painful arthritis associated with Lyme disease is linked to the fact that the immune system inappropriately attacks the body's joints. But why does the arthritis continue in some patients long after the *Borrelia* bacteria have been killed by antibiotics? This mysteriously persistent arthritis may depend on a *Borrelia* protein that resembles a normal cellular protein. A protein found in the *Borrelia* cell wall has amino acid sequences similar to those found in a human protein that is present on the surface of many human cells.

T cells that recognize the foreign *Borrelia* protein on infected cells are induced to divide in a normal immune response, but those same T cells also recognize the similar human protein on healthy cells. Thus, when T cells that recognize the *Borrelia* protein move into the body's joints to clear the bacterial infection, they also attack and destroy normal cells. The death of more and more cells in the joint causes inflammation and arthritis pain. Even after the bacteria have been eliminated, these T cells remain, and they continue their attack on normal body cells, prolonging the arthritis. This painful consequence of inappropriate cell division reminds us that both the timing and the identities of the cells undergoing division must be carefully controlled in the body. Cancer, another dire consequence of inappropriate cell division, will be discussed in Chapter 11.

■ An effective immune response requires the division of specific T cells that recognize and destroy the pathogen. The similarity between a *Borrelia* protein and a human protein found on healthy cells causes T cells to recognize and destroy both infected cells and healthy cells. The inappropriate destruction of healthy cells in the joints causes persistent arthritis in patients even after *Borrelia* has been killed.

SUMMARY

Stages of the Cell Division Cycle

- Cell proliferation requires the division of a single parent cell into two daughter cells.

- Cells divide in a carefully controlled cell division cycle with two distinct stages: interphase and mitosis.

- Interphase prepares the cell for division. It consists of three phases: S, G_1, and G_2.

- The cell's DNA is duplicated during S phase. The cell increases in size and produces specific proteins needed for division during the G_1 and G_2 phases.

- Mitosis marks the end of the cell division cycle, itself ending with the physical division of the parent cell into two identical daughter cells.

Mitosis and Cytokinesis: From One Cell to Two Identical Cells

- DNA in the nucleus is packed into chromatin with the assistance of proteins. The chromatin is further packed into chromosomes.

- The specific number and shapes of chromosomes found in a species constitute that species' karyotype.

- During mitosis, the parent cell's DNA is separated such that each daughter cell receives a complete karyotype.

Anaphase II	Telophase II and cytokinesis
The second cell division takes place.	Four daughter cells form.

- The five stages of mitosis are prophase, prometaphase, metaphase, anaphase, and telophase.
- The physical division of the parent cell into two daughter cells occurs in cytokinesis.

Meiosis: Halving the Chromosome Number

- Meiosis produces daughter cells with half the number of chromosomes found in the parent cell.
- The only cells in the human body that are produced by meiosis are gametes.
- Gametes are haploid (*n*), containing only one chromosome from each homologous pair. Other cells in the body are diploid (*2n*), containing a complete karyotype.
- Meiosis consists of two divisions. Meiosis I is called the reduction division because it produces haploid daughter cells from a diploid parent cell. Meiosis II is similar to mitosis.

Highlight: Immune Cell Proliferation and Lyme Disease

- An effective immune response requires the proliferation of specific T cells that can recognize and destroy an invading pathogen.
- The bacterium *Borrelia* causes Lyme disease. The similarity between a *Borrelia* protein and a human protein found on healthy cells causes T cells that recognize and destroy both kinds of cells to proliferate.
- The inappropriate destruction of healthy cells in the joints by T cells causes persistent arthritis in patients even after *Borrelia* has been killed.

KEY TERMS

anaphase p. 179	homologous chromosomes p. 178
bivalent p. 183	interphase p. 175
cell division cycle p. 174	karyotype p. 177
centromere p. 179	kinetochore p. 179
centrosome p. 179	meiosis p. 181
chromatid p. 179	meiosis I p. 183
chromatin p. 177	meiosis II p. 183
chromosome p. 177	metaphase p. 179
cytokinesis p. 180	mitosis p. 175
differentiation p. 176	mitotic spindle p. 179
diploid p. 183	prometaphase p. 179
DNA segregation p. 177	prophase p. 179
fertilization p. 182	S phase p. 177
G_0 phase p. 177	sex chromosome p. 179
G_1 phase p. 177	stem cell p. 176
G_2 phase p. 177	telophase p. 180
gamete p. 181	zygote p. 182
haploid p. 183	

CHAPTER REVIEW

Self-Quiz

1. Cell division in the skin is important because
   a. it helps block the harmful effects of ultraviolet light.
   b. the dead cells lost at the surface because of wear and tear must be replaced.
   c. it allows the animal to grow larger.
   d. all cells must divide or die.

2. DNA segregation is an essential feature of mitosis because
   a. each daughter cell must receive a complete set of chromosomes.
   b. it is necessary for the physical separation of the daughter cells.
   c. only one daughter cell should receive the parental DNA.
   d. it ensures that the parent cell is large enough to divide.

3. Which of the following statements is true?
   a. The cell lies dormant during interphase of the cell division cycle.
   b. The key event of S phase is the synthesis of proteins required for mitosis.
   c. The cell increases in size during the G_0 phase.
   d. The cell increases in size during the G_1 and G_2 phases.

4. Which of the following statements is *not* true?
   a. DNA is packed into chromatin with the help of proteins.
   b. Chromosomes are packed into chromatin with the help of proteins.
   c. Chromosomes are visible in the microscope only during mitosis or meiosis.
   d. Each species is characterized by a particular number of chromosomes.

5. Which of the following correctly represents the order of the phases in the cell division cycle?
   a. Mitosis, S phase, G_1 phase, G_2 phase
   b. G_2 phase, G_1 phase, mitosis, S phase
   c. S phase, mitosis, G_2 phase, G_1 phase
   d. G_1 phase, S phase, G_2 phase, mitosis

6. Cytokinesis occurs
   a. at the end of prophase.
   b. just before telophase.
   c. at the end of meiosis I.
   d. at the end of G_1 phase.

Review Questions

1. Horses have a karyotype of 64 chromosomes. How many chromosomes would a horse cell undergoing mitosis have? How many would a horse cell undergoing meiosis II have?

2. What essential role does the mitotic spindle play in mitosis?

3. Spindle microtubules attach to chromosomes in different ways during mitosis and meiosis I. Describe the effect of these differing attachments on the daughter cells produced by mitosis versus those produced by meiosis I.

4. Why is meiosis so important for the production of gametes? How would the offspring of sexual reproduction be affected if gametes were produced by mitosis instead of meiosis?

10   **The Daily Globe**

Spare Parts for Your Brain

To the Editor:

The human brain is a delicate and precious part of the body. But the tendency of adult brain cells not to divide means that any brain cells that die are not replaced. When the brain is damaged, either by physical injury or the effects of aging, there is no way to replace the cells that are lost, so the chances of recovering full brain function are limited.

Today, several research laboratories, including my own, have isolated stem cells that can be induced to divide and form new brain cells if given the right chemical cues. We first discovered these stem cells in the brains of fetal mice, and we have tested their ability to divide and repair brain tissue. These stem cells, when reintroduced into the brains of adult mice, divide and form nerve cells that coexist with the older brain cells of the mice.

This exciting breakthrough will mean that we probably will be able to repair brain damage in the future by administering the appropriate stem cells to the site of an injury. There they will respond to the normal chemical signals of the brain and divide to form new cells that will replace those lost to the injury. Since brain cells also die with advancing age, resulting in reduced cognitive ability, this approach may even allow us to maintain our mental faculties far into old age.

However, we can learn only so much using stem cells isolated from fetal mice. We must study stem cells from human fetal tissue to really understand how brain regeneration might occur in humans. A ban on research using human fetal tissue is therefore a major obstacle to our ultimate goal of limiting the ravages of disease and aging. Thus, I urge your readers to ask Congress not to ban research on human fetal tissue. Instead, Congress should support such research and help us develop the strict ethical guidelines that will allow us to improve human health with a clear conscience.

Yours sincerely,
Dr. Joseph Stemple
Division of Organ Transplantation
and Tissue Repair
New York Medical Institute

Evaluating "The News"

1. If you were to compare the brain cells of a human fetus with those of an adult, which would have more cells in G_0? How does this account for the potential importance of fetal tissue as a source for brain stem cells?

2. Some would argue that our pursuit of a prolonged life span has clouded our ethical judgment. In the hunt for human stem cells, do the possible benefits to our health justify research on tissues from aborted human fetuses?

3. If we do allow research on human fetal tissue, what restrictions do you think should be imposed?

chapter *11* Cancer: Cell Division Out of Control

Eugene Von Bruenchenhein, *Untitled no. 659*, 1957.

Turning a Virus into a Cancer Treatment

Cancer is the ultimate insult to the cooperative functioning of the cells in a multicellular organism. Cancerous tumors are nothing more than stubborn rebel colonies of cells that selfishly ignore the laws of cell-to-cell coordination that keep multicellular organisms alive. The cells that form a tumor divide with wild abandon, often failing to adopt the structures and behaviors required for the particular organ or tissue they are part of. At worst, these aggressive cells break out of the tumor and go on to establish other similarly rebellious colonies in other parts of the body. This is the malignant form of cancer that everyone fears, and it accounts for more than 500,000 deaths in the United States each year. The National Cancer Institute estimates that the collective price tag for the various forms of cancer is $107 billion per year—$37 billion for direct medical costs, $11 billion for lost productivity, and $59 billion for costs due to individual deaths.

Almost 25 years ago, President Richard Nixon declared a war on cancer in the United States by making anticancer research a high priority. Since then, some major victories have been won, thanks to improvements in radiation and drug therapies. Whereas in the early twentieth century very few individuals survived bouts of cancer, today roughly 40 percent of patients are alive 5 years after treatment is begun. Nevertheless, the war against cancer is far from over, and the need for powerful new treatments that can stop tumor growth and eliminate cancerous cells is as urgent as ever.

One of the most inventive potential methods of locating and destroying cancerous cells depends on the assistance of an unlikely ally—a virus. The adenovirus, which infects mammals, takes over the biochemical machinery of specific cells to produce more viruses. In the process, the virus kills the infected cells, often causing respiratory tract disease.

To harness this destructive power for the benefit of individuals with cancer, researchers have mutat-

■ MAIN MESSAGE

Cancer is characterized by cells that have lost their ability to control cell division.

ed the adenovirus so that it successfully infects only cancerous cells. These mutant viruses are unable to multiply, but once they infect a cancerous cell, they are still able to destroy it. In the future, mutant adenoviruses could be administered to tumor sites, where they would infect and kill only the cancerous cells, leaving healthy cells untouched. Although it is still too early to tell whether this kind of therapy will work in human patients, it represents a new potential solution to the problem of selectively eliminating cancerous cells.

This remarkable effort to tame a virus and turn it into an anticancer weapon depends on understanding how viruses take control of the cells they infect. New discoveries in this area have made it clear that many viruses have ways to bypass the normal controls that limit cell division, and have revealed that some viruses play an important role in causing cancer. In this chapter we discuss some of the ways in which cell division is normally controlled, and how the loss of these controls can lead to cancer, before looking at how this knowledge has been used to develop anticancer viruses.

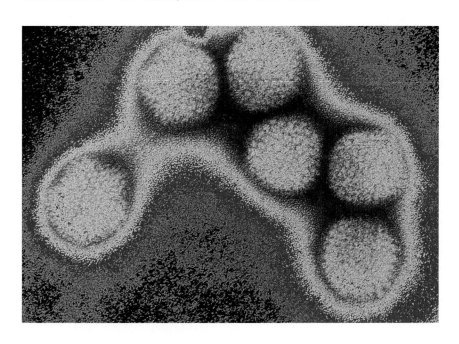

Adenoviruses: Potential Weapons against Cancer?

■ KEY CONCEPTS

1. Cancer is a group of diseases caused by rapid and inappropriate cell division. The resulting cell proliferation can form a tumor from which cancerous cells may spread, invading other tissues in the body.

2. All multicellular organisms have ways of regulating the proliferation of different types of cells. When a cell loses its ability to respond to these regulatory signals, it divides uncontrollably, leading to cancer.

3. Genes that promote cell division in response to normal growth signals are called proto-oncogenes. Overactive mutant versions of these genes can lead to excessive cell division and cancer.

4. Genes that inhibit cell division under normal conditions are called tumor suppressor genes. The complete loss of tumor suppressor activity can also lead to excessive cell division and cancer.

5. When and how often a cell divides depends on the balance between the promoting effects of proto-oncogenes and the inhibiting effects of tumor suppressor genes.

6. Most cancers are caused by a combination of environmental factors. A minority of cancers can be traced to inherited genetic defects.

To achieve a high level of organization, a multicellular organism must have a means of controlling and coordinating the behavior of individual cells. Any large community that does not have rules quickly falls into chaos, and the same is true of a multicellular organism. Therefore, both the metabolic activities and the frequency of division of every cell are closely regulated. As we saw in Chapter 9, signaling molecules called growth factors promote cell division. These positive growth signals work by activating cellular proteins required for the cell division cycle. In effect, such positive growth signals define one set of rules that cells obey by promoting cell division when and where it is needed.

More than a decade ago, cell division was thought to be controlled exclusively by positive signals that promote the cell division cycle. Today we know that whether or not a cell divides is not determined solely by positive signals. The proper functioning of a multicellular organism also depends on negative signaling molecules that can counterbalance positive growth signals and halt the cell division cycle. The life of every cell is therefore managed by a delicate interplay of positive and negative signals, both of which directly affect multiple proteins inside the cell.

Because each multicellular organism is a cooperative community of cells, the failure of just one cell to maintain the delicate balance between opposing positive and negative signals can have serious consequences. One of these consequences is **cancer**: a group of diseases caused by rapid and inappropriate cell division (Figure 11.1). In this chapter we see how the rules that govern cell division are enforced inside each cell, and how the failure of these control systems results in cancer.

Positive Growth Regulators: Promoting Cell Division

Our understanding of how cells respond to signals that promote cell division began with observations of cancer in animals. Perhaps one of the best ways to study the effects of a particular control system is to discover what happens when it is no longer working. This principle was applied in the first decade of the twentieth century by the biologist Peyton Rous, who studied cancerous tumors called sarcomas in chickens.

Rous discovered that he could grind up sarcomas and extract an unidentified substance that, when injected into healthy chickens, caused cancer. He knew that the extract contained no bacteria because it was carefully filtered, so the cause of the cancer had to be something much smaller—something that could pass through the filter. His work led to the discovery of the first animal tumor virus, which was named the Rous sarcoma virus in honor of its discoverer and the type of tumor from which it was obtained (Figure 11.2).

Some viruses can cause cancer

Viruses, as we saw in Chapter 1, are tiny assemblages of either RNA or DNA surrounded by protein. The nucleic acids found inside a virus contain genes that are necessary for the viral life cycle. However, viruses are more than 100 times smaller than the average animal

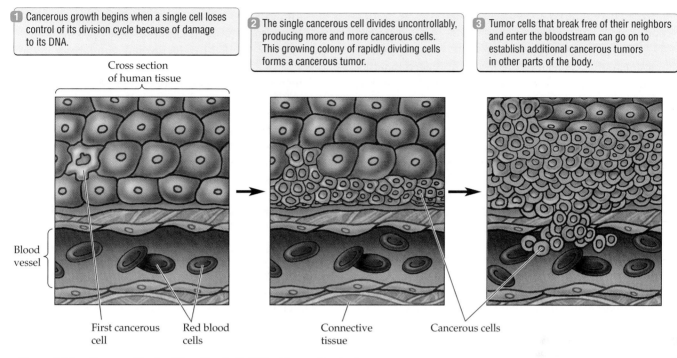

❶ Cancerous growth begins when a single cell loses control of its division cycle because of damage to its DNA.

❷ The single cancerous cell divides uncontrollably, producing more and more cancerous cells. This growing colony of rapidly dividing cells forms a cancerous tumor.

❸ Tumor cells that break free of their neighbors and enter the bloodstream can go on to establish additional cancerous tumors in other parts of the body.

Cross section of human tissue

Blood vessel

First cancerous cell

Red blood cells

Connective tissue

Cancerous cells

Figure 11.1   Cancers Start with a Single Cell That Loses Control

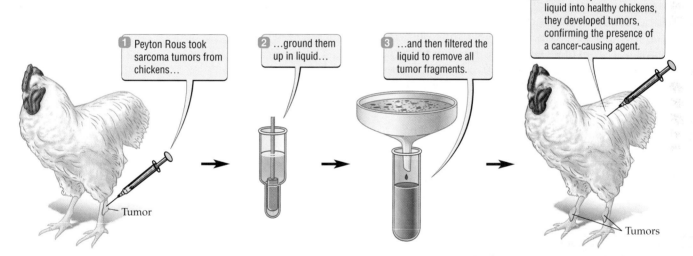

❶ Peyton Rous took sarcoma tumors from chickens…

❷ …ground them up in liquid…

❸ …and then filtered the liquid to remove all tumor fragments.

❹ When he injected the filtered liquid into healthy chickens, they developed tumors, confirming the presence of a cancer-causing agent.

Tumor

Tumors

11.2   The Rous Sarcoma Virus Causes Cancer in Chickens

cell. They are parasites that can multiply only by infecting cells and using the biochemical machinery of the infected cell for their own purposes.

The discovery that a virus could cause cancer in animals was a major breakthrough in our understanding of this type of disease, but it took many more decades before scientists discovered how the Rous sarcoma virus derails the normal controls that regulate cell division. The solution to this mystery came with the discovery of a particular strain of Rous sarcoma virus that could multiply in cells without causing them to divide rapid-

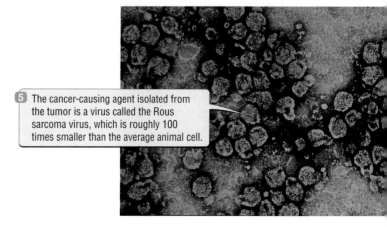

❺ The cancer-causing agent isolated from the tumor is a virus called the Rous sarcoma virus, which is roughly 100 times smaller than the average animal cell.

ly. Biologists could then compare the viral genes found in this virus with those found in cancer-causing strains. The virus that did not cause cancerous cell division was missing a single gene. Further research showed that this viral gene was responsible for destroying the internal controls of the host cell.

If you had to make an educated guess at the identity of the cancer-causing protein produced by the viral gene, what would you expect? As we saw in Chapter 9, growth factors activate many signal cascades inside the cell, which in turn promote cell division. Perhaps the cancer-causing protein somehow affects these signaling events. This is in fact the case, since the cancer-causing agent produced by the viral gene is a very active protein kinase.

Protein kinases activate their target proteins by adding phosphate groups to them, as we saw in Chapter 9. Under normal conditions, these activation events are counterbalanced by other enzymes that remove the phosphates, effectively turning the target proteins off. When an overactive viral protein kinase acts on the same target proteins, however, there is no way to turn the signal cascade off. It is a clear case of too much of a good thing. The avalanche of enzymatic reactions that drive the cell toward mitosis roars out of control, leading to a cell that just keeps dividing. Since the Rous sarcoma virus inserts all of its genes into the DNA of the infected cell, all the daughter cells also receive the cancer-causing gene. Eventually, the growing colony of rapidly dividing cells forms a tumor.

Oncogenes play an important role in cancer development

The overactive protein kinase gene that is found in the Rous sarcoma virus is named *Src* in honor of the virus. (By convention, the names of genes are always given in italic type, and the names of proteins are given in roman type.) *Src* is just one of several **oncogenes** (*onco*, "bulky mass"), or cancer-causing genes, found in viruses. What came as a real surprise was the discovery that the *Src* oncogene is not unique to the Rous sarcoma virus. Instead, it is an altered version of a gene normally found in the genetic material of host cells.

As we saw above, viruses like the Rous sarcoma virus multiply by becoming part of the infected cell's DNA. At some point in evolutionary history, a mutant protein kinase gene may have been picked up from an abnormal host cell and plugged into the genetic material of the virus. A **mutation**, as we will see in Chapter 12, is a change in the DNA sequence of a gene. Mutations can alter the characteristics of the protein produced by the gene, either increasing or decreasing its activity or

its ability to function. In the case of *Src*, the mutation results in the production of an overactive kinase that cannot be controlled like its normal counterpart in the cell.

The realization that the *Src* oncogene has a normal, controllable counterpart in host cells was an important step in identifying the cellular genes that regulate cell division. These normal cellular genes are called **proto-oncogenes** (*proto*, "first") because they are the predecessors of the viral oncogenes. Today scores of proto-oncogenes are known, most of which were first identified as oncogenic mutants in tumor viruses. Although most tumor viruses cause cancer only in animals such as chickens, mice, and cats, all the proto-oncogenes identified in these animal host cells are also found in human cells.

We now know that oncogenes play a major role in human cancers. Instead of being brought into the cell by an infecting virus, most human oncogenes come from mutations of proto-oncogenes caused by environmental factors such as chemical pollutants. In rare cases, oncogenes can also be inherited.

11.1 Proto-oncogenes produce many of the proteins in signal cascades

Not all proto-oncogenes produce protein kinases, as the predecessor of *Src* does. Not surprisingly, nearly any gene that produces a protein used in a cell division signal cascade can be a proto-oncogene (Figure 11.3). In other words, mutant versions of genes for cell surface receptors, kinases, and even signaling proteins such as growth factors can function as oncogenes. Since cell division depends on a complex and interconnected cascade of protein activities, any player in this scheme that misbehaves and overactivates its target can cause uncontrolled cell division.

Let's consider an example of an oncogene that produces a cell surface receptor. Epidermal growth factor (EGF) promotes cell division by binding to a receptor kinase embedded in the plasma membrane of the target cell. As described in Chapter 9, the binding of a growth factor to the external portion of its receptor activates the enzymatic portion of the receptor on the inside of the cell. This enzymatic activity initiates the signal cascade that leads ultimately to cell division.

In one type of blood cell cancer found in chickens, a gene from a tumor virus produces a mutant form of the EGF receptor, called ErbB, that is missing the external portion needed to bind the growth factor. This mutant receptor does not need to bind EGF to become enzymatically active, so it is always turned on (Figure 11.4). Just like the overactive kinase produced by *Src*, the over-

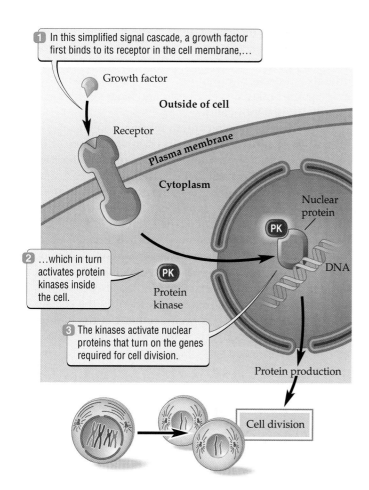

1. In this simplified signal cascade, a growth factor first binds to its receptor in the cell membrane,…

Growth factor

Outside of cell

Receptor

Plasma membrane

Cytoplasm

Nuclear protein

PK

DNA

2. …which in turn activates protein kinases inside the cell.

PK

Protein kinase

3. The kinases activate nuclear proteins that turn on the genes required for cell division.

Protein production

Cell division

Figure 11.3 Proto-Oncogenes Produce Proteins Found throughout the Cell
Most proteins in a signal cascade such as the one shown here can cause cancer if the genes that produce them (proto-oncogenes) are mutated to become oncogenes. The conversion of a proto-oncogene into an oncogene often means that the protein it produces will be overactive, causing the cell to divide uncontrollably.

active EGF receptor turns on too many target proteins over too long a time period, leading to uncontrolled cell division and cancer.

■ Oncogenes (overactive mutant versions of proto-oncogenes) promote excessive cell division, which can lead to cancer. Proto-oncogenes produce many of the protein components of growth factor signal cascades.

Negative Growth Regulators: Inhibiting Cell Division

The previous section might lead one to think that oncogenes are the sole villains responsible for the rampant cell growth that leads to cancer. Although this may be true in some cases, such as with *Src* in chickens, usually something else must also go wrong before cancer can occur. Normal cells have internal safeguards that must be overcome before the controls on cell division are totally removed. These safeguards consist of a family of proteins called **tumor suppressors** because their normal activities were first discovered to stop tumor growth. Tumor suppressor genes are therefore negative growth regulators that stop cells from dividing by opposing the action of proto-oncogenes.

Whether or not a normal cell divides depends on the activities of both proto-oncogenes and tumor suppressor genes. For a cell to divide, proto-oncogenes must be activated

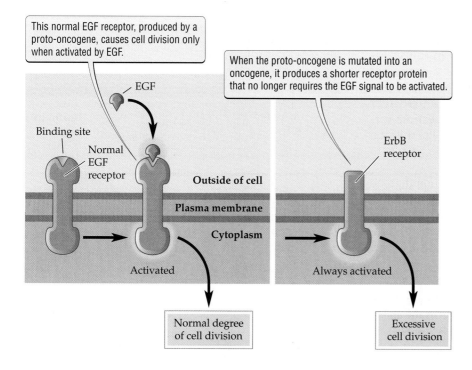

This normal EGF receptor, produced by a proto-oncogene, causes cell division only when activated by EGF.

When the proto-oncogene is mutated into an oncogene, it produces a shorter receptor protein that no longer requires the EGF signal to be activated.

EGF

Binding site

Normal EGF receptor

Outside of cell

Plasma membrane

Cytoplasm

ErbB receptor

Activated

Always activated

Normal degree of cell division

Excessive cell division

Figure 11.4 The *ErbB* Oncogene Causes Cancer by Producing an Altered Receptor Protein
The *ErbB* oncogene produces a receptor protein that can no longer be regulated by EGF and thus is always activated.

to promote the process, and tumor suppressor genes must be inactivated to allow the process to happen. Because controlling the timing and extent of cell division is so important, cells have these two counterbalancing control systems, which must be in agreement before a cell can divide.

How do tumor suppressor genes oppose the activity of proto-oncogenes under normal circumstances? The answer once again lies in the cascade of enzymatic events that lead to cell division.

Tumor suppressors block specific steps of growth factor signal cascades

In the same way that proto-oncogenes induce cell division by activating the components of a growth factor signal cascade, tumor suppressors block cell division by inactivating some of the same components. A well-known example of tumor suppressor activity was discovered during a study of a rare childhood cancer known as retinoblastoma. As the name indicates, retinoblastoma (*retino*, "net"; *blastoma*, "bud") is a cancer that forms in the retina of the eye, and it often leads to blindness (Figure 11.5). Retinoblastoma strikes one in every 15,000 children born in the United States and accounts for about 4 percent of childhood cancers.

A major breakthrough in our understanding of retinoblastoma came with the observation that some patients have visible abnormalities in chromosome 13. This important discovery was made by examining the karyotype of cancerous cells removed from retinoblastoma patients.

As we saw in Chapter 10, every species has a characteristic set of chromosomes, termed the karyotype, that can be seen under the microscope. In cancerous cells from some retinoblastoma patients, a portion of chromosome 13 appeared to be missing, hinting that the cancer might be caused by the absence of a particular gene. Today we know that the gene in question normally produces a protein called Rb. Thus, the missing *Rb* gene, and the resulting lack of Rb protein, results in retinoblastoma.

What can we conclude from the observation that the absence of the *Rb* gene leads to cancer? For one thing, this is not the effect we would expect an oncogene to have. An oncogene causes cancer by producing an overactive pro-

tein that pushes the cell to divide. The *Rb* gene has the opposite effect, since its *absence* promotes cell division. A simple explanation would be that the Rb protein normally inhibits a process required for cell division. When it is missing, the brakes on cell division no longer work, and cells divide uncontrollably.

As predicted, the Rb protein inhibits a key process in the cell's preparations for division. It binds to and inactivates regulatory proteins that are required for the cell's response to growth factor signals. When cells are stimulated to divide by growth factors under normal conditions, the resulting signal cascade involves not just the activation of proto-oncogenes, but also the inactivation of tumor suppressor genes such as *Rb*.

Given what you know about cell signaling, it should not be surprising that the Rb protein is inhibited by a protein kinase, which is activated by the growth factor signal cascade. True to its enzymatic identity, the kinase phosphorylates the Rb protein, causing it to change shape and release its target protein, which can then activate the genes needed for cell division (Figure 11.6). This

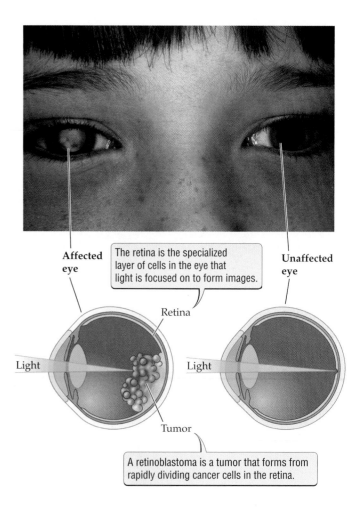

Affected eye

The retina is the specialized layer of cells in the eye that light is focused on to form images.

Unaffected eye

Retina

Light

Light

Tumor

A retinoblastoma is a tumor that forms from rapidly dividing cancer cells in the retina.

Figure 11.5 Retinoblastoma
This child has a visible retinoblastoma in her right eye. A retinoblastoma both blocks the light and destroys the ability of retinal cells to respond to light, frequently resulting in blindness.

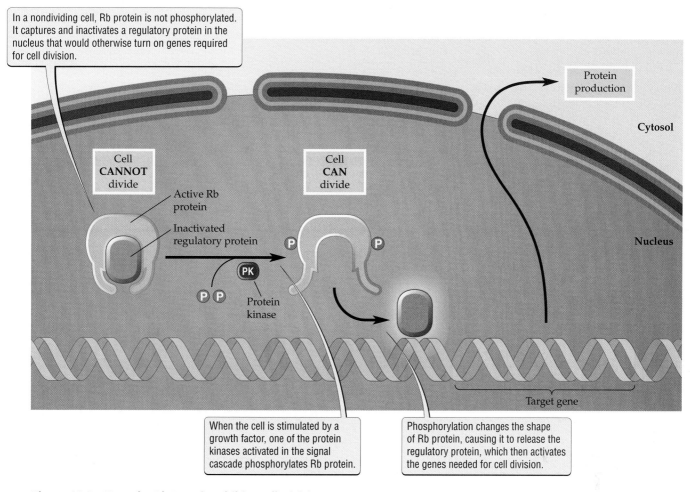

In a nondividing cell, Rb protein is not phosphorylated. It captures and inactivates a regulatory protein in the nucleus that would otherwise turn on genes required for cell division.

Protein production

Cytosol

Cell **CANNOT** divide

Cell **CAN** divide

Active Rb protein

Inactivated regulatory protein

Nucleus

Protein kinase

When the cell is stimulated by a growth factor, one of the protein kinases activated in the signal cascade phosphorylates Rb protein.

Phosphorylation changes the shape of Rb protein, causing it to release the regulatory protein, which then activates the genes needed for cell division.

Target gene

Figure 11.6   How the Rb Protein Inhibits Cell Division

example shows that the phosphorylation resulting from growth factor signals acts to turn on some proteins and turn off others, thus confirming the balance of positive and negative regulatory controls that must come into play before a cell can divide.

Both copies of a tumor suppressor gene must be mutated to cause cancer

The differences in how oncogenes and tumor suppressor genes function highlight differences in the kinds of genetic mutations that can lead to cancer. Since chromosomes exist in pairs, and the two chromosomes in each pair have the same set of genes, there are also two copies of each gene—one contributed by each parent (see Chapter 10). For an oncogene to promote cancer, only one copy of the matching proto-oncogene must be mutated to an oncogenic form. As long as one copy of

the gene is producing an overactive protein, this excessive activity can push the cell to divide.

In contrast, for a tumor suppressor gene to promote cancer, both copies of the gene must be mutated to an inactive form. In other words, if only one copy were inactivated, the other copy might still produce enough tumor suppressor protein to inhibit cell division. Therefore, complete loss of this negative control mechanism—meaning that no tumor suppressor protein is being made—requires that both copies of a tumor suppressor gene be inactivated (Figure 11.7).

■ When cells divide, proto-oncogenes must be activated to promote the process, and tumor suppressor genes must be turned off to allow the process to happen. Whereas overactivation of just one copy of a proto-oncogene can cause inappropriate cell division, both copies of a tumor suppressor gene must be inactivated to have the same effect.

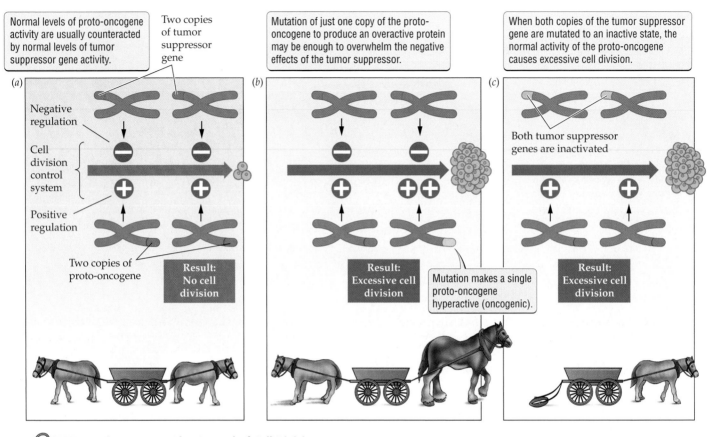

Normal levels of proto-oncogene activity are usually counteracted by normal levels of tumor suppressor gene activity.

Two copies of tumor suppressor gene

Mutation of just one copy of the proto-oncogene to produce an overactive protein may be enough to overwhelm the negative effects of the tumor suppressor.

When both copies of the tumor suppressor gene are mutated to an inactive state, the normal activity of the proto-oncogene causes excessive cell division.

(a)

Negative regulation

Cell division control system

Positive regulation

Two copies of proto-oncogene

Result: No cell division

(b)

Result: Excessive cell division

Mutation makes a single proto-oncogene hyperactive (oncogenic).

(c)

Both tumor suppressor genes are inactivated

Result: Excessive cell division

Figure 11.7 The Control of Cell Division by Proto-oncogenes and Tumor Suppressor Genes
Whether or not a cell divides depends on the balance between proto-oncogene and tumor suppressor gene activity. (a) A normal cell that is not dividing can be compared to a cart attached to two ponies pulling in opposite directions. Because the ponies are of equal size and strength, the cart remains stationary. Likewise, the activities of proto-oncogenes, which promote cell division, are counterbalanced by the activities of tumor suppressor genes, which inhibit cell division, and no cell proliferation occurs. (b) The mutation of one copy of a proto-oncogene to an oncogene is like substituting a workhorse for one of the ponies. The workhorse is larger and stronger than the pony; hence it can pull the cart to one side. In a cell, the result is inappropriate cell division. (c) When both copies of a tumor suppressor gene are inactivated, the result is similar to completely eliminating one pony from the cart: The remaining pony can pull the cart to one side. Again, the result in a cell is inappropriate cell division.

A Series of Chance Events Can Cause Cancer

Most human cancers involve more than the mutation of one proto-oncogene or the complete inactivation of one tumor suppressor gene. Indeed, the complex series of events involved in cell division means that both inherited and environmental factors can come into play. As we'll see in Unit 3, there are a number of factors in our environment that can cause changes in a cell's DNA. Only mutations found in the gametes or the gamete-producing cells, however, can be passed on to offspring. Only about 1 to 5 percent of all cancer cases can be traced exclusively to an inherited genetic defect. The remaining majority of cases involve either a combination of inherited and environmental factors or environmental factors alone.

In some ways this is good news, since many of the environmental factors that cause cancer are related to our lifestyles and behaviors, and we have the power to try to limit our exposure to those factors. To prevent cancer, we must try both to reduce the likelihood of dangerous mutations accumulating in the DNA of our cells and to understand the genetic characteristics that may lead to cancer later in life. In this section we will see just how important it is to lower the likelihood of mutations occurring in our genes.

Cancer is a multi-step process

Cancer is a group of diseases that is likely to affect all our lives eventually (Table 11.1). An American male has a 50 percent chance of developing cancer or dying from

11.1   Selected Human Cancers in the United States

Type of cancer	New cases in 2001[a]	Deaths in 2001[a]
Breast cancer The second leading cause of cancer deaths in women	192,200	40,600
Colon cancer The number of new cases is leveling off as a result of early detection and polyp removal	98,200	48,100
Leukemia Often thought of as a childhood disease, this cancer of white blood cells affects more than 10 times as many adults as children every year	31,500	21,500
Lung cancer Accounts for 28 percent of all cancer deaths and kills more women than breast cancer does	169,500	157,400
Ovarian cancer Accounts for 4 percent of all cancers in women	23,400	13,900
Prostate cancer The second leading cause of cancer deaths in men	198,100	31,500
Skin cancer Melanoma accounts for 80 percent of skin cancer deaths and has been on the rise since 1973	1,000,000	9,800

[a]Estimated numbers courtesy of the National Center for Health Statistics (NCHS) at the Centers for Disease Control and Prevention, with additional information from the American Cancer Society.

it. American women fare slightly better, with a 33 percent chance of developing cancer or dying from it. In the United States, one in four deaths is due to cancer, and more than 8 million Americans alive today have been diagnosed with cancer and are either cured or are undergoing treatment.

Given such a high incidence of cancer, you might think that only one or two mutations are sufficient to cause the disease. However, careful study of human cancers shows that several cellular safeguards have to fail before a cancerous tumor can form. The unlucky string of failures that produces a cancerous tumor includes both the mutation of proto-oncogenes and the loss of tumor suppressor activity.

Consider cancer of the colon (large intestine), which is diagnosed in more than 98,000 individuals each year in the United States. In many cases of colon cancer, tumor cells contain at least one overactive oncogene and several completely inactive tumor suppressor genes. In fact, since the mutations in different genes that lead to colon cancer usually occur over a period of years, the gradual accumulation of these mutations can be linked with the stepwise progression of the cancer.

Let's look at the step-by-step sequence of chance mutations that might lead to colon cancer. In most cases, the first sign of colon cancer is a relatively harmless, or **benign**, growth described as a polyp (Figure 11.8). The cells that make up the polyp are undergoing division at an inappropriate rate. These cells are the descendants of a single cell in the lining of the colon that has suffered one or more mutations.

In many large polyps, the cells contain mutations that inactivate both copies of a tumor suppressor gene, together with a single mutation that transforms a proto-

oncogene into an oncogene. The complete loss of one tumor suppressor's activity, combined with the presence of an overactive protein, is enough to allow inappropriate cell division. However, most such polyps do not spread to other tissues and can be safely removed surgically.

The progression from a benign polyp to a **malignant** tumor—that is, one that can spread throughout the body with life-threatening consequences—depends on the inactivation of additional tumor suppressor genes. In many colon tumors, the start of true malignancy coincides with the loss of a part of chromosome 18 that contains at least two important tumor suppressor genes. This complete loss of two additional tumor suppressors results in a far more aggressive and rapid multiplication of the cancerous cells, greatly increasing the chance that they will spread to other tissues.

One of the last key events in the unlucky slide to full malignancy is the complete inactivation of yet another tumor suppressor gene, named *p53*. For reasons that are not entirely clear, loss of the p53 protein seems to remove all controls on cell division, allowing the cancerous cells to break free of the original tumor and travel through the bloodstream to other parts of the body. At this point the cancerous cells are entirely resistant to both regulatory and defensive signals from the body, and the worst possible scenario—a malignant tumor—has come true.

Cancer is often related to lifestyle choices

The relative contributions of inherited and environmental factors to an individual's cancer risk have been debated for decades. In recent years, large-scale studies have tried to settle this issue by tracking cancer incidence

Figure 11.8   Colon Cancer Is a Multi-step Process
The sequential mutation of several genes that produce positive and negative growth regulators coincides with the progression from a benign polyp to a malignant cancer.

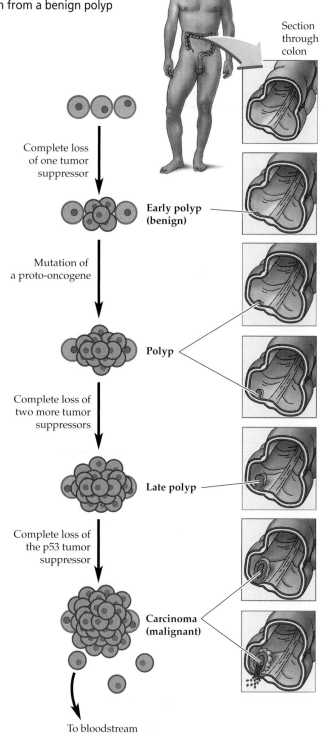

in thousands of pairs of identical twins, who share the same genetic makeup. If inherited genetic defects are more important than environmental factors in causing cancer, then one would expect to see a very similar incidence of cancer in both twins. On the other hand, if environmental factors play a greater role, one would expect to see significant differences in cancer incidence due to differences in the twins' adult environments or habits. In one Scandinavian study that tracked over 44,000 pairs of twins, the most important contributor by far to individual cancer risk was the environment, which included lifestyles and behaviors.

The contribution made by environmental factors to the vast majority of cancers confirms that changes in our personal behavior can reduce our risk. Let's consider one cause of cancer that is amenable to behavioral change: tobacco use.

Since 1982, cigarette smoking has been recognized as the single leading cause of cancer mortality in the United States. This acquired behavior is a major cause of lung and oral cavity cancer, and it contributes to a wide range of other cancers, including those of the kidney, stomach, and bladder. There are close to 50 million American smokers, and it is not surprising that tobacco use accounts for one in five deaths in the United States. Among the thousands of chemical compounds that have been identified in tobacco smoke, over 40 have been confirmed to be carcinogens. A **carcinogen** is a physical, chemical, or biological agent that causes cancer.

Polycyclic aromatic hydrocarbons, or PAHs, are an important class of carcinogens found in tobacco smoke. These organic compounds can bind to DNA, forming a physical complex known as an adduct. Adducts cause mistakes in DNA synthesis (see Chapter 14), which introduce mutations into the DNA sequence. PAHs tend to form adducts at several sites on the *p53* gene in the lung cells of smokers. The resulting mutations in the *p53* gene prevent the production of functional p53 protein in the affected cells. In the same way that the inactivation of *p53* contributes to colon cancer, a similar loss in lung cells allow them to divide uncontrollably, leading to lung cancer. Furthermore, the formation of adducts due to PAHs is not restricted to lung cells. The white blood cells of smokers also show PAH-related genetic damage, which can contribute to other forms of cancer.

The good news is that stopping smoking can dramatically reduce an individual's cancer risk. People who quit smoking before the age of 50 reduce their risk of dying in the subsequent 15 years by half. Regardless of age, people who quit smoking live longer than those who continue to smoke. While nicotine, which is the addictive drug in tobacco, makes quitting smoking difficult, all should find inspiration in the fact that one in five Americans is a former smoker.

> ■ Most cases of cancer involve either a combination of inherited and environmental factors or environmental factors alone. Colon cancer requires mutations that alter the activities of several different genes, resulting in at least one oncogene and several completely inactive tumor suppressor genes. Behaviors such as tobacco smoking increase cancer risk by promoting mutations.

■ HIGHLIGHT

Making the Most of Losing p53

Although the connection between loss of the tumor suppressor protein p53 and cancers of the colon and lungs emphasizes this protein's importance, it only hints at the protein's broad range of different activities. The p53 protein is perhaps most famous for its multiple roles in guarding the integrity of the cell. It not only prevents the cell from dividing at inappropriate times, but also halts cell division when there is evidence of DNA damage that could result in harmful mutations. This protection gives cells the opportunity to repair the damage. If the repair process fails, p53 then goes so far as to induce a cascade of enzymatic reactions that kills the cell. In other words, if the cell's DNA is too badly damaged to repair, the cell commits suicide, rather than passing on mutations that could potentially harm the entire organism.

Given the important guardian functions of the p53 protein, it is not surprising that more than half of all cancers involve a complete loss of p53 activity in tumor cells. The number goes as high as 80 percent in some types of cancer, such as colon cancer. Although the loss of cell division control in each case of cancer involves mutations in a variable roster of several genes, the loss of p53 activity is a factor in most cancers. Thus, the absence of p53 activity might be used as a recognizable characteristic of cancerous cells.

Returning to the proposed cancer therapy introduced at the start of this chapter, the first hint that the adenovirus could be used against cancerous cells came with the discovery that the virus must inactivate the p53 protein in order to multiply. The same mechanism that enables p53 to halt cell division also stops the adenovirus

DNA in an infected cell from being used to make viral proteins. To avoid this defensive measure, an adenovirus gene produces a protein that binds to and disables the p53 protein, thereby allowing the virus to use the cell to make components for new viruses. In other words, the adenovirus can function effectively only in cells that have no active p53 protein. Since p53 also happens to be the tumor suppressor that is most often absent in cancerous cells, an important connection can be forged between the virus and cancerous cells that lack p53 activity.

In a clever turn of events, biologists have mutated the adenovirus such that it no longer produces the protein that disables p53. This mutant virus can multiply only in cells that already lack p53 activity, resulting in the selective killing of these cells. Since the only cells in the body that are likely to lack p53 activity are cancerous cells, this virus works like a smart bomb that seeks out its target and destroys it. Early tests in humans have shown that injection of the mutant virus can reduce the size of some tumors, but only time will tell if this inventive application of our understanding of cancer will help those with the disease.

> ■ Adenoviruses successfully infect cells by inactivating the p53 tumor suppressor protein. Mutant adenoviruses that cannot inactivate p53 can infect only cells that already lack this protein. These mutant adenoviruses can therefore be used to selectively infect and destroy cancerous cells, which usually lack p53 activity.

SUMMARY

Positive Growth Regulators: Promoting Cell Division

- Viruses can multiply only by inserting their genes into a host cell's DNA and using the cell's biochemical machinery.
- The first gene found to promote cell division was isolated from the Rous sarcoma virus. This gene, the *Src* oncogene, causes cancer in chickens.
- Oncogenes (overactive mutant versions of proto-oncogenes) promote excessive cell division, which leads to cancer.
- Proto-oncogenes produce many of the protein components of growth factor signal cascades.

Negative Growth Regulators: Inhibiting Cell Division

- Tumor suppressor genes stop cells from dividing by opposing the action of proto-oncogenes.
- For a cell to divide, proto-oncogenes must be activated to promote the process, and tumor suppressor genes must be inactivated to allow the process to happen.

- The Rb tumor suppressor protein inhibits regulatory proteins required for growth factor signal cascades. A complete lack of Rb activity can cause the eye cancer retinoblastoma.

- Mutation of one copy of a proto-oncogene may be sufficient to cause inappropriate cell division. In contrast, both copies of a tumor suppressor gene must be inactivated to cause inappropriate cell division.

A Series of Chance Events Can Cause Cancer

- Most cases of cancer involve either a combination of inherited and environmental factors or environmental factors alone.

- Cancer usually requires mutations that alter the activities of several different genes. In colon cancer, tumor cells contain at least one oncogene and several completely inactive tumor suppressor genes.

- The inactivation of both copies of the *p53* tumor suppressor gene is a key step in the transition from benign growth to malignant cancer.

- Tobacco smoke contains carcinogens that inactivate the *p53* gene, leading to lung cancer.

Highlight: Making the Most of Losing p53

- The tumor suppressor protein p53 has several roles in guarding the integrity of the cell. It prevents the cell from dividing at inappropriate times, such as when DNA damage could result in harmful mutations.

- The adenovirus must inactivate the p53 protein in order to multiply in cells. Biologists have produced a mutant adenovirus that can no longer disable p53, and thus can multiply only in cells that already lack p53 activity. Since cancerous cells frequently lack p53 activity, the mutant virus multiplies only in those cells, thus selectively killing cancerous cells.

KEY TERMS

benign p. 197	oncogene p. 192
cancer p. 190	*p53* p. 197
carcinogen p. 198	proto-oncogene p. 192
malignant p. 197	tumor suppressor p. 193
mutation p. 192	virus p. 190

CHAPTER REVIEW

Self-Quiz

1. Viruses are different from cells because
   a. they do not contain nucleic acids.
   b. they do not have genes.
   c. they cannot multiply on their own.
   d. they are larger than most cells.

2. The *Src* oncogene
   a. was first discovered in a virus that causes cancer in chickens.
   b. is found in all healthy cells.
   c. inhibits cell division.
   d. can be found only in cancer-causing viruses.

3. Proto-oncogenes
   a. are not related to oncogenes.
   b. can be mutated to become oncogenes.
   c. inhibit signal cascades.
   d. are found only in viruses.

4. Which of the following statements is correct?
   a. Tumor suppressors are able to halt cell division only in cancer cells.
   b. Tumor suppressors inhibit proteins that promote cell division.
   c. Only one type of tumor suppressor is known.
   d. Proto-oncogenes and tumor suppressors have similar effects on the cell.

5. Which of the following is least likely to be produced by a proto-oncogene?
   a. actin cytoskeletal protein
   b. growth factor receptor
   c. protein kinase
   d. growth factor

6. A benign tumor is more likely to become malignant when
   a. the tumor cells no longer respond to growth factor signals.
   b. the rate of cell division slows down.
   c. it reaches a certain size.
   d. several tumor suppressor genes are inactivated by mutations.

Review Questions

1. Describe one way in which p53 prevents the passing on of harmful mutations from one cell to its daughter cells.

2. When Peyton Rous isolated the first cancer-causing virus, why was it so important for him to filter the extract made from the cancerous tumors?

3. On the basis of what you know about how malignant cancers develop, why is it so important to reduce your long-term exposure to chemicals that damage DNA?

4. How does the normal regulation of Rb activity confirm the role of phosphorylation in signal pathways?

11

The Daily Globe

Cell Phones as a Health Risk?

MELBOURNE, AUSTRALIA—While millions of Americans work and socialize using cell phones, a new study on mice suggests that these increasingly popular devices, which many people already cannot live without, may in fact increase a user's chances of developing cancer.

Researchers in Australia reported today in the journal *Medical Research* that mice exposed to high-frequency microwaves like those produced by cell phones developed twice as many cancers as mice that were not exposed to such radiation. Furthermore, the mice that did receive the radiation were exposed to the equivalent of only 1 hour of cell phone use per day.

ATIO, the country's largest cell phone producer, called the new study ridiculous, stating: "There have been and will be no scientific evidence to indicate that cell phones are a hazard to human health."

Dr. Wade Garland, a geneticist at Melbourne Provincial College in Australia, said the new study indicates that the high-frequency microwaves produced by cell phones might be causing cancer by damaging DNA in cells, just as ultraviolet light and X-ray radiation do. Dr. Janice Owen of The University of Boston, however, argued that cell phones probably cannot damage DNA, since the microwaves they produce lack sufficient energy. However, other researchers interviewed argued that these microwaves might make DNA vibrate, making it more prone to damage by other factors, or that microwaves might even alter the physical properties of plasma membranes, thereby changing the behavior of cells.

"There's still more work to do," said Dr. Garland. "But until we know more, I'm not going to use my cell phone again."

Evaluating "The News"

1. Why does DNA damage increase the risk of cancer?

2. Given the unknown risks of extended cell phone use, do you think telecommunications companies should encourage use of a cell phone as your primary means of calling?

3. Even if cell phone use has only a small risk of causing cancer, would this risk make you less likely to make cell phone calls?

4. Some people argue that so many factors in the environment already contribute to cancer that it doesn't matter if cell phone use is added to the list. Given what you know about the multi-step process that leads to cancer, would you agree with this lack of concern?

Paul Giovanopoulos, *Man 2/Man 1*, 1987.

UNIT 3

Genetics

chapter **12** **Patterns of Inheritance**

Jean Dubuffet, *Le Strabique*, 1953.

The Lost Princess

*I*n the early hours of July 17, 1918, the Russian royal family was awakened and taken to the basement of a house in the industrial city of Ekaterinburg. Told they were to be photographed, the Tsarina, Alexandra, and her young son, Alexis, who suffered from hemophilia, were seated in chairs. The rest of the family—Tsar Nicholas II and his four daughters, Olga, Tatiana, Maria, and Anastasia—stood behind Alexandra and Alexis, as did the family physician, cook, maid, and valet. Suddenly, 11 men burst into the room, each with an assigned victim, and began firing with revolvers. In a brutal act that brought to an end the Romanov dynasty of pre-Communist Russia, the seven members of the royal family, and their four servants, were killed.

Or were they? In 1920, a woman was pulled freezing from a Berlin canal. At first known simply as

Grand Duchess Anastasia of Russia, Fourth Daughter of the Tsar

"Fraulein Unbekannt," or "Miss Unknown," and later as Anna Anderson, she claimed that she was the Grand Duchess Anastasia. Her knowledge of minute details of life at the Russian imperial court convinced many that she was indeed Anastasia, the youngest daughter of Nicholas and Alexandra. Others, troubled by her inability to speak Russian and her bouts of erratic behavior, thought she was a pretender. Anna Anderson herself never doubted that she was Anastasia, a conviction she held to her death in 1984, at the age of 83.

Was Anna Anderson really Anastasia? Ultimately, the mystery was solved with a combination of careful detective work and genetic analyses (see p. 217). In the genetic analyses, investigators used basic principles of genetics to determine whether Anna Anderson could have been the lost princess.

What are the rules that govern how characteristics are inherited? How could such information be used to determine whether Anna Anderson was a member of the Russian royal family? To answer these and many other questions about inherited characteristics, we must understand the principles of genetics, the scientific study of genes.

■ KEY CONCEPTS

1. Genetics is the study of genes, which are the basic units of inheritance.

2. Organisms contain two copies of each gene, one inherited from each parent. If the two copies of a gene are identical, the individual is homozygous for that gene. If the two copies of a gene are different, the individual is heterozygous for the gene.

3. Alternative versions of genes are called alleles. Different alleles of a gene cause hereditary differences among organisms.

4. During meiosis, alleles separate equally into sex cells or gametes. With some exceptions (see Chapter 13), the separation of alleles for one gene is independent of the separation of alleles for other genes.

5. The genetic makeup of an organism is its genotype. The phenotype is the observable characteristics of an organism, such as its physical appearance, behavior, and pattern of development.

6. Most aspects of an organism's phenotype are determined by many genes that interact with each other and with the environment.

Genetics is an extremely useful field of biology. For example, the past few years have witnessed an impressive series of discoveries about the genetic basis of human disease. These discoveries have the potential to improve our understanding and treatment of conditions ranging from cancer to Alzheimer's disease to diabetes—diseases that touch all our lives. In addition, genetic principles are used routinely in plant and animal breeding, which lead ultimately to the foods we eat. The use of genetic techniques has also become common in criminal investigations and in paternity lawsuits. Genetic methods have even been used in "biological computers" capable of solving complex computational problems.

Throughout this unit, we will explore the many practical applications of genetics. To do so, however, we will need to understand how genetic traits are passed from one generation to the next. The transmission of genetic traits is the subject of this chapter, which focuses on basic principles of inheritance and some important extensions of those principles. We begin our journey into the study of inheritance with an overview of the field of genetics and a look at some of the concepts that are important in that field.

Genetics: An Overview

Humans have used the principles of inheritance for thousands of years. Knowing that offspring resemble their parents, for example, people allowed only those organisms with desirable traits, such as large grains in wheat, to reproduce. Over time, people used this method to domesticate animals and to develop agricultural crops from wild plant species. As a field of science, however, genetics did not begin until 1866, the year that Gregor Mendel (Figure 12.1) published his landmark paper on inheritance in pea plants. Prior to Mendel's work, many facts about inheritance were known, but no one had organized those facts by describing and testing a hypothesis that could explain how traits are passed from parent to offspring.

Mendel changed all that. His experiments led him to propose that the inherited characteristics of organisms are controlled by hereditary factors—now known as genes—and that one factor for each characteristic is inherited from each parent. Although he did not use the word "gene," Mendel was the first to propose the concept of the gene as the basic unit of inheritance. The emphasis that Mendel placed on genes continues today. In fact, we define **genetics** as the scientific study of genes.

In the more than 100 years since Mendel's work, we have learned a great deal about genes, especially about their physical and chemical properties. We now know that genes are located on chromosomes. Structurally, each gene is a segment of DNA within the long DNA molecule of the chromosome. As we will see in this unit, most genes contain instructions for the synthesis of a single protein product. Finally, most of the trillions of cells in our bodies contain exactly the same set of genes. Each cell contains two copies of every gene, one inherited from each parent (Figure 12.2). The exception to this rule is our sperm and egg cells—our gametes—which have only one of the usual two copies of each gene.

Figure 12.1 Mendel and the Monastery Where He Performed His Experiments
Gregor Mendel was a monk at the monastery shown in this photograph. For many years it was believed that Mendel had performed his experiments behind the fence visible here in front of the monastery. Staff members at a museum devoted to Mendel recently discovered that Mendel's garden actually was located in the foreground of this photograph.

■ Genetics is the study of genes. Inherited characteristics are controlled by genes. Genes are made of DNA and are located on chromosomes. Most genes contain instructions for a single protein product.

Essential Concepts in Genetics

By the end of this chapter, you will know how genetic traits are inherited, and you will be able to use that information to understand how genetic analyses helped solve the mystery of Anna Anderson's identity. You will also

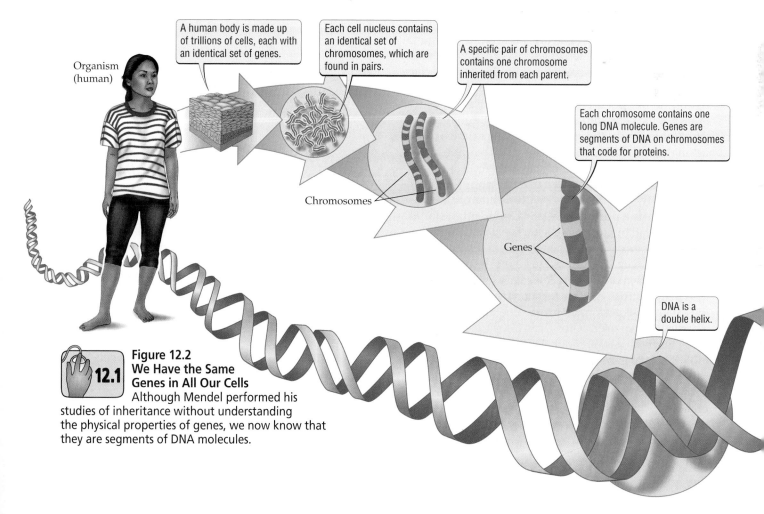

Organism (human)

A human body is made up of trillions of cells, each with an identical set of genes.

Each cell nucleus contains an identical set of chromosomes, which are found in pairs.

A specific pair of chromosomes contains one chromosome inherited from each parent.

Each chromosome contains one long DNA molecule. Genes are segments of DNA on chromosomes that code for proteins.

Chromosomes

Genes

DNA is a double helix.

Figure 12.2 We Have the Same Genes in All Our Cells
Although Mendel performed his studies of inheritance without understanding the physical properties of genes, we now know that they are segments of DNA molecules.

be able to predict the chance that a newborn baby will inherit a certain trait from its parents, such as an inherited disease. To understand the material in this chapter, however, you must first become very familiar with some key genetic concepts, which we introduce here.

Organisms differ in many traits, or **characters**. A character is a feature of an organism such as its height, flower color, or the chemical structure of a protein. Many characters are determined at least in part by **genes**, which are individual units of genetic information for specific characters. The gene is the basic unit of inheritance.

As we saw in Chapter 10, the body cells of most plants and animals are diploid—that is, they contain two copies of each chromosome type. Therefore, they contain two copies of each gene, one inherited from each parent. These two copies are not necessarily identical, however. Such alternative versions of a gene are called **alleles**. The different alleles of a gene are often denoted by uppercase and lowercase letters—for example, _A_ and _a_. No individual has more than two alleles for a particular gene, but among a large number of individuals of the same species a single gene can be represented by many more than two alleles. An individual that carries two copies of the same allele, such as an _AA_ or an _aa_ individual, is called a **homozygote**. An individual that carries one copy of each of two different alleles, such as an _Aa_ individual, is called a **heterozygote**.

The genetic makeup of an organism is called its **genotype**; for example, a heterozygote has genotype _Aa_. The **phenotype** of an organism is its observable physical characteristics. Aspects of an organism's phenotype include its appearance (for example, flower color), behavior (for example, the courtship display of a bird), and biochemistry (for example, the amount of a gene's protein product in the body). Two individuals with the same phenotype can have different genotypes (Figure 12.3).

An allele that determines the phenotype of an organism even when it is paired with a different allele is referred to as a **dominant** allele. Dominant alleles are often denoted by uppercase letters, such as _A_. An allele that does not express its phenotypic effect when paired with a dominant allele is said to be **recessive**. Recessive alleles are often denoted by lowercase letters, such as _a_.

A **genetic cross**, or cross for short, is a controlled mating experiment performed to examine the inheritance of a particular character. "Cross" can also be used as a verb, as in "individuals of genotype _AA_ were crossed with individuals of genotype _aa_." The parent generation of a genetic cross is called the **P generation**. The first generation of offspring in a genetic cross is called the **F$_1$ generation** ("F" is for "filial," a word that refers to a son or daughter). The second generation of a cross is called the **F$_2$ generation**.

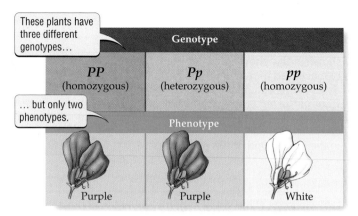

Figure 12.3 Genotype and Phenotype
Flower color in peas is controlled by a gene with two alleles (_P_ and _p_), resulting in three genotypes (_PP_, _Pp_, and _pp_). Although there are three genotypes, there are only two phenotypes (purple flowers and white flowers), since genotypes _PP_ and _Pp_ both produce purple flowers.

Definitions of these important genetics terms are collected in Table 12.1. Study these terms carefully, and refer to them as needed throughout the chapter.

■ The following are some key genetics terms: character, gene, allele, homozygote, heterozygote, genotype, phenotype, dominant, recessive, genetic cross, P generation, F$_1$ generation, F$_2$ generation.

12.2 _Gene Mutations: The Source of New Alleles_

Different alleles of a gene are responsible for hereditary differences among organisms. Although a single individual has at most two different alleles for any given gene, when we examine the genotypes of many individuals, we may find that a particular gene has many different alleles. For example, in human populations, many proteins can be found in three or more forms, each of which is produced by a different allele.

New alleles arise by **mutation**, which we can define briefly here as any change in the DNA that makes up a gene (see Chapter 15 for a more detailed discussion). Many mutations are harmful, and many have little effect on the organism. A few mutations are beneficial. Even when they are beneficial, however, mutations occur at random with respect to their usefulness. There is no evidence that specific mutations occur because they are needed.

12.1   *Key Terms in Genetics*

Term	Definition
Character	A feature of an organism, such as height, flower color, or the chemical structure of a protein.
Gene	An individual unit of genetic information for a specific character. Genes are located on chromosomes and are the basic functional unit of inheritance.
Allele	One of two or more alternative versions of a gene.
Homozygote	An individual that carries two copies of the same allele (for example, an *AA* or *aa* individual).
Heterozygote	An individual that carries one copy of each of two different alleles (for example, an *Aa* individual).
Genotype	The genetic makeup of an organism.
Phenotype	The physical characteristics of an organism.
Dominant allele	An allele that determines the phenotype of an organism even when paired with a different (recessive) allele.
Recessive allele	An allele that does not have a phenotypic effect when paired with a dominant allele.
Genetic cross	Controlled mating experiment, usually performed to examine the inheritance of a particular character.
P generation	Parent generation of a genetic cross.
F_1 generation	First generation of offspring in a genetic cross.
F_2 generation	Second generation of offspring in a genetic cross.

Mutations can happen at any time and in any cell of the body. In multicellular organisms, however, only mutations that occur in the cells that produce gametes, or in the gametes themselves, can be passed on to offspring.

■ In a population of many individuals, a particular gene may have one to many alleles. Different alleles cause hereditary differences among organisms. New alleles arise by mutation, a change to the DNA that makes up the gene. Many mutations are harmful, many have little effect, and a few are beneficial.

12.3   *Basic Patterns of Inheritance*

Now that we've defined some key genetic concepts and discussed how mutations produce new alleles, we are ready to explore how genes are transmitted from parents to offspring. Prior to Mendel, many people argued that the characters of both parents were blended in their offspring, much as paint colors blend when they are mixed together. According to this theory, which was known as the theory of blending inheritance, offspring should be intermediate in phenotype to their two parents, and it should not be possible to recover characters from previous generations. Thus, if a white-flowered plant were mated with a red-flowered plant, the offspring should have pink flowers, and the original flower colors of white and red should not be seen in later generations.

Many observations do not match these predictions, however. The features of offspring often are not intermediate to those of their parents, and it is common for characters to skip a generation (for example, for a child to have blue eyes like one of its grandparents, but unlike its brown-eyed parents). How can such observations be explained? Gregor Mendel answered this question with a series of experiments on pea plants.

Mendel's experiments

During eight years of investigation, Mendel conducted experiments on inheritance in pea plants. His results led him to reject the theory of blending inheritance. Mendel proposed instead that for each character, offspring inherit two separate units of genetic information (genes), one from each parent.

Peas are an excellent organism for studying inheritance. Ordinarily, peas self-fertilize; that is, a given pea plant contains both male and female reproductive organs, and it fertilizes itself. But because peas also can be mated experimentally, Mendel was able to perform carefully controlled genetic crosses. In addition, peas have varieties that breed true for easy-to-measure characters, such as the color and shape of seeds. When a plant of a **true-breeding variety** is self-fertilized, all of its offspring have the same phenotype as the parent. For example, when self-fertilized, a variety that breeds true for yellow seeds produces only offspring with yellow seeds. Mendel used only true-breeding varieties.

In his experiments, Mendel observed inherited characters for three generations. For example, he crossed plants that bred true for purple flowers with plants that bred true for white flowers (Figure 12.4). Mendel then allowed the F₁ plants (the first generation of offspring) to self-fertilize, thereby producing the F₂ generation.

Inherited characters are determined by genes

According to the theory of blending inheritance, the cross shown in Figure 12.4 should have yielded F₁ generation plants bearing flowers of intermediate color. Instead, all the F₁ plants had purple flowers. Furthermore, when the F₁ plants self-fertilized, about 25 percent of the F₂ offspring had white flowers. Thus, the white-flowered character skipped a generation, something that should not happen under blending inheritance.

Mendel studied seven characters in peas, and his results for each of these characters were similar to those shown in Figure 12.4. These results led him to propose a new theory of inheritance. Mendel's theory is referred to as the "particulate" theory of inheritance because it proposes that genes behave like separate units, or particles, not like colors of paints that blend together. Using modern terminology, Mendel's theory can be summarized as follows:

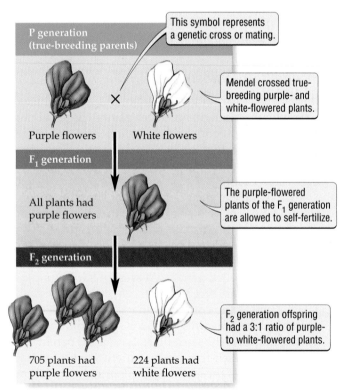

Figure 12.4 Three Generations in One of Mendel's Experiments

[figure labels]

P generation (true-breeding parents)

This symbol represents a genetic cross or mating.

Mendel crossed true-breeding purple- and white-flowered plants.

Purple flowers White flowers

F₁ generation

All plants had purple flowers

The purple-flowered plants of the F₁ generation are allowed to self-fertilize.

F₂ generation

705 plants had purple flowers   224 plants had white flowers

F₂ generation offspring had a 3:1 ratio of purple- to white-flowered plants.

1. *Alternative versions of genes cause variation in inherited characters.* For example, pea plants have one version of a certain gene that causes flowers to be purple, and another version of the same gene that causes flowers to be white. These alternative versions of a gene are known as alleles.

2. *Offspring inherit one copy of a gene from each parent.* In his analysis of crosses like that in Figure 12.4, Mendel reasoned that for white flowers to reappear in the F₂ generation, the F₁ plants must have had two copies of the flower color gene (one copy that caused white flowers and one copy that caused purple flowers). Mendel was right: Most cells in the adult organism contain one maternal and one paternal copy of each of their many genes (see Figure 12.2). The only exceptions to this rule are the gametes—the egg or sperm cells produced by meiosis (see Chapter 10). Gametes contain only one copy of each gene. Two copies of each gene are restored when two gametes, an egg and a sperm, fuse to form a zygote, the first cell of a new offspring.

3. *An allele is dominant if it determines the phenotype of an organism even when paired with a different allele.* Thus, if allele A is dominant over allele a, AA and Aa individuals will have the same phenotype. An allele is recessive if it has no phenotypic effect when paired with a dominant allele. For example, let's call the allele for purple flower color P and the allele for white flower color p. Plants that breed true for purple flowers have two copies of the P allele (i.e., they are of genotype PP), since otherwise they would occasionally produce white flowers. Similarly, plants that breed true for white flowers have two copies of the p allele (genotype pp). Thus, the F₁ plants in Figure 12.4 must have genotype Pp; that is, they must have received a P allele from the PP parent with purple flowers, and a p allele from the pp parent with white flowers. Since all the F₁ plants had purple flowers, the P allele is dominant and the p allele is recessive.

4. *The two copies of a gene separate during meiosis and end up in different gametes.* As already mentioned, each gamete receives only one copy of each gene. If an organism has two identical alleles for a particular character, as Mendel's homozygous true-breeding varieties did, all the gametes will contain that allele. However, if the organism has two different alleles, like an individual of genotype Pp, then 50 percent of the gametes will receive one of the alleles and 50 percent of the gametes will receive the other allele.

5. *Gametes fuse without regard to which allele they carry.* When gametes fuse to form a zygote, they do so

randomly with respect to the alleles they carry for a particular gene. As we'll see, this element of randomness allows us to use a simple method to determine the chance that offspring will have a particular genotype.

> ■ Mendel's experiments on pea plants led him to propose a particulate theory of inheritance, which states that (1) alleles cause variation in inherited characters, (2) offspring inherit one copy of a gene from each parent, (3) alleles can be dominant or recessive, (4) the two copies of a gene separate into different gametes, and (5) gametes fuse without regard to which allele they carry.

Mendel's Laws

Mendel summarized the results of his experiments in two laws: the law of equal segregation and the law of independent assortment. Let's take a look at each of Mendel's laws and how he developed them.

Mendel's first law: Equal segregation

The **law of equal segregation** states that the two copies of a gene separate during meiosis and end up in different gametes. This law can be used to predict how a single character will be inherited. As an illustration, let's revisit the experiment shown in Figure 12.4. In that experiment, Mendel crossed plants that bred true for purple flowers (genotype *PP*) with individuals that bred true for white flowers (genotype *pp*). This cross produced an F_1 generation composed entirely of heterozygotes (individuals with genotype *Pp*). According to the law of equal segregation, when the F_1 plants reproduced, 50 percent of the pollen (sperm) should have contained the *P* allele, and the other 50 percent the *p* allele. The same is true for the eggs.

Reginald Punnett

We can represent the equal separation of alleles by a **Punnett square**, a method first used in 1905 by the British geneticist Reginald Punnett (Figure 12.5). In a Punnett square, all possible male gametes are listed on one side of the square, and all possible female gametes are listed on the perpendicular side of the square. Regardless of whether it has a *P* or a *p* allele, each sperm

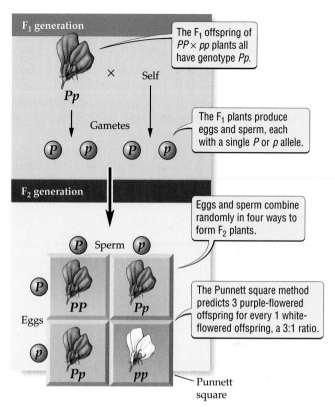

Figure 12.5 The Punnett Square Method
The Punnett square method can be used to represent the equal segregation of alleles and to predict the outcome of a genetic cross.

has an equal chance of fusing with an egg that has a *P* allele or an egg that has a *p* allele. Thus, the four combinations of alleles shown within the Punnett square are all equally likely.

Using the Punnett square method, we can predict that, on average, ¼ of the F_2 generation should have genotype *PP*, ½ should have genotype *Pp*, and ¼ should have genotype *pp*. Because the allele for purple flowers (*P*) is dominant, plants with *PP* or *Pp* genotypes have purple flowers, while *pp* genotypes have white flowers. Thus, we predict that, on average, ¾ (75 percent) of the F_2 generation should have purple flowers and ¼ (25 percent) should have white flowers. This prediction is very close to Mendel's actual results for the F_2 generation: 705 (76 percent) had purple flowers and 224 (24 percent) had white flowers.

Mendel's second law: Independent assortment

Mendel also performed experiments in which he simultaneously tracked the inheritance of two characters. For example, pea seeds can have a round or wrinkled shape, and they can be yellow or green. Two different genes control these aspects of the plant's phenotype. With respect to seed shape, Mendel determined that the allele for round seeds (denoted *R*) was dominant to the allele for wrinkled seeds (*r*). With respect to seed color, he determined that the allele for yellow seeds (*Y*) was dominant to the allele for green seeds (*y*).

What would happen if round, yellow-seeded individuals of genotype *RRYY* were crossed with wrinkled, green-seeded individuals of genotype *rryy*? As might be expected, when Mendel performed this experiment, all of the resulting F_1 plants had genotype *RrYy* and hence round, yellow seeds (Figure 12.6).

But would there be any relationship between the inheritance of seed color and the inheritance of seed shape? To find out, Mendel crossed *RrYy* plants with each other. He obtained the following results in the F_2 generation: approximately $9/16$ of the seeds were round and yellow, $3/16$ were round and green, $3/16$ were wrinkled and yellow, and $1/16$ were wrinkled and green (a 9:3:3:1 ratio). As shown in Figure 12.6, Mendel's results were similar to what should have happened if the alleles for these two characters were inherited independently of each other.

Mendel made similar crosses for various combinations of the seven characters he studied. His results led him to propose the **law of independent assortment**, which states that when gametes form, the separation of alleles for one gene is independent of the separation of alleles for other genes. (There are some exceptions to this law, however; we will find out why in Chapter 13.)

When developing both his first and second laws, it was important that Mendel observed the results of his genetic crosses in large numbers of offspring, for reasons we explore in the box on page 213.

> ■ Mendel proposed two laws of inheritance based on the results of his experiments. The law of equal segregation states that the two copies of a gene separate during meiosis and end up in different gametes. The law of independent assortment states that when gametes form, the separation of alleles for one gene is independent of the separation of alleles for other genes. The Punnett square method considers all the possible combinations of gametes to predict the outcome of a genetic cross.

Hypothesis: Independent inheritance

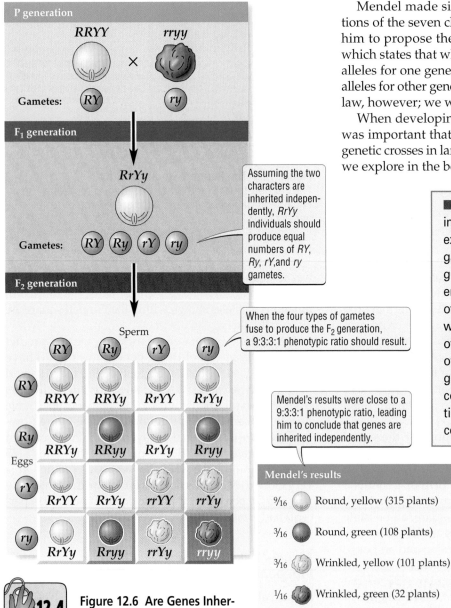

Assuming the two characters are inherited independently, *RrYy* individuals should produce equal numbers of *RY*, *Ry*, *rY*, and *ry* gametes.

When the four types of gametes fuse to produce the F_2 generation, a 9:3:3:1 phenotypic ratio should result.

Mendel's results were close to a 9:3:3:1 phenotypic ratio, leading him to conclude that genes are inherited independently.

Mendel's results

$9/16$ Round, yellow (315 plants)

$3/16$ Round, green (108 plants)

$3/16$ Wrinkled, yellow (101 plants)

$1/16$ Wrinkled, green (32 plants)

12.4

Figure 12.6 Are Genes Inherited Independently?

■ THE SCIENTIFIC PROCESS

12.6 *Tossing Coins and Crossing Plants: Probability and Mendel's Experiments*

Mendel produced many plants in the course of his experiments. That aspect of his work was important, for it allowed him to identify the simple rules that determine the inheritance of flower color and other characters in peas.

The basic principles of probability explain why Mendel had to produce large numbers of plants. The probability of an event is the chance that the event will occur. For example, there is a probability of 0.5 that a fair coin will turn up "heads" when it is tossed. A probability of 0.5 is the same thing as a 50 percent chance.

As an illustration, consider a hypothetical coin-tossing experiment. If a fair coin is tossed only a few times, the observed percentage of heads can differ greatly from 50 percent. For example, if you tossed a coin only 10 times, it would not be unusual to get 70 percent (7) heads. However, if you tossed a coin 10,000 times, it would be very unusual to get 70 percent (7000) heads. If you got such a result, you would (and should) suspect that the coin was not fair after all.

Each toss of a coin is an independent event, in the sense that the outcome of one toss does not affect the outcome of another toss. For a series of independent events, we can esti-

mate the probability of each event from the results. For example, suppose we tossed a coin 10,000 times and got 5046 heads. From these results, it would be reasonable to estimate the chance of getting heads on the next toss as 50 percent, the percentage we expect from a fair coin. When only a small number of events are observed, our estimates of the underlying probabilities are less likely to be accurate, as when we toss a coin only a few times.

How does all this relate to Mendel's experiments? When Mendel crossed heterozygous plants (for example, *Pp* individuals) with each other, the offspring always had a phenotypic ratio close to 3:1. The reason for this consistent ratio is that the chances of getting individuals with *PP*, *Pp*, and *pp* genotypes are 25 percent, 50 percent, and 25 percent, respectively (see Figure 12.5). Because the 25 percent *PP* individuals and the 50 percent *Pp* individuals have the same phenotype, 75 percent of the individuals should look alike, thus giving a 3:1 phenotypic ratio.

The Punnett square method predicts the percentages of *PP*, *Pp*, and *pp* offspring that Mendel should have observed. The method assumes that all sperm cells and all egg cells have

an equal chance of achieving fertilization. In reality, some sperm or egg cells do not achieve fertilization. With large numbers of offspring, however, the assumption that all sperm and egg cells have an equal chance of achieving fertilization is not too far off, because on average, the successes or failures of the different types of gametes tend to balance one another. But if Mendel had used only small numbers of offspring in his experiments, his results probably would have differed greatly from a 3:1 ratio. If that had been the case, he might not have discovered his two fundamental laws: the law of equal segregation and the law of independent assortment.

In general, it is important to understand that the chance of obtaining an offspring with a particular genotype is just that, a chance. If there is a ¼ chance that an offspring will be a homozygous recessive (*pp*) individual, that means that, on average, 25 percent of the offspring will have genotype *pp*. But it does not mean that if four offspring are produced, one will always have genotype *pp*. That may happen, but often, none of the offspring will have genotype *pp*, and in some cases more than one will have genotype *pp*.

Extensions of Mendel's Laws

Mendel's laws describe how genes are passed from parents to offspring. In some cases—such as the seven true-breeding characters of pea plants that Mendel studied—these laws allow accurate prediction of patterns of inheritance. In other cases, however, they do not. To account for these special cases, extensions of Mendel's

laws have been developed. These extensions supplement, rather than invalidate, Mendel's laws. Even when Mendel's laws do not accurately predict observed patterns of inheritance, the genes in question are inherited according to those laws. As we'll see, what differs is how the genes affect the phenotype of the organism, not how the genes are inherited.

Many alleles do not show complete dominance

For dominance to be complete, a single copy of the dominant allele must be enough to produce the maximum phenotypic effect; for example, one *P* allele ensures that even a *Pp* pea plant has purple flowers. But often dominance is not complete. There are many examples of a lack of complete dominance in plants. In snapdragons, for example, when a true-breeding variety with red flowers (*AA*) is crossed with a true-breeding variety with white flowers (*aa*), the heterozygous offspring (*Aa*) have pink flowers. Animals also can show a lack of complete dominance, as in the coat color of horses (Figure 12.7).

The colors of snapdragons and horses illustrate **incomplete dominance**: The heterozygotes are intermediate in phenotype between the two homozygotes. Although incomplete dominance superficially resembles blending inheritance, it is really just an extension of Mendelian inheritance. For example, if two heterozygous snapdragons (*Aa*) are crossed, on average, ¼ of the offspring will have red flowers (genotype *AA*), ½ will have pink flowers (genotype *Aa*), and ¼ will have white flowers (genotype *aa*). Work this out for yourself using the Punnett square method. You will see that Mendel's laws still apply; the main difference is that the heterozygotes (*Aa*) look different from *AA* individuals. Because later generations can return to the original flower colors of red and white—something that cannot occur under blending inheritance—this example shows that incomplete dominance is very different from blending inheritance.

Snapdragon

Alleles for one gene can alter the effects of another gene

A particular phenotype often depends on more than one gene. In such cases the genes are said to interact because the phenotypic effect of each gene depends partly on its own function and partly on the function of other genes. Such gene interactions are common in all types of organisms.

Coat color in mammals provides an example of gene interaction. In mice and many other mammals, for example, a gene that controls the type of pigment produced has a dominant allele (*B*) that produces black fur and a recessive allele (*b*) that produces brown fur. But the effects of the pigment alleles (*B* and *b*) can be eliminated completely, depending on which alleles are present at another gene that interacts with the pigment gene and determines whether any pigment will be produced at all.

To illustrate this point, if a mouse has genotype *cc* at the gene that interacts with the pigment gene, it produces no pigment, regardless of which alleles it has for the pigment gene (Figure 12.8). A lack of pigment causes mice to have white fur. Thus, although we would expect *BB* mice to be black and *bb* mice to be brown, both *BBcc* and *bbcc* mice actually have white fur because they have the *cc* genotype at the gene that interacts with the pigment gene.

The environment can alter the effects of a gene

The effects of many genes depend on environmental conditions, such as temperature, amount of sunlight, or concentration of salt. An allele for coat color in

Figure 12.7 Incomplete Dominance in Horses

Chestnut (sorrel), genotype *CC*

Palominos have genotype *Cc* and are intermediate in color to genotype *CC* and *cc*.

Palomino, genotype *Cc*

Cremello, genotype *cc*

Figure 12.8 Gene Interactions
Gene interactions are very common, as illustrated here by the effects of the *c* allele of a gene that interacts with a pigment gene in mice.

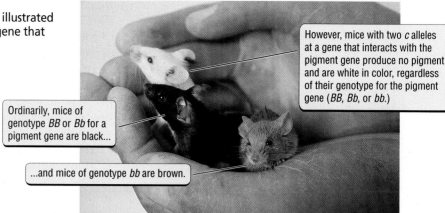

However, mice with two *c* alleles at a gene that interacts with the pigment gene produce no pigment and are white in color, regardless of their genotype for the pigment gene (*BB*, *Bb*, or *bb*.)

Ordinarily, mice of genotype *BB* or *Bb* for a pigment gene are black...

...and mice of genotype *bb* are brown.

Siamese cats (Figure 12.9), for example, is sensitive to temperature. This allele leads to the production of dark pigment only at low temperatures. Because a cat's extremities tend to be colder than the rest of its body, the paws, nose, ears, and tails of Siamese cats tend to be dark. If a patch of light fur is shaved from the body of a Siamese cat and covered with an ice pack, when the fur grows back, it will be dark. Similarly, if dark fur is shaved from the tail and allowed to grow back under warm conditions, it will be light-colored.

Chemicals, nutrition, sunlight, and many other environmental factors also can alter the effects of genes. In plants, for example, genetically identical individuals (clones) grown in different environments often differ greatly in many aspects of their phenotype, including their height and the number of flowers they produce. Thus, plants on a windswept mountainside may be short and have few flowers, while clones of the same plants grown in a warm, protected valley are tall with many flowers.

Most characters are determined by multiple genes

Mendel studied characters that were under simple genetic control: A single gene determined the phenotype for each of the characters he studied. Most characters, however, are determined by the action of more than one gene. For example, skin color, running speed, and body size in humans are controlled by multiple genes, as are height, flowering time, and seed number in plants. Let's look at one of these examples, the inheritance of skin color in humans, in more detail.

A dark pigment, melanin, determines a person's skin color. Many of the differences among people in the amount of melanin in the skin are controlled by three genes. Each of these genes affects skin color equally. The

skin colors that result from these three genes vary considerably, as shown in Figure 12.10. Differences between genotypes are then smoothed over by suntans, causing the skin color of humans to vary nearly continuously from light to dark.

Figure 12.9 The Environment Can Alter the Effects of Genes
Coat color in Siamese cats is controlled by an allele that produces dark pigment (as on the nose, tail, paws, and ears) only at low temperatures.

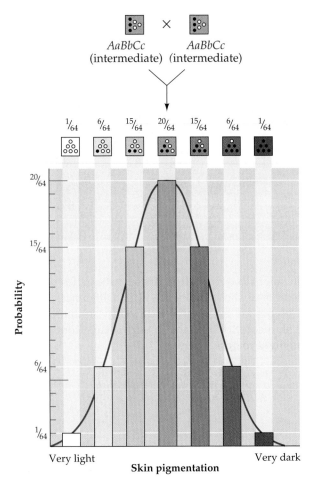

Figure 12.10 Three Genes Produce a Wide Range of Skin Color in Humans
Humans differ with respect to the amount of melanin, a dark pigment, in the skin. For each of the three genes shown here, the allele denoted by a capital letter leads to the production of a relatively large amount of melanin. Thus, individuals of genotype *aabbcc* will tend to have very light skin, while individuals of genotype *AABBCC* tend to have very dark skin. None of the alleles of these three genes is dominant. Alleles that lead to light skin (*a, b, c*) are represented by open circles, while alleles that lead to dark skin (*A, B, C*) are represented by solid circles. The graph illustrates the range of skin color phenotypes that could occur in the children of two parents with genotype *AaBbCc* and, hence, an intermediate phenotype. Additional variation in skin color would result from different levels of tanning.

Putting it all together

Patterns of inheritance are determined by genes that are passed from parent to offspring according to the simple rules summarized in Mendel's laws. Some characters are controlled by one gene and are little affected by environmental conditions. For such characters, such as seed shape and flower color in pea plants, it is possible to pre-

dict the phenotypes of offspring just from knowledge of the alleles the parents have for a single gene.

Many other characters, however, are influenced by sets of genes that interact with one another and with the environment. For such characters, the relationship between genotype and phenotype is more complex. A given gene does not act in isolation; rather, its effect depends on its own function, the function of other genes with which it interacts, and the environment (Figure 12.11).

Many human diseases, for example, including heart disease, cancer, alcoholism, and diabetes, are strongly influenced by multiple genes and by many different environmental factors, such as smoking, diet, and overall mental and physical health. To predict the phenotypes of offspring for such characters requires a detailed understanding of how genes and the environment influence the final product of the genes, the phenotype. Such prediction is a challenging and important task. We could reduce the death rate from heart disease and cancer, for example, if we knew how specific genes interacted with the environment to cause these diseases. We still have much to learn about how such interactions cause heart disease and cancer, but recent developments in genetics (see Chapters 16 through 18) suggest promising leads for future improvements in our ability to predict such disease phenotypes.

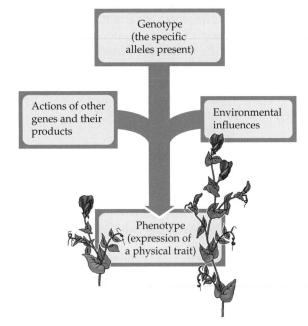

Figure 12.11 From Genotype to Phenotype: The Big Picture
The effect of a gene on an organism's phenotype depends on the gene's own function, the function of other genes with which it interacts, and the environment.

■ Exceptions to Mendel's laws occur for several reasons: (1) many alleles do not show complete dominance; (2) alleles for one gene can alter the effects of another gene; (3) the environment can alter the effects of a gene; (4) most characters are determined by multiple genes. Even when Mendel's laws do not accurately predict observed patterns of inheritance, the genes in question are inherited according to Mendel's laws.

■ HIGHLIGHT

Solving the Mystery of the Lost Princess

Tsar Nicholas II abdicated in 1917, ending over three centuries of rule by the House of Romanov. While alive, Tsar Nicholas and his wife, the Tsarina Alexandra, served as a rallying point for people opposed to the new Communist revolutionary government. Even after death, the royal family was viewed as a threat, and so Communist officials spread misleading information about them. Tsar Nicholas, they said, had been shot, but his family had been moved to a place where they were safe from the turmoil of civil war. The resulting uncertainly over the fate and location of Alexandra and her daughters set the stage for the legend of Anastasia, the lost princess.

The mystery of Anna Anderson was finally solved in 1994. The secret grave of Tsar Nicholas and his family was unearthed in 1989, and the remains were carefully analyzed. Investigators electronically superimposed photographs of the skulls on archive photographs of the family, and they compared skeletal measurements with clothing known to have belonged to the Tsar and his daughters. They also matched the platinum dental work on one skull with the Tsarina's dental records. All these and other tests yielded a match: The skeletons seemed to be those of the Russian royal family. Finally, to make their case airtight, the investigators turned to the ultimate arbitrator, genetic analyses.

When DNA obtained from the skeletons was compared to DNA obtained from relatives of the Russian royal family (including the Tsar's brother, who died in 1899), the results showed conclusively that the skeletal remains were those of the Russian royal family. However, two sets of bones were missing, those of Prince Alexis and one of the two princesses, either Maria or Anastasia. Could it be that Anastasia had escaped and that Anna Anderson was who she claimed to be?

Here, too, DNA analyses provided the answer: The genetic data revealed that the Tsar and Tsarina were not Anna Anderson's parents. For example, for one region

of DNA, five alleles (A, B, C, D, and E, all of which are capitalized since none is dominant) were found among all the persons who were analyzed. The Tsar had genotype *AB*, the Tsarina had genotype *BC*, and Anna Anderson had genotype *DE*. According to Mendel's laws, for Anna Anderson to have been the daughter of the Tsar and Tsarina, and hence the Grand Duchess Anastasia, she should have had one of the following genotypes: *AB*, *AC*, *BB*, or *BC*. Anna Anderson's actual genotype (*DE*) was not consistent with the Tsar and Tsarina being her parents. Three other regions of DNA yielded similar results, indicating that Anna Anderson was not the lost princess.

■ DNA analyses solved a long-standing mystery, revealing that Anna Anderson was not the Russian royal daughter Anastasia.

SUMMARY

Genetics: An Overview

■ Genetics is the scientific study of genes.

■ Genes are made of DNA and are located on chromosomes.

■ Most genes contain instructions for a single protein product.

■ Inherited characters are controlled by genes. One copy of each gene is inherited from each parent.

Essential Concepts in Genetics

■ Key concepts in genetics include the following: character, gene, allele, homozygote, heterozygote, genotype, phenotype, dominant, recessive, genetic cross, P generation, F_1 generation, F_2 generation.

Mutations: The Source of New Alleles

■ Different alleles of a gene cause hereditary differences among organisms.

■ In a population of many individuals, a particular gene may have one to many alleles.

■ New alleles arise by mutation.

■ Mutations occur at random with respect to their usefulness.

■ Many mutations are harmful, many have little effect, and a few are beneficial.

Basic Patterns of Inheritance

■ Mendel's experiments on pea plants led him to propose a particulate theory of inheritance, summarized as follows: (1) Alleles of genes cause variation in inherited characters. (2) Offspring inherit one copy of a gene from each parent. (3) Alleles can be dominant or recessive. (4) The two copies of a gene separate into different gametes. (5) Gametes fuse without regard to which allele they carry.

Mendel's Laws

- Mendel summarized the results of his experiments in two laws: the law of equal segregation and the law of independent assortment.

- The law of equal segregation states that the two copies of a gene end up in different gametes.

- The law of independent assortment states that when gametes form, the segregation of alleles for one gene is independent of the segregation of alleles for other genes.

- The Punnett square method considers all possible combinations of gametes to predict the outcome of a genetic cross.

Extensions of Mendel's Laws

- For some characters, Mendel's laws do not predict patterns of inheritance accurately.

- Reasons for these departures from Mendelian inheritance patterns are as follows: (1) many alleles do not show complete dominance; (2) alleles for one gene can alter the effects of another gene; (3) the environment can alter the effect of a gene; (4) most characters are determined by multiple genes.

- Even when Mendel's laws do not accurately predict observed patterns of inheritance, the genes in question are inherited according to Mendel's laws. What differs is how the genes affect the phenotype, not how they are inherited.

Highlight: Solving the Mystery of the Lost Princess

- DNA analyses solved a long-standing mystery, indicating that Anna Anderson was not the Russian royal princess Anastasia.

KEY TERMS

allele p. 208	incomplete dominance p. 214
character p. 208	
dominant p. 208	law of equal segregation p. 211
F₁ generation p. 208	
F₂ generation p. 208	law of independent assortment p. 212
gene p. 208	mutation p. 208
genetic cross p. 208	P generation p. 208
genetics p. 206	phenotype p. 208
genotype p. 208	Punnett square p. 211
heterozygote p. 208	recessive p. 208
homozygote p. 208	true-breeding variety p. 209

CHAPTER REVIEW

Self-Quiz

1. Alternative versions of a gene for a given character are called
   a. alleles.
   b. heterozygotes.
   c. genotypes.
   d. phenotypes.

2. If an allele for long hair (L) is dominant to an allele for short hair (l), then a cross of $Ll \times ll$ should yield
   a. ¼ short-haired offspring.
   b. ¾ short-haired offspring.
   c. ½ short-haired offspring.
   d. all offspring with intermediate hair length.

3. If A and a are two alleles of the same gene, then individuals of genotype Aa are
   a. homozygous.
   b. heterozygous.
   c. dominant.
   d. recessive.

4. Genes
   a. are the basic units of inheritance.
   b. are located on chromosomes and composed of DNA.
   c. usually contain instructions for a single protein product.
   d. all of the above

5. Coat color in horses shows incomplete dominance. CC individuals have a chestnut color, Cc individuals have a palomino color, and cc individuals have a cremello color (see Figure 12.7). What is the predicted phenotypic ratio of chestnut:palomino:cremello if Cc individuals are crossed with other Cc individuals?
   a. 3:1
   b. 2:1:1
   c. 9:3:1
   d. 1:2:1p

Review Questions

1. For flower color in peas, the allele for purple flowers (P) is dominant to the allele for white flowers (p). A purple-flowered plant, therefore, could be of genotype PP or Pp. What genetic cross could you make to determine the genotype of a purple-flowered plant? Explain how your cross enables you to do this.

2. Many lethal human genetic disorders are caused by a recessive allele, whereas relatively few are caused by a dominant allele. Why might dominant alleles for lethal human diseases be uncommon? (*Hint:* Solve Sample Genetics Problems 4 and 5 below and use the results to guide your answer to this question.)

3. Referring to the box on page 213, explain in your own words why it was important that Mendel observed large numbers of offspring in his genetic crosses.

chapter *13* **Chromosomes and Human Genetics**

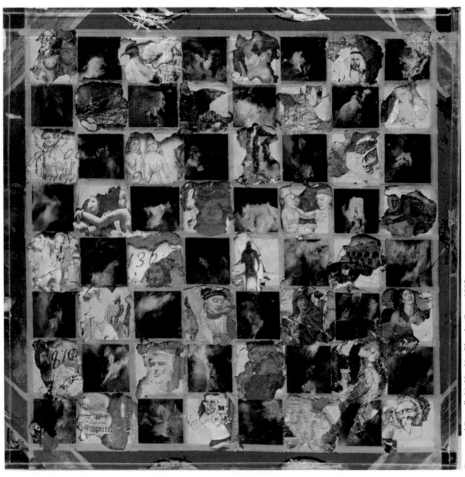

Rosamond Purcell, *It's How You Play the Game*, 1992.

A Horrible Dance

As a child in the mid-1800s, George Huntington went with his father, a medical doctor, as he visited patients in rural Long Island, New York. On one such trip, the young boy and his father saw two women by the roadside, a mother and daughter, who were bowing and twisting uncontrollably, their faces contorted in a series of strange grimaces. Huntington's father paused to speak with them, then left them, continuing on his rounds.

For any child, such an encounter would be unforgettable. For the young Huntington, it was that and more. Like his father and grandfather before him, George Huntington became a doctor, and in 1872 he

The Gene for Huntington's Disease
The gene for Huntington's disease is located at one end of chromosome 4 (shown by the white arrows).

described and named the disease that had plagued the two women. He called it "hereditary chorea" (the word *chorea* comes from the Greek word for "dance").

Huntington described hereditary chorea as an inherited disease, one that destroyed the nervous system and caused jerky, involuntary movements of the body and face. The disease had no cure. Eventually, it killed its victims, but first it reduced them to a quivering wreck of their former selves. It also caused memory loss, severe depression, mood shifts, personality changes, and intellectual deterioration.

If a parent had hereditary chorea, it did not necessarily strike all the children of that parent. The children who did get it usually showed no symptoms until they were in their 30s, 40s, or 50s. Thus, hereditary chorea was like a genetic time bomb. In Huntington's words, the combination of the terrible symptoms of the disease and its late and uncertain onset caused "those in whose veins the seeds of the disease are known to exist [to speak of it] with a kind of horror."

Hereditary chorea is now known as Huntington's disease, in honor of George Huntington. There still is no cure, but researchers have identified a mutation in a gene on chromosome 4 that causes the disease, and they have isolated that gene. With the gene in hand, the quest continues for further understanding of, and ideally, an effective treatment for, Huntington's disease.

■ KEY CONCEPTS

1. A gene is a region of DNA within the DNA of a chromosome. Each gene has a specific location on the chromosome.

2. In humans, males have one X and one Y chromosome, and females have two X chromosomes. A specific gene on the Y chromosome is required for human embryos to develop as males.

3. Unless they are located far from each other, genes on the same chromosome tend to be inherited together, or linked. Genes on different chromosomes are not linked.

4. The homologous chromosomes that pair during meiosis can exchange genes in a process called crossing-over.

5. The genotypes of offspring can be different from that of either parent as a result of crossing-over, the random distribution of maternal and paternal chromosomes into gametes, and fertilization.

6. Many inherited genetic disorders in humans are caused by mutations of single genes. A far smaller number of human genetic disorders are caused by abnormalities in chromosome number or structure.

Humans are afflicted by many types of genetic disorders, with effects that range in severity from mild to deadly. For some of these conditions, such as some forms of breast cancer, an understanding of the genetic basis of the disorder has contributed to effective means of treatment. For others, such as Huntington's disease, the search for successful treatments is still under way.

Although we emphasize human genetic disorders in this chapter, we first continue our discussion of basic genetic principles. In Chapter 12, we described Mendel's discovery that inherited characters are determined by genes. We begin this chapter with a second foundation of modern genetics, the chromosome theory of inheritance. We then explain how gender (sex) is determined in humans and other organisms, and how new combinations of genes different from those of either parent can occur in offspring. This information about chromosomes, gender determination, and new gene combinations provides helpful background material for the discussion of human genetic disorders that follows.

The Role of Chromosomes in Inheritance

When Mendel developed his particulate theory of inheritance (see Chapter 12), he had no idea what the physical properties of his "particles" were. By 1882, studies using microscopes had revealed that threadlike structures—the chromosomes—exist inside of dividing cells. The German biologist August Weismann hypothesized that the number of chromosomes was first reduced by half during the formation of sperm and egg cells, then restored to its full number during fertilization. In 1887, meiosis was discovered (see Chapter 10), thus confirming Weismann's hypothesis. Weismann also suggested

that the hereditary material was located on chromosomes, but no experimental data supported or refuted this idea.

Genes are located on chromosomes

The idea that genes are located on chromosomes is known as the **chromosome theory of inheritance**. Much experimental evidence now supports this theory. In fact, modern genetic techniques allow us to pinpoint where particular genes are located on a chromosome, as in the photograph of the gene for Huntington's disease at the beginning of this chapter.

How are chromosomes, DNA, and genes related? As we learned in Chapter 10, chromosomes are made up of a single DNA molecule and many proteins; the proteins provide structural support for the DNA. Each gene on a chromosome is a region of DNA within the long strand of DNA in the chromosome. The physical location of a gene on a chromosome is called a **locus** (plural loci) (Figure 13.1). Chromosomes that pair during meiosis and that have the same gene loci and structure are called **homologous chromosomes**. One member of each pair is inherited from the mother, and the other is inherited from the father. Thus, at each gene locus, there are two alleles, one inherited from each parent. Each chromosome contains many gene loci, and these loci have a physical relationship to each other: Some genes are located near each other on the chromosome, others far away from each other (see Figure 13.1).

■ Chromosomes are composed of a single DNA molecule and many proteins. A gene is a region of DNA within the DNA of the chromosome. The physical location of a gene on a chromosome is called a locus.

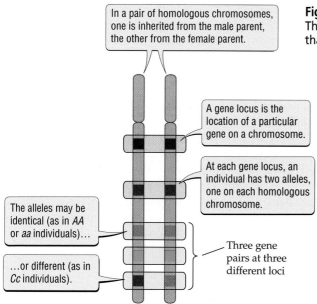

Figure 13.1 Genes Are Located on Chromosomes
The genes shown here take up a larger portion of the chromosome than they would if they were drawn to scale.

In a pair of homologous chromosomes, one is inherited from the male parent, the other from the female parent.

A gene locus is the location of a particular gene on a chromosome.

At each gene locus, an individual has two alleles, one on each homologous chromosome.

The alleles may be identical (as in *AA* or *aa* individuals)…

…or different (as in *Cc* individuals).

Three gene pairs at three different loci

izes the egg, the child will be a girl; if a sperm carrying a Y chromosome fertilizes the egg, the child will be a boy (see Figure 13.2).

Compared with the X chromosome, the Y chromosome has few genes. It does, however, carry one important gene: a gene that functions as a master switch, committing the gender of the developing embryo to "male." In the absence of this gene, a human embryo develops as a female, but when this gene is present, it develops as a male.

Occasionally, XY individuals develop as females, and XX individuals develop as males. In most cases, this occurs because an XY female is missing the por-

Autosomes and Sex Chromosomes

In most chromosome pairs, as we saw in Chapter 10, the maternal and paternal copies are exactly alike in terms of length, shape, and the set of genes they carry. But in humans and many other organisms, this is not true of the chromosomes that determine the gender (sex) of the organism. In humans, for example, males have one X chromosome and one Y chromosome, whereas females have two X chromosomes (Figure 13.2). The Y chromosome in humans is much smaller than the X chromosome, and it does not contain the same set of genes. In other organisms, such as birds, butterflies, and some fish, males have two identical chromosomes, which we denote ZZ, whereas females have one Z chromosome and one W chromosome. Chromosomes that determine gender are called **sex chromosomes**; all other chromosomes are called **autosomes**.

Sex determination in humans

Because human females have two copies of the X chromosome, all the gametes (egg cells) they produce contain one X chromosome. In males, however, half of the gametes (sperm) produced contain an X chromosome and half contain a Y chromosome. The sex chromosome carried by the sperm therefore determines the gender of the child. If a sperm carrying an X chromosome fertil-

Figure 13.2 Sex Determination in Humans
Human females have two X chromosomes, while human males have one X and one Y chromosome.

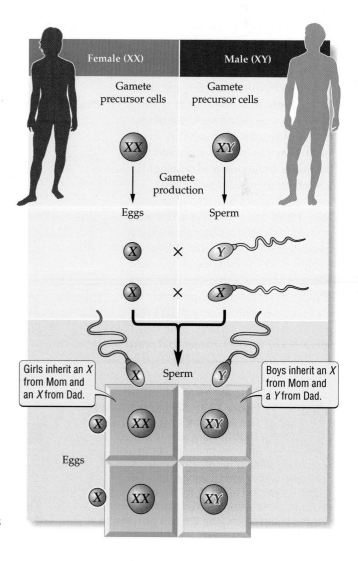

Female (XX) Male (XY)

Gamete precursor cells

Gamete precursor cells

Gamete production

Eggs Sperm

Girls inherit an *X* from Mom and an *X* from Dad.

Boys inherit an *X* from Mom and a *Y* from Dad.

Sperm

Eggs

tion of the Y chromosome that contains the sex-determining gene (and hence cannot develop as a male), or similarly, because an XX male has that portion of the Y chromosome stuck to one of the X chromosomes (and hence develops as a male).

> ■ Chromosomes that determine gender are called sex chromosomes. All other chromosomes are called autosomes. Human males have one X and one Y chromosome. Human females have two X chromosomes. A specific gene on the Y chromosome is required for human embryos to develop as males.

13.1 Linkage and Crossing-Over
Exceptions to the law of independent assortment

As we saw in Chapter 12, the results of Mendel's experiments indicated that genes are inherited independently of one another. These results led him to propose the law of independent assortment, which stated that the alleles for one gene separate into gametes independently of alleles for other genes. Early in the twentieth century, however, results from several laboratories indicated that certain genes were often inherited together, and thus that their behavior contradicted the law of independent assortment.

Thomas Hunt Morgan discovered some genes that were inherited together in his research on fruit flies, which began in 1909 at Columbia University in New York City. In one experiment, Morgan crossed a true-breeding variety of fruit fly that had a gray body (*G*) and wings of normal length (*W*) with a true-breeding variety that had a black body (*g*) and wings that were greatly reduced in length (*w*). That is,

he crossed *GGWW* flies with *ggww* flies to obtain flies of genotype *GgWw* in the F₁ generation. He then mated those *GgWw* flies with *ggww* flies, as Figure 13.3 shows. Morgan's results were very different from the results he expected based on the law of independent assortment. What had happened?

Thomas Hunt Morgan

Morgan concluded that the genes for body color and wing length must be located on the same chromosome. Because they were on the same chromosome, they did not assort independently during meiosis; instead, they were inherited together. Genes that are located on the same chromosome and that do not assort independently are said to be **genet-**

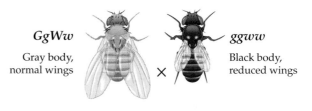

GgWw
Gray body, normal wings

×

ggww
Black body, reduced wings

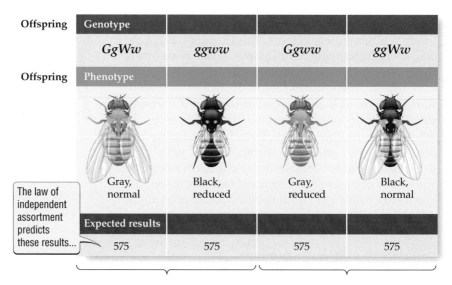

Offspring	Genotype			
	GgWw	*ggww*	*Ggww*	*ggWw*
Offspring	Phenotype			
	Gray, normal	Black, reduced	Gray, reduced	Black, normal
Expected results				
	575	575	575	575

The law of independent assortment predicts these results...

...but these results were observed.

Parental phenotypes | Nonparental phenotypes

Actual results			
965	944	206	185

> Conclusion: These two genes do not assort independently. They are linked together on the same chromosome.

Figure 13.3 Some Alleles Do Not Assort Independently Thomas Hunt Morgan found that the gene for body color (dominant allele *G* for gray, recessive allele *g* for black) was linked to the gene for wing length (dominant allele *W* for normal length, recessive allele *w* for reduced length). This linkage occurred because the two genes were on the same chromosome.

ically linked; as discussed at the end of this section, some genes located far from one another on a chromosome are not linked. Genes located on different chromosomes are not genetically linked.

Because organisms have many more genes than chromosomes, many of their genes are inherited together, or linked. For example, humans have an estimated 35,000 genes, which are located on 23 pairs of homologous chromosomes. Thus, on average, our 23 chromosomes have 35,000/23, or 1521, genes per chromosome.

Crossing-over disrupts genetic linkage

If the linkage between two genes on a chromosome was complete, all offspring would be of a parental type; that is, all of them would have a genotype that matches that of one of their parents. For example, in the cross shown in Figure 13.3, if the same genes were completely linked, half of the offspring would have had genotype *GgWw*, and the other half would have had genotype *ggww*. Since many of the offspring did have these two parental genotypes, the two genes clearly were linked. But how can the appearance of some nonparental—*Ggww* and *ggWw*—genotypes be explained?

To explain the appearance of these genotypes, Morgan suggested that genes are physically exchanged between homologous chromosomes during meiosis. This exchange of genes is called **crossing-over**. To make this concept more concrete, imagine that the two chromosomes illustrated in Figure 13.4 come from one of your cells. You inherited one of these chromosomes from your father, the other from your mother. In crossing-

over, part of the chromosome inherited from one parent may be exchanged with the corresponding DNA inherited from the other parent (see Figure 13.4). By physically exchanging pieces of homologous chromosomes , crossing-over combines alleles inherited from one parent with those inherited from the other. This exchange makes possible the formation of gametes with nonparental genotypes, which in turn can lead to the formation of offspring with nonparental genotypes, such as the *Ggww* and *ggWw* offspring shown in Figure 13.3.

Crossing-over can be compared to the cutting of a string. Two points that are far apart on the string will be separated from each other in most cuts of the string, whereas points that are very close to each other will rarely be separated. Similarly, genes that are far from each other on a chromosome are more likely to be separated by crossing-over than are genes that are close to each other. In fact, two genes on the same chromosome that are very far from each other may be separated by crossing-over so often that they are not linked. Such genes assort independently even though they are located on the same chromosome. Among the traits that Mendel studied in pea plants, we now know that the genes for flower color and seed color are on the same chromosome, but are so far apart that they are not linked. Thus, the law of independent assortment holds for these genes.

> ■ Genes that are located on the same chromosome and that do not assort independently are genetically linked. Crossing-over, the exchange of genes between chromosomes, disrupts the linkage between genes.

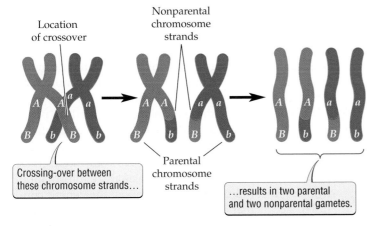

Figure 13.4   Crossing-Over Disrupts the Linkage between Genes
In the case shown here, crossing-over occurs at a point on the chromosome between two linked genes. As a result, half of the gametes have a parental genotype (either *AB* or *ab*), and the other half have a nonparental genotype (either *Ab* or *aB*).

13.3 Origins of Genetic Differences among Individuals

Inheritance is both a stable and a variable process. It is stable in that genetic information is transmitted accurately from one generation to the next. Despite this stability, offspring are never exact replicas of their parents, even in species, such as bacteria and dandelions, that reproduce asexually. In organisms that reproduce sexually, one reason offspring differ from their parents and from one another (and from other individuals of the same species) is that different offspring inherit different combinations of alleles. Genetic differences among individuals can affect whether a person has a genetic disorder and, in some cases, how severe that disorder is. Furthermore, as we will see in Unit 4, these different combinations of alleles produce genetic differences among individuals on which evolution can act.

How do genetic differences among individuals arise? First, as we discussed in Chapter 12 (see p. 208), new alleles can be formed by mutation. Once formed, those alleles are shuffled or arranged in new ways by crossing-over, independent assortment of chromosomes, and fertilization.

Let's examine how crossing-over typically causes a group of offspring to have a wide range of different genotypes. Every time meiosis occurs, crossing-over produces some "new" chromosomes. These chromosomes are new in the sense that they contain some alleles inherited from one parent and other alleles inherited from the other parent. By exchanging alleles between chromosomes, crossing-over causes some offspring to have a nonparental genotype, that is, a genotype that differs from the genotype of either parent (see Figures 13.3 and 13.4).

During meiosis, the maternal and paternal chromosomes are distributed randomly into gametes. This **independent assortment of chromosomes** also produces new combinations of alleles. Independent assortment happens because the orientation of the maternal and paternal chromosomes varies at random when the chromosomes line up at the metaphase plate during meiosis (see Chapter 10), and hence the chromosomes are shuffled into new combinations in the gametes. This process has great potential for producing new combinations of alleles. In humans, for example, the 23 pairs of homologous chromosomes can be arranged on the spindle fibers in 2^{23}, or 8,388,608, different ways. Of these 8,388,608 ways of arranging the chromosomes, only two are the combination originally inherited from the parents. Thus, like crossing-over, the independent assortment of chromosomes can cause the formation of nonparental genotypes.

Finally, fertilization has the potential to add a tremendous amount of genetic variation to that already produced by crossing-over and the independent assortment of chromosomes. In the previous paragraph we saw that each particular gamete represents one of over 8 million different gametes that could have been formed in meiosis. Thus, each sperm cell represents one of over 8 million sperm that could have been formed; similarly, each egg represents one of over 8 million possible egg cells. As a result, even without considering the variation caused by crossing-over, there are over 64 trillion (8 million possible sperm multiplied by 8 million possible eggs) different offspring that could be formed each time a human sperm fertilizes a human egg. Thus, when fertilization occurs and a particular sperm unites with a particular egg, a truly unique individual with a unique set of alleles is formed.

> ■ The independent assortment of chromosomes is the random distribution of maternal and paternal chromosomes into gametes during meiosis. Crossing-over, the independent assortment of chromosomes, and fertilization cause offspring to differ genetically from one another and from their parents. Such genetic differences underlie human genetic disorders and provide the genetic variation on which evolution can act.

Human Genetic Disorders

The topics covered thus far in this chapter have included the chromosome theory of inheritance, how gender is determined in humans and other organisms, linkage and crossing-over, and the independent assortment of chromosomes. Let's see how this information is being applied to the study of genetic disorders in humans. The application of genetic principles to human genetic disorders has been a powerful and useful approach, one that has led to greater understanding and improved treatment of these disorders.

Many of us know someone who has suffered from a genetic disorder, such as a hereditary form of cancer, heart disease, or one of the many other diseases caused by gene mutations. Because human genetic disorders are so widespread, it is important to study them, since such studies could lead to the prevention or cure of much human suffering. But the study of human genetic disorders faces daunting problems. Unlike fruit flies and the other organisms often studied by geneticists, we humans have a long generation time, we select our own mates, and we decide when and whether to have children. Understanding the inheritance of human genetic disorders also can be difficult because our families are small (see the box in Chapter 12 on p. 213).

One approach to these problems is to analyze pedigrees. A **pedigree** is a chart that shows genetic relationships among family members over two or more generations of a family's history (Figure 13.5). Pedigrees provide geneticists with a way to analyze information from many families so as to learn about the genetic control of a particular disease. The pedigree shown in Figure 13.5, for example, shows the inheritance of a lung disease called cystic fibrosis, which usually causes death before the age of 30. Individuals 2 and 3 in generation III had cystic fibrosis, but their parents did not. The pedigree in Figure 13.5 indicates that the allele that causes

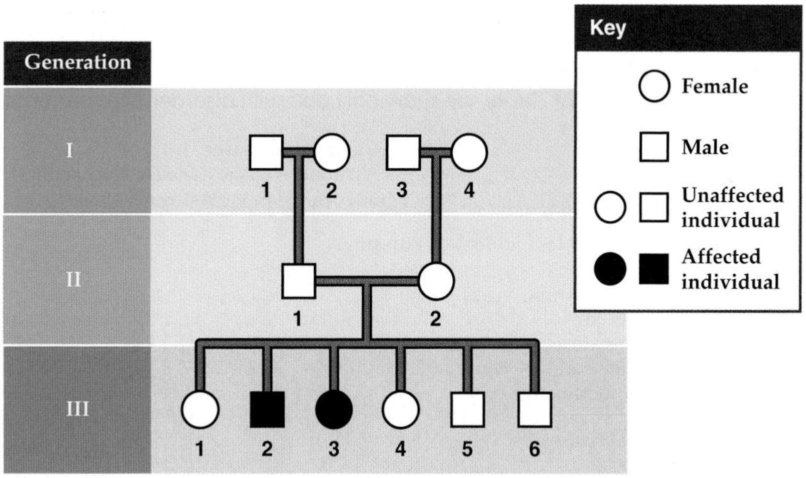

Figure 13.5 Human Pedigree Analysis
This pedigree of cystic fibrosis, a recessive genetic disorder in humans, illustrates symbols commonly used by geneticists. The Roman numerals at left identify different generations. Numbers listed below the symbols are used to identify individuals of a given generation. Individuals 1 and 2 in generation II each were carriers of cystic fibrosis (i.e., each had genotype *Aa*, where the recessive *a* allele causes cystic fibrosis).

cystic fibrosis is not dominant, for if it were, one of the parents of the affected individuals would have had the disease.

Humans suffer from a variety of genetic disorders. Some of these disorders result from new mutations that occur in the cells of an individual sometime during his or her life. Most cancers fall into this category (see Chapter 11). Other genetic disorders are passed down from parent to child. These inherited genetic disorders can be caused by mutations of individual genes (Figure 13.6) or by abnormalities in chromosome number or structure.

In the remainder of this chapter we focus on inherited genetic disorders that have relatively simple causes: those caused by mutations of a single gene or by chromosomal abnormalities. We organize our discussion of single-gene genetic disorders by whether the gene is located on an autosome or a sex chromosome. As you read this material, however, it is important to bear in mind that the tendency to develop some diseases, such as heart disease, diabetes, and some cancers, is caused by interactions among multiple genes and the environment. For most diseases caused by multiple genes, the identity of the genes involved and how they lead to disease is poorly understood. In Chapters 16–18, we will revisit human genetic disorders as we describe exciting new approaches for understanding and treating genetically based human diseases, including those caused by multiple genes.

> ■ Humans suffer from a variety of genetic disorders. Pedigrees, which show genetic relationships among family members over two or more generations of a family's history, provide a useful way to study human genetic disorders. Some inherited genetic disorders are caused by mutations of a single gene; others are caused by abnormalities in chromosome number or structure.

Autosomal Inheritance of Single-Gene Mutations

Several thousand human genetic disorders are inherited as recessive characters. Most of these, such as cystic fibrosis, sickle-cell anemia (see Figure 15.10 and Chapter 17), Tay-Sachs disease, and PKU (phenylketonuria) (see Figure 13.6), are caused by recessive mutations of genes located on autosomes. Recessive genetic disorders range in severity from those that are lethal to those with relatively mild effects. Tay-Sachs disease, for example, is a recessive genetic disorder in which a crucial enzyme does not work properly, causing lipids to accumulate in brain cells. As a result, the brain begins to deteriorate during a child's first year of life, causing death within a few years. At the other end of the severity spectrum, albino skin color in humans is controlled by a complex set of genes, including a recessive allele similar to that which produces a white coat color in mice and other mammals (see Figure 12.8).

For genetic disorders caused by a recessive allele (*a*), the only individuals who actually get the disease are those who have two copies of the disease-causing allele (*aa*). Most commonly, when a child inherits a recessive genetic disorder, both parents are heterozygous for the disorder; that is, they both have genotype *Aa*. Because the *A* allele is dominant and does not cause the disease, *Aa* individuals are said to be **carriers**: They carry the disease-causing allele (*a*), but do not get the disease.

If two carriers of a recessive genetic disorder have children, the patterns of inheritance are the same as for any recessive character: On average, ¼ of the children have genotype *AA*, ½ have genotype *Aa*, and ¼ have genotype *aa*. Thus, as shown in Figure 13.7, each child has a 25 percent chance of not carrying the disease-caus-

Figure 13.6   Examples of Single Genes That Cause Inherited Genetic Disorders
Mutations of single genes that cause genetic disorders are found on the X chromosome and on each of the 22 autosomes in humans. Thousands of single-gene genetic disorders are known; for clarity, we show only one such disorder per chromosome.

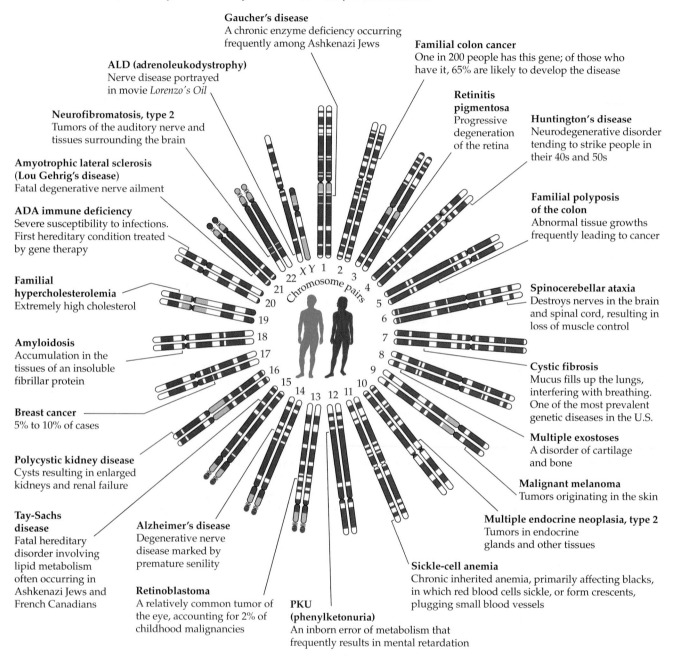

Gaucher's disease
A chronic enzyme deficiency occurring frequently among Ashkenazi Jews

Familial colon cancer
One in 200 people has this gene; of those who have it, 65% are likely to develop the disease

ALD (adrenoleukodystrophy)
Nerve disease portrayed in movie *Lorenzo's Oil*

Retinitis pigmentosa
Progressive degeneration of the retina

Huntington's disease
Neurodegenerative disorder tending to strike people in their 40s and 50s

Neurofibromatosis, type 2
Tumors of the auditory nerve and tissues surrounding the brain

Amyotrophic lateral sclerosis (Lou Gehrig's disease)
Fatal degenerative nerve ailment

Familial polyposis of the colon
Abnormal tissue growths frequently leading to cancer

ADA immune deficiency
Severe susceptibility to infections. First hereditary condition treated by gene therapy

Familial hypercholesterolemia
Extremely high cholesterol

Spinocerebellar ataxia
Destroys nerves in the brain and spinal cord, resulting in loss of muscle control

Amyloidosis
Accumulation in the tissues of an insoluble fibrillar protein

Cystic fibrosis
Mucus fills up the lungs, interfering with breathing. One of the most prevalent genetic diseases in the U.S.

Breast cancer
5% to 10% of cases

Multiple exostoses
A disorder of cartilage and bone

Polycystic kidney disease
Cysts resulting in enlarged kidneys and renal failure

Malignant melanoma
Tumors originating in the skin

Tay-Sachs disease
Fatal hereditary disorder involving lipid metabolism often occurring in Ashkenazi Jews and French Canadians

Multiple endocrine neoplasia, type 2
Tumors in endocrine glands and other tissues

Alzheimer's disease
Degenerative nerve disease marked by premature senility

Retinoblastoma
A relatively common tumor of the eye, accounting for 2% of childhood malignancies

PKU (phenylketonuria)
An inborn error of metabolism that frequently results in mental retardation

Sickle-cell anemia
Chronic inherited anemia, primarily affecting blacks, in which red blood cells sickle, or form crescents, plugging small blood vessels

Chromosome pairs
22 X Y 1 2 3 4 5 6 7 8 9 10 11 12 13 14 15 16 17 18 19 20 21

ing allele (genotype *AA*), a 50 percent chance of being a carrier (genotype *Aa*), and a 25 percent chance of actually getting the disease (genotype *aa*).

These percentages identify one way in which lethal recessive disorders such as Tay-Sachs disease can persist in the human population: Although homozygous recessive individuals (with genotype *aa*) die long before they are old enough to have children, carriers (with genotype *Aa*) are not harmed by the disorder, and hence, on average, they pass the disease-causing allele to half of their children. In a sense, then, the *a* alleles can "hide" in the heterozygous carriers. Recessive genetic disorders also

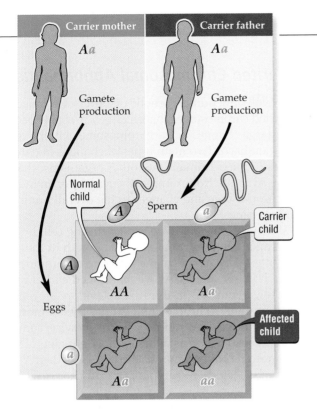

Carrier mother
Aa

Carrier father
Aa

Gamete production

Gamete production

Normal child

Sperm

A

a

Carrier child

Eggs

A

AA

Aa

a

Aa

Affected child

aa

Figure 13.7   Inheritance of Autosomal Recessive Disorders
The patterns of inheritance for a human autosomal recessive genetic disorder are the same as for any recessive character (compare this figure with the pattern shown by Mendel's pea plants in Figure 12.6). Recessive disease-causing alleles are colored red and denoted *a*. Dominant, normal alleles are denoted by *A*. Here, a carrier male (genotype *Aa*) mates with a carrier female (genotype *Aa*).

remain in the human population because new mutations generate new copies of the disease-causing alleles.

A dominant allele (*A*) that causes a genetic disorder cannot "hide" in the same way that a recessive allele can. In this case, *AA* and *Aa* individuals get the genetic disorder; only *aa* individuals are symptom-free. For a dominant genetic disorder with serious negative effects, individuals that have the *A* allele tend to survive or reproduce poorly; hence few of them pass the allele on to their children. In particular, most lethal dominant genetic disorders are not common and remain in the population primarily because new disease-causing alleles are generated by mutation.

Huntington's disease, which was described at the beginning of this chapter, illustrates another way in which dominant lethal alleles persist. The symptoms of Huntington's disease begin relatively late in life, often after victims of the disease have had children. Because the allele that causes the disease can be passed on to the next generation before the victim dies, the disease is more common than it would be if it persisted by mutation alone.

■ In the case of a genetic disorder caused by a recessive allele (*a*) of a gene on an autosome, only homozygous *aa* individuals have the disorder; heterozygous (*Aa*) individuals are unaffected carriers. Dominant genetic disorders affect both *AA* and *Aa* individuals. Lethal dominant genetic disorders can remain in the population because the disease symptoms begin late in life or because new disease-causing alleles are generated by mutation.

Sex-Linked Inheritance of Single-Gene Mutations

Roughly 1100 of the estimated 35,000 human genes are located on the sex chromosomes. Such genes are said to be **sex-linked**. Of the 1100 sex-linked genes in humans, about 1000 are located on the X chromosome, and 95 are located on the much smaller Y chromosome. Although there are no well-documented cases of disease genes located on the Y chromosome, X chromosomes do contain genes known to cause human genetic disorders (see Figure 13.6). Genes on the X chromosome, whether or not they cause a genetic disorder, are said to be **X-linked**.

Because males inherit only one X chromosome, genes on sex chromosomes have different patterns of inheritance than do genes on autosomes. Consider how an X-linked recessive allele that causes a human genetic disorder is inherited (Figure 13.8). We label the recessive disease-causing allele *a*, and in the Punnett square we write this allele as X^a to emphasize the fact that it is on the X chromosome. Similarly, the dominant allele is labeled *A* and is written as X^A in the Punnett square. If a carrier female (with genotype $X^A X^a$) has children with a normal male (with genotype $X^A Y$), 50 percent of their male children, on average, will get the disease. This result differs greatly from what would happen if the same disease-causing allele (*a*) were on an autosome: In that case, none of the children, male or female, would get the disease.

For recessive X-linked genetic disorders, males of genotype $X^a Y$ get the disease because the Y chromosome does not have a copy of the gene, and hence a dominant *A* allele cannot mask the effects of the *a* allele. In general, males are much more likely than females to be afflicted with X-linked disorders. To get the disorder, males have to inherit only a single copy of the disease-causing allele, whereas females must inherit two copies of the allele (see Figure 13.8). In contrast, both genders are equally likely to be affected by autosomal recessive disorders.

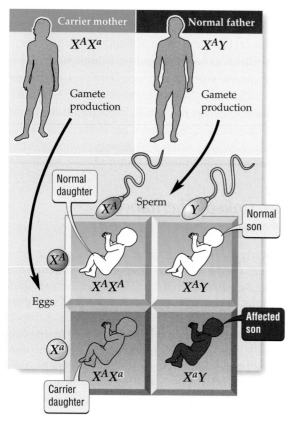

Carrier mother X^AX^a

Normal father X^AY

Gamete production

Gamete production

Normal daughter

Sperm

X^A

Y

Normal son

Eggs

X^A

X^AX^A

X^AY

X^a

Carrier daughter

X^AX^a

X^aY

Affected son

13.5

Figure 13.8 Inheritance of X-Linked Recessive Disorders
The recessive disease-causing allele (*a*) is located on the X chromosome and so is denoted by X^a. The dominant, normal allele (*A*) is also located on the X chromosome and is denoted by X^A.

X-linked genetic disorders in humans include hemophilia, a serious disorder in which minor cuts and bruises can cause a person to bleed to death, and Duchenne's muscular dystrophy, a lethal disorder that causes the muscles to waste away, usually leading to death in the early 20s. Both of these X-linked disorders are caused by recessive alleles. An example of a dominant X-linked disorder is one with a daunting name—congenital generalized hypertrichosis, or CGH. It is discussed in the box on p. 234.

■ Because males inherit only one X chromosome, genes on sex chromosomes have different patterns of inheritance than genes on autosomes. Males are more likely than females to have recessive X-linked genetic disorders because they need inherit only one copy of the disease-causing allele to be affected.

Inherited Chromosomal Abnormalities

Relatively few human genetic disorders are caused by inherited chromosomal abnormalities, probably because most large changes in the chromosomes kill the developing embryo. Two main types of changes in chromosomes occur in humans and other organisms: changes in the structure (for example, the length) of an individual chromosome and changes in the overall number of chromosomes.

When chromosomes are copied during cell division, breakages can occur that alter the length of the chromosome. Such changes in the structure of chromosomes have dramatic effects. As we mentioned earlier in this chapter, some XY individuals develop as females because their Y chromosome is missing a region that contains the sex-determining gene. Likewise, some XX individuals develop as males because a region of the Y chromosome has become attached to one of their X chromosomes. Another example is cri du chat syndrome, which develops when a child inherits a chromosome 5 that is missing a particular region (Figure 13.9). *Cri du chat* is French for "cry of the cat," a name that describes the characteristic mewing sound made by infants suffering from this condition. Individuals with this syndrome grow slowly and tend to have severe mental retardation, small heads, and low-set ears.

Unusual numbers of chromosomes—such as one or three copies instead of the normal two—can be produced when chromosomes fail to separate properly during meiosis. Most changes in chromosome number result in the death of the embryo. It is estimated that at least 20 percent of human pregnancies are spontaneously aborted, largely as a result of changes in chromosome number.

There is only one case in which a person who inherits the wrong number of autosomes can reach adulthood: Down's syndrome. Individuals with this genetic disorder have three copies of chromosome 21, the smallest autosome in humans. They tend to be short and mentally retarded, and to have defects of the heart, kidneys, and digestive tract. They also tend to be cheerful and affectionate, to enjoy music and dance, and to have a flair for mimicry. Live births also can result when an infant has three copies of chromosome 13, 15, or 18. Such children have severe birth defects, however, and they rarely live beyond their first year.

Compared with autosomes, changes in the number of sex chromosomes can have relatively minor effects. XXY males, for example, have a normal life span and normal intelligence, and they tend to be tall. However, they also have reduced fertility, and some have feminine characteristics, such as enlarged breasts.

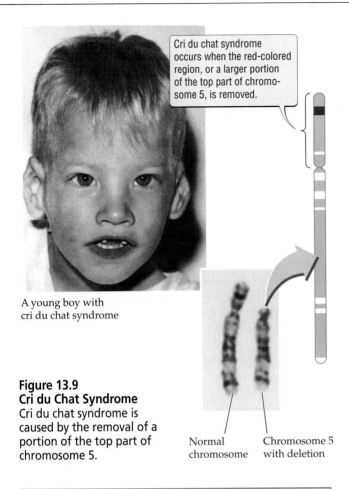

Cri du chat syndrome occurs when the red-colored region, or a larger portion of the top part of chromosome 5, is removed.

A young boy with cri du chat syndrome

Figure 13.9
Cri du Chat Syndrome
Cri du chat syndrome is caused by the removal of a portion of the top part of chromosome 5.

Normal chromosome   Chromosome 5 with deletion

■ Chromosomal abnormalities include changes in the structure of an individual chromosome and changes in the overall number of chromosomes. Changes in chromosome structure can have profound effects. Most changes to the number of autosomes result in the death of the developing embryo. Changes to the number of sex chromosomes can have relatively minor effects.

HIGHLIGHT

Uncovering the Genetics of Huntington's Disease

George Huntington wrote the classic paper describing the disease named for him in 1872. From then until 1983, there was little hope for progress in treatment of the disease. It was known that the gene for the disease is on an autosome, and that the disease-causing allele is dominant (*A*). But there was no cure, and little helpful information could be given to potential victims of the disease. For example, by constructing a pedigree, a geneticist might learn that a person's father (who had Huntington's disease) was of genotype *Aa*. If the mother did not have the disease, and hence was of genotype *aa*, all the geneticist could say was that the person had a 50 percent chance of developing the disease (see Problem 3 in the sample genetics problems at the end of the chapter).

In 1983, however, this situation changed dramatically when researchers discovered that the gene for Huntington's disease is located on a portion of chromosome 4. This discovery was made possible by pedigree analyses that looked for patterns of genes inherited together. These analyses indicated that the gene for Huntington's disease (HD) was linked to other genes known to be located on chromosome 4. This discovery set off an intense effort to isolate the gene. Ten years later, the gene was found. To find it, researchers carefully determined which chromosome 4 genes were linked most closely to the HD gene, a process that eventually allowed them first to pinpoint the gene's location on chromosome 4, then to isolate the gene itself.

By isolating the HD gene, scientists were able to learn how the protein it produced in individuals with Huntington's disease differed from the protein produced in individuals without the disease. (We will return to the general topic of how genes control the production of proteins in Chapter 15.) The key point is that from these analyses, researchers learned the identity of an abnormal protein produced by the allele that causes Huntington's disease. This protein forms clumps in the brains of people with the disease; these clumps are correlated with, and may cause, the disease symptoms.

Dramatic new results within the past few years suggest that knowledge of the HD gene and its associated protein may help in the design of effective treatments. For example, in mice genetically engineered to have the human HD gene, scientists have developed ways to slow the progression of, or even reverse, the disease symptoms. There have been similarly exciting results in humans: In a human patient with Huntington's disease, cells lacking the disease-causing allele were transplanted into the patient's brain, where they survived and remained free of the clumps of the HD protein for 18 months. These results offer hope that brain repair may eventually be possible in the treatment of Huntington's disease.

In addition to providing clues to possible ways of treating the disease, isolation of the HD gene had another, immediate effect: It allowed scientists to design a diagnostic genetic test for the disease. With this test, a person whose parent had the disease can now know with near certainty whether or not he or she also will get the disease. In some cases, the genetic test offers hope. For example, if persons at risk want to have a family without the

■ THE SCIENTIFIC PROCESS

Tracing the Inheritance of a Disease Gene

People often construct family trees to trace their ancestry. In genetics, a pedigree (see p. 228) is used in much the same way, with the emphasis shifted to focus on the ancestry of alleles rather than the ancestry of people. To illustrate how pedigrees are used, we trace the inheritance of congenital generalized hypertrichosis, or CGH.

CGH is a rare genetic disorder that causes extreme hairiness of the face and upper body. It is caused by a mutation of a single gene. Let's work through the pedigree shown here to see if we can determine whether the CGH allele is dominant or recessive and whether it is X-linked. Let's assume we already know that indi-

vidual 1 in generation I and individuals 2 and 5 in generation II did not carry the disease allele.

First, the pedigree reveals that the disease is not an autosomal recessive disorder, for if it were, none of the three children in generation II would have CGH (given that individual 1 in generation I did not carry the disease allele). Similarly, the disease could not be an X-linked recessive disorder, for if it were, only the male children in generation II would have CGH. (Make sure you understand why this is so by completing a Punnett square for the cross $X^A Y \times X^a X^a$, where A is the normal allele and a is the allele for CGH.) Thus, the allele for CGH must be dominant. Is it X-linked?

The genetic characteristics of the ten children produced by individuals 4 and 5 of generation II suggest that the CGH allele is probably X-linked and dominant. If this is the case, then individual 4 of generation II should have genotype $X^A Y$, and his wife should have genotype $X^a X^a$. Assuming that is true, all of their female children, but none of their male children, should have CGH, exactly the pattern seen. Other pedigrees confirm that the condition is X-linked, making CGH an example of an X-linked dominant genetic disorder.

A Pedigree for Congenital Generalized Hypertrichosis (CGH)

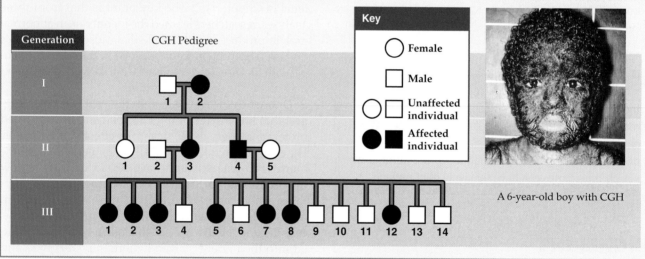

A 6-year-old boy with CGH

fear of passing a terrible disease on to their children, they can take the test and make an informed decision about having children. But the test poses an agonizing choice for these and all other individuals at risk: If they take the test, they may experience tremendous relief if the results show that they will not get the disease. Alternatively, they may experience a crushing burden and loss of hope if they find out they have a lethal disease with horrible symptoms that, as of now, cannot be cured. Given these alternatives, would you take the test?

■ Analysis of human pedigrees indicated that the gene that causes Huntington's disease (HD) is located on a portion of chromosome 4. By determining which genes were most closely linked to the HD gene, researchers pinpointed its location and then isolated the gene itself. The isolation of the HD gene has improved our understanding of the disease and has led to progress toward possible treatments as well as a diagnostic test.

SUMMARY

The Role of Chromosomes in Inheritance

- The chromosome theory of inheritance states that genes are located on chromosomes.
- Chromosomes are composed of a single DNA molecule and many proteins.
- A gene is a region of DNA within the DNA of the chromosome.
- The physical location of a gene on a chromosome is called a locus.

Autosomes and Sex Chromosomes

- Chromosomes that determine gender are called sex chromosomes; all other chromosomes are called autosomes.
- Human males have one X and one Y chromosome. Human females have two X chromosomes.
- A specific gene on the Y chromosome is required for human embryos to develop as males.

Linkage and Crossing-Over

- Genes that are located on the same chromosome and that do not assort independently are said to be genetically linked.
- Crossing-over, the exchange of genes between chromosomes, disrupts the linkage between genes.

Origins of Genetic Differences among Individuals

- In the independent assortment of chromosomes, maternal and paternal chromosomes are randomly distributed into gametes during meiosis.
- Crossing-over, the independent assortment of chromosomes, and fertilization cause offspring to differ genetically from one another and from their parents.
- Genetic differences among individuals underlie human genetic disorders and provide the genetic variation on which evolution can act.

Human Genetic Disorders

- Humans suffer from a variety of genetic disorders, including those caused by mutations of a single gene and those caused by abnormalities in chromosome number or structure.
- Pedigrees provide a useful way to study human genetic disorders.

Autosomal Inheritance of Single-Gene Mutations

- For genetic disorders caused by a recessive allele (*a*) of a gene on an autosome, only homozygous *aa* individuals get the disease. Heterozygous (*Aa*) individuals are merely carriers of the disease.
- Dominant autosomal genetic disorders affect both *AA* and *Aa* individuals.
- Lethal dominant genetic disorders can remain in the population because the disease symptoms begin late in life, as in Huntington's disease, or because new disease-causing alleles are generated by mutation.

Sex-Linked Inheritance of Single-Gene Mutations

- Because males inherit only one X chromosome, genes on sex chromosomes have different patterns of inheritance than do genes on autosomes.
- Males are more likely than females to have recessive X-linked genetic disorders because they need to inherit only one copy of the disease-causing allele to be affected, while females must inherit two copies to be affected. Both sexes are equally likely to get autosomal genetic disorders.

Inherited Chromosomal Abnormalities

- Chromosomal abnormalities include changes in chromosome structure and chromosome number.
- Changes in the structure of an individual chromosome can have profound effects.
- Changes in the number of autosomes in humans are usually lethal. Down's syndrome, which occurs in individuals with three copies of chromosome 21, is an exception to this rule.
- Changes to the number of sex chromosomes in humans can have relatively minor effects.

Highlight: Uncovering the Genetics of Huntington's Disease

- Analysis of human pedigrees indicated that the gene that causes Huntington's disease (HD) is located on a portion of chromosome 4.
- By determining which genes on chromosome 4 were most closely linked to the HD gene, researchers pinpointed the location of the HD gene and then isolated the gene itself.
- Isolation of the HD gene has improved our understanding of the disease and has led to the development of a diagnostic test, as well as progress toward possible treatments.

KEY TERMS

autosome p. 225	independent assortment of chromosomes p. 228
carrier p. 229	
chromosome theory of inheritance p. 224	locus p. 224
	pedigree p. 228
crossing-over p. 227	sex chromosome p. 225
genetic linkage p. 226	sex-linked p. 231
homologous chromosome p. 224	X-linked p. 231

CHAPTER REVIEW

Self-Quiz

1. Genes are
   a. located on chromosomes.
   b. composed of DNA.
   c. composed of both protein and DNA.
   d. both a and b

2. Which of the following is an autosomal dominant disorder whose symptoms begin late in life and which destroys the nervous system, resulting in death?
   a. Tay-Sachs disease
   b. Huntington's disease
   c. Down's syndrome
   d. Cri du chat syndrome

3. Crossing-over is
   a. more likely between genes that are close together on a chromosome.
   b. more likely between genes that are on different chromosomes.
   c. more likely between genes that are far apart on a chromosome.
   d. not related to the distance between genes.

4. Comparatively few human genetic disorders are caused by chromosomal abnormalities. One reason is that
   a. most chromosomal abnormalities have little effect.
   b. it is difficult to detect changes in the number or length of chromosomes.
   c. most chromosomal abnormalities result in spontaneous abortion of the embryo.
   d. it is not possible to change the length or number of chromosomes.

5. Nonparental genotypes can be produced by
   a. crossing-over and the independent assortment of chromosomes.
   b. linkage.
   c. autosomes.
   d. sex chromosomes.

Review Questions

1. Explain how nonparental genotypes are formed.

2. Do you think alleles that cause lethal dominant genetic disorders are likely to be as common in the human population as alleles that cause lethal recessive genetic disorders? Explain your answer.

3. Huntington's disease is a genetic disorder whose symptoms begin in middle age and which destroys the nervous system, eventually killing its victims. Although at present no cure is available, the gene that causes the disease has been isolated, and a genetic test has been developed that can tell potential victims with near certainty whether or not they will get the disease. If you were at risk for the disease, would you take the genetic test? More generally, what are the advantages and disadvantages of genetic tests that reveal whether or not a person has a particular genetic disorder?

Sample Genetics Problems

1. Human females have two X chromosomes; human males have one X chromosome and one Y chromosome.
   a. Do males inherit their X chromosome from their mother or their father?
   b. If a female has one copy of an X-linked recessive allele that causes a genetic disorder, does she have the disorder?
   c. If a male has one copy of an X-linked recessive allele that causes a genetic disorder, does he have the disorder?
   d. A female is a carrier of an X-linked recessive disorder. With respect to the disease-causing allele, how many types of gametes can she produce?
   e. Assume that a male with an X-linked recessive genetic disorder has children with a female who does not carry the disease-causing allele. Could any of their children have the genetic disorder? Could any of their children be carriers for the disorder?

2. Cystic fibrosis is a genetic disorder caused by a recessive allele, *a*; the disease-causing allele is located on an autosome. What are the chances that parents with the following genotypes will have a child with the disorder?
   a. $aa \times Aa$
   b. $Aa \times AA$
   c. $Aa \times Aa$
   d. $aa \times AA$

3. Huntington's disease is a genetic disorder caused by a dominant allele, *A*; the disease-causing allele is located on an autosome. What are the chances that parents with the following genotypes will have a child with Huntington's disease?
   a. $aa \times Aa$
   b. $Aa \times AA$
   c. $Aa \times Aa$
   d. $aa \times AA$

4. Hemophilia is a genetic disorder caused by a recessive allele, *a*; the disease-causing allele is located on the X chromosome. What are the chances that parents with the following genotypes will have a child with hemophilia?
   a. $X^A X^A \times X^a Y$
   b. $X^A X^a \times X^a Y$
   c. $X^A X^a \times X^A Y$
   d. $X^a X^a \times X^A Y$
   e. Do male and female children have the same chance of getting the disease?

5. Explain why the terms "homozygous" and "heterozygous" do not apply to X-linked traits in males.

6. Study the pedigree shown below. Is the disease-causing allele dominant or recessive? Is the disease-causing allele located on an autosome or on the X chromosome? What are the genotypes of individuals 1 and 2 in generation I?

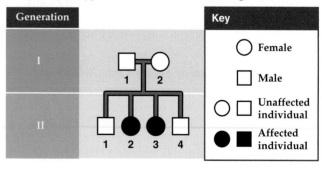

7. Study the pedigree shown on p. 237. Is the disease-causing allele dominant or recessive? Is the disease-causing allele located on an autosome or on the X chromosome? To answer this question, assume that individual 1 in genera-

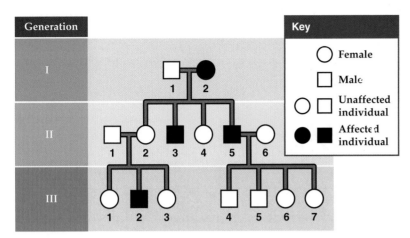

tion I and individuals 1 and 6 in generation II do not carry the disease-causing allele.

8. Consider the inheritance of two genes, one with alleles *A* or *a*, the other with alleles *B* or *b*. *AABB* individuals are crossed with *aabb* individuals to produce F_1 offspring, all of which have genotype *AaBb*. These *AaBb* F_1 offspring are then crossed with *aaBB* individuals. Construct Punnett squares and list the possible offspring genotypes that you would expect in the F_2 generation.
   a. if the two genes were completely linked.
   b. if the two genes were on different chromosomes.

13

The Daily Globe

Bringing Genes into the Courtroom

NEW YORK, NY—Scientists have recently completed a multibillion-dollar study of human DNA, which has enabled them to produce a highly detailed "map" of our chromosomes that can be used to pinpoint the locations of genes responsible for many disorders. Now a controversial study that links violent behavior to particular genes may represent the beginning of a new and largely unintended application of the human chromosome map: It soon may be possible for lawyers to

mount a "genetic defense," in which defendants admit to carrying out a crime but argue that they are not guilty because their genes made them do it. As more and more information about our DNA pours in from laboratories around the world, the possibility that defendants will use a genetic defense is expected to grow with time.

Such a possibility has not been lost on New York state judges, a group of whom attended a two-day workshop to learn more about ge-

netics in general and the genetic defense in particular. During the workshop, the judges heard a hypothetical case in which a man accused of murder claimed he was innocent by virtue of a genetic predisposition to violent behavior. As Judge Robin Denova later commented, "We are entering uncharted territory. Will defendants soon claim 'their genes made them do it' for a broad range of crimes? And if they do, how will the courts and society respond?"

Evaluating "The News"

1. Do you think people should be held responsible for actions "determined by their genes"? Is a plea of "innocent by virtue of genetics" different from a plea of "innocent by virtue of insanity"?

2. Suppose someone driving a car has a seizure that causes them to lose control of their car, crash into another car, and kill a mother and her two young children. In a wrongful death court case, the person who had the seizure is found to

carry genes that predisposed them to have seizures—a genetic condition of which the person had previously been unaware. In such a case, do you think an argument that "my genes made me do it" would be a convincing line of defense? Why or why not?

3. Suppose instead that a person under the influence of alcohol drives a car the wrong way on a highway, crashes into another car, and kills a mother and her two young

children. In the ensuing court case, the person who had driven while intoxicated is found to carry a putative alcoholism gene. Prior to the court case, the defendant was known to have a drinking problem, but had no prior history of drunken-driving arrests. In such a case, do you think an argument that "my genes made me do it" would be a convincing line of defense? Why or why not?

chapter *14* DNA

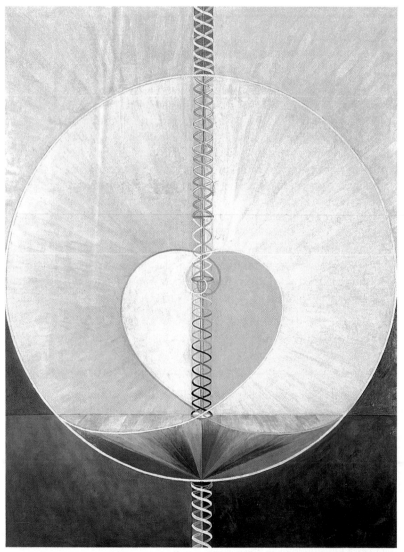

Hilma af Klint, *Group 9. Series UW. Dove no. 25*, 1915.

The Library of Life

The book you are reading has more than a million letters in it. How long do you think it would take you to copy these letters by hand, one by one? How many mistakes do you think you would make? Would you check your work for mistakes, and if so, how long do you think that would take?

Difficult as the job of copying all the letters in this book would be, it pales in comparison to the job your cells do each time they divide. When a cell divides, it must make a copy of all its genetic information. That information is stored in deoxyribonucleic acid, or DNA, the substance of which genes are made. The amount of information stored in your DNA is mind-boggling:

DNA is the genetic material, its three-dimensional structure consisting of two strands of nucleotides twisted into a double helix.

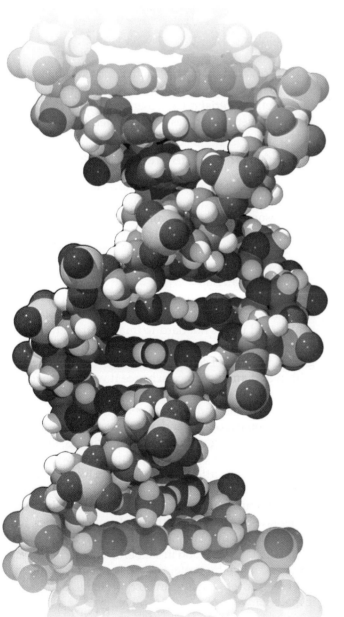

Whereas this book contains roughly a million letters, the DNA in each of your cells has the equivalent of 3,300,000,000 letters. If it were printed with the same font in the same size as the letters on this page, the information that is in your DNA would fill thousands of books similar in length to this one.

Your cells copy the phenomenal amount of information in your DNA in a matter of hours. Yet despite the speed with which they work, on average they make only one mistake for every billion "letters." How do they do this? What are the "letters" that make up the DNA molecule? More broadly, what is the structure of the DNA molecule, and what implications does this structure have for the processes of life?

A Model of the DNA Molecule

■ KEY CONCEPTS

1. Genes are composed of DNA.

2. Four nucleotides are the building blocks of DNA. Each of these nucleotides contains one of four nucleotide bases: adenine, cytosine, guanine, or thymine.

3. DNA consists of two strands of nucleotides twisted together into a spiral. The two strands are held together by hydrogen bonds that form between adenine and thymine and between cytosine and guanine.

4. Each strand of DNA has an enormous number of bases arranged one after another. The sequence of bases in DNA differs among species and among individuals within a species. These differences are the basis of inherited variation.

5. Because adenine pairs only with thymine and cytosine pairs only with guanine, each strand of DNA can serve as a template from which to duplicate the other strand.

6. DNA in cells is damaged constantly by factors such as heat energy and ultraviolet light. If this damage were not repaired, the organism would die.

We have looked at genes in several different ways in this unit. In Chapter 12, for example, genes were treated as abstract entities that control the inheritance of characters. In Chapter 13, we saw that genes are physically located on chromosomes. Knowing that genes are located on chromosomes makes them less abstract, but this knowledge leaves unanswered many fundamental questions: What are genes made of? When a cell divides to form two daughter cells, how is the information in the genes copied? How are errors in copying corrected, and how is damage to the cell's genetic material repaired?

To answer such questions, geneticists had to discover the substance of which genes are made, and they had to learn the physical structure of this substance. As they began this search, they were guided by three basic biological facts about the nature of the genetic material. First, the genetic material had to contain the information necessary for life. It had to contain, for example, the information needed to build the body of the organism and to control the complex metabolic reactions on which life depends. Second, the genetic material had to be composed of a substance that could be copied accurately. If this could not be done, reliable genetic information could not be passed from one generation to the next. Finally, the genetic material had to be variable. If the genetic material were not variable, there would be no genetic differences within or among species, something that scientists knew was not true.

Parallel to the search for the chemical composition and physical structure of genes was a search for the function of genes: How exactly did genes produce their effects? In Chapter 12, we learned that DNA is the genetic material. In this chapter, we describe how scientists discovered that genes are composed of DNA. We also discuss the physical structure of genes and how the genetic material is copied and repaired. In Chapter 15 we will see how genes produce their effects.

The Search for the Genetic Material

By the early 1900s, geneticists knew that genes control the inheritance of characters, that genes are located on chromosomes, and that chromosomes are composed of DNA and protein. With this knowledge in hand, the first step in the quest to understand the physical structure of genes was to determine whether DNA or protein was the genetic material.

Initially, most geneticists thought that protein was the more likely candidate. Proteins are large, complex molecules, and it was not hard to imagine that they could store the tremendous amount of information needed to govern the lives of cells. Proteins also vary considerably within and among species; hence it was reasonable to assume that they caused the inherited variation observed within and among species.

DNA, on the other hand, was initially judged a poor candidate for the genetic material, mainly because DNA was thought to be a small, simple molecule whose composition varied little among species. Over time, these ideas about DNA were discovered to be wrong. In fact, DNA molecules are large and vary tremendously in their nucleotide composition within and among species. Still, as we will see, the variation contained in DNA is more subtle than the variation in shape, electrical charge, and function shown by proteins, so it is not surprising that most researchers initially favored proteins as the genetic material.

Over a period of roughly 25 years (1928–1952), geneticists became convinced that DNA, not protein, was the genetic material. The results of three key experiments helped to cause this shift of opinion.

Harmless bacteria can be transformed into deadly bacteria

In 1928, a British medical officer named Frederick Griffith published an important paper on *Streptococcus pneumoniae*, a bacterium that causes pneumonia in humans and other mammals. Griffith was studying two genetic varieties, or strains, of *Streptococcus* to find a cure for pneumonia, which was a common cause of death at that time. The two strains, called strain S and strain R, were named after differences in their appearance. When the bacteria were grown on a petri dish, strain S produced colonies that appeared smooth, while strain R produced colonies that appeared rough.

Griffith conducted four experiments on these bacteria (Figure 14.1). First, when he injected bacteria of strain R into mice, the mice survived and did not develop pneumonia. Second, when he injected bacteria of strain S into mice, the mice developed pneumonia and died. In the third experiment, he injected heat-killed strain S bacteria into mice, and once again the mice survived. His plan was to test mice from the third experiment to see if they were resistant to later exposure to live strain S bacteria.

None of these results were particularly unusual: Griffith had simply shown that there were two strains of bacteria, one of which (strain S) killed mice and was itself killed and rendered harmless by heat. In the fourth experiment, however, Griffith mixed heat-killed bacteria of strain S with live bacteria of strain R. On the basis of the results from the first three experiments, he expected the mice to survive. Instead, the mice died, and Griffith recovered live bacteria of strain S from the blood of the dead mice.

In Griffith's fourth experiment, something had caused harmless strain R bacteria to change into deadly strain S bacteria. Griffith showed that the change was genetic: When they reproduced, the altered strain R bacteria produced strain S bacteria. Overall, the results of Griffith's fourth experiment suggested that genetic material from heat-killed strain S bacteria had somehow changed living strain R bacteria into strain S bacteria.

This remarkable and unexpected result stimulated an intensive hunt for the material that caused the change. We now know that the strain R bacteria had absorbed DNA from the heat-killed strain S bacteria, causing the genotype of the strain R bacteria to change. A change in the genotype of a cell or organism after exposure to the DNA of another genotype is called **transformation**.

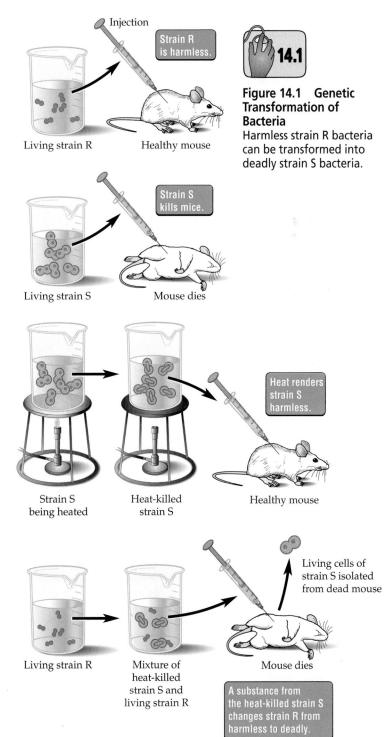

Figure 14.1 Genetic Transformation of Bacteria
Harmless strain R bacteria can be transformed into deadly strain S bacteria.

DNA can transform bacteria

For 10 years, the American Oswald Avery and his colleagues at Rockefeller University in New York struggled to identify the genetic material that had transformed the bacteria in Griffith's experiments. They isolated and tested different compounds from the bacteria. Only DNA was able to transform harmless strain R bacteria into deadly strain S bacteria. In 1944, Avery, Colin MacLeod, and Maclyn McCarty published a landmark paper that summarized their results. The paper created quite a stir.

In addition to showing that DNA transforms bacteria, Avery, MacLeod, and McCarty's paper led many biologists to a broader conclusion: that DNA, not protein, is the genetic material. As a leading DNA researcher later remarked, the paper stimulated an "avalanche" of new research on DNA. Some biologists remained skeptical, arguing, for example, that DNA was too simple a molecule to be the genetic material. However, the tide was definitely turning in favor of DNA.

The genetic instructions of viruses are contained in DNA

In Griffith's experiments, heat killed the strain S bacteria, but did not destroy its genetic material. Since most proteins are destroyed by heat, this result suggested that protein was not the genetic material. Then the work by

Avery, MacLeod, and McCarty provided very strong evidence that DNA was the genetic material. Additional proof came in 1952, when Alfred Hershey and Martha Chase published an elegant study on the genetic material of viruses.

Hershey and Chase studied a virus that consists only of DNA and a coat of protein (Figure 14.2). To reproduce, this virus attaches to the wall of a bacterium and injects its genetic material into the bacterium. The genetic material of the virus then takes over the bacterial cell, eventually killing it, but first causing it to produce many new viruses. Because the virus is composed only of protein and DNA, it provided an excellent experimental system in which to test whether DNA or protein was the genetic material.

Hershey and Chase demonstrated that only the DNA portion of the virus was injected into the bacterium (see Figure 14.2). This result indicated that DNA was respon-

Figure 14.2 DNA Is the Genetic Material
The Hershey–Chase experiments on viruses that infect bacteria used a radioactive labeling technique. Hershey and Chase knew that a virus injects its genetic material into a bacterial cell, where it directs the production of new viruses. They grew viruses in two different radioactive solutions that labeled either the viruses' DNA or their proteins. When Hershey and Chase found labeled DNA—but not labeled protein—inside bacterial cells, they concluded that DNA was the genetic material.

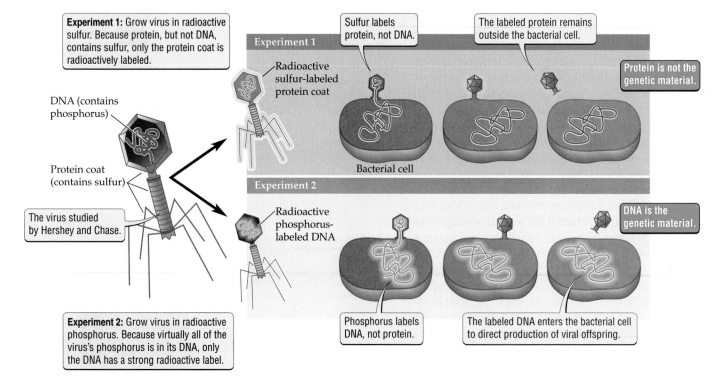

Experiment 1: Grow virus in radioactive sulfur. Because protein, but not DNA, contains sulfur, only the protein coat is radioactively labeled.

DNA (contains phosphorus)

Protein coat (contains sulfur)

The virus studied by Hershey and Chase.

Experiment 1

Radioactive sulfur-labeled protein coat

Sulfur labels protein, not DNA.

The labeled protein remains outside the bacterial cell.

Protein is not the genetic material.

Bacterial cell

Experiment 2

Radioactive phosphorus-labeled DNA

DNA is the genetic material.

Experiment 2: Grow virus in radioactive phosphorus. Because virtually all of the virus's phosphorus is in its DNA, only the DNA has a strong radioactive label.

Phosphorus labels DNA, not protein.

The labeled DNA enters the bacterial cell to direct production of viral offspring.

sible for taking over the bacterial cell and for causing the production of new viruses. These experiments convinced most remaining skeptics that DNA, not protein, was the genetic material.

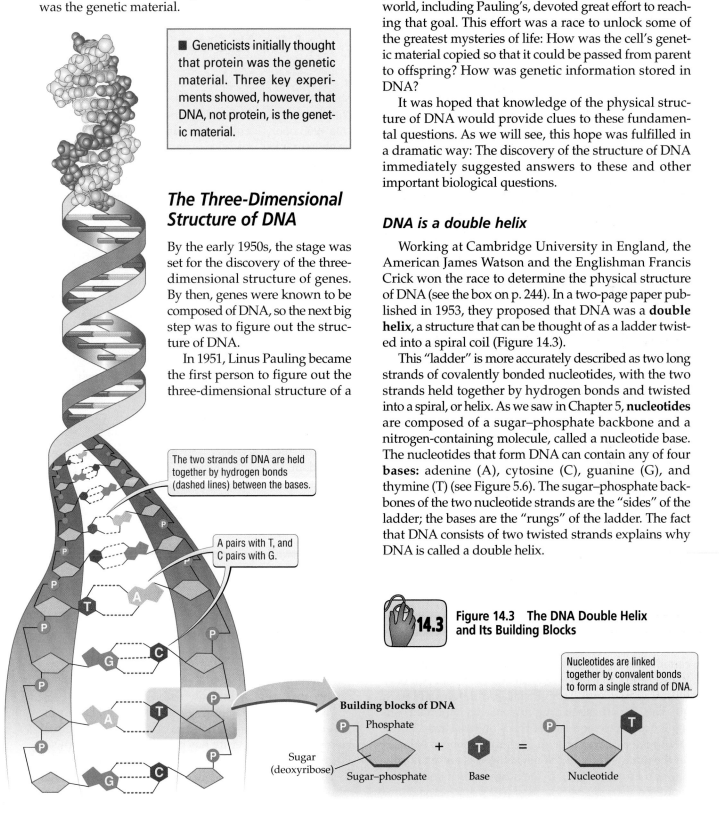

> ■ Geneticists initially thought that protein was the genetic material. Three key experiments showed, however, that DNA, not protein, is the genetic material.

The Three-Dimensional Structure of DNA

By the early 1950s, the stage was set for the discovery of the three-dimensional structure of genes. By then, genes were known to be composed of DNA, so the next big step was to figure out the structure of DNA.

In 1951, Linus Pauling became the first person to figure out the three-dimensional structure of a

protein. Pauling's success suggested that determining the three-dimensional structure of DNA should also be possible. Major research laboratories from around the world, including Pauling's, devoted great effort to reaching that goal. This effort was a race to unlock some of the greatest mysteries of life: How was the cell's genetic material copied so that it could be passed from parent to offspring? How was genetic information stored in DNA?

It was hoped that knowledge of the physical structure of DNA would provide clues to these fundamental questions. As we will see, this hope was fulfilled in a dramatic way: The discovery of the structure of DNA immediately suggested answers to these and other important biological questions.

DNA is a double helix

Working at Cambridge University in England, the American James Watson and the Englishman Francis Crick won the race to determine the physical structure of DNA (see the box on p. 244). In a two-page paper published in 1953, they proposed that DNA was a **double helix**, a structure that can be thought of as a ladder twisted into a spiral coil (Figure 14.3).

This "ladder" is more accurately described as two long strands of covalently bonded nucleotides, with the two strands held together by hydrogen bonds and twisted into a spiral, or helix. As we saw in Chapter 5, **nucleotides** are composed of a sugar–phosphate backbone and a nitrogen-containing molecule, called a nucleotide base. The nucleotides that form DNA can contain any of four **bases:** adenine (A), cytosine (C), guanine (G), and thymine (T) (see Figure 5.6). The sugar–phosphate backbones of the two nucleotide strands are the "sides" of the ladder; the bases are the "rungs" of the ladder. The fact that DNA consists of two twisted strands explains why DNA is called a double helix.

The two strands of DNA are held together by hydrogen bonds (dashed lines) between the bases.

A pairs with T, and C pairs with G.

14.3 Figure 14.3   The DNA Double Helix and Its Building Blocks

Nucleotides are linked together by convalent bonds to form a single strand of DNA.

Building blocks of DNA

Phosphate

Sugar (deoxyribose)

Sugar–phosphate   +   Base   =   Nucleotide

■ **THE SCIENTIFIC PROCESS**

Rosalind Franklin: Crucial Contributor to the Discovery of DNA's Structure

In 1962, James Watson and Francis Crick received the Nobel prize in physiology or medicine for determining the three-dimensional structure of DNA. They shared the prize with Maurice Wilkins, who also had worked to discover the structure of DNA. Missing from the 1962 Nobel ceremony was Rosalind Franklin, a gifted young scientist whose research provided Watson and Crick with critical data.

Rosalind Franklin was born in London in 1920. At St. Paul's Girls' School and Cambridge University, she studied physics and chemistry. After leaving Cambridge University in 1942, she worked in industry for several years. Next, from 1947 to 1950, she worked in Paris learning how to use X-rays to produce photographs of crystals. In 1951, she returned to Cambridge University to work on the structure of DNA. By that time, she was an expert at taking X-ray photographs of crystals, a procedure mastered by few people in the world.

Franklin's research at Cambridge provided several important clues for Watson and Crick. First, she took very clear X-ray photographs of the DNA fibers prepared by Wilkins. As described by Watson in his 1968 book, *The Double Helix*, these photographs provided essential information to Watson and Crick as they sought to determine the molecular structure of DNA. Franklin also demonstrated that phosphate groups are located on the outside of the DNA molecule, a key feature of Watson and Crick's description of the three-dimensional structure of DNA.

Rosalind Franklin died of cancer in 1958 at the age of 37. Nobel prizes cannot be awarded after the recipient's death, so no one will ever know whether or not Franklin would have been awarded a Nobel prize for her critical contributions to determining the structure of DNA. Unfortunately, at times during her scientific career, her work was not given the high regard that it fully deserved. In *The Double Helix*, Watson commented that he realized years too late what a struggle it was for a woman like Rosalind Franklin to be accepted by a scientific community that often failed to take female scientists seriously. Increasingly recognized for her scientific accomplishments, Franklin, because of her achievements as well as her courage in the face of such skepticism, is an excellent role model for scientists today, both male and female.

Rosalind Franklin

Watson and Crick's paper contained two key realizations. First, they recognized that there were two strands of nucleotides in DNA. Second, they realized that only certain combinations of bases could pair with each other. Watson and Crick proposed that adenine could pair only with thymine and that cytosine could pair only with guanine. These **base-pairing** rules had an important consequence: If the sequence of bases on one strand of the DNA molecule was known, then the sequence of bases on the other strand of the molecule was automatically known. For example, if one strand consisted of the sequence

ACCTAGGG

then the other strand had to have the sequence

TGGATCCC

Any other sequence would violate the rule that A pairs only with T and C pairs only with G.

We now know that the physical structure of DNA proposed by Watson and Crick is correct in all its essential elements. This structure has great explanatory power. For example, as we will see in the following section, the fact that adenine can pair only with thymine and that cytosine can pair only with guanine immediately sug-

Chapter 14 DNA 245

gested a simple way in which the DNA molecule could be copied, or replicated: The original strands could serve as templates on which to build the new strands.

Knowledge of the three-dimensional structure of DNA also indicated that DNA could be viewed as a long string of the four bases: A, C, G, and T. Although A had to pair with T and C had to pair with G, the four bases could be arranged in any order along a strand of DNA. The fact that each strand of DNA is composed of millions of these bases suggested that a tremendous amount of information could be contained in the order, or sequence, of the bases along the DNA molecule (see Chapter 15).

The sequence of bases in DNA differs among species and among individuals within a species (Figure 14.4). Differences in DNA sequence are the basis of inherited variation. Different alleles of a gene have different DNA

sequences. For example, people with a genetic disorder such as Huntington's disease or cystic fibrosis (see Chapter 13) inherit particular alleles that cause the disease. At the molecular level, one allele causes a disease and another allele does not because the two alleles have a different sequence of bases.

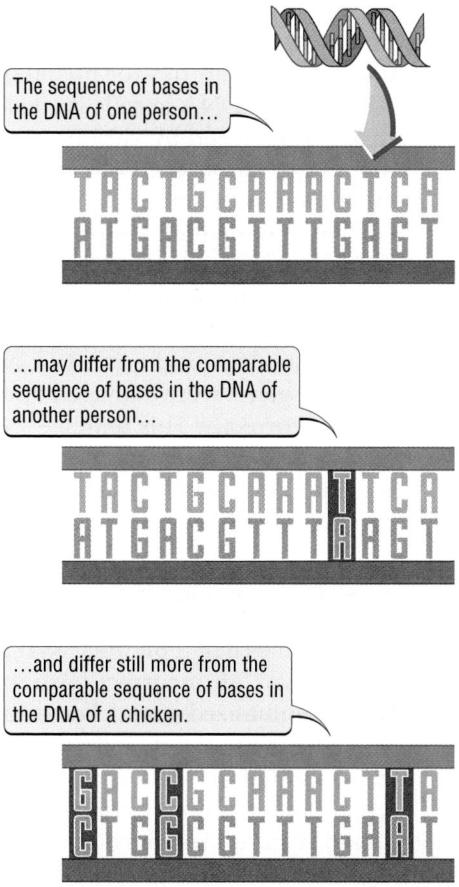

> ■ DNA is a double helix consisting of two long strands of covalently bonded nucleotides held together by hydrogen bonds between the nitrogen-containing bases adenine, cytosine, guanine, and thymine. Adenine pairs only with thymine; cytosine pairs only with guanine. Differences in the sequence of bases are the basis of inherited variation.

14.4 How DNA Is Replicated

As Watson and Crick noted in their historic 1953 paper, the structure of the DNA molecule suggested a simple way that the genetic material could be copied. They elaborated on this suggestion in a second paper, also published in 1953. Because A pairs only with T and C pairs only with G, each strand of DNA contains the information needed to duplicate the other strand. For this reason, Watson and Crick suggested that DNA replication—the duplication of a DNA molecule—could work in the following way (Figure 14.5):

1. The hydrogen bonds connecting the two strands could be broken.
2. Breaking the hydrogen bonds would cause the two strands to unwind and separate.
3. Each strand could then be used as a template for the construction of a new strand of DNA. We now know that the main enzyme involved in the replication of DNA is called DNA polymerase.
4. When this process was completed, there would be two copies of the DNA molecule, each composed of one "old" strand of DNA (from the parent DNA molecule), and one newly synthesized strand of DNA.

Five years later, other researchers confirmed that DNA replication produces DNA molecules composed of one old strand and one new strand, as predicted by Watson and Crick.

The Watson–Crick model of DNA replication is elegant and simple, but the mechanics of actually copying DNA are far from simple. More than a dozen enzymes and proteins are needed to unwind the DNA, to stabilize the separated strands of DNA, to start the replication process, to attach nucleotides to the correct positions

Figure 14.4 Variation in the Sequence of Bases in DNA
The sequence of bases in DNA differs among species and among individuals within a species. Here, the sequence of bases for a hypothetical region of DNA in two humans and a chicken is compared. Base pairs highlighted in red are different from the corresponding base pairs in the first human.

Figure 14.5 DNA Replication

In this overview of DNA replication, the parent DNA strands are blue, and the newly synthesized strands are yellow.

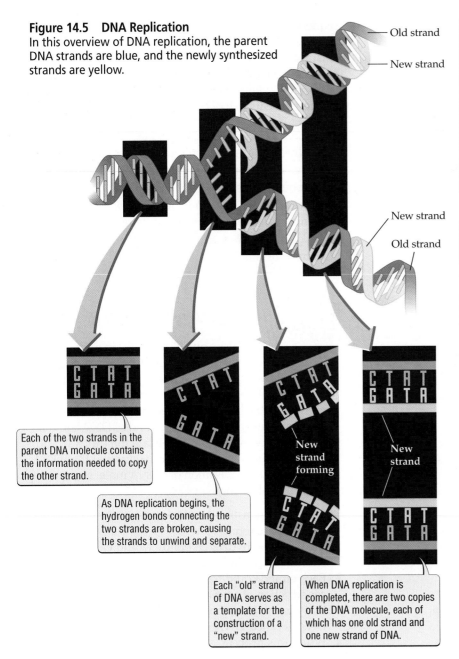

Old strand

New strand

New strand

Old strand

Each of the two strands in the parent DNA molecule contains the information needed to copy the other strand.

As DNA replication begins, the hydrogen bonds connecting the two strands are broken, causing the strands to unwind and separate.

New strand forming

New strand

Each "old" strand of DNA serves as a template for the construction of a "new" strand.

When DNA replication is completed, there are two copies of the DNA molecule, each of which has one old strand and one new strand of DNA.

■ Because A pairs only with T and C pairs only with G, each strand of DNA can serve as a template from which to copy the other strand. DNA replication produces two copies of the DNA molecule, each composed of one old strand from the parent DNA molecule and one newly synthesized strand.

14.5 Repairing Damaged DNA

When DNA is copied, there are many opportunities for mistakes to be made. In humans, for example, more than 3 billion nucleotides must be copied each time a cell divides, so there are over 3 billion opportunities for mistakes. In addition, the DNA in cells is constantly being damaged by various sources, including radiation, collisions with other molecules in the cell, and chemicals, many of which are produced by the cell itself. Failure to repair copying errors and damage to the DNA would be disastrous for the organism: Normal cell functions would grind to a halt if essential genes were damaged or replicated incorrectly. This damage would lead to the death of many cells and, ultimately, to the death of the organism.

Few mistakes are made in DNA replication

The enzymes that replicate DNA sometimes insert an incorrect base in the newly synthesized strand. For example, if a cytosine (C) were inserted across from an adenine (A) located on the template strand, an incorrect C–A pair bond would form instead of the correct T–A pair bond (Figure 14.6). Such mistakes are made about once in every 10,000 bases. Depending on the organism, one or more enzymes checks or "proofreads" the pair bonds as they form, reducing the error rate to about one in every 10 million bases.

When an incorrect base is added and escapes the proofreading mechanism, a **mismatch error** is said to occur. Mismatch errors are so named because the bases

on the template strand, and to join partly replicated fragments of DNA to one another.

Although DNA replication is a complex task, cells can copy DNA molecules containing billions of nucleotides in a matter of hours. This speed is achieved in part by starting the replication of the DNA molecule at thousands of different places at once. Despite their speed, cells make few mistakes when they copy their DNA. As we discuss in the following section, when a mistake does occur, DNA repair operations usually correct it.

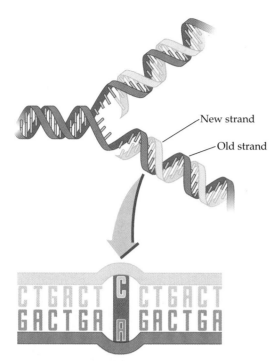

Figure 14.6   Mistakes Can Be Made in DNA Replication
Here, a cytosine (C) has incorrectly been inserted opposite an adenine (A). DNA repair enzymes almost always fix such problems before the DNA is replicated again.

There are three steps in **DNA repair**: The damaged DNA must be recognized, removed, and replaced (Figure 14.7). Different sets of repair proteins specialize in recognizing and removing different types of DNA damage. Once these first two steps have been accomplished, the final step is to add the correct sequence of bases to the damaged strand, replacing the nucleotides that were removed when the damaged section was cut out. This third step of the repair process is the same for

are not correctly matched with each other. Cells contain an additional set of repair proteins that specialize in fixing mismatch errors. These repair proteins fix 99 percent of the remaining errors, reducing the overall chance of an error to the incredibly low rate of one mistake in every billion bases.

On the rare occasions when a mismatch error is not corrected, the DNA sequence is changed. A change in the sequence of bases in DNA, is called a **mutation**. Thus, when a mistake in the copying process is not corrected, a mutation has occurred. Mutations result in the formation of new alleles, including those that cause cancer and other human genetic disorders, such as sickle-cell anemia and Huntington's disease.

Normal gene function depends on DNA repair

The DNA in cells constantly suffers mechanical, chemical, and radiation damage. Every day, the DNA in each of our cells is damaged thousands of times by heat energy and random chemical accidents. The vast majority of the damage to DNA is fixed by a complex set of repair proteins. Single-celled organisms such as yeast have more than 50 different repair proteins, and humans probably have even more.

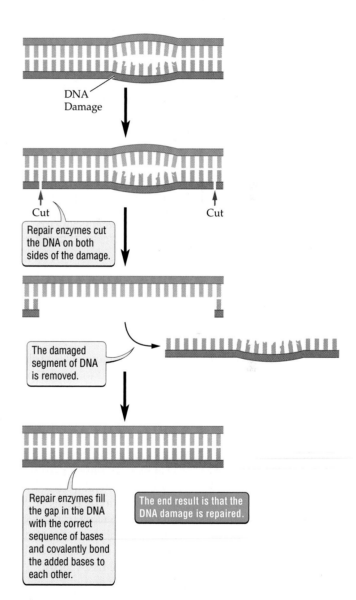

Figure 14.7   DNA Repair
Complex sets of DNA repair proteins work together to fix many types of damage to DNA. Some of these proteins bind to damaged DNA, thereby providing a molecular "tag" that indicates where such damage is located. Other proteins (enzymes) cut out the damaged DNA and replace it with newly synthesized DNA.

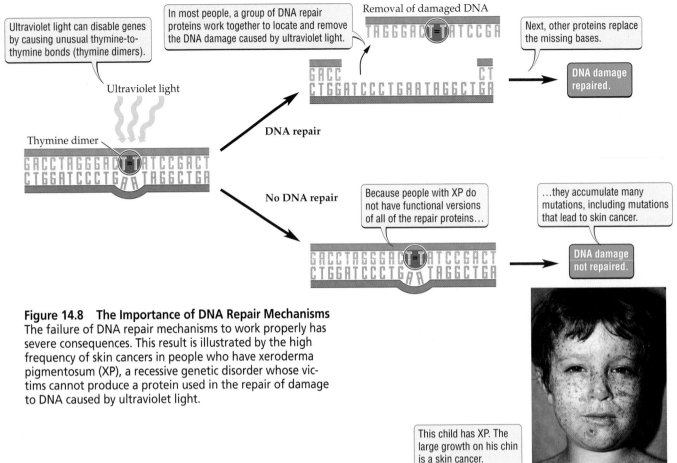

Ultraviolet light can disable genes by causing unusual thymine-to-thymine bonds (thymine dimers).

In most people, a group of DNA repair proteins work together to locate and remove the DNA damage caused by ultraviolet light.

Removal of damaged DNA

Next, other proteins replace the missing bases.

DNA damage repaired.

Ultraviolet light

DNA repair

Thymine dimer

No DNA repair

Because people with XP do not have functional versions of all of the repair proteins…

…they accumulate many mutations, including mutations that lead to skin cancer.

DNA damage not repaired.

Figure 14.8   The Importance of DNA Repair Mechanisms The failure of DNA repair mechanisms to work properly has severe consequences. This result is illustrated by the high frequency of skin cancers in people who have xeroderma pigmentosum (XP), a recessive genetic disorder whose victims cannot produce a protein used in the repair of damage to DNA caused by ultraviolet light.

This child has XP. The large growth on his chin is a skin cancer.

most types of DNA repair, including the correction of mismatch errors.

The importance of DNA repair mechanisms is highlighted by what happens when they fail to work. The child in Figure 14.8 has xeroderma pigmentosum (XP), a recessive genetic disorder in which even brief exposure to sunlight causes painful blisters. In XP, the disease-causing allele (*a*) produces a nonfunctional version of one of our many DNA repair proteins. The normal form of the protein that is disabled in XP functions in the repair of damage to DNA caused by ultraviolet light. The lack of this DNA repair protein makes individuals with XP highly susceptible to skin cancer. Several inherited tendencies to develop cancer, including some types of breast and colon cancer, also appear to be caused by defective versions of genes that control DNA repair.

■ On rare occasions, mistakes are made when DNA is replicated. In addition, DNA suffers mechanical, chemical, and radiation damage. Most of the mutations produced by errors in copying DNA or by damage to DNA are fixed by DNA repair proteins.

■ HIGHLIGHT

Sunburns and Parsnips

The medical literature has many reports of individuals whose skin mysteriously blistered and turned red in a manner resembling a sunburn, but who had not been overexposed to sunlight. Often doctors made the wrong diagnosis when they examined such patients, initially blaming the skin damage on factors such as child abuse, jellyfish stings, or a malignant form of cancer. Sometimes these cases of skin damage appeared in batches, as when workers at a particular grocery store or farm all developed similar symptoms. Other times they were isolated events: a child who was playing with limes at the beach, or a woman who drank a home remedy of celery broth (which, ironically, was meant to improve a skin condition).

The common thread in these examples is a class of chemicals called psoralens (pronounced SORE-uh-luns). These chemicals are found in plants such as celery, limes, and parsnips. Many plants contain chemicals that can cause rashes or other problems when they contact human skin. But psoralens are a skin irritant with a twist: If the plant were handled or eaten at night or in a room with no sunlight, it would have no effect.

Parsnips

Why do psoralens cause skin problems only in sunlight? The answer is found in the way they work. Sunlight contains a type of light known as ultra-violet, or UV, light. When in the presence of UV light, psoralens can insert themselves into the DNA double helix, bridging the two strands of the DNA molecule and disrupting its normal function. This damage can kill the cells whose DNA is under assault, causing changes in skin color and painful blisters. Such effects are rare and typically occur only in people that have handled a large number of psoralen-containing plants.

The attack of psoralens on DNA mimics the damage that UV light causes on its own. UV light damages DNA by causing unusual chemical bonds to form within the DNA molecule (see Figure 14.8). If the damage is severe enough, the cell dies. Sunburns result from massive damage to the DNA of skin cells caused by UV light. This damage causes the death and removal (by peeling) of many skin cells. Cells that survive try to repair the damage, but some damage remains, resulting in mutations of the DNA. If such a mutation occurs in certain genes, such as genes that control DNA repair or suppress the growth of tumors, a sunburn can be the first step on a path that leads ultimately to skin cancer.

Knowledge of what UV light does to your DNA suggests precautionary actions: While enjoying the outdoors, you should always protect yourself from overexposure to the sun. This protection is especially important if you have a higher-than-average risk of developing skin cancer. You should use special caution if you have any of the following characteristics: (1) fair skin; (2) a family history of cancer; (3) patches of skin that remain sore, dry, or scaly; or (4) moles that are frequently subject to rubbing or other forms of irritation. Skin cancer can strike people of all ages, so if you find a suspicious growth on your skin, such as those shown in Figure 14.9, consult a doctor immediately. Prompt action can save your life.

■ Chemicals called psoralens, found in certain plant species, can damage DNA in the presence of UV light. UV light also can damage DNA on its own. Overexposure to psoralens or UV light can damage DNA so greatly that the cell's DNA repair mechanisms fail and the cell dies. Mutations of DNA that result from sunburn can be the first step on the path to skin cancer.

Examples of skin cancer

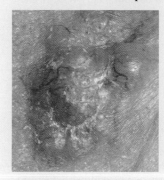 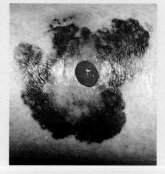

Basal cell carcinoma   Malignant melanoma

Figure 14.9   Sunburn Can Cause Cancer
Overexposure to ultraviolet light can cause DNA mutations that transform normal skin cells into cancer cells.

SUMMARY

The Search for the Genetic Material

■ Geneticists initially thought that protein was the genetic material. Three landmark experiments showed that this initial view was wrong and that DNA, not protein, is the genetic material.

■ The first experiment showed that harmless strain R bacteria could be transformed into deadly strain S bacteria when exposed to heat-killed strain S bacteria.

■ The second experiment showed that only the DNA from heat-killed strain S bacteria was able to transform strain R bacteria into strain S bacteria.

- The third experiment showed that the DNA of a virus, not its protein, was responsible for taking over a bacterial cell and producing new viruses.

The Three-Dimensional Structure of DNA

- In 1953, James Watson and Francis Crick determined that DNA is a double helix.

- The double helix is formed by two long strands of covalently bonded nucleotides. The two strands of nucleotides are held together by hydrogen bonds between the nitrogen-containing bases: adenine, cytosine, guanine, and thymine.

- An adenine on one strand pairs only with a thymine on the other strand. Similarly, a cytosine on one strand pairs only with a guanine on the other strand.

- These specific base-pairing rules indicate that each strand can serve as a template from which the other strand can be copied.

- The sequence of bases in DNA, which differs among species and among individuals within a species, is the basis of inherited variation.

How DNA Is Replicated

- Because A pairs only with T and C pairs only with G, each strand of DNA contains the information needed to duplicate the other strand.

- A complex set of enzymes and other proteins guides the replication of DNA. To replicate DNA, these enzymes must first break the hydrogen bonds connecting the two nucleotide strands.

- Breaking of the hydrogen bonds causes the two strands to unwind and separate. Each of these strands is then used as a template from which to build a new strand of DNA.

- DNA replication produces two copies of the DNA molecule, each composed of one old strand of DNA (from the parent DNA molecule) and one newly synthesized strand of DNA.

Repairing Damaged DNA

- On rare occasions, mistakes occur when DNA is copied. Mistakes in the copying process introduce mutations, which are changes in the sequence of bases in DNA.

- The DNA in each of our cells is altered thousands of times every day by mechanical, chemical, and radiation damage. If DNA damage in an organism's cells were not repaired, the cells, and ultimately the organism, would die.

- Mutations produced by replication errors or by damage to DNA are fixed by DNA repair proteins.

- Several inherited genetic disorders result from the failure of DNA repair proteins to work properly.

Highlight: Sunburns and Parsnips

- Certain plant species contain psoralens, which can cause painful blisters if the plant is extensively handled or eaten in the presence of UV light.

- In the presence of UV light, psoralens damage DNA. Acting on its own, UV light also can damage DNA. Overexposure to psoralens or UV light can damage DNA so greatly that the cell's DNA repair mechanisms fail and the cell dies.

- Mutations of DNA that result from sunburn can be the first step on the path to skin cancer.

KEY TERMS

base p. 243	double helix p. 243
base pairing p. 244	mismatch error p. 246
DNA polymerase p. 245	mutation p. 247
DNA repair p. 247	nucleotide p. 243
DNA replication p. 245	transformation p. 241

CHAPTER REVIEW

Self-Quiz

1. The base-pairing rules for DNA are
   a. any combination of bases is allowed.
   b. T pairs with C, A pairs with G.
   c. A pairs with T, C pairs with G.
   d. C pairs with A, T pairs with G.

2. DNA replication results in
   a. two DNA molecules, one with two old strands and one with two new strands.
   b. two DNA molecules, each of which has two new strands.
   c. two DNA molecules, each of which has one old strand and one new strand.
   d. none of the above

3. Experiments performed by Oswald Avery and colleagues showed that
   a. protein, not DNA, transformed bacteria.
   b. DNA, not protein, transformed bacteria.
   c. carbohydrates, not protein, transformed bacteria.
   d. both DNA and protein were able to transform bacteria.

4. The DNA of cells is damaged
   a. thousands of times per day.
   b. by collisions with other molecules, chemical accidents, and radiation.
   c. not very often and only by radiation.
   d. both a and b

5. The DNA of different species differs in
   a. the sequence of bases.
   b. the base-pairing rules.
   c. the number of nucleotide strands.
   d. the location of the sugar–phosphate backbone in the DNA molecule.

Review Questions

1. Explain why the structure of DNA proposed by Watson and Crick immediately suggested a way DNA could be replicated.

2. Explain how the following three things are related:
   a. the sequence of bases in DNA
   b. mutations
   c. alleles that cause human genetic disorders

3. Explain how a mutation that disables a DNA repair protein can lead to cancer.

14

The Daily Globe

Saving Our Children From Cancer

To the Editor:

When surveyed, most people say they fear cancer. So why have Americans chosen to ignore the biggest news of all about this frightening disease? According to scientists, the vast majority of cancers are not inherited, but are caused by mutations to your DNA that occur over the course of your lifetime. Scientists have also shown that diet is strongly linked to cancer, so it appears that many cancer-causing mutations can be avoided by simple changes in what you eat.

Yet despite this good news, Americans continue to eat a deadly diet—high in animal fat and salt and low in fiber, fruits, and vegetables. The result? In North America and Europe alone, nearly 400,000 people die each year from cancers that could have been prevented by changes in diet.

To save our children from having preventable cancers later in life, we should encourage them to have healthy eating habits. Unfortunately, one look at the menu of any public school cafeteria reveals a cancer-causing diet: hamburgers, french fries, pepperoni pizza. And with all the desserts available—cakes, cookies, and candy bars—what chance is there that these children will eat the lifesaving fruits and vegetables that they are served?

It's time for America to face the facts: We are using tax dollars to serve up cancer every day at our nation's schools. Americans for Healthy Living are calling for the immediate removal of all meat, nonfruit desserts, and other dangerous foods from public schools. We routinely remove asbestos from our schools because of the cancer risk it poses. Asbestos is linked to fewer cases of cancer than is diet, so it's time for the hamburgers to go, too.

Bob Pinter
Americans for Healthy Living

Evaluating "The News"

1. Is there a difference between removing asbestos from a school and removing hamburgers from a school?

2. There is overwhelming evidence that the link between diet and cancer is real. However, scientists still don't know exactly how what you eat translates into a risk of, or protection from, cancer. Should that uncertainty alter whether or not public policy—such as what children are given to eat at school—is changed?

3. Smoking causes 30 percent of all cancer deaths. The drinking of very hot beverages and overexposure to sunlight contribute a few more percentage points each to the total number of cancer deaths. Yet drinking hot beverages and sitting in the sun are allowed at schools across the country. Many high schools even provide areas where students can smoke. Should these activities be banned for sake of the health of the nation's children?

chapter 15 *From Gene to Protein*

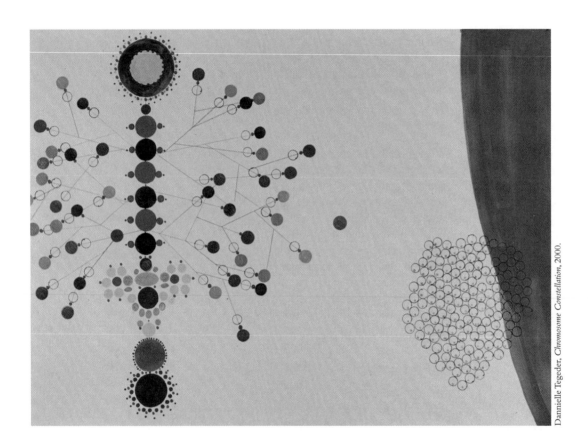

Dannielle Tegeder, *Chromosome Constellation*, 2000.

Finding the Messenger and Breaking the Code

We live in a global economy. The headquarters of a corporation may be located in one country—say, Germany—but the company's factories may be located elsewhere—say, the United States. Immediately there is a problem: Decisions made in Germany need to be communicated to employees in the United States. This problem is easy to solve: A message is sent from one location to the other. In addition, the message must be translated from German, the language in which the decision was made, to English, the language in which the decision must be implemented.

Eukaryotic cells face similar challenges. Genes work by controlling the production of proteins. Whereas genes are located in the nucleus of the cell, their protein products are made on ribosomes,

252

A Model of a tRNA Molecule
Transfer RNA molecules like this one read the genetic code.

which are located outside of the nucleus, in the cytoplasm. How does a gene control from a distance how a ribosome constructs a protein? Like our imaginary corporate headquarters, the gene does this by sending a message. What is the chemical messenger that carries the gene's instructions from the nucleus to the ribosome? And once the message reaches the ribosomes, how do the ribosomes "read" it?

This last question highlights another similarity between cells and our imaginary international corporation: To be effective, the information contained in genes must be translated from one "language" (that of DNA, which is based on its four nucleotide bases) to another (that of proteins, which is based on the 20 amino acids they contain). In the mid-1950s, biologists working on this problem realized that cells must have a "genetic code" that allows the instructions of the gene to be translated from the language of DNA to the language of proteins. The discovery of how cells do this was one of the crowning achievements of twentieth-century science: the breaking of the genetic code.

■ KEY CONCEPTS

1. Most genes contain instructions for building proteins. The DNA sequence of a gene encodes the amino acid sequence of its protein product.

2. A few genes encode RNA molecules as their final product.

3. Two steps are required to go from gene to protein: transcription and translation.

4. In eukaryotic cells, transcription occurs in the nucleus and produces a messenger RNA copy of the information stored in the gene.

5. Translation occurs in the cytoplasm and converts the sequence of bases in an mRNA molecule to the sequence of amino acids in a protein.

6. Mutations can alter the sequence of amino acids in a gene's protein product. Such changes, in turn, can alter the protein's function. Although changes in protein function are usually harmful, occasionally they benefit the organism.

Chapters 12 through 14 have described how genes are inherited, where they are located (on chromosomes), and what they are made of (DNA). But we have yet to describe how genes work. How do genes store the information needed to build their final products, proteins? How does the cell use that information? Knowing how genes work can help us understand how mutations produce new phenotypes, including disease phenotypes. We begin this chapter by discussing how genetic information is encoded in genes and how the cell uses that information to build proteins. We then discuss how a change to a gene can change an organism's phenotype.

15.1 Genes Encode Proteins

Genes work by controlling the production of proteins. For most genes, the relationship between gene and protein is direct: The gene contains instructions for how to build a particular protein. Some genes have an indirect relationship to proteins. Rather than encoding a particular protein, these genes specify how to build ribonucleic acid (RNA) molecules that help the cell construct proteins.

Proteins are essential to life. They are used by cells and organisms in many ways: Some provide structural support, others transport materials, still others defend against disease-causing organisms. In addition, the many chemical reactions on which life depends are controlled by a crucial group of proteins, the enzymes. Enzymes and other proteins influence so many features of the organism that they, along with the environment, determine the organism's phenotype.

Early clues that genes work by controlling the production of proteins came at the beginning of the twentieth century from the work of British physician Archibald

Garrod, who studied several inherited human metabolic disorders. In 1902, he argued that these disorders were caused by an inability of the body to produce specific enzymes. Garrod was particularly interested in alkaptonuria, a condition in which the urine of otherwise healthy infants turns black when exposed to air. He proposed that infants with alkaptonuria had a defective version of an enzyme that ordinarily would break down the substance that caused urine to turn black. Garrod did not stop there; he and his collaborator, William Bateson, went on to suggest that in general, genes worked by controlling the production of enzymes.

Genes contain information for the synthesis of RNA molecules

Garrod and Bateson were on the right track, but they were not entirely correct: Genes control the production of all proteins, not just enzymes. In addition, as mentioned above, a few genes do not directly specify proteins. Rather, these genes specify as their final product one of several RNA molecules used in the construction of proteins. Thus, directly and indirectly, genes control the production of proteins.

As we will see shortly, even genes that specify proteins make an RNA molecule as their initial product. Thus, modifying the definition in Chapter 12, we can redefine a **gene** as a sequence of DNA that contains information for the synthesis of one of several types of RNA molecules used to make proteins.

Three types of RNA are involved in the production of proteins

The nucleic acids DNA and RNA (see Chapter 5) play key roles in the construction of proteins. Several types of RNA, as well as many enzymes and other proteins, are required for the cell to make proteins. As already described, DNA controls the production of all these essential molecules, so DNA controls all aspects of protein production.

Cells use three types of RNA molecules to construct proteins: **ribosomal RNA** (abbreviated **rRNA**), **messenger RNA** (**mRNA**), and **transfer RNA** (**tRNA**). The function of each kind of RNA is defined in Table 15.1 and discussed in more detail in the sections that follow. But first we describe several general differences between the structure of RNA and the structure of DNA. Whereas DNA molecules are double-stranded, most RNA molecules are single-stranded. Overall, the structure of a single strand of RNA is similar to the structure of a single strand of DNA. RNA is composed of a long string of nucleotides covalently bonded together. Each nucleotide, in turn, is composed of a sugar–phosphate backbone and one of four nitrogen-containing bases (Figure 15.1). However, the nucleotides in RNA and DNA differ in two respects: (1) RNA uses the sugar ribose, whereas DNA uses the sugar deoxyribose; and (2) in RNA, the base uracil (U) replaces the DNA base thymine (T). The other three bases (A, C, and G) are the same in RNA and DNA.

■ Genes work by controlling the production of proteins. A gene contains information for the synthesis of one of several types of RNA molecules used to make proteins. Three types of RNA and many enzymes and other proteins are required for this process. RNA consists of a single strand of nucleotides, each composed of a sugar–phosphate backbone and one of four nitrogen-containing bases: A, C, G, or U.

How Genes Control the Production of Proteins

In both prokaryotes and eukaryotes, genes specify the production of proteins in two steps: transcription and translation. In transcription, an RNA molecule is made from the DNA sequence of a gene. If the gene specifies the production of rRNA or tRNA, its transcription produces those molecules as the final product. If the gene specifies a protein, however, its transcription produces an mRNA molecule, which in turn specifies the amino acid sequence of a protein. In translation,

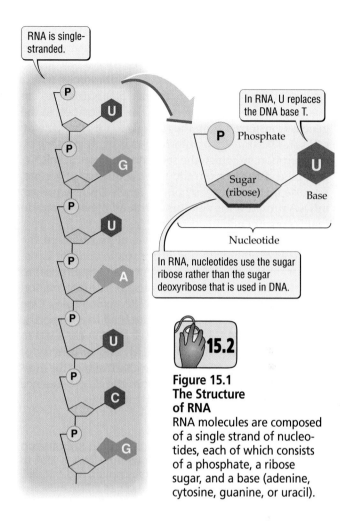

RNA is single-stranded.

In RNA, U replaces the DNA base T.

P Phosphate

Sugar (ribose)

Base

Nucleotide

In RNA, nucleotides use the sugar ribose rather than the sugar deoxyribose that is used in DNA.

15.2

Figure 15.1 The Structure of RNA
RNA molecules are composed of a single strand of nucleotides, each of which consists of a phosphate, a ribose sugar, and a base (adenine, cytosine, guanine, or uracil).

rRNA, mRNA, and tRNA molecules direct the synthesis of a given gene's protein product.

We will discuss transcription and translation in detail later in the chapter. First, however, let's consider how genes work from the perspective of information flow. We will describe the flow of genetic information in eukaryotes. Events are similar in prokaryotes except that, because they lack a nucleus, both genes and ribosomes are located in the cytoplasm.

15.1 RNA Molecules and Their Functions

Type of RNA molecule	Function
Ribosomal RNA (rRNA)	Major component of ribosomes, the molecular machines that make the covalent bonds that link amino acids together into a protein
Messenger RNA (mRNA)	Specifies the order of amino acids in a protein
Transfer RNA (tRNA)	Transfers the correct amino acid to the ribosome, on the basis of the information encoded in the mRNA

For a protein to be made, the information in the gene must be sent from the gene, which is located in the nucleus, to the site of protein synthesis. As we learned in Chapter 6, proteins are synthesized at ribosomes, which are located in the cytoplasm (Figure 15.2). Ribosomes are made up of rRNA and proteins. The information in the gene is transferred from the nucleus to the ribosome by mRNA. This transfer of information is made possible by transcription, in which the sequence of bases in mRNA is copied directly from the DNA sequence of a gene. Because it is a direct copy of the gene's DNA sequence, mRNA provides the ribosome with all the information that is contained in the gene.

Once the mRNA molecule arrives at the ribosome, the information on how to build the protein must be translated from the language of DNA (nucleotide bases) to the language of proteins (amino acids). The information is translated at the ribosomes by tRNA molecules. One portion of each tRNA molecule can bind to one specific sequence of mRNA, while another portion can bind to and carry one specific amino acid. The specificity of these binding rules is essential because it allows the message in mRNA to be translated into the exact sequence of amino acids that is called for by the gene.

Figure 15.2 The Flow of Genetic Information
Genetic information flows from DNA to RNA to protein in two steps, transcription and translation. Transcription produces an mRNA molecule, which then moves to the ribosome, where translation occurs and the protein is made. Different amino acids in the protein being constructed at the ribosome are represented by different colors. This illustration shows the flow of information in a eukaryotic cell.

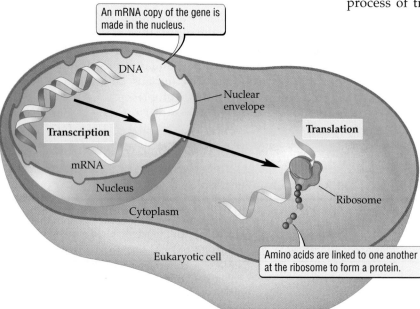

An mRNA copy of the gene is made in the nucleus.

DNA

Nuclear envelope

Transcription

Translation

mRNA

Nucleus

Ribosome

Cytoplasm

Eukaryotic cell

Amino acids are linked to one another at the ribosome to form a protein.

■ Two steps are required for the synthesis of proteins: transcription and translation. In transcription, an mRNA molecule is made using the DNA sequence of the gene. In translation, rRNA, mRNA, and tRNA molecules direct protein synthesis. Information for the synthesis of a protein flows from the gene, located in the nucleus, to the site of protein synthesis, the ribosome, located in the cytoplasm.

15.4 Transcription: From DNA to RNA

Transcription is the synthesis of an RNA molecule using the DNA sequence of a gene as a template. As we saw above, transcription produces all the kinds of RNA molecules we have mentioned; here, we focus on the transcription of mRNA from genes that encode proteins.

Transcription is somewhat similar to DNA replication in that one strand of DNA is used as a template from which a new strand—in this case, a strand of mRNA—is formed. However, transcription differs from DNA replication in three important ways. First, a different enzyme guides the process: Whereas the enzyme used in DNA replication is DNA polymerase, the key enzyme in transcription is RNA polymerase. Second, whereas the entire DNA molecule is duplicated in DNA replication, in transcription only the small portion of a DNA molecule that includes a particular gene is transcribed into mRNA. Finally, whereas the process of DNA replication produces a double-stranded DNA molecule, the process of transcription produces a single-stranded mRNA molecule.

Transcription begins when the RNA polymerase enzyme binds to a region of DNA that is called a **promoter**. Promoters contain specific base sequences to which RNA polymerase can bind. Once bound to the promoter, the **RNA polymerase** enzyme unwinds the DNA double helix and separates the two strands. Then the enzyme begins to construct an mRNA molecule (Figure 15.3).

As discussed earlier, RNA molecules consist of four bases: adenine (A), cytosine (C), guanine (G), and uracil (U). These four bases pair with the four bases in DNA according to specific rules: A in RNA pairs with T in DNA,

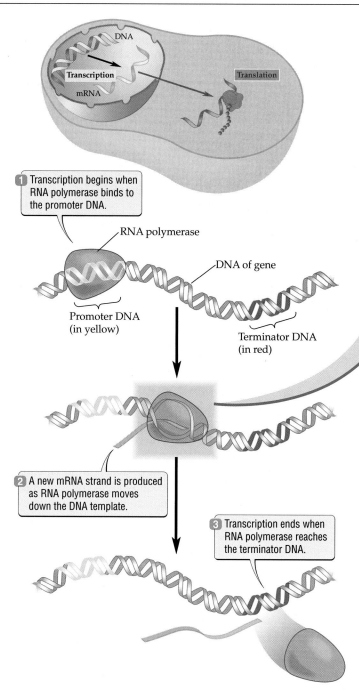

1 Transcription begins when RNA polymerase binds to the promoter DNA.

RNA polymerase

DNA of gene

Promoter DNA (in yellow)

Terminator DNA (in red)

2 A new mRNA strand is produced as RNA polymerase moves down the DNA template.

3 Transcription ends when RNA polymerase reaches the terminator DNA.

Figure 15.3 Overview of Transcription

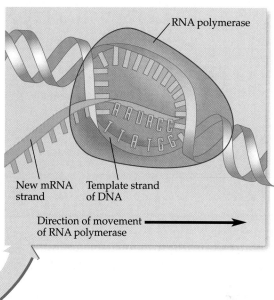

RNA polymerase

New mRNA strand

Template strand of DNA

Direction of movement of RNA polymerase

an mRNA molecule synthesized from this DNA template would have the sequence

AAUACCGUGGC

Synthesis of an mRNA molecule from the DNA template continues until the RNA polymerase enzyme reaches a sequence of bases called a terminator, at which point transcription ends and the newly formed mRNA molecule separates from its DNA template. The two strands of the DNA template then bond back to each other, ready to be used again when needed by the cell.

Messenger RNA molecules produced by transcription will carry the information from the gene to the ribosome, where the protein specified by the gene is built. First, however, the newly formed mRNA molecule must often be modified before it is used. In eukaryotes, most genes contain internal sequences of bases called **introns** that do not specify part of the protein encoded by the gene (Figure 15.4). Introns must be removed from the initial mRNA product if the protein encoded by the gene is to function properly.

C in RNA pairs with G in DNA, G in RNA pairs with C in DNA, and U in RNA pairs with A in DNA. These base-pairing rules determine the sequence of bases in the mRNA molecule that is made from a DNA template. For example, if the DNA sequence were

TTATGGCACCG

■ In transcription, an mRNA molecule is synthesized from a gene's DNA template. In eukaryotes, genes contain internal sequences of DNA called introns that must be removed from the initial mRNA product.

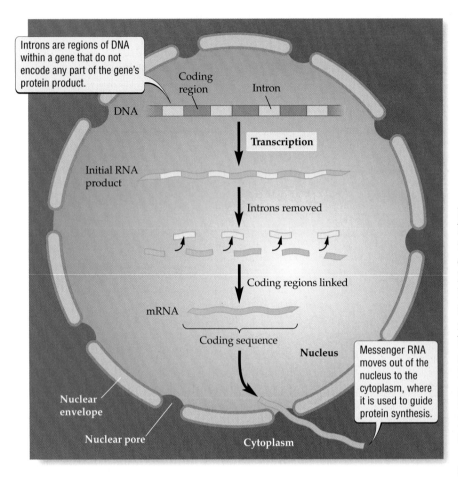

Introns are regions of DNA within a gene that do not encode any part of the gene's protein product.

Coding region

Intron

DNA

Transcription

Initial RNA product

Introns removed

Coding regions linked

mRNA

Coding sequence

Nucleus

Messenger RNA moves out of the nucleus to the cytoplasm, where it is used to guide protein synthesis.

Nuclear envelope

Nuclear pore

Cytoplasm

Figure 15.4 Removal of Introns by Eukaryotic Cells Before mRNA molecules can be used, enzymes in the nucleus must remove noncoding sequences (introns) and link the remaining coding sequences together.

The Genetic Code

Genes are composed of DNA and consist of a sequence of the four bases adenine, cytosine, guanine, and thymine. The information in a gene is encoded in its sequence of bases. As we learned in the previous section, the gene's DNA sequence is used during transcription as a template to produce an mRNA molecule. How does the sequence of bases in mRNA encode, or specify, the sequence of amino acids in a protein? From the late 1950s to the mid-1960s, geneticists worked feverishly to figure out the code that accomplishes this task. By 1966, the genetic code had been broken (Figure 15.5).

In the **genetic code**, an amino acid is specified by a nonoverlapping sequence of three nucleotide bases in an mRNA molecule. Each group of three bases in mRNA is called a **codon**. For example, as Figure 15.6 shows, if a portion of an mRNA molecule consisted of the sequence UUCACUCAG, the first codon (UUC) would specify one amino acid (phenylalanine), the next codon (ACU) would specify a second amino acid (threonine), and the last codon (CAG) would specify a third amino acid (glutamine). There are four possible bases at each of the three positions of a codon, so there are a total of 64 possible codons $(4 \times 4 \times 4 = 64)$.

When reading the code, the cell begins at a fixed starting point, called a start codon (usually the codon AUG), and ends at one of several stop codons (such as UGA or UAA) (see Figure 15.5). By beginning at a fixed point, the cell

Figure 15.5 The Genetic Code

UAA, UAG, and UGA do not code for an amino acid. Translation stops when these codons are reached.

Like arginine, most amino acids are specified by more than one codon.

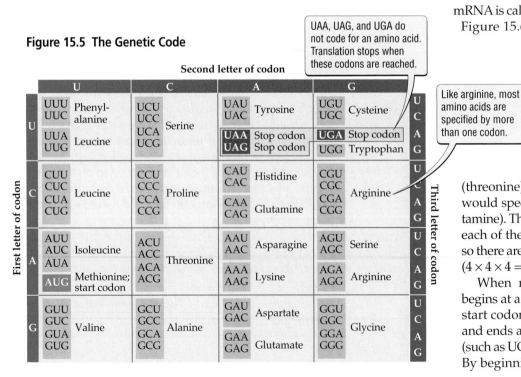

	U	C	A	G	
U	UUU UUC Phenyl-alanine UUA UUG Leucine	UCU UCC UCA UCG Serine	UAU UAC Tyrosine UAA Stop codon UAG Stop codon	UGU UGC Cysteine UGA Stop codon UGG Tryptophan	U C A G
C	CUU CUC CUA CUG Leucine	CCU CCC CCA CCG Proline	CAU CAC Histidine CAA CAG Glutamine	CGU CGC CGA CGG Arginine	U C A G
A	AUU AUC Isoleucine AUA AUG Methionine; start codon	ACU ACC ACA ACG Threonine	AAU AAC Asparagine AAA AAG Lysine	AGU AGC Serine AGA AGG Arginine	U C A G
G	GUU GUC GUA GUG Valine	GCU GCC GCA GCG Alanine	GAU GAC Aspartate GAA GAG Glutamate	GGU GGC GGA GGG Glycine	U C A G

Second letter of codon

First letter of codon

Third letter of codon

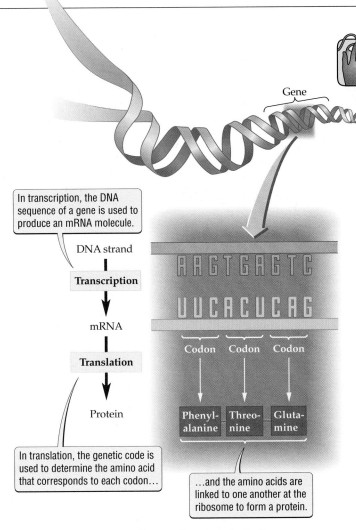

In transcription, the DNA sequence of a gene is used to produce an mRNA molecule.

DNA strand

Transcription

mRNA

Translation

Protein

In translation, the genetic code is used to determine the amino acid that corresponds to each codon…

Gene

A A G T G A G T C

U U C A C U C A G

Codon   Codon   Codon

Phenyl-alanine   Threo-nine   Gluta-mine

…and the amino acids are linked to one another at the ribosome to form a protein.

Figure 15.6 How Cells Use the Genetic Code

ensures that the message from the gene does not become scrambled. To see why this is important, use Figure 15.5 to determine the amino acid sequence that would results if the sequence UUCACUCAG in Figure 15.6 were read in codons that began with the second U, not the first.

■ The information in a gene is encoded in its sequence of bases. In the genetic code, an amino acid is specified by a codon, a nonoverlapping sequence of three nucleotide bases in an mRNA molecule. When reading the genetic code, the cell begins at a fixed starting point, thus ensuring that the message from the gene does not become scrambled.

Translation: From mRNA to Protein

15.8

The genetic code provides the cell with the equivalent of a dictionary with which to translate the language of DNA into the language of proteins. The conversion of a sequence of bases in mRNA to a sequence of amino acids in a protein is called **translation**. Translation is the second major step in the process by which genes specify proteins (see Figure 15.2). It occurs at the ribosomes, which are composed of several different sizes of rRNA molecules and more than 50 different proteins. The ribosomes are molecular machines that make the covalent bonds that link amino acids together into a protein. As a major component of ribosomes, rRNA plays a central role in protein synthesis (see Table 15.1).

Transfer RNA molecules also play a crucial role in the synthesis of proteins at ribosomes. There are several types of tRNA molecules, but they all have a similar structure with two binding sites (Figure 15.7). First, each tRNA molecule has a particular sequence of nucleotide bases, called an **anticodon**, that can bind to a particular mRNA codon. Each tRNA molecule also has a site that can bind to a particular amino acid. Each tRNA molecule can carry the specific amino acid that is called for by the mRNA codon to which it can bind. If this were not the case, the genetic code would not work.

Figure 15.7 Transfer RNA (tRNA)
Translation is accomplished by tRNA molecules, which bind to specific mRNA codons and to specific amino acids. Shown here are a computer model (at left) and a simplified illustration (at right) of a tRNA molecule. Similar regions in the computer model and simplified illustration of a tRNA molecule are drawn in matching colors. The site at which tRNA binds to an mRNA codon is called the anticodon.

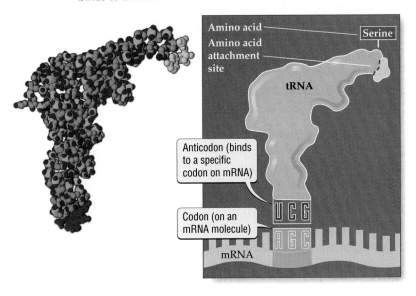

Amino acid

Amino acid attachment site

Serine

tRNA

Anticodon (binds to a specific codon on mRNA)

Codon (on an mRNA molecule)

U C G

A G C

mRNA

15.9   **Figure 15.8 Translation**

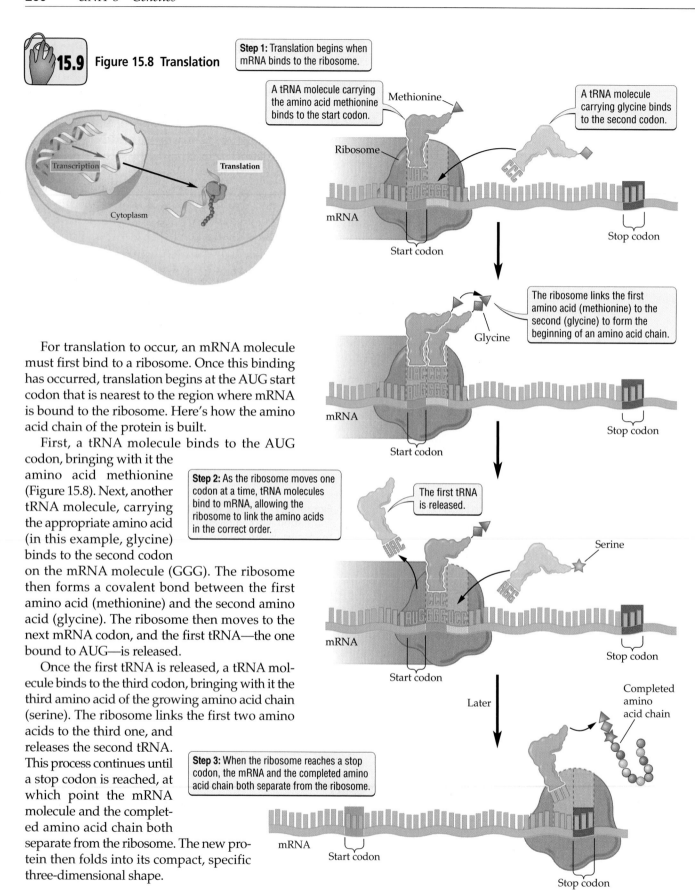

Step 1: Translation begins when mRNA binds to the ribosome.

A tRNA molecule carrying the amino acid methionine binds to the start codon.

Methionine

Ribosome

A tRNA molecule carrying glycine binds to the second codon.

mRNA

Start codon

Stop codon

The ribosome links the first amino acid (methionine) to the second (glycine) to form the beginning of an amino acid chain.

Glycine

mRNA

Start codon

Stop codon

Step 2: As the ribosome moves one codon at a time, tRNA molecules bind to mRNA, allowing the ribosome to link the amino acids in the correct order.

The first tRNA is released.

Serine

mRNA

Start codon

Stop codon

Later

Completed amino acid chain

Step 3: When the ribosome reaches a stop codon, the mRNA and the completed amino acid chain both separate from the ribosome.

mRNA

Start codon

Stop codon

Transcription

Translation

Cytoplasm

For translation to occur, an mRNA molecule must first bind to a ribosome. Once this binding has occurred, translation begins at the AUG start codon that is nearest to the region where mRNA is bound to the ribosome. Here's how the amino acid chain of the protein is built.

First, a tRNA molecule binds to the AUG codon, bringing with it the amino acid methionine (Figure 15.8). Next, another tRNA molecule, carrying the appropriate amino acid (in this example, glycine) binds to the second codon on the mRNA molecule (GGG). The ribosome then forms a covalent bond between the first amino acid (methionine) and the second amino acid (glycine). The ribosome then moves to the next mRNA codon, and the first tRNA—the one bound to AUG—is released.

Once the first tRNA is released, a tRNA molecule binds to the third codon, bringing with it the third amino acid of the growing amino acid chain (serine). The ribosome links the first two amino acids to the third one, and releases the second tRNA. This process continues until a stop codon is reached, at which point the mRNA molecule and the completed amino acid chain both separate from the ribosome. The new protein then folds into its compact, specific three-dimensional shape.

■ In translation, a sequence of bases in mRNA is converted into a sequence of amino acids in a protein. Transfer RNA molecules carry the specific amino acids called for by the mRNA to the ribosome, which links them together to form a protein molecule.

15.10 The Effect of Mutations on Protein Synthesis

In Chapter 13 we discussed two types of mutations: those that affect individual genes and those that alter the number or structure of chromosomes. Both types of mutations alter the DNA of a cell. In general, we can define a mutation as a change in either the sequence or the amount of DNA found in a cell. Mutations range in extent from a change in the identity of a single base pair to the addition or deletion of one or more chromosomes.

In this section we describe how mutations affect protein synthesis. Here we focus on mutations that occur in the portions of a gene that encode proteins, rather than mutations that occur in introns or mutations that affect entire chromosomes.

Many mutations alter a single base pair

Many mutations are changes in a single base pair of a gene's DNA sequence. There are three major types of such mutations: substitution, insertion, and deletion mutations. In a **substitution mutation**, one base is changed to another at a single position in the DNA sequence of the gene. In the substitution mutation shown in Figure 15.9, for example, the sequence of the gene is changed when a thymine (T) is replaced by a cytosine (C). As the figure shows, this particular change causes the substitution of one amino acid for another because the mRNA codons made from the DNA sequences TAA and CAA encode different amino acids (isoleucine and valine, respectively).

Not all substitution mutations lead to changes in the amino acid sequence of a protein. For example, although a change in the DNA sequence from GGG to GGA would alter the mRNA sequence from CCC to CCU, since both CCC and CCU code for the same amino acid, glycine (see Figure 15.5), this change would not alter the amino acid sequence of the protein. In such cases, the substitution mutation is said to be "silent" because it produces no change in the structure of the protein, and thus no change in the phenotype.

Insertion or **deletion mutations** occur when a single base is inserted into or deleted from a DNA sequence. Such mutations alter the identity of many of the amino acids in the encoded protein (see Figure 15.9), which usually prevent the protein from functioning properly. When an insertion or deletion mutation occurs, it is said to cause a **frameshift**. A genetic frameshift is similar to what happens if you accidentally record the answer to a question twice on the answer sheet of a multiple choice test: All the answers from that point forward are likely to be wrong, since each is an answer to the previous question. Similarly, a frameshift shifts all the codons by one base, scrambling the message.

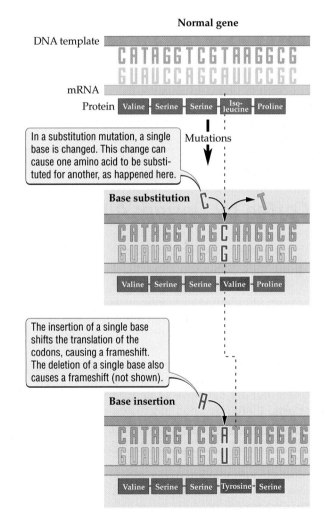

Figure 15.9 Effects of DNA Mutations
Changes to the sequence of bases in a gene can change the sequence of amino acids in the gene's protein product. DNA bases, RNA bases, and amino acids that change as a result of a mutation are shown in red.

Insertions or deletions of a series of bases are common

Mutations can alter more than one base pair of the DNA sequence of a gene. For example, from two to thousands of bases can be inserted within a DNA sequence. Unless the number of bases that are inserted is a multiple of three (the length of a complete codon), such an insertion will cause a frameshift, usually preventing the protein from functioning properly. Even if no frameshift occurs, the insertion of extra bases into a gene causes extra amino acids to be added to the protein being built, often destroying protein function and hence harming the organism.

Mutations can cause a change in protein function

Mutations alter the DNA sequence of a gene, which in turn alters the sequence of bases in an mRNA molecule made from the gene. Changes to the sequence of bases in mRNA can have a wide range of effects. If the change is a substitution of a single base and does not result in the substitution of one amino acid for another, the structure and function of the protein are not changed. Even if one or a few amino acids of the protein are changed, there may be little or no effect. For example, changes that do not alter the region of the protein that binds to its substrate may not affect protein function. However, amino acid changes that alter the binding region of a protein usually do change how the protein works (see the box below). Such changes often are harmful to the organism because they decrease or destroy protein function. On rare occasions, changes to the binding region of a protein benefit the organism by improving its efficiency, or by causing the protein to take on a new and useful function.

> ■ Many mutations are caused by the substitution, insertion, or deletion of a single base pair. Insertions or deletions of a series of bases are also common. Mutations can change the function of a gene's protein product.

■ BIOLOGY IN OUR LIVES

Coffee by Design: All the Taste, None of the Jolt

Java. The aroma of a hot cup of coffee, its taste, and the "kick start" it provides makes coffee a beverage of choice for many people. But coffee has a dark side, with side effects that include increased blood pressure, insomnia, anxiety, tremors, heart palpitations, and gastrointestinal disturbances. Because many people love coffee but are sensitive to its side effects, there has been increasing demand for decaffeinated coffee. Unfortunately, flavors and aromas can be lost in the decaffeination process, making decaf a less attractive choice than it otherwise would be. And decaf is not 100 percent caffeine-free, so for some people, even a single cup of decaf after dinner can mean a long night with the jitters.

Coffee lovers, take heart. Researchers have now isolated the gene

Decaf Anyone?
Genetic engineering techniques may soon produce a cup of decaf with as much flavor as regular coffee.

for caffeine synthase, an enzyme that controls the final two steps in the caffeine biosynthesis pathway in the coffee plant. With this gene in hand, it should now be possible to develop caffeine-free coffee that retains all of its flavor. For example, genetic engineering techniques could be used to add an intron to the caffeine synthase gene, thereby destroying the function of the enzyme it encodes and rendering the coffee plant unable to make caffeine. The coffee that resulted would be 100 percent caffeine-free and should keep all of its flavor, since caffeine is odorless and tasteless.

To some people, coffee without the kick will always be an oxymoron. But for people who are strongly affected by caffeine or who are trying to reduce their coffee consumption, the availability of a fully flavored decaf would be a welcome development. A cup of decaf could then be savored like a regular cup of coffee, but without the aftereffects. If a jumpstart were desired, there would still be the option of a cup of caffeinated coffee. And it would taste just as good as decaf.

Putting It Together: From Gene to Phenotype

Humans have approximately 35,000 genes. More than 99 percent of these genes code for the amino acid sequences of proteins; the rest specify RNA molecules such as tRNA or rRNA. Here we review the major steps in how cells go from gene to protein to phenotype, focusing on genes that encode proteins. However, it is important to remember that transcription—the first step in the process that leads from gene to protein—is similar in all genes, including the small percentage that specify tRNA and rRNA molecules. Translation does not occur for genes that specify tRNA or rRNA because the tRNA and rRNA molecules produced by such genes are the gene's final product.

Chromosomes contain many genes. Each gene on a chromosome is composed of DNA and consists of a sequence of the four bases adenosine (A), cytosine (C), guanine (G), and thymine (T). The particular sequence of bases in the DNA of the gene specifies the amino acid sequence of the gene's protein product.

As we have discussed in this chapter, the two major steps from gene to protein are transcription and trans-lation (see Figure 15.2). In transcription, the sequence of bases in the DNA of a gene is used as a template to produce an mRNA molecule. The cell then transports this mRNA molecule to a ribosome, where translation occurs. In translation, the sequence of bases in the mRNA molecule is used to synthesize the gene's protein product.

The proteins encoded by genes are essential to life. A mutation in a gene can alter the sequence of amino acids in the gene's protein product, and this change can disable or otherwise alter the function of the protein. When a critical protein is disabled, the entire organism may be harmed. For example, in people who suffer from the genetic disease sickle-cell anemia, a single base in the gene that encodes hemoglobin is altered (Figure 15.10). Hemoglobin is a protein that functions in the transport of oxygen by red blood cells. The red blood cells of people with sickle-cell anemia are curved and distorted under low oxygen conditions, such as those found in narrow blood vessels like our capillaries. The distorted red blood cells clog these narrow blood vessels, thereby leading to a wide range of serious effects, including heart and kidney failure.

In sickle-cell anemia, a gene mutation alters the gene's protein product, which in turn produces a change in the organism's phenotype. A similar chain of events occurs for other genes. Overall, the phenotype of an organism is determined by the organism's proteins and by the environment (see Chapter 12). Because genes control the production of proteins, genes play an important role in determining the phenotype of the organism.

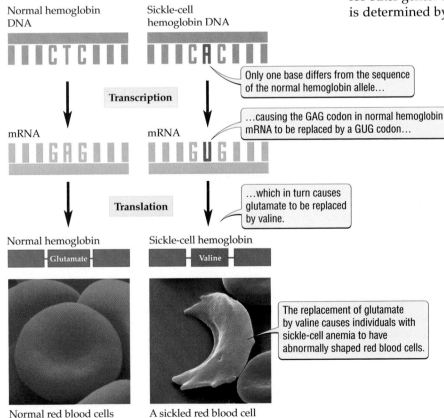

Normal hemoglobin DNA

Sickle-cell hemoglobin DNA

CTC

CAC

Only one base differs from the sequence of the normal hemoglobin allele...

Transcription

mRNA

mRNA

GAG

GUG

...causing the GAG codon in normal hemoglobin mRNA to be replaced by a GUG codon...

Translation

...which in turn causes glutamate to be replaced by valine.

Normal hemoglobin

Glutamate

Sickle-cell hemoglobin

Valine

The replacement of glutamate by valine causes individuals with sickle-cell anemia to have abnormally shaped red blood cells.

Normal red blood cells

A sickled red blood cell

■ Proteins are essential to life. In conjunction with the environment, proteins determine an organism's phenotype. Because genes control the production of proteins, they play a key role in determining an organism's phenotype.

Figure 15.10 A Small Genetic Change Can Have a Large Effect
Sickle-cell anemia is caused by a change in a single base. People with sickle-cell anemia usually die before they reach childbearing age.

■ HIGHLIGHT

The Evolving Genetic Code

The genetic code lies at the heart of the process by which genes make proteins: It provides the "dictionary" that cells use to translate the language of DNA into the language of proteins. As we have learned in this chapter, the genetic information stored in genes is first transcribed into mRNA molecules, then translated by tRNA molecules. Transfer RNA molecules are able to translate the genetic code because one portion of their structure binds to a specific mRNA codon while another portion binds to the specific amino acid called for by that mRNA codon.

We might expect the consequences of any change in the genetic code to be disastrous. For example, consider what would happen if a particular type of tRNA molecule began carrying the wrong amino acid. For every protein made by every cell in the body, whenever the codon recognized by that tRNA was translated, the wrong amino acid would be inserted into the protein that was being built. As a result, the function of many proteins would probably be destroyed, which would kill or severely harm the organism.

This line of reasoning suggests that once the genetic code reached its current form, early in the history of life, it should have remained "frozen" in place because any changes to it would be so harmful. This strong emphasis on the unchanging nature of the genetic code explains why it is often called a universal code, common to all life.

All organisms do have a similar genetic code, and many use the exact same genetic code, the one shown in Figure 15.5. But the genetic code is not really "frozen." Instead, it has slowly evolved, or changed over time. In six species of yeasts, for example, CUG codes for serine instead of leucine. In the mitochondria of many organisms, AUA codes for methionine instead of isoleucine. Such departures from the genetic code are not limited to yeasts and mitochondria; bacteria, protists, and even humans violate portions of the code. In total, 15 of the 64 codons—almost 25 percent—have changed their meaning at least once throughout the history of life.

How did the genetic code evolve without killing the organisms in which the changes occurred? One answer to this question is that much of the genetic code is "redundant" in the sense that several codons have the same meaning—that is, they call for the same amino acid (see Figure 15.5). In some cases, organisms do not use one of the redundant codons at all, nor do they produce the tRNA molecules that are usually associated with the unused codon. If an organism no longer uses a particular codon, that codon can change its meaning without damaging the organism.

Such changes appear to have happened repeatedly, albeit very slowly, during the history of life. Early in the history of life, all organisms probably did have the same genetic code. But since that time, the genetic code, like all other aspects of life, has evolved and continues to evolve.

> ■ The genetic code has evolved slowly over time. It can change without killing the organism because different codons call for the same amino acid, and in some cases, some of these redundant codons are not used.

SUMMARY

How Genes Work

- Genes work by controlling the production of proteins.

- Most genes encode proteins.

- A few genes do not encode proteins, but rather specify as their final product one of several RNA molecules that are used to synthesize proteins.

- Three types of RNA (rRNA, mRNA, and tRNA) and many enzymes and other proteins are required for the cell to make proteins.

- RNA consists of a single strand of nucleotides, each composed of a ribose sugar–phosphate backbone and one of four nitrogen-containing bases: adenine (A), cytosine (C), guanine (G), or uracil (U).

How Genes Control the Production of Proteins

- In both prokaryotes and eukaryotes, two steps are required for the synthesis of proteins: transcription and translation.

- In transcription, an RNA molecule is made using the DNA sequence of the gene.

- In translation, rRNA, mRNA, and tRNA molecules direct protein synthesis.

- In eukaryotes, information for the synthesis of a protein flows from the gene, located in the nucleus, to the site of protein synthesis, the ribosome, located in the cytoplasm.

Transcription: From DNA to RNA

- In transcription, an mRNA molecule is synthesized from a gene's DNA template.

- The mRNA molecule is constructed using the DNA sequence of the gene according to specific base-pairing rules: A in DNA pairs with U in RNA, T in DNA pairs with A in RNA, and C pairs with G.

■ In eukaryotes, genes contain internal sequences of DNA (introns) that must be removed from the initial mRNA product if the protein encoded by the gene is to function properly.

The Genetic Code

■ The information in a gene is encoded in its sequence of bases.

■ In the genetic code, an amino acid is specified by a nonoverlapping sequence of three nucleotide bases, called a codon, in an mRNA molecule.

■ When reading the genetic code, the cell begins at a fixed starting point, thus ensuring that the message from the gene does not become scrambled.

Translation: From mRNA to Protein

■ In translation, a sequence of bases in mRNA is converted into a sequence of amino acids in a protein.

■ Translation occurs at the ribosomes, which are composed of rRNA and more than 50 different proteins.

■ Ribosomes are molecular machines that make the covalent bonds that link amino acids together into a protein.

■ Transfer RNA molecules carry the specific amino acids called for by the mRNA to the ribosome.

The Effect of Mutations on Protein Synthesis

■ Many mutations are caused by the substitution, insertion, or deletion of a single base pair of a gene's DNA sequence.

■ Insertion or deletion of a single base pair causes a frameshift, which usually destroys the function of the gene's protein product.

■ Insertions or deletions of a series of bases are also common.

■ Mutations can change the function of a gene's protein product.

Putting It Together: From Gene to Phenotype

■ More than 99 percent of genes specify proteins.

■ Proteins are essential to life. In conjunction with the environment, proteins determine an organism's phenotype.

■ Because genes control the production of proteins, genes play a key role in determining an organism's phenotype.

Highlight: The Evolving Genetic Code

■ The genetic code has evolved slowly over time.

■ The genetic code can change without killing the organism because different codons call for the same amino acid and, in some cases, particular codons are not used.

KEY TERMS

anticodon p. 259	messenger RNA (mRNA) p. 255
codon p. 258	promoter p. 256
deletion mutation p. 261	ribosomal RNA (rRNA) p. 255
frameshift p. 261	RNA polymerase p. 256
gene p. 254	substitution mutation p. 261
genetic code p. 258	transcription p. 256
insertion mutation p. 261	transfer RNA (tRNA) p. 255
intron p. 257	translation p. 259

CHAPTER REVIEW

Self-Quiz

1. For genes that specify a protein, what molecule carries information from the gene to the ribosome?
   a. DNA
   b. mRNA
   c. tRNA
   d. rRNA

2. During translation, the nucleotide bases in mRNA are read _____ at a time to produce a protein.
   a. one
   b. two
   c. three
   d. four

3. Which molecule carries the amino acid called for by mRNA to the ribosome?
   a. rRNA
   b. tRNA
   c. codon
   d. DNA

4. In transcription, which of the following molecules is produced?
   a. mRNA
   b. rRNA
   c. tRNA
   d. all of the above

5. A portion of the DNA sequence of a gene has the nucleotide bases CGGATAGGGTAT. What is the sequence of amino acids specified by this DNA sequence?
   a. alanine-tyrosine-proline-isoleucine
   b. arginine-tyrosine-tryptophan-isoleucine
   c. arginine-isoleucine-glycine-tyrosine
   d. none of the above

6. What molecular machine makes the covalent bonds that link the amino acids of a protein together?
   a. tRNA
   b. mRNA
   c. rRNA
   d. ribosome

Review Questions

1. What is a gene?

2. What are the functions of the three types of RNA used by cells to make proteins: rRNA, tRNA, and mRNA?

3. Summarize the key steps in transcription and translation.

4. Why is it essential that tRNA be able to bind to both an amino acid and an mRNA codon?

5. Describe the flow of genetic information from gene to phenotype.

15

The Daily Globe

Gene Therapy Bounces Back

In the past 50 years, science has revealed that genes work by controlling the production of proteins. Our knowledge of how genes work has raised a tantalizing prospect: Could we reach inside our cells and repair genes that cause disease? Scientists in France appear to have done just that in a groundbreaking study on SCID-X1, a lethal human immune deficiency disease that leaves its victims vulnerable to many kinds of infections.

The French scientists used a technique known as gene therapy, which seeks to correct genetic disorders by fixing the genes that cause the disorder. This technique, which is still in the early stages of development, has recently fallen on hard times: It has produced few unqualified success stories, and a gene therapy patient died within the past few months, apparently as a result of the therapy. The lack of progress and the recent death have led many to argue that gene therapy is both a scientific failure and unsafe.

The new results from France may provide the boost gene therapy needs to bounce back. SCID-X1 is caused by a gene that produces a defective version of a protein, the usual version of which plays a critical role in the development of two types of white blood cells used in the human immune system, T cells and natural killer cells. The French scientists successfully inserted a corrected version of this gene into two patients, aged 8 and 11 months, suffering from SCID-X1. Three years later, the results are very encouraging: The numbers of T and natural killer cells in the patients are up to normal levels, suggesting that they will be able to live a healthy and disease-free life. That's good news for the patients and their families. And the news will also be welcomed by recently beleaguered gene therapy advocates, who argue that despite the inevitable setbacks associated with any new technology, gene therapy holds more potential than any other area of medicine to provide cures for human genetic disease.

Evaluating "The News"

1. Despite more than a decade of effort, gene therapy has produced few success stories. Should society continue to fund gene therapy research, which is relatively expensive but may, in the long run, allow us to fix otherwise incurable genetic diseases? Or would our money be spent more wisely on low-cost preventative measures that we know can reduce the frequency of AIDS, cancer, and other human diseases?

2. Gene therapy has been criticized for promising too much and delivering too little since its inception in the early 1990s. Is this a fair criticism for a technology in the early stages of development?

3. Many people think it is ethically wrong for people to use modern genetic techniques to modify the genes of organisms, including ourselves. The modification of genes is a key component of gene therapy. In general, do you think it is ethically acceptable for people to modify the genes of organisms? In particular, do you think gene therapy, which strives to cure human genetic diseases, is ethically acceptable? Why or why not? If you find gene therapy ethically acceptable, are there modifications of human genes that you would not find ethical?

chapter 16 Control of Gene Expression

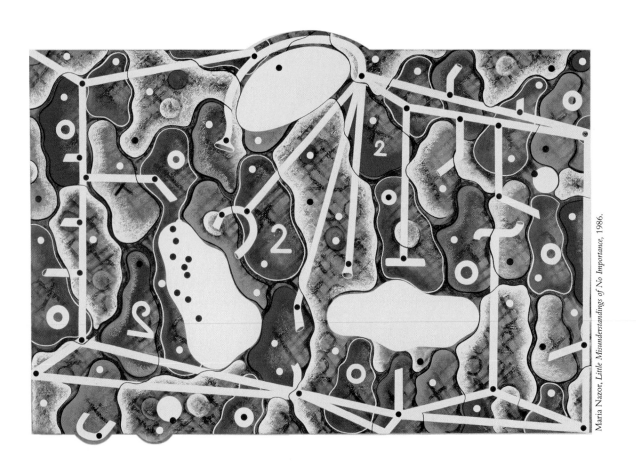

Maria Nazor, *Little Misunderstandings of No Importance*, 1986.

Greek Legends and One-Eyed Lambs

Among his many adventures, the Greek mythological hero Odysseus encountered (and outwitted) a Cyclops, a gigantic humanlike creature with great strength and a single large eye. The Cyclopes of legend have characteristics that resemble those caused by some rare genetic and developmental disorders. Lambs, mice, and humans occasionally are born with a single large eye, along with other abnormalities of the brain and face. Such individuals die soon after birth.

What causes an animal to be born with only one eye? Two causes are known, both of which relate to the function of genes. Certain master-switch genes guide the development of an organism by "turning on" a series of other crucial genes. Some one-eyed individuals have a defect in one of these master-

268

A Cyclops of Legend

switch genes. The defect prevents the master-switch gene from turning on other genes that control the normal development of the brain and face. Other one-eyed individuals were exposed as embryos to chemicals that prevented the master-switch gene's protein product from having its usual effect (which is to turn on the other crucial genes). In broad overview, whether it is due to a defective gene or exposure to chemicals, the formation of a single large eye results from the failure of cells to control how a series of crucial genes are turned on.

Deciphering the causes of developing a single eye brings us to one of the most exciting areas of modern genetics: the control of gene expression. To develop and function normally, organisms must express the right genes at the right place and time, producing just the right amount of each gene's product. This is a task of bewildering complexity, but each of us does it, many times, every day. How organisms control gene expression is the subject of this chapter.

■ KEY CONCEPTS

1. Eukaryotic DNA is organized by a complex packing system that allows cells to store an enormous amount of information in a small space.

2. Prokaryotes have relatively little DNA, most of which encodes proteins. Eukaryotes have more DNA, and more genes, than prokaryotes. Unlike prokaryotes, eukaryotes have large amounts of DNA that does not encode proteins.

3. Organisms turn genes on and off in response to short-term changes in food availability or other features of the environment. Organisms also control genes over long periods of time, as when different genes are expressed at different times during embryonic development.

4. In multicellular organisms, different cell types express different genes.

5. The main way in which cells control gene expression is to regulate transcription. Cells also control gene expression in other ways.

6. Genetics is moving from the study of single genes to the study of interactions among large numbers of genes. This shift has the potential to revolutionize our understanding of genes and gene expression, leading to dramatic improvements in the practice of medicine.

Figure 16.1   Different Cells Have the Same Genes
The nucleus of each cell type contains the same DNA. Although they have the same genes, cells can differ greatly in structure and function because different genes are active in different types of cells.

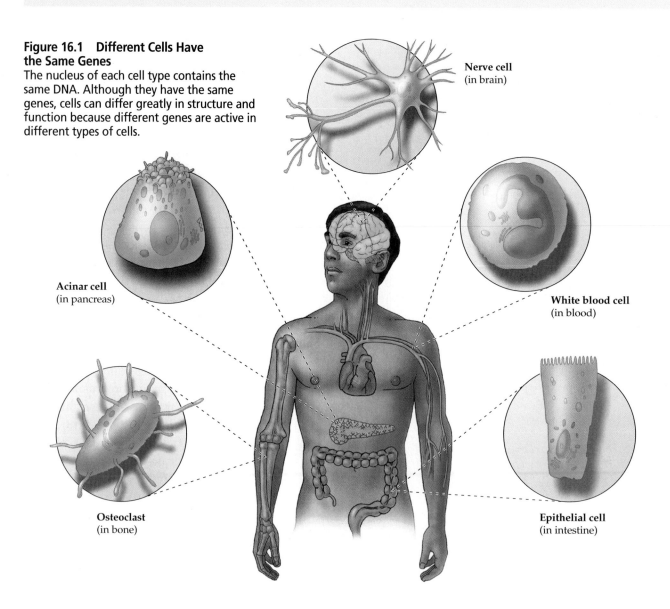

Nerve cell
(in brain)

Acinar cell
(in pancreas)

White blood cell
(in blood)

Osteoclast
(in bone)

Epithelial cell
(in intestine)

*E*ven though they all have the same genes, the various cells of a multicellular organism differ greatly in their structure and function (Figure 16.1). Since genes provide cells with the "blueprint" of life, how can cells with the same genes be so different? The answer lies in how the genes are used: Cells of different types differ in which of their genes are actually used to make proteins. As we will see, these differences relate to the cells' functions.

As we learned in Chapter 15, each gene contains instructions for the synthesis of a protein (or an RNA molecule). **Gene expression** is the synthesis of a gene's protein (or RNA) product. For different cells to use different genes, organisms must have ways to control which genes are expressed, and indeed they do. Organisms control where, when, and how much of a given gene's product is made. Factors that influence gene expression include cell type, the chemical environment of the cell, signals received from other cells in the organism, and signals received from the external environment. In this chapter we describe how cells control their own genes.

Before describing control of gene action, we need to describe how DNA is organized in chromosomes. To be expressed, the information in a gene must first be transcribed into an RNA molecule; thus, it is essential for the enzymes that guide transcription to be able to reach the gene. This task may sound simple, but it involves what may be the ultimate storage problem: How to store an enormous amount of information (its DNA) in a small space (the nucleus), yet still be able to retrieve each single piece of that information precisely when it is needed. We will see how cells solve this storage problem in the following section.

DNA Packing in Eukaryotes

Each chromosome in each of our cells is made up of one DNA molecule. These molecules hold a vast amount of genetic information. As we learned in Chapter 10, the haploid number of chromosomes in humans is 23; these chromo-

somes together contain about 3.3 billion base pairs of DNA. If the DNA from all 46 chromosomes in a human cell were stretched to its full length, it would be more than 2 meters long (taller than most of us). That is a huge amount of DNA, especially considering that it is packed into a nucleus that is only 0.000006 meter in diameter. In total, the amount of DNA in our bodies is staggering: The human body has about 10^{13} cells, each of which contains roughly 2 meters of DNA. Therefore, each of us has about 2×10^{13} meters of DNA in his or her body, a length that is more than 130 times the distance from Earth to the sun.

How can our cells pack such an enormous amount of DNA into such small spaces? The DNA molecule is very narrow (only 2 billionths of a meter, or 2 nanometers, wide), and it is highly packed and organized into a complex system. At the highest level of packing, a chromosome has its characteristic shape because its DNA is packed together very tightly (Figure 16.2). As we begin to "unpack" the chromosome, we see that each portion

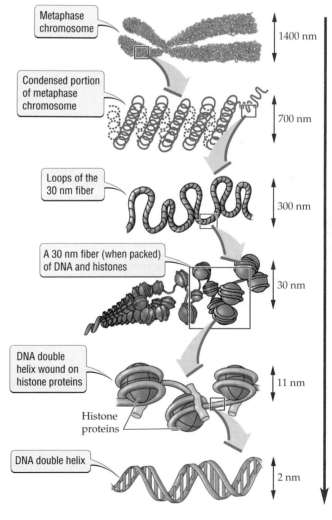

Figure 16.2
DNA Packing in Eukaryotes
The DNA of eukaryotes is highly organized by a complex packing system. This chromosome is shown at metaphase, the phase of the cell cycle at which DNA is most tightly packed.

Metaphase chromosome — 1400 nm

Condensed portion of metaphase chromosome — 700 nm

Loops of the 30 nm fiber — 300 nm

A 30 nm fiber (when packed) of DNA and histones — 30 nm

DNA double helix wound on histone proteins — 11 nm

Histone proteins

DNA double helix — 2 nm

Decreasing tightness of packing

of the chromosome consists of many tightly packed loops. Each loop, in turn, is composed of a fiber that is 30 nanometers wide. If we unpack that fiber, we see that it is made from many histone spools packed together tightly. Each of these histone spools consists of a segment of DNA wound around a "spool" composed of proteins called histones. Finally, if we unwind the DNA from the histone spools, we reach the level of the DNA double helix, which is 2 nanometers wide.

The packing scheme we've just described holds for eukaryotes during mitosis or meiosis, when the chromosomes are most condensed. The DNA of eukaryotes is much less tightly packed during interphase, when most gene expression occurs. During interphase, much of the DNA is folded into 300-nanometer loops of the 30-nanometer fibers, but some remains more tightly packed. Genes in the tightly packed regions are not expressed; here, the DNA is packed so tightly that the proteins necessary for transcription cannot reach it. Compared with eukaryotes, prokaryotic cells have much less DNA per cell, and their DNA packing system is less complex.

> ■ The DNA of eukaryotes is highly organized by a complex packing system. During interphase, when most gene expression occurs, the tightness of DNA packing is reduced. Genes in tightly packed regions are not expressed because the proteins necessary for transcription cannot reach them.

16.2 *Functional Organization of DNA*

We've just seen how the massive amount of DNA in a eukaryotic cell must be packaged to fit into the nucleus. Now let's turn to several related questions: How much DNA do different organisms have? What are the functions of different portions of an organism's DNA?

Compared with eukaryotes, prokaryotes have little DNA. A typical bacterium has several million base pairs of DNA, whereas most eukaryotes have hundreds of millions to billions of base pairs. In addition, prokaryotic DNA is organized by function in a straightforward way: The different genes that are needed for a given metabolic pathway usually are grouped together along the DNA molecule. In contrast, eukaryotic genes with related functions often are not grouped near one another on a chromosome, and may even be on different chromosomes. Finally, most of the DNA in prokaryotes encodes proteins, and prokaryotic genes rarely contain noncoding segments of DNA. Eukaryotic genes, in contrast, contain noncod-

ing DNA both within and between genes, as we will see shortly. Overall, then, prokaryotic DNA is streamlined and organized by function in a simple, direct way.

The total amount of DNA in prokaryotes varies from 0.6 to 30 million base pairs. The eukaryotes show much greater variation, from 12 million base pairs in yeast (a single-celled fungus) to over a trillion base pairs in a single-celled protist. Most animals with backbones have hundreds of millions to billions of base pairs of DNA. For example, puffer fish have 400 million base pairs, humans have 3.3 billion base pairs (the range among all mammals is 1.5 to 6.3 billion base pairs), and some salamanders have 90 billion base pairs.

Eukaryotes usually have far more DNA than prokaryotes. Why is this so? In part, the reason is that eukaryotes are more complex organisms than prokaryotes and hence need more genes to run their metabolic machinery. A typical prokaryote has about 2000 genes. Among the eukaryotes studied to date, the single-celled yeast *Saccharomyces cerevisiae* has 6000 genes, the nematode worm *Caenorhabditis elegans* has 19,100 genes, several plant species have an estimated 20,000 genes, and humans have an estimated 35,000 genes. However, the primary reason that eukaryotes have more DNA than prokaryotes is that in most

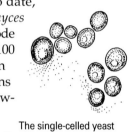

The single-celled yeast *Saccharomyces cerevisiae*

eukaryotes, only a small percentage of the DNA encodes proteins used by the organism. Instead, most of the DNA in eukaryotes is "superfluous." Thus, differences in the amount of DNA in prokaryotes and eukaryotes reflect not only more genes, but also the presence of considerable amounts of superfluous DNA, in eukaryotes.

Genes comprise only a small percentage of the DNA in eukaryotes

Scientists estimate that genes make up less than 1.5 percent of human DNA. The rest consists of various types of superfluous DNA. **Noncoding DNA**, which does not encode proteins or RNAs, includes introns and spacer DNA. Introns, as we saw in Chapter 15, are sequences of noncoding DNA within genes. **Spacer DNA** consists of stretches of DNA that separate genes (Figure 16.3). Eukaryotic DNA also contains many **transposons**, the mobile genetic elements discussed in the box on page 274. Although transposons often do encode proteins, those proteins are used by the transposon, not by the organism. Transposons may surround genes or may even be inserted in the middle of a gene. They make up an estimated 36 percent of human DNA and more than 50 percent of the 5.4 billion base pairs in corn.

Figure 16.3   The Composition of Eukaryotic DNA
Eukaryotic genes are surrounded by spacer DNA and transposons, as shown here for five human genes. Each of the two different types of transposons shown here is found in many copies throughout human DNA. Note that a transposon, colored red, has been inserted into one of the five human genes.

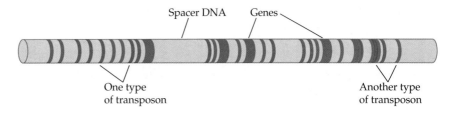

Spacer DNA   Genes

One type of transposon   Another type of transposon

■ Compared with eukaryotes, prokaryotes have less DNA and fewer genes. Genes constitute only a small portion of the DNA of many eukaryotes; the rest consists of noncoding DNA and transposons.

Patterns of Gene Expression

Gene expression—the synthesis of a gene's protein product—is the process by which the gene influences a cell or organism. At any given time, roughly 5 percent of the genes in a typical human cell are being actively used or expressed. The rest of the genes are not in use. The particular genes that are expressed by a cell can change over time, and different cells express different sets of genes. What determines which genes an organism expresses at a particular time or in a particular type of cell?

In prokaryotes, genes are turned on and off as a direct, short-term response to changing environmental conditions. Gene control in eukaryotes is much more complex. Eukaryotes have more genes, their DNA is more tightly packed, and they must control genes over both short and long periods of time. Here we provide a broad overview of patterns of gene expression. In the next section, "How Cells Control Gene Expression," we will see how cells turn genes on and off.

Organisms turn genes on and off in response to short-term environmental changes

Single-celled organisms such as bacteria face a big challenge: They are directly exposed to their environment, and they have no specialized cells to help them deal with changes in that environment. One way they meet this challenge is to express different genes as conditions change (Figure 16.4).

Bacteria respond to changes in nutrient availability, for example, by turning genes on or off. If the nutrient lactose (a sugar found in milk) is given to *E. coli* bacteria, within a matter of minutes the genes that encode the enzymes needed to break down lactose are activated. When the lactose is used up, the bacteria stop producing those enzymes. In effect, the bacteria specialize temporarily on an available resource. When that resource runs out, they switch to the next resource that becomes available (in Figure 16.4, the sugar arabinose). By producing the enzymes to process a particular food only when that food is available, bacteria do not waste energy and cellular resources making enzymes that are not needed.

Multicellular organisms also change which genes they express in response to short-term changes in the environment. For example, we humans change the genes we express when our blood sugar or blood pH levels change, allowing us to keep these levels from becoming too high or too low. Similarly, when exposed to high temperatures, humans, plants, and many other organisms turn on certain genes. Such heat-induced genes produce proteins that protect cells against heat damage.

Figure 16.4   Bacteria Express Different Genes as Food Sources Change

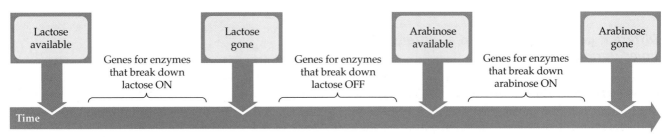

Lactose available   Lactose gone   Arabinose available   Arabinose gone

Genes for enzymes that break down lactose ON   Genes for enzymes that break down lactose OFF   Genes for enzymes that break down arabinose ON

Time

Barbara McClintock and Jumping Genes

Transposons, known informally as "jumping genes," are DNA sequences that can move from one position on a chromosome to another, or from one chromosome to another. Transposons were first discovered in corn by Barbara McClintock in 1948, a discovery that earned her the 1983 Nobel prize in physiology or medicine. They have since been found in many other organisms, ranging from bacteria to peas to humans.

Born in 1902, Barbara McClintock completed her undergraduate and graduate studies at Cornell University, where she was awarded a Ph.D. in 1927. Working at Cornell as an instructor, McClintock published an important paper in 1931, which proved that crossing-over leads to the formation of nonparental genotypes (see Chapter 13). This discovery established McClintock's reputation as one of the world's leading geneticists.

Despite the great respect in which McClintock was held, her discovery of transposons in 1948 was slow to be appreciated, in part because genes were thought to have fixed locations on chromosomes. It took roughly 20 years for the field of genetics to catch up with McClintock, but by 1967 she had begun to receive a series of awards for her work on transposons, culminating in the 1983 Nobel prize. McClintock continued to do research until her death in 1992, at the age of 90.

What role do transposons play in organisms? One effect of transposons relates to the fact that they often duplicate themselves when they move. Such duplication events serve to produce many copies of the transposon (see Figure 16.3) and to increase the total amount of DNA in the organism. Whether they duplicate themselves or simply move to a new location in the genome, many scientists view transposons as genetic parasites that spread through an organism's DNA. When a transposon moves from one location in the DNA to another, it may insert itself into a gene, causing a mutation that changes the gene's function. As with any gene mutation, mutations caused by transposons can be harmful. In contrast, other scientists argue that transposons may be advantageous. For example, transposons alter patterns of gene expression in corn and many other species. McClintock viewed transposon-caused changes in gene expression as a mechanism that organisms use to cope with changing environmental conditions.

Whether transposons are genetic parasites or are beneficial to organisms, their discovery helped change how scientists think about DNA. Biologists used to think of genes as fixed in position, like beads on a string. The discovery of transposons changed that thinking forever. DNA can move from one location to another, organisms can change their patterns of gene expression quickly, and over the course of evolution, genes can take on new functions. The world of genes is exciting and dynamic, and Barbara McClintock's discoveries played a major role in showing biologists just how dynamic that world can be.

Barbara McClintock

Different genes are expressed at different times during development

Turning the correct genes on and off in response to changing environmental conditions is a challenging task. But multicellular organisms must also coordinate an even more difficult operation: developing from a single-celled zygote into a large, complex organism. The control of gene expression and the timing of gene activity during embryonic development is a task of great complexity, and errors in the control process can result in death or deformity (Figure 16.5).

Head of a normal fruit fly

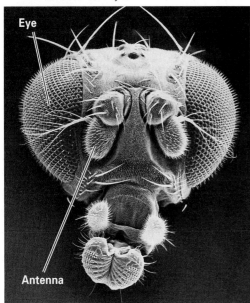

Eye

Antenna

Head of a developmental mutant

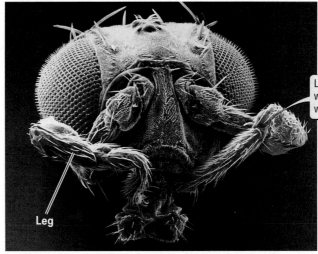

Legs have been produced where antennae normally would be located.

Leg

Figure 16.5   A Developmental Mutation with Bizarre Results
A mutation in a homeotic gene that controls development in fruit flies produces legs where antennae should be.

Organisms accomplish gene control during development through cascades of gene expression. In such a cascade, a series of genes are turned on one after another, like a series of dominoes in which one knocks over another, which in turn knocks over the next, and so on. In such a **gene cascade**, the protein products of certain genes interact with one another and with signals from the environment to turn on different sets of genes in different cells. The proteins produced by those newly activated genes then interact with one another and with the environment to turn on still more genes, and so on. Eventually, genes are expressed whose protein products alter the structure and function of cells, allowing cells to become specialized for particular tasks. (We will learn more about the process of development in Chapter 37.)

The master-switch genes described at the beginning of this chapter play a central role in the control of gene cascades. Each master-switch gene, or **homeotic gene**, controls the expression of a series of other genes whose proteins direct the development of the organism. Given this crucial role, it is not surprising that defective versions of homeotic genes can have striking phenotypic effects, such as those shown in Figure 16.5.

At different times, different homeotic genes are active in the body's different cell types. For example, a homeotic gene that coordinates the development of the eye may be active in cells that will give rise to the eyes. In other parts of the body, although this gene is present, it is not in use; instead, other homeotic genes are expressed. Finally, as the body changes during development, the homeotic genes expressed by cells also change.

Recently it was discovered that similar homeotic genes control development in organisms as different as fruit flies, mice, and humans (Figure 16.6). This similarity indicates that these genes are ancient. Homeotic genes first evolved hundreds of millions of years ago, and since then they have been used in similar ways by a wide variety of organisms.

Different cells express different genes

We have just described how different cells express different homeotic genes during development. The same holds true for many other genes: In general, different types of cells express different genes, in both developing embryos and adults.

Whether a cell expresses a particular gene depends on the function of the gene's product. Not surprisingly, a gene that encodes a specialized protein will be expressed only in cells that use or produce that protein. For example, red blood cells are the only cells in the human body that use the oxygen transport protein hemoglobin, and developing red blood cells are the only cells that express the gene for this protein (Figure 16.7). Similarly, the gene for crystallin, a protein that makes up the lens of the eye, is expressed only in developing eye lens cells. Finally, the gene for insulin, a hormone produced in the pancreas and used elsewhere in the body, is expressed only in pancreas cells.

Fruit fly

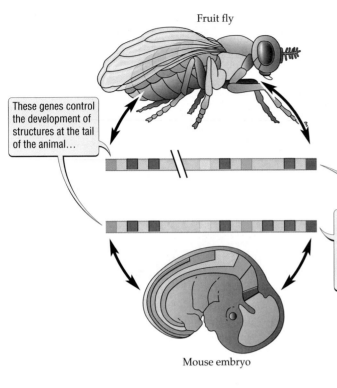

These genes control the development of structures at the tail of the animal...

...and these genes control structures in the head. Similar genes in between control the development of parts between the head and tail.

Mouse embryo

Figure 16.6 Homeotic Genes in Different Organisms Are Similar

Development is controlled by similar homeotic genes in very different organisms. The homeotic genes that control development in fruit flies and mice are arranged in a similar order on fruit fly and mouse chromosomes. Moving from tail to head, the body parts affected by these genes are positioned in the same order in which the genes are arranged on the chromosome. Similar homeotic genes and the structures whose development they control are matched here by color.

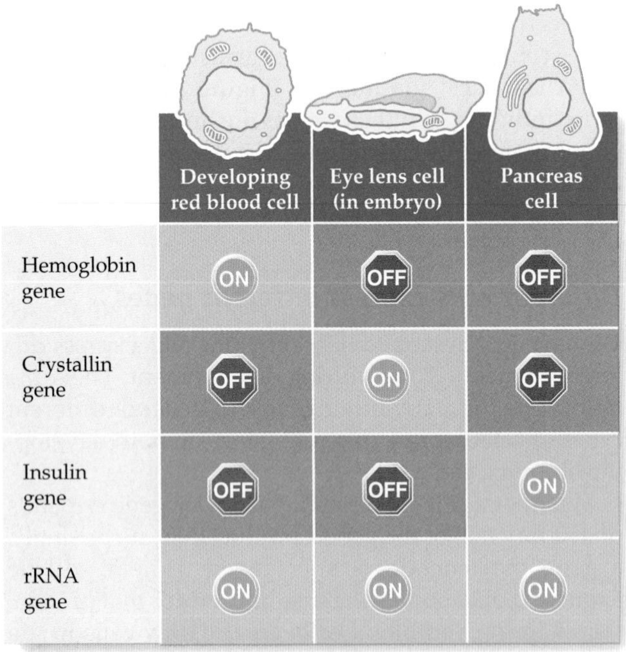

Some genes, known as **housekeeping genes**, have an essential role in the maintenance of cellular activities and are expressed by most cells in the body. Genes for ribosomal RNA (rRNA), for example, are expressed by most cells (see Figure 16.7). This is not surprising, since rRNA is a key component of ribosomes, which are the sites at which proteins are synthesized.

■ Both prokaryotes and eukaryotes turn genes on and off in response to short-term changes in environmental conditions. Multicellular organisms also control genes that influence development through gene cascades, which are regulated by homeotic genes. The different cells of multicellular organisms express different genes, in both developing embryos and adults.

	Developing red blood cell	Eye lens cell (in embryo)	Pancreas cell
Hemoglobin gene	ON	OFF	OFF
Crystallin gene	OFF	ON	OFF
Insulin gene	OFF	OFF	ON
rRNA gene	ON	ON	ON

Figure 16.7 Different Types of Cells Express Different Genes

16.3 Some genes, such as those encoding hemoglobin, crystallin, and insulin, are active only in cells that use or produce the protein encoded by the gene. Other "housekeeping" genes, such as the rRNA gene, are active in most types of cells. "OFF" signifies inactive genes; "ON," active genes.

How Cells Control Gene Expression

Cells receive signals that influence which genes they turn on and off. Some of these signals are sent from one cell to another, as when one cell releases a signaling molecule that alters gene expression in another cell (see Chapter 9). Cells also receive signals from other features of the organism's internal environment (for example, blood sugar level in humans) and external environment (for example, sunlight in plants). Overall, cells combine information from a variety of signals and use that information to determine which genes to express. In the discussion that follows, we describe how cells turn genes on and off.

Cells control the expression of most genes by controlling transcription

The most common way in which cells control gene expression is to turn the transcription of particular genes on or off. For example, the bacterium *E. coli* requires a supply of the amino acid tryptophan. If this amino acid is available in the environment, the bacterium absorbs it from the environment and does not waste cellular resources making it. But if tryptophan is not readily available, the bacterium expresses a series of five genes that together encode the enzymes used to make tryptophan.

E. coli controls these five genes in the following way. When tryptophan is present in the environment, it binds to a **repressor protein** in the bacterial cell, so called because it represses the expression of the tryptophan genes. When bound to tryptophan, the repressor protein can bind to a DNA sequence called an **operator**. The function of an operator is to control the transcription of a gene or group of genes—in this case, the group of five genes required to make tryptophan. When bound to the operator, the repressor protein prevents RNA polymerase from binding to the operator, thus blocking transcription of the five genes needed to produce tryptophan (Figure 16.8*a*). In the absence of tryptophan, the repressor protein cannot bind to the operator, and transcription of the genes occurs (Figure 16.8*b*). As a result, the cells do not make tryptophan when it is already present, but they do make it when tryptophan levels are low. This control of gene expression ensures that the cell does not waste precious resources producing tryptophan when it is readily available.

A few genes, such as those in *E. coli* that encode the tryptophan repressor protein, are always expressed at a low level; their transcription is not regulated. But nearly all genes in prokaryotes and eukaryotes are regulated. In general, the control of transcription has two essential elements, both of which are illustrated by tryptophan synthesis in *E. coli*. First, there are **regulatory DNA sequences**, such as the tryptophan operator, that can switch a gene on and off. Second, to switch genes on and off, regulatory DNA sequences must interact with **regulatory proteins** that signal whether a gene should be expressed. The repressor protein that binds to the tryptophan operator when tryptophan is present is an example of a regulatory protein. Together, regulatory DNA sequences and regulatory proteins turn genes on and off in both prokaryotes and eukaryotes.

In eukaryotes, the situation is considerably more complex than in prokaryotes: Eukaryotes make more different proteins, and often dozens of regulatory DNA sequences are involved in controlling gene expression. Despite this added complexity, the basic concepts are the same: Regulatory proteins bind to regulatory DNA sequences, which in turn promote or inhibit transcription.

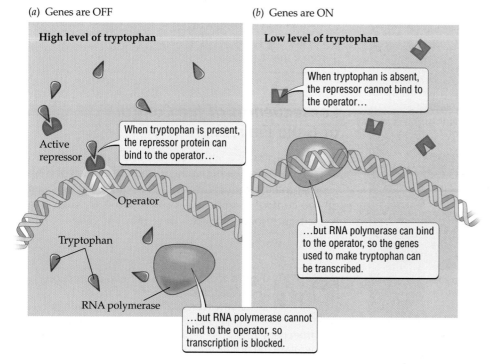

(*a*) Genes are OFF

High level of tryptophan

When tryptophan is present, the repressor protein can bind to the operator…

Active repressor

Operator

Tryptophan

RNA polymerase

…but RNA polymerase cannot bind to the operator, so transcription is blocked.

(*b*) Genes are ON

Low level of tryptophan

When tryptophan is absent, the repressor cannot bind to the operator…

…but RNA polymerase can bind to the operator, so the genes used to make tryptophan can be transcribed.

Figure 16.8 Repressor Proteins Turn Genes Off In the bacterium *E. coli*, a repressor protein binds to an operator to control the transcription of a group of genes that encode the enzymes needed to make tryptophan. (*a*) When tryptophan is present, it binds to a repressor protein, which then turns the genes off by binding to the operator. (*b*) When tryptophan is absent, the repressor protein cannot bind to the operator; thus RNA polymerase can bind to the operator, and the genes used to make tryptophan are turned on.

Cells also control gene expression in other ways

In addition to controlling transcription, cells can control gene expression at other key steps in the pathway from gene to protein (Figure 16.9):

- *Tightly packed DNA is not expressed.* Most gene expression occurs during the interphase portion of the cell cycle. At that time, the chromosomes are long and narrow, and their DNA is relatively loosely packed. Even during interphase, however, some DNA remains tightly packed. Genes in this tightly packed DNA are not transcribed, in part because the proteins necessary for transcription simply cannot reach them.

- *Cells regulate how quickly messenger RNA molecules are broken down.* Recall from Chapter 15 that when cells express a gene to make a protein, an mRNA molecule is transcribed from the gene and then used in translation to make the gene's protein product. Most mRNA molecules are broken down by cells a few minutes or hours after being made, although a few persist for days or weeks. The longer it takes to break down an mRNA molecule, the more protein can be made from that mRNA molecule. In some cases, cells determine how long mRNA molecules will persist by chemically modifying the mRNA.

- *Cells can inhibit translation.* Some proteins can bind to mRNA molecules and prevent their translation. This method of control is especially important for some long-lived mRNA molecules. It allows the cell to deactivate an mRNA molecule that otherwise might continue to produce a protein that is no longer being used.

- *Proteins can be regulated after translation.* Cells must modify or transport many proteins before they can be used; both modification and transport can be used to regulate the availability of a protein. Cells also can target certain proteins for destruction, thus controlling gene expression at the final step in the chain from gene to protein.

> ■ Cells control most genes by controlling transcription. To switch genes on and off, regulatory DNA sequences interact with regulatory proteins that signal whether a gene should be transcribed. Cells also control gene expression at other points in the pathway from gene to protein.

Consequences of the Control of Gene Expression

In this chapter, we set out to describe how cells and organisms control when, where, and how much of each gene's protein product is made. Thus, our focus has been on the control of gene expression. Now we turn to a related question: How does the control of gene expres-

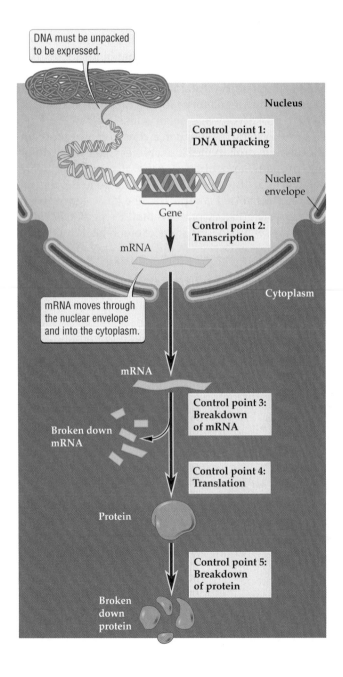

DNA must be unpacked to be expressed.

Nucleus

Control point 1: DNA unpacking

Nuclear envelope

Gene

Control point 2: Transcription

mRNA

mRNA moves through the nuclear envelope and into the cytoplasm.

Cytoplasm

mRNA

Control point 3: Breakdown of mRNA

Broken down mRNA

Control point 4: Translation

Protein

Control point 5: Breakdown of protein

Broken down protein

Figure 16.9 Control of Gene Expression in Eukaryotes
Eukaryotes can control gene expression in many ways. Each control point on the pathway from gene to protein represents a point at which cells can regulate the production of proteins.

sion relate to the phenotype of the organism? In many instances, there is a powerful and direct relationship. Consider, for example, the effect of homeotic genes on a developing organism. As you will recall, a homeotic gene controls the expression of a series of other genes that influence development. If the protein specified by the homeotic gene does not function properly, the expression of these genes will not be controlled properly, and extreme phenotypic effects may result (see p. 275 and Figure 16.5).

But the control of gene expression and the phenotype of the organism are not equivalent. Organisms turn genes on and off in response to signals from the environment, thus changing which proteins they produce. These proteins have the potential to influence the phenotype. However, as we discussed in Chapters 12 and 15, the phenotype of an organism results from interactions between its genetic makeup and its environment. As a result, the environment not only can influence which genes are expressed, it also can alter the effects of the proteins produced by genes that are expressed (see Figure 12.9). Thus, as summarized in Figure 12.11, the phenotype of an organism results from the combined effects of the organism's genotype (the specific alleles the organism has), the interactions among genes and their protein products (patterns of gene expression), and the environment.

> ■ The control of gene expression and the phenotype of the organism are not equivalent. Instead, the phenotype of the organism results from the effects of the organism's genotype, the interactions of genes and their protein products, and the environment.

■ HIGHLIGHT

Genome Biology

A revolution is brewing in the field of genetics—one that, in its effect on society, may ultimately rival the computer revolution. Like the computer revolution, technological advances are pushing this genetic revolution forward (see Chapter 17). However, the revolution in genetics is not centered on a particular new type of machine or technology. Instead, it is due to a shift in perspective, from the study of single genes to the study of whole genomes.

A **genome** is all the DNA of an organism, including its genes; in eukaryotes, the term "genome" refers to a haploid set of chromosomes, such as that found in a sperm or egg. The human genome, for example, consists

of all the DNA in our 23 chromosome types. At the molecular level, a genome can be described by its **DNA sequence**—that is, by the order in which the nucleotide bases adenine (A), cytosine (C), guanine (G), and thymine (T) are arranged throughout all of the organism's DNA. Genome biologists are currently striving to meet two major goals: (1) determining the DNA sequences of entire genomes and (2) understanding the expression and function of large numbers of genes.

How much progress has been made in determining the DNA sequences of entire genomes? The most exciting milestone in this process was the publication in 2001 of the first draft of the sequence of the human genome, the story of which is told in Chapter 18. As of this writing, the entire DNA sequences of 55 species of bacteria and three eukaryotes (the 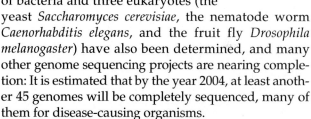 yeast *Saccharomyces cerevisiae*, the nematode worm *Caenorhabditis elegans*, and the fruit fly *Drosophila melanogaster*) have also been determined, and many other genome sequencing projects are nearing completion: It is estimated that by the year 2004, at least another 45 genomes will be completely sequenced, many of them for disease-causing organisms.

Three *Caenorhabditis elegans*

What can all these DNA sequence data—long lists of millions of As, Cs, Gs, and Ts—tell us? Let's illustrate the answer with several examples.

First, consider what has already been learned from analyses of the human DNA sequence. Such studies have uncovered more than 40 new disease genes, including genes that influence epilepsy, deafness, color blindness, and muscular dystrophy. Over the next few years, it is likely that hundreds more genes for human genetic disorders will be discovered. The initial analyses of the human genome also revealed a number of surprises. Perhaps the biggest single surprise was just how few genes it takes to make a person: roughly 35,000 genes, slightly more than double the number of genes that it takes to make the nematode worm, *C. elegans*. Other interesting facts about our genome include the discovery that our DNA contains hundreds of bacterial genes, and the as yet unexplained finding that males tend to be twice as likely as females to transmit mutations to their offspring.

Second, once the entire DNA sequence of an organism is known, we can learn a tremendous amount not only about that organism, but also about the general principles of genetics. For example, the DNA sequence of the budding yeast *Saccharomyces cerevisiae* was published in 1996. There are approximately 6000 genes in this yeast, which is a simple, single-celled eukaryote.

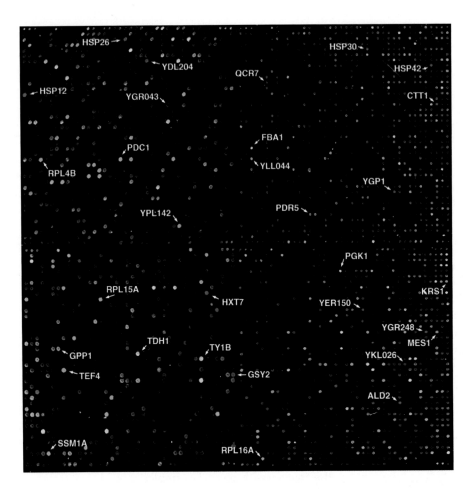

Figure 16.10   Results of a DNA Chip Experiment
In this experiment, yeast cells were provided with nutrients that caused them to switch metabolic pathways from fermentation to aerobic respiration. Each colored dot represents a particular gene. Genes colored red were turned on when the yeast cells switched to aerobic respiration; genes colored green were turned off when the yeast cells switched to aerobic respiration. Genes colored yellow were expressed at roughly equal levels during fermentation and aerobic respiration.

Once the sequence of these genes was known, scientists placed each of the yeast genes on a small glass surface (roughly 2 centimeters by 2 centimeters), producing a "DNA chip" that is packed with genetic information. Geneticists can use such DNA chips to monitor the activity of all the yeast genes at once (Figure 16.10). This approach lets biologists observe how a particular stage of development or a particular set of environmental conditions influences the activity of all the genes of an organism. Geneticists have long realized that many genes influence the metabolism and phenotype of an organism; now, for the first time, they can begin to study how.

Finally, genome sequence data can be used to design DNA chips that have the potential to alter the practice of medicine. For example, by surveying how individuals with different genetic characteristics respond to different drugs, doctors will be able to use DNA chips to decide which of several possible drugs would be likely to work best in any given patient. Similarly, DNA chips can be used to determine whether the cells of a breast cancer have a particular mutation (see Chapter 11): If the answer is yes, the most aggressive treatments should be used to treat the tumor; if the answer is no, less traumatic treatments could be used with equal effectiveness. DNA chips also could be mass-produced and used to identify the exact strain of bacteria causing a particular individual's sore throat, along with the antibiotics to which that bacterial strain is resistant (and which hence should not be prescribed by the physician). DNA chips could thus remove the guesswork from the diagnosis and treatment of both cancer and infectious diseases. By identifying the correct treatment, DNA chips have the potential to save lives and improve medical care for millions of people.

■ The science of genetics is shifting from the study of single genes to the study of whole genomes. New technological advances allow geneticists to monitor the activity of all the genes of an organism simultaneously, making it possible to study how large numbers of genes influence metabolism and phenotype. Genome biology also has the potential to revolutionize the diagnosis and treatment of human disease.

SUMMARY

DNA Packing in Eukaryotes

- The DNA molecule is very narrow, and in eukaryotes it is highly organized by a complex packing system. Together, these two features allow cells to pack an enormous amount of DNA into a very small space.

- The tightness of the DNA packing system is reduced during interphase, when most gene expression occurs.

- Genes in DNA regions that are tightly packed cannot be expressed, because the proteins necessary for transcription cannot reach them.

Functional Organization of DNA

- Compared with eukaryotes, prokaryotes have little DNA. Most prokaryotic DNA encodes proteins, and functionally related genes in prokaryotes are grouped together in the DNA.

- Eukaryotes have more genes than prokaryotes. In many eukaryotes, genes constitute only a small portion of the DNA; the rest consists of noncoding DNA and transposons. Eukaryotic genes with related functions usually are not located near each other.

Patterns of Gene Expression

- Both prokaryotes and eukaryotes turn genes on and off in response to short-term changes in environmental conditions.

- Multicellular eukaryotes also must regulate gene expression over long periods of time during embryonic development.

- Organisms control gene expression in development through gene cascades, which are regulated by homeotic genes.

- The different cells of multicellular organisms express different genes, in both developing embryos and adults.

How Cells Control Gene Expression

- Cells control most genes by controlling transcription.

- Transcription is controlled by regulatory DNA sequences that can switch genes on and off.

- To switch genes on and off, regulatory DNA sequences interact with regulatory proteins that signal whether a gene should be expressed.

- Cells also can control gene expression at other points on the pathway from gene to protein.

Consequences of the Control of Gene Expression

- The control of gene expression and the phenotype of the organism are not equivalent.

- The phenotype of the organism results from the combined effects of the organism's genotype, interactions among genes and their protein products, and the environment.

Highlight: Genome Biology

- The science of genetics is shifting from the study of single genes to the study of whole genomes.

- New technological advances allow geneticists to monitor the activity of all genes of an organism simultaneously, making it possible to study how large numbers of genes influence metabolism and phenotype.

- Genome biology has the potential to revolutionize the diagnosis and treatment of human disease.

KEY TERMS

DNA sequence p. 279	operator p. 277
gene cascade p. 275	regulatory DNA sequence p. 277
gene expression p. 271	
genome p. 279	regulatory protein p. 277
homeotic gene p. 275	repressor protein p. 277
housekeeping gene p. 276	spacer DNA p. 272
noncoding DNA p. 272	transposon p. 272

CHAPTER REVIEW

Self-Quiz

1. Select the term that is best defined as all the DNA of an organism.
   a. genes
   b. spacer DNA
   c. gene frequency
   d. genome

2. In prokaryotes and eukaryotes, gene expression is most often controlled by regulation of which of the following?
   a. the destruction of a gene's protein product
   b. how long mRNA remains intact
   c. transcription
   d. translation

3. Which of the following is a regulatory DNA sequence?
   a. repressor protein
   b. operator
   c. intron
   d. housekeeping gene

4. During development, different cells express different genes, leading to
   a. the formation of different cell types.
   b. gene mutation.
   c. DNA packing.
   d. the spread of transposons.

5. Assume that an organism has 20,000 genes and a large quantity of DNA, most of which is noncoding DNA. What kind of organism is it most likely to be?
   a. insect
   b. plant
   c. bacterium
   d. either a or b

Review Questions

1. Cell types in multicellular organisms often differ considerably in structure and in the metabolic tasks they perform, yet each cell of a multicellular organism has the same set of genes.
   a. Explain how cell types with the same genes can be so different in structure.
   b. Explain how cell types with the same genes can be so different in the metabolic tasks they perform.

2. Genes in eukaryotes are often separated by large amounts of noncoding DNA. How would you expect long stretches of noncoding DNA located between genes to influence the frequency of crossing over?

3. Describe the tryptophan operator in *E. coli*. Relate the way this operator works to the general way in which gene expression most often is controlled in prokaryotes and eukaryotes.

4. As outlined in the section "Highlight: Genome Biology" (see p. 279), there is an ongoing shift in genetics from the study of single genes to the study of interactions among large numbers of genes. What are the advantages and disadvantages of this approach?

16   𝕿𝖍𝖊 𝕯𝖆𝖎𝖑𝖞 𝕲𝖑𝖔𝖇𝖊

Does Genetic Engineering Threaten Monarch Butterflies?

AMES, IA—In recent years, geneticists have developed ways to alter the genes of crop plants to improve their resistance to insect pests. One such success story was the insertion into corn plants of a toxin-producing bacterial gene, known as the *Bt* gene. The toxin produced by the *Bt* gene protected corn from attack by the corn borer, an insect pest that causes millions of dollars of damage each year.

Now it seems that the *Bt* gene has a dark side: It may harm the caterpillar stage of monarch butterflies. The *Bt* toxin gets into corn pollen, which spreads by wind to coat the leaves of nearby plants, including the milkweed plants that are eaten by monarch caterpillars.

Do these pollen-coated leaves harm monarchs? Two recent studies indicate they may.

First, in a laboratory experiment in which researchers dusted milkweed leaves with corn pollen, 50 percent of monarch caterpillars that ate pollen with the *Bt* gene died. In a second study, researchers let nature do the dusting: They collected milkweed leaves from plants growing near cornfields and then fed those leaves to monarch caterpillars. Monarch caterpillars that ate leaves collected near fields of corn that had the *Bt* gene died at a significantly higher rate than caterpillars that ate leaves collected near corn that lacked the *Bt* gene. Taken together, these two studies suggest

that the *Bt* toxin could hit natural populations of these beautiful butterflies hard.

The *Bt* gene has already been approved for use in crop plants; when testing for unwanted side effects, no one considered that the *Bt* toxin might spread via pollen and harm other species. Some environmentalists have been quick to call for a stop in the use of the *Bt* gene until its potential negative effects can be more carefully examined. Researchers at the MonPont Company, which produced the corn with the *Bt* gene, have countered that corn pollen does not spread very far and thus is not likely to cause much of a threat.

Evaluating "The News"

1. Do you agree with the environmentalists who argue that more careful testing is needed before farmers use genetically modified crops? Why or why not?

2. Which do you think is more likely to have unintended negative

effects—the use of corn with the *Bt* gene or the traditional practice of spraying (often by airplane) pesticides on crop fields?

3. In principle, how could an understanding of the control of expression of the *Bt* gene be used to

remove the risk that the *Bt* toxin will spread in pollen and harm monarch caterpillars? In considering this question, recall that not all genes are expressed in all portions of a multicellular organism.

chapter # 17 # DNA Technology

Alexis Rockman, *The Farm*, 2000.

Glowing Bunnies and Food for Millions

Some jellyfish can produce flashes of light that may serve to ward off attacks from predators. Recently these same lights made the headlines: The gene that enables jellyfish to produce light was isolated and transferred to a rabbit, creating a piece of "living art" meant to confront people with a creature that was both lovable and alien.

To date, the rabbit, whose name is Alba, has never been shown in an art exhibit because the outcry over her creation caused her to be confined to the laboratory in which she was made.

While some artists use genes that cause organisms to glow in the dark to create controversial art exhibits, such genes are usually used in scientific

■ MAIN MESSAGE

DNA technology makes it possible to isolate genes, produce many copies of them, determine their sequence, and insert them into organisms.

research and in medical applications. A light-producing gene from a firefly, for example, can help doctors treat the lung-destroying disease tuberculosis (TB), which is caused by a bacterium. To achieve this, the firefly gene is inserted into TB bacteria sampled from a patient. The bacteria are then screened for resistance to different antibiotics. When exposed to an antibiotic, resistant bacteria are able to grow and express their genes, including the inserted firefly gene. As a result, resistant bacteria glow in the dark, whereas bacteria that are not resistant sicken and die. Doctors can use the results of this test to prescribe an antibiotic to which the bacteria are not resistant.

Techniques similar to those used to transfer light-producing genes are also being used to develop crop plant varieties that could improve the health and nutrition of millions of people. The basic principle is simple: New genes are transferred to crops such as rice or wheat, enabling them to produce essential nutrients that they otherwise could not make. Such genetically altered crops could save the lives of millions of undernourished children.

The creation of a glowing rabbit and the development of crops with improved nutritional value were both made possible by new innovations in DNA technology, the set of techniques used to manipulate DNA. These new techniques have led to many medical and commercial applications, from the isolation of disease-causing genes to the production of industrial lubricants. In this chapter we describe the techniques scientists use to manipulate DNA, and we discuss the applications and the risks of those techniques.

Alba, a Genetically Altered White Rabbit that Glows in the Dark

■ KEY CONCEPTS

1. Recent innovations in the techniques used to manipulate DNA have greatly increased our ability to isolate and study genes and to alter the DNA of organisms.

2. Restriction enzymes cut DNA molecules at specific target sequences. When used with gel electrophoresis, a technique that sorts the chopped pieces of DNA by size, restriction enzymes provide a powerful way to examine DNA sequence differences.

3. A gene is said to be cloned when geneticists isolate it and produce many copies of it. Once a gene is cloned, automated sequencing machines can quickly determine its DNA sequence.

4. Cloning and sequencing a gene can provide vital clues about its function, making these techniques critical to the study of genes that cause inherited genetic disorders.

5. In genetic engineering, a gene is isolated, modified, and inserted back into the same species or into a different organism. Expression of the transferred gene changes the performance of the genetically modified organism.

6. DNA technology provides many benefits, but its use also raises ethical dilemmas and poses risks to human society.

H uman beings have been manipulating the DNA of other organisms for thousands of years. This fact is well illustrated by the many differences we have sculpted between domesticated species and their wild ancestors. For example, because of genetic changes brought about through selective breeding, dog breeds differ greatly from one another and from their wild ancestor, the wolf (see the box in Chapter 21, p. 363). Similarly, due to selective breeding, food plants such as wheat and corn bear little resemblance to the wild species from which they arose.

Although we have a long history of altering the DNA of other organisms, the past 30 years have witnessed a huge increase in the power, precision, and speed with which we can make such changes. For the first time, we can now select a particular gene, produce many copies of it, and put it back into living organisms. In doing so, we can rapidly alter DNA in ways that would never happen naturally, as when the light-producing gene from a jellyfish was inserted into a rabbit.

This chapter discusses **DNA technology**, the set of techniques scientists use to manipulate DNA. We will see how scientists locate a gene, analyze its sequence, and insert it into a living organism. DNA technology is increasingly in the news, whether because of the millions of dollars made or lost in biotechnology stocks, the identification of new disease-causing genes, or the use of DNA fingerprinting in murder and rape trials. We begin this chapter by describing the methods of DNA technology, then turn to its practical applications, ethical issues, and risks.

Working with DNA: Techniques for DNA Manipulation

DNA is the genetic material of all organisms. Although the sequence of DNA varies greatly among species, the chemical structure of the DNA molecule (see Figure 14.3) is the same in all species. This consistency in the structure of DNA means that similar laboratory techniques can be used to analyze DNA from organisms that are as different as bacteria and people.

In this section we describe some of the basic methods of DNA technology. To illustrate how these methods can be used, we show how they can be applied to sickle-cell anemia, a lethal recessive genetic disorder in humans. Sickle-cell anemia is caused by a mutation that alters a single amino acid in hemoglobin, a protein that is involved in the transport of oxygen by red blood cells (see Figure 15.10). Individuals who have two copies of the disease-causing allele (genotype *ss*) suffer from many serious complications, including damage to the heart, lungs, kidneys, and brain. Most individuals with sickle-cell anemia die before they have children. Heterozygous individuals of genotype *Ss*, who carry the sickle-cell allele but usually have few or no symptoms, can pass the allele on to the next generation.

Key enzymes of DNA technology

Humans have 3.3 billion base pairs of DNA on our 23 unique chromosomes. Each of these chromosomes contains a DNA molecule so large that it is difficult for scientists to work with. Thus, after DNA has been extract-

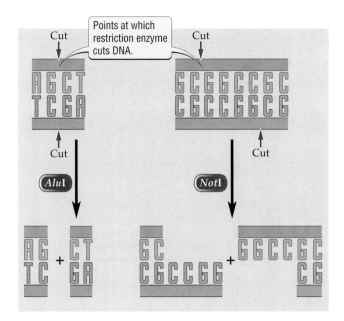

Figure 17.1 Restriction Enzymes Cut DNA at Specific Places
The restriction enzyme *Alu*I specifically cuts the DNA sequence AGCT (and no other), while the restriction enzyme *Not*I specifically cuts the DNA sequence GCGGCCGC (and no other).

ed from a person's cells, it must be broken into smaller pieces.

DNA can be broken into small pieces by **restriction enzymes**, which cut DNA at highly specific sites. For example, a restriction enzyme called AluI cuts DNA everywhere that its target sequence (AGCT) occurs, but nowhere else (Figure 17.1). There are many different restriction enzymes, each of which recognizes and cuts its own unique target sequence of DNA. Restriction enzymes were discovered in bacteria in the late 1960s; in nature, these enzymes protect bacteria against foreign DNA, such as the DNA of a virus.

When a restriction enzyme is used to chop up a person's DNA in a test tube, the specificity of the enzyme ensures that the same results are obtained at different times or from the DNA of different tissues (for example, skin and hair). Because restriction enzymes work on the DNA of all organisms, they are used in virtually all applications of DNA technology.

Ligases and DNA polymerases are two other important enzymes used in DNA technology. **Ligases** are enzymes that can connect two DNA fragments to each other, making it possible to insert a gene from one species into the DNA of another species. **DNA polymerases**, the key enzymes that cells use to replicate their DNA, can be used to make many copies of a gene or other DNA sequence in a test tube.

Gel electrophoresis

Once a DNA sample is cut into fragments by a restriction enzyme, researchers often use gel electrophoresis to help them see and analyze the fragments. In **gel electrophoresis**, DNA that has been chopped up by a restriction enzyme is placed into a depression (a "well") in a gelatin-like substance (a "gel") (Figure 17.2). An electrical current is then passed through the gel. Since DNA has a negative electrical charge, the electrical current causes the DNA to move toward the positive end of the gel. Large pieces of DNA (those with more base pairs) pass through the gel with more difficulty, and thus move more slowly, than small pieces. Because they move more slowly, large fragments of DNA do not travel as far as small fragments. Thus, rapidly moving small fragments will be found toward the bottom of the gel, while slowly moving large fragments will be located toward the

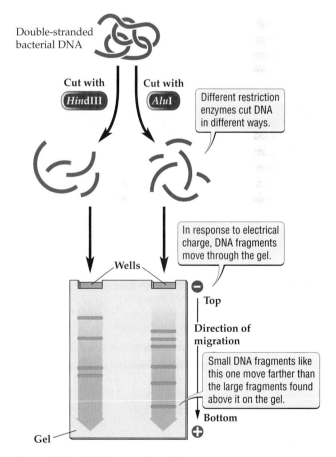

Figure 17.2 Gel Electrophoresis
When subjected to an electrical charge, DNA fragments move through a gel at different rates, depending on their size. Fragments found toward the bottom of the gel are smaller than fragments found toward the top of the gel.

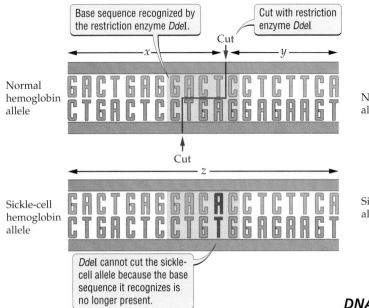

Normal hemoglobin allele

Base sequence recognized by the restriction enzyme *Dde*I.

Cut with restriction enzyme *Dde*I

Sickle-cell hemoglobin allele

*Dde*I cannot cut the sickle-cell allele because the base sequence it recognizes is no longer present.

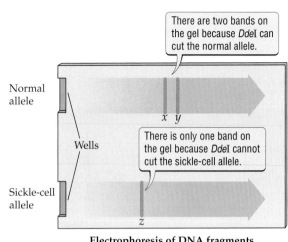

Normal allele

There are two bands on the gel because *Dde*I can cut the normal allele.

Wells

Sickle-cell allele

There is only one band on the gel because *Dde*I cannot cut the sickle-cell allele.

Electrophoresis of DNA fragments

17.2

Figure 17.3 A Restriction Enzyme and Gel Electrophoresis Can Be Used to Identify the Sickle-Cell Allele
The restriction enzyme *Dde*I cuts DNA every time it encounters the sequence GACTC (shown in yellow in the top panel). The single base-pair mutation that causes the disease sickle-cell anemia changes the DNA sequence of part of the hemoglobin gene from the normal GACTC to GACAC, thus preventing *Dde*I from recognizing and cutting the DNA at this location. As a result, the normal hemoglobin allele shows up as two bands on the gel, representing two fragments of different sizes, whereas the sickle-cell allele produces only one. The substitution mutation that causes sickle-cell anemia is highlighted in red.

DNA hybridization

Another way of testing for the sickle-cell allele is to use a DNA probe. A **DNA probe** is a short single-stranded sequence of DNA, usually tens to hundreds of bases long, that can pair with a particular gene or region on another strand of DNA. A probe can pair with another segment of DNA if the sequence of bases in the probe (e.g., CCTAGT) is complementary to the sequence of bases in the other segment of DNA (e.g., GGATCA). DNA probes are used in **DNA hybridization** experiments (Figure 17.4), which involve the pairing of DNA from two different sources.

Scientists can use DNA hybridization experiments to test whether a person has one, two, or no copies of the

top. The different-sized fragments are invisible to the human eye, so to be seen they must be stained or labeled by one of various methods.

By using restriction enzymes and gel electrophoresis together, we can examine differences in DNA sequences. For example, the restriction enzyme *Dde*I cuts the normal hemoglobin allele into two pieces, but it cannot cut the sickle-cell allele, providing a simple test for the disease allele (Figure 17.3).

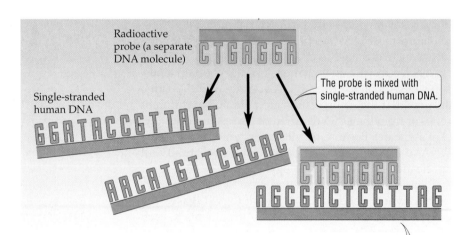

Radioactive probe (a separate DNA molecule)

Single-stranded human DNA

The probe is mixed with single-stranded human DNA.

DNA hybridization occurs when the probe can pair with a portion of the single-stranded human DNA.

Figure 17.4 DNA Hybridization
DNA hybridization occurs when DNA from two different sources can form complementary base pairs. Here, a DNA probe can bind to one of these single-stranded segments of human DNA, but not to the other two. In such experiments, the probe may be radioactively or chemically labeled to make it easier to identify the DNA segments to which it binds.

allele (*s*) that causes sickle-cell anemia. In such experiments, two 21-base-long DNA probes are used: one probe that can bind only to the sickle-cell allele, and a second probe that can bind only to the normal allele (*S*). If both probes can bind to a person's DNA, the person must have both alleles, and hence must be a heterozygous carrier with genotype *Ss*. However, if only the probe for the normal allele can bind, the person must have two copies of the normal allele (genotype *SS*). Similarly, if only the probe for the sickle-cell allele can bind, the person must have two copies of the sickle-cell allele (genotype *ss*).

DNA sequencing and synthesis

Both basic research and practical applications of modern genetics often depend on knowing the sequence of bases in a DNA fragment, a gene, or even an entire genome. DNA sequences can be determined by several methods, the most efficient of which rely on automated sequencing machines (Figure 17.5). One of these machines can sequence over a half million bases per day, thus making it possible to determine the sequence of a single gene quickly. DNA can also be sequenced without a sequencing machine by relatively slow, but still highly effective, methods.

Figure 17.5 A DNA Sequencing Machine
Automated DNA sequencing machines can determine DNA sequences rapidly. The machine shown here can sequence a total of over 700,000 bases per day.

The synthesis of probes and other DNA fragments can also be automated. DNA synthesis machines can rapidly produce DNA segments of a specified base sequence up to hundreds of bases long. For example, in less than an hour, a DNA synthesis machine can produce the two probes used to test for sickle-cell anemia, each of which is 21 bases long.

> ■ The methods of DNA technology can be used to study the DNA of all organisms. Scientists can manipulate DNA with a variety of laboratory techniques, including restriction enzymes, gel electrophoresis, DNA hybridization, and automated sequencing and synthesis.

Producing Many Copies of a Gene: DNA Cloning

A single copy of a gene is difficult to study. For this reason, geneticists may isolate a gene and make many copies of it—in other words, the gene may be **cloned**. Once a gene is cloned, it can be sequenced, transferred to other cells or organisms, or used as a probe in DNA hybridization experiments. In addition, the cloning and sequencing of a gene enable researchers to use the genetic code (see Figure 15.5) to determine the amino acid sequence of the gene's protein product. As occurred for Huntington's disease (see p. 233), knowledge of a gene's product can provide vital clues to the gene's function. For this reason, DNA cloning is a key step in the study of genes that cause inherited genetic disorders or cancers.

In the sections that follow, we'll describe two of the ways genes can be cloned: by the construction of DNA libraries or through the use of the polymerase chain reaction (PCR).

DNA libraries

A **DNA library** is a collection of DNA fragments from one organism that is stored in another host organism. For humans, a complete DNA library would contain tens to hundreds of thousands of DNA fragments, which collectively would include all the 3.3 billion bases in the human genome.

The concepts behind the formation of a DNA library are simple. First, the DNA is broken into pieces by a restriction enzyme. The fragments are then inserted into **vectors**, which are pieces of DNA that are used to transfer genes or other DNA frag-

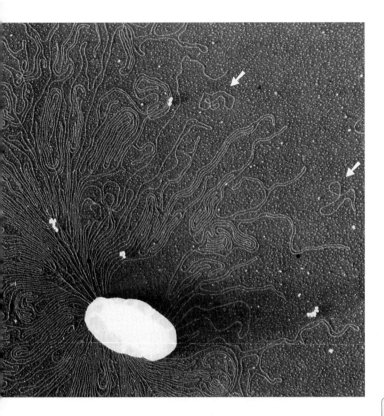

Figure 17.6 Plasmids
Plasmids are circular segments of DNA found naturally in bacteria. Plasmids are not part of, and are much smaller than, the single main chromosome of the bacterium. Here, a ruptured *E. coli* bacterium spills out its chromosome and several plasmids, two of which are indicated with arrows.

ments from one species to another. Common types of vectors include DNA from viruses and **plasmids**, which are small circular segments of DNA that are found naturally in bacteria (Figure 17.6). Vectors can move DNA fragments into a host organism, such as a bacterium.

If, for example, plasmids were used to construct a human DNA library, fragments of human DNA would be inserted into plasmid vectors and then mixed with bacteria under conditions that would cause some of the bacteria to take up a plasmid (Figure 17.7). Once the bacteria had taken up the plasmids, our library would be formed: We would have a large number of bacteria, each of which contained a single fragment of human DNA. Collectively, these bacteria would contain many fragments of the 3.3 billion bases in the human genome.

Bacteria reproduce rapidly, and as they reproduce, they make new copies of the inserted DNA fragments. Thus, we can use the bacteria in a DNA library to make multiple copies of a gene. In this process, a few bacterial cells from the library are grown on a small dish. Each

bacterial cell gives rise to a mass of cells, called a colony. Each cell in the colony contains the DNA fragment inserted into the bacterium that started the colony.

To produce many copies of a particular gene, we first must locate bacterial colonies that carry the gene of interest. Therefore, the colonies are tested, or "screened," by DNA hybridization to see if their DNA can pair with a probe for the gene of interest. Colonies whose DNA can pair with the probe contain all or part of the gene we seek. Bacteria from such colonies are then grown in a liquid broth, producing billions of bacterial cells. Each of these cells contains a plasmid that has the gene of interest in it. Thus, by screening a DNA library, we can isolate a particular gene from a large number of different human DNA fragments, and then produce many copies of that gene.

The polymerase chain reaction

In some cases, DNA can be cloned by the **polymerase chain reaction (PCR)**, a method that uses the DNA poly-

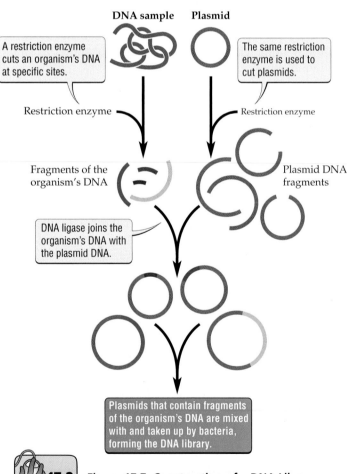

17.3   Figure 17.7 Construction of a DNA Library

merase enzyme to make billions of copies of a targeted sequence of DNA in just a few hours (Figure 17.8).

When PCR is used to clone a particular gene, two short segments of synthetic DNA, called primers, must be used. Each primer is designed to pair with one of the two ends of the gene of interest. The DNA polymerase enzyme then produces many copies of the sequence of DNA that is between the primers; that is, it produces many copies of the gene. To use PCR in gene cloning, scientists must know the DNA sequence of both ends of the gene; without this knowledge, they cannot synthesize the specific primers that will pair with the ends of the gene.

■ Genes can be cloned via construction of a DNA library or by the polymerase chain reaction. Once it has been cloned, a gene can be sequenced, transferred to other organisms, or used in DNA hybridization experiments.

17.5 *Applications of DNA Technology*

Now that we've discussed some of the methods used in DNA technology, let's consider some of the purposes for which those methods can be used. There are many important applications of DNA technology, recent examples of which include prenatal screening for genetic disorders and the use of genetic profiling to determine which of several drugs should work best in a given patient (see Chapter 18). Other examples include gene therapy, a topic addressed at the close of this chapter, and the potential use of DNA technology in the conservation of endangered species (see the box on p. 292). In this section we focus on two common uses of DNA technology: DNA fingerprinting and genetic engineering.

Figure 17.8 The Polymerase Chain Reaction
In PCR, short primers that can pair with the two ends of a gene of interest are mixed in a test tube with the organism's DNA, the enzyme DNA polymerase, and nucleotides (containing the bases A, C, G, or T). A machine then processes the mixture, going through a repeated series of steps in which the temperature is first raised, then lowered, as indicated. Billions of copies of a targeted DNA sequence can be made in this way in a few hours. The targeted DNA sequence (blue) and the primers (red) are color-coded here for reasons of clarity only; the colors do not indicate any biological differences in the DNA sequences.

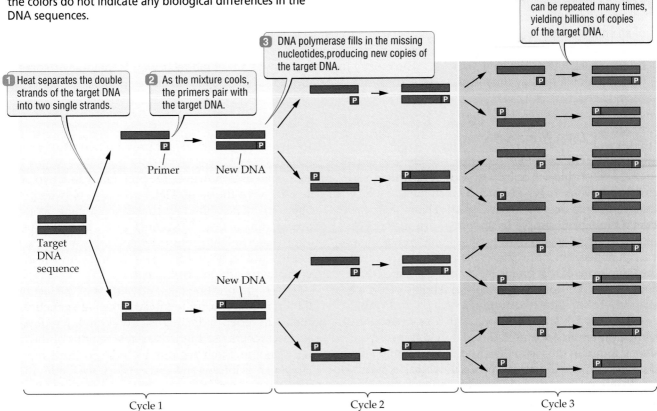

■ BIOLOGY IN OUR LIVES

Can DNA Technology Be Used to Save Endangered Species?

The rapidly growing human population has altered the global environment and has brought many species to the brink of extinction (see Chapter 4). Some scientists argue that we should harness the power of DNA technology to save endangered species, or even bring extinct species back to life.

How could DNA technology accomplish such goals? It is possible, for example, to use DNA technology to transfer genes to plant cells and then grow entire plants from such genetically modified cells. This approach could be used to rescue plant populations suffering from extreme reductions in genetic diversity caused by small population size (see Chapter 20). In principle, key alleles no longer found in natural populations could even be recovered from museum specimens and transferred to the cells of an endangered plant species.

But how can an extinct species be brought back to life? While *Jurassic Park* remains a fantasy (in part because DNA recovered from fossils is highly degraded), cells collected from endangered species could be frozen and then used to produce genetic copies, or clones (see p. 619), of the organism if the species were to become extinct.

Scientists are currently working on just such a project. In 2000, the last remaining bucardo, a type of Spanish mountain goat, was killed when a tree fell on it and crushed its skull. Fortunately, Spanish researchers had previously collected and frozen some of the last bucardo's cells, and plans are under way to create bucardo clones, effectively bringing an extinct species back to life. Plans have also been made to clone many other endangered species, such as the African bongo antelope, the Sumatran tiger, and the reluctant-to-reproduce giant panda.

The most important step we can take to save endangered species is to

Can this extinct species, a bucardo, be brought back to life?

set aside large areas of natural habitat and otherwise reduce the human impact on ecological communities. While the potential of genetic engineering efforts should not be used as an excuse to continue to destroy the natural habitat on which endangered species depend, the cloning of endangered species may provide a useful, high-tech way to help save some species from extinction.

17.6 *DNA fingerprinting*

The use of DNA analyses to identify individuals is called **DNA fingerprinting**. Scientists use DNA fingerprinting in much the same way that traditional fingerprints are used. A laboratory can take a biological sample of unknown origin, such as blood, tissue, or semen from a crime scene, and develop a DNA fingerprint, or profile, of the person from whom the sample came. That profile can then be compared with another profile—for example, that of a crime victim or suspect—to see if the two profiles match (Figure 17.9).

When two DNA profiles don't match, as when DNA from a rape suspect differs from DNA obtained from semen found in the victim, the results are definitive (in this case, proving that the suspect is innocent). A match, like that shown in Figure 17.9, provides evidence that a crime scene sample (for example, a drop of blood on a shirt) could have come from the tested victim or suspect.

However, a match does not provide definitive proof, since two people can have the same DNA profile. In most legal cases, however, the probability that two people will have the same DNA profile is between one chance in 100,000 and one chance in a billion. Widely used in criminal cases, DNA fingerprints have been used to convict criminals of murder and rape and to prove people's innocence, sometimes freeing convicts after years of wrongful imprisonment.

DNA fingerprinting takes advantage of the fact that all individuals (except identical twins) are genetically unique. In order to distinguish between different people, researchers use highly variable regions of the human genome. Such regions include portions of our DNA, such as introns and spacer DNA (see Figure 16.3), that do not encode proteins; these regions tend to vary greatly in size and base sequence between different people.

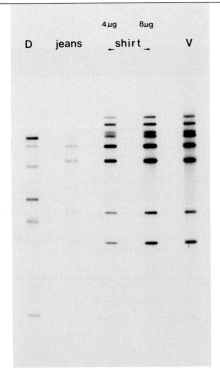

Figure 17.9 Catching the Bad Guys Using DNA
The DNA profile on the left comes from a RFLP analysis of a sample taken from the defendant (D) on trial for murdering the victim (V). The defendant's jeans and shirt were splattered with blood—the DNA from which matches the victim's.

DNA fingerprinting can be done in various ways, including **restriction fragment length polymorphism (RFLP) analysis** and PCR amplification. In RFLP analysis, restriction enzymes are used to cut a person's DNA into small pieces. Next, those fragments are sorted by size using gel electrophoresis. Finally, a DNA probe is used to identify the number and size of the fragments that can bind to the probe, enabling DNA profiles like those in Figure 17.9 to be formed.

To ensure that the chance that two individuals will have the same DNA profile is very low, usually several DNA probes are used, each of which binds to regions of the genome that are known to vary greatly between individuals. Similarly, PCR amplification can be carried out using primers from highly variable regions of the genome. By amplifying a number of these highly variable regions, technicians can produce a DNA fingerprint specific to each person.

Figure 17.10 Genetic Engineering of Plants via Plasmids
Plasmids can be used to insert a gene of interest into plant cells. In many plant species, adult plants can then be grown from the genetically modified cells, thus producing a genetically engineered plant.

Genetic engineering

All organisms share a similar, often identical, genetic code. For this reason, if a gene can be transferred from one species to another, it often can make a functional protein product in the new species. For example, the light-producing gene from a firefly mentioned at the beginning of this chapter has been transferred to and expressed in organisms as different as plants, mice, and bacteria. The deliberate transfer of a gene from one species to another is an example of genetic engineering. **Genetic engineering** is a three-step process in which a DNA sequence (often a gene) is isolated, modified, and inserted back into the same species or into a different species.

Firefly

Several techniques can be used to insert genes into organisms. We have already discussed how plasmids can be used to transfer a gene from humans or other organisms to bacteria. Plasmids also can be used to transfer genes to plant or animal cells (Figure 17.10).

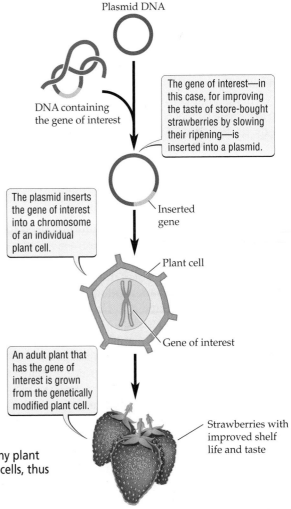

Plasmid DNA

DNA containing the gene of interest

The gene of interest—in this case, for improving the taste of store-bought strawberries by slowing their ripening—is inserted into a plasmid.

Inserted gene

The plasmid inserts the gene of interest into a chromosome of an individual plant cell.

Plant cell

Gene of interest

An adult plant that has the gene of interest is grown from the genetically modified plant cell.

Strawberries with improved shelf life and taste

17.1 *Products of Genetic Engineering*

Product	Method of production	Use
Protein		
Human insulin	*E. coli*	Treatment of diabetes
Human growth hormone	*E. coli*	Treatment of growth disorders
Taxol biosynthesis enzyme	*E. coli*	Treatment of ovarian cancer
Luciferase (from firefly)	Bacterial cells	Testing for antibiotic resistance
Human factor VIII	Mammalian cells	Treatment of hemophilia
ADA	Human cells	Treatment of ADA deficiency
DNA sequence		
Sickle-cell probe	DNA synthesis machine	Testing for sickle-cell anemia
BRCA1 probe	DNA synthesis machine	Testing for breast cancer mutations
HD probe	*E. coli*	Testing for Huntington's disease
M13 probe	*E. coli*, PCR	DNA fingerprinting in plants
33.6 and other probes	*E. coli*, PCR	DNA fingerprinting in humans

In some species, including many plants, and recently in mammals such as sheep, cows, and mice, genetically modified adults can be generated, or "cloned," from these altered cells. Other techniques for gene transfer include the use of viruses that infect cells with genes from other species and "gene guns" that fire microscopic pellets coated with the gene of interest into target cells.

Genetic engineering is commonly used to alter the performance of the genetically modified organism. Pea seeds, for example, have been given new genes that protect them against insect attack (Figure 17.11). Other crop plants have been genetically engineered for disease resistance, frost tolerance, and herbicide resistance (to allow the crops to survive the application of weed-killing chemicals).

Genetic engineering is also commonly used to produce many copies of a DNA sequence, a gene, or a gene's protein product (Table 17.1). For example, insulin, a human hormone used to treat millions of people suffering from diabetes, is mass-produced by *E. coli* bacteria that have been engineered to contain the human insulin gene. In addition, bacteria have been genetically engineered to produce large amounts of the cell surface proteins of particular disease-causing organisms. When injected into the body, these cell surface proteins stimulate the immune system to recognize the organism that normally carries the proteins. Hence, they can be used as a vaccine to protect us against future attack by that organism. Currently, a test is under way to evaluate the effectiveness of an AIDS vaccine that uses a genetically engineered cell surface protein from the AIDS virus.

■ DNA fingerprinting can be used for a wide variety of purposes, such as convicting criminals and freeing the innocent. In genetic engineering, a DNA sequence (often a gene) is isolated, modified, and inserted back into the same species or a different species. Genetic engineering can be used to alter the performance of the genetically modified organism or to produce many copies of a DNA fragment, a gene, or a gene's protein product.

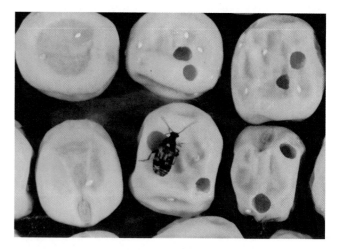

Figure 17.11 Genetically Engineered Protection from Insect Attack
The pea seeds on the left, which were genetically engineered to be protected against insect attack, are intact. The seeds on the right have holes caused by damage from weevils. A weevil is shown here.

Ethical Issues and Risks of DNA Technology

As we have seen, DNA technology provides many benefits to human society. At the same time, the immense power and scope of genetic engineering and other aspects of DNA technology raise ethical dilemmas and pose a variety of risks. For example, at a most basic level, what gives us the right to alter the DNA of other species? We typically do so for our own advantage, but is this an ethical thing to do? And if, say, most people could agree that there was no ethical conflict in altering the DNA of a bacterium or a virus, does that mean there is also no conflict associated with altering the DNA of a plant, a dog, a chimpanzee, or a person?

With respect to altering human DNA, how do we distinguish between acceptable and unacceptable uses of genetic engineering? If it is ethical to genetically engineer a human to prevent a horrible disease, is it also acceptable to make less critical changes? For example, if it were possible to do so, would it be ethical to alter the future intelligence, personality, looks, or sexual orientation of our children before birth? According to a March of Dimes survey, more than 40 percent of Americans would make such modifications if given the chance, but is it fair for parents to make such decisions on behalf of their children? We will return to such issues in Chapter 18, but be forewarned: Often there are no easy answers to questions like these.

The use of DNA technology also involves risks. For example, 12 of the world's 13 most important crop plants can mate and produce offspring with a wild plant species in some region where they are grown. If a crop plant is genetically engineered to be resistant to an herbicide, the potential exists that the resistance gene will be transferred (by mating) from the crop to the wild species. Thus, there is a risk that by engineering our crops to resist herbicides, we will unintentionally create "superweeds" that are resistant to the same herbicides. Similar risks exist for most efforts to alter the performance of an organism. In essence, such risks boil down to the problem that a gene that is good for humans when it is in one species (or one set of circumstances, such as an agricultural field), may be bad for us if it is in another species (or another set of circumstances, such as a more natural field environment).

Environmental or social costs may also be associated with genetic engineering. Engineering crops to be resistant to herbicides, for example, might promote the increased use of herbicides, many of which are harmful to the environment. Alternatively, a product might be environmentally safe yet still entail social costs. Consider bovine growth hormone (BGH), which is mass-produced by genetically engineered bacteria. Among other effects, BGH increases milk production in cattle. Before the introduction of genetically engineered BGH in the 1980s, milk surpluses already were common. The use of BGH by large corporate milk producers has further increased the amount of milk available. The resulting drop in milk prices threatens to drive small producers of milk—the traditional family farms—out of business. Are lower milk prices for consumers worth the social cost of driving small dairy farms into bankruptcy?

> ■ The use of DNA technology raises ethical questions and poses risks. There is a fine line between acceptable and unacceptable changes to the DNA of humans and other species. Benefits and risks must be considered carefully in evaluating the potential effect of any particular use of DNA technology on human society.

■ HIGHLIGHT

Human Gene Therapy

On September 14, 1990, 4-year-old Ashanthi DeSilva made medical history when she received intravenous fluid that contained genetically modified versions of her own white blood cells. She suffered from adenosine deaminase (ADA) deficiency, a genetic disorder that severely limits the ability of the body to fight disease and can make ordinary infections, such as colds or flu, lethal. This disorder is caused by a mutation to a single gene that is expressed in white blood cells, which play a central role in our ability to combat disease (see Chapter 32). Earlier, doctors had removed some of Ashanthi's white blood cells and added the normal ADA gene to them (Figure 17.12). Thus, her cells were genetically engineered in an attempt to fix a lethal genetic defect. A few months later, a similar experiment was performed on 8-year-old Cynthia Cutshall, who also suffered from ADA deficiency. Both girls responded very well to the treatment and now lead essentially normal lives.

The treatments received by Ashanthi DeSilva and Cynthia Cutshall were the first clinical gene therapy experiments ever performed. Human **gene therapy** seeks to correct genetic disorders by fixing the genes that cause them. The possibility of our curing even the worst of genetic diseases by reaching into our cells and repairing the mutations that caused them is a bold and captivating prospect.

17.8

Figure 17.12 Gene Therapy for ADA Deficiency
ADA deficiency, a lethal disorder caused by a genetic defect in white blood cells, was the first genetic disorder to be treated by gene therapy.

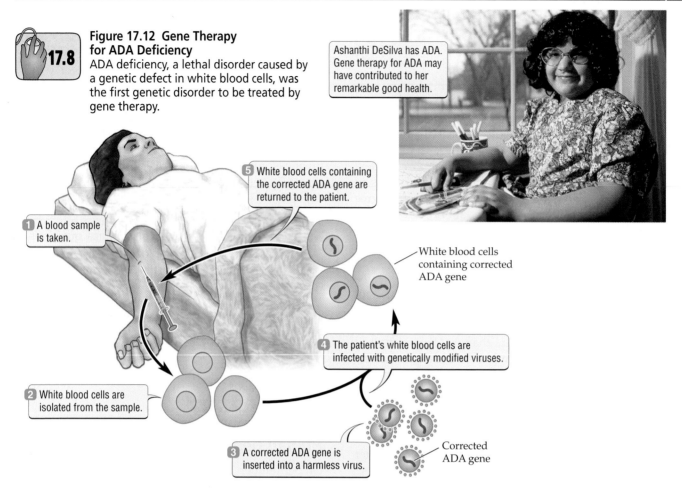

Ashanthi DeSilva has ADA. Gene therapy for ADA may have contributed to her remarkable good health.

5 White blood cells containing the corrected ADA gene are returned to the patient.

1 A blood sample is taken.

White blood cells containing corrected ADA gene

2 White blood cells are isolated from the sample.

4 The patient's white blood cells are infected with genetically modified viruses.

3 A corrected ADA gene is inserted into a harmless virus.

Corrected ADA gene

As such, gene therapy has attracted much media attention—some of it, unfortunately, bordering on hype. Take the cases of Ashanthi and Cynthia. In addition to gene therapy, both girls received other treatments for ADA deficiency. Hence, contrary to what some reports in the media might have us believe, their remarkable good health cannot be attributed to gene therapy alone. Overall, although there are now more than 400 gene therapy experiments in progress worldwide, there are few clear success stories. Why? Has gene therapy been "oversold"?

Concerns about gene therapy were driven home by the recent death of a young man in a gene therapy experiment. This tragic event raised new questions about human gene therapy and how best to guide its development. These concerns are important, yet at one level, scientists know that gene therapy works: There are many examples in which corrected genes have been transferred to a patient and expressed. And in one recent clinical study, conducted in France, gene therapy treatment alone brought the crippled immune systems of two patients, ages 8 and 11 months, to normal levels. As of this writing, the gene therapy treatment had completely eliminated the disease symptoms of these two patients for more than three years. Although these and other new results are very promising, whether gene therapy can achieve its ultimate goal of permanently fixing a genetic disease is not yet known.

For gene therapy to reach its full potential, formidable hurdles remain to be cleared—perhaps most importantly, how to deliver the engineered gene to the cells where it is needed. Harmless viruses are often used for this purpose, but to date viral delivery methods have had limited success. In part, the reason for this low rate of success is that the human body defends itself so well against viruses that the viruses often are destroyed before they deliver the corrected gene to where it is needed.

Viral methods for delivering engineered genes continue to be improved, and recently a novel approach that uses an artificial "hybrid" molecule (part DNA, part RNA) was found to be astonishingly effective at delivering an engineered gene to the liver cells of live laboratory rats. In this experiment, up to 60 percent of the rat liver cells received the engineered gene—far more than would be needed to cure many genetic disorders. Thus, although many challenges lie ahead, there is cause for excitement and hope.

■ Human gene therapy seeks to correct genetic disorders by fixing the genes that cause them. In some gene therapy experiments, the corrected gene has been transferred and expressed in the patient. However, many challenges remain, as illustrated by the recent death of a gene therapy patient and by the difficulty in delivering corrected genes to the cells that need them.

SUMMARY

Working with DNA: Techniques for DNA Manipulation

■ Scientists can manipulate DNA using a variety of laboratory techniques. Because the structure of DNA is the same in all organisms, these techniques work on all species.

■ Restriction enzymes break DNA into small pieces, and gel electrophoresis is used to separate the resulting DNA fragments by size.

■ DNA probes are used in DNA hybridization experiments to test for the presence of a particular allele or gene.

■ DNA can be sequenced and synthesized by automated machines.

Producing Many Copies of a Gene: DNA Cloning

■ A gene is cloned if it has been isolated and many copies of it can be made.

■ After being cloned, a gene can be sequenced, transferred to other organisms, or used in DNA hybridization experiments.

■ Genes can be cloned by constructing a DNA library or by using the polymerase chain reaction.

Applications of DNA Technology

■ Genetic engineering works because all organisms share a similar, often identical, genetic code.

■ In genetic engineering, a DNA sequence (often a gene) is isolated, modified, and inserted back into the same species or into a different species.

■ Genetic engineering is used to alter the performance of the genetically modified organism or to produce many copies of a DNA fragment, a gene, or a gene's protein product.

Ethical Issues and Risks of DNA Technology

■ The use of DNA technology raises ethical questions and poses risks. There is a fine line between acceptable and unacceptable changes to the DNA of humans and other species.

■ Benefits and risks must be considered carefully in evaluating the potential effect of any particular use of DNA technology on human society.

Highlight: Human Gene Therapy

■ Human gene therapy seeks to correct genetic disorders by fixing the genes that cause them.

■ In some gene therapy experiments, the corrected gene has been transferred and expressed in the patient.

■ Although results to date are encouraging, there are no clear examples in which gene therapy has permanently cured a genetic disease.

KEY TERMS

clone (of a gene) p. 289	genetic engineering p. 293
DNA fingerprinting p. 292	ligase p. 287
DNA hybridization p. 288	plasmid p. 290
DNA library p. 289	polymerase chain reaction (PCR) p. 290
DNA polymerase p. 287	
DNA probe p. 288	restriction enzyme p. 287
DNA technology p. 286	restriction fragment length polymorphism (RFLP) analysis p. 293
gel electrophoresis p. 287	
gene therapy p. 295	vector p. 289

CHAPTER REVIEW

Self-Quiz

1. Which of the following cuts DNA at highly specific target sequences?
   a. ligase
   b. DNA polymerase
   c. restriction enzyme
   d. RNA polymerase

2. A collection of an organism's DNA fragments stored in a host organism is called a
   a. DNA library.
   b. DNA restriction site.
   c. plasmid.
   d. DNA clone.

3. The pairing of DNA from two different sources is called
   a. DNA replication.
   b. DNA hybridization.
   c. genetic engineering.
   d. DNA cloning.

4. Genetic engineering
   a. can be used to make copies of a DNA sequence, a gene, or a gene's protein product.
   b. can provide benefit to people by altering the performance of the genetically modified organism.
   c. raises ethical questions and poses risks to society.
   d. all of the above

5. When DNA fragments are placed on an electrophoresis gel and subjected to an electrical current, the _____ fragments move the farthest.
   a. smallest
   b. largest
   c. PCR
   d. DNA library

Review Questions

1. a. Define DNA cloning and describe how it is done.

   b. Discuss the advantages of gene cloning.

2. When a DNA library is made, tens to hundreds of thousands of fragments of an organism's DNA are stored in a host organism. How is the library screened so that scientists can find a particular gene of interest?

3. Is it ethical to genetically alter the DNA of a bacterium? A single-celled yeast? A worm? A plant? A cat? A human? Give reasons for your answers.

4. If it were possible to do so, should it be legal to alter the DNA of a person to cure a genetic disorder, such as Huntington's disease, that causes great suffering and kills its victims?

5. Are there some changes to the DNA of humans that are not acceptable? Assuming you think some changes are not acceptable, what criteria would you use to draw the line between acceptable and unacceptable changes?

The Daily Globe

"Terminator Gene" Causes Riots in Developing World

NEW DELHI, INDIA. Small farmers in Mullampur rioted today, burning fields owned by the MonPont Company that contained genetically engineered plants. The farmers expressed fear that a "terminator gene" would spread from the agribusiness company's fields to the small plots on which they depend for food, making their crop plants sterile.

Seed companies have spent vast sums of money to develop new genetic varieties of crop plants. The companies "own" these new genetic varieties, and when they have sold their seed to farmers, they have insisted that the farmers agree not to replant any seed from the harvest. Thus, each year the farmers must start from scratch by buying more seed from the seed companies.

Despite this agreement, until now the possibility remained that some farmers could "cheat" by planting some of the seed they harvested. But the MonPont Company has now developed a new, genetically engineered variety of corn that gives the seed companies the final say. MonPont researchers have inserted a "terminator gene" that makes the next generation of corn plants sterile. Thus, if a farmer "cheats" and plants seed from his harvest, the plants produced from that seed will be sterile and will not produce any corn.

As word of the terminator gene has spread, riots like the one in Mullampur have spread as well. As one angry local farmer put it, "We are hungry enough already. We don't need MonPont to genetically engineer our crops to be sterile." A spokesperson for MonPont assured local farmers that the gene could not spread to their crop plants, but few farmers appeared to be in a mood to listen.

Evaluating "The News"

1. When seed companies sell seed for genetic varieties of crop plants that cost a lot of money to develop, is it reasonable for them to insist that farmers not replant any of the seed they harvest?

2. Farmers work hard to grow their crops. When the harvest is complete, do you think farmers should have the right to do whatever they want with the seeds they harvest, including saving some of them to plant as crops for the next year?

3. Roughly a billion of Earth's 6 billion people are malnourished. Given that so many people lack food, do you think the "terminator gene" is a wise use of genetic engineering?

Lawrence Berzon, *The Collector* (detail), 2000.

A Crystal Ball for Your Health

*D*espite the worldwide celebrations that greeted the start of the year 2000, the new millennium, based on the Gregorian calendar, actually began on January 1, 2001. A milestone whose timing is not debatable was the publishing of the first draft of the human genome sequence. This historical event, which occurred in February of 2001, was the fruit of the combined efforts of thousands of researchers over a period of 15 years. To many scientists, it represented the crowning achievement of twentieth-century biology, and many press conferences and articles have touted it as such.

How will this scientific achievement affect our lives? The simple answer is that knowing the entire sequence of the human genome amounts to knowing the blueprint that dictates every biological process in our bodies. Encoded in our more than 3 billion base pairs of DNA are variations in our individual genes that are likely to directly affect our future health. While all of us have genomes that are 99.9 percent identical with one another, the 0.1 percent of difference has great bearing on our susceptibility to certain diseases and even our overall life span. Thus, our DNA sequences will reveal not just how our bodies work in the present, but also how they are likely to function in the future.

For us to use this information to predict the health of individuals, there must be a simple way to identify relevant genetic variations.

Single base-pair differences, known as **single nucleotide polymorphisms (SNPs)**, are one important source of genomic variation. If one could associate the presence of a specific SNP or group of SNPs with susceptibility to a disease such as breast cancer, individuals that present this SNP pattern could be forewarned of oncoming disease far in advance, allowing timely therapeutic intervention. Naturally, the detailed matching of SNP profiles with disease susceptibility depends on knowledge of the human genome sequence as a basis for comparison.

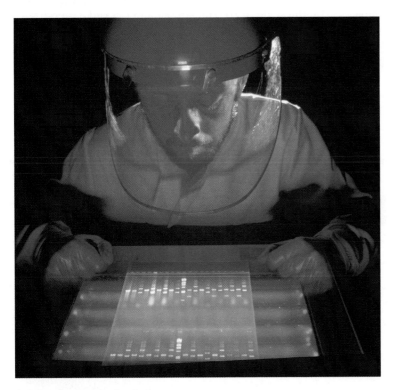

A Scientist Reads the Future

■ **MAIN MESSAGE**

The Human Genome Project has revolutionized biology and medicine while raising difficult ethical questions.

Even before the draft genome sequence was released in 2001, scientists in Great Britain began laying the foundation for a massive database of human SNP profiles. The database would rely on SNP profiles from hundreds of thousands of blood samples donated by adult volunteers. Physicians would refer these volunteers to the project, and their current health status would be recorded and matched with their SNP profile. In addition, the volunteers would be tracked over time so that changes in their health would also be matched with their SNP profile. Based on this database and others like it, specific SNPs will eventually be identified as indicators of disease susceptibility and even particular patterns of aging. The creation of such databases redefines the idea of health profiling, and creates new concerns about the possible effect on individual privacy and even health insurance coverage.

The SNP database is just one example of the tremendous promise and accompanying ethical problems that result from having access to our complete genetic blueprint. In this chapter, we explore the history that led up to the sequencing of the human genome, then return to the medical benefits and ethical challenges this new knowledge is likely to bring us.

■ KEY CONCEPTS

1. Genomics seeks to understand the structure and expression of entire genomes and how they change during evolution.

2. The quest to understand inheritance led to the discovery of DNA as the blueprint of life. Subsequent breakthroughs in DNA cloning and sequencing provided the tools necessary for the Human Genome Project.

3. The competition between the public and commercial efforts to sequence the human genome sped up the

process, but it also raised difficult ethical questions along the way.

4. Comparisons of the human genome with the genomes of other organisms reveal the commonality of many biological processes.

5. Knowing the human genome sequence will greatly enhance the ability of genetic screening to determine the likelihood that a person will contract a particular disease or respond poorly to a given drug.

*T*he twenty-first century has opened with a bang as far as biology is concerned. The completion of the first draft of the human genome sequence is expected to change not only the science of biology, but also our daily lives. In the immediate afterglow of this incredible achievement, nearly every branch of biology is reassessing its future in the brave new postgenomic world.

The successful sequencing of the human genome depended on several earlier efforts to sequence the genomes of other organisms. These efforts produced draft genome sequences for a bacterium in 1995, a budding yeast in 1996, a nematode worm in 1998, and a fruit fly in 2000 (Table 18.1). The success of these and other efforts gave birth to a new field of biology called **genomics**. In the earlier chapters of this unit, you learned that genetics is the study of genes and how they are expressed and transmitted in cells and organisms. The field of genomics builds on genetics, seeking to understand the structure and expression of entire genomes and how they change during evolution.

The study of genetics includes the methods used to analyze processes such as inheritance. Experiments like those performed by Mendel (see Chapter 12) remain important tools for uncovering the role played by specific genes in producing a particular trait or phenotype. Likewise, the study of genomics includes a related group of methods used to analyze the broad-scale expression of genes in whole genomes. As you read in Chapter 16, some of these genomic methods use DNA chips to monitor the expression levels of all the genes in a single cell.

Genetics can be further distinguished from genomics by the scale of the questions being asked. Genetics focuses on individual genes and how they function, either alone or together with a limited set of other genes, to control a phenotype. Genomics, on the other hand, takes a far more comprehensive view, monitoring the coordinated activities of all the genes in the genome. The expanded scale of the issues addressed by genomics has already had a profound effect on other fields in biology. Indeed, the enormous volume of data represented by complete genome sequences encourages biologists to

18.1 *A Selection of Sequenced Genomes*

Organism	Date	Estimated genome size (millions of base pairs)	Predicted number of genes
Bacterium *Haemophilus influenzae*	1995	1.8	1,740
Bacterium *Escherichia coli*	1997	4.6	3,240
Budding yeast *Saccharomyces cerevisiae*	1996	12	6,000
Nematode worm *Caenorhabditis elegans*	1998	97	19,100
Flowering plant *Arabidopsis thaliana*	2000	125	25,500
Fruit fly *Drosophila melanogaster*	2000	180	13,600
Human *Homo sapiens*	2001	3,200	30,000–40,000

study specific genes and their encoded proteins in the broad context of the whole cell or even the whole organism. For example, when a human cell undergoes division in response to a hormone signal, genomic analysis using DNA chips can now detect all the genes that are turned on and off. In effect, all of the players in the signal cascade can be studied as they work together to promote cell division (see Chapter 9), in much the same way as all the components of a clock can be observed working together to keep time and move the hour, minute, and second hands on the clock face.

Genomics allows biologists to integrate ideas in a way that was previously not possible. In the past, the ability to study thousands of genes at once was limited by the cost in terms of time, money, and human effort. Today, genomic methods have removed this barrier, but they have replaced it with a new one; namely, the huge amount of information they produce. The greatest challenge of our postgenomic era will not be how to sequence genomes or monitor genome-wide changes in transcription; it will be how to make sense of all that information so that it reveals something about biological processes.

The 2000-Year-Old Quest for the Human Genome

The focused effort to sequence the human genome spanned a mere 15 years, but the intellectual and technical breakthroughs that gave scientists the means of achieving this goal have spanned more than two millennia. Here we outline some of the critical discoveries and events that led up to the genome sequencing effort (Figure 18.1).

Around the fourth century BC, the ancient Greek philosophers were already formulating ideas to account for the remarkable phenomenon of heredity. Hippocrates, who is thought of as the father of medicine, believed that illness had a scientific basis and had nothing to do with spirits or gods. He even presaged key principles of heredity by noting that signs of disease could appear throughout a family, even over successive generations. Aristotle went one step further by declaring that both the mother and father contributed biological material to their offspring. However, he mistakenly believed that blood carried the biological information that determined the child's appearance, giving birth to the idea of inherited bloodlines. Despite this misconception about the nature of what was transmitted, Aristotle's notion of a transmissible blueprint laid the foundation for Gregor Mendel's work nearly two thousand years later.

From Mendel's peas to the double helix

The idea of inheritance remained a topic of philosophical debate well into the seventeenth century, when a Dutch scientist named Antonie van Leeuwenhoek invented a simple microscope powerful enough to reveal sperm cells. His groundbreaking observations led him to speculate that offspring arose from interacting cells contributed by both parents. It was not until the nineteenth century, however, that Mendel performed his revolutionary pea breeding experiments, which demonstrated that physical traits are passed to offspring by both parents. Mendel deduced that the factors responsible for passing on these traits occurred in pairs, with one contributed by each parent, and that either could play a dominant role in determining a particular trait. His laws of equal segregation and independent assortment described critical features of gene transmission (see Chapter 12) even though the role of genes as hereditary factors would not be confirmed for another 35 years.

The first half of the twentieth century witnessed a whirlwind of discoveries that grew out of Mendel's laws of inheritance. As described in Chapter 13, Morgan's work on fruit flies revealed that genes are arranged in linkage groups that are transmitted together. This discovery led to the realization that chromosomes carry linear arrays of genes. In 1928, Griffith's work on *Streptococcus* demonstrated that an unknown transforming factor could change harmless bacteria into deadly bacteria. By the early 1950s, Avery and colleagues, followed by Hershey and Chase, used bacteria and viruses to identify DNA as the genetic material (see Chapter 14). When Watson and Crick revealed the double-helical structure of DNA in 1953, all the parts fell into place, answering key questions on the replication, transmission, and coding capacity of the genetic material. Roughly 50 years of research had uncovered the molecular basis for heredity, establishing the DNA genome as the blueprint of life.

Understanding the physical nature and transmission of life's molecular blueprint, however, did not immediately reveal how the encoded instructions were used to create a living system. It took more than a decade following the discovery of the DNA double helix for geneticists and biochemists to crack the genetic code. This work revealed how a DNA sequence serves as a template for RNA, which in turn is translated into protein (see Chapter 15). For the first time in history, scientists knew how information encoded in the genome determined the physical makeup of living organisms.

Meaningful applications of this knowledge remained elusive, however, until the creation of ways to isolate and manipulate fragments of DNA. The gold rush to

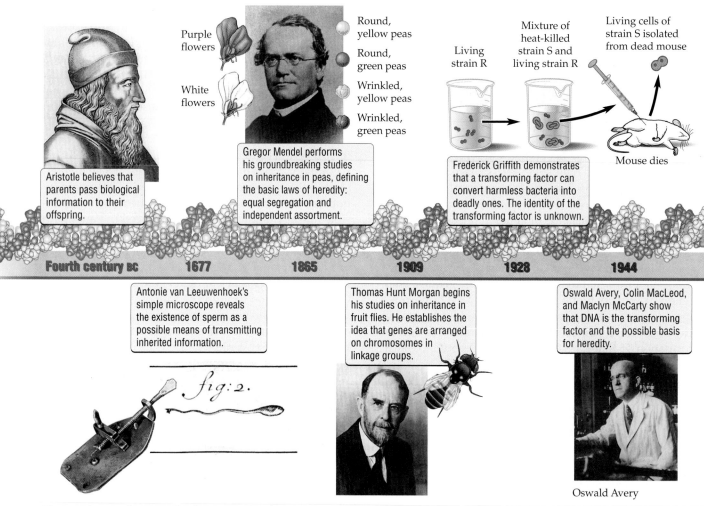

Purple
flowers

White
flowers

Round,
yellow peas

Round,
green peas

Wrinkled,
yellow peas

Wrinkled,
green peas

Living
strain R

Mixture of
heat-killed
strain S and
living strain R

Living cells of
strain S isolated
from dead mouse

Mouse dies

Aristotle believes that
parents pass biological
information to their
offspring.

Gregor Mendel performs
his groundbreaking studies
on inheritance in peas, defining
the basic laws of heredity:
equal segregation and
independent assortment.

Frederick Griffith demonstrates
that a transforming factor can
convert harmless bacteria into
deadly ones. The identity of the
transforming factor is unknown.

Fourth century BC   1677   1865   1909   1928   1944

Antonie van Leeuwenhoek's
simple microscope reveals
the existence of sperm as a
possible means of transmitting
inherited information.

fig:2.

Thomas Hunt Morgan begins
his studies on inheritance in
fruit flies. He establishes the
idea that genes are arranged
on chromosomes in
linkage groups.

Oswald Avery, Colin MacLeod,
and Maclyn McCarty show
that DNA is the transforming
factor and the possible basis
for heredity.

Oswald Avery

Figure 18.1   Milestones in the Quest for the Human Genome

develop such molecular tools began in the early 1970s, when the basic techniques of DNA cloning were first worked out. The subsequent revolution in DNA technology allowed genomes to be studied by breaking them up into fragments and inserting them into vectors (see Chapter 17). For the first time, scientists could manipulate and propagate fragments of DNA, which allowed them to study individual genes of interest.

The genomic revolution

Isolating and manipulating genes is one thing, but understanding the information they contain requires some way to read the genetic code. The first techniques for sequencing DNA were developed in 1977. Once these methods were sufficiently refined and easy to use, a single scientist could accurately sequence several hundred bases of DNA per day. The ability to read DNA sequences explicitly revealed the link between changes in a gene's sequence due to mutation and functional changes in its protein product. Furthermore, once a gene was identified and sequenced, scientists could compare it with closely related genes of known function from other organisms. As the number of sequenced genes from various organisms climbed during the 1980s, relationships between families of genes began to appear in the fledgling sequence databases. The genes that control cell division in yeast, for example, were found to be similar to human genes with the same function. Likewise, of the 289 human genes that are linked to a known disease, over 60 percent have counterparts in the fruit fly. This process of comparison across groups of related genes has uncovered many common mechanisms for solving biological problems, such as cell-to-cell signaling and metabolism. The realization that different organisms share significant numbers of related genes accelerated the pace of discovery of individual gene functions and enhanced the evolutionary view of biology.

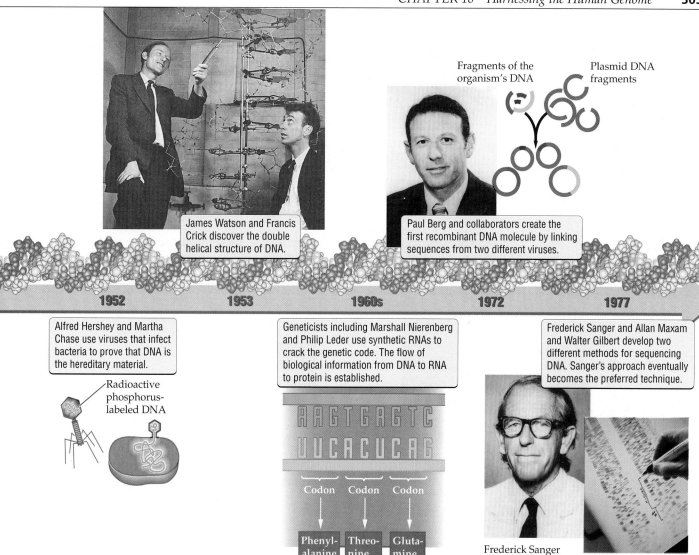

James Watson and Francis Crick discover the double helical structure of DNA.

Fragments of the organism's DNA

Plasmid DNA fragments

Paul Berg and collaborators create the first recombinant DNA molecule by linking sequences from two different viruses.

Alfred Hershey and Martha Chase use viruses that infect bacteria to prove that DNA is the hereditary material.

Radioactive phosphorus-labeled DNA

Geneticists including Marshall Nierenberg and Philip Leder use synthetic RNAs to crack the genetic code. The flow of biological information from DNA to RNA to protein is established.

Frederick Sanger and Allan Maxam and Walter Gilbert develop two different methods for sequencing DNA. Sanger's approach eventually becomes the preferred technique.

1952   1953   1960s   1972   1977

AAGTGAGTC
UUCACUCAG

Codon   Codon   Codon

Phenyl-alanine   Threo-nine   Gluta-mine

Frederick Sanger

With the technology to manipulate and sequence DNA in place, talk of sequencing the entire human genome began to circulate among scientists in the 1980s. The idea was initially greeted with skepticism due to the logistics of such a mammoth undertaking. At the time, one could obtain up to 500 base pairs of sequence from a single trial, which meant that it would take a multitude of scientists several decades to completely sequence the 3 billion base pairs of the human genome. Others argued that even if the goal could be attained, it would not be worth the effort, since coding sequences account for less than 2 percent of the genome. The remaining 98 percent of noncoding DNA would be a waste of time to sequence, since it would reveal nothing about the function of genes. On the other hand, not everyone agreed with this view, and many leading scientists pointed out that noncoding DNA includes the regulatory sequences that control when genes are transcribed. Thus, knowledge of these noncoding sequences would be essential to understanding the timing and extent of gene expression.

Arguments for the value of noncoding sequences prevailed, and in 1987, the U.S. Department of Energy commenced funding for a project to sequence the human genome. Coincidentally, during that same year, new technological developments both sped up and automated the DNA sequencing process. By 1990, the National Institutes for Health (NIH) and the Department of Energy had created an international consortium of sequencing laboratories or centers, collectively known as the **Human Genome Project (HGP)**. The HGP was originally scheduled to complete the human genome sequence by 2005. The DNA to be sequenced was obtained from several anonymous donors from diverse backgrounds. To ensure privacy, the identity of the donors was never associated with the DNA samples. Although human genomes are generally 99.9 percent identical, using DNA from several individuals would

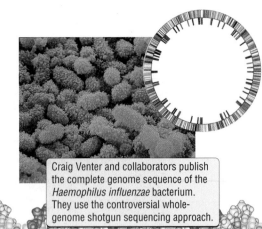

Charles DeLisi at the U.S. Department of Energy opens discussions with several scientists on a project to sequence the human genome.

NIH and the DOE establish the publicly funded Human Genome Project, under the directorship of James Watson. An international consortium of sequencing centers is scheduled to completely sequence the human genome by 2005.

Craig Venter and collaborators publish the complete genome sequence of the *Haemophilus influenzae* bacterium. They use the controversial whole-genome shotgun sequencing approach.

1985   **1987**   **1990**   **1992**   **1995**

Applied Biosystems Inc. markets the first automated DNA sequencing machine.

Watson resigns as director of the HGP due to disagreements on the patenting of genes. Francis Collins is appointed the new director of the HGP.

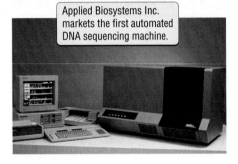

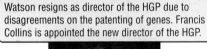

Figure 18.1 Milestones in the Quest for the Human Genome *(continued)*

avoid the possibility of sequencing a single genome with significant mutations and mistakenly using it as the genomic benchmark.

In the early stages of the HGP, the decision was made to sequence the smaller genomes of several model organisms as a rehearsal for tackling the human genome. Smaller sequencing projects would allow scientists to improve on existing methods and develop the computer programs necessary for knitting together and analyzing huge quantities of sequence data. The selected genomes came from a bacterium (*Escherichia coli*) with 4.6 million base pairs, budding yeast (*Saccharomyces cerevisiae*), with 12 million base pairs, and a nematode worm (*Caenorhabditis elegans*) with 97 million base pairs. Interestingly, however, the first complete genome sequence of a free-living organism was published in 1995 by a group that was not involved in the HGP. The 1.8 million-base-pair genome of the *Haemophilus influenzae* bacterium was completely sequenced in a collaborative effort spearheaded by Craig Venter, a former NIH scientist. Herein lay the seed of a controversy that continues to swirl around the Human Genome Project even today.

■ Early studies of inheritance led to the recognition of DNA as the molecular blueprint of life. The development of DNA cloning and sequencing techniques revealed the evolutionary relationships among genes from different organisms and paved the way for the sequencing of whole genomes.

Public Science versus the Commercial Bottom Line

From the earliest days of the HGP, there was vocal disagreement on how the genome should be sequenced and on who should control access to the resulting information. James Watson, the co-discoverer of the DNA double helix, was appointed head of the project and was instrumental in securing the first round of federal funding. Thanks to his scientific credentials and forceful personality, Watson was able to rally congressional support for what he described as an effort to understand what it means to be

Venter announces the formation of a commercial company, Celera Genomics, that will sequence the entire human genome by 2001. HGP scientists publish the complete genome sequence of the nematode worm *Caenorhabditis elegans*.

Celera and collaborators publish the genome sequence of the fruit fly, *Drosophila melanogaster*. Venter and Collins agree to collaborate. Disagreement over the data-release policy ends the short-lived collaboration between the HGP and Celera.

| 1996 | 1998 | 1999 | 2000 | 2001 |

The HGP publishes the complete genome sequence of budding yeast, *Saccharomyces cerevisiae*. They use the map-based sequencing approach.

Both the public and commercial sequencing efforts pass the 1 billion-base-pair milestone. HGP scientists publish the first complete human chromosome sequence, that of chromosome 22.

The Human Genome Project and Celera both publish complete draft sequences of the human genome in separate journals: HGP in *Nature*; Celera in *Science*.

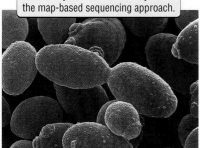

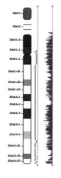

human. In addition, he astutely agreed to use some of the approved funds to address the social and ethical issues that would surely arise from the project.

At the time, the HGP was to begin by creating genetic maps of the various human chromosomes. These genetic maps would identify the locations of known DNA polymorphisms on each chromosome, thereby providing a low-resolution roadmap of the genome. The next step would be to sequence many DNA fragments from each chromosome and reassemble them using the genetic maps as a guide. This kind of **map-based sequencing** was viewed as a conservative way to ensure an accurate genome sequence.

The controversy began when Venter, who was then head of a large NIH sequencing laboratory, proposed an alternative sequencing method that focused exclusively on fragments of expressed genes. Furthermore, he announced that NIH would patent the resulting sequences even when the functions of the expressed gene fragments were unknown. This announcement outraged Watson and many other scientists, since such a practice could result in the wholesale patenting of the

human genome and limit the study of individual genes based on their function. It also seemed unfair to simply patent DNA sequences with no insight into their function. After all, the sequence of the human genome could be considered the property of all human beings.

Further contentious disagreements between the NIH and Watson on this and related issues ultimately led Watson to resign from the project in 1992. Francis Collins, a physician researcher at the University of Michigan, was subsequently appointed the new director of the HGP. Collins had already played a critical role in finding the genes responsible for Huntington's disease and cystic fibrosis, and he brought a strong emphasis on medical applications to the genome project.

Venter also left NIH and joined a new nonprofit organization, which, in collaboration with scientists at Johns Hopkins University, successfully sequenced the 1.8 million base pairs of the *Haemophilus influenzae* genome. This sequence, published in 1995, represented the first complete genome sequence of a free-living organism. Venter and his collaborators used a new approach known as **whole-genome shotgun sequencing**. Unlike

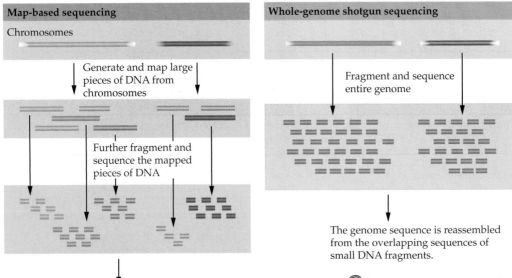

Map-based sequencing

Chromosomes

Generate and map large
pieces of DNA from
chromosomes

Further fragment and
sequence the mapped
pieces of DNA

Genome sequence is reassembled from the
overlapping sequences of small DNA fragments
that are placed in order according to the genetic maps.

Whole-genome shotgun sequencing

Fragment and sequence
entire genome

The genome sequence is reassembled
from the overlapping sequences of
small DNA fragments.

18.1

Figure 18.2 Alternative Methods for Sequencing Whole Genomes
The public Human Genome Project used the map-based sequencing method, while Celera's commercial effort used the whole-genome shotgun sequencing method.

the HGP approach based on genetic maps, whole-genome shotgun sequencing broke up the entire genome into many small fragments, which were then sequenced and reassembled by computer based strictly on their overlapping ends (Figure 18.2).

The HGP approach bore fruit the following year, when the complete genome sequence of the budding yeast, *Saccharomyces cerevisiae*, was published. At the time, some argued that the whole-genome shotgun approach would work only for relatively small bacterial genomes (1–5 million bases), since the size of a eukaryotic genome (12 million or more bases) would make reassembly of the genome very difficult. Furthermore, unlike bacterial genomes, the human genome contains long stretches of repetitive DNA sequences, which many scientists believed would confuse attempts to align the overlapping fragments in the correct register.

Despite these doubts about Venter's whole-genome shotgun approach, he surprised the world in 1998 by announcing the formation of a commercial company that planned to sequence the entire human genome in 3 years. The company was named Celera Genomics, and it had strong ties to the leading manufacturer of high-speed automated sequencing machines. In effect, Venter threw down the gauntlet and challenged the public sequencing consortium to a race. This challenge sent shock waves through the HGP community for several reasons. Many felt that Venter's method would not work on the 3-billion-base-pair human genome and might divert funding from conservative approaches that were

more likely to succeed. In addition, there were grave concerns about how accessible the resulting information would be if a commercial outfit succeeded in sequencing the genome first. This prospect raised the specter of a commercially patented human genome sequence that would exclude scientists who were unable or unwilling to pay for access to the data. These and other concerns galvanized the HGP community of scientists, and the race was on!

Over the next couple of years, a flurry of sequencing milestones was announced by both sides. In October of 1999, Venter announced that Celera had successfully sequenced 1 billion base pairs of the human genome. In response, Collins and the public consortium pointedly noted that no one had access to the sequence to check the veracity of Celera's claim. The HGP reached the 1 billion base milestone a month after Celera. In 2000, a collaborative effort between Celera and a publicly funded consortium of scientists published the genome sequence for the fruit fly, *Drosophila melanogaster*. This sequence was proof that the shotgun method could successfully tackle a large and complex genome (180 million base pairs). At about the same time, squabbling between the public and commercial teams reached a fever pitch, with each side issuing statements questioning the quality of the other's data. At the urging of several prominent scientific and political figures, a truce was called, which lasted just long enough for a collegial

White House ceremony and a few exchanges of flattering comments in the press. During this time, both sides made major gains in the sequencing effort, and it became clear that both would complete a draft sequence of the entire human genome by 2001.

The short-lived spirit of collaboration between the HGP and Celera came to an abrupt end over the question of how to publish the human genome sequence. Traditionally, all genomic sequences published in the scientific literature are also deposited in a free, unrestricted public database called GenBank. Venter wished to break with tradition and allow access to the Celera sequence only on a company website. Many prominent scientists, including Collins, cried foul, claiming this was contrary to the fundamental principle of sharing scientific data. The heated disagreement over this issue killed any chance of a joint publication. In February of 2001, two separate first drafts of the human genome sequence were published in different journals. The HGP published its sequence in the journal *Nature*, while Celera published its sequence in the journal *Science*.

Both the HGP's and Celera's publications presented complete drafts of the human genome, but that does not mean the genome is 100 percent sequenced. The human genome, like other eukaryotic genomes, contains stretches of repetitive DNA that are difficult to sequence. Fortunately, these regions tend to contain very few genes or regulatory sequences, so the loss of information is not great. The two published human genome sequences are approximately 83 percent complete in terms of total coverage, and over 90 percent complete in terms of coverage of gene-rich regions. Sequencing efforts are therefore continuing on all the eukaryotic genomes published so far.

It is unfortunate that the crowning achievement of twentieth-century biology was published in such a contentious manner. The Celera sequence is available only on the company website, and access is subject to several conditions, including restrictions on the number of base pairs that can be downloaded per week, their commercial use, and their distribution. In the tradition of GenBank, the HGP sequence is available without any restrictions. The existence of these two databases stands as a testament to the increasingly differing views of public versus commercial research. Few would deny that the commercial effort greatly accelerated the timetable for both projects, resulting in significant technological breakthroughs under the pressure of active competition. In addition, it is often said that commercial gain is one of the most powerful incentives for scientific research. What remains to be seen is whether future commercial efforts will dramatically alter the openly collaborative environment of biology.

■ The public Human Genome Project adopted a conservative sequencing approach based on creating genetic maps and using them as a guide to assemble sequenced DNA fragments. Celera's commercial effort used whole-genome shotgun sequencing, in which randomly generated DNA fragments were sequenced and reassembled based on their overlapping ends. The frequently contentious competition between the public and commercial efforts resulted in a faster sequencing of the human genome, but raised difficult questions about who would have access to the resulting information.

Previewing Our Blueprint

Now that we can browse our genetic blueprint, what do all those sequences of base pairs tell us? The sequencing of the human genome remains an ongoing process as scientists seek to fill in gaps of missing sequence, check for accuracy, and confirm their predictions of which sequences represent bona fide genes. Analysis of the human genome sequence is therefore in its infancy. However, some significant observations have already emerged.

For one thing, the human genome sequence reveals the number of genes required to encode all the physical complexities of the human organism. Prior to the first draft sequence, scientists had predicted that humans would have at least 100,000 different genes. The first big surprise from reviewing the sequence was a revised estimate of 30,000 to 40,000 human genes. Estimates of gene numbers are largely based on computer analyses of genome sequences, using programs that predict the beginnings and ends of previously unknown genes. Given that genes occupy only about 2 percent of the human genome, such analyses can be difficult, and they will have to continue for several more years before the estimate can be considered definitive. Consequently, the estimated number of human genes is expected to rise modestly due to improvements in both the quality of the sequence data and the computer programs used to make the predictions. Nevertheless, it appears that humans require only ten times the number of genes needed to construct a bacterium like *E. coli* (see Table 18.1).

Before you take offense at the implication that humans are not that much more complex than bacteria, let's consider some factors that determine the complexity of the two organisms in question. If one assumes that every gene can independently adopt two states, either on or off, then the number of different populations of RNA products that can be produced by a human genome with 30,000 genes is $2^{30,000}$. In comparison, the *E. coli* genome contains 3,240 genes, which would give a possible $2^{3,240}$ different RNA populations. This means that even based on such an oversimplified model of gene expression, the human genome can produce a vastly larger number of different RNA populations than that of *E. coli*. Thus, a tenfold increase in gene number produces far more than a tenfold increase in complexity.

A more accurate comparison of human and bacterial RNA populations would have to take into account the selective removal of introns from human RNA products (see Chapter 15). The variable removal of introns from eukaryotic RNAs means that a single gene can produce many different RNA transcripts. This greatly increases the number of different RNA transcripts produced by eukaryotic genomes such as ours. Bacterial genes do not have introns, which means that each gene only produces one RNA product. In addition, different mechanisms of regulating gene expression in humans versus bacteria further determine the complexity of their respective RNA populations (see Chapter 16).

Many years down the road, when the gene complements and RNA populations have been characterized for each genome, truly accurate comparisons of complexity across different organisms will be possible.

Many of the most intriguing observations stemming from the human genome sequence are based on comparisons with genome sequences from other organisms. These comparisons shed light on the common sets of genes that have been conserved during evolution. Perhaps the most striking patterns emerge when the protein products of these genes are compared across species. Most proteins can be categorized according to their functional domains. A domain is a segment of a given protein that has a specific function, be it enzymatic activity or the ability to bind to another protein. Interestingly, nine out of ten protein domains found in humans have counterparts in fruit flies and nematode worms. While the arrangements of protein domains are more complex in humans, their very existence in fruit flies and worms implies that certain protein activities and interactions were established early in evolution. On a broader scale, roughly 60 percent of human proteins are similar to proteins found in fruit flies and worms, and participate in the core processes required for life. Unsurprisingly, these conserved protein sets include DNA synthesis enzymes, transcription factors, metabolic enzymes, and many of the receptors and kinases involved in signal cascades.

Evolutionary conservation across species also means that well-understood processes in other organisms may yield insight into similar but less understood processes in humans. The regulation of biological clocks is just one essential process that is likely to involve similar proteins in different species. If you have ever flown on an airplane across several time zones, you have probably experienced a disturbance in your sleep patterns. This phenomenon, known as jet lag, is due to a disruption of your biological clock, not the airline food. In mammals, biological clocks dictate the daily cycling of many physiological events, including sleep. Earlier studies in the fruit fly, *Drosophila*, have uncovered many so-called clock genes that control the timing of the fly's activities. Comparisons between fruit fly clock gene sequences and the human genome have already yielded several previously unknown candidate genes that may play a role in the daily activity cycles of humans. Furthermore, the human genome sequence has revealed the chromosomal locations of the known human clock genes, at least one of which is responsible for an inherited sleep disorder. This kind of dialogue with research done on other organisms is therefore bound to play an important role in the ongoing analysis and interpretation of the human genome.

> ■ The blueprint for a human being requires an estimated 30,000 to 40,000 genes. Many of these genes have counterparts in other organisms and participate in the core processes necessary for life. Understanding how these genes function in other organisms sheds light on the functions of their human counterparts.

■ HIGHLIGHT

Health Care for You Alone

Long before the race to sequence the human genome began, scores of genes were already known as culprits in human disease. The link between mutations in a given gene and the likelihood of coming down with a particular disease formed the basis for genetic screening. The traditional notion of **genetic screening** was the process of looking for gene mutations in order to assess a person's future health risks. By the end of the twentieth century, people had the option of testing themselves or even their unborn fetuses for mutations linked to diseases such as breast cancer, cystic fibrosis, and Huntington's disease. With hundreds of tests for genetic diseases available, people were able to learn more about their children, their families, and themselves than they had ever thought possible.

However, many diseases, such as breast cancer, are not caused by mutations in just one or two genes. As discussed in Chapter 11, many malignant cancers require

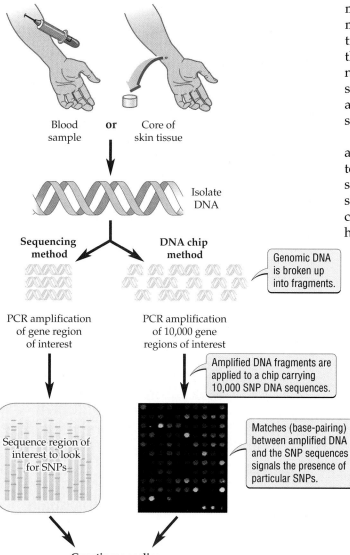

Blood sample   **or**   Core of skin tissue

Isolate DNA

Sequencing method

DNA chip method

Genomic DNA is broken up into fragments.

PCR amplification of gene region of interest

PCR amplification of 10,000 gene regions of interest

Amplified DNA fragments are applied to a chip carrying 10,000 SNP DNA sequences.

Sequence region of interest to look for SNPs

Matches (base-pairing) between amplified DNA and the SNP sequences signals the presence of particular SNPs.

Genetic counseling

18.2

Figure 18.3   From Tissue to Test Results
Using cells from tissue or body fluids, scientists can isolate a person's DNA and use a variety of methods to screen for single nucleotide polymorphisms (SNPs).

mutations in several different genes, resulting in a far more complex genetic profile than can be revealed by traditional genetic tests of a few genes. Knowledge of the human genome sequence and new genomic techniques are sure to exponentially increase the number of such variations in an individual's genome that can be analyzed, radically expanding the scope of genetic screening.

Single nucleotide polymorphisms, or SNPs, represent a powerful means of characterizing individual genomes today. By the end of the twentieth century, a few thousand SNPs were already known—a number that was simply not adequate to cover the entire genome. The completion of the draft sequence of the human genome has uncovered the identities and locations of over 1.4 million SNPs. This number of SNPs brings the variation in individual genomes into sharper focus by allowing them to be compared with a vastly more detailed SNP map of the human genome. Of course, there has to be a rapid and affordable way to detect thousands of SNPs in an individual DNA sample for this breakthrough to be medically relevant. Using traditional polymerase chain reaction (PCR) technology (see Chapter 17), it would take weeks and cost tens of thousands of dollars to create a detailed SNP profile for a single patient. However, the DNA chip technology introduced in Chapter 16 has recently been modified to allow the screening of 10,000 SNPs in less than an hour (Figure 18.3). Within a few years it is likely this technology will be further refined to allow the screening of 100,000 SNPs in a matter of hours. People will then have the option of viewing their genomic fingerprint and subjecting it to analysis.

Only a minority of SNPs fall within gene coding sequences, and only some of these will be found to affect the activity of the proteins produced by those genes. Nevertheless, a significant number of SNPs have already been identified that have dramatic effects on human health. The presence of a particular SNP in the apolipoprotein E (*apoE*) gene, for example, is associated with a risk of developing Alzheimer's disease. Alzheimer's disease (AD) is the leading cause of cognitive degeneration in the elderly, and is characterized by the formation of plaques of protein in the brain, leading to extensive cell death. The normal ApoE protein plays a role in delivering cholesterol to cells and is thought to help remodel the plasma membranes of brain cells. The exact role played by *apoE* in AD plaque formation is unclear, but clinical studies show that a particular *apoE* SNP correlates with an increased risk of AD. On the other hand, the presence of a different *apoE* SNP correlates with a reduced likelihood of AD. Thus, by determining the SNP profile for the *apoE* gene, a person can determine whether or not they are at high risk for AD later in life.

SNP profiles can also give us insight into how an individual will respond to drug therapy. The wide range of possible responses a patient may have to a particular drug often complicates the prescribing of safe and appropriate therapies. These variations in responses may be due to preexisting disease, the presence of other drugs, and even nutritional status. Genetic variations in drug-metabolizing enzymes can also have a significant effect on patient response. Several genes encoding enzymes that process drugs in the body have been identified. Some of these genes contain SNPs that give an indication of their likely activity. For example, the activity of the CYP2D6 enzyme is required for the activation of analgesic painkillers such as codeine. Nearly 10 percent of the population are homozygous for a SNP that renders this enzyme inactive. These unfortunate individuals fail to respond to codeine and do not get adequate pain relief from the drug. Similarly, certain plasma membrane proteins tend to pump anticancer drugs out of cells, resulting in the need for higher drug doses. SNP profiling of the genes that encode these proteins may identify variations that either decrease or increase their pumping activity, allowing physicians to prescribe drug doses tailored to the individual.

As the number of SNPs with known effects on gene function increases, the possibility of using an SNP profile to optimize an individual's drug regimen will move closer to reality. In the near future, patients will sit down with their physicians, review their SNP profile, and find out what it reveals about their susceptibility to several diseases and how well a particular drug therapy might work for them. These methods will result in a personalized health program designed to head off future problems while treating current illnesses as effectively as possible. In fact, submitting a blood sample for DNA chip analysis of hundreds of thousands of SNPs could become the first health-related activity of childhood.

Despite the exciting promise that this kind of detailed health profiling may hold, it will raise a number of ethical problems. Is it possible that such detailed biological profiles will be used to discriminate against individuals? Health insurance companies will be tempted to raise premiums for people whose profiles show a high susceptibility to disease down the road. Individuals with a high risk of serious illness might even find themselves denied health and life insurance. Furthermore, as a new field of behavioral genomics enters the stage, many believe that SNP profiling will reveal a person's sus-

ceptibility to syndromes such as alcoholism, schizophrenia, and clinical depression. The revelation that someone has a heightened susceptibility to drug addiction might cost them a job. Perhaps by understanding our genetic blueprint too well, we run the risk of losing sight of our humanity in an avalanche of predictions and genetic susceptibilities.

To prevent the potential misuse of SNP profiles, new guidelines for personal privacy must be established. To this end, about 5 percent of the budget for the Human Genome Project is devoted to studying the ethical, legal, and social issues surrounding the availability of genetic information. If we are to truly reap the benefits of this scientific achievement, addressing these issues is as important as filling gaps in the genome sequence.

> ■ The identification and mapping of single nucleotide polymorphisms (SNPs) has opened up the possibility of creating highly detailed individual genetic profiles. These SNP profiles could be used to predict a person's future susceptibility to disease and to tailor optimum drug treatments for that individual. Having such a detailed record of a person's health status raises difficult questions about personal privacy and the possibility of genome-based discrimination.

SUMMARY

The 2000-Year-Old Quest for the Human Genome

- The 2000-year-old realization that parents pass on traits to their offspring laid the foundation for Mendel's studies of inheritance.

- Mendel's laws of inheritance paved the way for the discovery of DNA and genes as the basis of heredity.

- Watson and Crick's discovery of the DNA double helix revealed the molecular basis for heredity, establishing the DNA genome as the blueprint for life.

- Breakthroughs in DNA cloning and sequencing techniques were necessary precursors to the idea of sequencing whole genomes.

- The Human Genome Project (HGP) was officially launched in 1990 as a publicly funded consortium of many sequencing centers.

Public Science versus the Commercial Bottom Line

- From the earliest days of the effort to sequence the human genome, there was controversy surrounding what methods would be used and who would control access to the resulting information.

- The public HGP focused on creating genetic maps of each chromosome, then sequencing DNA fragments from each chromosome, which were then reassembled using the genetic maps as a guide.

- The commercial Celera project used whole-genome shotgun sequencing, in which the entire genome was broken into many small fragments, which were then sequenced and reassembled based on their overlapping ends.

- Due to multiple disagreements between the public and commercial sequencing efforts, two complete draft sequences of the human genome were published separately in 2001.

- While competition from the commercial sequencing effort clearly accelerated the pace of both sequencing projects, freedom of access to the commercial data remains a controversial issue.

Previewing Our Blueprint

- The sequencing of the human genome remains an ongoing process as scientists seek to fill in gaps of missing sequence, to check for accuracy, and to confirm gene predictions.

- Genes account for approximately 2 percent of the human genome.

- The human genome is currently estimated to contain 30,000 to 40,000 genes.

- The 60 percent of human proteins that have counterparts in fruit flies and nematode worms participate in core processes of life such as metabolism and signal cascades.

- Well-understood gene functions in other organisms may yield insight into similar but less understood processes in humans.

Highlight: Health Care for You Alone

- Traditional genetic screening is the process of looking for genetic mutations in order to assess a person's future health risks.

- The identification of over 1.4 million SNPs in the human genome sequence exponentially increases the number of variations in an individual's genome that can be analyzed, radically expanding the scope of genetic screening.

- DNA chip technology will eventually allow screening of 100,000 SNPs in a matter of hours, giving every person the option of having their genetic profile subjected to medical analysis.

- A number of SNPs within genes have been identified that affect a person's susceptibility to illnesses such as Alzheimer's disease.

- The potential to create highly detailed genetic profiles with the power to predict susceptibility to disease and responses to drugs raises difficult questions about the limits of personal privacy and the possibility of genomic discrimination.

KEY TERMS

genetic screening p. 310	map-based sequencing p. 307
genomics p. 302	single nucleotide polymorphisms (SNPs) p. 301
Human Genome Project (HGP) p. 305	whole-genome shotgun sequencing p. 307

CHAPTER REVIEW

Self-Quiz

1. The field of genomics seeks to understand
   a. how individual genes affect human health.
   b. how organisms interact with their environment.
   c. the expression patterns of entire genomes.
   d. the inheritance of genes.

2. The idea that parents transmit heritable information to their offspring first arose
   a. among Greek philosophers around the fourth century BC.
   b. with Mendel's genetic experiments in the nineteenth century.
   c. with the discovery of DNA sequencing.
   d. among the scientists who discovered genes in the early twentieth century.

3. The noncoding regions of the human genome were sequenced because
   a. they form such a small percentage of the genome they would not have increased the cost.
   b. they contain important regulatory sequences that control gene expression.
   c. it is impossible to distinguish them from coding sequences.
   d. some people have a lot more of them than others.

4. In comparison to the genome of the fruit fly, the human genome has
   a. one hundred times the number of genes.
   b. two to three times the number of genes.
   c. half the number of genes.
   d. the same number of genes.

5. Whole-genome shotgun sequencing was originally thought to be unsuitable for the human genome because
   a. the method does not work on human DNA, regardless of its sequence.
   b. there are too many genes in the human genome.
   c. there are long stretches of repetitive DNA in the human genome.
   d. it only works on coding DNA.

6. Which of the following statements is *not* true?
   a. The public and commercial genome sequencing efforts closely collaborated with one another from the beginning.
   b. The public and commercial genome sequencing efforts placed different access restrictions on the human genome sequence.

c. The public and commercial genome sequencing efforts published their complete draft sequences in separate scientific journals.

d. The public and commercial genome sequencing efforts initially announced very different schedules for completion.

Review Questions

1. Why does the fact that so many human proteins have counterparts in other organisms make studying their functions in humans so much easier?

2. Why was it important for the Human Genome Project to sequence DNA samples from several volunteers instead of just one? What do you think the basis for selecting those volunteers should have been?

3. Do you view competition between publicly funded and commercial research as having a generally positive effect on scientific progress? Why or why not?

4. Do you think that SNPs in the noncoding sequences of the human genome are worth identifying? Why or why not?

18

The Daily Globe

Should Doctors Genetically Screen All Their Patients?

ALTONIA, LA—Last month, when Jason Schultze, a 6-year-old boy suffering from leukemia, was given a standard leukemia treatment—a drug known as a thiopurine—he nearly died. Jason nearly died not because of his leukemia, but because he had a genetic tendency that made the drug deadly for him to take—a tendency for which his doctors could have, but did not, test him.

"All this could have been avoided," said Jason's father, Mitch Schultze, "if they'd done a simple test that would've cost less than two hundred dollars."

Every year, 2.2 million Americans have adverse reactions to drugs, and 100,000 of these people die as a result. At the same time, the number of genetic screening tests available that can identify people who are genetically at risk of harmful or nonbeneficial reactions to drugs is growing. There are now genetic screening tests for the asthma medication albuterol (trade name Ventolin), the painkiller codeine, and the antidepressant Prozac, as well as others.

Jason is one of the 1 in 300 people who carry a gene that makes them unable to metabolize thiopurines. Rather than helping cure his leukemia, the drugs his doctors gave him accumulated in his body, reaching toxic levels and destroy-ing his bone marrow. An emergency bone marrow transplant saved Jason's life.

Despite their potential to save lives, genetic screening tests for drug reactions are still very uncommon—and, pharmaceutical industry representatives say, rightly so. "These kinds of adverse reactions are extremely rare," said John Smith, spokesman for United Pharmaceutical Industrial Corporations Inc., a trade association of pharmaceutical companies. "There is no reason to do widespread testing for these genetic defects or to frighten people into thinking that medicines that have been used safely for decades pose a high risk."

Evaluating "The News"

1. Should doctors begin genetically screening their patients before issuing them drug treatments? Why or why not? Are there any ethical issues to be considered?

2. Why might a pharmaceutical trade association be against widespread screening?

3. As researchers discover more and more genes that predict certain tendencies—for example, tendencies to obesity, to high blood pressure, to depression—should doctors be screening their patients for some or all of these genes to help them adopt the healthiest lifestyles possible? Why or why not?

Philip Taaffe, *Passage II*, 1998.

UNIT 4

Evolution

chapter *19* *How Evolution Works*

Mary Frank, *Natural History*, 1985.

A Journey Begins

The Galápagos Islands are isolated, encrusted with lava, and home to bizarre creatures found nowhere else on Earth. Life here offers odd twists on the usual: tortoises that reach giant size, land-dwelling lizards that take to the sea, and vampire finches that suck the blood of other birds.

The unusual species of the Galápagos Islands have long fascinated biologists. One such biologist was the 22-year-old Charles Darwin, who visited the Galápagos as part of a remarkable 5-year journey on the ship *Beagle*. The *Beagle* left England in 1831, sailed to South America, from there to the Galápagos Islands, and eventually back to England in 1836. Throughout this long voyage, Darwin collected specimens and made careful observations of the regions he visited and the organisms he found there.

Upon his return to England, Darwin consulted other scholars and thought deeply about what he

■ **MAIN MESSAGE**

Strong evidence for evolution is provided by fossils, features of existing organisms, continental drift, and direct observations of genetic change.

had seen on his journey. Early in 1837, he learned from taxonomists that many of the specimens he had collected in the Galápagos were entirely new species, and that many of these species were confined to a single island. In addition, where the organisms he collected from different islands turned out to be groups of new species, those species often shared unique characteristics, and thus appeared to be closely related. These and many other observations led Darwin to a bold conclusion: Species were not, as he and everyone else of his time had been taught, the unchanging result of separate acts of creation by God. Instead, species had descended with modification from ancestor species; that is, they had changed over

time, or evolved. Darwin's recognition in 1837 that species had changed over time was a key step in an intellectual journey that would blossom, 22 years later, into the publication of a book that shook the world: *The Origin of Species*.

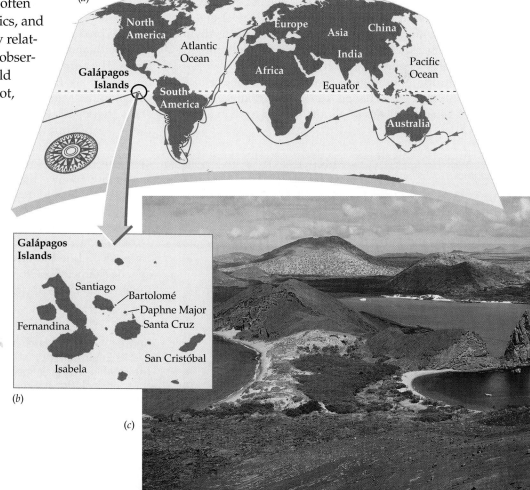

Charles Darwin's Voyage
(a) The course sailed by the *Beagle*. (b) The Galápagos Islands, located 1000 kilometers to the west of Ecuador. (c) Bartolomé, one of the Galápagos Islands.

■ KEY CONCEPTS

1. Biological evolution is change in the genetic characteristics of populations of organisms over time. For evolution to occur, there must be inherited differences among the individuals in a population.

2. Populations evolve when individuals with certain characters (such as large body size) leave more offspring than other individuals. The inherited characters of the individuals that leave more offspring become more common in the following generation.

3. Adaptations are features of an organism that improve its performance in its environment. Adaptations are products of natural selection, the process in which individuals with particular inherited characters survive and reproduce at a higher rate than other individuals because of those characters.

4. The great diversity of life on Earth has resulted from the repeated splitting of species into two or more species.

5. When one species splits into two, the two species that result share many features because they have evolved from a common ancestor.

6. The evidence that evolution has occurred is overwhelming. One strong line of evidence comes from the fossil record, which allows biologists to reconstruct the history of life on Earth and shows how new species arose from previous species.

E arth teems with organisms, many of which are exquisitely matched to their environments. The soaring flight of a hawk, the beauty and practicality of a flower, and the stunning camouflage of a caterpillar each provide a glimpse into the remarkable designs of organisms. How did organisms come to be as they are, seemingly engineered to match their surroundings (Figure 19.1)? What has caused the amazing diversity of life? And within this diversity, why do organisms share so many characteristics? Scientists who study evolution seek to answer questions such as these.

Evolution is biological change over time. In this context, "biological change" can refer to changes in the genetic characteristics of populations or to changes in the kinds of species living on Earth. In this chapter we provide an overview of evolution, the evidence for it, and its consequences for life on Earth.

19.1 Biological Evolution: The Sum of Genetic Changes

Defined broadly, evolution is descent with modification, often with an increase in the variety of the descendant forms. The term can be applied to organisms, cars, computers, or hats. In each of these cases, new items represent modified versions of previous items, and often several varieties arise where only one existed before. But there is an important and fundamental difference between biological evolution and, say, the evolution of hats: Hats change over time because of deliberate decisions made by their designers. As we will see throughout this chapter, biological evolution is not guided by a "designer" in nature, though humans can, and do, direct the course of evolution in some species.

Biological evolution can be defined in several ways. Here, we define **biological evolution** as change in the genetic characteristics of populations of organisms over time. We take this approach to emphasize how evolu-

Figure 19.1 The Match between an Organism and Its Environment
The dolphin is a whale species that evolved from a terrestrial ancestor. Evolution has changed its body form to fit its marine environment.

tion occurs: through changes in the inherited characters of organisms. In Chapter 20 we will look at the mechanisms of evolution and see what kinds of genetic changes they can produce.

Biological evolution also can be defined as the pattern of changes that occur over time. From this perspective, biological evolution is history; specifically, it is the history of the formation and extinction of species over time. Focusing on the pattern of evolution provides us with a grand view of the history of life on Earth, a view to which we will return in Chapter 22.

> ■ Biological evolution is change in the genetic characteristics of populations of organisms over time.

19.2 Mechanisms of Evolution

Evolutionary change requires that differences exist among the individuals within a population, and that those differences be inherited. Inherited differences among individuals depend ultimately on the formation of new alleles by gene mutation. As we learned in Unit 3, gene mutations occur at random and are not directed toward any goal. In sexually reproducing organisms, the genetic variation produced by mutations is increased greatly by crossing-over, the independent assortment of chromosomes (both of which occur during meiosis), and fertilization (see p. 228). As a result of these three processes, the genetic information in the parent generation is rearranged to produce new combinations of alleles in the offspring. Thus, even if no new mutations occur, offspring produced by sexual reproduction differ genetically from their parents and from one another.

The net effect of gene mutations, crossing-over, and other processes that rearrange the genetic information is that individuals within populations differ greatly in many inherited characters, including aspects of their morphology (form and structure), biochemistry, and behavior. These differences in the inherited characters of individuals in a population provide the raw material upon which evolution acts.

One way that populations evolve is when some individuals within a population survive and reproduce at a higher rate than other individuals. This causes the inherited characters of such individuals to become more common in the following generation. There are two main reasons why some individuals survive and reproduce at a higher rate than others: natural selection and genetic drift.

In many cases, there are consistent differences in the survival and reproduction of individuals within a pop-

ulation that are based on inherited characters. For example, deer that are large or fast—two traits that are usually at least in part under genetic control—may escape from predators more easily than deer that are small or slow. Thus, more of the large and fast deer are likely to survive and produce offspring, and their offspring will tend to be large and fast. The deer example illustrates **natural selection**, the process by which individuals that possess particular inherited characters survive and reproduce at a higher rate than other individuals because of those characters (Figure 19.2). Over time, natural selection can cause a population to evolve so that more and more individuals possess characters that enhance their survival and reproduction.

Chance events also can cause some individuals to leave more offspring than others. For example, when a windstorm causes trees to fall amid a population of forest wildflowers, some of the plants are crushed (and leave no offspring), while others survive (and leave offspring), by chance alone (Figure 19.3). Differences in reproduction or survival caused by such chance events

In regions where soot from industrial pollution darkens the bark of trees…

…dark moths are favored by natural selection because they are less conspicuous than light moths.

Figure 19.2   Natural Selection at Work
The peppered moth has two color forms, a light form and a dark form, both shown here. Moth color is an inherited character. Many peppered moths are eaten by birds, which are most successful at finding those moths that differ most in color from the bark of the trees on which they rest. Thus, in regions like that shown here, where industrial soot has darkened the bark of trees, dark moths are harder for birds to find, and so moth populations have evolved to consist mostly of dark moths. In regions where soot does not darken the bark of trees, light moths are harder for birds to find, and so moth populations have evolved to consist mostly of light moths.

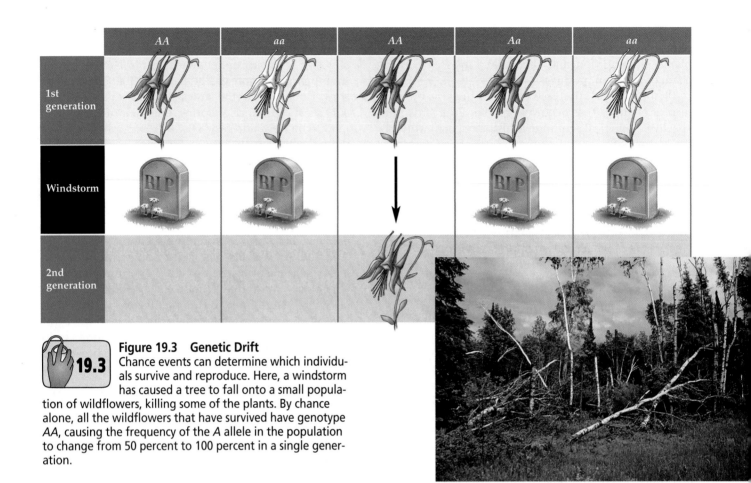

	AA	*aa*	*AA*	*Aa*	*aa*
1st generation					
Windstorm	R.I.P	R.I.P		R.I.P	R.I.P
2nd generation					

Figure 19.3   Genetic Drift
Chance events can determine which individuals survive and reproduce. Here, a windstorm has caused a tree to fall onto a small population of wildflowers, killing some of the plants. By chance alone, all the wildflowers that have survived have genotype *AA*, causing the frequency of the *A* allele in the population to change from 50 percent to 100 percent in a single generation.

can lead to **genetic drift**, a process in which the genetic makeup of a population changes at random over time, rather than being shaped in a nonrandom way by natural selection. Like natural selection, genetic drift can cause populations to evolve.

Although it is a simplification, we can summarize how evolution works as follows: Mutations and genetic rearrangements occur at random, thus generating genetically based differences in the characters of individuals, and these differences are then acted on by natural selection or by genetic drift to produce evolutionary change. In Chapter 20 we will examine natural selection, genetic drift, and other mechanisms that influence the evolution of populations.

> ■ Evolution can be summarized as a three-step process: (1) Mutations and genetic rearrangements occur at random. (2) These events generate inherited differences in the characters of individuals in populations. (3) These differences are then acted on by natural selection or genetic drift to produce evolutionary change.

19.4   Consequences of Evolution for Life on Earth

Life on Earth is distinguished by matches between organisms and their environments, by a great diversity of species, and by many puzzling examples of organisms that differ greatly in many respects, yet share certain key characteristics. Evolutionary biology seeks to explain all these features of life on Earth. This attempt to understand why the living world is the way it is motivates the three great themes of evolutionary biology: adaptation, the diversity of life, and the shared characteristics of life. In this section we describe the consequences of evolution for life on Earth, focusing on these three themes.

Adaptations result from natural selection

Some of the most striking features of the natural world are the complex designs of living things and the often remarkable ways in which they are suited to their environment (see Figure 19.1). These aspects of life are **adap-**

tations, which are defined as characteristics of an organism that improve the performance (that is, the reproductive success) of that organism in its environment.

Adaptations are the product of natural selection. Here's the reasoning behind this statement: If their reproduction were not restricted in some way, all organisms would reproduce so much that their populations would outstrip the limited resources available to them. Because organisms produce more offspring than can survive, the individuals in a population must struggle for existence. In this struggle, the individuals whose inherited characters provide the best match to their environments tend to leave more offspring than other individuals. This is natural selection in action. Over time, natural selection leads to the accumulation of favorable characters—adaptations—within the population.

The diversity of life results from the splitting of one species into two or more species

A major focus of evolutionary biology is the great diversity of life on Earth. Here, too, evolution provides a simple, clear explanation: The diversity of life is a result of the repeated splitting of one species into two or more species, a process called **speciation**.

Speciation can result from a variety of processes, as we will see in Chapter 21. One of the most important of these processes is adaptation to different environments. Consider two populations of a species that live in different environments and which are isolated from each other, as by a mountain or other barrier that prevents individuals from moving between the populations. Over time, natural selection may cause each population to become better adapted to its own particular environment, leading to changes in the genetic makeup of both populations. Eventually, these genetic changes may be so great that individuals from the two populations can no longer reproduce with each other. As we learned in Chapter 1, species are often defined in terms of reproduction: A species is a group of interbreeding populations whose members cannot reproduce with members of other such groups. Thus, evolution by natural selection can lead to the formation of new species.

Shared characteristics are due to common descent

The natural world is filled with puzzling examples of very different organisms that share certain characteristics. For example, the wing of a bat, the arm of a human, and the flipper of a whale all have five digits and contain the same kinds of bones (Figure 19.4*a*). Why do limbs that look so different and have such different func-

tions share the same set of bones? Surely if the best possible wing, arm, and flipper were designed from scratch, their bones would not be so similar. Likewise, why do we humans have reduced tailbones and the remnants of

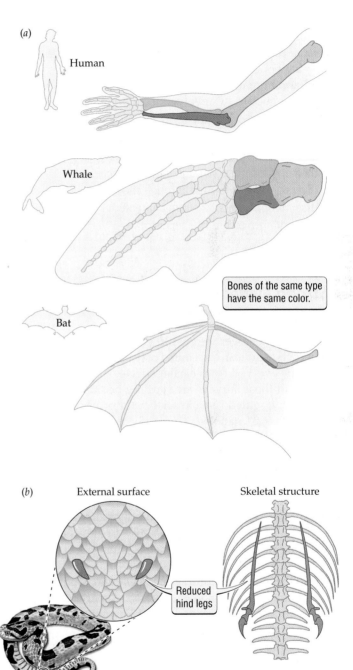

Figure 19.4   Shared Characteristics
(*a*) The human arm, whale flipper, and bat wing all have five digits and contain the same set of bones. (*b*) The rudimentary hind legs of a python snake, as seen in the skeletal structure and from the external surface of the snake.

muscles for moving a tail? Why do some snakes have rudimentary leg bones, but no legs (Figure 19.4*b*)?

Evolution answers these and many other questions about shared characteristics of life. Many similarities among organisms are due to common descent—that is, to the fact that the organisms share a common ancestor. When one species splits into two, the two species that result share many features because they have evolved from a common ancestor. Some snakes have rudimentary leg bones because they evolved from reptiles with legs, and humans have rudimentary bones and muscles for a tail because we evolved from organisms that had tails.

> ■ Evolution explains how organisms are matched with their environment (through natural selection), how the diversity of life is formed (through speciation), and why dissimilar organisms share characteristics (through descent from a common ancestor).

Strong Evidence Shows that Evolution Happened

Surveys taken over the past 10 years reveal that almost half of the adults in the United States do not believe that humans evolved from earlier species of animals. The results of these surveys are startling because evolution has been a settled issue in science for nearly 150 years. Scientists of all nations, races, and creeds agree that the evidence for evolution is very strong. In his landmark book, *The Origin of Species*, published in 1859, Charles Darwin argued convincingly that organisms were descended with modification from common ancestors—that is, that evolution had occurred. The scientific issue today is not whether evolution occurs, but how. To this question Darwin also offered an answer: He argued that natural selection was the principal cause of evolutionary change.

On the issue of natural selection Darwin was less successful in convincing other scientists, in part because at that time no one understood the underlying mechanisms of inheritance. For 60 years after the publication of *The Origin of Species*, many scientists thought Darwin was wrong to place so strong an emphasis on natural selection. However, the rediscovery of Gregor Mendel's work (see Chapter 12) and the understanding of genetics that resulted made it clear that natural selection could cause significant evolutionary change and, hence, that Darwin was at least partially correct.

Biologists still argue about the relative importance of natural selection and other mechanisms of evolution, but they do not dispute whether evolution occurs. Today's debate about the causes of evolution can be compared to a dispute over what caused World War I to progress as it did: Although we might argue over its causes, we all recognize that the war did indeed happen.

Why do scientists find the case for evolution so convincing? As we saw in Chapter 1, a scientific hypothesis must lead to predictions that can be tested, and hypotheses about evolution are no exception. Scientists have tested many predictions about evolution and have found them to be true. In this section we look at five compelling lines of evidence: fossils, traces of evolutionary history in existing organisms, continental drift, direct observations of genetic change in populations, and the formation of new species.

The fossil record strongly supports evolution

Fossils are the preserved remains of formerly living organisms. The fossil record allows biologists to reconstruct the history of life on Earth, and it provides some of the strongest evidence that species have evolved over time (Figure 19.5). As we saw in Chapter 2, the evolutionary relationships among organisms—their pattern of descent from a common ancestor—can often be determined by comparing their anatomical characteristics. This technique can be applied to fossils as well as to living organisms. The fossil record also contains excellent examples of how major new groups of organisms have arisen from previously existing organisms. We will discuss one of these examples, the evolution of mammals from reptiles, in Chapter 23. Fossils that show how new organisms evolved from ancestral organisms also exist for microorganisms, fish, amphibians, reptiles, birds, and humans.

A fossil trilobite

Another important line of evidence for evolution is the observation that the times at which organisms appear in the fossil record match predictions based on evolutionary patterns of descent. For example, evolutionary relationships among horses and their ancestors can be determined from fossilized anatomical data. On the basis of those evolutionary relationships and the age of fossils of horses thought to be ancestors to the modern horse, we can conclude that the modern horse (genus *Equus*) evolved relatively recently (about 5 million years ago) (see Figure 19.5). Thus, we can predict that *Equus* fossils should not be found in very old rocks, and thus far they have not.

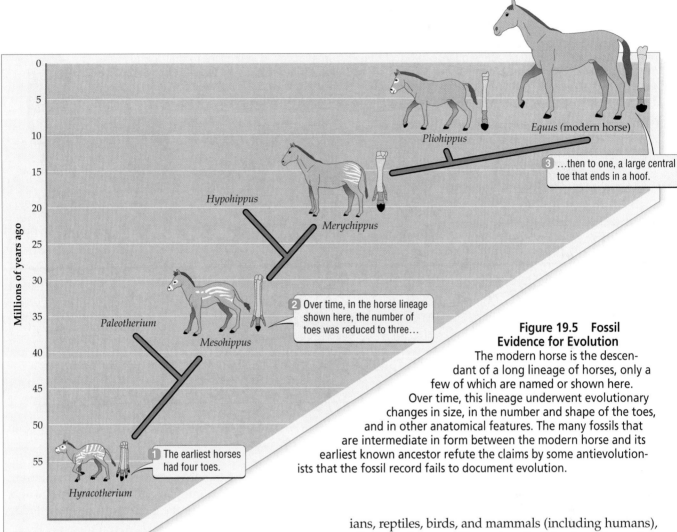

Millions of years ago

0
5
10
15
20
25
30
35
40
45
50
55

Equus (modern horse)

Pliohippus

3 ...then to one, a large central toe that ends in a hoof.

Hypohippus

Merychippus

Paleotherium

Mesohippus

2 Over time, in the horse lineage shown here, the number of toes was reduced to three...

Figure 19.5 Fossil Evidence for Evolution
The modern horse is the descendant of a long lineage of horses, only a few of which are named or shown here. Over time, this lineage underwent evolutionary changes in size, in the number and shape of the toes, and in other anatomical features. The many fossils that are intermediate in form between the modern horse and its earliest known ancestor refute the claims by some antievolutionists that the fossil record fails to document evolution.

1 The earliest horses had four toes.

Hyracotherium

Organisms contain evidence of their evolutionary history

A major prediction of evolution is that organisms should carry within themselves evidence of their evolutionary past—and they do. We described some examples of such evidence earlier in this chapter; namely, the reduced versions of ancestral organs found in some organisms (for example, the "legs" of a snake and the "tail" of a human) and the remarkable similarity in design of limbs that differ greatly in function (for example, the bat wing and the human arm).

Patterns of embryonic development provide further evidence of organisms' evolutionary past. For example, anteaters and some whales do not have teeth as adults, but as embryos they do. Why should the embryos of these organisms produce teeth and then reabsorb them? Or consider the embryos of fish, amphib-

ians, reptiles, birds, and mammals (including humans), all of which have gill pouches. In fish, the gill pouches develop into gills that the fish use to breathe under water, but why should the embryos of organisms that breathe air develop gill pouches?

Our understanding of evolution provides a common answer to these puzzles: Similarities in patterns of development are caused by descent from a common ancestor. Fossil evidence suggests that anteater and whale embryos have teeth because anteaters and whales evolved from organisms with teeth. Similarly, fossil evidence indicates that the first mammals and the first birds each evolved from a (different) group of reptiles, that the first reptiles evolved from a group of amphibians, and that the first amphibians evolved from a group of fish. Thus, it is likely that the embryos of air-breathing organisms such as humans, birds, lizards, and tree frogs all have gill pouches because all of these organisms share a common (fish) ancestor.

Finally, as discussed above, the evolutionary relationships among organisms—their patterns of descent

Figure 19.6   Independent Lines of Evidence Yield the Same Result
The pattern of evolutionary relationships among the animals shown here is based on the number of mutations required to change the amino acid sequence of cytochrome c from that found in humans to that found in the other organisms. Cytochrome c is an enzyme found in all eukaryotes that functions in aerobic respiration. The pattern of evolutionary relationships shown here, in which humans are most closely related to rhesus monkeys and least closely related to moths, matches the pattern derived from anatomical data.

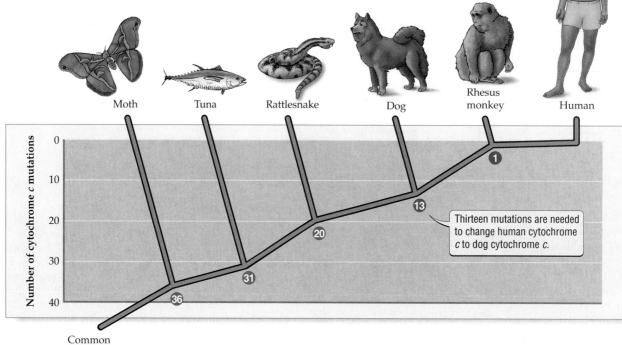

from a common ancestor—often can be determined from anatomical data. These patterns of descent can then be used to make predictions about the similarity of molecules such as DNA and proteins; in such studies, scientists examine proteins or DNA sequences that do not encode or have a function related to the anatomical characteristics originally used to determine the pattern of evolutionary relationships. Biologists have correctly predicted that the proteins and DNA of organisms that share a recent common ancestor should be more similar than the proteins and DNA of organisms that do not share a recent ancestor (Figure 19.6). The finding that the patterns based on these two different sources of data are usually the same provides strong evidence for evolution.

Continental drift and evolution explain the geographic locations of fossils

Earth's continents move over time, a process called **continental drift**. Each year, for example, the distance between South America and Africa increases by about 3 centimeters. But 240 million years ago, the land masses

of Earth drifted together to form one giant continent, called Pangea. Beginning about 200 million years ago, Pangea split up to form the continents we know today (see Figure 22.6).

We can use knowledge about evolution and continental drift to make predictions about the geographic locations where fossils will be found. For example, organisms that evolved when Pangea was intact could have moved relatively easily between what later became widely separated regions, such as Antarctica and India. For that reason, we can predict that their fossils should be found on most or all continents. In contrast, the fossils of species that evolved after the breakup of Pangea should be found on only one or a few continents (for example, the continent on which they evolved and any connected or nearby continents). These predictions have proved correct, and they provide another important line of evidence for evolution.

Direct observation reveals genetic changes within species

In thousands of studies, researchers observing populations in the wild or in the laboratory have seen them change genetically over time. Such observations provide direct, concrete evidence for evolution. Consider how humans have altered the crop species *Brassica oleracea* (Figure 19.7). By allowing only individuals with certain characters to breed—a process called **artificial selection**—humans have crafted enormous evolutionary changes within this species. The tremendous variation that humans have produced within dogs, ornamental flowers, and many other species illustrates the power of artificial selection to produce evolutionary change. Natural selection can produce evolutionary changes in a similar way, as shown by the often striking match of organisms to their environment (see Figure 19.1 and the Highlight on p. 328).

The formation of new species can be produced experimentally and observed in nature

Biologists have directly observed the formation of new species from previously existing species. The first experiment in which a new species was formed took place in the early 1900s, when the primrose *Primula kewensis* was produced. Scientists have also observed the formation of new species in nature. For example, two new species of salsify plants were discovered in Idaho and eastern Washington in 1950. Neither of the new species was found in this region or anywhere else in the world in 1920. Thus both of the new species appear to have evolved from previously existing species sometime between 1920 and 1950. The two new species continue to thrive, and one of them has become common since its discovery in 1950.

Salsify

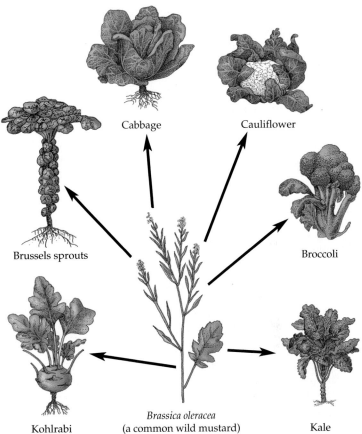

Brassica oleracea
(a common wild mustard)

Cabbage

Cauliflower

Broccoli

Brussels sprouts

Kohlrabi

Kale

Figure 19.7   Artificial Selection Produces Genetic Change
Humans have directed the evolution of *Brassica oleracea*, the wild mustard shown in the center of the figure, to produce such different crop plants as Brussels sprouts, cabbage, and broccoli. Despite their obvious differences, all the crop plants shown here are members of the same species, *B. oleracea*.

> ■ Scientists are convinced that evolution occurred because of (1) evidence from the fossil record; (2) the many cases in which organisms share characteristics because they share a common ancestor; (3) evidence from continental drift; (4) direct observations of genetic change in populations; and (5) the experimental and natural formation of new species.

The Impact of Evolutionary Thought

Before the concept of evolution was developed, adaptations were taken as evidence for the existence of a Creator. The many examples of organisms that seemed well designed for their environments implied to many people that there must be a supernatural designer—a God—much as the existence of a watch implies the existence of a watchmaker. In addition, there was a long tradition, dating from Plato, Aristotle, and other Greek philosophers, in which species were viewed as unchanging through time. Darwinian evolution shook these ideas to their very foundation. No longer could species be viewed as unchanging, nor could apparent design in nature be offered as logical proof that God existed, since evolution by natural selection provided an alternative, scientific explanation for the design of organisms.

The evolution of species and, even more, the argument that the design of organisms could be explained by natural selection were radical ideas in the mid-nineteenth century. These ideas not only revolutionized biology, but also had profound effects on other fields, ranging from literature to philosophy to economics.

The idea of Darwinian evolution also had a profound effect on religion. Evolution was viewed initially as a direct attack on Judeo-Christian religion, and this presumed attack prompted a spirited counterattack by many prominent members of the clergy. Today, however, most religious leaders and most scientists view evolution and religion as compatible but distinct fields of inquiry (but see the box on p. 329). The Catholic Church, for example, accepts that evolution explains the physical characteristics of humans, but maintains that religion is required to explain their spiritual characteristics. Similarly, although the vast majority of scientists accept the scientific evidence for evolution, many of those same scientists have religious beliefs. Overall, most scientists recognize that religious beliefs are up to the individual, and that science cannot answer questions regarding the existence of God or other matters of religious import.

The emergence of evolutionary thought has also had an effect on human technology and industry. For example, an understanding of evolution has proved essential as farmers and researchers have sought to prevent or slow the evolution of resistance to pesticides by insects and weeds. Information about the evolutionary relationships among organisms can also be used to increase the efficiency of searching for new antibiotics and other pharmaceuticals, food additives, pigments, and many other valuable products.

> ■ The concept of evolution has had a profound effect on the sciences, philosophy, religion, agriculture, and many other aspects of human society.

■ HIGHLIGHT

Evolution in Action

Since Darwin's time, the Galápagos Islands have provided a natural biological laboratory in which scientists have studied evolution. The climate of these islands, which are located off the west coast of South America, is usually hot and relatively wet from January to May, and cooler and dry for the rest of the year. But in 1977 the wet season never arrived: Very little rain fell the entire year. On Daphne Major, a small island near the center of the Galápagos Islands, the lack of rain withered the plants that lived there (Figure 19.8). Soon the effects were also felt by a seed-eating bird, the medium ground finch. During the drought, the number of these birds on Daphne Major plummeted from 1200 to 180.

The lack of rain not only caused many birds to die, but also caused evolutionary change in the medium ground finch population. One effect of the drought was that the seeds available to the finch were larger than normal, a condition that prevailed until the wet season of 1978. This occurred because most of the smaller seeds had already been eaten by the abundant finch population by the time the drought began, and few new seeds were formed during the drought. Large seeds can be difficult or impossible for birds with small beaks to crack open and eat. Thus finches with larger beaks had an edge, and many more large-beaked than small-beaked finches survived the

Figure 19.8 A Drought Results in Rapid Evolutionary Change
The 1977 drought on Daphne Major in the Galápagos Islands had a dramatic effect on the plant life there, thus setting the stage for natural selection to cause rapid evolutionary change in birds that depended on the plants for food.

■ BIOLOGY IN OUR LIVES

Creationism: A Scientific Alternative to Evolution?

The Roman Catholic Pope and religious leaders from a diverse set of creeds, including Muslims, Jews, Episcopalians, Methodists, Presbyterians, the United Church of Christ, and the Lutheran World Federation, have all expressed respect for science and for evolutionary biology as a part of science. However, a minority of Christian fundamentalists remain opposed to evolutionary biology. They exert pressure on the governments of many states not to allow evolution to be taught in secondary schools, or if it is taught, to give equal time to what they call creation science.

Creation science states that all species were created by God roughly 10,000 years ago and that they have not evolved since then. As scientific issues, we know these assertions are not correct: The scientific evidence indicates that (1) Earth is more than 4 billion years old, (2) life began about 3.5 billion years ago, and (3) evolution has occurred, continues to occur, and

is responsible for the great diversity of life on Earth.

Creation science explains the diversity of life on Earth in terms of the actions of a supernatural Creator. Such an explanation rings true to some people because their religious beliefs suggest that a supernatural being was directly or indirectly responsible for life on Earth. Science, however, is limited to natural, not supernatural, explanations. For this reason, most scientists do not think that creation science has a place in biology classrooms.

It is important for students—the future leaders of society—to understand evolution. For example, if medical doctors had no understanding of evolution, they would not realize that overuse of antibiotics has the disastrous effect of causing bacteria to evolve resistance to those antibiotics (see the Highlight in Chapter 20, p. 347). Although antibiotics have been overused, medical doctors are aware of the situation, they know why bacteria evolve resistance, and they are tak-

ing steps to fix the problem. An understanding of the evolutionary history of life is also important. For example, as was discussed in Chapter 2, information about evolutionary relationships among organisms can guide the search for useful pharmaceuticals and suggest ways to combat crop pests.

With respect to political decisions about whether evolution should be taught in secondary schools, if we prevent our students from learning what science has to offer, we run the risk that they will not be able to compete effectively in college classrooms or in today's global economy. When scientific understanding and nonscientific beliefs come into conflict, as illustrated by the conflict between evolutionary biology and creation science, the debates that result can be enlightening. Such debates, however, should not be used as an excuse to prevent our students from being taught according to our best and most current scientific understanding of how the world works.

drought to contribute offspring to future generations. As a result, the beak size of the medium ground finch population evolved toward a larger size.

As the research on the medium ground finch shows, we live in an ever-changing world in which species are constantly being shaped by evolutionary forces. In particular, this research shows how natural selection can cause a character to vary over time within a species. Natural selection and other evolutionary forces also can have much greater effects, such as causing entirely new species to form.

New species have evolved many times on the Galápagos Islands. These islands are home to many unique and unusual plant and animal species, including 13 species of finches (Figure 19.9). The Galápagos finches are closely related to one another, they are found nowhere else on Earth, and they exhibit many behav-

iors that are unusual for a finch. For example, whereas most finches are specialized to eat seeds, different Galápagos finches are specialized to feed on everything from seeds to insects to green leaves to blood.

Why do these small islands harbor so many unique but closely related species of finches? The answer seems to be that all the finches on the Galápagos Islands descended from a single species, most likely a finch that reached the islands (perhaps blown in a storm) from the nearest mainland, South America. Upon its arrival, this species found itself in a place where many kinds of birds, such as insect-feeding woodpeckers and warblers, were absent. Over time, natural selection favored birds that developed new ways (for a finch) to feed themselves. For example, while it is unusual for finches to feed on insects, the Galápagos finches that do so evolved to fill an ecological role that is usually taken by other birds,

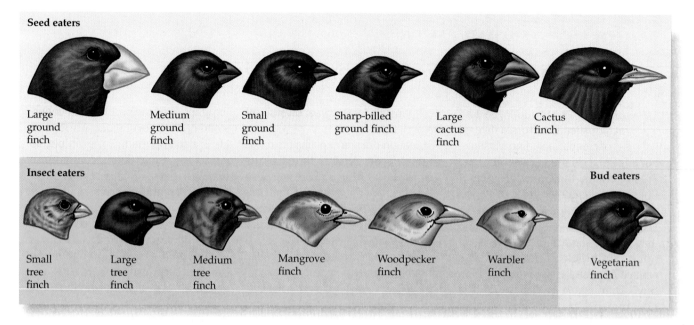

Figure 19.9   The Galápagos Finches
Thirteen unique species of finches evolved on the Galápagos Islands in isolation from other finches. Recent DNA evidence suggests that warbler finches may actually include birds from two different (but morphologically similar) bird species. If confirmed by additional data, such a result would imply that 14, not 13, unique species of finches evolved on the Galápagos Islands.

such as the absent woodpeckers and warblers. The end result of this process was that new species of finches evolved from the single species that originally colonized the islands. Thus, the odd behaviors of these newly evolved finch species make sense when viewed from the perspective of "evolution in action."

> ■ The Galápagos Islands represent a natural biological laboratory in which both small (changes in beak size over a few generations) and large (the formation of new species) evolutionary changes have been documented.

SUMMARY

Biological Evolution: The Sum of Genetic Changes

- Biological evolution is change in the genetic characteristics of populations of organisms over time.
- Biological evolution also can be defined as history—namely, the history of the formation and extinction of species over time.

Mechanisms of Evolution

- Gene mutations occur at random and are not directed toward any goal.
- Mutations and processes that rearrange genetic information cause individuals to differ morphologically, biochemically, and behaviorally.
- The differences among individuals in a population are acted on by natural selection or genetic drift to produce evolutionary change.

Consequences of Evolution for Life on Earth

- Life on Earth is characterized by adaptations that match organisms to their environment, by a great diversity of species, and by many puzzling examples in which otherwise dissimilar organisms share certain characteristics. Evolution explains all these features of life on Earth.
- Adaptations are characteristics of an organism that improve its performance in its environment. Adaptations result from natural selection.
- The diversity of life on Earth has resulted from the repeated splitting of one species into two or more species.
- The shared characteristics of organisms are due to descent from a common ancestor.

Strong Evidence Shows that Evolution Happened

- Organisms are descended with modification from common ancestors; that is, they have evolved.
- Scientists of all nations, races, and creeds agree that evolution has occurred.
- The fossil record provides clear evidence of the evolution of species over time. It also documents the evolution of

major new groups of organisms from previously existing organisms.

■ The extent to which organisms share characteristics is consistent with patterns of evolutionary relationships. For example, the proteins and DNA of organisms that share a recent common ancestor are more similar than the proteins and DNA of organisms that do not share a recent ancestor.

■ As predicted by our understanding of evolution and continental drift, fossils of organisms that evolved when the continents were still connected to each other have a wider geographic distribution than do fossils of more recently evolved organisms.

■ In thousands of studies, researchers have observed genetic changes in populations over time, providing direct evidence of small evolutionary changes.

■ Biologists have observed the evolution of new species from previously existing species.

The Impact of Evolutionary Thought

■ Darwin's ideas on evolution and natural selection revolutionized biology and had a profound effect on many other fields, including literature, economics, and religion.

■ Evolutionary biology has many technological applications.

Highlight: Evolution in Action

■ The Galápagos Islands provide a natural biological laboratory in which scientists have studied evolution.

■ On the Galápagos Islands, natural selection has caused rapid changes in the size of the beak of a seed-eating bird, the medium ground finch.

■ Large evolutionary changes have also taken place on the Galápagos Islands, including the evolution of many new and unusual plant and animal species.

KEY TERMS

adaptation p. 322	fossil p. 324
artificial selection p. 327	genetic drift p. 322
biological evolution p. 320	natural selection p. 321
continental drift p. 326	speciation p. 323

CHAPTER REVIEW

Self-Quiz

1. Which of the following provides evidence for evolution?
   a. direct observations of genetic changes in populations
   b. shared characteristics of organisms
   c. the fossil record
   d. all of the above

2. In natural selection,
   a. the genetic composition of the population changes at random over time.
   b. new mutations are generated over time.
   c. all individuals in a population are equally likely to contribute offspring to the next generation.
   d. individuals that possess particular inherited characters survive and reproduce at a higher rate than other individuals.

3. Adaptations
   a. are features of an organism that hinder its performance in its environment.
   b. are not common.
   c. result from natural selection.
   d. result from genetic drift.

4. The first mammals evolved 220 million years ago. The supercontinent Pangea began to break apart 200 million years ago. Therefore, fossils of the first mammals should be found
   a. on most or all of the current continents.
   b. only in Antarctica.
   c. on only one or a few continents.
   d. none of the above

5. The fact that the flipper of a whale and the arm of a human both have five digits and the same kinds of bones can be used to illustrate that
   a. genetic drift can cause the evolution of populations.
   b. organisms can share characteristics simply because they share a common ancestor.
   c. whales evolved from humans.
   d. humans evolved from whales.

Review Questions

1. How does evolution explain (a) adaptations, (b) the great diversity of species, and (c) the many examples in which otherwise dissimilar organisms share certain characteristics?

2. Why are scientists throughout the world convinced that evolution happened? Consider the five lines of evidence discussed in this chapter.

3. Although biologists agree that evolution occurs, they debate which mechanisms are most important in causing evolutionary change. Does this mean that the "theory" of evolution is wrong?

4. Genetic drift occurs when chance events cause some individuals in a population to contribute more offspring to the next generation than other individuals. Are such chance events likely to have a greater effect in small or in large populations? (*Hint:* Examine Figure 19.3. Consider whether the proportion of the *A* allele in the population would be likely to change from 50 percent to 100 percent if there were 1000 plants instead of 5 plants in the population.)

The Daily Globe

Genetically Engineered Corn Could Aid Farmers

AMES, IA—In a discovery that some are saying could save farmers millions of dollars in pesticide expenses and lost crops, researchers have produced a genetically engineered variety of corn that can knock out one of its most troublesome pests, the corn mealybug. This pest, which has evolved resistance to most pesticides and attacks millions of acres of crops each year, has been continually troublesome to farmers.

"We think this is going to revolutionize corn farming," said Ms. Carol Barnes, public relations director at the MonPont Company. The genetically engineered corn produces a toxin that prevents the larva, or immature form, of the mealybug from developing normally. Thus mealybugs are unable to grow and reproduce on the engineered corn.

But Dr. Joseph Purpurata, evolutionary biologist at Western Iowa University, urged caution. "The mealybug has been a tough foe in the past. If farmers plant only the engineered corn, the mealybug will evolve resistance to the toxin in a few years, and then we'll be right back to where we started." MonPont spokesperson Barnes agreed and pointed out that her company routinely asks farmers not to plant all of their fields with the engineered corn. As Ms. Barnes explained, "When farmers plant non-engineered corn, a large number of mealybugs survive and reproduce on that corn. Such mealybugs have many offspring, and most of their offspring are not resistant. In contrast, in fields of engineered corn, most mealybugs cannot reproduce and hence few (but resistant) offspring are produced. By planting non-engineered corn, farmers help ensure that each year there are large numbers of non-resistant offspring and few resistant offspring, thus keeping the overall resistance of the mealybug population to a minimum."

Organic farmers also have a stake in seeing that insects do not develop resistance. The toxin produced by the genetically engineered corn is also widely used by these farmers, who depend heavily on it, as it is one of the few powerful, naturally produced toxins available for fighting insect attack. "Our members use this stuff only when we're dealing with large numbers of mealybugs," said Jake Granola, coordinator of the Iowa Organic Farmers' Cooperative. "We use it in a restrained manner, whereas the engineered corn produces it constantly."

Many traditional farmers, however, are not convinced by these calls for restraint. As farmer Joe Henderson of Straight, Iowa, put it, "It's fine to worry about the future, but I need to feed my family now. I'm planting all my fields with the engineered corn."

Evaluating "The News"

1. Organic farmers use the toxin produced by the genetically engineered corn only when large numbers of mealybugs threaten crop production, whereas the engineered plant produces the toxin continuously. Which method is more likely to cause the insect to evolve resistance to the protein? (*Hint:* The evolution of resistance is due to natural selection. Consider whether it would be easier to select for resistance if the toxin was always in use or was only used sporadically.)

2. Organic farmers are insisting that traditional farmers do more to keep insects from becoming resistant to this potent toxin. Does one set of farmers have the right to dictate methods that might alter the yields and profits of another set of farmers? Why or why not?

3. It costs a lot of money to develop a genetically engineered crop. If farmers decide to plant all of their fields with the engineered corn and the mealybug rapidly evolves resistance, do you think MonPont or other companies will continue to develop crops that are genetically engineered to resist insect attack? Is there any way to coordinate the activities of individual farmers?

20 Evolution of Populations

Ismael Vargas, *Tiempos Idos Que Felices Llamo*, 1986.

Evolution and AIDS

Within the past 10 years, several advances in the battle against HIV (human immunodeficiency virus), the virus that causes AIDS, have been heralded in the media as breakthroughs that might lead to the end of that dread disease. The new findings have included exciting advances in clinical treatment and great strides in basic research.

With respect to clinical practice, new therapies have reduced the blood concentration of HIV in a majority of patients to undetectable levels. These new treatments rely on "triple drug cocktails," so called because they contain three different drugs. Two of these drugs inhibit the ability of HIV to transcribe its genetic material, which it does at an early stage of viral attack. The third drug inhibits the ability of HIV to assemble offspring viruses, which it does at a late stage of viral attack.

Complementing the development of treatments that reduce the concentration of HIV in the blood, new discoveries have shed light on how HIV first invades human cells. Researchers knew that HIV must bind to a particular human cell surface protein, called CD4, before it can infect human cells. But they also knew that additional (but unidentified) cell surface proteins were required for HIV to enter cells. In 1996, researchers identified two of these additional proteins, at least one of which had to be present (along with CD4) for HIV to invade human cells. This research offered hope for the design of new drugs to thwart the virus: If drugs could be developed that blocked the ability of HIV to bind to the two newly discovered proteins, they might prevent the virus from infecting human cells.

Promising new treatments for HIV had been developed before, however, and shortly thereafter, the virus evolved new ways to get around those treatments. Sadly, the same thing has happened in the past few years. First, in some patients given triple drug cocktails, new resistant strains of the virus appeared. Viral concentrations in the patients' blood rose, on average only 4 to 6 months after treatment began. Second, two new strains of the virus were discovered in early 2001. These strains entered human cells in an entirely new way: They did not bind to CD4 or to the two other recently discovered pro-

teins at all, but instead bound to a different cell surface protein, called CD8. The ability of HIV to alter the way it enters human cells may make it very hard to develop drugs that block HIV infection, since if we block one or more points of entry, the virus may simply shift to another point of entry. Overall, the recent changes in the virus send a sobering message: With remarkable speed and power, HIV evolves the ability to cope with our best efforts to defeat it.

A Model of HIV

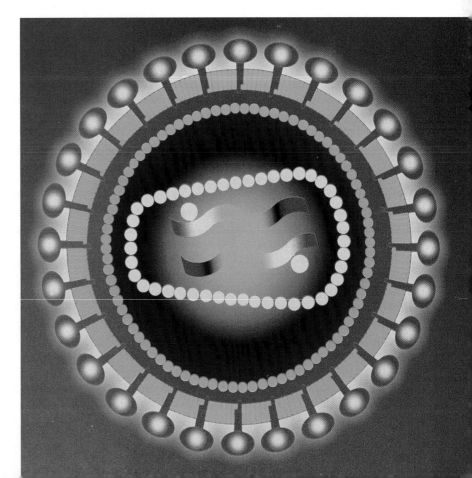

■ KEY CONCEPTS

1. Populations evolve when allele or genotype frequencies change from generation to generation.

2. Individuals within populations differ genetically in behavioral, morphological, physiological, and biochemical characters. This genetic variation provides the raw material on which evolution can work.

3. Mutations are the original source of genetic variation within populations. The genetic variation produced by mutations is increased by recombination.

4. Five factors can cause populations to evolve: mutation, nonrandom mating, gene flow, genetic drift, and natural selection.

5. Some populations can evolve very rapidly (in months to years). This potential for rapid evolution has serious implications for the evolution of pesticide resistance by insects and antibiotic resistance by bacteria.

In the nineteenth century, Charles Darwin wrote that evolution occurs too slowly to observe, but more recently, thousands of studies have documented the evolution of populations. Collectively, these studies indicate that Darwin had the basic ideas about evolution right, but he was wrong about the time required for evolution to occur. The recent studies show that populations can evolve very rapidly (within a few generations, or in months to years), and that new species can form in surprisingly short periods of time (one year to a few thousand years; see Chapter 21).

Evolution can be defined as a change in allele or genotype frequencies (percentages) in a population over generations, where a population is a group of individuals of a species that lives in the same area. For example, the frequency of a particular genotype (such as *aa*) in a population might change from 5 percent to 15 percent in several generations. Changes in allele or genotype frequencies result from a two-step process. First, as described in Chapter 13, mutation, crossing-over, independent assortment of chromosomes, and fertilization result in genetic variation among the individuals in a population. Second, the pattern of genetic variation in the population changes over time, as when an allele that once was common is replaced by another allele.

Changes in allele or genotype frequencies in a population over time are referred to as **microevolution**, so called because they represent the smallest scale at which evolution occurs. Evolution also occurs on a much larger scale, as when new species or entire new groups of organisms evolve (macroevolution). We will discuss the formation of new species in Chapter 21 and macroevolution in Chapter 22. In this chapter we describe the factors that cause allele and genotype frequencies in a population to change from generation to generation. Let's begin by defining two essential terms: genotype frequency and allele frequency.

20.1 *Key Definitions: Genotype and Allele Frequencies*

Genotype frequency refers to the proportion, or percentage, of a genotype in a population, and **allele frequency** refers to the proportion of an allele in a population.

To see how these proportions are calculated, let's imagine that wing color in a population of 1000 moths is caused by a single gene, which has two alleles: *W*, for orange wing color, and *w*, for white wing color. If the population contains 160 *WW* individuals, 480 *Ww* individuals, and 360 *ww* individuals, then we can obtain the frequencies for the three genotypes (*WW*, *Ww*, and *ww*) by dividing their numbers by the total number of individuals in the population (1000). Thus, the genotype frequencies for *WW*, *Ww*, and *ww* are 0.16, 0.48, and 0.36, respectively. Note that these genotype frequencies add up to 1.0, as they should because *WW*, *Ww*, and *ww* are the only three genotypes possible and a frequency of 1.0 is equivalent to 100 percent.

Allele frequencies can be computed by the following method, which we illustrate for the *W* allele. There are 1000 individuals in our moth population, each of which has two alleles of the wing color gene. Thus, the total number of alleles in the population equals 2000. There are 160 *WW* individuals, each of which carries two *W* alleles, for a total of 320 *W* alleles. *Ww* individuals have one *W* allele each, for a total of 480 *W* alleles, and *ww* individuals have no *W* alleles. Thus, there are 800 (320 + 480 + 0) *W* alleles in the population. Finally, we calculate the frequency of the *W* allele by dividing the number of *W* alleles by the total number of alleles in the population: (800 *W* alleles)/(2000 total alleles) = 0.4. Since the gene for wing color has only two alleles, *W* and *w*, the sum of their frequencies must equal 1.0. Hence, the frequency of the *w* allele is 1.0 − 0.4 = 0.6. We can check this value by performing a calculation for the *w* allele similar to the calculation we made for the *W* allele.

With definitions for genotype frequency and allele frequency in hand, we are ready to begin our study of the evolution of populations. We start with a discussion of genetic variation, the raw material on which evolution acts.

> ■ Genotype frequency and allele frequency are, respectively, the proportion of a genotype and of an allele in a population.

Genetic Variation: The Raw Material of Evolution

Genetic variation refers to genetic differences among the individuals of a population. Within a population, individuals often differ in behavioral, morphological (Figure 20.1), and physiological characters. As described in Unit 3, much of this variation is under genetic control. Organisms also vary greatly for biochemical characters that are under direct genetic control, such as the amino acid sequences of their proteins. The underlying cause of all genetic differences among individuals within a population is a difference in their DNA sequences. Genetic differences among individuals are important because they provide the raw material on which evolution can work.

Mutation and recombination create genetic variation

Mutations are changes in the sequence of an organism's DNA. As we learned in Chapter 12, mutations give rise to new alleles. Thus, mutations are the original source of all genetic variation. Mutations are thought to occur at random in the sense that (1) they are not directed toward any particular goal (that is, genes do not "know" when and how it might be beneficial for them to mutate), and (2) we cannot predict which copy of a gene in a population will mutate.

Even though the mechanisms that correct mistakes in DNA fix most errors in DNA replication (see Chapter 14), mutations occur regularly in all organisms. Humans, for example, have two copies each (one copy from each parent) of their 35,000 or so genes. On average, two or three of these 70,000 gene copies have mutations that make them different from those of either parent. Furthermore, the genetic variation that is initially present as a result of mutations is increased when fertilization, crossing-over, and independent assortment of chromosomes result in new combinations of alleles (see Chapter 13). These processes are known collectively as **recombination**.

Many mutations have little effect, many are harmful, and a few are beneficial. The effect of a mutation often depends on the environment in which it occurs. For example, certain mutations that provide houseflies with resistance to the pesticide DDT also reduce their rate of growth. Flies that grow slowly take longer to mature, and thus do not produce as many offspring in their lifetime as flies that grow at the normal rate. In the absence of DDT, such mutations are harmful, but when DDT is sprayed, these mutations provide an advantage great enough to offset the disadvantage of slow growth, and hence the mutant alleles spread throughout housefly populations.

Figure 20.1 Morphological Variation
The individuals within a population often vary greatly for many morphological characters, such as the color patterns of the starfish shown here.

■ Individuals within populations differ for many important characters, and much of this variation is under genetic control. The genetic variation among individuals in a population provides the raw material on which evolution can work.

When Populations Do Not Evolve

In this section we turn to a discussion of conditions under which populations do not evolve. Specifically, we discuss the Hardy–Weinberg equation, a simple formula that allows us to predict genotype frequencies in a hypothetical nonevolving population. This calculation provides a baseline against which real populations can be compared.

A population does not evolve when the following five conditions all hold:

1. There is no net change in allele frequencies due to mutation.
2. Members of the population mate randomly.
3. New alleles do not enter the population via immigrating individuals, seeds, or gametes (for example, pollen).
4. The population contains a large number of individuals.
5. Natural selection does not occur.

These five conditions do not all hold completely in most natural populations. However, many populations meet these conditions well enough that the Hardy–Weinberg equation is approximately correct, at least for some of the genes within the population.

To derive the Hardy–Weinberg equation, let's return to the hypothetical population of 1000 moths introduced earlier in this chapter. In that example, the dominant allele (W) for orange wing color had a frequency of 0.4, and the recessive allele (w) for white wings had a frequency of 0.6. What we seek to do now is predict the frequencies of *WW*, *Ww*, and *ww* genotypes in the next generation.

If mating among the individuals in the population is random, and if the other four conditions described on this page also hold, we can use a Punnett square approach (see Chapter 12) to predict the offspring that will be produced in the next generation (Figure 20.2). This approach is similar to mixing all the gametes in a bag and then randomly drawing one egg and one sperm to determine the genotype of each offspring. With such random drawing, the allele and genotype frequencies in our moth population do not change from one generation to the next (see Figure 20.2).

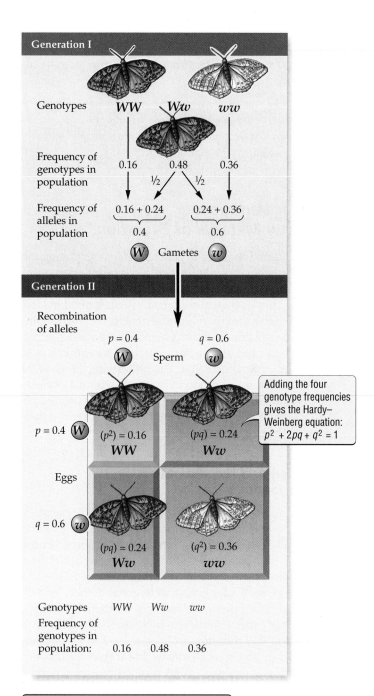

Adding the four genotype frequencies gives the Hardy–Weinberg equation: $p^2 + 2pq + q^2 = 1$

Conclusion:
Generation II genotype and allele frequencies = frequencies of generation I.

20.2

Figure 20.2 The Hardy–Weinberg Equation
When mating is random and certain other conditions are met, allele and genotype frequencies in a population do not change. p = frequency of the W allele, q = frequency of the w allele.

Because the *WW*, *Ww*, and *ww* genotypes are the only three types of zygotes that can be formed, the sum of their frequencies must equal 1. As shown in Figure 20.2, when we sum the frequencies of the three genotypes, we get the **Hardy–Weinberg equation**:

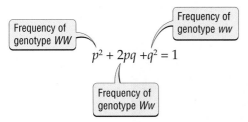

In this equation, the frequency of the *W* allele is labeled *p* and the frequency of the *w* allele is labeled *q*.

In our moth population, allele and genotype frequencies do not change from one generation to the next. Once the Hardy–Weinberg frequencies are reached, they remain constant over time, assuming the five conditions listed on page 338 continue to hold.

Populations can evolve when one or more of the five conditions for the Hardy-Weinberg equation are violated. The five major factors that violate these conditions are mutation, nonrandom mating, gene flow, genetic drift, and natural selection. The remainder of this chapter will look at each of these factors in turn. These factors can cause genotype frequencies in real populations to differ greatly from the predictions of the Hardy–Weinberg equation.

> ■ A population does not evolve when (1) mutation does not change allele frequencies, (2) mating is random, (3) new alleles do not enter the population, (4) the population is large, and (5) natural selection does not occur. When these five conditions hold or are approximately met, the Hardy–Weinberg equation can be used to calculate genotype frequencies in a population.

 ## 20.3 Mutation: The Source of Genetic Variation

By creating new alleles, mutation provides the raw material for evolution. In this sense, all evolutionary change depends ultimately on mutation. However, mutations occur so infrequently at each gene locus that they cause little direct change in the allele frequencies of populations. Allele frequencies in real populations often change rapidly, however, indicating that mutation, acting alone, is not responsible for most evolutionary change.

Although mutations have only a small direct effect on changes in allele frequencies, in some dramatic cases new mutations have stimulated the rapid evolution of populations. For example, the speed with which HIV has overcome clinical treatments is due in part to its high mutation rate (see the box on p. 341).

In another example, genetic evidence indicates that the resistance of the mosquito *Culex pipiens* to organophosphate pesticides was caused by a single mutation that occurred in the 1960s. Since that time, migrant mosquitoes have carried the initially rare mutant allele from its place of origin in Africa or Asia to North America and Europe. This mutant allele is highly advantageous to the mosquito: Individuals that lack the allele die when exposed to organophosphate pesticides. Thus, when the mutant allele is introduced into a population that is exposed to organophosphate pesticides, natural selection causes it to increase rapidly in frequency, leading to the evolution of resistance within the new population.

> ■ Mutations cause little direct change in allele frequencies, but they are the original source of all evolutionary change.

Nonrandom Mating: Changing Genotype Frequencies

Nonrandom mating occurs when certain subsets of individuals within a population are more likely to mate with each other than they are to mate with individuals selected at random from the population at large. Nonrandom mating can occur in a variety of ways. For example, in organisms as different as mice and pine trees, individuals usually mate with close neighbors (Figure 20.3). Another common form of nonrandom mating occurs when organisms select mates on the basis of shared characters, as in the preference some people show for mates of similar height or skin color. The underlying cause of nonrandom mating differs in these two examples. It is driven by geographic proximity in pine trees and mice and by active preferences in people. But the net effect is the same: Matings among members of the population are not random.

Nonrandom mating does not directly change allele frequencies, but it can change genotype frequencies. Hence, it can cause the evolution of populations, which refers to changes in either allele or genotype frequencies. To understand how nonrandom mating can cause genotype frequencies to change, consider an extreme form of nonrandom mating, self-fertilization in plants.

Roughly 40 percent of all plant species mate with themselves, a practice known as self-fertilization, or "selfing." When a homozygous *AA* or *aa* plant mates with itself, all the offspring inherit the genotype of their

House mice that live within a barn are more likely to mate with each other…

…than they are to mate with mice that live in a different barn.

Figure 20.3
Nonrandom Mating
In the house mouse, as in many other species, individuals mate with nearby individuals more often than they mate with distant individuals.

parent. However, when a heterozygous *Aa* plant mates with itself, one-fourth of the offspring, on average, have genotype *AA*, one-fourth *aa*, and one-half *Aa* (see Chapter 12). Thus, if the plants in a population of 100 heterozygotes each produce a single offspring by selfing, in the next generation we would expect the number of heterozygotes to drop from 100 to 50. Similarly, we would expect the number of both *AA* and *aa* homozygotes to increase from 0 to 25. As this example illustrates, over time selfing causes the frequency of heterozygotes to decrease and the frequency of homozygotes to increase.

■ Nonrandom mating does not alter allele frequencies directly, but it can alter genotype frequencies.

20.4 Gene Flow: Exchanging Alleles between Populations

Gene flow is the exchange of alleles between populations (Figure 20.4). Alleles can be exchanged when seeds or individuals move from one population to another. If the immi-

grants then reproduce, their alleles enter the collection of alleles in their new population, and gene flow has occurred. Gene flow can also occur when individuals in one population are fertilized by the gametes of individuals from another population (as, for example, in the case of windblown pollen from plants).

Because it consists of an exchange of alleles between one population and another, gene flow makes the genetic composition of different populations more similar to each other. In this way gene flow can counteract the effects of mutation, nonrandom mating, genetic drift, and natural selection, all of which can cause populations to become more different from one another. In some plant species, for example, neighboring populations live in very different environments, yet remain genetically similar. Natural selection, by favoring different alleles in the different environments, would tend to make the populations differ genetically. The lack of genetic difference between such populations appears to be due to gene flow, which occurs at a rate high enough to cause the populations to remain genetically similar despite the effects of natural selection.

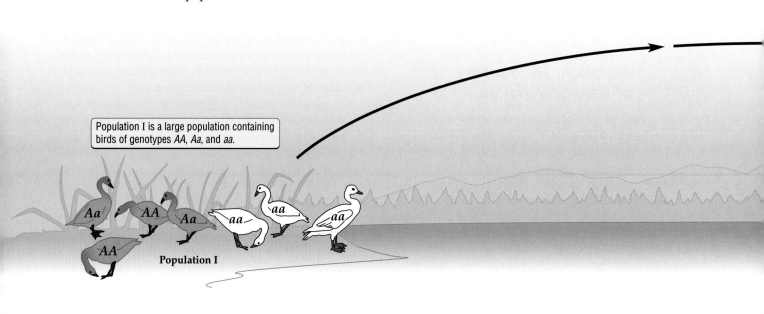

Population I is a large population containing birds of genotypes *AA*, *Aa*, and *aa*.

Population I

■ BIOLOGY IN OUR LIVES

Mutation and Resistance in HIV

The virus that causes AIDS has an exceptionally high mutation rate. Each time it reproduces itself, HIV produces one mutation for every 10,000 to 100,000 of its nucleotide bases. This mutation rate is considerably greater than that of most living organisms, which typically undergo one mutation for every 1 billion to 100 billion bases. HIV's high mutation rate may enable it to generate large numbers of new genetic variants rapidly, some of which can cope with changing environmental conditions (such as the threats posed by triple drug cocktails).

A high rate of mutation is not the only reason why HIV evolves resistance to clinical treatments. Often a 10-year interval passes between the time an individual is infected with HIV and the onset of the symptoms of full-blown AIDS. During this time, the concentration of HIV that is detectable in the blood remains low, but the virus is not just quietly waiting. Rather, HIV repeatedly attacks the cells of the infected person's immune system, producing many billions of viral offspring.

The fact that the concentration of HIV in the blood remains low even though the virus goes through many infective cycles implies that, prior to the onset of AIDS symptoms, huge numbers of the virus are being destroyed shortly after they are produced. Within a given patient, the high mutation rate of HIV and the high rate at which viruses are produced and destroyed causes the body to harbor many competing and different strains of HIV. Some of these strains of the virus may contain mutations that make it resistant to clinical treatments.

Overall, the characteristics of HIV make it a difficult and moving target: Different patients with AIDS are likely to harbor very different strains of HIV, and within a single infected individual, HIV can be expected to evolve resistance to new therapies extremely rapidly. There are many promising avenues for progress in our battle against HIV, but for now perhaps the best we can hope for is to gain an edge and then remain one step ahead of this remarkable and terrible virus.

Gene flow can also play a role similar to that of mutation by introducing new alleles into a population. The introduction of new alleles by gene flow can have a dramatic effect. For example, in the case of the mosquito *Culex pipiens*, discussed earlier, a new allele that made the mosquito resistant to organophosphate pesticides spread by gene flow across three continents. This spread of the mutated allele allowed billions of mosquitoes to survive the application of pesticides that otherwise would have killed them.

■ Gene flow can introduce new alleles into a population, thus providing new genetic variation. Gene flow makes the genetic composition of different populations more similar.

Figure 20.4 Gene Flow
New alleles can be introduced into populations in several ways, as when individuals move from one population to another.

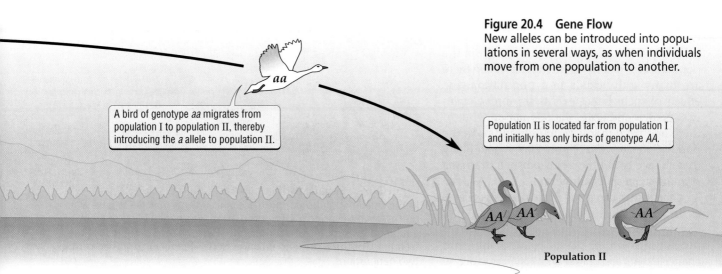

A bird of genotype *aa* migrates from population I to population II, thereby introducing the *a* allele to population II.

Population II is located far from population I and initially has only birds of genotype *AA*.

Population II

Genetic Drift: The Effects of Chance

As we learned in Chapter 19, chance events may determine which individuals contribute offspring to the next generation (see Figure 19.3). Chance events may cause alleles from the parent generation to be sampled at random for inclusion in the next generation. The process by which alleles are sampled at random over time is called **genetic drift**. This process leads to random changes in allele frequencies from generation to generation.

Genetic drift affects small populations

Genetic drift can have dramatic effects on small populations. If a population is very large, chance events are less likely to cause genotype and allele frequencies to change greatly from one generation to the next. Chance has less of an effect on large populations because when there are many individuals, random events usually have a similar effect on all the alleles and genotypes in the population. To help you understand this idea, consider what happens when you toss a coin. If the coin is tossed only a few times, the observed percentage of heads will often differ greatly from the expected 50 percent. But if the coin is tossed thousands of times, the observed percentage of heads will usually be very close to 50 percent.

In natural populations, the number of individuals in the population has an effect similar to the number of times the coin is tossed. Consider the small population of moths shown in Figure 20.5. By chance alone, some individuals leave offspring and others do not. In this example, such chance events have altered the genotype and allele frequencies of a gene with two alleles (*A* and *a*). The changes are so rapid that one of the alleles has been lost from the population in just two generations. When there are many individuals in a population, the effects of such chance events are spread evenly over all members of the population. Thus, if there had been many more individuals in the population shown in Figure 20.5, it is unlikely that chance could have caused such dramatic changes in so short a time.

Genetic drift can result from a variety of causes, including the random alignment of alleles during gamete formation (causing some alleles but not others to be passed to offspring), and chance events associated with the reproduction

and survival of individuals. Genetic drift can have a strong effect on the evolution of populations in several ways:

- Many evolutionary changes in small populations are due to genetic drift. Genetic drift also occurs in large populations, but in these cases its effects are more

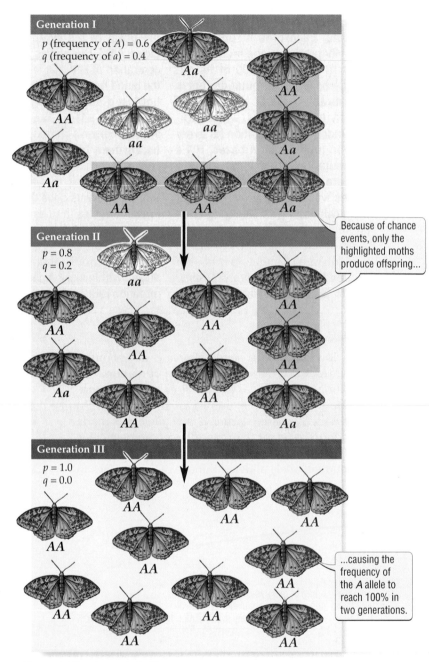

Figure 20.5   Genetic Drift
In this small population of moths, chance events may determine which moths leave offspring, without regard to which individuals are better equipped for survival or reproduction. Here, chance events cause the frequency of the *A* allele to increase from 60 percent to 100 percent in two generations.

easily overcome by natural selection and other evolutionary forces. Thus in large populations, genetic drift causes little change in allele or genotype frequencies over time.

- Genetic drift reduces genetic variation within populations because chance alone eventually causes one of the alleles to reach **fixation**, that is, to reach a frequency of 100 percent (see Figure 20.5). The fixation of alleles can happen rapidly in small populations, but in large populations it takes a long time.
- Genetic drift can lead to the fixation of alleles that are neutral, harmful, or beneficial. As emphasized in Chapter 19, only natural selection consistently leads to adaptive evolution.

Genetic bottlenecks can threaten the survival of populations

The importance of genetic drift in small populations has implications for the preservation of rare species. If the number of individuals in a population falls to very low levels, genetic drift may lead to a loss of genetic variation or to the fixation of harmful alleles, either of which may hasten the extinction of the species. When a drop in the size of a population causes low genetic variation or the fixation of harmful alleles, the population is said to have experienced a **genetic bottleneck**.

Genetic bottlenecks are thought to have occurred for the Florida panther, the northern elephant seal, and the African cheetah. However, each of these examples poses a problem: There is no way of knowing whether the observed low level of genetic variation in these animals really was caused by a decrease in population size (it could just be a natural feature of the organism).

Elephant seal

Recent studies on greater prairie chickens in Illinois avoided this problem by comparing the DNA of modern birds to the DNA of their prebottleneck ancestors, obtained from (nonliving) museum specimens. There were millions of greater prairie chickens in Illinois in the nineteenth century, but the conversion of prairies to farmland caused their numbers to drop to only 50 birds by 1993. This drop in numbers did cause a genetic bottleneck: The modern birds lacked 30 percent of the alleles found in the museum specimens, and they suffered poor reproductive success compared with prairie chicken populations that had not experienced a genetic bottleneck (Figure 20.6).

> ■ Genetic drift causes random changes in allele frequencies over time. It can have a large influence on the evolution of small populations, resulting in the loss of genetic variation and the fixation of harmful alleles.

Illinois

Prebottleneck (1820)

Postbottleneck (1993)

In 1820, the grasslands in which greater prairie chickens live covered most of Illinois.

In 1993, less than 1% of the grassland remained, and the birds could be found only in these two locations.

By 1993, only 50 greater prairie chickens remained in Illinois, causing both the number of alleles and the percentage of eggs that hatched to decrease.

	Illinois		Kansas	Minnesota	Nebraska
	Prebottleneck (1933)	Postbottleneck	No bottleneck		
Population size	25,000	50	750,000	4,000	75,000 – 200,000
No. of alleles at 6 genetic loci	31	22	35	32	35
Percentage of eggs that hatch	93	56	99	85	96

Figure 20.6 A Genetic Bottleneck
The Illinois population of greater prairie chickens dropped from 25,000 birds in 1933 to only 50 birds in 1993. This drop in population size caused a loss of genetic variation and a drop in the percentage of eggs that hatched. Here, the modern, postbottleneck Illinois population is compared with the 1933 prebottleneck Illinois population, as well as with populations in Kansas, Minnesota, and Nebraska that never experienced a bottleneck.

Natural Selection: The Effects of Advantageous Alleles

Natural selection is a process by which individuals with particular heritable characters survive and reproduce at a higher rate than other individuals in a population (see Chapter 19). Therefore, the alleles for the heritable characters favored by natural selection tend to become more common in the offspring generation than in the parent generation.

Even though natural selection may favor one allele over another, genetic drift, gene flow, or mutation may oppose the action of natural selection and prevent allele frequencies from changing. Hence, natural selection can, but does not necessarily, lead to evolutionary change. Although it does not always lead to evolutionary change, natural selection is the only evolutionary mechanism that consistently improves the reproductive success of the organism in its environment.

The research on the medium ground finch described in Chapter 19 shows that natural selection can cause characters to evolve rapidly in response to changes in the environment. Although natural selection often has a strong and rapid effect, it is powerless unless individuals within the population differ genetically. For example, if no insects in a population carry alleles for resistance to a particular pesticide, then natural selection cannot promote the evolution of resistance to that pesticide.

Types of natural selection

There are three major types of natural selection: directional selection, stabilizing selection, and disruptive selection. Despite the fact that the three types have different names, they all operate by the same mechanism: Individuals that possess certain forms of a heritable character tend to survive better and produce more offspring than individuals that possess other forms of that character.

In **directional selection** (Figure 20.7*a*), individuals with one extreme of a heritable phenotypic character have an advantage over other individuals in the population. For example, if large individuals produce more offspring than small individuals, then there will be directional selection for large body size.

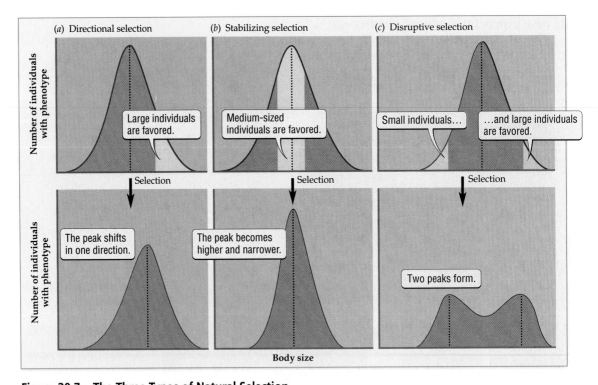

Figure 20.7   The Three Types of Natural Selection
Directional selection (*a*), stabilizing selection (*b*), and disruptive selection (*c*) affect body size differently. The graphs in the top row show the relative numbers of individuals with different body sizes in a population before selection. The phenotypes that are favored by selection are shown in yellow. The graphs in the bottom row show how each type of natural selection affects the distribution of body size in the population.

Directional selection has often caused increased resistance to pesticides in insects. In India, for example, the pesticide DDT was first applied in the late 1940s to control mosquitoes that spread malaria. Initially it was very effective, but in 1959 the world's first DDT-resistant mosquitoes appeared in India. The resistant mosquitoes increased rapidly in frequency, making DDT less and less effective. DDT-resistant mosquitoes are now found throughout the globe. As a consequence, when a local spraying program starts, mosquito populations evolve resistance in a few generations—that is, within months, not years (Figure 20.8). This problem is not restricted to DDT and mosquitoes: A decrease in the time required for the evolution of pesticide resistance has been found for many other pesticides and species of insects.

In **stabilizing selection** (Figure 20.7b), individuals with intermediate values of a heritable phenotypic character have an advantage over other individuals in the population. Birth weights in humans provide a classic example. Historically, light or heavy babies did not survive as well as babies of average weight (Figure 20.9). By the late 1980s, however, selection against small and large babies had virtually disappeared in some wealthy countries with advanced medical care, such as Italy, Japan, and the United States. This change was caused by advances in the care of very light premature babies and by increases in the use of cesarean deliveries for babies that are large relative to their mothers (and hence pose a risk of injury to mother and child).

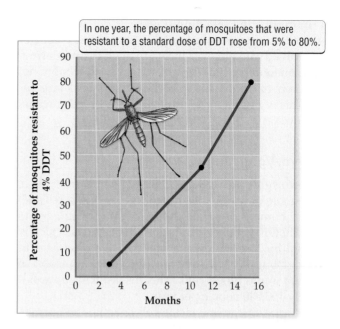

Figure 20.8 Directional Selection for Pesticide Resistance
Over a two-year period, mosquitoes were captured from a population at different times and sprayed with the pesticide DDT. Directional selection caused the population of mosquitoes to evolve resistance rapidly, thus limiting the effectiveness of the pesticide. Mosquitoes were considered resistant if they were not killed by a standard dose of DDT (4% DDT) in one hour.

Figure 20.9 Stabilizing Selection for Human Birth Weight
This graph is based on data for 13,700 babies born from 1935 to 1946 in a hospital in London.

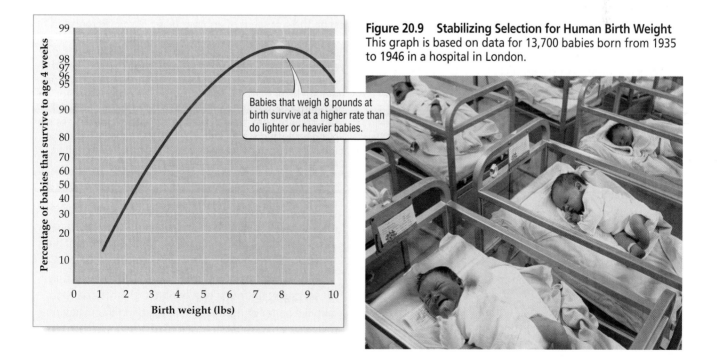

In **disruptive selection** (Figure 20.7c), individuals with either extreme of a heritable phenotypic character have an advantage over individuals with an intermediate phenotype. This type of selection is probably not common, but it appears to be the cause of beak size differences within a population of African seed crackers (Figure 20.10).

Heterozygote advantage

Heterozygote advantage occurs when heterozygotes leave more offspring than homozygotes. Sickle-cell anemia in humans illustrates heterozygote advantage. As described in Chapter 15, sickle-cell anemia is typically caused by a mutation that alters one amino acid in hemoglobin, a protein that functions in the transport of oxygen in red blood cells. This disease kills approximately 100,000 people per year.

In terms of causing sickle-cell anemia, the normal, nondisease hemoglobin allele (*S*) is dominant. Thus, only homozygous recessive individuals (*ss*) have the disease. The red blood cells of *ss* individuals are curved and distorted (sickle-shaped) when oxygen concentrations are low, otherwise they are fine. Victims of sickle-cell anemia suffer from many serious health problems, including stroke, heart failure, and kidney failure.

Individuals with sickle-cell anemia usually die before they are old enough to reproduce. Nevertheless, the *s* allele persists, reaching frequencies as high as 10 to 20 percent in regions of Africa where malaria is common (Figure 20.11). Why should this be? Heterozygous *Ss* individuals are more resistant to malaria than are homozygous (normal) *SS* individuals, especially as infants. Thus, the *s* allele is found at high frequencies in malaria-infested locations because it gives heterozygotes a survival advantage in such places. Due to the rarity of malaria in the United States, Americans of African descent have shown a drop in the frequency of the *s* allele in the last 200 years.

Notice that there is an important difference between heterozygote advantage and the three types of natural selection (directional, stabilizing, and disruptive selection): Whereas heterozygote advantage is defined in terms of the *genotype* of the organism, directional, stabilizing, and disruptive selection are defined in terms of the *phenotype* of the organism. If, as is often the case, heterozygotes are intermediate in phenotype to both homozygotes, then heterozygote advantage is an example of stabilizing selection. In some cases, heterozygotes may have an extreme phenotype, as when they are larger or produce more seeds than homozygotes. When heterozygotes have an extreme phenotype, heterozygote advantage is an example of directional selection.

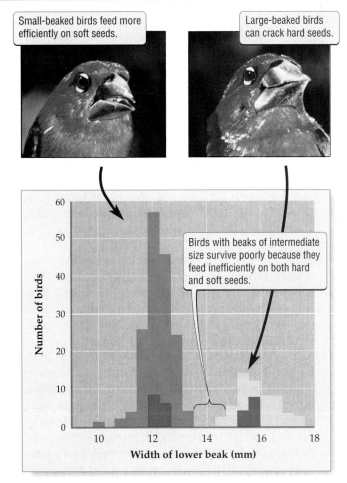

Small-beaked birds feed more efficiently on soft seeds.

Large-beaked birds can crack hard seeds.

Birds with beaks of intermediate size survive poorly because they feed inefficiently on both hard and soft seeds.

Figure 20.10   Disruptive Selection for Beak Size
In African seed crackers, large-beaked birds are the most efficient feeders on the hard seeds of one common food plant, while small-beaked birds are the most efficient feeders on the soft seeds of another plant. Such differences in feeding efficiencies may cause similar differences in survival. For example, among a group of young birds hatched in one year, only those with a small or large beak size survived the dry season, when seeds were scarce; all the birds with intermediate beak sizes died. Thus, natural selection favored both large-beaked and small-beaked birds over birds with intermediate beak sizes. Red bars indicate the beak sizes of young birds that survived the dry season, while yellow and blue bars indicate the beak sizes of young birds that died.

■ Natural selection is the only evolutionary mechanism that consistently favors alleles that improve the reproductive success of the organism in its environment. In each of the three types of natural selection (directional, stabilizing, and disruptive selection), individuals that possess certain forms of a heritable character tend to survive better or produce more offspring than individuals that possess other forms of that character.

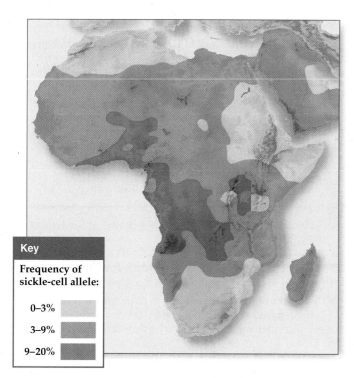

Figure 20.11 Frequency of the Sickle-Cell Allele
The sickle-cell allele is most common in regions where malaria is common.

■ HIGHLIGHT

20.6 *The Crisis in Antibiotic Resistance*

Thirty-five years ago, the U.S. surgeon general confidently testified to Congress that it was time to "close the book on infectious diseases." He was sadly mistaken. Today, one or more species of bacteria are resistant to every antibiotic now in use. Similarly, most viruses, fungi, and parasites that cause disease have evolved resistance to clinical treatments. Thus, HIV (see the box on p. 341) is not the exception, but rather the rule: Most pathogens have evolved resistance to our best efforts to destroy them, a compelling illustration of directional selection at work.

During the past 65 years, the bacteria that cause rheumatic fever, staph infections, pneumonia, strep throat, tuberculosis, typhoid fever, dysentery, gonorrhea, and meningitis all have evolved resistance to multiple antibiotics. This widespread resistance of bacteria to antibiotics has serious implications. Some biologists worry that we may enter a "postantibiotic" era, in which bacterial diseases ravage human populations on a scale not seen since the widespread introduction of antibiotics in the late 1930s. Before antibiotics were in common use,

even a seemingly mild infection, such as a boil on the face, was potentially lethal.

We have not entered a postantibiotic era yet, but there are early signs of a return to just such a nightmarish situation. For example, in 1941, pneumonia could be cured in several days if patients took 40,000 units of penicillin per day; today, however, despite excellent medical care and the administration of 24 million units of penicillin per day, a patient could die from complications of this disease.

Resistance to antibiotics evolves rapidly in bacteria partly because bacteria can transfer resistance genes within and among species with ease (see Figure 14.1). The rapid movement of resistance genes from one species to another is especially troubling: Genes for resistance that evolve in a relatively harmless species of bacteria can be transferred to a highly pathogenic species, creating the potential for a public health disaster.

To prevent this sort of doomsday scenario, what should we, as a society, do? Consider the costs and benefits of the following actions we could take:

- Devote greater resources to the study of the biology of pathogens, thereby improving our ability to design drugs or pest management strategies that attack weak points in their life cycles.
- Insist on more prudent use of antibiotics in human, plant, and animal health care. Medical doctors and agriculturists frequently use antibiotics inappropriately. For example, one U.S. study found that doctors prescribed antibiotics for 51 to 66 percent of adult patients with colds or bronchitis. Antibiotics do little for either of these conditions because they are usually caused by viruses, which are not affected by antibiotics. The indiscriminate use of antibiotics encourages the evolution of antibiotic resistance in the many species of bacteria that are normally found in our bodies. As we have seen, these resistant, nonpathogenic bacteria can then transfer genes for antibiotic resistance to other, harmful species of bacteria.
- Improve sanitation, thus decreasing the spread of resistant bacteria from one person to another. This action is of critical importance in hospitals, where the abundant use of antibiotics has led to the emergence of highly resistant strains of bacteria that can cause a variety of "hospital diseases," some of which can be lethal.

■ Bacteria evolve resistance to antibiotics rapidly. One or more species of bacteria are resistant to every antibiotic now in use. Genes for resistance that evolve in harmless bacteria can be transferred to highly pathogenic species.

SUMMARY

Key Definitions: Genotype and Allele Frequencies

- Genotype frequency is the proportion of a genotype in a population. Allele frequency is the proportion of an allele in a population.

Genetic Variation: The Raw Material of Evolution

- Individuals within populations differ in biochemical, physiological, morphological, and behavioral characters, many of which are under genetic control.
- Mutation and recombination result in genetic variation.
- Genetic variation provides the raw material on which evolution can work.

When Populations Do Not Evolve

- A population does not evolve when (1) mutation does not change allele frequencies, (2) mating is random, (3) the movement of individuals does not bring new alleles into the population, (4) the population is large, and (5) natural selection does not occur.
- When these five conditions hold, the Hardy–Weinberg equation can be used to calculate genotype frequencies in a population.

Mutation: The Source of Genetic Variation

- Mutation creates new alleles. Hence all evolutionary change depends ultimately on mutations.
- Mutations cause little direct change in allele frequencies over time.
- New mutations can stimulate the rapid evolution of populations by providing new genetic variation on which evolution can act.

Nonrandom Mating: Changing Genotype Frequencies

- Nonrandom mating does not alter allele frequencies directly, but it does alter genotype frequencies.
- Nonrandom mating leads to an increase in the frequency of homozygotes.

Gene Flow: Exchanging Alleles between Populations

- Gene flow makes the genetic composition of populations more similar.
- Gene flow can introduce new alleles into a population, providing new genetic variation on which evolution can work.

Genetic Drift: The Effects of Chance

- Genetic drift causes random changes in allele frequencies over time.
- Genetic drift can cause small populations to lose genetic variation.
- Genetic drift can cause the fixation of harmful, neutral, or beneficial alleles.
- In a genetic bottleneck, a drop in population size reduces genetic variation or causes the fixation of harmful alleles.

Natural Selection: The Effects of Advantageous Alleles

- In natural selection, individuals that possess certain forms of a heritable character tend to survive better or produce more offspring than individuals that possess other forms of that character.
- Natural selection is the only evolutionary mechanism that consistently favors alleles that improve the reproductive success of the organism in its environment.
- There are three forms of natural selection: directional, stabilizing, and disruptive selection.

Highlight: The Crisis in Antibiotic Resistance

- Bacteria evolve resistance to antibiotics rapidly. One or more species of bacteria are resistant to every antibiotic now in use.
- The indiscriminate use of antibiotics encourages the evolution of resistance in the many species of bacteria normally found in our bodies.
- Genes for resistance that evolve in harmless bacteria can be transferred to highly pathogenic species. Thus, it is essential that antibiotics be used only when needed.

KEY TERMS

allele frequency p. 336
directional selection p. 344
disruptive selection p. 346
fixation p. 343
gene flow p. 340
genetic bottleneck p. 343
genetic drift p. 342
genetic variation p. 337
genotype frequency p. 336
Hardy–Weinberg equation p. 339
heterozygote advantage p. 346
microevolution p. 336
mutation p. 337
natural selection p. 344
nonrandom mating p. 339
recombination p. 337
stabilizing selection p. 345

CHAPTER REVIEW

Self-Quiz

1. A population of birds has roughly 15 individuals. If allele frequencies were observed to change in a random way from year to year, which of the following would be the most likely cause of the observed changes in gene frequency?
   a. stabilizing selection
   b. disruptive selection
   c. genetic drift
   d. mutation

2. A study of a plant population finds that large individuals consistently survive at a higher rate than small individuals. Most likely, the evolutionary mechanism at work here is
   a. heterozygote advantage.
   b. directional selection.
   c. stabilizing selection.
   d. genetic drift.

3. Two large populations of a species that are found in neighboring locations with different environments are observed to become genetically more similar over time. Which evolutionary mechanism is the most likely cause of this trend?
   a. gene flow
   b. nonrandom mating
   c. natural selection
   d. genetic drift

4. Which of the following terms describes the situation in which individuals within a subset of a population are more likely to mate with each other than with individuals selected at random from the population at large?
   a. gene flow
   b. genetic drift
   c. nonrandom mating
   d. random mating

5. Assume that individuals of genotype *Aa* are intermediate in size and that they leave more offspring than either *AA* or *aa* individuals. This situation is an example of
   a. heterozygote advantage.
   b. disruptive selection.
   c. stabilizing selection.
   d. both a and c

6. A population of toads has 280 individuals of genotype *AA*, 80 individuals of genotype *Aa*, and 60 individuals of genotype *aa*. What is the frequency of the *a* allele?
   a. 0.24
   b. 0.33
   c. 0.14
   d. 0.07

Review Questions

1. Using your own words, define the following terms: nonrandom mating, gene flow, genetic drift, and natural selection.

2. To prevent a small population of a plant or animal species from going extinct, some individuals from a large population of the same species can be moved from the large to the small population. In terms of the evolutionary mechanisms discussed in this chapter, what are potential benefits and drawbacks of transferring individuals from one population to another? Do you think biologists and concerned citizens should take such actions?

3. Reconsider the toads in Question 6 of the Self-Quiz. How do the numbers of toads with genotypes *AA*, *Aa*, and *aa* compare to the numbers you would expect based on the Hardy–Weinberg equation? Discuss factors that could cause any differences you find.

The Daily Globe

Lessons from Mad Cow and Other Killers

To the Editor:

Have you ever wondered where new killer diseases come from? Recent headlines on Mad Cow disease illustrate the answer: That disease, and all of recent history's most lethal diseases, such as tuberculosis, smallpox, and bubonic plague, have spread from animal hosts to us. Among the many diseases that strike animals, those that kill people usually come from domesticated animals. We usually get our diseases from domesticated animals (as opposed to wild animals) because we are in close contact with the animals we keep as pets or for food, and hence, we are continually exposed to the germs they carry.

Because so many lethal human diseases originate as animal diseases, animal care practices are very important. For example, Mad Cow disease reached epidemic propor-

tions in cattle—and then spread to people—because people began to include parts of dead cattle in cattle feed, thus providing a new and highly efficient way for the disease agent to spread from one cow to another.

And it is not just what we feed domesticated animals that matters. Antibiotics are routinely given to chickens and other domesticated animals that are not sick. The purpose of this practice is to increase their rate of growth by keeping them healthy. However, giving antibiotics to animals that are not sick makes it much more likely that bacteria carried by these animals will build resistance to the antibiotics, which in turn makes it more likely that diseases that spread from domesticated animals to us will arrive in a highly resistant, and hence deadly, form.

Hundreds of scientific studies show that disease organisms rapidly evolve resistance to our best efforts to kill them. Because diseases jump from animals to humans on a regular basis, we risk disaster if we continue practices—such as the indiscriminate use of antibiotics—that favor the evolution of resistance in our domesticated animals. Overall, our understanding of how diseases evolve and spread from animals to people provides a simple lesson: We should follow the same rules when we care for our animals as when we care for ourselves. If we do otherwise, we run the risk that we will foster and facilitate the spread of yet another new and terrible disease.

Julia S. Owens
Dean, State University School of
Agriculture

Evaluating "The News"

1. How do you think agricultural policy should balance short-term economic gain (for example, the use of antibiotics to ensure rapid growth in domestic animals) against long-term public health (for example, discouraging the routine use of antibiotics in animals, thereby reducing the risk of antibiotic-

resistant bacteria infecting humans)?

2. If we chose to do so, how could our society stop the routine use of antibiotics as an agricultural practice?

3. If antibiotics were not used routinely, it is likely that domesticated animals such as chickens would grow

more slowly (because they got sick more often) and hence that farmers would show reduced profits. Farming is already a difficult business, with many farmers losing money each year. Should farmers bear the cost of reduced profits if they were no longer allowed to use antibiotics on a routine basis?

chapter *21* Adaptation and Speciation

Suzanne Stryk, *Tree of Life*, 1997.

Cichlid Mysteries

The surface of Earth changes slowly but dramatically over time. Islands rise from the sea, mountains are thrust up to divide once-continuous land masses, new lakes form and old ones disappear, and entire continents come together, then break apart. Such changes alter the environments in which species live, setting the stage for grand natural experiments in evolution.

No evolutionary experiment has been more wondrous than that of the cichlid fish of Lake Victoria, the largest of the Great Lakes of East Africa. Lake Victoria first formed about 750,000 years ago, but geologic evidence indicates that the lake may have been completely dry as recently as 12,000 years ago. Whenever a lake forms or refills with water, it may be colonized by one or more species of fish, some of which then evolve to form new

■ MAIN MESSAGE

Adaptive evolution causes species to adjust to environmental change and helps to generate the great diversity of life.

species unique to the new lake. Such a sequence of events has happened many times in lakes around the world, but nowhere more spectacularly than in Lake Victoria.

A Lake Victoria Cichlid

Until recently, Lake Victoria harbored over 500 species of cichlids, a greater number of fish species than found in all the lakes and rivers of Europe combined. The cichlids are a diverse and colorful group of fish, well known to the aquarium trade. Genetic evidence suggests that virtually all of the cichlids of Lake Victoria evolved in that lake, which brings us to the first of our "cichlid mysteries": How did so many species form in the lake, perhaps in as little time as 12,000 years?

The mystery deepens when we realize that many of the Lake Victoria cichlid species differ considerably from one another in color, jaw structure, and feeding specialization, yet genetically all the species in the lake are extremely closely related to one another. What evolutionary forces have caused some of these closely related species to differ so much? Furthermore, in the past 30 years, roughly 200 Lake Victoria cichlid species have disappeared. Some, we know, were driven to extinction by an introduced predatory fish species, the Nile perch. But many species rarely eaten by Nile perch also have vanished. What drove these species to extinction? We'll return to these cichlid mysteries at the close of this chapter.

■ KEY CONCEPTS

1. Adaptations are features of organisms that improve the performance of those organisms in their environments. Adaptations result from natural selection.

2. The process by which natural selection improves the match between an organism and its environment over time is called adaptive evolution. Adaptive evolution causes organisms to adjust to environmental change, sometimes over short periods of time (months to years).

3. A species is a group of interbreeding natural populations that is reproductively isolated from other such groups.

4. Speciation, the process by which one species splits to form two or more species, is usually a by-product of genetic differences between populations that are caused by other factors (for example, natural selection or genetic drift).

5. Speciation often occurs when populations of a species become geographically isolated. Such isolation limits gene flow between the populations, which makes the evolution of reproductive isolation more likely.

The three great themes of evolutionary biology are adaptation, the diversity of life, and the shared characteristics of life (see Chapter 19). In this chapter we return to two of these themes, adaptation and biodiversity. We will examine the characteristics of adaptations and discuss how they are shaped by natural selection. Then we will focus on speciation, the process that generates the diversity of life.

21.1 Adaptation: Adjusting to Environmental Challenges

An **adaptation** is a feature of an organism that improves the performance (that is, the reproductive success) of the organism in its environment. Many adaptations result in what appears to be a remarkably well-designed match between the organism and its environment. As we saw in Chapter 19, however, adaptations do not result from any intentional "design." Instead, they result from natural selection: Individuals with heritable characters that allow them to survive and reproduce better than other individuals replace those with less favorable characters (see Chapter 20). This process, which improves the match between organisms and their environment over time, is called **adaptive evolution**.

There are many different types of adaptations

Weaver ants construct nests of living leaves by the concerted actions of many individual ants, some of which draw the edges of leaves together while others weave them in place by moving silk-spinning larvae (immature ants) back and forth over the seam of the two leaves. These actions are not the result of conscious planning on the part of the ants; rather, they illustrate how a simple evolutionary mechanism—natural selection—can produce a complex behavioral adaptation (cooperative nest building).

Weaver ants

In another example of an adaptation, the caterpillars (the larval or immature stage) of a certain moth species differ in shape depending on whether they feed on the flowers or the leaves of their food plant, the oak tree. Caterpillars that feed on flowers resemble oak flowers; those that feed on leaves resemble oak twigs (Figure 21.1). The larvae develop so as to match whatever background they feed on, making them more difficult for predators to locate.

Natural selection has also shaped some astonishing adaptations that facilitate reproduction. The flowers of some orchid species, for example, use chemical attractants and appearance to mimic female wasps, thereby attracting male wasps and fooling them into attempting to mate with the flowers. In the course of these attempts, the insects become coated with pollen, which they then transfer from one plant to another.

All adaptations share certain key characteristics

Although there are literally millions of examples of adaptations, the few examples given in the previous section illustrate the most important characteristics of adaptations:

1. Adaptations have the appearance of having been designed to match the organism to its environment.

2. Adaptations are often complex, as exemplified by the nest building behavior of weaver ants.

(a)

(b)

Figure 21.1   Caterpillars that Match Their Environments
Caterpillars of the moth *Nemoria arizonaria* differ in shape depending on their diet. (*a*) Caterpillars that hatch in the spring resemble the oak flowers on which they feed. (*b*) Caterpillars that hatch in the summer eat leaves and resemble oak twigs. Experiments have demonstrated that chemicals in the leaves control the switch that determines whether caterpillars will mimic flowers or twigs.

3. Adaptations help the organism accomplish important functions, such as feeding, regulation of body chemistry, defense against predators, and reproduction.

Populations can adjust rapidly to environmental change

Male guppies in the mountain streams of Trinidad and Venezuela have bright and variable colors that serve to attract females. But the bright colors that help the males succeed in attracting mates also make them easier for visually hunting predators to find. How do guppy populations evolve in response to such conflicting pressures?

Field observations show that guppies from streams where few predators lurk have bright colors, but guppies from streams with more predators are drab in comparison. This match between organism and environment evolves very rapidly: When guppies are experimentally transferred to a different environment (from an area with few predators to an area with many predators, or vice versa), their color patterns evolve to match the new conditions within 10 to 15 generations (14 to 23 months).

The ability to evolve rapidly in response to changing environmental conditions is not limited to guppies. For example, in soapberry bug populations, beak length evolves rapidly to match the size of the fruits upon which these insects feed (Figure 21.2). Similarly, beak sizes in the medium ground finch evolve from year to year to match the size of the seeds on which they depend for food (see Figure 19.8). In addition, as we saw in Chapter 20, viruses, bacteria, and insects can evolve resistance to our best efforts to kill them in only a few months or years. Collectively, these examples illustrate an important point: Evolution by natural selection can improve the adaptations of organisms in short periods of time.

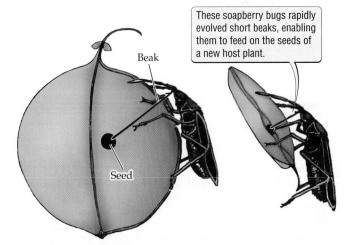

These soapberry bugs rapidly evolved short beaks, enabling them to feed on the seeds of a new host plant.

Beak

Seed

Figure 21.2   Rapid Evolution in an Insect
In Florida, soapberry bugs traditionally fed on seeds within the large, round fruit of a native species of vine, the balloon vine (shown at left). Over the past 30–50 years, some populations of soapberry bugs have evolved short beak lengths, enabling them to feed on seeds within the fruit of an introduced tree species, the golden rain tree (shown at right).

■ Adaptations result in an apparent match between an organism and its environment. They are often complex, and they help organisms accomplish important functions such as food capture, reproduction, and defense against predators. Natural selection can improve adaptations in short periods of time (months to years).

Adaptation Does Not Craft Perfect Organisms

As impressive as the adaptations we see in nature may be, no organism matches its environment perfectly. In many cases, genetic or developmental limitations or ecological trade-offs prevent further improvements in an organism's adaptation. In this section we look at these barriers to perfection.

Genetic limitations

For the quality of an adaptation to increase over time, there must be genetic variation for traits that can enhance the match between the organism and its environment. In some cases, the absence of such genetic variation places a direct limit on the ability of natural selection to cause adaptive evolution. For example, the mosquito *Culex pipiens* is now resistant to organophosphate pesticides, but this resistance is based on a single mutation that occurred in the 1960s (see Chapter 20). Before this mutation occurred, adaptation to these pesticides was not possible, and billions of mosquitoes were killed because their populations lacked genetic variation for resistance to the pesticides.

Developmental limitations

As we saw in Chapter 16, changes in genes that control development can have dramatic effects on the phenotype. Such a change in the developmental program of an organism often influences more than one part of the phenotype. Thus, changes to genes that control development can have many effects, some of which may be advantageous and others of which may harm or kill the organism.

The multiple effects of developmental genes can limit the ability of the organism to evolve in certain directions, which in turn limits what can be achieved by adaptive evolution. For example, the larval stages of some insects, such as beetles and moths, never have wings or well-developed eyes, two important adaptations that the adult forms of these insects have (Figure 21.3). Beetle and moth larvae have a wide range of lifestyles, so wings or well-developed eyes probably would benefit many of these larvae. Since the adult and larval forms of these insects have exactly the same genetic instructions, yet only the adults have wings or eyes, it is likely that turning on or expressing the genes that control the production of wings or eyes would have other, extremely harmful effects in larvae. Thus, the lack of wings and eyes in beetle and moth larvae probably results from developmental limitations.

Figure 21.3 An Apparent Developmental Limitation
The larval form of this moth (*a*) does not have wings or eyes, two important adaptations that the adult form (*b*) has.

(*a*)

Wings or eyes might benefit some moth larvae.

Since larvae and adults have the same genetic instructions, developmental limitations probably prevent the larvae from having wings or eyes.

(*b*)

Ecological trade-offs

To survive and reproduce, organisms must perform many functions, such as finding food and mates, avoiding predators, and surviving the challenges posed by the physical environment. Within the realm of what is genetically and developmentally possible, natural selection increases the overall ability of the organism to survive and reproduce. However, the many and often conflicting demands that organisms face causes trade-offs or compromises in their ability to perform important functions.

High levels of reproduction, for example, often are associated with decreased longevity. This apparent trade-off may be due to relatively subtle costs of reproduction: Resources directed toward reproduction are not available for other uses, such as storing energy to help the organism survive a cold winter. In red deer, for example, females that reproduced the previous spring have a higher rate of death during winter than do females that did not reproduce. But costs associated with reproduction can sometimes be immediate and very dramatic, as illustrated by the mating calls of the túngara frog (Figure 21.4). In general, the widespread existence of trade-offs between reproduction and other important functions ensures that organisms are not perfect, for the simple reason that it is not possible to be the best at all things at once.

■ Adaptation does not craft perfect organisms. Adaptive evolution can be limited by genetic constraints, developmental constraints, and ecological trade-offs.

What Are Species?

Before we discuss the processes that have generated the great diversity of species on Earth, we must first define what a species is. In a practical sense, species are usually defined in terms of morphology; that is, two groups of organisms are classified as members of different species if they look sufficiently different. All of us use such a definition: for example, we distinguish bald eagles from other birds by how they look (Figure 21.5) Most species can be identified by morphological characteristics. However, a morphological definition of species does not tell us what makes one species different from another, nor does it always work well. What holds a species together and causes it to be phenotypically different from other species?

How we answer this question can be important, for the definition of a species that we use can have practical effects. For example, the definition we choose can determine whether or not a particular organism is classified (in a legal sense) as a rare or endangered species. For the organism, the legal protection provided by such a classification can make the difference between recovery and extinction. For people, such a classification might limit some forms of economic development (such as construction projects on land used by the rare species) and stimulate others (such as revenues from tourists that visit an area to see the rare species).

Figure 21.4 Does Love or Death Await?
Male túngara frogs face an ecological trade-off: The same calls that are most successful at attracting females also make it easy for predatory bats to locate calling males.

(a)

(b)

Figure 21.5   Members of a Species Look Alike
Bald eagles that live in Alaska (a) look the same as bald
eagles that live in Colorado (b). Although these birds live far
apart, they remain phenotypically similar.

Species are reproductively isolated from each other

In most cases, members of different species cannot reproduce with each other. When barriers to reproduction
exist between species, the species are said to be **reproductively isolated** from each other. Barriers to reproduction can act before (prezygotic) or after (postzygot-
ic) the formation of a zygote (Table 21.1). While there are
a wide range of barriers to reproduction, they all have

the same net effect: few or no genes are exchanged
between species. This restriction ensures that the members of a species share a common and exclusive set of
genes and alleles. Because members of a species share
an exclusive set of genes and alleles, they remain phenotypically similar to one another but different from
members of other species.

Species, then, can be defined in terms of reproductive
isolation: A **species** is a group of interbreeding natural
populations that is reproductively isolated from other
such groups. The phrase "interbreeding natural populations" is meant to include populations that could interbreed if they were in contact with one another, but do

21.1 *Barriers that Can Reproductively Isolate Two Species in the Same Geographic Region*

Type of barrier	Example	Effect
Prezygotic barriers		
Ecological isolation	The two species breed in different portions of their habitat, at different seasons, or at different times of the day	Mating is prevented
Behavioral isolation	The two species respond poorly to each other's courtship displays	Mating is prevented
Gametic isolation	The gametes of the two species are incompatible	Fertilization is prevented
Postzygotic barriers		
Zygote death	Zygotes fail to develop properly, and die before birth	No offspring are produced
Hybrid performance	Hybrids survive poorly or reproduce poorly	Hybrids are not successful

not because they have no opportunity to do so (for example, because they are too far away from one another).

A definition of species based on reproductive isolation has important limitations. For example, it is of no use when defining fossil species, since no information can be obtained about whether or not two fossil forms were reproductively isolated from each other. Instead, fossil species are distinguished on the basis of morphology. Nor does such a definition apply to organisms, such as bacteria and dandelions, that reproduce mainly by asexual means.

Our definition of species also fails to work well for the many plant and animal species that mate in nature to produce fertile offspring. Species that interbreed in nature are said to **hybridize**, and their offspring are called **hybrids**. Although they can reproduce with each other, species that hybridize often look different from

each other (Figure 21.6) or are distinct ecologically (for example, they are usually found in different environments, or they differ in how they perform important biological functions, such as obtaining food).

Despite the limitations of such a definition, most biologists define species on the basis of reproductive isolation, and this is the definition we will use here. Many alternative definitions of species exist, but there are problems with those definitions as well. It is perhaps best to think of our definition of a species as a simple conceptual definition, but to recognize that the reality of a species in nature can be considerably more complicated.

■ A species is a group of interbreeding natural populations that is reproductively isolated from other such groups.

Figure 21.6   Some Species Interbreed yet Remain Different
(*a*) The Gambels oak and Gray oak can mate to produce fertile hybrids. However, the species remain phenotypically different, as is evident from their leaves. (*b*) Hybrid individuals, with varying degrees of fall color, are found throughout the hillside shown in this photograph. The two oak species are found nearby, one at lower elevations (Gray oak), the other at higher elevations (Gambels oak).

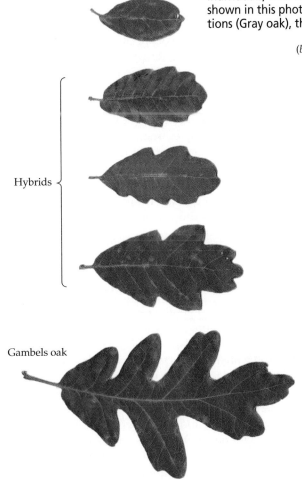

(*a*) Gray oak

Hybrids

Gambels oak

(*b*)

21.2 Speciation: Generating Biodiversity

The tremendous diversity of life on Earth is caused by **speciation**, the process in which one species splits to form two or more species that are reproductively isolated from one another. The study of speciation is fundamental to understanding the diversity of life on Earth.

How do new species form? The crucial event in the formation of new species is the evolution of reproductive isolation, which requires that populations diverge from one another, at least with respect to their ability to reproduce. But populations within a species are connected by gene flow, which tends to keep them similar to one another (see Chapter 20). How does reproductive isolation develop within a species, whose members interbreed and therefore share a common set of genes and alleles? Throughout this section, we discuss how reproductive isolation originates, despite ongoing gene flow between populations.

Speciation can be explained by the same mechanisms that cause the evolution of populations

Speciation is usually considered a secondary consequence of the evolution of populations. In essence, populations evolve genetic differences from one another—for whatever reason—and some of these genetic differences have the accidental by-product of causing partial or total reproductive isolation.

As we saw in Chapter 20, natural selection can cause populations to diverge genetically, as when populations located in different environments face different selection pressures. Over time, these different selection pressures can cause the populations to differ genetically from one another. But the divergence of populations does not have to be due to natural selection. Populations can also diverge as a result of mutation, nonrandom mating, and genetic drift (to review these three terms, see Chapter 20). In contrast, gene flow always operates to *prevent* the genetic divergence of populations. Thus, for populations to accumulate genetic differences, the factors that promote divergence must have a greater effect than does the amount of ongoing gene flow.

Speciation often results from geographic isolation

Many species are formed when populations of a single species become separated, or **geographically isolated**, from one another. This process can begin, for example, when a newly formed geographic barrier, such as a river

or a mountain chain, isolates two populations of a single species. Alternatively, geographic isolation can occur when a few members of a species colonize a region that is difficult to reach, such as an island located far outside the usual geographic range of the species.

The distance required for geographic isolation varies tremendously with the species in question. Populations of squirrels and other rodents that live on opposite sides of the Grand Canyon have diverged considerably, whereas populations of birds—whose members can cross the canyon relatively easily—have not. In general, geographic isolation is said to occur whenever populations are separated by a distance that is great enough to limit gene flow.

However they arise, geographically isolated populations are connected by little or no gene flow. For this reason, mutation, nonrandom mating, genetic drift, and natural selection can more easily cause the populations to diverge genetically from one another. If the populations remain isolated for long enough periods of time, they can evolve into new species (Figure 21.7). The formation of new species from populations that are geographically isolated from one another is called **allopatric speciation**.

Much evidence indicates that geographic isolation can lead to speciation. For example, in many groups of organisms, the number of species is greatest in regions where strong geographic barriers increase the potential for geographic isolation. Examples include species that live in mountainous regions, on island chains, or in a series of small, isolated lakes. A second line of evidence is found in some cases when individuals from populations that live at the extreme ends of a species' geographic range cannot reproduce with each other, even though they can reproduce with individuals from intermediate portions of the species' range. Finally, when populations are separated from one another in the laboratory, they often show considerable reproductive isolation from one another.

Speciation can occur without geographic isolation

There is a greater potential for gene flow between populations whose geographic ranges overlap or are adjacent to one another than between populations that are geographically isolated from one another. Thus it can be difficult for speciation to occur in the absence of geographic isolation. Nevertheless, it has long been known that plants can form new species in the absence of geographic isolation, and recent work has provided convincing evidence that animals can as well. The formation of new species in the absence of geographic isolation is called **sympatric speciation**.

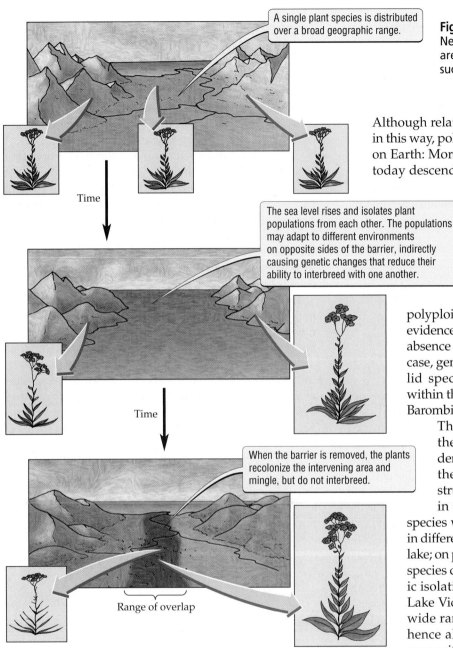

A single plant species is distributed over a broad geographic range.

Time

The sea level rises and isolates plant populations from each other. The populations may adapt to different environments on opposite sides of the barrier, indirectly causing genetic changes that reduce their ability to interbreed with one another.

Time

When the barrier is removed, the plants recolonize the intervening area and mingle, but do not interbreed.

Range of overlap

Figure 21.7 Allopatric Speciation
New species can form when populations are separated by a geographic barrier, such as a rising sea.

Although relatively few species originate directly in this way, polyploidy has had a large effect on life on Earth: More than half of all plant species alive today descended from species that originated by polyploidy. A few animal species also appear to have originated by polyploidy, including several species of lizards and fish, and one mammal (an Argentine rat).

Evidence is mounting that sympatric speciation can occur in animals by means other than polyploidy. For example, there is compelling evidence that new fish species can form in the absence of geographic isolation. In one such case, genetic data indicate that 9 and 11 cichlid species have originated, respectively, within the confines of Lake Bermin and Lake Barombi Mbo, two small lakes in West Africa.

The formation of new species within these lakes provides very strong evidence for sympatric speciation because these lakes are small and simple in structure. As a result, populations within these lakes cannot evolve into new species while living apart from one another in different environmental subportions of the lake; on page 362, we discuss how new cichlid species can form in the absence of geographic isolation. In contrast, large lakes such as Lake Victoria (see p. 352) usually provide a wide range of different environments, and hence allow geographic isolation to occur even within the confines of a single lake.

Strong evidence in support of sympatric speciation has also been found in other animals, such as Pacific salmon, rough periwinkle snails, and a variety of insect species. For example, researchers think that North American populations of the apple maggot fly, *Rhagoletis pomonella*, are in the process of diverging into new species, even though their geographic ranges overlap. Historically, *Rhagoletis* usually ate native hawthorn fruits, but in the mid-nineteenth century these flies were first recorded as pests on apples, an introduced nonnative species. *Rhagoletis* populations that feed on apples are

Rhagoletis pomonella

In plants, rapid chromosomal changes can cause sympatric speciation. New plant species can form in a single generation as a result of **polyploidy**, a condition in which an individual has more than two sets of chromosomes, usually due to the failure of chromosomes to separate during meiosis (see Chapter 10). Polyploidy also can occur when a hybrid (formed by a mating between individuals of two different species) spontaneously doubles its chromosome number. A doubling of the chromosomes leads to reproductive isolation because the chromosome numbers in the gametes of the new polyploid no longer match those of either of its parents.

now genetically distinct from populations that feed on hawthorns; members of these populations also mate at different times and usually lay their eggs only on the fruit of their particular food plant. As a result, there is little gene flow between fly populations that feed on apples and fly populations that feed on hawthorns. In addition, researchers have identified alleles that benefit flies that feed on one host plant, but are detrimental to flies that feed on the other host plant. Thus, natural selection operating on these alleles acts to limit whatever gene flow does occur. Over time, the ongoing research on *Rhagoletis* may well provide a dramatic case history of sympatric speciation.

> ■ Speciation is usually a by-product of the evolutionary divergence of populations. Speciation often occurs in geographic isolation, but new species also can form in the absence of geographic isolation.

Rates of Speciation

When speciation is caused by polyploidy or other types of rapid chromosomal change, new species form in a single generation. New species also appear to have formed with extraordinary speed in the case of some cichlid fish. Genetic analyses indicate that the 500 species of cichlids in Lake Victoria, the large East African lake discussed in the introduction to this chapter (see p. 352), descended from just two ancestor species. Recent estimates for the length of time during which the lake has remained filled with water range from 200,000 years to a mere 12,000 years. Thus, in perhaps as little time as 12,000 years, 500 fish species evolved from two species.

In many—perhaps most—cases, speciation occurs more slowly. Among fish, the time required for speciation has been estimated to range from 1500 years (in Lake Victoria cichlids) to over 9 million years (in characins, a group that includes carp and piranha). In other groups of organisms, including fruit flies, snapping shrimp, and birds, the time required for speciation has been estimated to range from 600,000 to 3 million years. Furthermore, some populations can be geographically isolated for long periods of time without evolving reproductive isolation. North American and European sycamore trees, for example, have been separated for more than 20 million years, yet the two populations remain morphologically similar and can breed with each other.

> ■ Speciation occurs rapidly in some cases, but requires hundreds of thousands to millions of years in other cases.

Implications of Adaptation and Speciation

Adaptation and speciation are, respectively, the means by which organisms adjust to the challenges posed by new or changing environments and the means by which the diversity of life has come into being. Both, therefore, are critical to understanding how evolution works.

Adaptation and speciation are also very important from an applied perspective. For example, to combat rapidly evolving pathogens—such as HIV, the virus that causes AIDS—we must have a detailed understanding of the new adaptations that enable them to overcome our best efforts to kill them (see Chapter 20). Speciation has long been of practical importance to humans, as our development of domesticated crop and animal species readily attests (see the box on p. 363).

In addition, understanding the often relatively slow pace of speciation gives us a strong incentive to stop the ongoing extinctions of species (see Chapters 4 and 45). Speciation can require hundreds of thousands to millions of years, yet humans are driving species extinct in decades to hundreds of years. If we continue to drive species extinct at the present rate, it will take millions of years before the speciation process can replace the large number of species that are currently being lost.

> ■ Adaptation is the means by which species adjust to challenges posed by new or changing environments. Speciation is the means by which the diversity of life has come into being. Adaptation and speciation also influence such practical matters as how we fight disease and develop new domesticated species.

■ HIGHLIGHT

Rapid Speciation in Lake Victoria Cichlids

As we saw earlier in this chapter, the Lake Victoria cichlids formed new species at a very rapid rate, resulting in an evolutionary expansion unmatched by any other group of vertebrates. The rapid speciation of these fish brings us back to one of the "cichlid mysteries" discussed in the introduction to this chapter: How did so many species form in so short a time? The answer hinges on two key aspects of cichlid biology.

First, cichlids in Lake Victoria use color as a basis for mate choice: Females prefer to mate with males of a particular color. Within Lake Victoria, there are many pairs of ecologically similar cichlid species that differ from each other in color, but little else. In each of these pairs

■ BIOLOGY IN OUR LIVES

The Origin of Dogs, Corn, and Cows

Humans can—and do—direct the course of evolution. For thousands of years we have domesticated wild species and controlled their breeding, thus molding their evolution to suit our own ends. Wheat, rice, corn, cows, chickens, and dogs all were derived from wild species. Among animal species, the evidence suggests that cattle were domesticated three times, all from the now-extinct wild ox, while chickens were domesticated once, from a southern Asian jungle fowl. Dogs also were domesticated only a few times, from gray wolves. Thus, the remarkable diversity of dogs represents the effects of artificial selection on a small number of lineages of domesticated wolves.

Among plant species, wheat and corn illustrate two different ways in which crop species have evolved. Wheat evolved from two polyploid speciation events, the first leading to the production of emmer wheat (*Triticum turgidum*), the second to the formation of bread wheat (*T. aestivum*). Corn, on the other hand, evolved directly from a single plant species, teosinte. Teosinte and corn have very different forms. Whereas teosinte has long side branches, each tipped with male reproductive structures, corn has short side branches, each tipped with female reproductive structures (see Figure 1.10). As a result, corn produces more seeds and thus more food. Recent genetic studies indicate that many of the differences in the form of corn and teosinte are controlled by differences in the regulation of a single gene. Early Central American farmers probably spotted a teosinte plant with mutations that altered the regulation of this gene and, by selectively breeding this mutant, went on to guide the evolution of corn from teosinte.

Gray wolf
(Common ancestor)

Europe · North America · China · India

European toy dogs · Terriers · Mastiffs · Arctic spitz · Native American dogs · Feral dogs · Chow chow · Oriental toy dog · Sight hounds · Scent hounds

Pug · Bulldog · St. Bernard · Great Dane · Asian pariah dogs · Dingo · Afghan hound · Borzoi · Gun dogs · Spaniels

of species, the males of one species tend to be blue, while the males of the other species are red or yellow. Researchers have discovered that cichlids are most sensitive to blue and to red/yellow light. In normal light, females of a given species show a strong preference for mating with males of their own species. But when researchers experimentally changed the quality of the light so that the females could not see the difference between blue and red/yellow males, the female preference for mating with males of their own species broke down. This research suggests that female preferences for mating with males of a certain color may cause reproductive isolation between cichlid populations, thus setting the stage for the formation of new species.

In addition, Lake Victoria cichlids have unusual jaws that can be modified relatively easily over the course of evolution to specialize on new food items. This feature of their biology causes the cichlids in the lake to vary greatly in form and feeding behavior (Figure 21.8). How do their easily modified jaws relate to speciation? Once female mate choice causes two populations to begin to become reproductively isolated from each other, if those populations then specialize on different sources of food, it becomes increasingly likely that they will continue to diverge and form new species.

A final cichlid mystery remains: We saw that predation by the introduced Nile perch has driven many species extinct (see p. 352), but what has caused species not eaten by the Nile perch to vanish? Here, too, the answer may be related to female mate choice. As the experiments described above showed, female cichlids need to be able to see the colors of males to be able to distinguish males of their own species from males of closely related species. However, pollution due to human activities has caused the water of Lake Victoria to become cloudy. Because the water is cloudy, females cannot distinguish the color of potential mates, and hence reproductive barriers between species break down. As reproductive barriers lose their effectiveness, species that once were distinct can interbreed freely and become more similar to each other. If continued long enough, such interbreeding can "reverse" the speciation process and cause species to go extinct. Remember that cichlid speciation, too, depends on the ability of females to recognize differences in the color of males; when cloudy water impairs that ability, new species cannot form. Thus, pollution from human activities appears to have two profound effects: It halts the formation of new cichlid species while simultaneously causing existing species to go extinct. That pollution must be reduced if we are not to destroy an amazing evolutionary experiment, the cichlids of Lake Victoria.

Figure 21.8 Lake Victoria Cichlids
The four species shown here illustrate some of the differences in morphology and feeding behavior found in Lake Victoria cichlids.

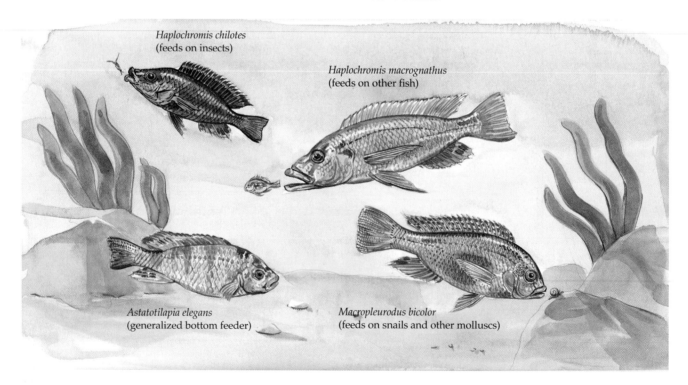

Haplochromis chilotes
(feeds on insects)

Haplochromis macrognathus
(feeds on other fish)

Astatotilapia elegans
(generalized bottom feeder)

Macropleurodus bicolor
(feeds on snails and other molluscs)

■ The rapid formation of cichlid species in Lake Victoria may have resulted from the interplay of two factors: (1) female preference for males of a certain color, which can promote reproductive isolation, and (2) the easily modified jaws of these fishes, which allow populations that are partially reproductively isolated to specialize on new sources of food.

SUMMARY

Adaptation: Adjusting to Environmental Challenges

■ Adaptations result in an apparent match between organisms and their environment.

■ Adaptations are often complex.

■ Adaptations help organisms accomplish important functions, such as mate attraction and predator avoidance.

■ Adaptations can be improved in short periods of time (months to years).

Adaptation Does Not Craft Perfect Organisms

■ Adaptive evolution can be limited by genetic constraints, developmental constraints, and ecological trade-offs.

What Are Species?

■ A species is a group of interbreeding natural populations that is reproductively isolated from other such groups.

■ This definition of species in terms of reproductive isolation has important limitations. It does not apply to fossil species, to organisms that reproduce mainly by asexual means, or to organisms that hybridize extensively in nature.

Speciation: Generating Biodiversity

■ The crucial event in the formation of a new species is the evolution of reproductive isolation.

■ Speciation usually occurs as an accidental by-product when natural selection, genetic drift, or mutation cause populations to diverge genetically from one another.

■ Speciation usually occurs when populations are geographically isolated from one another.

■ Speciation can occur without geographic isolation.

Rates of Speciation

■ Speciation occurs rapidly in some cases, but it requires hundreds of thousands to millions of years in other cases.

Implications of Adaptation and Speciation

■ Adaptations are the means by which organisms adjust to challenges posed by new or changing environments.

■ Speciation is the means by which the diversity of life has come into being.

■ Adaptation and speciation influence such practical matters as how we fight diseases and develop new crop species.

Highlight: Rapid Speciation in Lake Victoria Cichlids

■ Female cichlids prefer to mate with males of a certain color, and this preference can promote reproductive isolation between cichlid populations.

■ Because the jaws of Lake Victoria cichlids can be modified easily, populations that are partially reproductively isolated, and hence represent potential new species, can specialize on new sources of food.

■ These two features of cichlid biology, female mate choice and the ease with which the jaws can be modified, may have driven the rapid formation of cichlid species in Lake Victoria.

KEY TERMS

adaptation p. 354	polyploidy p. 361
adaptive evolution p. 354	reproductive isolation p. 358
allopatric speciation p. 360	speciation p. 360
geographic isolation p. 360	species p. 358
hybrid p. 359	sympatric speciation p. 360
hybridize p. 359	

CHAPTER REVIEW

Self-Quiz

1. Species whose geographic ranges overlap but that do not interbreed in nature are said to be
   a. geographically isolated.
   b. reproductively isolated.
   c. influenced by genetic drift.
   d. hybrids.

2. Which of the following evolutionary mechanisms acts to slow down or prevent the evolution of reproductive isolation?
   a. natural selection
   b. gene flow
   c. mutation
   d. genetic drift

3. Speciation usually occurs
   a. when populations are not geographically isolated.
   b. by genetic drift.
   c. when populations are geographically isolated.
   d. suddenly.

4. The time required for populations to diverge to form new species
   a. varies from a single generation to millions of years.
   b. is always greater in plants than in animals.
   c. is never less than 100,000 years.
   d. is rarely more than 1000 years.

5. Adaptations
   a. have the appearance of a close match between the organism and its environment.
   b. are often complex.
   c. help the organism accomplish important functions.
   d. all of the above

Review Questions

1. Should species that look different and are ecologically distinct, such as the oaks in Figure 21.6, be classified as one species or two? These oak species hybridize in nature. Should species that hybridize in nature be called one species or two?

2. Hundreds of new species of cichlids evolved within the confines of Lake Victoria, but some of these species live in different habitats within the lake and rarely encounter one another. For such species, would you consider them to have evolved with or without geographic isolation?

3. Imagine that a species that is legally classified as rare and endangered is discovered to hybridize with a more common species. Since the two species interbreed in nature, should they be considered a single species? Since one of the two species is common, should the rare species no longer be legally classified as rare and endangered?

4. Discuss the implications of adaptation for human medical practice.

21   𝕿𝖍𝖊 𝕯𝖆𝖎𝖑𝖞 𝕲𝖑𝖔𝖇𝖊

Protected Plant Causes Development Project to Grind to a Halt

RED MOUNTAIN, NM—Two months ago, things were looking good for this small town. Construction was about to begin on a new tourist resort, the town's first. The construction jobs would have provided badly needed work, and once built, the resort was expected to attract tourist dollars to this scenic, but economically depressed, area.

Those hopes were dashed when environmentalists announced that a small population of indigo trumpet flowers had been found on the site of the future resort. This small wildflower is protected as a rare and endangered species by the fed-eral Endangered Species Act (ESA). Private landowners are pro-hibited from harming species pro-tected by the ESA. Since the indigo trumpet flower probably would have been harmed by development at the resort site, as environmental-ist Steven Cooper, of the non-governmental agency Nature De-fense, says, "Either the resort or the plant had to go. We're glad that we've saved the plant."

Cooper believes the plant evolved at the site very recently from a common species of trumpet flower, the crested trumpet flower. As he explains, "We've surveyed that area several times in the past 15 years. We found the indigo trumpet flower for the first time a few years ago, and it is not known to live any-where else. So it seems to have evolved there very recently. Know-ing exactly when and where the species originated makes it unique."

Cooper's enthusiasm for the rare plant is not shared by many of the locals, however. As one woman, who preferred to remain anony-mous, said, "I only wish we had known the plants were there. We could have dug all of them up and saved our town. What right do these outsiders have to ruin our future?"

Evaluating "The News"

1. Should we protect rare species? Why or why not?

2. The indigo trumpet flower is thought to have evolved from a common species of trumpet flower, the crested trumpet flower. The two species are very similar to each other. Since the indigo trumpet flower is not all that different from another plant species, is it really worth the effort to save it?

3. Should an endangered species be valued more than the economic well-being of a town? Why or why not?

4. Are environmental and develop-ment interests necessarily in con-flict? Could the town of Red Moun-tain turn the discovery of the indigo trumpet flower into an economic benefit? How might the town be able to protect the plant, yet still build the resort?

chapter 22 The Evolutionary History of Life

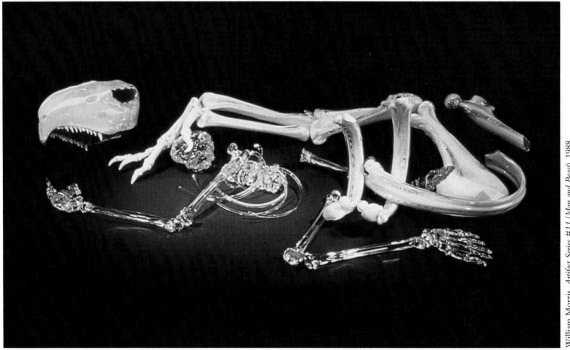

William Morris, *Artifact Series #11 (Man and Beast)*, 1988.

Puzzling Fossils in a Frozen Wasteland

A ntarctica is a crystal desert, a land in which heat and liquid water are very scarce. Few organisms can survive the extreme cold and lack of available water, and most of those that can are small and live near the sea. The entire continent has only two species of flowering plants (bundle grass and Antarctic pink), and its largest terrestrial animal is a fly 2 millimeters long.

In the interior of the continent, the organisms are even smaller: In most places the only living things are microscopic bacteria, algae, and protists, including some that survive in a state of suspended animation (frozen but alive in the ice). Some of the interior valleys are mostly free of ice, and hence seem a little less forbidding. But these valleys are so dry and cold that they support no visible life. There, the only organisms found on land are photosynthetic bacteria and lichens that spend their entire lives in a narrow zone just under the translucent surface of certain types of rocks.

■ MAIN MESSAGE

Over long periods of time there have been major changes in the kinds of organisms that have dominated life on Earth.

Despite the nearly lifeless appearance of this continent, fossils reveal that Antarctica used to differ greatly from today's frozen landscape. Where life now maintains an uncertain foothold, the land once was bordered by tropical reefs and later was covered with forests. At different times, ferns, freshwater fish, large amphibians, aquatic beetles, and trees as tall as 22 meters thrived in what is now the harshest of environments. Dinosaurs once roamed these lands, and millions of years later mammals and reptiles were pursued by the terror-bird, a fast, flightless bird that stood 3.5 meters tall.

Early explorers and scientists were amazed when they discovered these fossils of ancient life forms in Antarctica. The fossils showed that life used to be rich and abundant where now it barely exists. And so these scientists and explorers were left with a simple question: What happened?

Antarctic Landscape, with Photo Inset of One of Just Two Species of Flowering Plant Found on the Continent

■ KEY CONCEPTS

1. The fossil record documents the history of life on Earth and provides clear evidence of evolution.

2. Early photosynthetic organisms released oxygen to the atmosphere as a waste product, thereby setting the stage for the evolution of the first eukaryotes, followed later by multicellular organisms.

3. An astonishing increase in animal diversity occurred 530 million years ago, when large forms of most of the major living animal phyla appeared suddenly in the fossil record.

4. The colonization of land by the first plants (descendants of green algae) and animals (millipedes and spiders) marked the beginning of another major increase in the diversity of life.

5. The history of life can be summarized by the rise and fall of major groups of protists, plants, and animals. This history has been greatly influenced by continental drift, mass extinctions, and evolutionary radiations.

Earth abounds with life. About 1.5 million species have been described, millions of species have been collected and await formal description, and millions more await discovery (estimates for the total number of species on Earth range from 3 to 30 million). Even so, the species alive today represent far less than 1 percent of all the species that have ever lived.

In previous chapters of this unit we have discussed how biodiversity arose, focusing on the mechanisms that drive the evolution of populations and lead to the formation of new species (speciation). The evolution of populations, referred to as microevolution, is the smallest scale at which evolution occurs. In this chapter, we broaden our scope to discuss large-scale evolutionary changes, or macroevolution.

Macroevolution refers to the rise and fall of major taxonomic groups of organisms—that is, groups above the species level (see Chapter 2). The study of macroevolution focuses on evolutionary expansions that bring new groups to prominence and large-scale extinctions that greatly alter the diversity of life on Earth. Macroevolution emphasizes the pattern of evolutionary change over time. As we saw in Chapter 19, an emphasis on pattern leads us to define evolution as history—specifically, the history of the formation and extinction of species over time. Let's look at how that history is documented in the fossil record.

The Fossil Record: Guide to the Past

Fossils are the preserved remains or impressions of past organisms (Figure 22.1). In many fossils, the body parts of dead organisms are replaced with rock; thus, the original structure is maintained, but with completely new material. Fossils are usually found in sedimentary rock (rock that consists of layers of hardened sediments), but

also can be formed in a few other situations. Insects, for example, have been found in amber, the fossilized sap of a tree (see Figure 22.1*e*), and many mammals, including mammoths and a 5000-year-old man (see the photograph on p. 19), have been found in melting glaciers.

The fossil record documents the history of life and is central to the study of evolution. Historically, fossils provided the first compelling evidence that past organisms were unlike living forms, that many forms had disappeared completely from Earth, and that life had evolved through time.

As mentioned above, fossils are often found in sedimentary rocks. The relative depth or distance from the surface of Earth at which fossils are found is referred to as their "order" in the fossil record; usually, older fossils are in deeper rocks. The order in which organisms appear in the fossil record agrees with our understanding of evolution based on other evidence, thus providing strong support for evolution. For example, analyses of the morphology, DNA sequences, and other characteristics of living organisms indicate that bony fish gave rise to amphibians, which later gave rise to reptiles, which still later gave rise to mammals. This is exactly the order in which fossils from these groups appear in the fossil record. The fossil record also provides excellent examples of the evolution of major new groups of organisms, such as the evolution of mammals from reptiles (see Chapter 23).

Although many fossils have been found, the fossil record is not complete. Most organisms decompose rapidly after death; hence, very few form fossils. Even if an organism is preserved initially as a fossil, a variety of common geologic processes (including erosion and extreme heat or pressure) can destroy the rock in which it is embedded. Finally, fossils can be difficult to find.

(a)

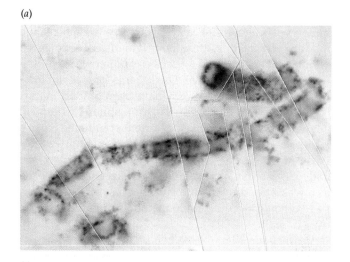

(b)

(c)

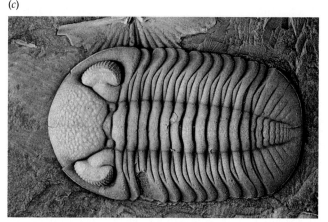

(d)

Figure 22.1 Fossils through the Ages
(a) The oldest fossils are of bacteria, such as this 3.5-billion-year-old fossil found in Western Australia. (b) Soft-bodied animals such as the ones that left these fossils dominated life on Earth 600 million years ago (mya). (c) A fossil of a trilobite that lived in the Devonian period (410 to 355 mya). Note the rows of lenses on each eye. (d) This leaf of a 300-million-year-old seed fern was found near Washington, DC. The fossil formed during the Carboniferous period (355 to 290 mya). The great forests of this period led to the formation of the fossil fuels (oil, coal, and natural gas) that we use today as sources of energy. (e) This 20-million-year-old termite is preserved in amber, the fossilized sap of a tree.

(e)

Given the unusual circumstances that must occur for a fossil to form, remain intact, and be discovered, a species could evolve, thrive for millions of years, and become extinct without our ever finding evidence of its existence in the fossil record.

Although it is not complete, the fossil record shows clearly that there have been great changes in the groups of organisms that have dominated life on Earth over time. These changes have been caused by the extinction of some groups and the expansion of other groups. In the discussion that follows we describe the broad patterns in the history of life revealed by the fossil record, and we discuss the factors that cause these patterns.

■ The fossil record documents the history of life on Earth and provides strong evidence that life has evolved through time. The fossil record shows that past organisms were unlike living organisms, that many species have gone extinct, and that there have been great changes in the dominant groups of organisms over time.

22.1 The History of Life on Earth

Figure 22.2 provides a sweeping overview of the history of life on Earth; study it carefully. The sections that follow focus on three of the main events in the history of life: the origin of cellular organisms, the beginning of multicellular life, and the colonization of the land.

The first single-celled organisms arose at least 3.5 billion years ago

Our solar system and Earth formed 4.6 billion years ago. The oldest known rocks on Earth (3.8 billion years old) contain carbon deposits that hint at life. The first solid evidence for life, however, comes from fossils that resemble present-day bacteria (see Figure 22.1a). These 3.5-billion-year-old fossils are fully formed cellular organisms, suggesting that the earliest forms of life arose before that time, probably between 4 and 3.5 billion years ago. (See Chapter 1 for more discussion of the origin of life.)

Eukaryotes first appear in the fossil record at about 2.1 billion years ago; these early eukaryotes resemble some modern algae. Thus, after the origin of prokaryotes 3.5 billion years ago, it took well over a billion years for the first eukaryotes to evolve. During this long period, the evolution of eukaryotes may have been limited in part by low levels of oxygen in the atmosphere. Chemical analyses of very old rocks indicate that initially, Earth's atmosphere contained almost no oxygen. Shortly after life began, however, some groups of bacteria evolved the ability to conduct photosynthesis, which releases oxygen as a waste product. As a result, photosynthesis caused the oxygen concentration in the atmosphere to increase over time (Figure 22.3).

Prokaryotes that use oxygen absorb it directly across their plasma membranes. These aerobic organisms require a concentration of oxygen in the atmosphere that is at least 1 percent of current levels. Eukaryotes, on the other hand, are larger than most prokaryotes. Because of their size, eukaryotes depend on aerobic respiration, which provides more energy per unit of food than do metabolic reactions (such as fermentation) that do not use oxygen (see Chapter 8). In addition, oxygen absorbed across a plasma membrane spreads more slowly through a large cell than through a small cell. Overall, because of their relatively large size, eukaryotic cells cannot get enough

Figure 22.2 The History of Life on Earth The history of life can be divided into 12 major geologic time periods, beginning with the Pre-cambrian (3500 to 545 mya) and extending to the Quaternary (1.8 mya to present).

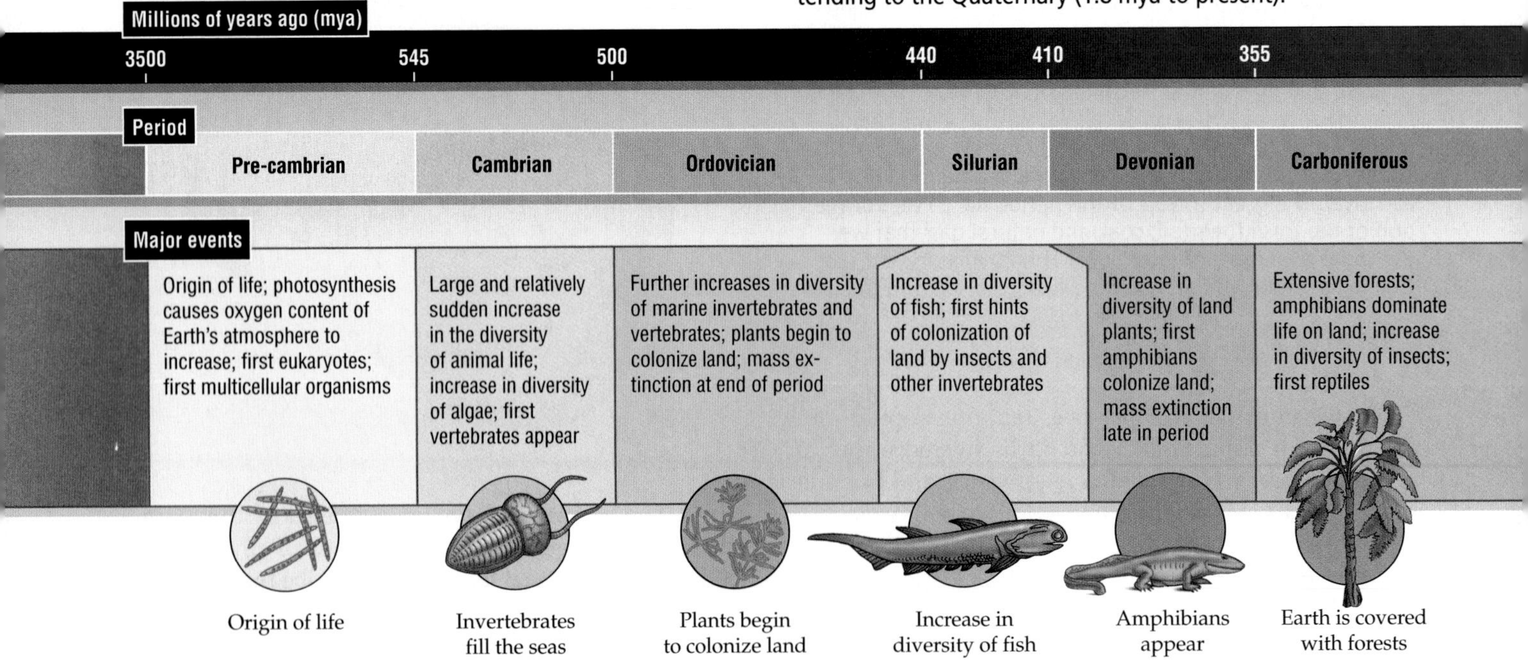

Millions of years ago (mya)

3500	545	500	440	410	355

Period

Pre-cambrian	Cambrian	Ordovician	Silurian	Devonian	Carboniferous

Major events

| Origin of life; photosynthesis causes oxygen content of Earth's atmosphere to increase; first eukaryotes; first multicellular organisms | Large and relatively sudden increase in the diversity of animal life; increase in diversity of algae; first vertebrates appear | Further increases in diversity of marine invertebrates and vertebrates; plants begin to colonize land; mass extinction at end of period | Increase in diversity of fish; first hints of colonization of land by insects and other invertebrates | Increase in diversity of land plants; first amphibians colonize land; mass extinction late in period | Extensive forests; amphibians dominate life on land; increase in diversity of insects; first reptiles |

| Origin of life | Invertebrates fill the seas | Plants begin to colonize land | Increase in diversity of fish | Amphibians appear | Earth is covered with forests |

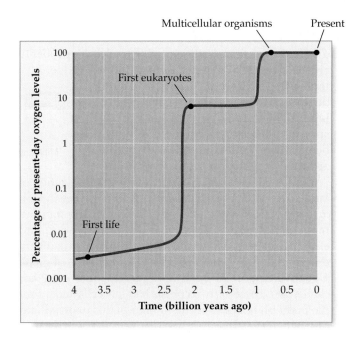

Figure 22.3   Oxygen on the Rise
The release of oxygen as a waste product by photosynthetic organisms has caused its concentration in Earth's atmosphere to increase greatly in the last 3 to 4 billion years.

increased, many early prokaryotes went extinct or became restricted to environments that lacked oxygen. Because it drove many early organisms extinct while simultaneously setting the stage for the origin of multicellular eukaryotes, the biologically driven increase in the oxygen concentration of the atmosphere was one of the most important events in the history of life on Earth.

Multicellular life evolved about 650 million years ago

All early forms of life evolved in water. About 650 million years ago (mya), there was an increase in the number of organisms appearing in the fossil record. At that time, much of Earth was covered by shallow seas, which were filled with protists, small multicellular animals, and algae, collectively known as plankton.

By 600 mya, larger soft-bodied multicellular animals had evolved (see Figure 22.1*b*). These animals were flat and appear to have crawled or stood upright on the seafloor, probably feeding on living plankton or their remains. No evidence indicates that any of these animals preyed on the others. Many of these early multicellular animals may have belonged to groups of organisms that are no longer found on Earth.

oxygen unless the atmospheric concentration of oxygen is at least 2–3 percent of current levels. Once these levels were reached, about 2.1 billion years ago, the first eukaryotes evolved (see Figure 22.3). As oxygen levels continued to increase, the evolution of larger and more complex multicellular organisms became possible.

Oxygen was toxic to many of the early forms of life. Thus, as the oxygen concentration in the atmosphere

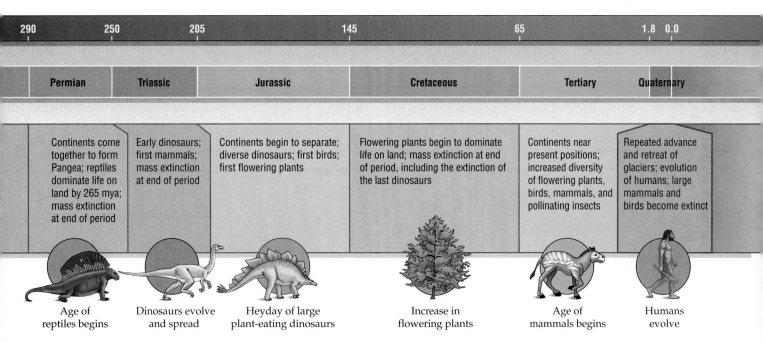

The early to middle Cambrian period (530 mya) witnessed an astonishing burst of evolutionary activity. In a dramatic increase in the diversity of life that is known as the **Cambrian explosion**, large forms of most of the major living animal phyla, as well as other phyla that have since become extinct, appeared suddenly in the fossil record. (The word "explosion" here refers to a rapid increase in the number and diversity of species, not to a physical explosion.) The Cambrian explosion lasted only 5 to 10 million years, a blink of the eye in geologic terms (compare this time span with the 1.4 billion years it took for eukaryotes to evolve from prokaryotes).

The Cambrian explosion was one of the most spectacular events in the evolutionary history of life. It changed the face of life on Earth: From a world of relatively simple, slow-moving, soft-bodied scavengers and herbivores (plant-eaters), suddenly there emerged a world that was filled with large, mobile predators (animals that kill other animals for food) and herbivores with hard body coverings for defense against these predators (Figure 22.4).

Figure 22.4   Before and After the Cambrian Explosion
The Cambrian explosion greatly altered the history of life on Earth.

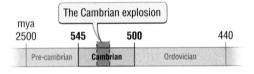

Colonization of the land followed the Cambrian explosion

The first organisms to live on dry land were probably crusts of bacteria living close to the water's edge. A more extensive colonization of land did not begin until the late Ordovician period (about 450 mya), at which time plant spores and burrows believed to have been formed by millipedes (a type of arthropod) appear in the fossil record.

Millipede

Because life first evolved in water, the colonization of land posed enormous challenges. As we will see in Unit 5, many of the functions basic to life, including support, movement, reproduction, and the conservation and exchange of ions, water, and heat, must be handled very differently on land than in water. Descendants of green algae were some of the first organisms to meet these challenges. These early colonists had few cells and a simple body plan, but from them land plants evolved and diversified greatly.

As new groups of land plants arose, they evolved a series of key evolutionary innovations, including waterproofing, stems with efficient transport mechanisms, structural support tissues (wood), leaves and roots of various kinds, seeds, the tree growth form, and specialized sexual organs (see Chapter 3). These and other important changes allowed plants to cope with the transition to life on land. Waterproofing, stems with efficient transport

Before the Cambrian explosion, Earth was dominated by soft-bodied scavengers and herbivores.

After the Cambrian explosion, life was dominated by more complex animals, including predators and well-defended herbivores.

mechanisms, and roots, for example, were important features that helped plants acquire and conserve water while living on dry land.

The key innovations that made life on land possible for plants took roughly 120 million years to evolve. Taken together, these evolutionary changes represent an unparalleled episode in the history of plant life, and nothing like it has occurred before or since. By the end of the Devonian period (345 mya), Earth was covered with plants. Like plants today, the plants of the Devonian included low-lying spreading species, short upright species, shrubs, and trees.

Although there are hints of land animals as early as 450 mya, the first definite fossils of terrestrial animals are of spiders and millipedes that date from about 410 mya. Many of the early animal colonists on land were predators; others, such as millipedes, fed on plant material and decaying matter. Insects, which are currently the most diverse group of terrestrial animals, first appeared roughly 400 mya, and they played a dominant role on land by 350 mya.

The first vertebrates to colonize land were amphibians, the earliest fossils of which date to about 365 mya. Early amphibians resembled, and probably descended from, lobe-finned fish (Figure 22.5). Amphibians were the most abundant organisms on land for about 100 million years. In the late Permian period, the reptiles, which had evolved from a group of reptile-like amphibians, rose to become the most common vertebrate group. Reptiles were the first group of vertebrates that could complete their entire life cycles on land. Earlier vertebrates were unable to reproduce without returning to open water (for example, to lay eggs), and so remained semiaquatic. Thus reptiles were the first vertebrates that could fully exploit the available opportunities for terrestrial life.

Reptiles, including the dinosaurs, dominated life on land for 200 million years (265 mya to 65 mya), and remain important today. Mammals, the group that currently dominates life on land, evolved from reptiles roughly 220 mya, as we will see in Chapter 23. The origin of mammals from reptiles is beautifully documented in the fossil record and provides an excellent example of macroevolution. Since the dinosaurs went extinct, about 65 mya, the mammals have diversified greatly. The increase in the diversity of mammals was influenced in part by continental drift, the subject we turn to next.

> ■ Three major events in the history of life were the evolution of the first single-celled prokaryotes (3.5 billion years ago) and eukaryotes (2.1 billion years ago), the great expansion of multicellular organisms that took place in the Cambrian explosion (530 to 525 mya), and the colonization of land by plants and invertebrates (450 to 410 mya), followed later by vertebrates (365 mya).

22.4 The Effect of Continental Drift

The enormous size of the continents may cause us to think of them as immovable. But this notion is not correct. The continents move slowly relative to one another, and over hundreds of millions of years they travel considerable distances (Figure 22.6). This movement of the continents over time is called **continental drift**. The continents can be thought of as plates of solid matter that "float" on the surface of Earth's mantle, a hot layer of semisolid rock.

Two forces cause the continental plates to move: First, hot plumes of liquid rock rise to the surface and push the continents apart (Figure 22.7). This process can cause the seafloor to spread, as it is doing between North America and Europe,

Millions of years ago (mya)

| 2500 | 545 | 500 | 440 | **410** | **355** | | 290 | 250 |

| Pre-cambrian | Cambrian | Ordovician | Silurian | Devonian | Carboniferous | Permian | Triassic |

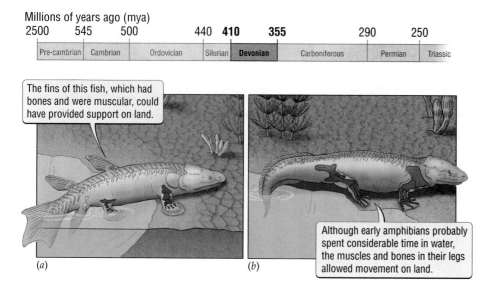

The fins of this fish, which had bones and were muscular, could have provided support on land.

Although early amphibians probably spent considerable time in water, the muscles and bones in their legs allowed movement on land.

(a)   (b)

Figure 22.5 The First Amphibians
(*a*) Amphibians probably descended from a lobe-finned fish, such as the one shown here. (*b*) This early amphibian was reconstructed from a 365-million-year-old (late Devonian) fossil.

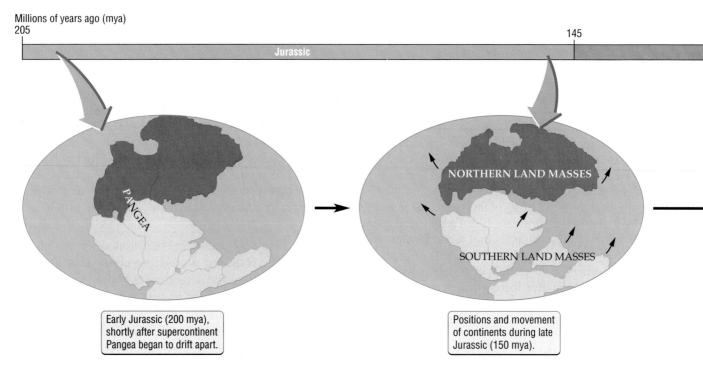

Millions of years ago (mya)
205

Jurassic

145

PANGEA

Early Jurassic (200 mya), shortly after supercontinent Pangea began to drift apart.

NORTHERN LAND MASSES

SOUTHERN LAND MASSES

Positions and movement of continents during late Jurassic (150 mya).

Figure 22.6   Movement of the Continents over Time
The continents move over time, as these snapshots of the break up of the supercontinent Pangea show. Earlier movements of the continents had led to the gradual formation of Pangea, a process that was complete by 250 mya.

which are separating at a rate of 2.5 centimeters per year. This process also can cause bodies of land to break apart, as is currently happening in Iceland and East Africa. Second, where two continental plates collide, one can sink into the mantle below the other. This sinking action pulls the rest of the continental plate along with it, eventually causing the sinking plate to melt.

22.5

Figure 22.7   The Causes of Continental Drift
Two forces cause continental drift: hot plumes of liquid rock that push apart continental plates, and collisions that force one plate under another.

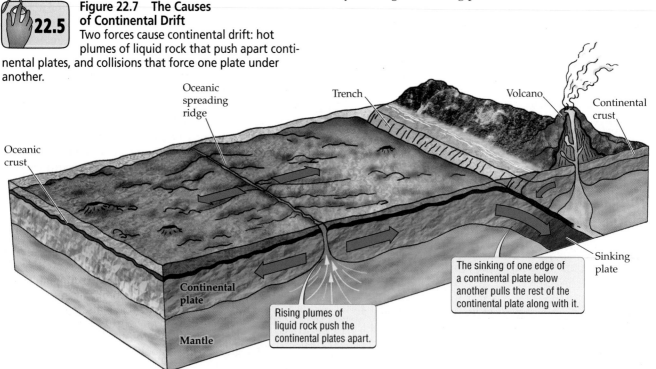

Oceanic spreading ridge

Trench

Volcano

Continental crust

Oceanic crust

Continental plate

Mantle

Rising plumes of liquid rock push the continental plates apart.

The sinking of one edge of a continental plate below another pulls the rest of the continental plate along with it.

Sinking plate

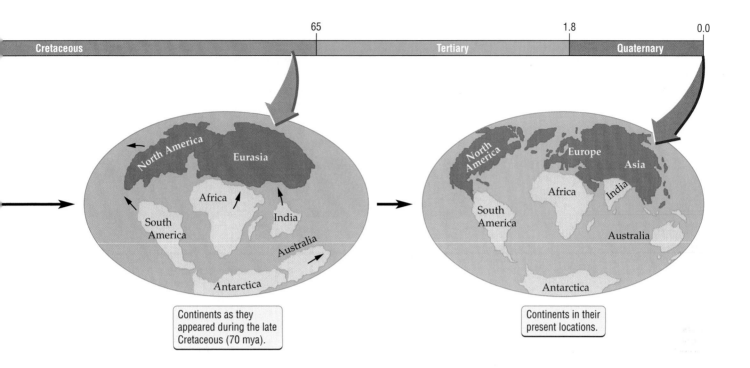

Continents as they appeared during the late Cretaceous (70 mya).

Continents in their present locations.

Patterns of continental drift—most notably the breakup of the ancient supercontinent Pangea—have had a tremendous effect on the history of life. Pangea began to break apart early in the Jurassic period (about 200 mya), ultimately separating into the continents we know today (see Figure 22.6). As the continents drifted apart, populations that once were connected by land became isolated from one another. As we learned in Chapter 21, geographic isolation reduces gene flow and thereby promotes speciation. The separation of the continents was geographic isolation on a grand scale, and it led to the formation of many new species. Among mammals, for example, kangaroos, koalas, and other marsupials that are unique to Australia evolved in geographic isolation on that continent, which broke apart from Antarctica and South America about 40 mya.

Continental drift also affects climate, which has a profound effect on organisms. Shifts in the position of the continents alter ocean currents, and these currents have a major influence on the global climate. At various times, changes in the global climate caused by the movement of the continents have led to the extinction of many species. In the next section we'll look at the evolutionary effect of such large-scale extinction events, which are known as mass extinctions. Moreover, geographic isolation has led to the formation of many morphologically similar, but distantly related plant species living in the world's deserts, which are widely scattered across the globe.

> ■ Continental drift has affected the history of life on Earth. The separation of the continents during the past 200 million years has promoted the evolution of many new species. Changes in the climate caused by the movement of the continents have led to the extinction of many species.

Mass Extinctions: Worldwide Losses of Species

As the fossil record shows, species have gone extinct throughout the long history of life. The rate at which species have gone extinct has varied over time, from low to very high. At the upper end of this scale, there have been five **mass extinctions**, periods of time during which great numbers of species went extinct throughout most of Earth. Each of these five biological upheavals left a permanent mark on the history of life (Figure 22.8). The causes of the five mass extinctions are difficult to determine, but are thought to include such factors as climate change, explosive volcanism, asteroid impacts, changes in the composition of marine and atmospheric gases, and changes in sea levels. In addition to the five mass extinctions revealed by the fossil record, we may be entering a sixth, human-caused mass extinction today (see Chapters 4 and 45).

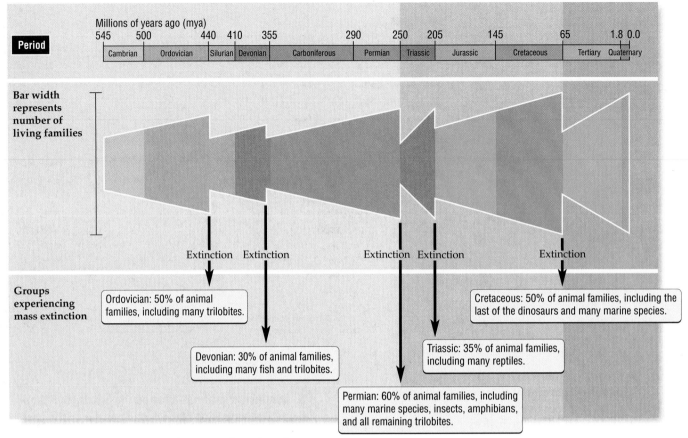

Figure 22.8   The Five Mass Extinctions of Earth's History

Of the five previous mass extinctions, the largest occurred during the Permian period, 250 mya. This mass extinction radically altered life in the oceans. Among marine invertebrates, an estimated 50 to 63 percent of the existing families, 82 percent of the genera, and 95 percent of the species became extinct. The Permian mass extinction was also highly destructive on land, removing 62 percent of the families, bringing the reign of the amphibians to a close, and causing the only major extinction of insects in their 390-million-year history (8 of 27 orders of insects became extinct).

Although not as severe as the Permian extinction, each of the other mass extinctions also had a profound effect on the diversity of life (see Figure 22.8). The best-studied mass extinction occurred at the end of the Cretaceous period, 65 mya. At that time, half of the marine invertebrate species perished, as did many families of terrestrial plants and animals, including the dinosaurs. The Cretaceous mass extinction was probably caused at least in part by the collision of an asteroid with Earth. A 65-million-year-old crater 10 kilometers wide was

recently found buried in sediments off the Yucatán coast of Mexico. An asteroid of this size would have caused great clouds of dust to hurtle into the atmosphere, blocking sunlight around the globe for months to years, thus causing temperatures to drop drastically and driving many species extinct.

The effects of mass extinctions on the diversity of life are twofold. First, entire groups of organisms perish in a mass extinction. The loss of some groups, but not others, greatly alters the subsequent course of evolution and the history of life. If the dinosaurs had been spared, for example, and our early primate ancestors had become extinct 65 mya instead, humans would not exist, and the world would be a very different place.

Second, the extinction of one or more dominant groups of organisms can provide new ecological and evolutionary opportunities for groups of organisms that previously were of relatively minor importance, thus dramatically altering the course of evolution. We discuss this second effect of mass extinctions in the following section.

■ There have been five mass extinctions during the history of life on Earth. Mass extinctions have two effects on the diversity of life. First, the loss of some groups and the survival of others greatly alters the subsequent course of evolution. Second, the extinction of a dominant group of organisms can provide new opportunities for other groups.

Evolutionary Radiations: Increases in the Diversity of Life

After each of the five mass extinctions, some of the surviving groups of organisms diversified to replace those that had become extinct. These bursts of evolution, which lasted 1 to 7 million years each, were just as important for the future course of evolution as the extinctions themselves. The diversification of a group of organisms to form new species and higher taxonomic groups is called an **evolutionary radiation**. In this section we look at four factors that promote evolutionary radiations: release from competition, key evolutionary innovations, ecological interactions, and low amounts of gene flow.

Release from competition

Evolutionary radiation can occur when species are released from competition. The first mammals, for example, evolved in the Triassic period, about 220 mya. Fossil and genetic evidence suggests that several of the orders of living mammals diverged from one another about 100 to 85 mya, well before the extinction of the dinosaurs. But most of the major radiations within these and other groups of mammals did not occur until after the dinosaurs went extinct (65 mya). Many dinosaurs were large and fierce; thus competition with dinosaurs may have prevented the mammals from expanding to fill new ecological roles, such as those of large herbivores or large predators. After the dinosaurs became extinct, some land mammals reached enormous sizes, such as the extinct Beast of Baluchistan, which was over three times as large as an elephant (see p. 429).

On a much shorter time scale, the spectacular radiations of organisms that colonize newly formed islands or lakes illustrate how migration to a place free from the usual competitor species can also promote evolutionary radiation. The Galápagos finches (see Figure 19.9) and the cichlids of Lake Victoria (see Figure 21.8) are examples of this phenomenon.

Key evolutionary innovations

A group may diversify greatly if it acquires a new adaptation that lets it use its environment in new ways. There are many examples of such evolutionary radiations in the fossil record, as in the Cambrian explosion, the radiation of land plants (see Chapter 3), and the radiations that followed the evolution of flight in insects, birds (Figure 22.9), and bats.

Ecological interactions

Species depend on one another for food and other resources, as we will see in Chapter 42. For this reason, diversification within one group of organisms can promote diversification within other groups. For example, if one group radiates because it has evolved a new way to use its environment, its diversification can stimulate the evolutionary radiation of another group that depends on the newly expanded group. Such was the case with the radiation of land plants, which was followed by a radiation of insects that fed on those plants.

Low amounts of gene flow

Lack of gene flow promotes evolutionary radiation because when there is little exchange of alleles between populations, the populations diverge more rapidly, a condition that favors the formation of new species (see Chapter 21). On a broad geographic scale, the separation of the continents over the past 200 million years isolated many previously connected populations. Many of these geographically isolated populations diverged from one another to form new species, resulting in a large increase in the diversity of life.

■ Evolutionary radiations can be promoted by release from competition, key evolutionary innovations, ecological interactions, and low amounts of gene flow.

Overview of the Evolutionary History of Life

The history of life on Earth can be summarized by the rise and fall of major groups of protists, plants, and animals (see Figure 22.2). These broad patterns in the history of life are caused by the extinction or decline of some groups and the origin or expansion of other groups. Taken together, mass extinctions and evolu-

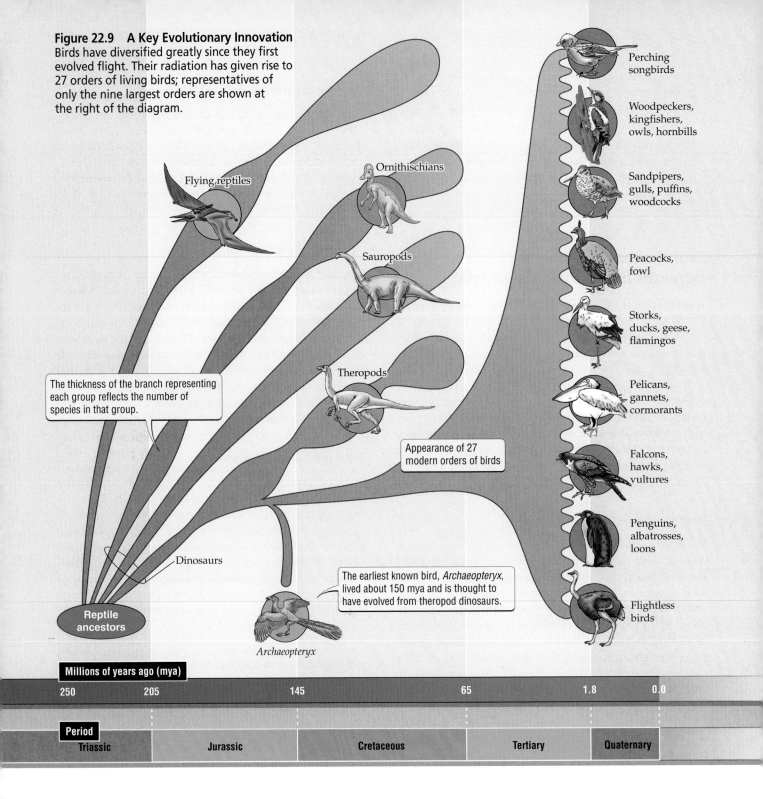

Figure 22.9 A Key Evolutionary Innovation
Birds have diversified greatly since they first evolved flight. Their radiation has given rise to 27 orders of living birds; representatives of only the nine largest orders are shown at the right of the diagram.

Flying reptiles

Ornithischians

Sauropods

Theropods

The thickness of the branch representing each group reflects the number of species in that group.

Appearance of 27 modern orders of birds

Dinosaurs

The earliest known bird, *Archaeopteryx*, lived about 150 mya and is thought to have evolved from theropod dinosaurs.

Reptile ancestors

Archaeopteryx

Perching songbirds

Woodpeckers, kingfishers, owls, hornbills

Sandpipers, gulls, puffins, woodcocks

Peacocks, fowl

Storks, ducks, geese, flamingos

Pelicans, gannets, cormorants

Falcons, hawks, vultures

Penguins, albatrosses, loons

Flightless birds

Millions of years ago (mya)

| 250 | 205 | 145 | 65 | 1.8 | 0.0 |

Period

| Triassic | Jurassic | Cretaceous | Tertiary | Quaternary |

tionary radiations have been largely responsible for shaping macroevolution.

In addition to offering this broad view of the evolutionary history of life, we want to emphasize two important related concepts: the increase in biodiversity over time, and the difference between the evolution of populations (microevolution) and the evolution of higher taxonomic groups (macroevolution). We close this chapter by briefly summarizing these two points.

The diversity of life has increased over time

Despite the severity of the five mass extinctions, the diversity of life has increased over time, especially dur-

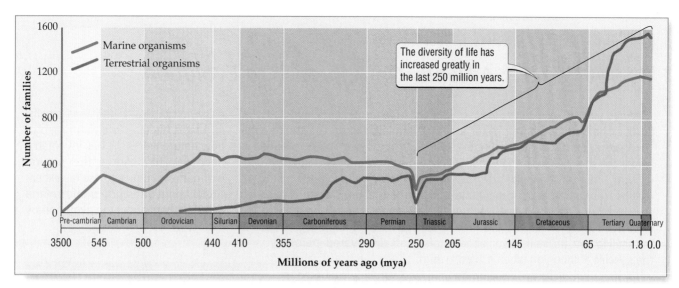

Figure 22.10 Biodiversity Has Increased Over Time
The diversity of life in the oceans and on land has increased dramatically over time, most notably during the past 250 million years.

ing the last 250 million years (Figure 22.10). As described on page 377, a major cause of this increase in biodiversity has been the movement of the continents during the past 200 million years. Other important factors include release from competition following the mass extinctions, the evolution of key innovations (such as the ability to live on land), and ecological interactions that have promoted diversity.

Macroevolution differs from the evolution of populations

As we saw in Chapter 21, the often exquisite adaptations of organisms result from natural selection. Evolution by natural selection is a short-term process: Adaptations are shaped by natural selection to match the organism's current environment. Can natural selection provide a complete explanation of the evolution of higher taxonomic groups? The answer is no, in part because natural selection cannot predict future changes to the environment; hence there is no reason to suppose that adaptations that are currently advantageous will remain so if the environment changes.

Furthermore, mass extinctions can remove entire groups of organisms, seemingly at random—even those that possess highly advantageous adaptations. A group of predatory gastropods (snails and their relatives), for example, went extinct in the Triassic mass extinction, shortly after they had evolved the ability to drill through the shells of other gastropods. The ability to drill through

shells had opened up a major new way of life for these organisms (see the box on p. 382). If these predatory gastropods had not gone extinct, they probably would have thrived and formed many new species that had the ability to drill through shells (as did another group of shell-drilling gastropods that evolved 120 million years later). As this example shows, even species that have highly beneficial adaptations don't always win.

Overall, broad patterns in the history of life cannot be predicted solely from an understanding of the evolution of populations. To have a full understanding of the history of life on Earth, we must also understand factors such as mass extinctions, evolutionary radiations, and continental drift, all of which can have a tremendous effect on evolution above the species level.

> ■ Evolutionary radiations and mass extinctions have been responsible for shaping the rise and fall of major groups of organisms. Biodiversity has increased over time. The evolution of higher taxonomic groups differs from the evolution of populations.

■ HIGHLIGHT

When Antarctica Was Green

Antarctic fossils of tropical organisms, dinosaurs, and forests are vivid testimony to the fact that we live in a dynamic world. These fossils reveal great changes over time, ranging as they do from Cambrian marine organisms to early land plants to birds and mammals. The

■ THE SCIENTIFIC PROCESS

Geerat Vermeij: A Hands-On Approach to Evolution

Give Geerat Vermeij the fossil of a shell and watch a master at work. As he runs his fingers across the shell, he sees what most of us would miss: evidence of a narrow escape from a predator, or a tell-tale clue to what killed the creature that once lived in the shell. As he returns the shell to you, he'll talk to you about the species, telling you when it lived, what its enemies were like, and its position in a sweeping view of the evolutionary history of life. He sees so much in a few short moments of examining the shell. And yet, literally, he sees nothing at all, for he is blind.

In over 30 years of studying fossil shells, Dr. Vermeij (pronounced "Ver-MAY") has used his hands to examine how the history of life in the oceans was influenced by ongoing battles between predator and prey. His results show that over long periods of time, predators such as crabs and shell-drilling snails evolved increasingly sophisticated ways to crush, pry open, or drill through the shells of their prey. And the prey responded in kind, developing tougher, better-defended shells. Dr. Vermeij realized that this evolutionary "arms race" had proceeded in a single direction for hundreds of millions of years, shaping the history of life in the oceans. His hands revealed a fundamental story about life in the oceans, a story that no sighted biologist had discovered before him.

Dr. Vermeij has worked in challenging environments throughout the world (such as snake-filled swamps and shark-infested oceans), published many groundbreaking papers and books, and won numerous awards. The fact that he is blind is in some ways irrelevant—he is a great scientist by any standard. But his blindness is also basic to what he has accomplished. He has a legendary ability to use senses other than vision to explore the outdoors, typically reaching a deeper, fuller understanding of a place than would be achieved by a biologist who can see. As he listens, smells, and feels his way to a rich grasp of how nature works, he has invented a new way to observe the natural world. He continues to use his exceptional talent and gifts to inform and delight others about the evolutionary history of life on Earth.

Geerat Vermeij

very different organisms that have lived in Antarctica at different times illustrate the broad changes in the history of life described in this chapter, such as the Cambrian explosion, the colonization of land, and the periods of domination by amphibians, reptiles, and mammals.

The Antarctic fossils also show the striking contrast between the diverse life forms that once lived in Antarctica and the few that live there today. The small number of organisms that now live in Antarctica is a consequence of continental drift. First, as the continents broke apart, Earth's climate grew colder. This process was heightened in Antarctica, which experienced an ever-colder climate as it moved toward its present position over the South Pole. Once Antarctica separated from Australia and South America, about 40 mya, the organisms on the continent were trapped there. Thus, as the climate in Antarctica became increasingly cold, most species perished.

The movement of Antarctica to its present position may have contained within it not only the seeds of destruction, but those of creation as well. As Antarctica and the rest of the continents drifted apart over the last 40 million years, they altered the flow of ocean currents. The rerouting of ocean currents contributed to the formation of the Antarctic ice cap and produced the largest differences in temperature between the poles and the Tropics that Earth has ever known. The wide range of new habitats that resulted from these temperature differences helped set the stage for evolutionary radiations in many organisms, including humans.

■ The small number of species that currently live in Antarctica is a consequence of continental drift. Continental drift isolated Antarctica from other continents and caused most species that lived there to perish as the climate became increasingly cold.

SUMMARY

The Fossil Record: Guide to the Past

■ The fossil record documents the history of life on Earth.

■ Fossils reveal that past organisms were unlike living organisms, that many species have gone extinct, and that there have been great changes in the dominant groups of organisms over time.

■ The order in which organisms appear in the fossil record is consistent with our understanding of evolution.

■ Although the fossil record is not complete, it provides excellent examples of the evolution of major new groups of organisms.

The History of Life on Earth

■ The first single-celled organisms resembled bacteria and evolved 3.5 billion years ago.

■ Shortly after life began, some groups of bacteria evolved the ability to conduct photosynthesis, which releases oxygen as a waste product.

■ The release of oxygen by photosynthetic bacteria caused oxygen concentrations in the atmosphere to increase. As a result, the evolution of single-celled eukaryotes (2.1 billion years ago) became possible.

■ Life on Earth changed dramatically during the Cambrian explosion (530 mya), when large predators and well-defended herbivores suddenly appeared.

■ The land was colonized by plants and invertebrates (450 to 400 mya), followed later by the vertebrates (365 mya).

The Effect of Continental Drift

■ Continental drift has had a profound effect on the history of life on Earth.

■ The separation of the continents over the past 200 million years has led to geographic isolation on a grand scale, promoting the evolution of many new species.

■ At different times, changes in the climate caused by the movement of the continents have led to the extinction of many species.

Mass Extinctions: Worldwide Losses of Species

■ There have been five mass extinctions during the history of life on Earth.

■ The extinction of some groups and the survival of others greatly alters the subsequent course of evolution.

■ The extinction of a dominant group of organisms can provide new opportunities for other groups.

Evolutionary Radiations: Increases in the Diversity of Life

■ After each of the five mass extinctions, other groups of organisms diversified to replace those that had become extinct.

■ Each of these great evolutionary radiations took 1 to 7 million years and forever altered the history of life on Earth.

■ Evolutionary radiations are promoted by release from competition, key evolutionary innovations, ecological interactions, and low amounts of gene flow.

Overview of the Evolutionary History of Life

■ Evolutionary radiations and mass extinctions are primarily responsible for the rise and fall of major groups of organisms that characterize the history of life on Earth.

■ The diversity of life has increased over time due to continental drift, release from competition after mass extinctions, the evolution of key innovations, and ecological interactions.

■ Macroevolution differs from the evolution of populations. Mass extinctions, evolutionary radiations, and continental drift have all had large effects on macroevolution.

Highlight: When Antarctica Was Green

■ The diversity of organisms that once lived in Antarctica, as indicated by fossils, contrasts sharply with the few organisms that live there today.

■ The large changes in the organisms living in Antarctica over time illustrate the rise and fall of different groups of dominant organisms during the history of life on Earth.

■ The low number of species that currently live in Antarctica is a consequence of continental drift, which isolated Antarctica from other continents and caused most species that lived there to perish as the climate became increasingly cold.

KEY TERMS

Cambrian explosion p. 374	fossil p. 370
continental drift p. 375	macroevolution p. 370
evolutionary radiation p. 379	mass extinction p. 377

CHAPTER REVIEW

Self-Quiz

1. Continental drift
   a. can occur when liquid rock rises to the surface and pushes the continents apart.
   b. no longer occurs today.
   c. has led to the geographic isolation of many populations, thus promoting speciation.
   d. both a and c

2. The fossil record
   a. documents the history of life.
   b. provides examples of the evolution of major new groups of organisms.
   c. is not complete.
   d. all of the above

3. Mass extinctions
   a. are all caused by asteroid impacts.
   b. are periods of time in which many species went extinct throughout Earth.
   c. have little lasting impact on the history of life.
   d. are usually recovered from within 100,000 years or so.

4. The Cambrian explosion
   a. caused a spectacular increase in the diversity and complexity of animal life.
   b. caused a mass extinction.
   c. was the time during which all living animal phyla suddenly appeared.
   d. had few consequences for the later evolution of life.

5. The history of life shows that
   a. biodiversity has remained constant for about 400 million years.
   b. extinctions have little effect on biodiversity.
   c. macroevolution is greatly influenced by mass extinctions and evolutionary radiations.
   d. macroevolution can be understand solely in terms of the evolution of populations.

Review Questions

1. Mass extinctions can remove entire groups of organisms, seemingly at random—even groups that possess highly advantageous adaptations. How can this be?

2. Evidence from the fossil record indicates that it usually takes 1 to 7 million years for an evolutionary radiation to replace the species lost during a mass extinction. Discuss this observation in light of your understanding of the speciation process (see Chapter 21). What does it suggest about the consequences of the losses of species that are occurring today?

3. Is macroevolution fundamentally different from microevolution? Can macroevolutionary patterns be explained solely in terms of microevolutionary processes? Does any evolutionary mechanism that we have studied link macroevolution and microevolution?

22

The Daily Globe

Appreciate That Sparrow (It's a Dinosaur)

TUCSON, AZ—Next time you see a sparrow in flight, or a roadrunner dashing across the desert, look closely: They may be dinosaurs.

The idea that birds might have evolved from dinosaurs has been around for over a hundred and thirty years. Although a few biologists are not convinced, recent analyses of the bones of dinosaurs and birds support this idea, revealing that the roadrunner or sparrow in your backyard is actually a not-too-distant cousin of the fearsome *Tyrannosaurus rex*.

It may seem like a big leap from *T. rex* to a sparrow, but spectacular fossil discoveries from China and Spain have made that big leap seem more like a small hop. Small dinosaurs, closely related to the group to which *T. rex* belongs, were recently discovered to have feathers, a feature once thought restricted to birds. Dinosaurs also share other features that biologists had thought were shared only by birds, including wishbones, sideways-flexing wrists, hollow bones, and nesting behavior. Taken together, these recent discoveries suggest that the dinosaurs did not go extinct after all.

Many dinosaurs evolved to be big, as *T. rex* and the fossils of *Apatosaurus* and other large herbivores readily attest. But birds evolved rapidly in the opposite direction, toward smaller size. Now look again at that roadrunner. If you watch it closely, it's not hard to imagine it as a small dinosaur, in hot pursuit of its prey. As you watch the roadrunner, remember that it is closely related to *T. rex*. And with that thought in mind, let's just hope that bird evolution does not take a sharp U-turn anytime soon, returning birds to the large size of their terrible and awe-inspiring ancestors.

Evaluating "The News"

1. Does thinking of birds as dinosaurs change how you look at birds?

2. Do you think birds really might evolve to be large, like many dinosaurs? Why or why not?

3. It can cost considerable sums of money to discover, excavate, analyze, and display the bones of dinosaurs. In a world plagued by social and environmental problems, should taxpayer money be spent on learning more about the dinosaurs? Does it matter whether birds are dinosaurs?

4. In general, should we direct our educational efforts toward only those subjects that yield material or economic gain? Or should we direct our efforts toward all forms of knowledge? What do you think the effect would be if society ceased to support types of knowledge, expression, and belief that address questions about who we are, where we came from, or why we are here?

23 Human Evolution

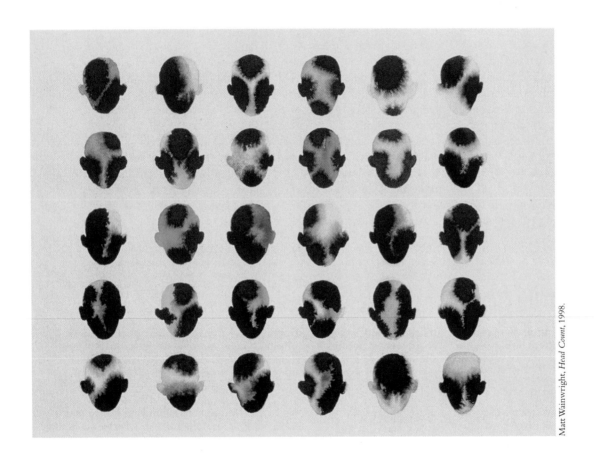

Matt Wainwright, *Head Count,* 1998.

The Neandertals

Neandertals loom large in our imagination, having been the stuff of many movies, books, and magazine articles. Their name alone calls to mind words such as "brutish," "sub-human," and "caveman." The image these words convey is not accurate. Neandertals were not sub-human; they may even have been early members of our own species, *Homo sapiens.*

Even so, if a group of Neandertals were to stroll down the street today, they certainly would attract attention. Neandertals had large, arching ridges above their eyes, a low, sloping forehead, no chin, and a face that, compared with ours, looked as if it were pulled forward. Without shirts, they would be even more striking. Slightly shorter than humans of today, Neandertals had thick necks, were heavily

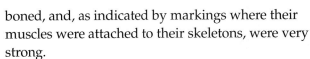

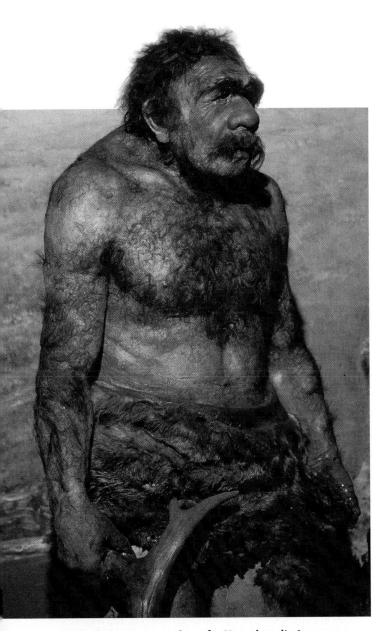

An Artist's Interpretation of a Neandertal's Appearance

boned, and, as indicated by markings where their muscles were attached to their skeletons, were very strong.

Neandertals take their name from the Neander Valley in eastern Germany (*tal* comes from the German for "valley"). Fossilized bones discovered there in 1856 were thought by some people to be those of a bear. Others, however, argued that they were the remains of an ancient human. Known as the Neandertal Man, these remains became the subject of great debate: Were the bones really human? Finally, after similar fossils were found in other places, scientists became convinced that the bones were indeed of human origin. Collectively, the Neandertal fossils shook our understanding of ourselves, for they provided dramatic proof that different forms of humans once existed.

The Neandertals hold a central place in the study of human evolution. Nevertheless, their relationship to modern humans has long been a mystery. They first appeared in Europe 200,000 years ago, where they thrived until they were replaced by anatomically modern humans, about 28,000 years ago. In both physical appearance and culture, they were like us, yet not like us.

What caused the disappearance of the Neandertals? Were they our direct ancestors? Or were they an evolutionary dead end, an experiment that failed and contributed little or nothing to the genetic constitution of modern humans?

■ KEY CONCEPTS

1. Humans are primates, which in turn are mammals. Mammals evolved from reptile-like ancestors 220 million years ago. The first primates split off from other mammals more than 65 million years ago.

2. Among primates, humans and our humanlike ancestors are hominids. Hominids diverged from other primates 5 to 6 million years ago.

3. The earliest hominids walked upright, but otherwise resembled the apes from which they descended. Over time, hominid brain size increased greatly.

4. Early members of our genus, *Homo*, used stone tools 2.5 million years ago. Toolmaking technology improved slowly for over 2 million years, then changed more rapidly over the past 300,000 years.

5. Modern humans probably arose in Africa 100,000 to 200,000 years ago.

6. The human evolutionary tree is "bushy," consisting of multiple species and many side branches at different points in time.

Human beings—*Homo sapiens*—are animals, classified as members of the chordate phylum, the mammal class, the primate order, and the hominid family. Humans share with all other mammals certain distinguishing characteristics, including body hair (which provides insulation in many species) and the feeding of the young with milk produced by mammary glands. As we'll see in this chapter, humans also share more specific characteristics with other primates and with other members of the hominid family (which contains humans and our humanlike ancestors).

As animals, we are descended from earlier animals. Most recently, the branch of the evolutionary tree leading to humans split from the branch leading to chimpanzees 5 to 6 million years ago (see Figure 2.3). Although chimpanzees are our closest living relatives, we did not evolve directly from them, as is sometimes misleadingly stated. Rather, the different evolutionary lineages leading to humans and to chimpanzees both originated from a common ape ancestor. In this chapter, we describe the origin of mammals and primates, then focus on the evolution of humans and our immediate ancestors.

Evolution of Mammals: From A Minor Role to Dominance

Mammals evolved from reptiles. More specifically, mammals arose from the cynodonts, the end of a long lineage of mammal-like reptiles. Living mammals differ from living reptiles in many respects, including the way they move (Figure 23.1), the nature of their teeth, and the structure of their jaws. It is difficult to draw the line between mammals and reptiles in the fossil record, however. Some fossil species are excellent intermediates between what we now call reptiles and what we now call mammals, providing a beautiful illustration of an evolutionary shift from one major group of organisms (the reptiles) to another (the mammals).

There were many species of mammal-like reptiles in the early Triassic period, 245 million years ago (mya). By 200 mya, however, the mammal-like reptiles had declined as other reptiles, most notably the dinosaurs,

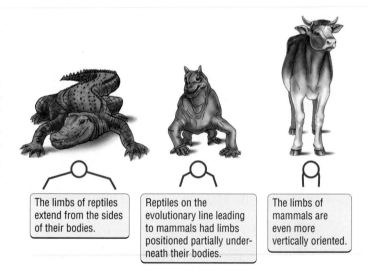

The limbs of reptiles extend from the sides of their bodies.

Reptiles on the evolutionary line leading to mammals had limbs positioned partially underneath their bodies.

The limbs of mammals are even more vertically oriented.

Figure 23.1 From Reptiles to Mammals
The legs of most living reptiles stick out to the side of their bodies, causing them to have a sprawling gait as compared with the upright gait of mammals. Over time, the legs of mammal-like reptiles became positioned under the body, leading eventually to the vertical orientation of the legs in living mammals.

came to dominate Earth. Although the mammal-like reptiles became extinct, they left behind the first mammals as their descendants. The earliest mammals were small, rodent-sized organisms that evolved about 220 mya, roughly the same time as the first dinosaurs.

Throughout the long reign of the dinosaurs, the mammals remained small. They appear to have been nocturnal (active at night), because they had large eye sockets, a characteristic that is found in many living nocturnal organisms and that appears to improve vision in low-light conditions. By being nocturnal and small, early mammals may have been to dinosaurs what a mouse is to a lion: hard to notice and too small to eat. Following the mass extinction that killed off the dinosaurs 65 mya, the mammals radiated greatly to include many new forms that were large and active by day.

A fundamental event in the evolution of mammals was an increase in the speed and duration of brain growth during embryonic development. The resulting increase in the size of the adult brain had many effects, including an increase in the ability to learn and in the ability to respond appropriately to new situations. This increase in brain size was especially prominent in the order to which humans belong: the primates.

■ Mammals evolved from reptiles about 220 million years ago. Throughout the long reign of the dinosaurs, mammals remained small and nocturnal. Following the extinction of the dinosaurs, the mammals radiated to include many species that were large and active by day.

The Origin and Evolution of Primates

Primates are thought to have originated more than 65 mya from small nocturnal mammals, similar to tree shrews, that ate insects and lived in trees. However, the fossil evidence of primate origins is sketchy, and the first definite primate fossils are roughly 50 million years old. These early primates resembled modern lemurs and lorises. Living primates include lemurs, tarsiers, monkeys, and humans and other apes (Figure 23.2).

The first higher primates, a group that includes New World monkeys, Old World monkeys, and apes, arose 37 to 45 mya. The earliest primates had small bodies and probably ate insects or fruit. By 35 mya, the group had diversified greatly. One branch gave rise to the **hominoids**, a group of primates whose living members

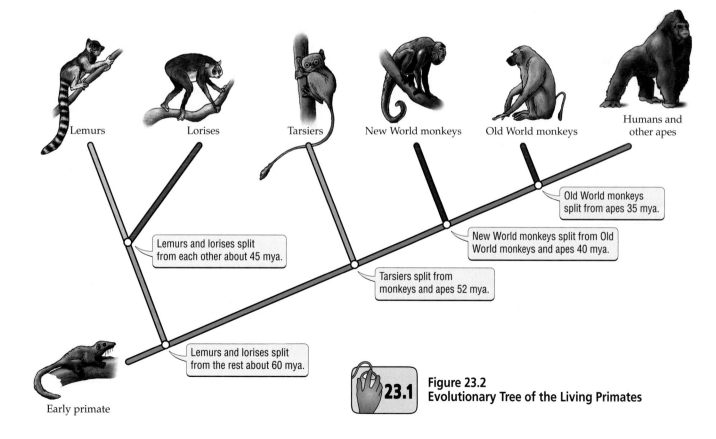

Lemurs   Lorises   Tarsiers   New World monkeys   Old World monkeys   Humans and other apes

Old World monkeys split from apes 35 mya.

New World monkeys split from Old World monkeys and apes 40 mya.

Tarsiers split from monkeys and apes 52 mya.

Lemurs and lorises split from each other about 45 mya.

Lemurs and lorises split from the rest about 60 mya.

Early primate

23.1

Figure 23.2
Evolutionary Tree of the Living Primates

include the gibbons and the great apes (orangutans, gorillas, chimpanzees, and humans).

The early hominoids radiated to form many new species and higher taxonomic groups. Between 23 and 5 mya, 30 genera of hominoids evolved, most of which are now extinct. One branch of the hominoids gave rise to the **hominids**—that is, humans and our now extinct, humanlike ancestors.

Evolutionary trends in primates

Primates share numerous characteristics, including flexible shoulder and elbow joints, five functional fingers and toes, thumbs and big toes that are opposable (that is, they can be placed opposite other fingers or toes), flat nails (instead of claws), forward-facing eyes, short snouts, and brains that are large in relation to body size. Many of these traits appear to be adaptations to life in trees. Three characteristics became increasingly well developed during primate evolution: limb mobility and grasping ability, daytime vision, and brain size.

- *Limb mobility and grasping ability.* The limbs of animals that run on four legs are often adapted for speed and stability, as in horses. The limbs of primates, on the other hand, became adapted for greater mobility. The evolution of flexible shoulder and elbow joints improved the ability of primates to climb trees and swing from branch to branch. Primates also evolved greater grasping ability in their hands and feet, largely through the development of opposable big toes and thumbs. The freedom of movement of their limbs and the grasping ability of their feet and hands were essential to primates as they moved through trees.

- *Daytime vision.* Early primates were nocturnal. They had relatively long snouts, as well as eyes located on the sides of their heads. During primate evolution, the position of the eyes moved forward, resulting in greater overlap in their fields of vision and improving depth perception. Simultaneously, the snout shortened, further improving the forward-facing vision of primates. Primates also evolved an increased ability to distinguish the color and brightness of light, both of which were useful for daytime vision. In general, these evolutionary changes in primate vision would have benefited organisms living in trees. Improved depth perception, for example, made it less likely that a primate moving from branch to branch would miscalculate and fall.

- *Brain size.* Perhaps the most notable feature of primate evolution is the tendency toward larger brain size. In general, the size of an animal's brain relative to the size of its body provides a crude measure of intelli-

gence. Increases in brain size among the primates were linked to a greater emphasis on learning, as well as to the more complex social behaviors shown by higher primates. In many primate lineages, parents raised fewer offspring but invested more effort in each of them. For such primates, the period of time during which offspring depended on and learned from their parents increased.

> ■ Primates originated about 65 mya. Primates have flexible elbow and shoulder joints, forward-facing eyes, and opposable big thumbs and toes. These traits make them well adapted for life in trees. Three characteristics became increasingly well developed during primate evolution: limb mobility and grasping ability, daytime vision, and brain size.

23.2 *Hominid Evolution: The Switch to Walking Upright*

Hominids—humans and our now extinct humanlike ancestors—are characterized by large brains, an upright walking posture, and complex toolmaking behaviors. Of these traits, our intelligence and toolmaking abilities, and our associated culture, are central to what it means to be human. In an evolutionary sense, however, the increases in our intelligence and toolmaking abilities were secondary changes: They occurred relatively late in our evolutionary history and resulted from the general trend toward large brain size in primates. Instead, the first big step in human evolution was the switch from moving on four legs to walking upright on two legs, a change that occurred long before hominids evolved large brains.

The switch to walking upright required a drastic reorganization of primate anatomy, especially of the hip bones. Walking upright also brought other important changes, including a shift in how the skull is oriented on the spinal cord, a change in the position of the big toe and the loss of its opposability, the development of a pronounced heel and arch on the foot, a lengthening of the legs, and a change in the angle of weight support from the hip to the knee (Figure 23.3).

It is not necessary—or even feasible—to walk upright in a tree, and for an organism that lived primarily in trees, the loss of an opposable big toe would be a handicap. On the ground, however, walking upright would have provided several advantages, including freeing the hands for carrying objects or using tools and improving the line of sight (that is, being able to see over nearby

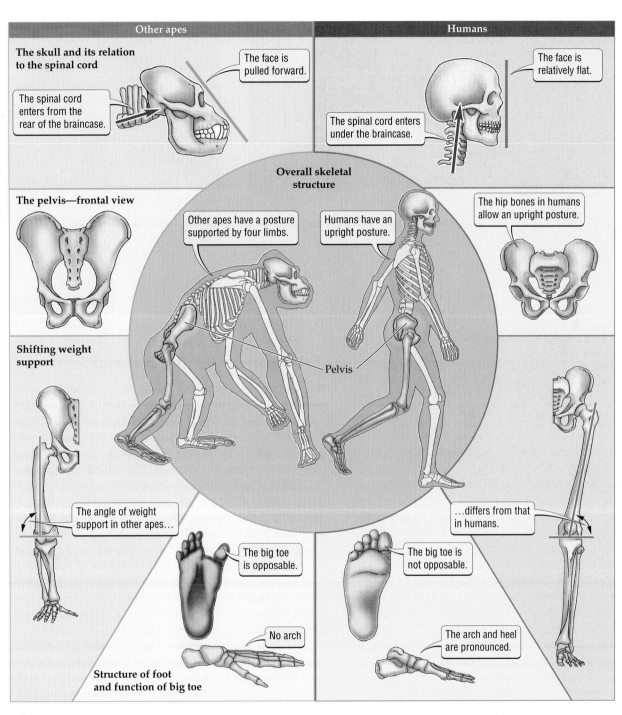

 Figure 23.3   Evolutionary Differences between Humans and Other Apes

objects). Thus, it is likely that the evolution of an upright posture was linked to a switch from life in trees to life on the ground, a switch that probably occurred between 8 and 5 mya.

The switch to life on the ground was probably not sudden or complete. The skeletal structure of the oldest fossil hominids (from 4.4 mya) indicates that they walked upright. However, foot bones and fossilized footprints from 3 to 3.5 mya show that the hominids living at that time still had partially opposable big toes (Figure 23.4), suggesting they may have continued to use trees some of the time.

Foot bones of early hominid

Partially opposable big toe

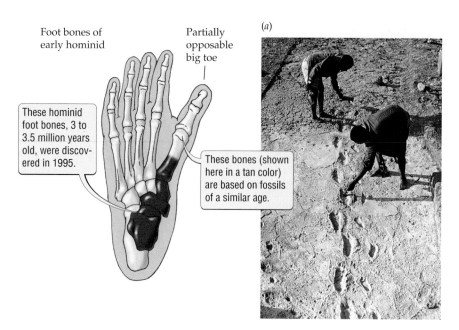

These hominid foot bones, 3 to 3.5 million years old, were discovered in 1995.

These bones (shown here in a tan color) are based on fossils of a similar age.

(a)

(b)

Figure 23.4   Early Hominids Had Partially Opposable Big Toes
(a) Footprints of two early hominids (3 to 3.5 million years old) walking upright, side by side. (b) Chimpanzees, the closest living relatives of humans, have fully opposable big toes.

The earliest known hominids are *Ardipithecus ramidus* (5.8 to 4.4 mya) and several *Australopithecus* species, including *Australopithecus afarensis* (3.9 to 3.0 mya). Although each of these organisms is thought to have walked upright, the size of their brains (measured by the volume of the braincase) was relatively small, and their skulls and teeth were more similar to those of other apes than to those of humans (compare the skull of *A. afarensis* with that of *Homo sapiens* in Figure 23.5).

The *Australopithecus* species may have used simple tools, as do living chimpanzees and other animals, but no direct evidence to support this assumption has been found. This lack of evidence is not surprising: If the tools used by early hominids, like those used by chimpanzees, consisted of items such as sticks, blades of grass, or rocks, they would not be identifiable as "tools" in the fossil record.

> ■ The first big step in human evolution was the switch to walking upright, which was linked to a change from life in trees to life on the ground. Fossil evidence indicates that early hominids walked upright, but had relatively small brains, as well as jaws and skulls that differed greatly from those of living humans.

Evolution in the Genus Homo

The first members of the genus *Homo*, to which modern humans belong, originated in Africa 2 to 3 million years ago. The earliest *Homo* fossil fragments date from 2.4 mya. More complete early *Homo* fossils exist for the period from 1.9 to 1.6 mya; these fossils have been given the species name *Homo habilis*. The oldest *H. habilis* fossils resemble those of *Australopithecus africanus*, the species from which *H. habilis* may have evolved. In more recent *H. habilis* fossils, the face is not pulled forward as much and the skull is more rounded. In these and other ways, more recent *H. habilis* specimens have features that are intermediate between those of *A. africanus* and *Homo erectus*, a species that evolved after *H. habilis*. Thus, *H. habilis* fossils provide an excellent record of the evolutionary shift from ancestral (*Australopithecus*) to more recent (*H. erectus*) characteristics.

H. habilis made a variety of simple stone tools such as choppers, scrapers, and handaxes. These tools are known as Oldowan technology, named after Olduvai Gorge, Tanzania, a site at which many important hominid fossils have been found. Oldowan tools were first made about 2.5 mya. Oldowan tools were replaced in Africa and southwestern Asia by similar, but technically more advanced tools about 1.4 mya, but they persisted in Europe and eastern Asia until 1 mya.

The more advanced tools were probably made by *Homo erectus*, a species that first appeared about 1.8 mya (Figure 23.6). Taller and more robust than *H. habilis*, *H. erectus* also had a larger brain and a skull more like that of modern humans (see Figure 23.5). It is likely that by

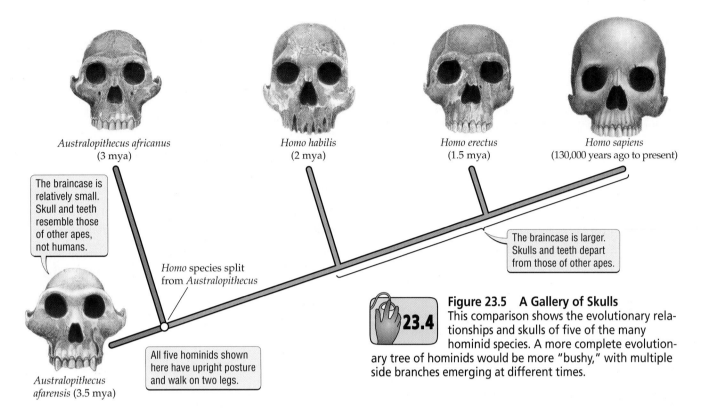

Australopithecus africanus
(3 mya)

Homo habilis
(2 mya)

Homo erectus
(1.5 mya)

Homo sapiens
(130,000 years ago to present)

The braincase is relatively small. Skull and teeth resemble those of other apes, not humans.

Homo species split from *Australopithecus*

The braincase is larger. Skulls and teeth depart from those of other apes.

All five hominids shown here have upright posture and walk on two legs.

Australopithecus afarensis (3.5 mya)

23.4

Figure 23.5   A Gallery of Skulls
This comparison shows the evolutionary relationships and skulls of five of the many hominid species. A more complete evolutionary tree of hominids would be more "bushy," with multiple side branches emerging at different times.

500,000 years ago *H. erectus* could use, but not necessarily make, fire, and they probably hunted large species of game animals. The evidence to support the latter conclusion includes the remarkable discovery in Germany of three 400,000-year-old spears, each about 2 meters long and designed for throwing with a forward center of gravity (like a modern javelin).

It was long thought that *H. habilis* gave rise to *H. erectus*, which then spread from Africa about 1 mya and later evolved into *Homo sapiens*. This simple picture has become more complicated with recent fossil discoveries. Some evidence now suggests that the fossils labeled *H. habilis* are from two different species, and there is debate over which of these species gave rise to *H. erectus*. In addition, it now appears that *H. erectus* or an earlier form of *Homo* migrated from Africa much earlier than previously thought. *Homo* fossils dating from 1.9 to 1.7 mya have been found in Java, the Central Asian republic of Georgia, and China.

Overall, current research on *H. habilis*, *H. erectus*, and other early *Homo* species indicates that there were more species of *Homo* than once thought, and that several of these species existed in the same places and times. More research and evidence will be necessary before general agreement is reached regarding the number of early *Homo* species and their evolutionary relationships.

(a)

(b)

Figure 23.6   Tools from Long Ago
Toolmaking technology changed relatively little from 2.5 million to 400,000 years ago. The stone chopper (a) and stone handaxe (b) both date to 700,000 years ago.

■ The first members of the genus *Homo*, to which modern humans belong, originated in Africa 2 to 3 mya. By 400,000 to 500,000 years ago, our ancestors used fire and hunted large game animals. The number of early *Homo* species and their evolutionary relationships are the subject of ongoing debate.

The Origin and Spread of Modern Humans

Fossils with features that are intermediate between *Homo erectus* and *Homo sapiens* appear in the fossil record from 400,000 to 130,000 years ago. Known as archaic (meaning "old" or "early") *H. sapiens*, these fossils have been found in Africa, China, Java, and Europe. These ancestors of living humans developed new tools and new ways of making tools, used new foods, built complex shelters, and controlled the use of fire.

Early populations of archaic *H. sapiens* gave rise to both Neandertals (an advanced type of archaic *H. sapiens*) and anatomically modern humans. With that point in mind, let's return to a question posed in the chapter introduction: Were Neandertals direct ancestors of modern humans? In two recent studies, DNA extracted from Neandertal fossils was sequenced and compared with the DNA of modern humans. The results of these studies suggest that Neandertals were not direct ancestors of modern humans (Figure 23.7) and that there was relatively little interbreeding between the two groups. Some scientists dispute this conclusion, however, arguing that there is fossil evidence of such interbreeding, as we will see shortly.

The oldest fossils of anatomically modern humans, dating from 130,000 years ago, have been found in Africa. More recent fossils of anatomically modern humans have been found in such places as Israel (dating from 115,000 years ago), China (60,000 years ago), Australia (56,000 years ago), and the Americas (13,000 years ago).

There has been considerable controversy over the origin of anatomically modern humans. Two conflicting hypotheses have been proposed: the out-of-Africa hypothesis and the multiregional hypothesis. According to the **out-of-Africa hypothesis** (Figure 23.8a), modern humans first evolved in Africa sometime within the past 200,000 years. They then spread from Africa to the rest of the world, completely replacing archaic *H. sapiens*, including advanced forms such as the Neandertals. In contrast, the **multiregional hypothesis** (Figure 23.8b) proposes that modern humans evolved over time from *H. erectus* populations located throughout the world. According to this hypothesis, regional differences among human

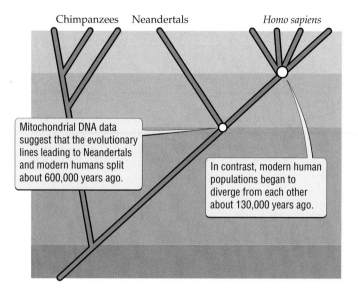

Figure 23.7 DNA Evidence from Fossils
DNA data from humans and from two Neandertal fossils suggest that Neandertals were not the direct ancestors of modern humans.

populations developed early, but worldwide gene flow caused these different populations to evolve modern characteristics simultaneously and to remain a single species.

Which of these hypotheses is correct? Let's consider some of the evidence. According to the multiregional hypothesis, when different populations of early humans came into contact, extensive gene flow should have caused them to become more similar to one another. Thus, we would not expect different types of early humans to coexist in the same geographic region, yet remain distinct for long periods of time. But in fact Neandertals and more modern humans coexisted in western Asia for about 80,000 years. Even as recently as 25,000–30,000 years ago, *H. sapiens* may have shared some parts of their range with *H. erectus* (*H. erectus* fossils from Java have been dated to 25,000 years ago). These findings call into question the extensive gene flow assumed by the multiregional hypothesis.

The best fossil evidence for the shift from archaic to modern *H. sapiens* comes from Africa, providing some support for the out-of-Africa hypothesis. Recent analyses of human genome data are also consistent with the out-of-Africa hypothesis. However, the "complete replacement" part of the hypothesis may not be correct. For example, fossils have been found that some scientists interpret as showing a mix of Neandertal and modern human characteristics. Similarly, some genetic studies indicate that genes from ancient *H. sapiens* populations outside Africa may have contributed to the

Figure 23.8 The Origin of Anatomically Modern Humans (*a*) According to the out-of-Africa hypothesis, anatomically modern humans originated in Africa within the past 200,000 years, then migrated to Europe and Asia, replacing *Homo erectus* and early *Homo sapiens* populations. (*b*) In contrast, the multiregional hypothesis proposes that populations of *H. erectus* in Africa, Asia, and Europe evolved simultaneously into anatomically modern humans.

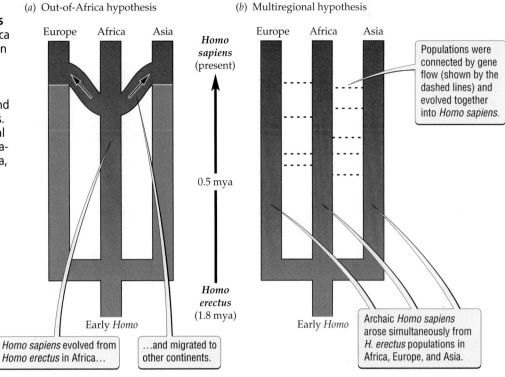

(*a*) Out-of-Africa hypothesis

Europe   Africa   Asia

Homo sapiens evolved from *Homo erectus* in Africa...

...and migrated to other continents.

(*b*) Multiregional hypothesis

Europe   Africa   Asia

Populations were connected by gene flow (shown by the dashed lines) and evolved together into *Homo sapiens*.

Archaic *Homo sapiens* arose simultaneously from *H. erectus* populations in Africa, Europe, and Asia.

Homo sapiens (present)

0.5 mya

Homo erectus (1.8 mya)

Early *Homo*

genetic makeup of modern humans, thus suggesting that limited interbreeding did take place.

In summary, many scientists think that anatomically modern humans arose in Africa and spread from there to other parts of the world. However, the origin of modern humans continues to be debated; this debate is especially active concerning the extent to which early *Homo sapiens* interbred with, and hence did not completely replace, more ancient *Homo* populations.

> ■ Fossils with a mix of features intermediate between *Homo erectus* and *Homo sapiens* appear in the fossil record 130,000 to 400,000 years ago. The oldest fossils of anatomically modern humans, dating from 130,000 years ago, have been found in Africa. Based on current genetic and fossil evidence, most scientists think that anatomically modern humans arose in Africa and spread from there to the rest of the world.

From Stone Tools to Agriculture

Homo toolmaking technology showed relatively little change from the time of origin of the first stone tools, roughly 2.5 million years ago, to 400,000 years ago (see Figure 23.6). However, over the past 300,000 years, a rich set of new tools was developed, first in Africa (Figure 23.9), then in other parts of the world. The archaeological record shows that these tools did not appear suddenly together, but rather were developed over a long period of time and a broad geographic area.

By 12,000 years ago, the tools used by human populations in different regions of the world were much more sophisticated than the tools used by early hominids. The increasing sophistication of human tools may have contributed to the origin of agriculture. Consider the following hypothetical scenario: Before the development of agriculture, human populations were small, mobile, and dependent on hunting and gathering (collecting edible plants and seeds, insects, and small vertebrates). By 12,000 years ago, our technology and culture may have developed to the point at which people could gather food more efficiently, allowing them to live in villages that shifted in location relatively infrequently. As villages grew in size and remained in one place for longer periods of time, it would be increasingly likely that seeds collected from the wild would "escape" and grow in disturbed areas (such as refuse disposal areas) near the village. Agriculture may have begun gradually as people first harvested and later deliberately cultivated the seeds of these plants.

However they began, the first agricultural societies were located in two parts of Eurasia, the eastern Medi-

(a)

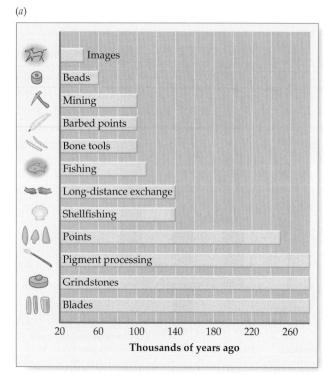

(b)

Figure 23.9 Advanced Stone Age Tools and Art
(*a*) New tools and art forms developed over the past 300,000 years in Africa. (*b*) Spearhead and figurine of a bison (40,000–10,000 years ago).

terranean region (about 10,500 years ago) and China (about 10,000 years ago). Agriculture also developed independently in other parts of the world, but considerably later (about 5500 years ago in the Americas). Early Eurasian farmers domesticated plants such as wheat, barley, peas, and rice and animals such as sheep, goats, pigs, and cattle.

Why did agriculture begin first in Eurasia and not in other regions of the world, such as Africa, the Americas, or Australia? The answer may hinge on several features of the environment. First, the number of potential domestic plant and animal species is much larger in Eurasia than in other regions of the world. For example, 32 of the world's 56 large-seeded grass species are located in the Mediterranean region, compared with 11 in the Americas, 4 in sub-Saharan Africa, and 2 in Australia. Thus, the world's prize grasses, from which crops such as wheat, barley, and rice were developed, were overwhelmingly concentrated in the Mediterranean, giving early farmers in that region a large head start. In addition, the wild ancestors of all five of the world's major domesticated animals (sheep, goats, cattle, pigs, and horses) were from Eurasia, again giving early Eurasian farmers a major advantage.

The development of agriculture had enormous consequences for human societies. Farming led to increased population sizes and a resulting increase in the complexity of human societies. As societies became larger and more complex, technology and other aspects of culture such as art and music expanded greatly, as did our capacity to conduct warfare. Finally, because people in agricultural societies lived in dense populations and in close contact with animals, humans were struck by new infectious diseases that spread from animals to humans; all of the major lethal infectious diseases, including smallpox, tuberculosis, and the plague, originated in this way.

■ Beginning about 300,000 years ago, a rich set of new tools was developed first in Africa, then in other parts of the world. The first agricultural societies arose 10,500 to 10,000 years ago in the eastern Mediterranean region and in China. The development of agriculture had a profound effect on human life, leading to an increase in the size and complexity of human societies, to great strides in technology and culture, and to the origin and spread of new lethal diseases.

Overview of Human Evolution

The fossil record covers 4 to 5 million years of hominid evolution. We conclude this chapter by emphasizing four important points that have emerged from the fossil evidence:

1. From 5 million to roughly 30,000 years ago, there was more than one hominid species alive. As recently as 25,000–30,000 years ago, *Homo sapiens* may have shared parts of their range with Neandertals and *Homo erectus*. Thus, hominid evolution did not consist of a single evolutionary line that began with *Ardipithecus ramidus* and ended with anatomically modern humans. Rather, the human evolutionary tree is "bushy," consisting of multiple species and many side branches at different times.

2. Some hominid traits evolved toward a modern condition earlier than others. For example, hominids were walking upright on two feet by 4.4 mya, but did not evolve exceptionally large brains until much later (Table 23.1).

3. Brain size increased greatly from early hominids to *Homo sapiens*, then decreased slightly in recent times (see Table 23.1).

4. Early *Homo* first made tools about 2.5 mya. Toolmaking technology improved slowly for more than 2 million years, then improved more rapidly over the past 300,000 years.

HIGHLIGHT

The Evolutionary Future of Humans

Now that we have examined our evolutionary past, what can it tell us about our evolutionary future? Will we go the way of science fiction stories and evolve toward beings with huge brains? Such a direction appears unlikely, since for the last 75,000 years human brain size appears to have been decreasing, not increasing (see Table 23.1). More realistically, what changes can we expect as a result of genetic drift, gene flow, and natural selection? (To review these concepts, see Chapter 20.)

Before the development of agriculture about 10,500 years ago, human populations were small and widely scattered. On a large geographic scale, these populations were isolated from one another by geographic barriers. For example, 30,000 years ago, the chance of an African meeting an Australian was virtually nil. Early human populations were probably isolated on a much smaller geographic scale as well, as illustrated by the fact that before the twentieth century, people from nearby valleys in New Guinea spoke different languages and had little contact with one another.

Translated into evolutionary terms, the conditions of early human populations were exactly those under which genetic drift should be important: Population sizes were small, and there was probably little gene flow among populations. Thus, we would predict that genetic drift has played a major role in causing genetic differences among human populations.

Some evidence supports this claim: Analyses of the rate of evolution of the skulls and teeth of modern humans indicate that genetic drift was a more important factor in their evolution than natural selection. Because the human population is now large and mobile, however, genetic drift is less likely to play a major role in future human evolution. Instead, high rates of gene flow

23.1 *Key Characteristics of Hominids, Chimpanzees, and Gorillas*				
Species	**Dates (thousands of years ago)**	**Body weight (kilograms)**	**Brain size (cubic centimeters)**	**Relative brain size (EQ)[a]**
Homo sapiens	Present	58	1349	5.3
Homo sapiens	35–10	65	1492	5.4
Neandertal	75–35	76	1498	4.8
Late *Homo erectus*	600–400	68	1090	3.8
Early *H. erectus*	1800–600	60	885	3.4
Homo habilis	2400–1600	42	631	3.3
Australopithecus africanus	3000–2300	36	470	2.7
Australopithecus afarensis	4000–2800	37	420	2.4
Chimpanzee	Present	45	395	2.0
Gorilla	Present	105	505	1.7

[a]The encephalization quotient (EQ) is the ratio of the actual brain size to the average brain size for a mammal of that body weight.

among human populations could substantially reduce their differences over time (Figure 23.10).

What about the role of natural selection? Our technological advances have removed many of the selection pressures our ancestors faced. Thus, there is now relatively little selection pressure on many characters (such as poor vision) that might have been greatly disadvantageous at previous times in our evolutionary history. This does not mean that natural selection will be powerless in the future. For example, infectious diseases take a terrible toll each year in human death and suffering; hence there remains strong selection pressure for the evolution of increased disease resistance in human populations.

In many ways, however, humans have now stepped outside the evolutionary framework described in this unit. In a sense, we have taken control of our own evolutionary future. We constantly modify both ourselves (at present, mostly in an indirect way through our use of tools and machines) and our environment to suit our needs. We rely on our culture—our technology, languages, institutions, and traditions—to adapt to our world. Furthermore, we transmit that culture within and between generations, changing it rapidly to meet new challenges. Cultural change proceeds at a far more rapid pace than biological evolution. This fast pace of change provides us with both hope and peril: Will we direct our future wisely?

■ Early human populations were small and isolated, conditions that favor evolution by genetic drift. Because the human population is now large and mobile, gene flow could lead to a blurring of existing differences among populations. Technological developments have removed many of the selection pressures faced by our ancestors. To a large extent, we now rely on rapid cultural change, rather than natural selection, to meet new challenges posed by our environment.

SUMMARY

Evolution of Mammals: From a Minor Role to Dominance

- Humans are apes, which are a type of primate. Primates are a type of mammal.
- Mammals evolved from mammal-like reptile ancestors about 220 mya.
- Throughout the long reign of the dinosaurs, mammals remained small and nocturnal.
- Following the extinction of the dinosaurs, the mammals radiated greatly to include many species that were large and active by day.

The Origin and Evolution of Primates

- Primates are thought to have originated more than 65 mya.
- Primates have flexible elbow and shoulder joints, opposable big thumbs and toes, and forward-facing eyes. These characteristics make primates well adapted for life in trees.
- Three characteristics became increasingly well developed during primate evolution: limb mobility and grasping ability, daytime vision, and brain size.

Hominid Evolution: The Switch to Walking Upright

- The first big step in human evolution was the switch to walking upright.
- The evolution of an upright posture probably was linked to a change from life in trees to life on the ground, a switch that probably occurred between 8 and 5 mya.

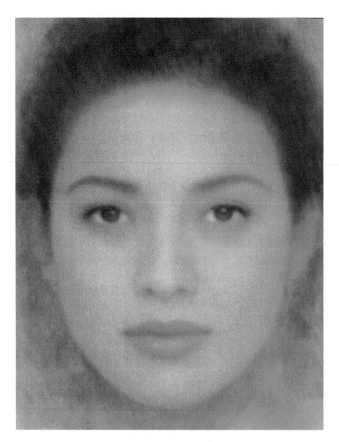

Figure 23.10   Gene Flow in Our Future
Gene flow among human populations could reduce some of the features that distinguish different groups of people. Future humans might look something like this computer composite image, which was formed from photographs of eight Afro-Caribbean models, eight Caucasian models, and eight Japanese models.

- Though they walked upright, early hominids had relatively small brains, and their jaws and skulls were more similar to those of nonhuman apes than to those of modern humans.

Evolution in the Genus Homo

- The first members of the genus *Homo*, to which modern humans belong, originated in Africa 2 to 3 mya.

- *H. habilis* fossils (from 1.9 to 1.6 mya) provide an excellent record of the evolutionary shift from ancestral (*Australopithecus*) to more recent (*Homo erectus*) characteristics.

- By 400,000–500,000 years ago, *Homo* species were using fire and hunting large game animals.

- The number of early *Homo* species and their evolutionary relationships are the subject of ongoing debate.

The Origin and Spread of Modern Humans

- Fossils with features intermediate between *Homo erectus* and *Homo sapiens* appear in the fossil record from 400,000 to 130,000 years ago.

- The oldest fossils of anatomically modern humans, dating from 130,000 years ago, have been found in Africa.

- Although the origin of anatomically modern humans is still the subject of a lively debate, genetic and fossil evidence suggests that they arose in Africa and spread from there to the rest of the world, but may have occasionally interbred with more ancient *Homo* populations.

From Stone Tools to Agriculture

- Beginning 300,000 years ago, a rich set of new tools was developed first in Africa, then in other parts of the world.

- The first agricultural societies arose 10,500 to 10,000 years ago in the eastern Mediterranean region and in China.

- The development of agriculture affected many aspects of human life, leading for example, to an increase in the size and complexity of human societies, to great strides in technology and other aspects of human culture, and to the origin and spread of new types of lethal diseases.

Overview of Human Evolution

- The human evolutionary tree is "bushy," consisting of multiple species and many side branches at different points in time.

- Some hominid traits evolved toward a modern condition earlier than others; for example, hominids evolved an upright walking posture long before they evolved large brain size.

- Brain size (volume of the braincase) increased greatly from early hominids to *Homo sapiens*.

- Early *Homo* first made tools 2.5 mya. Toolmaking technology improved slowly for more than 2 million years, then improved more rapidly over the past 300,000 years.

Highlight: The Evolutionary Future of Humans

- Early human populations were small and had little gene flow among them, conditions that favor evolution by genetic drift.

- Because the human population is now large and mobile, gene flow could lead to a blurring of existing differences among populations.

- With respect to natural selection, technological developments have removed many of the selection pressures faced by our ancestors.

- To a large extent, we now rely on rapid cultural change to meet new challenges posed by our environment.

KEY TERMS

hominid p. 390	out-of-Africa hypothesis p. 394
hominoid p. 389	
multiregional hypothesis p. 394	primate p. 389

CHAPTER REVIEW

Self-Quiz

1. An early (about 5 to 8 mya) and crucial step in human evolution was
   a. the development of large brains.
   b. a sudden and dramatic improvement in toolmaking technology.
   c. the switch to walking upright.
   d. genetic changes that improved spoken language.

2. Which of the following sentences is *not* true?
   a. A single evolutionary line led from *Ardipithecus ramidus* to modern humans.
   b. Some hominid traits evolved more rapidly than others.
   c. Brain size increased greatly from early hominids to *Homo sapiens*.
   d. Toolmaking technology improved greatly over the past 300,000 years.

3. Fossils of anatomically modern humans and Neandertals coexisted for thousands of years in many parts of the world, yet remained different. Such data are not what you would predict based on the
   a. out-of-Africa hypothesis.
   b. multiregional hypothesis.
   c. genetic drift hypothesis.
   d. Oldowan hypothesis.

4. Which of the following features do humans lack that other primates have?
   a. forward-facing eyes
   b. short snouts
   c. flexible shoulder and elbow joints
   d. opposable big toes

5. About 5 to 6 mya,
   a. humans evolved directly from chimpanzees.

b. the evolutionary line leading to humans split from that leading to monkeys.

c. the evolutionary lines leading to humans and chimpanzees both originated from a common ape ancestor.

d. primates first diverged from other mammals.

Review Questions

1. Gene flow has the potential to make living human populations far more similar than they are today. Would this change represent a net loss or gain for human societies? What factors would operate to prevent gene flow from making human populations more similar? (*Hint:* Refer to Chapter 20.)

2. Humans are so closely related to chimpanzees that some have referred to modern humans as the "fourth chimpanzee." Why do you think we label ourselves members of the genus *Homo*, instead of placing ourselves in the genus *Pan*, along with the other three species of chimpanzees?

3. How did agriculture develop, and what effect did it have on human societies?

23   **The Daily Globe**

Scientists Hunt for Genes that Make Us Human

RESEARCH TRIANGLE PARK, NC— Scientists are hot on the trail of the genes that make us human. Identifying these genes might appear to be an easy task, since we seem to differ greatly from our closest living relative, the chimpanzee. But the differences are not as great as you might think: Chimpanzees share many characteristics with humans, including tool use, a capacity for symbolic communication, a sense of self-awareness, the performance of deliberate acts of deception, and the passing on of social customs from generation to generation. And at the genetic level, the difference between chimpanzees and humans is downright minuscule: 98–99 percent of human DNA is identical to chimpanzee DNA.

Although we share many similarities with chimpanzees, we also differ from them in many and obvious ways. The extent to which humans make tools, solve problems, communicate using language, exterminate members of our own and other species, and create art, music, and other forms of culture are all unique to our species. How can the 1–2 percent difference in DNA between humans and chimpanzees have caused this enormous gulf between us?

That question drives the research of scientists who seek to use the recently completed human genome sequence to find the genes that make us human. Their goal is to understand the genetic basis of the differences between humans and chimpanzees. Scientists have begun to identify small differences in the gene sequences of the two species, looking for what may turn out to be a small handful of genes.

But other scientists question this research. How, they ask, will the results be used? If it proves possible, for example, to identify the genes that enable people to use language, will those genes be used in an attempt to design a talking chimpanzee? Of perhaps even greater concern, once we have identified the genes that make us human, will we succumb to the urge to redesign ourselves?

Evaluating "The News"

1. Do you think we should seek to understand the genetic basis of what makes us human?

2. The identification of the genes that make us human could be of considerable medical importance, allowing, for example, the design of new treatments that target genes that influence our susceptibility to infectious disease or to memory loss. Do such potential benefits outweigh the potential ethical dilemmas associated with creating designer chimpanzees or designer humans?

3. Assuming that it proves possible to identify the genes that make us human, do you think use of those genes should be regulated in any way? If not, why not? If so, how and by what mechanism should the use of those genes be controlled?

Robert Kushner, *Chrysanthemum Brocade*, 1994.

Form and Function

24

An Overview of the Form and Function of Life

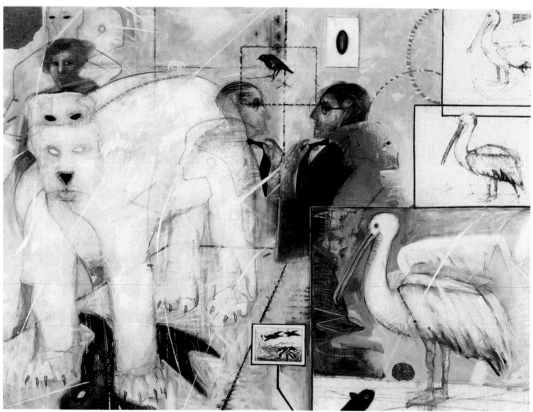

James G. Davis, *Museum of Natural History I*, 1986.

Pollution Detectives

Lichens grow all around us. Although you can easily overlook most lichens, you may have noticed some of the more spectacular kinds, hanging like gray-green beards from tree branches or forming colorful crusts on gravestones. Lichens are worth getting to know both for their bizarre biology and for what they have to teach us about our environment.

Lichens do not belong to any one of the six kingdoms of life described in Unit 1 of this book. Instead, each kind of lichen results from a remarkable interaction between a fungus species and a photosynthetic protist or bacteria species. The fungus forms the visible body of the lichen. When moistened, the fungal tissue becomes transparent and reveals a green layer of single-celled photosynthetic organisms living just

■ MAIN MESSAGE

All life shares a set of functions essential to survival, including the ability to regulate what enters and leaves cells.

These Orange and Gray Lichens Growing on a Gravestone Arise from the Interaction between Single-Celled Photosynthesizers and Fungi

beneath the surface. The fungus attaches to the surface of rocks or plants. It provides a protective environment for the photosynthetic organisms and exchanges mineral nutrients and water with them. The photosynthetic organisms, in turn, convert sunlight and carbon dioxide into sugars, which they share with the fungus. When the fungus and the photosynthetic organism are grown separately in the laboratory, neither develops into anything that looks or functions like a lichen. Lichen development depends entirely on the intimate interplay between the two component organisms.

Apart from their role as biological oddities, lichens have long been recognized for their ability to detect air pollution. Lichens show extreme sensitivity to pollutants in the air and typically are among the first living things to die as air quality drops. By monitoring the fates of lichens around factories and cities, we have access to an early warning system for environmental problems.

In this chapter, we introduce some of the functions basic to all life. We return to lichens at the end of the chapter to consider how some of the basic features of the biology of the fungus and the photosynthetic organism contribute to making lichens such excellent pollution detectives.

■ KEY CONCEPTS

1. The plasma membrane acts as a selective filter that determines which materials enter and leave a cell.

2. The exchange of materials between living organisms and their environment depends on a combination of passive transport, which requires no energy input, and active transport, which does require energy.

3. The survival of any organism depends on its ability to carry out certain basic functions, which include, most conspicuously, the exchange of materials with the environment and reproduction.

4. The cell is the basic unit of life. Cells are arranged in an organized fashion to form complex multicellular plants, animals, and other organisms.

One of the themes of Unit 5 is that all living things deal with the challenges of survival and reproduction in biologically related ways. All prokaryotes and both single-celled and multicellular eukaryotes share several basic features that define them as living organisms. As outlined in Chapter 1, living things consist of highly organized cells that respond to and capture energy from their surroundings, reproduce and develop, and evolve from one generation to the next. When the chemical factories inside cells are working properly, organisms of all sorts can carry out a shared set of tasks that begins with the gathering of energy and culminates in the production of offspring. A carefully controlled exchange of materials between cells and their environment creates conditions inside the cell that can support the chemical reactions of life. We begin this chapter by discussing how organisms control the conditions inside their cells; next, we survey the things that all living organisms must do to survive and reproduce.

Although similar in some basic and important ways, organisms vary in form in ways that influence how they function. Plants and animals, for example, which have followed separate evolutionary paths for perhaps a billion years, carry out their shared set of tasks in ways that most of us would recognize as distinctively plantlike or animal-like. Accordingly, this chapter closes with a look at some of the essential differences in form among the six kingdoms introduced in Chapter 2.

24.1   *How Do Organisms Control Conditions Inside Their Cells?*

Most of the chemical reactions that sustain life cannot take place outside the cells of which all organisms are built. Cells manage to maintain suitable conditions for the chemistry of life only by carefully controlling the uptake of materials from and loss of materials to their environment (Figure 24.1). Thus, all organisms must have a way of moving materials into and out of their cells, as well as a way of controlling which materials can enter or leave. In this section we discuss how the plasma membrane of the cell controls the exchange of materials between the cell and the environment in which it lives.

Figure 24.1   Life Depends on the Exchange of Materials between Organisms and Their Environment
Marathon runners lose water and salt to the environment by sweating and gain water by drinking. Every breath brings oxygen to working cells and takes away carbon dioxide that the cells release as a waste product.

Everything that enters or leaves a cell crosses the plasma membrane

The plasma membrane, as we saw in Chapter 6, separates the inside of a cell from the outside environment. The structure of the plasma membrane is as universal a feature of life on Earth as is the DNA-based genetic code. A double layer of lipids, or **phospholipid bilayer**, provides the framework for the plasma membrane (Figure 24.2). Although some biologically important materials can pass directly through the phospholipid bilayer, most cannot. Embedded in the phospholipid bilayer are proteins, which together typically make up more than half the weight of the plasma membrane. Many of these proteins span the plasma membrane and provide a path by which materials can enter or leave cells. Together, the phospholipid bilayer and its associated proteins act as a selective filter, controlling which materials enter and leave the cell (see the box on p. 409).

Cells can move materials across the plasma membrane either with or without the expenditure of energy

Two general rules can help us understand how materials move into and out of cells:

1. Molecules move down **concentration gradients**—that is, from areas of abundance to areas of scarcity—and they do so **passively** (with no energy input).
2. Molecules can move up concentration gradients—that is, from areas of scarcity to areas of relative abundance—but must do so **actively** (with energy input).

We can use the physical example of a ball moving down or up a hill, in which the ball represents a chemical and the hill represents a concentration gradient (Figure 24.3). The ball rolls downhill on its own, but it cannot roll uphill unless we provide the energy needed to push it.

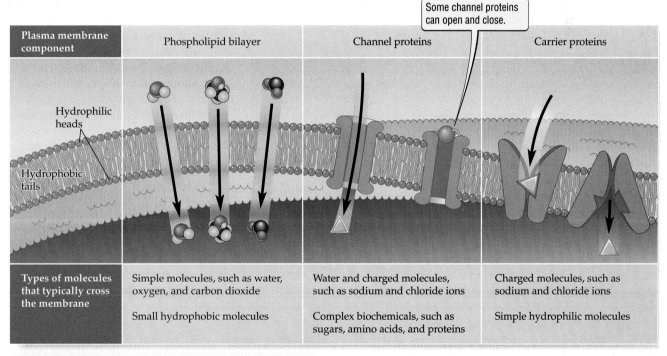

Plasma membrane component	Phospholipid bilayer	Channel proteins	Carrier proteins
	Hydrophilic heads / Hydrophobic tails	Some channel proteins can open and close.	
Types of molecules that typically cross the membrane	Simple molecules, such as water, oxygen, and carbon dioxide Small hydrophobic molecules	Water and charged molecules, such as sodium and chloride ions Complex biochemicals, such as sugars, amino acids, and proteins	Charged molecules, such as sodium and chloride ions Simple hydrophilic molecules

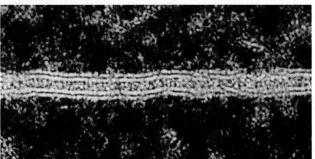

Figure 24.2 The Plasma Membrane Controls What Enters and Leaves a Cell
The organization of the plasma membrane determines what materials enter and leave a cell. The basic structure of the plasma membrane consists of a double layer of phospholipid molecules with hydrophilic "heads" and hydrophobic "tails." This phospholipid bilayer is clearly visible in the electron micrograph at left. Proteins that span the plasma membrane play an important role in the movement of materials into and out of cells. Different kinds of biologically important molecules cross the membrane by different routes.

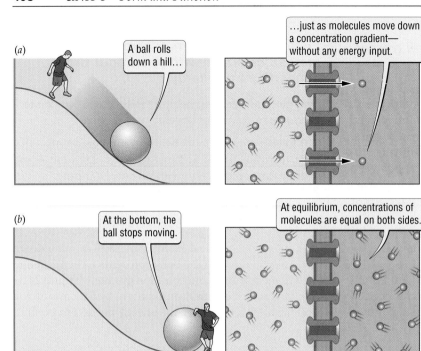

(a)

A ball rolls down a hill…

…just as molecules move down a concentration gradient— without any energy input.

(b)

At the bottom, the ball stops moving.

At equilibrium, concentrations of molecules are equal on both sides.

(c)

We must work to move the ball back uphill…

… just as a cell must work to move molecules up a concentration gradient.

Energy

Energy

24.2

Figure 24.3   Active versus Passive Movement of Molecules
Materials can move into and out of organisms either passively (without an input of energy) or actively (with an input of energy). (*a*) Molecules can move passively through a membrane from areas of high concentration to areas of low concentration. (*b*) When the concentration of molecules is equal on both sides of a membrane, the molecules have no tendency to move. (*c*) Energy is required to move materials from areas of low concentration to areas of high concentration.

wastes. Active transport, which requires energy, plays an essential role in allowing organisms to move the raw materials for their biochemical reactions from the environment, where they exist at low concentrations, into cells, where they exist at high concentrations. Without active transport, organisms could not survive. As we will see throughout this unit, however, the same diffusion process that allows color to spread through a pitcher of water also plays a key role in the transfer of some of the most basic molecules, such as water, oxygen, and carbon dioxide, into and out of cells.

Imagine that you empty a packet of Kool-Aid powder into a pitcher of water. You can watch the coloring agents in the powder gradually spread from the area of high concentration—the powder—throughout the water in the pitcher. This will happen even if you do not expend any energy to stir the mixture. The coloring agent will continue to **diffuse**, or spread passively, until it is evenly distributed throughout the water. Once the color is distributed evenly, the concentration differences that make diffusion possible have disappeared and diffusion stops.

In contrast, concentrated drink powder will never form spontaneously in a pitcher of Kool-Aid, no matter how long you watch. The formation of concentrated powder requires that the chemicals in Kool-Aid move up a concentration gradient from relatively low concentrations in the drink to high concentrations in the powder. We can re-form the powder only by adding energy—for example, as heat—to evaporate the water.

Organisms rely heavily on both passive and active transport to take up and get rid of nutrients, gases, and

Small molecules can diffuse through the phospholipid bilayer

Materials that can cross the phospholipid bilayer of the plasma membrane do so strictly by moving passively down concentration gradients. Water, oxygen, and carbon dioxide usually enter and leave cells in this way (see Figure 24.2). All these molecules are small and simple, consisting of just a few atoms each. In addition, some simple hydrophobic (*hydro*, "water"; *phobic*, "fearing"; see Chapter 5) molecules can pass through the hydrophobic core of the phospholipid bilayer. (Many of the early pesticides, such as DDT, worked effectively precisely because they could easily get into the cells of insect pests in this way.)

At the same time, the phospholipid bilayer acts as a barrier to the movement of most kinds of biologically important molecules. Even nutrients such as the simplest sugars and amino acids, for example, consist of more than 20 atoms each, and thus are both too large and too hydrophilic (*philic*, "loving") to diffuse through the hydrophobic core of the phospholipid bilayer. Various ions (small atoms or molecules carrying a positive or

■ BIOLOGY IN OUR LIVES

Dogsleds and Diphtheria

Every winter, dozens of dog teams head into the bitterly cold Alaskan winter to start the Iditarod, a grueling dogsled race that covers over 1500 kilometers along a route connecting Anchorage with the isolated town of Nome. More than simply a race, the Iditarod commemorates a dramatic 1925 dogsled relay that brought life-saving medicine to Nome, where an epidemic of diphtheria had broken out. Today most people receive immunizations against diphtheria, but in the early 1900s it ranked among the most feared of childhood diseases. This disease affected mostly children under the age of 10, causing up to half of its young victims to suffocate when swelling of tissues blocked their throats. The desperate race to save the lives of Nome's children made the headlines in newspapers around the world. Today, a statue in New York City's Central Park still honors Balto, the lead dog on the sled that carried the medicine into Nome just five and a half days after it left Anchorage.

In addition to regulating which materials enter and leave a cell, the plasma membrane protects the cell from invaders. Diphtheria results from a three-stage attack on the plasma membranes of human cells by diphtheria toxin, a protein produced inside the bacterium *Corynebacterium diphtheriae*. Diphtheria toxin has three distinct sections: the A and B domains at the two ends of the protein, and the T domain in the middle. In the first stage of its attack, the B domain of diphtheria toxin binds to a receptor protein embedded in the plasma membrane, which causes a small bubble, or vesicle, of plasma membrane to pinch off. As a result,

the diphtheria toxin is trapped inside a vesicle inside the human cell under attack. The second stage of the attack starts as the hydrophobic T domain embeds itself in the vesicle's hydrophobic phospholipid bilayer. A change in the shape of the embedded T domain carries the A domain across the plasma membrane and into the cell. In the third stage the A domain breaks off inside the cell and stops

protein synthesis. A single A domain can shut down protein synthesis in a cell, killing it.

As an ironic twist on the way in which *C. diphtheria* attacks its victims, only *C. diphtheria* cells infected by a virus can produce the diphtheria toxin on which their success as a disease depends. The virus, not the bacterium, carries the DNA that codes for diphtheria toxin.

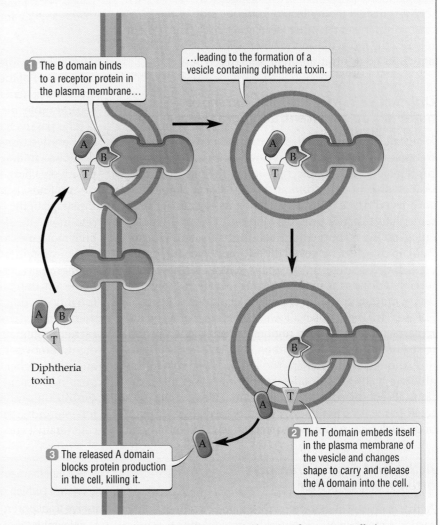

1 The B domain binds to a receptor protein in the plasma membrane…

…leading to the formation of a vesicle containing diphtheria toxin.

Diphtheria toxin

2 The T domain embeds itself in the plasma membrane of the vesicle and changes shape to carry and release the A domain into the cell.

3 The released A domain blocks protein production in the cell, killing it.

Diphtheria Toxin Penetrates the Plasma Membrane of Human Cells in a Three-Stage Process

negative charge) are also too hydrophilic to penetrate the phospholipid bilayer. As we will see, the proteins embedded in the membrane provide the only path by which these molecules can cross the plasma membrane.

Channel and passive carrier proteins allow molecules to cross the plasma membrane passively

Two types of plasma membrane proteins allow large and hydrophilic molecules to cross the plasma membrane passively. **Channel proteins** form openings through the phospholipid bilayer that allow hydrophilic molecules of the right size and charge to move through the plasma membrane down a concentration gradient (see Figure 24.2). **Passive carrier proteins** can bind to a particular molecule that fits into the folds of the protein. When this happens, the protein changes its shape in such a way that it transfers the molecule from one side of the plasma membrane to the other (see Figure 24.2). The protein releases the molecule on the other side of the membrane only if its concentration there is relatively low.

Only active carrier proteins can move materials up a concentration gradient

Molecules can cross a plasma membrane up a concentration gradient only by active transport. **Active carrier proteins** can move molecules across the plasma membrane using energy from an energy storage molecule such as ATP (see Chapter 5). Like passive carrier proteins, active carrier proteins bind only to certain molecules having a shape that fits into the protein (Figure 24.4). In this case, however, the addition of energy causes a shape change in the active carrier protein that forcibly releases the molecule being transferred, regardless of the concentration of that molecule near the site of release. This mechanism allows active transport proteins to carry molecules from regions of low concentration to regions of high concentration.

Although active transport plays a critical role in the ability of cells to maintain appropriate internal conditions, its energy cost can be substantial. For example, 30 to 40 percent of the energy used by a resting human body fuels active transport across plasma membranes.

Active and passive transport often work together

Active carrier proteins often work together with passive carrier proteins to move molecules up concentration gradients. In these situations, an active carrier pumps positively charged ions—typically sodium ions in animals and hydrogen ions in other organisms—against a con-

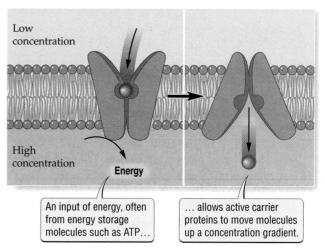

Low concentration

High concentration

Energy

An input of energy, often from energy storage molecules such as ATP…

… allows active carrier proteins to move molecules up a concentration gradient.

24.3

Figure 24.4   Active Carrier Proteins
Active carrier proteins use energy to move materials from areas of low concentration to areas of high concentration.

centration gradient so that ion concentrations build up on one side of the membrane. These ions then move passively back across the membrane, carrying with them a second molecule.

The passive carrier proteins involved in this type of cooperative transfer differ from the simple ones already described in that they bind two different molecules: the positively charged ion and a second molecule. Such carrier proteins usually bind first to the ion, which changes the shape of the carrier protein so that it can bind the second molecule. Binding to the second molecule transfers both molecules to the opposite side of the membrane. The carrier protein releases the ion readily on the side of the membrane where its concentration is low. Upon release of the ion, the carrier protein returns to its original shape, forcibly ejecting the second molecule on the side of the membrane where its concentration is high. Although the carrier protein moves the ion down a concentration gradient, it moves the other molecule against a gradient to an area of higher concentration. This indirect active transport mechanism allows a single power source—a concentration gradient in one kind of positively charged ion—to move many different molecules into or out of cells against concentration gradients.

■ The plasma membrane acts as a selective filter that controls which materials can enter and leave cells. Passive transport requires no energy, but can move materials only down concentration gradients. Active transport can move materials up concentration gradients, but requires energy.

The Functions Essential to Life

Although organisms make their living in an overwhelming variety of ways, we can identify certain functions that are essential to the success of all organisms (Figure 24.5). Here, and elsewhere in this unit, we sometimes consider a function separately from the others, but keep in mind that the various functions depend heavily on one another. In this section we merely introduce the various functions; the remaining chapters in this unit cover each function in much greater detail.

The transfer of materials into and out of organisms makes life possible

Much of what any organism does to accumulate the resources it needs to stay alive and reproduce depends on its ability to transfer the materials it needs into its body and move the wastes it produces out of its body. Organisms take up nutrients, such as the sugars and amino acids animals get from their food and the mineral nutrients that plants extract from the soil (Chapter 28).

Other essential molecules, such as oxygen and carbon dioxide, move into and out of the body as gases and therefore require different kinds of exchange structures (Chapter 29). The ability of organisms to maintain a relatively constant and appropriate environment inside their cells (see Chapter 31) depends on the constant transfer of materials into and out of organisms.

Many common functions speed the transfer of materials

Several functions that are carried out by most, though not all, organisms make it much easier for organisms to accumulate resources. Most organisms can support their bodies against the pull of gravity, allowing them to maintain shapes that determine how they interact with their environment (Chapter 26). Many organisms can move from one place to another, an ability that gives them some control over where they look for resources (Chapter 27). An essential part of survival is defense against being eaten oneself or becoming diseased (Chapter 32). Finally, to coordinate these various functions so that they take place in the right way and under the right conditions, organisms must have a way of sensing what is happening in their environment (Chapter 34) and of responding to that information (Chapter 35).

Reproduction measures an organism's success

Individuals that accumulate enough resources can eventually reproduce—that is, they can make genetic copies of themselves (Chapter 36). Reproduction measures how successfully individuals accumulate resources: Individuals that accumulate resources quickly produce more offspring more rapidly than do those that accumulate resources more slowly. Thinking back to our discussion of natural selection (see Chapter 20), recall that individuals that produce the most offspring the most quickly generally contribute the most genes to the next generation. Thus reproduction connects individual success with evolutionary processes.

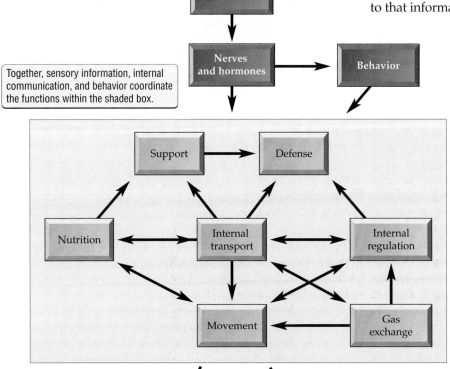

Together, sensory information, internal communication, and behavior coordinate the functions within the shaded box.

Reproduction and development depend on the functions within the shaded box.

Figure 24.5   The Functions Basic to Life
All organisms must carry out a number of interrelated functions to survive and reproduce. The functions listed here correspond to the topics discussed in Chapters 26–37 of this unit.

Multicellular organisms face challenges not faced by single-celled organisms

Large multicellular plants, animals, and fungi face some special problems that single-celled organisms do not. First, multicellular organisms must develop from a single-celled fertilized egg into an adult organism consisting of huge numbers of precisely organized cells (Chapter 37). Second, once a nutrient or gas, for example, enters the body of a multicellular organism, it has not necessarily reached the cells where it will be used. Cells in the human liver or heart, for example, lie at a distance from the lungs and intestines that absorb oxygen and nutrients, respectively. Thus, most multicellular organisms have a way of rapidly distributing materials from the place where they enter the body to the cells that use them (Chapter 30). Finally, multicellular organisms must coordinate the functions of the individual cells. This coordination requires internal communication systems that allow the exchange of information among cells throughout the body (Chapters 33 and 34).

> ■ All organisms, regardless of their form or way of making a living, must carry out a series of interrelated functions.

How Do Differences in Form Lead to Differences in Function?

All life consists of basic units called cells, within which all of the functions basic to life take place (see Chapter 6). Successful survival and reproduction require that the cell or cells that make up an organism meet a number of challenges faced by all life. Each of the 1.5 million or so described species on Earth, however, has its cells arranged in a unique way. The human body consists of many trillions of cells, each of which shares certain essential features with the cells of all other living organisms. Like a human, a tomato plant consists of a huge number of cells, but they are arranged to allow a lifestyle dramatically different from our own. In contrast, an *Escherichia coli* bacterium in the human digestive system thrives in its environment while consisting of but a single cell.

In the last section of this chapter we consider how the diversity of form influences how organisms deal with the basic tasks that confront all life. Differences in the complexity of organization, the support of individual cells, and the source of energy define many of the differences among the six kingdoms that we will encounter in this unit.

Multicellular organisms can devote cells to a specialized function

Whereas most archaeans, bacteria, and protists consist of a single cell, plants, fungi, and animals each consist of many cells. In single-celled organisms, all the functions essential to life occur in one all-purpose cell (Figure 24.6a). All of the vast number of cells in a multicellular plant, fungus, or animal are similar in design and size to the single-celled protists from which they evolved. Because of this essential similarity, the individual cells of multicellular organisms cannot do anything that the single cell of a protist cannot. In multicellular organisms, however, each cell is specialized to perform just a few of the many functions needed to keep the organism alive. As a result, the individual cells of plants and animals can often perform their particular specialties more effectively than the all-purpose cell of a single-celled organism can. On the other hand, because individual plant or animal cells depend on other cells in the body to carry out the functions that they cannot, they could not survive as individual cells.

The single cell that makes up a bacterium or protist is surrounded by a physical environment over which it has little control. In contrast, most cells in most multicellular organisms lie inside the body, surrounded by other cells. This means that multicellular organisms have much more control over the environment in which their cells function, which reduces the difficulties faced by their cells in controlling what crosses their plasma membrane.

Differences among multicellular organisms arise from how they get energy and maintain the shape of their cells

Among the multicellular organisms, plants obtain their energy from sunlight, whereas animals and fungi rely on other organisms for their energy. Plants, as a result, can get most of the raw materials that they need from their physical environment. Animals and fungi, in contrast, must take what they need from other organisms.

Plants and fungi have rigid cell walls that hold the shape of individual cells. Animals lack cell walls, which allows their cells to change shape easily. This seemingly minor difference makes possible the muscles that allow animals to move to find food and mates and to escape danger. Plants and fungi cannot move, so they must find food and mates and deal with danger in different ways than animals. On the other hand, because of their flexible cells, animals require complex skeletons to provide their bodies with support and shape, and require the rapid communication made possible by nervous systems to control their movements.

(a)

Single-celled organisms must perform all the functions essential to life within one all-purpose cell.

Organism

(b)

Plants consist of three basic organs...

...that differ mainly in the arrangement of the three different tissue systems: dermal (red), ground (green), and vascular (yellow).

Organism   **Organs**

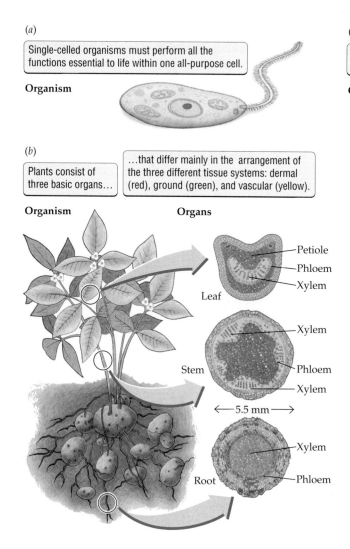

Petiole
Phloem
Xylem
Leaf

Xylem
Stem
Phloem
Xylem
← 5.5 mm →

Xylem
Root
Phloem

(c)

The organs of animals, which are devoted to specialized functions...

Organism   **Organ**

...are formed from tissues made up of different cell types.

System

Tissue

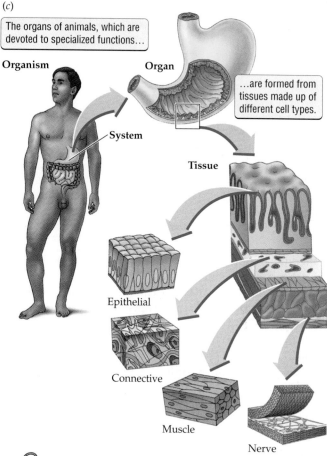

Epithelial

Connective

Muscle

Nerve

24.4

Figure 24.6 The Organization of Life
(a) Single-celled organisms such as bacteria and protists must perform all the basic functions of life within a single, all-purpose cell. The bodies of multicellular plants (b) and animals (c) consist of organs that work together, each carrying out one or a few functions. These organs are made up of tissues, which in turn are made up of cells that are specialized for particular functions.

The cells in multicellular organisms are highly organized

The millions of cells that make up the body of a typical plant or animal are highly organized (Figure 24.6b and c). Think of the human body: We are not random masses of specialized cells. Our bodies contain recognizable **organs**, such as eyes and lungs, each consisting of a collection of cells that work together to perform a shared function. Perhaps the most surprising feature of the organization of plant and animal bodies is how few basically different cell types form the many different specialized organs that make up a functioning organism. Animals consist of four distinctly different cell types, plants of ten types.

In animals, cells of the same basic type are gathered into **tissues**, which consist of many similar cells. The cells in a tissue produce a **matrix** (plural matrices) of chemicals, which surrounds them and glues them together. Each organ, in turn, consists of a characteristic arrangement

of several tissues that together perform a specific task. Our intestines, for example, are made up of the four animal tissues (Figure 24.6c), working together to absorb nutrients from food that we have digested. Plant leaves consist of three tissue systems (Figure 24.6b), working together to convert the sun's energy into sugars.

Biologists recognize four basic cell types in animals

Although more than 200 different kinds of animal cells have been identified, they all fall into just four broad classes whose members share the same basic function and appearance (Figure 24.6c). **Epithelial cells** form the surfaces of animals and line their body cavities. A matrix

24.1   *The Organization of Plant Cell Types into Tissues and Tissue Systems*

Cell types	Tissue	Tissue system	Function
Parenchyma, fiber, sclereid	Epidermis	Dermal	Protection, absorption
Parenchyma, fiber, sclereid	Periderm	Dermal	Protection
Tracheid, vessel, parenchyma, fiber, sclereid	Xylem	Vascular	Water and nutrient transport, support
Sieve, sieve tube, albuminous, companion, parenchyma, fiber, sclereid	Phloem	Vascular	Transport of photosynthetic products
Parenchyma	Parenchyma	Ground	Photosynthesis, storage, respiration, support
Collenchyma	Collenchyma	Ground	Support
Fiber, sclereid	Sclerenchyma	Ground	Support

called the basal membrane holds epithelial cells tightly together to form epithelial tissue. The epithelial tissues that form the surface of the skin act as a barrier between the inside of the body and the outside environment. In contrast, the epithelial tissues that line the lungs and intestines control how gases and nutrients enter and leave the body.

Connective cells perform many functions, including, most prominently, producing the matrix that connects and supports other cells in the body. Unlike epithelial cells, which are tightly packed, connective cells typically lie loosely packed in a matrix that occupies more space than the cells themselves. Connective cells form the bones that support the body and provide the elastic connections that allow the skin to bend and stretch. Both epithelial and connective cells replace themselves by dividing throughout the life of the organism.

In contrast, the more narrowly specialized muscle and nerve cells divide only slowly or stop dividing altogether shortly after birth. **Muscle cells** can contract, giving animals their unique ability to move. **Nerve cells** are highly specialized for transmitting signals from one part of the body to another, allowing an animal to coordinate the function of its various body parts.

Biologists recognize ten basic types of plant cells arranged into seven tissues and three tissue systems

Plant biologists use a system of grouping the diversity of plant cells that differs from that used for animal cells. In plants, ten structurally and functionally distinct cell types combine to form seven types of tissues (Table 24.1). Whereas animal biologists define a tissue as consisting of a single cell type, plant biologists define tissues as consisting of certain combinations of the ten basic plant cell types. As is true of animal tissues, however, each plant tissue fulfills a well-defined function.

We can further group the seven plant tissues into three **tissue systems**, each of which consists of several tissues grouped together to serve a particular function (Table 24.1). The **dermal tissue system** forms the outer surface of the plant. Like the epithelial tissue of animals, the dermal tissue system protects the plant (on the aboveground stem and leaves) and absorbs nutrients (on the roots). The **vascular tissue system** transports water and mineral nutrients from the roots to the leaves, and moves sugars produced in the leaves by photosynthesis to other parts of the plant. The **ground tissue system** plays an important role in supporting the plant, and it houses the sites of photosynthesis and nutrient storage.

The concept of tissue systems helps us understand plant structure because each of the three basic plant organs—roots, stems, and leaves—represents a different arrangement of the three tissue systems (see Figure 24.6*b*).

■ The cell is the basic functional unit of all organisms. In single-celled organisms one cell must serve many functions. In multicellular organisms individual cells can specialize. Cells of similar types combine to form tissues, and the various tissues, in turn, are arranged into organs that perform particular functions.

■ HIGHLIGHT

Why Are Lichens So Sensitive to Air Pollution?

The city of Sudbury, in the Canadian province of Ontario, is surrounded by some of the world's richest nickel deposits. Large smelter operations convert the ore into 2 million kilograms of nickel each year. Large amounts of sulfur dioxide (SO_2) are released into the air each year as a by-product of the smelting process. SO_2 reacts with water to produce sulfuric acid, which leads to acid rain. The high SO_2 concentrations in the air around Sudbury have given the city an unenviable reputation as a biological wasteland.

Lichens were one of the first groups of organisms to succumb to the SO_2 pollution around Sudbury (Figure 24.7). Unlike most plants, the fungal species in lichens absorb nutrients from the atmosphere and from precipitation extremely efficiently by both passive and active transport. In unpolluted environments, this ability to take up nutrients from the air and rain allows the fungus to gather the chemical nutrients needed by the photosynthetic organisms. In the area surrounding Sudbury, however, it also ensured that the lichens would rapidly accumulate SO_2. Inside the lichen, the SO_2 combined with water to form sulfuric acid. The sulfuric acid broke down the phospholipid bilayer of the plasma membranes, creating leaky cells. The leaky membranes disrupted the exchange of nutrients between the single-celled photosynthesizers and the associated fungus.

Since 1970, when air pollution around Sudbury was at its worst, a number of events have reduced the amount of SO_2 deposited in the surrounding countryside. Improved SO_2 recovery at the smelters for the production of marketable sulfuric acid has reduced the SO_2 emitted from the chimneys. The construction of the world's tallest chimney at one of Sudbury's three smelters has reduced SO_2 fallout by causing the chemical to travel farther from Sudbury before settling to the ground. Finally, one of the other two smelters has closed. Together, these events have reduced the SO_2 falling on the landscape around Sudbury from 2.5 million tons to less than 1 million tons each year. In response, the lichen abundance and diversity around Sudbury has slowly begun to increase (see Figure 24.7).

> ■ The fungal component of lichens efficiently absorbs nutrients from the air and rain as a way of providing the photosynthesizing organism with nutrients. In polluted air, this efficient nutrient-gathering mechanism leads lichens to accumulate toxic quantities of sulfur dioxide, which break down the plasma membranes of cells, disrupting the exchange of nutrients between the fungus and the photosynthesizing organisms.

(a) Before giant chimneys (1968)

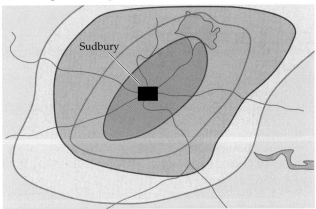

(b) After giant chimneys (1990)

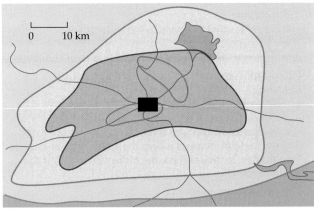

Lowest abundance and diversity of lichens   Highest abundance and diversity of lichens

Figure 24.7 Giant Chimneys Near Sudbury, Ontario, Release Sulfur Dioxide into the Air
Lichens growing on trees around Sudbury died when they took up SO_2 from the air. (a) The building of the 380-meter-tall chimney (below) in the 1970s reduced SO_2 pollution around Sudbury by ensuring that the SO_2 did not reach ground level near the city. (b) As local SO_2 levels decreased, the lichens around Sudbury recovered somewhat.

SUMMARY

How Do Organisms Control Conditions Inside Their Cells?

- The plasma membrane, which consists of a phospholipid bilayer containing embedded proteins, controls the movement of materials into and out of cells.

- Passive movement from areas of high concentration to areas of low concentration provides an energy-free way of moving materials.

- Active transport makes it possible to move materials from areas of low concentration to areas of high concentration, but it requires energy.

- Small, simple molecules such as oxygen and carbon dioxide cross the phospholipid bilayer passively by diffusion.

- Channel and passive carrier proteins selectively allow certain charged ions and large molecules such as sugars and amino acids to cross the plasma membrane passively.

- Only active carrier proteins can transport materials up a concentration gradient, but they consume a great deal of energy in doing so.

- Active carriers of positively charged ions often act together with passive transport mechanisms to move molecules indirectly against a concentration gradient.

The Functions Essential to Life

- All organisms, regardless of their form or way of making a living, must carry out a series of interrelated functions.

- Functions that transfer materials into and out of organisms make life possible.

- Reproduction links the success of individuals to evolutionary processes.

- Multicellular organisms carry out various functions related to the development of multiple cells and to the transport of materials and information among those cells.

How Do Differences in Form Lead to Differences in Function?

- The cell is the basic unit of life.

- Multicellular organisms can devote individual cells to specialized tasks, and can provide their cells with a more protected environment than is possible in single-celled organisms.

- The rigid cell walls of plants and fungi keep them from moving around like animals, which have cells that lack cell walls.

- Photosynthesizing plants obtain their energy from the sun. Fungi and animals rely on other organisms for energy.

- In multicellular organisms, cells are organized into tissues, and tissues are grouped into organs that serve specific functions.

- All animal organs consist of various arrangements of four basic cell types. All plant organs consist of various arrangements of ten basic cell types.

Highlight: Why Are Lichens So Sensitive to Air Pollution?

- The fungal component of lichens efficiently absorbs nutrients from air and rain to provide the photosynthesizing organisms with the nutrients that they need.

- In polluted air, this efficient nutrient-gathering mechanism causes lichens to accumulate toxic quantities of sulfur dioxide, which acts by breaking down the plasma membranes of cells.

KEY TERMS

active carrier protein p. 410
active transport p. 407
channel protein p. 410
concentration gradient p. 407
connective cell p. 414
dermal tissue system p. 414
diffusion p. 408
epithelial cell p. 413
ground tissue system p. 414
matrix p. 413
muscle cell p. 414
nerve cell p. 414
organ p. 413
passive carrier protein p. 410
passive transport p. 407
phospholipid bilayer p. 407
tissue p. 413
tissue system p. 414
vascular tissue system p. 414

CHAPTER REVIEW

Self-Quiz

1. The single cell of a bacterium
   a. typically carries out many more functions than each cell in the human body does.
   b. can survive on its own, whereas the individual cells of a human being cannot.
   c. lacks a plasma membrane.
   d. both a and b

2. Which of the following gives reproduction a unique place among the basic functions of life?
   a. It is the only function carried out by all organisms.
   b. It is the only function without which an organism could not survive.
   c. It connects what happens to individuals with evolutionary processes.
   d. It does not depend on any of the other basic functions.

3. Passive movement of a molecule involves which of the following?
   a. movement down a concentration gradient
   b. the expenditure of energy
   c. movement of molecules unimportant to living organisms
   d. movement from areas where the molecule is at low concentration to areas where the molecule is at high concentration

4. The plasma membrane of a cell
   a. plays an important role in determining which materials the cell can take up.
   b. is found only in the cells of multicellular plants and animals.

c. consists of pure lipids arranged in a characteristic bilayer.

d. both a and b

5. Active carrier proteins do which of the following?

a. carry molecules up concentration gradients

b. work together with passive carrier proteins to transport molecules across plasma membranes

c. use energy storage molecules such as ATP to transport molecules across plasma membranes

d. all of the above

Review Questions

1. When we wash with soap, we kill bacteria on our hands by breaking open their plasma membranes. Why does destruction of the plasma membrane kill a bacterium?

2. One of the main points of this chapter is that despite their outward differences, all living things have many features in common. In light of what you learned about evolution in Unit 4, do the shared features of life come as a surprise? Why or why not?

3. The various functions that make life possible all interact closely with one another. Although we will discuss some of these interactions in the chapters to follow, can you think of any examples from your own experience of interactions between the functions described in this chapter?

4. Consider how it is possible to create an organism as complex as a tree from a very small number of different kinds of plant cells.

5. Compare the challenges faced by the single cell of a bacterium to those faced by an individual cell in the human body.

---

24 ## The Daily Globe

Improper Formula Preparation Puts Babies at Risk

Portland, OR—The Portland Regional Health Authority (PRHA) unveiled a campaign yesterday that will warn new parents of the risks associated with improper preparation of baby formula.

A recent spate of hospitalizations of infants highlights the importance of following manufacturers' instructions in preparing formula. According to PRHA spokesperson Cathy Olay, five babies were hospitalized for severe dehydration in the past month alone after having been fed concentrated baby formula at full strength.

"Concentrated baby formula must be diluted with water before being fed to infants," warned Dr. Beverly Ridge of the Breastfeeding Clinic at Stoneyview Hospital. "When a baby is fed undiluted concentrate, the infant actually loses water from its body as a result of its attempts to dilute its meal."

"We were shocked," said Susan Lake, mother of an infant who became extremely dehydrated due to drinking concentrated formula. "We never thought that giving our baby a liquid formula would cause her to get so sick."

Most cases of dehydration occur when parents unwittingly switch from ready-to-serve formula to the concentrate, which often comes in very similar packaging. Ms. Olay emphasized the importance of checking the label carefully to make sure of the contents before feeding it to an infant, with potentially tragic results.

Evaluating "The News"

1. The parent interviewed in the article expressed shock that an improperly mixed baby formula could lead to dehydration. How can babies lose water while drinking a liquid baby formula that consists mostly of water?

2. We use many concentrated or powdered drink mixes not only for infants, but also as adults for weight loss and for rehydration during exercise. Explain why it is important that we make these drinks the correct strength.

3. Rats and mice are often used as test animals in nutritional studies. Both of these animals can survive on diets containing much less water than is necessary in the diet of human babies. Would tests with rats or mice reveal the hazards of using a too-concentrated infant formula? why or why not?

chapter 25

Size and Complexity

Roberto Marquez, *Theory of King David*, 1999.

Movie Monsters

On a remote, fog-enshrouded island in the Pacific Ocean lives a Stone Age tribe that worships a giant ape. They keep the ape from harming them by occasionally sacrificing a virgin to him. An adventurous film crew from New York City intrudes into this peaceful setting and decides that a giant ape is just the thing to make them rich. With considerable effort, they capture the ape and return to civilization.

This is the idea behind the famous 1933 movie *King Kong*. Kong (the ape) doesn't particularly like the idea of being put on public display, so he escapes. The cast and the audience quickly discover that a large, angry ape can have a wild night on the town even by New York City standards: Subways derail, people die, and the city ordinance against climbing skyscrapers is broken.

Kong belongs to a long tradition of mythical creatures based on real animals that become monstrous by virtue of huge size and foul tempers. Large size makes such monsters terrifying because with large size comes great strength. Thus, as a giant gorilla, Kong can knock out dinosaurs, lay waste to an entire village, and trash the New York City public transportation system. Moreover, weapons such as spears and guns that work at a human scale do little more than annoy something as big as Kong.

This fearful and shallow view of the effect of large size on monsters such as Kong ignores the complex way in which size affects the biology of organisms. As we will see in this chapter, large organisms are not simply magnified versions of small organisms. By understanding how biology changes with size, we can get an idea of how Kong might function if he really existed.

King Kong atop the Empire State Building

■ KEY CONCEPTS

1. Living things span a tremendous range of size and complexity.

2. As the dimensions of an organism increase, volume increases faster than surface area.

3. Large organisms have less surface area relative to volume than small organisms, and this relationship affects how they exchange materials with their environment.

4. Large size stems mainly from an increase in the number of cells, rather than from an increase in the size of cells.

5. Complexity probably evolved through the specialization of cells or organs that were present in multiple copies.

6. Specialization of cells increases the efficiency with which they can carry out their functions, but it also increases their dependence on other cells in the organism.

*A*dult humans typically weigh between 100 and 200 pounds, stand approximately 5 ft to 6½ feet tall, and as we saw in Chapter 24, consist of many trillions of cells arranged into tissues and organs specialized to carry out well-defined functions. Amid the incredible diversity of life, humans stand out as large, complex organisms. As we will see, simply understanding the implications of our size and complexity can tell us a great deal about what we are and how we function.

In this chapter we consider some of the implications of size and complexity. After surveying the range of the size and complexity of life, we consider how size affects biology and how scientists use the relationship between size and function to understand the biology of various organisms. Finally, we consider the relationship between size and complexity in an attempt to understand the evolution of large, complex organisms from small, simple ones.

25.1 An Overview of Size and Complexity

The variation in the size and complexity of living things is astonishing (Table 25.1). To provide one example, it would take an incredible 100,000,000,000,000,000 individuals of *Mycoplasma*, one of the smallest bacteria known, to weigh as much as a single 200-pound person. We humans, in fact, rank among the largest of the animals (see the box on p. 422). In turn, it would take 10,000 200-pound people to weigh as much as a giant sequoia tree, one of the largest land plants. Whereas we need a powerful microscope to see *Mycoplasma*, the underground parts of some individuals of the fungus *Armillaria bulbosa* spread over an area larger than 10 city blocks.

Life encompasses an equally impressive range of complexity. Whereas a bacterium consists of a single cell with a relatively simple internal structure, complex animals and plants consist of many millions of cells, each containing well-defined organelles, organized into highly structured tissues and organs.

Mushrooms produced by the fungus *A. bulbosa*

Although size and complexity often go hand in hand, they do not mean the same thing (Figure 25.1). Large organisms consist of many cells, and complex organisms consist of many different kinds of cells organized into tissues and organs. As we will see later in this chapter, multicellular organisms have had more opportunities for the evolution of complexity than have single-celled organisms. Nonetheless, large organisms may be simple and small organisms may be complex. Impressively large marine sponges, for example, may consist of several cell types, but these cells are not arranged into tissues or organs; a sponge functions essentially as a col-

25.1 *Some of the Smallest and Largest Living Things*		
Organism	**Type of organism**	**Weight (grams)**
Mycoplasma	Bacterium	0.000000000001
Giant squid	Mollusc	450,000
King Kong	Movie character	13,950,000
Baluchitherium	Extinct land mammal	14,250,000
Blue whale	Aquatic mammal	100,000,000
Armillaria	Fungus	110,000,000
Giant sequoia	Plant	1,000,000,000

(a)

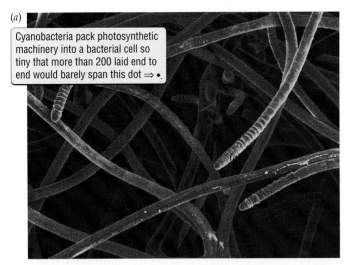

Cyanobacteria pack photosynthetic machinery into a bacterial cell so tiny that more than 200 laid end to end would barely span this dot ⟹ •.

(b)

Large sponges may contain just a few cell types and lack tissues.

(c)

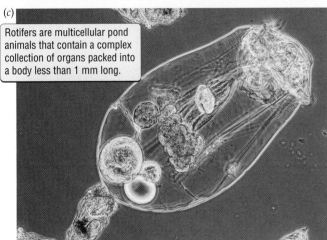

Rotifers are multicellular pond animals that contain a complex collection of organs packed into a body less than 1 mm long.

(d)

Individual aspens can cover a large area. The yellow patch may consist of a single plant that includes hundreds of trunks but a single root system.

Figure 25.1 Living Things Vary Tremendously in Size and Complexity

lection of independent cells (Figure 25.1*b*). In contrast, microscopic aquatic animals called rotifers consist of many different kinds of cells arranged into distinct tissues and organs (Figure 25.1*c*).

As we look more deeply into the importance of size and complexity, keep in mind that large, complex organisms are not necessarily better or evolutionarily more successful than small or simple organisms. Consider that *Yersinia pestis*, the simple, single-celled bacterium that causes black plague, managed to kill up to half the population of large, complex humans in medieval Europe.

■ Life on Earth spans a tremendous range of size and complexity. Complexity does not necessarily depend on size, and neither size nor complexity reflects how successful an organism is.

25.2 Surface Area and Volume: Why Size Matters

We begin our consideration of the biological importance of size by looking at what happens when we change the size of an imaginary creature. This example will lead us to a biologically important conclusion about how organisms function: Large organisms face greater challenges than small organisms in exchanging materials with their environment.

Changes in size affect various body measurements differently

To understand how size affects the biology of organisms, let's consider two forms of an imaginary cubical crea-

■ BIOLOGY IN OUR LIVES

Appreciating the Little Things in Life

Humans have a warped perspective on life. In a rough survey of the lengths of almost 1 million animals, only 0.02 percent, or 1 in 5000, had body lengths of 1 meter or more. Almost 75 percent of the animals in this survey were smaller than 1 centimeter. In other words, we are really big animals! Even among plants, which we tend to regard as relatively large, humans would find themselves in the tallest 25 percent of species.

We perceive the world from the perspective of a large organism. Small organisms—in other words, almost all other organisms—perceive a very different world, in which they face problems that differ greatly from our own. Imagine, for example, a world where the air consists of a syrupy fluid through which you have to fight your way. Imagine a world where you can become trapped in a dewdrop and where you can walk on water. To bacteria, protists, and small animals, this is exactly how the world appears. Most of us have seen an insect trapped in a tiny droplet of water and

wondered why it could not escape. To us these observations may seem strange, but they reflect the reality of life at a small size.

To small organisms, the world appears much more varied. A single human head, for example, provides a variety of distinctly different habitats for small organisms: Lice live on our scalp; mites live at the bases of our eyelashes; a variety of bacteria inhabit the oily skin on our forehead, nose, and chin. There is even some suggestion that for smaller animals

the world is physically bigger, because surfaces and edges that appear smooth to us appear textured to them. What to us is a single step across a flat carpet may be a lengthy series of climbs up huge mountains to a microscopic dust mite.

In understanding how other animals make their livings, we would do well to remember the words of the British biologist Richard Dawkins: "To a first approximation all animals are smaller than us."

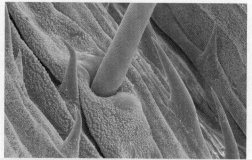

Small Organisms See a Very Different World
A blade of grass as it appears to humans (left); and the surface of a blade of grass (right) maginified 120 times, showing sharp spines and a hair protruding.

ture, which has a conveniently simple geometry (Figure 25.2). We can easily express the size of the cubical creature by its length, by its volume, or by its surface area. What do those measurements tell us about the creature's biology?

Imagine a small cubical creature that has a length of 1 centimeter. Its length determines, for example, the size of the hole into which it can fit or how high it can reach for food. The small cubical creature has a volume of 1 cubic centimeter. Volume tells us how much tissue there is, and because this tissue consists of metabolizing cells, it also tells us about how much food, water, and oxygen or carbon dioxide the creature needs. Volume relates closely to the weight of the organism, so much so that weight and volume are often used interchangeably. Finally, our small cubical

creature has a surface area of 6 square centimeters (six sides of 1 square centimeter each). This measurement represents the surface across which the creature can absorb the nutrients and gases that it needs, and across which it can dispose of its wastes.

Now imagine that we discover a second species of cubical creature whose cells have multiplied to give a length twice that of the smaller species. Doubling the length of each side leads to several important changes in the creature's biology (see Figure 25.2). Not surprisingly, as we double the cubical creature's length, we also increase its volume and its surface area. The larger species has a volume, and therefore a weight, that is eight times that of the smaller species (2 cm × 2 cm × 2 cm = 8 cm^3). At the same time, the surface area of the creature increases to four times that

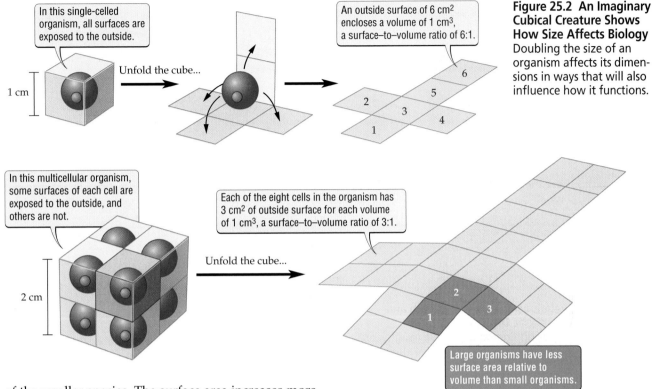

In this single-celled organism, all surfaces are exposed to the outside.

1 cm

Unfold the cube...

An outside surface of 6 cm² encloses a volume of 1 cm³, a surface–to–volume ratio of 6:1.

6
5
2
3
4
1

Figure 25.2 An Imaginary Cubical Creature Shows How Size Affects Biology Doubling the size of an organism affects its dimensions in ways that will also influence how it functions.

In this multicellular organism, some surfaces of each cell are exposed to the outside, and others are not.

2 cm

Each of the eight cells in the organism has 3 cm² of outside surface for each volume of 1 cm³, a surface–to–volume ratio of 3:1.

Unfold the cube...

2
1
3

Large organisms have less surface area relative to volume than small organisms.

of the smaller species. The surface area increases more than the length (length two times, surface area four times), but less than the volume (eight times).

Thus, as the length of the cubical creature doubles, its volume and surface area more than double. As a result, the relationships among length, volume, and surface area differ between small and large species. Because length, volume, and surface area all affect the biology of an organism, we should expect the small species to function differently from the large species.

The surface area–to–volume ratio decreases as size increases

Of the various relationships among different body measurements, the one that we will encounter repeatedly in the next several chapters is the **surface area–to–volume ratio**:

$$\frac{\text{Surface area}}{\text{Volume}}$$

The surface area–to–volume ratio tells us how easily an organism can move materials across its surface relative to the demand for those materials by the cells that make up its volume.

In our example, our small cubical creature has twice as big a surface area–to–volume ratio (6 square centimeters of surface area per cubic centimeter of volume) as our large cubical creature (3 square centimeters of sur-

face area per cubic centimeter of volume) (see Figure 25.2). Thus, the small cubical creature has more surface area across which it can supply each cubic centimeter of metabolizing tissue with nutrients, water, and gases than does the large species.

The larger cubical creature, therefore, must somehow compensate for the greater difficulty it faces in supplying its cells. On the other hand, the large species should have an easier time keeping valuable materials inside its body. Terrestrial organisms, for example, need to prevent scarce water from diffusing out of their bodies. With less surface area across which water can diffuse relative to the volume of the cells in which the water is stored, the large cubical creature should lose its body water relatively more slowly than the small species. Even though they look the same, the small and large species must function differently.

Does the cubical creature model work for real organisms? We selected a cubical creature as our example because cubes have such simple relationships among length, volume, and surface area. Surprisingly, the relationships hold up remarkably well for real organisms having complex shapes. For photosynthetic organisms as different as single-celled bacteria, single-celled algae, and terrestrial plants, we find that as length doubles, weight, which measures the same thing as volume, increases eight times, or the increase in length cubed (Figure 25.3). This

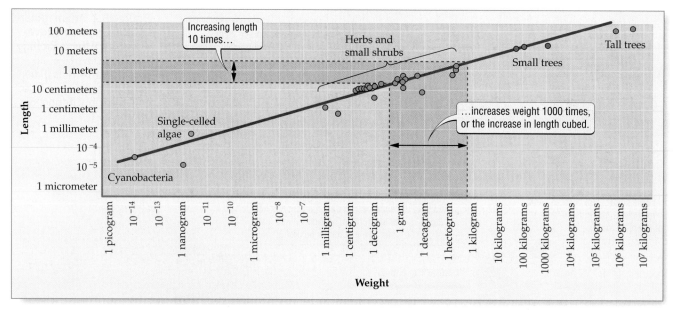

Figure 25.3 The Relationship between Length and Weight for Photosynthetic Organisms
In spite of the dramatic differences in shape among cyanobacteria, single-celled algae, and plants, they all show the same relationship between weight and length. Furthermore, the relationship between weight and length for these organisms as a group (shown by the solid line) is the same as it is for the imaginary cubical creatures in Figure 25.2. (Note that each interval in the graph represents ten times the previous hatch mark; this convention allows us to show the incredible range of size among these organisms in a single graph.)

is the same geometric relationship that we found for our cubical creatures.

> ■ The relationships among length, volume, and surface area change as size changes. Large organisms have less surface area relative to their volume than do small organisms. As a result, large organisms function differently from small organisms.

25.3 *Using Body Size to Understand Biology*

In the chapters that follow, we will often refer to relationships between body size and a biological feature such as the surface area–to–volume ratio. Such relationships are called **allometric relationships**. In most discussions of allometric relationships, body size is given in terms of the weight of the organism because weight is more easily measured than volume. In this section we introduce some examples of the different ways in which allometric relationships can help us understand how organisms function.

Allometric relationships can reveal general patterns in biological features and functions

Among mammals, sleep shows some interesting patterns. Cats, for example, seem to spend their entire lives sleeping. Whereas most humans sleep only 8 hours, an average cat sleeps 14.5 hours each day. A broader look at patterns of sleep among mammals, however, reveals that the time devoted to sleep decreases predictably with body size (Figure 25.4). From this allometric relationship, we can conclude that large mammals spend less time sleeping each day than do small mammals. In other words, cat naps reflect the small size of cats, not their laziness.

In contrast, the relationship between brain size and body size in mammals reveals that not all brains are created equal. As body size increases, brain size generally increases in a predictable manner (Figure 25.5). Most of the points representing individual species in Figure 25.5 cluster close to the line representing the general pattern, indicating that most mammals have brain sizes close to those predicted for their body size. The points representing humans and dolphins, however, lie well above the line. These data suggest that for their weight, humans and dolphins have exceptionally large brains. This finding has led scientists to wonder what special selective pressures might favor the evolution of large brain size in these two mammals living in very different environments.

Figure 25.4 How Much Mammals Sleep Depends on Their Size

Cats sleep much more than humans not because they are lazy, but because they are smaller than we are. Plotting the relationship of sleeping time to body size for a number of mammals reveals an allometric relationship for this function. The orange zone indicates where most of the points representing mammals lie.

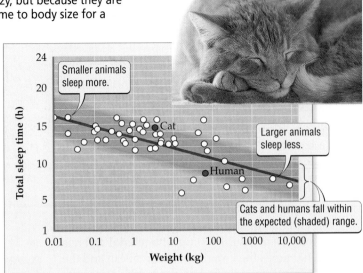

Allometric relationships have important practical applications

The relationship between body size and function affects our everyday lives. When doctors or veterinarians prescribe drugs, for example, they recommend a dosage that considers both the weight of the patient and how quickly the drug passes through the patient's body. Large animals, such as humans, clear drugs relatively slowly, whereas small animals clear drugs rapidly. Thus a small animal such as a cat would receive a higher dose of a particular drug relative to its body weight than we would, because the cat's cells would otherwise clear the drug from its body before it had a chance to take effect.

The sad tale of Tusko the elephant illustrates the disastrous consequences of ignoring the allometric relationship between body size and the drug clearance rate. In a stunningly misguided effort to study an elephant behavior in which normally calm males become extremely aggressive, some psychologists decided to mimic the behavior by giving Tusko the hallucinogenic drug LSD. The researchers estimated the dose of LSD they gave to Tusko based on laboratory experiments involving 5-kilogram cats. A dose of 0.15 milligrams per kilogram of cat body weight caused a definite change in cat behavior.

Based on the cat studies, the psychologists gave Tusko what they thought was a safe dose: 0.1 milligrams per kilogram (or 300 milligrams of LSD for a 3000-kilogram elephant). Within 2 hours, Tusko had overdosed and died. Unfortunately, these researchers did not take into account the allometric relationship between size and drug clearance rates. Because of their much larger size, elephants remove LSD from their bodies much more slowly than cats. Tusko should have received no more than 0.03 milligrams per kilogram to avoid overdosing.

Figure 25.5 Brain Size Increases as Body Size Increases

Humans and dolphins have exceptionally large brains for mammals of their size. The orange zone indicates where most of the points representing mammals lie.

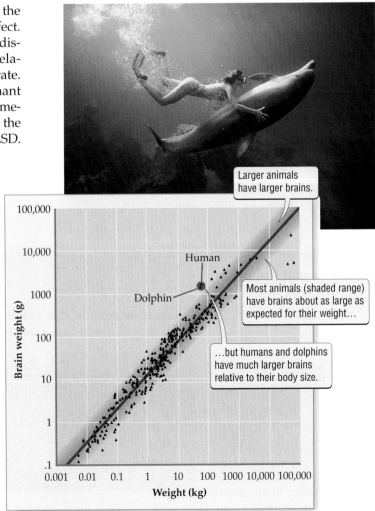

■ Allometric relationships, which describe how a biological feature or function changes in relation to body size, are useful tools in understanding the biology of organisms.

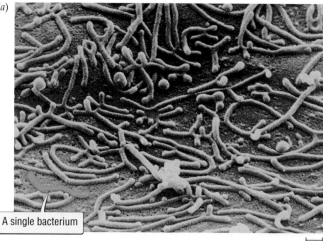

(a)

A single bacterium

├─┤
0.0001mm

The Evolution of Large Size

Large organisms consist of many cells. Given that large size could have evolved either by increases in cell size or by increases in the number of cells, why is it that the latter seems to be the general approach?

We can start our search for an answer by considering single-celled organisms. Bacteria and single-celled algae both show great variation in cell size (Figure 25.6*a* and *b*). The largest bacterial cell has about 100 times the volume of the smallest, and the largest single-celled alga has about 1000 times the volume of the smallest. Cells, however, cannot simply evolve an ever-larger size. They must remain small enough so that their surface area–to–volume ratio is large enough to allow movement of enough materials across the plasma membrane to meet their needs. Thus, when natural selection favored much larger size in bacteria or single-celled protists, rather than evolving larger cells, these organisms formed collections of cells (Figure 25.6*c*).

By having many small cells rather than fewer, larger ones, organisms can increase in size while still accommodating the fact that the cell must exchange materials with its environment. In mammals, cell size does not vary much with body size, suggesting that elephants are bigger than humans entirely

(b)

├─┤
0.001mm

(d)

(c)

Figure 25.6 Variation in Cell Size and Number
The size of living things depends on both cell size and number of cells. Single-celled organisms such as (*a*) small bacteria and (*b*) large algae can vary a thousand-fold in size. (*c*) Alternatively, when cells remain together, organisms can reach larger effective sizes, as in the multicellular green alga *Volvox*. (*d*) The size of mammals such as elephants depends largely on the number of cells they contain, not on cell size.

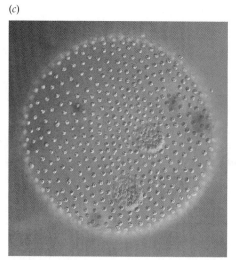

├─┤
0.1mm

├──────────┤
1000mm

because elephants have many more cells than we do (Figure 25.6*d*).

At the same time, however, increasing in size by increasing cell number rather than cell size leaves many cells far from the body surface. The cells inside the organism, therefore, still face the challenges of obtaining the nutrients and gases needed for metabolism and getting rid of wastes. As we will see in the chapters to come, many features found in multicellular organisms address this problem of getting materials to and from the internal cells.

■ Large size evolved mainly by increases in the number, rather than in the size, of the cells making up organisms. Cell size had to remain relatively small so that cells could rapidly import and export the materials essential for life.

Complexity and Its Implications

In this section we briefly consider three aspects of the complexity that has arisen in most large multicellular organisms: its evolution, its advantages, and its problems.

Complexity evolves from repeated units

How did complexity evolve? As Chapter 24 pointed out, plants and animals consist of relatively few cell types arranged in complex ways to form tissues and organs. A key step in the evolution of complexity seems to be the presence of duplicate structures. Once multicellularity evolved, perhaps because of selective advantages associated with larger size, it may have provided the raw materials for the evolution of complexity by providing duplicate cells (Figure 25.7). If an organism has multiple cells, some of those cells can become specialized to perform a particular function without threatening the well-being of the whole organism, which can rely on other cells to do the things the specialized cells can no longer do. On a larger scale, if an organism has duplicate organs or other repeated units, some of those units can also become specialized for a particular purpose without threatening the well-being of the organism. Both plants and animals have evolved complexity through the specialization of repeated units in this way (Figure 25.8).

The advantage of complexity is the specialization of cells

Specialization is an almost universal feature of large multicellular organisms. A structure that is specialized for

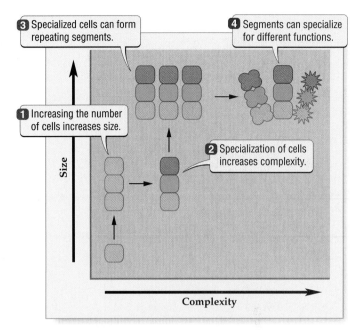

Figure 25.7 The Evolution of Complexity
This schematic diagram shows the relationship between size and complexity and a possible pathway for the evolution of complexity. The vertical axis shows that larger organisms have more cells; the horizontal axis shows that more complex organisms have more kinds of cells.

carrying out a particular function can usually perform that function more efficiently than could an all-purpose structure. For example, the epithelial cells that line the human small intestine include two specialized types: secretory cells that pump ions into the gut and absorptive cells that absorb ions from the gut. Each of these epithelial cells has many carrier proteins suited to either releasing ions or absorbing them. In contrast, the single cell of a protist would need many of both types of carrier proteins in its plasma membrane to perform both functions. Presumably, the specialized secretory cell of the human small intestine can transport ions more quickly than an unspecialized protist cell.

Complexity requires coordination among cells

The disadvantage of complexity is that it increases the dependence of each cell on other cells in the organism. Complexity generates two closely related problems:

1. The activities of the specialized cells must be coordinated.
2. Some of the cells must give up the chance to reproduce.

If an organism consists of multiple interdependent parts, then the activity of each part must be coordinat-

(a)

The aboveground portion of a plant consists of many stem/leaf units.

Flowers are modified stem/leaf units.

The potatoes we eat are underground stems modified for storage.

Below ground, plants consist of many root units.

(b)

Each segment of a shrimp represents a repeating unit bearing a pair of legs.

The legs of the last body segment have been modified to form a fin for swimming.

The antennae are modified legs associated with head segments.

Some legs have been modified for walking.

Figure 25.8 Complexity Evolves through the Specialization of Repeated Units
Both plants and animals have evolved complexity through the repetition and modification of a basic structure. (*a*) The flowers of flowering plants, such as this potato plant, evolved as specialized leaves. (*b*) The diverse legs of shrimps evolved by specialization of a single basic leg design.

ed with that of the other(s). Plants, for example, must regulate their root growth so that they meet the demands of their aboveground parts for mineral nutrients and water. At the same time, the roots must not grow so much that they use up the supply of sugars and proteins produced in the leaves. Complexity therefore creates a need for systems of communication and coordination among cells. The hormones that serve as chemical mes-

sengers in plants and animals and the unique nervous system of animals play this role.

A more fundamental problem related to complexity is that some cells must give up the option of reproduction to support the cells that do reproduce. The members of one odd group, the cellular slime molds, demonstrate this sacrifice clearly. For most of their lives, cellular slime molds live as individual cells, feeding on rotting material in the soil. When the food runs out, however, the cells release a chemical signal into the soil. The signal stimulates surrounding cells to collect into a multicellular group, called a slug, that acts like a single individual. Some of the cells in the slug form a stalk, atop which sit a few cells that form a reproductive structure. Only the cells in the reproductive structure reproduce; those in the stalk produce no spores. In much the same way, our gut cells give up the chance of giving rise to a new human so that a few cells in our sex organs can have the chance to reproduce.

Because natural selection favors genotypes that leave behind as many copies of themselves as possible, why should the stalk cells of the cellular slime mold slug give up their chance to reproduce? The answer seems to be that cooperation leads to greater overall reproductive success for a collection of genetically similar cells. Only slime mold cells that share similar genes can come together to form a slug, so the offspring produced by the few cells that reproduce genetically resemble the stalk cells that have given up reproduction. In much the same way, most of the cells in our body do not contribute their genes to any offspring produced, but instead support the few cells in our reproductive system that do.

■ Complexity probably evolved through the specialization of duplicate cells or organs into structures that could carry out particular functions more efficiently. Complexity carries with it the problems of coordination among cells and the sacrifice of some cells so that others can reproduce.

■ HIGHLIGHT

The Collapse of Kong

How can we use allometric relationships such as those discussed in this chapter to predict the biology of King Kong? From Kong's approximate height of 7.2 meters (the height of the model used in the movie) and the relationship between body length and weight, we arrive at an estimated weight of 13,950 kilograms. Although big,

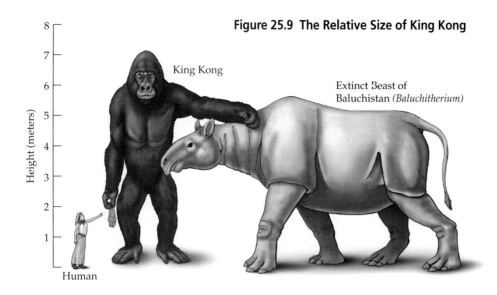

Figure 25.9 The Relative Size of King Kong

King Kong

Extinct Beast of
Baluchistan (*Baluchitherium*)

Human

Kong was actually not outside the range of mammal size: He weighed about as much as two elephants, and less than the extinct Beast of Baluchistan (at 20,000 kilograms, the biggest mammal ever to roam the land) (Figure 25.9).

Table 25.2 makes some predictions about the biology of Kong based on his size. We would expect Kong to be challenging to maintain in captivity: He would eat a great deal, and he would produce a lot of waste. Once loose, he would probably sleep little, move quickly, and cover a lot of territory in a single day. In addition, Kong should live a long time, yet produce relatively few offspring. Compared with humans, Kong should have a similarly sized heart and a small brain relative to his overall weight. Table 25.2 provides no indication of why Kong should have the foul temper that he did, however. We should not expect the numbers in Table 25.2 to be too precise, but they should give us a good idea of the general patterns.

In many ways, Kong represents a biologically possible creature. At least one feature of Kong's natural history, however, suggests that he could not have existed: The island on which he lived was far too small to support a population of giant gorillas. Kong alone would require about 2210 square kilometers of land to satisfy his immense appetite, but his island covered only about 15 to 20 square kilometers, as estimated from a map shown in the movie. Because Kong would not be immortal and because, as a mammal, he would need at least one mate to produce offspring, the existence of Kong implies the existence of several other giant gorillas. Supporting even a minimum population of Kong, Kongette, and two Konglets would require an island much larger than that depicted in the movie.

> ■ Using allometric relationships, we can make predictions about what the biology of an imaginary creature such as King Kong would have been based on its size. Although he falls within the range of possible sizes for a mammal, the island on which Kong lived was too small to have supported a population of gigantic apes.

25.2 *Comparative Biology of King Kong and Humans*

Trait	Value for trait (relative to weight where appropriate)	
	King Kong	**Human**
Metabolic rate (kcal/kg/d)	6	29
Home range size (km²)	2210	—
Organ sizes (percentage of body weight)		
Brain	0.1	2
Gut	16	15
Skeleton	28	13
Testes	0.01	0.07
Heart	0.4	0.5
Urine production (l/d)	77	2
Life span (yr)	149	100
Reproductive rate (offspring/yr)	0.3	1.3
Speed (km/h)	54	37
Daily sleep time (h)	3.9	8

SUMMARY

An Overview of Size and Complexity

■ Living organisms span a tremendous range of size, from microscopic organisms to organisms that weigh thousands of times more than humans.

■ Living organisms also span a tremendous range of complexity, from single-celled bacteria with a simple internal structure to organisms composed of many millions of specialized cells, each with well-defined organelles, organized into tissues and organs.

■ Neither size nor complexity reflects the success of an organism.

Surface Area and Volume: Why Size Matters

■ As size increases, surface area increases more rapidly than length, but less rapidly than volume.

■ Volume determines how much food, water, and oxygen or carbon dioxide an organism needs. Surface area determines the amount of surface available to the organism for exchanging these materials with its environment.

■ Large organisms must find ways to compensate for having less surface area relative to their volume than small organisms do. Therefore, large organisms are not simple magnifications of small organisms.

Using Body Size to Understand Biology

■ Allometric relationships are relationships between body size and a biological feature or function.

■ Allometric relationships allow us to account for the effects of size as we work with a variety of species, as well as to identify species with unique characteristics that are not simply a result of their size.

The Evolution of Large Size

■ Large size evolved mainly by increases in the number, rather than in the size, of the cells making up an organism. Larger organisms do not necessarily have larger cells.

■ The relationship between surface area and volume limits cell size. Cell size has to remain relatively small so that cells can import and export sufficient materials across the plasma membrane.

Complexity and Its Implications

■ Complexity refers to the arrangement of different specialized cell types into tissues and organs.

■ Complexity probably evolved through the specialization of duplicate cells or organs into structures that could carry out particular functions more efficiently.

■ A cost associated with complexity is that specialized cells are interdependent and require communications systems that allow them to coordinate their functions. In addition, many specialized cells are not directly involved in reproduction.

Highlight: The Collapse of Kong

■ Using allometric relationships, we can make predictions about the biology of the fictitious giant ape, King Kong, based on his size.

■ Because Kong's size falls within the range of sizes reported for mammals, it is conceivable that he could exist. The island on which Kong lived, however, was too small to support such a large creature.

KEY TERMS

allometric relationship p. 424

surface area–to–volume ratio p. 423

CHAPTER REVIEW

Self-Quiz

1. Among living organisms, large size arises primarily from
   a. more cells.
   b. larger cells.
   c. both fewer and larger cells.
   d. increasingly complex cells.

2. Which of the following statements about the size and complexity of living organisms is true?
   a. Organisms fall within a narrow range of sizes.
   b. Only animals ever evolved to reach truly large sizes.
   c. Because of their simplicity and small size, single-celled organisms cannot survive in a world filled with large, complex organisms.
   d. Organisms vary greatly in both size and complexity.

3. As size increases,
   a. surface area increases more slowly than volume.
   b. length increases more rapidly than surface area.
   c. length increases more rapidly than volume.
   d. surface area increases more rapidly than volume.

4. The psychologists experimenting with Tusko the elephant could have prevented his death if they had taken into account that
   a. elephants have more cells than cats.
   b. weight affects the rate at which drugs pass through an animal's body.
   c. elephants sleep less than cats.
   d. elephants have more surface area relative to their volume or weight than cats do.

5. Cell specialization leads to
   a. increased efficiency.
   b. the need for coordination and communication between cells.
   c. both a and b.
   d. neither a nor b.

Review Questions

1. On a visit to the pet store, you encounter two different species of parrots: tiny budgies that weigh less than 25 grams and larger lovebirds that weigh five times as much. As a potential bird owner, you start thinking about what it would involve to own these birds. How would the biologies of the budgies and the lovebirds differ? Think about how much they would eat, how much of a mess they would make, how much space they would need, and so on.

2. Explain why relationships between surface area and volume limit cell size.

3. Why might cells or organs need to be duplicated in a species before being able to evolve specialized functions?

The Daily Globe

Greenland Well Yields A New Life Form

COPENHAGEN, DENMARK—Danish scientists Reinhardt Møbjerg Kristensen and Peter Funch have discovered a completely new kind of animal in a sample of well water. The animal was found in samples taken in 1994 from a well in Isunngua on Disco Island in Greenland.

The 0.1-millimeter-long freshwater organism represents a completely new animal phylum, given the name Micrognathozoa. This is only the fourth time in the past century that a new phylum has been described.

Dr. Kristensen, who has been involved in the discovery of two other new phyla, noted that the animal, named *Limnognathia maerski*, shared some characteristics with three well-known phyla of animals familiar to science.

The most remarkable feature of the tiny animal is a set of complex jaws. *Limnognathia maerski* appears to use its jaws to scrape the bacteria and algae it eats from underwater mosses growing in wells that freeze over in winter.

Evaluating "The News"

1. How is it possible that after all of the effort put into understanding our natural world, we can still discover not just new species, but new phyla?

2. Does it surprise you that the newly discovered phylum consists of a species less than 1 millimeter long? Why or why not?

3. The world had existed without knowledge of the phylum Micrognathozoa up until the year 2000. Construct an argument in favor of or against funding research of this sort.

4. Although *Limnognathia maerski* is tiny, it possesses complex jaws. Is it reasonable to conclude from organisms such as *L. maerski* that small organisms are no simpler than large organisms?

chapter 26 Support and Shape

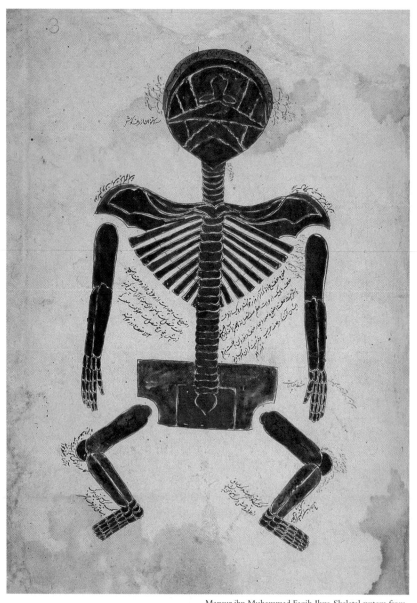

Mansur ibn Muhammad Fagih Ilyas, Skeletal system from *Five Anatomical Figures*, late fourteenth century.

Leaping Cats and Humans

*L*ife in the city can be tough. Humans must deal with polluted air, violent crime, and speeding taxis. As our companions, cats face their own hazards in a landscape of vertical buildings and hard pavement. For example, the legendary curiosity of cats, combined with the promise of better hunting grounds, lures many cats through open windows or over balcony railings. Unfortunately, these adventures can turn into a quick trip that ends abruptly on the streets and sidewalks several stories down.

For humans, such falls usually end in disaster: The death rate increases rapidly as people fall from ever greater heights up to the seventh story, at which point the chances of dying level off at almost

■ MAIN MESSAGE

Support systems depend on structures that can resist compression and tension.

a hundred percent. In sharp contrast, information on 115 cats admitted to animal hospitals in New York City reveals two amazing facts. Although these cats fell an average distance of 5.5 stories, only 11 of the cats died from their injuries. Even more incredibly, cats stood a better chance of surviving a fall from *above* the fifth story than from *below* the fifth story. One cat that fell from the thirty-second story suffered only minor injuries.

The secret to the survival of these skydiving cats lies in the ability of a cat's skeleton, which is primarily responsible for supporting its body, to handle the tremendous forces it experiences as it hits the pavement. We will explain why cats survive such falls after we review some of the principles that have affected the evolution of support systems.

Does the Ability of Cats to Land on Their Feet Contribute to Their Proverbial Nine Lives?

■ KEY CONCEPTS

1. The human skeleton relies primarily on bone for support, but cartilage and hydrostatic structures play a role as well.

2. Animals rely on specialized tissues that provide support, but also accommodate movement.

3. Single-celled organisms, plants, and fungi rely for support on cell walls associated with each cell.

4. Support structures must resist both compression and tension to maintain their shape.

5. The biologically produced materials, minerals, and hydrostatic skeletons that make up support systems each offer unique advantages and disadvantages.

6. The support systems of aquatic organisms need not provide as much support as those of terrestrial organisms.

7. Joints consist of combinations of stiff and flexible materials that allow animals to move.

We could not survive without a skeleton. Our skeleton gives us our shape, and our shape defines how we move, how we eat, how we breathe, how we protect ourselves, how we give birth, and of course, how we survive falls. Our skeleton is an amazingly complex system of support that holds our bodies in a human shape. At the same time, it allows us to change our shape in a controlled way as we move.

We begin this chapter by surveying the variety of support systems found in living things. Against this backdrop of diversity, we consider the shared selection pressures that have influenced the evolution of all support systems and which have led to some surprisingly general patterns in their form and function. We then consider the unique ability of animal skeletons to provide support while at the same time allowing movement. With all of this information in hand, we return to the question of why cats can survive falls that are fatal to humans.

The Human Skeleton

We will use our own skeleton as a starting point in our overview of support systems because it is the one most familiar to us. By understanding some basic features of our own skeleton, we will have a point from which we can begin to explore some general patterns in the way support systems work.

Bone is the major support tissue in the human body

Humans have an internal skeleton that supports the body and gives it shape (Figure 26.1a). The major component of the skeleton is **bone**. Bone tissue consists of connective cells (see Chapter 24) that surround themselves with a nonliving matrix made largely of calcium, which we accumulate from our environment. Bone is stiff, meaning that it does not bend easily, so it is well suited to providing a framework for the body. Bone supports our legs and arms and forms our rib cage and skull.

Connective tissues and hydrostatic skeletons provide additional support

Other connective tissues supplement the support provided by bone. **Cartilage** consists of connective cells that produce a matrix consisting mostly of a protein called **collagen**. Cartilage supports the nose and the ears, among other body parts. Compared with bone, the cartilage in the bridge of your nose is less stiff and bends more easily, but it is strong and resists breaking.

Our skin and other flexible connective tissues, including the ligaments and tendons that we will describe later in this chapter, help to hold the more rigid bones and cartilage together. Like cartilage, these tissues contain collagen.

The final component of the human support system is less obvious, but can be seen in the tongue and in the female's clitoris or male's penis. Human tongues, penises, and clitorises lack bone or cartilage, yet can become quite stiff when we wish. These parts of our bodies are examples of **hydrostatic skeletons**. Hydrostatic skeletons become stiff when fluid under pressure pushes against a membrane, as we will explain more fully below.

> ■ The human skeleton consists of several different types of support structures derived from specialized connective tissues: bone, cartilage, skin, ligaments, tendons, and hydrostatic skeletons.

(a)

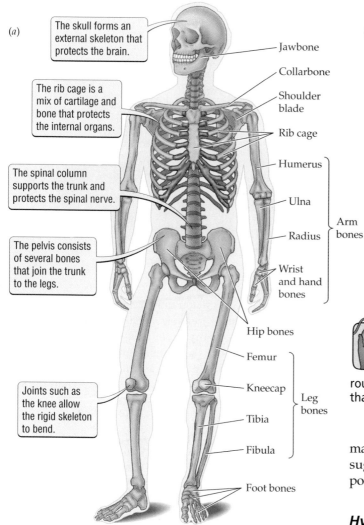

The skull forms an external skeleton that protects the brain.

The rib cage is a mix of cartilage and bone that protects the internal organs.

The spinal column supports the trunk and protects the spinal nerve.

The pelvis consists of several bones that join the trunk to the legs.

Joints such as the knee allow the rigid skeleton to bend.

Jawbone

Collarbone

Shoulder blade

Rib cage

Humerus

Ulna

Radius

} Arm bones

Wrist and hand bones

Hip bones

Femur

Kneecap

Tibia

Fibula

} Leg bones

Foot bones

(b)

26.1

Figure 26.1   Animal Support Systems
Animals have specialized tissues that are devoted to supporting their bodies. (*a*) Humans have a rigid internal skeleton surrounded by soft tissues. (*b*) Clams have a hard external shell that supports and protects the soft tissues inside.

An Overview of Animal Skeletons

Animal skeletons share a number of features that set them apart from the support systems of other organisms and allow animals to move.

Animals rely on specialized support tissues

Although animals have evolved a diversity of support systems, all of them, including humans, rely on specialized tissues to support the rest of the body. These specialized support tissues consist of connective cells that produce an extensive nonliving matrix around themselves. This matrix can be made up of minerals from the environment, like the bones, or biologically produced materials, like the collagen in cartilage. The minerals that form rocklike matrices typically contain calcium, as is true of our bones and the shells of clams (Figure 26.1*b*). Among animals, the most widespread biologically produced matrices are collagen, the protein found in cartilage and

many other animal support tissues, and **chitin**, a complex sugar that forms the basis of support tissues in arthropods, (such as insects and lobsters) and many other phyla.

Hydrostatic skeletons are the main source of support for soft-bodied animals

Many animals, including terrestrial and marine worms, caterpillars, and octopuses, are like our tongues in that they lack an obvious means of support (Figure 26.2*a*). These organisms gain their support from hydrostatic skeletons, which provide support not by being made of an inherently stiff material, but through an interaction between the skin and a fluid under pressure.

Balloons illustrate how hydrostatic skeletons work. When you inflate a balloon, you increase the pressure with which the air inside pushes against the skin of the balloon. As you blow more air into the balloon, the pressure inside increases, and the balloon becomes stiffer. Balloons highlight two important features of hydrostatic skeletons: First, as pressure increases, stiffness increases. Second, it takes energy to create the pressure on which the hydrostatic skeleton depends.

The hydrostatic skeletons that support earthworms and tongues work basically like balloons. Most animals use water instead of air as the fluid, since living organisms consist largely of water. The skin contains connec-

(a)
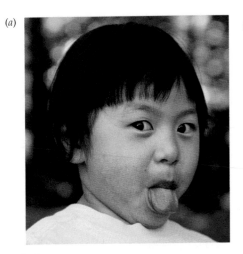

(b)

Figure 26.2 Some Hydrostatic Skeletons
Hydrostatic skeletons provide support without requiring any rigid materials. (*a*) The human tongue and (*b*) a young plant shoot both rely on hydrostatic pressure for support. When plant tissues dry out, they wilt and lose their shape.

tive cells that typically produce a collagen or chitin matrix. Rather than increasing the pressure on the skin by increasing the amount of water in the hydrostatic skeleton, animals contract muscles in the skin to decrease the volume of water it can hold. You can do the same thing with a partially deflated balloon by pinching off some of the balloon to force the air into a smaller volume. Generating stiffness in a hydrostatic skeleton requires muscle contraction, which requires energy.

Internal or external skeletons?

Over evolutionary time, two distinct ways of arranging stiff support tissues have arisen in the animal kingdom (see Figure 26.1). Some animals, including ourselves and most other vertebrates, have an **internal skeleton**, in which the support tissues lie inside the body surrounded by soft tissues. Other animals, such as lobsters and clams, have **external skeletons** that surround their soft tissues.

You can get an idea of some of the selective advantages that might favor the evolution of internal or external skeletons by imagining a seafood dinner. It takes much less effort to pull the meat off the internal skeleton of a fish than it does to crunch your way through the shell that forms the external skeleton of a lobster. Thus, external skeletons provide protection in addition to support. If you think about your skull, you will realize that it forms a protective external skeleton that simultaneously supports and protects your brain. At the end of

your meal, however, you may notice that the remaining lobster skeleton is heavier than the fish bones. This observation reflects the lower weight of internal as compared with external skeletons relative to the amount of support offered (Table 26.1). As we shall see later in this chapter, weight considerations have played an important role in the evolution of skeletons.

Joints allow animal skeletons to bend

Animals have a unique ability to move because they have muscle tissue that can contract or shorten (as we will see in detail in Chapter 27) and because their cells, which lack cell walls, can change shape relatively freely compared with those of other organisms. The stiff skeletons that maintain their shape, however, would make movement impossible for animals were it not for their design. **Joints** are breaks in the stiff skeletal system that allow the skeleton to bend in specific ways (Figure 26.3). Skeletons without joints—for example, snail shells—severely limit movement. Later in this chapter, we will see how joints combine stiff and flexible components to

26.1	*External Skeletons Weigh More than Internal Ones*	
Animal	**Skeleton type**	**Percentage of body weight devoted to skeleton**
Clam	External shell	30
Spider	External chitin	10
Mammal	Internal bone	4
Bony fish	Internal bone	3

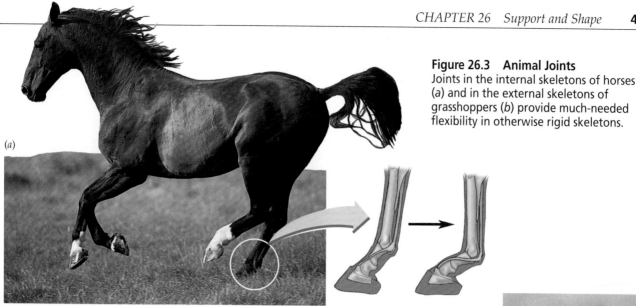

Figure 26.3 Animal Joints
Joints in the internal skeletons of horses (*a*) and in the external skeletons of grasshoppers (*b*) provide much-needed flexibility in otherwise rigid skeletons.

(a)

(b)

allow animals to bend, while at the same time providing the stiffness needed to maintain their shape.

■ Animal support systems depend on specialized tissues that provide stiffness, while at the same time accommodating the changes in shape needed for movement. The support tissues of animals can lie within or can surround their soft tissues.

The Support Systems of Plants, Fungi, and Single-celled Organisms

Many single-celled organisms, plants, and fungi support themselves differently from animals. These organisms all rely on a cell wall (which animal cells lack) for support.

Single-celled organisms are supported by their cell walls

The cells of most bacteria and many protists produce a **cell wall** that lies outside the plasma membrane. The bacterial cell wall usually contains murein, a compound made up of sugars and amino acids. The cell walls of protists may consist of a complex carbohydrate called **cellulose** or of minerals concentrated from the environment.

In a single-celled organism, the cell wall may either contribute to a hydrostatic skeleton or be inherently stiff. When the cell wall acts as a hydrostatic skeleton it resists

stretching as pressure inside the cell increases. The organism "inflates" the cell to generate pressure by allowing water to diffuse into the cell down a concentration gradient. Creating the concentration gradient requires active transport of materials into the cell (see Chapter 24) and thus requires energy.

Multicellular plants and fungi also rely on cell walls for support

The support systems of plants and fungi represent an elaboration of the cell wall support system of the protists from which they evolved. Each cell in a plant produces a matrix containing cellulose, and each cell in a fungus

Figure 26.4 Support by Cell Walls
Plants, fungi and many single-celled organisms obtain support from cell walls associated with individual cells.

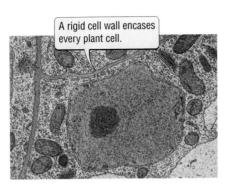

A rigid cell wall encases every plant cell.

In woody plant tissues, the thick cell walls may consist of many layers.

produces a matrix containing chitin, to form the cell wall (Figure 26.4). Both groups rely on a combination of hydrostatic skeletons and inherently rigid cell walls for stiffness.

We can see this approach most clearly in plants. The young, growing parts of plants have thin cell walls and rely on hydrostatic skeletons for support. If the plant is allowed to dry out, these young parts lose their shape. A shortage of water in these plant tissues makes it impossible to keep enough water in the cells to maintain hydrostatic pressure (Figure 26.2*b*). The older stems, in contrast, do not wilt even when they dry out. In plant parts that have completed their growth, the cell walls are stiffened by the addition of a chemical called **lignin** to the cellulose matrix. Wood consists of lignified plant tissue that remains stiff even when dry.

The support systems of plants and fungi depend on the cell walls associated with each cell, rather than on specialized skeletal tissues, as in animals. Thus, the cells of plants and fungi cannot change shape as easily as animal cells. The inability of individual cells to change shape severely limits the ability of plants and fungi to move, as we will see in Chapter 27.

> ■ The support systems of single-celled bacteria and protists, plants, and fungi all rely on cell walls associated with each cell. This strategy greatly restricts the mobility of multicellular plants and fungi.

How Do Support Systems Support Organisms?

Two physical stresses, compression and tension, tend to change the shape of things. The function of supporting an organism is essentially the function of resisting both **compression**, which squeezes molecules closer together and shortens things, and **tension**, which pulls molecules apart and lengthens things.

When a tree bends in a strong wind, tension stretches the wood on the outside of the curve in the trunk, whereas compression squeezes the wood on the inside of the curve (Figure 26.5*a*). How easily wood stretches

and compresses in the wind determines how much the tree bends. Stiff wood stretches and compresses little and, therefore, bends little, whereas flexible wood stretches and compresses easily, and bends more.

If the tension generated by bending exceeds the tensile strength of the wood, which is the maximum tension that it can handle before the molecules in trunk pull apart, a crack will form on the outer edge of the curve in the trunk. Of course, from the tree's perspective, this is bad: If the crack passes all the way through the trunk, the tree breaks (Figure 26.5*b*). Similarly, if the weight of the tree or the bending of the trunk generate more compression than the wood can handle, the trunk breaks. The tensile and compressive strength of a support system depends both on the materials of which it is made and the arrangement of those materials.

> ■ Support systems must resist changes in shape caused by tension, which pulls molecules apart, and compression, which squeezes them together.

 ## 26.2 The Materials of which Skeletons Are Made

In our survey of support systems, we have encountered three different classes of materials: biological materials produced by the organism, minerals accumulated from the environment, and hydrostatic skeletons. Let's briefly compare their strengths and weaknesses as components of support systems (Table 26.2).

Biologically produced materials are light and resist tension well

Cellulose, murein, collagen, and chitin represent the most widespread biologically produced support mate-

(a)

Figure 26.5   Compression, Tension, and Tensile Strength in Support Systems
Successful support systems resist breaking when subjected to tension and compression. (*a*) When wind blows against a tree, it causes both compression and tension in the tree trunk. (*b*) Trees, such as palm trees, whose flexible wood resists these forces well can survive hurricane-force winds. (*c*) Other trees, such as pines, with stiffer wood break under the same conditions.

The outer edge of the curve is stretched, creating tension, which tends to pull things apart.

Tension   Compression

The inner edge of the curve is squeezed together, creating compression, which tends to crush things.

rials. Biologically produced support molecules typically weigh little and resist tension exceptionally well. The cellulose in wood, for example, can give it a tensile strength greater than that of a steel cable of similar weight. These materials, however, do not provide as much inherent stiffness as mineral skeletons do. Their stiffness generally comes from **cross-links**, chemical bonds that form between adjacent molecules. The cross-links allow these materials to resist compression more effectively. We can see this if we look at a caterpillar: The stiff case protecting the head consists of cross-linked chitin, whereas a flexible skin made of chitin lacking cross-links cov-

(b)

The ability to resist compression and tension…

(c)

…determines how trees respond to the stress of hurricane-force winds.

26.2 *Properties of Materials Used in Support Systems*

	Material		
Property	**Biological**	**Mineral**	**Hydrostatic skeleton**
Stiffness	Moderately stiff, if cross-linked	Stiffest	Variable, but not very stiff
Resistance to forces	Resists tension	Resists compression	Resists neither tension nor compression
Weight per volume	Light	Heavy	Light

ers the rest of the body. Similarly, lignin in plants stiffens cell walls by increasing their resistance to compression.

Minerals accumulated from the environment provide stiffness

Organisms of all kingdoms use minerals accumulated from the environment in their support structures. As mentioned above, these minerals typically incorporate calcium, but some organisms, such as grasses (among plants) and sponges (among animals), accumulate other minerals, such as silica (of which glass is made). These minerals are often abundant and easily accumulated from the environment. Mineral support systems resist compression very well, and they are inherently stiff.

Mineral support systems suffer from two drawbacks, however. A mineral skeleton weighs up to twice as much as a similar volume of biologically produced material, and mineral skeletons tend to have low tensile strengths, which makes them vulnerable to breaking under tension. In response to selection pressures favoring lightweight support systems, organisms with mineral-based support systems tend to evolve skeletons that contain as little of the mineral as is needed to safely support them. We consider this trade-off in more detail below. In response to selection for support systems that resist breakage, virtually all organisms with mineral-based support systems mix some tension-resistant biologically produced materials with the minerals. The mineral matrix produced by our bone cells, for example, includes tension-resistant collagen.

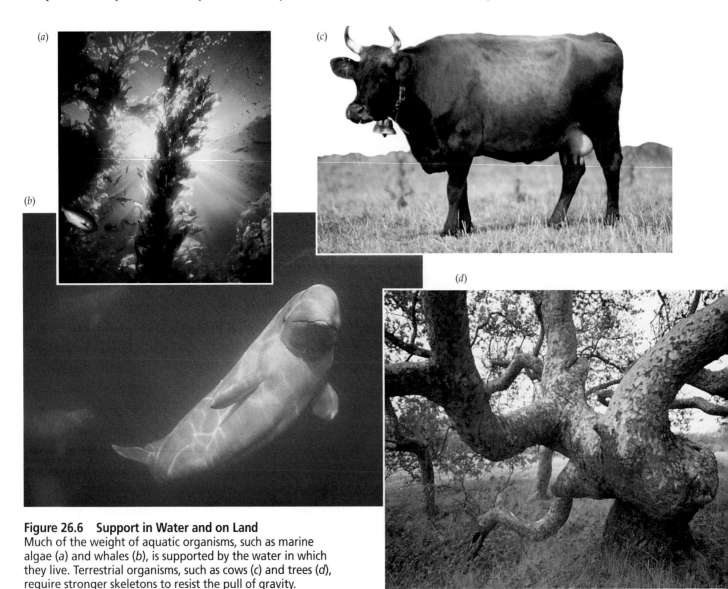

Figure 26.6   Support in Water and on Land
Much of the weight of aquatic organisms, such as marine algae (a) and whales (b), is supported by the water in which they live. Terrestrial organisms, such as cows (c) and trees (d), require stronger skeletons to resist the pull of gravity.

Hydrostatic skeletons offer relatively low stiffness but great flexibility

Even the most high-pressure hydrostatic skeletons never attain the stiffness of support systems based on biologically produced or mineral materials. Thus, organisms that rely on hydrostatic skeletons cannot maintain their shape as well when under compression or tension. On the other hand, hydrostatic skeletons offer the advantages of light weight and of a control over stiffness that is not possible with either biologically produced or mineral support systems. Animals control the pressure in, and hence the stiffness of, their hydrostatic skeletons by contracting muscles. Other organisms accomplish this by regulating the movement of water into or out of their cells. By controlling hydrostatic pressure, organisms can change the stiffness of their hydrostatic skeletons, which allows them to change their shape. A little experimenting with your tongue will convince you that you can make it quite stiff or quite limp.

> ■ Biologically produced support materials provide light weight and high tensile strength, but relatively little stiffness. Mineral support materials offer stiffness and resistance to compression, but are relatively heavy and prone to breaking under tension. Hydrostatic skeletons provide changeable, but relatively low, stiffness and light weight.

The Evolution of Skeletons in Water and on Land

Life in water and life on land place very different demands on the support systems of organisms. The water in which aquatic organisms live supports them, so that they lead almost weightless existences. On land, however, the support system of an organism must support its full weight against the pull of gravity.

Accordingly, aquatic organisms tend to have support systems that are less stiff than those of terrestrial organisms (Figure 26.6). Hydrostatic support systems, which generally provide less stiffness than mineral or biologically produced ones, are found much more often in aquatic habitats than on land. The low-pressure hydrostatic skeleton that provides shape to a jellyfish in water is useless on land: When a jellyfish washes ashore, it collapses into a shapeless blob on the beach. Even aquatic animals that rely on mineral skeletons, such as fish, often have more delicate ones than terrestrial organisms of similar weight.

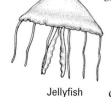
Jellyfish

Terrestrial organisms need stiff skeletons to support their weight, but the support tissues themselves add weight to the organism. In these organisms, body weight and skeleton weight show an allometric relationship (see Chapter 25): As the weight of an organism increases, the weight of its support tissues increases even more rapidly. For example, the shrew, a tiny mammal that weighs about 8 grams, has a skeleton that makes up 4.6 percent of its body weight, whereas a 7000-kilogram elephant has a skeleton that makes up 24 percent of its body weight (Figure 26.7). This difference is visible in the much stouter appearance of elephant bones as compared

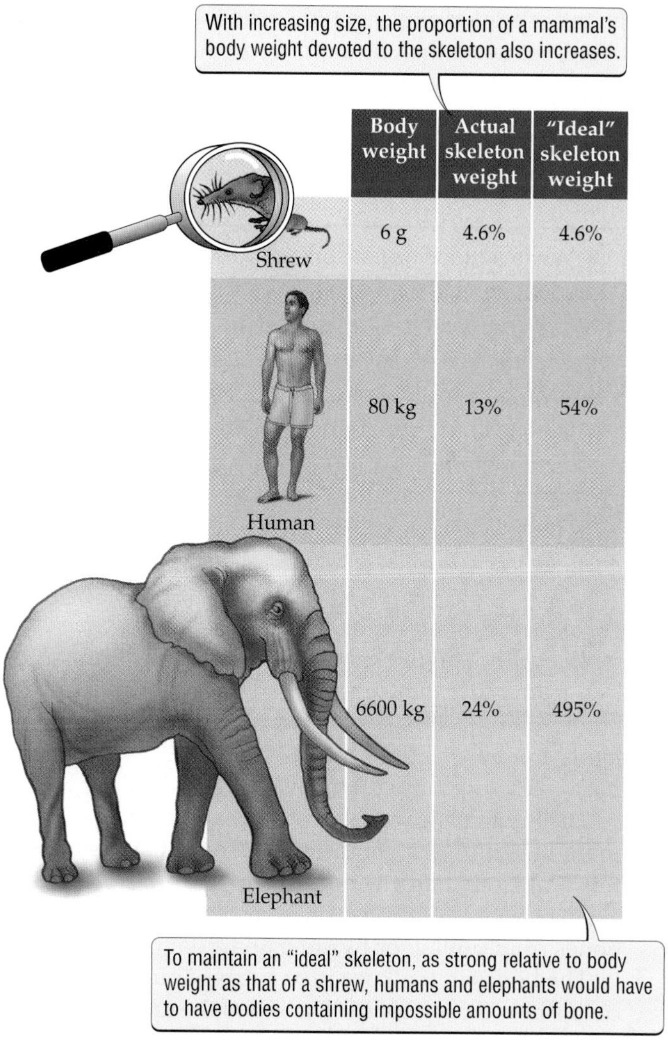

> With increasing size, the proportion of a mammal's body weight devoted to the skeleton also increases.

	Body weight	Actual skeleton weight	"Ideal" skeleton weight
Shrew	6 g	4.6%	4.6%
Human	80 kg	13%	54%
Elephant	6600 kg	24%	495%

> To maintain an "ideal" skeleton, as strong relative to body weight as that of a shrew, humans and elephants would have to have bodies containing impossible amounts of bone.

Figure 26.7 Skeleton Weight Reflects Body Weight
Large organisms must devote more of their body weight to skeletal materials than small organisms. To have skeletons proportionately as strong as those of shrews, elephants would have to devote more than a hundred percent of their actual body weight to bone weight—which, of course, is impossible.

with the equivalent bones in shrews (Figure 26.8*a*). This allometric relationship sets up a vicious circle: Elephants need a heavier skeleton in part to support their heavier skeleton. Furthermore, it means that elephants can devote less body weight to other tissues. It appears that elephants have cut corners to keep their skeletons from being even heavier than they already are (see the box on p. 445). Considering the properties of bone, we can estimate that for an elephant's skeleton to be as strong relative to its body weight as is that of a shrew, the elephant's skeleton alone would have to weigh almost five times more than its entire body does in reality (see Figure 26.7). Because an elephant's skeleton actually weighs much less than that, its skeleton is much less strong relative to its body weight than is a shrew's.

■ While aquatic organisms need relatively little support, terrestrial organisms need stiff skeletons that can support the organism's entire body weight against the pull of gravity. Large terrestrial organisms have relatively weaker support systems than small ones because of the prohibitive weight of support materials.

(a)

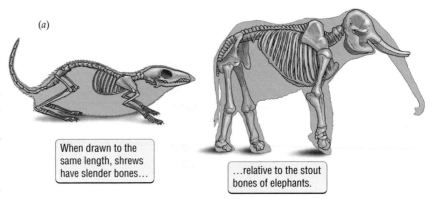

When drawn to the same length, shrews have slender bones...

...relative to the stout bones of elephants.

(b)

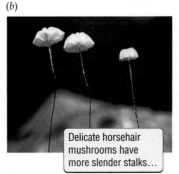

Delicate horsehair mushrooms have more slender stalks...

...when depicted at the same size as large fly agaric mushrooms.

Figure 26.8   Heavy Organisms Need Thick Support Structures
(*a*) Heavy elephants have bones that are thicker relative to their legs than those of tiny shrews. (*b*) Note the similar pattern in the thin stalk of the small mushroom on the left as compared with the thick stalk of the large one. In both pairs of illustrations, the scale of the pictures has been adjusted to make the organisms appear equal in body length.

The Human Knee

We return now to joints, using the human knee as a model. Although the knee differs in details of construction from other joints in our bodies and from joints in other animals, it makes a good general model for how joints combine flexible and stiff materials to allow animals to move while still maintaining their shape.

The human knee consists of bone and cartilage. These materials can resist compressive forces that may reach ten times our body weight during exercise. Various arrangements of collagen surround the knee to keep the bones from pulling or twisting apart (Figure 26.9).

The bones in the knee joint support our weight and define how the leg bends. The end of the upper leg bone, the femur, rides in a pair of grooves at the end of the tibia, which is the larger of our two lower leg bones. This arrangement allows the leg to swing forward and back like a hinge, but not from side to side, thereby providing both motion and stability. The knee must also allow for the slight twisting of the tibia relative to the femur that takes place when we walk.

Ligaments are connective tissues that attach bone to bone. Two pairs of ligaments, one on the front and back of the knee and a second on the sides of the knee, connect the femur to the lower leg bones and prevent sliding along the joint surface when the knee bends, keeping the bones in their proper place. Because the collagen of the ligaments can stretch slightly, it allows some bending and twisting motion in the knee. **Tendons**, which attach muscle to bone, represent another form of connective tissue. They help to hold the knee together by connecting the upper leg muscles to the bones of the lower leg.

Wherever two moving parts of a joint rub against each other, wear can destroy bone and friction can waste energy. Layers of cartilage in the knee cushion the points at which the femur meets the tibia, and sacs containing lubricating **synovial fluid** reduce friction between them. As a result, the femur and tibia slide past each other more easily than a skate slides over ice.

26.3

Figure 26.9
The Human Knee
The human knee represents a good example of how animal skeletons balance rigidity and flexibility. The deep view in this diagram indicates what surgeons might see as they operate.

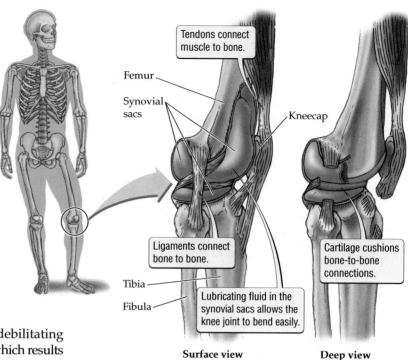

Tendons connect muscle to bone.

Femur

Synovial sacs

Kneecap

Ligaments connect bone to bone.

Tibia

Fibula

Lubricating fluid in the synovial sacs allows the knee joint to bend easily.

Cartilage cushions bone-to-bone connections.

Surface view   **Deep view**

When all of these components—bone, cartilage, ligaments, tendons, and synovial fluid—work together, we have a joint that can control and withstand motion over decades. When one of these components fails, however, we face potentially serious medical problems (Table 26.3).

Various kinds of arthritis involve damage to the cartilage and bone in the knee joint, and can make bending difficult and painful. Tears in the cartilage and ligaments of the knee represent common and debilitating sports injuries (Figure 26.10). Torn cartilage, which results most often from rapidly speeding up or slowing down while turning, can lead to increased wear on the knee joint. Unless the tears are repaired surgically, the remaining cartilage and underlying bone may become permanently damaged, leading to arthritis. Ligaments tear when they are stretched to more than one and a half times their normal length. Tears can result from a hard blow to the front or sides of the knee or from severe twisting of the lower leg relative to the femur. Torn ligaments cannot hold the femur in place relative to the tibia and fibula, leading to movement in the knee joint that may make walking impossible. Like torn cartilage, torn ligaments require surgery.

■ The human knee illustrates many of the basic features of the joints that allow animals to move while still maintaining their shape. Flexible materials around the joint help control the movement of the stiff skeletal components relative to one another.

■ HIGHLIGHT

Landing Cats and Humans

How can cats survive falls that are fatal to people? Three potential explanations have been proposed (Figure 26.11).

First, cats have stronger skeletons relative to their weight than humans do. Both cats and humans have stiff

Figure 26.10 Damage to the Knee Joint Can Severely Restrict Mobility

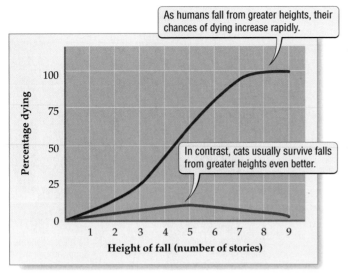

As humans fall from greater heights, their chances of dying increase rapidly.

In contrast, cats usually survive falls from greater heights even better.

Percentage dying

Height of fall (number of stories)

Figure 26.11 Survival of Falling Cats and Humans
Statistics collected by doctors and veterinarians show that cats survive falls from high places much better than humans do. Note that cats are much less likely to die from a fall than are humans, and that the mortality rate for cats actually decreases for falls from heights greater than the fifth story.

internal skeletons made of collagen-reinforced bone. Recall that the weight of bone increases more rapidly than the body weight it can support (see Figure 26.7). The conflict between the advantages of a stronger skeleton and the disadvantages of a heavier skeleton has led to an evolutionary trade-off: Humans have sacrificed skeletal strength to reduce weight. We must pay for the benefit of a tolerably light skeleton by looking carefully before we leap, whereas smaller and more strongly constructed cats can get away with less caution.

Second, cats have an exceptional ability to soften their landing. Perhaps because falls are a normal part of life for tree-climbing creatures, natural selection strongly favors cats that instinctively rotate their bodies as they fall so that they land feet first. Humans, in contrast, may land either arms or feet first. By landing as they do, cats spread the force of impact over four relatively stronger limbs, whereas humans concentrate a larger force over only two, relatively weaker limbs.

Third, cats may even relax a bit during a fall and prepare for a crash landing by bending their legs so that the ligaments and tendons associated with the joints can act as springs. We, quite understandably, tend not to relax during a free fall. By using their legs as springs, cats further reduce the force of impact by spreading it over a slightly greater period of time. Tense humans suffer the full force of impact virtually instantaneously. While the difference between decelerating from a fall to a dead stop in 0.002 seconds instead of 0.001 seconds may seem trivial, doubling the time to decelerate nonetheless halves the force of impact. For similar reasons, a padded dashboard in a car or a bicycle helmet can dramatically reduce the injuries suffered in a collision. The decline in deaths when cats fall from heights greater than the fifth story may reflect the increased time that these high-diving cats have to prepare properly for impact.

Because cats naturally climb and really do let curiosity have a good shot at killing them, it would be tempting to conclude that the ability of cats to survive falls from high places represents a marvelous adaptation to their way of being. In part, the cat's amazing ability to survive falls does seem to represent an adaptation to life in the trees. Equally important, however, may be the relatively greater strength of cat leg bones, which is an incidental consequence of their size rather than an adaptation.

■ Cats probably survive falls better than humans do because their skeletons are relatively stronger. In addition, cats may benefit from their ability to prepare for landing by relaxing, bending their joints, and landing on all four feet.

26.3 *Some Ailments of the Knee Joint*

Problem	Bone	Cartilage	Ligaments	Synovial fluid
Osteoarthritis	X	X		
Rheumatoid arthritis	X	X		X
Torn cartilage		X		
Torn ligament			X	
Bursitis				X

■ THE SCIENTIFIC PROCESS

For the Extinct Elephant Bird, Was Incubation a Crushing Disaster?

The eggs of birds, when not serving as part of a nutritious breakfast, provide a wonderful example of conflicting selection pressures at work. The shell of an egg, for example, acts both as an external skeleton and as a filter through which materials enter and leave the egg. The shell must be weak enough that a tiny chick can break it open when ready to hatch, yet strong enough that it can support the weight of an incubating adult bird. The shell must be thin and porous enough to allow in the oxygen that the developing chick needs, yet it must resist the loss of water so that the developing chick does not dry out.

The largest bird in the world, the 450-kilogram elephant bird of Madagascar, laid the biggest eggs in the world, weighing 9 kilograms. Sadly, the elephant bird went extinct in the seventeenth century, leaving almost no records of its natural history in its wake. We can use what we know about the eggs of other bird species, however, to reconstruct some of the selection pressures that shaped the nesting biology of this remarkable bird.

Among living birds, large species have eggshells that are thin relative to egg weight. The relatively thinner eggshells of large eggs appear to compensate for their smaller surface area–to–volume ratio as compared with small eggs. In other words, a thick shell on a large egg would keep the developing chick from getting enough oxygen.

Shell thickness, however, directly determines how much force an egg can tolerate before it breaks. We know that the eggs of small birds can support a weight many times that of the parent bird. The eggs of large birds, however, are relatively fragile. An egg of the ostrich, the largest living bird species, cannot quite support the weight of a single adult. Ostriches deal with this problem by laying 10 or more eggs in a nest, so that each egg supports just a fraction of the parent's weight.

Birds as big as the elephant bird, however, must have faced tremendous problems during incubation. A single elephant bird egg would break under a weight 1/200 that of an adult bird. In other words, an elephant bird nest would have to contain more than 200 eggs before a parent could safely settle in. Because the elephant bird is extinct, we may never know how this species managed to incubate its eggs.

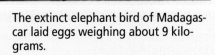

The extinct elephant bird of Madagascar laid eggs weighing about 9 kilograms.

SUMMARY

The Human Skeleton

- Bone is the major support tissue in the human body.
- Cartilage and hydrostatic skeletons provide additional support.

An Overview of Animal Skeletons

- All animal support systems depend on specialized tissues that provide stiffness. Animal skeletons also accommodate the changes in shape needed for movement by including joints.
- Soft-bodied animals are supported mainly by hydrostatic skeletons.
- Support tissues that are external to the soft tissues offer greater protection.

- Support tissues that are internal to the soft tissues weigh less than external skeletons.

The Support Systems of Plants, Fungi, and Single-Celled Organisms

- The support systems of single-celled bacteria and protists and of multicellular plants and fungi rely on cell walls that surround each cell.
- Because each cell has a cell wall associated with it, this type of support system greatly restricts the mobility of multicellular plants and fungi.

How Do Support Systems Support Organisms?

- Support systems must resist changes in their shape caused by tension, which pulls molecules apart, and compression, which squeezes them together.

- If the tensile or compressive strength of a support system is exceeded, it will break.

The Materials of which Skeletons Are Made

- Biologically produced support materials, such as cellulose, murein, collagen, and chitin, provide light weight and high tensile strength, but relatively little stiffness.

- Mineral support materials offer stiffness and resistance to compression, but they are relatively heavy and do not offer much tensile strength.

- Hydrostatic skeletons provide changeable but relatively low stiffness and light weight.

The Evolution of Skeletons in Water and on Land

- Because water supports much of their weight, aquatic organisms can rely on hydrostatic skeletons or relatively delicate mineral skeletons for support.

- Terrestrial organisms need stiff skeletons that can support the organism's entire body weight against the pull of gravity.

- Large terrestrial organisms have relatively weaker support systems than small ones because of the prohibitive weight of support materials.

The Human Knee

- The human knee illustrates many of the basic features of the joints that allow animals to move while still maintaining their shape.

- Flexible ligaments and tendons around the joint help control the movement of the two bones—the femur and the tibia—relative to each other.

- To work efficiently and to withstand a lifetime of movement, the joints requires friction-reducing components such as cartilage and synovial fluid.

Highlight: Landing Cats and Humans

- Cats can survive falls better than humans probably because their skeletons are relatively stronger than those of humans.

- Cats may also benefit from their apparent ability to prepare for landing by relaxing, bending their joints, and landing on all four feet.

KEY TERMS

bone p. 434	hydrostatic skeleton p. 434
cartilage p. 434	internal skeleton p. 436
cell wall p. 437	joint p. 436
cellulose p. 437	ligament p. 442
chitin p. 435	lignin p. 438
collagen p. 434	synovial fluid p. 442
compression p. 438	tendon p. 442
cross-link p. 439	tension p. 438
external skeleton p. 436	

CHAPTER REVIEW

Self-Quiz

1. An important advantage of hydrostatic skeletons is that they
   a. allow the organism to vary its stiffness.
   b. resist compression better than any other kind of skeleton.
   c. resist tension better than any other kind of skeleton.
   d. all of the above.

2. Collagen and chitin are
   a. biologically produced materials.
   b. connective tissues.
   c. hydrostatic skeletons.
   d. minerals.

3. Which of the following is true of both plants and fungi?
   a. They rely on external skeletons.
   b. They rely on skeletons made of minerals.
   c. Their support comes from cell walls.
   d. Their inadequate support systems force them to live in aquatic environments.

4. Large mammals have bones that are relatively heavier than the bones of small mammals because
   a. their limbs are longer.
   b. their weight is greater.
   c. they eat less.
   d. their muscles are stronger.

5. Ligaments
   a. connect muscles to other muscles.
   b. connect muscles to bones.
   c. connect bones to other bones.
   d. all of the above

Review Questions

1. Recall that plants do not add lignin to the cell walls of growing plant parts. Why might this be?

2. Why do most organisms combine support structures that contain matrices of biologically produced materials with structures based on mineral matrices?

3. Under what circumstances could using a hydrostatic skeleton be advantageous?

The Daily Globe

Knee Injury Fuels Debate

BOSTON, MA—Less than 2 weeks after the U.S. victory in the Women's World Cup, soccer star Sarah Domore, one of the most popular and visible players on the U.S. team, tore her anterior cruciate ligament (ACL) in an exhibition game against up-and-coming college players. The injury, though not career-threatening, will require surgery and is expected to sideline Domore for at least 6 months.

Given Domore's immense popularity, some people are wondering whether her ACL injury will serve as a warning—or even a deterrent—to young women interested in soccer. Female athletes are more susceptible than males to ACL injuries: In some studies, females suffered such injuries at rates four to five times higher than men. These injuries are especially common in sports that require many sudden stops and turns, such as soccer, basketball, and volleyball, but why women are at higher risk for ACL tears is not clear.

Physical differences may account for some of the increased risk to women. For instance, the wider pelvis of women results in a different angle of the bones at the knee joint, and monthly changes in hormones appear to affect the flexibility of connective tissue. However, weight lifting and training routines that strengthen the hamstring muscles, along with new jumping techniques, may help protect women from some of these injuries.

Although many athletic trainers and orthopedic specialists suggest that such education and exercise programs can reduce the risk of injury to acceptable levels, not everyone agrees. Some family doctors and parents are trying to urge schools not to support soccer and other sports, such as basketball and volleyball, that pose a risk of injury to women and girls, but instead to provide better funding for less risky sports, such as softball and track and field. However, given the success of the U.S. women's soccer team and professional leagues such as the Women's National Basketball Association, they may face an uphill battle.

Evaluating "The News"

1. Think about other familiar sports in which players are sometimes injured. What types of tissues seem to be most injury-prone, and why do you think this is so?

2. Do schools that support sports with taxpayer dollars have a responsibility to support only the safest sports? Should schools that support women's soccer be involved in educational campaigns about how to minimize ACL injuries, or does that responsibility fall on someone else? If so, whose responsibility is it?

3. If you had a daughter, how would you decide which sport(s) you would encourage her to play?

chapter 27 Movement

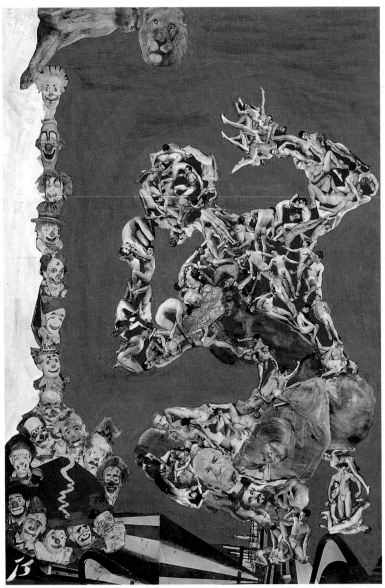

Jess, *The Mouse's Tale*, 1951.

Speeding Sperm and Racing Runners

The question of how fast we can run has motivated competitions since the beginning of organized athletics. Florence Griffith-Joyner has sprinted 100 meters in a world record of 10.49 seconds. At the other extreme of distance, world record holder Catherine Ndereba has run the 42.2-kilometer marathon in 2 hours, 18 minutes, and 47 seconds. The excitement of a sprinter like Griffith-Joyner comes from seeing her reach maximum possible speeds over short distances (10.2 meters per second), whereas the challenge for a long-distance runner such as Ndereba lies in her ability to sustain a lower speed (5.0 meters per second) over a much longer distance.

Humans stage races for the entertainment they provide, but remarkably similar tests of speed and endurance serve a more serious purpose in the biological realm. Mammalian sperm swim as quickly as possible toward unfertilized eggs moving down the oviducts of the female reproductive system (see Chapter 36). From a starting field of millions of sperm released in a single ejaculation, only one will succeed in fertilizing each egg—probably the one that is the first to reach it. Researchers have found that sperm

similar factors determine how fast and how far they can travel. Even though the muscles that propel runners and the tiny tail-like flagella that propel sperm work differently, we will see in this chapter that they are responses to similar challenges.

Moving Toward the Finish Line

behave much like track stars. Mouse sperm sprint down the short reproductive tract of female mice at 0.2 millimeters per second. In the much longer reproductive tract of human females, however, human sperm pace themselves at about half this speed as they swim the sperm equivalent of a marathon.

Comparisons between runners and sperm that travel short and long distances illustrate an important point: Although athletes and sperm may seem to have little in common,

■ KEY CONCEPTS

1. The interaction of actin and myosin filaments causes muscle fibers to contract.

2. Muscles can only contract.

3. Muscles can increase in strength by increasing cross-sectional area, and can increase in speed by lengthening or including more fast muscle fibers.

4. Animals can move by walking, flying, or swimming.

5. Drag resists the forward motion of an organism.

6. Single-celled prokaryotes and eukaryotes propel themselves by hairlike structures call cilia or flagella.

7. Plants and fungi rely on animal dispersers or on wind and water currents to move their pollen, seeds, and spores.

The ability to move through the environment provides an almost endless list of potential advantages to an organism. Consider how different your own life would be if you could not move: You could not explore your surroundings, you could not get to the refrigerator to forage for food, you could not search for a mate, and you could not escape an underheated room for a warmer one.

Although movement undoubtedly benefits organisms in all six kingdoms of life, we saw in Chapter 26 that animals, more than any other group, have the ability to move through their world. The unique combination of specialized support tissues, cells that lack rigid cell walls, and muscle tissues that can contract makes it possible for the animals to exploit the potential of movement more effectively than any other kingdom (Figure 27.1a).

In this chapter we focus on how animals move. We begin by discussing the form and function of the muscle tissue that serves as a biological motor. We then consider how natural selection has changed the arrangement of muscle tissue, the interactions between muscle and support tissue, and the shape of animal bodies to allow animals to move in a variety of ways. Next we look at some of the unique ways in which bacteria and protists move (Figure 27.1b). Fi-

(a)

(b)

(c)

Figure 27.1 Movement Is Essential for Animals and Plants
Movement plays an important part in the lives of all organisms. (*a*) Cheetahs pursue their antelope prey in the hope of a meal, while the antelopes flee for their lives. (*b*) Bacteria can swim toward and away from chemical cues using hairlike flagella. (*c*) Plant tendrils cling to objects that they encounter in their environment.

nally, we see how plants and fungi have overcome some of the limitations that come with their lack of muscle tissue (Figure 27.1*c*).

27.1 Animal Muscles as Biological Motors

Muscle tissue, which is unique to animals, plays a leading role in their ability to move. Muscle tissue specifically does one invaluable thing: it contracts. Before turning to the many ways in which animals convert these simple contractions into useful motion, let's examine the structure of muscle tissue and see how muscles contract.

Muscle tissue has a structure that reflects its function

Most of us are familiar with muscle tissue because when we eat fish or chicken or beef, we are eating muscles. Most features that relate to the role of muscle tissue as a biological motor, however, remain hidden to the unaided eye. Each muscle—for example, the biceps in our upper arm (Figure 27.2)—consists of many basic units of muscle tissue, called **muscle fibers**. Muscle fibers are unique in that they arise from the fusion of many mus-

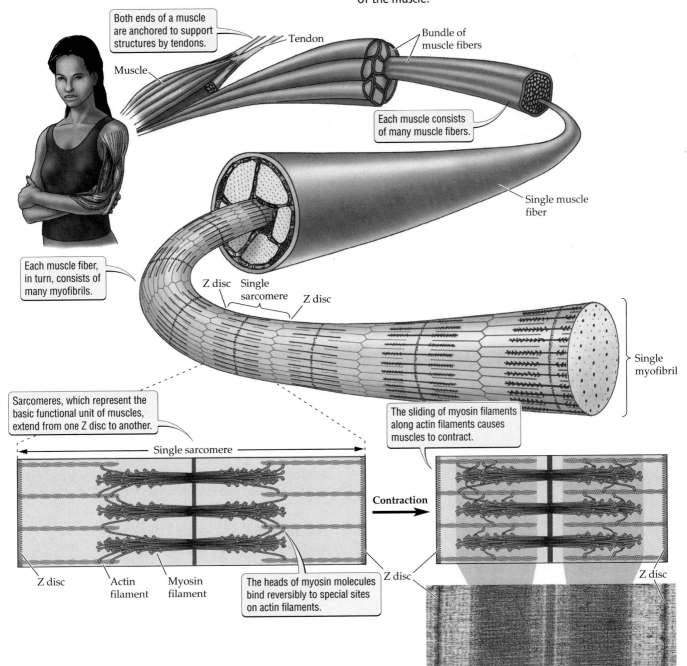

Figure 27.2 The Fine Structure of the Biceps Muscle
Muscle contraction ultimately depends on the sliding of myosin filaments along actin filaments to shorten the length of the muscle.

Both ends of a muscle are anchored to support structures by tendons.

Tendon

Muscle

Bundle of muscle fibers

Each muscle consists of many muscle fibers.

Single muscle fiber

Each muscle fiber, in turn, consists of many myofibrils.

Z disc   Single sarcomere   Z disc

Single myofibril

Sarcomeres, which represent the basic functional unit of muscles, extend from one Z disc to another.

The sliding of myosin filaments along actin filaments causes muscles to contract.

Single sarcomere

Contraction

Z disc   Actin filament   Myosin filament

The heads of myosin molecules bind reversibly to special sites on actin filaments.

Z disc   Z disc

cle cells, so that each fiber contains many nuclei. Each fiber, in turn, is packed with smaller **myofibrils** (*myo*, "muscle"). The myofibrils consist of units, called **sarcomeres**, which do the actual work of contraction. Sarcomeres are visible as stripes when seen through a microscope. At extremely high magnification we can see each sarcomere clearly. Each sarcomere extends between two **Z discs**, which are visible as dark lines.

Each sarcomere is based on a very specific arrangement of two proteins, **actin** and **myosin**, that allow muscles to contract. The Z discs represent the anchor point for actin filaments, each of which consists of two actin molecules. The actin filaments extend toward the center of each sarcomere from the Z discs. Between the actin filaments lie myosin filaments, each of which consists of many myosin molecules. As we will see shortly, the myosin filaments can use energy from ATP to pull the two Z discs closer together, causing the muscle to contract. We see muscles bulge when they contract because all of the actin and myosin filaments must squeeze into a shorter length of muscle.

Actin and myosin interact to cause muscle contraction

When you lift a weight, actin and myosin filaments slide past each other in the sarcomeres of the biceps muscle of your arm. If you could watch your biceps at high magnification as you lifted the weight, you would see the two Z discs defining each sarcomere pull closer together (see Figure 27.2). An even closer look would reveal that each myosin molecule ends in a protruding head that can bind to specific sites on the adjacent actin filaments. As you lift the weight, each myosin head repeatedly attaches to a binding site on the actin filament, changes its shape so as to pull the myosin filament closer to the Z disc, and then releases the binding site before reattaching to another binding site even closer to the Z disc. In this way, the myosin heads "walk" their way from binding site to binding site along the actin filaments, pulling the Z discs closer together and contracting the muscle. Each "step" of each myosin head requires the energy released by converting one ATP molecule to one ADP molecule (see Chapter 7).

Muscles must work in pairs

Because the myosin heads can only pull the Z discs together, not push them apart, a contracted muscle cannot stretch itself out again. Thus, we find muscles arranged in pairs, so that the contracting muscle stretches the other muscle back to its starting position following a contraction. In the upper arm, for example, we have

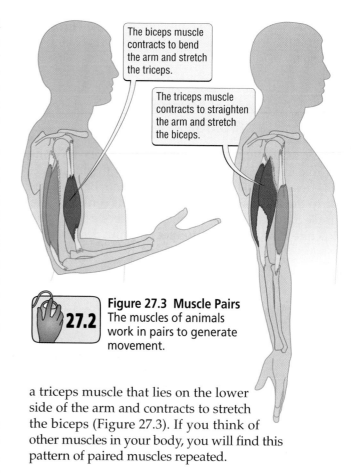

27.2

Figure 27.3 Muscle Pairs
The muscles of animals work in pairs to generate movement.

The biceps muscle contracts to bend the arm and stretch the triceps.

The triceps muscle contracts to straighten the arm and stretch the biceps.

a triceps muscle that lies on the lower side of the arm and contracts to stretch the biceps (Figure 27.3). If you think of other muscles in your body, you will find this pattern of paired muscles repeated.

> ■ Muscle fibers consist of sarcomeres that contract through the action of actin and myosin filaments. Myosin molecules "walk" along the actin filaments to pull the Z discs of a sarcomere closer together. Muscles must work in pairs because individual muscles can only contract.

27.3 How Animals Convert Muscle Contraction into Motion

As we have seen, muscle is a remarkable tissue that is specialized for contraction. Within the animal kingdom, mechanisms have evolved to convert simple muscle contractions into the wide range of motions of which most animals are capable. Let's consider how animals can vary the strength and speed with which their muscles contract.

The strength of muscles depends on the amount of muscle tissue

All animals have essentially identical muscle tissue, which contains identical actin and myosin proteins

Figure 27.4 Exercise Can Increase Muscle Size
Weight lifting increases muscle strength by stimulating the production of additional muscle fibers, which increases muscle bulk.

arranged in much the same way. It should not surprise us, therefore, that pieces of muscle tissue *containing similar numbers of muscle fibers* can lift about the same weight, regardless of whether they come from an elephant or an earthworm.

Differences in strength between species and between muscles within an individual depend almost entirely on the cross-sectional area of the muscles. The greater the cross-sectional area, the more muscle fibers can lie side by side and contract together. Elephants are stronger than earthworms because elephants have muscles with bigger cross-sectional areas.

Exercise can change muscle strength. If you train with weights, your muscles ini-

tially increase in strength without increasing in cross-sectional area. The number of muscle fibers in your muscles remains the same, but a greater proportion of them contract together at any one time. Thus the initial increase in strength comes about by using muscle fibers more efficiently. With continued weight training, the number of fibers in the muscle increases, leading to the greater muscle bulk for which bodybuilders strive (Figure 27.4).

The speed of muscle contraction depends on the length and type of muscle

Just as animals have muscles that can generate different forces, they also have muscles that contract at different speeds. Like differences in strength, these differences in speed arise not because the muscle tissues differ in any fundamental way, but because they differ in their arrangement.

One way to increase the speed with which a muscle can contract is to increase its length. When fully contracted, a muscle shortens to about 60 percent of its resting length. Thus a 10-centimeter-long muscle in your arm shortens by 4 centimeters to a 6-centimeter length when fully contracted. In the same amount of time, a 30-centimeter-long muscle in your thigh shortens by 12 centimeters to an 18-centimeter length when fully contracted. This means that the longer leg muscle contracts three times the distance in the same amount of time as the shorter arm muscle.

Many groups of animals also have two classes of muscles, which differ in how their fibers respond to the nerve signals that tell them to contract (Table 27.1). Muscle fibers that contract quickly, such as those that generate the explosive power of a runner sprinting from the starting blocks, are called **fast muscle fibers**. Fast muscle fibers contract completely in an all-or-nothing

27.1 A Comparison of Fast and Slow Muscle Fibers		
	Type of muscle fiber	
Characteristic	**Fast**	**Slow**
Speed of contraction	Fast	Slow
Force of contraction	Weak	Powerful
Length of contraction	Brief	Sustained
Response of sarcomeres	Either no contraction or complete contraction	Partial contraction possible
Source of ATP	Fermentation	Aerobic respiration
Human example	The quadriceps muscle in the thigh	The gluteus maximus muscle in the buttocks

response to signals from nerves. In contrast, **slow muscle fibers** respond to signals from nerves by contracting in small steps (see Table 27.1). Slow muscle fibers are used most often for activities that are sustained over long periods of time, as during the slower pace of a marathon.

Fast muscles generally sacrifice force for speed, whereas slow muscles sacrifice speed for force. Fast muscles contract so rapidly that they quickly use up the available oxygen and have to rely on fermentation, the form of metabolism that takes place when oxygen runs out (see Chapter 7). The slow muscle fibers, on the other hand, rely on aerobic respiration. To ensure a steady oxygen supply, slow muscle fibers contain a pigment called **myoglobin** that helps capture oxygen for the muscle. We can usually distinguish slow muscle tissue visually by its darker color, which reflects its myoglobin content. Thus the white meat of a turkey's breast consists mostly of fast muscle fibers used to power flight, whereas the dark meat of the legs consists mostly of slow muscle fibers that must contract for long periods with enough force to hold up the body of the bird.

Muscles and skeletons work together to control strength and speed of movement

The speed or strength of the movement that results from a muscle contraction also depends on the arrangement of muscle and skeleton to form **lever** systems. All of us have used mechanical levers of some sort (Figure 27.5a). In each case, the lever system had a **fulcrum**, which served as a pivot point for stiff lever arms. You applied a force to the **power arm**, which moved the **load arm** that did the work you wanted done. Lever systems show a clear trade-off between the force they generate at the tip of the load arm and the speed with which the tip of the load arm moves: Large force means slow movement, and high speed means low force.

The relative length of the power and load arms determines how much speed or force a lever system gener-

Figure 27.5 Lever Systems
Lever systems allow us to use our muscles more effectively. (a) In a lever system, force applied to the power arm (red) causes the lever to pivot about the fulcrum and do work at the load arm (blue). (b) A human arm working with a hammer to pull a stubborn nail illustrates a biological and a mechanical lever system working together, and also reveals a common difference between the two. The mechanical lever system represented by the hammer has a fulcrum (Δ) where the head of the hammer pivots on the board. A long power arm extends the length of the handle, and a much shorter load arm runs from the fulcrum to where the claw of the hammer grips the nail. In the arm, the elbow joint forms the fulcrum (Δ) of the biological lever system. The biceps muscle attaches to bones in the forearm to define the power arm, and the entire length of the forearm, from the elbow to the hand, forms the load arm. In this case, the load arm is longer than the power arm.

(a)

Work—lifting an object—is done at the end of the load arm.

Our muscles apply force on the power arm.

The lever pivots on the fulcrum.

Power arm

Load arm — Fulcrum

Because the power arm is twice as long as the load arm, we can use this lever to lift an object twice as heavy as the force we apply.

Although the power and load arms are differently arranged, this lever system is equivalent to the one above because the power arm is still twice as long as the load arm.

(b)

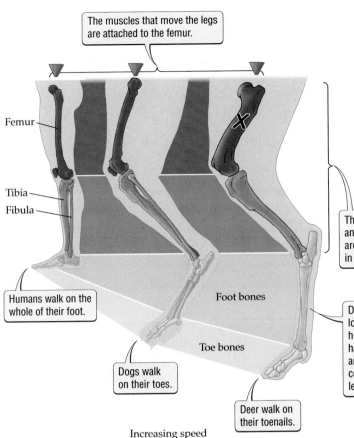

The muscles that move the legs are attached to the femur.

Femur

Tibia
Fibula

The lengths of the femur and the lower leg bones are proportionally similar in the three animals.

Humans walk on the whole of their foot.

Foot bones

Dogs walk on their toes.

Toe bones

Dogs and deer have longer load arms than humans because they have longer foot bones, and the foot bones contribute more to leg length.

Deer walk on their toenails.

Increasing speed

Figure 27.6 Adaptations of the Lever System in Mammalian Legs
Human legs lend themselves less well to high-speed running than do the legs of most other animals. All of our foot and toe bones touch the ground when we run, meaning that only the upper (red) and lower (orange) leg bones contribute to leg length. With increased leg length, the length of the load arm (the full length of the leg) relative to the power arm (the distance from the fulcrum at the tip of the femur to the X marking where the muscle attaches) increases, decreasing the proportion of the force generated by muscle contraction that is applied where the foot touches the ground, but increasing the speed of the foot movement relative to the speed of muscle contraction.

ates. We tend to use mechanical levers to increase the force that we can generate from a given muscle contraction. For example, we use the claw of a hammer to extract nails too firmly stuck in wood to pull out with our bare hands. In this case, the load arm is shorter than power arm (Figure 27.5*b*). Thus, the tip of the load arm moves a much shorter distance than the power arm, and the force that you apply to the power arm is "concentrated" over the short distance that the load arm moves. Conversely, if the load arm of a lever is longer than the power arm, the force is reduced and the speed is increased at the tip of the load arm as compared with that of the power arm. In most animal body parts, the load arm is longer than the power arm. Thus the force generated at the tip of the load arm is usually less than that generated by the muscle.

The arrangement of the legs of mammals with different lifestyles illustrates how natural selection can act on lever systems to emphasize either force or speed (Figure 27.6). In humans, only the upper and lower leg bones contribute to the length of the leg, which represents the length of the load arm. In contrast, some of the bones that form part of the foot in humans add to leg length in dogs,

and in speedy runners such as deer, all of the foot and toe bones contribute to making longer legs. The longer the leg, the longer the load arm, and thus the farther a muscle contraction moves the foot in a given amount of time. The human leg, with its relatively short load arm, works well for steady walking over long distances (the high force generated at the foot relative to the force generated by the muscles that move the leg means that our leg muscles do not have to work so hard). Dog legs are suited to running over moderate distances, and the long legs of deer let them run very rapidly over short distances. (The low force generated at the foot relative to the force generated by the muscles that move the leg means that a deer's leg muscles must generate a lot of force for each stride, which rapidly tires the animal.)

■ The strength of a muscle depends on the number of muscle fibers packed into it. The speed of muscle contraction depends on the length of a muscle and the response of its muscle fibers to nerve signals. Lever systems that arise out of the interaction of muscle and skeleton influence both the strength and speed with which organisms move.

Animal Locomotion

As animals move through their environment, they convert the contractions of their muscles into **thrust** that propels them forward. Inevitably, however, forces collectively called **drag** resist the animals' motion.

Animals can move by running, flying, or swimming

Among animals we find three main modes of locomotion: running, swimming and flying (Figure 27.7). Animals use energy to generate thrust with their muscles. Running animals, such as humans, use this energy to push off against the ground with their legs. Flying and swimming animals move through a fluid environment that requires a different way of generating thrust. Flying animals have evolved wings and swimming animals have evolved fins that provide broad surfaces with which to push off against air and water. These various modes of locomotion exact different energetic costs, which we consider in the box on page 457.

Drag resists forward motion

"Drag" is a blanket term that includes two very different forces that resist motion, which we will call pressure drag and friction drag.

To understand **pressure drag**, imagine that you are riding a bicycle, and a breeze begins to blow on your face as you set off. The breeze makes you work harder, as anyone who has pedaled a bicycle into a head wind can appreciate. What you feel as wind is air colliding with your face to create a zone of high pressure in front of you, whereas your hair streams back into a zone of low pressure behind you. High pressure in front combined with low pressure behind creates the pressure drag that pushes you backward while you do your best to pedal forward. The faster you go, the stronger the breeze feels, because the pressure difference from front to back increases. The resistance offered by pressure drag increases very quickly as speed increases, so fast-moving fish and birds tend to suffer the most pressure drag.

Friction drag, on the other hand, occurs when air molecules stick to you and to one another as you rush by. In other words, as you pedal down the road, you move not only yourself, but extra baggage in the form of the air molecules that you carry along. This friction drag can become a serious burden to an animal moving through a **viscous** (that is, "syrupy") and heavy fluid such as water. If you ride your bicycle through a deep puddle, you can feel the much greater friction drag offered by the relatively syrupy water compared with that offered by air. Friction drag increases most significantly as the total surface area of the organism increases and as the viscosity of the surrounding fluid increases. Friction drag is of greatest importance to organisms with a large surface area and to those that live in water. Small swimming animals, especially, must spend a lot of their energy overcoming friction drag.

(a)

(b)

(c)

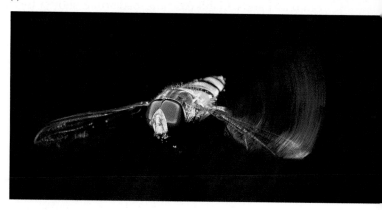

Figure 27.7 Modes of Locomotion in Animals
(a) Running animals generate thrust by pushing against the ground with their legs. (b) Swimming organisms push off against the surrounding water with paddle-like flippers or fins. (c) Flying animals such as this hoverfly use the broad surfaces of their wings to push off against air.

■ BIOLOGY IN OUR LIVES

The Energetic Costs of Running, Swimming, and Flying

How much energy do animals spend to move from here to there? Scientists have calculated the energy used for movement by animals that have different weights and use different means of locomotion. Unsurprisingly, heavy animals use more energy to move a given distance than lighter animals. A graph of these calculations, however, reveals two less obvious patterns.

First, all the lines in the graph slope downward from left to right, indicating that heavy animals spend less energy to move a given body weight than do light animals. Thus, a cat uses less than ten times as much energy to run a given distance than a squirrel weighing one-tenth as much. Why do heavy animals move more efficiently than light animals? Perhaps it is because large animals have relatively less surface area than small ones (see Chapter 25), which subjects large animals to less drag.

Second, the red line representing runners lies above the green line representing fliers, indicating that run-

ners spend more energy moving a given distance than do equally heavy fliers. Swimmers (blue) use the least energy. Why should this be? Half of the energy spent on running is "wasted" in the movement of the body as it bobs up and down and in changing the momentum of the legs as they alternate between moving forward

and backward. Bicycling, which involves fewer changes in momentum, is more efficient than running. Fliers use less energy to change momentum than runners, but spend energy to keep themselves aloft. Swimmers lose little energy to changes in momentum, and they receive support from the surrounding water.

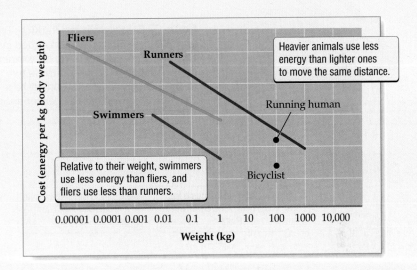

The Relative Cost of Different Modes of Locomotion

Natural selection favors organisms that minimize drag

Because organisms must use energy to generate thrust, drag represents a waste of valuable resources. If an organism reduces drag, then it frees up energy for other, more productive uses. Because they have different causes, it should come as no surprise that minimizing pressure drag and friction drag requires different solutions. The most effective way of minimizing pressure drag seems to be a streamlined shape, which makes it easy for water or air to flow smoothly around the organism. A streamlined shape has evolved independently in many aquatic organisms and birds (Figure 27.8).

Organisms reduce friction drag by having non-stick surfaces that allow air or water molecules to slip off eas-

ily. The smooth, slimy surface that makes many aquatic animals so unappealing to touch effectively reduces friction drag.

> ■ Drag resists forward motion. Pressure drag increases as the speed of an organism increases, whereas friction drag increases as the organism's surface area or the viscosity of the surrounding fluid increases.

How Protists and Bacteria Move

Although protists, bacteria, and archaeans lack muscle tissue, many of them prove quite adept at moving

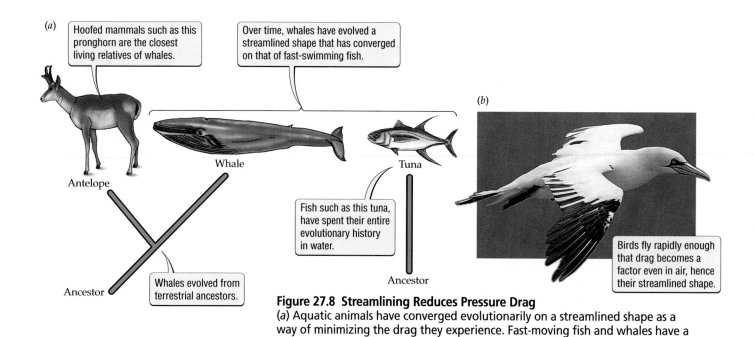

(a)

Hoofed mammals such as this pronghorn are the closest living relatives of whales.

Over time, whales have evolved a streamlined shape that has converged on that of fast-swimming fish.

Antelope

Whale

Tuna

(b)

Fish such as this tuna, have spent their entire evolutionary history in water.

Ancestor

Whales evolved from terrestrial ancestors.

Ancestor

Birds fly rapidly enough that drag becomes a factor even in air, hence their streamlined shape.

Figure 27.8 Streamlining Reduces Pressure Drag
(a) Aquatic animals have converged evolutionarily on a streamlined shape as a way of minimizing the drag they experience. Fast-moving fish and whales have a similar shape, even though these two groups evolved from very different ancestors. (b) Birds such as this gannet have also evolved a streamlined shape to reduce the pressure drag that they experience during rapid flight.

(a)

(b)

Cilia generate motion in much the same way as our arms do when we swim the backstroke.

Waves pass down the length of a eukaryotic flagellum to generate motion.

Power stroke

Recovery stroke

27.4

(c)

(d) **1** With energy supplied by ATP, the heads of the left microtubule pair walk down the right microtubule pair,...

2 ...causing the flagellum or cilium to bend to the right when seen from the side.

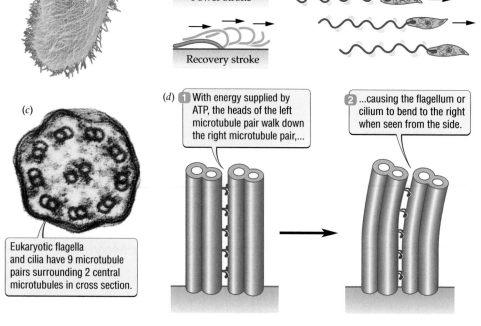

Eukaryotic flagella and cilia have 9 microtubule pairs surrounding 2 central microtubules in cross section.

Figure 27.9 Cilia and Eukaryotic Flagella
Many eukaryotic organisms, especially single-celled ones, use cilia or flagella to generate thrust. (a) This protist is covered with cilia. (b) Although cilia look like flagella, they generate thrust by different kinds of bending motions. (c) The cilia and flagella of eukaryotes all share a characteristic arrangement of paired microtubules. (d) These microtubules slide past one another, much as myosin slides along actin, causing the cilium to bend.

through their environment. These organisms use a variety of different means of propelling themselves, but the use of hairlike flagella and cilia predominates.

How eukaryotes other than animals move

Many single-celled protists, as well as the sperm produced by some plants and all animals, propel themselves using hairlike structures called **cilia** (singular cilium) or **eukaryotic flagella** (singular flagellum). Cilia move much as our arms do when swimming the breaststroke, whereas eukaryotic flagella beat in a wavelike pattern (Figure 27.9b). The bending of cilia and flagella depends on the sliding of one member of a pair of tiny tubules relative to the other in a manner reminiscent of the sliding of actin and myosin filaments in muscles (Figure 27.9d). A protein that makes up arms projecting out from these microtubules, called dynein, shares with myosin the ability to "walk" up an adjacent microtubule using energy from ATP.

Bacterial flagella have rotary motors

Although bacteria use a variety of interesting means of propelling themselves, the most widespread of these are **bacterial flagella**. A unique rotary motor rotates the stiff, corkscrew-like flagellum of a bacterium like a propeller. This rotary motor is a truly remarkable structure that differs from anything found in eukaryotes (Figure 27.10). It rotates in response to hydrogen ions that diffuse into the bacterium from the environment. The bacterium uses active carrier proteins to pump hydrogen ions out of the cell to create a hydrogen ion gradient (see Chapter 24). It may take as many as 100 hydrogen ions diffusing back into the cell to rotate the flagellum once, yet these bacterial propellers may rotate as rapidly as 1000 times in a second.

■ Protists propel themselves using eukaryotic flagella and cilia, which move by means of microtubules that slide past each other, much as actin and myosin do. Bacteria rely on a unique rotary motor connected to a bacterial flagellum.

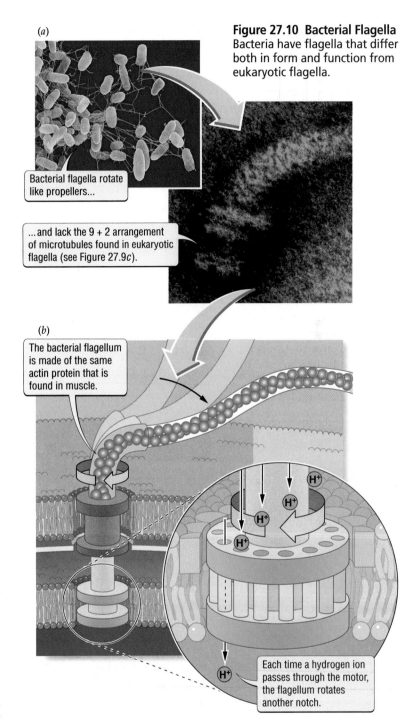

Figure 27.10 Bacterial Flagella Bacteria have flagella that differ both in form and function from eukaryotic flagella.

Bacterial flagella rotate like propellers...

...and lack the 9 + 2 arrangement of microtubules found in eukaryotic flagella (see Figure 27.9c).

The bacterial flagellum is made of the same actin protein that is found in muscle.

Each time a hydrogen ion passes through the motor, the flagellum rotates another notch.

How Plants and Fungi Get from Here to There

As we saw in Chapter 26, multicellular plants and fungi have rigid cell walls that prevent their tissues from changing shape in the way needed for motion. Nonetheless, plants and fungi still need to explore their environment to find food and mates.

Plants and fungi can move through their immediate surroundings by growing. Plant roots and stems, for example, can grow by several millimeters a day (some bamboo plants can grow over a meter in a day), enough to allow them to reach nutrients in the soil or sunlight. The hyphae with which fungi absorb nutrients can grow through their food at a similar speed.

Reproducing with distant mates and dispersing to colonize new habitats pose a more formidable challenge. Many plants and fungi can move their seeds or spores several meters in a matter of seconds (Figure 27.11a). They manage to generate the thrust needed to launch these reproductive structures by slowly letting pressure build up in their reproductive organs. A rapid release of this pressure sends the seed or spore flying.

Much more widespread is the use of animals or wind and water currents to transport pollen, seeds, or spores. Ani-

mals can carry these structures either inside their bodies, by swallowing them, or by having them stick to the body surface. Plants and fungi attract animal dispersers by offering food as bait (Figure 27.11b). Wind and water currents can carry pollen, seeds, and spores over thousands of kilometers. Wind-dispersed structures usually have a shape that slows the speed at which they fall. Slowly falling pollen, for example, can travel on even very light winds. Tiny size and elaborate hairs or wings increase the surface area of pollen, which in turn increases the drag that resists falling through air. Thus wind-dispersed structures actually increase drag to ensure that they stay aloft as long as possible.

(a)

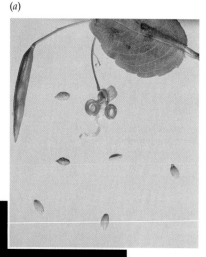

(b)

(c)

(d)

Figure 27.11 How Plants and Fungi Get Around
Plants and fungi move in a diversity of ways. (*a*) Some plants, like this touch-me-not, explosively release their seeds, sending them several meters. (*b*) Other plants use nectar rewards in their flowers to lure animals that can disperse their pollen. (*c*) Wind currents can carry pollen, spores, and seeds over tremendously long distances. Raindrops hitting a puffball fungus send tiny spores into the breeze. (*d*) The elaborate feathery structures on a dandelion seed act as a parachute to keep the seed aloft.

■ Plants and fungi must explore their environment despite their inability to swim, walk, or fly. They find nutrients by growing, and they rely on animals or on wind or water currents to disperse their pollen, seeds, and spores.

■ Runners spend a lot of energy in generating an up-and-down motion, and swimming sperm spend a lot of energy overcoming drag. These energetic costs force a trade-off between speed and distance.

■ HIGHLIGHT

Racing Runners and Speeding Sperm Revisited

Sprinters and marathoners and their sperm counterparts face a similar challenge: how to best allocate energy resources for generating thrust while overcoming drag. Both runners and sperm rely on motors based on molecules that use energy to slide past each other to generate the power needed for movement. Runners have at their disposal a body in which the muscles and rigid bones have combined to form intricate lever systems that control how the power generated by the muscles translates into motion. In addition, athletes can, through training, influence the force that their muscles can generate. Sperm have only one chance to reach an egg, and they must rely on a much simpler flagellum to propel themselves.

Runners and sperm face very different problems when it comes to minimizing drag. Humans are relatively large and fast, and therefore live in a universe dominated by pressure drag rather than friction drag. Even sprinters, however, barely reach speeds at which pressure drag becomes significant as they move through air. Instead, the major source of energy wastage for runners lies in unnecessary up and down movement of the body (see the box on p. 457), something that proper running technique can minimize. Small, slow-swimming sperm inhabit a watery universe dominated by friction drag even greater than that in aquatic environments, because the fluids in the reproductive tract have a greater viscosity than water. For sperm, pressure drag represents a minor problem as compared with friction drag.

In spite of the functional and morphological gulf separating runners and sperm, the differences between them diminish when viewed on the common scale of the energetic cost of locomotion. Sperm that move short distances can, like sprinters, afford to consume a lot of energy quickly to attain as high a speed as possible. On the other hand, sperm that swim long distances to meet up with an egg must, like marathon runners, conserve their energy so that it lasts the whole trip.

SUMMARY

Animal Muscles as Biological Motors

■ The muscle tissue that is unique to animals gives them an ability to move that other multicellular organisms lack.

■ Muscle fibers are made up of sarcomeres, which contain actin and myosin.

■ Sarcomeres contract when myosin heads "walk" along actin filaments by binding to successive sites along the filaments. This action pulls the Z discs anchored to the ends of the actin filaments closer together.

■ Muscles always work in pairs because individual muscles can only contract.

How Animals Convert Muscle Contraction into Motion

■ Muscles increase in strength when the number of muscle fibers within them, and therefore their cross-sectional area, increases.

■ Long muscles contract faster than short muscles.

■ Fast muscle fibers differ from slow muscle fibers in the way they respond to nerve signals and in the type of metabolism on which they rely.

■ Muscles and skeletons work together as lever systems, in which the relative lengths of the load and power arms help determine the force generated by muscle contraction.

Animal Locomotion

■ Animals move by running, flying, or swimming.

■ Two types of drag resist forward motion. Pressure drag increases with increased speed; friction drag increases with increased surface area or increased fluid viscosity.

■ Streamlined shapes reduce pressure drag by changing the flow of air or water around an organism.

■ Smooth surfaces reduce friction drag by allowing a viscous fluid to slip off an organism.

How Protists and Bacteria Move

■ Protists move by using cilia or eukaryotic flagella. The bending of these hairlike structures results from a pair of microtubules that slide past each other, much like actin and myosin filaments do in animal muscles.

■ Bacteria move by using rotary motors attached to bacterial flagella, which rotate like a propeller.

How Plants and Fungi Get from Here to There

■ Plants and fungi must compensate for their inability to swim, walk, or fly. They explore their environment by growing, and they rely on animals, wind, or water currents to disperse their pollen, seeds, and spores.

Highlight: Racing Runners and Speeding Sperm Revisited

■ Both runners and sperm travel either quickly or far, but not both.

■ Running consumes a lot of energy in generating an up-and-down motion, and swimming sperm spend a lot of energy overcoming drag. These energetic costs force the trade-off between speed and distance.

KEY TERMS

actin p. 452	myofibril p. 452
bacterial flagellum p. 459	myoglobin p. 454
cilium p. 459	myosin p. 452
drag p. 455	power arm p. 454
eukaryotic flagellum p. 459	pressure drag p. 456
fast muscle fiber p. 453	sarcomere p. 452
friction drag p. 456	slow muscle fiber p. 454
fulcrum p. 454	thrust p. 455
lever p. 454	viscous p. 456
load arm p. 454	Z disc p. 452
muscle fiber p. 451	

CHAPTER REVIEW

Self-Quiz

1. Which of the following is true of animal muscles?
   a. The muscle tissue in strong animals has different kinds of filaments in it than the muscle tissue in weak animals.
   b. Animal muscles can only contract, or shorten.
   c. Muscle contraction requires no energy.
   d. both a and b

2. From which of the following are grasshoppers most likely to get their great jumping power?
   a. leg muscles with a large cross-sectional area
   b. leg muscles that contain extra nuclei
   c. short leg muscles
   d. leg muscles that contain actin instead of myosin

3. To increase the power produced by a lever system, you should
   a. make the power arm thicker.
   b. shorten the load arm.
   c. lengthen the load arm.
   d. make the load arm thicker.

4. Swimming is energetically cheaper than flying because
   a. water helps support an organism.
   b. water generates more thrust.
   c. there is no pressure drag in water.
   d. there is less friction drag in water.

5. The inner workings of eukaryotic flagella are most similar to those of
   a. bacterial flagella.
   b. plant growth.
   c. lever systems.
   d. animal muscles.

Review Questions

1. Describe how myosin and actin interact to make a sarcomere contract.

2. How can animals affect the speed and strength of muscle contraction if they all share essentially similar muscle tissue?

3. Consider the following pairs of organisms:
   a. a walking elephant and a swimming whale
   b. a fast-swimming penguin and a fast-flying falcon (about the same size as a penguin)
   In each pair, which organism would you expect to face more resistance from pressure drag, and why? Similarly, which organism would face more friction drag?

4. How do fungi manage to spread so quickly in our refrigerators to spoil our food, when they lack muscles?

The Daily Globe

Bike Paths Are the Answer

To the Editor:
I am writing in response to the Letter to the Editor that appeared in your newspaper last week condemning the expenditure of our taxes on maintaining bicycle paths in our city. The writer argued that tax money would be much better spent on roads used by everyone rather than on "luxuries" used by "a relative few." I find this view not only narrow-minded but wrong.

Our city should be proud of having an excellent system of over 150 kilometers of bike paths. These paths are more than just scenic byways along our rivers and through our parks. They provide an important commuter route connecting many of our neighborhoods with the downtown core.

People constantly complain about the parking problems downtown, and the pollution problems that have come with the dramatic increase in car traffic around the city over the past several years. Bicycles provide an excellent solution to these problems and more. Bicycles do not require parking places and cause no pollution, yet provide a fast and energy-free way of getting around. They make no noise, and they do not damage the environment.

Perhaps we should spend more, not less, valuable tax money on extending our bicycle paths to even more parts of our city.

Janice C. Shum

Evaluating "The News"

1. The author of this letter claims that bicycles provide an energy-free mode of transportation. Biologically, why is this claim not reasonable?

2. In what ways does the source of energy used to fuel the bicycle rider (food) also take an environmental toll?

3. Imagine that you are a member of the city council faced with letters like this one and like the earlier letter favoring roads for cars over paths for bicycles. How would you evaluate the relative costs and benefits to the environment of roads versus bicycle paths?

chapter 28 Nutrition

Rene Magritte, *L'Ile aux Tresors*, 1942.

The Rise of Agriculture and Malnutrition

The transition from a hunting and gathering way of life to an agricultural one represented a milestone in the development of human civilization, as we saw in Chapter 23. Early hunter-gatherer cultures probably ate equal proportions of animal and plant foods. In contrast, agricultural societies depend mostly on plants, especially cereals such as corn and wheat and root crops such as potatoes and cassava. Cereals and root crops supply energy; they store well, so that they can be available when wild food is scarce; and they can support more people in a given area than can animal-based foods. By stimulating tremendous growth in human populations, agriculture created the conditions that led to cities, art, and industry.

For all its advantages, agriculture has an unexpected side effect: It can lead to poor nutrition. Agriculture arose independently in three regions—

the Middle East, southeastern Asia, and Central America—from which it spread to other regions. In each case, the rise of agriculture coincided with an increase in diet-related diseases, as evidenced by the skeletons unearthed by archaeologists.

We see evidence of agriculture-related malnutrition at archaeological sites throughout the Americas, where agriculture traditionally depended almost completely on plant, rather than animal, crops. Around AD 1000, for example, the hunter-gatherers of the Ohio River Valley of the United States adopted the cultivation of first corn and then beans from peoples to the south. The adoption of agriculture coincided with a decline in the importance of wild game as food. It also coincided with an increase in

■ MAIN MESSAGE

Whereas producers can obtain all their nutritional needs from the physical environment, consumers must get their nutrients by feeding on other organisms.

the size and number of farming communities, indicating that farming increased the population capacity of the Ohio River Valley. Evidence from skeletal remains, however, suggests that these farmers were short, faced frequent food shortages, suffered nutrient deficiencies, and had rotten teeth compared with the hunter-gatherers that came before them.

In this chapter we consider how organisms acquire the nutrients they need. After introducing the basics of nutrition, we return to the question of why an improved ability to produce food led, paradoxically, to malnutrition among early Americans.

Indian Burial Mounds Provide Clues to the Effect of Agriculture on Human Nutrition

■ KEY CONCEPTS

1. All life depends on a relatively small number of nutrients.

2. Whereas producers photosynthesize and respire, consumers can only respire.

3. Plants use energy to absorb scarce nutrients from the soil into their roots.

4. Animal guts process food in a logical sequence of mechanical and chemical breakdown, absorption, and waste production.

5. Without extensive processing, consumers cannot absorb the nutrients locked in the complex organic molecules of their food.

6. Rapid absorption of nutrients depends on having large absorptive surfaces.

7. The details of the design of nutrient processing systems reflect the nutrients on which organisms rely.

Some humans may live to eat, but all living things must eat to live. All living organisms must get nutrients from their environment. Nutrients provide the raw materials needed for all the functions essential to life. Although all life relies on a surprisingly small and similar set of nutrients, several features of the biology of organisms influence how they accumulate those nutrients. One of these features affects the nutrition of organisms profoundly: Whereas producers can accumulate all the nutrients they need from their physical surroundings and capture energy from sunlight, consumers must feed on other organisms to get energy and nutrients. The ability to take food inside the body before absorbing nutrients and the ability to move also shape the nutritional biology of organisms.

In this chapter we survey the different ways in which living things obtain nutrients. After reviewing the nutrients basic to life and the differences between producers and consumers, we look at the structures used by plants and animals to process nutrients. We then consider some general features of the ways in which organisms process and absorb nutrients. We conclude by considering some of the adaptations of producers and consumers for gathering scarce nutrients.

The Elements of Nutrition

Life depends on relatively few of the known chemical elements (Figure 28.1). Depending on the kingdom to which they belong, organisms need between 9 and 11 elements in relatively large amounts. These elements, known as **macronutrients**, make up over 99 percent of the body weight of an organism. Of the macronutrients, carbon, hydrogen, and oxygen alone make up about 93 percent of the weight of organisms. All of the complex organic compounds that characterize life, including carbohydrates, proteins, fats, and nucleotides, depend on

a framework of carbon, hydrogen, and oxygen (see Chapter 5). These organic compounds not only make up the structure of the organism, but also store energy in the chemical bonds that hold them together. By breaking these chemical bonds, organisms can release the energy that they need to survive (see Chapter 7).

The remaining macronutrients together make up about 6 percent of the weight of organisms. These macronutrients, such as the nitrogen needed to make proteins, are required for structures or functions that occur throughout an organism.

The other essential elements, called **micronutrients**, make up less than 1 percent of an organism's body weight. Although micronutrients occur in tiny amounts, organisms cannot survive without them. The human body, for example, contains an amount of iodine that would easily fit into a quarter teaspoon, the smallest measuring spoon commonly used in the kitchen. Without this tiny quantity of iodine, however, the body could not make some hormones that regulate how we burn energy and how the body grows. Unlike the macronutrients, most of which are required by all organisms, micronutrients often reflect the peculiarities of the biology of particular groups of organisms. Cobalt is an essential micronutrient for the nitrogen-fixing bacteria described in the box in Chapter 29 (p. 485), and grasses and tiny single-celled protists called diatoms need silicon to produce their support structures.

Diatom

■ All life depends on a limited number of chemical elements, some of which are needed in relatively large amounts and other in tiny amounts. The macronutrients carbon, hydrogen, and oxygen make up the bulk of an organism's body weight.

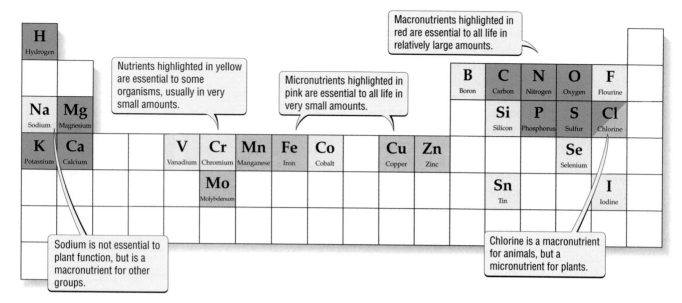

Macronutrients highlighted in red are essential to all life in relatively large amounts.

Nutrients highlighted in yellow are essential to some organisms, usually in very small amounts.

Micronutrients highlighted in pink are essential to all life in very small amounts.

Sodium is not essential to plant function, but is a macronutrient for other groups.

Chlorine is a macronutrient for animals, but a micronutrient for plants.

Figure 28.1 The Elements Essential to Life

This figure highlights the chemical elements that play an essential role in life. The arrangement of the elements follows that of the periodic table used by chemists. The one- or two-letter abbreviation, along with the full name of the element, is provided only for elements that are essential to at least some living organisms. The blank squares represent common chemical elements not known to play an essential part in the biology of organisms.

28.1 Comparing Producers and Consumers

All organisms must acquire nutrients and store energy. In addition, all organisms must respire to release the energy that allows them to organize the nutrients they acquire into the complex organic molecules and structures that make life possible. In spite of these basic similarities, life has evolved two distinctly different approaches to dealing with energy and nutrients.

Producers, which include bacteria, protists, and plants, can get all of their

Figure 28.2
Producers and Consumers

Producers use photosynthesis to capture the energy of sunlight and store it in the chemical bonds of sugar molecules. Through respiration, they can then release the energy the sugars contain. Consumers can only respire; they must therefore obtain organic compounds as sources of energy from their food.

nutrients and energy from their physical surroundings. Producers are defined by the ability to photosynthesize: They harness the sun's energy to convert carbon dioxide and water into storable chemical energy in the form of organic sugar molecules (Figure 28.2). As described in Chapter 8, producers respire to convert the energy stored in sugars into chemical bonds in ATP, which acts as a widely accepted energy currency in the chemistry of life. ATP provides producers with the energy required to combine the sugars produced by photosynthesis with nutrients from the surrounding soil or water to manufacture the full range of organic compounds they need. We will consider photosynthesis in more detail in Chapter 29.

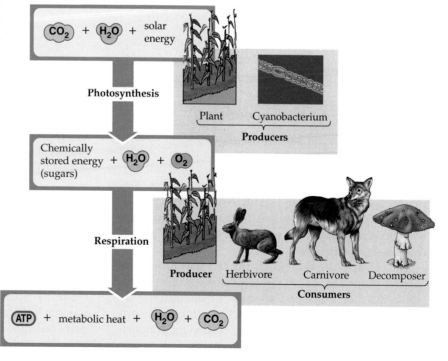

Consumers, such as animals and fungi, as well as many archaeans, bacteria, and protists, depend absolutely on nutrients in organic form. In other words, consumers must obtain nutrients by consuming other organisms, either living or dead. Consumers cannot photosynthesize, and must therefore break down the carbon-based compounds made by other organisms to produce their own ATP (see Figure 28.2). In addition to using organic compounds as their primary energy source, consumers also depend heavily on organic compounds as a source of the raw materials from which they synthesize their own carbohydrates, fats, proteins, and nucleotides. Instead of assembling proteins, for example, by combining sugars with nitrogen obtained from the physical environment, consumers typically construct proteins out of ready-made amino acids from their food.

We can classify consumers according to the form in which they obtain their organic nutrients. **Herbivores**, which include many animals and fungi, eat living plant tissues. **Carnivores** are consumers, primarily animals, that eat living animals. **Decomposers** feed on the tissues of dead organisms and include many single-celled organisms, fungi, and animals.

> ■ Producers can photosynthesize and respire. Consumers can only respire to release the energy in organic nutrients.

Nutritional Requirements of Producers and Consumers

A close look at the nutritional labels on packages of human and plant foods reveals much about the differences in how producers and consumers obtain nutrients and energy.

Producers rely on mineral nutrients

The label on a typical package of plant food, or fertilizer, reveals that it provides plants with three mineral macronutrients—nitrogen, phosphorus, and potassium—and an assortment of mineral micronutrients (Figure 28.3*a*). We do not feed fertilizer directly to plants; instead, we mix it into the soil. In the soil, the minerals in the fertilizer dissolve in water. Plant roots then absorb the dissolved minerals.

The fertilizer label conceals the fact that producers such as plants can absorb mineral nutrients only in certain forms. Plants take up almost all of their nitrogen, for example, as nitrate ions. Similarly, producers absorb phosphorus as phosphate ions and potassium as potas-

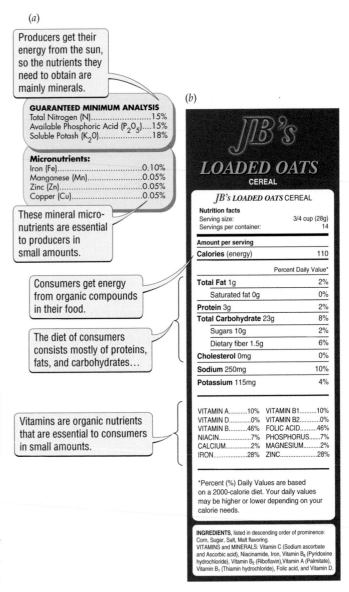

Figure 28.3 Nutritional Labels Can Tell Us a Great Deal
(*a*) This label from a fertilizer for house plants shows the percentages of the essential plant nutrients—nitrogen (N), phosphorus (P), and potassium (K)—contained by the fertilizer. (*b*) This label from a cereal box indicates the important sources of energy and nutrients for humans. We obtain energy from carbohydrates, fats, and proteins.

sium ions. These absorbable ions share several features: They carry a slight electrical charge, they dissolve readily in water, and they can cross the cell membranes of producers.

The label in Figure 28.3*a* comes from a fertilizer that contains no organic compounds. Commonly used organic fertilizers such as compost or manure improve plant nutrition not by providing plants with ready-made carbohydrates or proteins, but by increasing the ability of soil to hold water and mineral nutrients. Organic fertilizers contain relatively small amounts of the mineral nutrients that plants need. Instead, the spongelike texture of rotting organic material soaks up water in which mineral nutrients can dissolve. In addition, organic particles act as magnets that hold many of the mineral nutrients in the soil, keeping them from being washed out of reach of the plant roots.

Note, too, that the plant fertilizer provides no energy. Remember that photosynthesis allows plants to capture the energy of the sun to convert carbon dioxide into sugars (see Figure 28.2).

Consumers rely on organic compounds for both nutrients and energy

The nutritional labels on foods intended for consumers, such as humans, differ greatly from those on fertilizers intended for producers. The label on a cereal box lists carbohydrates, fats, proteins, and vitamins, which come from the grass seeds used to make the cereal (Figure 28.3*b*). A few of the nutrients in cereal come in a mineral form (an example is salt, which provides sodium and chlorine).

The breakdown of carbohydrates and fats by respiration provides consumers (such as humans) with most of the energy they need to survive. The organic compounds in our food also serve as building blocks that we reassemble into our own organic compounds. Proteins provide as much energy per unit of weight when broken down as do carbohydrates, but they play a much more important role as the source of the amino acids that we use to build our own proteins. Although humans can synthesize some amino acids from sugars, others, called **essential amino acids**, must be obtained from our food (Figure 28.4). Whereas we can get the eight essential amino acids by eating animal foods, we must include a variety of plants in our diet to get all eight.

Vitamins are a type of nutrient unique to consumers. We can think of **vitamins** as organic micronutrients: Consumers need them in only tiny amounts, but they either

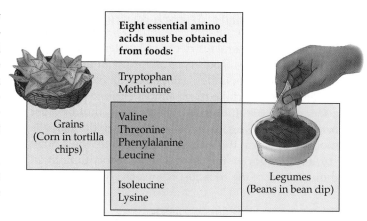

Figure 28.4   Essential Amino Acids in the Human Diet
Humans cannot synthesize eight of the twenty amino acids we use to build proteins, so we must obtain them from our food. To get all eight from plants in our diet, we must combine different grains and vegetables.

Eight essential amino acids must be obtained from foods:

Tryptophan
Methionine
Valine
Threonine
Phenylalanine
Leucine
Isoleucine
Lysine

Grains
(Corn in tortilla chips)

Legumes
(Beans in bean dip)

cannot make enough of these substances to meet their needs or cannot make them at all. Thus consumers must get vitamins in their food. As with mineral micronutrients, consumer species may differ greatly in which vitamins they need. Humans require two broad classes of vitamins: water-soluble vitamins, which help speed essential chemical reactions in the body, and fat-soluble vitamins, which play a variety of roles (Table 28.1). Insufficient vitamins in the diet lead to deficiency diseases such as pellagra (see the Highlight on p. 477). Although vitamin supplements are widely available commercially, there is little evidence of a benefit from taking more than the recommended daily dose of a vitamin: The human body quickly rids itself of excess water-soluble vitamins, and fat-soluble vitamins may build up in the body to toxic levels.

■ Producers require mineral nutrients. Consumers require organic nutrients to provide them with energy as well as with building blocks for their own organic compounds.

28.1 *Vitamins in the Human Diet*

Vitamin type	Examples	Main function
Water-soluble	Thiamin, riboflavin, pyridoxine, B_{12}, biotin, vitamin C[a], folates, niacin, pantothenic acid	Act with enzymes to speed metabolic reactions, or act as raw materials for chemicals that do so
Fat-soluble	Carotene, vitamin D, vitamin E, vitamin K	Produce visual pigment; calcium uptake in bone formation; protect fats from chemical breakdown, produce clotting agent in blood

[a]Vitamin C is a vitamin only for primates, including humans, and a few other animals, such as guinea pigs. All other animals can make the vitamin C they need.

How Plant Roots Absorb Nutrients

Plants absorb mineral nutrients and water from the soil though their roots. Roots typically grow in the form of long, branching cylinders (Figure 28.5). Nutrient and water absorption take place almost entirely in a small zone, often less than a centimeter long, that lies just behind the growing tip of the root. Here, epidermal cells modified to form threadlike **root hairs** provide a huge surface area for water and nutrient absorption. Even though most of the root absorbs relatively little, typical crop plants still manage to place about a square centimeter of absorptive root surface in each cubic centimeter of soil volume.

Plants must spend energy to absorb nutrients into their roots. Because nutrient concentrations in roots may be ten to a thousand times greater than in the surrounding soil water, plants rely on active transport by carrier proteins in the plasma membranes of their root cells (see Chapter 24) to absorb nutrients. Barley, a grain used in making beer, illustrates the cost of nutrient absorption: Almost 20 percent of the energy used by barley roots fuels the active absorption of scarce, nitrogen-containing ions from the soil. In contrast, water diffuses into the root passively, flowing between cells (see Figure 28.5).

Although plants cannot move as animals can, they must constantly seek out new sources of nutrients and water. Plants take up some nutrients, notably phosphorus and potassium (not coincidentally, two of the nutrients found in standard fertilizers) more rapidly than they are replenished in the soil water surrounding the root. As a result, plants quickly deplete the soil around the root hairs of these essential nutrients. The growth of roots allows plants to seek new soil rich in nutrients and water, but costs the plant both nutrients and resources. The narrow, tubelike shape of roots helps reduce the cost of their growth by providing a maximum of length and surface area with a minimum of tissue (Figure 28.6a). To further increase the efficiency with which they locate mineral nutrients, plants often concentrate root growth in patches of soil containing abundant resources (Figure 28.6b). Fungi, which also grow through their nutrient source, deploy absorptive structures called **hyphae** (singular hypha) in a pattern strikingly like that of plant roots (Figure 28.6c).

To reach the rest of the plant, nutrients must move from the root surface to the core of the root to reach the **xylem** tissue, which carries water and nutrients from the roots to the photosynthesizing leaves (see Chapter 30 and Figure 28.5). The tubelike cells that make up the xylem lie in bundles at the center of the root, surrounded by a wall of cells called the endodermis (*endo*, "inside"; *dermis*, "skin"). Once inside a root cell, nutrients move toward the xylem bundles through minute pores that connect the root cells to one another. The root core is encircled by a band of waxy material that seals the gaps between the cells of the endodermis, forcing everything that passes into the root core where the xylem lies to move *through* the endodermal

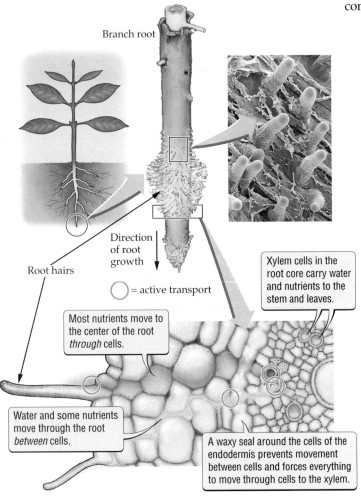

Branch root

Root hairs

Direction of root growth

○ = active transport

Most nutrients move to the center of the root *through* cells.

Water and some nutrients move through the root *between* cells.

Xylem cells in the root core carry water and nutrients to the stem and leaves.

A waxy seal around the cells of the endodermis prevents movement between cells and forces everything to move through cells to the xylem.

Figure 28.5   Plant Roots Absorb Mineral Nutrients
The branched root systems of plants allow them to absorb mineral nutrients from a large volume of soil. Just behind the growing root tip lie epidermal cells modified to form delicate root hairs. The root hairs provide a large surface area across which the plant can absorb mineral nutrients. Carrier proteins in the plasma membrane of the root hairs transport nutrients into the root epidermis. From here they pass from one cell into the next until they reach xylem tubes in the center of the root. Water and a few nutrients move through the outer layers of the root between cells, but are forced to pass into cells when they reach the waterproof waxy layer surrounding the root core.

(a)

Figure 28.6   Plant Roots and Fungal Hyphae Forage for Nutrients by Growing (*a*) The cylindrical shape of roots allows plants to produce great root length and surface area at relatively little cost. (*b*) Plants explore the soil for water and nutrients through root growth. (*c*) Plant roots grow to fill the soil in a pattern remarkably like that of fungal hyphae, which face a similar challenge in growing through the organic matter from which they absorb nutrients.

Roots grow through the soil that provides plants with their mineral nutrients.

(c)

Seen from above, the growth pattern of these plant roots looks remarkably like…

…the growth pattern of these fungal hyphae.

(b)

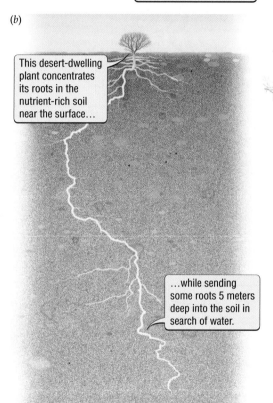

This desert-dwelling plant concentrates its roots in the nutrient-rich soil near the surface…

…while sending some roots 5 meters deep into the soil in search of water.

28.2 How Animal Guts Process Nutrients

The digestive system, or **gut**, of animals differs in important ways from the roots of plants. In most animals, including humans, the gut forms a tube that runs through the organism, which means that animals bring food inside the body before they absorb nutrients. As consumers, animals encounter nutrients in a more complex form than plants do. Therefore, animals must do a lot of physical and chemical work to convert their food into molecules that they can absorb.

Animals have a remarkably uniform gut design

Considering the diversity of form and diet among animals, their digestive systems follow a strikingly similar pattern (Figure 28.7*a*). The human digestive system shows this basic pattern well (Figure 28.7*b*). The human gut carries a stream of food through the body from the mouth to the anus by means of muscular contractions. This one-way flow sets up a sort of food-processing assembly line, along which the various portions of the gut perform specialized functions.

The mouths of many animals have structures that, like human teeth, allow the animal to break large pieces of food into small ones. While chewing, humans mix their relatively dry food with moisture in the form of saliva to smooth its passage through the esophagus, which carries food to the **stomach**. Contractions of the muscular wall

cells. This arrangement plays an important role in root function because it gives plants much more control over both the kinds and amounts of molecules that enter the xylem. Cells that lie next to the xylem cells provide additional control over nutrient uptake by actively and selectively pumping ions into the xylem.

■ Roots absorb nutrients from the soil water through root hairs concentrated in a narrow zone near their tip. Because nutrients occur in the soil at much lower concentrations than they do in roots, plants rely on closely controlled active transport to absorb needed nutrients. Plants use energy to produce root tissue and to transport nutrients into roots.

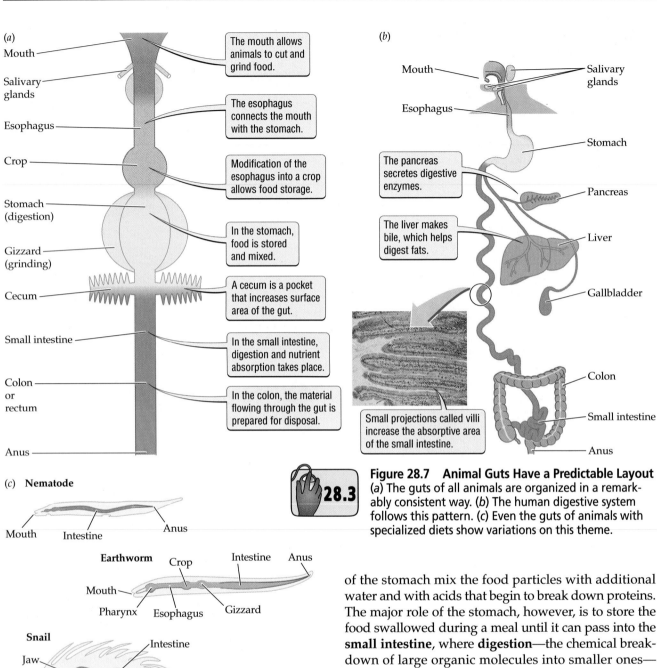

(a)

Mouth — The mouth allows animals to cut and grind food.

Salivary glands

Esophagus — The esophagus connects the mouth with the stomach.

Crop — Modification of the esophagus into a crop allows food storage.

Stomach (digestion) — In the stomach, food is stored and mixed.

Gizzard (grinding)

Cecum — A cecum is a pocket that increases surface area of the gut.

Small intestine — In the small intestine, digestion and nutrient absorption takes place.

Colon or rectum — In the colon, the material flowing through the gut is prepared for disposal.

Anus

(c)   **Nematode**

Mouth   Intestine   Anus

Earthworm   Crop   Intestine   Anus

Mouth

Pharynx   Esophagus   Gizzard

Snail   Intestine

Jaw

Mouth   Liver   Stomach   Anus

Cockroach   Crop   Gizzard

Esophagus

Anus
Rectum
Jaws   Salivary glands   Intestine

(b)

Mouth — Salivary glands

Esophagus

The pancreas secretes digestive enzymes. — Stomach

The liver makes bile, which helps digest fats. — Pancreas

Liver

Gallbladder

Colon

Small projections called villi increase the absorptive area of the small intestine. — Small intestine

Anus

28.3

Figure 28.7   Animal Guts Have a Predictable Layout
(*a*) The guts of all animals are organized in a remarkably consistent way. (*b*) The human digestive system follows this pattern. (*c*) Even the guts of animals with specialized diets show variations on this theme.

of the stomach mix the food particles with additional water and with acids that begin to break down proteins. The major role of the stomach, however, is to store the food swallowed during a meal until it can pass into the **small intestine**, where **digestion**—the chemical breakdown of large organic molecules into smaller ones—begins in earnest. In the small intestine, enzymes produced by the pancreas and the liver are mixed with the food stream to digest carbohydrates, fats, and proteins. The small intestine also plays a central role in the absorption of nutrients. Before the food stream passes out of the body through the anus, the **colon** removes additional nutrients and water from it, leaving waste, or feces.

The functions of the various parts of the human gut follow a logical sequence of food breakdown, nutrient absorption, and waste disposal. We should not be surprised, therefore, to learn that other animals have guts laid out in much the same sequence (Figure 28.7*c*). Parts of animal guts that share a function with part of the human gut have often been given the same name, even if their evolutionary origins differ greatly.

Animals can move to their food or wait for their food to come to them

Muscles allow animals to move in search of their food. By moving, **active foragers** can feed on food that moves (as when ospreys dive into water to catch fish) as well as food that does not move (as when herbivorous bison seek out patches of grass) (Figure 28.8*a,b*). Some animals, however, wait for their food to come to them. For example, web-spinning spiders, which choose not to move in search of prey, and barnacles, which cannot move, must wait for food to come to them. Such organisms are called **sit-and-wait foragers**. They spend little energy on locomotion, and they depend on food that moves on its own or that rides on air or water currents. Many sit-and-wait foragers build elaborate traps that increase the chances that food will find them (Figure 28.8*c*).

Barnacles

■ Animals typically have a tubelike gut arranged to carry out a sequence of functions: physical and chemical breakdown of food, absorption of nutrients, and waste formation. Muscles allow animals to forage actively for food.

Some General Patterns in the Breakdown and Absorption of Nutrients

After you eat a bowl of cereal, your body has much to do before you can absorb and use the nutrients it contains. Entire flakes of cereal obviously cannot pass from your gut into your body. Even after it is thoroughly chewed, the cereal contains organic compounds that are too big and complex to cross your intestinal wall.

Because consumers—and, to a lesser extent, producers—cannot absorb most nutrients in the form in which they occur in the environment, they must process those nutrients first. Absorption requires that the nutrients cross a plasma membrane, which, as we have seen in Chapter 24, controls what enters and leaves cells. Not surprisingly, the absorptive surfaces of single-celled organisms, fungal hyphae, animal guts, and plant roots selectively let some substances pass while excluding others. Nutrients, therefore, must be converted from the form in which the organism encounters them in the environment into a form that can cross the plasma membrane.

Figure 28.8 How Animals Forage for Food
Animals can forage actively by seeking out their food (*a*, *b*), or they can sit and wait for food to come to them (*c*).

(*a*)

Herbivores such as these stationary bison must move to their food.

(*b*)

Carnivorous ospreys dive into the water to catch fish.

(*c*)

The web increases the chance that the waiting spider will catch prey.

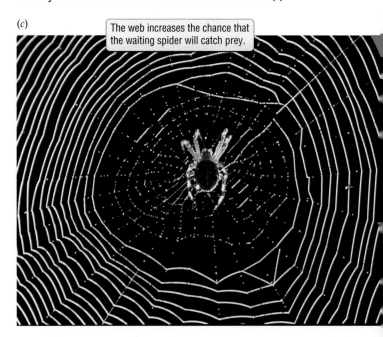

Many animals break food into small pieces to speed digestion

Animals use muscles in combination with hard surfaces to mechanically break large food particles into smaller ones. The physical breakup of large food items into small pieces provides an important benefit beyond allowing the animal to swallow the food: Relative to their volume, smaller food particles expose more of their surface area to acids and enzymes, which greatly speeds their digestion compared with that of large food particles.

Humans and many other animals use hard teeth or jaws in their mouths to crush, rip, tear, or grind large food particles into small ones (see Figure 28.7a). Other animals, including birds and earthworms, have parts of their guts modified into muscular **gizzards**, which grind food against rocks or sand collected from the environment. Other consumers, including animals that lack jaws, rely entirely on chemical digestion to break down food, and as a result, may have to feed on food that comes in small particles.

Consumers and producers digest complex molecules into simpler ones that they can absorb

All organisms must convert complex nutrient-containing molecules into simpler molecules that can cross the plasma membrane. The human gut, for example, can absorb carbohydrates only as simple sugars, but the carbohydrates in cereal come mostly in the form of starches and fiber, which are large molecules consisting of many simple sugar units bonded together (Figure 28.9). Similarly, the human gut can absorb the amino acids needed to build proteins only as individual units, but the amino acids in cereal come packaged as proteins containing hundreds of units. Fats represent a special challenge. Not only do they come as molecules too large to cross the plasma membrane, but they dissolve poorly in the watery mass of food moving through the gut, making absorption difficult. Other consumers, whether single-celled bacteria or multicellular fungi, must also break down

the complex organic compounds they encounter in their food before absorbing them. Even the roots of plants, which rely on simpler inorganic nutrients, can absorb only a small fraction of the nutrients in the soil without further processing.

Plants improve the availability of minerals in the soil by releasing various substances into the water surrounding the root hairs. Weak acids released into the soil water by root cells help separate mineral ions from soil particles. Some bacteria and fungi living in the soil produce chemicals that have a similar effect on insoluble forms of mineral nutrients, particularly iron, making them more available to plant roots.

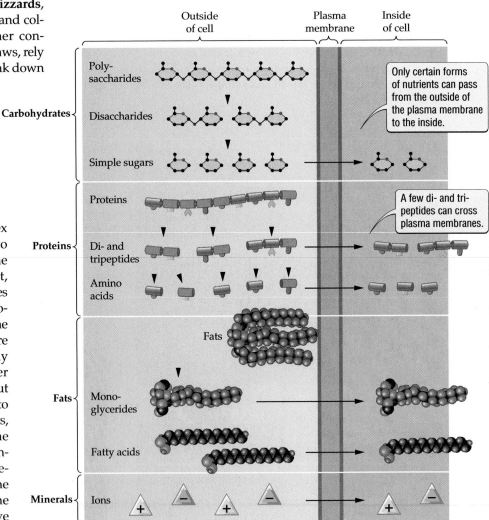

Figure 28.9   The Absorbable Forms of Nutrients
Most consumers can absorb only the simple breakdown products of the carbohydrates, proteins, and fats they encounter in their environment. Many minerals can cross membranes only in their ionic forms.

Compared with producers, consumers rely much more heavily on chemicals to break down food because they get their energy and nutrients from complex organic molecules. They use acids and a variety of enzymes to convert these organic molecules into simpler, easily absorbed forms. In contrast to acids, which break down a wide array of chemical compounds, each digestive enzyme tends to break a specific type of chemical bond. Rennin, for example, is an enzyme that helps newborn mammals digest milk proteins. Humans exploit rennin to curdle milk as a first step in producing cheese. Rennin specifically cuts the bonds joining two hydrophobic amino acids, while leaving the bonds joining other combinations of amino acids untouched.

Humans produce enzymes that can digest most of the nutrients in our cereal. Our saliva contains amylases, which break starches into sugars, and the stomach produces strong acids that contribute to the chemical breakdown of proteins. The pancreas releases several enzymes into the small intestine. These enzymes include a second amylase; proteases, which break proteins into their constituent amino acids; and lipases, which break fats into fatty acids and monoglycerides.

The digestion of fats poses a particular problem because fats dissolve poorly in water. Because they are hydrophobic, they form globules that are too big for the cells lining the small intestine to absorb. These globules must be broken down and made to mix more evenly with the watery contents of the gut. The human liver produces a substance called **bile**, which acts like dish detergent to put a hydrophilic coating on small fat droplets, causing the large globules to break into tiny droplets on which the lipases can work more easily.

When an organism lacks the enzymes needed to break down nutrients into absorbable units, those nutrients are not accessible. Consider the enzyme lactase, which breaks the disaccharide lactose into simple sugars (Figure 28.10). Human infants, like all other young mammals, need lactase when nursing because lactose is the sugar in their mothers' milk. Until humans began dairy farming, however, adults rarely encountered lactose in their diets. Accordingly, most humans stop producing lactase as they mature. Only individuals whose ancestors came from Europe or from a few other, scattered areas of the world, commonly continue to produce lactase as adults. The ability of adults in these populations to digest milk probably arose by chance mutation, but it opened up the possibility of using animal milk as a rich source of protein. In adults of other populations, bacteria living in the gut break down the undigested lactose, and the by-products of this bacterial digestion lead to diarrhea and painful gas. Bacterially altered milk products, such as cheese and yogurt, are generally diges-

tible because the bacteria used in their production convert much of the lactose into a form we can digest.

Absorption depends on the surface area across which nutrients can move

How rapidly an organism can take up nutrients depends on the surface area available for absorption. Because nutrient demand depends mostly on the volume of the body, tiny single-celled organisms, with their high surface area–to–volume ratios, need little modification to absorb enough nutrients from the environment. Multicellular plants, fungi, and animals, however, often need specialized absorptive surfaces that increase their surface area enough to provide their greater volume of metabolizing cells with an adequate supply of nutrients.

Plant roots, fungal hyphae, and the small intestines of animal guts are all organs specialized for nutrient absorption. The shape of each organ maximizes its surface area–to–volume ratio. Each of these organs consists of

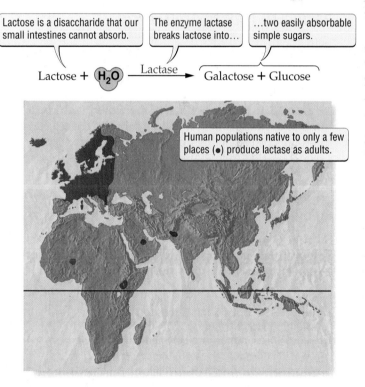

Figure 28.10 Lactase and Milk in the Human Diet
Lactase is the enzyme that digests lactose, the major sugar in milk. After being weaned, most people do not produce lactase. Adults from populations native to only a few places (red areas on the map) produce this enzyme. In areas where virtually no one produces lactase as an adult (green areas on the map), most milk products—for example, cheese and yogurt—are treated with bacteria or fungi that break down lactose.

long, narrow cylinders with branches or folds that maximize surface area while minimizing volume. Folded inside the human belly is a small intestine that has a diameter of 4 centimeters and a length about 1.5 times that of the body. The inside surface of the small intestine is covered with tiny fingerlike projections that increase the actual absorptive surface area to 300 square meters, roughly the equivalent of a tennis court (see Figure 28.7*b*). Root hairs have an even more spectacular effect on the absorptive surface area of plants. Rye plants (which produce the grain used in making rye bread) can pack about 650 square meters of root surface area into a volume similar to that occupied by the human small intestine.

> ■ Most animals process food by mechanically breaking it into small pieces before digesting it. Chemical digestion converts large molecules into forms that the organism can absorb. Absorptive surfaces usually have features that increase their surface area.

When Nutrients Are Scarce

Although all plants share similar root designs and all animals share similar gut designs, the details of these systems reflect the unique features of a particular species' biology. Let's look at one group of plants and one group of animals to illustrate some of the ways in which the process of nutrient absorption responds to the selection pressures placed on organisms by nutrient scarcity.

Plants growing in nutrient-poor environments have specialized features that allow them to get and hold onto nutrients

Peat bogs develop where abundant moisture and cool temperatures allow a particular kind of moss, called peat moss, to flourish. As they die, the remains of the peat mosses build up to form a spongelike blanket many meters thick. The roots of plants growing on top of the bog cannot reach the mineral soil underneath the peat moss blanket. As a result, plants growing on peat bogs have access only to the small amounts of mineral nutrients that come dissolved in rain or snow.

Among the most conspicuous plants in peat bogs are members of the heath family, which include blueberries and cranberries. Below ground, members of this family engage in mutually beneficial relationships with fungi, known as **mycorrhizae** (*myco*, "fungus"; *rhizae*, "roots") (Figure 28.11*a*). The mycorrhizal fungi live inside the root cells of the plant, but send threadlike hyphae out into the soil. The hyphae act like extremely efficient root hairs that pull scarce nutrients out of the bog. The fungus transports these nutrients into the root, where they become available to the plant. In return, the plant provides the fungus with the products of photosynthesis that the fungus, as a consumer, needs. Above ground, these plants typically have leathery leaves that may per-

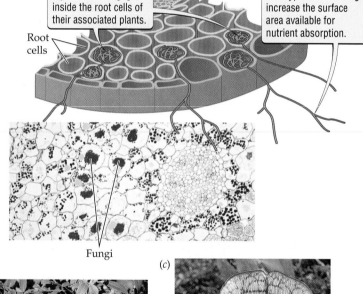

(*a*) Mycorrhizal fungi grow inside the root cells of their associated plants.

The hyphae of the fungi increase the surface area available for nutrient absorption.

Root cells

Fungi

Figure 28.11   How Plants Survive in Habitats Containing Few Nutrients
(*a*) Mycorrhizal interactions between fungi and the roots of members of the heath family greatly increase the surface area for mineral absorption by the plant. The fungi obtain proteins and carbohydrates from the plant in return. (*b*) Plants growing where nutrients are scarce hold onto nutrients tightly, often by producing tough, leathery leaves that survive for more than one growing season. (*c*) In extreme cases, plants turn to animals for nutrients. This pitcher plant does not use the tissues of its prey as food, but only as a source of nitrogen and other nutrients that most plants get in a mineral form.

(*b*)

(*c*)

A few remarkable carnivorous plants have traps, like the pitcher that gives the insect-eating pitcher plant its name.

sist for more than a year even in cold climates (Figure 28.11*b*). These thick, persistent leaves allow the plants to hold onto their hard-earned nutrients for a much longer time than do plants that drop their leaves each year.

Bogs also harbor plants that go to even greater lengths to get nutrients; namely, carnivorous plants that capture and digest insects. Carnivorous plants have leaves that are modified into ingenious insect traps. Unlike true consumers, however, carnivorous plants use their prey only as a source of nutrients, and still rely on photosynthesis as a way of getting energy.

Herbivores have specialized guts that allow them to extract nutrients from protein-poor plant tissues

Herbivorous animals that feed on plant tissues have a much harder time getting nutrients than do carnivorous animals that eat other animals. Plant tissues contain much less protein and much more indigestible material than do animal tissues. Thus herbivores must eat more food than carnivores, and they must put more effort into breaking down their food. We can highlight some of the evolutionary modifications of the animal digestive system for herbivory by comparing two similarly sized mammals: a carnivorous dog and an herbivorous sheep.

Several parts of the digestive systems of dogs and of sheep reflect the differences in their diet (Figure 28.12). Dogs have bladelike teeth suited for slicing meat into chunks small enough to swallow. Sheep have broad teeth that literally grind tough plant tissues into small pieces. The simple stomachs of dogs produce enzymes that partially break down meat. The sheep's stomach has evolved into a complex, four-chambered structure that contains a thriving population of bacteria, fungi, and protists. As outlined in the box on page 478, plant tissues contain abundant cellulose, which animals cannot digest. The microorganisms in the stomach break down the cellulose, making the nutrients that it contains available to the sheep.

Dogs have a relatively short small intestine that provides only a relatively small surface area for absorption. Sheep, in contrast, have a small intestine that is six times as long as that of a similarly sized dog. This design provides a huge surface area over which the sheep can absorb the scarce protein in its diet. The elaborate gut of the sheep is what allows it to survive on a diet that could not support a dog.

> ■ The basic design of the root system in plants and the gut in animals has been modified by evolution in response to the particular challenges associated with surviving where nutrients are scarce.

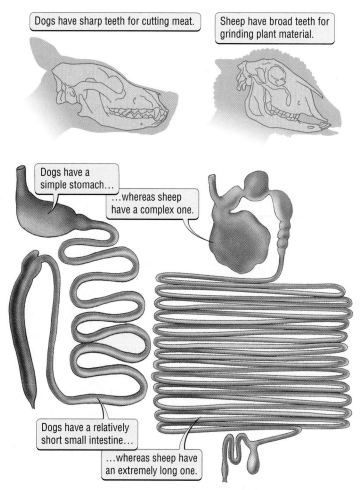

Figure 28.12 Herbivores and Carnivores Compared
Although of similar body length, carnivorous dogs and herbivorous sheep have remarkably different digestive tracts that reflect their different diets. Sheep skulls have broad teeth suited to grinding tough, fibrous grass; dog skulls have slicing teeth suited to cutting meat into bite-sized pieces. Dogs have relatively simple stomachs and short small intestines that reflect the easily digestible nature of their food. Sheep, however, have evolved four-chambered stomachs and long small intestines that allow them to digest the difficult-to-digest grass in their diet and then to absorb most of the nutrients that are released.

■ HIGHLIGHT

Agriculture and Malnutrition Revisited

The puzzling effect of agriculture on human populations—the simultaneous increase in population numbers and malnutrition—stems from an increasing reliance on plants as a source of nutrients in agricultural societies. The high carbohydrate content of grains, such as the corn that served as the staple food of agricultural populations

■ BIOLOGY IN OUR LIVES

The Case of the Missing Enzyme

Animals that feed on plants face a serious problem: They cannot digest much of the food they eat. Much of the carbon in plants is locked up in the complex carbohydrate cellulose, which forms the cell walls of the plant. Some bacteria, protists, and fungi produce cellulase, an enzyme that digests cellulose into simple sugars. Animals, however, do not produce this enzyme. How herbivores handle cellulose matters a great deal to humans because the animals that are most important in agriculture (for example, cattle, sheep, and goats) and many important insect pests (for example, leaf-cutter ants) thrive on plant tissues rich in cellulose.

Animals that feed on foods rich in cellulose enter into a mutualistic relationship with microorganisms that can produce cellulase. These animals depend on fungi, protists, or bacteria to digest cellulose for them. The microorganisms, for their part, use the animals to provide them with a well-maintained home and to gather food for them (tiny organisms cannot forage widely for food).

The complex, four-chambered stomach of a cow, for example, is really an elaborate fermentation chamber that provides ideal conditions for a complex community of cellulase-producing fungi, protists, and bacteria. These cellulase-producing organisms allow cattle to convert the cellulose in the grass they eat into absorbable sugars. Tropical leaf-cutter ants, in contrast, maintain their fungal cellulose-digesting partners outside of their bodies. The ants bring carefully selected leaf fragments to an underground fungus "garden," which they tend with great care. By eating the fungus that grows in their garden, leaf-cutter ants indirectly get at the nutrients in the abundant foliage of tropical forests. These animal–microorganism teams represent combinations of mobile foraging units (the animals) and digestive units (the fungi, protists, and bacteria) that make it possible for consumers to process the tremendous quantities of cellulose that plants produce annually.

Digestive Partnerships
Cows and leaf-cutter ants rely on microorganisms to digest the cellulose in the plant tissues on which they subsist. The cows and ants mechanically break down the food, and the microbes, which produce the enzyme cellulase, digest the cellulose in the food.

in the Americas, makes them rich in energy. A diet of corn could support more people on a given area of land than the protein-rich, but energy-poor, diets of hunter-gatherer societies. In addition, the easily digestible starches in grains provided a potential energy source for infants too young to chew meat. Thus crops such as corn allowed mothers to wean infants at an earlier age than is possible among hunter-gatherers. This ability, in turn, decreased the minimum time between pregnancies and increased the population growth rate.

Although it is rich in energy, corn is poor in several essential nutrients. The overall protein content of corn is well below that of animal tissues, and corn proteins contain relatively little of two essential amino acids (lysine and tryptophan) that humans cannot make on their own. In addition, humans cannot easily digest the forms of iron and the vitamin niacin found in corn.

Protein deficiency most strongly affects young children, in whom it stunts growth. Only by eating about 0.8 kilogram (2 pounds) of corn tortillas each day could

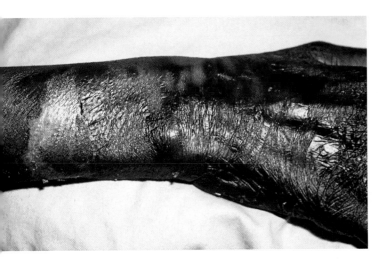

Figure 28.13 The Effect of Corn as a Staple in the Human Diet
Pellagra results from a deficiency in the vitamin niacin that often occurs among people who depend heavily on corn for their diet.

a growing child get enough protein. Not only would this amount of food be a lot for a child to eat, but it would provide many more calories than needed.

A shortage of niacin and tryptophan leads to a deficiency disease called pellagra, the symptoms of which include skin irritations, diarrhea, and dementia (Figure 28.13). A shortage of absorbable iron leads to anemia because of inadequate production of red blood cells. As a final insult, carbohydrate-rich diets cause dental decay and cavities.

Cultural approaches can overcome some of the nutritional deficiencies of agricultural diets. The addition of lysine-rich beans to the diet improves the quality of protein in a corn-based diet (see Figure 28.4). Furthermore, aboriginal cultures in Central and North America commonly treat corn with a chemical called lye to form the corn meal used in making tortillas and to produce hominy. The treatment of corn with lye makes niacin and iron more accessible, thereby reducing the incidence of both pellagra and anemia. In such ways, ancient and contemporary agricultural societies have managed to reap the energetic benefits of plant-based diets while avoiding some of the nutritional pitfalls.

> ■ The development of agricultural societies may have been associated with malnutrition because some crop plants lack important nutrients. Later modifications to agricultural diets and food preparation techniques helped overcome these difficulties.

SUMMARY

The Elements of Nutrition

- Life depends on a small subset of the known chemical elements.
- Macronutrients are chemical elements that are needed in relatively large amounts. They include carbon, hydrogen, and oxygen, which form the backbone of most organic compounds, and various other elements that are needed for structures or functions that occur throughout an organism.
- Micronutrients are chemical elements that are needed only in tiny quantities, but are essential to survival.

Comparing Producers and Consumers

- Producers use photosynthesis to capture the sun's energy in the form of sugar molecules. They get all of their nutrients from the physical environment.
- Consumers—herbivores, carnivores, and decomposers—must eat other organisms to acquire energy. They get most of their nutrients from organic compounds as well.

Nutritional Requirements of Producers and Consumers

- Producers rely on mineral ions dissolved in water for their nutrients. These nutrients do not provide producers with energy.
- Consumers rely on organic nutrients for both chemical building blocks and energy. Respiration releases the energy in carbohydrates, fats, and proteins; proteins are used mostly as a source of amino acids. Vitamins are essential organic micronutrients.

How Plant Roots Absorb Nutrients

- Plant roots absorb mineral nutrients from the soil water primarily at a small portion of the root near the root tip, which is covered with epidermal cells specialized to form root hairs.
- Plants must absorb nutrients actively because nutrient concentrations inside the root are much higher than in the surrounding soil.
- Plant roots grow through the soil to reach soil not yet depleted of nutrients.
- The active transport of nutrients into the root from the soil and then into the xylem gives the plant control over the kinds and quantities of nutrients and water entering the plant.

How Animal Guts Process Nutrients

- The guts of most animals share a remarkably similar design, which follows from a logical sequence of events in the processing of food as it flows from the mouth to the anus.
- Animals break food down physically, digest it, absorb the nutrients, and finally form waste for disposal.
- Animals can forage by moving actively in search of food or by sitting and waiting for food to come to them.

Some General Patterns in the Breakdown and Absorption of Nutrients

- Organisms must process most nutrients they encounter before they can absorb them. This is especially true of consumers, which cannot absorb the large, complex organic molecules in their food.

- Animals use muscles and hard surfaces to break large pieces of food into smaller ones that they can digest more rapidly.

- Digestion breaks large organic molecules into simpler ones that can be absorbed. Enzymes play a critical role in the digestion of food by consumers.

- Absorptive surfaces are extensively modified to increase their surface area and, therefore, the rate of absorption.

When Nutrients Are Scarce

- Mycorrhizal associations between plant roots and fungi greatly increase the ability of plants to absorb nutrients.

- Plants growing in nutrient-poor soil hold onto scarce nutrients more efficiently than do plants growing on nutrient-rich soil. In extreme cases, plants resort to carnivory to get nutrients.

- Herbivores have evolved grinding teeth to break up tough plant tissues. Their long small intestines are specialized for absorbing proteins, which are in short supply in their diet. The stomachs of many herbivores are specialized to accommodate microorganisms that help break down plant tissue.

Highlight: Agriculture and Malnutrition Revisited

- The development of agricultural societies may have been associated with malnutrition because some crop plants lack important nutrients or contain them in an indigestible form.

- Later modifications to agricultural diets and food preparation techniques helped overcome these difficulties.

KEY TERMS

active forager p. 473	hypha p. 470
bile p. 475	macronutrient p. 466
carnivore p. 468	micronutrient p. 466
colon p. 472	mycorrhiza p. 476
consumer p. 468	producer p. 467
decomposer p. 468	root hair p. 470
digestion p. 472	sit-and-wait foragers p. 473
essential amino acid p. 469	small intestine p. 472
gizzard p. 474	stomach p. 471
gut p. 471	vitamin p. 469
herbivore p. 468	xylem p. 470

CHAPTER REVIEW

Self Quiz

1. All organic molecules
   a. occur only in producers.
   b. always contain water.
   c. are easily absorbed by consumers.
   d. contain the elements carbon, hydrogen, and oxygen.

2. Organisms that can both respire and photosynthesize are
   a. very rare.
   b. producers.
   c. consumers.
   d. fungi.

3. Macronutrients include which of the following elements?
   a. vitamins
   b. flourine
   c. potassium
   d. cobalt

4. Digestion is necessary because
   a. nutrients generally occur in a form that cannot be absorbed by an organism.
   b. nutrients must be converted into an organic form before they can be used.
   c. it allows organisms to chew food more easily.
   d. both b and c

5. The guts of herbivores such as kangaroos
   a. process food in an assembly-line fashion, much as we do.
   b. have relatively short small intestines.
   c. allow them to break down the large amounts of protein in the animals on which they feed.
   d. none of the above

Review Questions

1. Why does it matter that all nutrients absorbed by plants are actively transported into and out of root cells?

2. The digestive system of humans is intermediate between that of strict carnivores and that of strict herbivores. How might you expect your digestive system to be different if we had fed exclusively on plant tissues throughout our evolutionary history?

3. Which features of fungi mentioned in this chapter suit them for life as consumers? From your knowledge that fungi lack muscles and from what we know about other consumers, such as animals, how do you think fungi go about breaking down and absorbing their food?

Fertilizers Threaten Gulf Fisheries

GALVESTON, TX—Fishermen based on the Gulf Coast met in Galveston today to discuss their growing dismay over their shrinking catches. They voiced particular concern over growing "dead zones" on once productive fishing grounds.

A series of studies presented at the meeting documented the appearance of virtually lifeless zones that can cover tens of thousands of square miles in the Gulf. The reports placed the blame for the dead zones on fertilizers carried into the Gulf by rivers draining farmlands.

"I've fished these waters for 30 years and my daddy fished them before me," said J. C. Hollings of Freeport. "We used to fill our holds out there in places where I can't catch a thing today."

A statement released by the group indicated that Gulf fisheries were being threatened by a chain of events that starts far from the coast. According to the report, when rivers dump fertilizers from agricultural runoff into the Gulf, tiny organisms called algae undergo a "bloom." The fish that normally eat the algae cannot control the bloom, which ends in a massive die-off. Bacteria that feed on the dead algae use up all of the oxygen in the water, which suffocates the fish, crabs, and shrimp.

"The fish can't live without oxygen, and we can't live without the fish," said Hollings. Others at the meeting echoed his fears, some going so far as to wonder about the future of fisheries along the Gulf Coast.

The Gulf fisheries pump millions of dollars annually into the Texas economy, and their collapse would be likely to hit hard in many communities. "We understand that farmers need to grow their crops, but we need to make a living, too," said Winston Tranh, who owns a small boatyard near Galveston.

Reducing fertilizers in runoff will not come cheaply. Changes in farm practices, including how much and where fertilizer is applied, would require investment in expensive new equipment for many farms. Another proposed solution, setting aside buffer zones around fields to catch fertilizer before it reaches rivers, would take valuable land out of cultivation.

Evaluating "The News"

1. Dead zones are an example of what some might call too much of a good thing—fertilizers. What kinds of economic impacts could dead zones in coastal waters have?

2. Having benefited in your lifetime from the ability of farmers to produce food cheaply by adding large amounts of nutrients to the environment, how much more would you be willing to pay for vegetables to protect the environment? Would you be willing to pay more even if most other people were still buying cheap vegetables grown in an environmentally damaging way?

3. In the dead zones, the animals that need oxygen die, but the bacteria, which do not require oxygen, thrive. Crabs and shrimp, for example, are dying, but they are being replaced by many more bacteria. Do conservationists really need to be worried about the changes caused by excess nutrients in the water? Why should we view the fate of bacteria as any less important than that of crabs or shrimp?

chapter *29* Gas Exchange

Vivian Torrence, Greek Air, 1989.

In High Places

When hiking high in the Rocky Mountains, the Andes, or the Himalayas, you may find yourself short of breath. Shortness of breath is one of the first signs of the oxygen deprivation that humans experience at high altitudes. The amount of oxygen that we absorb from each liter of air that we inhale decreases by half for every 5500 meters that we climb (see Figure 29.2). Physiological changes in the body resulting from decreasing oxygen availability start to appear at about 3000 meters, and the symptoms of what is called mountain sickness become more severe as we climb higher. From shortness of breath, the symptoms progress to headaches, fatigue, and loss of appetite, and in severe cases, to nausea, heart palpitations, and hallucinations. At 7100 meters, where each breath of air provides only one-third as much oxygen as at sea level, most people lose consciousness.

Although the consequences of reduced oxygen availability at high altitudes can be severe, humans can adapt to these conditions. The highest permanent human habitations lie at 5700 meters in the Andes of South America. Several mountain climbers have reached the top of Mount Everest (8848 meters) without using oxygen tanks.

■ MAIN MESSAGE

Gas exchange in plants and animals depends on the diffusion of oxygen and carbon dioxide into and out of the body.

Other animals and plants can also function at high altitudes; the record is held by an unfortunate griffon vulture that met its end in an encounter with a transport plane flying over Africa at 11,300 meters.

In this chapter we consider how organisms move the gases they need— namely, oxygen and carbon dioxide—into and out of their bodies. Once we understand the basic principles of gas exchange, we can return to considering how humans and other organisms survive in environments where these essential gases are in short supply.

High Altitude, Little Oxygen
Residents of the Andes have adapted to low oxygen availability; the griffon vulture of Africa holds the high altitude record for animals.

■ KEY CONCEPTS

1. Oxygen and carbon dioxide enter and leave organisms exclusively by diffusion.

2. Land-dwelling and water-dwelling organisms face very different challenges in exchanging gases.

3. Specialized gas exchange surfaces provide a large surface area in a small space.

4. Whereas plants and insects transport gases directly to metabolizing cells through air spaces, many animals transport gases in their blood.

5. Water loss across the gas exchange surfaces is one of the most serious problems faced by land-dwelling organisms.

6. Where low carbon dioxide concentrations reduce the efficiency of photosynthesis, producers resort to modifications of the basic photosynthetic pathway.

More than 95 percent of the weight of living things consists of four macronutrients: carbon, hydrogen, oxygen, and nitrogen (see Chapter 28). Of these elements, organisms regularly take in carbon, oxygen, and in a few important cases, nitrogen as gases. Almost all organisms exchange carbon dioxide and oxygen with their environment. Carbon dioxide provides the carbon from which photosynthesizing producers build the organic compounds basic to life. Both producers and consumers use oxygen when they break down organic compounds to release energy (see Chapter 8). Organisms of one very special group, the nitrogen-fixing bacteria, play an essential role in biological systems by taking up gaseous nitrogen, which most organisms cannot use, and converting it into biologically available forms (see the box on p. 485).

At first glance, corn plants and humans differ greatly when it comes to gas exchange. As producers, corn plants absorb carbon dioxide from the air, which they use to make carbohydrates, and they give off oxygen as a waste product. As consumers, humans absorb oxygen from the air, which we use to release energy from the foods we eat, and we give off carbon dioxide as a waste product. The leaves of corn plants, which absorb carbon dioxide and release oxygen, seem to have little in common with our lungs, which absorb oxygen and release carbon dioxide. Nonetheless, both corn plants and humans have common gas exchange problems to solve.

In many important ways, humans resemble corn plants in how we exchange gases with our environment. Both corn plants and humans must process a lot of air to get the gases they need. Corn plants must process about 2500 liters of air to get enough carbon dioxide to produce a quarter teaspoon of sugar, and when humans exercise intensively, they consume the oxygen in 3000 liters of air each hour.

Whereas single-celled organisms simply exchange gases across their plasma membrane, large multicellu-lar organisms such as humans and corn plants face two problems that complicate gas exchange. First, their large size requires that they transport oxygen and carbon dioxide to or from the individual cells where these gases are used. Second, their small surface area–to–volume ratio (see Chapter 25) requires that they have specialized structures that speed gas exchange. Thus, humans have **lungs**, which pack almost 100 square meters of exchange surface into the chest, and corn plants have intercellular air spaces that form an intricate system of tubes inside their leaves. In addition, as terrestrial organisms, humans and corn plants must both reduce water loss across their gas exchange surfaces, a problem evolution has dealt with by moving the exchange surface inside the organism.

In this chapter we focus on how oxygen and carbon dioxide move into and out of organisms, and we see why organisms as different as corn plants and humans have evolved functionally similar gas exchange structures. We begin by describing how gases enter and leave organisms and how the availability of carbon dioxide and oxygen differs in water and on land. We then turn to some basic biological questions: What adaptations increase the rate at which organisms can absorb gases? How do plants and animals transport gases to and from the cells that use them? How does gas exchange influence photosynthesis in plants? Throughout the chapter we refer to the different adaptations gas exchange requires on land and in water.

29.1 Oxygen and Carbon Dioxide Enter Organisms by Diffusion

Given that organisms need to exchange so much oxygen and carbon dioxide, it may come as a bit of a surprise to learn that these gases move into and out of all living things solely by the passive process of diffusion (see

■ BIOLOGY IN OUR LIVES

Nitrogen, Nitrogen Everywhere...

Like sailors stranded at sea, surrounded by water but unable to drink it, organisms struggle to obtain nitrogen with which to make amino acids while they are bathed by air or water containing vast quantities of unusable nitrogen. Only a few organisms can use N₂, the gaseous form of nitrogen that makes up 78 percent of Earth's atmosphere. These nitrogen-fixing bacteria convert N₂ into forms of nitrogen that other organisms can use.

Some of these remarkable bacteria live in the soil, but many form close relationships with plants. The most famous of these are bacteria of the genus *Rhizobium*, which live in the roots of familiar garden vegetables such as beans and peas. *Rhizobium* receives carbohydrates from its plant host, and the plant benefits from the nitrogen fixed by the bacteria.

Farmers and gardeners will immediately recognize how valuable a service these bacteria perform. Every year they spend tens of millions of dollars on chemical fertilizers to provide plants with enough nitrogen to allow vigorous growth. Nitrogen-fixing bacteria, on the other hand, make nitrogen at no cost to us, and play an essential part in making nitrogen available to all life.

The lumps on these roots are nodules containing large numbers of *Rhizobium* bacteria.

The dark areas in this electron micrograph are individual *Rhizobium* bacteria.

Nitrogen-fixing bacteria of the genus *Rhizobium* form close, mutually beneficial relationships with certain plants.

Chapter 24). No known membrane proteins actively carry oxygen or carbon dioxide across plasma membranes. Because the exchange of oxygen and carbon dioxide depends on diffusion, three basic rules apply to gas exchange in all organisms.

Rule 1: The higher the concentration of a gas in the environment compared with that inside an organism, the faster the gas will diffuse into the organism.

Two biologically important points follow from the diffusion of gases down a concentration gradient: First, the concentration of a gas inside a metabolizing cell must be lower than that in the environment from which it is absorbed. Second, organisms that live in habitats where a gas occurs at a high concentration can maintain a higher rate of diffusion of that gas into their bodies than can organisms living in habitats with low concentrations of the gas.

Rule 2: The more surface area available for diffusion, the faster the gas will diffuse.

The relationship between surface area and diffusion rate also has two important biological consequences. First, single-celled and other small organisms face fewer problems with gas exchange than large organisms because smaller organisms have higher surface area–to–volume ratios (see Chapter 25). Second, large multicellular organisms depend on specialized gas exchange structures that maximize the area of the exchange surface. Our lungs, the intercellular air spaces of plants, and the gills of aquatic animals are examples of such structures (Figure 29.1).

Rule 3: The shorter the distance over which a gas must diffuse, the faster diffusion can supply an organism with that gas.

Oxygen and carbon dioxide cannot diffuse over long distances fast enough to meet the metabolic needs of large organisms. Thus, multicellular organisms must either have a shape that brings all their cells within a few millimeters of the air or water around them or have a way of transporting gases within their bodies. Plants have flat, thin leaves in which all cells lie near a gas exchange surface, and humans have a circulatory system that carries oxygen from our lungs to our metabolizing cells.

The much greater density of water as compared with air (a liter of water weighs 1000 times more than a liter of air) also has profound implications for gas exchange. Consider how humans absorb oxygen compared with how a trout does. We use about 2500 times more oxygen per hour than the trout, both because we are much bigger and because we use energy much more quickly. To meet our oxygen demands, we breathe about 3 kilograms of air in an hour. The trout must meet its much lower demand for oxygen by passing 150 kilograms of heavy, oxygen-poor water over its gas exchange surfaces in the same amount of time.

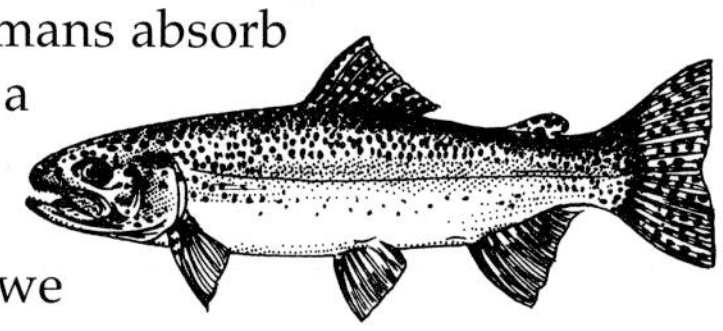

Trout

■ Oxygen and carbon dioxide pass into and out of organisms by diffusion. Large concentration gradients, large surface areas, and short diffusion distances all help speed diffusion.

29.3 Oxygen and Carbon Dioxide in the Environment

The concentrations of oxygen and carbon dioxide found in air and water set upper limits on how large the concentration differences of these gases between the environment and organisms can be. Therefore, we turn our attention next to the distribution of these gases in the physical environment.

1 Whereas single-celled organisms, such as diatoms, can exchange gases across their plasma membranes,...

SINGLE-CELLED ORGANISMS   Diatoms

Earthworm

2 ...multicellular organisms usually need specialized structures for gas exchange.

MULTICELLULAR ORGANISMS

3 Multicellular organisms with a large surface area relative to their body volume, such as earthworms or algae, simply exchange gases across their outer layer of cells.

4 Most multicellular organisms, however, need specialized gas exchange surfaces to compensate for low surface–to–volume ratios or harsh environments.

The problems of gas exchange on land differ from those in water

The circumstances under which aquatic organisms exchange gases with their environment differ from those in which terrestrial organisms exchange gases with their environment (Figure 29.2). Oxygen occurs at much higher concentrations in air than in water. Carbon dioxide, on the other hand, occurs in similar concentrations in both aquatic and terrestrial habitats.

In addition to these differences in the abundance of oxygen and carbon dioxide, air and water differ physically in ways that affect the uptake of gases. Oxygen and carbon dioxide diffuse thousands of times more slowly in water than in air. In still water, therefore, organisms can easily use up the supply of oxygen or carbon dioxide in the surrounding water.

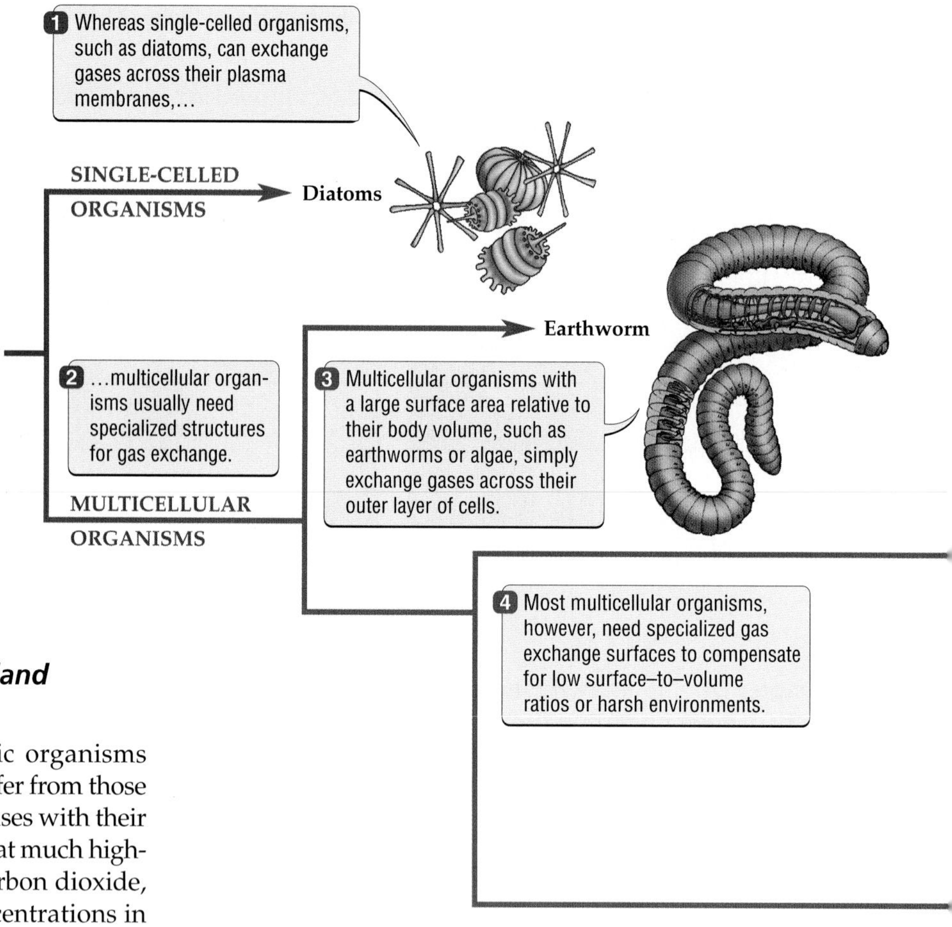

Figure 29.1   Gas Exchange Structures
The structures involved in gas exchange show clear patterns depending on the form of the organism and its environment.

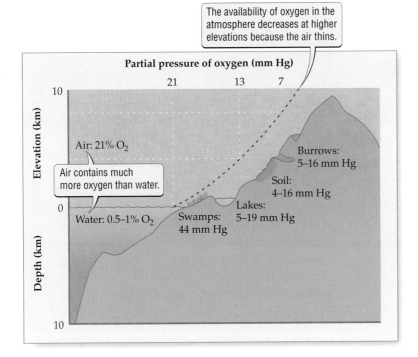

The availability of oxygen in the atmosphere decreases at higher elevations because the air thins.

Air: 21% O₂

Air contains much more oxygen than water.

Water: 0.5–1% O₂

Burrows: 5–16 mm Hg

Soil: 4–16 mm Hg

Lakes: 5–19 mm Hg

Swamps: 44 mm Hg

Figure 29.2 Availability of Oxygen in the Environment
The availability of oxygen (O₂) differs greatly among habitats. The partial pressure exerted by oxygen (indicated in mm Hg—the height in millimeters to which it can lift a column of the heavy liquid mercury) indicates its availability: the greater the pressure, the greater the availability to organisms.

If we breathed water instead of air, our lungs would have to move about 350,000 kilograms of water in an hour. People drown not because water contains no oxygen, but because they cannot possibly pump enough heavy water through their lungs fast enough to meet their oxygen demand.

Oxygen diffuses into animals more rapidly than carbon dioxide diffuses into plants

Oxygen-absorbing consumers generally face fewer problems in exchanging gases than do carbon dioxide–absorbing producers. Oxygen is found at a concentration almost 1000 times

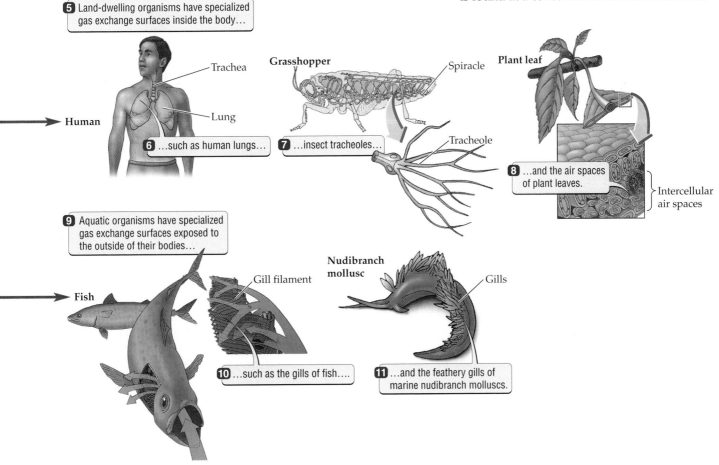

5 Land-dwelling organisms have specialized gas exchange surfaces inside the body…

Human — Trachea — Lung

6 …such as human lungs…

Grasshopper — Spiracle

7 …insect tracheoles… — Tracheole

Plant leaf

8 …and the air spaces of plant leaves. — Intercellular air spaces

9 Aquatic organisms have specialized gas exchange surfaces exposed to the outside of their bodies…

Fish — Gill filament

10 …such as the gills of fish….

Nudibranch mollusc — Gills

11 …and the feathery gills of marine nudibranch molluscs.

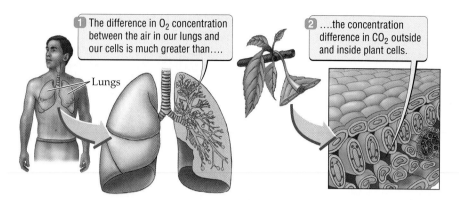

1 The difference in O_2 concentration between the air in our lungs and our cells is much greater than....

Lungs

2 ....the concentration difference in CO_2 outside and inside plant cells.

Figure 29.3 Gas Exchange in Plants and Animals
The difference in carbon dioxide (CO_2) concentration between plant leaves and the surrounding air is much smaller than the difference in oxygen (O_2) concentration between our bodies and the air we take into our lungs. Thus, carbon dioxide diffuses into plants much more slowly than oxygen diffuses into the human body.

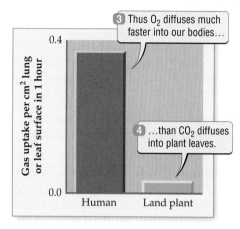

3 Thus O_2 diffuses much faster into our bodies...

4 ...than CO_2 diffuses into plant leaves.

that of carbon dioxide in air, and at a concentration almost 25 times greater than carbon dioxide in water. Rule 1 in the previous section tells us that oxygen should therefore diffuse into our bodies many times faster than carbon dioxide diffuses into the crop plants on which we depend (Figure 29.3). In fact, many plants photosynthesize more rapidly if supplied with air that is artificially enriched with carbon dioxide.

> ■ Air is richer in oxygen than water is. Carbon dioxide occurs in both environments in similar concentrations, but those concentrations are much lower than concentrations of oxygen in either environment. As a result, oxygen diffuses into animals more quickly than carbon dioxide diffuses into plants.

29.4 Gas Exchange Structures

As rule 2 above makes clear, the part of the body surface used for gas exchange must provide enough surface area to supply the organism's needs as rapidly as possible. This requirement presents little problem for very small organisms, but poses special challenges for large ones.

Many small organisms lack specialized gas exchange structures

Single-celled organisms can carry out gas exchange, like all functions that require a large surface area–to–volume ratio, with little modification (see Figure 29.1). Most bacteria and protists, whether aquatic or terrestrial and whether producers or consumers, simply exchange oxygen and carbon dioxide across their plasma membranes.

Some small multicellular organisms can do without specialized gas exchange surfaces as well. Because of the higher oxygen concentrations in air than in water, land-dwelling organisms can reach a greater size than can aquatic organisms before needing specialized gas exchange surfaces. The 9-centimeter-long red-backed salamander, one of the most common salamanders in eastern North America, absorbs all the oxygen it needs across its skin; it has no lungs.

Red-backed salamander

The gas exchange structures of aquatic and terrestrial organisms differ

Most organisms with bodies more than a few millimeters in diameter need a specialized gas exchange surface to meet their gas exchange needs. We call the specialized gas exchange surfaces of aquatic animals **gills**. Gills have evolved several times among aquatic animals, but in all cases they follow rule 2 in providing a large surface area in a small space. Gills typically consist of finely folded sheets of thin epithelial tissue. Gills may lie on the surface of the organism, as in nudibranch molluscs, or in a more protected internal location, as in fish (see Figure 29.1).

In contrast, most terrestrial organisms have internal gas exchange surfaces. The lungs of humans and other land-dwelling vertebrates are a familiar example. Our lungs are elastic sacs that allow us to pump air into and out of the body. Inside the lungs, we exchange gases in tiny sacs called **alveoli** (singular alveolus) that lie at the ends of a multitude of tiny tubular branches (Figure

**Figure 29.4
Human Lungs**
The structure of our lungs speeds the diffusion of oxygen and carbon dioxide into and out of our bodies by providing a large surface area for gas exchange.

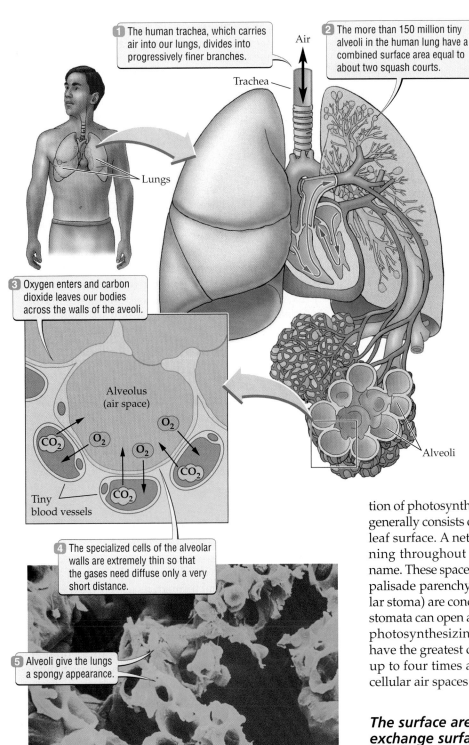

Insects have evolved a distinctly different sort of internal gas exchange structure (see Figure 29.1). The **tracheoles** of insects are a set of tiny tubes that carry oxygen from openings called **spiracles** on the outside of the body to all the respiring cells inside.

The sheetlike shape of plant leaves suits them both to capturing as much sunlight as possible and to bringing cells close to the leaf surface (Figure 29.5). The **epidermis** of a leaf forms a waterproof outer layer. Inside the leaf lie two types of ground tissue (see Chapter 25): **palisade parenchyma**, which generally consists of cells neatly arranged near the upper leaf surface, where they can intercept the light needed for their primary function of photosynthesis, and **spongy parenchyma**, which generally consists of loosely arranged cells near the lower leaf surface. A network of **intercellular air spaces** running throughout the spongy parenchyma gives it its name. These spaces allow carbon dioxide to diffuse to the palisade parenchyma. Openings called **stomata** (singular stoma) are concentrated on the lower epidermis. The stomata can open and close to regulate gas exchange. The photosynthesizing palisade parenchyma cells, which have the greatest demand for carbon dioxide, may have up to four times as much surface area along the intercellular air spaces as the spongy parenchyma cells.

The surface area provided by the gas exchange surface matches the needs of the organism

Evidence from both animals and plants suggests that natural selection has led to gas exchange surface areas that match the metabolic demands of the organism. The lungs of active mammals, such as dogs, provide a large alveolar area compared with those of less active mammals, such as cows. Among fish, the amount of gill sur-

29.4). Although they are individually tiny, the combined surface area of the 150 million alveoli in a typical human lung provides a gas exchange surface 90 times the surface area of our skin.

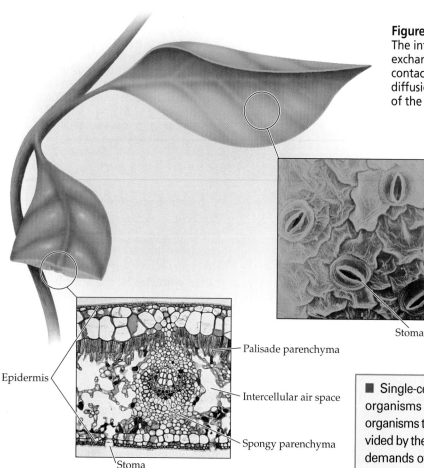

Figure 29.5   The Internal Structure of Plant Leaves
The internal structure of leaves creates a large gas exchange surface where the intercellular air space contacts the parenchyma cells. Stomata regulate the diffusion of carbon dioxide and oxygen into and out of the leaf.

Stoma

Epidermis
Palisade parenchyma
Intercellular air space
Spongy parenchyma
Stoma

would also create other problems. The same gill surface area that lets oxygen enter a fish may also, for example, increase the drag forces that resist forward motion (see Chapter 27) or allow valuable nutrients to be lost from the body by diffusion. In terrestrial organisms, as we will see later in this chapter, gas exchange surfaces allow precious water to escape.

■ Single-celled organisms and some small multicellular organisms lack specialized gas exchange structures. In organisms that have such structures, the surface area provided by the gas exchange surface matches the metabolic demands of the organism.

face area relative to body weight is much greater in active species than in slow-swimming ones (Figure 29.6).

Among plants, the leaves of the aquatic water milfoil take on different forms depending on whether they grow in the water or in the air (Figure 29.7). The feathery shape of the underwater leaves provides a large surface area for gas exchange, which allows them to absorb carbon dioxide even at the low concentrations found in water. Carbon dioxide concentrations around submerged leaves often drop to low levels because the plant uses carbon dioxide more quickly than it can be replaced by the slow diffusion of carbon dioxide through water. Above the water, the water milfoil can produce sheet-like leaves well suited to collecting sunlight because the rapid diffusion of carbon dioxide in air places fewer limits on leaf shape.

A noteworthy feature of the match between gas exchange surface area and demand is not so much that organisms with large demands have large gas exchange surfaces, but that organisms with small demands have small gas exchange surfaces. Although an excess surface area for gas exchange in organisms with small demand would speed the supply of gases to their tissues, it

How Plants and Animals Transport Gases to Metabolizing Cells

After oxygen or carbon dioxide enters the body of a multicellular plant or animal through its gas exchange structures, the gas must still reach the respiring or photosynthesizing cells. As rule 2 above makes clear, diffusion allows for the rapid movement of materials only over very short distances. It works too slowly to adequately supply most metabolizing cells at distances much more than a millimeter from the gas exchange surface. Thus, plants and animals must transport the gases they need to their individual cells.

Plants and insects bring gases to their cells before absorbing them

Plants and insects have evolved similar approaches to supplying their cells with gases: Both groups bring the atmosphere to within a fraction of a millimeter of their cells through a system of air passages. In effect, plants and insects transport gases to their cells before absorbing them.

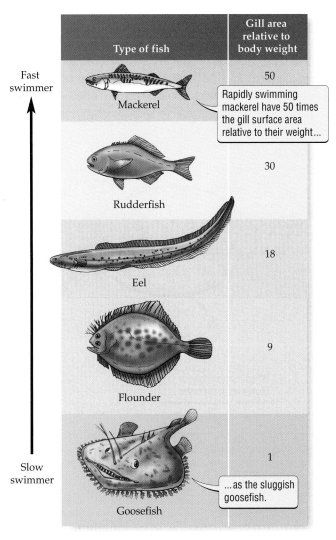

Figure 29.6 The Surface Area of Fish Gills Varies
Actively swimming fish have more gill surface area relative to their body weight than do sluggish bottom-dwelling fish. This difference suggests that natural selection favors gas exchange surface areas large enough to meet the gas demands of the organism without being so extensive as to cause other problems.

Figure 29.7 Adaptations of Leaves to Carbon Dioxide Availability
The different leaf shapes of the water milfoil above and below the water's surface reflect the different selective pressures faced by plants in terrestrial and aquatic habitats.

Oxygen and carbon dioxide move through the intercellular air spaces of leaves and the tracheoles of insects by diffusion, which works only because gases diffuse rapidly in air. Even so, diffusion through the intercellular air spaces and tracheoles is slow enough that it requires leaves to have a flat shape and insects to be small. If an insect were ever to grow as big as suggested in some monster movies, it would pose no threat to anyone: It would simply drop dead from lack of oxygen.

Humans and many other animals absorb oxygen before transporting it

Humans absorb oxygen across the surfaces of the alveoli in our lungs. After absorbing oxygen, we, like many other animals, use our circulatory system (see Chapter 30) to move it rapidly throughout our bodies.

Unfortunately, the liquid portion of animal blood, called **plasma**, is poorly suited to the transport of oxygen. It cannot absorb enough oxygen to meet the demands of most organisms. The plasma of human blood can carry only about 3 milligrams of oxygen per liter, or roughly one-third that of fresh water—much too low to meet our high metabolic demand.

Oxygen-binding pigments carried in the plasma greatly increase the capacity of blood to carry oxygen. Oxygen-binding pigments appear to have evolved independently in many different groups of animals, but they all share a similar basic design: a complex protein associated with a heme group that usually contains iron and that binds reversibly to oxygen. We call these molecules "pigments" because they have distinctive colors that change depending on whether they are bound to oxygen.

Hemoglobin is an oxygen-binding pigment found in many animal species, including humans. Hemoglobin changes color from blue when it is not bound to oxygen to red when it is. It does not float freely in the plasma, but is packaged inside red blood cells. Each hemoglobin molecule can carry up to four oxygen molecules, thus greatly increasing the oxygen-carrying capacity of the blood from the 3 milligrams per liter that can be dissolved in the plasma to 270 milligrams per liter (Figure 29.8).

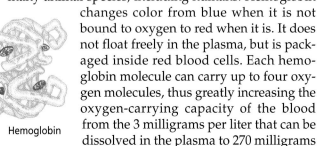

Hemoglobin

Hemoglobin is essential to our ability to supply our cells with oxygen. Iron deficiency anemia results when we do not get enough iron in our diet. Without iron, the human body cannot make enough hemoglobin; thus the oxygen-carrying capacity of blood is reduced, leading to chronic fatigue.

The transport of oxygen in the blood of organisms with oxygen-binding pigments involves several steps. These steps are illustrated in Figure 29.9. When oxygen enters the blood at the gas exchange surface, it first diffuses into and dissolves in the blood plasma. The oxygen-binding pigment then picks up the oxygen from the plasma. Because many animals, including humans, carry the oxygen-binding pigment in red blood cells, picking up oxygen requires the diffusion of oxygen across the plasma membrane of the red blood cell to reach the pigment.

In the body away from the gas exchange surface, this sequence is reversed. In respiring tissue, the oxygen-binding pigment releases oxygen into the blood plasma. Oxygen then diffuses through the blood plasma and across the plasma membrane into the respiring cell. At any time, most of the oxygen in the blood is bound to oxygen-binding pigments, with only a small fraction dissolved in the plasma.

■ Plants and insects allow gases to diffuse directly to each metabolizing cell through air spaces. Many animals, in contrast, transport oxygen in their blood. Oxygen-binding pigments such as hemoglobin greatly increase the amount of oxygen that animal blood can carry.

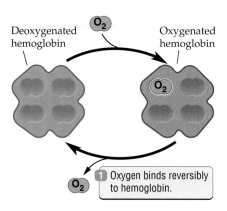

1 Oxygen binds reversibly to hemoglobin.

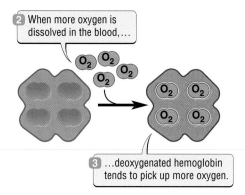

2 When more oxygen is dissolved in the blood,...

3 ...deoxygenated hemoglobin tends to pick up more oxygen.

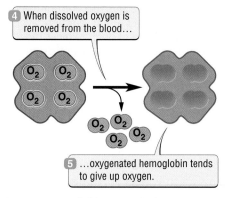

4 When dissolved oxygen is removed from the blood...

5 ...oxygenated hemoglobin tends to give up oxygen.

Figure 29.8   Hemoglobin Can Bind to and Release Oxygen
Hemoglobin is a complex molecule that allows our blood to carry much more oxygen than it could otherwise. Hemoglobin picks up or releases oxygen molecules depending on the amount of oxygen dissolved in the blood plasma around it.

Figure 29.9   How Hemoglobin Picks Up and Delivers Oxygen

In humans, hemoglobin picks up oxygen in the lungs, where the concentration of oxygen dissolved in the blood plasma is relatively high. Hemoglobin releases oxygen as it passes through respiring tissue, where the concentration of oxygen in the blood plasma is relatively low. Most of the oxygen in our blood at any time is bound to hemoglobin rather than dissolved in the blood plasma itself.

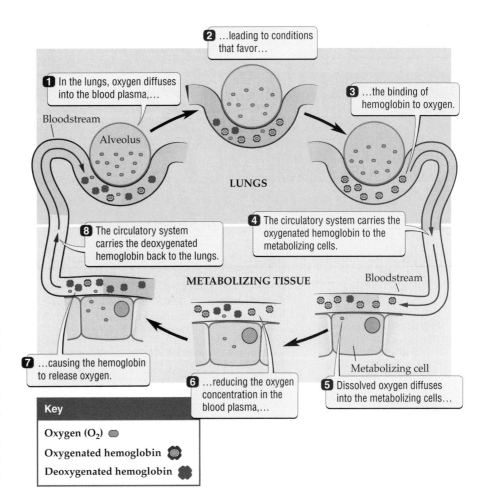

❷ ...leading to conditions that favor...

❶ In the lungs, oxygen diffuses into the blood plasma,...

Bloodstream

Alveolus

❸ ...the binding of hemoglobin to oxygen.

LUNGS

❹ The circulatory system carries the oxygenated hemoglobin to the metabolizing cells.

❽ The circulatory system carries the deoxygenated hemoglobin back to the lungs.

METABOLIZING TISSUE

Bloodstream

❼ ...causing the hemoglobin to release oxygen.

Metabolizing cell

❻ ...reducing the oxygen concentration in the blood plasma,...

❺ Dissolved oxygen diffuses into the metabolizing cells...

Key

Oxygen (O_2)

Oxygenated hemoglobin

Deoxygenated hemoglobin

Gas Exchange on Land

Terrestrial organisms must cope with a problem that inevitably accompanies gas exchange on land: Any surface that allows the diffusion of oxygen and carbon dioxide into an organism also allows the diffusion of water out of the organism. Thus, for terrestrial organisms, gas exchange means water loss. This problem probably represented one of the major obstacles to the evolution of terrestrial life.

Terrestrial organisms represent an evolutionary compromise between the conflicting demands of water conservation and gas exchange. Most of the outer surfaces of terrestrial creatures are waterproof, and these surfaces are therefore useless for gas exchange. The gas exchange structures of virtually all terrestrial organisms lie inside rather than outside the body. By moving their gas exchange surfaces inside their bodies, terrestrial organisms gain more control over what crosses those surfaces.

Organisms that live in particularly challenging environments may have even more elaborate strategies for avoiding water loss. Kangaroo rats, for example, live in a dry desert environment and must be particularly careful to conserve water. This animal takes advantage of the fact that cool air cannot hold as much moisture as warm air. When the kangaroo rat inhales the dry desert air, it picks up moisture from the tissue that lines the complexly folded nasal passage (Figure 29.10). The inhaled air warms to body temperature and picks up even more water in the lungs. The kangaroo rat recovers much of this water, however, when it exhales the air through its cool nasal passages. Much as moisture condenses on the outside of a glass of ice water on a hot, humid day, moisture in the warm exhaled air condenses out in the cool nasal passages and is reabsorbed.

> ■ In terrestrial organisms, surfaces that allow the diffusion of oxygen and carbon dioxide necessarily allow the diffusion of water. Gas exchange is thus a major source of water loss for terrestrial organisms.

Gas Exchange and Photosynthesis in Plants

The relative availability of oxygen and carbon dioxide to producers determines how efficiently they can photosynthesize. In hot, dry habitats, the need to prevent water loss may reduce carbon dioxide concentrations in the leaves. Some producers have evolved ways of solving this problem by altering the photosynthetic pathway.

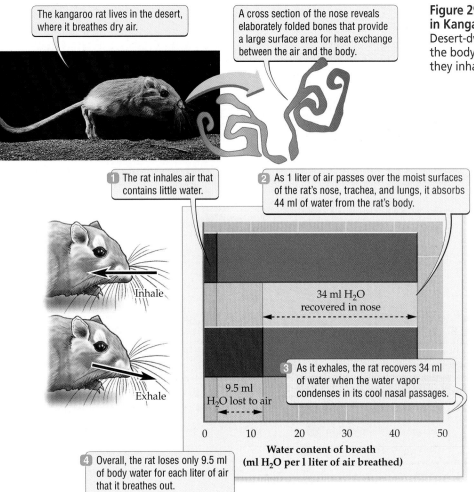

The kangaroo rat lives in the desert, where it breathes dry air.

A cross section of the nose reveals elaborately folded bones that provide a large surface area for heat exchange between the air and the body.

Figure 29.10 Water Conservation in Kangaroo Rats
Desert-dwelling kangaroo rats recover most of the body water that is taken up by the dry air they inhale.

1 The rat inhales air that contains little water.

2 As 1 liter of air passes over the moist surfaces of the rat's nose, trachea, and lungs, it absorbs 44 ml of water from the rat's body.

Inhale

Exhale

34 ml H₂O recovered in nose

3 As it exhales, the rat recovers 34 ml of water when the water vapor condenses in its cool nasal passages.

9.5 ml H₂O lost to air

4 Overall, the rat loses only 9.5 ml of body water for each liter of air that it breathes out.

Water content of breath
(ml H₂O per l liter of air breathed)

0   10   20   30   40   50

The rarity of carbon dioxide in the atmosphere causes problems for producers

The presence of too much oxygen and too little carbon dioxide can interfere with photosynthesis. Before we see how this happens, let's quickly review the process of photosynthesis as presented in Chapter 8. Recall that during the dark reactions of photosynthesis, carbon dioxide combines with a five-carbon molecule in the presence of an enzyme called **rubisco** and energy captured by the light reactions. This reaction produces a three-carbon compound called glyceraldehyde 3-phosphate (see Figure 8.6). This photosynthetic pathway is generally called C_3 **photosynthesis** after its three-carbon end product. Producers then use glyceraldehyde 3-phosphate to build six-carbon sugars, such as glucose.

Rubisco, which binds carbon dioxide as it combines with the five-carbon compound during photosynthesis, can have a second, unexpected role. When oxygen concentrations are high relative to carbon dioxide concentrations, rubisco may bind with oxygen, rather than carbon dioxide. When this happens, it combines the oxygen with the five-carbon molecule, causing the resulting five-carbon compound to split into three-carbon glyceraldehyde 3-phosphate and a two-carbon compound that eventually breaks down to release carbon dioxide. This reaction is called **photorespiration** because, like respiration, it uses oxygen and releases carbon dioxide, but unlike respiration, it occurs only in the presence of light. Photorespiration undoes the work of photosynthesis, releasing up to half of the carbon fixed by photosynthesis.

Whether rubisco binds carbon dioxide to catalyze the dark reactions of photosynthesis or binds oxygen to catalyze photorespiration depends on the relative concentration of carbon dioxide and oxygen in photosynthetic cells. Rubisco strongly favors carbon dioxide over oxygen, ensuring that it usually catalyzes the dark reactions of photosynthesis. Keep in mind, however, that carbon dioxide is 1000 times less abundant than oxygen in the atmosphere. Thus, even with the much greater affinity of rubisco for carbon dioxide than oxygen, photorespiration still takes place.

Two situations often cause carbon dioxide levels in the body of a producer to fall below those in the atmosphere:

1. Rapid photosynthesis in bright light, which can deplete carbon dioxide more rapidly than it can be supplied by diffusion through the intercellular air spaces
2. Closing of the stomata to conserve water under dry or hot conditions, which cuts off the supply of carbon dioxide to the photosynthesizing cells

C_4 and CAM photosynthesis allow producers to use carbon dioxide more efficiently under sunny or dry conditions

Although 85 percent of plants use C_3 photosynthesis, many plants, and even some single-celled algae, resort to one of two variations on this basic photosynthetic pathway to reduce water loss in sunny, dry environments. In such environments, plants face two conflicting pressures: They must ensure that enough carbon dioxide enters the leaves to prevent photorespiration, but they must keep their stomata closed to prevent dangerous water loss.

Many economically important plants, including corn and sugarcane, use C_4 **photosynthesis**. This process sep- arates photosynthesis into two steps, which take place in separate locations in the leaf. Carbon is initially fixed in parenchyma cells not as a three-carbon, but as a four-carbon compound (Figure 29.11a). This four-carbon compound then moves out of the parenchyma cells into specialized cells called **bundle sheath cells**. In the bundle sheath cells, C_4 plants use energy to split the four-carbon compound into a three-carbon compound and carbon dioxide. The carbon dioxide then combines with a five-carbon molecule in the presence of rubisco to form three-carbon glyceraldehyde 3-phosphate, just as it does in C_3 photosynthesis. C_4 photosynthesis works at low carbon dioxide levels because the enzyme that catalyzes carbon fixation in the parenchyma cells, **PEP carboxylase**, favors carbon dioxide over oxygen much more strongly than does rubisco. Thus, C_4 plants can avoid photorespiration when carbon dioxide concentrations in the leaf drop. This ability to photosynthesize at low carbon dioxide concentrations allows them to keep their stomata closed more of the time to reduce water loss.

Many plants that live in extremely dry conditions, such as deserts, use **CAM photosynthesis** to avoid photorespiration (Figure 29.11b). CAM plants work much like C_4 plants, except that the two steps of photosynthesis are separated in time rather than in space. CAM plants open their stomata to take up carbon dioxide primarily at night, when temperatures are lower and humidity is higher. Using PEP carboxylase, the plant immediately captures carbon in a four-carbon compound, which it then stores. During the day, the plant converts PEP carboxylase to an inactive form, allowing rubisco to convert the carbon dioxide released from the four-carbon compound into glyceraldehyde 3-phosphate using energy from the sun. Because they have stored carbon dioxide in a four-carbon compound, CAM plants

(a)

(b) Transport tissue

Parenchyma   Bundle sheath cell

(c)

Catalyzed by PEP carboxylase

CO_2

Energy

5C

3C

4C

Sugar

Catalyzed by rubisco

4C

3C

Energy

CO_2

(d)

Figure 29.11 C_4 and CAM Photosynthesis Reduce Water Loss in Plants
(a) The leaves of C_4 plants, such as corn, have a specialized arrangement of cells that allows the plant to spatially separate the initial fixation of carbon as a four-carbon compound in photosynthesis and its subsequent transfer to glyceraldehyde 3-phosphate (b, c). This allows the plant to tolerate low concentrations of carbon dioxide in the leaves, letting it keep its stomata closed to conserve water. (d) CAM plants are common in desert habitats. Many CAM plants, like this barrel cactus, store water in their bodies as well as altering the photosynthetic pathway to conserve water.

can keep their stomata closed during the day to reduce water loss.

It is important to keep in mind that C_4 and CAM photosynthesis protect plants from photorespiration at a cost: It takes twice the energy to fix each carbon molecule in C_4 as in C_3 photosynthesis. Nonetheless, in hot, dry, open habitats such as grasslands and deserts, the benefits of avoiding photorespiration may outweigh these costs, and C_4 or CAM species may outnumber C_3 species.

> ■ Photorespiration at low carbon dioxide concentrations can result in the loss of much of the carbon fixed during photosynthesis. The C_4 and CAM photosynthetic pathways reduce water loss by avoiding photorespiration at low carbon dioxide concentrations, which allows the plants to keep their stomata closed.

■ HIGHLIGHT

Gas Exchange in High Places

Human populations that live at high altitudes, where each breath of air provides relatively little oxygen, have several genetically based features that allow them to live in such environments. People who live permanently at high altitudes tend to have large lungs. In addition, the hemoglobin concentration in their blood is higher, often lending a red cast to the skin. Their hemoglobin itself picks up oxygen more readily from the blood plasma, which allows their blood to leave their lungs 100 percent saturated with oxygen, despite the smaller difference in oxygen concentration between air and blood. In contrast, when individuals who are native to lowland areas visit high altitudes, the hemoglobin in the blood leaving their lungs is typically not fully oxygenated.

Without prior exposure to high-altitude environments, our initial response to thin mountain air is an emergency response: an increased breathing rate accompanied by an increased heart rate. In essence, our bodies respond as if we were running at a lower altitude. The increased breathing rate maintains as steep an oxygen gradient between the air and the body as possible, and the rapidly pumping heart quickly carries oxygenated blood to the metabolizing cells.

As athletes and mountaineers know, the human body can adjust to a low-oxygen environment if we give it time. One recommendation is to allow one day of adjustment time for each 600 meters we wish to go above 2000 meters. With time spent at high altitudes, the body

adjusts by increasing the hemoglobin concentration in the blood and by *decreasing* the ability of the hemoglobin to bind oxygen. The former strategy increases the oxygen-carrying capacity of our blood, and the latter allows us to unload oxygen from the blood more quickly at the site of metabolism. Note that the decreased attraction of hemoglobin to oxygen is just the opposite of the property found in the hemoglobin of humans evolutionarily adapted to high altitudes, thus illustrating an interesting difference between a short-term physiological response and a long-term evolutionary response.

> ■ Peoples native to high altitudes have evolved adaptations such as large lungs and high hemoglobin concentrations in the blood. Individuals not used to high altitudes can compensate for low oxygen availability by breathing more rapidly, increasing blood flow, and over time, increasing the hemoglobin concentration in the blood.

SUMMARY

Oxygen and Carbon Dioxide Enter Organisms by Diffusion

- Oxygen and carbon dioxide enter and leave organisms exclusively by diffusion.

- Steeper concentration gradients cause gases moving from the environment into an organism to diffuse more quickly.

- The more surface area available for diffusion, the faster a gas diffuses.

- The shorter the distance across which a gas must diffuse, the faster the gas diffuses.

Oxygen and Carbon Dioxide in the Environment

- The oxygen concentration in air is higher than that in water.

- Concentrations of carbon dioxide are similar in air and in water.

- Oxygen is much more abundant than carbon dioxide in both air and water. As a result, oxygen diffuses more quickly into animals than carbon dioxide does into plants.

Gas Exchange Structures

- Small organisms, because of their large surface area–to–volume ratios, generally do not need specialized gas exchange structures.

- Aquatic animals more than a few millimeters in diameter rely on gills for gas exchange.

- Terrestrial organisms rely on internal structures, such as lungs (in vertebrates), tracheoles (in insects), and intercellular air spaces (in plants), for gas exchange.

- The relative surface areas of gas exchange structures in different organisms match the relative metabolic demands of those organisms.

How Plants and Animals Transport Gases to Metabolizing Cells

- Because gases diffuse slowly across long distances, multicellular organisms must transport gases from their specialized gas exchange structures to other tissues in their bodies.
- Plants and insects transport gases to their cells directly through an extensive system of air spaces.
- Many animals absorb the gases taken up by their gas exchange structures into their blood, which then transports the gases throughout their bodies.
- Specialized oxygen-binding pigments, such as the hemoglobin in human red blood cells, improve the blood's ability to transport oxygen.

Gas Exchange on Land

- The gas exchange surfaces of terrestrial organisms can allow significant water loss.
- Terrestrial organisms have evolved internal gas exchange systems as a means of conserving water.

Gas Exchange and Photosynthesis in Plants

- Plants must keep their stomata open enough to get sufficient carbon dioxide, but keep them closed enough to prevent excessive water loss.
- When concentrations of carbon dioxide around the site of photosynthesis are low, photorespiration can cause producers to release much of the carbon fixed during photosynthesis.
- The C_4 and CAM photosynthetic pathways, found commonly in plants of dry, sunny habitats, reduce photorespiration, but increase the energetic cost of fixing each molecule of carbon.

Highlight: Gas Exchange in High Places

- When a person first encounters the oxygen-poor air of high altitudes, the body's immediate response is to increase the heart and breathing rates.
- If the same individual stays at high altitudes for a few days, the body adjusts by increasing the hemoglobin content of the blood and reducing the attraction of the hemoglobin to oxygen.
- Human populations that have adapted to high altitudes through natural selection have hemoglobin with a higher attraction to oxygen, as well as larger lungs and increased hemoglobin concentrations.

KEY TERMS

alveolus p. 488	oxygen-binding pigment p. 492
bundle sheath cells p. 495	palisade parenchyma p. 489
C_3 photosynthesis p. 494	PEP carboxylase p. 495
C_4 photosynthesis p. 495	photorespiration p. 494
CAM photosynthesis p. 495	plasma p. 492
epidermis p. 489	rubisco p. 494
gill p. 488	spiracle p. 489
hemoglobin p. 492	spongy parenchyma p. 489
intercellular air space p. 489	stoma p. 489
lung p. 484	tracheole p. 489

CHAPTER REVIEW

Self-Quiz

1. The rate at which oxygen moves from the alveoli of our lungs into our blood
   a. depends on the difference in oxygen concentration between the alveoli and the blood.
   b. depends on the color of the alveoli.
   c. depends on the availability of energy to transport gases across the membrane.
   d. none of the above

2. The spiracles of insects function similarly to the
   a. lungs of humans.
   b. gills of fish.
   c. air filters of cars.
   d. stomata of plants.

3. The reason why single-celled organisms do not use specialized gas exchange structures is that their bodies
   a. are not large enough to accommodate them.
   b. have a large surface area–to–volume ratio.
   c. are not complex enough to need them.
   d. have a slow metabolism.

4. Oxygen is transported through the human body
   a. as a dissolved gas in the blood plasma.
   b. bound to hemoglobin molecules in red blood cells.
   c. through intercellular air spaces.
   d. by diffusion.

5. An important advantage of C_4 and CAM photosynthesis over C_3 photosynthesis is that C_4 and CAM photosynthesis
   a. increase the energy lost to photorespiration.
   b. allow plants to grow better in wet terrestrial environments such as swamps and rainforests.
   c. use less energy to fix each carbon dioxide molecule in a sugar.
   d. allow plants to lose less water while photosynthesizing in dry environments.

Review Questions

1. Compare carbon dioxide uptake by terrestrial producers with oxygen uptake by terrestrial consumers.

2. Our oxygen demand changes dramatically depending on how active we are. Can you think of changes that take place in the way your body works as you begin to exercise? Think of how these changes help provide more oxygen to the metabolizing cells in an active body.

3. What advantage might air-breathing aquatic mammals such as dolphins have over a hypothetical aquatic mammal that depended on gills for gas exchange?

---

29   The Daily Globe

Ask The Garden Doctor

Dear Garden Doctor:

I keep reading about how humans are causing the carbon dioxide levels in our atmosphere to rise, and it seems as if the media is presenting this as a bad thing. But don't plants use carbon dioxide to grow? Shouldn't this actually help our gardens and agricultural fields?
Confused in Calgary

Dear Confused:

Don't worry; you're not the only one who's having a tough time sorting this one out. Scientists are still debating the overall effects of more carbon dioxide in our air. As you point out, carbon dioxide levels are rising, and plants need carbon dioxide to photosynthesize. In fact, we know that plants can increase their photosynthetic rates when exposed to higher levels of carbon dioxide. That's exactly why many farmers pump carbon dioxide into greenhouses filled with crop plants.

There is much more that we don't know, however. Because plant growth depends on nutrients in the soil and water in addition to carbon dioxide, we don't know if increasing carbon dioxide will automatically increase plant growth. We also don't know if increased carbon dioxide will lead to the type of growth that we want. Research seems to show that increasing carbon dioxide in the air will lead primarily to additional growth of leaves, stems, and roots. This would benefit some of the plants in our gardens, such as potatoes and foundation plantings, but would do little to help produce more flowers, seeds, or fruits.

The greatest effect of the changes in the atmosphere may be on which plants we can grow in our gardens. Changes in atmospheric carbon dioxide will cause changes in temperature and rainfall (especially here in Canada, and not always in the way that we'd expect!). This will change the choices of plants and vegetables available to us as we pore over spring seed catalogues.

The bottom line is that nobody knows exactly what the effects of the increasing carbon dioxide concentrations in the atmosphere will be. Given the multitude of factors interacting in this complex global phenomenon, it's no wonder that the scientific community sometimes seems a little confused when trying to predict the effects on our crops and gardens.

Evaluating "The News"

1. If we do not know the actual effects, good or bad, of increasing atmospheric carbon dioxide on our crops and garden plants, how worried should home gardeners—or, for that matter, farmers—be?

2. If the effects of increasing atmospheric carbon dioxide where you live seem to benefit plant growth, should you encourage increases in carbon dioxide?

3. Would increasing carbon dioxide concentrations affect C_3 plants (such as tomatoes) and C_4 plants (such as corn) differently? Is it likely that farmers growing different kinds of crops might view increases in carbon dioxide differently?

4. What should you do when it appears that even scientists are confused, as The Garden Doctor asserts? How can our society respond to issues when we are faced with conflicting scientific results or an incomplete understanding of a natural phenomenon?

chapter *30* *Internal Transport*

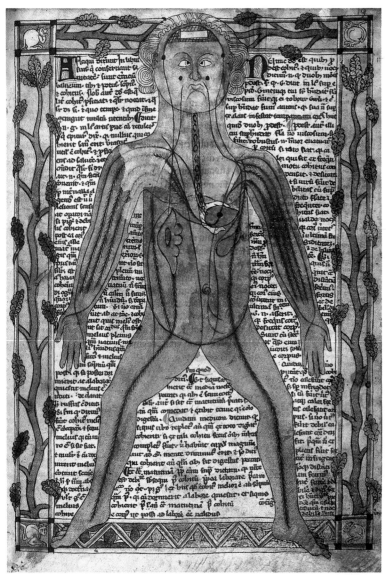

Circulatory System, late thirteenth-century manuscript.

High Pressure on the African Savanna

The human heart generates the pressure that drives blood through the blood vessels of the circulatory system. As part of a routine checkup, doctors measure blood pressure to assess the health of the circulatory system. A blood pressure of about 120/80 millimeters of mercury (mm Hg) in a person at rest usually indicates a heart that is working normally. The two numbers refer to the maximum and minimum pressures produced, respectively, as the blood pulses through blood vessels in the arm. The unit of measurement used, millimeters of mercury, refers to the height to which a pressure can lift a column of the heavy liquid mercury: A larger value indicates a larger pressure.

Blood pressures higher than 120/80 indicate an overworked heart. In resting adults, blood pressures of 160/90 or greater warn of potential problems. About 20 percent of North American adults have high blood pressure: Their hearts are working too hard to push blood through their bodies, thus increasing their risk of heart disease.

■ MAIN MESSAGE

Most multicellular organisms need internal transport systems to move materials to and from their various tissues.

Scientists have turned to the East African savanna in their efforts to understand how the human body might respond to high blood pressure. Here giraffes move gracefully across sweeping grasslands, munching on flat-topped acacia trees. Though apparently safe from the poor diets and stressful lifestyles often blamed for high blood pressure in humans, giraffes live with blood pressures of 260/160. Why does the lifestyle of a giraffe lead to blood pressures that would threaten human lives?

The stately acacia trees on which the giraffes browse face even greater challenges. They must somehow lift sap—the plant's equivalent of blood—to a height of 15 meters or more above the soil surface. Lifting sap to such heights requires an astonishing pressure of 8000 mm Hg, enough to burst a human heart.

This chapter considers how organisms generate the pressures needed to move blood or sap through their bodies. At the end of the chapter, we consider how and why giraffes and acacia trees maintain such high pressures.

Giraffes and Acacias on the East African Savanna

■ KEY CONCEPTS

1. Diffusion is too slow to transport materials throughout the bodies of multicellular organisms.

2. The human circulatory system, like that of most other animals, depends on a muscular heart that contracts to circulate blood through blood vessels.

3. Hearts move blood by creating pressure within the circulatory system.

4. Blood vessels must transport blood and allow for the exchange of materials between the blood and the surrounding tissue.

5. In the phloem tissue of plants, osmotic pressure moves the products of photosynthesis throughout the plant body.

6. The xylem tissue of plants moves water and mineral nutrients from the roots to the leaves using the pulling power generated by the evaporation of water from leaf surfaces.

Multicellular organisms must move materials from certain parts of their bodies to other parts. Nutrients and gases must be moved from sites of production or uptake to sites of use. Coordinated activities such as growth and the regulation of water concentration and temperature also depend on the transfer of signaling molecules and other materials within the body.

In this chapter we consider how multicellular animals and plants transport materials throughout their bodies. Internal transport systems must balance the demands of moving materials and of exchanging those materials with the tissues they supply. As animals evolved from their single-celled ancestors, they developed internal transport systems based on muscular hearts that circulate blood through their bodies. In plants, the evolution of internal transport systems has followed a completely different path. Plants transport nutrients and water in noncirculating sap that flows through their bodies in response to diffusion and evaporation.

materials essential to life. Over short distances (well under 1 millimeter)—across a plasma membrane, within a cell, or through the soil water next to a root tip— dissolved materials can diffuse within a fraction of a second. However, even over distances that we humans would consider tiny—for example, the thickness of a leaf or an ant's body—it takes several seconds to many minutes for dissolved materials to diffuse (Figure 30.1).

At the scale of very large organisms, such as the giraffe and acacia tree mentioned at the beginning of this chapter, transport by diffusion would take years. Thus, organisms more than a few cells thick cannot rely on diffusion to transport materials throughout their bodies. Instead, they need an internal transport system to move materials from one part of their body to another.

> ■ Diffusion works too slowly for effective transport of materials in most multicellular animals and plants.

Who Needs Internal Transport?

In discussing nutrition and gas exchange in Chapters 28 and 29, we focused on the role of diffusion as a way of moving the

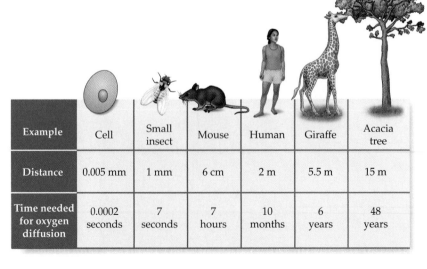

Figure 30.1   The Relationship between Body Size and Diffusion Time
As body size increases, the time that it takes for materials to move from one end of the organism to the other by diffusion increases greatly. The times shown represent the time that it takes oxygen to diffuse various distances through water, the main component of living things.

Example	Cell	Small insect	Mouse	Human	Giraffe	Acacia tree
Distance	0.005 mm	1 mm	6 cm	2 m	5.5 m	15 m
Time needed for oxygen diffusion	0.0002 seconds	7 seconds	7 hours	10 months	6 years	48 years

The Human Circulatory System

How do large multicellular animals move materials inside their bodies? Let's look first at our own internal transport system as an example. The human circulatory system—a muscular heart that circulates blood through a complex series of loops formed by blood vessels—has many of the basic elements of the various circulatory systems found in the animal kingdom (Figure 30.2).

Blood flows to all parts of the human body through blood vessels

In the human circulatory system, tubular blood vessels form closed loops through which blood moves from the heart to the lungs or metabolizing cells and then back to the heart. These vessels carry a rich mixture of minerals, gases, organic nutrients, and products of metabolism

dissolved in the blood. In a typical adult, 5 to 6 liters of blood circulate continuously through the blood vessels.

The human body contains three kinds of blood vessels:

1. Thick-walled **arteries** carry blood away from the heart for distribution to the body.
2. Tiny thin-walled **capillaries**, less than 0.01 millimeter in diameter, allow the exchange of materials between the blood and the surrounding tissue.
3. Thin-walled **veins** collect blood from the body for return to the heart.

The capillaries connect arteries to veins to make a **closed circulatory system** (Figure 30.3a).

Because diffusion works well only over short distances, nearly all cells in the body lie within 0.03 millimeter of a blood vessel (less than one-third the thickness of this page). A piece of muscle tissue the size of a pencil tip may contain more than a thousand capillaries. That is why blood oozes from a tiny cut made anywhere on the body. Clearly, carrying blood so close to all the billions of cells in the body requires an extensive system of blood vessels. If laid end to end, the approximately 17 billion blood vessels in the human body would stretch an incredible 80,000 kilometers, or twice around the world.

In a pattern that we have already seen in other parts of multicellular organisms, the structures involved in the exchange of materials—in this case, the capillaries—provide an astonishingly large surface area. Taken together, the individually tiny capillaries make up most of the length of the circulatory system and provide a surface area equal to that of the floor space in three large houses.

The human heart generates pressure that propels blood through the vessels

Contraction of the muscular heart pumps blood through our blood vessels. The human heart has four cham-

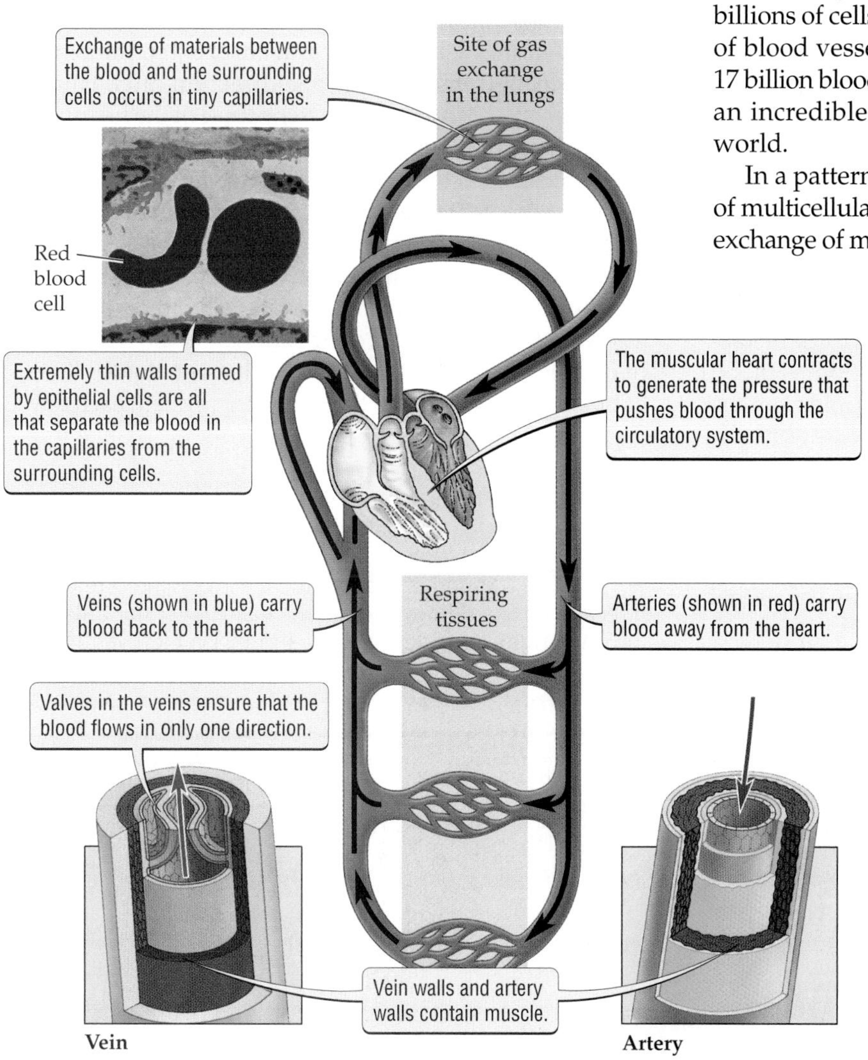

Exchange of materials between the blood and the surrounding cells occurs in tiny capillaries.

Site of gas exchange in the lungs

Red blood cell

Extremely thin walls formed by epithelial cells are all that separate the blood in the capillaries from the surrounding cells.

The muscular heart contracts to generate the pressure that pushes blood through the circulatory system.

Veins (shown in blue) carry blood back to the heart.

Respiring tissues

Arteries (shown in red) carry blood away from the heart.

Valves in the veins ensure that the blood flows in only one direction.

Vein walls and artery walls contain muscle.

Vein

Artery

Figure 30.2   The Human Circulatory System The human circulatory system, like those of all other mammals, depends on a heart to push blood through closed loops formed by blood vessels. In humans, one loop carries blood to and from the lungs for gas exchange; a second loop carries blood to and from metabolizing tissues in the body.

(a) Closed circulatory systems

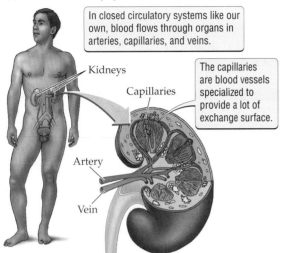

In closed circulatory systems like our own, blood flows through organs in arteries, capillaries, and veins.

The capillaries are blood vessels specialized to provide a lot of exchange surface.

Kidneys

Capillaries

Artery

Vein

(b) Open circulatory systems

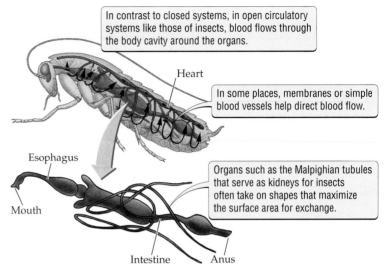

In contrast to closed systems, in open circulatory systems like those of insects, blood flows through the body cavity around the organs.

Heart

In some places, membranes or simple blood vessels help direct blood flow.

Esophagus

Mouth

Organs such as the Malpighian tubules that serve as kidneys for insects often take on shapes that maximize the surface area for exchange.

Intestine   Anus

Figure 30.3   Closed and Open Circulatory Systems
Two main types of circulatory systems are found in the animal kingdom. (a) In closed circulatory systems, blood flows through organs (such as the human kidney shown here). (b) In open circulatory systems, blood flows around organs. Those organs (such as the Malpighian tubules of insects, which are equivalent to our kidneys) often have shapes that maximize the surface area available for exchange.

bers: a left atrium and ventricle and a right atrium and ventricle (Figure 30.4). Each side of the heart acts as a separate pump. The larger left side pumps blood through the loops of blood vessels that serve our metabolizing tissues, and the smaller right side pumps blood through a shorter loop that carries blood to and from the gas exchange surfaces in our lungs.

On each side of the heart, contraction of the smaller, thinner-walled **atrium** (plural atria) pumps blood into the corresponding **ventricle**, which has a thicker, more muscular wall. The ventricle pumps blood out of the heart and through the blood vessels. One-way valves that

separate the atrium from the ventricle, and the ventricle from the artery leading out of it, ensure that the blood flows through the heart in one direction only. Heard through a stethoscope, the *lub-dub* sound of the heartbeat is actually the sound of these valves snapping shut. The *lub* sound is the closing of the valve separating the atrium from the ventricle, and the *dub* sound is the closing of the valve separating the ventricle from the artery.

Our hearts beat without fail—we hope—about 75 times each minute, which amounts to 3 billion beats in a 70-year lifetime. At rest, our hearts pump all 5 to 6 liters of our blood each minute—the equivalent of 7000 liters of blood moving through the circulatory system each day.

Figure 30.4   The Human Heart
30.3 One-way valves in the heart direct the flow of blood and make the familiar *lub-dub* sound of the heartbeat. Arrows indicate the direction of blood flow through the heart. RA = right atrium; RV = right ventricle; LA = left atrium; LV = left ventricle.

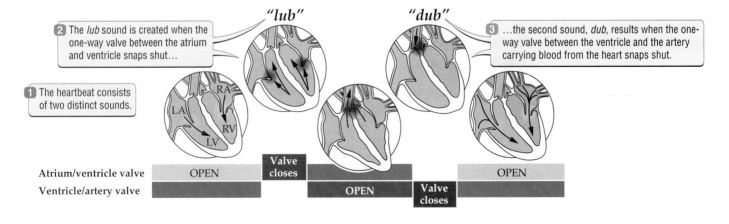

"lub"   "dub"

2 The *lub* sound is created when the one-way valve between the atrium and ventricle snaps shut...

3 ...the second sound, *dub*, results when the one-way valve between the ventricle and the artery carrying blood from the heart snaps shut.

1 The heartbeat consists of two distinct sounds.

RA
LA
RV
LV

Atrium/ventricle valve   OPEN   Valve closes   OPEN
Ventricle/artery valve   OPEN   Valve closes

Each contraction of the heart generates a sharp increase in pressure that pushes blood through the blood vessels (Figure 30.5). The blood pressure measured in a doctor's office reflects the pressure changes in the arteries leading to the body from the left ventricle. If your blood pressure reading is 120/80, for example, contraction of your left ventricle generates 120 mm Hg of pressure, followed by a drop to 80 mm Hg when the ventricle relaxes and refills. Similar measurements made in the arteries leading from the smaller right ventricle into the short loop running through your lungs would reveal lower pressures, ranging from 8 to 25 mm Hg.

By the time the blood enters the capillaries, the pressure has dropped to a steady 35 mm Hg. When the blood leaves the capillaries, the pressure is only 10 mm Hg. The drop in pressure as the blood progresses through the circulatory system results from friction drag (see Chapter 27) between the flowing blood and the blood vessel walls.

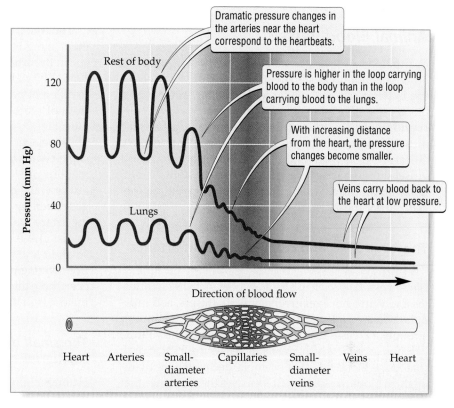

Figure 30.5 Pressure Changes in the Human Circulatory System
Both the pressure and the changes in pressure in human blood vessels decrease as the blood moves away from the heart and through the circulatory system.

The human circulatory system responds to the changing needs of the body

When we leave the comfort of the couch and set off on a bicycle ride, the nutrient and, in particular, oxygen demands of our muscles increase greatly. The blood vessels and heart respond to this increased demand with a coordinated set of changes in heart rate, blood flow, and blood distribution (Figure 30.6). The heart rate doubles, thus increasing the blood pressure in the arteries to about 180 mm Hg. The resulting increase in blood pressure tends to stretch the blood vessels, allowing them to carry more blood. The blood vessels supplying the muscles involved in riding the bike carry more blood, whereas muscles in the walls of those supplying tissues not involved in riding the bike, such as those of the digestive system, contract, so that they carry less blood. As a result of these changes, the supply of blood to the active muscles may almost triple.

Figure 30.6 The Response of the Human Circulatory System to Exercise
The human circulatory system responds to increased activity by increasing the flow of blood to the active muscles (measured in liters per minute).

■ The human circulatory system consists of a muscular heart that contracts to push blood through closed loops consisting of blood vessels. The arteries and veins transport blood to and from the millions of tiny capillaries, which exchange materials with the metabolizing tissues.

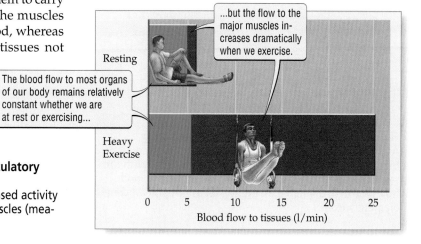

Animal Blood Vessels

In addition to moving materials throughout an animal's body, the blood vessels must also pick up and deliver the materials that they carry. The same features of a blood vessel that allow it to exchange materials with surrounding cells, such as thin walls and a large surface area–to–volume ratio, suit it poorly to blood transport. To carry blood with a minimum of resistance, a blood vessel should have a small surface area–to–volume ratio to minimize drag. Thus, different blood vessels are specialized for these two functions.

The large sizes and complex walls of arteries and veins suit them to transporting blood

Blood vessels specialized for transport, such as our arteries and veins, have relatively large diameters, and muscular, elastic walls. Decreasing blood vessel diameter strongly increases the surface area relative to the volume of blood carried, which greatly increases the friction drag that resists blood flow. For example, decreasing the diameter of a blood vessel by half increases the resistance that the blood faces by 40 times and forces the heart to work 40 times harder to maintain the same blood flow (Figure 30.7). To reduce the amount of work the heart must do to transport blood, the vessels that carry blood the farthest have the largest diameter, and the small-diameter capillaries carry blood over only short distances.

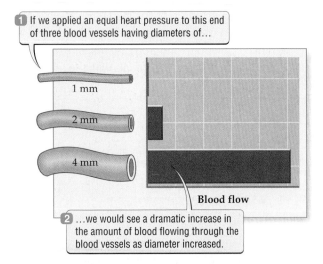

1 If we applied an equal heart pressure to this end of three blood vessels having diameters of…

1 mm

2 mm

4 mm

Blood flow

2 …we would see a dramatic increase in the amount of blood flowing through the blood vessels as diameter increased.

Figure 30.7   Blood Vessel Size and Friction Drag
As the diameter of a blood vessel increases, the amount of blood that it can transport increases dramatically. With the same pressure from the heart, 256 times as much blood flows through a 4-mm-diameter vessel as through a 1-mm-diameter vessel.

The thick, complex walls of arteries and veins equip them to regulate blood distribution. Layers of muscle tissue in the artery walls can contract to reduce blood flow. This ability allows animals to control the flow of blood to different parts of the body, much as we can control the flow of water in our homes by opening and closing faucets.

The elasticity of the artery walls also allows them to stretch when the heart increases blood pressure. When we feel our pulse in an artery, we are actually feeling the artery walls bulging in response to a surge of pressure generated by the contraction of the heart. The ability of the artery walls to stretch reduces the pressure changes as distance from the heart increases, thus protecting the smallest and most delicate blood vessels, the capillaries, from damaging pressure changes (see Figure 30.5). However, these thick elastic walls make it difficult to exchange materials in the blood with the surrounding tissues.

The small diameters and thin walls of capillaries suit them to exchanging materials

Human circulatory systems, like those of vertebrates, generally exchange materials across the tiny capillaries that connect the arteries with the veins. Such closed circulatory systems give the animal great control over where the blood flows and where exchange takes place (see Figure 30.3*a*).

In comparison to the arteries and veins, capillaries have extremely thin, porous walls, formed exclusively of epithelial cells, across which materials diffuse easily. The small diameter of the capillaries increases the surface area for exchange relative to the volume of blood passing through them. However, because of the dramatic increase in friction drag that accompanies a decrease in diameter, capillaries strongly resist blood flow. Thus, organisms with capillaries must have strong hearts that generate enough pressure to push blood through their capillaries.

Insects and some molluscs, such as snails and clams, have taken a different approach from the vertebrates to minimize friction drag. These creatures have **open circulatory systems**, in which blood vessels leading from the heart direct blood into spaces between the organs (Figure 30.3*b*). Blood makes its way back to the heart by flowing through these spaces. Animals with open circulatory systems need not generate pressures as high as those whose blood flows through vessels. However, they have relatively little control over how their blood circulates, and their blood returns to the heart more slowly than in closed circulatory systems. The organs of animals with open circulatory systems are

Snail

often shaped so as to create a large surface area–to–volume ratio, which facilitates the exchange of materials with the blood that bathes them.

> ■ The arteries and veins that carry blood to and from the capillaries have muscular walls and large diameters. These features allow them to regulate blood distribution and to carry blood over relatively long distances with a minimum of resistance. Capillaries have thin walls and large surface area–to–volume ratios, features that facilitate exchange with the surrounding tissues.

30.4 Animal Hearts

Like human hearts, the hearts of other animals rely on muscles to pump blood through the circulatory system. The work that hearts do must both lift the blood against the pull of gravity and overcome the friction drag generated as blood flows past blood vessel walls.

Animal hearts use muscles to generate one-way flow

Animal hearts come in several forms, but all of them use the muscle tissue that is unique to animals to generate pressure, and all of them use one-way valves to direct blood flow. Unlike the other muscles in the human body, heart muscle stimulates its own contraction. Thus the heart can continue to beat regularly even when removed from the body.

The human heart is an example of the most widespread type of heart, the chambered heart (Figure 30.8a). All chambered hearts consist of at least two chambers—an atrium and a ventricle—separated from each other and from the blood vessels by one-way valves. Working alone, the ventricle could fill to only about 75 percent of its capacity. The smaller atrium provides the force needed to fill the ventricle completely, thus increasing the amount of blood moved with each heartbeat.

Some animals also have hearts without chambers. A simple type of pump found in some animals consists of a blood vessel that contains one-way valves and is surrounded by muscle tissue (Figure 30.8b). In many animals, such as humans and sharks, this sort of pump supplements a chambered heart. Contraction of the surrounding muscle tissue during normal activity

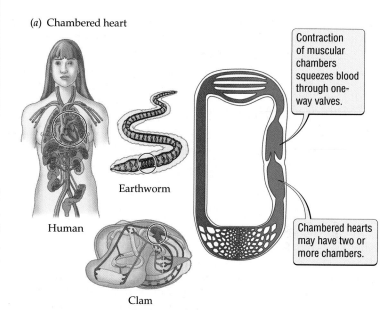

(a) Chambered heart

Human

Earthworm

Clam

> Contraction of muscular chambers squeezes blood through one-way valves.

> Chambered hearts may have two or more chambers.

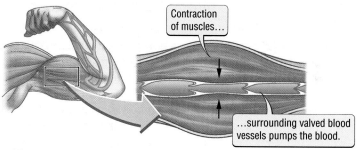

(b) Blood vessels surrounded by muscle tissue

> Contraction of muscles…

> …surrounding valved blood vessels pumps the blood.

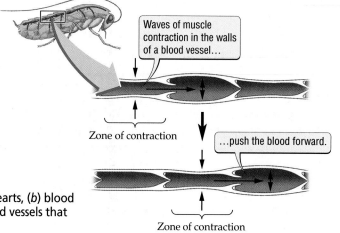

(c) Muscular blood vessels

> Waves of muscle contraction in the walls of a blood vessel…

Zone of contraction

> …push the blood forward.

Zone of contraction

Figure 30.8 The Hearts of Animals
Animal hearts come in three basic types: (a) chambered hearts, (b) blood vessels surrounded by muscle tissue, and (c) muscular blood vessels that undergo waves of contraction.

squeezes the vessel, forcing blood or body fluid out of the squeezed portion. The one-way valves ensure that the squeezing of the vessel forces the fluid to move in only one direction. The veins in our legs work in this way to ease the task of lifting the blood back up to the heart.

Another kind of heart used by some animals, such as insects, consists of a muscular blood vessel in which a wave of muscular contraction pushes the blood ahead of itself, much as food is pushed through the gut (see Chapter 28) (Figure 30.8c).

Animal hearts must overcome friction drag and gravity

The pressure generated by animal hearts accomplishes two things:

1. It overcomes the friction drag that blood encounters as it flows past the walls of the blood vessels.
2. It lifts blood against the pull of gravity.

As already mentioned, anything that decreases the diameter of the blood vessels dramatically increases the pressure that the heart must generate. Blockages of blood vessels can also increase resistance to blood flow, and therefore the pressure needed to maintain adequate blood flow to the tissues.

In humans, diets high in saturated fats encourage the development of fatty deposits called **plaques** in the blood vessels, resulting in a condition called **atherosclerosis**. As these deposits build up, the diameter of the blood vessels decreases. The heart must therefore work harder to force blood through the narrowed arteries. Over many years, this continued high blood pressure can cause changes in the heart that reduce its ability to pump blood. These changes can lead to heart failure. More dramatically, if the plaques develop in one of the major blood vessels supplying the heart itself, they can reduce the supply of oxygen and nutrients to the heart enough to kill heart muscle, causing a **heart attack**.

The heart must also provide the force needed to lift blood to the highest part of the body against the pull of gravity (see the box below). Of the 7000 liters of blood that the human heart pumps through the circulatory system each day, roughly 1000 liters go to the brain. Because

■ BIOLOGY IN OUR LIVES

How Do You Make a Sea Serpent Faint?

The answer to this question is simple: Grab it (very carefully) by its neck and hold it with its head up. Far from being mythical creatures, sea serpents are alive today, in the form of sea snakes that live in the waters of the Pacific Ocean. Like their close relatives, the cobras, sea snakes produce potent poisons with which they subdue their prey.

To understand why a sea snake passes out when held vertically, we need to understand how the environment has shaped the evolution of its circulatory system. Sea snakes spend most of their lives swimming in a horizontal position. Therefore, their hearts usually do not have to lift blood much more than a centimeter. Sea snake hearts need to generate only enough pressure to overcome the friction drag their blood encounters as it flows through their blood vessels. Thus, sea snakes can get by with hearts that generate pressures of between 15 and 30 mm Hg. If you hold a sea snake vertically, with its head pointing upward, its heart cannot generate enough

Sea Snakes Have Weak Hearts Compared with Snakes that Climb Trees

pressure to pump blood to its brain. The result is an unconscious snake. Terrestrial snakes that climb trees, in contrast, have hearts that generate over three times as much pressure.

a liter of blood weighs about 1 kilogram, each day our hearts must do the work necessary to lift 1000 kilograms of blood the 30- to 40-centimeter distance from our hearts to our heads. The extinct dinosaur *Barosaurus*, which had to lift blood the length of its extraordinary 10-meter-long neck, would have needed a heart that could generate a pressure of 646 mm Hg to lift blood to its head.

> ■ Animal hearts produce the pressure needed to overcome the friction drag that resists the flow of blood past blood vessel walls and to lift blood against the pull of gravity.

Internal Transport in Plants: A Different Approach

Although the accomplishments of the human circulatory system, or that of organisms such as *Barosaurus*, may seem impressive, they pale in comparison to what the internal transport systems of trees must do. Against the pull of gravity, trees must lift water and dissolved nutrients from their roots to their leaves, which may be 20 meters or more above the ground. To manage this, a 20-meter-tall tree must generate a pressure difference between its roots and its uppermost leaves of at least 3500 mm Hg, or about 20 times the pressure generated by the human heart during exercise. Now consider the world's tallest living tree, a 114-meter-tall giant sequoia growing in California: It must generate an astonishing pressure difference of 16,500 mm Hg to supply its uppermost leaves with water and nutrients.

The internal transport system that plants use to move materials over long distances and to great heights bears little resemblance to the circulatory systems of animals. The blood equivalent of plants, called **sap**, moves through the interior of cylindrical cells that form the equivalent of blood vessels. In contrast to animal blood, sap does not circulate, so we refer to the internal transport system of plants not as a circulatory system, but by the more general term **vascular system**. Perhaps the most puzzling aspect of the plant vascular system lies in its lack of an obvious heart equivalent, even though we know that plants must have a way of generating pressures even greater than those achieved by animals.

> ■ Plants must generate great pressures to lift sap from the soil to the heights commonly attained by their uppermost leaves.

The Structure of Plant Vascular Tissues

An individual plant depends on two distinctly different vascular systems: the phloem and the xylem (Figure 30.9). **Phloem** transports sap containing the dissolved products of photosynthesis throughout the plant. **Xylem** carries sap consisting of water and dissolved mineral nutrients from the roots to the leaves.

Phloem tissue is made up of living cells. These cells are connected to one another by pores, which allow materials to move directly from the cytoplasm of one cell to the cytoplasm of the adjacent cell. The phloem is the means by which sugars produced by the leaves are moved to nonphotosynthesizing plant parts, such as roots or buds, that use those sugars as an energy source. When dissolved in phloem sap, sugars can move up and down the plant through the living cells that make up the phloem.

A major advance in the evolution of land plants was the development of a distinct vascular tissue, the xylem, through which water and mineral nutrients could be moved from the soil up to the leaves. Water and dissolved minerals move in only one direction—upward—through the xylem cells. Unlike phloem cells, functional xylem cells are dead. They consist of empty cylinders arranged end to end. Connections between the xylem cells allow a continuous column of water to form between the roots and the leaves. As we will see, this thin column of water is critical to the movement of xylem sap. Only plants with xylem tissue have evolved a tall, upright form, suggesting the importance of xylem tissue in the lives of land plants.

Plants differ from animals in that they move sap through the insides of cells, rather than through multicellular organs such as blood vessels. A single vascular cell cannot extend from the roots to the leaves in even the smallest of plants. Instead, the vascular cells are placed end to end to form vascular tissue. Pores that connect the cells allow sap to flow from one cell to the next. An important consequence of this strategy is that the diameter of cells limits the diameter of plant vessels. Typically, the cells of the plant vascular system have diameters ranging from 0.02 millimeter to a maximum of 0.5 millimeter, placing them in the range of capillaries and the smallest arteries and veins of animals. In a pattern that parallels that of animal circulatory systems, however, exchange of materials usually takes place in the vascular cells that have the smallest diameter. The vascular cells with the largest diameter, although tiny compared with many arteries and veins, serve mainly for transport, not for exchange. The small diameter of the transport vessels means that the vascular systems of plants offer very high resistance to sap flow.

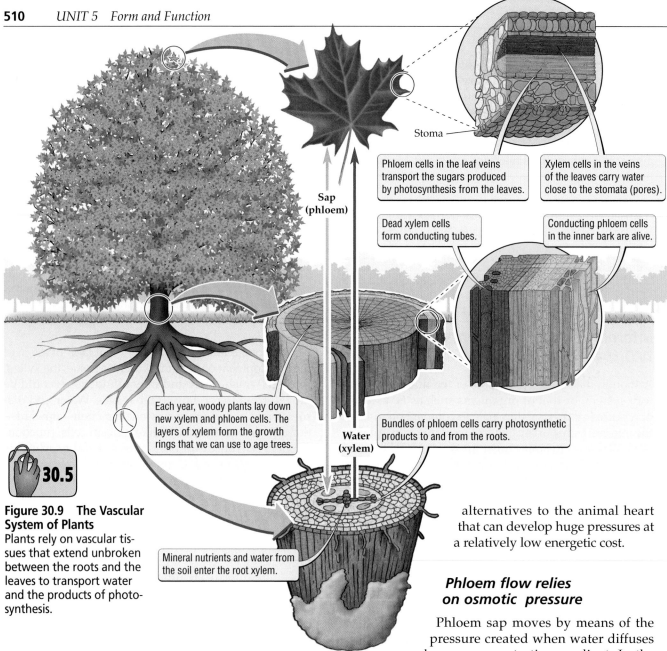

30.5

Figure 30.9   The Vascular System of Plants
Plants rely on vascular tissues that extend unbroken between the roots and the leaves to transport water and the products of photosynthesis.

Sap (phloem)

Phloem cells in the leaf veins transport the sugars produced by photosynthesis from the leaves.

Xylem cells in the veins of the leaves carry water close to the stomata (pores).

Stoma

Dead xylem cells form conducting tubes.

Conducting phloem cells in the inner bark are alive.

Each year, woody plants lay down new xylem and phloem cells. The layers of xylem form the growth rings that we can use to age trees.

Water (xylem)

Bundles of phloem cells carry photosynthetic products to and from the roots.

Mineral nutrients and water from the soil enter the root xylem.

■ Plants rely on two separate vascular systems for internal transport. Phloem transports the products of photosynthesis throughout the plant, and xylem transports water and mineral nutrients from the roots to the leaves.

How Do Plants Move Sap?
30.6

Given the heights to which plants must lift sap and the drag generated by movement through their narrow phloem and xylem cells, muscular pumps like those of animals almost certainly could not generate the necessary pressure. Instead, plants have evolved clever alternatives to the animal heart that can develop huge pressures at a relatively low energetic cost.

Phloem flow relies on osmotic pressure

Phloem sap moves by means of the pressure created when water diffuses down a concentration gradient. In the leaves, specialized **companion cells** associated with the phloem use energy to pump the sugars produced by photosynthesis into the phloem against a concentration gradient (Figure 30.10). The active carrier proteins in the plasma membranes of companion cells can build up sugar concentrations of 10 to 30 percent in the phloem. Such high sugar concentrations result in low water concentrations in the phloem cells, causing water to diffuse from the surrounding leaf tissue into the sugar-rich phloem cells. The water creates pressure as it accumulates in the phloem cells of the photosynthesizing leaf.

In nonphotosynthesizing plant structures, which may be distant from the leaves, cells take up sugars from nearby phloem tissue. Once the sugars have been removed

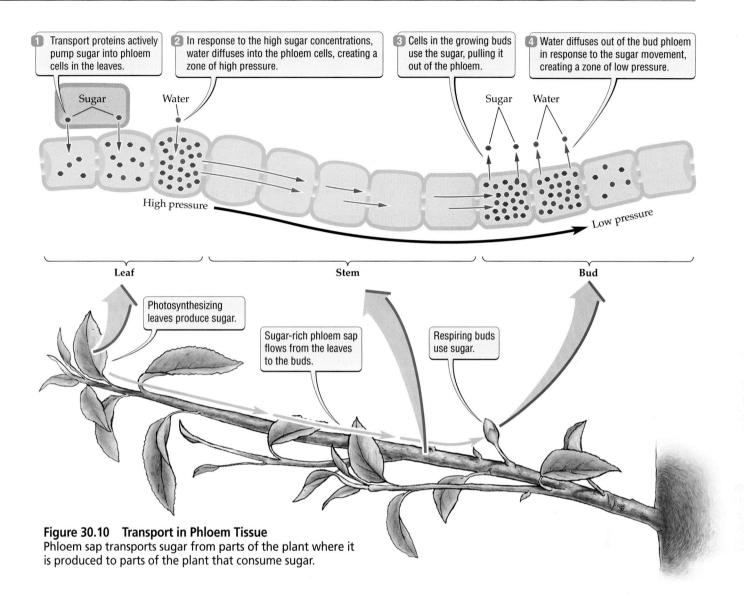

Figure 30.10 Transport in Phloem Tissue
Phloem sap transports sugar from parts of the plant where it is produced to parts of the plant that consume sugar.

from the phloem sap, water diffuses from the now sugar-poor phloem cells into the sugar-rich surrounding cells. The diffusion of water out of the phloem cells in the non-photosynthesizing tissues leads to a region of relatively low pressure there.

The difference in pressure between sugar-rich and sugar-poor phloem cells can easily reach 700 mm Hg, much greater than the pressure that the human heart generates. This pressure difference causes the phloem sap to flow from the leaves to all the tissues that consume sugars as an energy source.

Xylem flow relies on evaporation of water from the leaves

The remarkable pump that lifts xylem sap represents an amazingly efficient evolutionary solution to the prob-

lem of lifting sap from the roots up to the leaves. This mechanism works at almost no energetic cost to the plant because it relies on solar energy to evaporate water from the leaf surface (Figure 30.11). To understand how an evaporation-driven pump can pull sap 100 meters up a tree trunk, we must understand three key points:

1. Evaporation can generate very high tension at very little cost to the plant. Water evaporates readily from wet laundry, for example, even on a humid summer day. As the laundry dries, water is pulled off the fibers of the clothes into the air. We can measure the strength of this pull, or **tension**, in the same units that we measure pressure. Even in humid air, evaporation generates a tension of more than −20,000 mm Hg (when a force pulls rather than pushes, we report a negative pressure—hence the minus sign—but it has the same

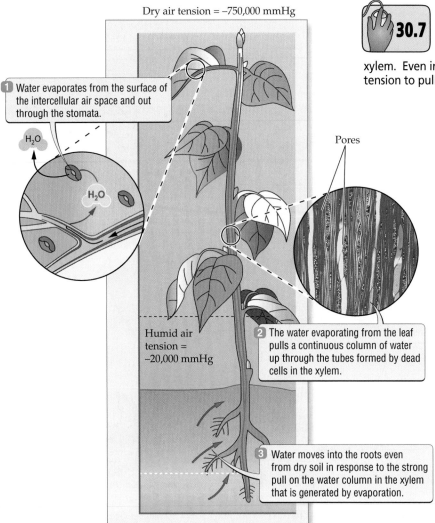

Dry air tension = –750,000 mmHg

1 Water evaporates from the surface of the intercellular air space and out through the stomata.

H₂O

H₂O

Pores

Humid air tension = –20,000 mmHg

2 The water evaporating from the leaf pulls a continuous column of water up through the tubes formed by dead cells in the xylem.

3 Water moves into the roots even from dry soil in response to the strong pull on the water column in the xylem that is generated by evaporation.

Figure 30.11   Transport in Xylem Tissue
Evaporation from the intercellular air spaces of the leaves pulls a continuous column of water up from the soil through the xylem. Even in humid air, evaporation generates enough tension to pull sap to the uppermost leaves of tall trees.

have a hard time thinking of water as a cable. By some calculations, however, the molecules in a column of water hold together so tightly that they can withstand greater tensions than a steel cable of equal diameter. Measured values in plants fall well below this, but are still over 20 times what is needed to withstand the tension generated in the tallest sequoia.

Thus, as water evaporates from the leaves, a "cable" of water is drawn up from the roots, transporting dissolved nutrients and replacing the water lost to evaporation. Anything that breaks the continuous column of water, such as an air bubble, breaks the cable and prevents the lifting of that particular water column. Both drought and freezing temperatures promote air bubble formation, often forcing plants to stop xylem transport during dry or cold seasons.

■ Phloem tissue relies on differences in pressure to move phloem sap. Evaporation from the leaves pulls a continuous column of water upward through the xylem cells. Both mechanisms generate pressures far greater than those generated by animal hearts.

effect as a push). In dry air, the tension generated can be many times greater. Moreover, evaporation costs the plant almost nothing. In a sense, the sun, rather than the plant, pays the high cost of moving sap through small-diameter vessels.

2. The tension generated by the evaporation of water from the intercellular air spaces of the leaves pulls a column of xylem sap up from the roots. In contrast, the pump of the phloem and the muscular hearts of animals push fluid. We can easily see that water is pulled up the xylem by the leaves, rather than pushed up to the leaves from the roots, by considering cut flowers: Flowers in a vase lack roots, yet they continue to draw water out of the vase.

3. For evaporation to pull a column of xylem sap up from the roots, the xylem sap must act as a thin cable strong enough to withstand very high tensions—as much as 16,500 mm Hg in a giant sequoia. Most of us would

■ HIGHLIGHT

Acacias and Giraffes Revisited

Acacia trees and giraffes have internal transport systems that can lift sap or blood to great heights. Acacias lift water 15 meters or more from the dry soil of the savanna to their leaves. Although no known animal could manage this feat, the xylem tissue of the acacia easily meets this challenge. In the relatively dry air of the African savanna, evaporation can generate a tension of as much as −525,000 mm Hg, far more than the acacia needs to lift water to its leaves.

When the roots cannot supply the leaves with water quickly enough, the tension generated by the dry air could break the water columns in the xylem and threaten the life of the acacia. When such dangerous conditions arise during the growing season, acacias respond by closing their stomata (see Chapter 29). In this manner, the acacia slows evaporation from its leaves. However, closing the stomata also cuts off the carbon dioxide supply to the intercellular air spaces of the leaves, preventing photosynthesis. During the seven-month dry season, therefore, closing the stomata is not a reasonable option, so the acacias simply shed their leaves.

Giraffes have hearts that, for a mammal, can generate exceptionally high blood pressures. The giraffe needs these high pressures to lift a column of blood that extends over 2 meters from its heart to its brain. Although the giraffe's heart generates pressures that could kill a human, by the time the blood reaches the brain, its pressures fall within the normal range for humans.

Capillaries in the lower parts of a giraffe's body experience pressures so great (260 mm Hg) that in a human, they would cause blood plasma to leak out of capillaries into the spaces between cells, causing potentially dangerous swelling. To compensate, giraffe capillaries have thicker walls than human capillaries. In addition, the tight skin on giraffe legs places enough pressure on the fluid surrounding the capillaries to stop any leaks.

Now consider what happens when a giraffe lowers its head to drink water. As the head reaches ground level, the heart, which is still generating enough pressure to lift blood to a raised head, increases blood pressure in the lowered head to the same level as those recorded in giraffe feet. The capillaries in the giraffe's head lack the features that protect the capillaries in its legs from leaking. Moreover, leakage from capillaries supplying the brain could easily lead to a fatal increase in pressure inside the skull. When a giraffe drinks, therefore, it reduces the distance of the heart above the ground by spreading its front legs. This behavior reduces the pressure of the blood in the head so that it never reaches the 260 mm Hg measured in the feet of a standing giraffe. In addition, closely spaced one-way valves in the long neck veins near the brain prevent a backflow of blood from the veins to the brain, further reducing the pressure on the capillaries of the brain. Finally, some evidence suggests that the pressure of the fluid surrounding the capillaries in the head may prevent leaks from capillaries in much the same way as fluid surrounding the capillaries of the legs does.

Drinking giraffe

■ The internal transport systems of acacias and giraffes reflect the challenges faced by these two tall organisms. Acacias limit the water loss that accompanies xylem transport by closing stomata or shedding their leaves. Giraffes have a circulatory system suited to tolerating the high blood pressures that their hearts must generate to lift blood to their brains.

SUMMARY

Who Needs Internal Transport?

■ Diffusion can move materials rapidly across very short distances, but it is too slow to move materials across distances of more than a few cell layers.

■ Multicellular organisms need internal transport systems to supplement diffusion in distributing materials throughout their bodies.

The Human Circulatory System

■ The human heart pumps blood—which contains a mixture of dissolved minerals, gases, nutrients, and metabolic products—through closed loops of blood vessels.

■ There are three kinds of blood vessels: arteries, which transport blood away from the heart; capillaries, which exchange materials between the blood and surrounding tissues; and veins, which carry blood back to the heart.

■ Billions of tiny capillaries spread throughout the tissues so that diffusion can efficiently move materials the short distances between the cells and the nearest blood vessel.

■ The heart consists of two separate pumps: the right atrium and ventricle, which work together to pump blood to the lungs, and the left atrium and ventricle, which work together to pump blood to the metabolizing tissues.

■ One-way valves that separate the atria from the ventricles and the ventricles from the arteries maintain a one-way flow of blood through the circulatory system.

Animal Blood Vessels

■ Arteries and veins have a large diameter and a low surface area–to–volume ratio to minimize friction drag. These characteristics allow for the efficient transport of blood through the body.

■ With their small diameter, high surface area–to–volume ratio, and thin, porous walls, capillaries encourage the exchange of materials between the blood and surrounding tissues.

■ Because the high surface area–to–volume ratio of capillaries greatly increases friction drag, animals with closed circulatory systems must have strong hearts.

■ Open circulatory systems, which direct blood flow around tissues rather than through them, avoid the friction drag associated with capillaries, but they also distribute materials to tissues less precisely and more slowly.

Animal Hearts

- Chambered hearts have atria, which forcefully fill ventricles, which then pump blood through vessels. Blood can also be pumped by the contraction of muscles surrounding vessels that contain one-way valves, or by muscular vessels.

- The muscular hearts of animals must create sufficient pressure to overcome the friction drag associated with the flow of blood past blood vessel walls and to lift blood against the pull of gravity.

Internal Transport in Plants: A Different Approach

- Some plants must generate tremendous pressures to move sap from their roots to the heights commonly attained by their uppermost leaves.

The Structure of Plant Vascular Tissues

- Plants move sap through the insides of cells rather than through multicellular vessels, as in animals.

- Phloem, composed of living cells lined up end to end, transports sugars produced by photosynthesis up and down the plant from photosynthesizing leaves to nonphotosynthesizing tissues.

- Xylem, composed of dead cells lined up end to end, transports water and dissolved minerals from the roots to the leaves.

How Do Plants Move Sap?

- Phloem sap moves through the plant by means of osmotic pressure created as the diffusion of water follows sugars that are actively pumped from photosynthesizing tissues into the phloem and from the phloem into nonphotosynthesizing tissues.

- Evaporation of water from the intercellular air spaces of the leaf generates the tension needed to pull xylem sap from the roots up to the leaves.

Highlight: Acacias and Giraffes Revisited

- The internal transport systems of acacias and giraffes reflect the challenges faced by these two tall organisms.

- Acacias control the evaporative water loss that accompanies xylem transport by closing their stomata or shedding their leaves.

- Giraffes have a circulatory system suited to tolerating the high blood pressures that their hearts must generate to lift blood to their brains.

KEY TERMS

artery p. 503	phloem p. 509
atherosclerosis p. 508	plaque p. 508
atrium p. 504	sap p. 509
capillary p. 503	tension p. 511
closed circulatory system p. 503	vascular system p. 509
companion cell p. 510	vein p. 503
heart attack p. 508	ventricle p. 504
open circulatory system p. 506	xylem p. 509

CHAPTER REVIEW

Self-Quiz

1. In which of the following ways does the circulatory system of animals differ from the vascular system of plants?
   a. The circulatory system of animals forms loops, whereas the sap of plants does not circulate.
   b. The sap of plants flows through cells, whereas the blood of animals flows through multicellular organs called blood vessels.
   c. The internal transport system of plants can lift fluids to greater heights than the internal transport system of animals.
   d. all of the above

2. Houseflies and cockroaches have an internal transport system that works at relatively low pressures and in which the blood flows around, rather than through, the organs. We would call such an internal transport system
   a. an open circulatory system.
   b. a capillary.
   c. xylem tissue.
   d. a chambered pump.

3. As the diameter of a vessel increases, the friction drag that must be overcome by the heart
   a. remains constant.
   b. increases.
   c. decreases.
   d. increases and then decreases.

4. Xylem sap is
   a. pulled through the xylem by osmotic pressure.
   b. pushed through the xylem by a heart.
   c. pulled through the xylem by evaporation.
   d. pushed through the xylem by companion cells.

5. Which of the following statements about the flow of sap through phloem is *false*?
   a. Sap moves through phloem under pressure.
   b. Sap moves through phloem in one direction only: from the roots to the leaves.
   c. Phloem sap moves through the inside of living cells.
   d. Sugars move into phloem sap by active transport.

Review Questions

1. Describe and explain the variation in blood pressure throughout an individual's circulatory system during a single heartbeat.

2. After running across campus, how would your circulatory system adjust to the lower demand for oxygen and energy once you arrived in class and sat down?

3. Discuss the advantages and disadvantages of a closed circulatory system.

4. Why do you think that plants never evolved an internal transport system more like our own? Think of the two things that a plant's internal transport system must accomplish and of the limitations under which evolution by natural selection works.

30

The Daily Globe

Irrigation Efficiency Increases

BISMARCK, ND—Statistics provided by the North Dakota Department of Agriculture indicate that farmers are getting a greater increase in crop yields than they have in the past for every dollar spent on irrigation systems. Local farmers and officials see this as a major step forward for agriculture in this relatively rain-poor region.

In announcing the study results, North Dakota Secretary of Agriculture Bob Johanson indicated that the increase in efficiency benefits not only farmers and consumers,

but also the environment. "By increasing irrigation efficiency," said Johanson, "we not only reduce the production costs of growing crops, but we also conserve water, which is such a precious resource in our state."

A number of economically important local crops, including corn and canola, depend on irrigation. The costs of installing irrigation equipment, and of water rights, contribute significantly to the cost of producing these crops. Recently, concern has also arisen that irriga-

tion is using up the aquifer, or underground water supply, on which most local farms depend. The shrinking of the aquifer has given rise to fears that water is being used for irrigation faster than it is being replaced by rainfall.

Canola farmer Leon Upchyk commented after Johanson's announcement that "this is the first bit of good news that [farmers] here have had in a long time. It tells us that the investments we've made in upgrading our irrigation equipment are beginning to pay off."

Evaluating "The News"

1. Why do farmers spend money to irrigate their fields? Think of the various ways in which plants use water that have been introduced in this chapter and preceding chapters.

2. Plants that are native to the North Dakota prairie can survive without

irrigation in that environment. How do you think they manage to do this? (*Hint:* See Chapter 29.)

3. Those in favor of irrigation in North Dakota point out that it allows farmers to grow crops on otherwise less productive land,

whereas those opposed to irrigation argue that valuable underground water reserves are being used up to support crops that are more easily grown elsewhere. How do you feel about the use of irrigation?

chapter *31* Maintaining the Internal Environment

Roy de Forest, *The Importance of Being Earnest*, 1964.

The Giant Guardians of the Sonoran Desert

You've seen them whizzing by as Wile E. Coyote chases the Road-runner, and they appear in more Western movies than John Wayne and Clint Eastwood combined. Saguaro cacti symbolize the American West. These massive plants can grow 10 meters tall in a lifetime that may span three centuries, towering above the shrubs that often huddle at their base. Although the film industry gives the impression that they sprout up wherever cowboys roam, saguaros grow only in the Sonoran Desert of the southwestern United States and northwestern Mexico. Here, saguaros bake in blistering 45°C summer heat and endure winter temperatures that can drop below freezing. What little rain falls on the Sonoran Desert falls mostly during a few weeks in the spring and again in the late summer. Thus, saguaros must cope with the chronic lack of water that characterizes deserts.

Saguaros look different from plants that live in the forests and grasslands of the world. They even look different from most other plants in the Sonoran Desert. As members of the cactus family,

■ MAIN MESSAGE

Organisms must regulate their internal concentrations of water and solutes and keep their internal temperatures within a livable range.

saguaros lack photosynthesizing leaves. Instead, their photosynthetic cells are located in the light green "skin" that covers most of the stem. Compared with those of other cacti, however, saguaro stems are not only tall, but massive. Patches of sharp spines protect the pronounced pleats that run the length of the stem.

How does this unique plant survive in the Sonoran Desert environment? In this chapter, we consider how organisms respond to environmental variations in water availability and temperature. With this background, we consider how the form and function of the saguaro allow it to survive in a hot, dry desert environment. We focus especially on the close relationship between water and temperature regulation in the lives of organisms, including saguaros.

Saguaro Cacti Dominate the Hot, Dry Sonoran Desert of North America

Sonoran Desert

■ KEY CONCEPTS

1. Most organisms function well under only a narrow range of internal conditions.

2. Most organisms manage to keep their internal environment within the narrow range of conditions under which they function best.

3. Organisms gain and lose water by diffusion across the surfaces that they use to exchange gases or nutrients.

4. Organisms exchange heat with their environment by conduction and radiation. Evaporation allows organisms to lose heat. Heat generated as a by-product of metabolism contributes to heat gain in some organisms.

5. The ways in which organisms gain and lose water and heat depend strongly on the environment in which they live.

Living organisms are distinguished from their non-living physical environment by their ability to create an environment inside their cells that supports the unique and essential chemistry of life. Two of the most important components of that internal environment are the contents of the water and temperature inside cells. Life depends on the ability of organisms to maintain a distinct environment inside their cells in the face of a different and ever-changing physical environment. We call the process of maintaining appropriate and constant conditions inside cells **homeostasis** (*homeo*, "similar"; *stasis*, "constant").

Temperature and water availability vary greatly over Earth's surface, often reaching extremes that living cells cannot tolerate. In addition, water availability and temperature can change dramatically over time. The saguaro cactus, for example, faces swelteringly hot afternoons followed by surprisingly chilly evenings in the Sonoran Desert, and it survives months of drought broken only by two brief rainy periods each year. To tolerate both the extremes in environment encountered in different locations and the variation that a single individual may experience over time, organisms have evolved a tremendous diversity of homeostatic mechanisms.

After using a human example to underscore the importance of homeostasis, we present an overview of the ways in which organisms regulate their internal environment. We then turn specifically to how organisms regulate the water content of their bodies. Similarly, we review when and how organisms regulate their internal temperatures. We close by discussing how water

and temperature regulation interact to allow saguaro cacti to survive the harsh Sonoran desert environment.

When Homeostasis Fails: Heat Stroke in Humans

Heat stroke in humans illustrates the importance of homeostasis. Normally, our core body temperature stays within the narrow range of 36.5°C to 37.5°C, no matter how hot or cold the air around us gets. Heat stroke occurs when body temperature rises above 41°C (Figure 31.1). Its symptoms include dizziness, nausea, confusion, loss of muscle control, and in severe cases, death.

Figure 31.1 Heat Stroke
People who cannot keep cool on a hot day can suffer from heat stroke. During 1995, more Americans died of overheating than from any other climate-related cause. In Chicago alone, 465 people died of heat-related causes during the heat wave that hit the city during July of that year.

What causes the body temperature to rise to this dangerous level? Under very hot conditions, our bodies overheat when we can no longer lose enough heat by sweating. Normally, when the body temperature begins to rise above normal, we start to sweat. As the water in the sweat evaporates, it takes with it a surprising amount of heat. A heavily sweating person can get rid of 900,000 calories in an hour—enough heat to raise the body temperature by over 12°C. In heat stroke, the sweating response fails, either because the body has already lost so much water that the sweating mechanism shuts down to conserve water (the main cause of heat stroke in young people) or because old age or certain prescription drugs (the main cause of heat stroke in the elderly) impair the sweating response.

> ■ Heat stroke threatens humans when a failure of the sweating mechanism allows the body temperature to exceed normal temperatures by over 4°C.

Cells Function Best under Predictable Internal Conditions

Life thrives in a wide range of environments, from parched deserts to freshwater lakes. It survives in the –70°C cold of Siberian winters and in the boiling temperatures of 100°C hot springs. Most living organisms, however, function best within much narrower limits.

Life depends on liquid water

The chemical reactions on which life depends take place inside water-filled cells. Without enough liquid water, cells die. More than half of the weight of most organisms consists of water. Sixty percent of the weight of the human body, for example, consists of water. The loss of 5 percent of our body water reduces our ability to do physical work by one-third, and the loss of more than 15 percent can kill us.

Liquid water allows the materials needed for life to dissolve and move to the sites where they are needed. Materials dissolved in water are referred to as **solutes**. Too much water dilutes the concentration of solutes in cells; too little water causes proteins such as enzymes to lose their three-dimensional shape and, therefore, their function.

Most organisms need temperatures somewhere between 0°C and 40°C to remain active

At 0°C, water freezes. Freezing has two important consequences for a living cell: First, materials that are frozen in ice cannot move through the organism. Second, when water freezes, it expands, which may rip apart the plasma membranes of cells (Figure 31.2). At the other end of the scale, as temperatures rise above 40°C, the enzymes that mediate chemical reactions in cells begin to unfold and lose the three-dimensional shape that is so essential to their function. A handful of species can survive at

1 Most living things function best when their internal body temperatures lie between 10 and 40 °C.

45°C

0°C

°F   °C
114   44
110   42
106   40
102   38
98   36
94   34
90   32
86   30
82   28
78   26
74   24

2 At temperatures over 40°C, proteins begin to break down.

3 Some organisms, such as humans, can survive only within a very narrow range of internal temperatures.

4 Other organisms, such as sugar maples, can survive over a broad range of internal temperatures, but still function best at relatively high internal temperatures.

5 Because water expands when it freezes, ice formation can destroy cells.

Normal cells   Frozen cells

Figure 31.2   The Temperature of Life
Most living things require a body temperature that falls between 0°C and 40°C. Although most organisms can survive over a range of temperatures, they usually perform best within a relatively narrow range of temperatures.

100°C because their enzymes hold their shape at these high temperatures (see the Highlight in Chapter 5).

Most organisms function best under a narrow range of conditions

Within the broad limits set for life by these properties of water and proteins, most organisms function best within a relatively narrow range of conditions. As described above, humans function best within narrow ranges of temperature and water content. Even organisms that can tolerate a wide variation in internal conditions still function best under a relatively narrow range of conditions.

For example, during a typical growing season, the photosynthetic cells in the leaves of a sugar maple tree experience temperatures that range from near freezing to well above 30°C. Photosynthesis, however, reaches maximum rates between 20°C and 25°C, the temperature at which photosynthetic enzymes work best (see Figure 31.2). Similarly, brine shrimp, sold commercially as fish food or as pets under the more appealing name "sea monkeys," live in extremely salty water and can survive almost complete dehydration. If their water content drops below 40 percent, however, brine shrimp become dormant, and only at 46 percent to 60 percent water content do they metabolize at full capacity.

Brine shrimp

> ■ Organisms can survive only when the water content and temperature of their cells fall within a limited range of values. They function well within an even narrower range of internal conditions.

Some Basic Features of Homeostasis

In spite of the importance of maintaining conditions inside cells within a narrow range, many organisms cannot respond quickly or effectively to changes in their environment. In addition, because water and temperature regulation are closely linked, organisms must often make trade-offs between maintaining an ideal internal temperature and an ideal level of solutes in cell water.

Gains must equal losses

To maintain constant internal conditions, gains of water and heat must equal losses of water and heat. When gains do not equal losses, the water content and temperature of the body change. To understand how organisms regulate their internal conditions, we must there-

fore understand how water and heat enter and leave organisms. We will see below how organisms exchange heat with their environment. Here, we look at the major routes by which water and solutes move into and out of organisms (see Figure 31.7).

Single-celled organisms exchange water and solutes with their environment across their plasma membranes. As we saw in Chapter 24, water moves passively into and out of cells across the phospholipid bilayer and through channel proteins. The channel and carrier proteins that allow mineral nutrients to cross the plasma membrane give the organism considerable control over their movement. Single-celled eukaryotes have specialized organelles called contractile vacuoles that allow them to move water and minerals out of their bodies.

Amoeba showing a vacuole

Multicellular plants, animals, and fungi control the movement of water and solutes by limiting their exchange to specialized organs that make up only a small fraction of their surface area. Gas exchange structures such as lungs, gills, and intercellular air spaces represent important routes of water loss and, in aquatic habitats, water uptake. Plant roots and animal guts represent important water absorption structures. Animals lose both water and solutes in the urine and feces they produce.

Some organisms cannot regulate

Whereas some organisms maintain constant internal conditions in the face of changes in their environment, others allow their internal conditions to change as their surroundings change (Figure 31.3). The more effectively an organism can regulate its internal environment, the more easily it can keep its internal conditions near the ideal for its enzymes. For example, our ability to regulate our internal temperature allows humans to be equally active on warm days and on cool nights. Butterflies, in contrast, cannot regulate their body temperatures as closely, and must become inactive during cool nights.

Many organisms that cannot regulate their internal environment live in unvarying habitats. Naked mole-rats burrow in the soils of African grasslands, where temperatures remain constantly warm throughout the year (see Figure 31.3). They are unusual among mammals in having little control over their body temperature. For much the same reason, most ocean-dwelling plants and algae have little control over water entering or leaving their bodies. The solute concentration of ocean water changes little from day to day and resembles the concentration of solutes inside cells. As a result, these marine organisms would benefit little from the energet-

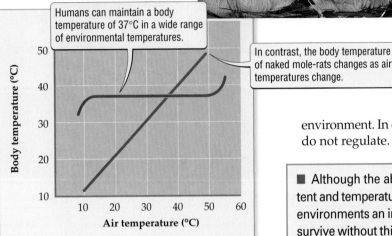

Humans can maintain a body temperature of 37°C in a wide range of environmental temperatures.

In contrast, the body temperature of naked mole-rats changes as air temperatures change.

Figure 31.3 Organisms Differ in How They Respond to Environmental Changes
Humans maintain a constant body temperature around 37°C, even though they inhabit environments that can have temperatures ranging from –70°C, at Verkhoyansk in Siberia, to 58°C, at Tripoli in northern Africa. Other animals, such as the naked mole-rat, have body temperatures that mirror those of their surroundings.

environment. In other words, tiny organisms typically do not regulate.

■ Although the ability to regulate internal water content and temperature gives organisms living in variable environments an important advantage, organisms can survive without this ability. Large organisms can maintain constant internal environments more easily than small organisms can.

ically costly mechanisms for regulating water content that make life possible in other habitats.

Conditions inside large organisms change more slowly than conditions inside small organisms

Heat and water enter and leave an organism across its surface. As we saw in Chapter 25, large organisms generally have less surface area relative to their volume than small organisms do. From this observation, we can draw three conclusions:

1. Small organisms can exchange water, solutes, or heat with their environment more easily than large organisms can.

2. Simply by virtue of their size, large organisms can maintain a constant internal temperature and water content more easily than small organisms can. The body temperature of a small lizard, for example, changes quickly as its environment heats up or cools down, but the temperature of a large alligator or crocodile remains remarkably constant over the course of a day.

3. Tiny organisms exchange heat, water, and solutes with their environment so quickly that their internal conditions cannot differ much from those of their

How Organisms Regulate Water Content and Solute Concentrations

31.1

The water inside living organisms is not pure, but exists as a rich chemical soup of mineral nutrients and organic compounds. As we will see, the materials dissolved in water affect how it works as a medium for life.

Water enters and leaves organisms passively

Because there are no known carrier proteins that can move water actively across plasma membranes, water moves into and out of organisms passively by diffusion (see Chapter 24). We saw in Chapters 29 and 30 that water evaporates readily from gas exchange surfaces such as lungs and the intercellular air spaces of leaves. For terrestrial organisms, this evaporation represents an important water loss.

Both terrestrial and aquatic organisms lose or gain water as it diffuses down a concentration gradient. Thus water diffuses across a membrane from the side where the solute concentration is low to the side where the solute

Figure 31.4   Water Moves Into and Out of Cells by Diffusion Differences in the concentration of dissolved materials between the water outside and inside cells determine how water moves into and out of cells.

Neither gains nor loses water	Loses water	Gains water
When solute concentrations outside the cell equal concentrations inside the cell, the cell neither gains nor loses water.	When solute concentrations outside the cell exceed those inside the cell, the cell shrinks as it loses water to its environment.	When solute concentrations outside the cell are lower than those inside the cell, the cell swells as it gains water from its environment.

Plant cells

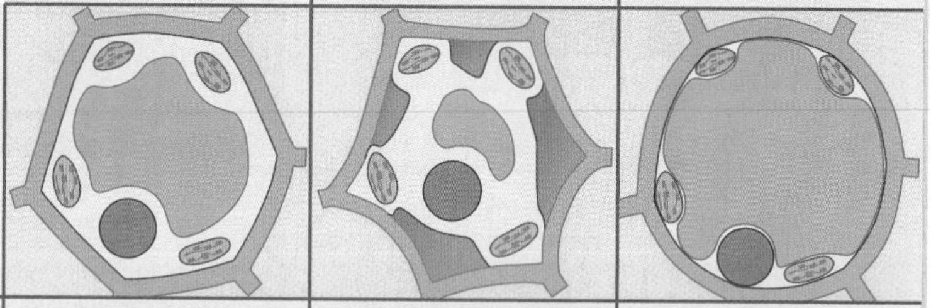

Animal cells

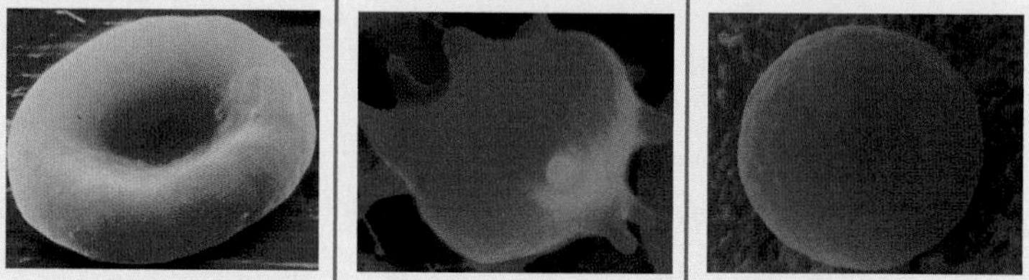

Normal red blood cell	Shrunken red blood cell	Bloated red blood cell

Examples	Most marine organisms	Terrestrial organisms and marine vertebrates	All freshwater organisms

concentration is high (see Chapter 24) (Figure 31.4). Whenever the solute concentration in a cell or an organism exceeds that in its environment, water diffuses in. Alternatively, whenever the solute concentration in the cell or organism is lower than that in its environment, water diffuses out. It is important to understand that water diffuses in response to the concentration gradient of all the dissolved materials combined, not in response to the concentrations of individual chemicals. Diffusion of water into or out of cells can cause them to shrink or expand.

Metabolism constantly changes the content of the water inside organisms

Metabolizing cells constantly use and produce the materials that are dissolved in cell water. Organisms must accommodate this constant change in the composition and concentration of the solution inside their cells. In most cases, organisms need to make only minor adjustments to maintain the overall solute concentration inside their cells.

Metabolic wastes, however, can pose greater problems than other solutes. The metabolism of proteins, for example, produces **ammonia**, the same chemical found in many household cleaners. If a high concentration of ammonia builds up in a cell, it can poison the cell. Ammonia diffuses easily out of single-celled organisms, and multicellular plants can use it as a source of nitrogen. Animals, however, cannot use ammonia, so it is dumped from the cells into the circulating blood. Ammonia in the blood continues to pose a potential danger to the animal, so it must be removed from the body. Some animals flush ammonia from their bodies, usually with the loss of much water; others use energy to convert it into less dangerous molecules. Aquatic animals can afford the water loss needed to simply flush ammonia from their bodies. Some terrestrial animals, including humans, that can get plenty of water to drink convert ammonia into a nontoxic compound called urea, which they eliminate in liq-

uid urine. Other terrestrial animals, such as birds and insects, conserve precious water by converting ammonia into white crystals of uric acid, which they then pass using relatively little water (Figure 31.5).

Most animals have specialized kidneys that regulate water content and solute concentrations

The **kidneys** found in many groups of animals regulate their bodies' supply of water and solutes, and help them dispose of metabolic wastes (Figure 31.6). Although the designs of kidneys differ greatly among species, they all function in basically the same way. They first filter body water and a diversity of small molecules, both useful and not, through a porous membrane under pressure, and then actively recover the useful molecules that have passed through the membrane.

Consider our own kidneys. In the human body, the pressure generated by the heart forces an amazing 180 liters of water a day out of the blood through the porous walls of capillaries in the kidneys. After this filtration, only large molecules such as proteins and carbohydrates remain in the blood. The water that passes into the tubules of the kidney carries valuable sugars, amino

acids, and minerals along with metabolic wastes. Active transport of solutes, however, lets us reabsorb almost all the water and valuable solutes back into the blood, leaving only excess solutes, wastes, and a small amount of water to leave the body as urine. Of the 180 liters of water that the human kidneys filter in a typical day, for example, only about 2 liters (or 1 percent) leaves the body as concentrated urine, containing mostly wastes and excess solutes. If we do not drink enough water, our kidneys may recover even more of the water to produce an even more concentrated urine.

■ Water moves passively into and out of cells in response to gradients of solute concentration. Metabolism constantly disrupts the water balance of water and solutes by producing and using chemicals dissolved in body water. Animals rely on specialized kidneys to regulate the amount and solute content of body water and to remove toxic by-products of metabolism.

Figure 31.5   Birds Change Toxic Ammonia Into Harmless Uric Acid
The white color associated with bird droppings, such as these deposits made by seabirds, comes from uric acid. Birds and other animals, including reptiles and insects, use energy to convert the toxic ammonia produced by protein metabolism into white crystals of uric acid.

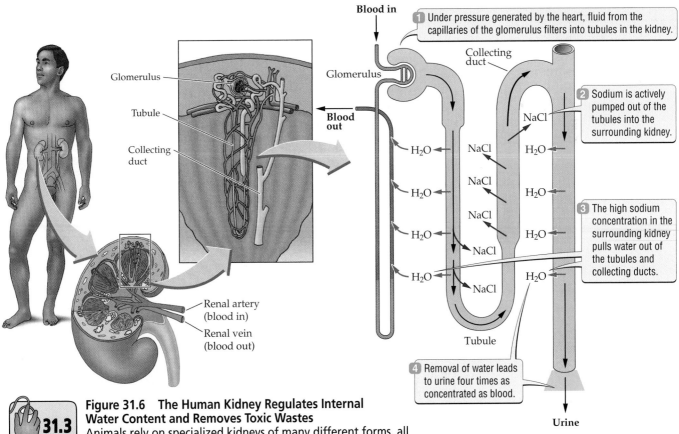

Figure 31.6   The Human Kidney Regulates Internal Water Content and Removes Toxic Wastes
Animals rely on specialized kidneys of many different forms, all of which, like the mammalian kidney, remove the toxic by-products of protein metabolism and assist in regulating internal water content.

The Environment Affects Regulation of Internal Water Content

The tendency of an organism to gain or lose water depends on its physical environment. Some of the problems organisms face in maintaining their water content in different environments, and some of their solutions to those problems, are summarized in Figure 31.7.

Marine organisms usually have internal solute concentrations that closely match those in the surrounding seawater. As a result, most marine organisms neither gain nor lose water. Notable exceptions to this pattern are marine fish, which, for reasons that remain unclear, have lower concentrations of solutes in their body fluids than are found in seawater. Marine fish tend to lose water to the ocean as it diffuses out of their bodies into

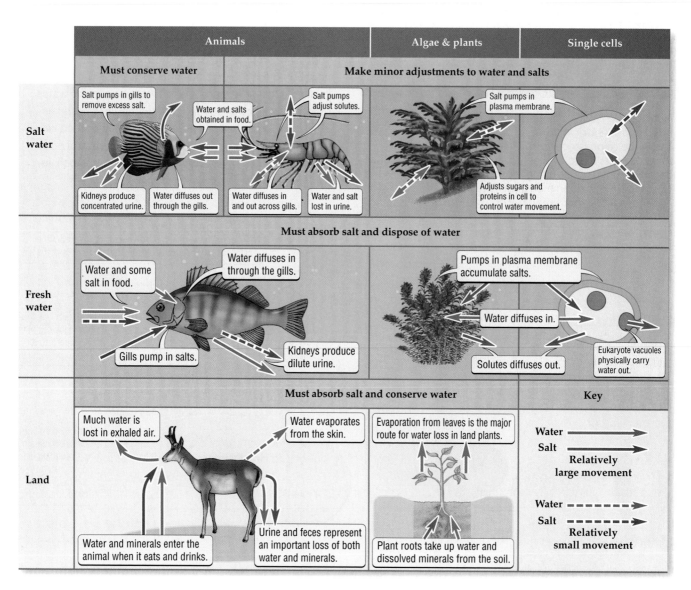

Figure 31.7   How Organisms Regulate Their Internal Water Content Depends on Their Environment
Single-celled organisms, animals, and plants meet the challenges of water content regulation posed by different environments in ways that reflect their basic form. Individual cells within multicellular organisms respond to changes in the surrounding body water in much the same way as single-celled organisms respond to their physical environment.

the more concentrated seawater. At the same time, they must get rid of excess solutes that diffuse in.

Freshwater organisms have a problem opposite to that of marine fish: They always contain much higher solute concentrations than the surrounding water. Freshwater organisms, therefore, tend to gain water from their surroundings as it moves into their bodies down a concentration gradient, and they must constantly replace lost solutes. Terrestrial plants place their absorptive roots, and animals place their absorptive guts, in contact with fresh water when it is available, and they lose a lot of water through evaporation. This means that terrestrial organisms must conserve scarce mineral solutes and somehow limit water losses.

The form of an organism determines the way in which it can respond to a tendency to gain or lose water or solutes (Figure 31.7). Single-celled organisms face potentially severe water regulation problems because their high surface area–to–volume ratio makes it difficult to keep the content of their internal water different from that of the environmental water. Whereas most marine microorganisms need only fine-tune their internal solute concentrations, freshwater microorganisms rely heavily on active carrier proteins in their cell membranes to pull scarce solutes from their environment, and, where possible, remove excess water from the cell.

Individual cells within the bodies of multicellular organisms regulate their internal water content in much the same way as single-celled organisms do. In addition, most multicellular organisms buffer the water inside their cells from the environment by surrounding them with carefully regulated body water. A waterproofing tissue covers most of the surface of multicellular organisms and restricts water and solute exchange to specialized structures (Figure 31.7). Animals take in water and solutes with their food, and lose them in their urine and solid wastes. All animals lose or gain water across their gas exchange surfaces (see Chapter 29). The gills that aquatic animals use to exchange gases, in addition, also play an important role in taking up and disposing of solutes. Land plants take up water and solutes through their roots (see Chapter 28), and unavoidably lose water from their leaves during gas exchange and in drawing sap upward through their xylem tissues (see Chapters 29 and 30).

> ■ The tendency of an organism to gain or lose water depends on its physical environment. Single-celled organisms, like the individual cells of multicellular organisms, regulate their water content by controlling internal solute concentrations. Multicellular organisms, in addition, regulate the solute concentration of the body water that surrounds their cells.

31.5 How Organisms Gain and Lose Heat

In addition to regulating their internal concentrations of water and solutes, many organisms must regulate their body temperature to survive and maintain high activity levels. A first step toward understanding temperature regulation is to understand that heat enters and leaves an organism in several different forms: conductive heat, radiant heat, evaporative heat, and metabolic heat (Figure 31.8). Because organisms in all but the hottest environments function best at internal temperatures higher than those of their environment, organisms that can manage heat losses and gains usually do so to increase their internal temperature.

The exchange of heat by conduction and radiation depends on environmental conditions

Organisms obtain heat from their environment in two ways: through the absorption of light, primarily from the sun, to obtain **radiant heat**, and through the **conduction** of heat by direct contact with warm objects. How effectively an organism gains radiant and conductive heat depends on the difference in temperature between the surface of the organism and the heat source. We warm up more quickly at the beach by lying on hot sand than by lying on warm sand because the larger temperature difference between the hot sand and our skin speeds the conduction of heat. When an organism has a surface temperature warmer than that of the environment, the conductive heat flow reverses to carry heat out of the organism. The surface area that an organism presents to the heat source also influences the rate of radiant or conductive heat exchange. When we warm ourselves in the radiant heat of the sun at the beach, we face the sun fully to place as much of our skin as possible in full sunlight. On the other hand, when we go to bed in a cold room, we curl into a ball that minimizes our surface area and keeps us from losing too much heat.

Although similar rules govern the exchange of radiant and conductive heat, organisms regulate the exchange of these two forms of heat differently. Although almost all objects in the environment, both living and not, give off some radiant heat, the sun emits much more radiant heat—mostly as visible light—than anything on Earth. During the day, radiant energy from the sun provides the major heat source for organisms and for the environments in which they live. At night, without the sun's light, the environment, and organisms that depend on environmental heat sources, slowly cool. Objects that have dark surfaces absorb sunlight better than objects

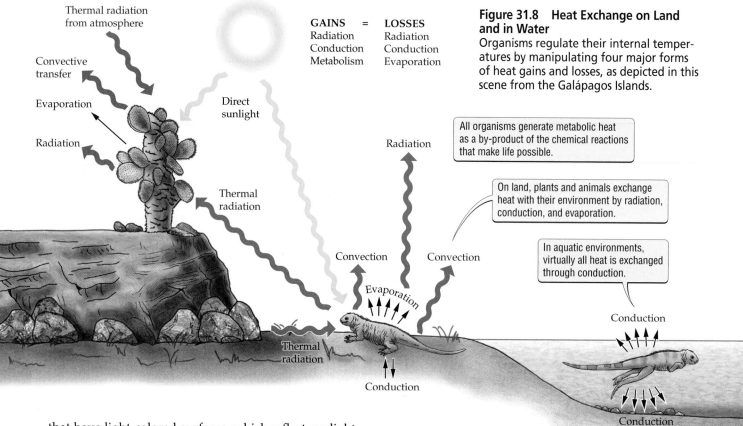

GAINS = LOSSES
Radiation   Radiation
Conduction   Conduction
Metabolism   Evaporation

Figure 31.8 Heat Exchange on Land and in Water
Organisms regulate their internal temperatures by manipulating four major forms of heat gains and losses, as depicted in this scene from the Galápagos Islands.

All organisms generate metabolic heat as a by-product of the chemical reactions that make life possible.

On land, plants and animals exchange heat with their environment by radiation, conduction, and evaporation.

In aquatic environments, virtually all heat is exchanged through conduction.

that have light-colored surfaces, which reflect sunlight (Figure 31.9). Whereas light-colored clothing keeps us comfortable on a sunny summer day, dark clothing keeps us warm on a sunny winter day.

Heat conduction also depends on the materials involved. Solids and liquids conduct heat much more quickly than air does. We can remain comfortable indefinitely in 20°C air, but we start to shiver with cold if we stay in a 20°C swimming pool for too long, because water conducts heat from the body 25 times as quickly as air. Biologically, this difference matters a great deal, because it means that aquatic organisms lose or gain conductive heat much more quickly than land animals do.

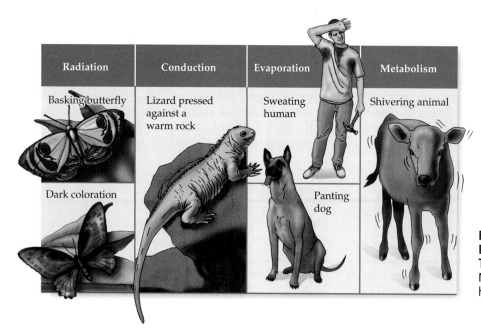

Figure 31.9 The Forms and Behaviors of Animals Influence Their Heat Gains and Losses
Many features of animals influence how they lose and gain heat.

As a result, whereas aquatic organisms can keep their body temperature different from that of their environment only with great effort, terrestrial organisms can do so much more easily.

Convection maintains a large temperature difference between the surface of an organism and its environment

As we have just seen, radiant and conductive heat exchange increase as the difference in temperature between the *surface* of an organism and a heat source increases. As an organism exchanges heat with its environment, the temperature difference decreases and the heat exchange proceeds more slowly. For example, hot water in a bathtub quickly feels less hot as your skin surface heats up and the water adjacent to your skin cools down. You can once again set up a large temperature difference by wiggling your legs to replace the water next to your skin with hotter water that has not yet lost heat to your body. This physical movement of heat that results from the movement of air or water is called **convection**.

Organisms can manipulate convection to speed or slow conductive and radiant heat exchange. Anything that prevents the physical movement of air or water near the surface of an organism reduces convection-aided heat exchange. For example, fur, feathers, and the synthetic fillings in sleeping bags trap a layer of still air next to the body, slowing heat loss to a cold environment. In contrast, organisms that lack feathers or fur can easily lose heat to, or gain heat from, their environment. Similarly, the circulation of blood can transfer heat to or from the skin convectively. When we begin to overheat, we flush bright red as the capillaries in our skin expand to carry heated blood from the body core to the skin surface.

Evaporation allows organisms to lose heat in hot terrestrial environments

When sweat evaporates from our skin, we feel cooler. Evaporating water changes from a liquid to a gas, and this transition from liquid water to water vapor requires a lot of energy. It takes roughly six times as much energy to change liquid water into vapor as it does to heat the same amount of water from near freezing to near boiling. When water evaporates from an organism, it takes most of this energy from body heat (see Figure 31.9).

Biologically, the most important feature of evaporation is that it provides a way of losing heat that does not depend on the temperature of the organism relative to its environment. An organism can lose conductive or radi-

ant heat only if the temperature of the environment is lower than the body temperature, but no such limitation exists for evaporative heat loss. In hot environments, such as deserts, where the environmental temperature often exceeds safe internal temperatures, heat loss through the evaporation of precious water often makes survival possible. On the other hand, the unavoidable evaporation of water during gas exchange often represents an important loss of heat in cold terrestrial environments.

Metabolic heat contributes to temperature regulation in a few groups of organisms

Metabolic heat is an inevitable by-product of the chemical reactions essential to life. The amount of metabolic heat most organisms produce represents only a tiny fraction of their total heat gain. In most plants, for example, metabolic heat production amounts to less than 1 percent of the total heat gain. Only a few groups of organisms generate enough metabolic heat for it to contribute in an important way to their heat gain. These organisms, known as **endotherms** (*endo*, "inside"; *therm*, "heat"), can maintain a high body temperature even in cold weather, when conductive heat sources are unavailable, and at night, when the radiant heat of the sun disappears. Birds and mammals are endotherms, and some scientists believe that the ancestors of birds—the dinosaurs—were endotherms as well (see the box on p. 529). Maintaining a constant high body temperature regardless of temperatures in the environment is rarely possible for **ectotherms** (*ecto*, "outside"), which depend on environmental heat sources to maintain their body temperature.

As endotherms, mammals and birds can generate enough metabolic heat to remain active despite varying temperatures in their environment. This advantage comes at a large cost, however: Both at rest and during activity, the metabolic rate of an endotherm is about 10 times that of an ectotherm of the same size. As a result, endothermic animals need to eat much more food than ectotherms do. To put this difference into perspective, if your house was guarded by a 50-kilogram endothermic dog, you would have to feed it roughly 500 kilograms of dog chow each year, whereas if you relied on a 50-kilogram ectothermic alligator for the same service, you would have to provide a mere 50 kg of alligator chow (Figure 31.10).

> ■ Organisms gain or lose heat through radiant heat exchange, conductive exchange of heat with the environment, evaporative heat loss, and metabolic heat production.

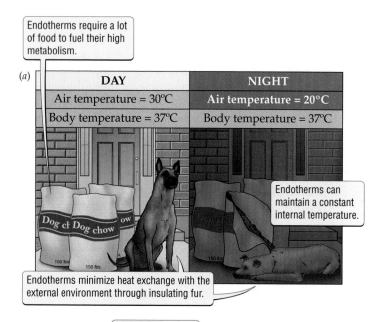

(a)

Endotherms require a lot of food to fuel their high metabolism.

DAY	**NIGHT**
Air temperature = 30°C	Air temperature = 20°C
Body temperature = 37°C	Body temperature = 37°C

Endotherms can maintain a constant internal temperature.

Endotherms minimize heat exchange with the external environment through insulating fur.

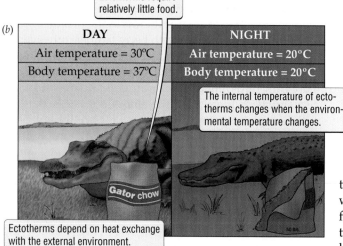

(b)

Ectotherms require relatively little food.

DAY	**NIGHT**
Air temperature = 30°C	Air temperature = 20°C
Body temperature = 37°C	Body temperature = 20°C

The internal temperature of ectotherms changes when the environmental temperature changes.

Ectotherms depend on heat exchange with the external environment.

Figure 31.10 Heat Exchange in Endotherms Versus Ectotherms
(a) Endothermic dogs gain independence from environmental heat sources by using a lot of energy to generate metabolic heat. (b) Ectothermic alligators use much less energy to regulate their internal temperature by relying almost completely on environmental heat sources.

How Organisms Regulate Body Temperature: Two Examples

Because conduction carries heat so rapidly to and from organisms that live in water, we find most organisms that can regulate their body temperature independently of their environment on land. The ability of terrestri-al organisms to use radiant, conductive, evaporative, and metabolic heat to regulate their internal temperature depends both on the characteristics of the organism and on the nature of its external environment. We will use two case studies to illustrate how organisms manipulate heat exchange.

Ectothermic marine iguanas use heat gained on land to sustain them while feeding under water

Marine iguanas, which feed in the cold ocean waters surrounding the Galápagos Islands (see Figure 31.8), illustrate how an ectotherm can take advantage of environmental heat sources to maintain a relatively constant internal temperature. Marine iguanas function best at body temperatures of about 37°C. Because they quickly lose heat by conduction to the relatively cool (20°C) ocean waters in which they feed, they need to warm up on land. Their black color and their habit of positioning themselves to present as much surface area to the sun's rays as possible (a behavior called basking) maximize their radiant heat gain. In addition, they gain heat conductively by pressing their bellies against the islands' black volcanic rock, which the sun can heat to temperatures of over 40°C. While basking, iguanas increase their heart rate and the flow of blood through the skin to help carry heat from the skin to the tissues inside the body.

After about an hour of basking, when their internal temperature has reached 37°C, the iguanas dive into the water. In the water, they slow their heart rate and the flow of blood to their skin. This strategy reduces heat transfer from the body core to the skin surface. Nonetheless, heat is conducted rapidly from the body into the cool water, allowing iguanas only about 30 minutes to feed before their falling body temperatures force them to return to land.

Endothermic chickadees must reduce heat loss to survive cold winters

Winter represents an especially stressful period in the lives of endotherms. Consider the chickadees that commonly visit North American bird feeders in winter. These small birds must survive winter temperatures of –40°C (almost 80°C lower than their core body temperature) at a time of year when food is scarce. The small size of chickadees means that they can lose heat over a large surface area relative to their body volume.

Chickadee

■ THE SCIENTIFIC PROCESS

Did Dinosaurs Blow Hot or Cold?

The early view of dinosaurs as large versions of modern-day reptiles—awkward, splay-legged, slow-moving, and ectothermic—has slowly evolved into a very different picture. Fossil bones and footprints clearly show that dinosaurs walked with their legs under their bodies, like birds and mammals, rather than like modern reptiles (see Figure 23.1). Careful interpretation of the distributions of fossil bones and nests suggests that many dinosaurs, like many modern mammals, lived in complex social groups. Biologists have even begun to think that, like birds and mammals, dinosaurs may have had carefully regulated body temperatures warmer than those of their environment.

Metabolically heated dinosaurs could have maintained body temperatures higher than those in their environment even without good sources of radiant or conductive heat. In addi-tion, the high metabolic rate associated with metabolic heating seems to go hand in hand with a greater ability to sustain vigorous activity. The advantages of metabolically heated dinosaurs would come at a massive energetic expense, however. As mentioned in this chapter, maintaining a constantly high body temperature, as mammals and birds do, requires 10 times more energy than does reliance on environmental heat sources. In cold environments, the cost could rise even higher. Interestingly, fossil evidence suggests that the amount of prey available per predatory dinosaur more closely resembles the pattern we see today for endothermic predatory birds and mammals than it does for ectothermic lizards.

How might being metabolically, rather than environmentally, heated have affected dinosaurs' appearance? Metabolic heating takes so much energy that endothermic dinosaurs would have had to conserve heat. Metabolically heated mammals and birds have fur or feathers. Ectothermic modern lizards and the artists' reconstructions of dinosaurs that we see in museums, on the other hand, have naked, scaly hides that encourage heat exchange with the air. Recent fossils suggest that some small dinosaurs had feathers even though they could not fly. At the same time, other fossils indicate that dinosaurs lacked the specialized nose bones that reduce evaporative heat loss in modern birds and mammals (see Figure 29.10). Thus, the question of endothermic dinosaurs is still far from being resolved, but as we develop a clearer picture of dinosaur biology, we may have to redo the murals that back many museum displays.

Dinosaurs May Have Been Endotherms

How do chickadees survive the bitter winter cold? They develop a greater feather mass in winter, which allows them to trap a thicker layer of still air next to their bodies. As temperatures drop, chickadees fluff out their feathers to further increase the thickness of the still layer of air, reducing convective heat loss. When not feeding greedily, chickadees nestle among the branches of conifers, which protect them from the wind. Chickadees also decrease their surface area by tucking their unfeathered legs against the body and huddling into as round a shape as possible. Tucking the bill under the wing reduces the evaporative heat (and water) lost with each breath. When such actions are not enough, chickadees begin to shiver by contracting their muscles rapidly. These muscle contractions do no useful work, but do generate metabolic heat.

Chickadees cannot sustain this energetically expensive shivering for the whole of a long winter night, however. When their energy reserves begin to run out, chickadees may have to lower their body temperature by up to 12°C, which saves about 20 percent over the metabolic cost of maintaining a normal body temperature. This strategy may make the difference between surviving to morning or dying in their sleep.

> ■ Ectotherms manipulate the exchange of radiant and conductive heat with the environment to keep their body temperatures as constant as possible. Endotherms reduce the amount of energy spent on generating metabolic heat by minimizing the exchange of heat with a relatively cold environment.

■ HIGHLIGHT

The Saguaro, Master of Extremes

The saguaro cactus must survive the brutal temperature fluctuations characteristic of its Sonoran Desert home. In addition, the saguaro may have to wait for months between rains, but when the rains arrive, they often come as torrential downpours. Unlike animals, which can seek shelter from extreme temperatures or move to find water, saguaros, like other desert plants, must deal with the full harshness of their desert environment. How does the saguaro juggle the conflicting demands of water and temperature regulation during dry spells that can last months or years?

Saguaros reduce evaporative water loss by not having leaves and by exchanging gases only at night. Like those of all cacti, the photosynthetic cells of saguaros lie in the stem, beneath a thick, waterproof epidermis. In addition, cacti belong to a select group of plants that use CAM photosynthesis (see Chapter 29), which allows them to carry out photosynthesis without simultaneous gas exchange. They open their stomata to collect and store carbon dioxide at night, when low temperatures and high humidity reduce evaporative water loss. During the hot, dry day,

they can use the stored carbon dioxide to photosynthesize without having to open their stomata.

Saguaros have two features that let them store tremendous amounts of water during the brief periods of heavy rain. Their long roots spread out near the soil surface, so that when the rain does finally pour down, they can absorb it rapidly. The vertical pleats that run the length of the trunk allow the cactus to expand as its cells accumulate water, greatly increasing the saguaro's capacity for water storage.

How do saguaro cacti keep from burning up during the day without using evaporative cooling? The pleats on the trunk that aid in water storage also allow sunlight to hit the saguaro's trunk at an angle, which spreads the radiant heat over a larger surface area (see Chapter 39). Moreover, the tons of water stored in the trunk of a large saguaro ensure that its internal temperature increases relatively little, even a scorchingly hot day. The large amount of radiant energy needed to heat a bathtub full of water raises the temperature of a 1000-kilogram saguaro by only 3°C. When the desert cools at night, the cactus can easily lose the heat stored by day.

> ■ CAM photosynthesis and a lack of leaves allow saguaro cacti to reduce water loss in their desert environment. An extensive root system and a specialized trunk allow them to take up and store water during brief rainy periods. A pleated trunk surface and a large body slow the rate of radiant heat gain during the day.

SUMMARY

When Homeostasis Fails: Heat Stroke in Humans

- Heat stroke occurs when the failure of the sweating mechanism that normally cools the body leads to an increase in body temperature to more than 4°C over normal.

Cells Function Best under Predictable Internal Conditions

- Life depends on liquid water, which typically makes up over half of an organism's weight.

- Most cells cannot survive temperatures lower than freezing (0°C), which can destroy cell membranes, or temperatures higher than 40°C, which cause proteins to lose their three-dimensional shape.

- Most organisms function best under an even more narrowly defined range of water content and temperature.

Some Basic Features of Homeostasis

- Gains must equal losses to maintain constant internal conditions.

- Not all organisms can regulate their internal water content or temperature. Although the ability to regulate their internal environment gives organisms more control over their performance, organisms that live in constant, favorable environments may gain no advantage from having this ability.

- Large organisms can maintain internal conditions different from those of their environment more easily than small organisms can.

How Organisms Regulate Water Content and Solute Concentrations

- Water diffuses across membranes in response to the relative concentration of solutes inside and outside an organism.

- Metabolism constantly adds solutes to and removes solutes from body water. In addition, metabolism may produce toxic by-products.

- Animals rely on organs called kidneys to regulate the solute content of their body water and to remove ammonia, the toxic by-product of protein metabolism, from the body.

The Environment Affects Regulation of Internal Water Content

- With the notable exception of fish, most marine organisms maintain internal solute concentrations that match those of the ocean around them, so that water tends neither to move into nor out of their bodies.

- Marine fish must replace water, which they tend to lose to the surrounding seawater, while disposing of excess solutes that diffuse into their bodies.

- Freshwater organisms must remove water that tends to diffuse into their bodies while conserving relatively scarce solutes.

- Terrestrial organisms must conserve both water and solutes.

How Organisms Gain and Lose Heat

- Radiant heat from the sun and the conduction of heat by direct contact with warmed objects in the environment represent the two most important environmental heat sources for most organisms.

- Convective movement of air or water can speed radiant or conductive heat exchange by maintaining a large temperature difference between the surface of an organism and the environment.

- Organisms can lose heat evaporatively even when the temperature of the environment is greater than their internal temperature.

- Heat produced as a by-product of metabolism is important to endotherms, including birds and mammals, which spend a great deal of energy to maintain a constant high body temperature.

How Organisms Regulate Body Temperature: Two Examples

- Aquatic organisms rarely maintain an internal temperature that differs from that of the surrounding water.

- Ectothermic organisms manipulate radiant and conductive heat sources to maintain optimal body temperatures.

- Endothermic organisms reduce heat losses to a relatively cool environment to conserve energetically expensive metabolic heat.

Highlight: The Saguaro, Master of Extremes

- Saguaro cacti must maintain their internal water content and temperature while living in an exceptionally hot, dry desert environment.

- CAM photosynthesis and a lack of leaves reduce evaporative water loss, while a large root system and water storage tissues in the stem allow saguaros to store water from rare, but torrential, rainfalls.

- The pleated surface and the large mass of the saguaro trunk help prevent damaging temperature increases in full sunlight.

KEY TERMS

ammonia p. 522	homeostasis p. 518
conduction p. 525	kidney p. 522
convection p. 527	metabolic heat p. 527
ectotherm p. 527	radiant heat p. 525
endotherm p. 527	solutes p. 519

CHAPTER REVIEW

Self-Quiz

1. Which of the following typically happens when water inside cells freezes?
   a. The cells function at full capacity.
   b. Animal cells begin to produce ammonia.
   c. The plasma membranes are ripped apart.
   d. Saguaro cacti shed their leaves.

2. Plants lose water to their environment
   a. during gas exchange.
   b. during transport of xylem sap.
   c. from the surfaces inside their leaves.
   d. all of the above

3. Marine fish tend to
   a. gain water through their gills.
   b. maintain a solute concentration equal to that of seawater.
   c. lose water through their gills.
   d. take up excess water.

4. The primary heat source for endothermic organisms is
   a. conductive heat.
   b. radiant heat.
   c. evaporative heat.
   d. metabolic heat.

5. Fur and feathers
   a. reduce heat loss in endothermic animals.
   b. reduce convective heat transfer.
   c. reduce heat loss by trapping air, which conducts heat poorly.
   d. all of the above

Review Questions

1. How does an organism's metabolism affect its heat gains and losses? Try to think not only of the direct effects of metabolism, but also of the indirect ones.

2. Under what circumstances do organisms tend not to regulate their internal water content or temperature?

3. Humans are endotherms. What advantages do we have over ectotherms? What disadvantages do we face?

---

31   **The Daily Globe**

Alcohol and Sports Do Not Mix

To the Editor:

I am writing to cancel my subscription to *The Daily Globe*. In last week's Sunday magazine, articles criticizing the poor examples that athletes set when they are arrested for using illegal drugs were comfortably sandwiched between pages showing healthy, active people enjoying alcohol as they enjoy sports. The subliminal message that you were sending is clear: Alcohol consumption goes hand in hand with sports.

Your readers should be informed that alcohol reduces coordination to a dangerous degree in many sports. But there are other problems with combining alcohol and exercise. You should be writing articles explaining these more subtle interactions to your readers rather than displaying advertisements that mislead them about the relationship between sports and alcohol.

One common and dangerous problem people face during exercise, especially in hot climates or at high elevations, is dehydration. Although drinking a cold beer may seem to quench thirst, it actually robs the body of precious water. All forms of alcohol interfere with a hormone that signals the kidneys to reabsorb water, and as a result excess water is lost in the urine. This water loss can still be felt hours later as the brain becomes dehydrated and the drinker feels hung over.

In cold-weather sports, alcohol is risky for a different reason. Although shots of schnapps or brandy can give us a feeling of warmth on a cold winter day, they actually cause us to lose body heat at a faster rate because the alcohol causes the blood vessels near the skin to expand, carrying vital heat away from the core of the body. Thus, alcohol consumption can dangerously increase the rate of heat loss in cold weather.

I encourage you to rethink your decision to accept money from advertisers that promote the dangerous combination of alcohol and sports. As long as you're taking their money, you won't be getting any more of mine.

John Singleton
Chicago, IL

Evaluating "The News"

1. Can you think of any examples of advertisements in which alcohol is depicted prominently in scenes involving exercise? What effect do you think such ads have on people?

2. Can you think of any other products that are advertised in a way that might distort the underlying biology involved?

3. If you were the owner of a magazine and you felt responsible for making a profit so that you could pay your employees, as well as providing your readers with accurate information, what would your policy be concerning the scientific accuracy of advertisements?

chapter *32* ***Defense***

AIDS in Africa

*A*IDS (acquired immunodeficiency syndrome) is caused by the human immunodeficiency virus, more commonly known as HIV. AIDS first came to the world's attention two decades ago, when doctors in the United States began to notice that gay men were falling victim to several previously rare diseases, including a skin cancer called Kaposi's sarcoma and an unusual kind of pneumonia, or lung infection. These diseases, it became clear, attacked AIDS victims when their immune systems, which provide the main defense against disease, failed. AIDS invariably killed its victims.

The number of cases of this frightening new disease rapidly increased in North America and Europe. It claimed tens of thousands of lives each year, mostly among gay men, intravenous drug users, and people requiring frequent blood transfusions. The common bond connecting these three groups was regular contact with the body fluids of others: gay men during sex, drug users by exchanging dirty needles, and those needing blood transfusions by receiving contaminated blood.

As sobering as the view of AIDS in North America and Europe may be, a global view proves even more fright-

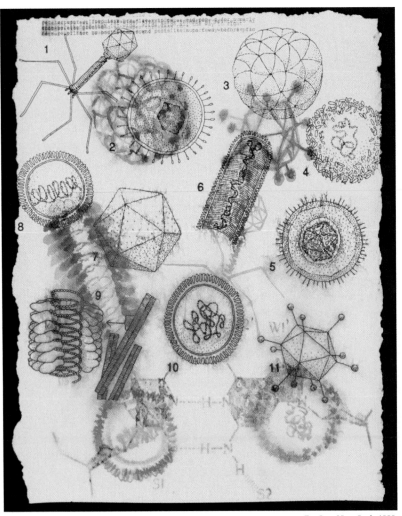

Eva Lee, *Virus Study*, 1999.

Defense against parasites, carni-vores, and herbivores is critical to the well-being of organisms.

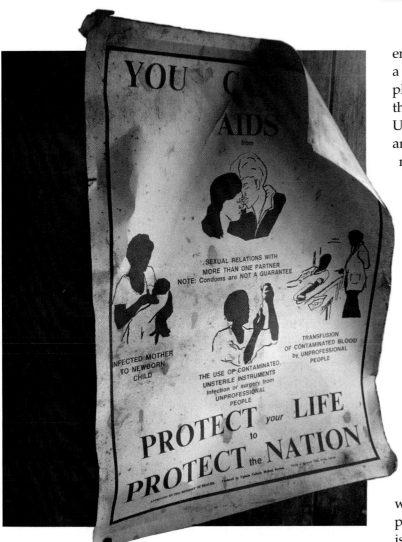

AIDS in Africa Affects Millions of Men, Women, and Children

ening. In Africa, the spread of AIDS has followed a very different pattern. Well over 10 million people in Africa carry HIV. HIV infects one-tenth of the population of the central African country of Uganda. The victims come from all walks of life and all classes of society, and they include both males and females and both adults and children.

The spread of AIDS in Africa probably stems from two main causes. First, insufficient funds force hospitals to reuse inadequately sterilized needles and blood transfusion equipment. Second, the social upheaval that followed the withdrawal of the European colonial powers from Africa during the 1960s created a dangerous mix of conditions. The frequent use of prostitutes by men in labor camps promoted the spread of sexually transmitted diseases, which often led to open sores that allowed for contact between the blood of partners during sex. Regular displacements of individuals and whole ethnic groups also sped the spread of disease. When AIDS established itself in Africa, it spread quickly and widely. The eventual toll of AIDS in Africa is likely to be staggering.

What makes AIDS so deadly? In this chapter, we review the various ways in which organisms protect themselves against attack by other organisms. We then return to the subject of AIDS to consider how HIV gets past our defenses against disease.

■ KEY CONCEPTS

1. Parasites attack a single host from within. Carnivores and herbivores attack multiple victims from the outside.

2. The outer surface of an organism forms the first line of defense against many enemies.

3. All organisms can distinguish their own cells from those of other organisms, a key requirement in fending off invading parasites.

4. Most organisms use chemicals to destroy invaders. Animals, in addition, use mobile defensive cells that consume the invading organisms.

5. The immune systems of vertebrates can distinguish both between host cells and invading parasites (in a nonspecific immune response) and among different kinds of invading parasites (in a specific immune response that depends on unique cells called lymphocytes).

6. Organisms defend themselves against carnivores and herbivores by avoiding discovery, by using defensive behaviors, or by making themselves unpleasant to eat or toxic through physical or chemical means.

We often downplay the importance of parasites and carnivores. Thanks to modern medicine and technology, humans live largely sheltered from the ravages of disease and in an environment stripped of carnivores that can kill us. As a result, we often live into our 70s and beyond. During the Roman Empire, however, life expectancy ranged from 22 to 47 years, and in medieval England only one in five individuals lived to see their twentieth birthday. Infectious diseases caused most of these early deaths. Death records from London for a single week in 1661 show that whereas 39 people died of old age, at least 205 died of disease (Figure 32.1). The recent emergence of new diseases such as AIDS and the evolution of antibiotic-resistant forms (see Chapter 20) of old killers such as tuberculosis (called "consumption" in Figure 32.1) reminds us that even humans must defend themselves against other organisms.

For most organisms, carnivores or herbivores, as well as diseases and parasites, are major causes of death. In this chapter we consider how organisms respond to attacks from other organisms of all kinds. We first consider defenses against parasites that invade organisms from within. We then turn to defenses against herbivores and carnivores that attack organisms from outside.

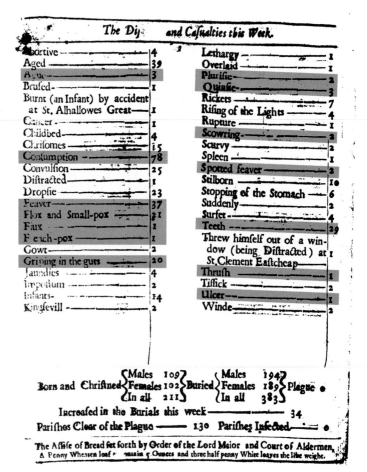

Figure 32.1   Bring Out Your Dead!
The major cause of death in seventeenth-century London was disease, not old age. This record of births and causes of death made during a single week in 1661 shows that slightly more than 50 percent of the people who died succumbed to disease (shaded in orange). In addition, many of the deaths of infants and the aged may well have resulted from disease. (Note that the letter that looks like ſ, when in the middle of a word, reads as the modern letter *s*.)

The Basics of Defense

Effective defense depends on avoiding or reducing damage done to a victim organism by an enemy organism. Victims defend against parasites, herbivores, and carnivores in ways that depend on the biologies of these different types of enemies.

Parasites differ from carnivores and herbivores

Organisms must defend themselves against parasites on the one hand and carnivores or herbivores on the other. **Parasites**, which include disease-causing organisms and viruses, feed and reproduce inside their victims, usually called **hosts**. Therefore, they are generally much smaller than their hosts (Figure 32.2). An individual parasite often spends its entire life associated with a single host, which may remain infected for days or months. Individual herbivores and carnivores, in contrast, may feed on many victims over their lifetime and typically attack them from

the outside. **Carnivores** are almost always larger than their animal prey and generally kill them in a matter of minutes. **Herbivores** are often smaller than the plants they eat, and, if so, they may damage a single food plant over days or weeks without ever killing it.

An overview of defensive options

We can break the attack on a victim by an enemy into three steps that apply equally to parasites and to carnivores and herbivores (Figure 32.3). First, enemies must find a victim. Although many carnivores and herbivores are mobile and can actively seek out their victims, most parasites depend on wind, water currents, or animals to carry them to their victims. Second, enemies must break through the outer surface of the victim's body to attack it. Therefore, anything that a victim can do to make this surface harder to penetrate provides protection. Once the enemy breaks through the outer surface, it can harm its victim.

The survival of a victim once the enemy penetrates its outer surface depends on how quickly it responds to the attack. Permanent defenses, which are always present, can act on an enemy almost instantaneously. Maintaining a permanent defense exacts a high cost, however, because the organism must devote resources to the defense even when it is not under attack. On-demand

Figure 32.2 Organisms Face Enemies with a Diversity of Biologies
A parasite, such as the protist (pink with blue areas inside) that causes malaria (*a*) or a tapeworm (*b*), can spend a lifetime spanning months or even years living inside its host. In contrast, carnivores such as lions (*c*) and herbivores such as this caterpillar (*d*) attack their food from the outside.

Figure 32.3 Defensive Responses to an Enemy Attack
Attacks on a victim by an enemy, and the victim's responses, can be broken into three steps that apply equally to parasites and to carnivores and herbivores.

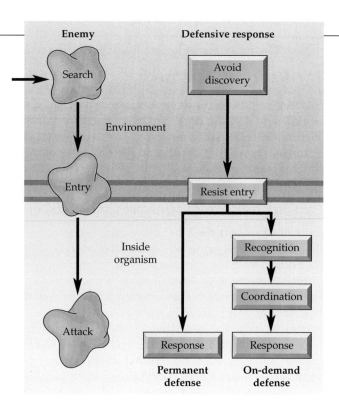

defenses use few resources except when needed, and therefore are less costly. They work well, however, only when an organism can detect an attack and use this information quickly to activate the defense. Thus, on-demand defenses may provide better defense against slowly acting parasites and herbivores than against the rapid attacks of carnivores.

■ Whereas parasites usually live inside a single host much larger than themselves, carnivores and herbivores attack one or more individuals from the outside. Victims can reduce their vulnerability by making themselves difficult to find and by thwarting attempts to penetrate their outer surface. Once its outer surface has been penetrated, the victim must limit damage by responding to the attack as quickly as possible.

Keeping Parasites Out

Although many of the interactions between parasites and their hosts take place inside the host, several characteristics of the host can reduce the chances that parasites will ever get in.

Dense host populations encourage parasites

During medieval plagues, people often fled disease-ridden cities for the relative safety of the countryside. The high density of people in cities greatly increases the chances that a parasite will spread from one person to another. Most parasites cannot survive long outside their hosts. Thus, the more potential victims that are crowded closely together, as in cities, the greater the chance that a parasite can cross the hostile environment separating two hosts quickly enough to survive the journey. For example, the body lice that carry the bacterium that causes typhus—a disease that caused millions of people to literally rot to death during World War I—can move only between hosts that come into close contact with each other. Cholera, which devastated nineteenth-century Europe, spreads in contaminated drinking water and food, which tends

The human louse that transmits typhus

to occur where humans live in dense concentrations. By living at lower densities, organisms can greatly reduce the ability of parasites to spread.

The body wall keeps parasites out

Before parasites can cause any harm, they must enter the host. The cell walls of bacteria protect them against attack by viruses and other bacteria. The dermal tissue system of plants defends them from attack by viruses, bacteria, and fungi. The human body wall—made up of the epithelial tissues of the skin, lungs, and gut wall—keeps out potentially dangerous microorganisms.

Many important parasites of animals and plants rely on insects to penetrate the body wall of the host. Mosquitoes, for example, help the organisms that cause malaria get into humans by biting through the skin to take a blood meal. Similarly, beetles spread Dutch elm disease, which threatens to drive North American elm trees to extinction. The beetles chew through the protective bark, introducing the fungus that causes the disease into the tree's vascular tissues.

Elm tree

Wounds allow parasites to enter an organism

Many parasites take advantage of wounds to enter their hosts. For example, the bacterium *Clostridium tetani*,

(a)

(b)

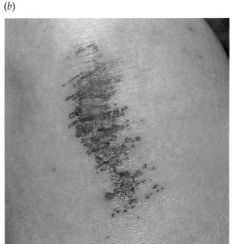

Figure 32.4 Closing Wounds Prevents Infection
(*a*) The resins produced by pines help the trees quickly seal wounds that might otherwise allow disease organisms to enter. (*b*) The scabs that seal wounds to human skin function in much the same way to keep out disease organisms.

which causes the disease tetanus, occurs commonly in soil and on other surfaces worldwide. Although we constantly come in contact with *C. tetani*, it becomes a threat only if it enters a poorly oxygenated break in the skin, such as a deep puncture wound. Away from oxygen, *C. tetani* can multiply and produce a toxic protein that causes severe and often fatal muscle spasms. In countries with poor sanitary conditions, *C. tetani* kills more than half a million babies annually by entering their bodies through the cut made at birth to sever the umbilical cord.

To reduce the risk posed by open wounds, plants and animals quickly seal the damaged area (Figure 32.4). Plants produce substances called resins and gums that help seal the wound quickly, thereby preventing entry by disease organisms. Animals with a circulatory system usually have specialized cells in the blood that quickly seal severed blood vessels. Insects seal wounds by gluing blood cells together, and mammals form **blood clots** consisting of sticky blood cells caught in a mesh formed from proteins that normally float freely in the blood. The growth of new tissues then repairs the wound more permanently.

> ■ Parasites spread between hosts more slowly in populations of widely scattered individuals than in dense populations. The body wall of a host helps keep parasites out. Most multicellular organisms can quickly seal breaks in the body wall to prevent the entry of parasites.

 The Vertebrate Immune System: Consequences of a Cut

A cut in our skin opens a route that allows disease-causing parasites to invade our bodies. Unless we can kill or isolate those parasites, they may multiply rapidly, stealing our resources and possibly, as *C. tetani* does, killing us in the process. In this section we introduce the **immune system**, a sophisticated defense against parasites found only in vertebrate animals. Although it is limited to vertebrates, the immune system illustrates many mechanisms that are basic to the internal defense systems of all organisms. For this reason, and because of its relevance to humans, we consider the vertebrate immune system as an introduction to the ways in which organisms defend themselves against parasites.

The immune system defends vertebrates against parasites

In our war against parasites, humans, like all vertebrates, rely on an extremely effective immune system (Figure 32.5*a*). The human immune system consists of an assortment of defensive proteins and trillions of **white blood cells** of several kinds, all of which help destroy microscopic invaders. The white blood cells and defensive proteins can move out of the vessels of the circulatory system into the **lymph**, the body fluid that surrounds our cells. From there the white blood cells collect in a second network of vessels, called the **lymphatic system**. The ducts of the lymphatic system carry the white blood cells back to the blood vessels. Along the lymphatic ducts lie pockets called lymph nodes containing huge numbers of white blood cells, which destroy bacteria and viruses roaming the body.

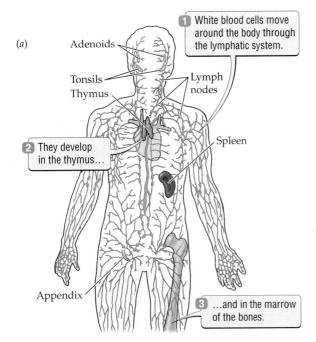

(a)

Adenoids

Tonsils
Thymus

Lymph nodes

1 White blood cells move around the body through the lymphatic system.

2 They develop in the thymus…

Spleen

Appendix

3 …and in the marrow of the bones.

(b)

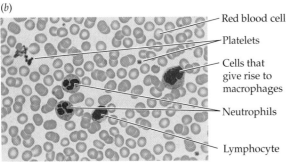

Red blood cell

Platelets

Cells that give rise to macrophages

Neutrophils

Lymphocyte

32.2

Figure 32.5
The Vertebrate Immune System
(a) The various white blood cells in the human circulatory system form in the marrow of the bones and in the thymus. They move through a specialized system of ducts called the lymphatic system (shown in green), and they accumulate in various organs that defend various points at which parasites can enter the body. The labeled structures in this figure house large concentrations of white blood cells and are often associated with places at which disease organisms could easily enter the body. (b) The blood contains several different kinds of specialized cells that enable the body to combat parasites.

We present the events that follow a deep cut under two different scenarios: (1) a cut that allows the tetanus-causing bacterium, *Clostridium tetani*, to enter the victim's body for the first time, and (2) a cut that leads to a second exposure to *C. tetani*, either because the victim survived an earlier infection or because they received a tetanus vaccination.

Scenario 1:
The body defends against a new invader

The first sign that the body has joined battle against an invader is **inflammation**, a characteristic swelling and reddening of the area around the cut. A sequence of defensive responses follows this initial reaction:

1. The tissues damaged by the cut release chemicals that stimulate the body's blood clotting substances. Invading *C. tetani* cannot easily penetrate the clot, which slows its spread by isolating the infected tissue.

2. A series of blood proteins known as **complement** (so called because they complement, or work together with, white blood cells) circulates to the wound, where they bind to the plasma membranes of cells such as *C. tetani*, marking them as invaders. Without distinguishing its own cells from foreign cells, no organism can defend itself effectively against invaders. The complement that marks invading cells allows white blood cells to distinguish them from the body's own cells. In addition to marking it, complement kills *C. tetani* by punching holes in its plasma membrane.

3. The proteins involved in blood clotting activate certain white blood cells that are stationed permanently just below the skin surface. Activation changes these cells into large **macrophage** (*macro*, "big"; *phage*, "eater") cells (Figure 32.5b), which start to prowl the wounded tissue much as predatory protists might. These macrophages bind to, then surround and digest, *C. tetani* or damaged host cells, a process called **phagocytosis**. The complement that marks the surfaces of parasites helps the macrophages hold onto the invaders that they devour. The macrophages also release chemical signals that result in a rise in the body temperature (see the box on p. 541).

4. Complement also attracts **neutrophils**, a second kind of phagocytic white blood cell, to the wound (Figure 32.5b). Each small neutrophil destroys fewer parasites than a large macrophage, and neutrophils cannot surround parasites bigger than bacteria, but they rapidly accumulate in huge numbers.

5. Over the next several days, more macrophages make their way to the wound from a population circulating in the blood. Although macrophages accumulate more slowly than neutrophils, they eventually become the dominant white blood cell type at the wound. Together the macrophages and neutrophils constitute a **nonspecific response** to parasites in that they indiscriminately attack anything other than healthy host cells.

■ BIOLOGY IN OUR LIVES

I've Got Fever

Grasshoppers Get Fevers, Too.

When we fell sick on a school day as children, our parents took our temperature to make sure we weren't faking it. We accept that a fever goes hand in hand with sickness. Medical research has begun to show, however, that fever is much more than a reliable symptom of an infection: Fever represents a coordinated attempt by the body to fight off the parasites that cause disease.

Fever differs from the rise in body temperature we experience when we overheat during exercise. When we overheat, our body does what it can to bring the temperature back down to the normal 37°C (see Chapter 31), which we can think of as the body's "target temperature." When we have a fever, however, the body raises the target temperature to a new, higher level. If you have a fever of 39°C, for example, your body will work to keep its temperature at 39°C. When we shiver with the chills while sick, we experience our temperature regulation mechanisms working to maintain the fever. Trying to cool down a fever by placing wet cloth on our skin simply makes the body work harder to maintain the fever temperature.

The immune system regulates fever. When macrophages destroy invading parasites, they release proteins called interleukins into the body fluids. These proteins circulate to the temperature control center in the brain, where they reset the body's target temperature to a higher level. This higher temperature may increase the activity of neutrophils and lymphocytes, weaken disease-causing parasites, and increase our survival rate.

A growing body of research indicates that fever occurs in all sorts of animals other than humans. Virtually all classes of vertebrates, insects and crustaceans, and earthworms and their relatives show evidence of fevers related to infections. Most of these organisms rely on environmental sources of heat to regulate their temperature, and some lack the vertebrate immune system. All, however, use the strategy of raising their body temperature in an attempt to fight off an attack by disease-causing parasites.

---

6. While the macrophages and neutrophils do their work, a third kind of white blood cell, called a **lymphocyte** (Figure 32.5*b*), builds in numbers. Unlike complement, which recognizes only that an invading cell is foreign, lymphocytes can distinguish between different kinds of parasites. Thus, lymphocytes recognize *C. tetani* as different from other invaders, and allow the body to mount a **specific response** to this organism. On first exposure, however, lymphocytes specific to *C. tetani* accumulate slowly, and they usually contribute little to a successful defense.

The success of the immune response to *C. tetani* depends on how quickly macrophages and neutrophils arrive at the wound compared with how quickly *C. tetani* multiplies. If steps 1 through 5 proceed slowly relative to the rate at which *C. tetani* multiplies, the victim loses the race, becomes ill, and may die. If the victim survives, however, the lymphocytes that played a minor role in the initial encounter with *C. tetani* will tip the balance strongly in that person's favor in any future encounter with that parasite.

Scenario 2:
The immune system responds much more effectively to a second exposure to an invader

The immune system mounts a much faster and more dramatic response to parasites when it encounters them a second time (Figure 32.6). For example, after our first exposure to *C. tetani*, we can survive an exposure more than 100,000 times as severe as the one that would have killed us the first time around; thus we are essentially immune to tetanus. The key to our ability to "learn" from

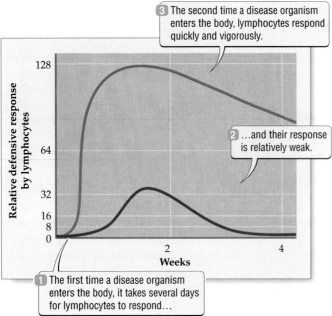

3 The second time a disease organism enters the body, lymphocytes respond quickly and vigorously.

2 ...and their response is relatively weak.

1 The first time a disease organism enters the body, it takes several days for lymphocytes to respond...

Figure 32.6   Initial versus Subsequent Responses by the Vertebrate Immune System
Lymphocytes in the immune system defend against specific parasites more quickly and vigorously in second and later attacks than in the original encounter.

sure to the *C. tetani* antigens, the *C. tetani*–specific memory lymphocytes rapidly multiply and mature into active lymphocytes that can defend against this bacterium. The vast numbers of these specific lymphocytes dramatically speed up and increase the effectiveness of all the steps of the immune response.

Among the lymphocytes, we can recognize three functionally different types:

experience with a particular parasite such as *C. tetani* lies with the lymphocytes that recognize a specific parasite.

How can lymphocytes tell *C. tetani* from a cold virus or the protist that causes malaria? The cell surface of each kind of parasite that invades the body carries unique **antigens**, molecules that have a shape that characterizes that kind of parasite in much the same way that fingerprints characterize individual people. At the same time, individual lymphocyte cells in the human body have a wide variety of differently shaped proteins embedded in their plasma membranes (Figure 32.7). Only a few of the lymphocytes bear membrane proteins that can bind to *C. tetani* antigens. When exposed to *C. tetani*, those few matching lymphocytes multiply rapidly and spread throughout the body, acting as a sort of memory of the parasite. A body equipped with large numbers of lymphocytes that recognize a particular parasite can much more effectively fight off a second exposure to that parasite.

In the case of attacks by viruses, lymphocytes respond not directly to the virus, but rather to human cells infected by the virus. Viruses leave behind bits of their protein coats on the plasma membrane of infected cells. These protein bits identify infected host cells just as antigens identify invading bacteria.

The much more effective response to a second exposure to *C. tetani* comes about because the lymphocytes that recognized the *C. tetani* antigens have remained in the victim's body in relatively large numbers since the initial exposure to that bacterium. Upon a second expo-

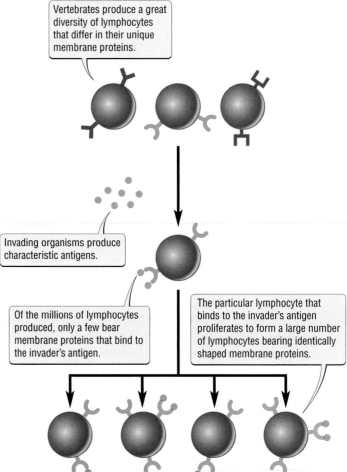

Vertebrates produce a great diversity of lymphocytes that differ in their unique membrane proteins.

Invading organisms produce characteristic antigens.

Of the millions of lymphocytes produced, only a few bear membrane proteins that bind to the invader's antigen.

The particular lymphocyte that binds to the invader's antigen proliferates to form a large number of lymphocytes bearing identically shaped membrane proteins.

Figure 32.7   How the Human Body Recognizes Specific Parasites
Lymphocytes that bear membrane proteins capable of binding to antigens that characterize invading organisms multiply rapidly to give rise to identical lymphocytes, all of which bind specifically to that invading organism.

1. Active **B lymphocytes** produce proteins called **antibodies** that circulate freely in the blood (Figure 32.8). The antibodies target specific invading parasites—in this case, *C. tetani*. When antibodies bind to an invading cell, they enhance the effect of the complement system already described, they can cause invading cells to clump together, and they help the complement, macrophages, and neutrophils bind to and destroy the invading cells. Antibodies can also bind directly to viruses and to toxins produced by invaders, which then cannot penetrate cells.
2. Active **helper T lymphocytes** help macrophages rapidly bind to and phagocytose parasites, and they stimulate the production of parasite-specific lymphocytes.
3. Active **killer T lymphocytes,** acting with macrophages and neutrophils, destroy any of the victim's cells that viruses have damaged or infected.

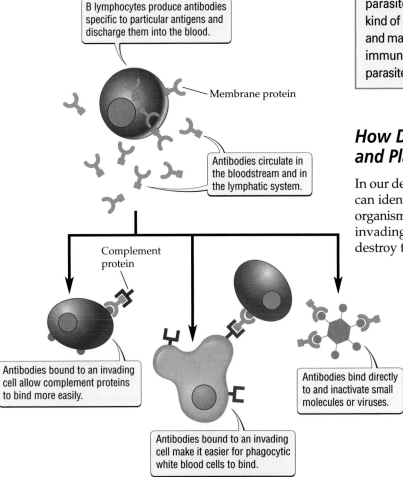

B lymphocytes produce antibodies specific to particular antigens and discharge them into the blood.

Membrane protein

Antibodies circulate in the bloodstream and in the lymphatic system.

Complement protein

Antibodies bound to an invading cell allow complement proteins to bind more easily.

Antibodies bound to an invading cell make it easier for phagocytic white blood cells to bind.

Antibodies bind directly to and inactivate small molecules or viruses.

32.5

Figure 32.8   How Antibodies Work
Antibodies can bind to invading cells (shown in blue) either to stimulate the binding of macrophages and neutrophils or to stimulate the complement system. Antibodies can also bind to and neutralize toxins and viruses.

Vaccines stimulate the immune system to provide protection

Preventative medicine relies heavily on **vaccines**, such as the tetanus shots many of us have had, to prime the immune system so that it is ready to defend the body in the event of a potentially dangerous exposure to a parasite. Vaccination involves the injection of killed or harmless forms of a parasite. It gives the immune system a chance to produce memory lymphocytes that recognize the antigens produced by that parasite without exposing us to the risk of the disease itself.

■ When exposed to a parasite, vertebrates first use nonspecific macrophages and neutrophils to destroy the invaders. Memory lymphocytes, which recognize a particular parasite, multiply after the first exposure to a parasite. In response to a later invasion by the same kind of parasite, memory lymphocytes rapidly multiply and mature into active lymphocytes. Vaccines prime the immune system so that it is ready to defend against a parasite when it invades.

How Do Bacteria, Invertebrates, and Plants Fight Parasites?

In our description of the vertebrate immune system, we can identify some functions basic to self-defense: First, organisms must distinguish between their own cells and invading parasites. Second, organisms must isolate or destroy those parasites.

Self-recognition is the first step in fending off invading parasites

In all organisms that have been investigated, the key to distinguishing self from nonself lies in proteins, complex carbohydrates, or nucleotides associated with the invading parasites. In contrast to the lymphocytes of vertebrates, the defense mechanisms used by most other organisms allow them to identify an invading parasite as foreign, but not to distinguish between different kinds of parasites. All animals use specialized mobile phagocytic cells similar to vertebrate macrophages and neutrophils as a cornerstone of their internal defense systems, but invertebrates rely solely on these nonspecific defenses.

Plants rely on the chemicals produced uniquely by parasites to signal an invasion. For example, fungi, many of which are important plant parasites, have cell walls made of chitin and other complex sugars that are absent from plant cell walls. The presence of these substances in a plant signals the presence of a foreign invader. Invertebrate animals recognize parasites much as vertebrates do: through a combination of marking foreign objects with proteins (as complement and antibodies do in vertebrates) and the direct binding of defensive cells to the invaders in response to those proteins.

Hosts limit the amount of damage that parasites cause

Single-celled bacteria and protists can tolerate little damage from invaders, which for them most often are viruses. They cannot isolate attacking viruses, and therefore must either avoid detection by viruses or kill them. Bacteria can hide from viruses by changing their plasma membrane proteins so that a virus cannot attach to them and initiate infection. Alternatively, bacteria may fend off a viral attack by producing proteins that stop replication of the viral genes, or by using restriction enzymes to destroy the nucleotides injected by the virus. Although researchers now use restriction enzymes as an important research tool (see Chapter 7), they evolved as a defense that protected bacteria against viruses.

Plants lack the specialized defensive cells that most animals have; they rely instead on a chemical defense against parasites. Plants typically respond to invasion with a **hypersensitive response** that isolates and kills the plant cells on which the parasite depends. In response to chemical signals that indicate invasion by parasites, plants seal off an area around the infection by reinforcing their cell walls. Then, the infected plant cells release chemicals that trigger a genetically controlled "suicide program." The result is a patch of dead tissue that isolates or kills the invading parasites (Figure 32.9). Recent evidence suggests that some plants acquire immunity to certain viral invaders after an initial attack, providing a response similar to that provided by the vertebrate immune system.

■ To mount an effective defense against parasites, organisms must distinguish self from nonself. Invertebrate animals rely on specialized mobile cells similar to the nonspecific defenses of vertebrates. Bacteria destroy attacking viruses by hiding from them or by destroying them. Plants isolate and destroy infected tissue along with the parasites it contains.

The circle of dead tissue isolates the diseased part of the leaf from its surroundings, dooming the organism that caused the response.

Figure 32.9   The Hypersensitive Response of Plants
These dead spots on a leaf are the result of the hypersensitive response of the plant to an invading parasite.

Avoiding Attack by Carnivores and Herbivores

Carnivores and many herbivores inflict damage on their victims much more quickly than parasites do. Thus effective defenses against carnivores and herbivores must prevent attack altogether rather than simply limiting the amount of damage done.

Avoiding discovery by carnivores and herbivores is a first line of defense

Searching carnivores and herbivores must first find their food, then they must identify what they have encountered as food. Just as they do in defending themselves against parasites, some organisms avoid attack by scattering themselves so widely that it becomes difficult for carnivores and herbivores to find them. If an insect species, for example, is rare enough, carnivorous birds, which often focus on the most abundant prey available, may ignore it. Many weedy plants (such as crabgrass in our lawns or the ragweed that may invade untended gardens) normally grow in isolated patches, where they remain for only a few years before colonizing new patches (Figure 32.10a). This lifestyle may help them avoid herbivores. Conversely, by growing crop plants in large fields that contain no other plant species, we humans have made it impossible for them to avoid

(b)

(a)

The tiger swallowtail caterpillar fools predators by imitating a snake.

(c)

Figure 32.10   Hide and Seek
(a) Herbivores cannot easily find a single food plant in the middle of a large field of other plants. Many animals avoid being recognized as food (b) by blending into their background or (c) by resembling something dangerous.

encountering herbivores. As a result, herbivorous pests easily find our crops and do a great deal of damage.

Many organisms avoid detection by making it hard for enemies to recognize them as food. The coloration or shape of many organisms lets them blend into their surroundings. Ground-dwelling birds blend into the leaf litter of their habitat, and many insects closely mimic the sticks, leaves, and flowers on which they live. The stripes and spots of many animals break up their shape so that they match the speckling effect of sun and shade (Figure 32.10b). Even plants can take on shapes that allow them to blend into the background of their habitat.

In other cases, organisms look like something they are not. For example, caterpillars of some butterflies resemble the head of a snake (Figure 32.10c). Many insects have large eyespots on their wings, which make them look much more threatening than they really are. The hognose snake, a harmless species, rattles its tail against dried leaves to sound like a poisonous rattlesnake.

Physical defenses prevent attack following discovery

A variety of organisms use biological suits of armor to fend off attacks (Figure 32.11a). Familiar examples among animals include the many different external skeletons that simultaneously provide support and pro-

tection (see Chapter 26), such as the shells of turtles and clams. Coral animals and many algae produce a mineral-based skeleton. Many land plants have cell walls reinforced with so much indigestible lignin and cellulose that herbivores have trouble chewing through their leaves. This defense not only makes herbivores work harder, but has the added bonus of reducing the nutritional quality of the plants, which makes them even less attractive as food (see Chapter 28). Some animals build or borrow their suits of armor. The prairie dog's burrows, the empty snail shells inhabited by hermit crabs, and the leaf shelters built by many caterpillars all reduce the vulnerability of these animals to their enemies.

Hermit crab in a snail shell

Alternatively, many organisms have surfaces that make them dangerous to attack (Figure 32.11b). Slow-moving porcupines have fearsome quills, as many dog owners know only too well. Although less imposing, the long hairs on many caterpillars can make them difficult for birds to swallow. A walk through the desert or through a patch of stinging nettles reveals painfully that plants, too, come equipped with an impressive arsenal of sharp weapons. Even the innocent-looking hairs on the surfaces of many leaves can create serious problems for the small insects trying to eat them.

The hard shell of a turtle protects it against predators.

Sharp cactus spines discourage many herbivores.

Figure 32.11   Biological Fortifications
Organisms can make themselves difficult to eat by protecting themselves with armor or spines.

Behavioral defenses provide animals with additional ways to prevent attacks

The combination of muscles (see Chapter 27) and nervous systems (see Chapter 34) gives animals a variety of behavioral defenses that can prevent attacks by carnivores. Animals can physically fight off attackers. Even some animals that we would not consider dangerous can seriously hurt or even kill carnivores. A healthy adult moose, for example, which is normally a peaceful herbivore, can kill attacking wolves. Animals can also avoid becoming someone else's meal through speed. Antelope, deer, and horses rank among the fastest of land mammals; their speed allows them to escape pursuing carnivores (Figure 32.12*a*).

Getting together with other animals also provides many defensive benefits. A group can be more watchful while feeding than an individual can (Figure 32.12*b*). A solitary individual cannot eat and watch for carnivores at the same time; it must either eat less or make itself more vulnerable to carnivores. In a group of 20 animals, however, each individual would need to watch for approaching carnivores only 3 minutes of each hour to provide full coverage. In addition, simply by belong-

Figure 32.12   Animals Can Use Behavioral Defenses
(*a*) These impalas are fleeing an approaching group of lions. (*b*) Groups of animals, like these flamingos, always contain a few individuals on the lookout for approaching carnivores.

(*a*)

(*b*)

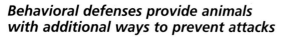

ing to a group, an animal can place other bodies between itself and a carnivore. When a flock of birds or a herd of mammals scatters at the first sign of danger (see Figure 32.12*a*), the resulting chaos often confuses carnivores enough to allow the intended victim to escape. In addition, groups of organisms can fend off attacks more effectively than individuals. We find the fury of a single honeybee annoying, but an angry hive can drive us away.

> ■ Organisms use a variety of defenses to avoid being attacked by carnivores and herbivores. Scattered distributions, hiding, and looking like something else reduce the chances of discovery by an enemy. Armor, spines and hairs, and defensive behaviors can prevent successful attacks once the organism is discovered.

Plants Can Limit Damage Following Discovery by Herbivores

Once under attack, the victims must limit the amount of damage they suffer. Successful attacks by carnivores usually result in a quick death, giving their prey little chance to respond. In contrast, attacks on large plants by relatively small herbivores—for example, a caterpillar feeding on a cabbage plant or a deer browsing on a tree—generally cause damage, but not immediate death. As a result, plants have evolved ways of limiting damage by herbivores. These defenses usually involve chemicals, including many of the same chemicals involved in the hypersensitive response of plants.

Many defensive chemicals are not part of the basic metabolism of the organism that produces them, and are therefore known as **secondary chemicals**. Researchers have identified tens of thousands of these secondary chemicals in plants. Many secondary chemicals act as poisons that disrupt the metabolism of herbivores. Nicotine, the notoriously addictive chemical in cigarette smoke, protects tobacco plants extremely well against herbivore attacks. A few drops of pure nicotine will kill a human, and it is also sold as a potent insecticide. Similarly, the ornamental foxglove commonly grown in flower gardens produces two toxins that disrupt heart function in vertebrates. Like many secondary chemicals that evolved as plant defenses, however, the foxglove toxins also provide us with important medical benefits. In 1785 the British physician and botanist William Withering reported that foxglove toxin,

used in controlled doses, could strengthen the heartbeat of patients with weak hearts. Sold as digitalis, it remains an important weapon in the fight against heart disease.

Like defenses against parasites, the chemicals used for defense against herbivores may be present at all times or may be available only on demand (see Figure 32.3). For example, cabbage plants produce mustard oils (which give cooking cabbage its distinctive smell) at relatively constant levels. Nicotine, in contrast, appears in tobacco leaves at much higher levels following an attack. The chemical signals that trigger the on-demand defenses against herbivores include many of the same signals that coordinate the defensive response to plant parasites, suggesting a close relationship between these two types of defense (see Chapter 33).

Some herbivores have hijacked plant secondary chemicals for their own fight against their enemies. Monarch butterflies defend themselves with chemicals closely related to those found in foxgloves (Figure 32.13). The butterflies cannot make these toxins themselves;

Foxglove

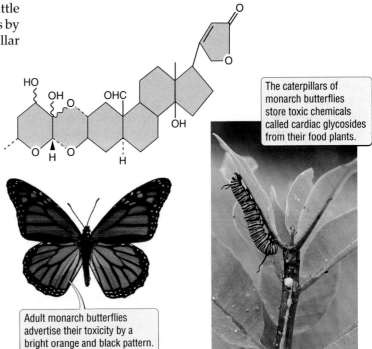

The caterpillars of monarch butterflies store toxic chemicals called cardiac glycosides from their food plants.

Adult monarch butterflies advertise their toxicity by a bright orange and black pattern.

Figure 32.13 Secondary Chemicals Defend Milkweeds and Monarch Butterflies
The milkweed plant produces toxic secondary chemicals called cardiac glycosides that disrupt heart function in vertebrates and have a foul taste. Monarch butterfly caterpillars store cardiac glycosides from the milkweeds they eat to protect themselves against their own attackers.

instead, they obtain the toxins from their food plant, the common milkweed, and store them in their own tissues.

> ■ Plants rely heavily on toxic secondary chemicals to reduce the amount of damage that herbivores do. Defenses against herbivores show similarities to defenses against plant parasites.

■ HIGHLIGHT

32.6 *Why Is AIDS So Deadly?*

Why is AIDS such a deadly disease? As we mentioned at the beginning of this chapter, HIV, the virus thatcauses AIDS, kills its host indirectly by weakening the immune system. HIV attacks two components of the human immune system: the macrophages, which phagocytose infected cells, and the helper T lymphocytes, which stimulate the activity of antibody-producing B lymphocytes and killer T lymphocytes.

Evidence suggests that the human immune system does a good job initially of keeping HIV at bay. When activated, killer T lymphocytes recognize and destroy helper T lymphocytes infected by the virus. By destroying infected helper T lymphocytes, the killer T lymphocytes greatly slow the rate at which HIV reproduces and spreads, and they allow replacement of the helper T lymphocytes lost to the virus. For up to 15 years, the immune system fights a valiant battle against HIV. Eventually, however, the inroads of HIV into the macrophage and helper T lymphocyte numbers weaken the immune system, and the symptoms of AIDS begin to appear.

Ongoing evolution of HIV inside the host's body seems to play an important part in its ultimate victory over the immune system. There is increasing evidence that the HIV population inside an infected individual evolves quickly following infection of a host (see Chapter 20). The antigens on the protein coat of HIV change rapidly over time. Thus, lymphocytes that fit the HIV antigen types that dominated at the time of infection may not recognize the most common antigens in the HIV population in the body a few years later. The lymphocytes do seem to have an increasingly difficult time identifying HIV as the infection progresses, leading to an increasingly less effective immune response.

Eventually HIV destroys helper T lymphocytes more quickly than they can be replaced. The downfall of the helper T lymphocytes leads to a decrease in the effectiveness of the B lymphocytes and the killer T lymphocytes. Ultimately, the victim's immune system collapses,

and infection by normally harmless diseases leads to death.

> ■ HIV gradually weakens the immune system by infecting the macrophages and helper T cells that play a central role in helping us fend off disease.

SUMMARY

The Basics of Defense
- Most parasites feed inside a single host for their entire lifetime.
- Carnivores and many herbivores attack their hosts from the outside, and often feed on one and often more individuals over their lifetime.
- To defend themselves against parasites, carnivores, or herbivores, potential victims must (1) reduce the chances of being found by an enemy, (2) stop the enemy from penetrating their outer surface, or (3) respond effectively to enemies that penetrate their outer surface.

Keeping Parasites Out
- Parasites cannot spread among widely scattered hosts as easily as they can among densely crowded hosts.
- The body wall—either the cell wall of a single-celled organism or the internal and external surface tissues of a multicellular organism—helps keep parasites out of hosts.
- Because a wound can allow parasites to enter a host's body, plants and animals quickly seal wounds.

The Vertebrate Immune System: Consequences of a Cut
- The immune system of vertebrates uses specialized proteins and white blood cells to defend against parasites.
- The components of the immune system move through blood vessels and through the ducts and nodes of the lymphatic system.
- The human immune system first reacts to a cut with inflammation and blood clotting, which physically block the passage of invaders into the body. Complement proteins then mark any invading cells for destruction.
- In the nonspecific immune response, two kinds of specialized white blood cells, macrophages and neutrophils, digest the parasites and any of the host's tissue that has been damaged.
- White blood cells known as lymphocytes, which can distinguish among different kinds of parasites, constitute the specific immune response. If the victim is faced with the same kind of parasite a second time, memory lymphocytes help the body mount a faster and more powerful defense.
- There are three functionally different types of lymphocytes: B lymphocytes, which produce antibodies; helper T lymphocytes, which make macrophages more vigorous

and stimulate the production of B and killer T lympho-cytes; and killer T lymphocytes, which help macrophages and neutrophils destroy damaged or infected cells.

■ Vaccines prime the lymphocytes so that they are ready to defend against certain kinds of parasites if they invade the body.

How Do Bacteria, Invertebrates, and Plants Fight Parasites?

■ All organisms achieve the critical step of distinguishing self from nonself by recognizing proteins or carbohydrates associated with invading parasites.

■ Invertebrate animals depend on specialized mobile cells to destroy invading parasites in a nonspecific response.

■ Bacteria can limit the damage done by viruses by hiding from them or by destroying them.

■ Plants use a hypersensitive response to isolate and destroy infected tissue along with the parasites it contains.

Avoiding Attack by Carnivores and Herbivores

■ Effective protection against carnivores and herbivores depends on reducing the chance of being discovered. Living in low-density populations, hiding, and looking like something else reduce an organism's chances of discovery.

■ Protective surfaces, such as armor or spines, prevent the enemy from attacking successfully.

■ Animals can use a variety of behaviors to avoid attack by carnivores.

Plants Can Limit Damage Following Discovery

■ Plants can effectively protect themselves against damage done by relatively small herbivores.

■ Plants produce a tremendous variety of secondary chemicals that are toxic to herbivores or interfere with their feeding.

■ Whereas some secondary chemicals are always present in the plant, the plant produces others only on demand.

■ Some herbivores appropriate chemicals in their food plants for their own defenses against carnivores.

Highlight: Why Is AIDS So Deadly?

■ HIV gradually weakens the immune system by destroying macrophages and helper T lymphocytes.

KEY TERMS

antibody p. 543
antigen p. 542
B lymphocyte p. 543
blood clot p. 539
carnivore p. 537
complement p. 540
helper T lymphocyte p. 543
herbivore p. 537
host p. 537
hypersensitive response p. 544
immune system p. 539
inflammation p. 540
killer T lymphocyte p. 543

lymph p. 539
lymphatic system p. 539
lymphocyte p. 541
macrophage p. 540
neutrophil p. 540
nonspecific response p. 540
parasite p. 537
phagocytosis p. 540
secondary chemical p. 547
specific response p. 541
vaccine p. 543
white blood cell p. 539

CHAPTER REVIEW

Self-Quiz

1. Parasites resemble small herbivores in which of the following ways?
   a. They attack their host from the inside.
   b. They both produce killer T lymphocytes.
   c. They often do damage to their victim over a long period of time.
   d. They are both much more likely to encounter widely scattered victims than densely crowded ones.

2. Which of the following statements about antibodies is true?
   a. They are produced by B lymphocytes.
   b. They are produced by parasites.
   c. They play a critical role in the hypersensitive response that allows plants to isolate invading parasites.
   d. all of the above

3. What role do macrophages play in the vertebrate immune system?
   a. They help form blood clots.
   b. They mark invading cells.
   c. They remember previous infections.
   d. They digest and destroy invaders.

4. A significant difference between the defensive systems of vertebrates and those of other organisms is that only vertebrates
   a. have specialized cells that destroy invading cells.
   b. isolate infected tissues with a hypersensitive response.
   c. have a well-developed and stronger response to later infections by the same parasite.
   d. can recognize foreign cells.

5. Which of the following features of plants could help protect them against herbivores?
   a. secondary chemicals
   b. hairs covering their leaves
   c. growing as widely scattered individuals
   d. all of the above

Review Questions

1. Why is self-recognition important to all organisms?

2. Explain how vaccines help protect us from disease. Include in your answer a description of how the three types of lymphocytes function.

3. Compare the way in which humans protect themselves against a bacterial attack with the way in which a tobacco plant protects itself against herbivorous insects seeking to feed on it.

32   **The Daily Globe**

Should Your Child Be Vaccinated?

BURNABY, BC—Experts from around the world gathered at the Convention Centre last weekend for a workshop intended to assess the effectiveness of government-sponsored vaccination programs. The Vaccination Action Committee workshop focused on the growing resistance among parents to mandatory vaccination programs for their children.

"Vaccines have become victims of their own success," explained Dr. Francis Stein of World Health International. "Once a vaccination program almost wipes out a disease, many parents no longer recognize the importance of vaccinations. They don't realize the harm that these diseases can do. They have the attitude, why should my child get vaccinated for a disease that no longer exists?"

Dr. David Mento of St. Peter's Hospital in Vancouver noted that the growth of World Wide Web access in wealthier countries has spread fears of the harmful side effects of vaccinations. "Websites blame vaccinations for a host of poorly understood medical conditions, such as autism or sudden infant death syndrome (SIDS), when in fact we have no reliable evidence connecting vaccination with these conditions in spite of constant monitoring. People simply assume that if they see it on the Web, it must be true."

The Burnaby area was hit by an outbreak of measles in 1997 in spite of decades of federally sponsored vaccination programs. Because measles is so rare in Canada, the overall chance of an unvaccinated person catching the disease is small. On the other hand, when the disease is introduced from countries where it is more common, it spreads quickly among unvaccinated people.

Warren Pace of Health Canada pointed out that when a disease such as measles becomes rare, the small risks associated with vaccination seem much more significant than the risk of catching the disease. A one in a million risk of contracting the brain disease encephalitis from the mumps–measles–rubella (MMR) vaccine seems significant when we forget that that one or two of every thousand measles victims contract encephalitis as a complication of getting measles. "When you realize this," Pace cautioned, "you can appreciate just how safe the MMR vaccine is."

Evaluating "The News"

1. Do you think that a government has a right to demand that children be vaccinated against potentially dangerous infectious diseases before being allowed to attend school? Would you want your children going to a school that allowed unvaccinated children to attend?

2. How would you, as a parent, balance the risks to your children associated with vaccination against the risks of catching the disease? Where would you look for information to help you make your decision wisely?

3. Do you think that the governments of wealthier nations should help fund expensive vaccination programs in countries that lack the resources to support such programs? Can you see any benefits to the wealthy countries of doing so?

chapter *33* Hormones

Roberto Juarez, *Kether Place*, 2001.

From Caterpillar to Butterfly

Butterflies live a double life. They hatch from eggs as wingless, wormlike caterpillars devoted completely to eating. Armed with strong, scissorlike jaws, caterpillars munch their way through plant tissues as they accumulate the proteins and carbohydrates that they need to grow.

Caterpillars have short, stubby legs better suited to holding onto a plant than to moving between plants. Indeed, they often complete their life as caterpillars on a single host plant.

As adults, butterflies devote themselves to reproduction. Brilliantly colored wings allow them to search widely for and identify potential mates. Once mated, females search their surroundings for good places to lay eggs, while males look for more females to mate with. Finding mates and laying eggs both require that butterflies fly long distances. Butterflies get the energy needed for flight by using strawlike mouthparts to suck up the sugary nectar produced by flowers. Unlike the plant tissues on which caterpillars feed, nectar contains few nutrients other than those needed to supply energy.

When you compare a caterpillar with an adult butterfly, it is hard to imagine that the two forms bear any relationship to each other. From most perspectives, a caterpillar differs more from the butterfly into which it develops than one species does from another. By having two very different body forms during their development, butterflies can essentially pack two specialized "species" into one organism: (1) a caterpillar stage, well suited for eating and growing, and (2) an adult stage, well suited for mating and egg laying. Compare this life cycle with that of humans, who must deal with the differ-

■ MAIN MESSAGE

All multicellular organisms use signaling molecules called hormones to coordinate the functions necessary for life.

ent demands of feeding and reproduction using a single, all-purpose body form.

Butterflies undergo the amazing transformation from caterpillar to adult during an inconspicuous pupal stage. During the pupal stage, the tissues of the caterpillar are rearranged completely to form adult body parts. This dramatic change, known as **metamorphosis**, depends on the coordination of a complex sequence of breaking down and building up of tissues. In this chapter we explore the role of signaling molecules called hormones in coordinating the biology of multicellular organisms. We see how hormones coordinate several functions in plants and animals, including humans, before returning specifically to how they regulate the metamorphosis of butterflies.

The Metamorphosis of a Caterpillar into a Butterfly

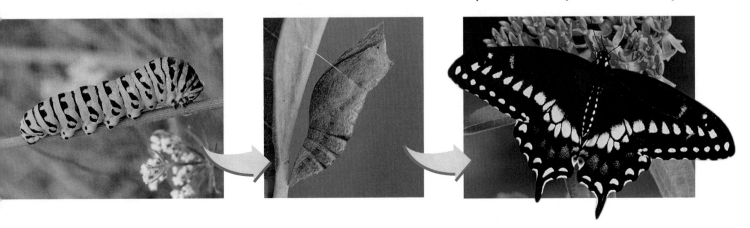

■ KEY CONCEPTS

1. Hormones allow distant cells within an organism to communicate and coordinate their activities.

2. A single hormone may have several different effects depending on the target tissue on which it acts.

3. The internal transport system carries hormones from sites of production to sites of action.

4. Several hormones often work together to coordinate a process or activity.

5. Hormones control growth and development in plants.

Multicellular plants and animals must coordinate the functioning of their many specialized tissues and organs. Just as a group of people cannot effectively complete a task without constantly communicating, the many cells within a plant or animal need to communicate to allow the individual to function effectively. Unlike people, cells cannot send memos or e-mail messages to one another. Instead, they rely on signaling molecules to let one part of the organism know what the others are doing.

In this chapter we discuss **hormones**, signaling molecules that circulate through the body. Both plants and animals rely on hormones (humans, in fact, often rely on commercially available hormones; see the box on p. 555). Animals also rely on the nervous system, a much faster means of signaling, which we will discuss in Chapter 34; however, the functions of hormones are often coordinated with those of the nervous system.

We begin by providing an overview of how hormones work. We then build on this general perspective by introducing examples that illustrate how both plants and animals use hormones to coordinate their many essential functions. We close by returning to consider how several hormones work together to coordinate butterfly metamorphosis.

How Hormones Work

Before we discuss specific examples of how hormones allow multicellular organisms to coordinate their many functions, let's first look at how a chemical can carry a message from one cell to another. In the process, we review some of what we learned in Chapter 9 about how signaling molecules function in communication between cells.

Hormones trigger a response in target cells

A hormone released by one cell causes a response in another cell, called a **target cell** (Figure 33.1). The tar-

get cells for a particular hormone may lie in several different kinds of tissue and may respond to the hormone in different ways. For example, **testosterone**, a hormone that influences the development of many features associated with maleness in humans, controls the development of the male sex organs in developing babies, stimulates growth in cells throughout the body, stimulates the production of sperm by target cells in the testes (see Chapter 36), stimulates the production of facial hair in target cells in the skin of the face, and causes behavioral changes through interactions with target cells in the brain. As we will see later in this chapter, organisms can get by with fewer different hormones by using them for multiple functions in this way.

As we saw in Chapter 9, hormones can act on target cells in two different ways:

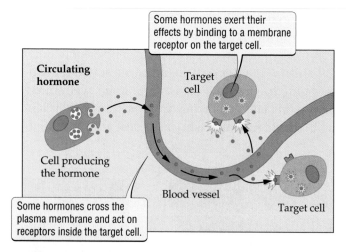

Figure 33.1   Hormones Allow Cells to Communicate with One Another
Plants and animals use signaling molecules called hormones to coordinate the diverse activities of their specialized cells. Hormones released by one cell travel through the internal transport system to cause a reaction in target cells in other parts of the organism.

■ BIOLOGY IN OUR LIVES

Steroids on Wheels

In 1998, scandal rocked the Tour de France, the world's premier bicycle race, when race officials expelled some of the top teams after discovering that their racing equipment included ample supplies of a steroid hormone called erythropoietin, or EPO.

The Tour de France places incredible demands on riders. The 2001 Tour, for example, ran for 20 days, extended over 3000 kilometers in France and Belgium, and crossed two major mountain ranges. As in many sponsor-driven professional sports, cyclists find themselves under intense pressure to win this high-profile event.

EPO is the drug of choice among cyclists who decide to use drugs to give themselves a competitive edge. The human body naturally produces EPO as a way to coordinate red blood cell production. For a long-distance athlete, more red blood cells mean that the circulatory system can carry more oxygen to rapidly metabolizing muscle cells (see Chapter 30). In a normal adult male, red blood cells make up 40 to 50 percent of blood volume; treatment with EPO can increase this proportion to 60 percent. For a bicycle racer, this increase can translate into over a minute in time saved for every 10 kilometers traveled.

Identifying athletes who abuse EPO poses some special challenges. First of all, the human body normally produces EPO, so we can expect to find at

What Price Victory?

least some EPO in anyone's blood. Second, the effects of EPO resemble those of acceptable training practices. For example, the bodies of athletes who train at high altitudes compensate for the low oxygen content of the air by producing more red blood cells (see Chapter 29). A recently developed test, however, can detect a genetically engineered version of EPO that is readily available to athletes. The Union Cycliste Internationale, which oversees many major races (including the Tour de France), has also decided to keep cyclists out of races if their red blood cells make up over 50 percent of their blood volume.

EPO may seem like a relatively harmless, if unethical, way of improving athletic performance. Nothing could be further from the truth. The body uses EPO as part of its system for regulating the red blood cell supply. If an athlete injects EPO into a body adequately supplied with red blood cells, it throws the circulatory system out of balance. An increase in the number of red blood cells in a volume of blood increases the drag that resists blood flow, forcing the heart to work harder (see Chapter 30). For a cyclist's heart already pushed to its limits, the thicker blood can prove too much to handle. As a result, several otherwise fit cyclists suffer heart attacks each racing season. The continued use of EPO in spite of the associated scandal and severe health risks indicates that some athletes will pay any price for victory.

1. Some hormones, particularly steroid hormones, can pass through the plasma membrane of a target cell and act inside that cell (see Figure 9.5).
2. Some hormones cannot pass through the plasma membrane of a target cell, and affect the target cell by acting on membrane receptors at the target cell's surface (see Figure 9.7).

Most hormones move through the internal transport system to act on distant cells

Hormones often act on target cells located far from the cells that produced them. To reduce their travel time, most hormones travel through the internal transport system of the organism: the circulatory system of animals or the vascular system of plants (see Chapter 30). Hormones can travel only as fast as the internal fluids move, which means that they take several seconds or more to move from the producing cells to the target cells. A few seconds may not seem like a long time, but it does mean that hormones cannot coordinate activities that require almost instantaneous responses. Only the nervous systems of animals can coordinate activities, such as muscle contraction and movement, that require an effectively instantaneous transfer of information (see Chapter 34). Instead, hormones coordinate functions that take place over time scales of seconds to months (Figure 33.2).

To survive their journey through the internal transport system, hormones must resist chemical breakdown. Rapid breakdown of hormones, on the other hand, allows organisms to prevent the chemical message from hanging around in the body and continuing to affect target cells for longer than intended.

Target cells amplify the hormonal signal

Organisms usually produce hormones in tiny amounts. One common plant hormone, for example, is found in the shoots of pineapple plants at concentrations of 0.000006 grams per 1000 grams of plant tissue, equivalent to the weight of a needle in 20 typical haystacks. How can such minuscule doses of a hormone regulate the function of a plant or animal? As we saw in Chapter 9, when a hormone binds to receptor in the target cell, it sets in motion a cascade of events that greatly amplifies the signal. As a result, the binding of a single hormone molecule may activate thousands of proteins in the target cell.

■ Hormones produced by one cell act on another, target cell. Hormones travel from the cells that produce them to distant target cells through the internal transport system. A cascade of biochemical events that follows the binding of a hormone to a target cell greatly amplifies the hormonal signal.

33.2 A Few Plant Hormones Control Plant Growth and Development

Plants face special challenges in coping with the demands of their environment. As we saw in Chapter 27, plants cannot run from a predator or move to a new or more favorable habitat to find water or food, as animals can. Instead, they often cope with unfavorable environmental conditions through growth responses. Plants do not have nervous systems; instead, all of their cellular processes, including growth, are coordinated by hormones. A surprisingly small number of hormones coordinate the growth and development of flowering plants: Auxin, cytokinin, and gibberellin promote growth and development, and abscisic acid and ethylene suppress growth and development (Table 33.1).

Hormones help coordinate the sprouting of seeds

The plant embryo inside a seed does not simply sprout as soon as the seed hits the ground. Instead, it waits until environmental conditions, such as temperature and water availability, can support its growth. When conditions are right—for example, when the seed has absorbed enough water—the embryo releases a hormone called gibberellin. Gibberellin stimulates the production of enzymes that break down storage compounds in the seed, releasing stored nutrients. These nutrients fuel the developing seedling until it can photosynthesize well enough to support itself (Figure 33.3).

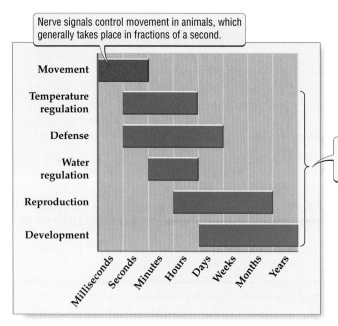

Figure 33.2 Hormones Act More Slowly than the Nervous System
The activities that organisms must coordinate occur over time scales that range from fractions of a second to many years. In animals, relatively slow-moving hormones tend to coordinate longer-term activities (blue bars). The nervous system coordinates activities that must occur rapidly, such as the muscular contractions of animals (red bar).

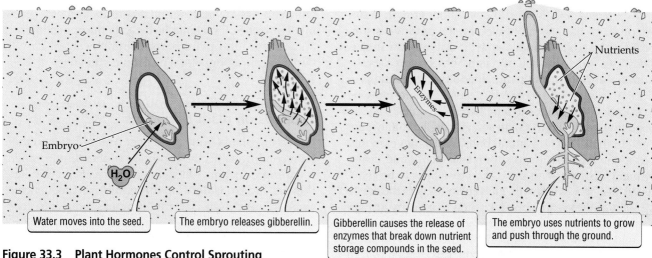

Embryo

H₂O

Nutrients

Enzymes

| Water moves into the seed. | The embryo releases gibberellin. | Gibberellin causes the release of enzymes that break down nutrient storage compounds in the seed. | The embryo uses nutrients to grow and push through the ground. |

Figure 33.3 Plant Hormones Control Sprouting
A hormone called gibberellin plays an important part in the early development of plants.

Plant hormones control shoot and root growth

Once the seedling becomes established, three hormones—auxin, gibberellin, and cytokinin—stimulate its growth. Together, they coordinate the upward growth of shoots toward the sunlight with the growth of roots through the soil to reach mineral nutrients (Figure 33.4). The three hormones are produced by cells in different plant tissues, and each hormone has different effects on target tissues in different plant parts. The regulation of plant growth by these three hormones illustrates two general points about how hormones work:

1. Individual hormones affect different target tissues differently.

2. Several hormones often work together to allow fine control in coordinating a process or activity.

33.1 *Plant Hormones and the Functions They Regulate*

Hormones	Major functions
Auxin	Promotes cell elongation
	Promotes growth at the top of the plant; suppresses branching
	Promotes vascular tissue development
	Regulates the response to light
Gibberellin	Releases enzymes to feed the plant embryo
	Stimulates stem elongation
	Works with auxin to stimulate fruit development
Cytokinin	Stimulates root growth (cell division)
	Delays leaf drop
	Promotes branching (counteracts the effects of auxin in suppressing branching)
	With auxin, establishes balance between leaf and root mass
Abscisic acid	Promotes leaf drop and plant dormancy
	Inhibits growth (and activity of auxin and cytokinin)
	Promotes nutrient storage in seeds
	Promotes water conservation by closing stomata
Ethylene	Stimulates fruit ripening
	Promotes dormancy and leaf drop
	Signals site of injury or attack
	Activates defense mechanisms

Auxin and gibberellin stimulate shoot growth in different ways. Both hormones loosen the rigid walls supporting plant cells, thus allowing the cells to grow in size. They also stimulate cell division, which increases the number of cells in the plant. Shade increases the production of auxin and gibberellin, causing plants growing in poorly lit conditions to lengthen more quickly and to bend toward the light. Auxin and gibberellin differ, however, in their effect on lateral buds, which develop into branches. Whereas auxin suppresses lateral bud development, causing all of the plant's resources to be channeled into upward growth, gibberellin triggers the development of these same buds, leading to a bushy growth form. Interestingly, auxin and gibberellin have an effect on root growth almost opposite to their effect on shoot growth. Auxin inhibits root growth, but stimulates root branching, and gibberellin inhibits root branching.

The third growth-stimulating hormone, cytokinin, acts on whole plants as a sort of reverse auxin. Root tips produce cytokinin, which stimulates root growth by encouraging cell division while inhibiting root branching. Damage to a root tip lowers cytokinin production,

which allows auxin to stimulate production of replacement branches. Cytokinin has exactly the opposite effect on aboveground plant parts: It inhibits shoot growth while encouraging lateral bud development.

The timing of flower and seed production depends on plant hormones

Flowering plants flower and bear fruit only when their own condition and the condition of their environment favor reproduction. Once flowering has been initiated, flowers and fruits develop through a fixed sequence of stages (Figure 33.5). Several hormones interact to coordinate timing and development in flowering plant reproduction. In some plants, environmental cues, such as temperature or day length, trigger the production of gibberellin, which in turn triggers the production of flowers. Seeds developing within fruits produce auxin, which stimulates growth of the surrounding fruit.

Two plant hormones that we have not yet introduced, ethylene and abscisic acid, control fruit ripening. Low concentrations of ethylene gas in the plant or in the air set in motion the series of changes that make ripe fruit so much more appealing to consumers than unripe fruit: Cell walls weaken to soften the fruit, starches and oils are metabolized into the sugars that sweeten the taste of the fruit, and colorful pigments replace the green chloro-

(a)
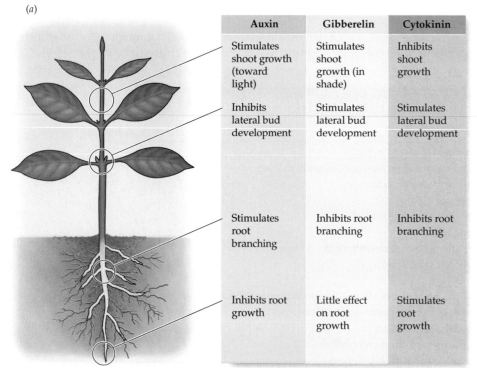

	Auxin	Gibberelin	Cytokinin
	Stimulates shoot growth (toward light)	Stimulates shoot growth (in shade)	Inhibits shoot growth
	Inhibits lateral bud development	Stimulates lateral bud development	Stimulates lateral bud development
	Stimulates root branching	Inhibits root branching	Inhibits root branching
	Inhibits root growth	Little effect on root growth	Stimulates root growth

(b)
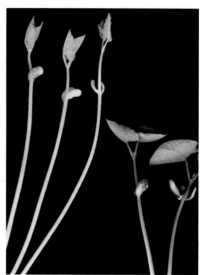

Figure 33.4   Three Plant Hormones Work Together to Stimulate Growth of Plant Roots and Shoots
(a) The hormones auxin, gibberellin, and cytokinin are produced in different plant tissues. Each hormone has a characteristic combination of effects on target tissues in various plant parts. Together, these three hormones coordinate a plant's growth toward sunlight and mineral nutrients. (b) Plants growing in low light grow tall and spindly.

(a)

(b)

Figure 33.5 Hormones Control the Timing of Events in Plant Reproduction
Hormones control the progress of raspberries from flowering to the production of ripe, red fruits.

phyll to signal ripeness. Commercial fruit producers exploit this phenomenon by picking and shipping unripe fruit, which resists bruising, then ripening it for market by exposing it to ethylene gas. Abscisic acid stimulates changes in fruit stems that allow humans and other fruit-eating animals to pull the fruit off the plant more easily.

Hormonally controlled dormancy allows plants to weather tough times

We have already seen (in Chapters 30 and 31) that some plants become dormant, shutting down their energy-requiring activities, when environmental conditions become too extreme. The same hormones that coordinate fruit ripening also play a lead role in regulating dorman-

cy. The tulip bulbs that we plant in our gardens in the fall, for example, are not seeds, but plants that have become dormant to avoid winter's freezing temperatures. During these tough times, ethylene and abscisic acid slow the plant's growth so that the plant can conserve energy until spring arrives. Other plants, such as the acacias described in Chapter 30, merely shed their leaves at the approach of the cold of winter or the drought of a dry season (Figure 33.6). The shed leaves undergo a well-organized "suicide program" closely controlled by ethylene and abscisic acid. Only after the plant has extracted and stored chlorophyll

Figure 33.6 Hormones Help Plants Survive Stressful Seasons
Every autumn, in a sequence of events controlled by hormones, the leaves of sugar maples first change to a brilliant orange and gold, then fall off. The color change reflects the breaking down and absorption of the green chlorophyll to reveal other pigments present in the leaves.

and other valuable materials from the leaves do they drop from the plant.

> ■ Plant hormones balance growth and reproductive processes with strategies for coping with stressful environmental conditions. Auxin, cytokinin, and gibberellin generally contribute to growth and development. Abscisic acid and ethylene generally suppress growth and other energy-requiring activities.

Hormones Coordinate Plant Defenses

The defensive response of a plant to attacks by both parasites and herbivores requires communication between the cells under attack and the surrounding tissues. Two classes of chemicals signal an attack by plant parasites: distinctive chemicals released by the parasite and chemicals produced by the plant itself in response to the attack. Proteins embedded in plant cell membranes may respond to either or both of these warning signals, triggering the hypersensitive response described in Chapter 32.

In addition to organizing the death of the cells under attack, the hypersensitive response triggers the production of a number of chemical alarm signals, which move out of the dying cells. These chemical messages, which include salicylic acid (familiar to us as aspirin) and the hormone ethylene, induce the hypersensitive response in healthy cells surrounding the ones under attack. In this way, the plant establishes a zone of dead cells around the infected cells in which the parasites cannot survive. Plants produce these same chemical alarm signals in response to an attack by herbivores, and they can cause changes in more distant plant tissues, such as other leaves, that make them more resistant to attack by parasites or herbivores.

> ■ Chemical messages coordinate the defensive response of plants to parasites and herbivores.

Hormones Regulate the Internal Environment of Animals

In animals, as in plants, hormones play a central role in coordinating biological activities. In animals, hormones are produced in groups of cells within tissues, as they are in plants, as well as in specialized organs called **glands**. These sites of hormone production together make up the **endocrine system** of the animal (Figure 33.7). The circulatory system carries the hormones throughout the

animal's body. The endocrine system works closely with the animal's nervous system, which we will introduce in Chapter 34. Here, we use the role of hormones in coordinating nutrient availability to emphasize that in animals, just as in plants, hormones affect different target tissues differently, and several hormones interact to coordinate a function.

Hormones regulate glucose levels

Glucose is one of the most important sugars in animal nutrition. It fuels glycolysis and cellular respiration, which produce the energy storage molecule ATP (see Chapters 7 and 8). Most animals, including humans, obtain glucose in their food. Glucose is absorbed from digested food in the small intestine, from which it travels through the circulatory system to metabolizing cells. Animals, including

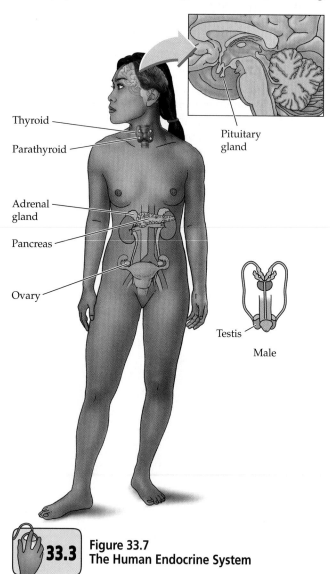

33.3   **Figure 33.7**
The Human Endocrine System

humans, must store excess glucose until it is needed, and they must be able to use it quickly. Three hormones—insulin, glucagon, and adrenaline—coordinate the storage and release of glucose in humans.

The pancreas, in addition to producing digestive enzymes (see Chapter 28), produces and releases two of the hormones that coordinate glucose storage and use. The pancreas releases insulin after a meal, when the blood contains more glucose than the body can use. Insulin acts on target cells in the liver, which respond by absorbing glucose from the blood. These storage cells assemble individual glucose molecules into long chains to form a storage carbohydrate called glycogen. Insulin also acts on other target cells to promote the storage of fatty acids from digested fats and amino acids from digested proteins. Insulin ensures that we do not lose excess nutrients by eliminating them from the body.

The pancreas also produces and releases glucagon, a hormone that has an effect opposite to that of insulin. When blood glucose falls to low levels, the pancreas releases glucagon, which stimulates the storage cells in the liver to convert stored glycogen into glucose and release it into the bloodstream.

The adrenal gland releases adrenaline, the third hormone involved in regulating glucose availability, in response to signals from the brain that warn of stress or danger (see Figure 9.6). The adrenaline circulates in the blood to target cells in the liver. Like glucagon, adrenaline stimulates glycogen breakdown in these storage cells to release glucose. In this way, glucose becomes available to fuel a rapid response to danger. (We will encounter adrenaline again in Chapter 34, where we will discuss its role as a neurotransmitter that carries information from one nerve cell to another.)

We always need to have some glucose in our blood, but too much of it can be harmful. In a condition called diabetes mellitus, the pancreas produces either too little insulin or a defective form of insulin that cannot bind to receptor proteins on the surface of the target cells. As a result, not enough insulin can be moved from the blood into storage cells. Over time, high levels of glucose in the blood damage small blood vessels, causing poor circulation and slowing the healing of injured tissues. Damage to blood vessels in the eyes can cause vision problems (in severe cases, blindness). Thanks to our understanding of the biology of insulin and our ability to produce the hormone in commercial quantities through genetic engineering techniques (see Chapter 17), most people who have diabetes can now monitor and manage their own glucose levels by modifying their diet and injecting additional insulin as needed (Figure 33.8).

Hormones regulate the concentration and distribution of mineral nutrients

In addition to regulating energy-supplying nutrients such as glucose, hormones coordinate the use of mineral nutrients by animals. Calcium in a mineral form, as we saw in Chapter 26, provides the framework for the internal and external support structures of many animals (see Chapter 26). Two hormones, calcitonin and parathyroid hormone, work together to regulate the amount of calcium circulating in the human bloodstream.

An excess of calcium in the blood triggers the release of calcitonin by the thyroid gland. Calcitonin removes calcium from the blood by promoting its storage in bones and by stimulating the kidneys to unload excess calcium into the urine. When there is too little calcium in the blood, the parathyroid gland releases parathyroid hormone, which

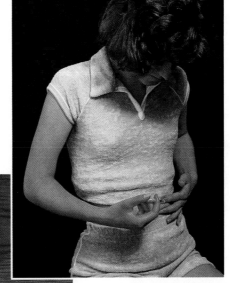

(a)

(b)

Figure 33.8 Compensating for a Defective Pancreas
(a) People affected by diabetes must monitor and coordinate their glucose use, a function normally performed by the pancreas. (b) When blood glucose drops below normal levels, diabetics can inject insulin, a hormone that their pancreas cannot produce.

stimulates the release of calcium from bones and prevents the kidneys from releasing it into the urine. Parathyroid hormone also stimulates special cells that dissolve some of the calcium stored in bones, releasing it back into the blood. In this way, calcitonin and parathyroid hormone maintain proper levels of blood calcium, while storing the excess and strengthening bones.

> ■ Several hormones act together to coordinate the availability of essential organic and inorganic nutrients such as glucose and calcium.

 ## 33.4 *Hormones Coordinate Human Reproduction*

Animals rely on hormones to coordinate their reproduction. In humans, hormones influence both long-term and short-term aspects of sexual development, pregnancy, and birth. The role of hormones in human reproduction illustrates the by now familiar points that a single hormone causes different responses in different target tissues and that several hormones usually work together to coordinate a particular function. In addition, human sex hormones illustrate the interaction between the endocrine and nervous systems in animals and the ability of hormones to regulate long-term and cyclical events. (We will learn more about the processes of human reproduction and development in Chapters 36 through 39.)

Hormones play a role in sexual development before birth

Hormones influence human sexual development even before birth. As we saw in Chapter 13, the presence or absence of a Y chromosome determines gender in humans, but it is through the action of hormones that a fetus develops into either a male or a female. If the fetus has a Y chromosome, male reproductive organs, called testes, begin to form just 4 to 6 weeks after fertilization of the egg. In the absence of a Y chromosome, female reproductive organs, called ovaries, develop. By the seventh week, each of these developing organs is producing steroid hormones specific to its gender. These sex hormones activate genes in their target cells to begin the process of sexual development (Figure 33.9).

The testes secrete three hormones that together coordinate the development of male reproductive structures. Testosterone and a second, closely similar hormone direct

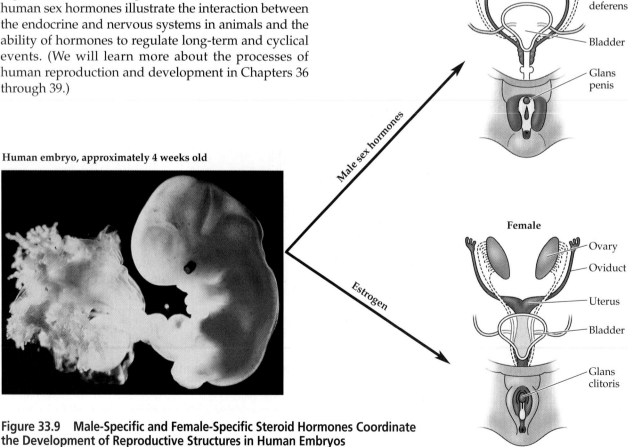

Human embryo, approximately 4 weeks old

Figure 33.9   Male-Specific and Female-Specific Steroid Hormones Coordinate the Development of Reproductive Structures in Human Embryos

the development of internal reproductive structures such as the sperm ducts and prostate gland; the third hormone directs the development of external structures such as the penis. The development of all female reproductive structures, both internal and external, falls under the control of a single steroid hormone, called **estrogen**.

Hormones coordinate sexual maturation at puberty

At an age of about 10 to 12 years, young humans make the transition to sexually mature adults capable of reproduction. During this transition, which is called puberty, the levels of sex hormones produced by our testes or ovaries rise markedly. Moreover, a structure within the brain,

called the **hypothalamus** (which is part of the nervous system), activates the production of two other hormones by the **pituitary gland**, which lies at the base of the brain. These two hormones, called **gonadotropins**, coordinate the development of sperm in males and eggs in females.

Gonadotropins also help maintain the functioning of the reproductive organs and glands throughout the individual's lifetime. In males, gonadotropins further stimulate the production of sperm in the presence of testosterone. At birth, the ovaries of a female already contain a lifetime supply of immature egg cells. These egg cells resume development upon the release of gonadotropins at puberty.

Hormones coordinate the menstrual cycle of human females

In humans, sperm is produced continuously by mature males. Human females, however, do not produce mature eggs continuously. Instead, individual eggs mature and are released in a 28-day **menstrual cycle**. Figure 33.10 summarizes the fluctuations in hormone levels that drive this cycle.

During the first 2 weeks of the menstrual cycle, estrogen levels rise steadily as a single egg completes its development in the ovary. The rising levels of estrogen that accompany egg maturation stimulate the development of a thickened lining in the uterus, the muscular organ that will receive a fertilized egg and support it during pregnancy.

When the egg has completed its development, a massive and sudden release of gonadotropins by the pituitary gland triggers the release of the egg from the ovary. This spike in gonadotropin levels also triggers the release of an additional hormone, progesterone, by cells in the ovary. Once the egg has been released, the gonadotropin and estrogen levels drop suddenly, and progesterone becomes the dominant hormone. Progesterone prepares the lining of the uterus to receive a fertilized egg: The

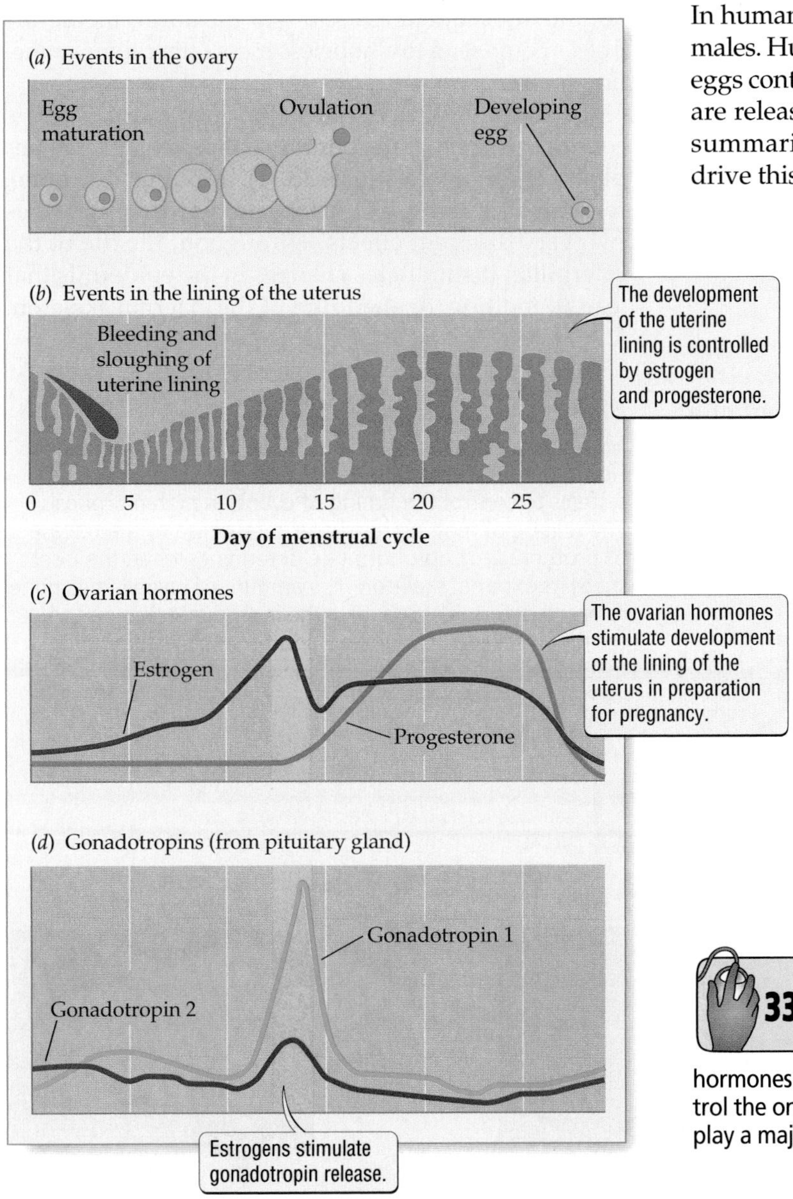

(a) Events in the ovary

Egg maturation   Ovulation   Developing egg

(b) Events in the lining of the uterus

Bleeding and sloughing of uterine lining

The development of the uterine lining is controlled by estrogen and progesterone.

Day of menstrual cycle

(c) Ovarian hormones

Estrogen

Progesterone

The ovarian hormones stimulate development of the lining of the uterus in preparation for pregnancy.

(d) Gonadotropins (from pituitary gland)

Gonadotropin 1

Gonadotropin 2

Estrogens stimulate gonadotropin release.

33.5

Figure 33.10 Hormonal Control of the Menstrual Cycle in Humans
The menstrual cycle of mature human females depends on the sequential release of several hormones over a 28-day cycle. Estrogen and progesterone control the onset of a menstrual cycle, and the two gonadotropins play a major role in stimulating the release of eggs.

lining thickens steadily, and new blood vessels develop that can supply a growing embryo.

If the egg is not fertilized, no embryo develops, and progesterone levels drop about 12 days after the egg's release. Without the high levels of supporting progesterone, the new blood vessels and thickened lining of the uterus do not last long. They separate from the uterus, and strong muscular contractions expel the unneeded tissues. Altogether, a menstrual discharge involves the release of about 2 to 6 tablespoons of blood and uterine lining over several days. The onset of menstrual discharge marks the beginning of a new menstrual cycle as estrogen levels rise and a new egg begins to mature.

> ■ In humans, an interacting set of sex hormones coordinates both the long-term development of male and female reproductive structures and, in females, the events that mark the shorter-term, recurring events of the menstrual cycle.

■ HIGHLIGHT

How Hormones Coordinate the Metamorphosis from Caterpillar to Butterfly

As we will see in Chapter 37, the development of a human or an oak tree from a fertilized egg to maturity involves the precisely timed turning off and on of specific genes in specific parts of the body. Imagine, then,

the complex orchestration needed to control the development of not just one, but several physically different body forms during the lifetime of a butterfly. Butterflies, like other animals, rely on hormones to orchestrate their development, providing a clear example of how a few different hormones can act together to control even tremendously complex events.

In a very real sense, adult butterflies are different animals than the caterpillars from which they arise. Adults develop from clusters of cells called imaginal discs that reside within the growing caterpillar. During the pupal stage in which caterpillars metamorphose into adults, the structures of the caterpillar body are disassembled. The cells of the imaginal discs begin to divide and grow to form adult structures, scavenging materials from the broken-down cells of the caterpillar. Clearly, without careful coordination of the disassembly of the caterpillar and subsequent assembly of the adult, metamorphosis could lead to a hopeless mess rather than a functioning butterfly.

Considering the complexity of metamorphosis, its control by a mere three hormones is astonishingly simple and elegant (Figure 33.11). Molting hormone, released by structures called the prothoracic glands, has two very different effects. Throughout the life of the caterpillar, it stimulates changes in the epidermis that lead to molting, or shedding of the external skeleton.

Figure 33.11   Three Hormones Interact to Coordinate Metamorphosis in Butterflies
Hormone-producing cells link the nervous and endocrine systems to control the timing of events in metamorphosis. The release of molting hormone in response to the release of prothoracicotropic hormone determines when the insect sheds its external skeleton. The amount of juvenile hormone released determines whether the individual takes on the form of a caterpillar, pupa, or adult following the molt.

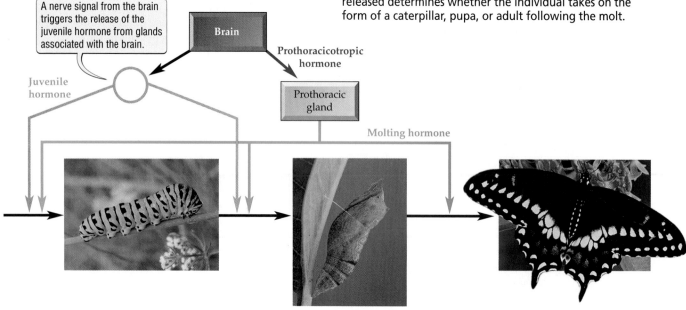

A nerve signal from the brain triggers the release of the juvenile hormone from glands associated with the brain.

Brain

Prothoracicotropic hormone

Juvenile hormone

Prothoracic gland

Molting hormone

Molting hormone also promotes development of the imaginal discs, leading to an adult insect. A second hormone, called juvenile hormone, released by a gland closely associated with the insect brain, controls the outcome of the molt. High juvenile hormone concentrations in the body fluid before a molt lead to a larger caterpillar, intermediate concentrations of juvenile hormone lead to a pupa, and an absence of juvenile hormone leads to an adult.

The timing of events during metamorphosis depends both on signals from the nervous system and on hormones. When signals from the nervous system indicate that a caterpillar has reached the proper stage of development, the brain initiates metamorphosis by stopping the release of juvenile hormone. The timing of molting hormone release depends on a more complex series of events. The brain produces a chemical message called prothoracicotropic hormone, which stimulates the prothoracic gland to release molting hormone. The hormone-producing cells in the gland release the hormone when two conditions are met:

1. Information reaching the brain indicates that the release of prothoracicotropic hormone will cause the molt to the pupal stage to take place at a relatively safe time of day.
2. The level of juvenile hormone indicates to target cells in the brain that the caterpillar has reached a suitable stage in development to metamorphose.

Butterflies can efficiently coordinate a process as complex as metamorphosis by using a combination of interacting chemical messages, by having different target tissues respond differently to each hormone, and, as is true of most animals, by integrating the function of hormones with that of the nervous system.

> ■ The interaction of juvenile hormone, molting hormone, and prothoracicotropic hormone with information supplied by the nervous system allows butterflies to coordinate key events in the metamorphosis from caterpillar to adult.

SUMMARY

How Hormones Work

- The release of a hormone by one cell triggers a response in a target cell. A single hormone may affect many different kinds of target cells, causing a different response in each.
- Hormones typically move through the internal transport system to reach distant target cells. They tend to control the slower, longer-term functions of an organism.

- Target cells amplify the hormonal signal, which means that organisms need to produce only tiny amounts of hormones.

A Small Number of Plant Hormones Control Plant Growth and Development

- Because plants lack a nervous system, they use hormones to coordinate and regulate processes throughout their bodies.
- Five major hormones interact to control all aspects of plant growth and development. Auxin, cytokinin, and gibberellin generally contribute to growth and development. Abscisic acid and ethylene usually suppress growth and other energy-requiring activities.
- Each plant hormone has a variety of effects, depending on the target cells it affects and the stage of plant development.

Chemical Messages Coordinate Plant Defenses

- Chemical messages coordinate the defensive response of plants to parasites and herbivores.

Hormones Regulate the Internal Environment of Animals

- Together, the endocrine system and the nervous system coordinate and regulate the life processes of animals.
- Animals have distinct tissues called glands that produce and release hormones.
- As in plants, several hormones typically act together to coordinate functions such as regulating the availability of essential nutrients.

Hormones in Human Reproduction

- Hormones coordinate human reproduction.
- Several steroid hormones coordinate the long-term process of development of male and female characteristics in humans before birth.
- Two steroid hormones involved in coordinating sexual development before birth—testosterone in males and estrogen in females—interact with a third group of hormones, the gonadotropins, to trigger and coordinate the production of sperm in males and eggs in females once an individual becomes sexually mature.
- In sexually mature females, estrogen and progesterone interact with gonadotropins to regulate the short-term events of the menstrual cycle.

Highlight: How Hormones Coordinate the Metamorphosis from Caterpillar to Butterfly

- During metamorphosis, butterflies undergo a complete reorganization from a caterpillar body form suited for feeding and growing to an adult body form suited for flight and reproduction.
- Three hormones interact to coordinate metamorphosis: juvenile hormone, molting hormone, and prothoracicotropic hormone.

■ Butterflies can coordinate a process as complex as metamorphosis with just three hormones by using them in combination, by having different target tissues respond differently to each hormone, and, as is true of most animals, by integrating the function of hormones with that of the nervous system.

KEY TERMS

endocrine system p. 560

estrogen p. 563

gland p. 560

gonadotropins p. 563

hormone p. 554

hypothalamus p. 563

menstrual cycle p. 563

metamorphosis p. 553

pituitary gland p. 563

target cell p. 554

testosterone p. 554

CHAPTER REVIEW

Self-Quiz

1. Which of the following is true of target cells?
   a. Target cells occur only in the endocrine system of animals.
   b. Target cells amplify the chemical message carried by hormones.
   c. All target cells have receptor proteins on their plasma membrane that bind a specific hormone.
   d. both b and c.

2. The hormones that regulate growth in plants
   a. have no other functions.
   b. all work in the same way.
   c. also occur in animals.
   d. produce a balance between growth and root.

3. Hormones
   a. affect all target cells similarly.
   b. interact with the nervous system in plants.
   c. typically act at very low concentrations.
   d. none of the above

4. The use of hormones by animals differs in which of the following ways from the use of hormones by plants?
   a. Animals do not use hormones.
   b. Animals typically rely on the interaction between several hormones to coordinate a function.
   c. Animal hormones do not move through the internal transport system.
   d. none of the above

5. Estrogen has which of the following effects on target cells in humans?
   a. It coordinates development of the ovaries in the developing embryo.
   b. It coordinates the production of mature eggs at puberty.
   c. It stimulates thickening of the uterus as it prepares to receive a fertilized egg.
   d. all of the above

Review Questions

1. Explain how hormones interact to control the growth of plants.

2. Describe the monthly cycling of hormones in adult human females.

3. In medicine, we use commercially available hormones to compensate for the body's inability to produce a particular hormone, as in the use of insulin by diabetics. In athletics, on the other hand, commercially available hormones are used to supplement otherwise normal levels of hormones. Using what you now know about how hormones work, discuss the similarities and differences in how these two uses of hormones affect the functioning of the body.

33

The Daily Globe

Hormone-Treated Beef Is Safe

To the Editor:

The European Union's stubborn refusal to listen to the international scientific community will continue to cost American beef producers and processors nearly $500 million annually.

In 1989, the EU imposed a ban on imports of American beef because of safety concerns over residual hormones in meat. Now they are refusing to abide by the decision of two international organizations, the World Trade Organization and a joint committee organized by the World Health Organization and the United Nations, who concluded that the EU's ban was overly restrictive.

By giving our cattle hormones such as estrogen, progesterone,

and testosterone, we can produce more flavorful, tender meat at a lower cost. We are not creating a hazardous food; just ask any of the millions of consumers of American beef, 90 percent of which is produced with hormone supplementation.

The EU contends that some of the hormones used to treat cattle could cause cancer in human consumers. Women taking these hormones for birth control or to relieve some of the symptoms of menopause have probably heard from their doctors about the risks of breast and uterine cancer associated with hormone supplements. However, even without such supplements, women's bodies circulate

much higher concentrations of female sex hormones at certain times of the month than they will ever get from eating beef with trace residues of these hormones. Without even eating beef, both men and women consume estrogen-like substances that occur naturally in foods such as yams, peas, dairy products, wheat germ, and soybean oil.

The EU's continuing ban is motivated by politics, not science. The only way to fight the EU's decision is to use a political weapon: trade sanctions that will force it to open European markets to our beef.

William Herdstrom II
Executive Vice President
Organization of American Beef
Producers

Evaluating "The News"

1. Could cultural differences between Europeans and Americans, such as the ways they value scientific information or institutions, affect how they assess the health risks associated with eating agricultural products that have been grown with hormone supplements?

2. Do you find the argument presented in this letter convincing? Why or why not? Are there any other points of view that you would particularly want to hear before deciding how you stand on this issue?

3. Does the argument that women have, at certain times, higher levels of female sex hormones in their

bodies than can be gotten from a meal of beef mean that hormone-treated beef is safe to eat? Why or why not? If different risk factors for cancer can add up to increasing risk, how safe is hormone-treated beef for women? What does this mean for men eating hormone-treated beef?

chapter *34* Nervous and Sensory Systems

Sharon Ellis, *Afternoon*, 1998.

Missing Limbs and Persistent Pain

How can an amputated limb still itch or cause pain? Many amputees suffer from persistent pain that seems to come from the missing limb. The pain is not imagined, and it is not simply a matter of remembering previous pain.

For some, the problem is mild and can be as subtle as an itch. For others, though, it is a severe pain that seems to occur in empty space, but is all too real. Phantom pain is a well-documented problem in amputees. Sometimes physical sensations, such as arthritis pain, that affected the limb before its loss continue after its amputation. In some cases, time lessens the severity and the frequency of the painful sensations.

Amputees Can Feel Missing Limbs

In a related problem, called reflex sympathetic dystrophy syndrome, pain occurs in an area of former injury, even after all the damage has healed completely. This painful condition is difficult to relieve. In some desperate cases, nerves have been cut in an effort to stop the pain, but the pain still does not lessen or go away. What could cause the painful sensations of reflex sympathetic dystrophy syndrome and phantom pain to persist in otherwise healthy tissues?

In this chapter we explore how the nervous systems of animals gather information, transmit information to the brain, and interpret and respond to information. We then return to a consideration of phantom pain and reflex sympathetic dystrophy syndrome.

■ KEY CONCEPTS

1. Neurons transmit information rapidly in the form of electrical signals called action potentials.

2. A variety of sensory cells generate action potentials in response to chemicals, mechanical stimuli, and light in the environment.

3. The central nervous system allows animals to integrate and respond to sensory information.

*A*ll multicellular organisms must coordinate the functions of their various cells and tissues. As we saw in Chapter 33, plants and animals rely heavily on signaling molecules called hormones to transmit information between cells. Animals have a second, high-speed internal communication system called the **nervous system**. The nervous system consists of specialized cells that can transmit information much faster than hormones traveling through the circulatory system (see Figure 33.2). These nerve cells, or **neurons**, can carry messages throughout the body in a fraction of a second. Animals need this almost instantaneous form of communication to coordinate the rapid contraction of their muscles, which play such an essential role in their ability to run, swim, and fly (see Chapter 27).

Having a high-speed nervous system also gives animals an unequaled ability to gather information about the environment in which they live. Many of the internal adjustments that organisms must make depend on external conditions. To gather information about those conditions, they rely on sensory structures that detect chemicals, pressure, and light. Although all organisms need to gather this sort of information, the nervous systems of animals allow them to collect and process much more complex sensory information than other organisms do.

In this chapter we first consider how the nervous system can transmit information so quickly. We then introduce sensory structures, focusing on those of animals. We finish by discussing how the arrangement of neurons into a nervous system affects the ability of animals to process sensory information and generate an appropriate response.

34.1 How Neurons Create and Transmit Rapid Electrical Signals

Neurons, like muscle cells, are found only in animals. In this section we look at how neurons work and how they communicate with one another in an interconnected nervous system.

Neurons and nerves

Neurons receive a signal from another cell or from the environment, transmit the signal rapidly over a long distance, and then pass the signal on to another cell. Neurons interact with other neurons, with muscles, and with hormone-producing cells.

The form of a neuron reflects its unique function (Figure 34.1). The cell body of a neuron contains all the basic components of a eukaryotic cell. Branched **dendrites** at one end of the neuron receive signals from adjacent cells. These signals travel to the **axon**, a wirelike extension that carries them to other cells, sometimes over a long distance. Although we think of cells as small, human neurons can have axons extending up to a meter in length. In vertebrates, helper cells associated with neurons produce an insulating sheath of a fatty material called **myelin** that encases the axon. As we will see, myelinated axons can carry signals more rapidly than unmyelinated ones. At the end of the axon, axon terminals transmit signals to other cells. The axons of many individual neurons may be bundled together to form major communication pathways called **nerves**.

Axons transmit self-strengthening electrical signals in one direction

The secret to the rapid transmission of information by neurons lies in electrical signals that travel along the axon. We are most familiar with electrical signals in the form of electrical currents. When an electrical current flows through a wire, negatively charged particles called electrons move from an area of negative charge to an area of positive charge (Figure 34.2a). When you switch on a flashlight, for example, electrons flow from the negative end of the battery to the positive end. Electrical currents can move rapidly (for example, electrical currents can carry your voice almost instantaneously over hundreds of kilometers of telephone line), but have the disadvantage of weakening over distance. In metal wires, the current weakens only a little even over long distances, but in biological materials, including axons, it

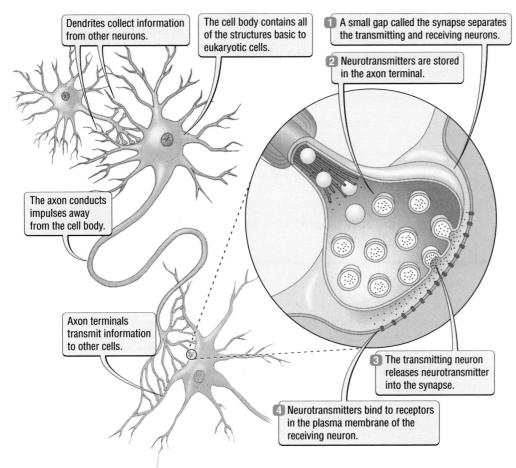

Dendrites collect information from other neurons.

The cell body contains all of the structures basic to eukaryotic cells.

The axon conducts impulses away from the cell body.

Axon terminals transmit information to other cells.

1 A small gap called the synapse separates the transmitting and receiving neurons.

2 Neurotransmitters are stored in the axon terminal.

3 The transmitting neuron releases neurotransmitter into the synapse.

4 Neurotransmitters bind to receptors in the plasma membrane of the receiving neuron.

34.2

Figure 34.1 Neurons Carry Information Rapidly from One Cell to Another Neurons rapidly transmit signals in the form of action potentials. Where neurons meet, electrical action potentials are converted into chemical signals in the form of neurotransmitters, which diffuse across a small gap (the synapse) between the neurons.

weakens by half over distances as small as a few millimeters.

To solve the problem of rapidly weakening signals, axons transmit information as a special kind of self-strengthening electrical signal called an **action potential**. These signals depend on the rapid movement of positively charged sodium ions across the plasma membrane of the axon. When the axon is not transmitting an action potential, active carrier proteins in the plasma membrane pump positively charged sodium ions from the axon into the surrounding fluid, creating a relatively negative charge inside the axon (Figure 34.2*b*). An action potential begins when, in response to a signal of some kind, the charge difference across a portion of the plasma membrane decreases to a critical level—a phenomenon called **depolarization**. Channel proteins in the depolarized stretch of the plasma membrane open to let sodium ions flow rapidly into the axon from the surrounding fluid. The rapid entry of sodium ions reverses the charge difference across the plasma membrane, so that the inside of the affected portion of the axon takes on a relatively positive charge. This reversal of charge depolarizes the next segment of axon plasma membrane, triggering a wave of reversal of charge that

remains consistently strong as it moves down the axon.

Because action potentials involve the movement of ions across plasma membranes, they move more slowly than the flow of electrons in an electrical current. Whereas we notice no delays in a telephone conversation carried by an electrical current between two cities 1000 kilometers apart, it would take an action potential traveling at a typical speed of 5 meters per second more than 2 days to cover the same distance. For most animals, however, an action potential moving at this clip travels the length of the body in a fraction of a second.

The myelinated axons of vertebrates transmit action potentials especially rapidly

Among vertebrates and in a few invertebrates, the axons of most neurons are surrounded by a myelin sheath, which greatly speeds signal transmission. The myelin sheath covers segments of the axon, leaving unmyelinated openings between the myelinated zones. Action potentials can begin only in the unmyelinated openings, because only here can sodium ions move across the plasma membrane. When an action potential begins at one unmyelinated opening, the axon interior there depolarizes, just as in the action potentials described earlier. The myelin sheath, however, prevents the action potential from moving along the axon. Instead, an electrical current flows between the depolarized opening and the next opening along the axon (Figure 34.2*c*). As a result, the axon depolarizes at the next opening. Because the electrical current moves along the axon much more quickly than an action potential would, the signal jumps

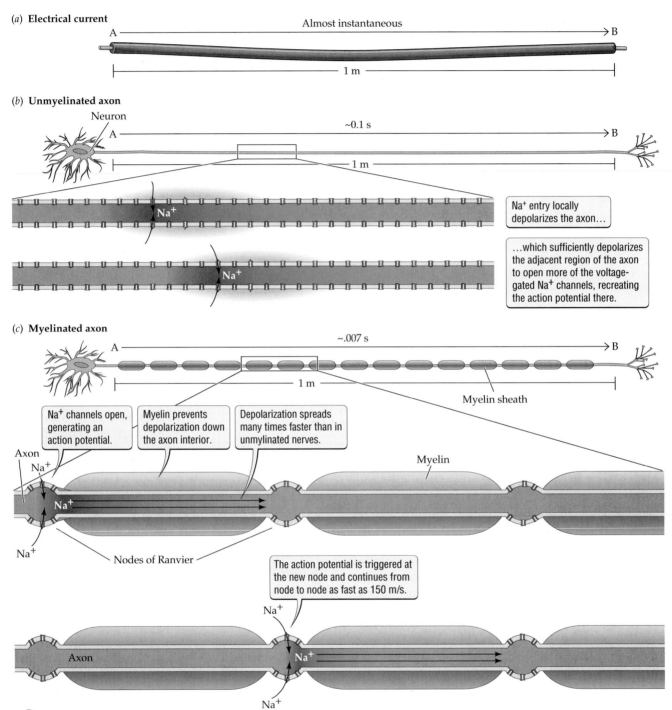

(a) **Electrical current**

A ————————————— Almost instantaneous ————————————→ B

|———————————————— 1 m ————————————|

(b) **Unmyelinated axon**

Neuron

A ————————————— ~0.1 s ————————————→ B

|———————————————— 1 m ————————————|

Na⁺

Na⁺ entry locally depolarizes the axon...

Na⁺

...which sufficiently depolarizes the adjacent region of the axon to open more of the voltage-gated Na⁺ channels, recreating the action potential there.

(c) **Myelinated axon**

A ————————————— ~.007 s ————————————→ B

|———————————————— 1 m ————————————|

Myelin sheath

Na⁺ channels open, generating an action potential.

Myelin prevents depolarization down the axon interior.

Depolarization spreads many times faster than in unmylinated nerves.

Myelin

Axon

Na⁺

Na⁺

Na⁺

Nodes of Ranvier

The action potential is triggered at the new node and continues from node to node as fast as 150 m/s.

Na⁺

Axon

Na⁺

Na⁺

34.3

Figure 34.2   How Neurons Transmit Signals
(a) Electrical wires rapidly carry currents that flow from areas of negative charge to area of positive charge. (b) Unmyelinated axons carry self-strengthening action potentials that rely on changes in the charge difference across their plasma membranes. (c) Myelinated axons carry signals by a combination of rapid electrical currents and self-strengthening action potentials.

rapidly from one unmyelinated opening to the next along the myelinated axon.

Organisms with myelinated axons benefit from the best features of signal transmission by action potential and by electrical current. The action potentials that form in the unmyelinated openings maintain the strength of the signal as it travels down the axon. The current that flows between the unmyelinated openings transmits the signal many times more rapidly than would be possible by action potential alone: Whereas completely unmyelinated nerves transmit action potentials at a maximum speed of about 30 meters per second, myelinated nerves allow action potentials to jump along at up to 150 meters per second.

A defective myelin sheath can lead to serious problems in humans. Diseases that affect the myelin sheath, such as multiple sclerosis and Guillain-Barré syndrome, can disrupt vision, speech, balance, and muscular coordination, and may eventually kill their victims.

Neurotransmitters transmit signals between adjacent cells

The exchange of information between a neuron and an adjacent cell takes place across a minute gap less than a millionth of a millimeter in width, called a **synapse** (see Figure 34.1). Synapses separate the axon terminals of a neuron from the adjacent cells that receive its signals. In a few instances, action potentials pass directly from one cell to the next. In most cases, however, signaling molecules called **neurotransmitters** bridge the synapse. When an action potential traveling down an axon reaches the end of the axon, it causes the axon terminals to release neurotransmitters into the synapse. Like hormones, neurotransmitters diffuse across the synapse to the target cell and bind to matching receptor proteins in the plasma membrane of the target cell. When enough of the appropriate neurotransmitter binds to the membrane of the target neuron, it depolarizes to trigger an action potential, which begins its rapid journey down the axon of the target neuron.

Given that the speed with which the nervous system can transmit information represents a major advantage over the endocrine system, it may seem odd that neurons depend on relatively slow signaling molecules to transmit information between cells. We should recall, however, that molecules can diffuse over tiny distances in fractions of a second (see Figure 30.1). In addition, long axons ensure that action potentials have to cross only a few synapses to reach their destinations.

The connections between neurons form an essential component of nervous systems because it is here that neurons exchange information. Some neurons in humans may interact with as many as 10,000 other neurons in a dense tangle of dendrites and axon terminals. Neurons both produce and respond to many different neurotransmitters (Table 34.1). The type of receptor that interacts with a certain neurotransmitter determines the response that follows. The diversity of receptor types on the target cell membrane determines which events can

34.1 Functions of Common Human Neurotransmitters	
Neurotransmitter	**Major functions**
Acetylcholine	Muscle control
Dopamine	Muscle activity
GABA (gamma-aminobutyric acid)	Assists muscle coordination by inhibiting counterproductive or unneeded neurons
Adrenaline	Maintains state of alertness, readiness
Serotonin	Temperature regulation, sensory transmission, sleep
Melatonin	Day–night cycles, such as sleep regulation
Enkephalins and endorphins	Inhibits transmission of signals from pain sensors, pain perception
Substance P	Regulates transmission of signals from pain sensors

follow neurotransmitter binding. Thus, one neurotransmitter can bind several different receptor types and trigger different responses. Acetylcholine, for example, is the major neurotransmitter responsible for signaling skeletal muscles to contract, yet in heart muscle it signals relaxation.

> ■ Neurons transmit information down axons as self-strengthening electrical signals called action potentials. A myelin sheath surrounds some axons, enhancing the speed of action potential transmission. Neurotransmitters diffuse rapidly across the narrow synapses that separate neurons from their target cells.

34.5 Sensory Structures Allow Organisms to Respond to Their Environment

All life—from single-celled bacteria to plants to animals—must have ways of gathering information about its surroundings. Perceiving such things as the time of day, the temperature, or the presence of enemies or food makes it possible for organisms to coordinate their activities effectively. Although all organisms gather such sensory information, none do so in as sophisticated a way as animals. When you walk down the street, for example, you sense and interpret constantly changing patterns of light hitting millions of light-sensitive cells in each of your eyes, and your brain compares the input from your two eyes to make sense of what you see. The speed with which neurons transmit information makes it possible for animals to deal with rapid changes in their environment and to compare information collected at the same time at different points. For this reason, we focus here on how animals sense their environment, although we also mention examples of sensation by other organisms.

In looking at how animals sense their surroundings, we should keep in mind three general patterns:

1. The sensory structures of animals all convert some property of the environment into action potentials.
2. Having more cells sensitive to a particular feature of the environment allows an animal to detect smaller changes in that feature.
3. Paired patches of cells sensitive to a stimulus of a particular type make it possible for an animal to locate the source of the stimulus.

Chemicals provide organisms with essential sensory information

The ability to sense the presence of chemicals allows organisms from bacteria to animals to identify and find important features of their environment. Bacteria use chemicals dissolved in water to locate suitable food and habitat (see Chapter 35); the airborne gas ethylene influences the rate of fruit ripening in flowering plants (see Chapter 33); and animals use chemicals to identify food, detect predators, and find mates. Chemicals in the environment can provide remarkably detailed information. A mother bat, for example, uses her sense of smell to identify the unique combination of chemicals coming from her baby among the million or so other bats that may roost in the same cave. Male silkworm moths can detect bombykol, a chemical produced by female silkworm moths as a way of attracting mates, at concentrations as low as 200 molecules in a cubic centimeter of air. Even humans, who have a rather low sensitivity to tastes and smells, can detect small amounts of important chemicals. Bitter tastes, for example, often warn of toxic secondary chemicals in plants that we eat (see Chapter 32), and we can taste quinine, the chemical that gives tonic water its bitter taste, at concentrations 2500 times lower than those we need to detect table sugar.

Chemical sensors respond to the presence of chemicals by triggering an action potential. These sensors may be modified neurons or specialized sensory cells associated closely with neurons. Although diverse in their form, all chemical sensors depend on receptor proteins in their plasma membrane that bind specifically to certain chemicals, much as the receptors of target cells bind

specifically to certain hormones. When the matching chemical binds to a receptor protein, the sensory neuron or cell depolarizes to set in motion an action potential.

Some chemicals "fool" the sensory cells, leading us to taste or smell something that is not really there. The sweetener saccharine, for example, binds to sugar-specific receptor proteins to make us sense a sweet taste. Toothpaste and artichokes, on the other hand, contain chemicals that block the functioning of some taste sensors, leading to distorted tastes—try drinking some orange juice right after brushing your teeth.

Humans and many other land animals have distinct senses of smell and taste. Smell involves detecting airborne chemicals that often reach the sensory cells at low concentrations, such as the bombykol that attracts male silkworm moths. Taste, on the other hand, involves detecting chemicals dissolved in water at relatively high concentrations. When we taste something, however, we combine information from our senses of taste and smell. Professional food tasters (they do exist!) focus equally on smell and taste when evaluating a product. For aquatic creatures, the distinction between smell and taste becomes even muddier, since all chemicals reach aquatic animals dissolved in water.

In animals, chemical sensors are concentrated in those parts of the body most likely to be exposed to airborne or dissolved chemicals (Figure 34.3). Thus, in humans, smell sensors crowd the lining of the nose, through which the air we breathe moves in and out, and taste sensors fill some 4000 taste buds arranged on the surface of the tongue, over which all food passes. Similarly, male silkworm moths have about 50,000 sensors specific to bombykol arranged on paired feathery antennae that sift the air, and female silkworm moths have taste sensors on their legs that allow them to taste plant leaves as they search for the right place to lay their eggs. Fish have dense arrays of chemical sensors in their paired nostrils, their mouths, and spread over the surface of their skin.

The senses of touch and hearing interpret mechanical stimuli

Animals and many other organisms collect information about mechanical stimuli for many of the same reasons that they use the more widespread chemical-detecting sense. The sense of touch informs us about our immediate environment: Is it hot or cold? Smooth or rough? Is something hurting us? Is something crawling on our skin? Perhaps more importantly, touch sensors inform an organism about the position of its body: Are we standing or sitting? In what position is our arm? Which way is up? How fast are we moving?

(a)

Molecules that we sense as taste bind to membrane receptors on the sensory cells.

Taste sensor

Taste buds are located on the surface of the tongue.

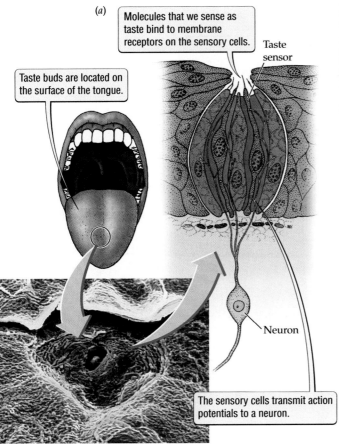

Neuron

The sensory cells transmit action potentials to a neuron.

A taste bud

Figure 34.3 Taste in Humans and Fish
(*a*) Taste buds are groups of sensory and supporting cells that respond to chemicals that enter the mouth and throat as we break up food with our teeth. (*b*) Catfish have chemical sensors inside the mouth, inside the nostrils, and scattered over the body surface

(b)

these vibrating air molecules create alternating zones of high pressure and low pressure that radiate out in all directions. Sound, then, consists of extremely rapid pressure changes that occur many times in a second. The number of these changes per second determines the pitch, or frequency, of the sound. The lowest frequency sounds that humans can hear involve about 20 changes from high to low pressure per second, and the highest frequency sounds

Many different kinds of sensory cells respond to different mechanical stimuli (Figure 34.4). Like chemical sensors, mechanical sensors include both neurons and specialized cells associated with neurons. All trigger an action potential in response to a particular kind of mechanical stimulus. For example, some neurons that are sensitive to pressure depolarize when their plasma membrane is stretched. Neurons specialized to detect slight pressures are often associated with a hair, which works like a lever to convert a small amount of pressure applied to the end of the hair into a greater pressure on the sensory neuron. For this reason, your relatively hairy forearm can easily detect a slight breeze that you barely feel on the back of your hand. The number of mechanical sensory cells varies over the surface of an animal. Human fingers, which we use to manipulate objects in our environment, have more sensory cells sensitive to mechanical stimuli than any other part of our body surface.

Although we think of hearing as a distinct sense, it is really no more than an ability to detect minute changes in pressure. We hear the vibrations of engines or vocal cords or guitar strings as sound. When a vibrating surface or string causes adjacent air molecules to vibrate,

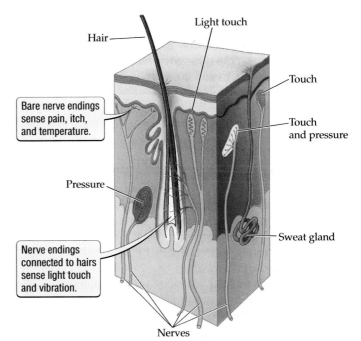

Light touch

Hair

Touch

Bare nerve endings sense pain, itch, and temperature.

Touch and pressure

Pressure

Sweat gland

Nerve endings connected to hairs sense light touch and vibration.

Nerves

34.6

Figure 34.4 Touch Sensors in Human Skin
The skin contains many different types of neurons and sensory cells that provide detailed information about mechanical stimuli.

(a)

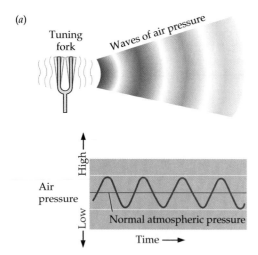

Figure 34.5   The Human Ear Detects Minute Pressure Changes
(a) Sound caused by a vibrating object consists of rapidly changing zones of high and low air pressure. (b) The construction of the human ear allows it to convert tiny and rapid changes in air pressure into movements that pressure-sensitive cells can detect. The outer ear acts as a funnel that collects sound over a large area and concentrates it onto the small area of the eardrum. The eardrum converts changes in air pressure into physical movements, which three tiny bones in the middle ear transmit to the cochlea of the inner ear. In the cochlea, patches of sensory cells distinguish different sound frequencies.

(b)

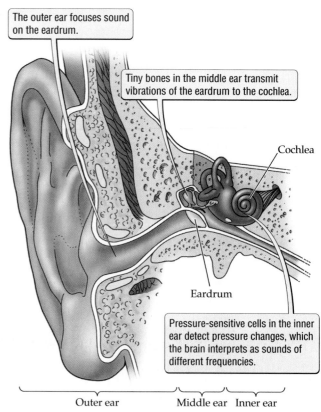

The outer ear focuses sound on the eardrum.

Tiny bones in the middle ear transmit vibrations of the eardrum to the cochlea.

Cochlea

Eardrum

Pressure-sensitive cells in the inner ear detect pressure changes, which the brain interprets as sounds of different frequencies.

Outer ear   Middle ear   Inner ear

involve around 20,000 such pressure changes each second (Figure 34.5a). The loudness of sound reflects the distance over which the air molecules vibrate, which in turn determines the difference between the lowest and highest pressure. Even for painfully loud sounds, however, air molecules vibrate over a distance of only a fraction of a millimeter. The challenge of detecting sound, which relatively few animals have managed to meet, lies in arranging pressure-sensitive cells so that they can detect these tiny and rapid movements of air.

The human ear illustrates the essential features of the various ear designs used by animals to convert tiny, rapid changes in air pressure into a detectable signal: a funnel form combined with thin, delicate membranes (Figure 34.5b). The ear magnifies the pressure changes by concentrating sound collected over a large area onto a small area of membrane. Our outer ear, like that of most animals that can hear, has a funnel shape that collects sound over an area about 20 times that of the eardrum. The eardrum is a delicate membrane that vibrates quickly in response to rapid changes in air pressure. By doing so, it converts these changes into physical movements, which the three tiniest bones in the human body transmit to a second membrane. This second membrane has a smaller surface area than the

eardrum, continuing the funneling effect and creating relatively strong vibrations in the fluid-filled cochlea of the inner ear. Pressure-sensitive hairs arranged along the cochlea convert the resulting changes in the shape of the cochlea wall into action potentials that the brain interprets as sound. Not all animals that can hear can distinguish different frequencies of sound. The structure of the cochlea, which is found only in vertebrate ears, allows us to recognize high and low sounds: Sounds having different frequencies stretch the wall of the cochlea at different points, so that pressure-sensitive hairs trigger an action potential corresponding to the frequency of the sound.

Many organisms can distinguish light from dark, but few can form images

Many organisms, including bacteria, plants, fungi (see the box on page 577), and animals, can distinguish light from dark, but only a relatively few animals can distinguish images. To detect light, an organism needs only a light-sensitive pigment. If the change in the shape of the pigment molecule in response to light triggers either a signaling molecule or an action potential, the organism can respond to light. The clustering of sensory cells containing light-sensitive pigments increases the chances

■ BIOLOGY IN OUR LIVES

Eye on the Pie

Most of us avoid the life that lives on the end products of cow metabolism: cow pies. By doing so, we miss out on some of life's little wonders.

For a small fungus called *Pilobolus*, however, cow pies are the stuff of life. *Pilobolus* feeds exclusively on the droppings of grazing mammals. The moment they emerge from the tail end of a cow, droppings quickly begin to dry from the outside in, and they eventually disappear as a host of competing detritivores feasts on their nutrients. To succeed, *Pilobolus* must colonize the freshest possible cow pies, both to allow its feeding structures, the hyphae, time to penetrate the outside of the dropping before it crusts over and to get a head start on the competition. *Pilobolus* manages this feat by passing its spores, from which new individuals grow, through the cow along with its food. In this way *Pilobolus* is there when the cow pie hits the ground.

To accomplish this, *Pilobolus* must place its spores on plants growing in open areas where cows like to graze, rather than in sheltered locations. Whereas typical mushrooms drop their spores passively into a breeze, which spreads them all willy-nilly through the surroundings, *Pilobolus* uses a light-guided cannon that blasts sticky spores specifically into open areas. At the heart of this spore-dispersal mechanism is perhaps the most sophisticated eye found outside the animal kingdom. At the tip of the capsule containing its spores, *Pilobolus* produces a patch of cells that contain light-sensitive pigments. To increase its sensitivity to light, a clear lens focuses light on the patch of sensory cells. By keeping this eye aimed toward bright light, *Pilobolus* orients its capsule toward an open

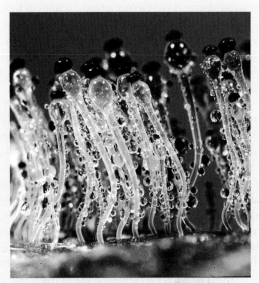

The Light-Guided Cannon of *Pilobolus*
This tiny fungus uses a sophisticated eye to direct its spores toward suitable habitat.

area. When the spores mature, the capsule explodes to launch the spores several meters into prime grazing territory. The unprepossessing lifestyle of *Pilobolus* reminds us that not only animals sense their environment.

---

that the animal will be able to detect low levels of light. A **lens** concentrates light in much the same way that ears concentrate sound (think of the magnifying glass that some of us, in our misguided youth, used to toast innocent ants), providing a second way of increasing sensitivity to light (Figure 34.6). The simple ability to distin-

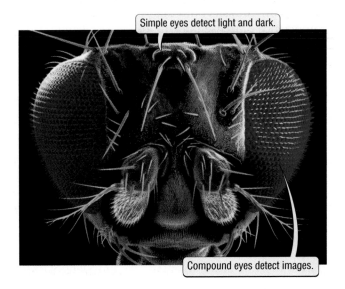

Simple eyes detect light and dark.

Compound eyes detect images.

Figure 34.6   The Many Eyes of Insects
Insects such as this fly have two kinds of eyes. Simple eyes distinguish light and dark by clustering light-sensitive cells together and using lenses to increase their sensitivity. Compound eyes represent an alternative approach to the image-forming eyes of vertebrates such as humans. Instead of having a single lens focusing light onto a retina consisting of many light-sensitive cells, the compound eye of insects consist of many units, each having a lens that focuses light onto a few light-sensitive cells. In the brain, the combined information from thousands of such units forms an image.

guish light from dark can provide an organism with information about its position and the time of day. If animals, with their rapidly acting nervous systems, have two or more clusters of light-sensitive cells on the body surface, they can detect motion by following changes over time in which light-sensitive areas are detecting light. Thus, even extraordinarily simple light-detecting structures can provide a great deal of information.

Human eyes go well beyond merely responding to light and dark to form images of objects in their environment. The great variety of image-forming eyes found in animals have two features in common: first, a way of capturing the image—that is, a biological equivalent of the film we use in cameras—and second, a way of ensuring that all of the light reaching the eye from a single point on an object goes to a single point on the biological "film." A patch of light-sensitive cells that is used to capture images is called a **retina**. Separate neurons carry information from one or a few of the light-sensing cells to the brain. Think of a newspaper photograph, which under slight magnification resolves into a field of black and white dots: Each neuron carries information for one of those "dots." The more densely packed the independent light-sensing units are, the crisper the image (compare the pattern of dots in relatively low resolution newspaper photographs with those in glossy magazines). Because about a million axons lead from each of our eyes to the brain, humans can form sharp images. In the center of our retina lies a zone of especially tightly packed light-sensing units that allows us to form especially sharp images. As you read this page, you line up your eyes so that the word that you are reading projects onto this zone. To see the importance of dense packing of light-sensing units, look at a particular word and then, without moving your eyes, try to make out surrounding words.

The lens also contributes to the sharpness of an image. Its characteristic shape allows it to bend light so that all of the light coming from a particular point on an object is focused on a particular point on the retina (Figure 34.7). One of the most complex tasks facing an image-forming eye lies in adjusting the focus of the lens to accommodate objects at different distances from the eye: The lens must bend light from nearby objects more sharply than light from distant objects to maintain focus. The lens of the human eye changes its shape to focus images on the retina. Some other animals either move the entire retina nearer to or farther from a lens of fixed

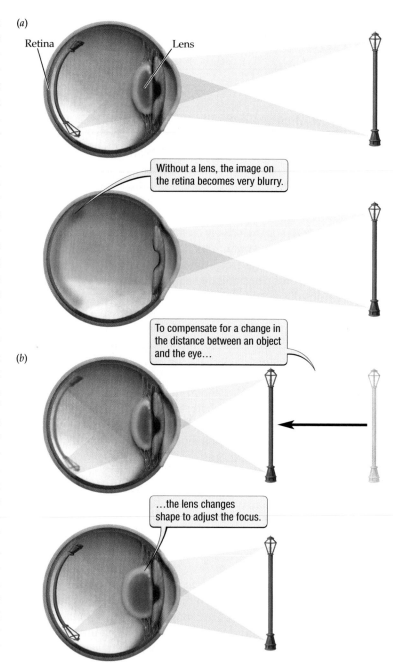

(a)

Retina   Lens

Without a lens, the image on the retina becomes very blurry.

(b)

To compensate for a change in the distance between an object and the eye…

…the lens changes shape to adjust the focus.

Figure 34.7 How the Human Eye Forms Sharp Images (a) Lenses bring images into focus on a layer of light-sensitive cells in the retina. (b) As the distance to an object changes, the eye must adjust to keep the image focused on the retina.

shape, as in jumping spiders, or move the lens nearer to or farther from the retina, as in octopuses. Whatever the mechanism, focusing requires precise muscular control and a nervous system that can continually monitor the quality of an image.

■ Sensory information lets organisms identify and find important features in their environment. Sensory cells respond to a particular type of stimulus and convert that information into action potentials. Whereas sensory structures that can detect chemicals are widespread among living organisms, structures that can detect mechanical stimuli and light occur most conspicuously among animals.

34.7 Making Sense of Action Potentials in the Central Nervous System

We have introduced the basic features of the neurons that make up the nervous system and the sensory structures that feed information into the nervous system, but we have yet to consider how the nervous system itself works. In other words, how does the nervous system convert sensory information into an appropriate response? The central nervous system of animals is made up of complex interconnections among neurons that convert sensory input into meaningful information that can trigger an appropriate reaction. The central nervous system, and especially the brain, converts information into action.

Sensory information can move through the nervous system by various routes, from simple circuits involving only a few neurons to complicated pathways involving vast numbers of neurons in different areas of the brain and central nervous system. As a rule, the more complicated the required response to information, the more processing it requires. In this section we focus on our own highly developed central nervous system to explore some of its basic functions, looking at how and where we process information and how its processing affects the way we perceive our world. In Chapter 35, we will look at the output that results from this processing: the complex behaviors of animals.

Figure 34.8 Organization of the Human Central Nervous System
Sensory information (shown in red) travels to the spinal cord, where it may be processed and acted upon or, more often, sent on to the brain for additional integration and interpretation. Certain parts of the brain are devoted to the processing of certain kinds of information. Commands from the brain (shown in blue) leave the spinal cord by a slightly different route. In the body, these commands may stimulate or inhibit organs, muscles, or glands, and may override the actions triggered by reflex arcs or spinal cord processing.

The organization of the central nervous system

Neurons carry action potentials to and from the **central nervous system**, the component of the nervous system devoted to the exchange of information among neurons. Even animals with simple nervous systems, such as jellyfish, have structures called **ganglia** (singular ganglion), consisting of densely packed neuron cell bodies, that allow the exchange of signals between many different neurons. In animals that move, the forward end usually houses a dominant ganglion or a more complex structure, called a **brain**, which serves as the major clearinghouse for information (Figure 34.8). The forward end of a mobile animal also contains most of the sensory

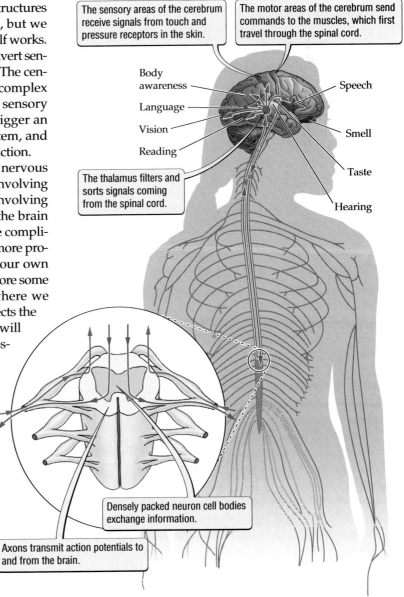

The sensory areas of the cerebrum receive signals from touch and pressure receptors in the skin.

The motor areas of the cerebrum send commands to the muscles, which first travel through the spinal cord.

Body awareness

Language

Vision

Reading

Speech

Smell

Taste

Hearing

The thalamus filters and sorts signals coming from the spinal cord.

Densely packed neuron cell bodies exchange information.

Axons transmit action potentials to and from the brain.

structures because it encounters new stimuli first. Placing the brain near the sensory structures reduces the time it takes for the brain to receive sensory information. Animals that do not move, such as clams, or do not have an end that is consistently at the front, such as jellyfish, often lack brains.

A large diameter nerve or collection of nerves that carries information between the brain and the hindmost part of the animal forms the final component of a typical central nervous system. Vertebrates, including humans, have a specialized spinal cord that serves this purpose. In contrast to the large nerves used by other animals, the spinal cord includes large concentrations of dendrites and axon terminals that allow information exchange.

The reflex arc is an example of simple information processing

When immediate action can help an organism avoid injury, **reflex arcs** allow a rapid response to signs of trouble. In a reflex arc, information from a sensory neuron passes directly to a neuron associated with a muscle, without much processing by the central nervous system (see Figure 9.2). The advantage of reflex arcs is that they process sensory inputs more quickly than they can travel to the brain for interpretation. When your finger touches a hot stove, you do not want to waste time processing information from various sensory structures to figure out just how much pain you are in. You need to get your finger off the burner quickly.

The spinal cord and the brain handle more complex processing

The **spinal cord**—a collection of nerve tissue about as thick as a finger—is an organized collection of neuron cell bodies and axons. Dense collections of neuron cell bodies, dendrites, and axon terminals exchange information, and large bundles of axons carry this information to and from the brain (see Figure 34.8). The spinal cord of vertebrates forms an intermediate site of information processing that acts as a buffer between the brain and the sensory neurons.

In contrast to the exchange of information between a few neurons in a reflex arc, the brain receives information in the form of action potentials from millions of sensory neurons, and it can direct a response to that information to millions of neurons associated with muscles. The human brain consists of huge numbers of neurons arranged to allow the efficient exchange of information. Although the brain requires more time to process information than a reflex arc does, its vast num-ber of neurons can sort and interpret a bewildering array of incoming information.

Organization of the human brain

Ultimately, most information ends up in the brain. Far from being a homogeneous mass of neurons, the brain has distinct portions that handle specific kinds of information in specific ways. One of the major relay areas is the **thalamus**, a small central switching area in the brain (see Figure 34.8). In fact, the thalamus is a critical filter, saving the brain from processing all the stimuli that constantly bombard our sensors. The thalamus determines which of the sensory signals coming from the spinal cord should go to the conscious perception centers in the cerebrum and which should go to the other portions of the brain that act on sensory inputs without our being aware of it. The act of swatting away a fly, for example, might require an unthinking reflex the first time. After the tenth fly, however, the information might go to the cerebrum for more careful or strategic contemplation.

The **cerebrum** forms the most conspicuous portion of the human brain (but not in the brains of many other vertebrates). Large bundles of axons carry information between the left and right halves of the cerebrum. The outer layer of the cerebrum, called the cerebral cortex, looks heavily wrinkled. If the thalamus sends signals to the cerebrum, they usually arrive in a specific part of this wrinkled outer portion that corresponds to the area of the body that received the original stimulus (Figure 34.9). These sensory signals become part of our conscious awareness. Responses to these signals by our muscles, which also arise in well-defined portions of the cerebrum, are also those of which we are aware.

As we saw in Chapter 33, the brain also has close connections with the endocrine system. Its close communication with the endocrine system plays an important role in the regulation of internal functions. Neurons from a portion of the brain called the hypothalamus, for example, directly stimulate the release of hormones from the pituitary gland, an important endocrine gland that coordinates many important long-term processes involved in growth, regulation of water and solute content, and reproduction.

■ The human central nervous system integrates and interprets sensory information to allow appropriate responses to environmental stimuli. The central nervous system consists of the spinal cord and the brain. Reflex arcs let the body bypass the central nervous system in situations requiring especially rapid responses.

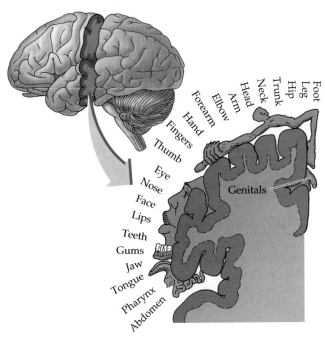

Figure 34.9 How the Brain "Sees" the Body
Different parts of the cerebral cortex are devoted to processing sensory information received from different parts of the body. The amount of brain tissue devoted to processing this sensory information does not necessarily correspond to the size of the body part. The body parts shown here look distorted because they are drawn to indicate the amount of brain area devoted to the sensory information they send to the brain. For example, you can see that more of the brain is devoted to receiving sensory information from the foot, a relatively small body part, than from the leg, a much larger body part.

■ HIGHLIGHT

Perception and Pain

When we encounter light, sound, smells, or any other kind of stimulus in the environment, we receive one type of signal, but we *perceive* something completely different in each case. Our eyes, for example, do not perceive the world as consisting of thousands of points of color, even though images leave the retina that way. The brain assembles the raw information from the sensory neurons into our perception of the world around us. Signals may be shared among several areas of the cerebrum for further interpretation.

Pain is a special case of perception. The perception of pain protects an animal by encouraging it to stop a potentially damaging behavior. The sole function of certain sensory cells and neurons appears to be to notify the brain of painful stimuli, especially in the skin. In addition, the role of some neurotransmitters appears to be limited to communicating pain messages. Pain would be much easier to manage if it were a simple matter of knocking out the pain sensors or pain transmitters. However, the perception of pain seems to be a complicated result of processing in the brain.

Conditions such as reflex sympathetic dystrophy syndrome and phantom pain remind us of the complex ways in which the human central nervous system processes information. The fact that cutting the nerves does not cure reflex sympathetic dystrophy syndrome provides an additional clue that sensory cells are not solely responsible. Even without receiving any information from sensory cells in the amputated body part, our brain fills in information that it thinks should be there.

Even as a surgical patient, you may experience phantoms. Many patients who have been anesthetized using a chemical that blocks action potential transmission nonetheless continue to "feel" the anesthetized limb. Interestingly, the phantom limb often is perceived to have a different position than the anesthetized limb has. When the patient looks at the anesthetized limb, the phantom limb appears to jump into the real limb as the brain tries to reconcile what it sees with what it thinks it feels.

Phantom limb pain provides another example of the brain's role in perception. Sensory pathways in the brain that once gave it useful information about the status of a limb fail to reorganize after an amputation. Until these pathways change, a particular part of the brain still has neurons reporting a problem. Because the brain cannot filter out the false information, the pain persists.

Modern advances in prosthetic technology (the technology of artificial limbs) take advantage of electrical activity in the remaining muscles, as well as electrical signals coming from brain itself. These tiny electrical currents can drive some of the prosthetic devices that help amputees reclaim some of the lost function once handled by a missing limb. These devices are being developed to manage tasks that require both strength and delicate coordination.

For some amputees, the sensory pathways adjust over time, and the phantom sensations diminish. For others who must endure them, phantom pain and reflex sympathetic dystrophy syndrome provide a lifelong reminder of the complexities of the human nervous system.

> ■ Our perception of the world comes from the brain's integration and interpretation of incoming information, as in pain perception.

SUMMARY

How Neurons Create and Transmit Rapid Electrical Signals

- Animals have unique cells called neurons that rapidly transmit electrical signals along specialized extensions called axons.

- Neurons interact with other neurons, with muscles, and with hormone-producing cells.

- Neurons transmit information in the form of a self-strengthening electrical signal called an action potential.

- A myelin sheath that surrounds most vertebrate axons limits where action potentials can begin, causing the action potential to move along myelinated axons much more rapidly than it does along unmyelinated axons.

- Neurons transmit information to other cells across narrow gaps called synapses. When an action potential reaches an axon terminal, it releases signaling molecules called neurotransmitters into the synapse. The neurotransmitters cross the synapse by diffusion to trigger an action potential or other response in the target cell.

Sensory Structures Allow Organisms to Respond to Their Environment

- Many organisms can detect environmental stimuli—chemicals, mechanical stimuli, or light—with specialized sensors that convert the stimuli into action potentials.

- Most organisms can detect chemicals in their environment through specific receptor proteins embedded in the plasma membrane of sensory cells. These chemical-detecting cells are concentrated in those parts of the organism most likely to encounter chemicals.

- On land, the sense of smell involves detecting low concentrations of airborne chemicals, and the sense of taste involves detecting high concentrations of dissolved chemicals. This distinction breaks down in aquatic organisms that come in contact only with chemicals dissolved in water.

- Mechanical stimuli inform organisms about the condition of their immediate surroundings, the location and identity of distant objects, and the position of their own bodies.

- To detect sounds, or hear, animals must have ears that can magnify tiny, rapid air pressure changes to the point that they can be detected by pressure-sensitive cells.

- Organisms rely on light-sensitive pigments to detect light.

- The simple ability to distinguish light from dark supplies organisms with valuable information about their position, the time of day, and the presence of moving objects in their environment.

- To see images, as the human eye can, organisms need a retina containing many light-sensitive cells (which acts as the film in a camera does) and a lens that can focus light coming from an object to form a sharp image on the retina.

Making Sense of Action Potentials in the Central Nervous System

- The central nervous system of most animals includes a brain, located near sensory structures in the head, and a major nerve or spinal cord, which rapidly carries information between the brain and all the parts of an animal's body.

- The central nervous system integrates sensory information and sends out a response in the form of commands to the muscles. The greater the number of connections between neurons in the central nervous system, the more sophisticated the processing of information and the response.

- When immediate action is required, simple reflex arcs bypass the brain.

- The brain handles signals that require more complex processing. The thalamus filters incoming signals and sends those that need further processing to the appropriate part of the brain. Specific areas of the cerebrum are involved in the conscious processing of information.

Highlight: Perception and Pain

- The perception of pain is a complex process involving the integration of information in the brain.

- Even after an injury has healed or a limb has been amputated, a person may perceive pain because of continued stimulation of sensory pathways in the brain.

KEY TERMS

action potential p. 571	nerve p. 570
axon p. 570	nervous system p. 570
brain p. 579	neuron p. 570
central nervous system p. 579	neurotransmitter p. 573
cerebrum p. 580	reflex arc p. 580
dendrite p. 570	retina p. 578
depolarization p. 571	spinal cord p. 580
ganglion p. 579	synapse p. 573
lens p. 577	thalamus p. 580
myelin p. 570	

CHAPTER REVIEW

Self-Quiz

1. To respond to light, an organism must have
   a. light-sensitive pigments.
   b. a lens.
   c. temperature sensors.
   d. a complex brain.

2. Which of the following senses is found most widely among living organisms?
   a. The ability to detect light
   b. The ability to detect sound
   c. The ability to detect chemicals
   d. The ability to feel pain as a warning of injury

3. An action potential traveling along an axon
   a. moves rapidly in both directions.
   b. moves faster than a neurotransmitter.
   c. is slowed by myelin.
   d. travels through the blood.

4. Which of the following comments applies to the brains of most animals?
   a. Within the brain, neurons exchange information with one another.
   b. Brains usually lie as near as possible to the important sensory structures in an animal.
   c. Brains send action potentials to the hindmost portion of the animal by means of major nerves.
   d. all of the above

5. Which of the following is true of action potentials?
   a. They travel more slowly than electrical currents.
   b. They can form only where a myelin sheath surrounds the nerve.
   c. They rapidly weaken as they travel down an axon.
   d. all of the above

Review Questions

1. Contrast the advantages and disadvantages of a reflex arc response with those of a response processed by the brain.

2. Describe the movement of an action potential down the axon of one neuron and across the synapse to the next neuron.

3. Not all animals have a well-developed ability to detect light. What are some of the advantages and disadvantages of using light as a way of finding out about the environment?

The Daily Globe

34

A Whale of a Problem

Woods Hole, MA—For years, many species of whales were threatened by overhunting, but international cooperation has reduced that threat significantly. Now marine biologists are warning that humans may be responsible for a much more subtle problem: noise pollution that could harm many marine mammals, especially whales and dolphins.

Although humans live in a world that is dominated by visual perception, other species sense a very different world. In the watery depths, highly social species such as humpback whales and bottlenose dolphins rely on their hearing to communicate with one another, coordinating cooperative activities as diverse as escap-

ing predators and reproducing. Some species, such as the beluga whale of Arctic waters, also rely on their hearing to locate and capture their prey. The effects of noise pollution on such species could be both immediate and long-term: immediate if a single event disrupts their behavior temporarily, and long-term if exposure to dangerous levels of noise damages their hearing irreversibly.

One of the main sources of noise pollution in the ocean is ship engines. Ironically, some people have observed that the immense popularity of whale-watching boat tours, which allow people to come very close to whales, may add to the problem. Another source of noise is military activity. For in-

stance, a test of a powerful sonar system for detecting submarines was blamed for the deaths of four humpback whales in 1997. Conservationists found temporary relief for the whales in 1998 when they won their lawsuit demanding a halt to the sonar testing while the effects of the sonar system were being investigated.

"Given the wide-ranging, migratory populations of many marine mammals," said Dr. Olivia Peach, scientist for the Marine Mammal Research Institute, "any successful attempt to reduce this new threat will require international cooperation, public education, and a commitment to increasing our understanding of life in the sea."

Evaluating "The News"

1. Can you think of any other examples of situations that might be described as sensory pollution, in which something could affect the ability of a person or other animal to detect sensory stimuli?

2. Is sensory pollution more or less dangerous than other kinds of pollution that can cause a specific disease?

3. What are the positive and negative effects of ecotourism businesses such as whale watching, swimming with dolphins, jeep rides through

beautiful but delicate desert habitats—activities that allow people to have an "up close and personal" experience with threatened or rare organisms? Given your assessment, are these activities valuable enough to continue, or should more of them be banned?

chapter 35 *Behavior*

Bright Butterflies

Imagine you are walking through a forest. You encounter a large butterfly with iridescent blue wings rimmed with yellow spots. It flies slowly, and every now and then it lands on a plant. Clearly, this is a butterfly on a mission. Curious about what that mission might be, you follow the butterfly. You discover that the butterfly usually does no more than briefly land on a plant. Every now and then, however, it lays a cluster of three to five orange eggs on the bottom of a leaf. Knowing what you do about herbivorous insects, you realize that this butterfly must be a female searching for plants on which her young can feed when they hatch. The food plants chosen by this particular butterfly have small, rounded leaves and seem to grow abundantly on the forest floor.

Confident that you've figured things out, you follow another of the pretty blue butterflies. Like the first butterfly, this butterfly alights on many plants before finding one she considers a suitable place to lay her eggs. As you follow the second butterfly, however, you notice two things. First, when she finally lays eggs, she does so on a plant that looks comple-

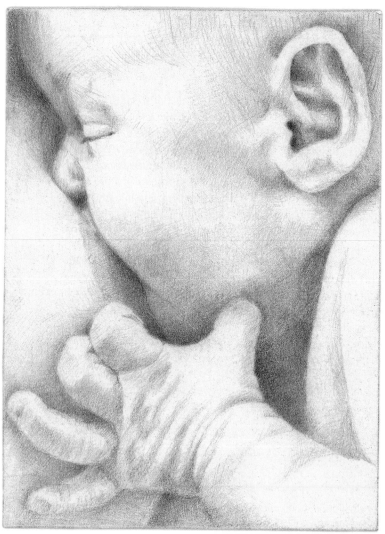

Anique Taylor, *Bliss*, 1989.

584

ly different from the plant selected by the first butterfly—an inconspicuous plant with long, narrow leaves. Second, most of the other plants on which she lands also bear long, narrow leaves.

Perhaps the two outwardly identical butterflies belong to different species that survive on different foods. Following a third blue butterfly, however, shows this not to be the case. The third female initially behaves like the first one—landing mostly on plants with rounded leaves and laying clusters of eggs on some of these. At one point, however, she happens to land on one of the narrow-leaved plants favored by the second female—and promptly lays some eggs. Thereafter, this third female switches from landing on broad-leaved plants to landing on narrow-leaved plants.

The preference for landing on broad-leaved or narrow-leaved plants by this butterfly species, the pipe vine swallowtail of the southern United States, is a response of individuals to their environment as they search for their food plants, two different species of pipe vines. Although this behavior is not a physical feature like wings for flying or eyes for seeing, the way in which pipe vine swallowtails search for food plants makes their survival possible. In this chapter we look at how behaviors contribute to survival. At the end of the chapter, we revisit food plant selection by pipe vine swallowtails to learn why these butterflies do what they do.

Female Pipevine Swallowtail Butterflies Carefully Choose Leaves on Which to Lay Eggs

■ KEY CONCEPTS

1. Behavior results from interactions among the senses, coordinating mechanisms, and some means of generating motion.

2. Animals rely more heavily on behavior than do members of other kingdoms.

3. Behavior can be fixed from birth or learned based on the organism's experience.

4. Both fixed and learned behaviors can evolve.

5. Communication allows two or more individuals to coordinate their activities.

6. Behavioral interactions within groups of organisms allow them to defend themselves more effectively and gain access to resources that are not available to individuals.

The passage at the beginning of this chapter describes the plant-searching behavior of the pipe vine swallowtail butterfly. A **behavior** is a movement made by one organism in response to another organism or the physical environment. This broad definition encompasses behaviors that we recognize, such as plant searching by butterflies and courtship by humans, as well as more subtle behaviors. Behavior allows organisms, especially animals, to respond quickly and flexibly to an ever-changing environment (see the box on p. 587). Moreover, behavior allows organisms to coordinate their own activities with those of other organisms.

We begin this chapter by looking at the behavioral interaction between a mother and her nursing child as a way of emphasizing the relationship between behavior and the hormonal and nervous systems described in Chapters 33 and 34. We then make the point that although members of all the kingdoms of life use behavior, animals rely on behavior most heavily. The subsequent discussion of behavior focuses on how it works in animals.

Breast-Feeding in Humans as an Example of Behavior

Behavior is a crucial part of human life. To understand the function and importance of behavior, let's look at the first behavioral interaction in human lives: that between newborns and their mothers during breast-feeding (Figure 35.1).

When a mother brings her breast to the mouth of her child for the first time, the child searches for the nipple with its mouth in an attempt to get milk. This behavior, called the **rooting reflex**, is stimulated by a touch against the cheek or mouth of an infant. The touch triggers a signal that is carried to the infant's brain by its nervous sys-

tem. The infant's brain, in turn, coordinates nursing behavior by stimulating contractions in its mouth muscles.

The baby's nursing behavior stimulates a less obvious behavioral response in the mother: the **milk-letdown reflex**. Although you have probably forgotten, for the first 30 to 60 seconds that you nursed, you received no milk. The reason for the delay is that the milk storage cells in a mother's breasts release milk only in response to nursing. Pressure-sensitive cells around the

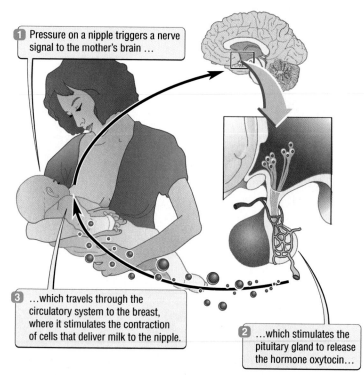

1 Pressure on a nipple triggers a nerve signal to the mother's brain …

2 …which stimulates the pituitary gland to release the hormone oxytocin…

3 …which travels through the circulatory system to the breast, where it stimulates the contraction of cells that deliver milk to the nipple.

Figure 35.1   The Rooting and Milk-Letdown Reflexes of Humans
Nursing by an infant stimulates the contraction of cells in the breast to release milk into the nipple, in a behavior coordinated by both the nervous system and hormones.

■ BIOLOGY IN OUR LIVES

The Long Journey Home

Each year in rivers all along North America's West Coast various salmon species fight their way upstream in search of their ancestral breeding grounds. Breeding marks the end of a remarkable life cycle for these fish. Salmon hatch in freshwater streams. For the first year or two of their lives they feed on small aquatic animals in sheltered streams located near their hatching site. The young salmon then follow the currents downstream to larger and larger rivers until they reach the Pacific Ocean. For most of the remainder of their lives, these freshwater fish switch over to a marine way of life. Schools of salmon feed as predators hundreds of kilometers out to sea.

After several years, those individuals that have survived and grown to breeding size return to freshwater rivers. Salmon do not return to just any river to breed—they return over immense distances specifically to spawn and die in the stream in which they themselves hatched. The mystery of how salmon manage to navigate so accurately through featureless oceans ended when scientists discovered that they follow their noses. Each stream carries dissolved in it a characteristic combination of chemicals that identifies it, just as a fingerprint identifies an individual human. Salmon follow this scent home.

With the damming of many rivers along the West Coast, the strong homing instinct of salmon placed them in jeopardy. If a population of salmon could not reach their breeding stream because a dam blocked the way, then that population disappeared. Without any salmon breeding in that stream, no new salmon hatched that responded to the smell of that stream. Even when dam designs were improved to include a way for salmon to continue their journey upstream, there were no salmon left that recognized that stream as home.

Fortunately for the recovery of salmon, they learn the scent of their stream, rather than having it programmed in their genes. This learning takes place during the early, freshwater phase of their lives, and by the time they enter the ocean, they are programmed to recognize their hatching stream as home. Thus, it has proved possible to remove newly hatched salmon from hatcheries based along populated streams and introduce them into empty streams. These young salmon learn their adopted stream's odor and, with some luck, become the founders of a successful population.

Salmon Fight Their Way Upriver to Their Ancestral Breeding Stream

nipple send signals to the mother's brain, which trigger the release of a hormone called oxytocin from the pituitary gland at the base of the brain (see Figure 35.1). Oxytocin circulates in the blood to target cells in both breasts. These cells respond by contracting, which forces milk out of the storage cells and into tubes leading to the nipples. Over weeks, as her breast-feeding experience increases, a mother's milk-letdown reflex changes in an intriguing way: Instead of releasing milk in response to nursing, she releases milk in response to the hungry cries indicating that her baby needs to feed.

Behaviors give organisms the flexibility to respond rapidly to a changing environment. During breast-feeding, the rooting and milk-letdown reflexes ensure that a baby tries to feed only when a breast is available and that a mother releases milk only when her hungry baby is nearby. The baby's mouth and the mother's breast are structures that allow babies to feed, and the behavioral interaction between mother and baby allows effective use of these structures. The switch from the touch of the baby's lips to the baby's cry as the stimulus that triggers the milk-letdown reflex emphasizes the flexibility of behavioral responses. Both behaviors depend on sensory information and on the hormones and action potentials that coordinate functions in multicellular organisms.

> ■ The behavioral interaction between a mother and child during breast-feeding coordinates the feeding behavior of the child with milk release by the mother. Sensory stimulation that triggers action potentials and hormone release controls the behaviors.

Who Behaves?

Although we typically think of behavior as something animals do, other organisms behave, too. As we have seen in the preceding chapters, members of other kingdoms have the ability to gather information about their environment and to move in response to that information. In this section we present two examples of behavior among organisms other than animals, before focusing on animal behavior for the remainder of the chapter.

Bacteria can move toward food

Bacteria can move toward or away from objects in their environment. Such **orientation behavior** depends on simple mechanisms that emphasize the role of coordinating systems in generating behavior.

Bacteria orient toward food by following two simple rules (Figure 35.2):

1. If the concentration of chemicals diffusing from the food source increases, continue in the same direction.
2. If the concentration of chemicals diffusing from the food source decreases, turn.

Bacteria move in a straight line by rotating their flagella in one direction (see Figure 27.10), and they turn by rotating the same flagella in the opposite direction. In practical terms, then, bacteria move toward food by rotating their flagella in one direction in response to increasing chemical concentrations and in the reverse direction in response to decreasing chemical concentrations. The requirements for coordinating such an orientation behavior are simple: (1) a stimulus (the chemical signal

from the food), (2) an ability to compare the strength of a signal at two different points (to determine whether it is increasing or decreasing), and (3) a rule about how to respond to a change in signal strength (rotate flagella clockwise or counterclockwise).

Plants can open and close their flowers in response to temperature

If you watch flowers closely, you will notice that many of them open and close according to the time of day or the weather. These movements generally cause flowers

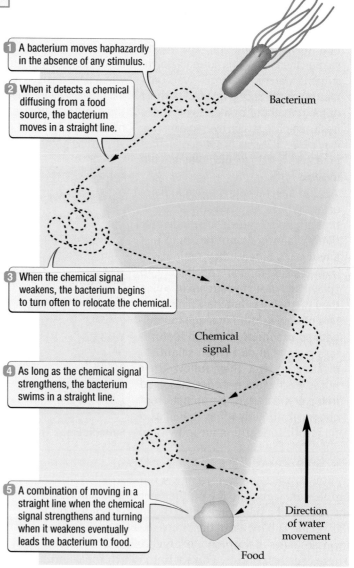

1. A bacterium moves haphazardly in the absence of any stimulus.

2. When it detects a chemical diffusing from a food source, the bacterium moves in a straight line.

Bacterium

3. When the chemical signal weakens, the bacterium begins to turn often to relocate the chemical.

Chemical signal

4. As long as the chemical signal strengthens, the bacterium swims in a straight line.

5. A combination of moving in a straight line when the chemical signal strengthens and turning when it weakens eventually leads the bacterium to food.

Direction of water movement

Food

Figure 35.2 Orientation Behavior of Bacteria
Bacteria can move toward food by combining a few very simple behaviors.

to open when they are most likely to be visited by the animals that transfer their pollen between plants (see Chapter 27).

Tulip flowers open at warm temperatures when insects are active and close when temperatures cool and insects stop visiting (Figure 35.3). When the temperature increases, the cells on the *upper* surface of the petals briefly increase their rate of division, while the rate of cell division on the lower surface of the petals remains constant. The lengthening of the upper petal surface relative to the lower one causes the flower to open. When the temperature drops, the cells on the *lower* side of the petal briefly increase their rate of division relative to those on the upper side, and the flower closes. Change in temperature is the stimulus, and different rates of cell division on the upper and lower surfaces of the petals coordinate the opening and closing of the flower.

Animals rely heavily on behavior

The nerve and muscle tissues of animals suit them particularly well to coordinating sensory information with movement to generate behavior. Neurons can relay signals more rapidly than hormones can, and the collection of neurons into a brain gives animals a unique ability to

process sensory information. Muscles give animals an unparalleled ability to move. Not surprisingly, therefore, behavior plays a more prominent role in animals than it does in the members of any other kingdom. For this reason, the rest of this chapter focuses on animal behavior.

> ■ Animals rely more heavily on behavior than do members of other kingdoms. Nonetheless, simple behaviors play an important part in the lives of members of all kingdoms.

35.3 Fixed and Learned Behaviors in Animals

Behavior patterns can either be fixed features of an animal from birth or be developed through the animal's experiences. For instance, males of most bird species sing a distinctive song that they use to court females. Males of some species know their courtship songs from birth, but others must learn them. Male cowbird chicks that are raised in captivity, and thus never hear an adult male cowbird sing, still give a flawless rendition of their species' courtship song when they mature. In contrast, male white-crowned sparrows raised in captivity fail miserably in their attempts to sing a song attractive to females. To learn their courtship song, white-crowned sparrow males must hear other males of their species sing.

White-crowned sparrow

Animals display fixed behaviors in response to simple stimuli

In **fixed behaviors**, stimulus A leads predictably to response B. Many examples of such behaviors are familiar to us. For instance, newborn babies display the rooting reflex, and most first-time mothers display the milk-letdown reflex without having learned how. An abandoned barn kitten that we rescue somehow knows to cover its droppings when first introduced to a litter box. Butterfly caterpillars know which food plants to eat and which to avoid from the moment they hatch.

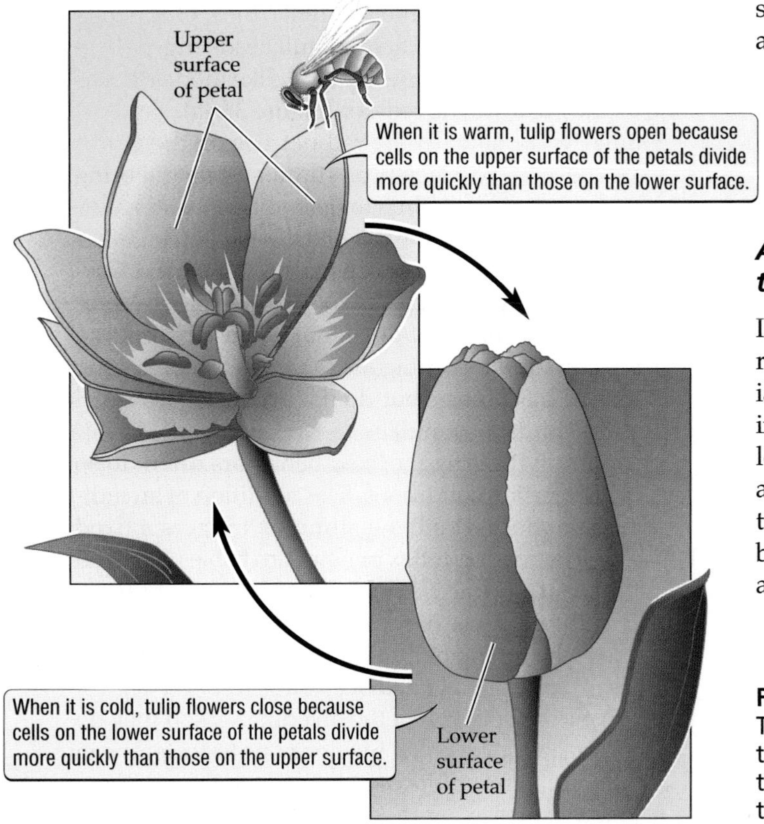

Upper surface of petal

When it is warm, tulip flowers open because cells on the upper surface of the petals divide more quickly than those on the lower surface.

When it is cold, tulip flowers close because cells on the lower surface of the petals divide more quickly than those on the upper surface.

Lower surface of petal

Figure 35.3 How Tulips Open and Close Their Flowers
The cell division rates on the upper and lower surfaces of tulip petals are affected differently by temperature, causing the flower to open at warm temperatures and close at cool temperatures.

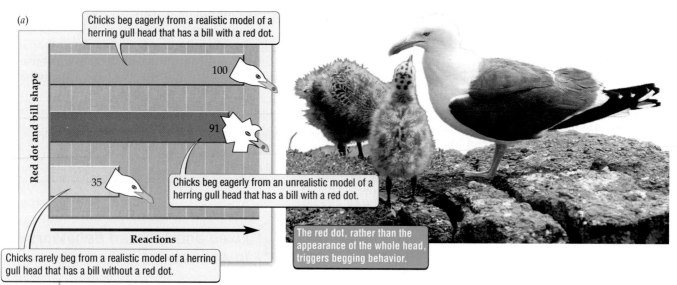

(a)

Chicks beg eagerly from a realistic model of a herring gull head that has a bill with a red dot.

Red dot and bill shape

100

91

35

Reactions

Chicks beg eagerly from an unrealistic model of a herring gull head that has a bill with a red dot.

Chicks rarely beg from a realistic model of a herring gull head that has a bill without a red dot.

The red dot, rather than the appearance of the whole head, triggers begging behavior.

Figure 35.4 Simple Stimuli Trigger Fixed Behaviors
(*a*) A herring gull chick will beg as eagerly from an unrealistic model of a gull head that bears the critical red dot (the trigger for begging) as they will from the real thing. (*b*) Red-winged blackbirds flash red wing patches at each other during territorial displays.

(b)

The red patches on the wings of male red-winged blackbirds play a key role in territorial behavior.

Cowbird

Fixed behaviors allow organisms to behave appropriately when they have no chance to learn by experience or when the risks associated with the wrong behavior are great. A baby that had to learn to nurse, for example, might not learn in time to keep from starving. The cowbirds we mentioned cannot learn their courtship song from their fathers because female cowbirds have the odd habit of laying their eggs in the nests of other bird species. As a result, cowbird chicks grow up in the care of adoptive parents of a different species that sing different courtship songs.

Simple stimuli often trigger fixed behaviors. Herring gull chicks, for example, beg eagerly for food when their parents return to the nest. In a famous series of experi-

ments, Nobel prize–winning scientist Niko Tinbergen showed that the chicks aim their begging behavior at a conspicuous red dot on their parents' bills. Oddly colored or shaped models of herring gull heads trigger begging behavior just as effectively as lifelike models do, as long as they feature the red dot (Figure 35.4*a*).

Similarly, the sight of the red patches on the wings of male red-winged blackbirds stimulates neighboring male red-winged blackbirds to defend their nesting territories vigorously. It is the red of the wing patch, rather than the overall appearance of a male red-winged blackbird, that stimulates this behavior (Figure 35.4*b*). A person can cause a frenzy of territorial defense by putting on a red shirt and marching into a marsh full of nesting red-winged blackbirds (but do not do this, or you will disrupt the birds' nesting!).

The stimuli that trigger fixed behaviors often do so only under certain conditions. The condition of an individual may affect whether a stimulus triggers a fixed behavior. The rooting reflex of newborn babies is a fixed behavior, but the baby exhibits it only when hungry. Similarly, the response of an organism to a stimulus may depend on the timing of the stimulus. During the breeding season, male red-winged blackbirds respond aggressively to red, but outside of the breeding season they do not. The greater level of the sex hormone testosterone in the male birds during the breeding season causes them to respond aggressively to red wing patches.

Learned behaviors provide animals with flexible responses

In **learned behaviors**, the response to a stimulus depends on the animal's past experiences. For instance, we can train a dog to sit in response to the command "Sit" or to a whistle. A dog trained to the oral command sits in response to "Sit"; a dog trained to the whistle also sits, but in response to a different stimulus. The two dogs learn to respond appropriately to their different environments.

Much of our own behavior is learned rather than fixed. Mothers learn to release milk in response to a baby's cry rather than to nursing, and students learn to answer questions properly on biology exams in response to information gained from lectures and textbooks. Most other animals also modify their behavior in light of their previous experiences.

The feeding behavior of rats illustrates the advantages of learned behavior. One aspect of the biology of rats that has made them so successful is that they thrive on the ever-changing variety of garbage that humans discard. However, we provide rats with many potentially poisonous food items as well as edible ones. Because of the diversity of their diet, rats cannot have a set of fixed rules about what to eat and what to avoid. Instead, rats learn to avoid poisonous foods.

When they encounter a food for the first time, rats sample only an amount small enough not to kill them if the food proves toxic. If the food sample sickens them, they avoid that food in the future. If the food causes no harm, they add it to their learned list of good things to eat. By learning what to avoid based on their experiences, rats can cope with new and unexpected kinds of food, something they could not do with inflexible fixed behaviors.

■ Animals display fixed behaviors in response to a particular stimulus without any previous experience. Fixed behaviors allow animals to function when there is no opportunity to learn from other members of their species or when the risks associated with making a mistake are great. Learned behaviors allow organisms to incorporate experience into their responses and to deal with unpredictable situations.

The Evolution of Animal Behavior

Although behaviors may lack the concreteness of structural characters, and although learned behaviors, in particular, depend strongly on environmental factors, behaviors have a genetic basis. For this reason, natural selection can act on behaviors just as it does on physical characters.

Genes control fixed behaviors

We can easily imagine the genetic control of fixed behaviors. One of the clearest examples of the direct genetic control of a fixed behavior involves the nest-cleaning behavior of honeybees (Figure 35.5). Honeybees sometimes fall victim to a contagious bacterial disease that kills larvae (immature bees) in the hive. The hives of some genetic varieties, or genotypes (see Chapter 12), of honeybees rarely suffer from the disease. These "resistant" genotypes do become infected, but they reduce the spread of the bacteria by quickly removing infected larvae from the hive. Nest cleaning involves two behaviors, each controlled by a different gene: (1) cutting open cells containing larvae killed by the bacteria,

A *UuRr* × *ur* cross yields the following phenotypes:

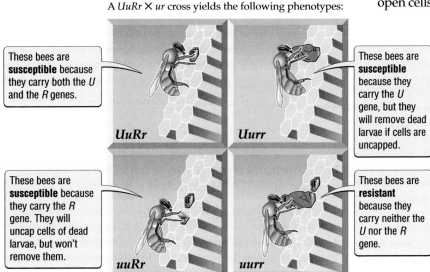

These bees are **susceptible** because they carry both the *U* and the *R* genes.

UuRr

These bees are **susceptible** because they carry the *U* gene, but they will remove dead larvae if cells are uncapped.

Uurr

These bees are **susceptible** because they carry the *R* gene. They will uncap cells of dead larvae, but won't remove them.

uuRr

These bees are **resistant** because they carry neither the *U* nor the *R* gene.

uurr

Figure 35.5 Genes Control the Nest-Cleaning Behavior of Honeybees The results of a cross between a *UuRr* female bee and a *ur* male bee reveal that nest-cleaning behavior has a genetic basis. (Male honeybees are haploid, hence the two genes instead of four.) Bees that carry either the *U* gene or the *R* gene will be susceptible to disease because they will not work to keep their hive disease-free: Bees with the *U* gene do not uncap cells containing dead larvae, while bees with the *R* gene do not remove the dead larvae from their cells.

and (2) removing the dead bodies from the hive. "Susceptible" genotypes of honeybees neither cut open cells nor dispose of dead larvae. Genetic crosses like those described in Chapter 12 have revealed that the nest-cleaning behavior of honeybees is under simple genetic control.

More typically, genes influence behavior, but interactions between genes and the environment ultimately determine the behavioral patterns shown by an individual. Careful studies of identical and fraternal twins in humans, for example, have revealed that schizophrenia—a mental illness in which a person hallucinates, feels persecuted, and behaves strangely—has a strong genetic component.

Identical twins arise from a single fertilized egg that splits into two genetically identical individuals. Fraternal twins result when two sperm simultaneously fertilize two different eggs, leading to twins that are no more similar genetically than are typical brothers and sisters. If genes control schizophrenia, then we would expect that either both members of a pair of identical twins should have schizophrenia or both should be healthy. Fraternal twins should less frequently both have schizophrenia because they differ genetically.

Researchers have found that it is indeed much more common for both members of a pair of identical twins to have schizophrenia than for both members of a pair of fraternal twins to have the disorder (Figure 35.6). At the same time, however, if one member of a pair of identical twins has schizophrenia, the other may still be healthy, even though he or she carries all the same genes.

This observation indicates that environmental influences play a role in determining whether a person with the genes for schizophrenia actually develops the disease.

Genes also influence learned behaviors

Learned behaviors, by definition, are strongly influenced by the environment: What animals learn depends on what they experience during their lifetimes. Nonetheless, differences in what, how much, and when animals can learn reflect genetic differences in their learning capacity.

Individual laboratory rats, for example, learn to negotiate mazes at different rates. By mating fast learners with other fast learners, researchers can produce offspring that also learn to negotiate mazes quickly. Mating slow learners with other slow learners produces rats that have a difficult time learning to find their way through a maze.

Some species of birds, such as Clark's nutcrackers, excel at remembering where they have stored food. These birds have brains that differ structurally, and presumably genetically, from those of closely related species that have an average ability to find stored food. Researchers have even identified genes that seem to control the ability to learn the songs that are characteristic of a bird species.

Clark's nutcracker burying a seed

> ■ Both fixed and learned behaviors have a genetic component and can therefore evolve.

Communication Allows Behavioral Interactions between Animals

Communication is a kind of behavior that allows one individual to exchange information with another, thereby making it possible for the animal to coordinate its activities with those of other individuals. By communicating with one another, animals in groups can do things that no individual could do on its own, such as fend off large predators or build pyramids.

Communication is the production of signals by one animal that stimulate a response in another

The communication behavior of animals varies widely in its complexity. At the simple end of the spectrum, the release of a chemical, called a **pheromone**, by one individual informs others of its identity and location. The release of bombykol by a female silkworm moth to

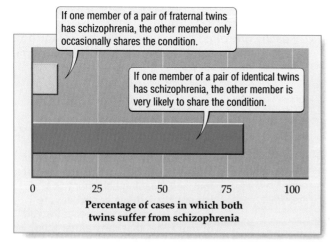

If one member of a pair of fraternal twins has schizophrenia, the other member only occasionally shares the condition.

If one member of a pair of identical twins has schizophrenia, the other member is very likely to share the condition.

Percentage of cases in which both twins suffer from schizophrenia

Figure 35.6   The Genetics of Schizophrenia
Schizophrenia is a genetically based mental disorder in humans. However, the environment clearly affects the expression of the genes causing schizophrenia, because even among genetically identical twins there are many cases in which only one twin develops the condition.

attract a male (see Chapter 34) is an example of this simple form of communication. Probably the most widespread means of communication, pheromones are used not only by animals, but also by bacteria, fungi, protists, and plants.

At the complex end of the spectrum lie human **languages**, consisting of thousands of words that represent everything from objects to actions to abstract ideas, and the dance language used by honeybees to communicate the location of food to other members of their hive (Figure 35.7). Communication includes just about any type of signal that other animals can sense: sound, visual signals, odors, electrical pulses, touch, and tastes.

Why do animals communicate?

Animals most often communicate to identify themselves, to avoid conflict, and to coordinate their activities. Self-identification is probably the most common function of communication. An animal uses a variety of signals to inform other members of its species of its sex, its physical condition, and its location. When a dog marks a fire hydrant or a post with its scent, the scent communicates the dog's species, sex, breeding condition, health, and status to other dogs. The mating songs and behaviors of male birds, frogs, and crickets tell the females they court about their species, their location, and their potential quality as mates (Figure 35.8a).

A second important function of communication in animals is to avoid potentially harmful conflicts. Physical conflict over food or mates can lead to serious injury for both the winner and the loser. To reduce the risk of injury in such encounters, many species communicate their fighting ability through ritual displays. When male red-winged blackbirds flash red wing patches at each other, for example, they signal their quality as fighters to other males. This ritual allows the poorer fighter to back down without engaging in a potentially dangerous fight. By communicating in this way, animals can resolve contests without actually coming to blows.

Sheep, antelope, and their relatives bear horns that vary greatly in shape—from stout, straight weapons capable of doing great damage to impressive but ineffective weapons. The species with the most lethal horns usually engage in displays that communicate their suitability as mates rather than in actual fights, whereas the species with less lethal horns more often engage in physical combat (Figure 35.8b).

Animals that live in groups communicate to coordinate their behaviors to accomplish a shared task. Humans do this when they work as a team (Figure 35.8c). Wolf packs and lion prides communicate when they hunt for food. The dance language of honeybees (see Figure 35.7) allows the members of a hive to coordinate their foraging activity.

Language may be a uniquely human trait

Humans rely heavily on spoken, written, or sign language for communication. Much of our human identity depends on our ability to express complex and abstract ideas through language. There is no strong evidence of language among other animals. Many birds and mammals can produce a variety of sounds, each of which conveys a particular message, but do not assemble the sounds into ideas. Research on the ability of chimpanzees to communicate indicates that they can string together symbols provided by humans to express

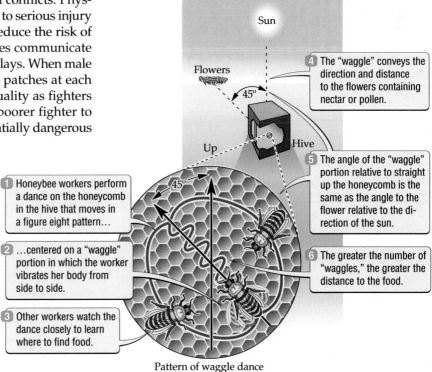

Sun

Flowers

45°

Up

Hive

4 The "waggle" conveys the direction and distance to the flowers containing nectar or pollen.

5 The angle of the "waggle" portion relative to straight up the honeycomb is the same as the angle to the flower relative to the direction of the sun.

6 The greater the number of "waggles," the greater the distance to the food.

45°

1 Honeybee workers perform a dance on the honeycomb in the hive that moves in a figure eight pattern...

2 ...centered on a "waggle" portion in which the worker vibrates her body from side to side.

3 Other workers watch the dance closely to learn where to find food.

Pattern of waggle dance

35.4

Figure 35.7 Honeybees Communicate by Dancing in Their Hive Bees use a dance language to communicate complex information about the position of distant sources of food to other worker bees in their hive.

(a)

(b)

Figure 35.8 Why Animals Communicate (c) (a) A singing male frog advertises to listening females its species, its location, and its quality as a mate. (b) Male bighorn sheep communicate their quality by engaging in head-butting contests. (c) Humans communicate to coordinate their activities.

abstract ideas, but there is no indication that chimpanzees use such sophisticated communication in nature.

■ Communication allows animals to coordinate their activities with those of other animals.

Social Behavior in Animals

Many animals besides humans live in closely interacting groups. The behavioral interactions among members of a group offer them advantages that are not available to solitary organisms. In this section we look at some of the advantages of group living, and we consider some of the reasons why individuals might make sacrifices for the sake of the group as a whole.

Group living offers many benefits

The many benefits of group living can more than compensate for the increased competition for resources that results from living with other animals. Groups can find some foods better than individuals can. Groups can stir

up prey more effectively than individuals can, and unsuccessful hunters can watch and learn while others in their group find food. By working together, members of a group can get foods that are not available to individuals. The hunting of large mammals by groups of cooperating wolves or humans illustrates such cooperative behavior well (Figure 35.9a).

Living in groups also gives animals two effective defenses against predators that are not available to single individuals, as we saw in Chapter 32. First, groups of animals contain more eyes, ears, and noses, making early detection of an approaching predator more likely. Second, the individuals on the edge of a group are more likely to be attacked than those on the inside, offering the lucky ones in the middle some protection. The individuals on the inside of a herd of gazelles, for example, benefit because a lion is most likely to kill an individual at the edge of the group. Moreover, all group members benefit because at least one is likely to warn of approaching lions before they are close enough to attack (Figure 35.9b).

Figure 35.9   The Benefits of Group Life
(*a*) As a group, wolves can hunt larger prey than they can as individuals. (*b*) Some Thomson's gazelles watch for predators. (*c*) Family groups of scrub jays work together to feed young, defend nests, and hold onto valuable breeding territories.

(*a*)

(*b*)

(*c*)

In some cases, living in a group gives animals access to a scarce resource. Groups of Florida scrub jays include some members that do not breed, but instead help care for the offspring of a breeding pair. The nonbreeding birds appear to benefit because belonging to the group puts them on a sort of "waiting list" to inherit the territory from the breeding pair. The chance that a solitary scrub jay can find and hold control over a high-quality breeding territory is so small that it pays to put off breeding for several years while remaining on this waiting list (Figure 35.9*c*).

Individuals living in groups often act in ways that benefit other group members more than themselves

Animals that live in groups often do things that help other members of their group survive or reproduce while decreasing their own chances of doing so. Such **altruistic** behavior seemed for many years to contradict Darwin's idea that only traits that improve individual reproductive success will spread through a population (see Chapter 20). The scrub jays mentioned in the previous section provide a good example of altruistic behavior: Nonbreeding members of a group help the breeding pair raise their young. This behavior improves the reproductive success of the breeding pair, while presumably plac-

ing the nonbreeding individuals at risk as they forage for food and defend the territory against intruders.

In general, social groups consist of closely related individuals. A pride of lions, for example, centers around a group of closely related females. Similarly, among scrub jays, the helpers most often turn out to be older offspring of the breeding pair, so the young birds they are helping to raise are their younger siblings. In these cases, the individuals that benefit from altruistic behavior carry many of the same genes the altruist does, so the altruistic behavior often helps the spread of many of the same genes that the altruist carries.

Do social insects represent the ultimate in group living?

Social insects such as ants, bees, and termites include some of the most successful species on Earth. Although they evolved independently of one another, these three groups share some remarkable traits.

Each species lives in large colonies containing individuals belonging to distinct groups that serve distinct functions within the colony. The closely related workers work together to build complex nests (Figure 35.10), forage widely for food, and defend the colony effectively against predators in ways that individuals never could. The queen spends her life producing massive numbers of eggs while being tended carefully by the workers. The workers in a hive are offspring of the queen and therefore are closely related to her. The workers represent the extreme in altruism in that they are sterile; thus, they give up their own opportunity to reproduce in order to help their mother do so.

> ■ The social behavior of animals living in groups allows them to obtain otherwise unavailable foods, avoid predators more effectively, and defend scarce resources. Social behavior often involves altruism, in which an individual reduces its own chances of reproducing by aiding other members of its group.

Figure 35.10 Cooperation Creates Termite Mounds
The giant termite mound on which this cheetah sits was built through the cooperative effort of millions of closely related, sterile termite workers. Such workers make up most of the population of each colony.

■ HIGHLIGHT

Why Pipe Vine Swallowtails Learn

We do not usually think of insects as intelligent animals. The switching of the pipe vine swallowtail's preference for landing on broad-leaved or narrow-leaved pipe vines, however, is a remarkably complex mechanism that allows female butterflies to respond rapidly to changes in food plant availability. These butterflies can search effectively for only one leaf shape at a time. Just as we have much more success in looking for something once we know what the object of our search looks like, butterflies that search exclusively for either broad-leaved or narrow-leaved plants find pipe vines faster than do individuals looking for both types of plants at once.

The relative abundance of broad-leaved and narrow-leaved pipe vines changes from place to place within the forest. A female butterfly looking for narrow-leaved plants where broad-leaved plants predominate finds pipe vines on which to lay eggs at a much lower rate than a female searching for the predominant pipe vine species. The ability of a female to change the leaf shape she seeks to match the changing forest floor over which she flies allows her to lay more eggs with less time spent searching.

The very fact that female pipe vine swallowtails focus on leaf shape in their search reflects a second kind of learning. The true indicator of the quality of a plant as a place to lay eggs is its species. Females can tell for sure whether a plant is a pipe vine only by landing on it and tasting the leaves. Because landing takes time, females learn to associate leaf shape with their food plant species, and thus can rule out some plant species without landing. In this way, females take the time to land on a plant only when there is a good chance the plant is a pipe vine.

The next time you see a butterfly, remember some of the remarkable and sophisticated behaviors that these animals use to find quickly what they seek.

> ■ The ability of pipe vine swallowtails to change their searching behavior depending on the relative abundance of broad-leaved and narrow-leaved pipe vines reduces the time needed to find plants on which to lay eggs.

SUMMARY

Breast-Feeding in Humans as an Example of Behavior

- Sensory stimulation initiates responses, including action potentials and the release of hormones, that stimulate muscle contractions, which allow breast-feeding to occur.

■ The behavioral interaction between infant and mother coordinates their responses so that the infant tries to feed only when milk is available and the mother releases milk only when her hungry baby is nearby.

Who Behaves?

■ Bacteria exhibit orientation behavior that allows them to move toward food or other stimuli in their environment.

■ Plants can also respond to stimuli in their environment, as by opening and closing their flowers in response to temperature.

■ The nerve and muscle tissues of animals allow them to rely heavily on behavior.

Fixed and Learned Behaviors in Animals

■ Animals display fixed behaviors in response to simple stimuli. Fixed behaviors can enhance an organism's survival when there is no opportunity to learn how to respond or when the costs of a mistake are great.

■ Learned behaviors, in which the response depends on past experience, provide animals with flexible responses.

The Evolution of Animal Behavior

■ Because both fixed and learned behaviors have a genetic basis, all behaviors can evolve.

■ Many behaviors, both fixed and learned, can be influenced by the environment.

■ Learning is one example of how environmental influences can affect behaviors that have a genetic basis, but the capacity to learn may also be under genetic control.

Communication Allows Behavioral Interactions between Animals

■ Communication is the production of signals by one animal to stimulate a response in another.

■ Animals communicate to identify themselves, to avoid conflict, and to coordinate their activities.

■ Language may be a uniquely human trait.

Social Behavior in Animals

■ Living in groups can enable animals to avoid predation and to gain access to resources, including food and breeding territories, more effectively.

■ Altruistic behavior can evolve within groups of closely related animals.

Highlight: Why Pipe Vine Swallowtails Learn

■ The ability of pipe vine swallowtails to change their searching behavior depending on the relative abundance of broad-leaved and narrow-leaved pipe vines reduces the time needed to find sites for egg laying.

KEY TERMS

altruism p. 595

behavior p. 586

communication p. 592

fixed behavior p. 589

language p. 593

learned behavior p. 591

milk-letdown reflex p. 586

orientation behavior p. 588

pheromone p. 592

rooting reflex p. 586

CHAPTER REVIEW

Self-Quiz

1. The rooting reflex
   a. occurs in response to milk production.
   b. is an example of a fixed behavior.
   c. occurs in response to a visual stimulus.
   d. is an example of a learned behavior.

2. Behavior
   a. occurs only in animals.
   b. always involves communication.
   c. allows organisms to evolve.
   d. allows organisms to respond quickly to changes in their environment.

3. Learning
   a. occurs only in humans.
   b. overrides all genetic control of behavior.
   c. depends on an organism's past experience.
   d. requires communication.

4. Which of the following is *not* an example of altruistic behavior?
   a. A butterfly lays an egg.
   b. A scrub jay chases a snake away from its parents' nest.
   c. A sterile worker ant defends the nursery of its anthill from invading ants.
   d. A lioness regurgitates food for her sister's cubs.

5. Animals that live in groups
   a. have no fixed behaviors.
   b. usually rely on language for communication.
   c. can reap benefits that compensate for increased competition among group members.
   d. have no genetically controlled behaviors.

Review Questions

1. What tissues and organs allow animals to rely so heavily on behavior, and why?

2. Design an experiment to test the hypothesis that a fear of heights is something that a person is born with. How would knowing whether a fear of heights was genetically programmed or learned influence the way we view this condition?

3. What kinds of fixed behaviors might be advantageous to animals that live in a group, and why?

4. Describe the three ways in which communication can benefit organisms, including examples not given in this chapter.

35 $\qquad$ **The Daily Globe**

Protesters Pick on New Breed Restrictions

CINCINNATI, OH—In response to several recent attacks by dogs, the City Council today passed a new regulation banning the ownership of pit bulls and Rottweilers within the city limits.

In the past 18 months, at least six people in Cincinnati have been seriously injured in attacks by dogs, and Joshua Corona, a 4-year-old boy, was mauled to death by a neighbor's Rottweiler. Four of the nonlethal injuries were caused by pit bulls or pit bull crosses, leading to the concern over this dog breed.

Outside City Hall, about 20 dog enthusiasts gathered to protest the new regulation. John Sykes, a Rottweiler breeder and trainer, complained that the regulation unfairly targets the breed rather than the behavior. "Sure, Rottweilers are big, scary-looking dogs. But when raised in a loving environment and trained by a conscientious owner, they can become wonderful family pets. The problem is not the breed; it's irresponsible owners who encourage ferocious behavior and don't keep their dogs under control."

Leading her pit bull "Chucky" on a short leash, 20-year-old Melissa Whitsen held a sign that said, "A dog is this woman's best friend." She bought Chucky 3 years ago, when she first moved out on her own. "As a woman, living alone, I wanted both the companionship and the security that a dog like this could give me." She credits Chucky with scaring away an intruder just three nights ago.

Across town, Joshua's father, Michael Corona, applauded the Council's decision. He explained that his son had been playing in his own backyard when the attack occurred, and he stated, "These dogs have been bred to be guard dogs, which makes them ferocious and strong. They have no business in residential areas where kids and families are trying to lead normal lives."

After the protesters had dispersed, the City Council defended its decision in a short statement prepared for the press, noting that the regulation was modeled on similar laws passed by cities in neighboring states.

Evaluating "The News"

1. Consider the wild ancestors of domestic dogs. Do you think the behaviors we now consider fierce in domestic dogs would ever have developed without human intervention? Why or why not?

2. Explain how you think dog attacks of the sort described in this article could most effectively be reduced. In other words, if you lived in Cincinnati, would you have supported the council's decision or another alternative? Why?

3. How do you think these sorts of regulations can (or should) deal with the complicating factor of mixed-breed dogs?

chapter 36 Reproduction

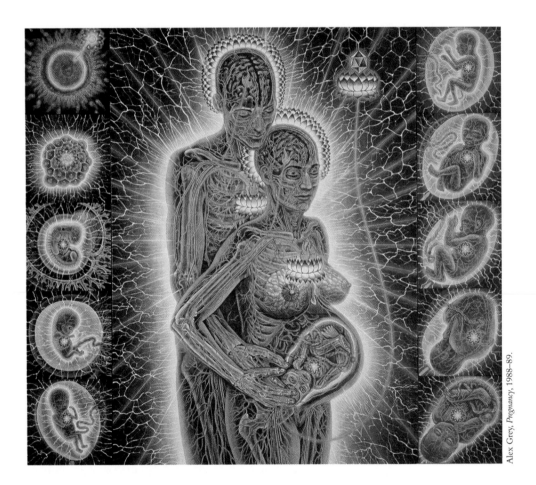

Alex Grey, *Pregnancy*, 1988–89.

Gender-Bending Plants

We generally assume that maleness or female-ness is an easily identified characteristic of an individual. Most humans live their entire lives as either a sperm-producing male or an egg-producing female. For many organisms, however, human notions of gender have little relevance.

Jack-in-the-pulpits are plants whose name comes from the unique shape of their flowers: A modified leaf shaped like a church pulpit surrounds a thick standing object, which is actually a spike of tiny, densely crowded flowers. These plants grow in moist forests throughout eastern North America. A

600

A Change in Size Means a Change in Gender
Jack-in-the-pulpits switch from male to female or from female to male depending on their size.

close look at jack-in-the-pulpits reveals that some are males, producing only powdery yellow pollen, and others are females, which eventually produce a cluster of bright red berries.

Among jack-in-the-pulpits, however, all is not as simple as it seems. This year's male may have been last year's female, and vice versa. The gender of a jack-in-the-pulpit depends on the amount of energy the plant manages to store in its roots, rather than on the presence or absence of genes on a Y chromosome, as in humans (see Chapter 13). The amount of energy the plant can store in its roots in summer determines how large the aboveground parts can grow the following year. Jack-in-the-pulpits switch from being male when they are small to being female when they can grow large. This year's males, if they store enough energy, may produce female flowers next year. In contrast, females that store little energy turn into small, male plants the next year.

In this chapter we explore some of the basic patterns of reproduction found among living things. Once we understand some of the fundamental features of reproduction, we can see what lies behind the odd reproductive life of jack-in-the-pulpits.

■ KEY CONCEPTS

1. Sexual reproduction involves combining the genes of two individuals to form a new, genetically unique individual.

2. Both sexual and asexual reproduction offer important advantages.

3. Prokaryotes reproduce asexually, and they combine genes from two individuals sporadically and by a variety of means.

4. Eukaryotes can reproduce both sexually and asexually. In sexual reproduction, the male and the female each contribute half of the genes to each of their offspring.

5. Both animals and plants can improve their reproductive success by moving close to a potential mate before releasing eggs or sperm.

6. Plants and animals can also improve their reproductive success by selecting mates carefully to increase the likelihood that the genes contributed by their mates will produce high-quality offspring.

*I*f life has a purpose, it is to make more of itself. Thus **reproduction**, the making of offspring, characterizes all life. Like many other organisms, humans reproduce **sexually**, meaning that reproduction involves a recombination of the genes from two parents. As we saw in Chapter 20, the genetic rearrangements that take place during sexual reproduction maintain the genetic variation needed for evolution by natural selection. Many organisms, however, reproduce **asexually**; that is, a single parent produces offspring that are genetically identical copies of itself.

Is sexual reproduction better than asexual reproduction? The answer seems to depend on the environment in which an organism lives (Figure 36.1). Sexual reproduction leads to offspring that differ genetically from their parents. As a result, sexual reproduction may work best when offspring must survive in a habitat that differs from the one in which their parents were successful. Thus, the species that stand to benefit most from sexual reproduction either have young that move away from their parents to live in slightly different habitats, or have long-lived individuals whose habitat changes from the time of their parents' lives to the time when their own offspring develop. In contrast, the recombination of genes that characterizes sexual reproduction may be a disadvantage for species in which the offspring develop under conditions identical to those in which their parents thrived.

Because an organism's evolutionary success hinges on its reproductive success, it devotes much of its time and energy to producing as many offspring as possible

These sexually produced seeds that dot the outside of a strawberry may be carried by animals to distant habitats and may not sprout for many years.

These strawberry plants produced asexually on runners are developing close to the plant that produced them.

Figure 36.1 Sexual and Asexual Reproduction
Asexually produced offspring usually end up living in habitat similar to that of their parents. Sexually produced offspring, by moving away from their parents or by living for a long time, often experience habitat that is different from that of their parents.

and ensuring that those offspring will themselves survive to reproduce. In a sense, the function of all the structures and processes described so far in Unit 5 is to allow the organism to accumulate the resources it needs to reproduce.

We begin this overview of reproduction by comparing and contrasting reproduction and genetic recombination, and the relationship between these two processes, in bacteria and in eukaryotic organisms. For the rest of the chapter, we focus on reproduction in animals and flowering plants, the two groups of eukaryotes most familiar to us. We describe reproduction in animals and plants to introduce the basic reproductive structures, to give an idea of the frequency of sexual compared with asexual reproduction in these groups, and to show the different evolutionary directions in which animals and plants have taken the basic pattern of eukaryotic sexual reproduction. We then look at how animals and plants deal with two important challenges associated with eukaryotic sexual reproduction: ensuring that eggs and sperm can get together, and selecting high-quality mates with which to recombine genetic material.

How Bacteria and Eukaryotes Reproduce and Recombine Genes

The relationship between reproduction and the recombination of genes in prokaryotes such as bacteria differs greatly from that relationship in eukaryotes such as animals and plants. Three outstanding features distinguish the recombination of the genes of two prokaryotic individuals from that of eukaryotes:

1. In eukaryotes, genes recombine during reproduction; in prokaryotes, the two processes occur separately.
2. In eukaryotes, two parents contribute genes equally to their offspring; in prokaryotes, the transfer of genes is unequal.
3. In eukaryotes, the combination of genetic material from two eukaryotic parents is highly predictable; the transfer of genes in prokaryotes is often sporadic.

Prokaryotes transfer genes in one direction in a process separate from reproduction

Prokaryotes such as bacteria reproduce when one cell divides asexually into two independent cells (Figure 36.2a). Reproduction in prokaryotes does not involve the combining of genes from two individuals.

Prokaryotes do occasionally transfer genes, however, and they do so in two general ways. In **transformation**,

bacterial individuals, called recipients, take up fragments of double-stranded DNA released into the environment by other bacteria, called donors (Figure 36.2b). Donor bacteria release their DNA into the environment when they die or in response to chemicals produced by other bacteria. We saw an example of bacterial transformation in Chapter 14 (see Figure 14.1) Transformation almost certainly evolved as a way of introducing genetic variation into bacterial DNA. Only bacterial cells that have the proper binding proteins in their plasma membranes can take up DNA released into the environment. Once inside the recipient, a single strand of a donor DNA fragment becomes incorporated into the recipient's genome. Thus, in transformation, the recipient becomes a single, new, genetically different individual, clearly showing the distinction between gene transfer and reproduction in prokaryotes. The relative contributions of the donor and recipient bacteria to the transformed individual depend on how much donor DNA is incorporated into the recipient's genome.

A second means of gene transfer involves **plasmids** (small, circular pieces of DNA) or viruses. Usually when viruses or plasmids transfer from one host bacterium to another, only viral or plasmid DNA enters the new host. Occasionally, however, some of original host's DNA is accidentally transferred in the process. Although almost certainly not a mechanism that evolved to recombine bacterial DNA, these accidental transfers of bacterial DNA by viruses and plasmids can play an important role in the transfer of genetic information between bacteria. Biotechnologists use viruses and plasmids as tools to insert potentially useful genes into bacteria, and even into animals and plants (see Chapter 17).

Both eukaryotic parents contribute genes during sexual reproduction

In eukaryotes, the combining of genes from two parents to form a genetically unique offspring always takes place during sexual reproduction. Eukaryotic species that reproduce sexually produce individuals with two distinct genders: males and females. The two genders produce different kinds of sex cells, or **gametes**. **Males** produce specialized gametes called sperm. **Sperm** can move, and they contain little beyond chromosomes and the cellular machinery needed for them to move. **Females** produce gametes called eggs. **Eggs** are typically much larger than sperm and generally cannot move on their own. Eukaryotic organisms produce gametes through meiosis, in which diploid ($2n$) cells (containing two paired copies of each chromosome) give rise to haploid (n) gametes (containing only one member of each chromosome pair). Thus sperm and eggs are hap-

(a)

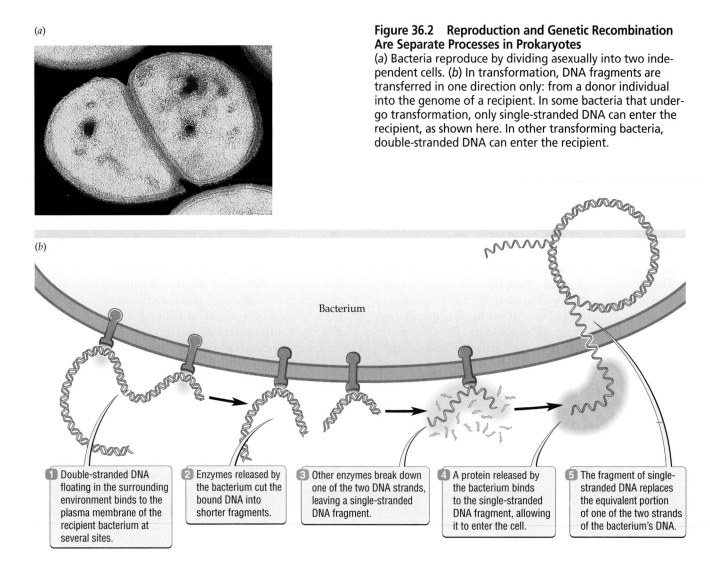

Figure 36.2 Reproduction and Genetic Recombination Are Separate Processes in Prokaryotes
(a) Bacteria reproduce by dividing asexually into two independent cells. (b) In transformation, DNA fragments are transferred in one direction only: from a donor individual into the genome of a recipient. In some bacteria that undergo transformation, only single-stranded DNA can enter the recipient, as shown here. In other transforming bacteria, double-stranded DNA can enter the recipient.

(b)

Bacterium

1 Double-stranded DNA floating in the surrounding environment binds to the plasma membrane of the recipient bacterium at several sites.

2 Enzymes released by the bacterium cut the bound DNA into shorter fragments.

3 Other enzymes break down one of the two DNA strands, leaving a single-stranded DNA fragment.

4 A protein released by the bacterium binds to the single-stranded DNA fragment, allowing it to enter the cell.

5 The fragment of single-stranded DNA replaces the equivalent portion of one of the two strands of the bacterium's DNA.

loid, containing only half of the genetic information carried by the diploid individual that produced them. (You may find it helpful to review the discussion of meiosis in Chapter 10.)

A haploid sperm and a haploid egg fuse to form a diploid cell called a **zygote** during **fertilization**. Eggs of many eukaryotes release chemical pheromones that allow sperm to find them. Flowering plants seem to physically guide the sperm to the egg, as we will describe later in the chapter. Chemical interactions between the sperm and the egg allow the sperm to enter the egg. Once inside the egg, the sperm's nucleus separates from the rest of the sperm and fuses with the egg's nucleus to form a genetically unique zygote. One member of each pair of chromosomes in the zygote comes from the male, and one comes from the female. The organelles and other cellular machinery of the fertilized egg, however, come almost entirely from the female.

■ In prokaryotes, genes are transferred in a process separate from reproduction. Prokaryotes transfer genes in one direction only: from a donor to a recipient. In eukaryotes, the recombination of genes occurs during sexual reproduction. Sexual reproduction involves a combining of genetic material, of which half comes from the female and half from the male.

36.1 *Animal Reproduction*

Having seen the dramatic differences between prokaryotic and eukaryotic reproduction, we now turn to the different ways of reproducing that have evolved within the eukaryotes. To set the stage for a more general discussion of asexual and sexual reproduction and of male and female genders in animals, we first briefly describe reproduction in our own species.

Human reproduction illustrates many basic features of sexual reproduction in animals

Sexual reproduction in humans provides a good example of sexual reproduction in animals. As we all know, humans come in two genders: male and female. Males produce haploid sperm (containing 23 chromosomes) by meiosis in specialized organs called **testes** (singular testis). Human sperm can swim through fluid by means of a flagellum (see Chapter 27). Females produce haploid eggs (containing 23 chromosomes) by meiosis in organs called **ovaries** (see Chapter 33).

The female usually releases one egg from her ovaries each month, which moves down a tube called the **oviduct** (Figure 36.3). During intercourse, the male's **penis** directs almost 300 million tiny sperm into the **vagina** of the female.

Massive numbers of sperm swim up the oviduct in response to a pheromone released by the ovary. The female aids the progress of the swimming sperm by a combination of muscular contractions of the reproductive tract and the release of chemicals that stimulate sperm to swim. However, only a few hundred sperm reach the egg in the oviduct. Only one of these sperm can successfully fuse with the egg because changes in the egg immediately after fusion prevent entry by other sperm. Fusion of the nucleus of the egg with that of the sperm produces a diploid zygote containing 46 chromosomes (23 pairs) that come equally from both parents.

Humans, like most organisms, form zygotes only following a sequence of activities that increase the chances that a sperm will encounter an egg and that their fusion will result in a zygote that develops into a successful new individual. During courtship, we spend an amazing amount of energy, time, and effort on finding and impressing a potential mate (see Chapter 35). Careful mate selection is important to

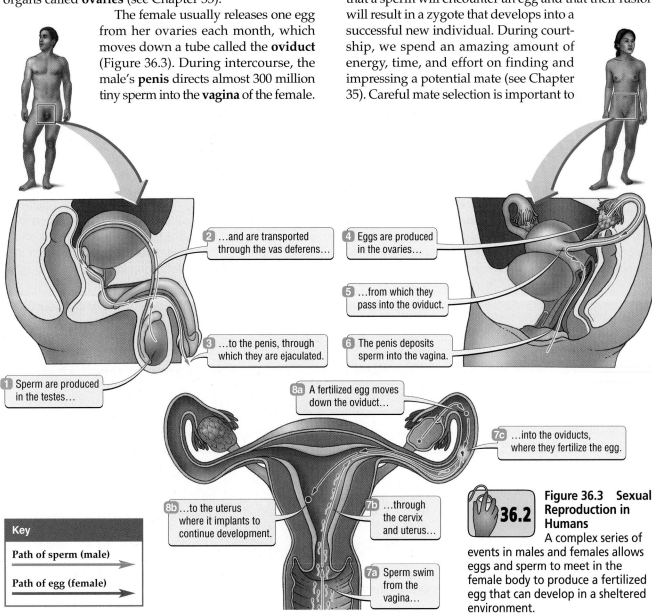

Figure 36.3 Sexual Reproduction in Humans
A complex series of events in males and females allows eggs and sperm to meet in the female body to produce a fertilized egg that can develop in a sheltered environment.

avoid some of the hazards involved in sexual reproduction (see the box below), and it is also important in ensuring that our offspring will survive and reproduce, as we will see below. Once we manage to win someone's heart, we consummate the relationship by having sexual intercourse, during which the male releases sperm directly into the female's body. As a zygote develops into an adult, it receives a great deal of care, first inside its mother's body (see Chapter 37), then outside (see Chapter 38). This parental care greatly increases the chances that it will survive to become a successful adult.

Many animals reproduce asexually

Many animal species reproduce asexually, including some members of all animal phyla. Asexually reproducing animals have lifestyles as different as those of oceangoing jellyfish and desert-dwelling lizards. The off-

spring that result from asexual reproduction get all of their genes from a single parent.

Although some animals rely exclusively on asexual reproduction, most asexually reproducing species switch between sexual and asexual reproduction depending on environmental conditions. Aphids, the small insects that sometimes infest our greenhouses or garden plants, usually reproduce asexually. The crowds of aphids that we see sucking sap from a plant in the summer are all genetically identical copies of a female that originally colonized that plant. Aphids rely on sexual reproduction, however, when approaching winter drives them from their summer food plant (Figure 36.4).

Animals may be male, female, or both

Gender in animals is more variable than our human perspective might lead us to think. As already noted,

■ BIOLOGY IN OUR LIVES

Sexually Transmitted Diseases: The Dark Side of Sex

AIDS, gonorrhea, syphilis, chlamydia, and herpes are some of the best-known human sexually transmitted diseases (STDs). As the name implies, these diseases are transmitted during sexual contact between two individuals. STDs include some of the most common and serious infectious diseases among humans: In the United States alone, 12 million people contract an STD each year, half of the population contracts at least one STD by the age of 35, and the annual cost related to STDs (other than AIDS) is over $10 billion.

The success of any STD, from the perspective of the disease organism, lies in its use of sexual contact as a way of spreading from one host to another. The most obvious way to reduce the risk of sexually transmitted disease is to abstain from sex. Avoiding sex, however, prevents an individual from contributing genes to the next generation. Thus, STDs have found the perfect way of spreading,

since sexual intercourse is essential to our reproductive success.

Efforts to fight the spread of STDs focus on three approaches: (1) reducing the number of sexual partners that an individual has; (2) promoting

safe sex involving the use of condoms and other barriers (see Chapter 38) to eliminate the exchange of bodily fluids between sex partners; and (3) rapid detection and treatment of STDs following infection.

Sexually Transmitted Diseases May Affect Over Half the United States Population during Their Lifetimes

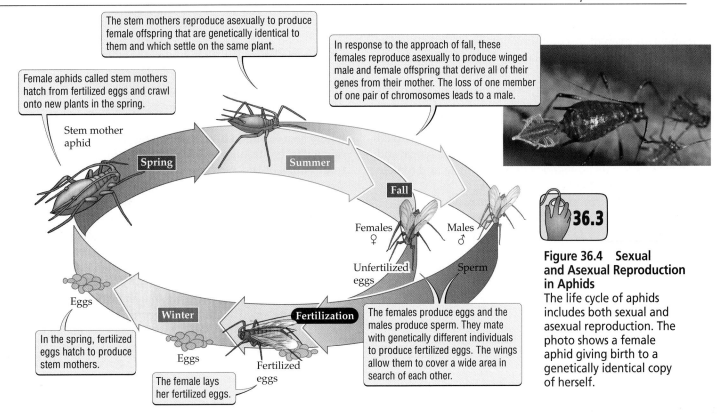

The stem mothers reproduce asexually to produce female offspring that are genetically identical to them and which settle on the same plant.

Female aphids called stem mothers hatch from fertilized eggs and crawl onto new plants in the spring.

In response to the approach of fall, these females reproduce asexually to produce winged male and female offspring that derive all of their genes from their mother. The loss of one member of one pair of chromosomes leads to a male.

Stem mother aphid

Spring

Summer

Fall

Females ♀   Males ♂

Unfertilized eggs   Sperm

Eggs

Winter   Fertilization

In the spring, fertilized eggs hatch to produce stem mothers.

Eggs

Fertilized eggs

The female lays her fertilized eggs.

The females produce eggs and the males produce sperm. They mate with genetically different individuals to produce fertilized eggs. The wings allow them to cover a wide area in search of each other.

36.3

Figure 36.4 Sexual and Asexual Reproduction in Aphids
The life cycle of aphids includes both sexual and asexual reproduction. The photo shows a female aphid giving birth to a genetically identical copy of herself.

humans may be either sperm-producing males or egg-producing females. The majority of animal phyla, however, include many species made up of individuals that produce both functional testes and functional ovaries and are therefore both female and male. We call such individuals **hermaphrodites**.

Hermaphrodites live in a wide variety of habitats and even include members of our own phylum (Chordata). Probably the most familiar hermaphrodite is the common earthworm (Figure 36.5). Mature earthworms have functional testes and ovaries at the same time, but many other hermaphrodites, including many fish species, change gender depending on their size (as the jack-in-the-pulpit plants described at the beginning of this chapter do).

A common misconception about hermaphrodites is that they fertilize their own eggs. Most hermaphrodites, even those that have functional testes and ovaries at the same time, must still mate with another individual. Nonetheless, they have an advantage over animals that are male only or female only in that they can mate with any individual they encounter.

■ Among animals, males produce haploid sperm in their testes and females produce haploid eggs in their ovaries. Many animals reproduce asexually instead of, or more commonly in addition to, reproducing sexually. Hermaphrodites have both functional testes and functional ovaries.

Plant Reproduction

Plants share with animals most of the basic features of their reproductive biology. In plants, as in animals, a sperm fertilizes an egg to form a new individual containing pairs of chromosomes that come equally from the male and female parents. Asexual reproduction is,

Figure 36.5 The Familiar Earthworm Is a Hermaphrodite
Each earthworm produces both eggs and sperm. For this reason, all earthworms of the same species are potential mates. In contrast, in animals that have separate males and females, only members of the opposite gender are potential mates.

however, even more common in plants than in animals. And although individual plants may be exclusively male or exclusively female, most are hermaphroditic.

Plant evolution has generated some unique variations on the basic eukaryotic life cycle. One important difference between plant and animal reproduction is that whereas animals produce sperm and eggs in specialized organs (the testes and ovaries), plants produce sperm and eggs in specialized *individuals*, called male and female **gametophytes** (*gameto*, "sex cell"; *phyte*, "plant." Some gametophytes produce sperm; others produce eggs.

A second unique aspect of plant reproduction arises because plants, unlike most animals, cannot walk, fly, or swim to find a mate. Flowering plants have overcome this problem by relying heavily on animals and wind to bring their gametes together.

Plants produce gametes in separate individuals

Because ferns illustrate the process so well, we begin our discussion of sexual reproduction in plants by describing the process in ferns. The ferns that we most commonly see (Figure 36.6) consist of typical diploid eukaryotic cells containing paired chromosomes. On the undersides of their leaves, ferns produce specialized structures that release dustlike haploid **spores**.

Although spores may seem like the equivalent of sperm or eggs, they differ in one critical way: Unlike sperm or eggs, spores do not fuse with other spores to form diploid embryos. Instead, the spores released by a fern usually sprout to form small, independent plants consisting of haploid cells.

These small, haploid plants are the gametophytes, and they produce small swimming sperm or large stationary eggs. The fern sperm swim through water to fuse with an egg. As in animals, fertilization results in a genetically unique, diploid zygote. The zygote develops into a diploid, spore-producing fern to start the cycle again.

Figure 36.6 Reproduction in Ferns
Ferns have distinct spore-forming and gamete-forming stages in their life cycle.

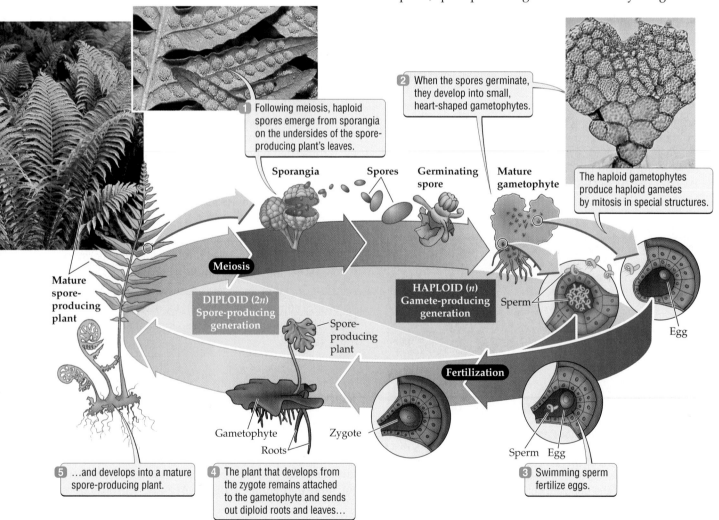

1 Following meiosis, haploid spores emerge from sporangia on the undersides of the spore-producing plant's leaves.

2 When the spores germinate, they develop into small, heart-shaped gametophytes.

The haploid gametophytes produce haploid gametes by mitosis in special structures.

Sporangia Spores Germinating spore Mature gametophyte

Meiosis

DIPLOID (2n) Spore-producing generation

HAPLOID (n) Gamete-producing generation

Sperm

Mature spore-producing plant

Spore-producing plant

Egg

Fertilization

Gametophyte

Roots

Zygote

Sperm Egg

3 Swimming sperm fertilize eggs.

5 ...and develops into a mature spore-producing plant.

4 The plant that develops from the zygote remains attached to the gametophyte and sends out diploid roots and leaves...

Flowering plants follow the basic plant model in their sexual reproduction

For those of us who have grown flowering plants in our homes or gardens, the reproductive cycle just described for ferns may seem to have little in common with flower and seed production. A closer look, however, reveals a similar reproductive cycle in flowering plants (Figure 36.7). The plants that we think of as apple trees, for example, consist of diploid cells, just as the spore-producing stage of ferns do. The flowers of the apple tree that represent its reproductive structures produce neither sperm nor eggs. Instead, like the spore-producing structures on fern leaves, flowers produce haploid spores that do not fuse with other spores.

Unlike ferns, however, flowering plants do not release these spores. Instead, the spores remain in the flower and develop into simple nonphotosynthesizing gameto-

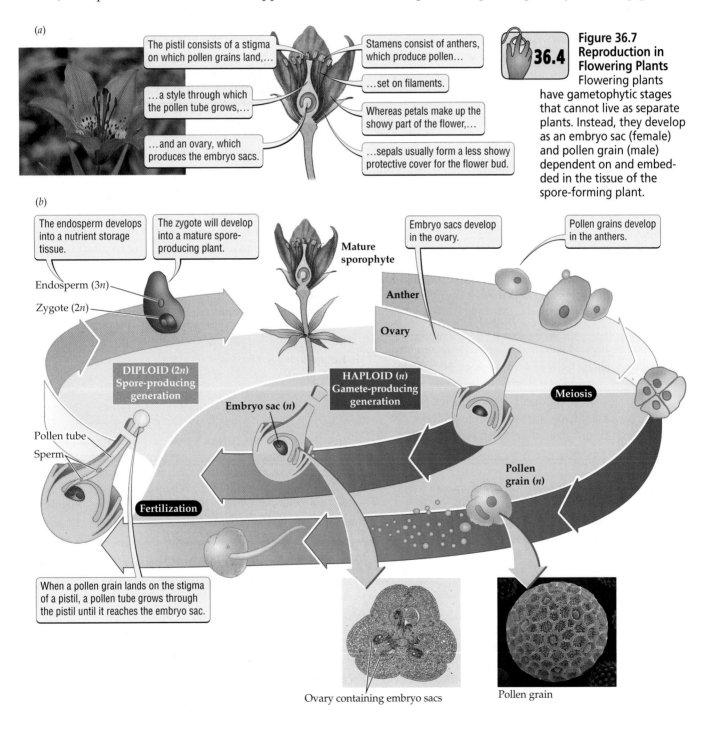

(a)

The pistil consists of a stigma on which pollen grains land,…

…a style through which the pollen tube grows,…

…and an ovary, which produces the embryo sacs.

Stamens consist of anthers, which produce pollen…

…set on filaments.

Whereas petals make up the showy part of the flower,…

…sepals usually form a less showy protective cover for the flower bud.

36.4

Figure 36.7 Reproduction in Flowering Plants Flowering plants have gametophytic stages that cannot live as separate plants. Instead, they develop as an embryo sac (female) and pollen grain (male) dependent on and embedded in the tissue of the spore-forming plant.

(b)

The endosperm develops into a nutrient storage tissue.

The zygote will develop into a mature spore-producing plant.

Embryo sacs develop in the ovary.

Pollen grains develop in the anthers.

Endosperm (3n)

Zygote (2n)

Mature sporophyte

Anther

Ovary

DIPLOID (2n) Spore-producing generation

HAPLOID (n) Gamete-producing generation

Meiosis

Embryo sac (n)

Pollen tube

Sperm

Pollen grain (n)

Fertilization

When a pollen grain lands on the stigma of a pistil, a pollen tube grows through the pistil until it reaches the embryo sac.

Ovary containing embryo sacs

Pollen grain

phytes that live as a sort of parasite on the flower tissues while they produce sperm and eggs. **Pollen grains**, which develop in the **stamens** of the flower, are the male gametophytes. The female gametophytes, called **embryo sacs**, are embedded in the **pistil** of their spore-producing parent (Figure 36.7*a*).

Fertilization of the eggs produced by the embryo sac requires that the pollen grain somehow move from the stamens of its parent plant to another flower. It also requires that the pollen grains release their sperm in such a way that they can penetrate the tissue of the pistil surrounding the embryo sac. The shape of the mature pollen grain allows it to ride wind or water currents, or to stick to the body of a mobile animal, so that it can move to other flowers, as we will see below.

When a pollen grain reaches a flower of its species, it may catch on a sticky surface of the pistil, called the **stigma**. If the stigma of the plant on which it lands is of the right species, then the pollen grain produces a **pollen tube**. The pollen tube grows through the tissue of the pistil to the embryo sac (Figure 36.7*b*). Two sperm cells move through the pollen tube and enter the embryo sac. Inside the embryo sac, one of the sperm cells fertilizes the egg cell. The resulting zygote will develop into an embryo and eventually into a mature plant. The other sperm fuses with two other cells of the embryo sac to form the **endosperm**, a nutrient storage tissue that has three copies of each gene and makes up most of the developing seed.

> ■ In plants, haploid sperm and haploid eggs are produced by separate haploid individuals called gametophytes. In flowering plants, the male gametophyte (pollen grain), which produces sperm, and the female gametophyte (embryo sac), which produces eggs, live in the tissues of the diploid plant that produced them.

Ensuring that Sperm Meet Eggs

For sexually reproducing organisms, fertilization is only the last step in a complex sequence of events. Before fertilization can occur, sperm and eggs must be brought close together. Finding a mate allows the male's sperm to be released close to the female's eggs. This reduces the distance that the short-lived sperm must travel before reaching their goal. Simply finding a mate, however, is not the end of the story. As we'll see in the following section, careful selection of a mate increases the chances that the individual chosen is of sufficiently high genetic quality and that the resulting offspring stand a good chance of surviving.

Decreasing the distance between sperm and egg increases fertilization success

Males and females must be close to each other before releasing sperm or eggs, or the chances of fertilization fall to near zero. Under ideal conditions, human sperm can swim about 10 meters in a day. In the enclosed space of the female reproductive system, this rate of travel is more than adequate, but if they were released into the ocean, the chances of the sperm reaching an egg would become very small.

In addition, as sperm and eggs move away from their sources, their concentration rapidly decreases. To fertilize human eggs in test tubes during in vitro fertilization (see Chapter 38), 60,000 sperm are needed to give a reasonable chance of fertilization. Clearly it is important that males and females either seek one another out or live together in sufficiently dense populations that sperm and eggs need not travel far.

Mobile animals actively seek out mates

Finding a potential mate may seem less challenging to humans than the other component of mating behavior—choosing a high-quality mate—because we live in towns and cities and need not travel far to find a member of the opposite gender. For many organisms, however, encountering a potential mate is difficult. The chance that slowly moving animals, which include most small animals (see Chapter 27), or animals of a species whose members live widely scattered will bump into a potential mate is relatively small.

Animals can find other individuals of their species by using signals of various kinds (Figure 36.8). The most common signals are sounds and chemicals, both of which can travel great distances. The chirps and croaks produced by male birds, crickets, and frogs all guide females to eager mates. Sex pheromones released by female moths can draw in males from distances of many kilometers. Some animals, such as fireflies and deep-sea fish, produce flashes of light to signal mates through their dark environments.

Plants and immobile animals need other means of sperm transport

Plants and those aquatic animal species, such as corals, that cannot move as adults must nonetheless transport their sperm close enough to an egg that the sperm have a chance of encountering the egg. Immobile animals and plants commonly rely on wind or water currents to bring sperm to eggs. Allergy sufferers know that in the spring and fall the air is filled with pollen grains carrying plant

Figure 36.8 Finding a Mate
Animals use diverse signals to find and evaluate potential mates.

sperm. Sperm and eggs released in water go where the currents scatter them. Typically, organisms such as corals in a coral reef or grasses in a meadow that rely on water or wind to transport their eggs or sperm must live close together to give sperm a reasonable chance of encountering an egg.

The reproductive organs we call flowers originally evolved as a way of overcoming the limitations of water and wind as pollen dispersers. Flowers attract and manipulate the movements of mobile insects, birds, and even mammals so that they carry pollen grains from the stamens of one flower to the stigmas of other flowers of the same species (Figure 36.9). Plants bribe pollinating animals to visit their flowers with foods such as sugary nectar or protein-rich pollen. The

Figure 36.9 Pollinators Provide Stationary Flowering Plants with a Way of Getting Sperm to Eggs
The spectacular colors, shapes, and smells of flowers, in combination with rewards such as nectar, convince pollinating animals to visit several flowers of the same species, transferring pollen in the process.

(a)

Aquatic animals such as these salmon may release vast quantities of sperm and eggs into the water, where fertilization takes place.

Figure 36.10   Sperm Transfer in Animals and Plants

(b)

Male land animals typically transfer sperm directly into the female's body to avoid exposing it to the environment.

(c)

The sperm of flowering plants never face the external environment. They remain inside the pollen grain…

…until they move through the pollen tube to the embryo sac.

attraction of a pollinator species to a particular combination of flower color, shape, and smell increases the chance that the pollinator will carry the pollen to another flower of the same species. If you follow a honeybee as it moves through a field, for example, you will see that it visits only a few of the available flower types. With each visit to a flower, it deposits and picks up pollen caught in the hairs covering its body.

The environment determines where gametes must be released

The environment strongly limits the length of time that sperm and unfertilized eggs can survive outside the organism. If released onto land, neither eggs nor sperm will survive more than a few minutes, and the lack of water makes it impossible for the sperm to move. In ocean water, however, sperm can survive because the environment resembles the organism's body fluids. Thus many aquatic organisms simply release their sperm into the water surrounding a female (Figure 36.10*a*). Land-dwelling organisms almost always release their sperm directly into the female's body. **Copulation** (Figure 36.10*b*) is the means by which terrestrial animal males release sperm directly into a female's reproductive system (think of human sexual intercourse). Flowering plants release their sperm only after the pollen grain encasing the sperm lands on the stigma of a flower (Figure 36.10*c*).

■ Animals and plants move sperm close to eggs to increase the likelihood of fertilization. Mating behavior brings mobile male and female animals close together, but plants and immobile animals rely on wind, water, or pollinating animals to bring sperm close to eggs.

Choosing a Mate

Humans try to select their mates carefully. Only after considerable thought do most of us decide with whom we want to produce children. A few animal species, including humans, select mates for their ability to provide parental care for the offspring. Of much more general concern, however, is the fact that sexual reproduction involves recombining one's genetic information with that of another individual. If an organism selects its mate poorly, it risks combining its own genes with inferior ones to produce offspring with a poor chance of surviving and reproducing.

Males and females select mates differently

Because males and females produce different kinds of gametes, they choose their mates differently. Whereas each sperm costs a male relatively little in the way of resources, each egg takes up much more of the female's resources. This simple observation has two important consequences:

The traits that males use to convince females of their suitability as mates can reach extremes, as in the fantastic tail displays that peacocks use to attract peahens.

Figure 36.11 Some Females Judge Males by Their Displays
Careful selection of mates by females may increase their offspring's chances of success. Male courtship displays may allow females to choose the male with the best genes.

Females of the tiny spring peeper frog judge males by their ability to sing loudly and long.

1. Males can produce many more sperm than females can produce eggs.
2. A female stands to lose more if one of her eggs results in offspring with a poor chance of survival than does the male that fertilized that egg.

As a result, females tend to choose mates carefully to ensure that each offspring has as great a chance of success as possible, and males tend to mate with as many females as possible.

Females try to select high-quality males

Scientists who study mating behavior have typically focused on the elaborate contests among males for the chance to mate with females (Figure 36.11), such as those described in Chapter 35. Females play an equally active role in choosing their mates. Females actively accept some males and reject others. Because females often invest so much of their resources in their offspring, they stand to lose a great deal if they mate with a poor-quality male. The competitive mating displays in which males engage often indicate the quality of their genes, as shown by the observation that the offspring of winners often prove more successful than the offspring of losers.

In plants, too, competition plays an important part in mate selection. When pollen grains land on a stigma, they release sperm cells into pollen tubes that grow through the tissues of the pistil, which guide them to the embryo sac (see Figure 36.10c). The faster a pollen tube grows, the more likely its sperm will reach an unfertilized embryo sac. Several recent studies have shown that

seeds that result from pollen that produces rapidly growing pollen tubes produce more vigorous seedlings than do seeds that result from pollen that produces slowly growing pollen tubes.

■ Selection of a high-quality mate increases the chances that the recombination of genetic information from both parents will produce successful offspring.

■ HIGHLIGHT

Gender and Resource Allocation

Why do jack-in-the-pulpits change genders? Each large seed that a female plant produces costs a female much more of her resources than does each small sperm produced by a male. Only large jack-in-the-pulpits have sufficient resources to produce mature seeds, but small, resource-poor plants can produce at least some sperm-containing pollen. Thus, jack-in-the-pulpits use maleness as a fallback position that gives them a chance to reproduce even following years during which they have failed to store much energy.

Although the flexible gender of jack-in-the-pulpits may seem a bit over the edge to humans, other plants and animals do much the same sort of thing. Among fish we find several examples of changes in the gender of individuals. In freshwater sunfish and some marine coral reef fish, individuals switch irreversibly from female to male when they become large enough. In this case, males are successful only if they reach a size that allows them to compete with other males to hold hotly contested breeding territories. Small males will never hold a territory, and will therefore never obtain a mate. Small females, on the other hand, can produce at least some eggs. In other words, the logic for these fish mirrors that for the jack-in-the-pulpit even though the outcome is the reverse.

Because reproductive success translates so directly into evolutionary success, the way in which an organism uses its resources to mate and produce offspring matters a great deal. In addition, because males and females produce different kinds of gametes and often contribute to other aspects of reproduction in equally different ways, we would expect the two sexes to differ in how they use their resources. Organisms, such as humans, that are either always male or always female must devote all of their resources to either male or female reproduction. Organisms, such as jack-in-the-pulpits or the fish mentioned above, that can allocate resources to either male or female aspects of reproduction seem to do so based on the available resources.

Typical plants, which produce stamens and pistils at the same time, can also divide their resources between the male and female function according to the way in which each sex contributes to reproduction as well as the resources available. The scarlet gilia, for example,

produces flowers containing both stamens and pistils (Figure 36.12). Experiments have shown that pollen from scarlet gilia plants that devote almost all of their resources to stamens fathers only a few more seeds than that of individuals that devote only a small fraction of their resources to stamens. In contrast, scarlet gilia plants that devote almost all of their resources to pistils produce many more seeds than individuals that devote only a small fraction of their resources to pistils. In terms of reproductive success, therefore, a scarlet gilia stands to gain the most by devoting a lot of its resources to pistils and relatively little to stamens. In nature, scarlet gilia plants allocate only 30 percent of their resources to stamens and the remaining 70 percent to pistils.

> ■ How organisms allocate resources to male and female aspects of reproduction strongly influences their reproductive success.

SUMMARY

How Bacteria and Eukaryotes Reproduce and Recombine Genes

- Prokaryotes reproduce asexually when one parent cell divides to produce two new cells, each identical to the original cell.

- Prokaryotes transfer genetic information from one individual to another sporadically, in several ways, including transformation, and sometimes as a plasmid or virus moves from one host cell to another.

- In eukaryotes genetic material is recombined during reproduction. Haploid gametes are formed by meiosis, which distributes one chromosome from each pair into each sperm in males and each egg in females. Fertilization combines the genetic information from the sperm and egg to form a diploid zygote.

Animal Reproduction

- Human males release many millions of sperm from the penis into the female's vagina. Only a few hundred of these sperm manage to reach an egg, and only one of these can fertilize it.

- Human females produce approximately one egg each month, which moves from an ovary into the oviduct, where it may be fertilized.

- Unlike humans, which rely solely on sexual reproduction, some animals reproduce only asexually; others can reproduce either asexually or sexually.

- Many animals are hermaphrodites, which produce both eggs and sperm and are able to reproduce with any other individual of their species.

Figure 36.12   The Pistil-Packing Flowers of the Scarlet Gilia
By investing more heavily in pistils than in stamens, scarlet gilia maximizes its reproductive success.

Plant Reproduction

- Diploid plants use meiosis to produce spores, which grow into haploid gametophytes. These gametophytes then produce either sperm or eggs.

- Whereas fern gametophytes develop independently of the diploid plant, the gametophytes of flowering plants remain attached to the diploid plant as they develop.

- In flowering plants, the male gametophyte, a pollen grain, is carried to the stigma of another flower. The sperm move out of the pollen grain and through a pollen tube to reach the female gametophyte, the embryo sac. One sperm fertilizes the egg to form a zygote, and a second sperm fuses with two other cells to form the endosperm.

Ensuring that Sperm Meet Eggs

- To succeed at fertilization, sperm and eggs must be near each other.

- Animals use sound, pheromones, and even flashes of light as signals to search for and move toward potential mates.

- Immobile marine animals that use water currents and plants that use wind to carry sperm to eggs often must live near many individuals of the same species. Some flowering plants use animals to carry pollen to other flowers of the same species.

- In the ocean, many organisms release gametes into the water. On land, most males release sperm inside the female's body.

Choosing a Mate

- Females choose mates more carefully than males because eggs require a greater investment of resources than sperm and are in more limited supply.

- Competition between males and mate choice by females each plays a part in determining the identity of the male parent and therefore the quality of the genes contributed to the offspring.

Highlight: Gender and Resource Allocation

- How organisms allocate resources to male and female aspects of reproduction strongly influences their reproductive success.

KEY TERMS

asexual reproduction p. 602	pistil p. 610
copulation p. 612	plasmid p. 603
egg p. 603	pollen grain p. 610
embryo sac p. 610	pollen tube p. 610
endosperm p. 610	reproduction p. 602
female p. 603	sexual reproduction p. 602
fertilization p. 604	sperm p. 603
gamete p. 603	spore p. 608
gametophyte p. 608	stamen p. 610
hermaphrodite p. 607	stigma p. 610
male p. 603	testis p. 605
ovary p. 605	transformation p. 603
oviduct p. 605	vagina p. 605
penis p. 605	zygote p. 604

CHAPTER REVIEW

Self-Quiz

1. Which of the following organisms do *not* depend on reproduction to exchange genetic information?
   a. animals
   b. plants
   c. bacteria
   d. fungi

2. Human eggs are produced in the
   a. oviduct.
   b. testes.
   c. embryo sac.
   d. ovary.

3. Like animals, plants produce
   a. many more sperm than eggs.
   b. a few more sperm than eggs.
   c. equal numbers of sperm and eggs.
   d. fewer sperm than eggs.

4. Asexual reproduction produces offspring that are
   a. genetically identical to their parents.
   b. genetically identical to their siblings.
   c. none of the above
   d. both a and b

5. Which of the following is true about how female animals choose their mates?
   a. Females tend to choose mates more carefully than males do.
   b. Females tend to choose males that produce abundant pollen.
   c. Females tend to choose males that reproduce asexually.
   d. all of the above

Review Questions

1. Contrast the production of eggs and sperm in humans with the production of eggs and sperm in flowering plants.

2. What is it about the reproduction of certain flowering plants that allows them to live in more diverse communities, with many different plant species, than some wind-pollinated plants, which tend to live among many individuals of their own species?

3. How might the structure of the female part of a flower be modified to increase the intensity of the competition between male pollen tubes as they race toward the embryo sac, and how might this increased competition benefit the female parent?

4. How does habitat affect the reproduction of plants and animals?

36   𝕿𝖍𝖊 𝕯𝖆𝖎𝖑𝖞 𝕲𝖑𝖔𝖇𝖊

Disappearing Bugs Threaten Wildflowers and Crops

ITHACA, NY—Researchers studying bees and other pollinating animals say that a critical link in the life cycle of many flowering plants is being broken. The flowering plants that brighten our gardens and the crops that are the foundation of our agricultural production depend on pollinators—bees, butterflies, and even birds and bats—to carry pollen from plant to plant. As our wild lands have been replaced by agricultural and urban areas, the numbers of native insects that work for us as pollinators have dropped dramatically.

Nearly half of the pollination of agricultural plants has been taken over by managed and wild European honeybees, an introduced species. These honeybees also pollinate some wildflowers, although scien-

tists say they don't yet know what percentage of wildflowers now depend on honeybees. Unfortunately, even the number of managed honeybees in the United States has declined by more than 25 percent since the early 1990s. Researchers estimate that losses of honeybees have resulted in 5.7 billion dollars worth of damage to agricultural production around the world each year.

Although there are many reasons for the decline in honeybees and other insect pollinators, including introduced mites, diseases, introduced competitors such as the Africanized honeybee, and bad weather, Dr. Pauline Nation of the University of West Idaho points to two reasons for the dramatic drop in pollinators that every person can try to prevent: habitat destruction

and the misuse of pesticides. The Environmental Protection Agency now requires some pesticides to carry warning labels that recommend against their application near blooming crops or weeds, but these warnings have not reduced the effects on pollinators. Dr. Nation noted that pesticides continue to be used inappropriately.

Burt Umbleby of the Ithaca Metro Gardening Club suggests that local gardeners may be able to help reverse the pollinator loss by growing plants that provide nectar for insects. He added that many gardening books now list plants that are particularly beneficial to pollinators.

"We're part of the problem," said Mr. Umbleby, "and now gardeners need to become part of the solution."

Evaluating "The News"

1. In home gardens and commercial farms, reproduction—the successful pollination of flowers—is the key to successful harvests for most plants. Pesticides are also useful to farmers for protecting plants from herbivores so they can yield a good harvest. What are farmers to do? In deciding whether to use pesticides, how can they weigh the detriment

to pollinators against the protection of plants?

2. Is there any way to incorporate the aesthetic values associated with flowers into economic arguments about the costs and benefits of pesticide use?

3. If legislators are going to attempt to promote conservation to bring back

endangered plants that require pollinators, should conservation legislation focus on saving insects and other animal pollinators, plant species, specific environments, or interactions among species and their environments? How would you conserve an interaction?

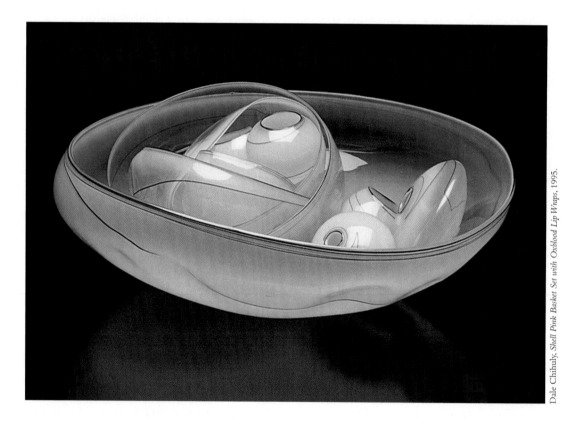

Dale Chihuly, *Shell Pink Basket Set with Oxblood Lip Wraps*, 1995.

chapter 37 Development

Cloning Carrots and Lambs

*I*n 1958, in a groundbreaking experiment that sounds more like an adventure in cooking than science, researchers managed to clone a new plant from a single mature plant cell. F. C. Steward and his colleagues placed tiny pieces of phloem tissue taken from a carrot root in a large container filled with coconut milk and stirred gently. The phloem cells divided as they circulated through the coconut milk to form lumps of callus, a type of

poorly organized tissue that typically forms in plants after wounding. The coconut milk, which is actually the unique, liquid endosperm, or nutritive tissue, of coconut seeds, provided the nutrients needed to support the dividing cells. The constant movement of the callus in the stirred coconut milk broke off individual callus cells. These individual cells continued to divide, but instead of forming another mass of disorganized callus tissue, they dif-

618

ferentiated to form a recognizable root and a mass of cells that eventually formed the stems and leaves of a new carrot plant. To understand just how amazing this event was, imagine putting a few cells from your little finger into a big vat full of mother's milk … and returning a few weeks later to find embryonic copies of yourself floating around in the vat.

In 1996, almost 40 years after the work of Steward and his colleagues, a group of scientists working in Scotland used a single cell taken from the udder of a sheep to produce a now famous lamb called Dolly. Cloning a sheep proved considerably

■ MAIN MESSAGE

The coordinated action of genes directs the development of a fertilized egg into a multicellular organism with specialized cells, tissues, and organs.

more complicated than cloning a carrot. The scientists first had to remove the haploid nucleus from an unfertilized sheep egg. They then inserted another nucleus, taken from an udder cell, into the egg. The egg now had a diploid nucleus containing two copies of each sheep chromosome, just as a fertilized egg would. The egg was then implanted in another female sheep and allowed to develop. An apparently healthy lamb, Dolly, was the result. In the years since Dolly was produced, cloning technology has advanced to the point at which we have now successfully cloned a human embryo.

The cloning of plants and animals raises interesting questions about how complex multicellular organisms can develop from the single cell of a fertilized egg. If we can grow whole carrots and sheep from a single cell, what prevents a phloem cell or an udder cell from growing into a whole new organism in nature? Conversely, what determines that a cell becomes a phloem cell rather than a xylem cell, or an udder cell rather than a brain cell? Why is it easier to clone carrots than to clone sheep? In this chapter, we explore these questions, which go right to the heart of how organisms control their development and apply equally to all multicellular organisms.

A Clone and Her Offspring

Dolly the sheep (left) is the first mammal to have been cloned from a mature cell. She has produced a lamb, Bonnie (right), by normal reproduction.

■ KEY CONCEPTS

1. The various organs that make up most animals develop in predictable ways from three distinct layers of cells.

2. All plants form three tissue types that are distinct from the tissues of animals early in development.

3. During development, the potential form and function of cells become more limited.

4. Chemicals called morphogens play a central role in coordinating how the genes in each cell are activated and inactivated.

5. Developmental changes have played an important part in generating evolutionary changes.

As described in Chapter 36, we humans start life as a fertilized egg containing equal genetic contributions from our mother and our father. This single cell multiplies and develops into a complex individual consisting of many different cell types organized into tissues and organs that interact in a coordinated way.

Our transformation from a fertilized egg into what we are today involved much more than an increase in the number of our cells. It required the precisely controlled differentiation of one cell into trillions of cells, each specialized for specific functions. During our development, these cells appeared not only in a carefully defined order, but also at carefully defined places in our bodies.

In Unit 3 we learned that DNA contains the blueprint for building an organism. However, because each cell in the organism contains exactly the same DNA, the differences among cells in their structure or function must stem from differences in how the body interprets its DNA blueprint as it develops.

In this chapter we investigate how a developing multicellular organism converts the information contained in its DNA into many different and interacting tissues. To provide a foundation for our discussion, we first contrast patterns of development in plants and animals. We then discover that in both groups, the development of different cell and tissue types depends on a closely regulated sequence of activation and inactivation of genes. Finally, we explore the evolutionary role of changes in the timing of developmental events.

37.1 *Animal Development*

For all their outward differences in size, shape, and complexity, all animals share remarkably similar developmental patterns. We begin this chapter by looking at the development of animals, using human development as an example to illustrate important steps in the process.

Rapid cell division characterizes early animal development

In the first stage of animal development, the single-celled zygote divides again and again to form a multicellular embryo. Never again in its life will an animal's cells divide as rapidly as they do right after fertilization. In part, the rapidity of cell division in the early embryo results because, unlike later cell division, it involves mainly subdivision of the cells rather than cell growth. At this point, cells increase in number while decreasing in size. For example, even after many cell divisions, a frog embryo containing tens of thousands of cells is still about the same size as the original zygote (Figure 37.1). At this early stage, the embryo, which has the shape of a hollow sphere, is called a **blastula**, and usually consists of cells that look very much alike.

A few features of the development of mammals are unique and noteworthy. Mammals, including humans, develop in a protected environment inside their mother's body, in an organ called the **uterus**. Although mammalian cells divide without growing early in development, they divide more slowly than in other animals. In the week that it takes a human egg to travel from the site of fertilization in the oviduct to the uterus (see Chapter 36), it divides only four times, so that it consists of just 16 cells.

A second difference is that the mammalian blastula is not just a simple, hollow sphere, like that of most other animals, but one that contains two distinct cell types. The outer layer of the blastula will become the developing embryo's portion of the **placenta**. This uniquely mammalian structure consists of tissues from both the mother's uterus and the embryo and permits the transfer of nutrients and wastes between the mother and the embryo. Only the inner layer of cells, which is clustered at one end of the blastula, becomes the actual embryo.

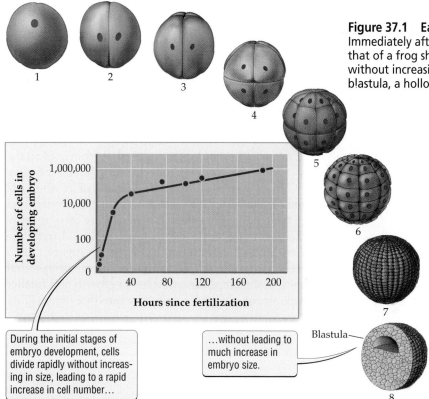

Figure 37.1 Early Development of Animals
Immediately after fertilization, most animal zygotes, like that of a frog shown here, develop rapidly as cells divide without increasing in size. The rapidly dividing cells form a blastula, a hollow sphere consisting of thousands of cells.

During the initial stages of embryo development, cells divide rapidly without increasing in size, leading to a rapid increase in cell number…

…without leading to much increase in embryo size.

Blastula

derm, and ectoderm are found in most animals and represent the first clear evidence that the developmental possibilities of a particular cell becomes more and more limited as development progresses (see the box on p. 622).

Initially, all three germ layers lie on the outside of the embryo (Figure 37.2). However, if the endoderm is to give rise to the lining of the gut and the mesoderm to muscles and skeletal tissues, these germ layers must move to more appropriate positions inside the embryo. This remarkable rearrangement takes place during **gastrulation**, a complex sequence of events in which the germ layers move relative to one another so that the endoderm lies inside the embryo and is surrounded by mesoderm. As a result, the ectoderm forms the entire outer layer of the embryo. In humans, the formation of the germ layers and their rearrangement during gastrulation take place during the 2 weeks following implantation of the embryo in the uterine wall.

The second stage of animal development produces three germ layers

During the second stage of animal development, two critical events take place. First, depending on their position within the blastula, cells in the embryo form three **germ layers**, which follow three distinct developmental paths to give rise to different tissues in the mature animal. The **endoderm** generally gives rise to the epithelial lining of the gut and lungs; the **mesoderm** gives rise to muscles and skeletal structures; and the **ectoderm** gives rise to nerves and the outer covering of the animal (Table 37.1). The endoderm, meso-

37.2

Figure 37.2 Animals Produce Three Germ Layers with Developmentally Distinct Fates
In spite of the difference in patterns of early cell division, all animals have recognizable endoderm, mesoderm, and ectoderm cells once the stage of rapid cell division ends. In the process of gastrulation, these three germ layers move to their proper positions in the embryo.

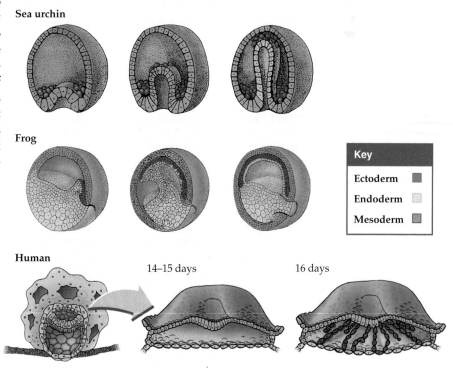

Sea urchin

Frog

Human

14–15 days   16 days

Key

Ectoderm

Endoderm

Mesoderm

37.1 *The Three Germ Layers and Their Fates*

Germ layer	Corresponding adult structure
Endoderm	Gut, liver, lungs
Mesoderm	Skeleton of limbs and body, muscles, reproductive structures, kidneys, blood, blood vessels, heart, inner layer of skin
Ectoderm	Skull, nerves and brain, outer skin, teeth

Cell proliferation, differentiation, and rearrangement generate the mature body

After gastrulation, animals undergo a period of rapid differentiation and rearrangement of cells in the germ layers into the various organs needed to make up a functioning individual. For example, just 3 weeks into our own 40-week-long development (Figure 37.3), the human embryo has an identifiable heart. By the eighth week of development, the developing human, now called a fetus, is 2.5 centimeters long, the head is clearly identifiable, the endoderm has differentiated into an identifiable liver, the mesoderm has differentiated into red blood cells and kidneys, and the beginnings of all the organs found in an adult are present.

By 12 weeks (the end of the first trimester), the fetus is 8 centimeters long, the fingers and toes have nails, the external genitalia are recognizably male or female, and the gut has developed from the endoderm to the point that it can absorb sugars.

Following the rapid differentiation of the first trimester, the emphasis switches to growth and further development of already existing organs (see Figure 37.3).

■ THE SCIENTIFIC PROCESS

Why Worms?

Scientific research often depends on organisms that themselves are of no particular interest. Fruit flies, for example, play a key role in genetic research, as mice do in medical research. Many studies of animal development rely on a particularly unimpressive experimental subject, the nematode worm *Caenorhabditis elegans*. This tiny, transparent worm (it is much shorter than its name) lives inconspicuously in the soil, where it feeds on bacteria. From birth to death it lives but 3 weeks. *C. elegans* causes no diseases, does not affect any crops, and has no economic use. Why, then, has *C. elegans* become a star in the realm of developmental biology?

Like most successful laboratory organisms, *C. elegans* lends itself to easy laboratory culture: Place these worms on a petri dish containing a good bacterial culture (or a commercially available nematode food) and they thrive. The short life span of *C. elegans* means that experiments can

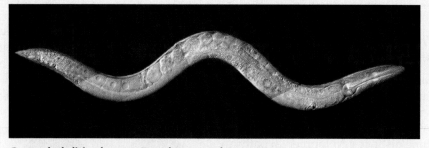

***Caenorhabditis elegans*, Developmental Superstar**

be completed quickly, allowing researchers to crank out publications.

What has placed *C. elegans* firmly on the path to biological stardom, however, is that its development has been remarkably thoroughly mapped. An adult *C. elegans* has just under a thousand cells. Developmental biologists have documented exactly how a single-celled *C. elegans* zygote divides and divides again to give rise to each of those adult cells. In other words, we know the complete family tree for every cell in an adult *C. elegans*. Moreover, the position of each cell at each stage of development has also been documented.

More recently, in 1998, scientists finished sequencing the entire genome of *C. elegans*, which consists of just under 20,000 genes. Given the importance of the activation and inactivation of genes in regulating development, this latest advance will only increase the stature of *C. elegans*. Researchers have already begun to determine which genes are active and which are inactive at various stages in its development. Even though the development of *C. elegans* itself might not particularly interest us, the general developmental principles that we can uncover using this little worm certainly do.

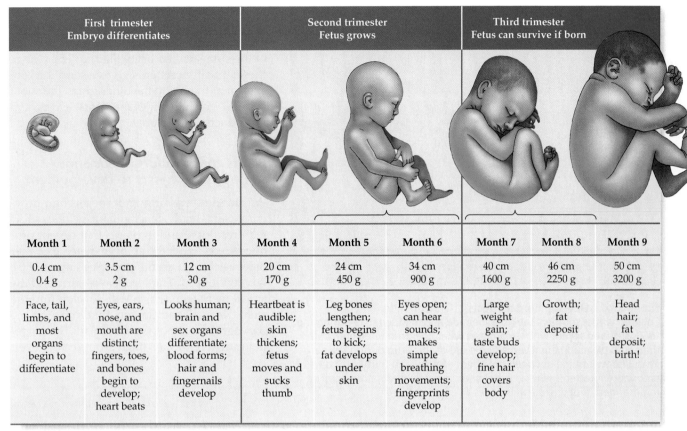

	First trimester Embryo differentiates		Second trimester Fetus grows			Third trimester Fetus can survive if born		
Month 1	**Month 2**	**Month 3**	**Month 4**	**Month 5**	**Month 6**	**Month 7**	**Month 8**	**Month 9**
0.4 cm 0.4 g	3.5 cm 2 g	12 cm 30 g	20 cm 170 g	24 cm 450 g	34 cm 900 g	40 cm 1600 g	46 cm 2250 g	50 cm 3200 g
Face, tail, limbs, and most organs begin to differentiate	Eyes, ears, nose, and mouth are distinct; fingers, toes, and bones begin to develop; heart beats	Looks human; brain and sex organs differentiate; blood forms; hair and fingernails develop	Heartbeat is audible; skin thickens; fetus moves and sucks thumb	Leg bones lengthen; fetus begins to kick; fat develops under skin	Eyes open; can hear sounds; makes simple breathing movements; fingerprints develop	Large weight gain; taste buds develop; fine hair covers body	Growth; fat deposit	Head hair; fat deposit; birth!

37.3

Figure 37.3 Human Development
Rapid differentiation of organs marks the early stages of human development, whereas rapid weight gain marks the later stages. The three trimesters, or three-month stages, commonly used to divide up human pregnancy correspond roughly to important developmental events. Improvements in medical technology now allow even some babies born in the middle of the second trimester to survive.

The fetus grows from the size of a large mouse to the typical birth weight of over 3 kilograms. Organs that formed by the end of the first trimester continue to develop, so that at birth the fetus can survive outside the uterus. We discuss development following birth in Chapter 38.

Genetic or environmental factors that disrupt the normal sequence of events, particularly during the rapid differentiation of the first 12 weeks of development, can lead to serious problems. In humans, most miscarriages take place during the critical first trimester of pregnancy because any developmental problems in the embryo are most likely to be expressed at this time.

To understand just how sensitive early development can be to even brief disruptions, consider the effect of thalidomide, a drug that was prescribed as a sedative in the 1950s. Thalidomide interferes with the development of limb bones. If taken by a pregnant woman between 24 and 35 days after fertilization, when the embryo normally develops its arms and legs, the drug can lead to devastating limb deformities (Figure 37.4).

■ Animal development has several identifiable stages. After fertilization, cells divide rapidly, without increasing in size, to form a blastula. The cells of the blastula then differentiate into three germ layers, each with a distinct developmental fate. Gastrulation moves the cells in each germ layer into appropriate positions. After gastrulation, animal organs differentiate quickly.

Plant Development

Like animals, plants begin life as a fertilized egg, and this single cell gives rise to a diversity of cell types carefully organized into the various tissues that make up a mature plant. In this section we survey the patterns of plant development to give us a basis for comparison with animal development. At the same time, we examine the mechanisms that control development.

Figure 37.4   Developmental Effects of Thalidomide
Children whose mothers took thalidomide as a mild sedative during the 1950s suffered severe limb deformities when the drug interfered with their limb development. Even a single use of this drug during the narrow window of time between the twenty-fourth and thirty-fifth days of pregnancy interferes with a critical phase of bone elongation, leading to deformed, flipperlike appendages.

Plants begin development inside seeds or spores

The plant zygote begins development surrounded by a protective spore (for example, in ferns) or seed (for example, in flowering plants). As in animal development, distinctive structures appear during the first stages of plant development. The first divisions of the zygote result in a mass of undifferentiated cells, all of which continue to divide. Soon, however, two important events take place within the developing embryo.

The cells of the embryo differentiate into three distinct types: (1) **protoderm** (*proto*, "first"; *derm*, "skin"), which develops from the outer layer of embryo cells, and (2) **ground meristem** and (3) **procambium**, both of which develop from the inner cells of the embryo (Figure 37.5). We can trace all the tissues in a mature plant back to one of these three embryonic tissues.

The second important event is the restriction of cell division and differentiation to zones called **apical meristems** that lie at the two tips of the elongated embryo (Figure 37.5). One of these apical meristems gives rise to the root system of the plant; the other forms the shoot system, which includes the stems, leaves, and reproductive structures.

Plants that produce seeds show some special twists to development

Within the seeds of gymnosperms and flowering plants, early development includes a few features not found in plants that do not produce seeds. First, the embryo develops a pair of leaves, called **seed leaves** (see Figure 37.5), within the seed. The seed leaves surround the apical meristem, which will develop into the stems, leaves, and flowers that make up the aboveground shoot system. In addition, special nutritive tissues form around the devel-

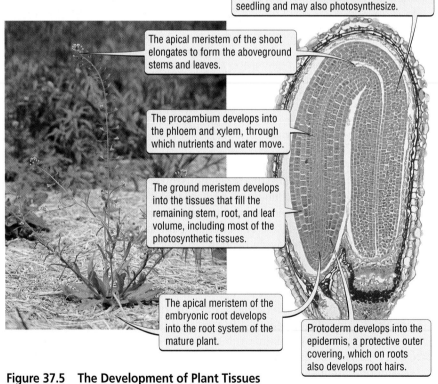

Seed leaves develop into the first pair of leaves, which often store nutrients for the germinating seedling and may also photosynthesize.

The apical meristem of the shoot elongates to form the aboveground stems and leaves.

The procambium develops into the phloem and xylem, through which nutrients and water move.

The ground meristem develops into the tissues that fill the remaining stem, root, and leaf volume, including most of the photosynthetic tissues.

The apical meristem of the embryonic root develops into the root system of the mature plant.

Protoderm develops into the epidermis, a protective outer covering, which on roots also develops root hairs.

Figure 37.5   The Development of Plant Tissues
Plant embryos produce three distinct tissue types—protoderm, ground meristem, and procambium—each of which gives rise to a predictable set of tissues in the mature plant.

oping embryo (Figure 37.6). In gymnosperms, these nutritive tissues form from tissues of the female gametophyte that produced the egg; in flowering plants, they form from the endosperm (see Chapter 36). In many flowering plants, such as corn and coconuts, the endosperm makes up much of the volume of the seed. Finally, the seed develops a hard external coat that protects it from the environment.

The endosperm and seed leaves produced early in the development of seed-producing plants allow them to enter a dormant stage in which development stops. For most plants, the transition between an embryo developing within a seed or spore and life as an independent, photosynthesizing plant is the most vulnerable stage of life. The well-stocked, well-protected, and well-developed embryos of seed-producing plants can lie dormant within the seed and resume development and growth only when conditions in their harsh terrestrial environment allow (see Chapter 33). In contrast, plants such as mosses and ferns cannot interrupt development at this vulnerable stage and must continue developing regardless of the condition of their environment.

The procambium, ground meristem, and protoderm differentiate to form plant organs

As in animals, each of the three basic cell types in the plant embryo gives rise to specific tissues in the mature plant. The protoderm gives rise to the outer covering of the plant. The procambium gives rise to the tissues that make up the internal transport system of the plant, which consists of water-conducting xylem and nutrient-conducting phloem (see Chapter 30). The ground meristem gives rise to the photosynthetic tissues that make up the bulk of the leaves and to the nonphotosynthetic tissue that lies between the xylem and phloem in the stems and roots.

As with animals, the way in which these three basic cell types differentiate produces the remarkable diversity of form and function in plants. Because plant cells have a rigid cell wall, however, they cannot move around during development as animal cells can. Instead, the shape of plant structures arises from patterns in cell division, cell enlargement, and cell death.

Plant development depends heavily on the production of identical modules

A universal feature of plant development is its reliance on repeated units, or **modules**, of a relatively small number of different structures. The aboveground portion of a plant consists of stem-and-leaf units repeated many times (Figure 37.7a). Each stem-and-leaf unit, including the specialized ones that make up flowers, contains developmentally flexible meristem cells, which can differentiate into any of the tissues in a stem or a leaf. Likewise, each root contains meristem cells at the tip that can give rise to any of the tissues in a root.

When we look at a plant branch from above, we can see that each stem-and-leaf unit is rotated at a constant angle from the one beneath it to generate a predictable spiral arrangement (Figure 37.7b). Thus plants appear to control the spatial arrangement of their cells and tissues very precisely during growth and development.

Because of their modular growth, plants often continue to develop new tissues throughout their lives. Modular development gives plants tremendous flexibility in their growth form. Even when the dandelions in your lawn are repeatedly mowed, for example, the meristem in the undamaged modules missed by the mower's blades can still produce seeds. On the other hand, nonmodular animals, such as humans, reach a distinct developmental end point.

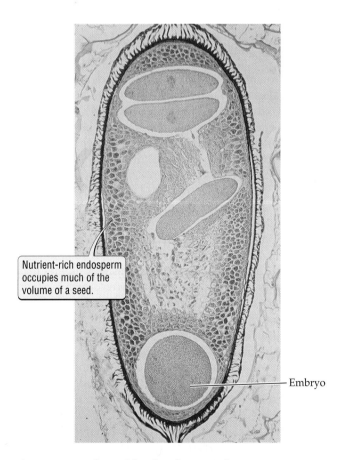

Nutrient-rich endosperm occupies much of the volume of a seed.

Embryo

Figure 37.6 The Inside of a Plant Seed
Plant seeds contain developing embryos and nutritive tissues encased in a protective outer covering. The micrograph has been tinted purple to make structures more prominent.

(a)

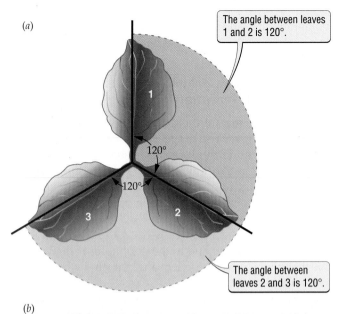

The angle between leaves 1 and 2 is 120°.

120°

120°

The angle between leaves 2 and 3 is 120°.

(b)

Figure 37.7 Modular Development in Plants
(a) A view down the shoot of a plant shows that each repeated stem-and-leaf unit is rotated by a constant angle relative to the one below it. In this instance, the angle is 120°. (b) This rotation of stem-and-leaf units gives rise to complex spiral patterns. Note that the single sunflower on the right actually consists of many small flowers, each of which sits at the tip of a stem.

■ In early plant development, three distinct cell types form: protoderm, ground meristem, and procambium. In seed-producing plants, the nutritive tissues and protective coverings of seeds allow the embryo to remain dormant until conditions allow for the growth of seedlings. Plants develop as a series of repeated modules, each of which contains developmentally flexible meristem cells.

37.4 What Controls Development?

The preceding descriptions of animal and plant development are simplified overviews of how the diversity of form and function develops in the cells, tissues, and organs of multicellular organisms. It should be clear from these descriptions that the development of a fertilized egg into a mature organism requires precise control over the fates of the cells as they divide. The different cell types that are arranged into tissues in an organism differ not because they contain different genetic information, but because they express the genetic information that they contain differently (see Chapter 16).

One of the major advances in biology in the past few decades has been a rapid growth in our understanding of how the expression of the genetic information in cells is controlled during development. We focus on two important aspects of this regulation: the activation and inactivation of genes within cells during development and the factors that control which genes are activated or inactivated.

The genes that are activated and inactivated during development determine cell fate

The reading of the DNA blueprint in each cell proceeds in two stages: transcription, in which the DNA message is converted into RNA in the nucleus, and translation, in which the RNA message crosses into the cytoplasm and directs protein production (see Chapter 15). If transcription of certain genes is either prevented or promoted, or if translation is altered, the identical genetic information in each cell may be expressed differently.

Humans and other mammals, for example, produce different kinds of oxygen-binding hemoglobin pigments (see Chapter 29) at different stages of development: embryonic hemoglobin at 0 to 8 weeks, fetal hemoglobin at 8 to 12 weeks, and then progressively greater proportions of adult hemoglobin at 12 weeks and beyond. The embryonic and fetal hemoglobins have a greater tendency to bind to oxygen than adult hemoglobin does, allowing the developing baby to pull oxygen from its mother's blood by way of the placenta. The genes that encode the embryonic, fetal, and adult hemoglobin proteins are associated with many DNA sequences that, when bound by regulatory proteins (such as the morphogens that we will describe shortly), allow transcription of the DNA (see Chapter 16). At certain stages during development, transcription of the appropriate hemoglobin genes is activated in specific cells, while the other hemoglobin genes remain inactive.

In other cases, cells transcribe a particular gene, but the processing of the resulting messenger RNA differs in cells with different fates. An example is the gene in sea urchins that controls the production of the protein actin, one of the molecules in muscle tissue. Actin is produced specifically in the ectoderm of sea urchins early in development. Although ectoderm, mesoderm, and endoderm cells all transcribe the actin gene into RNA, only in the ectoderm cells is the transcribed RNA converted into a piece of messenger RNA (mRNA) that can move out of the nucleus into the cytoplasm to produce actin.

Another way in which organisms can regulate gene expression during development is to cut up the transcribed RNA in different ways to make different mRNA products. For example, male and female fruit flies form different mRNAs from the transcription of a single gene that they both express. The resulting proteins, which determine gender in the developing flies, play an important part in controlling the sexual development of these organisms.

Cell fate becomes increasingly narrowly defined during development

As organisms develop, the range of potential fates of each cell in their bodies becomes more and more limited. The single cell of the zygote has the potential to give rise to any cell type in the mature organism. But by the time cells have differentiated into endoderm, mesoderm, and ectoderm in animals or into protoderm, ground tissue, and procambium in plants, each cell faces a more restricted future.

In the embryo, the potential fates of a cell depend on the particular genes in its DNA that can still be activated. In animals, the fate of a cell usually becomes fixed during development. In plants, the fate of a cell seems to depend somewhat on a continued interaction with surrounding tissues. Even leaf cells specialized for photosynthesis or xylem cells specialized for transporting water often retain the ability to produce all the cell types in the plant.

As development proceeds, then, the range of cell types that each cell can produce narrows. This growing limitation creates a problem: For multicellular organisms to reproduce, they must retain some cells that can produce all the structures needed in their offspring. In plants, this problem does not exist, because reproductive structures develop from cells in the undifferentiated meristems that lie at the tip of each stem-and-leaf unit. Animals, however, must overcome this problem by setting aside a few cells early in development to form the **germ line**. These unique cells, which eventually end up in the egg-producing ovaries or sperm-producing testes, remain unspecialized and separate from the other cells.

Specific molecules activate or inactivate genes in the developing organism

Genes are generally switched on or off by the presence of certain proteins called **morphogens**. One method by which morphogens can work is to spread outward from a cell that produces them, decreasing in concentration with increasing distance from the source. Cells with the proper receptor proteins in their plasma membrane can respond to binding by a morphogen by activating or inactivating certain genes. In this way, genes can be expressed at specific locations in the developing organism.

The role of morphogens in controlling development has been studied extensively in fruit flies. In the embryonic fruit fly we find that patterns in the concentration of several morphogens provide developing cells with information about their position relative to the head and tail (Figure 37.8a). By experimentally introducing these morphogens into a developing embryo, researchers can disrupt the morphogen patterns and, as a result, the developmental pattern of the embryo. If a morphogen that identifies the head end is injected into the tail end of the embryo, for example, a two-headed embryo can be created (Figure 37.8b).

Morphogens act on a family of homeotic genes (see Chapter 16), called Hox genes, that distinguish the various segments in the body of the fruit fly. The Hox genes appear to regulate other genes that control the form of a particular segment. Hox genes and morphogens that resemble those in fruit flies play a major role in directing head-to-tail differences in the development of tissues in all animals, including humans (see Figure 16.6).

Morphogens can also simply act on adjacent tissues to provide those cells with information on the identity of the surrounding cells. In birds, for example, the interaction between mesoderm-derived and ectoderm-derived tissues in the skin determines the form of the skin (Figure 37.9). Chicken ectoderm from the wing develops into wing feathers when placed next to mesoderm from the wing, but it develops into scales and claws when placed next to mesoderm from the foot. Chemical interactions between the mesoderm and ectoderm tissue determine which genes in the ectoderm tissue are activated.

Hormones also help direct development. Sex hormones, for example, influence the development of reproductive structures in animals, as we saw in Chapter 33. Hormones play a particularly important role in the ongoing growth and development of plant tissues, as we also saw in Chapter 33. Plants produce auxin in the apical meristem tissue of their shoots. Auxin affects the differentiation of plant tissue into phloem and xylem: At low concentrations it leads to phloem only; at higher

(a)

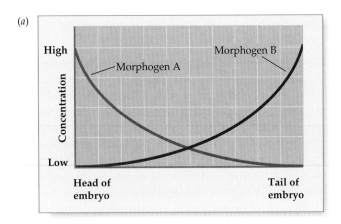

Figure 37.8 Morphogens Often Control Development
(a) Patterns of increasing or decreasing morphogen concentration provide cells with information about their position relative to the head and tail ends of a developing fruit fly embryo. (b) Experimental manipulation of morphogen concentration gradients can dramatically disrupt development. By injecting morphogen A, which identifies the head end of the embryo, into the tail end, cells near the tail can be "fooled" into developing into head tissues.

(b)

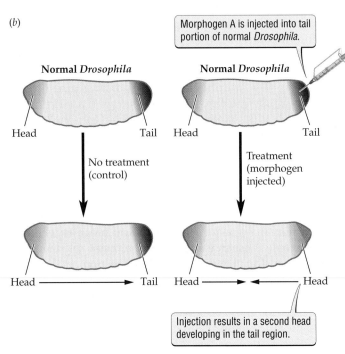

concentrations it leads to both phloem and xylem. In addition, auxin produced by the apical meristem stops the development of other meristems in other stem-and-leaf units (Figure 37.10). A decrease in the amount of auxin that reaches the other meristems, either as a result of removal of the apical meristem or because of long distances separating the two, allows the other meristems to develop.

The environment can influence development

The developmental patterns of many organisms depend not only on genetically programmed events, but also on environmental influences. In Chapter 36 we saw, for example, that the sex of a jack-in-the-pulpit plant depends on its ability to accumulate resources. Many plants trigger the development of flowers in response to the length of days relative to nights. Some plants, like the poinsettias that brighten up our homes in the middle of winter, flower only when days become progressively shorter relative to nights. Others, like the spinach in our vegetable gardens, flower only when days become progressively longer relative to nights.

Animals, too, change their developmental patterns in response to environmental influences. Gender in many turtles and in alligators depends on the temperature of the nest in which the eggs develop: Turtle eggs from cool nests tend to develop into males, and alligator eggs from cool

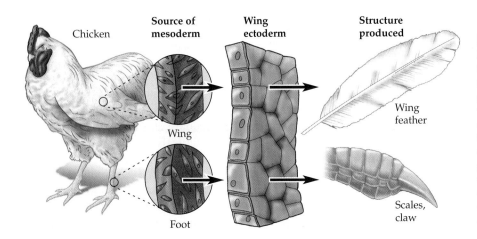

Figure 37.9 Contact between Cell Layers Can Influence Development
Cells in the ectoderm-derived layers of chicken skin develop differently depending on the source (wing or foot) of the mesoderm-derived layers of chicken skin they are adjacent to. When placed next to ectoderm from the wing, mesoderm from the wing stimulates the production of feathers typical of wings. When placed next to the same wing ectoderm, mesoderm from the foot stimulates the production of scales and claws typical of feet.

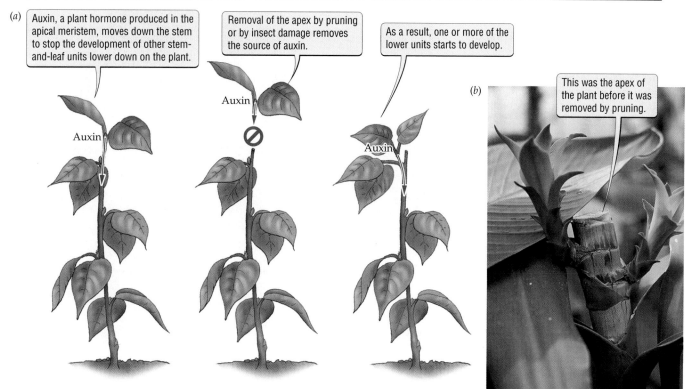

(a) Auxin, a plant hormone produced in the apical meristem, moves down the stem to stop the development of other stem-and-leaf units lower down on the plant.

Removal of the apex by pruning or by insect damage removes the source of auxin.

As a result, one or more of the lower units starts to develop.

Auxin

Auxin

Auxin

(b)

This was the apex of the plant before it was removed by pruning.

Figure 37.10 Hormonal Control of Plant Development
(*a*) The plant hormone auxin, produced in the apical meristem of plants, stops the development of other meristems, leading to a phenomenon known as apical dominance. (*b*) Removal of the apical meristem allows the other meristems to develop. That is why we prune plants: Pruning leads to bushier, fuller growth.

nests tend to develop into females. Many aquatic invertebrates develop a spiny or bristly form when exposed to chemicals released by nearby predators in their environment. The spines and bristles make it harder for the predators to eat their prey. Aphids (see Figure 36.4) show several environmental influences on their development: Scarcity of food and high densities of other aphids nearby stimulate the development of winged forms, and temperature and day length determine whether newly formed individuals reproduce sexually or asexually.

■ Cells and tissues become specialized as a result of the regulation of gene expression within cells, not because of genetic differences between cells. Unspecialized cells (meristems in plants and germ lines in animals) are required for reproduction. Morphogens coordinate development by influencing the expression of homeotic genes. Hormones and the environment also influence development.

Development and Evolution

Changes in development that may seem small can cause dramatic changes in the form and function of mature organisms. Because developmental processes have such a strong effect, rather simple genetic changes that affect those processes can lead to complex changes in their outcome. As a result, developmental changes have played an important role in the evolution of life. In this section we discuss three examples of the role of development in evolutionary change.

Changes in flower shape may arise from changes in a single gene

Shape plays an important role in helping pollinating animals identify flowers (see Chapter 36). We can distinguish two broad groups of flower shapes: radially symmetrical (such as daisy and dandelion flowers) and bilaterally symmetrical (such as snapdragon flowers) (Figure 37.11).

Experiments indicate that in snapdragons, a single gene that is expressed only in the uppermost portions of the flower meristem creates the bilateral symmetry. Where it is expressed, this gene leads to reduced petal growth and to a reduction in the number of nearby flower parts. Mutant snapdragons that have an inactive form of this gene produce radially symmetrical flowers with six petals instead of the usual five. Pollinators that visit normal, bilaterally symmetrical snapdragon flow-

Figure 37.11   Development and the Evolution of Flower Shape
Simple changes in the genes that control flower development may be all that separate radially symmetrical flowers such as buttercups from bilaterally symmetrical flowers such as snapdragons.

A single gene expressed only in the upper part of the developing flower slows growth there, creating bilateral symmetry.

Radially symmetrical flowers, such as this buttercup, look similar regardless of which side is up.

Bilaterally symmetrical flowers, like those of snapdragons, have a left side that is the mirror image of the right side, and a definite top and bottom.

ers might not recognize these radially symmetrical snapdragon flowers as worth visiting. Thus a simple genetic change leads to a far-reaching change in flower characteristics that could affect the plant's reproduction.

Changes in the expression of a single gene could dramatically change the shape of a chicken's foot

Chicken feet have four distinct toes. Chick embryos, however, have webbed feet more like those of an adult duck than those of an adult chicken. The disappearance of the webbing between the chick embryo's toes seems to depend on a single gene. This particular gene, when active in a cell, causes the cell to die. During development, this gene is activated specifically in the cells of the webbing, but not in the toes (Figure 37.12). A chick embryo that had a mutant form of this gene that failed to be activated in the webbing might well grow up to have ducklike, rather than chickenlike, feet. We can see how a minor genetic change—a mutation in a single gene—could lead to a dramatic morphological change, the development of a completely different kind of foot.

Humans and chimpanzees differ in the timing of developmental events

Chimpanzees are among our closest relatives (see Chapter 23). Outwardly, adult humans look little like adult chimpanzees. Adult chimpanzees have jaws that protrude more than human jaws, and their skulls are much less domed than human skulls (Figure 37.13). The heads of baby chimpanzees, on the other hand, have a shape much like those of human babies, and newborn chimpanzees share with adult humans a relatively flat face and domed skull. During human

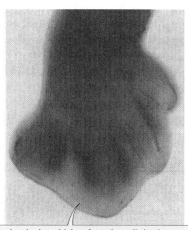

Chimpanzee

Figure 37.12   Why Chickens Don't Have Duck Feet
Adult chickens have feet with separate toes, but early in their development, a webbing that is similar to the webbing of an adult duck's feet connects their toes. In chickens, the webbing disappears during development when a gene is activated in the cells of the embryonic webbing that triggers the deaths of those cells.

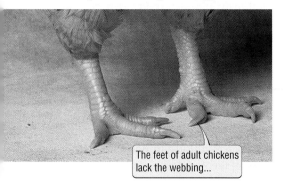
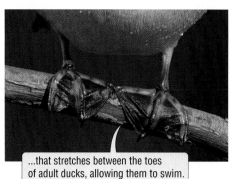
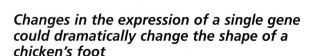

The feet of adult chickens lack the webbing...

...that stretches between the toes of adult ducks, allowing them to swim.

In developing chicken feet, the cell-death gene is active in the webbing between the embryonic toes, as indicated by the dark areas.

development, the shape of the skull changes little, but during chimpanzee development, skull shape changes greatly (Figure 37.13).

The difference in skull shape between adult humans and adult chimpanzees may simply reflect differences in the timing of the development of their bodies relative to the rate of sexual maturation that defines adulthood. Evidence suggests that, compared with the chimpanzee, the rate at which the human skull develops has slowed relative to the rate at which the reproductive organs develop. In other words, humans gain sexual maturity while still in a young animal's body. Here again, the effects of developmental changes are pronounced and complex, but the causes may be relatively simple.

> ■ Small developmental changes can have profound evolutionary effects. A change in a single gene regulating development can dramatically alter the form or function of an organism.

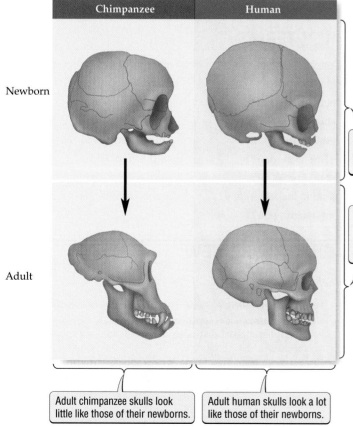

Newborn chimpanzees and humans have very similar skull shapes.

The skulls of adult chimpanzees and humans differ greatly in shape.

Adult chimpanzee skulls look little like those of their newborns.

Adult human skulls look a lot like those of their newborns.

Figure 37.13   Evolution of the Human Skull
The skulls of adult chimpanzees differ in shape from the skulls of newborns much more than is the case in humans.

■ HIGHLIGHT

Why Can We Clone Carrots and Sheep?

In most cases, cloning depends on techniques that humans have devised to bypass the factors that normally determine the fates of plant or animal cells during development. When the phloem cells of a carrot are isolated, for example, they no longer receive information from neighboring cells, and they no longer come under the influence of hormones released elsewhere in the plant. Similarly, the sheep's udder cell that gave rise to Dolly, once inserted into a sheep's egg, was cut off from all the morphogens that normally control its fate.

Plants typically commit their cells less firmly to a specific fate during development than animals do. The modular growth form of plants means that cells capable of developing into any plant tissue are scattered all over the plant. In animals, such cells are isolated in germ lines. We often take advantage of the developmental flexibility of plants to clone houseplants. If you place African violet leaves in moist soil, for example, they will readily take root to produce a new African violet plant.

Cloning sheep or humans requires much more elaborate methods than those needed to clone most plants. Success in cloning mammals has up to now required the insertion of a diploid nucleus into the cytoplasm of a nucleus-free egg. Egg cytoplasm seems to provide the conditions necessary to restore some developmental flexibility to mammalian nuclei taken from specialized cells of mature organisms. To clone Dolly, the Scottish researchers who produced her found that they had to deprive an udder cell of nutrients for 5 days for the cell to recover some developmental flexibility.

Stem cells are one kind of cell that has developmental flexibility that is lacking in most human cells. Because they can develop into any of several different kinds of cells, stem cells offer a way to replace damaged tissues. Stem cells offer hope of repairing damage to brain and heart tissues, which consist of cells that are notoriously bad at replacing themselves. While stem cells with limited potential to develop into a variety of different cell types occur throughout the human body, human embryos left over from in vitro fertilization (see Chapter 38) currently represent the most important source for stem cells capable of becoming a great variety of cell types. However, the use of human embryos to provide stem cells clearly raises issues about the ethics of harvesting developing humans for medical purposes.

■ Cloning mammals takes extensive manipulation, and to date it has been successful only when a nucleus from a mature cell has been placed into an unfertilized egg. Plants, with their many meristems and less restrictive cell fates, can be cloned more easily.

SUMMARY

Animal Development

■ Early development of animal embryos involves rapid cell division, with little, if any, cell growth or differentiation.

■ After a period of rapid cell division, which results in a hollow blastula, the three germ layers (endoderm, mesoderm, and ectoderm) form.

■ During gastrulation, the germ layers move into their appropriate positions within the embryo.

■ After gastrulation, the germ layers differentiate into specialized organs.

Plant Development

■ After a period of rapid cell division, plant embryos form three basic cell types: protoderm, ground meristem, and procambium.

■ The development of nutritive tissues and a protective covering around the embryos of seed-producing plants allows them to remain dormant until conditions favor the resumption of growth and development.

■ The rigid cell walls of plant cells prevent cell movement during development. Plants take shape by controlling patterns of cell division, cell enlargement, and cell death.

■ The meristem tissue in each repeated plant module enables plants to continue to grow throughout their lives, even after a significant portion of the plant has been damaged.

What Controls Development?

■ Cell and tissue specialization are a result of the regulation of gene expression within cells, not of genetic differences between cells.

■ Gene expression can be regulated by control of the transcription of DNA to make RNA, by control of the "editing" or cutting of RNA, or by regulation of the translation of mRNA to form proteins.

■ As cells become more specialized during development, their ability to produce other cell types becomes more limited. The need for reproduction, however, requires that some cell types—meristems in plants and germ lines in animals—retain the ability to produce cells that are not locked into a specific fate.

■ Patterns in morphogen concentrations inform cells of their position within the developing embryo and influence the expression of homeotic genes, which in turn influence the expression of many genes that influence structure and function.

■ Hormones also influence patterns of development.

■ The internal systems that regulate development respond to internal environmental cues, such as the amount of stored resources available, as well as external environmental, cues such as temperature, day length, or even the chemicals produced by potential predators and competitors.

Development and Evolution

■ Changes in the expression of a small number of genes that influence development can have profound effects on the resulting organism.

■ Changes in the rate of growth of one structure in relation to another can dramatically influence structure and function.

■ Changes in the pattern of programmed cell death during development can influence structure and function.

Highlight: Why Can We Clone Carrots and Sheep?

■ The fates of mature plant cells appear to be much less restricted than those of mature animal cells, allowing plants to be cloned much more easily.

■ The fates of mature mammal cells are so limited that, to date, mammals can be cloned only by placing the cell nucleus into an unfertilized egg.

■ Human stem cells can develop into many different kinds of cells, making them medically important as a potential way of replacing damaged tissues.

KEY TERMS

apical meristem p. 624	module p. 625
blastula p. 620	morphogen p. 627
ectoderm p. 621	placenta p. 620
endoderm p. 621	procambium p. 624
gastrulation p. 621	protoderm p. 624
germ layer p. 621	seed leaf p. 624
germ line p. 627	stem cells p. 631
ground meristem p. 624	uterus p. 620
mesoderm p. 621	

CHAPTER REVIEW

Self-Quiz

1. Which of the following events characterizes the early stages of human development?
   a. the appearance of distinct endoderm, mesoderm, and ectoderm cells
   b. a complex rearrangement of cells that leaves the ectoderm to form the outer layer of the developing embryo
   c. rapid differentiation of the various organs
   d. all of the above

2. Even after the top half of a plant has been grazed, it can continue to grow because
   a. gastrulation has not occurred.
   b. cell division accelerates.
   c. developmental genes have not been lost.
   d. the ungrazed modules have undamaged meristems.

3. Development can be regulated by control of which of the following processes?
   a. translation
   b. transcription
   c. editing of RNA
   d. all of the above

4. Which of the following differences between plant and animal cells allow gastrulation in animals and prevent gastrulation in plants?
   a. the presence of cell walls in plants and the absence of cell walls in animals
   b. the absence of cell walls in plants and the presence of cell walls in animals
   c. the difference in plasma membranes between plants and animals
   d. hormonal differences between plants and animals

5. The development of form and function is influenced by
   a. cell death.
   b. the rate of cell growth.
   c. both a and b
   d. none of the above

Review Questions

1. Why do you think that the fates of cells become increasingly narrowly defined as development progresses? What problems might arise if all cells in a mature plant or animal could become any other kind of cell?

2. What type of information does a cell need to appropriately regulate its growth and specialization? How does a cell obtain this information?

3. Can you think of reasons why many biologists believe that developmental changes have played a particularly important role in the evolution of plants and animals?

37

The Daily Globe

Parliament OKs Research on Human Embryos

LONDON—The British Parliament has voted in favor of continuing to allow research on human embryos.

Those in favor of the change argued that it could help scientists find cures for many diseases that are presently untreatable. So-called stem cells obtained from human embryos have the potential to develop into any of the many cell types found in the human body. Researchers believe that these embryonic cells will revolutionize the treatment of degenerative diseases, such as Alzheimer's and Parkinson's, by providing a way to replace the cells these disorders have destroyed.

Public health minister Yvette Cooper made an impassioned plea in support of stem cell research. She noted that the research could hold "the key to healing within the human body," giving hope to those suffering from many presently incurable degenerative diseases.

The British government specifically voted to allow stem cells to be taken from embryos at a very early stage of development. Currently, most embryos used to obtain stem cells are left over from in vitro fertilizations carried out in fertility clinics. The recent successful cloning of a human embryo, however, raises the possibility of producing large numbers of embryos specifically for medical purposes.

The decision has raised ethical concerns about how the sanctity of life will be affected. As Peter Garrett, President of the anti-abortion group LIFE, commented: "Once you open the floodgates on the production of human cloned embryos, you are setting up the preconditions for full pregnancy cloning," which results in functioning individuals.

Opposition health secretary Liam Fox disagreed on moral grounds with the use of stem cells obtained from human embryos. He felt that it was unrealistic to think that embryo research could be stopped, however, and therefore favored tough rules to help define its moral boundaries.

Evaluating "The News"

1. Imagine that one of your parents suffered from Alzheimer's disease. How might that influence your view of the use of human embryos in research aimed at curing the disease?

2. What are the arguments against the use of human embryos in research?

3. Secretary Liam Fox favored "tough rules" defining the "moral boundaries" of research involving human embryos. Can you describe some rules that you would want to see imposed on this kind of research?

chapter 38 *From Birth to Death*

Helen Lundeberg, *Double Portrait of the Artist in Time*, 1935.

Too Many People?

Humans are a remarkable species. The 6.2 billion humans on Earth outnumber all other large animals. Each day millions of babies are added to our rapidly growing population. Our use of tools has given us a unique ability to shape the environment to our needs. Because of this, humans have managed to settle and survive in just about every terrestrial habitat. Our technological talents have also allowed us to fend off many of the things that ultimately kill most animals: predators, disease, and starvation. By any measures of a species' success, humans are successful.

Our success has not been without its dark side, however. The very population growth that signifies biological success has damaged both human societies and their environment. Throughout history, human populations have repeatedly grown beyond their ability to support themselves. Overpopulation leads to famine. Famine, in turn, creates conditions in which war and disease thrive, and the deadly trio of famine, war, and disease ravages human society. In China, which has excellent records stretching back over a thousand years, famine, war, and disease contributed to the downfall of each of the many dynasties that has ruled that land. As we will see in more detail in Unit 6,

ever-increasing human populations also take their toll on the environment on which we depend, with potentially disastrous results for our continued survival.

In this final chapter of Unit 5, we look at the biological changes that take place as humans pass from childhood into adulthood and on into old age. By understanding how we mature, reproduce, and age, we can gain insight into both why we humans have achieved such remarkable reproductive success and what we might do to keep our population growth under control.

Humans Have Spread into Even the Most Hostile of Terrestrial Environments

■ KEY CONCEPTS

1. Humans have an extreme version of the lengthy development and complex social systems typical of primates.

2. Humans undergo an extended period of physical and behavioral development during childhood.

3. Human reproduction is relatively inefficient and risky.

4. Few animals other than humans live for a long time after they have stopped reproducing.

Many of the social issues that concern us from day to day have their roots in human biology. Everything we do ultimately depends on the basic biological functions that we must carry out to survive and reproduce. Like any other animal, we must eat, breathe, resist disease, and regulate our internal environment. On the other hand, how we survive and reproduce depends on some of the uniquely human characteristics introduced in this unit and in Chapter 23.

In this final chapter of the Form and Function unit, we look at how human biology contributes to our success as a species and to some of the problems that we face as individuals. We begin by introducing some of the unique features of human biology. Then we look at the human life cycle as a way of tying together many aspects of human biology. We can divide our lives into three important and biologically distinct stages:

1. The period from birth to sexual maturity, during which we undergo a highly organized sequence of developmental changes
2. The period from sexual maturity to menopause, during which we reproduce
3. The period after reproduction, during which we continue to contribute to society even as our bodies begin to function less well

How Human Lives Differ from Those of Other Animals

In many ways, human biology represents an extreme version of primate biology. **Primates** include the apes (which include humans), monkeys, tarsiers, and lemurs (see Figure 23.2). Compared with other animals, primates tend to have a long life span, have a long juvenile stage, invest a lot of time and energy in each offspring, produce relatively few offspring each year, have large brains, and live in complex societies. Within the primates, apes show these patterns most strongly. In other words, apes live longer

Eastern tarsier

than other primates, they invest more energy in their offspring, and they have larger brains. In turn, within the apes, humans show these patterns most strongly.

Humans develop slowly

As a general rule, larger animals live longer. Using allometric relationships of the kind introduced in Chapter 25, however, we find that humans live much longer than we would expect based on the allometric relationship between body weight and life span for mammals in general (Figure 38.1). A typical mammal having the same body weight as a human lives to a maximum age of about 20 years, and a typical primate having the same body weight as a human lives about 40 years. Humans, in contrast, can expect to live 80 years or more. Interestingly, the age at which human females stop reproducing (roughly 50 years) does not differ much from the expected maximum age for primates; it is our unique and lengthy life after reproduction that sets us apart from other primates (see Figure 38.2).

Our long life span affects all aspects of our development. Among the primates, as life span increases, so does the length of each developmental stage (Figure 38.2). Our childhood, the nonreproductive period during which we depend on our parents, lasts about 12 years, and in Western cultures it may take an additional 10 years or so before we attain true independence. The adult phase of life, during which we reproduce, extends over 3 decades in females and even longer in males. Whereas most animals begin to reproduce when they reach about 10 percent of their full adult weight, humans do not begin to reproduce until we reach about 60 percent of our full adult weight. Furthermore, humans, unlike most other animals, continue to live long after they have stopped reproducing.

Humans produce relatively few offspring, but devote a lot of attention to each one

Humans, like most primates, produce few offspring over their lifetimes. Even at a time when people tended to have big families, few mothers produced more than 20

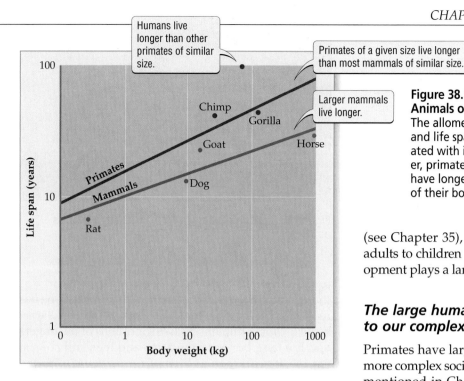

Figure 38.1   Humans Live Longer than Other Animals of Similar Size
The allometric relationship between body weight and life span shows an increase in life span associated with increasing size among mammals. However, primates as a group, and humans in particular, have longer life spans than predicted for mammals of their body weight.

(see Chapter 35), the information passed down from adults to children during this extended period of development plays a large part in a child's success.

The large human brain may be related to our complex social system

Primates have larger brains for their body weight and more complex social behaviors than other animals do. As mentioned in Chapters 23 and 25, human brains are much larger than expected for our body weight, even when compared with those of other primates (see Figure

children in a lifetime. Three factors contribute to this low reproductive output:

1. Humans tend to have just one baby at a time. Only 1 to 2 percent of births result in twins or triplets.
2. Often several years pass between births. While breast-feeding, mothers generally do not mature eggs. In many cultures, mothers breast-feed their children for 2 or more years after birth.
3. Humans do not start reproducing until they reach puberty in their early to mid-teens. Most animals reproduce in their first or second year of life, and even other slowly developing apes, such as chimpanzees, reproduce by the time they reach 9 years of age.

Humans lavish an incredible amount of time and energy on their children. Because humans, like other apes, depend so heavily on learned, rather than fixed, behaviors

Figure 38.2   Long Human Life Spans Go Hand in Hand with Slow Development
Among the primates, longer life spans correspond to a longer period spent in each stage of development. Only humans, however, survive for many years after they have stopped reproducing.

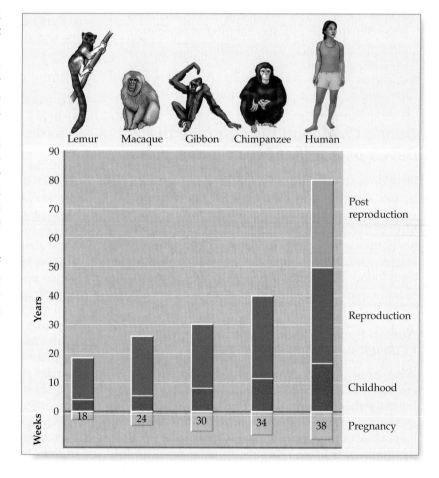

25.5). As we saw in Chapter 23, human brains increased beyond the norm for primates relatively recently in our evolutionary history. Over the past 1.5 million years the size of the human brain has increased from only slightly larger than expected for a primate of our size to the remarkable 1350 milliliters in volume of modern humans.

Most of this increase in brain size has resulted from an increase in the volume of the **cerebral cortex**, the portion of the brain that makes sense of and responds to sensory information. The cerebral cortex may have proved especially useful to humans who lived in groups and hunted together, and who were beginning to develop language and make tools. Large parts of the human cerebral cortex deal with the understanding and production of speech and with signals received from and sent to the hands (see Figures 34.8 and 34.9). In contrast, our closest relatives, the chimpanzees, have only small portions of their much smaller cerebral cortex devoted to these functions. Communication with others through language and the development of technology have been two of the hallmarks of the development of human societies.

> ■ All aspects of the human life cycle unfold slowly. Human parents produce few children and devote a lot of time and resources to each one. The exceptionally large human brain probably contributes to the complexity of our social system by making language and technology possible.

During Childhood We Continue the Development that Began in the Uterus

Development does not stop at birth. In humans, in particular, development continues for many years. We spend about a quarter of our lives reaching full size. Most of this development takes place during our childhood, the stage of our lives before we become sexually mature. The developmental changes that take place after birth follow a sequence that is every bit as predictable as the sequence of events before birth (see Chapter 37).

Human babies are born unusually early in development

When a human baby is born, it is ill-equipped to survive on its own (Figure 38.3). Compared with other animals, humans enter the world early in their development. In spite of spending a longer time developing in the uterus before birth, humans are born at an earlier stage of their development than any other ape. Whereas the newborns of most mammals have brains of almost adult size, new-

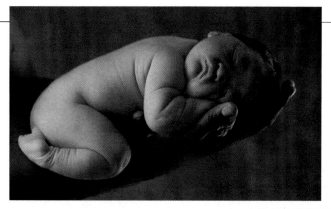

Figure 38.3   Human Babies Are Unusually Helpless at Birth

born humans have brains only one-quarter as large as those of adults. In addition, in most mammals, bone has replaced cartilage in almost all portions of the skeleton by birth, but in humans, growing portions of the fingers and the long bones of the arms and legs still consist entirely of cartilage at birth. Comparisons with the development of other apes suggest that humans would have to have 21-month-long pregnancies (instead of the 9-month-long pregnancies that we actually have) to reach the stage of development typical of chimpanzees and gorillas at birth. When we discuss the birth process in humans later in this chapter, we will see why we enter the world so early in development.

Human development continues after birth

Our lengthy childhood stems at least in part from our birth in a relatively undeveloped condition. In many ways, the development of a human baby during the first years of life represents a continuation of fetal development. Most mammals develop rapidly in the uterus compared with their development after birth. In the first years following birth, human babies continue to increase in size at a rate comparable to that of a fetus, rather than the slower rate of typical infant development. We triple our weight and grow to 1.5 times our birth length by the time we reach our first birthday. During this first year of life, as much as 40 percent of a baby's energy intake is used for growth. Following the first year, the growth rate slows to a weight gain of 2 to 3 kilograms per year, requiring a more modest 3 percent of the baby's energy intake. Males grow bigger than females mostly because they continue to grow until their late teens, while females stop growing in their mid-teens.

The human brain continues to increase in weight at the fetal rate until the age of 2 years. This continued rapid growth after birth probably plays an important part in the development of the exceptionally large human brain. The human brain reaches its mature weight by the time a child is 10 years old.

Behavior, including language, develops in a predictable way during childhood

Behavioral development is the most complex and probably the most important aspect of development after birth. Anyone who has watched a baby grow up will realize that its coordination increases by leaps and bounds during early childhood (see Figure 38.4). Babies progress from sitting upright to actions such as walking and throwing that require more coordination. At the same time, they become progressively better at delicate manipulations using their hands. The unparalleled ability of humans to use their hands to manipulate objects has played a central role in our ability to make and use tools (see Chapter 23).

In addition to skills requiring coordination, children learn social behaviors and language, which allow them to function in human society. The ability to use language to communicate sets humans apart from all other organisms (see Chapter 35). Our skill at using language to express our thoughts continues to develop throughout our lives.

Extensive studies show that children learn specific movements at specific times in their lives (Figure 38.4).

Figure 38.4   Humans Develop Behaviorally during Childhood
Children learn progressively more sophisticated behaviors as they grow up. These learned behaviors include large-scale movements of the body, fine manipulations using the hands, social skills, and language. Children begin showing various behaviors at remarkably predictable ages. This chart is a much simplified version of charts used by doctors to monitor the behavioral development of children. The orange part of each bar shows the typical range of ages over which children first show the behavior. The red zone indicates *slightly* delayed behavioral development.

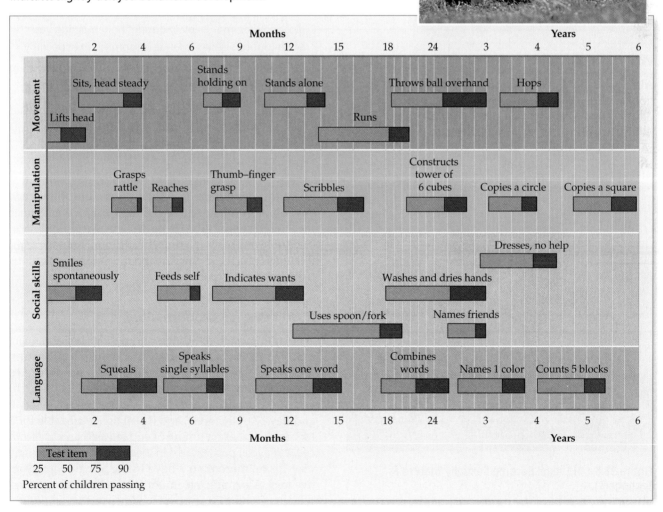

Logically, we would expect a child to master sitting before walking. Children learn social behaviors and language in a similarly fixed, but less obviously logical, order. For example, children begin to correctly identify and name their mother or father between the ages of 6 and 10 months. They do not start using "me" and "you" until they are between 18 and 28 months old. Similarly, children do not start using prepositions, such as "on" or "in" or "out," until they reach the age of 23 to 34 months, well after they have begun using simple sentences.

The behavioral development of children parallels physical changes in the brain. At birth the human brain contains its full complement of neurons, although it weighs only 25 percent of its adult weight. In the year following birth the number of synapses connecting neurons in the brain more than doubles. Between the ages of 1 and 12, however, the number of synapses drops to about 75 percent of its maximum value. The synapses lost are those that are not used. After the age of 12, the number of synapses connecting neurons changes relatively little. Both the creation and the loss of synapses are physical manifestations of learning, suggesting that up to the age of 12 children learn faster than at any other time in their lives.

We reach sexual maturity in our teenage years

Although growth may continue until the age of 20, childhood ends at **puberty**, the period during which humans develop into sexually mature individuals (Figure 38.5). At the beginning of puberty, the release of hormones

Figure 38.5   Humans Become Sexually Mature as Teenagers

called **gonadotropins** by the pituitary restarts a developmental sequence that stopped before birth (see Chapter 33). In females, puberty typically begins at the age of 8 to 13 years. The gonadotropins trigger the maturation of eggs that have been present in the body since birth. In addition, they stimulate the release of **estrogen** by the ovaries. Estrogen increases the sex drive and controls the development of breasts, pubic and armpit hair, and the reproductive tract.

In males, puberty begins slightly later than in females, between the ages of 10 and 15 years. The release of gonadotropins leads to the production of sperm cells and stimulates the testes to produce the male sex hormone, **testosterone**. Testosterone controls the final development of the penis and triggers the growth of hair on the face, in the pubic region, and in the armpits. It also triggers a growth spurt that includes a gain in both height and muscle mass, as well as a deepening of the voice. As an unwanted side effect, increased testosterone levels can also contribute to acne.

> ■ Humans are less developed at birth than other primates. They grow rapidly during the first two years of life, and then more slowly during the remaining years of childhood. Behavioral development is especially rapid up to the age of 12. At puberty, hormones end childhood by triggering the maturation of the reproductive organs.

38.1 The Reproductive Years

Although more than 6 billion people live on Earth, with hundreds of thousands more added each day, humans are not quite the reproductive machines they might appear to be. To begin with, we typically produce only one baby at a time. In addition, human reproduction is much less efficient and much more hazardous than we might expect.

It isn't easy to produce a baby

The rapid growth of human populations belies the fact that producing a baby is by no means easy. Mature females release a single egg into the oviduct only once in each 28-day-long menstrual cycle (see Figures 33.10 and 36.3). Typically, the egg is released between day 9 and day 17 of the cycle, and it remains fertilizable for 2 to 3 days. The exact timing of egg release can vary from cycle to cycle, particularly in teenagers. Sperm can survive for no more than 2 days in the female reproductive tract. Assuming maximum survival of both eggs and sperm, this leaves a variable 5-day window in each 28-day cycle during which intercourse can lead to the fertilization of an egg.

If a sperm successfully fertilizes the egg, the developing embryo should begin to divide as it travels through the oviduct to the uterus and implants itself in the thick wall of the uterus. About 15 percent of human zygotes never divide, however, usually because of a genetic problem. Another 15 percent begin to divide, but never find their way to the uterus. A final 25 percent divide and reach the uterus, but fail to implant themselves in the uterine wall. In none of these situations would there be any outward signs of anything wrong. Thus, only 45 percent of fertilized eggs ever result in a **pregnancy**, in which the embryo implants itself successfully in the uterine wall.

Because of these various obstacles to successful fertilization and implantation, even young, healthy couples who are doing their best to get pregnant have a success rate of only 25 to 30 percent per month. Over an entire year of trying, the success rate for such couples increases to 80 percent. After 2 years of trying, the success rate reaches 90 percent. About 15 percent of couples are infertile and cannot produce a pregnancy no matter how hard they try.

Not all pregnancies result in babies, however. Roughly one-quarter of the embryos that become implanted in the uterine wall do not survive and are lost from the uterus. Most of these miscarriages occur during the first 3 months of development, during which the cells of the embryo differentiate into the various tissues and organs (see Chapter 37). Most miscarriages are thought to arise from genetic problems. Many of them, especially the earliest ones, can easily go unnoticed. During the last 3 months of pregnancy and during birth, complications lead to the deaths of less than 1 percent of fetuses. While this percentage may seem small, it roughly equals the chances of dying during the first 40 years of life.

In spite of all the reproductive failures, mostly unnoticed, that are part of normal human reproduction, humans remain a reproductive success. As we will see in Chapter 41, as long as a species produces more offspring than there are parents (in other words, more than two per pair of parents), its population will grow.

New reproductive technologies can help solve infertility problems

Recent advances in reproductive technology have made it possible for infertile couples to have children. Since the first successful fertilization of a human egg outside the body in 1978, **in vitro fertilization** (*vitro*, "glass"; commonly referred to as IVF) has advanced to the point at which doctors can overcome many infertility-related problems.

In IVF, injections of hormones stimulate the female's ovaries to mature up to twenty eggs instead of the usual one in a particular month. Doctors collect these eggs surgically just before they are released into the oviducts. Some of the eggs are then placed in a test tube with a sample of about 60,000 sperm from the male (Figure 38.6). Under these conditions, 70 to 80 percent of the eggs typically become fertilized. Two days later, if all goes well, two to three healthy-looking embryos are selected and placed directly into the uterus of the female. IVF leads to pregnancy in about 50 percent of attempts.

In some cases, the problem lies specifically with the sperm. If the male's sperm cannot swim well enough to reach the egg, a sperm can be inserted directly into the egg. This process further increases the success rate of IVF.

Although reproductive technologies such as IVF make it possible for some infertile couples to have children, they also raise some difficult issues. For example, during IVF, doctors usually collect more eggs than they fertilize, and they usually fertilize more eggs than they place in the uterus. What do we do with the extra eggs and the extra embryos? Is it ethical to use them for medical research (see Chapter 37)?

Why is childbirth so risky?

One of the riskiest stages of our lives and those of our mothers occurs at birth. Before the development of sterile medical procedures, the deaths of babies, mothers, or both during childbirth were very common.

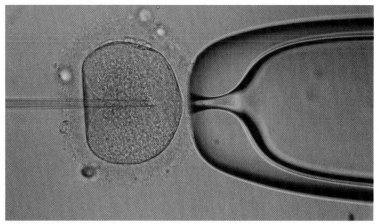

 Figure 38.6   In Vitro Fertilization Can Solve Many Cases of Infertility

The difficulty of birth in humans stems from the large size of the skull, which surrounds and protects the large brain. During birth, humans, like all other mammals, must leave the mother's body through a birth canal that passes through her hipbones. The skull is a baby's largest rigid structure, and it determines whether the baby can fit through this opening. Whereas the skull of a baby chimpanzee passes easily through its mother's hipbones during birth, it is much harder to fit the large skull of a human baby through the birth canal (Figure 38.7a).

To emerge from its mother's uterus, a human baby first has to bend its neck and then rotate its head as it passes through her hipbones (Figure 38.7b). At birth, the diameter of a human baby's skull actually exceeds the maximum diameter of the space between the hipbones. Thus, in addition to the baby's bending and turning, both the baby's skull and the mother's birth canal must change shape for the baby to squeeze through. At around the time of birth, the ovaries of pregnant moth-ers release a hormone called relaxin, which enables the ligaments that hold the hipbones together to stretch a bit. Without this stretching, the baby's head could not fit through the hipbones. A side effect of relaxin production, however, is that the mother often suffers from joint pain because relaxin affects the ligaments that hold bones together in other joints of the body as well.

The difficulty associated with passing a baby's skull through its mother's birth canal may have been an important factor leading to the early stage of development at which humans are born. If human fetuses remained in the uterus any longer than they currently do, the skull surrounding the rapidly growing brain would grow too big to fit through the hipbones.

> ■ The chances of successfully fertilizing an egg and then carrying a pregnancy to birth are surprisingly low. New reproductive technologies such as in vitro fertilization can help couples overcome fertility problems. Human birth is risky because the skull of an infant fits through its mother's birth canal only with difficulty.

(a)

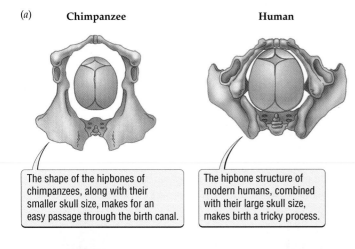

Chimpanzee   **Human**

The shape of the hipbones of chimpanzees, along with their smaller skull size, makes for an easy passage through the birth canal.

The hipbone structure of modern humans, combined with their large skull size, makes birth a tricky process.

Figure 38.7   Birth in Humans is a Complex Affair
(a) A rounder passage through the hipbones and a smaller skull allow a chimpanzee infant to pass easily through its mother's birth canal. In contrast, the skull of a human baby fits through its mother's birth canal only if it rotates to the side. (b) During birth, a human baby must tilt and rotate its head to pass through its mother's hipbones.

(b)

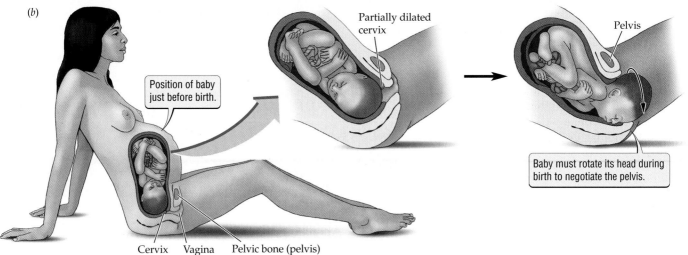

Partially dilated cervix

Pelvis

Position of baby just before birth.

Baby must rotate its head during birth to negotiate the pelvis.

Cervix   Vagina   Pelvic bone (pelvis)

Old Age: Life after Reproduction

Most species rarely live beyond their reproductive years. Even among the other apes, which have biologies similar to our own, no species has a significant life after reproduction. Let's consider why postreproductive life is so rare.

Menopause marks the end of reproduction

Both males and females reproduce less effectively as they age. Females show a clear drop in their ability to fertilize eggs and bear children when they pass 40 years of age (Table 38.1). If eggs from younger women are implanted into women over 40, the pregnancy rate equals that for the age group of the women who donated the eggs. This finding suggests that as women age, it is their eggs that decline in quality. Down's syndrome, a genetic disease in which a person has three copies of chromosome 21 (see Chapter 18), occurs much more commonly in children born to mothers over the age of 40 than in those with mothers younger than 40. Human females reach **menopause**, or the end of their reproductive lives, around the age of 50.

Similar patterns have been found for sperm produced by older and younger men. Children fathered by older men tend to live less long than children fathered by the same men at a younger age, suggesting that the genetic quality of sperm declines during life. In addition, as men age, they produce fewer sperm, decreasing the chances that they will fertilize an egg. Males do not undergo the clearly identifiable menopause characteristic of females. Instead, their ability to produce sperm and their sex drive slowly decrease as they age.

Because eggs and the cells that produce sperm are present in our bodies from birth, they accumulate harmful mutations over time. The older these cells are, the more likely they are to accumulate lethal mutations caused by various environmental factors.

38.1 The Effect of Age on the Reproductive Success of Females

Age (years)	Cycles that result in a pregnancy (%)	Cycles that result in an embryo that fails to complete development (%)
Under 30	29.0	14.9
30–35	19.8	16.5
35–40	17.1	22.4
40–45	12.8	33.2

Why is old age so rare?

With the notable exception of humans, most species rarely live beyond their reproductive years. We see this pattern because when an organism stops reproducing, natural selection no longer favors traits that keep the organism alive. In other words, any gene that keeps us alive while we are reproducing will tend to let us leave behind more offspring; therefore, such a gene should become more common in the population. On the other hand, any gene that keeps us alive after we stop reproducing will not affect the number of offspring we leave behind, so these genes will not become more common. Likewise, any gene—such as the one causing Huntington's disease—that harms us only after we have stopped reproducing will not be removed from the population by natural selection (see Chapter 13).

Thus we find in older people a host of health problems that are expressed after their reproductive lives have ended. Skeletons become weaker with age. Diseases such as Alzheimer's affect brain function. The ability of the heart to pump blood declines with age. Skin cancers appear after a lifetime of exposure to ultraviolet radiation from the sun (see Chapter 14). The immune system no longer protects us from disease as effectively as it did when we were younger.

Humans differ from other animals in having developed a sophisticated medical technology that can help us combat the diseases that in an earlier time would have killed us. Archaeological evidence suggests that prehistoric humans rarely lived longer than 50 years. As recently as 1890, citizens of England could expect to live an average of 45 years. Modern medicine has reduced the toll taken by infectious diseases that formerly killed people early in life, such as smallpox, influenza, and diphtheria (see Chapter 32). To a lesser extent, it has made inroads into the diseases we associate with old age by allowing us to patch up the damage done by heart disease, lung disease, and ailments of the skeletal system, among others. Today, life expectancies in Western nations approach 80 years, some 30 years beyond the end of a female's reproductive life. Thus medical advances have made it possible for many more people to survive well into their postreproductive years.

Can we live forever?

For as long as there has been recorded history, people have looked for a way to live forever. The philosopher's stone that was sought by medieval alchemists was thought to be one way of attaining immortality (Figure 38.8). Although

Figure 38.8   Alchemists Tried to Find the Philosopher's Stone, which They Thought Held the Key to Immortality

> ■ The end of reproduction comes at the age of about 50 years in women and later in men. Natural selection does not favor the spread of genes that prolong life after reproduction; as a result, the postreproductive body tends to function less well. New research is beginning to provide some insights into the maximum length of the human life span.

the philosopher's stone was never found, scientists are beginning to understand some of the factors that affect how long organisms live.

The upper limit to the human life span has often been placed at 120 years. This value reflects the oldest documented person in the world, Jean Calmet of France, who died at the age of 122 years. Recent evidence, however, suggests that the upper limit to the human life span may prove less fixed and much higher than we had suspected. Over the past century, the average age at which the oldest few members of the population died has increased by 13 months every 10 years. Based on this trend, we should expect at least a few people born in 2000 to survive to the ripe old age of 130. Perhaps most significantly, the rate at which the maximum age has increased shows no sign of slowing down. These observations suggest that the actual limit to the human life span may lie well beyond even 130 years.

We are just beginning to identify some of the factors that may determine how quickly animals age. Chemicals that prevent oxidation reactions (chemical reactions similar to the one that causes iron to rust) can extend the life span of the worm *Caenorhabditis elegans* (see Chapter 37) by 50 percent. In addition, a number of experiments have shown that keeping animals deprived of food consistently increases the life span. Although few people would look forward to going hungry all the time so that they could live longer, these experiments do give us our first glimpse into the mysteries of aging.

■ HIGHLIGHT

Birth Control and Human Population Growth

In almost every way, humans would seem to be poorly equipped to reproduce and spread throughout the world the way we have. We usually produce only one baby at a time, we often take 20 years or more to produce our first baby, and we invest over 12 years in teaching each baby the skills needed to survive. Nonetheless, our interest in reproduction, combined with our ability to control the diseases that would otherwise kill many of us, has allowed the human population to grow by many billions in the last century alone.

A variety of methods of **birth control**, which allow us to go through the motions of reproducing without actually producing any babies, form the front line in our attempts to rein in our population (Table 38.2). At one extreme lie birth control methods such as vasectomies, surgical operations in which the tubes carrying sperm from the testes are cut. Such operations are not reliably reversible, but almost never fail to prevent unwanted pregnancies. At the other extreme lie behavioral approaches to birth control such as the rhythm method. It costs no money to have intercourse only when the female member of a couple is not ovulating (releasing an egg). On the other hand, because the timing of ovulation is irregular at best and therefore hard to predict, the rhythm method is notoriously unreliable.

"The Pill," a mixture of synthetic female sex hormones, provides a commonly used middle path. The Pill provides effective birth control by changing hormone levels during the menstrual cycle to disrupt ovulation. Although the birth control pill can have a number of unpleasant side effects, including increased risk of heart disease and breast cancer, it does lower the chances of getting several different cancers of the female reproductive system.

38.2 *A Comparison of Commonly Used Methods of Birth Control*

Birth control method	Failure rate[a]	Advantages	Disadvantages
Surgical operations			
Vasectomy	0.02	Permanent	Permanent, risk of surgical complications
Tubal ligation	0.13	Permanent	Permanent, risk of surgical complications
Chemicals			
Estrogen + progestin pill	0.32	Protects against some cancers	Must take pills regularly and in sequence, some side effects
Progestin-only pill	1.2	Easier to use than E + P pill	Must take pills regularly
Spermicide	21	Protects against some sexually transmitted diseases	High failure rate, must use properly
IUD (Intrauterine device)	1.5	Always there	Can cause infections, may cause some discomfort or bleeding
Physical barriers			
Diaphragm	21	Some protection against STDs	Requires fitting by doctor, must put it in ahead of time
Condom	12	Protects against STDs, convenient, inexpensive	Must use properly
Behavioral			
Withdrawal	19	Free	High failure rate, requires self control
Rhythm	20	Free, no religious objections	High failure rate, requires regular menstrual cycles

[a]Number of failures under typical use if used by 100 women for 1 year

As a way of reining in the increase in world population, birth control works only if combined with an effective publicity campaign. In many countries, however, any attempt to promote birth control meets with stiff opposition from cultural traditions and religious groups.

■ Birth control methods range from effective but irreversible operations to inexpensive but ineffective behavioral methods. The birth control pill, which prevents ovulation by altering hormone levels during the menstrual cycle, has proved to be one of the most popular and reliable means of birth control. As a means of population control, birth control works only if combined with a publicity campaign.

SUMMARY

How Human Lives Differ from Those of Other Animals

■ The human life cycle unfolds slowly, even compared with those of other slowly developing primates.

■ Human parents produce relatively few children during their lifetimes, and devote a lot of time and resources to each child they produce.

■ The large human brain probably contributes to the complexity of our social system by making language and technology possible.

During Childhood We Continue the Development that Began in the Uterus

■ Humans are less developed at birth than other primates are.

■ Human children grow rapidly during the first 2 of years of life, then more slowly until the early teens in females and the mid-teens in males.

■ Much of childhood is devoted to behavioral development. The behavioral changes that take place rapidly during the first 12 years of life match changes in brain structure that occur during that time.

■ Childhood ends at puberty, when hormones trigger the maturation of the reproductive organs and a variety of gender-specific features.

The Reproductive Years

■ The chances of successfully fertilizing an egg and then carrying a pregnancy to birth are surprisingly low.

■ In vitro fertilization helps many otherwise infertile couples produce offspring.

■ Human birth is difficult and risky because the skull of an infant fits through its mother's birth canal only with difficulty.

Old Age: Life After Reproduction

- Menopause, which comes at the age of about 50 years, marks the end of reproduction for women. In men, reproduction ends later and more gradually.

- The postreproductive body tends to function less well because natural selection does not favor the spread of genes that prolong life after reproduction.

- New research is beginning to shed light on how long humans might be able to live and what we might need to do to live that long.

Highlight: Birth Control and Human Population Growth

- Birth control methods range from effective but irreversible operations to inexpensive and ineffective behavioral methods.

- The birth control pill, which prevents ovulation by altering hormone levels during the menstrual cycle, has proved to be one of the most popular and reliable means of birth control.

KEY TERMS

birth control p. 644

cerebral cortex p. 638

estrogen p. 640

gonadotropins p. 640

in vitro fertilization p. 641

menopause p. 643

pregnancy p. 641

primate p. 636

puberty p. 640

testosterone p. 640

CHAPTER REVIEW

Self-Quiz

1. Which of the following statements is true of the basic primate life cycle?
   a. Primates develop more slowly than humans.
   b. Compared with other animals, primates develop slowly.
   c. Primates have a very short reproductive life compared with most other animals.
   d. Primates produce more offspring during their lifetimes than most other animals.

2. Why are human babies born in such an undeveloped condition relative to other primates?
   a. Human parents devote little time and energy to each of their offspring.
   b. Otherwise their heads would be too big to pass through the birth canal.
   c. It helps to reduce their growth rate immediately after birth to a typical infant rate of development.
   d. It allows human babies to be more independent at birth than other primates.

3. Which of the following is true of behavioral development in human children?
   a. Behavioral development has no connection to the development of structures in the human brain.
   b. Behavioral development is not very important in humans.
   c. Behavioral development in humans is completed by the time babies are born.
   d. Behaviors are learned in a fixed and predictable order.

4. What is puberty?
   a. It is the structure through which all human and chimpanzee babies must fit at birth.
   b. It is the year immediately following birth in humans.
   c. It is the point in development at which human reproductive structures become functional.
   d. It is the part of the brain that processes information related to language.

5. Which of the following statements about life after reproduction is true of humans?
   a. Humans, like all animals, die shortly after they stop reproducing.
   b. Males have shorter postreproductive lives than females.
   c. Most humans live for about 50 years after their reproductive life ends.
   d. Postreproductive life begins at puberty.

Review Questions

1. The number of people using in vitro fertilization and other fertility treatments has increased over the past decade. Researchers believe that this trend is related to an increase in the age at which many women are choosing to have their first children. Explain how these two trends are related.

2. The small size and the shape of the passage through the hipbones in human females lie at the root of many problems encountered during birth. The shape of the hipbones in humans allows us to walk upright. What would happen if the size and shape of the birth canal were determined only by the challenges posed by childbirth? Is natural selection likely to favor such a change in human hipbones? Explain your conclusion.

3. Why is the chance of getting pregnant during any given menstrual cycle so low?

38

The Daily Globe

Have We Found the Fountain of Youth?

Manchester, UK—Scientists at Manchester University have reported the finding of a miracle drug that can extend life by 50 to 100 percent. The chemical, when applied to a microscopic worm called *Caenorhabditis elegans*, kept the worms alive much longer than a second group of worms that did not receive the treatment.

The lead researcher, Gordon Lithgow of the School of Biological Science, said he was "amazed by what we were seeing down the microscope as these experiments progressed. As the untreated worms began to die, the drug-treated worms were swimming around, full of life."

The wonder drug is an antioxidant that prevents oxidation from damaging cells. Karl Ehrlenmeyer, a chemist at Cosmetico, compared free-radical oxygens to a sort of chemical punk hanging around on a street corner looking for something to destroy. "Many of our skin creams contain antioxidants that help reduce the effects of aging."

According to Lithgow, the chemicals used in the new study are much more effective than the antioxidants used in skin cream or found in such familiar foods as carrots and oranges. Lithgow suggested that the development of similar drugs for humans might force us to "reconsider aging as an inevitability."

Evaluating "The News"

1. If the antioxidants used in this study worked as effectively on humans as they did on the worms, how might this change the way that we organize our lives?

2. Most Western countries are faced with an aging population. Can you think of some effects that extending life by another 50 years might have on our society?

3. Do you favor increasing the length of time that we live? Explain why you think it is or is not a good idea.

Kit Williams, *The Death of Spring*, 1980.

UNIT 6

Interactions with the Environment

chapter *39* *The Biosphere*

Dale Chihuly, *Sky Blue Soft Cylinder with Golden Yellow Lip Wrap*, 1993.

El Niño

Throughout much of the world something was seriously amiss. Off the western coast of South America, the fish on which both seabirds and local fishermen depended were gone. Thousands of seabirds starved to death. Farther north, along the coast of California, underwater "forests" dominated by long strands of brown algae called kelp were destroyed or heavily damaged by storms. Still farther north, the Canadian and

Alaskan coasts experienced high sea levels and unusual amounts of rainfall.

On the other side of the ocean, the sea level dropped throughout the entire western Pacific, killing huge numbers of animals that lived in coral reefs. Reefs were also hit hard in the eastern Pacific, where 50 to 98 percent of the corals of all species died. Lands separated by thousands of kilometers experienced unusual and violent swings in the

■ MAIN MESSAGE

Global patterns of air and water circulation can cause events in one part of the world to affect ecosystems all over the world.

weather: Tropical rainforests in Borneo were ravaged by catastrophic drought and fires, and deserts in Peru were flooded by torrential rains. The dramatic changes in the weather led to crop failures and disease outbreaks: Corn yields dropped drastically in southern Africa, Australian wheat fields were destroyed, and cholera ravaged parts of South America.

This list of disasters sounds like the beginning of a doomsday science fiction movie. But these events—and many others like them—actually happened in 1982 and 1983, and similar events occurred in 1997 and 1998. These catastrophes were not caused by alien invaders, or even by humans. They were set in motion by natural changes in wind patterns and water currents, changes known collectively as El Niño events.

In this chapter we examine the shaping of Earth's climate by its wind patterns and water currents. El Niño events are just one example of the profound effects these air and water circulation patterns can have on all life throughout the world.

A Rainforest Burns in Brazil during the 1997–98 El Niño

■ KEY CONCEPTS

1. Ecologists study interactions between organisms and their environment. All ecological interactions occur in the biosphere, which consists of all the living organisms on Earth together with the environments in which they live.

2. Climate has a major effect on the biosphere. Climate is determined by incoming solar radiation, global movements of air and water, and major features of Earth's surface.

3. The biosphere can be divided into major terrestrial and aquatic life zones called biomes.

4. Terrestrial biomes cover large geographic regions and are usually named for the dominant plants that live there. The locations of terrestrial biomes are determined by climate and by the actions of humans.

5. Aquatic biomes cover about 75 percent of Earth's surface. They are usually characterized by physical conditions of the environment. Aquatic biomes are heavily influenced by the surrounding terrestrial biomes, by climate, and by the actions of humans.

There are millions of known species on Earth, and many millions more yet to be discovered. Each of these organisms lives in a characteristic place, or habitat. Collectively, these organisms and their habitats make up the **biosphere**, which consists of all living organisms on Earth together with the environments in which they live. The biosphere is very complex. It includes grasslands, deserts, tropical rainforests, streams, and lakes, to name just a few of its many parts. It also includes the bottoms of our feet and the bacteria that grow there, as well as hydrothermal vents on the bottom of the ocean and the bacteria that grow there.

In this chapter we discuss Earth's climate and its importance to the biosphere. We also describe terrestrial and aquatic biomes, the major life zones into which the biosphere can be divided. First, however, we provide a brief overview of **ecology**, the scientific study of interactions between organisms and their environment, which is the subject of Unit 6.

Ecology: Studying Interactions between Organisms and Their Environment

All organisms interact with their environment. These interactions go both ways: Organisms affect their environment (as when a beaver builds a dam that blocks the flow of a stream and creates a pond or lake), and the environment affects organisms (as when an extended drought limits the growth of plant species on which the beaver depends for food). Ecologists study interactions in nature at several different levels of the biological hierarchy (see Chapter 1): individual organisms, groups of individuals of a species (populations), groups of different species (communities), and **ecosystems**, a term that refers to all the different species in an area plus the environment in which they live. Collectively, all ecosystems on Earth make up the biosphere.

Ranging as it does from the study of individual organisms to the study of the biosphere, ecology is a broad and complex subject. It is also an important area of applied biology. As we will learn in this unit, humans are changing the biosphere, often in ways that are not intended or that have unexpected consequences. A major goal of ecology is to document and understand the consequences of human actions for life on Earth.

To give just one example, chlorofluorocarbons, synthetic chemicals used as refrigerants, in aerosol sprays, and in foam manufacture, have created a hole in the ozone layer of the atmosphere—a potentially dangerous side effect of these technologies that people did not intend or expect (see Figure 45.3). The ozone hole is potentially dangerous because ozone in the atmosphere prevents much of the sun's ultraviolet light from reaching Earth, thus protecting organisms from DNA mutations caused by exposure to ultraviolet light (see Figure 14.9). Ecologists seek to understand both how the ozone hole was formed and its consequences for life on Earth.

In the chapters of this unit, our study of ecology will take us from individual organisms, to populations, to interactions among organisms, to communities, to ecosystems, and finally, to global change caused by humans. But all ecological interactions, at whatever level they occur, take place in the biosphere. Thus, by beginning our study of ecology with an overview of the biosphere, we will be able to use the material in this chapter to understand the different levels (individuals to ecosystems) at which ecology can be studied.

■ Ecologists study interactions between organisms and their environment at different levels, ranging from individuals to ecosystems. A major goal of ecology is to document and understand how human actions are affecting life on earth.

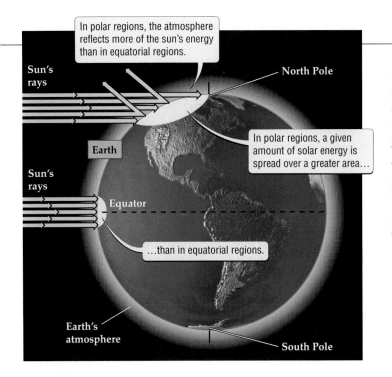

Figure 39.1 Solar Radiation Strongly Influences Climate
Sunlight strikes Earth less directly in polar regions than in equatorial regions. This difference has two main effects: First, rays from the sun travel a greater distance through the atmosphere in polar regions than they do near the equator. The atmosphere reflects, absorbs, and scatters sunlight, so the increased distance that sunlight travels through the atmosphere means that less solar energy reaches Earth's surface. Second, for a given amount of solar energy that reaches Earth's surface, that energy is spread over a larger area in polar regions than in equatorial regions. Together, these two effects cause polar regions to receive only 40 percent as much solar radiation as equatorial regions do.

the sections that follow, we'll see how climate influences the biosphere.

Incoming solar radiation shapes climate

Tropical areas near Earth's equator are much warmer than polar regions. The reason for this difference in temperature is that sunlight strikes Earth directly at the equator, but at a slanted angle near the North and South Poles. As a result, polar regions receive only 40 percent of the solar radiation that reaches the Tropics, making them much colder than the Tropics (Figure 39.1).

The amount of solar radiation received by regions outside the Tropics varies greatly during the year, giving rise to the seasons (Figure 39.2). The large seasonal variation

39.1 Climate

Weather refers to the temperature, precipitation (rainfall and snowfall), wind speed, humidity, cloud cover, and other physical conditions of the lower atmosphere at a specific place over a short period of time. Weather, as we all know, changes quickly and is hard to predict. But **climate**, the prevailing weather conditions experienced in an area over relatively long periods of time (30 years or more), is predictable. Climate has a major effect on ecological interactions. In fact, organisms in the biosphere are more strongly influenced by climate than by any other feature of the environment. On land, for example, whether a particular region is desert, grassland, or tropical rainforest depends primarily on such features of climate as temperature and precipitation.

In this section we consider the factors that shape climate: the amount of incoming solar radiation, the movement of air and water, and the major features of Earth's surface. In

Figure 39.2 The Tilt of Earth Causes the Seasons
In the Northern Hemisphere, winter officially begins on December 21, when days are at their shortest and the Northern Hemisphere is maximally tilted away from the sun. The reverse occurs on June 21, the first day of summer, when days in the Northern Hemisphere are at their longest and the Northern Hemisphere is maximally tilted toward the sun.

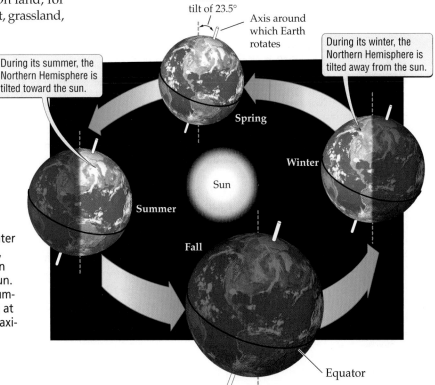

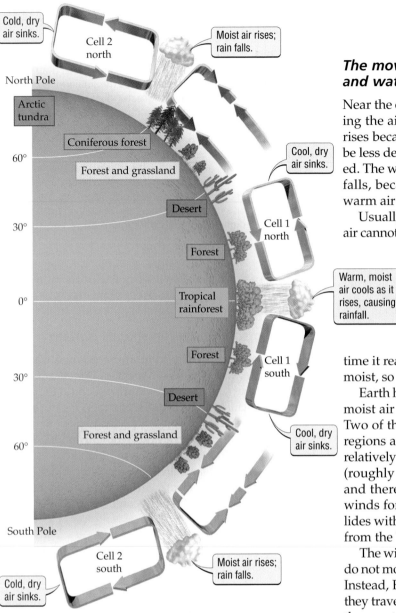

Cold, dry air sinks.

Cell 2 north

Moist air rises; rain falls.

North Pole

Arctic tundra

Coniferous forest

60°

Forest and grassland

Cool, dry air sinks.

Desert

30°

Cell 1 north

Forest

0°

Tropical rainforest

Warm, moist air cools as it rises, causing rainfall.

Forest

Cell 1 south

30°

Desert

Forest and grassland

Cool, dry air sinks.

60°

South Pole

Cell 2 south

Moist air rises; rain falls.

Cold, dry air sinks.

39.3

Figure 39.3 Earth Has Four Giant Convection Cells
Two giant convection cells are located in the Northern Hemisphere and two are in the Southern Hemisphere. In each of the four convection cells, relatively warm, moist air rises, cools, and then releases moisture as rain or snow. The cool, dry air then sinks to Earth and flows back toward the region where the warm air is rising.

in solar radiation occurs because Earth is tilted 23.5 degrees on its axis. As Earth revolves around the sun, the Northern Hemisphere is tilted toward the sun in June (and hence receives more energy) and away from the sun in December (and hence receives less energy). Careful examination of Figure 39.2 will show why it is winter in the Southern Hemisphere (for example, Chile) when it is summer in the Northern Hemisphere (for example, Canada).

The movement of air and water shapes climate

Near the equator, intense sunlight heats moist air, causing the air to rise from the surface of Earth. Warm air rises because heat causes it to expand and therefore to be less dense, or lighter, than air that has not been heated. The warm, moist air cools as it rises. As a result, rain falls, because cool air cannot hold as much water as warm air can (Figure 39.3).

Usually, cool air sinks. In this case, however, the cool air cannot sink immediately because of the warm air that is rising beneath it. Instead, the cool air moves to the north and south, tending to sink back to Earth at about 30° latitude. The cool air warms as it descends, which allows it to hold more water. As the air flows back toward the equator, it absorbs moisture from Earth's surface. By the time it reaches the equator, the air is once more hot and moist, so it rises to begin the cycle again.

Earth has four **giant convection cells** in which warm, moist air rises and cool, dry air sinks (see Figure 39.3). Two of the four convection cells are located in tropical regions and two in polar regions, where they generate relatively consistent wind patterns. In temperate regions (roughly 30° to 60° latitude), winds are more variable and there are no stable convection cells. The variable winds form when cool, dry air from polar regions collides with warm, moist air that moves toward the poles from the Tropics.

The winds produced by the four giant convection cells do not move straight north or straight south (Figure 39.4). Instead, Earth's rotation causes these winds to curve as they travel near Earth's surface. Winds that travel toward the equator, for example, curve to the west. When winds curve to the west, they blow from the east; hence such winds are called easterlies ("from the east"). Similarly, winds that travel toward the poles blow from the west and are called westerlies. Thus, at any given location, the winds usually blow from a consistent direction; these patterns of air movement are known as prevailing winds. In southern Canada and in much of the United States, for example, winds blow mostly from the west; thus storms in these regions usually move from west to east.

Ocean currents also have a major effect on climate. The rotation of Earth, differences in water temperatures between the poles and the Tropics, and the directions of prevailing winds all contribute to the formation of ocean currents. In the Northern Hemisphere, ocean currents tend to run clockwise between the continents; in the Southern Hemisphere, they tend to run counterclockwise (Figure 39.5).

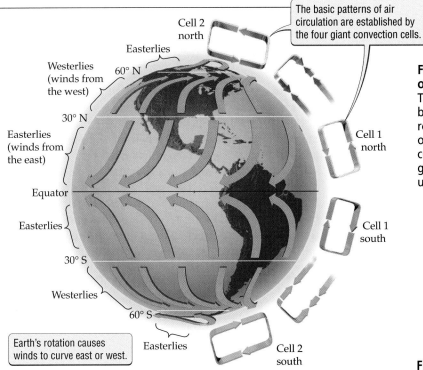

Cell 2 north

Easterlies

Westerlies (winds from the west)

60° N

30° N

Easterlies (winds from the east)

Equator

Easterlies

30° S

Westerlies

60° S

Easterlies

Cell 1 north

Cell 1 south

Cell 2 south

The basic patterns of air circulation are established by the four giant convection cells.

Earth's rotation causes winds to curve east or west.

Figure 39.4   Global Patterns of Air Circulation

The four giant convection cells determine the basic pattern of air circulation on Earth. Earth's rotation then causes winds to curve to the east or to the west. The direction in which they curve depends on their location, but for any given geographic region on Earth, the winds usually blow from a consistent direction.

Figure 39.5   The World's Major Ocean Currents

Cold ocean currents are colored blue; warm ocean currents are colored red.

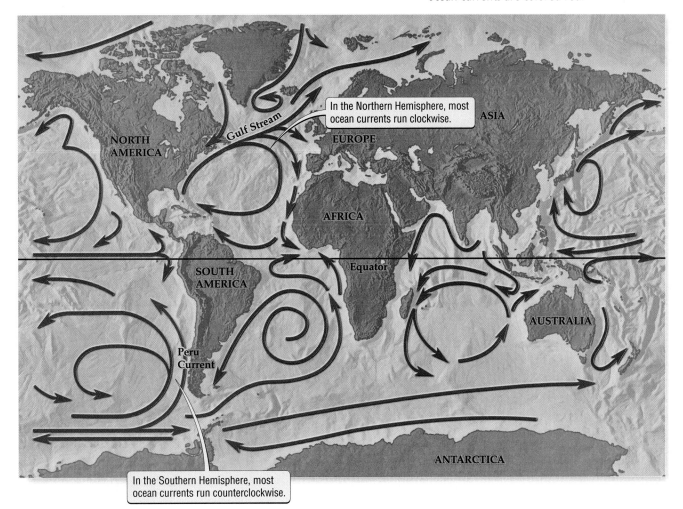

ASIA

NORTH AMERICA

Gulf Stream

EUROPE

In the Northern Hemisphere, most ocean currents run clockwise.

AFRICA

Equator

SOUTH AMERICA

AUSTRALIA

Peru Current

ANTARCTICA

In the Southern Hemisphere, most ocean currents run counterclockwise.

Ocean currents carry a huge amount of water and can have a great influence on regional climates. The Gulf Stream, for example, moves 25 times the water carried by all the world's rivers combined. Without the warming effect of the water carried by this current, countries such as Great Britain and Norway would have a subarctic to arctic climate. Overall, the Gulf Stream causes cities in Europe to be much warmer than cities at similar latitudes in North America, as illustrated by Rome versus Boston, Paris versus Montréal, and Stockholm versus Fort-Chimo (a town of 1400 people in Québec, Canada).

The major features of Earth's surface also shape climate

Heat is absorbed and released more slowly by water than by land. Therefore, because they retain heat comparatively well, oceans and large lakes moderate the climate of the surrounding lands. Mountains also can have a large effect on a region's climate. For example, mountains often cause a **rain shadow** effect, in which little precipitation falls on the side of the mountain that faces away from the prevailing winds (Figure 39.6). In the Sierra Nevada of North America, five times as much precipitation falls on the west side of the mountains (which faces toward winds that blow in from the ocean) as on the east side of the mountains. Mountain ranges in northern Mexico, South America, Asia, and Europe also create rain shadows.

The formation of deserts

Deserts are regions characterized by very low precipitation. Although most deserts are located in very hot places, deserts are also found in cooler climates. The formation of deserts illustrates how the climate of a region can be affected by a combination of incoming solar radiation, air and water currents, and major features of Earth's surface.

Many of the world's largest deserts are located at about 30° latitude (see Figure 39.7). Deserts are found at these latitudes for two reasons: First, the dry air that descends at these locations contains little moisture and thus brings little precipitation (see Figure 39.3). Second, these latitudes are relatively close to the equator, so they are hot. Of course, hot regions with little precipitation are well suited for the formation of deserts.

Other deserts, especially those in the temperate zone, are formed mainly or in part by rain shadows created by mountains. Rain shadows helped form the Mojave Desert and the Great Basin (a high-altitude, relatively cold desert) in North America, as well as the Gobi Desert of Asia. Finally, some of the driest deserts in the world, such as the Atacama (South America) and Namib (southern Africa) deserts, occur in places where prevailing winds blowing from the sea pass over cold ocean currents. The air cools as the winds pass over the cold water, causing rain to fall before the winds reach the shore.

39.4

Figure 39.6   Rain Shadow Effect
The side of a high mountain that faces the prevailing winds (the windward side) receives more precipitation than the side of the mountain that faces away from the prevailing winds (the leeward side). The leeward side is thus said to be in a rain shadow.

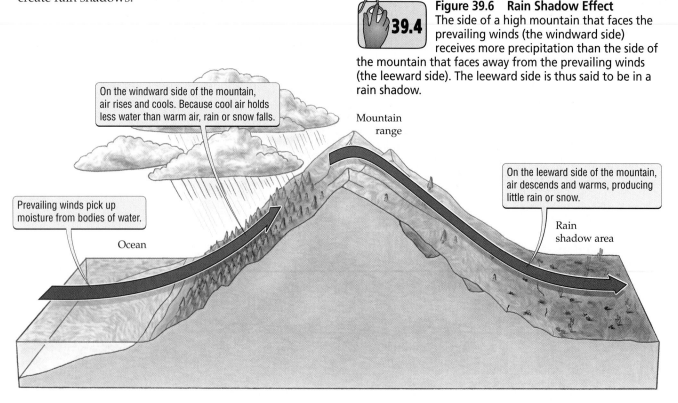

On the windward side of the mountain, air rises and cools. Because cool air holds less water than warm air, rain or snow falls.

Prevailing winds pick up moisture from bodies of water.

Ocean

Mountain range

On the leeward side of the mountain, air descends and warms, producing little rain or snow.

Rain shadow area

When the cool, dry winds reach the shore, they warm again, bringing little rain and causing extremely arid conditions; the Atacama Desert, for example, receives only 0.6 centimeters of precipitation per year.

> ■ The climate of a region is shaped by three major factors: the amount of incoming solar radiation, global patterns of air and water movement, and major features of Earth's surface.

Terrestrial Biomes

Now that we've discussed the factors that influence climate, let's turn to the effects of climate on the biosphere. The biosphere can be divided into several major terrestrial and aquatic life zones called **biomes**. It is climate that determines what biomes can exist at a particular location.

Terrestrial biomes, such as grasslands and tropical forests, cover large geographic areas and are usually named after the dominant vegetation of the region. Earth's terrestrial environment can be divided into seven major biomes: tropical forest, temperate forest, grassland, chaparral, desert, boreal forest, and tundra (which includes the tops of high mountains) (Figure 39.7). Figure 39.8 illustrates each of these biomes with a representative photograph. Keep in mind, however, that maps and photographs of a biome can give the misleading impression that the entire biome is the same. To produce a worldwide map of terrestrial biomes like the one shown in Figure 39.7, regions that actually are very different must be lumped together into a single biome. If this were not done, the map would be so complicated that it would not be worth much.

Consider grasslands, for example. The grassland biome includes arid grasslands (characterized by short, drought-resistant grasses) in southwestern North America, tallgrass prairie (characterized by tall grasses and many wildflowers) in north central North America, and savanna (grasslands dotted with occasional trees) in Africa. What is true of grasslands is true of all biomes: Both the species found in the biome and the ecological conditions of the biome can vary greatly from place to place.

The location of terrestrial biomes is determined by climate and human actions

Climate is the single most important factor controlling the potential or natural location of terrestrial biomes. The climate of a region—most importantly, the temperature and the amount and timing of precipitation—allows some species to thrive and prevents other species from living there. Overall, the effect of temperature and moisture on species causes particular biomes to be found under a consistent set of conditions (Figure 39.9).

Climate can exclude species from a region directly or indirectly. Species that cannot tolerate the climate of a

Figure 39.7   Major Terrestrial Biomes
This map shows the potential distribution of major terrestrial biomes on Earth. The actual distribution of biomes is heavily influenced by human activities.

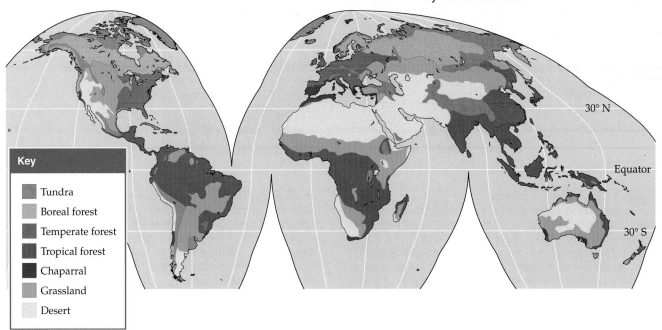

Key
- Tundra
- Boreal forest
- Temperate forest
- Tropical forest
- Chaparral
- Grassland
- Desert

(a) Tropical forest

(b) Temperate forest

(d) Chaparral

(e) Desert

(f) Boreal forest

Figure 39.8 Terrestrial Biomes

(a) Tropical forests form in warm, rainy regions and are dominated by a rich diversity of trees, vines, and shrubs. (b) Temperate forests are dominated by trees and shrubs that grow in regions with cold winters and moist, warm summers. (c) Grasslands are common throughout the world and are dominated by grasses and many different types of wildflowers. They often occur in relatively dry regions with cold winters and hot summers. (d) Chaparral is characterized by shrubs and small, nonwoody plants that grow in regions with mild summers and winters and low to moderate amounts of precipitation. (e) Deserts form in regions with low precipitation, usually 25 centimeters per year or less. (f) Boreal forests are dominated by coniferous trees that grow in northern or high-altitude regions with cold, dry winters and mild, humid summers. (g) Tundra is found at high latitudes and high elevations and is dominated by low-growing shrubs and nonwoody plants that can tolerate extreme cold.

(c) Grassland

(g) Tundra

region are directly excluded from that region. Species that can tolerate the climate but are outperformed by other organisms that are better adapted to the climate are excluded indirectly.

Although climate places limits on where biomes can be found, the actual extent or distribution of biomes in the world today is very strongly influenced by people (see the box on p. 660). We will return to the effects of humans on natural biomes when we discuss global change in Chapter 45.

> ■ There are seven major terrestrial biomes. Climate can exclude a species from an area directly or indirectly. Although climate is the most important factor controlling the potential location of terrestrial biomes, their actual location is heavily influenced by people.

Aquatic Biomes

Life evolved in water billions of years ago, and aquatic ecosystems cover about 75 percent of Earth's surface. There are eight major aquatic biomes: river, lake, wetland, estuary, intertidal zone, coral reef, ocean, and benthic zone (Figure 39.10). Unlike terrestrial biomes, aquatic biomes are usually characterized by physical conditions of the environment, such as salt content, water temperature, water depth, and the speed of water flow.

As with terrestrial biomes, the photographs in Figure 39.10 capture only a small portion of the diversity of aquatic biomes. Lake biomes, for example, include bodies of water that range in size from

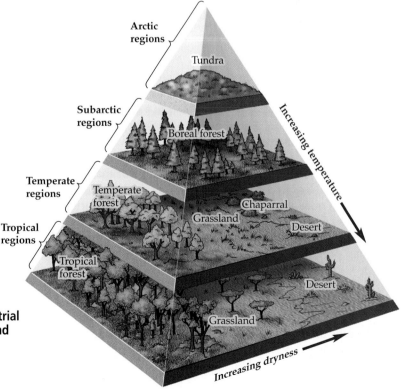

39.6

Figure 39.9   The Location of Terrestrial Biomes Depends on Temperature and Precipitation

■ BIOLOGY IN OUR LIVES

What Is a "Natural" Biome?

Locate the region where you live in Figure 39.7. Does the biome shown on the map match your actual surroundings? For many of you, the answer is "No." Figure 39.7 shows *potential* biomes—the vegetation types that could thrive given the climate of each particular region. But in reality, people have converted major portions of these potential biomes to urban (housing and industry) and agricultural areas. Vast regions of North America no longer have their original biomes, as illustrated by the almost complete conversion of the original grasslands (prairies) to urban and agricultural areas.

In some parts of the world, humans have altered the landscape so thoroughly and over such long periods of time that we now consider the altered landscape to be the "natural" one. Consider the moors of England and Scotland, famous from Sherlock Holmes stories and travel advertisements. Although many peo-

ple view the moors as a beautiful natural landscape, regions now covered with moors once were covered with oak woodlands. In fact, the moors are completely dependent on humans for their existence—they were produced when people cut the trees down and then used the land for grazing.

Moors are just one example of what is actually a common phenom-

enon: In many instances, actions by people have created or maintained landscapes that we now find beautiful and natural. Are landscapes that are dependent on the actions of people natural? What exactly is "nature?" Is it a pristine state that does not include people? Or does "nature" include—at least to some degree—people and our effects on the world?

small ponds to very large lakes. In addition, very different species can be found in two areas of an aquatic biome. For example, algae may thrive in a lake that contains large amounts of nitrogen and phosphorus. When the algae die, their remains are consumed by bacteria. As the bacteria grow and reproduce, they may use up so much of the oxygen dissolved in the water that fish in the lake die. In contrast, a lake containing little nitrogen and phosphorus may have few algae, few bacteria, high oxygen levels, and many species of fish. Thus, although both areas are lakes, the species that live in them differ tremendously.

Aquatic biomes are influenced by terrestrial biomes, climate, and human actions

Aquatic biomes, especially lakes, rivers, wetlands, and the coastal portions of marine biomes, are heavily influ-

Figure 39.10   Aquatic Biomes
(*a*) Rivers are relatively narrow bodies of fresh water that move continuously in a single direction. (*b*) Lakes are standing bodies of fresh water of variable size, ranging from a few square meters to thousands of square kilometers. (*c*) Wetlands are characterized by shallow waters that flow slowly over lands that border rivers, lakes, or ocean waters. (*d*) Estuaries are tidal ecosystems where rivers flow into the ocean. (*e*) Intertidal zones are found in coastal areas where the tides rise and fall on a daily basis, periodically submerging a portion of the shore. (*f*) Coral reefs form in warm, shallow waters located in the Tropics; they are named after the corals on which many of the reef's other organisms depend. (*g*) Oceans cover the majority of Earth's surface. They include a shallow layer (100 to 200 meters deep) in which photosynthesis can occur as well as deeper ocean waters that little light can penetrate. (*h*) Benthic zones, located on the bottom surfaces of other aquatic biomes, are home to a wide variety of organisms.

(a) River

(b) Lake

(c) Wetland

(d) Estuary

(g) Ocean

(e) Intertidal zone

(f) Coral reef

(h) Benthic zone

A deep-ocean-water fish

enced by the terrestrial biomes that they border or through which their water flows. For example, high and low points of the land determine the locations of lakes and the speed and direction of water flow. In addition, when water drains from a terrestrial biome into an aquatic biome, it brings with it dissolved nutrients (such as nitrogen or phosphorus) that were part of the terrestrial biome. Because nutrients are available only in low amounts in many aquatic biomes, nutrients imported from the surrounding terrestrial biome can have a significant effect. For example, the addition of nutrients from a farm field to a lake can cause algae to thrive and fish to die, as described above.

Aquatic biomes also are strongly influenced by climate. In temperate regions, for example, seasonal changes in temperature cause the oxygen-rich water near the top of a lake to sink in the fall and the spring, bringing oxygen to the bottom of the lake. In tropical regions, seasonal differences in temperature are not great enough to cause a similar mixing of water from the top and bottom of a lake. This lack of mixing causes the deep waters of tropical lakes to have low oxygen levels and relatively few forms of life.

Climate also has important effects in the open ocean. For example, climate helps determine the temperature, sea level, and salt content of the world's oceans. As we saw at the beginning of this chapter, the physical conditions of the ocean have dramatic effects on the organisms that live there; thus, climate has a powerful effect on marine life.

Finally, as with terrestrial biomes, the actions of humans strongly affect aquatic biomes. Portions of some aquatic biomes, such as wetlands and estuaries, are often destroyed to allow for development projects. Rivers, wetlands, lakes, and coastal marine biomes are negatively affected by pollution in most parts of the world. Aquatic biomes also suffer when humans destroy or modify the terrestrial biomes in which they are situated. For example, when forests are cleared for timber or to make room for agriculture, the rate of soil erosion can increase dramatically because trees are no longer there to hold the soil in place. Increased erosion can cause streams and rivers to become clogged with silt, which harms or kills invertebrates, fish, and many other species.

■ There are eight major aquatic biomes. Usually characterized by the physical conditions of the environment, aquatic biomes are strongly influenced by the surrounding terrestrial biomes, by climate, and by human actions.

■ HIGHLIGHT

One World

Spanish conquistadors are said to have cooled their water flasks in the unusually cold waters found off the coast of Peru. These cold waters are brought to the region from the south by the Peru Current (see Figure 39.5). The low sea surface temperatures of this region are also caused by cold, nutrient-rich waters that rise from the ocean depths to the surface. These waters provide food and suitable temperatures for enormous numbers of small marine organisms collectively known as plank-

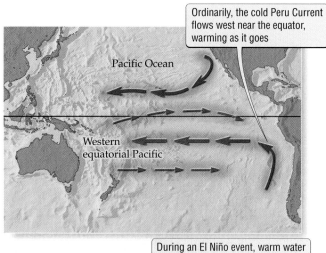

Ordinarily, the cold Peru Current flows west near the equator, warming as it goes

Pacific Ocean

Western equatorial Pacific

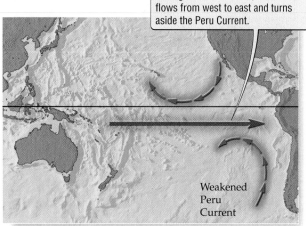

During an El Niño event, warm water flows from west to east and turns aside the Peru Current.

Weakened Peru Current

Figure 39.11   El Niño Events
During an El Niño event, winds from the west push warm surface water from the western Pacific to the eastern Pacific. The resulting changes in sea surface temperatures cause additional changes in ocean currents, wind patterns, sea levels, and patterns of precipitation throughout the world. Cold ocean currents are colored blue; warm ocean currents are colored red.

ton, which support a rich fishery. Before it collapsed in 1972 (during an El Niño event), the Peruvian anchovy fishery was the largest fishery in the world; its 1970 catch for this single species of fish equaled almost 20 percent of the total world harvest of fish.

Every year, the Peruvian fishing season ends when a current of warm water, known by local fishermen as El Niño (literally, "the child"; so named because it often appears close to Christmas), pushes south along the coast of Ecuador. Usually, this warm water has a temporary and local effect. But once every 2 to 10 years, the warm water that ends the fishing season is just the beginning of an **El Niño event**: a series of changes in the weather that triggers floods, fires, torrential rains, droughts, disease, crop failures and more, as described at the beginning of this chapter (Figure 39.11).

During a strong El Niño event, winds that usually blow from the coast of South America to the west shift in direction and blow to the east. As these winds blow east, they drive warm water from the western Pacific eastward. The arrival of warm water from the western Pacific changes the usual pattern of sea surface temperatures, leading to changes in ocean currents (including a weakening of the cold Peru Current), wind systems, sea levels, and patterns of precipitation over much of the world. These and other drastic changes in the weather cause the tremendous disruptions to natural systems described at the beginning of the chapter and shown in Figure 39.12.

Interesting in and of themselves, El Niño events also illustrate an important general point: Global patterns of air and water circulation can cause events that occur in one part of the world to affect ecosystems all over the world. As we will see in the upcoming chapters, it can be difficult to understand the ecological interactions that occur within a local area. In many cases, however, what

Figure 39.12   El Niño Events Have Large and Varied Effects throughout the World

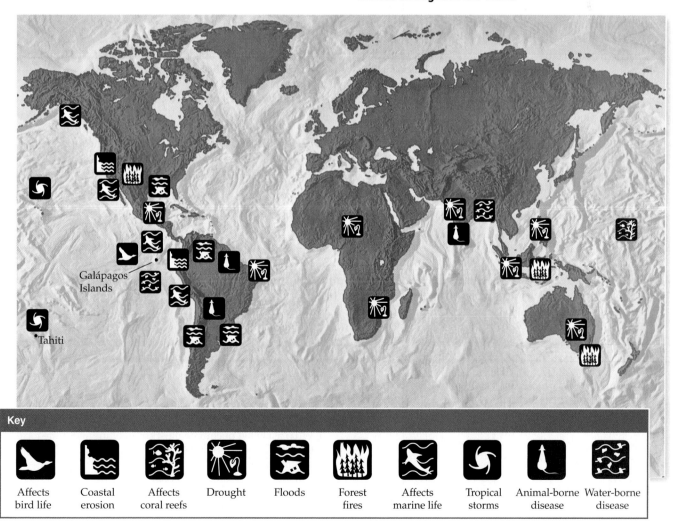

ecologists (and policymakers) really need to understand are both local interactions and the effects of distant events. Such an understanding requires a detailed understanding of the biosphere—a monumental, complex task. To anyone who argues that there are no major scientific frontiers left to explore, the biosphere looms large as proof that they are wrong.

> ■ Once every 2 to 10 years, a stronger than usual warm water current flowing along the west coast of South America signals the onset of an El Niño event, a series of changes that alter weather patterns and affect ecosystems all over the world.

SUMMARY

Ecology: Studying Interactions between Organisms and Their Environment

■ Ecologists study interactions between organisms and their environment at different levels, ranging from individuals to ecosystems.

■ A major goal of ecology is to document and understand the consequences for life on Earth of human actions that are changing the biosphere.

Climate

■ Climate depends on incoming solar radiation. Tropical regions are much warmer than polar regions because sunlight strikes Earth directly at the equator but at a slanted angle near the poles.

■ Climate is strongly influenced by four giant convection cells that generate relatively consistent wind patterns over much of Earth.

■ Ocean currents carry an enormous amount of water and can have a large effect on regional climates.

■ Regional climates are also affected by major features of Earth's surface. For example, mountains can create rain shadows, thereby contributing to the formation of deserts.

Terrestrial Biomes

■ There are seven major terrestrial biomes: tropical forest, temperate forest, grassland, chaparral, desert, boreal forest, and tundra.

■ Climate is the most important factor controlling the potential location of terrestrial biomes. Climate can exclude a species from an area directly or indirectly.

■ The actual location of terrestrial biomes is heavily influenced by human activities.

Aquatic Biomes

■ There are eight major aquatic biomes: river, lake, wetland, estuary, intertidal zone, coral reef, ocean, and benthic zone.

■ Aquatic biomes are usually characterized by physical conditions of the environment, such as temperature, salt content, and water movement.

■ Aquatic biomes are strongly influenced by the surrounding terrestrial biomes, by climate, and by human actions.

Highlight: One World

■ El Niño events illustrate how global patterns of air and water circulation can cause events that occur in one part of the world to have far-reaching consequences.

KEY TERMS

biome p. 657

biosphere p. 652

climate p. 653

ecology p. 652

ecosystem p. 652

El Niño event p. 663

giant convection cell p. 654

rain shadow p. 656

weather p. 653

CHAPTER REVIEW

Self-Quiz

1. Which of the following is *not* a level of the biological hierarchy commonly studied by ecologists?
   a. ecosystem
   b. individual
   c. organelle
   d. population

2. Which of the following can contribute to the formation of a desert?
   a. winds that pass over warm ocean currents
   b. cool, dry air that sinks toward Earth at about 30° latitude
   c. rain shadows
   d. both b and c

3. London, England, and Winnipeg, Canada, are located at a similar latitude, yet London is much warmer than Winnipeg. Why?
   a. The Gulf Stream warms Europe.
   b. The Peru Current keeps Canada cold.
   c. Easterlies occur in London, westerlies in Winnipeg.
   d. There is a rain shadow effect in Winnipeg.

4. The biosphere consists of
   a. all organisms on Earth only.
   b. only the environments in which organisms live.
   c. all organisms on Earth and the environments in which they live.
   d. none of the above

5. What aspect(s) of climate most strongly influence the locations of terrestrial biomes?
   a. rain shadows
   b. temperature and precipitation
   c. only temperature
   d. only precipitation

Review Questions

1. Discuss the difference between potential and actual locations of biomes. What factors control the potential and actual locations of biomes?

2. Explain in your own words how global patterns of air and water circulation can cause local events to have far-reaching ecological consequences. Give an example that shows how local ecological interactions can be altered by distant events.

3. Explain how climate can exclude species from a region in both terrestrial and aquatic biomes.

4. To explore on your own: Why is it colder at the top of a high mountain (which is closer to the sun) than it is at the base of the mountain (which is farther from the sun)?

---

39   **The Daily Globe**

Editorial: Protect our Biosphere, Buy a Gas-Electric Hybrid Car

Americans are addicted to driving. We love our cars and we love to hit the open road. The problem is that with each mile we drive, we add carbon dioxide (CO_2) to the air. This is a problem because carbon dioxide is a greenhouse gas, meaning that it works like the windows of your car: It lets in sunlight but traps heat. Scientists warn that if we don't reduce the amount of CO_2 that we add to the air, the temperature of the planet may rise considerably, wreaking havoc on the biosphere's natural and agricultural ecosystems.

Transportation produces about 25 percent of global CO_2 emissions, mostly from cars and trucks. Given our addiction to driving, how much can we drive without risking large changes to the world's climate? Unfortunately, CO_2 emissions add up quickly. For example, if you own a typical SUV and drive more than 4 kilometers (2.5 miles) per day, you're adding more than your fair share of CO_2 to the air (where a person's "fair share" of CO_2 emissions is the amount that each person can add to the air without having global emissions of CO_2 disrupt the world's climate).

So what should we do? You could wait about 10 years for car engines that use new technologies (such as fuel cells) to become available. But if you want to make a difference sooner rather than later, Toyota and Honda already sell hybrid electric vehicles (HEVs) that get terrific gas mileage (up to 68 mpg) and qualify as "super low emission vehicles." These cars inflict minimal damage to the environment, in a sense allowing us to have our cake and eat it too.

Other car companies, including Ford, GM, DaimlerChrysler, and Fiat, have HEVs in production or design. Thus, for many of us, a "dream HEV" is available now or in the pipeline. HEVs look great, they have good power, and they are convenient to use (you don't plug them in, they recharge while you drive). HEVs also save you money and help save the global environment. What's not to like?

Evaluating "The News"

1. Another way to reduce greenhouse gas emissions would be by using various forms of mass transit. Why do you think car pooling and public transportation are little used in the United States?

2. In terms of yearly driving cost and effect on the environment, how well do you think your car would compare to an HEV? *Hint:* You can compare your car to two HEVs, the Honda Insight or the Toyota Prius, at the website http://209.10.107.169/tailpipetally/

3. Do you think people who buy cars and trucks that emit high levels of greenhouse gases should be penalized? For example, should a "global warming tax" be imposed on the sale of SUVs and other vehicles that emit large amounts of CO_2?

40

Why Organisms Live Where They Do

Erik La Gattuta, *Encouraging the Coelacanth*, 2000.

Life at the Extremes

The biosphere harbors millions of species, many of which live in environments that also serve as homes for people. But other species live in what to humans are extreme environments, very different from places in which people can or do live.

Some of these extreme environments, such as the deep regions of caves that never see the light of day, seem strange and inhospitable to us. But such places are home to insects, snails, crayfish, salamanders, and fish. Many of the species that live in caves have unusual characteristics. The grotto salamander, for example, spends its entire life deep in caves. At birth, these salamanders are dark in color and have functional eyes. However, as they grow, grotto salamanders lose their color and their eyelids fuse shut; in some specimens, the eyes appear to have sunk inside the head. In general, many cave-dwelling species are blind, have no color, and rely on sense organs other than eyes to navigate through their dark world.

Although deep caves can be a difficult place to live, other environments are even more extreme. Consider, for example, the abyssal zone, that region

Where organisms live is determined by history, the presence of suitable habitat, and dispersal.

of the ocean where the waters are more than 2000 meters deep. No sunlight reaches the abyssal zone, and the organisms that live there are subjected to pressures that would kill a person. Yet life abounds, as species that range from bacteria to worms to crabs to bizarre fish with few bones and transparent bodies make their home in the abyssal zone (see Figure 39.10*h*).

It is not hard to understand why organisms that inhabit extreme environments live where they do: Such creatures have unique features that allow them to thrive under conditions that would kill most other species. But what about species that live in environments that to us, at least, seem more hospitable? Why do we find such organisms in one place but not in another, seemingly similar, area? These questions bring us to the central issue addressed in this chapter: why organisms live where they do.

A Grotto Salamander

■ KEY CONCEPTS

1. The area where a species can be found is referred to as its distribution or range.

2. Species are distributed unevenly throughout the environment, in part because the environment varies greatly from place to place and in part for a variety of biological reasons.

3. The distribution of a species is determined by history (where the species evolved and continental drift), the presence of suitable habitat, and dispersal.

4. The distributions of species change naturally with time as a result of migration, climate change, and loss or gain of suitable habitat.

5. Humans are causing large changes in the distributions of species, both by destroying or degrading suitable habitat and by transporting species to new geographic regions.

*E*cology is a young science, having begun only about 110 years ago. Throughout the brief history of this science, ecologists have sought to understand why organisms live where they do. This information is of fundamental scientific interest because it enriches our knowledge of the world around us.

Knowledge of why organisms live where they do also helps us solve practical problems. To control a pest species, for example, we need to understand what features of the environment cause the pest to thrive and what features cause it not to thrive. With this knowledge, we can modify the environment so that it is less suitable for the pest species. On a much broader scale, human activities are causing the decline and extinction of species throughout the world. We need to understand why species live where they do so that we can improve their chances for survival while still allowing for sustainable forms of economic development.

Life is not distributed evenly on Earth: Any area, even an entire continent, contains only a small fraction of all the species on Earth. In this chapter we explore some of the reasons why a species is found in one region but not another, and we discuss some of the practical applications of such knowledge.

The Distributions of Species: Where Organisms Live

The area over which we can find a species is called its **distribution** or **range**. Many species have a small range. Some tropical plants, for example, are confined to very small areas (such as a single hillside), and the Organ Mountain primrose is found in only 11 canyons of the Organ Mountains, a small mountain chain in southern New Mexico. Other species, such as coyotes, live over most of one continent, while still others, such as wolves, live on restricted portions of several continents. Relatively few species are found on all or most of the world's continents. Notable exceptions include humans, Norway rats, and the bacterium *Escherichia coli*, which lives in the intestinal tracts of reptiles, birds, and mammals (including humans) and thus is found wherever its host organisms are found.

The place in which a species lives is called its **habitat**. The habitat of a species has a characteristic set of environmental conditions. These conditions may include both physical components (for example, the amount of sunlight) and biotic (living) components (for example, the other species present) of the environment. Consider a forest and an adjacent meadow in eastern North America (Figure 40.1). Some species of plants, such as starflower, thrive only in the relatively cool, low-light environment found on the forest floor. Other plant species, such as New England aster, are found only in more open areas. Still other plants, including young sugar maple trees and the wildflower rough-leaved goldenrod, are found in both forest and open habitat.

The distribution of a species includes the area it occupies during all of its life stages. We know so little about many organisms that we do not understand their distributions very well. This lack of knowledge is particularly true for insects, fungi, and plants that are hard to find or have life stages that are hard to study. We may know, for example, under what conditions the adult organism lives, yet have no idea about where or how other life stages live. The distributions of animals that migrate also can be hard to discover because they may include geographic regions separated by thousands of kilometers. We also know very little about the distributions of some large organisms, such as whales, and some small but striking organisms, such as the monarch butterfly (see the box on p. 670).

Figure 40.1 A Boundary between Forest and Meadow

New England aster

Starflower

Species Have Patchy Distributions

The natural world is patchy; that is, physical conditions, resources, and organisms vary greatly from place to place. Relatively few species in nature have a **regular distribution** in which they are positioned in an even fashion throughout their habitat (Figure 40.2). Instead, most organisms have a **random distribution** (in which they are positioned in an unpredictable fashion) or a **clumped distribution** (in which they are grouped together).

The patchy distribution of organisms is very clear if you take a bird's-eye view of the world. Viewed from above, landscapes are composed of patches: Here there may be a patch of aspen trees, there a meadow with pine seedlings. Trees are not unique in this regard: Over large areas, all species have a patchy distribution.

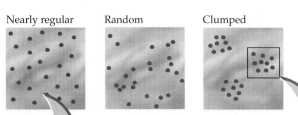

■ Some species have small ranges; others are found throughout one or a few continents. Relatively few species live on most or all of the world's continents. The habitat of a species has a characteristic set of environmental conditions, which include both physical and biotic components of the environment.

Nearly regular   Random   Clumped

Figure 40.2 Distribution of Organisms
A few organisms, such as creosote bush in the Mojave Desert, have a nearly regular distribution (probably in this case because of intense competition for limited water supplies). However, most organisms have either a clumped or a random distribution.

■ THE SCIENTIFIC PROCESS

Monarch Mysteries

Individual monarch butterflies are beautiful, and it is a breathtaking sight when thousands of them are found gathered together in eastern North America during their annual migration. Monarch caterpillars feed almost exclusively on milkweed plants. In the spring, large numbers of monarchs migrate north to the Gulf Coast states of eastern North America, lay eggs on milkweed, and die. The offspring of these butterflies migrate farther north, and several more generations of butterflies are born and spend their lives in the summer breeding range, where milkweed is abundant. In autumn, a final generation of butterflies migrates south, where they spend the winter.

Observations of the monarch migration gave rise to two great mysteries. First, although millions of butterflies migrate south from eastern North America each autumn, for a long time no one knew where they went. Scientists began trying to answer this question in 1857. It took almost 120 years. Finally, in January of 1975, biologists discovered that the monarchs that migrate from eastern North America spend the winter (as adults) in just a few spots in a mountainous area west of Mexico City. Compared with sites up north, the overwintering sites have few milkweed plants. Because butterflies that journey north have access to more milkweed plants —and hence, more food—it is likely that they can produce more offspring than could a "stay-at-home" butterfly that did not migrate north.

The second mystery has yet to be solved: How do the monarchs know where to go? Remember that as the monarchs migrate north, the adult butterflies lay their eggs on milkweed, and they die shortly thereafter. No individual monarch makes the entire journey. How the final generation can find its way back to the mountains of Mexico in autumn remains an unanswered and fascinating question.

Overwintering monarchs

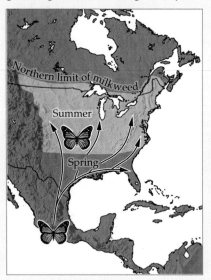

In the spring, monarch butterflies migrate north from Mexico to breed throughout much of eastern North America. Because monarch caterpillars depend on milkweed plants for food, the northern limit of the monarch's summer breeding range (shown in light blue) closely matches the northern limit of milkweeds. (This diagram shows only the migration patterns of the eastern North American population of monarchs; the smaller populations of monarchs found west of the Rockies overwinter along the Pacific coast of California.)

Why do organisms have patchy distributions? First, the underlying physical conditions often differ greatly from one location to the next. Factors such as the amount of nitrogen in the soil (which plants need to grow), the temperature, and the moisture available to organisms can vary greatly, even over distances as short as a few centimeters or meters. Variations in the physical environment can affect where organisms live, causing the distributions of organisms to match the patchy distributions of underlying physical conditions. For example, the shrub *Clematis fremontii* lives only on dry, rocky outcrops in Missouri. Few parts of the state have such outcrops. Within the few regions that do, the outcrops are found in clumps, and hence so are the shrubs.

Organisms may also have patchy distributions for biological reasons that are not related directly to the physical environment. For example, the vast majority of seeds produced by plants land close to the parent plant, thus causing plants of a given species to be clumped together. Animals that eat plants may have a patchy distribution for various reasons, the simplest of which is that the plants on which they feed also have a patchy distribution.

The shrub *Clematis fremontii*

> ■ Over large geographic areas, all organisms have patchy distributions. The patchy distributions of organisms often match the patchy distributions of underlying physical conditions. Organisms may also have patchy distributions for biological reasons.

Polar bears live where land meets the Arctic Ocean, which is shown in yellow. They hunt on ice packs far out into the sea.

**Figure 40.3
The Range of
Polar Bears**

40.1 *Factors that Determine the Distributions of Species*

The distribution of a species is determined mainly by one or more of the following factors: history, the presence of suitable habitat, and dispersal.

History: Evolution and continental drift

Past events have a profound effect on where organisms live today. Why, for example, are polar bears found in the Arctic (Figure 40.3), but not in Antarctica? Polar bears hunt on ice packs and eat seals, both of which abound in Antarctica. In part, the answer to this question lies in an accident of history: Polar bears evolved in the Arctic, and the tropical regions in between appear to have prevented them from reaching Antarctica. As the polar bear example shows, organisms do not necessarily live everywhere they are well suited to live.

In the early part of the twentieth century, the German scientist Alfred Wegener made the seemingly far-fetched suggestion that the continents move slowly over time (see Figure 22.6). At first, geologists hotly disagreed, but in the 1960s Wegener's suggestion was shown to be correct. Some biologists were early supporters of Wegener's theory because it helped them explain the curious distributions of many groups of organisms. For example, the southern beech tree is found today in Australia, New Zealand, and the southern parts of South America (Figure 40.4). The southern beech has heavy seeds and

is not able to spread long distances across ocean waters. So how did this tree species come to live in regions separated by vast stretches of ocean? The answer is simple: The southern beech evolved more than 65 million years ago, when South America, New Zealand, and Australia were connected to one another. The movement of the continents during the past 50 million years brought the tree to its current locations.

The presence of suitable habitat

Good and poor places to live exist for all species, and the distributions of species are limited by the presence of suitable habitat. This concept sounds simple, but what makes for "suitable habitat" can be complex. Let's consider how the physical environment, the biotic environment, interaction between the physical and the biotic environments, and disturbance influence what is suitable habitat for a species.

1. *Physical environment.* The survival and growth of organisms is limited by physical conditions. For example, some plants, such as the starflower of Figure 40.1, can live only in regions with relatively low light; such species are not found in open habitats. Likewise, cold-intolerant plants (Figure 40.5) and animals cannot live in regions where the weather gets very cold.

2. *Biotic environment.* Organisms interact with, and often require the presence of, other organisms. An animal that feeds on only one type of plant cannot live where that plant is absent, and a plant that requires a spe-

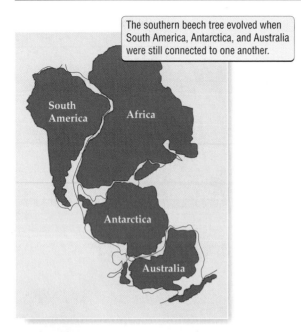

The southern beech tree evolved when South America, Antarctica, and Australia were still connected to one another.

Figure 40.4   How Did It Get There?

The modern distribution of the southern beech tree resulted when once-connected regions in South America, Australia, and New Zealand were separated by continental drift. Fossil pollen of the southern beech has been found at the two sites in Antarctica marked X.

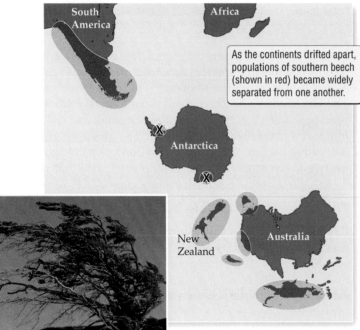

As the continents drifted apart, populations of southern beech (shown in red) became widely separated from one another.

together to determine the distributions of organisms. The barnacle *Balanus balanoides*, for example, requires certain physical conditions: It cannot survive where summer air temperatures are above 25°C, and it cannot reproduce where winter air temperatures do not remain below 10°C for 20 days or more. On the Pacific coast of North America, air temperatures are such that this barnacle should be found 1600 kilometers farther south than it currently is (Figure 40.6). *B. balanoides* is absent from the region shown in purple in Figure 40.6 because competition from other species of barnacles prevents it from living in what would otherwise be suitable habitat. To the north, as temperatures become increasingly low, a point is reached at which *B. balanoides* outcompetes the other barnacles and maintains healthy populations. Thus, the physi-

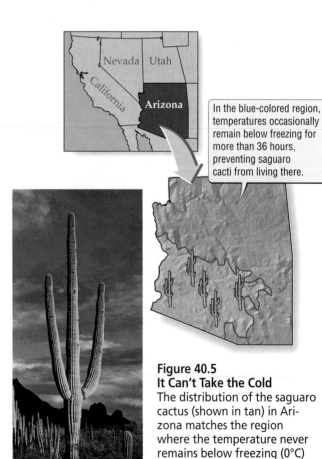

In the blue-colored region, temperatures occasionally remain below freezing for more than 36 hours, preventing saguaro cacti from living there.

Figure 40.5
It Can't Take the Cold
The distribution of the saguaro cactus (shown in tan) in Arizona matches the region where the temperature never remains below freezing (0°C) for more than 36 hours.

cific insect to pollinate its flowers cannot reproduce without that insect. In general, species that depend completely on one or a few other species for growth, reproduction, or survival cannot live where the species on which they depend are absent. Organisms also can be excluded from an area by predators (organisms that kill other organisms for food) or by competitors, either of which can greatly reduce their survival or reproduction (see Chapter 42).

3. *Interaction between the physical and the biotic environments.* The physical and biotic environments often act

Figure 40.6   Competition Limits a Species' Range
The barnacle *Balanus balanoides* attaches to rocks in the intertidal zone, as shown in this photograph taken along the coast of Alaska. The temperatures are suitable for *B. balanoides* in the region shaded purple on the map, but competition with other barnacles prevents it from living there. It is found only in the region shaded red.

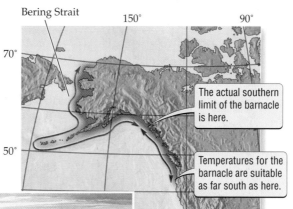

Bering Strait

The actual southern limit of the barnacle is here.

Temperatures for the barnacle are suitable as far south as here.

cal and biotic environments interact to determine where this barnacle is found.

4. *Disturbance.* The distributions of some organisms depend on regular disturbances. A **disturbance** is defined as an event that kills or damages organisms, creating opportunities for other organisms to grow and reproduce. Many species, for example, persist in an area only if there are periodic fires. For instance, fire causes the cones of some pine trees to open, thereby releasing the seeds and allowing the tree to reproduce. If humans prevent fires, such species are replaced by species that are not as tolerant of fire but that are superior competitors in the absence of fire. Floods, windstorms, and droughts are other forms of disturbance that can exclude some species from an area but give others an advantage.

Dispersal

Dispersal occurs when individuals travel long distances away from other members of their species. Dispersal may be active, as when organisms walk, fly, or swim.

Dispersal may also be passive, as when organisms are transported by water, wind, or other organisms.

Many species are not found in areas of suitable habitat at least in part because they cannot disperse—or have not yet dispersed—to that habitat. Although polar bears are known to travel more than 1000 kilometers in a year, they have never dispersed from the Arctic, where they evolved, to Antarctica, a region in which they would probably thrive (see Figure 40.3). Similarly, plant species often are not found in regions where they can grow, presumably because they have not been able to reach those areas by seed dispersal.

The failure of species to disperse to some areas of suitable habitat is especially clear on islands. The Hawaiian Islands, for example, have only one native mammal species, a bat, which was able to fly to the islands. No other mammals were able to disperse to Hawaii on their own, although cats, pigs, wild dogs, rats, goats, mongooses, and other mammals now thrive in Hawaii following their introduction to the islands by people.

Mongoose

■ The distributions of organisms are strongly influenced by history, including where the organisms evolved and continental drift. Dispersal and the presence of suitable habitat also have a great impact on where organisms live. The presence of suitable habitat is determined by the physical environment, biotic environment, interaction between the physical and the biotic environments, and disturbance.

40.2   Changes in the Distributions of Species

The distributions of species can change over time. If a species migrates on its own or is introduced by humans into new habitat, it may spread through a new region, thereby expanding its range. Starlings, for example, spread throughout North America after they were

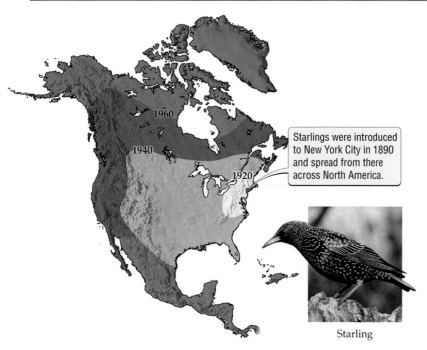

Starlings were introduced to New York City in 1890 and spread from there across North America.

Starling

Figure 40.7   From Shakespeare's *Henry IV* to North America
Because they were mentioned in Shakespeare's *Henry IV*, starlings were introduced to New York City by a group of Shakespeare enthusiasts who wanted all birds mentioned by the Bard to live in North America. Unfortunately, as they spread across the continent, starlings displaced native bird species such as bluebirds. Each year, starlings also cause extensive damage to wheat and some fruit crops. The dates printed on the map show the edge of the starling's range at different points in time.

brought by people from Europe (the region in which starlings were found naturally) to New York City (Figure 40.7).

Starlings, in fact, represent just one example of a general problem: Throughout the world, people have greatly altered the distributions of species by accidentally and deliberately introducing species to locations where they had not previously been found. For example, as people travel from one corner of the globe to another, they carry with them disease-causing organisms such as viruses, bacteria, and fungi, thereby enabling the rapid spread of diseases that once would have been confined to small geographic regions. Examples of such diseases include AIDS and West Nile fever, the latter of which first reached North America in 1999 (in New York) and has since spread to other regions of the United States.

The distributions of species also may change in response to relatively gradual climate change or losses or gains of suitable habitat. Let's explore each of these two factors in more detail.

Climate change can alter the distributions of species

A region's climate sets limits on the species that can live there (see Chapter 39). The climate of Earth, however, is not constant. What we experience today as "normal" climate actually is warmer than what has been typical for the past 400,000 years. Over even longer periods of time, the climate of North America has changed drastically (Figure 40.8). As the climate has changed, the plant and animal species that live in North America have also

changed. Fossil evidence indicates that 35 million years ago, the areas of southwestern North America that are now deserts were covered with tropical forests. The rapid global warming that is resulting from human activities is changing the distributions of species today, as we will see in Chapter 45.

Loss of habitat reduces the range of a species

The distributions of species change naturally with losses or gains of suitable habitat over time. Humans, however, are now destroying or degrading the habitat of other species at a rate that far exceeds what occurs naturally (Figure 40.9; see also Chapter 45). For any species, as its habitat is lost, its range is reduced. Species that are adapted to living in only one habitat type are hit particularly hard; such species can decline dramatically or go extinct when their habitat is reduced or destroyed. Species of desert pupfish, for example, are in danger of extinction throughout the southwestern United States because their unique habitat—salty desert ponds—is being degraded and destroyed.

Desert pupfish

■ The distribution of a species can increase when the species migrates or is introduced by humans to a new geographic region. The distribution of a species also can change in response to gradual changes in climate or to rapid changes caused by humans, such as the degradation or destruction of natural habitat.

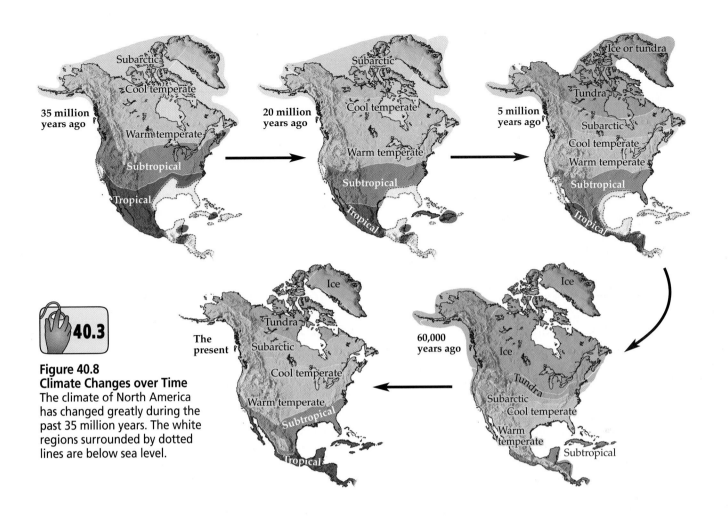

40.3

Figure 40.8
Climate Changes over Time
The climate of North America
has changed greatly during the
past 35 million years. The white
regions surrounded by dotted
lines are below sea level.

A stand of old-growth forest
in the Pacific northwest.

Old-growth
forest is
shown in red.

1620

Each dot
represents
25,000 acres
of old-growth
forest.

1920

Figure 40.9   The Destruction of Natural Habitat
From 1620 to 1920, vast regions of old-growth (original) forest in the United States were cut down for lumber and to
make room for agriculture, housing, and industry.

Using Knowledge about the Distributions of Species

Knowledge about the distributions of species has many practical applications. If biologists understand what makes a certain habitat suitable for a rare plant or for a species of economic importance, they can use that information to ensure that the habitat remains suitable for the species of interest. For example, people burn prairies periodically to increase the performance of rare prairie plants and decrease the performance of nonnative trees and shrubs. Similarly, knowing why pest species live where they do can guide efforts to control them, thereby saving considerable time and money.

Information about why species live where they do also can be used to provide early warning of a wide range of environmental problems. For example, many of the cave-dwelling aquatic organisms described at the beginning of this chapter have very specific habitat requirements, and they also are known to be highly sensitive to pollutants from agricultural and industrial sources. As a result, declines in the numbers of these organisms can provide an early indication that pollutants are entering groundwater supplies. In this and many similar examples, organisms with specific habitat requirements can serve as "biological indicators" for the quality of their habitat, thereby providing information that otherwise would be very hard to obtain.

As humans alter natural habitats throughout the world, it is becoming increasingly urgent that we improve our understanding of why organisms live where they do. For example, to understand how many acres of old-growth forest are required for populations of spotted owls to survive in the northwestern United States, biologists must know what makes that habitat suitable for the owl and how far the owl can disperse (see Chapter 41). Since the presence of suitable habitat and dispersal are key factors in determining the distribution of the spotted owl, biologists must know why the owl lives where it does if they are to improve its chance for survival.

Spotted owl

Finally, knowledge about why organisms live where they do can have far-reaching consequences. In the case of the spotted owl, such knowledge has influenced political decisions about how much logging can be allowed in the remaining stands of the old-growth forest that the owls need to survive. In general, if we are to make accurate predictions about the effects of humans on natural environments—thus enabling us to make intelligent decisions about what actions we should and should not take—we must understand why species live where they do.

■ Knowledge about why organisms live where they do can be used to protect rare species and species of economic importance; such knowledge also can be used to control pest species and to provide early warning of environmental problems. To understand the effects we are having on the natural environment, we must understand why organisms live where they do.

■ HIGHLIGHT

Coping with Life in Caves

Imagine the dilemma facing fish living in a region where the water table is falling rapidly, causing streams to run dry. Such fish depend absolutely on a resource that is disappearing. Matters would seem bleak indeed for fish in such a situation.

However, a stream that is dry on the surface may still flow underground. Fish sometimes remain in a stream as it runs from the surface to a region below ground, such as a pool in a cave or a system of underground streams. For a fish living in a place where surface waters were drying up, a new and risky existence below ground might offer a far greater chance of survival than would life as usual above ground.

Organisms can reach caves and underground streams by following the water table as it drops, or by accident. Undoubtedly, many organisms that end up in extreme environments such as caves fail to survive. But sometimes such colonists survive to start a new population. How do members of such a population cope with their new and challenging environment?

Initially, life might pose quite a struggle for the colonists and their immediate descendants. But, as we learned in Unit 4, populations often evolve in response to changing environmental conditions. Consider, for example, the blind cave-dwelling fish *Astyanax fasciatus*. All fish have lateral line systems, which consist of pressure-sensitive cells that provide information about the movements of other fish in the water. Compared with members of the same species that still live above ground and still rely on vision for navigation, cave-dwelling *Astyanax fasciatus* have evolved an extremely well developed lateral line system that allows them to move through the water without bumping into objects or one another, despite the fact that they have no eyes.

What is true of *Astyanax fasciatus* is true of many species that live in extreme environments: Such species possess special adaptations that allow them to thrive in an otherwise challenging environment. In fact, populations in extreme environments can evolve to the point at which they are well suited for their new home, but ill suited for life in the environment of their ancestors. Should that happen, the environment of their ancestors may actually be far more "extreme" to them than the place where they now live.

■ Populations that live in extreme environments evolve over time. As a result, they often possess special adaptations that allow them to thrive in an otherwise challenging environment.

SUMMARY

The Distributions of Species: Where Organisms Live

■ Some species have small ranges; others are found throughout one or several continents.

■ Few species are found on most or all of the world's continents.

■ The habitat of a species has a characteristic set of environmental conditions, which include both physical and biotic components of the environment.

Species Have Patchy Distributions

■ Over large geographic areas, all organisms have patchy distributions.

■ The patchy distributions of organisms often match the patchy distributions of underlying physical conditions.

■ Organisms may have patchy distributions for biological reasons.

Factors That Determine the Distributions of Species

■ The distributions of organisms are strongly influenced by history, including where the organisms evolved and continental drift.

■ Dispersal and the presence of suitable habitat have a great impact on where organisms live.

■ The presence of suitable habitat for an organism is determined by the physical environment, the biotic environment, interaction between the physical and the biotic environments, and disturbance.

Changes in the Distributions of Species

■ The distribution of a species can increase when the species migrates on its own or is introduced by humans to a new geographic region.

■ The distribution of a species can change in response to gradual changes in climate or to rapid changes caused by humans, such as the degradation or destruction of natural habitat.

Using Knowledge about the Distributions of Species

■ Knowledge about why organisms live where they do can be used to control pest species, protect rare species and species of economic importance, and provide early warning for a wide range of environmental problems.

■ To understand the effect of humans on the natural environment, we must understand why organisms live where they do.

Highlight: Coping with Life in Caves

■ Populations that live in caves or other extreme environments evolve over time. As a result, they possess special adaptations that allow them to thrive in an otherwise challenging environment.

KEY TERMS

clumped distribution p. 669   habitat p. 668
dispersal p. 673   random distribution p. 669
distribution p. 668   range p. 668
disturbance p. 673   regular distribution p. 669

CHAPTER REVIEW

Self-Quiz

1. The distribution of saguaro cacti shows how _____ can limit the distribution of an organism.
   a. temperature
   b. predation
   c. sunlight
   d. water

2. Plant seeds usually do not travel far. The short distances that seeds travel tend to cause offspring to have a _____ distribution.
   a. random
   b. nearly regular
   c. clumped
   d. proximate

3. If you were to go back in time 100 million years and then briefly examine what Knoxville, Tennessee, was like every 10 million years, what would you find?
   a. Climate and organisms changed little over time.
   b. The climate changed, but the organisms did not.
   c. The organisms changed, but the climate did not.
   d. Both climate and organisms changed over time.

4. Why don't polar bears live in Antarctica?
   a. There is no food for them in Antarctica.
   b. They evolved in the Arctic and have not dispersed to Antarctica.
   c. There are not enough ice packs in Antarctica.
   d. They used to live in Antarctica, but they went extinct.

5. Features of the environment that can make a habitat suitable for a species include
   a. temperature
   b. precipitation
   c. disturbance (such as fire)
   d. all of the above

Review Questions

1. Describe the factors that make habitat suitable for a species.

2. Explain why species are not distributed evenly throughout the environment.

3. Where a species evolved can influence its distribution. Do you think that *when* a species evolved might also be important? Relate your answer to the effect of continental drift and dispersal on the distributions of species.

4. How would you determine whether dispersal, the presence of suitable habitat, or history was most important in limiting the distribution of a particular species?

5. Describe the changes that have occurred in the distribution of tundra and tropical forest in North America over the last 35 million years.

The Daily Globe

40

Are Species Trying to Outrun Global Warming?

A number of years ago scientists warned that global warming might force plant and animal species to migrate north or up the sides of mountains in search of cooler conditions. New results show that these predicted migrations may have begun: Dozens of plant and animal species have migrated north or upward in elevation, in some cases at rates too slow to keep up with the changing climate.

The list of species on the move includes 19 species of butterflies and 55 species of birds. Land plants and a variety of sea creatures also seem to be feeling the heat. On land, 10 plant species have migrated up the sides of the European Alps; unfortunately, these plants are migrating at less than half the speed needed for them to keep up with climate change. Off the coast of California, reef fish and the microscopic plankton on which many marine organisms feed also have moved to new locations. Overall, the new studies show that many species have begun to migrate, seemingly in response to global warming.

What can we expect from future movements of species as the climate continues to warm? Species that move quickly should be fine, but species that migrate slowly may go extinct if global warming outstrips their ability to find a cooler place to live.

Trouble may be brewing for people as well. Like other organisms, insects that carry human diseases are on the move. Recent increases in temperature have allowed the mosquitoes that transmit malaria, dengue fever, and yellow fever to thrive at higher elevations than they once could. As these insects spread to new places, they bring the diseases they carry with them. In the words of Dr. Janet Malcolm of the United Health Organization, "If the planet continues to warm at current rates, many diseases will spread to new locations, causing untold human suffering."

Evaluating "The News"

1. Arctic species such as the polar bear may be at particular risk from global warming: As the climate warms, they may not have the option of moving farther north. Does it matter whether such species go extinct? If not, why not? If so, what actions would you be willing to take to prevent their extinction?

2. If human actions cause global climate change (see Chapter 45), do we have an ethical responsibility to help affected species find new places to live? Why or why not?

3. The most severe effects of global warming may occur in countries that are not major sources of the atmospheric pollution that causes it. Who should bear the costs associated with global warming? The country in which a particular effect occurs, or the countries that are most responsible for the pollution that causes global warming?

41 Growth of Populations

Rosamond Purcell, *European Moles*, 1992.

The Tragedy of Easter Island

*I*magine standing at the edge of a cliff on Easter Island, looking into the long-abandoned quarry of Rano Raraku. Scattered about the grassy slopes of the quarry lie hundreds of huge, eerie statues carved from stone hundreds of years ago. The scene is beautiful, yet also ghostly and disturbing. Some of the statues stand upright but unfinished; they look almost as if the artists dropped their tools in midstroke. Others are complete, but lie fallen at odd angles. Hundreds more statues are scattered

along the coast of Easter Island. Who carved these statues? Why were so many left unfinished? What happened to the people who made them?

The mystery deepens when we consider where Easter Island is and what it looks like today. Extremely isolated, the island is a small, barren grassland with little water and little potential for agriculture. How could such a remote and forbidding place support a civilization capable of carving, moving, and maintaining these enormous stone statues?

No population can continue to increase in size for an indefinite period of time.

The answers to these questions provide a scary lesson for people today. Easter Island was not always a barren grassland; at one time most of the island was covered by forest. According to archaeological evidence, no humans lived on the island until about AD 400. At that time, about 50 Polynesians arrived in large canoes, bringing with them crops and animals with which to support themselves. These people developed a well-organized society capable of sophisticated technological feats,

such as moving 15- to 20-ton stone statues long distances without the aid of wheels (they rolled the statues on logs).

By the year 1500, the population had grown to about 7000 people. By this time, however, virtually all the trees on the island had been cut down to clear land for agriculture and to provide the logs used to roll the statues from one place to another. The cutting of trees and other forms of environmental destruction caused increased soil erosion and decreased crop production, leading to mass starvation.

With no large trees remaining on the island, the people could not build canoes to escape the ever-worsening conditions. The society collapsed and sank into warfare, cannibalism, and living in caves for protection. The population crashed, and even 400 years later (in 1900), there were only 2000 people on the island, less than one-third the number that had lived there in 1500.

What caused the events on Easter Island? In essence, the number of people and the patterns of resource use on the island increased above the level that the land could support. And that's the scary part: Many scientists now think that Earth's human population is at or near the level the planet can support. Is the story of Easter Island a sneak preview of what will happen to the whole human race?

Easter Island
Located 3700 kilometers west of the coast of Chile and 4200 kilometers east of the nearest major population center in Polynesia (Tahiti), Easter Island is small (166 square kilometers) and extremely isolated.

Easter Island

■ KEY CONCEPTS

1. A population is a group of interacting individuals of a single species located within a particular area.

2. Populations increase in size when birth and immigration rates exceed death and emigration rates, and they decrease when the reverse is true.

3. A population that increases by a constant proportion from one generation to the next exhibits exponential growth.

4. Eventually, the growth of populations is limited by factors such as lack of space, food shortages, predators, and disease.

5. The world's human population is increasing exponentially. Rapid human population growth cannot continue indefinitely; either we will limit our own growth, or the environment will do it for us.

Ecology is the study of interactions between organisms and their environment. Two important questions ecologists ask are, Where do organisms live (and why)? and how many organisms live there (and why)? The answers to such questions not only provide insight into the natural world, but are essential for the solution of real-world problems, such as the protection of rare species or the control of pest species. We addressed the first question in Chapter 40, where we described the distribution of organisms. In this chapter we address the second question as we focus on the abundance of organisms, emphasizing the growth of populations over time. We begin by defining what populations are; then we describe examples of how populations grow over time. Finally, we consider the limits to growth that are faced by all populations, including the human population.

What Are Populations?

A **population** is a group of interacting individuals of a single species located within a particular area. This definition sounds simple to use, and in some cases it is. The human population on Easter Island, for example, consists of all the people who live on the island.

Ecologists usually describe the number of individuals in a population by the **population size** (the total number of individuals in the population) or by the **population density** (the number of individuals per unit of area). To return to the Easter Island example, in the year 1500, the population size was 7000 people and the population density was 42 people per square kilometer (7000/166 = 42, where 166 square kilometers is the area of the island)—a higher density than the 30 people per square kilometer that lived in the United States in 1999.

The Easter Island example is an easy one: Islands have well-defined boundaries, and human individuals are relatively easy to count. But often it is more diffi-

cult to determine the size or density of a population. For example, suppose a farmer wants to know whether the aphid population that damages his or her crops is increasing or decreasing (Figure 41.1). Aphids are small and hard to count. More importantly, it is not obvious how the aphid population should be defined. What do we mean by "a particular area" in this case? Aphids can produce winged forms that can fly considerable distances, so should only the aphids in the farmer's field be counted? What about the aphids in the next field over?

In general, what constitutes a population often is not as clear-cut as in the Easter Island example. Overall, the area appropriate for defining a particular population depends on the questions being asked and on aspects of the biology of the organism of interest, such as how far and how rapidly the organism moves.

> ■ A population is a group of interacting individuals of a single species located within a particular area. What constitutes an appropriate area depends on the questions of interest and the biology of the organism under study.

41.1 Changes in Population Size

All populations change in size over time—sometimes increasing, sometimes decreasing. In one year, abundant rainfall and plant growth may cause mouse populations to increase; in the next, drought and food shortages may cause mouse populations to decrease dramatically. Such changes in the population sizes of other organisms can have important consequences for people. For example, an increase in the number of deer mice, carriers of the deadly han-

Deer mouse

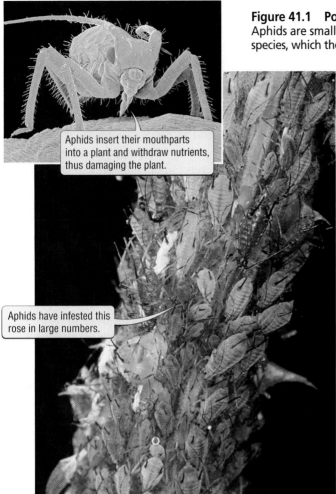

Figure 41.1 Populations of Aphids Can Cause Extensive Crop Damage
Aphids are small insects with sucking mouthparts. They are pests on many plant species, which they infest in large numbers.

Aphids insert their mouthparts into a plant and withdraw nutrients, thus damaging the plant.

Aphids have infested this rose in large numbers.

increases by a constant proportion from one generation to the next (Figure 41.2). When a population grows exponentially, the proportional increase is constant, but the numerical increase becomes larger each generation. For example, in terms of its proportional increase, the population in Figure 41.2, doubles in every generation. With respect to its numerical increase, however, the population increases by only 1 individual between generations 1 and 2, but by 16 individuals between generations 5 and 6. When plotted on a graph, an exponential growth pattern forms a J-shaped curve.

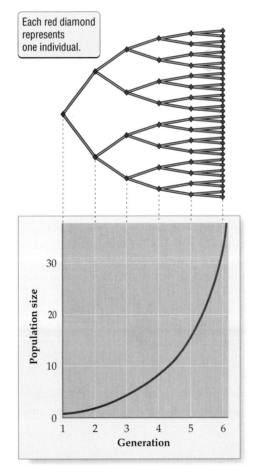

Each red diamond represents one individual.

Figure 41.2 Exponential Growth
In this hypothetical population, each individual produces two offspring, so the population increases by a constant proportion each generation (it doubles). The number of individuals added to the population increases each generation, causing the J-shaped curve that is characteristic of exponential growth.

tavirus, is thought to have been responsible for the 1993 outbreak of hantavirus in the southwestern United States.

Whether a population increases or decreases in size depends on the number of births and deaths in the population, as well as on the number of individuals that immigrate to (enter) or emigrate from (leave) the population. A population increases in size whenever the number of births and immigrations is greater than the number of deaths and emigrations. The environment plays a key role in the increase or decrease of a population because birth, death, immigration, and emigration rates all depend on the environment.

41.2 *Exponential growth*

Many organisms produce vast numbers of young, and if even a small fraction of those young survive to reproduce, the population can grow extremely rapidly. **Exponential growth** is an important type of rapid population growth that occurs when a population

The time it takes a population to double in size—the **doubling time**—can be used as a measure of how fast the population grows. It is nice when our bank accounts double rapidly, but when populations grow exponentially in nature, problems eventually result, as we will see in the following sections.

Exponential growth in an introduced species

It is not uncommon for populations to increase exponentially, at least initially, when they migrate to, or are introduced to, a new area. Consider the following tale of woe: In 1839, a rancher in Australia imported a species of *Opuntia* (prickly pear) cactus and used it as "living fence" (a thick wall of this cactus is nearly impossible for human or beast to cross). Unlike a real fence, however, the *Opuntia* cactus did not stay in one place: It spread rapidly throughout the landscape. As the cactus spread, whole fields were turned into "fence," crowding out cattle and destroying good rangeland.

In about 90 years, *Opuntia* cacti spread across eastern Australia, covering more than 243,000 square kilometers and causing great economic damage. All attempts at control failed until 1925, when scientists introduced a moth species, appropriately named *Cactoblastis cactorum*, whose caterpillars feed on the growing tips of the cactus. This moth killed billions of cacti, which successfully brought the cactus population under control (Figure 41.3). The moths' success, however, also caused their own numbers to plummet due to lack of food (both the cactus and the moth still exist in eastern Australia in low numbers). Overall, the number of *Opuntia* individuals

Cactoblastis cactorum moth

increased exponentially at first, then declined even more rapidly after introduction of the moth. Exponential growth has been observed for species other than *Opuntia* introduced by people to new areas, as well as for species that have expanded naturally to new areas.

■ All populations change in size over time. Populations increase when birth and immigration rates are greater than death and emigration rates, and they decrease when the reverse is true. Exponential growth occurs when a population increases by a constant proportion from one generation to the next.

41.3 Populations Cannot Increase without Limits

Two giant puffball mushrooms can produce up to 7000 billion offspring (Figure 41.4). If all of these offspring survived and reproduced, the descendants of two giant puffballs would weigh more than Earth in just two generations. Humans and *Opuntia* cacti have much longer doubling times than giant puffballs, but given enough time, they can also produce an astonishing number of descendants. Obviously, however, Earth is not covered

Figure 41.3 Blasting the Cactus
In Australia, the moth *Cactoblastis cactorum* was used to halt the exponential growth of populations of a nonnative cactus species. (*a*) A dense stand of *Opuntia* cactus 2 months before the release of the moth. (*b*) The same stand 3 years later, after the moth had killed the cacti by feeding on its growing tips.

(*a*)

(*b*)

Figure 41.4   Will They Overrun Earth?
Giant puffball mushrooms have the potential to produce 7000 billion offspring in a single generation. Large-sized giant puffballs weigh 40 to 50 kilograms each (a small one is shown here).

with giant puffballs, *Opuntia* cacti, or even humans. These examples illustrate an important general point: No population can increase in size indefinitely. Limits exist.

The reasons why populations cannot continue to increase are simple. Imagine that a few bacteria are placed in a closed jar that contains a source of food. The bacteria absorb the food and then divide, and their offspring do the same. The population of bacteria grows exponentially, and in short order there are billions of bacteria in the jar. Eventually, however, the food runs out and metabolic wastes build up. All the bacteria die.

This example may seem extreme because it involves a closed system: No new food is added, and the bacteria and the metabolic wastes cannot go anywhere. In many respects, though,

Figure 41.5   Carrying Capacity
A laboratory population of the single-celled protist *Paramecium caudatum* increases rapidly at first, then stabilizes at the maximum population size that can be supported indefinitely by its environment—that is, the carrying capacity. This growth pattern can be graphed as an S-shaped curve.

the real world is similar to a closed system: Space and nutrients, for example, exist in limited amounts. In the *Opuntia* example of the previous section, even if humans had not introduced the *Cactoblastis* moth, the cactus population could not have sustained exponential growth indefinitely. Eventually, the growth of the cactus population would have been limited by an environmental factor, such as a lack of suitable habitat.

The growth pattern of some populations can be represented by an S-shaped curve. Such populations grow rapidly at first, but then stabilize at the maximum population size that can be supported indefinitely by their environment, a level that is known as the **carrying capacity** (Figure 41.5). The growth rate of the population decreases as the population size nears the carrying capacity because resources such as food or water begin to be in short supply. At the carrying capacity, the population growth rate is zero.

In the 1930s, the Russian ecologist G. F. Gause found that laboratory populations of the protist *Paramecium* increased to a certain population size and remained there (see Figure 41.5). In these experiments, Gause added new nutrients to the protists' liquid medium at a steady rate

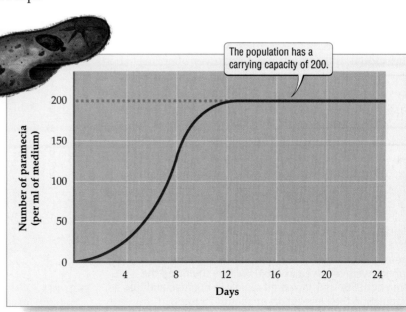

The population has a carrying capacity of 200.

and removed the old solution at a steady rate. At first, the *Paramecium* population increased rapidly in size. But as the population continued to increase, the paramecia used nutrients so rapidly that food began to be in short supply, slowing the growth of the population. Eventually the birth and death rates of the protists equaled each other, and the population size stabilized.

When birth and death rates are balanced, a population stops growing and stabilizes at the carrying capacity. Unlike a natural system, there was no immigration or emigration in Gause's experiments. In natural systems, populations reach and remain at a constant population size when

birth + immigration = death + emigration

for extended periods of time.

> ■ Because the environment contains a limited amount of space and resources, no population can continue to increase in size indefinitely. Some populations increase rapidly at first, then stabilize at the carrying capacity.

Factors that Limit Population Growth

Natural populations, like the laboratory populations of bacteria and *Paramecium* described in the preceding section, experience limits to their population growth (Figure 41.6). Their growth can be held in check by factors such as food shortages, lack of space, diseases, predators, habitat deterioration, weather, and natural disturbances. Let's consider how such factors limit the growth of natural populations.

Any area contains a limited amount of food and other essential resources. Thus, as the number of individuals in a population increases, there are fewer resources per individual. Individuals produce fewer offspring when they have fewer resources, causing the growth rate of the population to decrease. Similarly, when there are large numbers of individuals in a population, disease spreads more rapidly, and predators may pose a greater risk (since many predators prefer to hunt abundant sources of food) (see Chapter 32). Overall, when there are more individuals in a population, birth rates may

drop or death rates may increase, and either effect may limit the growth of the population.

Large populations also can cause habitat deterioration. For example, if a population exceeds the carrying capacity of its environment, it may damage that environment so severely that the population may decrease rapidly and the carrying capacity may be lowered (Figure 41.7). A lowered carrying capacity means that the habitat cannot support as many individuals as it once could.

The factors discussed so far in this section act more strongly as population size increases. When the number of individuals in a population increases, their density also increases (since density is the number of individuals in a population divided by the area in which the population lives). Such factors that limit the growth of a population more strongly as the density of the population increases are said to be **density-dependent**.

In other cases, populations are held in check by factors that are not affected by the density of the population, such as weather; such factors are said to be **density-independent**. Density-independent factors often prevent populations from reaching high densities in the first place. Year-to-year variation in weather, for example, may cause conditions to be suitable for rapid population growth only occasionally. Poor weather conditions may reduce the growth of a population directly (by

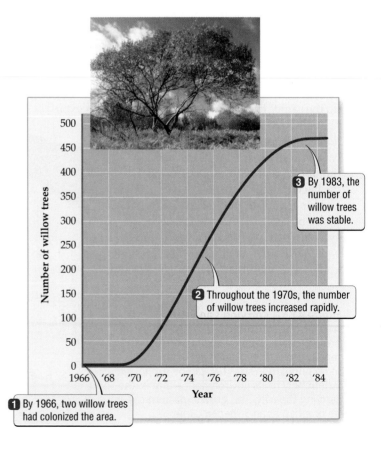

Figure 41.6 An S-Shaped Curve in a Natural Population
At a site in Australia, rabbits heavily grazed young willow trees, preventing willows from colonizing the area. The rabbits were removed in 1954. Willows colonized the area by 1966. They increased rapidly in number, and the population then leveled off at about 475 trees.

1 By 1966, two willow trees had colonized the area.

2 Throughout the 1970s, the number of willow trees increased rapidly.

3 By 1983, the number of willow trees was stable.

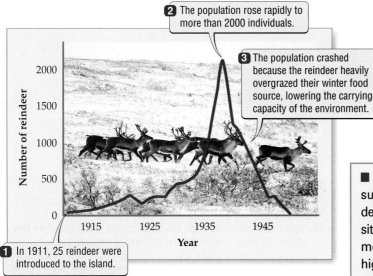

2 The population rose rapidly to more than 2000 individuals.

3 The population crashed because the reindeer heavily overgrazed their winter food source, lowering the carrying capacity of the environment.

1 In 1911, 25 reindeer were introduced to the island.

Figure 41.7   Boom and Bust
A reindeer population increased rapidly after it was introduced in 1911 to Saint Paul Island off the coast of Alaska. When the reindeer overgrazed the island, their population crashed. In 1950, only eight reindeer remained.

■ The growth of populations can be limited by factors such as food shortages, disease, predators, habitat deterioration, weather, and natural disturbances. Density-dependent factors limit the growth of a population more strongly when the density of the population is high. Density-independent factors limit the growth of populations without regard to their density.

freezing the eggs of an insect, for example) or indirectly (by decreasing the number of plants available as food to that insect). Natural disturbances such as fires and floods (see Chapter 40) also limit the growth of populations in a density-independent way. Finally, the effects of environmental pollutants such as DDT are density-independent; such pollutants can threaten natural populations with extinction (see the box below).

41.4 *Patterns of Population Growth*

Different populations may exhibit a number of different growth patterns over time. Under favorable conditions, the population size of any species increases rapidly. An initial period of rapid population growth can be seen in both J-shaped (see Figure 41.2)

■ THE SCIENTIFIC PROCESS

Back from the Brink of Extinction

Before being banned in 1972, the synthetic pesticide DDT was used widely in the United States for about 30 years. Since this chemical does not occur naturally, its use introduced an entirely new substance to ecosystems. The presence of DDT in ecosystems had undesirable and unanticipated effects. For example, birds with high levels of DDT in their bodies produced such fragile eggshells that they could not reproduce.

The bald eagle was one of many birds in the United States affected by DDT poisoning. By the early 1960s, only 400 breeding pairs of bald eagles remained in the lower 48 states, a huge drop from the estimated 100,000 breeding pairs present in 1800. The effect of DDT on these birds prompted a ban on its use within the United States. This ban has given populations of bald eagles and other birds, such as the peregrine falcon, the opportunity to bounce back from a perilous brush with extinction.

Today, there are more than 5800 breeding pairs of bald eagles in the lower 48 states. Although their numbers are still low, bald eagles represent one of many environmental success stories. When people recognize a problem and take decisive measures to fix it, as with the ban on DDT use, wild populations can often recover from the negative effects of human actions.

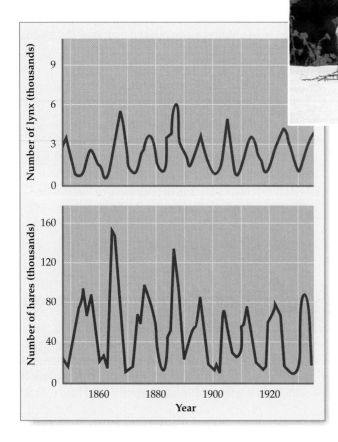

Figure 41.8 Population Cycles
In some cases, populations of more than one species cycle together. The Canada lynx depends on the snowshoe hare for food, so the number of lynx is strongly influenced by the number of hares. Experiments indicate that hare populations are limited by their food supply and by their lynx predators. (Numbers of lynx and hares were estimated from the number of furs sold by trappers to the Hudson's Bay Company, Canada.)

and S-shaped (see Figure 41.6) growth patterns. For a population with a J-shaped curve, rapid population growth may continue until resources are overused, causing the population size to drop dramatically (see Figure 41.7). In contrast, for a population with an S-shaped curve, the rate of population growth slows as the population size nears the carrying capacity. Predators, disease, and other factors may then keep the population near the carrying capacity for long periods of time.

As we have seen, populations change in size over time, increasing at some times and decreasing at others. Even populations that have an S-shaped growth pattern do not remain indefinitely at a single, stable population size; instead, they fluctuate slightly over time, yet remain close to the carrying capacity.

In some cases, the population sizes of two species change together in a tightly linked cycle (Figure 41.8). Populations of two species can cycle together when at least one of the species is very strongly influenced by the other. The Canada lynx, for example, depends on the snowshoe hare for food, so lynx populations increase

when hare populations increase and decrease when hare populations drop (Figure 41.8).

There are relatively few examples from nature in which the abundance of two species shows regular cycles like those of the hare and lynx. However, the populations of most species do rise and fall over time—just not as regularly as in Figure 41.8. Irregular fluctuations are far more common in nature than is a smooth rise to a stable maximum population size, as shown in Figure 41.6.

Finally, different populations of the same species may experience different patterns of growth. Understanding the reasons for these differences can provide critical information on how best to manage endangered or economically important species. For example, when forest managers needed to decide where and how much (if any) old-growth forest could be cut without harming the rare spotted owl, they gathered data on the birth rate of the owl, and on how much habitat each individual used, in different populations. They used these data to predict how the growth of spotted owl populations would be affected by the number and location of patches of the bird's preferred habitat, old-growth forest (Figure 41.9).

■ Different populations can exhibit different patterns of growth over time, including J-shaped curves, S-shaped curves, tightly linked cycles, and irregular fluctuations. Understanding why different populations have different patterns of growth can provide critical information on how best to manage endangered or economically important species.

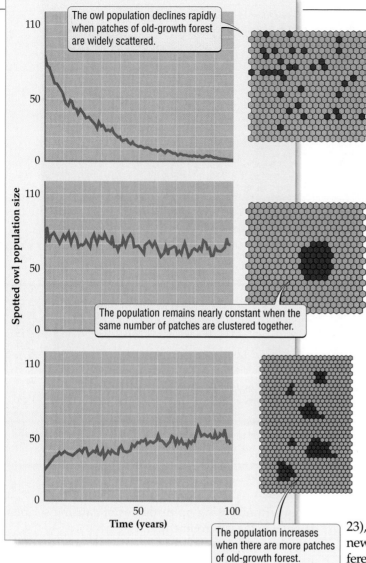

Figure 41.9   Same Species, Different Outcomes
Different populations of the endangered spotted owl are predicted to show different patterns of growth over time, depending on the arrangement and amount of their preferred habitat, old-growth forest.

41.5   *Human Population Growth: Beyond the Limits?*

The human population is growing today at a spectacular rate (Figure 41.10). It took more than 100,000 years for our population to reach 1 billion people, but now it increases by a billion people every 13 years. Our use of resources and our overall impact on the planet has increased even faster than our population size. From 1860 to 1991, the human population increased fourfold, but our energy consumption increased 93-fold.

The global human population passed the 6.2 billion mark in 2001. At present, the world's population increases by about 77 million people each year, or over 8500 people per hour. These numbers are all the more sobering when the following facts are considered:

- More than 1.3 billion people live in absolute poverty.
- 2 billion people lack basic health care or safe drinking water.
- More than 2 billion people have no sanitation services.
- Each year 15 million people, mostly children, die from hunger or hunger-related problems.

By the year 2025, the global human population is projected to increase to 8 billion people. Even if our birth rate dropped immediately from the current 2.8 children per female to a level that ultimately would allow the human population to replace itself, but not increase (about 2.1 children per female), the human population would continue to grow for at least another 60 years. The population will continue to increase long after birth rates drop because a huge number of existing children have not yet had children of their own.

How did the human population increase so rapidly, apparently escaping the limits to population growth described in this chapter? First, as our ancestors emigrated from Africa (see Chapter 23), they encountered and prospered in many kinds of new habitats. Few other species can thrive in places as different as grasslands, coastal environments, tropical forests, deserts, and arctic regions. Second, people increased the carrying capacity of the places where they lived. The development of agriculture, for example, allowed more people to be fed per unit of land area. Similarly, in the last 300 years, death rates have dropped as a result of improvements in medicine, sanitation, and food storage and transportation. However, birth rates did not drop at the same time; hence our population grew. Finally, our heavy use of fossil fuels and nitrogen fertilizers in the twentieth century led to great increases in crop yields.

Viewed broadly, human inventiveness and technology have allowed us to sidestep limits to population growth for some time. Ultimately, however, like all other populations, our population cannot continue to increase without limit.

> ■ The global human population is growing very rapidly, increasing by 1 billion people every 13 years. Resource use per person is also increasing very rapidly.

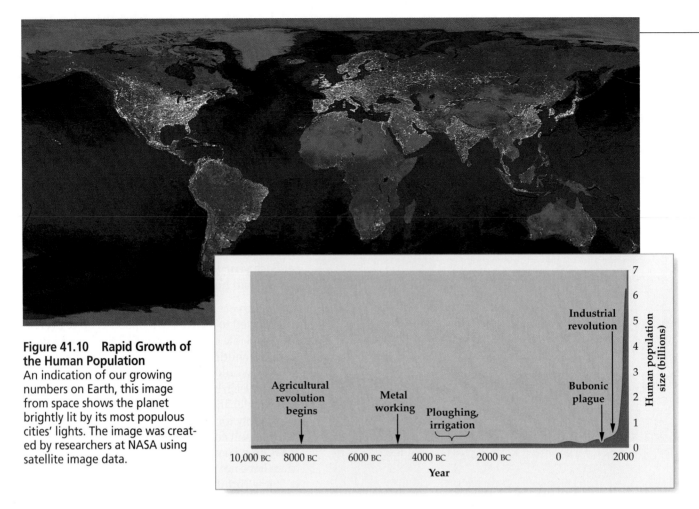

Figure 41.10 Rapid Growth of the Human Population
An indication of our growing numbers on Earth, this image from space shows the planet brightly lit by its most populous cities' lights. The image was created by researchers at NASA using satellite image data.

Chart labels: Agricultural revolution begins · Metal working · Ploughing, irrigation · Industrial revolution · Bubonic plague

Y-axis: Human population size (billions) — 0, 1, 2, 3, 4, 5, 6, 7

X-axis: Year — 10,000 BC, 8000 BC, 6000 BC, 4000 BC, 2000 BC, 0, 2000

■ HIGHLIGHT

What Does the Future Hold?

We began this chapter with a description of the rise and fall of the civilization on Easter Island. At least initially, the people who colonized Easter Island maintained a culturally rich and densely populated society. But their society did not persist. The people of Easter Island temporarily increased the carrying capacity of the island by cutting down the forest to create farm fields. Ultimately, however, cutting down the trees led to environmental deterioration, starvation, and the collapse of their civilization.

As on Easter Island, many of the problems facing humans today relate to population growth and environmental deterioration. More people means more environmental deterioration, which in turn makes it harder to feed the people we already have. Already much of Africa depends on imported food to prevent starvation, and cities in California persist only because of water imported from other states. In addition, our economy and culture, like those of Easter Island, are not based on the sustainable use of resources. The term **sustainable** describes an action or process that can continue indefi-

nitely without using up resources or causing serious damage to the environment.

Let's examine the implications of the nonsustainable use of resources for just one aspect of society, agriculture. In the past, humans responded to the need to feed more people by increasing the amount of land under cultivation (by cutting down forests and by irrigation) and by producing higher yields per unit of farmland. However, the best farmland is already taken, and the land that is easiest to irrigate is already irrigated. On a worldwide basis, the amount of farmland per person and the irrigated area per person are dropping. In addition, for a given amount of farmland, yields of major grains such as wheat are now increasing more slowly than the human population size.

Unless the trends described in the previous paragraph are reversed, there will be less food per person as our population continues to grow. In addition, as more people use more resources, agriculture will face growing challenges due to soil erosion, water shortages, and pollution from pesticides, fertilizers, and energy use. Overall, if we continue to have a nonsustainable impact on our environment, it will be very difficult to produce enough food to feed our growing population.

Will humans limit the growth and impact of our global population, or will the environment do it for us? There are some hopeful signs: The growth rate of the human population has slowed in recent years, and people throughout the world are conscious of the risks of environmental degradation. But much remains to be done. To limit the growth and impact of the human population, we must address the interrelated issues of population growth, poverty, unequal use of resources, environmental deterioration, and sustainable development. It is especially important for people who live in North America, Japan, and Europe to address such issues because people in these regions consume far more than their proportionate share of Earth's resources. For example, given the existing patterns of resource use in India and the United States, it would take more than 11 *billion* people living in India to consume the same amount of copper and petroleum that just 273 million Americans used in 1999.

Our hope for the future—for your future and the future of your children—lies in realistically assessing the problems we face, and then committing ourselves to take bold actions to address those problems (see Chapter 45). In the end, it is up to all of us to help ensure that humankind does not repeat on a grand scale the tragic lessons of Easter Island.

> ■ The rapid increases in human population size and resource use cannot continue indefinitely, since they are based on a nonsustainable use of resources. To prevent the environment from limiting the growth of the human population, people must act to address the interrelated problems of population growth, poverty, unequal use of resources, environmental deterioration, and sustainable development.

SUMMARY

What Are Populations?

- A population is a group of interacting individuals of a single species located within a particular area.

- What constitutes an appropriate area depends on the questions of interest and the biology of the organism under study.

Changes in Population Size

- All populations change in size over time.

- Populations increase when birth and immigration rates are greater than death and emigration rates, and they decrease when the reverse is true.

- A population grows exponentially when it increases by a constant proportion from one generation to the next.

- Populations may initially increase at an exponential rate when organisms are introduced to or migrate to a new area.

Populations Cannot Increase without Limits

- Because the environment contains a limited amount of space and resources, no population can continue to increase in size indefinitely.

- Some populations increase rapidly at first, then stabilize at the carrying capacity, the maximum population size that their environment can support.

Factors that Limit Population Growth

- The growth of populations can be limited by factors such as food shortages, disease, predators, habitat deterioration, weather, and natural disturbances.

- Density-dependent factors, such as food shortages and disease, limit the growth of a population more strongly when the density of the population is high.

- Density-independent factors, such as weather and natural disturbances, limit the growth of populations without regard to their density.

Patterns of Population Growth

- Different populations can exhibit different patterns of growth over time, including J-shaped curves, S-shaped curves, tightly linked cycles, and irregular fluctuations.

- Understanding why different populations have different patterns of growth can provide critical information on how best to manage endangered or economically important species.

Human Population Growth: Beyond the Limits?

- The global human population is growing very rapidly, increasing by 1 billion people every 13 years. Resource use per person is also increasing very rapidly.

Highlight: What Does the Future Hold?

- The rapid increases in human population size and resource use cannot continue indefinitely, since they are based on a nonsustainable use of resources.

- To prevent the environment from limiting the growth of the human population, people must act to address the interrelated problems of population growth, poverty, unequal use of resources, environmental deterioration, and sustainable development.

KEY TERMS

carrying capacity p. 685

density-dependent p. 686

density-independent p. 686

doubling time p. 684

exponential growth p. 683

population p. 682

population density p. 682

population size p. 682

sustainable p. 690

CHAPTER REVIEW

Self-Quiz

1. A population of plants has a density of 12 plants per square meter and covers an area of 100 square meters. What is the population size?
   a. 120
   b. 1200
   c. 12
   d. 0.12

2. A population that is growing exponentially increases
   a. by the same number of individuals each generation.
   b. by a constant proportion each generation.
   c. in some years and decreases in other years.
   d. none of the above

3. In a population that has an S-shaped growth curve, after an initial period of increase, the number of individuals
   a. continues to increase.
   b. drops rapidly.
   c. remains near the carrying capacity.
   d. cycles regularly.

4. The growth of populations can be limited by
   a. natural disturbances.
   b. weather.
   c. food shortages.
   d. all of the above

5. Factors that limit the growth of populations more strongly at high densities are said to be
   a. density-dependent.
   b. density-independent.
   c. exponential factors.
   d. sustainable.

Review Questions

1. Assume that a population grows exponentially, increasing by a constant proportion of 1.5 per year. Thus, if the population initially contains 100 individuals, it will contain 150 individuals in the next year. Graph the number of individuals in the population versus time for the next 5 years, starting with 150 individuals in the population.

2. a. What factors prevent populations from continuing to increase without limit?

   b. Why is it common for populations of species that enter a new region to grow exponentially for a period of time?

3. Describe the difference between density-dependent and density-independent factors that limit population growth. Give two examples of each.

4. Different populations of a species can have different patterns of population growth. Explain how an understanding of the causes of these different patterns can help managers protect rare species or control pest species.

5. List five specific actions that you can take to limit the growth or impact of the human population.

The Daily Globe

New Hope for Fighting Outbreaks of Deadly Disease

ATLANTA, GA—The huge floods of 1998, brought about by the strongest El Niño event of the twentieth century, not only caused a great amount of direct damage, but also created ideal conditions for disease outbreaks. Worldwide, waterborne diseases such as cholera and diseases spread by mosquitoes, such as malaria, dengue fever, and Rift Valley fever, hit hard, killing thousands of people.

Scientists at the Centers for Disease Control (CDC) say they have now learned enough about the ecology of Rift Valley fever, a disease that affects both cattle and people, to successfully predict outbreaks of this deadly disease. Like other studies in the young, growing field of ecological epidemiology, this newly published research provides a beacon of hope to humankind: We may soon be able to predict disease outbreaks months in advance, giving public health officials time to nip them in the bud.

The CDC scientists used historical data on sea surface temperatures, the amount of plant growth, and mosquito population sizes to predict the conditions under which mosquito populations would be likely to skyrocket. Armed with the ability to predict mosquito population increases, the scientists tested their approach and found that it successfully predicted past outbreaks of Rift Valley fever. The new approach could provide up to 5 months' advance warning, plenty of time for health officials to control mosquito populations and hence limit the spread of the disease.

As Dr. Roy West of the CDC put it, "Mosquitoes spread the disease from cattle to people, so we needed to be able to predict when mosquito populations would rise to dangerous levels. This new approach lets us do just that."

Evaluating "The News"

1. So far, the new method for predicting outbreaks of Rift Valley fever has relied on historical data on sea surface temperatures, plant growth, and disease outbreaks. What should scientists and health officials do next?

2. As the world's human population grows, why might disease outbreaks become increasingly common?

3. The control of diseases such as Rift Valley fever relies on knowledge about the ecology of populations and the effects of air and water currents on the biosphere. Is it more important to use public funds to deepen our understanding of basic ecology or to seek ways to predict and control specific diseases?

chapter 42 *Interactions among Organisms*

Melissa Miller, *Zebras and Hyenas*, 1985.

Lumbering Mantises and Gruesome Parasites

Praying mantises have been seen to walk to the edge of a river, throw themselves in, and drown shortly thereafter. If they are rescued from the water, they immediately throw themselves back in. What causes mantises to do this?

This bizarre behavior appears to be driven not by the mantises themselves, but by a parasitic worm. Less than a minute after the praying mantis lands in the water, a worm emerges from its anus. This worm attacks and infects terrestrial insects, such as praying mantises, but it also depends on an aquatic host for part of its life cycle. The worm has performed a neat trick: It has evolved the ability to cause its insect host to jump into the river, an act that kills the insect but increases the chance that the worm will eventually reach its aquatic host.

Moving from the bizarre to the gruesome, examine the accompanying photographs. The fungus that killed the ant first grew throughout its entire body, dissolving portions of its body and using them for food. Eventually, the fungus sprouted reproductive structures (indicated by arrows), which allowed it to spread and attack other ants. Fungi attack many other species, including crops such as the corn plant shown. Finally, the human example illustrates an important point: Hundreds of millions of people are disabled every year by protist, fungal, and animal attackers, each of which has unique and harmful effects on the human body.

Parasites, such as the worms that plague praying mantises and the fungi that riddle the bodies of ants, are organisms that live in or on other organisms (known as their hosts). They obtain nutrients from their hosts, often causing them harm, but not

■ MAIN MESSAGE

Interactions among organisms play a key role in determining the distribution and abundance of organisms.

immediate death. The effects of parasites on their hosts illustrate one important type of interaction among organisms: a relationship in which one species benefits and the other is harmed.

In this chapter we discover how ecological interactions such as parasitism help determine the distribution and abundance of organisms. At the end of the chapter, we return to parasites that alter their hosts' behavior, some of them in even more specific ways than those we have just seen.

Parasitic Relationships
(*a*) An ant has been killed by a fungus. (*b*) An ear of corn has been destroyed by a fungus known as corn smut. (*c*) A person has been infected with a protist that attacks the skin.

(*a*)

(*b*)

(*c*)

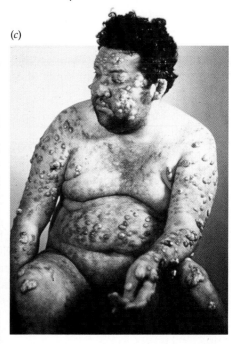

■ KEY CONCEPTS

1. Two species can interact to the benefit of both species. Such mutualisms evolve when the benefits of the interaction outweigh the costs for both species.

2. In consumer–victim interactions, one species in the interaction benefits (the consumer) while the other is harmed (the victim). Victims have evolved elaborate ways of defending themselves against consumers.

3. In competition, two species that share resources have a negative effect on each other. Competition can result in the evolution of greater differences between species.

4. Mutualisms, consumer–victim interactions, and competition help determine where organisms live and how abundant they are.

5. Interactions among organisms have a large effect on communities and ecosystems.

As we have seen throughout this unit, the focus of ecology is on interactions between organisms and their environment. An organism's environment includes the other organisms that live there. Thus, the subject of this chapter—interactions among organisms—is central to the very definition of ecology.

We have already seen how important interactions among organisms can be. For example, the climate sets broad limits on where organisms can live, but as we learned in Chapter 40, interactions with other organisms can prevent a species from living everywhere that it otherwise could (see Figure 40.6). Similarly, in Chapter 41, we saw how the moth *Cactoblastis cactorum*, which feeds on the cactus *Opuntia*, caused *Opuntia* populations to crash in Australia. Overall, interactions among organisms have an influence at all the levels at which ecology is studied.

The millions of species on Earth can interact in many different ways. In this chapter we classify interactions among organisms by whether the interaction is beneficial (+) or harmful (−) to each of the interacting species. In addition, we focus on the three most common and most important kinds of ecological interactions:

+/+ interactions, in which both species benefit (mutualisms)

+/− interactions, in which one species benefits and the other is harmed (consumer–victim interactions)

−/− interactions, in which both species are harmed (competition)

As we will see, each of these three types of interactions among organisms plays a key role in determining where organisms live and how abundant they are. We'll also discover how changes to interactions among organisms alter ecological communities.

42.1 Mutualisms

A **mutualism** is an interaction between two species in which both species benefit (a +/+ interaction). Mutualisms are common and important to life on Earth: Many species receive benefits from, and provide benefits to, other species. The benefits a pair of species receives in a mutualism increase the survival and reproduction of the interacting species.

Mutualisms can occur when two or more organisms of different species live together, an association known as **symbiosis**. Insects such as aphids and mealybugs, which feed on the nutrient-poor sap of plants, often have a mutualistic, symbiotic association with bacteria that live within their cells. The bacteria receive food and a home from the insects, and the insects receive nutrients that the bacteria (but not the insects) can synthesize from sugars in the plant sap. Such symbiotic associations can be amazingly complex. Scientists recently discovered that a second species of bacteria lives within the bacteria that live inside of citrus mealybug cells; it is not yet clear whether this second species benefits or harms the bacteria in which they live.

Citrus mealybug

The open question of whether the second species of bacteria benefits or harms its bacterial host illustrates an important point: Although many symbiotic associations benefit both of the organisms involved (and hence are examples of mutualisms), there are also many cases of symbiosis in which one species lives within another and harms it (a +/− interaction). Many parasites spend all or most of their lives within their host, yet they harm rather than benefit their host. In fact, by some estimates, nearly half of all animal species are parasites that harm the organism in which they live.

Types of mutualisms

Many types of mutualisms are found in nature. In **gut inhabitant mutualisms**, organisms such as bacteria that live in an animal's digestive tract receive food from their host and benefit the host by digesting foods such as wood or cellulose that the host otherwise could not use. In **seed dispersal mutualisms**, an animal such as a bird or coyote eats a fruit that contains plant seeds, then later defecates the seeds far from the parent plant. In this section we elaborate on two of the most common types of mutualisms: behavioral mutualisms and pollinator mutualisms.

Mutualisms in which each partner has evolved to alter its behavior to benefit the other species are called **behavioral mutualisms**. The relationship between certain shrimps and a fish called a goby is a good example of a behavioral mutualism (Figure 42.1). Shrimps of the genus *Alpheus* live in an environment that has plenty of food, but little shelter. They dig burrows in which to hide, but they see poorly and so are vulnerable to predators when they leave their burrows to feed. These shrimps have formed a fascinating relationship with gobies: The gobies act as "seeing-eye" fish, warning the shrimps of danger when the shrimps venture out of their burrows to eat. In return, the shrimps share their burrows with the gobies, thus providing the fish with a safe haven.

In **pollinator mutualisms**, an animal such as a honeybee transfers pollen (which contains male reproductive cells) from one plant to the female reproductive organs of another plant of the same species (see Chapter 36). These animals are known as pollinators, and without them the plants could not reproduce. To ensure that pollinators come to them, plants offer a food reward, such as pollen or nectar. Thus, both species benefit from the interaction. Pollina-

Honeybee

tor mutualisms are important in both natural and agricultural ecosystems. For example, the oranges we buy at the supermarket are available only because honeybees pollinated the flowers of orange trees, thus enabling the trees to produce their fruit.

Mutualists are in it for themselves

Although both species in a mutualism benefit from the relationship, what is good for one species may come at a cost to the other. For example, a species may use energy or increase its exposure to predators when it acts to benefit its mutualistic partner. From an evolutionary perspective, mutualisms evolve when the benefits of the interaction outweigh the costs for both species. But mutualisms are not cost-free, and the interests of the two species may be in conflict.

Consider the pollinator mutualism between the yucca plant and the yucca moth. Both the plant and its moth pollinator depend absolutely on each other. A female yucca moth collects pollen from yucca flowers, flies to another group of flowers, and lays her eggs at the base of the pistil (see Figure 36.7) of a newly opened flower. After she has laid her eggs, the female moth climbs up the pistil and deliberately places the pollen she collected earlier onto the stigma of the flower (Figure 42.2). When the moth larvae (singular larva; the immature form of the moth) hatch, they feed on the seeds of the yucca plant. Thus, the moth both pollinates the plant and eats some of its seeds.

In a cost-free situation for the plant, the moth would transport pollen, but would not destroy any of the plant's seeds. In a cost-free situation for the moth, the moth would produce as many larvae as possible, and they would consume many of the plant's seeds. In actuality, an evolutionary compromise has been reached: The moth usually lays only a few eggs per flower, and the plant tolerates the loss of a few of its seeds. Yucca plants have a defense mechanism that helps keep this compromise working: If a moth lays too many eggs in one of the plant's flowers, the plant can selectively abort the seeds in that flower, thereby killing the moth's eggs or larvae.

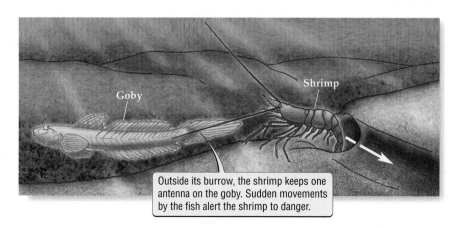

Outside its burrow, the shrimp keeps one antenna on the goby. Sudden movements by the fish alert the shrimp to danger.

Figure 42.1 A Behavioral Mutualism *Alpheus* shrimps build burrows for shelter, which they share with gobies. The fish provide an early-warning system to the nearly blind shrimps when they leave their burrows to feed.

**Figure 42.2
A Pollinator Mutualism**
The yucca and the yucca moth are dependent on each other for survival.

partner is found, the two species strongly influence each other's distribution and abundance. Second, a mutualism can have indirect effects on the distribution and abundance of species that are not part of the mutualism. Coral reefs, for example, are home to many different plant and animal species. The corals that build the reefs depend on their mutualisms with algae, and thus, indirectly, so do the many other species that live in coral reefs.

■ Mutualisms evolve when the benefits of the interaction are greater than its costs for both partners. Mutualisms help determine the distribution and abundance of the mutualist species as well as other species that depend directly or indirectly on the mutualist species.

Mutualisms are everywhere

Mutualisms are very common. Most of the plant species that dominate forests, deserts, grasslands, and other biomes are mutualists. For example, about 80 percent of plant species have mutualistic associations with fungi, called mycorrhizae (see Chapter 3). The fungi help the plant roots absorb nutrients and water from the soil, and the plant provides the fungi with products of photosynthesis.

As mentioned in the previous section, many animal species are involved with plants in pollinator mutualisms. Other examples of mutualisms involving animals include the spectacular reefs found in tropical oceans (see the box on p. 699). These reefs are built by corals (soft-bodied animals), most of which house photosynthetic algae—their mutualistic partners—inside their bodies. The corals provide the algae with a home and several essential nutrients, such as phosphorus, and the algae provide the corals with carbohydrates produced by photosynthesis.

Mutualisms can determine the distribution and abundance of species

Mutualisms can influence the distribution and abundance of organisms in two ways. First, because each species in a mutualism survives and reproduces better where its

42.2 Consumer–Victim Interactions

Consumer–victim interactions are interactions in which one species (the consumer) benefits and the other (the victim) is harmed (+/− interactions). The consumers in such interactions can be classified into four main groups:

1. **Predators** (also called carnivores) are consumers that kill their victims (called prey).
2. **Parasites** are consumers that both feed upon and live in or on their victims (which are called hosts).
3. **Pathogens** are disease-causing organisms.
4. **Herbivores** are consumers that eat plants.

These four major types of +/− interactions are very different from one another. For example, whereas predators (such as wolves) kill their prey immediately, herbivores (such as cows) and parasites (such as fleas) do not (see Chapter 32). Although the four types of consumer–victim interactions have obvious and important differences, in this section we look at some general principles that apply to all four.

■ BIOLOGY IN OUR LIVES
Working to Stop the Cyanide

Coral reefs are among the most stunningly beautiful and biologically diverse natural communities on Earth. Unfortunately, coral reefs are threatened by a wide range of human actions. Clear-cutting of forests on land causes increased erosion and hence silt deposition on the reefs. Industrial pollution and global warming are also having negative effects on reefs worldwide. Coral reefs in some parts of the world face still another lethal threat: exposure to cyanide used to catch fish for the marine aquarium trade.

In the Philippines and Indonesia, the source of 85 percent of the tropical fish that end up in saltwater aquariums, collectors often use cyanide to

stun coral reef fish, making them easier to catch. But some experts estimate that about half of the fish exposed to cyanide die immediately, and another 40 percent of the survivors die before they reach an aquarium. And the cyanide doesn't harm just fish: Cyanide concentrations much lower than those used for cyanide fishing were found to kill 10 different species of corals. In particular, the most important reef-building coral species were extremely sensitive to the effects of cyanide poisoning. Thus, cyanide fishing is putting the long-term health of the entire coral reef community at risk.

Because it is impossible to raise most marine tropical fish in captivi-

ty, the fate of the aquarium trade depends on preserving the reefs. To date, efforts by tropical fish exporting countries to halt cyanide fishing have met with little success. As a result, a different approach has been taken by an international nonprofit organization called the Marine Aquarium Council (MAC). This organization monitors the movement of fish every step of the way from coral reefs to local pet stores, and uses this information to label fish that are collected in an environmentally friendly manner. In addition, MAC teaches fish collectors in the exporting countries how to use hand nets as a viable alternative to cyanide fishing.

The initial results of these efforts are encouraging: Cyanide contamination of fish collected in the Philippines has dropped dramatically. Further progress will depend on the actions of people that buy marine tropical fish: If consumers refuse to buy fish collected by cyanide poisoning, MAC may well succeed in preventing cyanide from continuing to pose a threat to the world's coral reef communities.

The Home a Mutualism Built
The great diversity of life in tropical reefs depends on corals, which form the reefs in and around which many species live. The corals that build the reefs depend on a mutualism with algae.

Consumers can be a strong selective force

As we saw in Chapter 32, the presence of consumers in the environment has caused many species to evolve elaborate strategies to avoid being consumed. Many plants, for example, produce spines and toxic chemicals as defenses against herbivores. In some plants, the production of such defenses is directly stimulated by an

attack by herbivores: An individual cactus that has been partially eaten or grazed is much more likely to produce spines than is an individual that has not been grazed (Figure 42.3).

Many organisms have evolved bright colors or striking patterns that warn potential predators that they are heavily defended, usually by chemical means (Figure

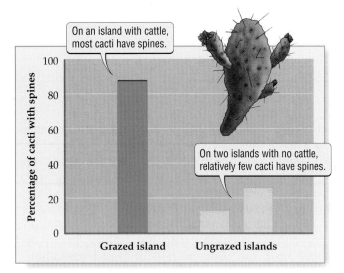

On an island with cattle, most cacti have spines.

On two islands with no cattle, relatively few cacti have spines.

Figure 42.3   A Cactus that Produces Spines when Grazed
On three islands off the coast of Australia, the percentage of cacti with spines is higher on the island that has cattle than on the two islands that do not. Field and laboratory experiments show that grazing by cattle directly stimulates the production of spines in this species of cactus.

42.4*a*). Such warning coloration can be highly effective. Blue jays, for example, quickly learn not to eat monarch butterflies, which are brightly colored and contain chemicals that make the birds very sick (Figure 42.4*b*). Other prey have evolved to avoid predators by being hard to find or hard to catch. And finally, animals have evolved molecular defenses (immune systems) to help them fight off the ravages of disease and parasitic infection. The many ways in which victim species have evolved to protect themselves against consumers indicate that consumers often apply strong selection pressure to their victims.

Consumers can alter the behavior of victims

The bizarre story of the praying mantises that jump to their deaths in rivers at the beginning of this chapter provides a dramatic example of how consumers can alter the behavior of their victims. But consumer–victim interactions can alter the behavior of victim species in more subtle ways as well.

Predators can be a driving force that causes animals to live or feed in groups. In some cases, several prey individuals acting together may be able to prevent predators from attacking (Figure 42.5). Large groups of prey also may be able to provide better warning of a predator's attack. Because more individuals can watch for predators, a large flock of wood pigeons detects the approach of a goshawk (a predatory bird) much sooner than a single pigeon does. The success rate of goshawk attacks drops from nearly 80 percent when these predators attack single pigeons to less than 10 percent when they attack flocks of more than 50 birds (Figure 42.6).

Consumers can restrict the distribution and abundance of victims

The American chestnut used to be a dominant tree species across much of eastern North America. In 1900, however, a fungus that causes a disease called chestnut blight was introduced into the New York City area. This fungus spread rapidly, killing most of the chestnut trees

(*a*)

(*b*)

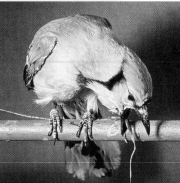

Figure 42.4   Warning Coloration
(*a*) The bright colors of this poison dart frog warn potential predators of the deadly chemicals it contains. (*b*) An inexperienced blue jay vomits after eating a monarch butterfly.

Figure 42.5 Come and Get Us
Although a single musk ox may be vulnerable to predators such as wolves, a group that forms a circle is a difficult target.

in eastern North America. Today the American chestnut survives throughout its former range only in isolated patches, primarily as sprouts that arise from the base of otherwise dead trunks.

The effect of chestnut blight on the American chestnut shows how a consumer (the fungus) can limit the distribution and abundance of a victim (the chestnut); in this case, a formerly dominant tree species was nearly eliminated from its entire range.

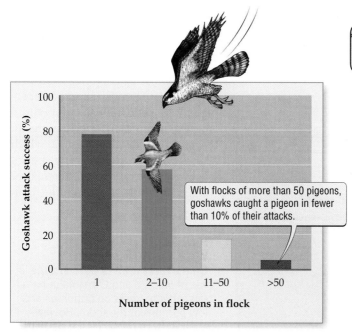

With flocks of more than 50 pigeons, goshawks caught a pigeon in fewer than 10% of their attacks.

Figure 42.6 Safety in Numbers
The success of goshawk attacks on wood pigeons decreases greatly when there are many pigeons in a flock.

Consumers can drive victims to extinction

Laboratory experiments with protists and with mites have shown that predators can drive their prey extinct. Consumer–victim interactions can drive victim species to extinction in natural systems as well. The effect of chestnut blight on the American chestnut provides one clear example: Although the chestnut tree is not extinct throughout its entire range, it has been driven to extinction in many local populations. Similarly, *Cactoblastis* moths drove many populations of the *Opuntia* cactus in Australia extinct (see Chapter 41).

In cases in which a consumer eats only one victim species, if the consumer drives a victim population extinct, the consumer must either locate a new population of victims or go extinct itself. This is exactly what happened to *Cactoblastis* in eastern Australia: The moth drove most populations of its cactus victim extinct, and now both species are found in low numbers.

> ■ Consumers can be a strong selective force, leading their victims to evolve various ways to avoid being consumed. Consumers can restrict the distribution and abundance of their victims, in some cases driving them to extinction.

42.3 Competition

In **competition**, each of two interacting species has a negative effect on the other (a –/– interaction). Competition is most likely when two species share an important resource, such as food or space, that is in short supply. When two species compete, each has a negative effect on the other because each uses resources (such as a source of food) that otherwise could have been used by its competitor. This is true even when one species is so superior a competitor that it ultimately drives the other species extinct: Until it actually becomes extinct, the inferior competitor continues to use some resources that could have been used by the superior competitor.

There are two main types of competition:

1. In **interference competition**, one organism directly excludes another from the use of a resource; for example, two species of birds may fight over the tree holes that they both use as nest sites.

2. In **exploitation competition**, species compete indirectly for a shared resource, each reducing the amount

of the resource available to the other; for example, two plant species may compete for a resource that is in short supply, such as nitrogen in the soil.

Evidence for competition

Competition between species is very common and often has important effects on natural populations. Along the coast of Scotland, the larvae of two species of barnacles, *Balanus balanoides* and *Chthamalus stellatus*, both settle on rocks on high and low portions of the shore. However, *Balanus* adults appear only on the lower portion of the shore, which is more frequently covered by water, and *Chthamalus* adults occur only on the higher portion of the shore, which is more frequently exposed to air.

In theory, the distribution of *Balanus* and *Chthamalus* could have been caused either by competition or by environmental factors. In an experimental study, however, ecologists discovered that *Chthamalus* could perform very well on low portions of the shore, but only if *Balanus* was removed (Figure 42.7). Hence, competition with *Balanus* ordinarily prevents *Chthamalus* from living low on the shoreline. The distribution of *Balanus*, on the other hand, depends mainly on physical factors: The increased heat and dryness found at higher levels of the shoreline prevent *Balanus* from surviving there.

In some cases, the resources shared by two species may be so readily available that little competition occurs. Competition among leaf-feeding insects, for example, is relatively uncommon. The reason is simple: A huge amount of leaf material is available for the insects to eat, and usually there are too few insects to cause their food

to be in short supply. As long as their food remains abundant, little competition occurs.

Competition can limit the distribution and abundance of species

A great deal of field evidence, such as the barnacle study described above, shows that competition can limit the distribution and abundance of species. A second example concerns wasps of the genus *Aphytis*. These wasps attack scale insects, which can cause serious damage to citrus trees. Female wasps lay eggs on a scale insect, and when the wasp larvae hatch, they pierce the scale insect's outer skeleton and then consume its body parts.

In 1948, the wasp *Aphytis lingnanensis* was released in southern California to curb the destruction of citrus trees caused by scale insects. A closely related wasp, *A. chrysomphali*, was already living in that region at the time. *A. lingnanensis* was released in the hope that it would provide better control of scale insects than *A. chrysomphali* did. *A. lingnanensis* proved to be a superior competitor (Figure 42.8), driving *A. chrysomphali* to extinction in most locations and, as hoped for, providing better control of scale insects.

Competition can increase the differences between species

As Charles Darwin realized when he formulated the theory of evolution by natural selection, competition between species can be intense when the two species are very similar in form. For example, birds whose beaks are similar in size eat seeds of similar sizes, and thus compete intensely, whereas birds whose beaks differ in size eat seeds of different sizes and compete less intensely. Intense competition between similar species may result in **character displacement**, in which the forms of the competing species evolve to become more different over time. By reducing the similarity in

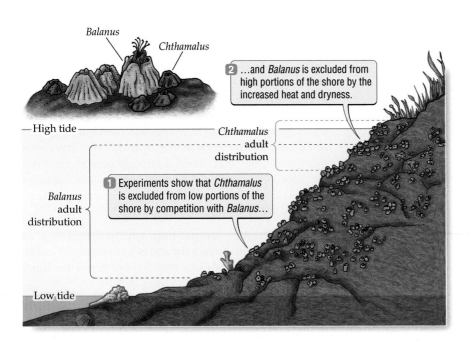

Figure 42.7
What Keeps Them Apart?
On the rocky coast of Scotland, the larvae of *Balanus* and *Chthamalus* barnacles settle on rocks on both high and low portions of the shore. However, adult *Balanus* barnacles are not found on high portions of the shore, and adult *Chthamalus* individuals are not found on low portions of the shore.

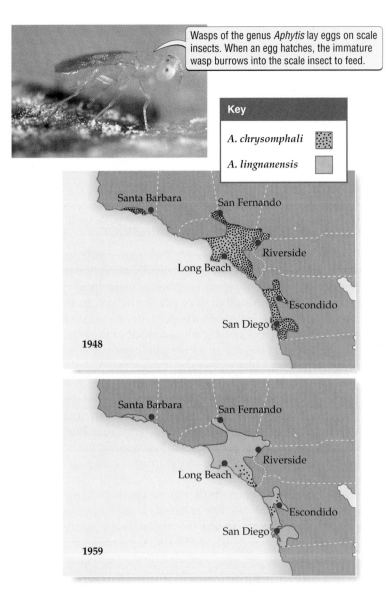

Wasps of the genus *Aphytis* lay eggs on scale insects. When an egg hatches, the immature wasp burrows into the scale insect to feed.

Key

A. chrysomphali

A. lingnanensis

Figure 42.8   A Superior Competitor Moves In
After being introduced to southern California in 1948, the wasp *Aphytis lingnanensis* rapidly drove its competitor, *A. chrysomphali*, extinct in most locations. Both species of wasps prey on scale insects that damage citrus crops such as lemons and oranges.

form between species, character displacement should reduce the intensity of competition. As discussed in Chapter 20 (see p. 344), however, species can evolve in this way only if their populations vary genetically for traits such as beak size on which natural selection can act.

Some evidence for character displacement comes from observations that the forms of two species are more different when they live together than when they live in separate places. In the Galápagos Islands, for example, the beak sizes of two species of Galápagos finches, and

hence the sizes of the seeds the birds eat, are more different on islands where both species live than on islands that have only one of the two species (Figure 42.9). Recent experiments with fish and lizards also suggest that character displacement is important in nature.

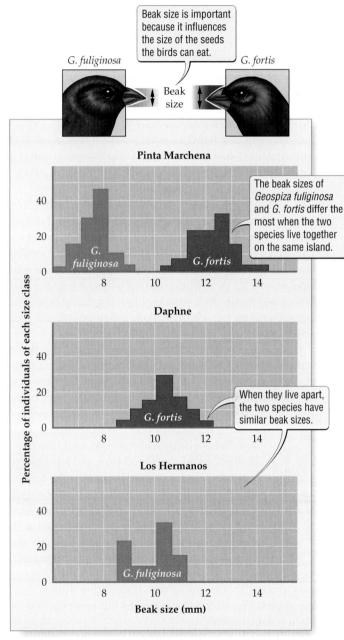

Beak size is important because it influences the size of the seeds the birds can eat.

G. fuliginosa   Beak size   *G. fortis*

The beak sizes of *Geospiza fuliginosa* and *G. fortis* differ the most when the two species live together on the same island.

When they live apart, the two species have similar beak sizes.

Figure 42.9   Character Displacement
In character displacement, competition for resources causes the competing species to become more different over time. Competition between two species of Galápagos finches, *Geospiza fuliginosa* and *G. fortis*, may be the driving force that causes the beak sizes of these birds to be more different when they live on the same island than when they live apart.

■ In interference competition, one species directly excludes another from the use of resources. In exploitation competition, species compete indirectly, each reducing the amount of a resource available to the other. Competition can affect the distribution and abundance of species, and can cause competitors to evolve greater differences over time.

els at which ecology can be studied: the individual organisms involved in an interaction, populations of those organisms, the communities in which those organisms live, and whole ecosystems.

■ Interactions among organisms affect individuals, populations, communities, and ecosystems.

Interactions among Organisms Shape Communities and Ecosystems

Throughout this chapter we have seen how interactions among organisms help determine their distributions and abundances. Interactions among organisms also have large effects on the communities and ecosystems in which those organisms live.

When dry grasslands are overgrazed by cattle, for example, grasses may become less abundant and desert shrubs may become more abundant (see Figure 43.7). These changes in the abundances of grasses and shrubs can change the physical environment. The rate of soil erosion may increase because shrubs do not stabilize soil as well as grasses do. Ultimately, if overgrazing is severe, the ecosystem can change from a dry grassland to a desert.

Changes in interactions among organisms can have complicated effects on natural communities. As an example, let's look at what happened when people removed dingoes from a particular region in Australia to prevent them from eating sheep. Dingoes are the largest (nonhuman) predators on the Australian continent. Where dingoes were removed, the population of their preferred prey, red kangaroos, increased dramatically (166-fold). The resulting increase in grazing by kangaroos changed the outcome of competition among plant species, causing some species to increase in abundance and others to decrease. Overall, the removal of dingoes resulted in (1) an increase in red kangaroos, (2) a decrease in plants consumed by red kangaroos, (3) changes in the competitive interactions among plants, and (4) changes in the composition of plant communities.

Dingo

In the two examples discussed in this section, a change in an interaction among organisms had a ripple effect, changing the abundances of populations, the community of species that lived in an area, and even, in the case of the dry grasslands, converting one ecosystem (dry grassland) to another (desert shrubland). In general, interactions among organisms can affect all the lev-

■ HIGHLIGHT

Parasites that Alter Host Behavior

Parasites affect their hosts in ways that range from merely annoying (fleas) to downright deadly (fungal parasites of ants). In addition, many parasites cause their victims to perform unusual or even bizarre behaviors that harm the victim but benefit the parasite. An example was provided by the worms we described at the beginning of this chapter, which cause praying mantises to throw themselves into rivers and drown. Similarly, the protist *Toxoplasma gondii* causes its rat host to become more curious and less fearful. Such changes make infected rats easier prey for cats, the other host of the protist. As these examples show, some parasites cause broad changes in their hosts' behavior, such as making the host less cautious or causing it to move from one habitat to another.

Other parasites cause much more specific changes in their hosts' behavior. For example, a parasitic wasp called *Hymenoepimecis* attacks the spider *Plesiometa argyra*. A female wasp stings the spider into temporary paralysis, then lays an egg on its body. The spider recovers quickly and builds normal webs for the next week or two (Figure 42.10*a*). During this period, the wasp egg hatches, and the wasp larva feeds by sucking body fluids from the spider. Then, one evening, the larva injects a chemical into the spider, causing the spider to spin a unique "cocoon web" (Figure 42.10*b*). Driven by the wasp's chemical attack, the spider performs many repetitions of one part of its normal web-building process, the other parts of which are suppressed. Thus, the wasp has evolved the ability to cause a very particular change in how the spider builds its web.

Why does the wasp alter spider behavior in this way? To see why, consider the end of the story. As soon as the spider finishes the cocoon web, the larva kills and consumes the spider. The larva then spins a cocoon of its own, in which it will complete its development. The larva uses the spider's cocoon web as a strong support from which to hang its own cocoon, thus protecting it from being swept away by heavy rains. In effect, the

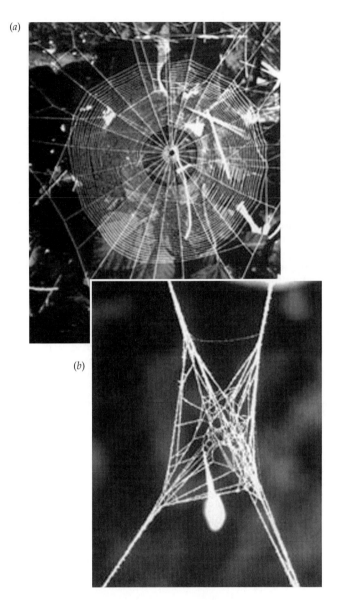

Figure 42.10 My Parasite Made Me Do It
(a) A typical web of the spider *Plesiometa argyra*. (b) A cocoon web, produced by a *P. argyra* spider infected with a wasp parasite that alters its web-spinning behavior. The wasp's cocoon can be seen hanging down from the center of the cocoon web.

wasp not only consumes the spider, but also forces the spider to build it a safe haven that protects it from potentially lethal torrential rains.

■ Many parasites cause relatively nonspecific changes in the behavior of their victims. Other parasites have evolved the ability to cause precise changes in host behavior.

SUMMARY

Mutualisms

■ A mutualism is an interaction between two species in which both species benefit (a +/+ interaction).

■ A symbiosis is an association of two species that live together. A symbiosis may or may not be mutualistic.

■ Mutualisms evolve when the benefits of the interaction are greater than its costs for both partners.

■ Mutualisms are very common in nature.

■ Mutualisms help determine the distribution and abundance of the mutualist species as well as other species that depend directly or indirectly on the mutualist species.

Consumer–Victim Interactions

■ Consumer–victim interactions are interactions in which one species (the consumer) benefits and the other (the victim) is harmed (+/− interactions).

■ Consumers include predators, parasites, pathogens, and herbivores.

■ Consumers can be a strong selective force, leading their victims to evolve various ways to avoid being consumed.

■ Consumers can restrict the distribution and abundance of their victims, in some cases driving their victims to extinction.

Competition

■ In competition, each of two interacting species has a negative effect on the other (a −/− interaction).

■ In interference competition, one species directly excludes another from the use of resources.

■ In exploitation competition, species compete indirectly, each reducing the amount of a resource available to the other.

■ Competition can have a strong effect on the distribution and abundance of species.

■ Competition can result in the evolution of greater differences between species.

Interactions among Organisms Shape Communities and Ecosystems

■ Interactions among organisms affect individuals, populations, communities, and ecosystems.

Highlight: Parasites that Alter Host Behavior

■ Many parasites cause relatively nonspecific changes in the behavior of their victims, such as making the victim less fearful or causing the victim to move from one habitat to another.

■ Other parasites have evolved the ability to cause precise changes in host behavior.

KEY TERMS

behavioral mutualism p. 697

character displacement p. 702

competition p. 701

consumer–victim interaction
    p. 698

exploitation competition p. 701

gut inhabitant mutualism p. 697

herbivore p. 698

interference competition p. 701

mutualism p. 696

parasite p. 698

pathogen p. 698

pollinator mutualism
    p. 697

predator p. 698

seed dispersal mutualism
    p. 697

symbiosis p. 696

CHAPTER REVIEW

Self-Quiz

1. Which of the following statements about consumers is true?
   a. Consumers cannot drive their victims extinct.
   b. Consumers are not important in natural communities.
   c. Consumers can apply strong selection pressure to their victims.
   d. Consumers cannot alter the behavior of their victims.

2. In what type of competition do species directly confront each other over the use of a shared resource?
   a. interference competition
   b. exploitation competition
   c. physical competition
   d. unstable competition

3. Interactions among species
   a. do not influence the distribution or abundance of organisms.
   b. are rarely beneficial to both species (that is, mutualism is not common).
   c. have a strong influence on communities and ecosystems.
   d. cannot drive species to extinction.

4. The advantages received by a partner in a mutualism can include
   a. food.
   b. protection.
   c. increased reproduction.
   d. all of the above

5. The shape of a fish's jaw influences what the fish can eat. Researchers found that the jaws of two fish species were more similar when they lived in separate lakes than when they lived together in the same lake. The increased difference in jaw structure when the fish live in the same lake is a potential example of
   a. warning coloration.
   b. character displacement.
   c. mutualism.
   d. a consumer–victim interaction.

Review Questions

1. A mutualism typically has costs for both of the species involved. Why, then, is mutualism so common?

2. How can a species that is an inferior competitor have a negative effect on a superior competitor?

3. Rabbits can eat many plants, but they prefer some plants over others. Assume that the rabbits in a grassland that contains many plant species prefer to eat a species of grass that happens to be a superior competitor. If the rabbits were removed from the region, which of the following do you think would be most likely to happen?

   a. The plant community would have fewer species

   b. The plant community would have more species

   c. The plant community would remain largely unchanged

   Explain your answer.

42 **The Daily Globe**

Can Wolves and People Live Together?

CATRON COUNTY, NM—Years after they were hunted to near extinction, the howl of the wolf once again can be heard in this remote region of New Mexico, a land of few people and fewer roads. Federal and state wildlife biologists recently reintroduced the wolf in an attempt to rescue this large predator from the brink of extinction.

Most local residents were strongly opposed to the reintroduction of the wolf. Some feared that wolves would harm one of the mainstays of the local economy, ranching, by preying on livestock. Many also resented the fact that federal and state wildlife biologists went ahead with reintroduction

plans despite the strong opposition of local people. The stage was set for ongoing conflict.

To some extent, that conflict has occurred. Several wolves have been shot on sight, and a local resident is currently on trial for one of these killings. Adding fuel to the fire, the wolves have killed a number of calves and sheep since their reintroduction. As a result, some people in the county argue that their worst fears have been realized, and tempers are running high over the perceived high-handedness of federal and state wildlife biologists.

But there also are signs of hope. Wildlife biologists know that the wolf reintroduction project cannot

succeed without the support of the local community. They are seeking that support door-to-door, in town halls, and in school classrooms. These efforts have begun to pay off. Some people opposed to the wolf reintroduction softened their position once they got to know the wildlife biologists and realized that they were willing, for example, to remove individual wolves that attack livestock.

As wildlife biologist Susan James put it, "Although the road may be rocky sometimes, people and wolves can live together. We're working hard to ensure that both people and wolves benefit from the return of the wolf to Catron County."

Evaluating "The News"

1. In what ways might people benefit from the return of the wolf to regions in which this species was hunted to extinction by people?

2. If an area provides excellent and otherwise hard-to-obtain habitat for

an endangered species, should that species be reintroduced to the area despite the objections of local people? Why or why not?

3. Do people have an ethical responsibility to help the recovery of species

driven to the brink of extinction by human actions? If not, why not? If so, how should the needs of endangered species be weighed against the needs of people?

chapter *43* Communities of Organisms

Jim Waid, *Daybreak*, 2000.

The Formation of a New Community

Whenever new habitat originates, as when an island rises out of the sea or a new lake is formed, it marks the beginning of a huge and exciting natural experiment. What organisms will first colonize the new habitat? How will those organisms interact and evolve over time?

Will the new habitat come to house unusual communities of organisms? Or will the communities of the new habitat be similar to other, nearby ecological communities?

The outcome of such grand experiments can be spectacular, especially when the new communities are located far from existing ones. Consider what has happened on the Hawaiian Islands, a remote chain of volcanic islands, the most recent of which (Hawaii) was formed about 600,000 years ago.

At first, the Hawaiian Islands were colonized by relatively few species. Over time, however, the few species that colonized the islands evolved to form many new species. As a result, the communities of organisms on Hawaii are very different from those anywhere else on Earth. Such a series of events is not restricted to Hawaii: Unusual communities often form when new habitat is located far from existing communities.

Today, many of the unique communities on Hawaii are threatened by species brought to the islands by people, as well as by habitat destruction and other forms of human impact. Is there something about island communities that makes them particularly vulnerable to human activities? What can be done, on Hawaii and elsewhere, to prevent human actions from having undesirable effects on ecological communities?

Pacific Ocean

Hawaii was formed 600,000 years ago.

A Natural Experiment
The Hawaiian Islands are part of a chain of volcanic islands that have risen from the sea over the past 70 million years. As newly formed islands were colonized by species from the mainland and as new species evolved on the islands, unique new ecological communities like this one were formed.

■ KEY CONCEPTS

1. A community is an association of populations of different species that live in the same area.

2. Food webs describe the feeding relationships within a community. Keystone species play a critical role in determining the types and abundances of species in a community.

3. All communities change over time. As species colonize new or disturbed habitat, they tend to replace one another in a directional and fairly predictable process called succession. Communities also change over time in response to changes in climate.

4. Communities can recover rapidly from some forms of natural and human-caused disturbance.

5. It may take thousands of years for communities to recover from other forms of human-caused disturbance.

A community is an association of populations of different species that live in the same area. There are many different types of communities, ranging from those found in grasslands and forests to those found in the digestive tract of a cow or deer (Figure 43.1). Most communities contain many species, and as we learned in Chapter 42, the interactions among those species can be complex. As we discover in this chapter, ecologists seek to understand how interactions among organisms influence natural communities.

Ecologists also seek to understand how human actions affect communities. At present, humans are having a profound effect on many kinds of ecological communities. When we cut down tropical forests, we destroy entire communities of organisms, and when we give antibiotics to a cow, we alter the community of microorganisms living in its digestive tract. To prevent our actions from having undesirable or unintended effects, we must understand how communities work and how they respond to both natural and human-caused disturbances.

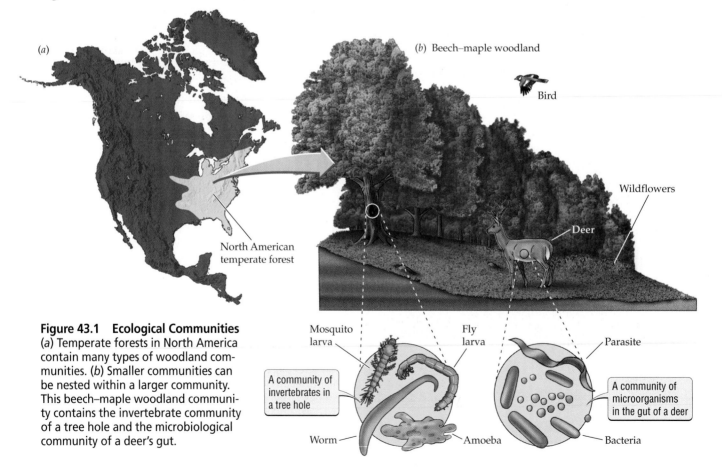

Figure 43.1   Ecological Communities
(*a*) Temperate forests in North America contain many types of woodland communities. (*b*) Smaller communities can be nested within a larger community. This beech–maple woodland community contains the invertebrate community of a tree hole and the microbiological community of a deer's gut.

In this chapter we describe some of the factors that influence what species are found in a community. We pay particular attention to how communities change over time and how they respond to disturbance, including disturbance caused by people. We begin by discussing the nature of ecological communities.

43.1 The Nature of Communities

Communities vary greatly in size and complexity, from the community of microorganisms that inhabits a small temporary pool of water, to the community of plants that lives on the floor of a forest, to a forest community that stretches for hundreds of kilometers. Communities also can be nested within each other, as Figure 43.1 shows. In general, just as in the study of populations (see Chapter 41), what constitutes a community depends on the organisms under study and the biological questions of interest.

Ecological communities are shaped by the individual species that live in them, by interactions among those species, and by interactions between those species and the physical environment. In this section we focus on how individual species and interactions among species affect communities. In the following section, we discuss how interactions between species and their environment influence communities.

Food webs consist of multiple food chains

One important aspect of a community is who eats whom. These feeding relationships can be described by **food chains**, each of which describes a single sequence of who eats whom in a community. The movement of food through a community can be summarized by connecting the different food chains to one another to form a **food web**, which describes the interconnected and overlapping food chains of a community (Figure 43.2).

Food webs and the ecological communities they describe are based on a foundation of producers. **Producers** are organisms that use energy from an external source, such as the sun, to produce their own food without having to eat other organisms or their remains. On land, plants, which harvest energy from the sun, are the major producers. In aquatic biomes, a wide range of organisms serve as producers, including phytoplankton (free-floating, photosynthetic microorganisms) in the oceans, algae in intertidal zones and lakes, and bacteria in deep-sea hydrothermal vents.

Consumers are organisms that obtain energy by eating all or parts of other organisms or their remains. Important groups of consumers include decomposers (see p. 727) and the pathogens, parasites, herbivores, and predators discussed in Chapter 42. **Primary consumers** are organisms, such as cows or grasshoppers, that eat producers. **Secondary consumers** are organisms, such

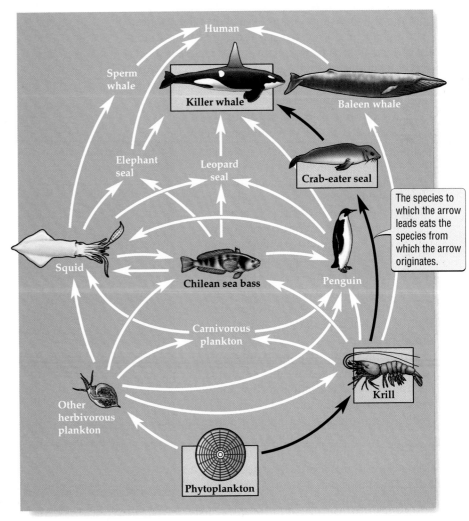

Figure 43.2 A Food Web
Food webs summarize the movement of food through a community. This figure is a simplified version of the food web in the Antarctic Ocean. Food webs are composed of many specific sequences of who eats whom, known as food chains. One of the food chains found in this food web is highlighted with black arrows and yellow boxes.

as humans or birds, that feed on primary consumers. This sequence of organisms eating organisms that eat other organisms can continue: A bird that eats a spider that ate a beetle that ate a plant is an example of a tertiary consumer.

Species interactions have profound effects on communities

Mutualism, consumer–victim interactions, and competition each influence the distribution and abundance of species found in a community, as shown by the coral reef, dingo, and barnacle examples discussed in Chapter 42. However, there are certain species that, relative to their own abundance, have a disproportionately large effect on the types and abundances of the other species in a community; these species are called **keystone species**.

In architecture, a keystone is the central, topmost stone that keeps an arch from collapsing. A keystone species serves a similar role with respect to the "architecture" of a food web: If a keystone species is removed, the entire food web will change drastically.

In an experiment conducted along the rocky Pacific coast of Washington, the ecologist Robert Paine demonstrated that the sea star *Pisaster ochraceus* is a keystone species in its intertidal zone community. He removed sea stars from one site and left an adjacent, undisturbed site as a control. He found that in the absence of the sea star, of the original 18 species in the community, all but mussels disappeared (Figure 43.3). When the sea stars were present, they ate the mussels, thereby keeping the number of mussels low enough so that the mussels did not crowd out the other species.

Organisms other than predators such as *Pisaster* can be keystone species. Plants such as fig trees, herbivores such as snow geese and elephants, and pathogens such as the distemper virus that kills lions have been found to function as keystone species; in addition, humans often function as keystone species. In general, the term "keystone species" can include any producer or consumer of relatively low abundance that has a large influence on

the interactions among organisms, and hence on the types and abundances of species, in its community. Although the most abundant or dominant species in a community (such as the corals in a coral reef or the mussels in Paine's intertidal zone) also have large effects on their communities, they are not considered keystone species because their abundance is not low.

In some cases, people remove a species from a community that turns out to be a keystone species, thus changing the community greatly. For example, when people removed rabbits from a region in England, they

When the sea star *Pisaster* was removed from a community experimentally, the number of species dropped from 18 to 1, a mussel.

Sea star not removed

Sea star removed

A *Pisaster* sea star feeding on a mussel

Figure 43.3   A Keystone Species
The sea star *Pisaster ochraceus* is a predator that feeds on mussels, thereby preventing the mussels from crowding out other species in the community.

unintentionally converted grasslands with a variety of plant species to grasslands containing just a few species of grasses. This change occurred because the rabbits had held the grasses in check. In the absence of rabbits, the grasses crowded out other plant species.

> ■ Communities can be described by food webs, which summarize who eats whom in the community. Relative to their abundance, keystone species have large effects on ecological communities because they alter the patterns of interaction among species, thus changing the types or abundances of species in the community.

43.2 Communities Change over Time

All communities change over time. The number of individuals of different species in a community often changes as the seasons change: Although they might be common in summer, for example, we would not find butterflies flying in a North Dakota field in the middle of winter. Similarly, every community shows year-to-year changes in the abundances of organisms, as we saw in Chapter 41.

In addition to such seasonal and yearly changes, communities show broad, directional changes in species composition over longer periods of time. These longer-term changes in communities are the subject of this section.

Succession establishes new communities and replaces disturbed communities

How do species come together to form a community? A community may begin when new habitat is created, as when a volcanic island rises out of the sea or when rock and soil are deposited by a retreating glacier. New communities also may form in regions that have been dis-turbed, as by a fire or hurricane. Some species arrive early in such new or disturbed habitat. These early colonists tend to be replaced later by other species, which in turn may be replaced by still other species. Earlier species may be replaced by later species because the replacement species are better able to grow and reproduce under the changing conditions of the area.

The process by which species in a community are replaced over time is called **succession**. In a given location, the order in which species will replace one another is fairly predictable (Figure 43.4). Such a sequence of species replacements sometimes ends in a **climax community** which, for a particular climate and soil type, is a community whose species are not replaced by other species. But in many—perhaps most—ecological communities, **disturbances** such as fires or windstorms occur so frequently that the community is constantly changing in response to a previous disturbance event, and a climax community never forms.

Primary succession is succession that occurs in newly created habitat, as when a glacier retreats or an island rises from the sea. In such a situation, the process begins with a habitat containing no species at all. The first species to colonize the new habitat usually have one of two advantages over other species: Either they disperse rapidly (and hence reach the new habitat first) or they are better able to grow and reproduce under the challenging conditions of the newly formed habitat.

In some cases of primary succession, the first species to colonize the area alter the habitat in ways that allow later-arriving species to thrive. In other cases, the early

Figure 43.4 Succession
When strong winds cause moving sand dunes to form at the southern end of Lake Michigan, succession often leads to a community dominated by black oak. Succession on such dunes occurs in three stages and forms black oak communities that have lasted up to 12,000 years. Under different local environmental conditions, succession in Michigan sand dunes can lead to the establishment of stable communities as different as grasslands, swamps, and sugar maple forests.

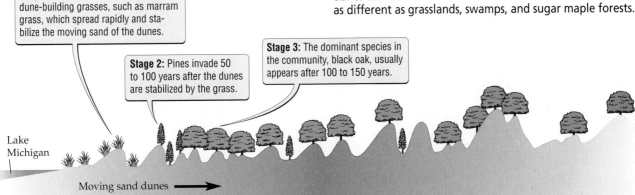

Stage 1: Bare sand is first colonized by dune-building grasses, such as marram grass, which spread rapidly and stabilize the moving sand of the dunes.

Stage 2: Pines invade 50 to 100 years after the dunes are stabilized by the grass.

Stage 3: The dominant species in the community, black oak, usually appears after 100 to 150 years.

Lake Michigan

Moving sand dunes →

colonists hinder the establishment of later species. An experimental study of primary succession in marine intertidal communities on the rocky coast near Santa Barbara, California, found that the first species of algae to colonize new habitats (concrete blocks placed there by the researchers) inhibited the establishment of the species that ultimately replaced them. In cases like this one, the early colonists lose their hold eventually because they are more susceptible to some particular feature of the environment, such as disturbance, grazing by herbivores, or extremes of heat or cold.

Secondary succession is the process by which communities recover from disturbance, as when a field ceases to be used for agriculture or when a forest grows back after a fire (Figure 43.5). In contrast to primary succession, habitats undergoing secondary succession often have well-developed soil that contains seeds from species that usually occur late in the successional process. The presence of such seeds in the soil can considerably shorten the time required for the later stages of succession to be reached.

Communities change as climate changes

Some groups of species stay together for long periods of time. For example, some plant communities in the southeastern United States and in China resemble an extensive community that once stretched across the northern parts of Asia, Europe, and North America. As the climate grew colder over the past 60 million years, plants in these communities migrated south, forming communities in Southeast Asia and southeastern North America that are similar in composition to the community from which they came.

Although groups of plants can remain together for millions of years, the community located in a particular place changes as the climate of that place changes. The climate at a given location can change over time for two reasons: global climate change and continental drift.

First, as we saw in Chapter 40, the climate of Earth changes over time. Historically, changes in the global climate have been due to relatively slow natural processes, such as the advance and retreat of glaciers. However, evi-

Figure 43.5   Secondary Succession
Forests in the United States grow back after they are cut by people, blown down by windstorms, or burned by fire. These photographs, taken in different locations, show a regrowth sequence following the large fire that struck Yellowstone National Park in 1988: (*a*) shortly after the fire in 1988; (*b*) 4 years later, Lodgepole pine saplings are growing in a stand of trees killed by the fire; (*c*) a mature Lodgepole pine forest.

(*a*)

(*b*)

(*c*)

dence is mounting that human activities are now causing rapid changes in the global climate (see Chapter 45).

Second, as the continents move slowly over time (see Figure 22.6), their climates change. To give a dramatic example of continental drift, 1 billion years ago Queensland, Australia, which is now located at 12° south latitude, was located near the North Pole. Roughly 400 million years ago, Queensland was at the equator. The species that thrive at the equator and in the Arctic are very different; thus continental drift has resulted in large changes in the communities of Queensland over time.

> ■ All communities change over time. Long-term directional changes in community composition have two main causes: succession and climate change. Primary succession occurs on newly created habitat, secondary succession occurs in communities recovering from moderate forms of disturbance. The climate at a given location can change because of global climate change or continental drift.

Communities Can Recover from Disturbance

Ecological communities are subject to many natural forms of disturbance, such as fires, floods, and windstorms. Following a disturbance, secondary succession can reestablish the previously existing community. Thus, communities can and do recover from some forms of disturbance. Depending on the community, the time required for recovery varies from years to decades or centuries.

Communities have been exposed over long periods of time to natural forms of disturbance. In contrast, humans may introduce entirely new forms of disturbance to ecological communities, such as the dumping of hot wastewater into a river. Humans also may alter the frequency of an otherwise natural form of disturbance—for example, causing a dramatic increase or decrease in the frequency of fires or floods.

Can communities recover from disturbance caused by humans? For some forms of human-caused disturbance, the answer is yes. Throughout the eastern United States, for example, there are many places where forests were cut down and used for farmland and then, years later, the farmland was abandoned. Second-growth forests have grown on these abandoned farms, often within 40 to 60 years after farming stopped.

Second-growth forests are not identical to the forests that were originally present. The sizes and abundances of tree species are different, and fewer plant species grow beneath the trees of a second-growth forest than beneath a forest that has never been cut down. However, previously logged, second-growth forests of the eastern United States already have recovered partially. If current trends continue, over the next several hundred years there will be fewer and fewer differences between such forests and the original forests.

In some cases, communities can also recover from pollution. Lake Washington is a large, clear lake in Seattle, Washington. As the city of Seattle grew, raw sewage was dumped into the lake. This practice declined after 1926 and was stopped by 1936. Beginning in 1941, treated sewage was discharged into Lake Washington from newly constructed sewage treatment plants.

A major effect of discharging sewage—treated or not—into Lake Washington was that phosphorus in the sewage added extra nutrients to the lake. As a result, the numbers of algae soared, decreasing the clarity of the water. As the algae reproduced and then died, their bodies provided an abundant food source for bacteria, whose populations also increased. Bacteria use oxygen when they consume dead algae, so concentrations of oxygen dissolved in the water decreased. The lowered oxygen concentrations killed invertebrates and fish.

By the early 1960s Lake Washington was highly degraded, so much so that it was referred to in the local press as Lake Stinko. From 1963 to 1968, the dumping of sewage into the lake was reduced, and virtually none was dumped after 1968. Once inputs of sewage were stopped, algae populations declined, oxygen concentrations increased, and Lake Washington returned to its former, clear state.

> ■ Communities can recover from some forms of natural and human-caused disturbance. Depending on the community, the time required for recovery varies from years to decades or centuries.

Humans Can Cause Long-Term Damage to Communities

A continent ages quickly once we come.

Ernest Hemingway

Communities do not always recover from disturbances caused by humans. Northern Michigan once was covered with a vast stretch of white-pine and red-pine forest. Between 1875 and 1900, nearly all of these trees were cut down, leaving only a few scattered patches of virgin forest. The loggers left behind large quantities of branches

(a)

(b)

(c)

Figure 43.6 Communities Do Not Always Recover from Disturbance
(*a*) A low-elevation Indonesian rainforest, photographed in 1980. (*b*) In 1982, this rainforest burned. As seen in this 1985 photograph, 3 years after the fire, few trees remained alive. (*c*) A second fire occurred in 1998, killing the remaining trees and converting the former rainforest to a grassland dominated by an introduced grass species.

and sticks, which provided fuel for fires of great intensity. In some locations, the pine forests of northern Michigan have never recovered from the combination of fire and logging.

A combination of logging and fire has also changed large regions of South America and the Pacific islands from tropical forest to grassland (Figure 43.6). Scientists estimate that it will take tropical forest communities hundreds to thousands of years to recover from such changes.

Finally, in some areas of the American Southwest, overgrazing by cattle has transformed dry grasslands into desert shrublands (Figure 43.7). How do cattle cause such large changes? Grazing and trampling by cattle decrease the abundance of grasses in the community. With less grass to cover the soil and hold it in place, the soil dries out and erodes more rapidly. Desert shrubs thrive under these new soil conditions, but grasses do not. These changes in soil characteristics can

Figure 43.7 Overgrazing Can Convert Grasslands into Deserts
(*a*) More than 200 years ago, large regions of the American Southwest were covered with dry grasslands. (*b*) Most of these grasslands have been converted to deserts, in large part because of overgrazing by cattle.

(a)

This dry grassland is in southern New Mexico.

(b)

This nearby, former dry grassland is now a desert.

make it very difficult to reestablish grasses, even when the cattle are removed.

It is interesting to compare how grazing and an atomic blast have affected dry grasslands over long periods of time. The first outdoor explosion of an atomic bomb occurred on July 16, 1945, at the Trinity site in New Mexico. Fifty years later, the dry grasslands that had been destroyed by the bomb blast (but which had never been grazed) had recovered. In contrast, nearby dry grasslands that had been heavily grazed (but not destroyed by the bomb blast) had not recovered. Thus, the plant community recovered more rapidly from the effects of a nuclear explosion than from the effects of grazing.

In the three examples discussed in this section—the pine forests of Michigan, the tropical forests of South America and the Pacific islands, and the grasslands of the American Southwest—humans have altered communities so greatly that it will take hundreds or thousands of years for those communities to recover. These long recovery times suggest that we need to carefully consider the long-term effects of our actions and practice restraint when we take actions that affect natural communities.

> ■ It can take hundreds to thousands of years for communities to recover from some forms of human-caused disturbance. It is important to consider these long-lasting effects when we take actions that affect natural communities.

43.3 *Disturbance, Community Change, and Human Values*

Change is a part of all communities. What, if anything, is special about human-caused changes in communities? What human values are affected by such changes?

Human actions have the power to change communities rapidly and dramatically across large geographic regions. Disturbances not caused by people, of course, can also occur rapidly and affect large geographic regions. For example, most scientists think that the impact of a large asteroid contributed to the sudden extinction of many species, including the last of the dinosaurs, 65 million years ago (see Chapter 22). However, changes in communities caused by humans are different in that we have direct control over whether or not we take the actions that cause such changes. Our unique ability to control our actions brings with it the responsibility to use our power to change communities wisely, a topic to which we will return in Chapter 45.

What human values are affected by human-caused community change? First, human actions that degrade or destroy ecological communities have ethical implications. When humans disrupt communities, our actions kill individual organisms, alter communities that may have persisted for thousands of years, and threaten species with extinction. According to results from surveys in North America and Europe, many people find such effects of human actions on individuals, communities, and species ethically unacceptable.

Second, human-caused community change often reduces the aesthetic value of a community. Tropical forests have unique aesthetic value to many people. When our actions cause tropical forests to be destroyed and replaced by introduced grasses (see Figure 43.6), we deprive current and future generations of experiencing the beauty of those forests.

Finally, in many cases, when humans change a community, we reduce its economic value. Economic value is lost, for example, when human actions convert grasslands to deserts, in part because there is more plant material to support grazing in grasslands than in deserts. In general, when our actions damage ecological communities, we often harm our own long-term economic interests as well. Unfortunately, it is common for people to take actions, such as allowing livestock to overgraze grasslands, that result in short-term economic benefit but cause long-term economic harm.

Communities change constantly in the face of both natural and human-caused disturbances. The challenge for human society is to manage Earth's changing communities while maintaining the diversity of their species and the ethical, aesthetic, and economic value that they provide for us. We describe one ecologist's thoughts on how we can successfully meet this challenge in the box on page 718.

> ■ Human-caused disturbances can change communities greatly. Our potential to control our actions gives us the responsibility to act in ways that do not reduce the ethical, aesthetic, and economic value of ecological communities.

■ HIGHLIGHT

Introduced Species and Community Change in the Hawaiian Islands

The Hawaiian Islands lie 4000 kilometers from the nearest continent. They are the most isolated chain of islands on Earth. Because the islands are so remote, entire groups

■ BIOLOGY IN OUR LIVES

Peter Vitousek Examines the Human Impact on Ecological Communities

Peter Vitousek

Peter Vitousek, an ecologist at Stanford University, grew up in Hawaii and performs research in the Hawaiian Islands. He has written several thought-provoking articles about the impact of humans on ecological communities. His time in Hawaii has given him firsthand knowledge of the ways in which humans have affected the unique ecological communities found there. That knowledge has contributed to the passion with which he seeks to understand and limit the negative effects of humans on all ecological communities.

As we have learned in this chapter, it can take thousands of years for ecological communities to recover from human actions. As we enter a new century, how can our growing population limit its negative effects on ecological communities, thereby maintaining the ethical, aesthetic, and economic value we get from them? In his scientific and other writings, Dr. Vitousek argues that we must take a multipronged approach.

First, we must reduce our impact on Earth by reducing the growth rate of our population and reducing the rate at which we use resources (see Chapters 41 and 45).

Second, we must recognize that humanity is a dominant force on Earth: No place on this planet is free from our impact. It is not enough to leave nature alone; our influence is too great for that. Instead, we must manage ecological communities actively if we are to maintain the value we receive from them.

The need to manage ecological communities implies a third need: We must devote far greater effort to understanding how ecological communities work. By understanding communities better, we will have a much better chance of managing them in ways that prevent the extinction of species while preserving the values they provide for us. In addition, as Dr. Vitousek and others have pointed out, to understand the natural world better is to enjoy it more and to want to take better care of it.

Finally, Dr. Vitousek argues that to maintain ecological communities and the value they provide to humans, we must all work together. Many scientists are better at figuring out what is going on and why than they are at seeing and implementing alternative ways of doing things. Thus, partnerships among scientists, economists, social scientists, governments, businesses, and private landowners will be essential as humanity strives to solve the complex environmental problems that we will face throughout the twenty-first century.

---

of organisms that live in most communities never reached Hawaii. For example, there are no native ants or snakes there, and there is only one native mammal (a bat, which was able to fly to the islands).

The few species that did reach the Hawaiian Islands found themselves in an environment that lacked most of the species from their previous communities. The sparsely occupied habitat and the lack of competitor species resulted in the evolution of many new species. The many different species of Hawaiian silversword plants found on the islands today (Figure 43.8), for example, all arose from a single ancestor. The new silversword species evolved very different forms, enabling them to live in a number of different habitats. Because other groups of plants and animals on the islands also

evolved many new species, the Hawaiian Islands have many unique natural communities.

Since the arrival of people in the islands about 1500 to 2000 years ago, Hawaii's unique biological communities have been threatened by habitat destruction, overhunting, and **introduced species** (species introduced by people to regions in which they would not otherwise be found). Of these threats, the effects of introduced species can be the easiest to overlook because such species often wreak their havoc quietly, behind the scenes. Introduced Argentine ants, for example, may drive native insects to extinction, but it can take years before even a trained biologist realizes what the introduced ants have done.

Island communities are particularly vulnerable to the effects of introduced species. Relatively few species colo-

Figure 43.8   Great Diversity from a Single Ancestor
Hawaiian silverswords are found only on the Hawaiian Islands. Genetic evidence indicates that this diverse group of plant species evolved from a single ancestor. Although the three silversword species shown here are closely related, they live in very different habitats and differ greatly from each other in form.

nize newly formed islands, and those species then evolve in isolation. For this reason, species on islands may be ill equipped to cope with new predators or competitors that are brought by people from the mainland. In addition, introduced species often arrive without the predator and competitor species that held their populations in check on the mainland. Thus, on islands, the potential exists for populations of introduced species to increase dramatically.

In some cases, introduced species can destroy entire communities. For example, if most of the native plants are not adapted to fire, an introduced species that alters the frequency or intensity of fire can have devastating effects. Consider what happened in Hawaii when beard grass was introduced.

Beard grass was brought to Hawaii by humans, and it invaded the seasonally dry woodlands of Hawaii Volcanoes National Park in the late 1960s. Before that time, fires occurred there, on average, every 5.3 years, and each fire burned only 0.25 hectare (1 hectare is about 2.5 acres). Since the introduction of beard grass, fires have occurred at a rate of more than one per year, and the average burn area of each fire has increased to more than 240 hectares.

Why did the introduction of beard grass increase the frequency and intensity of fires? As beard grass grows, it deposits a large amount of dry plant matter on the ground. This material catches fire easily, and the fires burn much hotter than they would in the absence of beard grass. Beard grass recovers well from large and hot fires, but the native trees and shrubs of the seasonally dry woodland do not. As a result, former woodlands have now been con-

verted to open pastures filled with beard grass and other, even more fire-prone, introduced grasses.

The Hawaiian dry woodland community has been destroyed, probably forever. Because there is no hope of restoring the native community, ecologists are now trying to construct a new community that is tolerant of fire yet contains native trees and shrubs. This is a difficult challenge, and it is uncertain whether the effort will succeed. If not, what was once woodland will remain indefinitely as open meadows filled with introduced grasses.

> ■ Island communities are vulnerable to the effects of introduced species in part because species on islands evolve in isolation and hence may not be able to cope with new predators or competitors. In the Hawaiian Islands, introduced species have caused the destruction of entire communities.

SUMMARY

The Nature of Communities

- Communities can be described by food webs, which summarize the interconnected food chains describing who eats whom in a community.
- A keystone species has a large effect on the food web of a community relative to its abundance.

- Keystone species alter the interactions between organisms in a community, thus changing the types or abundances of species in the community.

Communities Change over Time

- All communities change over time.

- Directional changes that occur over relatively long periods of time have two main causes: succession and climate change.

- Primary succession occurs in newly created habitat. Secondary succession occurs in communities recovering from moderate forms of disturbance.

- The climate at a given location can change because of global climate change or continental drift.

Communities Can Recover from Disturbance

- Communities can recover from some forms of natural and human-caused disturbances.

- Depending on the community, the time required for recovery varies from years to decades to centuries.

Humans Can Cause Long-Term Damage to Communities

- It can take hundreds to thousands of years for communities to recover from some forms of human-caused disturbance.

- It is important to consider these long-lasting effects when we take actions that affect natural communities.

Disturbance, Community Change, and Human Values

- Natural and human-caused disturbances can change communities greatly.

- Humans can control whether or not we perform actions that cause community change.

- Our potential to control our actions gives us the responsibility to act in ways that do not reduce the ethical, aesthetic, and economic value of ecological communities.

Highlight: Introduced Species and Community Change in the Hawaiian Islands

- Island communities are particularly vulnerable to introduced species, in part because species on islands evolve in isolation and hence may not be able to cope with new predators or competitors introduced by people.

- In the Hawaiian Islands, introduced species have caused the destruction of entire communities of organisms.

KEY TERMS

climax community p. 713	keystone species p. 712
community p. 710	primary consumer p. 711
consumer p. 711	primary succession p. 713
disturbance p. 713	producer p. 711
food chain p. 711	secondary consumer p. 711
food web p. 711	secondary succession p. 714
introduced species p. 718	succession p. 713

CHAPTER REVIEW

Self-Quiz

1. A species that has a large effect on a community relative to its abundance is called a
   a. predator.
   b. herbivore.
   c. keystone species.
   d. dominant species.

2. Organisms that can produce their own food from an external source of energy without having to eat other organisms are called
   a. suppliers.
   b. consumers.
   c. producers.
   d. keystone species.

3. Ecological communities
   a. cannot recover from disturbance.
   b. can recover from natural but not human-caused disturbance.
   c. can recover from all forms of disturbance.
   d. can recover from some, but not all, forms of natural and human-caused disturbance.

4. Which of the following was *not* caused by the introduction of beard grass to Hawaii?
   a. an increase in the growth of native trees and shrubs
   b. an increase in the frequency and intensity of fire
   c. the decline of native trees and shrubs
   d. the conversion of dry woodlands to grasslands

5. A directional process of species replacement over time in a community is called
   a. global climate change.
   b. succession.
   c. competition.
   d. community change.

Review Questions

1. Describe how each of the following factors influences ecological communities:
   a. species interactions
   b. disturbance
   c. climate change
   d. continental drift

2. What is the difference between primary and secondary succession?

3. Provide an example of how the presence or absence of a species in a community can alter a feature of the environment, such as the frequency of fire.

4. Do you think it is ethically acceptable for people to change natural communities so greatly that it takes thousands of years for the communities to recover? Why or why not?

5. Explain the difference between a keystone species and one of the most abundant or dominant species of a community.

43   𝕿𝖍𝖊 𝕭𝖆𝖎𝖑𝖞 𝕲𝖑𝖔𝖇𝖊

Strange Bedfellows?
Environmentalists and Ranchers Unite

ALBUQUERQUE, NM—Driving through the ranching country of the southwestern United States, one sees bumper stickers and signs with humorous but angry messages such as: "Hungry? Out of work? Eat an environmentalist." These messages are an expression of the long-standing conflict between ranchers who are trying to make a living and environmentalists who oppose the environmental damage caused by ranching. Now, however, the times are beginning to change, as environmentalists and ranchers are joining forces to save the grasslands that have so long supported ranchers and are so prized by conservationists.

"I grew up hating environmentalists," said Jake Simms, a rancher in Socorro County, New Mexico. "They were the enemy. But something had to be done. My land could hardly support cattle anymore."

As Mr. Simms and many other ranchers are discovering, while their great grandparents knew a Southwest in which grass stretched for miles and grew to above a person's knees, those days and those grasslands are gone, replaced by desert shrubs that provide little food for cattle. Regions once covered with black grama grass were heavily grazed over many years, then invaded by the desert shrub creosote bush, a plant that cattle don't like to eat.

With such changes occurring throughout his ranch, Mr. Simms, worried that he would have to give up ranching, turned to an unusual source for help: the Nature Defense Fund, a large environmental organization. Like an increasing number of ranchers, Mr. Simms found that he and the environmentalists had a common goal: improving the land.

The former foes have agreed to a plan in which ranchers remove cattle from environmentally sensitive areas, graze cattle only in numbers that the remaining grasslands can support, and devote a small portion of their land to a new industry, ecotourism.

"So far it seems to be working," said Ellen Deen, spokeswoman for the Nature Defense Fund. "Environmentalists and ranchers can cooperate to the benefit of both."

Evaluating "The News"

1. When the goals of environmentalists differ from those of ranchers, loggers, and others, the conflict is often portrayed as a simple choice of jobs versus the environment. If the economic activity in question is not sustainable, is the issue really as simple as this?

2. A major concern in our society is the economic bottom line. Such a focus leads people to maximize profits today. What might be some of the effects of such a strong emphasis on short-term economic gain?

3. Does the current generation have a responsibility to protect the environment for use by future generations? Or does the current generation have the right to use the land as they see fit, even if it harms the environment for generations to come?

chapter # 44 *Ecosystems*

David Ligare, *Landscape with Rainbow*, 1999.

Operation Cat Drop

In 1955, the World Health Organization staged an unusual mission in Borneo, Malaysia, called Operation Cat Drop, in which cats were parachuted into local villages. The story behind this strange event began roughly 15 years earlier, with research on a new pesticide called DDT.

In the late 1930s and early 1940s, it was discovered that DDT and other pesticides could be used to control insects that spread human diseases or ate crop plants. The new pesticides were an immediate success. Because they were cheap, easy to apply, and highly toxic to insects, most people viewed DDT and similar substances as a miracle of modern technology.

This view began to change in 1962, with the publication of Rachel Carson's *Silent Spring*. Carson questioned the widespread use of highly toxic pesticides, noting, "Future historians may well be

Ecosystems recycle materials and provide humans and other organisms with essential services.

amazed by our distorted sense of proportion. How could intelligent beings seek to control a few unwanted species by a method that contaminated the entire environment and brought the threat of disease and death even to their own kind?"

Carson was concerned because DDT is highly toxic to humans and other organisms, and because it tends to persist in the environment for long periods of time. Once DDT enters an organism's body, most of it remains there, stored in fatty tissue. When a predator eats the organism, it, too, becomes contaminated with DDT. Thus predators stored in their bodies most of the DDT that was contained in each of their prey. Because organisms toward the top of a food chain eat many smaller organisms, they collect high concentrations of DDT in their bodies.

To return to the parachuting cats, in 1955, malaria was very common in Borneo. To combat the disease, the World Health Organization sprayed the island with DDT to kill the mosquitoes that spread malaria. The disease was brought under control, but soon a

new health threat emerged: Villages were overrun with rats, and hence there was danger of an outbreak of other diseases, such as plague and typhus.

What had gone wrong? The DDT did kill the mosquitoes, but it also contaminated insects eaten by small lizards. Because the insects contained high concentrations of DDT, the DDT was even more concentrated in the lizards. When the lizards were eaten by the village cats, the cats died of DDT poisoning. With the cats out of the way, the rat population exploded, bringing with it the potential for dangerous diseases such as plague and typhus.

Although the cats delivered by Operation Cat Drop brought the rat population under control, this story demonstrates the need for caution whenever we add materials to the environment. In addition, the movement of DDT through the food chain (from insects to lizards to cats) provides an example of how natural systems recycle materials and follow a few simple rules, such as "everything goes somewhere" and "you can never do just one thing." By studying how materials like DDT move through natural systems, we can learn much about how natural systems work and why our actions can have surprising and unintended consequences.

A Forest in Malaysia

■ KEY CONCEPTS

1. An ecosystem consists of a community of organisms together with the physical environment in which those organisms live. Energy, materials, and organisms can move from one ecosystem to another.

2. Energy cannot be recycled. Energy captured by producers is lost at each step as it moves through a food chain. Because of the steady loss of energy, more energy is available toward the bottom than toward the top of a food chain.

3. Earth has a fixed amount of nutrients, the chemical elements needed to sustain life. If these nutrients were not recycled between organisms and the physical environment, life on Earth would cease.

4. Ecosystems provide humans with essential services, such as nutrient cycling, at no cost. Our civilization depends on these and many other free ecosystem services.

To survive, all organisms need energy to fuel their metabolism and materials to construct and maintain their bodies. Almost all life on Earth depends directly or indirectly on solar energy, the supply of which is renewed continually. Materials, on the other hand, such as the carbon, hydrogen, oxygen, and other elements of which we are made, are added to our planet in extremely small amounts (in the form of meteoric matter from outer space). Earth, therefore, has an essentially fixed amount of materials for organisms to use. This simple fact means that for life to persist, natural systems must recycle materials.

This chapter introduces ecosystem ecology, the study of how energy and materials flow through natural systems. We begin by discussing the capture and movement of energy in ecosystems. Next we consider how the chemical elements essential to life are cycled between organisms and the physical environment. At the close of the chapter, we discuss some of the effects humans are having on the cycling of materials in ecosystems.

How Ecosystems Function: An Overview

An **ecosystem** consists of a community of organisms together with the physical environment in which those organisms live. Like communities, ecosystems may be small or very large: A puddle teeming with protists is an ecosystem, as is the Atlantic Ocean. In fact, global patterns of air and water circulation (see Chapter 39) may be viewed as linking all the world's organisms into one giant ecosystem, the biosphere.

Ecosystem ecologists study how organisms capture energy and materials from the environment, transfer energy and materials to one another, and ultimately, return materials to the environment. Tracing these movements is a difficult task, in part because energy, materials, and organisms often move from one ecosystem to another.

The movement of energy and materials from one ecosystem to another forces ecosystem ecologists to measure as precisely as possible the amounts of energy and materials that enter and leave an ecosystem. Although this is not easy to do, a focus on energy and materials provides a powerful benefit: It allows very different ecosystems to be compared in a meaningful and standardized way. Measurements of the effects people have on the movement of energy and materials can be used, for example, to examine the relative effects of human actions on ecosystems as different as deserts and tropical rainforests.

The diagram in Figure 44.1 gives a broad overview of the movement of energy and materials through ecosystems. At each step in a food chain (see p. 711), a portion of the energy captured by producers is lost as metabolic heat, which is released as the inevitable by-product of the chemical breakdown of food. Because of this steady loss of heat, energy is not recycled, but exhibits a one-way flow through an ecosystem: It enters the ecosystem from the sun and leaves the ecosystem as metabolic heat. In contrast, nutrients—the materials required for life, such as carbon and nitrogen—are recycled between organisms and the nonliving environment. Nutrients are absorbed from the environment by producers, cycled among consumers for varying lengths of time, and eventually returned to the environment by decomposers, which, as described in Chapter 7, break down the dead bodies of organisms.

■ Energy and materials can move from one ecosystem to another. Energy is not recycled in ecosystems; rather, a portion of the energy captured by producers is lost as metabolic heat at each step in a food chain. Nutrients are recycled in ecosystems, moving between organisms and the nonliving environment.

Figure 44.1 How Ecosystems Work
At each step in a food chain, a portion of the energy captured by producers is lost as heat given off during the chemical breakdown of food by respiration (metabolic heat). Thus, energy flows through the ecosystem in a single direction and is not recycled. In contrast, nutrients cycle between organisms and the physical environment.

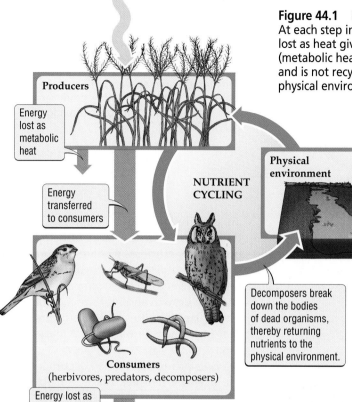

Producers

Energy lost as metabolic heat

Energy transferred to consumers

NUTRIENT CYCLING

Physical environment

Decomposers break down the bodies of dead organisms, thereby returning nutrients to the physical environment.

Consumers
(herbivores, predators, decomposers)

Energy lost as metabolic heat

ONE-WAY FLOW OF ENERGY THROUGH THE ECOSYSTEM

44.1 *Energy Capture in Ecosystems*

Most life on Earth depends directly or indirectly on the capture of solar energy by photosynthetic organisms such as plants. The energy captured by plants and other photosynthetic organisms is stored in their bodies in the form of chemical compounds, such as carbohydrates. Herbivores that eat plants, predators that eat herbivores, and decomposers that consume the remains of dead organisms all depend indirectly on the solar energy originally captured by plants.

The amount of energy that producers capture by photosynthesis, minus the amount they lose as heat in respiration, is called **net primary productivity**, or **NPP**. Although it is defined in terms of energy, ecologists usually measure NPP as the amount of new biomass produced by photosynthetic organisms during a specified period of time (the term "biomass" refers to the mass of organisms per unit of area). In a grassland ecosystem, for example, ecologists would estimate NPP by measuring the average amount of new grass and other plant matter produced in a square meter of area each year. Such NPP measurements based on biomass can be converted to units based on energy.

NPP is not distributed evenly across the globe. On land, NPP tends to decrease from the equator toward the poles (Figure 44.2a). This decrease occurs because the amount of solar radiation available to plants, the most important producers in terrestrial systems, is highest at the equator and lowest at the poles (see Chapter 39). But there are many exceptions to this general pattern. For example, there are large regions of very low NPP in northern Africa, central Asia, central Australia, and the southwestern portion of North America. Each of these regions is the site of one of the world's major deserts.

The low NPP in deserts emphasizes the fact that sunlight alone is not sufficient for NPP to be high; water is also essential. In addition to water and sunlight, productivity in terrestrial ecosystems can be limited by temperature and the availability of nutrients in the soil. On land, the most productive ecosystem types are tropical rainforests and cultivated land (Table 44.1). The least

44.1 *Net Primary Productivity (NPP) for Selected Ecosystem Types*

Ecosystem	Total area on Earth (10⁶ km²)	NPP (g per m² per year)	
		Range	Average
Terrestrial			
Tropical rainforest	17	1000–3500	2200
Cultivated land	14	100–3500	650
Temperate forest	12	600–2500	1240
Grassland	24	200–2000	790
Tundra and alpine communities	8	10–400	140
Desert	42	0–250	40
Aquatic			
Coral reef	0.6	500–4000	2500
Swamp and marsh	2	800–3500	2000
Estuary	1.4	200–3500	1500
Lake and stream	2	100–1500	250
Ocean upwelling zone	0.4	400–1000	500
Open ocean	332	2–400	125

(a)

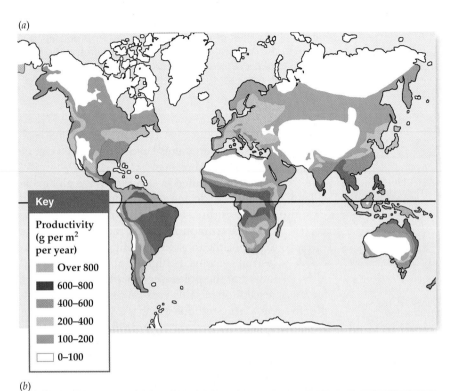

Key

Productivity (g per m² per year)

Over 800

600–800

400–600

200–400

100–200

0–100

(b)

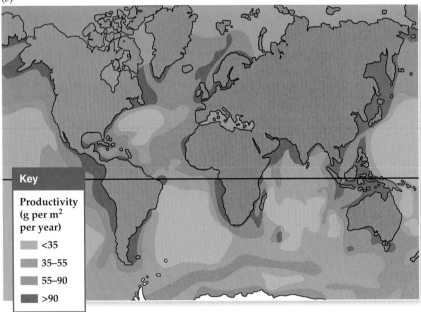

Key

Productivity (g per m² per year)

<35

35–55

55–90

>90

Figure 44.2 Net Primary Productivity Net primary productivity varies greatly in different parts of the world among both (a) terrestrial ecosystems and (b) marine ecosystems. Net primary productivity is measured as grams of new biomass made by plants or other producers each year in a square meter of area (g per m² per year).

are close to land. Streams and rivers that drain from the land into the ocean carry nutrients that are in short supply in the ocean. Addition of these nutrients to the ocean stimulates the growth and reproduction of phytoplankton, the small producers that form the foundation of aquatic food webs.

High productivity also occurs in the ocean wherever there are up-wellings of deep water. As we will see shortly, many important nutrients are stored in sediments at the bottom of the ocean, and upwellings can carry these nutrients to the surface. Upwellings can result in high productivity even at high latitudes, where both temperatures and solar radiation are low. Finally, the NPP in marine and other aquatic ecosystems can be strongly limited by sunlight and temperature.

Although the world's oceans generally have very low productivity, coral reefs and estuaries are among Earth's most productive ecosystems (see Table 44.1); in both of these ecosystems, currents and waves renew nutrients and the shallow waters allow photosynthesis. Some wetlands, such as swamps and marshes, also have high productivity.

productive terrestrial ecosystems are deserts and tundra (including alpine communities).

The global pattern of NPP in marine ecosystems (Figure 44.2b) is very different from that on land. There is little tendency for NPP to decrease from the equator to the poles. The open ocean has low productivity and is, in essence, a marine desert. The productivity of marine ecosystems is often high in regions of the ocean that

■ Most life on Earth depends on energy that is captured by photosynthetic organisms. Net primary productivity varies greatly among different ecosystem types and different parts of the world.

Energy Flow through Ecosystems

As described in Chapter 43, there are two major ways in which organisms capture energy. **Producers** get their energy from nonliving sources, such as the sun. The organisms at the bottom of a food web, such as plants, algae, and photosynthetic bacteria, are producers. **Consumers** get their energy by eating all or parts of other organisms or their remains. Consumers include **decomposers**, such as bacteria and fungi, that break down the dead bodies of organisms, as well as the herbivores, predators, parasites, and pathogens described in Chapter 42.

Energy pyramids

Energy from the sun is stored by plants and other producers in the form of chemical compounds. This energy can be transferred from organism to organism in a food chain. In the process, however, energy is steadily lost from the ecosystem, as described on page 724, and hence cannot be recycled. To illustrate this point, let's follow the fate of energy from the sun after it strikes the surface of a grassland.

A portion of the energy captured by grasses is transferred to the herbivores that eat the grasses, and then to the predators that eat the herbivores. The transfer of energy from grasses to herbivores to predators is not perfect, however. When a unit of energy is used by an organism to fuel its metabolism, for example, some of that energy is lost from the ecosystem as unrecoverable heat. Thus, energy moves through ecosystems in a single direction: As one proceeds up a food chain (for example, from grass to grasshopper to bird), portions of the energy originally captured by photosynthesis are steadily lost. Because of this steady loss of energy, more energy is available toward the bottom than toward the top of a food chain.

The amounts of energy available to organisms in an ecosystem can be represented by a pyramid. Each level of an energy pyramid corresponds to a step in a food chain and is called a **trophic level** (Figure 44.3). The grass–grasshopper–bird example given in the previous paragraph has three trophic levels: Grass is on the first trophic level, the grasshopper is on the second trophic level, and the bird is on the third trophic level. On average, roughly 10 percent of the energy at one trophic level is transferred to the next trophic level. The energy that is not transferred between trophic levels is either not consumed (for example, when we eat an apple, we eat only a small part of the apple tree) or not taken up by the body (for example, we cannot digest the cellulose that is contained in the apple), or lost as metabolic heat.

Secondary productivity

The rate of new biomass production by consumers is called **secondary productivity**. Because consumers depend on producers for both energy and materials, secondary productivity is highest in ecosystems with high net primary productivity. Tundra, for example, has a much lower NPP than grassland. For this reason, fewer herbivores per unit of area can be supported on tundra than on grassland.

New biomass made by plants and other producers is consumed either by herbivores or by decomposers. In some ecosystems, 80 percent of the biomass produced by plants is used directly by decomposers. Eventually, since

**Figure 44.3
An Idealized
Energy Pyramid**
On average, roughly 10 percent of the energy at each trophic level is transferred to the next trophic level. This figure shows the energy available at each trophic level for each 1,000,000 kilocalories of energy from the sun (1 kilocalorie equals 1000 calories).

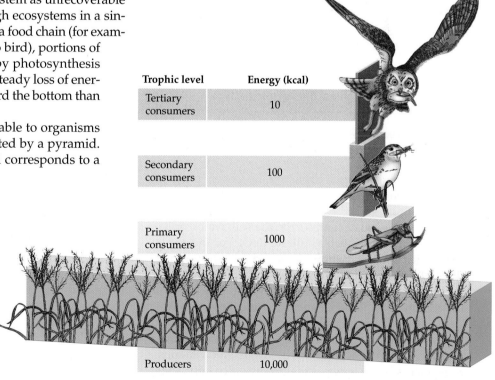

Trophic level	Energy (kcal)
Tertiary consumers	10
Secondary consumers	100
Primary consumers	1000
Producers	10,000

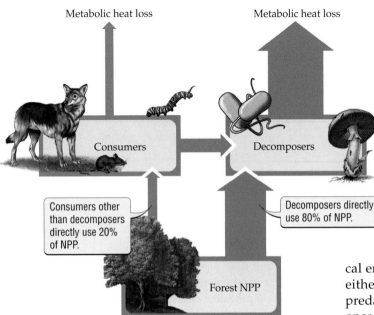

Forest NPP

Consumers

Decomposers

Metabolic heat loss

Metabolic heat loss

Consumers other than decomposers directly use 20% of NPP.

Decomposers directly use 80% of NPP.

Figure 44.4   Decomposers Rule
Decomposers such as bacteria and fungi directly use more than 50 percent of net primary productivity in ecosystems of all types. In this forest, 80 percent of NPP is used directly by decomposers, and the remaining 20 percent is used by other consumers (such as herbivores and predators).

all organisms die, all biomass made by producers, herbivores, and predators is consumed by decomposers (Figure 44.4). As we will see in the next section, decomposers are tremendously important to all life: They recycle nutrients, thereby ensuring that nutrients do not remain locked up in dead material once an organism dies.

■ The energy captured by producers can be transferred from organism to organism in a food chain. Once an organism uses energy to fuel its metabolism, that energy is lost from the ecosystem. Secondary productivity, the new biomass made by consumers, is highest in ecosystems with high net primary productivity.

Nutrient Cycles

Chemical elements such as carbon, hydrogen, oxygen, and nitrogen are used by organisms to construct their bodies (Figure 44.5). Producers obtain these and other essential chemical elements from the soil, water, or air in the form of ions such as nitrate (NO_3^-) or inorganic molecules such as carbon dioxide (CO_2). Consumers obtain them by eating producers or other consumers.

The essential elements required by producers are called **nutrients**. These elements include carbon, hydrogen, oxygen, nitrogen, phosphorus, sulfur, and many others. In discussing the biology of humans and other animals, the term "nutrients" is used to refer to vitamins, minerals, essential amino acids, and essential fatty acids

(see Chapter 28). In the ecosystem context of this chapter, however, we'll use the term "nutrients" to mean only the essential chemical elements required by producers.

Nutrients are cycled between organisms and the physical environment (see Figure 44.1). First, they are taken up from the physical environment by producers. They are then passed either directly or indirectly through herbivores and predators to decomposers. Decomposers break down once-living tissues into simple chemical components, thereby returning nutrients to the physical environment. Without decomposers, nutrients could not be repeatedly reused, and life would cease because all essential nutrients would remain locked up in the bodies of dead organisms.

Within the physical environment, a nutrient may move between air, water, soil, and rock before it is captured once again by a producer. Nutrients can cycle between organisms and the physical environment rapidly (in days to months) or very slowly: It may take many millions of years before an element that was once part of an organism is finally reabsorbed by a producer.

The cyclical movement of a nutrient between organisms and the physical environment is called a **nutrient cycle**. There are two main types of nutrient cycles on Earth: sedimentary and atmospheric. Both types of cycles are affected by human activities, as we'll see in the following section.

Sedimentary nutrient cycles

A nutrient that does not enter the atmosphere easily is said to have a **sedimentary cycle**. Such nutrients first cycle within terrestrial and aquatic ecosystems for variable periods of time; then they are deposited on the ocean bottom as sediments. Nutrients may remain in sediments, unavailable to most organisms, for hundreds of millions of years. Eventually, however, the bottom of the ocean is thrust up by geologic forces to become dry land, and once again the nutrients in the sediments may be available to organisms. Sedimentary nutrients usually cycle very slowly, so they are not replaced easily once they are lost from an ecosystem.

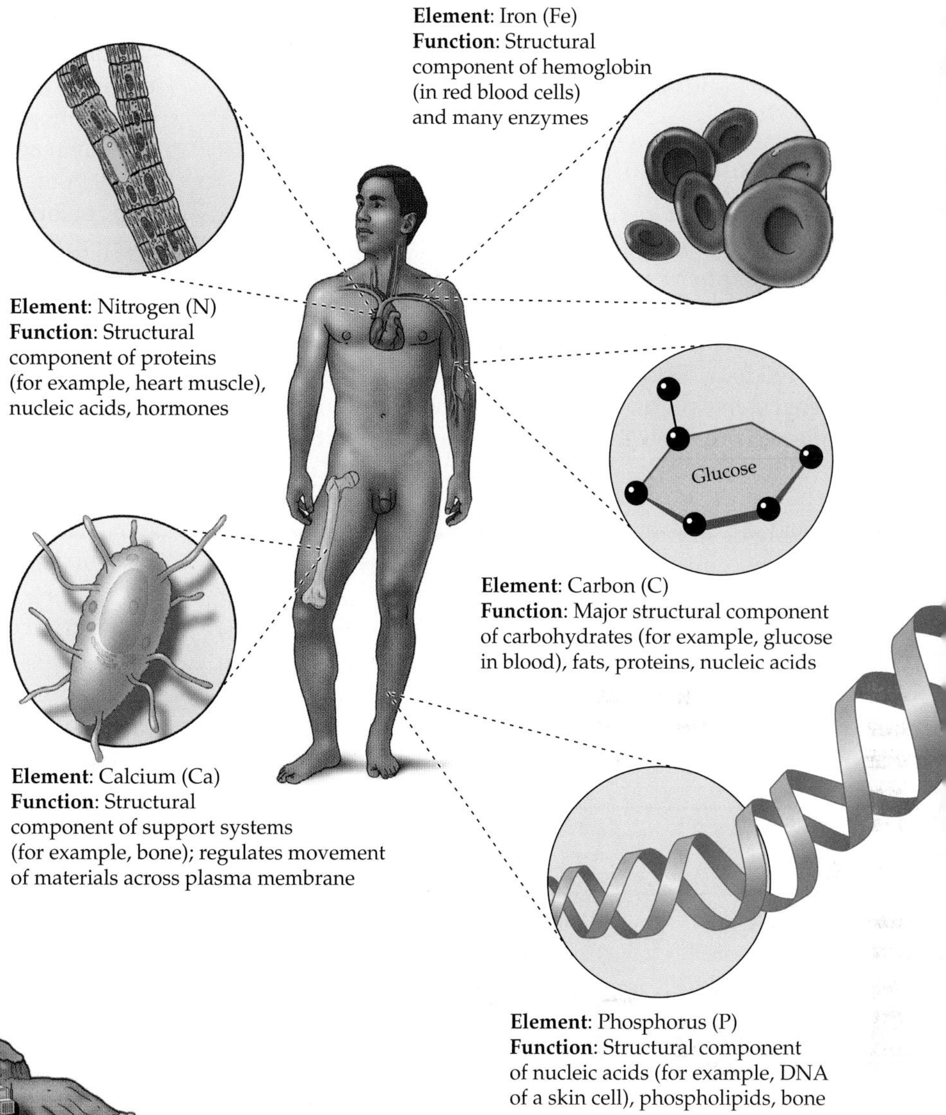
Element: Iron (Fe)
Function: Structural component of hemoglobin (in red blood cells) and many enzymes
Element: Nitrogen (N)
Function: Structural component of proteins (for example, heart muscle), nucleic acids, hormones
Glucose
Element: Carbon (C)
Function: Major structural component of carbohydrates (for example, glucose in blood), fats, proteins, nucleic acids
Element: Calcium (Ca)
Function: Structural component of support systems (for example, bone); regulates movement of materials across plasma membrane
Element: Phosphorus (P)
Function: Structural component of nucleic acids (for example, DNA of a skin cell), phospholipids, bone

Figure 44.5 Some of the Chemical Elements Required by Organisms

Phosphorus is one important nutrient that has a sedimentary cycle (Figure 44.6). Phosphorus is important to ecosystems because it strongly affects net primary productivity, especially in aquatic ecosystems. NPP usually increases, for example, when phosphorus is added to lakes. Such an increase in productivity can have undesirable effects: When phosphorus enters a lake in sewage or in water that drains from agricultural fields and contains fertilizer (agricultural runoff), the added phosphorus can cause populations of algae and other phytoplankton to increase dramatically. Such an increase in phytoplankton populations can lead eventually to the death of aquatic plants, fish, and invertebrates (see the story of Lake Washington on p. 715).

44.4

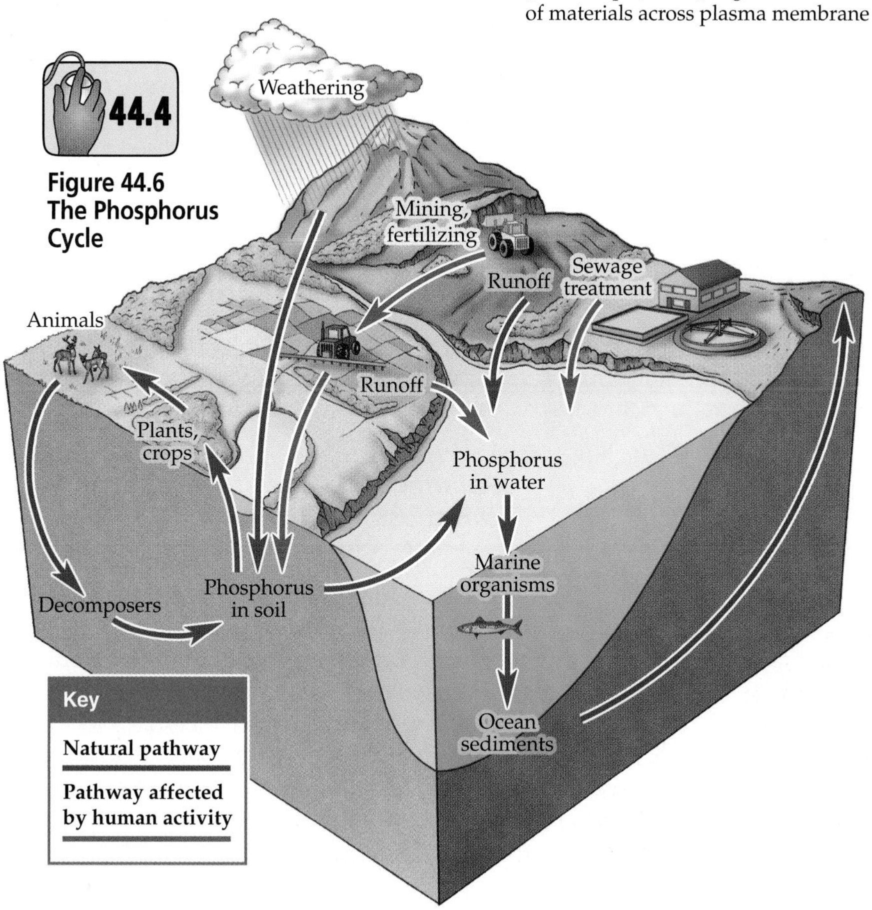
Weathering
Mining, fertilizing
Runoff
Sewage treatment
Animals
Runoff
Plants, crops
Phosphorus in water
Marine organisms
Decomposers
Phosphorus in soil
Ocean sediments
Key
Natural pathway
Pathway affected by human activity

Figure 44.6 The Phosphorus Cycle

The phosphorus cycle begins when a phosphorus atom is released from rock by weathering. Once this atom is absorbed by a producer, it may cycle within terrestrial ecosystems for years or centuries before it leaves them in a stream. From there, the phosphorus atom cycles within streams, rivers, or lakes, but within a few weeks to years it enters the ocean. Within the ocean, the atom cycles between deep and surface waters for an average of 100,000 years, after which it is deposited in sediments on the ocean bottom. It can remain trapped in these sediments for 100 million years or more. Finally, geologic forces cause the ocean floor to rise and become dry land, exposing the sediments to weathering, and the cycle begins again.

Atmospheric nutrient cycles

Nutrients that cycle between terrestrial ecosystems, aquatic ecosystems, and the atmosphere are said to have an **atmospheric cycle**. Because these nutrients enter the atmosphere easily, they can be transported by wind from one region of Earth to another. When nutrients are transported long distances in this way, their cycling in a local ecosystem, such as a lake in a remote region, may be affected by events that occur in distant parts of the globe.

Nitrogen, carbon, and sulfur are all important nutrients that have atmospheric cycles. Let's look at the cycling of sulfur as an example. There are three natural ways by which sulfur enters the atmosphere from terrestrial and aquatic ecosystems (Figure 44.7): in sea spray, as a metabolic by-product (the gas hydrogen sulfide, H_2S) released some types of bacteria, and, least importantly in terms of overall amount, as a result of volcanic activity. Human activities also cause sulfur to enter the atmosphere, as we will see shortly.

Sulfur enters terrestrial ecosystems through the weathering of rocks and as sulfate (SO_4^{2-}) that is lost from the atmosphere. Sulfur enters the ocean in stream runoff from land and, again, as sulfate lost from the atmosphere. Once in the ocean, sulfur cycles within marine ecosystems before being lost in sea spray or deposited in sediments. Compared with phosphorus, sulfur cycles through terrestrial and aquatic ecosystems quickly.

> ■ Nutrients are cycled between organisms and the physical environment. Decomposers return nutrients from the bodies of dead organisms to the physical environment. Nutrients such as phosphorus that do not enter the atmosphere easily are said to have sedimentary cycles. Nutrients such as carbon, nitrogen, and sulfur that enter the atmosphere easily are said to have atmospheric cycles.

Human Activities Can Alter Nutrient Cycles

Human activities can have major effects on nutrient cycles. Ecologists have shown, for example, that the clear-cutting of a forest, followed by spraying with herbicides to prevent regrowth, causes the forest to lose large amounts of NO_3^- (nitrate), an important source of nitrogen for plants (Figure 44.8). On a larger geographic scale, the nitrogen and phosphorus used to fertilize a farmer's fields can be carried by streams to a lake hundreds of kilometers away. The addition of nitrogen and phosphorus to the lake may disrupt local nutrient cycles and cause some species (for example, algae) to thrive at the expense of others (for example, fish and invertebrates).

Pollutants that contain sulfur can cause acid rain

When humans alter atmospheric nutrient cycles, the effects are often felt across international borders. Consider sulfur dioxide, which is released into the atmosphere when we burn fossil fuels such as oil and coal. The burning of fossil fuels has altered the sulfur cycle greatly: Annual human inputs of sulfur to the atmosphere are more than one and a half times the inputs from all natural sources combined.

44.5

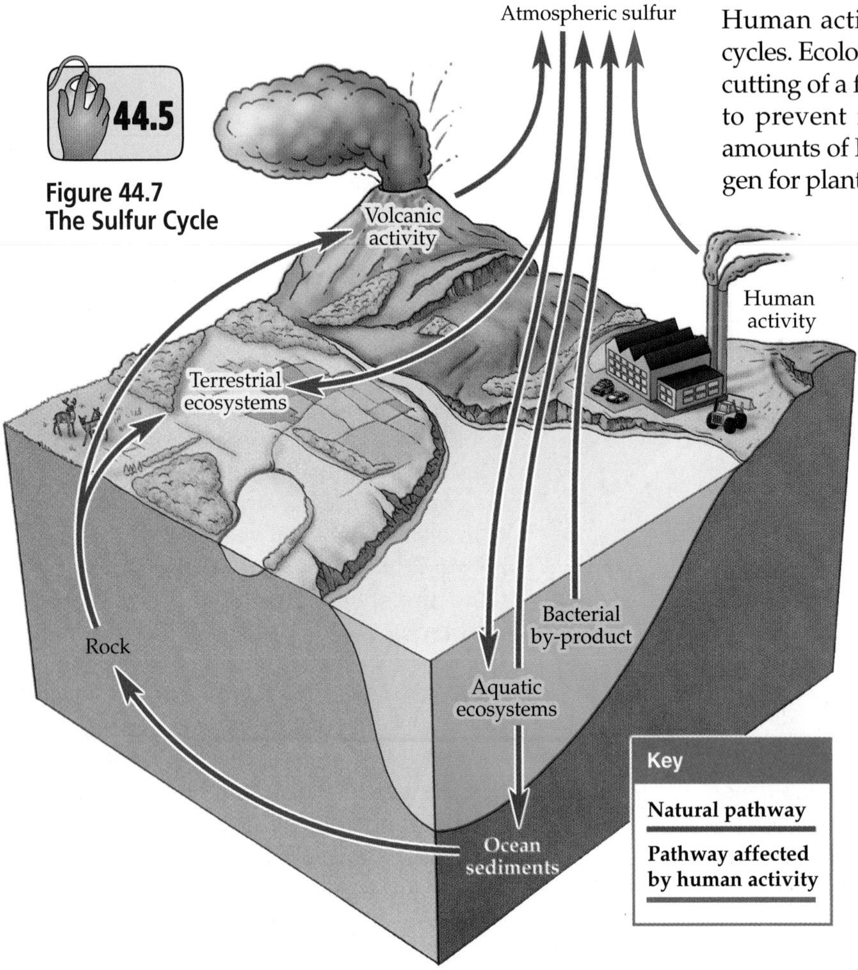

**Figure 44.7
The Sulfur Cycle**

Atmospheric sulfur

Volcanic activity

Human activity

Terrestrial ecosystems

Bacterial by-product

Rock

Aquatic ecosystems

Ocean sediments

Key

Natural pathway

Pathway affected by human activity

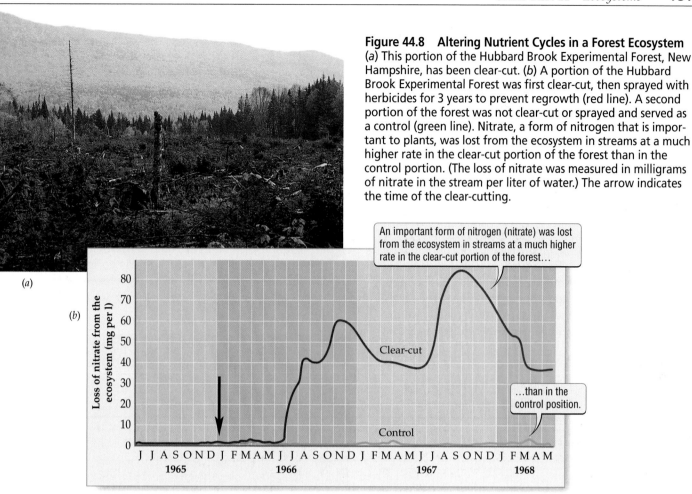

Figure 44.8 Altering Nutrient Cycles in a Forest Ecosystem
(*a*) This portion of the Hubbard Brook Experimental Forest, New Hampshire, has been clear-cut. (*b*) A portion of the Hubbard Brook Experimental Forest was first clear-cut, then sprayed with herbicides for 3 years to prevent regrowth (red line). A second portion of the forest was not clear-cut or sprayed and served as a control (green line). Nitrate, a form of nitrogen that is important to plants, was lost from the ecosystem in streams at a much higher rate in the clear-cut portion of the forest than in the control portion. (The loss of nitrate was measured in milligrams of nitrate in the stream per liter of water.) The arrow indicates the time of the clear-cutting.

Most human inputs of sulfur to the atmosphere come from heavily industrialized areas such as northern Europe and eastern North America. Once in the air, sulfur dioxide (SO_2) is dissolved in water and converted into sulfuric acid (H_2SO_4), which then returns to Earth in rainfall. Rainfall normally has a pH of 5.6, but sulfuric acid (as well as nitric acid, caused by nitrogen-containing pollutants) has caused the pH of rain to drop to values as low as 2 or 3 in the United States, Canada, Great Britain, and Scandinavia (see Chapter 5 to review pH). Rainfall with a low pH is called **acid rain**.

Acid rain can have devastating effects on human-made structures (such as statues) and on natural ecosystems. With respect to its effects on ecosystems, acid rain has drastically reduced fish populations in thousands of Scandinavian and Canadian lakes. Much of the acid rain that falls in these lakes is caused by sulfur dioxide pollution that originates in other countries (for example, Britain, Germany, and the United States). Acid rain has also caused extensive damage to forests in North America and Europe (Figure 44.9).

Figure 44.9 Acid Rain Damage
A spruce forest killed by acid rain in the Jizerske Mountains of Czechoslovakia.

The international nature of the acid rain problem has led nations to agree to reduce sulfur emissions. In the United States, sulfur emissions were cut by 40 percent between 1973 and 2001. Such reductions are a very positive first step, but the problems resulting from acid rain will be with us for a long time: Acid rain alters soil chemistry and thus has effects on ecosystems that will last for many decades after the pH in rainfall returns to normal levels.

Implications of nutrient cycles

The nutrients needed for life are cycled between air, water, soil, rock, and living organisms. Because many nutrients enter the atmosphere easily, human alteration of nutrient cycles can have local, regional, and global consequences. As ecologist Charles Krebs has written, humans "are all tied together to the fate of the earth. Because of the world scale of nutrient cycles, we are linked [to] communities all over the world. There are no more islands."

> ■ Human activities can alter nutrient cycles on local, regional, and global scales. Because many nutrients enter the atmosphere easily, actions in one part of the globe can affect ecological communities located in distant parts of the world.

■ HIGHLIGHT

Ecosystems Provide Essential Services at No Cost

In 1996 and 1997, floods struck the western United States with a fury. Damage amounted to billions of dollars in the states of Nevada, California, Oregon, and Washington (Figure 44.10a). Most news reports about the floods and associated mudslides (Figure 44.10b) said they were caused by unusually large amounts of rain and snowfall. Although the weather certainly had a major effect, it is only part of the story.

A huge flood is not always something that just happens, beyond our control. Human actions may help set the stage for flooding. By logging hillsides, diverting rivers, covering soil with concrete, and building on areas where rivers typically overflow (floodplains), we prevent ecosystems from responding as they normally would to heavy rainfall. Increasingly, the effects of our actions return to us in the form of devastating floods.

How do logging and other types of development alter the frequency and severity of floods? Trees hold soil in place with their roots, and a vast amount of water is removed from the soil by trees. Thus, by cutting down trees, logging increases both the rate of soil erosion (since the soil is no longer held in place by trees) and the amount of water in the soil; much of the extra water in the soil leaves the ecosystem in streams (stream runoff). The soil itself often can hold considerable water, and covering it with materials such as concrete and asphalt further increases runoff. Higher levels of ero-

(a)

(b)

Figure 44.10 Flood Devastation in the Pacific Northwest (a) Following heavy flooding, people use boats and float tubes to get from place to place in Oregon City. (b) Mudslides can kill people, contaminate stream ecosystems, and have undesirable aesthetic effects.

sion and stream runoff make both mudslides and floods more likely.

When we build dikes and levees and divert the flow of rivers, we seek to control rivers and to prevent floods. We often do this to protect homes or industrial areas located in what were once floodplains. But by preventing rivers from overflowing into floodplains, we reduce the ability of the ecosystem to handle periods of heavy rainfall. Floodplains function as huge sponges: When streams and rivers overflow, floodplains absorb excess water and prevent even more severe floods from occur-

ring farther downstream. By building on floodplains and attempting to control floods, we unintentionally make it much more likely that when a flood does occur, it will be a big one.

Floodplains provide us with a free service: They act as safety valves for major floods. Ecosystems provide us with many such essential services at no cost (Figure 44.11).Wetlands and estuaries, for example, have a great capacity to filter wastes from water and convert them to harmless substances. Throughout the world, humans take advantage of this fact by using wetlands and estuaries as cost-free waste treatment plants. Unfortunately, the filtering capacity of wetlands and estuaries can be overrun by large cities.

Other free ecosystem services include pollination by insects (essential for many crops), removal of pollutants from the atmosphere by forest ecosystems, maintenance of breeding grounds for shellfish and fish in marine ecosystems, prevention of soil erosion by plants, screening of dangerous ultraviolet light by the atmospheric ozone layer, moderation of the climate by the ocean, and, as we have seen throughout this chapter, nutrient cycling. These and other ecosystem services are essential to maintaining the productivity of ecological communities.

Too often we humans manipulate the environment for temporary advantage while neglecting to consider the full implications of our actions. Our civilization would collapse were it not for the many free and essential services provided to us by ecosystems. At present, we cannot duplicate with technology what ecosystems provide us for free (see the box on p. 734). Thus, when we destroy or degrade ecosystems, we do so at our own peril.

> ■ Human society depends on natural ecosystems for free and essential services such as nutrient cycling, flood control, and the filtering of pollutants from air and water.

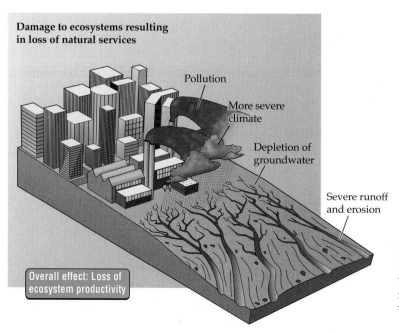

Natural services provided by ecosystems

Removal of pollutants; clean air and clear water

Climate moderation

Replacement of groundwater

Erosion control

Overall effect: Productive ecosystem

Damage to ecosystems resulting in loss of natural services

Pollution

More severe climate

Depletion of groundwater

Severe runoff and erosion

Overall effect: Loss of ecosystem productivity

Figure 44.11   Services Provided by Ecosystems
Ecosystems provide humans with many free services upon which our civilization depends. These same ecosystem services are also essential to maintaining the productivity of ecological communities.

■ BIOLOGY IN OUR LIVES

Ecosystem Design

Throughout the world, much effort is currently being devoted to the design and construction of ecosystems. One reason for these efforts is the desire to restore or replace ecosystems that have been degraded or destroyed by human activities. In the Netherlands, for example, where the land has been heavily modified by people for so long that few natural ecosystems remain, intensive work is under way to rebuild some of the original ecosystems. In many other countries, similar efforts are in progress to replace heavily damaged ecosystems, such as native prairie ecosystems in the United States.

People also seek to design and build ecosystems for economic reasons. For example, there is often a potential for conflict between developers who want to build homes or industrial parks and environmental activists who seek to prevent such development in order to protect natural ecosystems. Some argue that such conflict between developers and environmentalists could be avoided by playing a type of "zero sum" game: If an ecosystem is destroyed or degraded by development, a comparable ecosystem could be built to replace the damaged natural ecosystem.

In principle, such a zero sum game could work if people could build ecosystems with properties similar to those of natural ecosystems. Are we able to do this? Results to date indicate we still have much to learn. In a 2001 National Academy of Sciences report on wetland restoration, a team of scientists concluded that we can build ponds and cattail marshes, but we cannot replace fens, bogs, and many other complex, species-rich wetland ecosystems. Similar challenges face efforts to restore terrestrial ecosystems. Restored prairies, for example, can resemble native prairies in some respects after as few as 5 to 10 years. However, even 30 to 50 years after restoration work has begun, critical differences may remain in the way nutrients are cycled in restored versus native prairie ecosystems.

Ecosystem design is a difficult task. Most ecosystems are very complex, and we often know so little about fundamental ecosystem processes that it is hard to imagine we could replicate those processes in a human-built ecosystem. In addition, as we learned in Chapter 43, human actions can have long-term effects on natural communities, in some instances even shifting a community from one ecosystem type to another. In such cases, restoration of the original ecosystem may be impossible or—at best—may succeed only with prolonged, intensive, and costly management efforts.

A Restored Prairie Ecosystem

SUMMARY

How Ecosystems Function: An Overview

- Energy and materials can move from one ecosystem to another.
- Energy is not recycled in ecosystems. A portion of the energy captured by producers is lost as metabolic heat at each step in a food chain.
- Nutrients are recycled in ecosystems. They pass from the environment to producers to consumers, then back to the environment when decomposers break down the bodies of dead organisms.

Energy Capture in Ecosystems

- Most life on Earth depends on energy that is captured by photosynthetic organisms and stored in their bodies as chemical compounds.

- Net primary productivity (NPP) varies greatly among different ecosystem types and different parts of the world.

- On land, NPP tends to decrease from the equator toward the poles.

- In marine ecosystems, NPP is often high where the ocean borders land or where upwellings provide scarce nutrients to marine organisms.

Energy Flow through Ecosystems

- The energy captured by producers can be transferred from organism to organism in a food chain.

- Once an organism uses energy to fuel its metabolism, that energy is lost from the ecosystem.

- The amounts of energy available to organisms in an ecosystem can be represented in the form of a pyramid.

- Secondary productivity is highest in areas of high net primary productivity.

Nutrient Cycles

- Nutrients are cycled between organisms and the physical environment.

- Decomposers return nutrients from the bodies of dead organisms to the physical environment, thus allowing the cycling of nutrients, on which all life depends.

- Nutrients that do not enter the atmosphere easily have sedimentary cycles, which usually take a long time to complete.

- Nutrients that enter the atmosphere easily have atmospheric cycles, which occur relatively rapidly and can transfer nutrients between distant parts of the world.

Human Activities Can Alter Nutrient Cycles

- Human activities can alter nutrient cycles on local, regional, and global scales.

- Human inputs to the sulfur cycle exceed those from all natural sources combined, creating problems of international scope, such as acid rain.

- Because many nutrients enter the atmosphere easily, our actions in one part of the globe can affect ecological communities located in distant parts of the world.

Highlight:
Ecosystems Provide Essential Services at No Cost

- Human society depends on ecosystems for free and essential services such as nutrient cycling, flood control, and the filtering of pollutants from air and water.

KEY TERMS

acid rain p. 731	nutrient p. 728
atmospheric cycle p. 730	nutrient cycle p. 728
consumer p. 727	producer p. 727
decomposer p. 727	secondary productivity p. 727
ecosystem p. 724	sedimentary cycle p. 728
net primary productivity (NPP) p. 725	trophic level p. 727

CHAPTER REVIEW

Self-Quiz

1. The amount of energy captured by photosynthesis, minus the amount lost as heat in respiration, is
   a. secondary productivity.
   b. consumption efficiency.
   c. NPP.
   d. photosynthetic efficiency.

2. The movement of nutrients between organisms and the physical environment is called
   a. nutrient cycling.
   b. ecosystem services.
   c. NPP.
   d. a nutrient pyramid.

3. Free services provided to humans by ecosystems include
   a. control of atmospheric carbon dioxide concentration.
   b. prevention of soil erosion.
   c. filtering of pollutants from water and air.
   d. all of the above

4. Each step in a food chain is called a
   a. trophic level.
   b. consumer level.
   c. food web.
   d. producer.

5. What type of organisms consume 50 percent or more of the NPP in all ecosystems?
   a. herbivores
   b. decomposers
   c. producers
   d. predators

Review Questions

1. What prevents energy from being recycled in ecosystems?

2. What is the essential role of decomposers in ecosystems?

3. Explain why human alteration of nutrient cycles can have international effects.

4. Would it be in the self-interest of nations worldwide to form and abide by strict, enforceable international agreements on inputs to nutrient cycles?

5. Describe some key ecosystem services and discuss the extent to which human economic activity depends on such services.

44 **The Daily Globe**

New York City Buys Land to Save Water

NEW YORK, NY—This week, New York City officials announced the purchase of 1000 acres of streamside property in the Catskills. The largest purchase yet in the city's pioneering effort to preserve the drinking water that comes from this region, this latest acquisition has drawn attention around the country as municipalities of all sizes—from tens of thousands to hundreds of thousands—have begun looking for solutions to the decline in the quality of their drinking water.

For years, residents of New York City took their drinking water for granted, drawing it from the Catskills region of upstate New York. The water was kept pure by the root systems, soil microorganisms, and natural filtration processes of the region's forests. But in recent years, increasing amounts of sewage, fertilizers, and pesticides in the water have caused its quality to deteriorate.

When the city's drinking water no longer met Environmental Protection Agency (EPA) standards, New York City officials could have built a water treatment plant, as other cities have done, at a cost of 6 to 8 billion dollars, plus another 300 million dollars per year for its operation. Instead, for an estimated cost of 1 to 1.5 billion dollars, the city has embarked on an ambitious but simple plan: protect the environment so that natural ecosystems can once again supply the city with clean water. The city is buying land that borders rivers in the Catskills, protecting the land from development to minimize fertilizer and pesticide inputs into the water, and building sewage treatment plants for rural communities in the Catskills, thus decreasing sewage inputs.

"We're talking about a no-brainer here," said city water official Joe Marin, who has already been contacted by a number of cities hoping to follow New York's model. "For a one-time investment of less than 25 percent of the cost of building a treatment plant, we get clean water and we save huge annual expenses." In addition, the city protects the environment, in a case in which what's good for the ecosystem is also good for the bottom line.

Evaluating "The News"

1. The choice in New York City was clear: The city could save money and protect the environment at the same time. In other situations, protecting the environment may not be cost-effective in a strictly economic sense. Should society protect the environment regardless of cost? How can a balance between ecological and economic factors be reached?

2. Should individuals, companies, or cities that damage the ability of ecosystems to provide free services (such as pure drinking water) be charged for harming those ecosystems? Why or why not?

3. The profit motive can drive people to come up with creative solutions to complex problems, including environmental problems. One solution to New York City's problems would be to form a corporation to manage the city's water supply. This corporation would have the right to sell an ecosystem service—in this case, clean drinking water. Ownership of this right would enable the corporation to raise the money needed to improve New York City's drinking water supply. Driven by the profit motive, the corporation might develop new and cost-effective ways of protecting the ecosystem's ability to provide pure drinking water. Should such profit-driven approaches be used to protect the environment? Why or why not?

45

Global Change

Devastation on the High Seas

Nearly 75 percent of Earth's surface is covered with oceans. The oceans are so deep and so vast that many scientists once thought humans could never drive marine species extinct. They reasoned that no matter how much we overhunted a species or polluted local portions of its habitat, there would always be places where the species could thrive. Now it seems that even marine species are not safe from the impact of human actions.

Consider the white abalone's tale of woe. This large marine shellfish once was common along 1200 miles of the California coast. It lives on rocky reefs in relatively deep water (25 to 65 meters or deeper). The

738

fact that it lives in deep water protected it for a while: White abalone is delicious to eat, but people first hunted other species of abalone that live in shallower waters and hence are easier to find. When the shallow-water species became rare, fishermen turned to the white abalone. After only 9 years of commercial fishing, the fishery collapsed. This species, which once covered the seafloor with up to 10,000 individuals per hectare, is now on the verge of extinction.

And the white abalone is not alone. The barndoor skate, a large fish that once had a wide geo-graphic distribution, has also been hunted to near extinction. Overall, humans have had an enormous negative effect on fish populations worldwide. Recent studies indicate that 66 percent of the world's marine fisheries are in trouble due to overfishing.

Another sign of our effect on marine ecosystems is what fishermen are catching. In the past 45 years, the catch from fisheries across the globe has included fewer large carnivorous fish and more invertebrates and small fish that feed on plankton. Thus it appears we are altering the food webs of many ocean communities as well as reducing the populations of individual species.

The extinctions of marine species and the other effects of human actions on marine ecosystems are just a few of the changes we are making to the biosphere. Other examples include changes in the global sulfur cycle (see Chapter 44), effects on the location of biomes (see Chapter 39), and the extinctions or declines of many species worldwide (see Chapter 4). These and other worldwide environmental changes, referred to collectively as global change, are the subject of this chapter.

The White Abalone and Its California Coastal Habitat

■ KEY CONCEPTS

1. The effects of human actions on the world's lands and waters are thought to be the main causes of the current high rate of extinction of species.

2. Human inputs to the global nitrogen cycle now exceed those of all natural sources combined. If unchecked, these changes to the nitrogen cycle are expected to have negative effects on many ecosystems.

3. The concentration of carbon dioxide (CO_2) gas in the atmosphere is increasing at a dramatic rate, largely because of the burning of fossil fuels. Increased CO_2 levels

are expected to have large but hard-to-predict effects on ecosystems.

4. Increased concentrations of CO_2 and certain other gases in the atmosphere are predicted to cause a rise in Earth's temperatures. Most scientists think that such global warming is occurring, but its extent and consequences remain uncertain.

5. Because global change caused by humans is expected to have large, negative consequences for many species, including our own, we must learn to use Earth's ecosystems in a sustainable fashion.

Statements by politicians, talk show hosts, and others often give the impression that worldwide change to the environment, or **global change**, is a controversial topic. Such statements cause many in the general public to think that global change may not really be occurring, or cause them to wonder whether anything really needs to be done about it.

This impression of controversy is unfortunate. Contrary to what some reports in the media might lead us to believe, we know with certainty that global change is occurring. Invasions of nonnative species have increased worldwide (see examples in Chapters 40, 41, and 43), large losses of biodiversity have occurred (see Chapter 4), and pollution has altered ecosystems throughout the world (see Chapter 44). Each of these three examples illustrates an important type of global change that we know with certainty is happening today.

Although the examples of global change just mentioned are caused by people, the biosphere has always changed over time. The continents move, the climate changes, and succession and natural forms of disturbance change the composition of communities (see Chapters 39 and 43). Thus, even in the absence of human actions, we know that ecological communities face—and always have faced—global change.

In this chapter we describe how humans have influenced global change. We first discuss two types of global change that we know have occurred and that we know are caused by people: changes in land and water use and changes in the cycling of chemicals through ecosystems. As we will see, these and other kinds of human-caused global change are occurring at a rate and intensity that far exceeds natural rates of change. This observation leads us to discuss why it is

important for people to have a sustainable impact on Earth, and how we might achieve that goal.

45.1 Land and Water Transformation

Humans have a profound effect on the lands and waters of Earth.

Land transformation refers to physical and biotic changes that humans make to the land surface of Earth. Such changes include the destruction of natural habitat to allow for urban growth, agriculture, and resource use (as when a forest is clear-cut for lumber). Land transformation also includes many human activities that alter natural habitat to a lesser degree, as when we graze cattle on grasslands.

Similarly, **water transformation** refers to physical and biotic changes that humans make to the waters of our planet. For example, we have drastically altered the way water cycles through ecosystems. Humans now use more than half of the world's accessible fresh water, and we have altered the flow of nearly 70 percent of the world's rivers. Since water is essential to all life, our heavy use of the world's waters has many and far-reaching effects, including changing where water is found, and what species can survive, at a given location.

Evidence of land and water transformation

Aerial photos (or satellite data), changing urban boundaries, and local instances of destruction of natural habitats show how humanity is changing the face of Earth (Figure 45.1). Together, these and many other kinds of evidence show that land and water transformation is

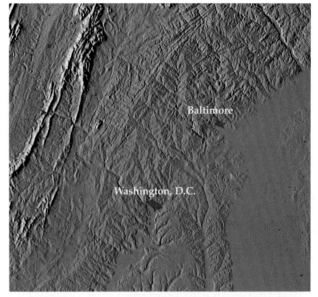

(c) 1850

1992

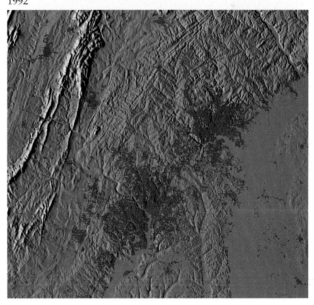

Figure 45.1 Examples of Land Transformation
(a) These forests in Washington State have been clear-cut.
(b) An open-pit copper mine in Arizona. (c) Change in the
urban boundaries (in red) of Baltimore, Maryland, and
Washington, DC, between 1850 and 1992.

occurring, is caused by human actions, and is global in
scope.

To estimate the total amount of land that has been
transformed by people, the effects of many different
human activities must be summed across every acre of
the world. This task may seem nearly impossible, but
with the aid of satellites and other new technologies, we
can now measure our total impact on Earth for the first
time in history. Although we are just beginning to determine the extent to which we have transformed the planet, one reasonable estimate is that humans have substantially altered one-third to one-half of Earth's land
surface. Thus, although we do not know the exact

amount of land we have transformed, we do know that
we have altered a large percentage of it.

In modifying the land and waters for our own use,
we have had a dramatic effect on many ecosystems. For
example, large regions of the United States were once
covered by wetlands. Many of these wetlands have been
drained so that the land could be used for agriculture or
other purposes. During the 200-year period beginning
in the 1780s, wetlands declined in every state in the United States (Figure 45.2), and they are still declining today.
Other examples of human effects on ecosystems include
the ongoing destruction of tropical rainforests and the

(*a*) Wetland distribution, circa 1780s

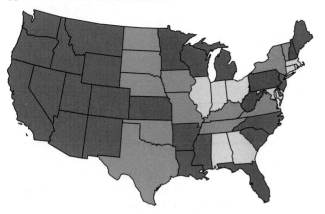

(*b*) Wetland distribution, circa 1980s

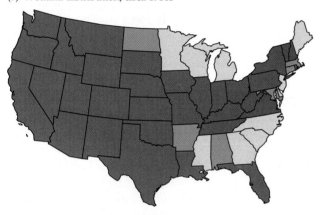

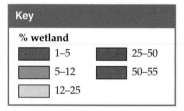

Key

% wetland

▓	1–5	▓	25–50
▓	5–12	▓	50–55
▓	12–25		

Figure 45.2   The Decline of Wetlands
Wetlands in the United States declined greatly over 200 years. The maps show the percentage of the land area of each state covered by wetlands (*a*) in the 1780s and (*b*) in the 1980s.

conversion of once vast grasslands in the American Midwest to croplands.

Consequences of land and water transformation

Many ecologists think that land transformation and water transformation are—and will remain for the immediate future—the two most important components of global change. There are several reasons why our transformation of the lands and waters of Earth is so important.

First, as humans alter the lands and waters to produce goods and services for an increasing number of people, we use a very large share of the world's resources. For example, estimates suggest that humans now control (directly and indirectly) more than 40 percent of the world's total net primary productivity on land (see Chapter 44 for a definition of net primary productivity). By controlling such a large portion of the world's land area and resources, humans have reduced the amount of land and resources available to other species, causing many species to go extinct (see Chapter 4). Water transformation has similar effects. When humans overfish or pollute Earth's waters, we may cause dramatic changes in the abundances and types of species found in the world's aquatic ecosystems (see p. 737 and Chapters 43 and 44).

The transformation of land and water has other effects as well. When a forest is cut down, for example, the local temperature may increase and the humidity may decrease. Such changes in the local climate can make it less likely that the forest will reappear. In addition, as we'll see shortly, the cutting and burning of forests increase the amount of carbon dioxide in the atmosphere, an aspect of global change that may alter the climate worldwide.

> ■ Human activities are changing the lands and waters of the entire planet. Land and water transformation has caused the extinction of species and has the potential to alter local and global climate.

Changes in the Chemistry of Earth

Life on Earth depends on, and is heavily influenced by, the cycling of nutrients in ecosystems, as we saw in Chapter 44. Net primary productivity often depends on the amount of nitrogen and phosphorus available to producers, and the amount of sulfuric acid in rainfall has many effects on ecological communities. The nitrogen and phosphorus that stimulate net primary productivity and the sulfur in acid rain are just two of many examples of naturally occurring chemicals that cycle through ecosystems.

Synthetic chemicals also cycle through ecosystems. The pesticide DDT, as we saw in Chapter 44, is one example of such a chemical. Another is chlorofluorocarbons (CFCs), chemical compounds used as coolants in refrigerators and in foam manufacture. Because CFCs are synthetic (made only by humans), there were no CFCs in the environment until recently. CFCs are not toxic, but their use and subsequent release into the atmosphere had a large and unexpected negative effect: They caused a decrease in the ozone layer of the atmos-

(a) 1979

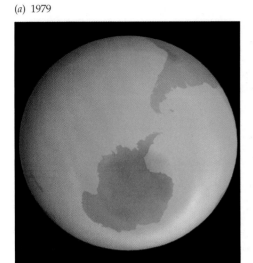

(b) 2000

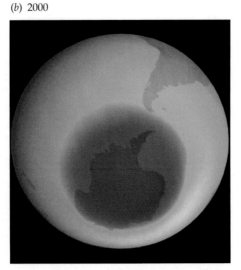

Figure 45.3
The Antarctic Ozone Hole
These satellite images show average ozone levels over Antarctica for the months of September 1979 (a) and September 2000 (b). Ozone levels declined slowly in the 1970s, then dropped dramatically in the 1980s. Thus, the September 1979 image represents near-normal conditions. In the September 2000 image, regions with the greatest ozone loss appear purple. At the time this book went to print, the largest ozone hole ever recorded occurred on September 16, 2000 and covered an area larger than all of North America.

phere (Figure 45.3). Because it shields the planet from harmful ultraviolet light, which can cause mutations in DNA, damage to the ozone layer poses a serious threat to life. Fortunately, the international community responded quickly to this threat, and treaties to halt the production of CFCs are now in place.

By adding synthetic and naturally occurring chemicals to the environment, humans have altered the way in which many chemicals cycle through ecosystems. In some cases, some of the harm caused by changes in chemical cycles has been undone (see the box on p. 687). In other cases, great challenges lie ahead. In the sections that follow, we consider two chemical cycles that human actions have changed dramatically: the global nitrogen cycle and the global carbon cycle.

■ Human activities are changing the way many chemicals are cycled through ecosystems.

 45.2 *Changes in the Global Nitrogen Cycle*

There is a large amount of nitrogen in Earth's atmosphere, where N_2 gas makes up 78 percent of the air we breathe. However, plants and most other organisms cannot use N_2 directly. Instead, the nitrogen in N_2 gas must be converted to other forms, such as ammonium (NH_4^+) or nitrate (NO_3^-), that can be used by plants and other producers. The conversion of N_2 to NH_4^+, called **nitrogen fixation**, is accomplished by several species of bacteria and, to a much lesser degree, by lightning. Once nitrogen is converted to NH_4^+, other bacteria can con-

vert it to NO_3^-. These two forms of nitrogen then cycle among plants, animals, and microorganisms. The amount of nitrogen that cycles among organisms is much smaller than the amount in the atmosphere.

Human technology is also capable of fixing nitrogen. In recent years, the amount of nitrogen fixed by human activities has exceeded the amount fixed by all natural processes combined (Figure 45.4). Much of this nitrogen fixation by humans is the result of industrial production of fertilizers. Other major sources of nitrogen fixation include fixation by car engines and fixation by bacteria that

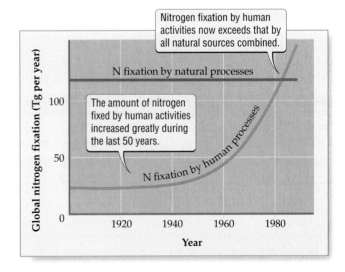

Figure 45.4 Human Effects on the Global Nitrogen Cycle
Nitrogen is fixed naturally by bacteria and by lightning at a rate of about 130 teragrams (Tg) per year (1 Tg = 10^{12} grams, or 1.1 million tons). Human activities such as the production of fertilizers now fix more nitrogen than all natural sources combined.

(a)

(b)

Figure 45.5 A Nitrogen Addition Experiment
Native grasslands in Minnesota often have 20 to 30 plant species per square meter. (*a*) This control plot received no added nitrogen, and lost no species from 1984 to 1994. (*b*) After the experimental addition of nitrogen to this plot, most of the native species disappeared and a nonnative species, European quackgrass, took over the plot.

have a mutualistic relationship with peas and other crop plants. The fact that human inputs of nitrogen to ecosystems are greater than natural inputs tells us that our activities have greatly changed the global nitrogen cycle.

The potential effects of changing the nitrogen cycle are far-reaching. When nitrogen is added to terrestrial communities, net primary productivity usually increases, but the number of species usually decreases (Figure 45.5). Ecosystems in the Netherlands receive more added nitrogen than those in any other country in the world. The addition of nitrogen to grasslands in the Netherlands that historically were poor in nitrogen has caused more than 50 percent of the species to be lost from some of these communities.

Similarly, when nitrogen is added to nitrogen-poor aquatic ecosystems, such as many ocean communities, productivity increases, but species are lost. In general, an increase in productivity caused by the addition of nitrogen is not necessarily a good thing for the ecosystem.

> ■ Nitrogen must be fixed before it can be used by producers. Human activities fix more nitrogen than all natural sources combined. The extra nitrogen fixed by human activities has altered the global nitrogen cycle, leading to increases in productivity that can cause the loss of species from ecosystems.

Changes in the Global Carbon Cycle

There are vast amounts of carbon in the biosphere, and this carbon cycles readily among organisms, soils, the atmosphere, and the ocean (Figure 45.6). We will focus here on one portion of the global carbon cycle that has been altered by human activities, the concentration of the gas carbon dioxide (CO_2) in the atmosphere.

Although CO_2 makes up less than 0.04 percent of Earth's atmosphere, its importance is far greater than its low concentration might suggest. As we have seen in earlier chapters, CO_2 is an essential raw material for photosynthesis, on which most life depends. CO_2 is also the most important of the greenhouse gases that contribute to global warming (see the discussion of global warming later in this chapter). Thus, scientists took notice in the early 1960s when new measurements showed that the concentration of CO_2 in the atmosphere was rising rapidly.

Atmospheric CO_2 levels have risen dramatically

Scientists have been measuring the concentration of CO_2 in the atmosphere since 1958. By measuring CO_2 concentrations in air bubbles trapped in ice for hundreds to hundreds of thousands of years, scientists have also estimated the concentration of CO_2 in both the recent and relatively distant past (Figure 45.7). At times for which direct measurements from the air and estimates

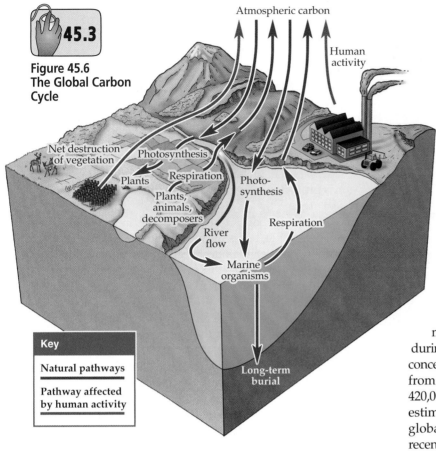

45.3

**Figure 45.6
The Global Carbon
Cycle**

Atmospheric carbon

Human
activity

Net destruction
of vegetation

Photosynthesis

Plants

Respiration

Plants,
animals,
decomposers

Photo-
synthesis

Respiration

River
flow

Marine
organisms

Long-term
burial

Key

Natural pathways

Pathway affected
by human activity

from ice bubbles have both been made, the two ways of measuring the concentration of CO_2 agree, giving us confidence that the ice bubble measurements are accurate. Both types of measurements show that CO_2 levels have risen greatly during the past 200 years.

Overall, about 75 percent of the current yearly increase in atmospheric CO_2 levels is due to the burning of fossil fuels. The cutting down and burning of forests is responsible for most of the remaining 25 percent.

The recent increase in CO_2 levels is striking for two reasons. First, the increase happened quickly: The concentration of CO_2 increased from 280 parts per million (ppm) to 370 ppm in roughly 200 years. Measurements from ice bubbles show that this rate of increase is greater than even the most sudden increase that occurred naturally during the past 420,000 years. Second, although the concentration of CO_2 in the atmosphere has ranged from about 200 ppm to 300 ppm during the past 420,000 years, CO_2 levels are now higher than those estimated for any time during this period. Thus, global CO_2 levels have changed very rapidly in recent years and have reached concentrations that are unmatched in the last 420,000 years.

Increased CO_2 concentrations have many biological effects

An increase in the concentration of CO_2 in the air can have large effects on plants (Figure 45.8). At least initially, many plants increase their rate of photosynthesis and use water more efficiently, and thus grow more rapidly, when more CO_2 is available. When CO_2 levels remain high, some plant species keep growing at higher rates, but others drop their growth rates over time. As CO_2 concentrations in the atmosphere rise, species that maintain rapid growth at high CO_2 levels might outcompete other species in their current ecological communities or invade new communities.

Differences in how individual species respond to higher CO_2 levels will probably cause changes to entire communities. However, it is difficult (at best) to predict exactly how communities will change under higher CO_2 levels. For example, as we will see in the following section, increased CO_2 levels in the atmosphere are likely to cause Earth's climate to warm. As both temperatures and CO_2 levels change, many different competitive interactions and many

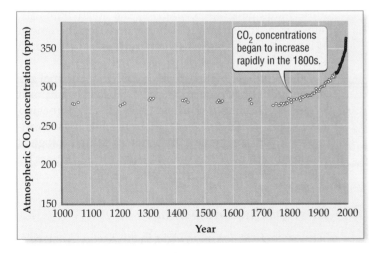

CO_2 concentrations began to increase rapidly in the 1800s.

Figure 45.7   Atmospheric CO_2 Levels Are Rising Rapidly
Atmospheric CO_2 levels (measured in parts per million, or ppm) have increased greatly in the past 200 years. The solid red circles show results from direct measurements of the concentration of CO_2 in the atmosphere. Open circles indicate CO_2 levels measured from bubbles of air trapped in ice.

(a)   (b)   (c)

Figure 45.8   High CO$_2$ Levels Can Increase Plant Size
This photograph shows three *Arabidopsis thaliana* plants of the same genotype grown under different CO$_2$ concentrations. Plants grew larger as the concentration of CO$_2$ was increased. CO$_2$ concentrations were (a) 200 ppm, a level similar to that found roughly 20,000 years ago; (b) 350 ppm, the level found in 1988; and (c) 700 ppm, a predicted future level.

different consumer–victim interactions may also change, but usually in ways that will not be known in advance. As we learned in Chapter 42, when interactions among species change, entire communities can change dramatically.

> ■ The burning of fossil fuels and the destruction of forests have caused the concentration of CO$_2$ in the atmosphere to increase greatly in the past 200 years. Increased CO$_2$ concentrations can alter the growth of plants in ways that will probably cause changes to many ecological communities.

45.4 Global Warming

Some gases in Earth's atmosphere, such as carbon dioxide (CO$_2$), methane (CH$_4$), and nitrous oxide (N$_2$O), absorb heat that radiates from Earth's surface to space. These gases are called **greenhouse gases** because they function much as the walls of a greenhouse or the windows of a car do: They let in sunlight, but trap heat. As the concentration of greenhouse gases in the atmosphere goes up, more heat should be trapped, thus raising temperatures on Earth.

Global temperatures appear to be rising

CO$_2$ is the most important of the greenhouse gases because so much of it enters the atmosphere. Scientists have predicted that current increases in atmospheric CO$_2$ concentrations will cause temperatures on Earth to rise. This aspect of global change, known as **global warming**, has proved controversial in both the media and the political arena.

We know that CO$_2$ concentrations in the atmosphere are increasing, but is the global climate getting warmer? Although the eight hottest years on record have occurred since 1990, there is so much year-to-year variation in the weather that it can be hard to show that the climate really is getting warmer. In 1995, however, the United Nations–sponsored Intergovernmental Panel on Climate Change concluded for the first time that our climate is warming (Figure 45.9). The panel also concluded that the increase in global temperatures is most likely due to human-caused increases in the concentration of CO$_2$ and other greenhouse gases in the atmosphere.

Since 1995, new statistical analyses have supported the panel's conclusion that recent rises in global temperatures represent a real trend, not just ordinary variation in the weather. In addition, results from new scientific studies suggest that recent temperature increases may have already changed ecosystems. For example, as temperatures have increased in Europe during the twentieth century, dozens of bird and butterfly species have shifted their geographic ranges to the north. Similarly, plants in northern latitudes have increased the length of their growing season as temperatures have warmed since 1980.

Other studies published since 1995 indicate that the warming that has occurred since 1950 has been caused largely by human activities. Overall, an increasing amount of scientific evidence suggests that global warming is already happening, that it is affecting ecological communities, and that it is caused at least in part by human activities, such as the use of fossil fuels.

What will the future bring?

Because there is no end in sight to the rise in CO$_2$ levels, the current trend of increasing global temperatures seems likely to continue. How will increased temperatures affect life on Earth? Not surprisingly, the effects will depend on how much global warming occurs and the rate at which it occurs.

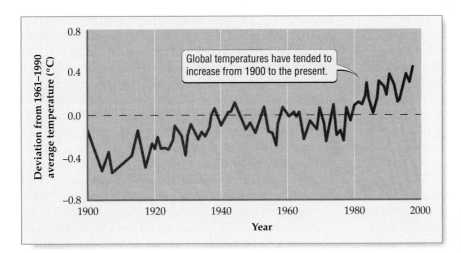

Figure 45.9   Global Temperatures Are on the Rise
Global temperatures are plotted here relative to the average temperature for the period 1961 to 1990. Portions of the curve below the dashed line represent lower-than-average temperatures; portions above the dashed line represent higher-than-average temperatures.

Scientists predict that 100 years from now, average temperatures on Earth will have risen by 1.4°C to 5.8°C. Selecting a temperature roughly in the middle of this range, a 3.5°C increase would probably have a very large effect on Earth's biomes (Figure 45.10). At global temperature increases of 3.5°C or higher, many species might go extinct simply because they were unable to migrate north fast enough to keep up with the changing climate. In addition, global temperature increases of 3°C or more by the year 2100 would be likely to have severe negative effects on the world's agricultural systems, especially since by then there probably will be 4 to 5 billion more people to feed (see Chapter 41).

Finally, at the high end of the climate change predictions (a 5.8°C increase), humanity may face a rise of over 5 meters in the global sea level. Such a rise would submerge many cities and

(a) Current climate

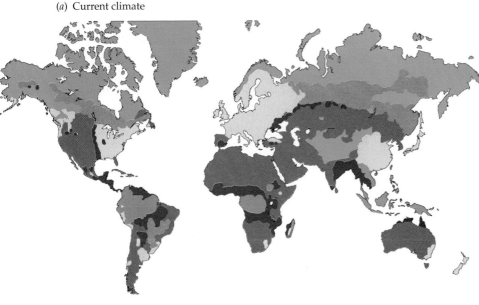

(b) Future climate (3.5°C increase in temperature)

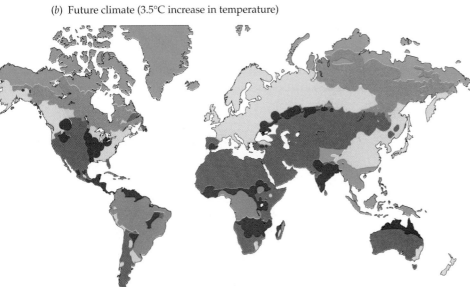

Figure 45.10   Biomes on the Move
If Earth's climate warms by 3.5°C, the distribution of forests, grasslands, deserts, and other biomes could be altered greatly by global climate change. (*a*) Current climate; (*b*) future climate (3.5°C increase in temperature).

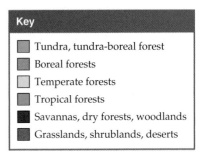

Key
Tundra, tundra-boreal forest
Boreal forests
Temperate forests
Tropical forests
Savannas, dry forests, woodlands
Grasslands, shrublands, deserts

even entire island nations. Even at the low end of the global warming predictions (a 1.4°C increase), the effects of increased global temperatures are likely to be considerable, ranging from negative effects on agriculture to an increased occurrence of severe weather (for example, floods, hurricanes, and extremes of heat and cold).

Estimates related to the timing, extent, and consequences of global warming are filled with uncertainty. This uncertainty puts us in the difficult position of choosing to act now, perhaps unnecessarily, or choosing to wait, perhaps until it is too late to do anything about the problem. Efforts to curb global warming will have social costs, but delays may have far greater costs. Given such uncertainties, what do you think we should do?

> ■ The global climate has warmed during the twentieth century, at least in part as a result of human activities. Although the amount of global warming that will occur in the twenty-first century is uncertain, if high-end predictions are correct, the social and economic costs will be extremely large.

A Message of Ecology

The science of ecology has many important and timely messages for humanity, such as the one we learned in Chapter 41: No population can continue to increase without limit. Although related, the message of this chapter is more complex. As one ecologist has written, "We are changing the world more rapidly than we are understanding it." In a very real sense, the world is in our hands. What we do to change it will determine our future and the future of all other species on Earth.

As we have seen in this chapter, human activities have caused global change and have had profound effects on life on Earth. In fact, many scientists are convinced that people are causing global change at a rate and intensity that is unmatched by natural patterns of change. Depending on the actions we take, global change has the potential to have even greater effects in the future.

As scientists, we believe that the main message provided by our knowledge of how much humans have changed the planet is that we must reduce the rate at which we alter Earth's ecosystems. This course of action is not only good for other species, it is in our own self-interest: As described in Chapter 44, our entire civilization depends on the many services that natural ecosystems provide to us at no cost. If we continue to ignore the effects of our actions on these natural systems, ultimately we will harm ourselves.

To reduce our effects on natural systems, we must limit the growth of the human population, and equally importantly, we must use Earth's resources more efficiently. Simply put, we must strive to have a **sustainable** impact on Earth—that is, our impact should be one that can continue indefinitely without using up resources or causing serious environmental damage. Overall, we must cease to alter natural systems at a rate that leads to short-term gain but long-term damage.

To achieve the goal of having a sustainable impact on the planet, we must anticipate the effects of our actions before they have disastrous consequences. No other species is capable of such forethought, but we are. Will we use that capability? Will we be bold enough, creative enough, and intelligent enough to take responsibility for our impact on Earth? Can we shift our worldview from one that seeks to dominate nature to one that recognizes the value and intrinsic worth of other species (Figure 45.11)—indeed, of the entire biosphere? As we will see in the following section, there is hope that the answers to these questions will be "Yes."

> ■ Knowledge of the impact of human activities on the planet suggests that humans should reduce the rate at which we alter Earth's ecosystems. To prevent human-caused global change from having a negative effect on ourselves and other species, humans must learn to have a sustainable impact on Earth.

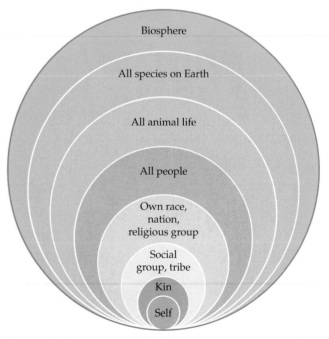

Figure 45.11   From Pure Selfishness to Concern for All
This figure illustrates an ethical sequence in which the individual expands his or her concern from a focus only on the self to a concern for the entire biosphere.

■ HIGHLIGHT

Building a Sustainable Society

To prevent serious harm to natural communities and to ourselves, people must learn to use Earth's ecosystems in a sustainable manner. Although building a sustainable society will require major changes to virtually all aspects of human life, there are many causes for hope. We close this book by discussing some of those sources of hope. To organize our discussion, we will focus on how five aspects of human society—education, individual action, research, government, and business—can contribute to the formation of a sustainable society.

Education. As many have written, people will save only what we love, and we love only what we know. In many human societies, the dominant view has been that a species or community is worth saving only to the extent that it provides direct benefits to people. The appreciation for natural communities that results from education provides a powerful alternative view: Communities are worth saving for their own sake. Because it is fundamental to how we view the world, education is central to all efforts to build a sustainable society. Education has already produced great changes: People are far more aware of environmental issues now than they were 30 or 40 years ago.

Individual action. Each day we make many choices that affect the environment. For example, with respect to what we buy, we can choose to purchase energy-efficient cars and appliances—or not. Similarly, we can refuse to buy throwaway products, such as disposable cameras—or not. Many people do make choices with the environment in mind, and as a result, "green" products that minimize environmental impacts are becoming increasingly common.

Research. Documenting what we are doing to the planet is a monumental task: We must understand our impact, acre by acre, across the entire globe. Now, for the first time in history, we have the research tools needed to accomplish this seemingly impossible task. Satellite images, for example, allow us to monitor Earth in unprecedented detail (Figure 45.12). As documented in this chapter, the preliminary results from such monitoring efforts show that we are causing serious damage to the world's ecosystems. Our newfound ability to recognize how much we are changing the planet provides a powerful new source of hope: We can use that information to motivate changes and to guide our efforts to build a sustainable society.

Government. Although individual actions can be very effective, government action also has an important role to play. Treaties among governments have helped reduce emissions of atmospheric pollutants that cross national boundaries, as illustrated by international agreements to

Figure 45.12 Measuring Earth's Vital Signs
This computer image of Earth was made using four different types of satellite data. Fires over land are shown in red. The large plume that extends from Africa over the Atlantic (and ranges in color from red to orange to yellow to green) was caused by the burning of vegetation and by windblown dust.

curb the production of CFCs that damage the ozone layer (see p. 743). Considerable potential remains for governments to take actions that would promote a sustainable society, as by initiating a major effort to develop new sources of energy, or by changing tax laws so that they would penalize polluters and provide incentives for environmentally friendly practices. For example, the German government provides tax incentives to families who choose to roof their homes with thatch roofs, which are made of renewable plant materials rather than asphalt-derived materials.

Business. When government mandates an outcome, such as lowering emissions of a pollutant to a specified level, but does not control how that outcome is achieved, businesses tend to develop efficient and cost-effective ways to solve the problem at hand. For example, when government regulations specified a level to which sulfur emissions had to be reduced, market forces led to rapid compliance with those regulations, at a much lower cost than originally predicted. In general, successful conversion to a sustainable economy will be an ongoing and complex process. As the growing sums of money corporations are investing in environmentally friendly technologies suggest, innovative companies that play a crucial role in this process have the potential to reap large profits.

> ■ The conversion to a sustainable society will be a complex process that requires fundamental change across many aspects of society. Important contributions to building a sustainable society can be made by education, individual action, research, government, and business.

SUMMARY

Land and Water Transformation

■ Human activities are changing the lands and waters of the entire planet.

■ Land and water transformation has caused the extinction of species and has the potential to alter local and global climate.

Changes in the Chemistry of Earth

■ Human activities are changing the way many chemicals, both natural and synthetic, are cycled through ecosystems.

Changes in the Global Nitrogen Cycle

■ Nitrogen must be fixed, or converted from N_2 gas, before it can be used by producers.

■ Human activities fix more nitrogen than all natural sources combined.

■ The extra nitrogen fixed by human activities has altered the global nitrogen cycle, leading to increases in productivity that can cause the loss of species from ecosystems.

Changes in the Global Carbon Cycle

■ The burning of fossil fuels and the destruction of forests have caused the concentration of CO_2 in the atmosphere to increase greatly in the past 200 years.

■ Increased CO_2 concentrations can alter the growth of plants in ways that will probably cause changes to many ecological communities.

Global Warming

■ Greenhouse gases in the atmosphere, such as CO_2, trap heat that radiates from Earth's surface. As the concentration of greenhouse gases goes up, temperatures on Earth are expected to rise.

■ The global climate has warmed during the twentieth century, at least in part as a result of human activities.

■ Although the amount of global warming that will occur in the twenty-first century is uncertain, if high-end predictions are correct, the social and economic costs will be extremely large.

A Message of Ecology

■ Knowledge of the impact of human activities on the planet suggests that humans should reduce the rate at which we alter Earth's ecosystems.

■ To prevent human-caused global change from having a negative effect on ourselves and other species, humans must learn to have a sustainable impact on Earth.

Highlight: Building a Sustainable Society

■ The conversion to a sustainable society will be a complex process that requires fundamental change across many aspects of society.

■ Important contributions to building a sustainable society can be made by education, individual action, research, government, and business.

KEY TERMS

global change p. 740   nitrogen fixation p. 743
global warming p. 746   sustainable p. 748
greenhouse gas p. 746   water transformation p. 740
land transformation p. 740

CHAPTER REVIEW

Self-Quiz

1. Which of the following do most ecologists think is mainly responsible for the current high rate of extinction of species?
a. increased CO_2 concentration in the atmosphere
b. global warming
c. hunting and fishing
d. land and water transformation

2. CO_2 absorbs some of the _____ that radiates from the surface of Earth to space.
a. ozone
b. heat
c. ultraviolet light
d. smog

3. The conversion of N_2 gas to a form of nitrogen that can be used by plants is called
a. nitrogen fixation.
b. nitrogen assimilation.
c. nitrogen cycling.
d. nitrogen uptake.

4. The concentration of CO_2 in the atmosphere is now about _____ ppm, a level that is over 30 percent higher than preindustrial levels.
a. 165
b. 265
c. 370
d. 465

5. Human-caused changes to the nitrogen cycle are expected to result in
a. an increase in acid rain.
b. an increase in the loss of species from ecosystems.
c. higher concentrations of a greenhouse gas.
d. all of the above

Review Questions

1. Summarize the major types of global change caused by humans.

2. Compare human-caused examples of global change to examples of global change not caused by people. What is different or unusual about human-caused global change?

3. What activities do you perform that are good for the biosphere? What activities do you perform that harm the biosphere?

4. What changes to human societies would have to be made for people to have a sustainable impact on Earth?

45

The Daily Globe

Society Taxes the Wrong Activities

To the Editor:

As a society, what do we want individuals and corporations to do? Have jobs, work hard, and create new and profitable businesses? Or do we want people and corporations to pollute the environment, destroy natural habitat, and contribute to global warming? If the way we are taxed is any guide, the answer is pollute, destroy, and warm the planet.

We tax people when they perform activities that benefit society, such as working hard at their jobs or creating new businesses. If the efforts of people or businesses result in their making more money, we tax them even more. In contrast, we charge few taxes to people or corporations that damage the environment. Emissions from the vehicles we drive contribute to global warming; pesticides and other chemicals we use alter the way materials are cycled through ecosystems; and activities such as mining or building housing developments—which we even subsidize!—destroy natural habitat.

Why does our society punish people (by taxation) for activities that benefit society? Why do we reward people (by not taxing them or by providing them with subsidies) for activities that damage the environment? When individuals and corporations damage the environment, society as a whole ultimately ends up sharing the cost of fixing the damage. Is it sensible for us to reward people for actions whose results we must all pay later to fix?

We are doing everything backward. We should lower the taxes we charge on income and profits from business and start to punish people and businesses that harm the environment by taxing them more heavily. How about high taxes on gasoline to help stop global warming? How about charging heavy taxes for environmentally harmful activities to pay for tomorrow's environmental repairs?

It is time to reward people who do what we say we want them to do and punish those who do the opposite.

Lisa Conracia

Evaluating "The News"

1. The author of this letter states that we punish people for socially beneficial activities and reward people for socially harmful activities. Do you think this is true? If so, why? If you disagree with the letter writer, explain why.

2. If gasoline were taxed heavily, do you think pressure from consumers would cause car manufacturers to respond by designing cars that got more miles per gallon? What do you think would happen if corporations were taxed heavily on the amount of pollution they released into the environment?

3. An assumption of this letter is that people are most likely to change their behavior if such changes can save them money. Do you agree with this assumption? Are there other ways that might be as effective or more effective in encouraging people to make changes that would benefit the global environment?

Suggested Readings

Unit 1 Diversity of Life

FOR FURTHER EXPLORATION

Futuyma, D. J. 1995. *Science on Trial: The Case for Evolution.* Sinauer Associates, Sunderland, MA. A leading evolutionary biologist summarizes the evidence for evolution and refutes the arguments made by creationists.

May, R. M. 1988. "How Many Species Are There on Earth?" *Science* 241: 1441–1449. Robert May, a distinguished English biologist, evaluates Terry Erwin's estimate of the number of species on Earth, which were based on fogging of canopies.

Mayr, E. 1997. *This is Biology: The Science of the Living World.* Harvard University Press, Cambridge, MA. A description of what biology is and how the science works by one of the most famous and respected biologists.

The New York Times Book of Science Literacy. Volume 2: *The Environment from Your Backyard to the Ocean Floor.* 1994. Times Books, New York. Collected articles from *The New York Times* on biodiversity and related environmental issues.

Pool, R. 1990. "Pushing the Envelope of Life." *Science* 247: 158–247. An exploration of what is known about Archaebacteria and their extreme lifestyles.

Takacs, D. 1997. *The Idea of Biodiversity.* Johns Hopkins University Press, Baltimore, MD. A discussion of the history of biodiversity and its meaning for biologists, written by a biologist–historian.

Wilson, E. O., ed. 1989. *Biodiversity.* National Academy Press, Washington, D.C. A discussion by a leading evolutionary biologist of the evolution of diversity.

FOR YOUR ENJOYMENT

Allmon, W. 2001. *Rock of Ages / Sands of Time.* The University of Chicago Press, Chicago. An intriguing book of paintings of fossils showing one page for every million years of the history of life.

Attenborough, D. 1995. *The Private Life of Plants.* Princeton University Press, Princeton, NJ. A popular natural history about plant "behavior."

Brodo, I. M., Sharnoff, S. D. and Sharnoff, S. 2001. *Lichens of North America.* Yale University Press, New Haven, CT. A beautifully photographed book describing scientific and folkloric knowledge of these species, each of which is a combination of two major groups or kingdoms, Protists and Fungi.

Gould, S. J. 1980. *The Panda's Thumb.* W. W. Norton and Co., New York. Essays by a leading evolutionary biologist on evolution and evolutionary relationships.

Raup, D. 1991. *Extinction: Bad Genes or Bad Luck?* W. W. Norton and Co., New York. In this book for lay audiences, renowned paleontologist David Raup examines why some organisms survive mass extinctions and others do not.

Watson, J. D. 1968. *The Double Helix: A Personal Account of the Discovery of the Structure of DNA.* Atheneum, New York. A lively account of the Nobel prize–winning discovery of the structure of DNA, written by one of its discoverers.

Wilson, E. O. 1991. "Rain Forest Canopy: The High Frontier." *National Geographic* 180: 78–107. A beautifully illustrated account of the diversity found in the highest branches of the tropical rainforest.

Unit 2 Cells: The Basic Units of Life

FOR FURTHER EXPLORATION

Atkins, P. W. 1987. *Molecules.* Scientific American Library, New York. Accurate and revealing renderings of important molecules.

Cooper, G. M. 1998. *The Cell: A Molecular Approach.* Sinauer Associates, Sunderland, MA. A concise introductory text on the molecular biology of cells.

Hall, D. O., and Rao, K. K. 1999. *Photosynthesis (Studies in Biology).* Cambridge University Press, Cambridge. An easily understandable and methodical treatment of photosynthesis.

Harold, F. M. 2001. *The Way of the Cell: Molecules, Organisms and the Order of Life.* Oxford University Press, New York. Thoughtful and witty discussion of what constitutes life, with a particular emphasis on single-cell microorganisms.

Needham, J. (ed.) 1970. *The Chemistry of Life*. Cambridge University Press, Cambridge. A selection of thoughtful essays on the history of biochemistry

Salway, J. G., and Salway, J. D. 1994. *Metabolism at a Glance*. Blackwell Science Inc., Cambridge, MA. A concise and complete overview of metabolism with clear diagrams.

Stossel, T. 1994. "The Machinery of Cell Crawling." *Scientific American*, September. Insightful discussion of how different cytoskeletal proteins contribute to cell movement.

Stryer, L. 1995. *Biochemistry*, 4th Ed. W. H. Freeman and Co., New York. Exceptionally well written and organized introductory text on biochemistry.

FOR YOUR ENJOYMENT

Angier, N. 1999. *Natural Obsessions: Striving to Unlock the Deepest Secrets of the Cancer Cell*. Houghton Mifflin Co., New York. Engaging account of what it is like to do biological research. Reveals the process of scientific inquiry as it unfolded for several important breakthroughs in molecular and cell biology.

Bloch, K. 1997. *Blondes in Venetian Paintings, the Nine-Banded Armadillo, and Other Essays in Biochemistry*. Yale University Press, New Haven. Colorful anecdotes by a Nobel prize–winning scientist on the chemical basis for diverse human and biological phenomena.

Emsley, J. 1998. *Molecules at an Exhibition: Portraits of Intriguing Materials in Everyday Life*. Oxford University Press, New York. Fascinating and entertaining anecdotes about the many chemical compounds encountered in daily life.

Galston, A. W. 1992. "Photosynthesis as a Basis for Life Support in Space." *Bioscience*, July/August. Provocative discussion of how photosynthetic organisms may be used to support artificial ecosystems in space.

Goodsell, D. S. 1998. *The Machinery of Life*. Springer Verlag, New York. Beautifully illustrated tour of the molecular structure of cells.

Loewenstein, W. R. 1999. *The Touchstone of Life: Molecular Information, Cell Communication, and the Foundations of Life*. Oxford University Press, New York. A lucid discussion of how the flow of information, molecular biology, and cell structure work together in living processes.

Rasmussen, H. 1991. *Cell Communication in Health and Disease: Readings from* Scientific American *Magazine*. W. H. Freeman and Co., New York. A collection of articles from *Scientific American* in which researchers examine the complexity of cell communication and its role in various human diseases such as atherosclerosis and diabetes.

Weinberg, R. A. 1998. *Racing to the Beginning of the Road: The Search for the Origin of Cancer*. W. H. Freeman and Co., New York. The fascinating history of cancer's origins, written by one of the most prominent researchers in cancer biology. Contains revealing renderings of important molecules.

Unit 3   Genetics

FOR FURTHER EXPLORATION

Carroll, S. B., Grenier, J. K. and Weatherbee, S. D. 2001. *From DNA to Diversity: Molecular Genetics and the Evolution of Animal Design*. Blackwell Science, Inc., Malden, MA. An exploration of how genes influence the shape and design of animals, including people.

Griffiths, A. J. F. et al. 2000. *An Introduction to Genetic Analysis*, 7th Ed. W. H. Freeman and Co., New York. A general genetics textbook.

"Making Gene Therapy Work." *Scientific American*, June 1997. A collection of articles outlining current progress in human gene therapy.

Mange, E. J. and A. P. Mange. 1999. *Basic Human Genetics*, 2nd Ed. Sinauer Associates, Sunderland, MA. An introduction to human genetics.

"Microchip Arrays Put DNA on the Spot." *Science*, October 16, 1998. A collection of news articles discussing the current and likely future impact of DNA chips.

Orel, V. 1996. *Gregor Mendel, the First Geneticist*. Oxford University Press, New York. A biography of the founder of the science of genetics.

Subramanian, S. 1995. *"The Story of Our Genes." Time* magazine, January 16. An overview piece on genetic screening and what our genes can tell us.

FOR YOUR ENJOYMENT

"GM foods: are they safe?" *Scientific American*, April 2001. A special report that explores the benefits and risks associated with our widespread use of genetically modified organisms in the food we eat.

Judson, H. F. 1996. *The Eighth Day of Creation: The Makers of the Revolution in Biology*. Cold Spring Harbor Laboratory Press, Cold Spring Harbor, NY. An engaging history of molecular genetics.

Keller, E. F. 1983. *A Feeling for the Organism: The Life and Work of Barbara McClintock*. W. H. Freeman and Co., San Francisco. The story of the life and work of a Nobel Prize–winning female scientist working in an all-male world of science.

Kitcher, P. 1996. "Junior Comes Out Perfect." *New York Times Magazine*, September 29. An article on the potential for producing the perfect baby through genetic screening.

Ridley, M. 2000. *Genome: The Autobiography of a Species in 23 Chapters.* HarperCollins, New York. A well-written and fascinating account of a single important gene from each of our chromosomes. The discussion includes a balanced narrative of the human genome projects.

Watson, J. D. 1968. *The Double Helix.* Atheneum, New York. A personal and controversial account of the Nobel Prize–winning discovery of the structure of DNA.

Unit 4 Evolution

FOR FURTHER EXPLORATION

Axelrod, R. 1984. *The Evolution of Cooperation.* Basic Books, New York. A far-reaching book on cooperation that shows how phenomena ranging from mutualism to trench warfare can be partially explained by an evolutionary perspective.

Cowen, R. 2000. *History of Life,* 3rd Ed. Blackwell Science, Inc., Malden, MA. An interesting and highly accessible survey of the history of life on Earth.

Darwin, C. 1859. *On the Origin of Species.* J. Murray, London. (reprint edition: 1964, Harvard University Press, Cambridge, MA.) The single most important book on evolution ever published. A landmark scientific work that revolutionized biology and had a large impact on many other areas of human society. With careful reading, this text is very clear to a nonscientific audience.

Freeman, S. and Herron, J. C. 2000. *Evolutionary Analysis,* 2nd Ed. Prentice Hall, Englewood Cliffs, NJ. A general textbook on evolution.

Futuyma, D. J. 1995. *Science on Trial: The Case for Evolution.* Sinauer Associates, Sunderland, MA. A leading evolutionary biologist summarizes the evidence for evolution and refutes the arguments made by creationists.

Slatkin, M. (ed.) 1995. *Exploring Evolutionary Biology: Readings from* American Scientist. Sinauer Associates, Sunderland, MA. An interesting collection of articles on evolution, originally published in the journal *American Scientist* and intended for both scientists and the general public.

FOR YOUR ENJOYMENT

Diamond, J. 1992. *The Third Chimpanzee: The Evolution and Future of the Human Animal.* HarperCollins, New York. A fascinating account of human evolution and the future of our species by one of the leading writers on science for the general public.

Johanson, D. and Shreeve, J. 1989. *Lucy's Child.* Avon Books, New York. The story of the discovery of the Australopithecus afarensis skeleton that became known in the popular press as Lucy.

Nesse, R. and Williams, G. 1995. *Why We Get Sick: The New Science of Darwinian Medicine.* Vintage Books, New York. A compelling argument for why both doctors and patients need to understand basic evolutionary principles.

Palumbi, S. R. 2001. *The Evolution Explosion: How Humans Cause Rapid Evolutionary Change.* W. W. Norton & Co., New York. A fascinating description of the many evolutionary changes caused by people and the effects such changes have on human society.

Weiner, J. 1994. *The Beak of the Finch: A Story of Evolution in our Time.* Knopf, New York. A Pulitzer Prize–winning account of evolution in the Galápagos finches.

Wilson, E. O. 1992. *The Diversity of Life.* W. W. Norton and Co., New York. A thought-provoking description of how the diversity of life evolved and how it is being threatened by human actions.

Unit 5 Form and Function

FOR FURTHER EXPLORATION

Bonner, J. T. 1980. *The Evolution of Culture in Animals.* Princeton University Press, Princeton, NJ. An interesting treatment of the evolution of cultural behavior.

Eisenberg, A., Murkoff, H. E. and Hathaway, S. E. 1996. *What to Expect When You're Expecting.* Workman Publishing, New York. This and other books on human pregnancy give a good overview of the steps in and timing of human reproduction and development.

Gordon, J. E. 1978. *Structures.* Pelican Books, London. Draws many parallels between the structure of plant and animal support systems and the way in which engineers assemble materials into bridges and buildings.

Lovejoy, C. O. 1988. "Evolution of human walking." *Scientific American* 259(5): 118–125. The evolution of walking and its relationship to the size of the birth canal in human females.

Olshansky, S. J., Carnes, B. A., and Butler, R. N. 2001. "If humans were built to last." *Scientific American* 284(3): 50–55. An interesting redesign of the human body to better accommodate the changes that come with aging.

Owen, J. 1980. *Feeding Strategy.* University of Chicago Press, Chicago. A broad and elementary review of nutrition among living things.

Ricklefs, R. E., and Finch, C. E. 1995. *Aging: A Natural History.* Scientific American Library, New York. An overview of the ecology and evolution of aging.

Schmidt-Nielsen, K. 1972. *How Animals Work.* Cambridge University Press, London. This book presents diverse and entertaining examples of how and why animals function the way that they do.

Schmidt-Nielsen, K. 1984. *Scaling: Why Is Animal Size So Important?* Cambridge University Press, New York. A readable but thorough overview of the effect of size on animal biology.

Vogel, S. 1988. *Life's Devices.* Princeton University Press, Princeton, NJ. A thorough overview of the mechanics of how organisms deal with their environment.

FOR YOUR ENJOYMENT

Andrews, M. 1976. *The Life that Lives on Man.* Taplinger Publishing Company, New York. A novel look at how various microbes, fungi, and animals make their living on the surfaces of our own bodies.

Bakker, R. T. 1986. *The Dinosaur Heresies.* Zebra Books, New York. A thorough presentation of the idea that dinosaurs were endotherms.

Bonner, J. T. 1980. *The Evolution of Culture in Animals.* Princeton University Press, Princeton, NJ. An interesting treatment of cultural evolution.

Gordon, J. E. 1978. *Structures.* Pelican Books, London. Draws many parallels between the structure of plant and animal support systems and the way in which engineers assemble materials into bridges and buildings.

Unit 6   Interactions with the Environment

FOR FURTHER EXPLORATION

Gates, D. M. 1993. *Climate Change and its Biological Consequences.* Sinauer Associates, Sunderland, MA. A clear description of past and present examples of climate change and how those changes affect life on Earth.

Kareiva, P. (ed.) 1998. *Exploring Ecology and its Applications: Readings from American Scientist.* Sinauer Associates, Sunderland, MA. An interesting collection of articles on ecology, originally published in the journal American Scientist and intended for both scientists and the general public.

Levin, S. A. 1999. *Fragile Dominion: Complexity and the Commons.* Perseus Books, Reading, MA. An assessment by a leading ecologist of natural ecosystems and how they are affected by people.

Newton, L. H. and Dillingham, C. K. 2002. *Watersheds 3: Ten Cases in Environmental Ethics.* Wadsworth Publishing Company, Belmont, CA. An excellent collection of environmental case studies, each of which provides readers with a clear description of the important biological and ethical considerations that relate to the issue at hand.

Primack, R. B. 1998. *Essentials of Conservation Biology,* 2nd Ed. Sinauer Associates, Sunderland, MA. A good introduction to the fast-developing field of conservation biology, a branch of science focused on the conservation of the diversity of life on Earth.

Ricklefs, R. E. and Miller, G. L. 1999. *Ecology,* 4th Ed. W. H. Freeman and Co., New York. A general ecology textbook.

FOR YOUR ENJOYMENT

Botkin, D. B. 1996. *Our Natural History: The Lessons of Lewis and Clark.* Perigee, New York. A provocative and interesting book that uses the journeys of Lewis and Clark as a springboard from which to discuss how ecological systems are always in a state of change and how endangered ecosystems can be saved.

Carson, R. 1962. *Silent Spring.* Houghton Mifflin Co., New York. An award-winning book that galvanized public interest in ecology and in the environmental movement.

Diamond, J. 1999. *Guns, Germs, and Steel: The Fates of Human Societies.* W. W. Norton & Co., New York. A leading scientist's riveting analysis of how the rise of human civilizations—and the dominance of some groups over others—was shaped by geography and by features of the natural environment.

Leopold, A. 1949. *A Sand County Almanac and Sketches Here and There.* Oxford University Press, New York. A collection of essays from one of the founders of the conservation movement.

McKibben, B. 1995. *Hope, Human and Wild: True Stories of Living Lightly on the Earth.* Little, Brown and Co., Boston. Beautifully written and penetrating essays that focus on successful community efforts to preserve wilderness and reverse environmental damage.

Orr, D. W. 1994. *Earth in Mind: On Education, Environment, and the Human Prospect.* Island Press, Washington, D.C. A thought-provoking collection of essays about causes of and solutions to global environmental problems.

Table of Metric–English Conversion

Common conversions

Length		To convert	Multiply by	To yield
nanometer (nm)	0.000000001 (10^{-9}) m	inches	2.54	centimeters
micrometer (μm)	0.000001 (10^{-6}) m	yards	0.91	meters
millimeter (mm)	0.001 (10^{-3}) m	miles	1.61	kilometers
centimeter (cm)	0.01 (10^{-2}) m			
meter (m)	—	centimeters	0.39	inches
kilometer (km)	1000 (10^{3}) m	meters	1.09	yards
		kilometers	0.62	miles

Weight (mass)				
nanogram (ng)	0.000000001 (10^{-9}) g	ounces	28.35	grams
microgram (μg)	0.000001 (10^{-6}) g	pounds	0.45	kilograms
milligram (mg)	0.001 (10^{-3}) g			
gram (g)	—	grams	0.035	ounces
kilogram (kg)	1000 (10^{3}) g	kilograms	2.20	pounds
metric ton (t)	1,000,000 (10^{6}) g (=10^{3}) kg			

Volume				
microliter (μl)	0.000001 (10^{-6}) l	fluid ounces	29.57	milliliters
milliliter (ml)	0.001 (10^{-3}) l	quarts	0.95	liters
liter (l)	—			
kiloliter (kl)	1000 (10^{3}) l	milliliters	0.034	fluid ounces
		liters	1.06	quarts

Temperature

degree Celcius (°C)   —

To convert Fahrenheit (°F) to Centigrade (°C):
$$°C = \tfrac{5}{9}(°F - 32°)$$

To convert Centigrade (°C) to Fahrenheit (°F):
$$°F = \tfrac{9}{5}\,°C + 32°$$

Answers to Self-Quiz Questions

Chapter 1
1. b
2. b
3. a
4. c
5. b
6. a

Chapter 2
1. a
2. c
3. d
4. c
5. a
6. d

Chapter 3
1. d
2. c
3. b
4. a
5. b
6. a

Chapter 4
1. c
2. c
3. b
4. c
5. d
6. b

Chapter 5
1. a
2. c
3. d
4. a
5. c
6. b

Chapter 6
1. d
2. a
3. c
4. b
5. a
6. b

Chapter 7
1. c
2. a
3. b
4. c
5. a
6. d

Chapter 8
1. b
2. d
3. a
4. a
5. a
6. c

Chapter 9
1. d
2. d
3. a
4. b
5. b
6. d

Chapter 10
1. b
2. a
3. d
4. b
5. d
6. c

Chapter 11
1. c
2. a
3. b
4. b
5. a
6. d

Chapter 12
1. a
2. c
3. b
4. d
5. d

Chapter 13
1. d
2. b
3. c
4. c
5. a

Chapter 14
1. c
2. c
3. b
4. d
5. a

Chapter 15
1. b
2. c
3. b
4. d
5. a
6. d

Chapter 16
1. d
2. c
3. b
4. a
5. d

Chapter 17
1. c
2. a
3. b
4. d
5. a

Chapter 18
1. c
2. a
3. b
4. b
5. c
6. a

Chapter 19
1. d
2. d
3. c
4. a
5. b

Chapter 20
1. c
2. b
3. a
4. c
5. d
6. a

Chapter 21
1. b
2. b
3. c
4. a
5. d

Chapter 22
1. d
2. d
3. b
4. a
5. c

Chapter 23
1. c
2. a
3. b
4. d
5. c

Chapter 24
1. d
2. c
3. a
4. a
5. d

Chapter 25
1. a
2. d
3. a
4. b
5. c

Chapter 26
1. a
2. a
3. c
4. b
5. c

Chapter 27
1. b
2. a
3. b
4. a
5. d

Chapter 28
1. d
2. b
3. c
4. a
5. a

Chapter 29
1. *a*
2. *d*
3. *b*
4. *b*
5. *d*

Chapter 30
1. *d*
2. *a*
3. *c*
4. *c*
5. *b*

Chapter 31
1. *c*
2. *d*
3. *c*
4. *d*
5. *d*

Chapter 32
1. *c*
2. *a*
3. *d*
4. *c*
5. *d*

Chapter 33
1. *b*
2. *d*
3. *c*
4. *d*
5. *d*

Chapter 34
1. *a*
2. *c*
3. *b*
4. *d*
5. *a*

Chapter 35
1. *b*
2. *d*
3. *c*
4. *a*
5. *c*

Chapter 36
1. *c*
2. *d*
3. *a*
4. *d*
5. *a*

Chapter 37
1. *d*
2. *d*
3. *d*
4. *a*
5. *c*

Chapter 38
1. *b*
2. *b*
3. *d*
4. *c*
5. *b*

Chapter 39
1. *c*
2. *d*
3. *a*
4. *c*
5. *b*

Chapter 40
1. *a*
2. *c*
3. *d*
4. *b*
5. *d*

Chapter 41
1. *b*
2. *b*
3. *c*
4. *d*
5. *a*

Chapter 42
1. *c*
2. *a*
3. *c*
4. *d*
5. *b*

Chapter 43
1. *c*
2. *c*
3. *d*
4. *a*
5. *b*

Chapter 44
1. *c*
2. *a*
3. *d*
4. *a*
5. *b*

Chapter 45
1. *d*
2. *b*
3. *a*
4. *c*
5. *d*

Answers to Review Questions

Chapter 1

1. Dr. Larson's honeysuckle work:

   Observations: Japanese honeysuckle spreads in the wild faster than the native coral honeysuckle. Vines need supports.

   Hypothesis: Japanese honeysuckle spreads faster than the native coral honeysuckle because it is better at finding supports, especially supports far from where it grows.

   Experiment: Grow both species of honeysuckle and place supports close to and far away from the plants.

   Further **observations** during the experiment: The Japanese species grew differently around a central axis, allowing it to produce roots faster. These observations led to another **hypothesis**: The difference in rotation patterns contributes to the rapid spreading of the Japanese species.

2. From smallest to largest, the elements of the biological hierarchy are: molecules, cells, tissues, organs, organ systems, individual organisms, populations, species, communities, ecosystems, biomes, biosphere.

3. Energy flows in one direction through biological systems from sunlight to producers (photosynthesizers such as plants), where it is converted to chemical energy, then to consumers (such as animals and fungi), which eat the producers and emit heat energy.

Chapter 2

1. Systematists compare characteristics of organisms, including structural characteristics, behaviors, and DNA. Systematists then can place an organism on the evolutionary tree of life with those organisms with which it has the greatest number of shared derived features.

2. A "real group" is one which contains all the descendants of a single common ancestor and only the descendants of that single common ancestor. If we were to make any single cut to the tree, all the branches that came off together from that single cut would constitute a real group. Many systematists feel that it is important only to name groupings of organisms that constitute "real groups" because they are based strictly on evolutionary relationships. Others, however, feel that doing away with groups like the reptiles would be too radical a move.

3. An evolutionary tree is like a hypothesis because both involve predictions that may be supported or rejected by the results of further testing.

Chapter 3

1. Factors that likely contributed to the success of the prokaryotes include their rapid rate of reproduction and their ability to obtain nutrients in very diverse ways.

2. *Giardia* is of interest to biologists concerned with evolution of eukaryotes because unlike most eukaryotic cells, it contains two nuclei but no mitochondria. As a result, *Giardia* provides evidence of one of the many evolutionary experiments in assembling eukaryotic cells from prokaryotic cells in the history of life. Slime molds are of interest because individual slime molds can exist as single-celled, then multicellular organisms, providing biologists a window into the transition that eukaryotes made from single-celled to multicellular living. Studying these two stages in the life of a slime mold may provide scientists with information about the evolution of multicellular organisms.

3. Sponges are the group in which the first specialized cells appeared. However, sponges are merely loose collections of such cells. The cnidaria were among the first animals to evolve true, differentiated tissues. Flatworms were among the first animals to evolve true organs.

4. While viruses exhibit some characteristics of living organisms, they are not placed into any existing kingdom. Viruses are not placed in any one kingdom because these highly degenerate organisms are thought to have evolved many times from many kingdoms.

Chapter 4

1. Terry Erwin fogged a single rainforest tree with insecticide and found more 1100 species of beetles living in its canopy. He estimated that 160 of these species were likely to be specialists. There are an estimated 50,000 or so tropical tree species. Thus if the tree species Erwin was studying were typical, beetle species in tropical trees should number 8,000,000 (50,000 × 160). Beetle species are thought to make up about 40 percent of all arthropod species. If that is the case, then the total number of arthropod species in tropical tree canopies should be 20 million. Many scientists believe that the total number of arthropod species in the canopy is double the number found in other parts of tropical forests, suggesting that there are another 10 million arthropod species in noncanopy tropical environments. Thus he estimated that the total number of arthropod species in the tropics should be 30 million. These indirect measures, which depend on other estimates, are difficult to do because they involve making a lot of assumptions about which scientists have little information.

2. Biodiversity has fluctuated ever since life began, declining rapidly during mass extinctions followed by slow recovery periods. Five previous mass extinctions (around 440, 350, 250, 206, and 65 million years ago) were not man-made, but are thought to have been caused by natural forces including climate changes, volcanic eruptions, changes in sea level, and atmospheric dust from an asteroid collision. Recovery periods lasted millions of years. Some scientists believe the current rapid expansion of human populations could lead to another such mass extinction.

3. As human populations grow, cities, suburbs and commercial areas expand and natural habitats are destroyed and degraded, driving out other species. Human population growth has increased pollution, further threatening biodiversity. Expansion of human populations has also increased the introduction of nonnative species, which is another threat to biodiversity.

Chapter 5

1. Polymers are complex molecules, often with attached functional groups. Combinations of these functional groups give polymers chemical properties not present in smaller molecules.

2. The pH of pure water should be 7. As a measurement, pH units represent the concentration of free hydrogen ions in water. In the presence of a base, the pH will be above 7, indicating that there are more hydroxyl ions than hydrogen ions; thus, the solution is basic. In the presence of an acid, the pH will be below 7, indicating more free hydrogen ions; thus, the solution is acidic. Pure water has equal amounts of hydrogen and hydroxyl ions and is thus neutral.

3. Each carbon atom has four bonding areas that can bond covalently to other carbons or to different atoms, creating large molecules containing hundreds of atoms.

4. Proteins are polymers of amino acids. Living organisms use proteins for physical structures such as hair and for enzymes to facilitate chemical reactions needed by the body.

Chapter 6

1. One function of proteins embedded in the plasma membrane is to form gateways that allow selective passage of ions and molecules in and out of the cell. This allows wastes to be removed and needed substances to enter. Another function is to recognize changes in the outside environment and to communicate with other cells so a quick response to change is possible.

2. Unlike lysosomes in animals cells, large vacuoles in plant cells can fill with water and push against the cell wall, giving it rigidity.

3. The enzyme is produced on ribosomes associated with the endoplasmic reticulum. A vesicle transports the enzyme to the Golgi apparatus, where it is packaged into a specialized vesicle called a lysosome. The lysosome is released into the cell. Since the enzyme is enclosed in a membrane or vesicle, it would not be manufactured on a free-floating ribosome.

4. Both microtubules and actin filaments can change length quickly. With microtubules, this allows movement of organelles within the cell. Movement of the cell itself occurs when actin filaments lengthen or shorten and change cell shape.

Chapter 7

1. If cells oxidized food molecules in one step, so much energy would be released that we would burst into flame. A step-by-step process allows some of the energy to be safely stored in chemical bonds and released when needed.

2. The second law of thermodynamics requires that the organization of the cell be compensated for by the transfer of disorder to the environment. Heat fulfills this requirement by increasing the disorderly and random movement of molecules in the environment.

3. Higher temperatures increase random molecular collisions while enzymes selectively bind to specific reactants and bring them closer together to facilitate a chemical reaction. This closeness lowers the energy of activation required for the reaction.

4. Specific organelles, such as mitochondria, increase the efficiency of catalysis by concentrating both the enzymes and substrates needed for a particular process. In mitochondria, free-floating enzymes are concentrated in the lumen, and other enzymes are physically arranged next to each other in the membrane. Both of these arrangements facilitate enzyme catalysis.

5. Enzymes in a detergent increase its efficiency by binding to the stains in clothing, bringing them in close contact with the detergent. The stain and detergent are the reactants.

Chapter 8

1. The transfer of electrons down an ETC produces a proton gradient in both chloroplasts and mitochondria. The protons move through a membrane channel known as the ATP synthase, which in turn catalyzes the production of ATP.

2. Sugar made in the leaves of a plant by photosynthesis is transported to the roots. Cellular respiration in the mitochondria of root cells converts this sugar to usable energy in the form of ATP.

3. O_2 is consumed when it picks up electrons and combines with protons to form water. Thus, O_2 is the last electron acceptor in the series of reactions that make up oxidative phosphorylation.

4. When protons pass through an ATP synthase channel, ADP is converted to ATP. Drugs that allow this channel to be bypassed will deter the production of ATP.

Chapter 9

1. The receptor could be located in the cytosol or the nucleus, since hydrophobic molecules, such as steroids, are able to pass through the lipid core of the plasma membrane.

2. Signal amplification is important to a cell's ability to communicate with other cells, because it activates many proteins at each step of the signal cascade, increasing the effect of the original signal and causing a faster response.

3. Phosphorylation by a kinase changes the shape of an enzyme, thus altering its activity. It activates some enzymes and inhibits others.

4. Adrenaline binds to and activates a receptor on the outside of a cell. This activates a G protein inside the cell by inducing it to exchange a guanine nucleotide (GDP) for another nucleotide (GTP) in the cytosol. This turns on an enzyme that converts ATP into another nucleotide (cAMP). The cAMP then activates a cascade of enzymes needed for the cell's response to the adrenaline.

Chapter 10

1. A horse cell undergoing mitosis would have 64 chromosomes. A horse cell undergoing meiosis II would have 32 chromosomes.

2. The mitotic spindle directs the movement of chromosomes during cell division.

3. Mitosis: Spindle microtubules from each centrosome attach to the duplicated centromere of a chromosome at the kinetochores. During anaphase, these microtubules pull the sister chromatids to opposite poles of the cell. Each sister chromatid forms a new chromosome in the daughter cells. Meiosis I: A single microtubule from opposite poles attaches to each homologous chromosome. During anaphase I, the homologous chromosomes (sister chromatids still joined) are pulled to opposite poles. This halves the number of chromosomes in each daughter cell.

4. Meiosis is important for the production of sex cells because it allows cells to maintain a constant chromosome number when fertilization occurs. If sex cells underwent mitosis instead, the gametes would contain a complete set of chromosomes and the zygote formed after fertilization would have twice the number of chromosomes as its parents. The developing embryo would not have the same karyotype as its parents and could not survive.

Chapter 11

1. p 53 prevents the passing on of harmful mutations by stopping a cell with DNA damage from dividing until the damage is repaired. If the damage is too severe, p53 kills the cell.

2. Peyton Rous filtered the extract from cancerous tumors to remove bacteria. If the chickens injected with the filtered extract developed cancer, Rous could rule out bacteria as the cause.

3. Malignancies develop over decades through an accumulation of several DNA mutations. Activation of at least one oncogene and deactivation of several tumor suppressors usually must occur first. Therefore, the less long-term exposure to chemical mutagens, the less risk of developing cancer.

4. If only one homologous chromosome is missing the *Rb* gene, some protein can still be produced and the disorder is less severe, usually involving a single tumor in one eye. A mutation in the remaining gene, however, causes multiple tumors in both eyes. If *Rb* were an oncogene, severe cancer would be produced with only one active gene. Therefore, *Rb* must be a tumor suppressor.

Chapter 12

1. The genotype of the unknown parent can be determined by crossing it with a with a pea plant that is homozygous recessive for this trait (*pp*). This parent can only provide recessive alleles, so if there are any progeny with the recessive phenotype (white flowers), the unknown parent must have been heterozygous (*Pp*) to provide the second recessive allele. If there are no progeny with white flowers, the unknown parent must have been homozygous (*PP*).

2. A lethal genetic disorder caused by a dominant allele would kill both the individuals with a homozygous dominant genotype as well as those with the heterozygous genotype. If the disorder kills people before they reach reproductive age, then the allele will rapidly be selected out of the population. A disorder caused by a lethal recessive allele would only kill individuals with the homozygous recessive genotype, so the allele could still be carried by heterozygous individuals and passed on to the next generation.

3. Probability refers to the chance that a certain event will occur. If a coin is only flipped a few times, for instance, then there is no way of knowing whether or not the results obtained were representative. However, if the coin is tossed many times, or if many pea plants are observed, then predictable patterns can be discerned.

Sample Genetics Problems

1. a. *A* and *a*

   b. *BC*, *Bc*, *bC*, and *bc*

   c. *Ac*

   d. *ABC*, *ABc*, *Abc*, *AbC*, *aBC*, *aBc*, *abC*, and *abc*

   e. *aBC* and *aBc*

2. a. genotype ratio: 1:1 phenotype ratio: 1:1

	A	*a*
a	*Aa*	*aa*

   b. genotype ratio: 1:0 phenotype ratio: 1:0

	B
b	*Bb*

   c. genotype ratio: 1:1 phenotype ratio: 1:1

	AB	*Ab*
ab	*AaBb*	*Aabb*

   d. genotype ratio: 1*BBCC*:1*BBCc*:2*BbCC*:2*BbCc*:1*bbCC*:1*bbCc* phenotype ratio: 6:2, reduced to 3:1

	BC	*Bc*	*bC*	*bc*
BC	*BBCC*	*BBCc*	*BbCC*	*BbCc*
bC	*BbCC*	*BbCc*	*bbCC*	*bbCc*

   e. genotype ratio:
   1*AABbCC*:2*AABbCc*:1*AABbcc*:1*AAbbCC*:2*AAbbCc*:1*AAbbcc*:1*AaBbCC*:2*AaBbCc*:1*AaBbcc*:1*AabbCC*:2*AabbCc*:1*Aabbcc*
   phenotype ratio: 6:2:6:2, reduced to 3:1:3:1

	ABC	ABc	AbC	Abc	aBC	aBc	abC	abc
AbC	AABbCC	AABbCc	AAbbCC	AAbbCc	AaBbCC	AaBbCc	AabbCC	AabbCc
Abc	AABbCc	AABbcc	AAbbCc	AAbbcc	AaBbCc	AaBbcc	AabbCc	Aabbcc

3.

	S	s
S	SS	Ss
s	Ss	ss

genotype ratio: 1SS:2Ss:1ss
phenotype ratio: 3 healthy:1 sickle-cell anemia

Each time two *Ss* individuals have a child there is a 25% chance that the child will have sickle-cell anemia.

4. a. *NN* and *Nn* individuals are normal; *nn* individuals are diseased.

b.

	N	n
N	NN	Nn
n	Nn	nn

genotype ratio: 1NN:2Nn:1nn phenotype ratio: 3 normal:1 diseased

c.

	N	n
N	NN	Nn

genotype ratio: 1:1 phenotype ratio: 2 healthy:0 diseased

5. a. *DD* and *Dd* individuals are diseased; *dd* individuals are normal.

b.

	D	d
D	DD	Dd
d	Dd	dd

genotype ratio: 1DD:2Dd:1dd phenotype ratio: 3 diseased:1 normal

c.

	D	d
D	DD	Dd

genotype ratio: 1:1 phenotype ratio: 2 diseased:0 healthy

6. The parents are most likely *BB* and *bb*. The white parent must be *bb*. The blue parent could potentially be *BB* or *Bb*, but if it were *Bb*, we would expect about half of the offspring to be white. Therefore, if the cross yields many offspring and all are blue, it is extremely likely that the blue parent's genotype is *BB*.

7. The yellow allele is dominant. Since each parent breeds true, that means each parent is homozygous. When a homozygous recessive parent is bred to a homozygous dominant parent, the F_1 generation will exhibit only the dominant phenotype. Therefore, the phenotype of the F_1 generation, yellow, is produced by the dominant allele.

Chapter 13

1. Nonparental genotypes are formed by crossing-over, independent assortment of chromosomes, and fertilization. Mutation can also lead to the creation of new, nonparental alleles.

2. It is unlikely that alleles that cause lethal dominant genetic disorders will be as common in the population as alleles that cause lethal recessive genetic disorders, because lethal dominant disorders affect both individuals with homozygous dominant genotypes and those with heterozygous genotypes. This would cause lethal dominant alleles to be removed from the population at a faster rate than lethal recessive alleles, which only cause the death of individuals with the homozygous recessive genotype. Since these dominant alleles are *lethal*, they will be selected out of the population more quickly, and not be "protected" by heterozygote carriers as a recessive allele would be.

3. Each person at risk for Huntington's disease must decide for themselves whether they want to know if they have the (dominant) disease allele, and hence later in life will develop a lethal disorder that has no cure. If someone had not yet had children and found he or she carried the allele, he or she would have the option to choose not to have children. Generally speaking, knowing whether a person has an allele that carries a genetic disorder may help determine whether that person should undergo certain treatments or lifestyle modifications. For instance, if parents learn that a baby has PKU, they could modify the baby's diet and avoid the disease; otherwise, the child's mental abilities would be affected. A limitation of genetic tests is that often the presence of a disease allele indicates a predisposition toward a disease, rather than the actual existence of a disease. This might lead to unnecessary preventative actions, such as a woman deciding to have a healthy breast removed because she fears a predisposition to breast cancer, or an insurance company increasing a person's premiums, without knowing whether the person would have actually contracted the disease.

Sample Genetics Problems

1. a. Males inherit their X chromosome from their mothers, since their Y chromosome must come from their fathers. Their mothers do not have a Y chromosome to give them, and they must have one in order to be male.

b. No, she does not have the disorder. If she has only one copy of the recessive allele, her other X chromosome must then have a copy of the dominant allele. She is a carrier, but she does not have the disorder herself.

c. Yes, he does have the disorder. The trait is X-linked, he has only one X chromosome, and that X chromosome carries the recessive disorder-causing allele. His Y chromosome does not carry an allele for this gene, so cannot contribute to the male's phenotype relative to this trait.

d. If the female is a carrier of an X-linked recessive disorder, her genotype is $X^D X^d$, where D = the dominant allele and *d* = the recessive, disorder-causing allele. This means she can produce two types of gametes relative to this trait: X^D and X^d. Only the X^d gamete carries the disease-causing allele.

e. None of their children will have the disorder, since the mother will always contribute a dominant non-disorder-causing allele to each child. However, all of the female children will be carriers, since their second X chromosome comes from their father, who only has one X chromosme to contribute, and it carries the disorder-causing allele.

2. a. 50% chance of *aa* cystic fibrosis genotype

	A	a
a	Aa	aa
a	Aa	aa

b. 0% chance of *aa* cystic fibrosis genotype

	A	A
A	AA	AA
a	Aa	Aa

c. 25% chance of *aa* cystic fibrosis genotype

	A	a
A	AA	Aa
a	Aa	aa

d. 0% chance of *aa* cystic fibrosis genotype

	A	A
a	Aa	Aa
a	Aa	Aa

3. a. 50% chance of Huntington's disease genotype, *Aa*

	A	a
a	Aa	aa
a	Aa	aa

b. 100% chance of Huntington's disease genotype, *AA* or *Aa*

	A	A
A	AA	AA
a	Aa	Aa

c. 75% chance of Huntington's disease genotype, *AA* or *Aa*

	A	a
A	AA	Aa
a	Aa	aa

d. 100% chance of Huntington's disease genotype, *Aa*

	A	A
a	Aa	Aa
a	Aa	Aa

4. a. 0% chance of a child with hemophilia

	X^a	Y
X^A	X^AX^a	X^AY
X^A	X^AX^a	X^AY

b. 50% chance of a child with hemophilia

	X^a	Y
X^A	X^AX^a	X^AY
X^a	X^aX^a	X^aY

c. 25% chance of a child with hemophilia

	X^A	Y
X^A	X^AX^A	X^AY
X^a	X^AX^a	X^aY

d. 50% chance of a child with hemophilia

	X^A	Y
X^a	X^AX^a	X^aY
X^a	X^AX^a	X^aY

e. No, male and female children do not have the same chance of getting the disease. Male children are more likely to have hemophilia since they do not possess a second allele for this trait to mask a recessive allele that they may inherit.

5. The terms "homozygous" and "heterozygous" refer to pairs of alleles for a given gene. Since a male has only one copy of any X-linked gene, it does not make sense to use these pair-related terms.

6. The disease-causing allele (*d*) is recessive and is carried by both parents, because although neither the mother nor the father expresses the trait in question, some of their children do. The disease-causing allele is located on an autosome. If it were on the X chromosome, the father would express the gene, since we have already determined that he must carry one recessive copy of the gene, and he would not have another copy of the gene to mask this recessive allele. Both individuals 1 and 2 of generation I have the genotype *Dd*.

7. The disease-causing allele is recessive. Neither individual number 1 nor number 2 in generation II exhibits the disease phenotype, yet they have a child who does. The disease-causing allele is located on the X chromosome. If individual 1 in generation II is not a carrier, as stated, then the only way he can have a child (individual 2, generation III) that expresses this recessive trait is if that child does not carry a second gene for the trait, which is true for males relative to X-linked traits. Otherwise, not being a carrier, he (individual 1, generation II) would have two copies of the dominant allele, and each of his children would express the dominant non-disease-causing phenotype no matter what allele they got from their mother. This is not what the pedigree shows for his offspring in generation III. His son inherited his Y chromosome from his father and happened to get his mother's recessive allele

for the trait, which is expressed since the son does not have a second allele to mask it. The daughters inherit a dominant non-disease-causing allele from their father, so it does not matter which of their mother's alleles they inherit. They will still not show the disease phenotype, as shown by the pedigree. In addition, if individual 6 in generation II is not a carrier as stated, then she will give a dominant non-disease-causing allele to every child, male or female, so none of them would express the disease phenotype, even though their father contributes a recessive disease-causing allele to each daughter (and no allele for this trait to his sons). This is consistent with the pedigree.

8. a. If the two genes are completely linked:

	AB	ab
aB	AaBB	aaBb

b. If the two genes are on different chromosomes:

	AB	Ab	aB	ab
aB	AaBB	AaBb	aaBB	aaBb

Chapter 14

1. The two strands of the double helix follow specific base-pairing rules. This means that when the two strands are separated, each can act as a template for a new DNA molecule. Each of the two new DNA molecules will have one original strand and one newly synthesized strand.

2. An allele is made up of a precise sequence of DNA bases. If this sequence is changed by mutation, a new allele has been produced. This new allele may no longer produce the correct gene product. These incorrectly made products can cause human genetic disorders. An example of mutation causing a genetic disorder is the defective hemoglobin molecule, which is caused by a single base mutation; this mutation is the cause of sickle-cell anemia.

3. Cancers are caused by uncontrolled cell division. If a mutation in a gene that controls cell division is not repaired due to a defective (mutated) repair protein, the cell may begin to divide rapidly and uncontrollably, thus causing cancer.

Chapter 15

1. A gene is a segment of DNA that codes for a particular molecule of protein or RNA. Although most genes code for proteins, there are genes that produce a final product of rRNA or tRNA.

2. rRNA is a component of ribosomes. Ribosomes make the bonds that link amino acids together to make proteins during the process of translation. During translation, tRNA transfers the correct amino acid to the ribosome. Each tRNA has a specific anticodon that matches with an mRNA codon that specifies the particular amino acid that should next be added to the growing protein. mRNA specifies the order of amino acids in a protein. During translation, tRNA will "read" the mRNA codons and transfer the correct amino acid onto the growing protein molecule.

3. Transcription: A DNA molecule is partially unwound so that the RNA polymerase enzyme can have access to the base pairs of a given gene. The RNA polymerase matches complimentary bases to the exposed DNA sequence, thus forming an RNA copy of the gene. When the RNA polymerase reaches a DNA terminator sequence, transcription ends, the new RNA molecule is released, and the DNA strands bond back to each other. Translation: mRNA molecules contain the information necessary to build proteins. A molecule of mRNA binds to a ribosome, and translation begins at the start codon nearest to where the mRNA is bound to the ribosome. A given tRNA binds to a specific mRNA codon and also to a specific amino acid. As the ribosome moves one codon at a time, different tRNAs bind to the mRNA codons and the ribosome links together the amino acids brought by the tRNAs. This is how the sequence of codons on mRNA is translated to a sequence of amino acids in a protein.

4. tRNA must be able to bind to both the mRNA codon and the amino acid to ensure that the amino acid being added to the protein chain is the correct one as coded by the mRNA codon.

5. A gene is a portion of a DNA molecule that contains information for the synthesis of a protein or RNA final product. That information is transcribed to a complimentary RNA molecule. In the case of RNAs that have protein products, the specific order of bases in the mRNA codes for the order of amino acids in the gene's protein product. The protein product is what determines a phenotype (for example, the presence of brown pigment in the iris of an eye).

Chapter 16

1. a. Although all the cells of a given organism contain the same DNA, different genes are expressed in different cells. The expression of different genes can affect cell structure.

b. Although all the cells of a given organism contain the same DNA, different genes are expressed in different cells. This allows different cells to perform various metabolic tasks.

2. Crossing-over occurs more often between genes that are located further apart. Thus, long sequences of noncoding DNA between genes should increase the frequency of crossing-over between the genes.

3. The tryptophan operator controls whether or not the gene that codes for tryptophan is transcribed. If tryptophan is present, a repressor protein is able to bind to the operator and prevent transcription, since the presence of tryptophan indicates that the cell does not need to waste energy by making more. If tryptophan is absent, the repressor is unable to bind to the operator, thus allowing the gene to be transcribed so that the tryptophan needed by the cell is produced. This example follows the general eukaryotic and prokaryotic pattern of gene control by regulation of transcription. Regulatory DNA sequences such as the tryptophan operator can switch genes on and off. In order to do this, the regulatory DNA must interact with regulatory proteins, such as the tryptophan repressor protein. Control of transcription determines whether or not the gene product is expressed.

4. The advantage of studying groups of genes together is that it gives a more accurate picture of how they interact in a living organism. The disadvantage is that studying several genes at once makes it harder to determine which genes are having what effect, or which combination of genes is having that effect.

Chapter 17

1. a. DNA cloning is the process of isolating a gene and making many copies of it. Two methods of cloning a gene are the construction of a DNA library and the use of the polymerase chain reaction (PCR) technique. To construct a gene library, the DNA of a given organism is digested into fragments by a restriction enzyme. These fragments are inserted into a vector that transfers the gene into bacteria. As these bacteria reproduce, they copy the inserted gene along with their own genes, thus cloning the gene. In PCR, short primers that can bond with a targeted region of an organism's DNA are mixed with the organism's DNA. Heat is used to separate the strands of the organism's DNA so that the primer can bond to each target area. DNA polymerase then fills in a complementary strand of DNA for each of the separated strands that have a primer attached. In this way, many copies of the gene can be made; thus, it is cloned.

b. The advantage of gene cloning is that it easier to study a gene once you have many copies of it. The cloned gene can then be sequenced or transferred to other organisms or used in various experiments.

2. To screen a DNA library for a particular gene, a probe for the gene of interest is needed. The DNA of the bacterial colonies that make up the library is tested to see if they can base-pair with the probe. If they can, they contain all or part of the gene of interest.

3. Some people argue that humans do not have the right to alter the DNA of other organisms and, hence, that all forms of genetic engineering are not ethical. However, other people think that it may be ethical to alter the DNA of all organisms, depending on the reasons for doing it. If altering the DNA of a child allowed that child to overcome a genetic disease and live a healthy life, many people would support such a use of DNA technology. If the genes were being altered merely for cosmetic reasons, many people would find such changes objectionable.

4. Many people would argue that altering the DNA of an individual suffering from a genetic disease should be legal. While altering a person's DNA is a serious matter, it could be argued that *not* to do so—thereby dooming the person with the disease to suffering and death—is ethically unacceptable.

5. However, since altering DNA in any way is a serious and not always predictable matter, limits should be placed on the kinds of traits that DNA technology is used to alter. One way to draw the line between "unacceptable" and "acceptable" changes to the DNA of humans would be to allow changes to genes that cause genetic disorders but not to genes that control cosmetic features.

Chapter 18

1. Information on the activities of proteins in other organisms is likely to yield useful predictions about the activities of their human counterparts. In addition, meaningful experiments such as genetic crosses are not possible in humans for ethical reasons, but can be performed in organisms such as yeast, by focusing on related genes.

2. In general, human genomes only differ by 0.01%, but there is always the possibility that some of these differences in an individual's genome may have significant effects on their health or metabolism. In order to establish a standard genome for reference, it is therefore important to sequence DNA from several individuals. This helps exclude any significant mutations in the genome of a single individual from being incorporated into the standard reference sequence. To represent the diversity of the human race, one could imagine selecting volunteers of both genders from various ethnic groups.

3. Competition often has a positive effect on the speed with which important scientific breakthroughs are made. There is little doubt that the timetable for the human genome sequencing effort was greatly accelerated by the competition between commercial and public concerns. However, for competition to remain a healthy driving force behind scientific research, it must not impede the open sharing of data and discussion of ideas. If this atmosphere of scientific openness is at odds with commercial interests, then competition could slow progress instead of accelerating it.

4. SNPs in the noncoding sequences of the human genome are worth identifying because they may affect the regulation of genes in the coding sequences. Variations in the SNPs found in these regulatory elements can therefore have as dramatic an effect on the activity of a protein as variations in the gene itself.

Chapter 19

1. Life on Earth is characterized by adaptations (a match between organisms and their environment), by a great diversity of species, and by many examples in which very different organisms share morphological or other characteristics. Each of these aspects of life on Earth can be explained by evolution: (a) Adaptations result from natural selection, (b) the diversity of life results from speciation, which occurs when one species splits to form two or more species, and (c) shared characteristics of life are due to common descent.

2. Overwhelming evidence indicates that evolution occurred and continues to occur. Support for evolution comes from five lines of evidence:

a. The fossil record provides clear evidence of the evolution of species over time and documents the evolution of major groups of organisms from previously existing organisms.

b. Organisms contain evidence of their evolutionary history. For example, when anatomical data alone are used to determine evolutionary relationships, scientists find that the proteins and DNA of closely related organisms are more similar than those of organisms that do not share a recent common ancestor. In this and many similar examples, the extent to which organisms share characteristics other than those used to determine evolutionary relationships is consistent with scientists' understanding of evolution.

c. Scientists' understanding of evolution and continental drift has allowed them to correctly predict the geographic distribution of certain fossils depending on whether the organisms evolved before or after the breakup of Pangea.

d. Scientists have gathered direct evidence of small evolutionary changes, in thousands of studies, by documenting genetic changes in populations over time.

e. Scientists have observed the evolution of new species from previously existing species.

3. In any area of science, new pieces of information are still being added to our knowledge of the subject. The debate between scientists as to which mechanisms are most important only means that evolution is not fully understood, not that it does not occur.

4. Genetic drift has a greater effect on smaller populations. If the plant population were larger, the likelihood that all plants of a certain type would die in a windstorm would be smaller, and the dramatic shift in the frequency of the A and a plant alleles would therefore also be less likely.

Chapter 20

1. Nonrandom mating occurs when an organism is more likely to mate with members of a portion of the population than with individuals chosen at random from the population at large. Gene flow is the exchange of genes between populations. Gene flow makes the genetic makeup of populations more similar to one another. Genetic drift is a process in which alleles are sampled at random over time. Genetic drift can have a variety of causes, such as chance events that cause some individuals to reproduce and other individuals not to reproduce. Natural selection is a process in which individuals with particular heritable characteristics survive and reproduce at a higher rate than other individuals.

2. Potential benefits include providing a larger population, which is therefore less susceptible to chance events and genetic drift, and providing an input of new alleles on which natural selection can operate. Drawbacks include the introduction of individuals with genotypes that are not well-matched to the local environmental conditions of the smaller population. Throughout time, some species have gone extinct locally or worldwide. This is a natural process. However, humans have increased the rate that populations and species have become extinct. If there are numerous other populations of a species, it may not be worth introducing new members to the smaller population. If the smaller population is one of the few populations of that species left, then it may be more important to introduce new individuals in an attempt to allow the population to recover and survive.

3. The genotype frequencies for the original population are:

$$AA: \frac{280}{280 + 80 + 60} = 0.67$$

$$Aa: \frac{80}{280 + 80 + 60} = 0.19$$

$$aa: \frac{60}{280 + 80 + 60} = 0.14$$

The allele frequencies for this population are:

Frequency of A allele =

$$p = \frac{2(280)+80}{2(280+80+60)} = 0.76$$

Frequency of a allele =

$$q = \frac{2(60)+80}{2(280+80+60)} = 0.24$$

Hardy–Weinberg predicts that the frequency of genotype AA should be

$$p^2 = (0.76)(0.76) = 0.58;$$

that the frequency of genotype Aa should be

$$2pq = (2)(0.76)(0.24) = 0.36;$$

and that the frequency of genotype aa should be

$$q^2 = (0.24)(0.24) = 0.06.$$

Note that the sum of the genotype frequencies,

$$p^2 + 2pq + q^2 = 1.0$$

These calculated genotype frequencies do not match those of the original population. This difference could be due to mutation, nonrandom mating, gene flow, a small population size, and/or natural selection.

Chapter 21

1. Classifying species by their inability to sexually reproduce with other species is a convenient definition, but there are many nonreproductive alleles that could cause these organisms to be different enough that they can be classified as separate species even though they can produce hybrids.

2. Some of the populations in Lake Victoria may have had so little contact with one another that they evolved into separate species in geographic isolation, despite the fact that they lived in the same lake. Other populations may have evolved into new species in the absence of geographic isolation.

3. Species that interbreed in nature may still be distinct species due to a host of nonreproductive alleles that may cause distinctions. For this reason, many people would argue that the rare species is separate from the common species and should remain classified as rare and endangered.

4. Many viral and bacterial pathogens rapidly evolve resistance to our best efforts to destroy them. Medical researchers seek a detailed understanding of how adaptations in pathogens provide such resistance. Armed with that knowledge, they then seek to develop new treatments that will be difficult for pathogens to become resistant to.

Chapter 22

1. A mass extinction may be associated with rapid environmental changes that have no relation to the conditions that favored a particular adaptation. Thus, organisms with wonderful adaptations can (and have) become extinct during mass extinction events.

2. Although speciation can happen within a single year, it often takes hundreds of thousands to millions of years to occur. Thus, it is not surprising that it usually takes one to seven million years for the number of species found in a region to rebound after a mass extinction event. The time required to recover from mass extinction events provides a powerful incentive for humans to halt the current, human-caused loss of species; otherwise, it will take millions of years for the number of species on Earth to recover.

3. Microevolution refers to the changes in allele or genotype frequencies within a population of organisms. This is fundamentally different than macroevolution, which deals with the rise and fall of entire groups of organisms, the large-scale extinction events that cause some of these changes, and the evolutionary radiations that follow extinction events. Macroevolutionary changes cannot be predicted solely from an understanding of the evolution of populations. Of the evolutionary processes studied in Unit 4, speciation occupies a "middle ground" between macroevolution and microevolution.

Chapter 23

1. Human prejudices might be decreased if individual humans looked more alike. However, a great diversity of cultural riches would be lost. Nonrandom mating, natural selection, and genetic drift all can prevent gene flow from making the genetic make-up of human populations more similar.

2. Classification systems are constructed so that we can better understand and speak about various organisms, but the separations we make between organisms are sometimes somewhat arbitrary. Although we classify ourselves in a separate genus from the three species of chimpanzees, a hypothetical observer from outer space might well classify humans and chimpanzees as members of the same genus.

3. Agriculture developed in the eastern Mediterranean region (ca. 10,500 years ago) and in China (ca. 10,000 years ago) when wheat, barley, rice, and other crops were developed from wild plant species. Agriculture transformed all aspects of human society. For example, farming led to increased population sizes, rapid improvements in technology, and the development and spread of infectious diseases.

Chapter 24

1. The plasma membrane serves two related functions essential to the survival of cells: 1) it selectively filters biologically important molecules entering and leaving the cell, and 2) it allows the cell to maintain an internal environment that is both suitable for life's chemistry and different from the external environment. Destroying the plasma membrane disrupts both of these functions, killing the bacterium.

2. Evolution teaches that all living things descended from a common ancestor. This idea is supported by evidence compiled from biogeography, fossil records, comparative anatomy, and molecular biology. Therefore, it comes as no surprise that different species share features.

3. There are too many examples of interactions between functions to list all of them here. The two following examples are familiar ones that illustrate the nature of such interactions: 1) When we eat, we rely on our eyes, nose, and tongue (senses) to identify food; muscles (movement) transport the food through our digestive system (nutrition). Once the food is digested and the nutrients are absorbed, the blood (internal transport) carries the nutrients to the metabolizing cells. 2) When we run, we rely on our eyes and sense of balance (senses) to keep us on our path. Muscles (movement) in our legs actually propel us. The oxygen and nutrients needed for activity are carried from our lungs (gas exchange) and guts (nutrition) to the muscles by our circulatory system (internal transport). Our nervous system helps to coordinate all of these activities (internal communication).

4. Although trees consist of relatively few cell types, these cell types are arranged into a variety of tissues that serve specific functions such as fluid conduction, support, photosynthesis, and active transport. These tissues combine in various ways into three tissue systems. Different arrangements of the three tissue systems lead to roots specialized for absorption, stems specialized for support and transport, and leaves specialized for photosynthesis.

5. A single-celled organism must be a generalist, performing all the functions needed to survive. On the other hand, cells in the multicellular human body are specialists. Although they can perform certain functions very efficiently, they must rely on other cells to meet some of their needs. This requires a speedy and coordinated system of communication and transport.

Chapter 25

1. Allometric relationships that relate the size of an organism to its form or function describe many aspects of an organism's biology. We will list just a few of the effects here; a careful search through the chapter will reveal more. The smaller budgies would have a larger surface area–to–volume ratio than the larger lovebirds. This means that all sorts of functions involving exchange, such as losing body heat and water, and absorbing nutrients, might be faster in the budgie than in the lovebird. Budgies would probably sleep longer than lovebirds. A sick budgie might require more medication relative to its body weight than would a sick lovebird. The lovebird would need more food and make more mess than a budgie, but would probably eat less than five times as much food and make less than five times as much waste. The budgie could probably get by in a smaller cage than the lovebird.

2. The size of a cell is limited by its ability to exchange with its environment a sufficient amount of material to support cell function. As cell size increases, the surface area across which exchange can take place decreases relative to the volume of the cell. Thus, cells that are too big cannot supply their needs

across the relatively small surface area provided by their plasma membrane.

3. If a single cell or organ performed several functions, the species could not survive if that cell or organ became modified to do just one thing. Duplication of cells and organs is necessary before specialization can occur so that the various functions can be divided amongthem.

Chapter 26

1. Lignin resists stretching under tension. Growth of the cell would cause the surrounding cell wall to stretch. If growing cells had cell walls rich in lignin, the cell wall would interfere with cell growth.

2. Biologically produced molecules and mineral materials have complementary properties that together produce effective support structures. Lightweight, biologically produced materials generally resist breakage very well under tension, but are often not very stiff and may not resist compression well. Heavy mineral materials, on the other hand, are stiff and resist compression, though they often break relatively easily under tension. Biologically produced and mineral materials combine to form support structures, such as our own bones, which resist bending, compression, and tension remarkably well.

3. Organisms that need relatively little support (such as many marine invertebrates) can get by with a hydrostat and can take advantage of the hydrostat's light weight. Hydrostats also provide a flexible support system without requiring joints, and they allow organisms to vary their stiffness. The hydrostat that supports earthworms takes advantage of both of these properties to allow the worm to move through the soil.

Chapter 27

1. In a microscopic view of muscle, a sarcomere is an area between two Z discs. Actin filaments extend from each Z disc toward the center of the sarcomere. Myosin filaments lie between actin filaments and have heads which, in the presence of ATP, can bind to the actin, change shape, and in the process, pull the Z discs closer together. This causes the sarcomere to shorten or contract.

2. The arrangement of stiff support structures into lever systems allows organisms to effectively increase or decrease the strength or speed of a muscle contraction. If the muscle that moves a limb is located near the joint that acts as the fulcrum in the lever system, the muscle contraction leads to relatively slow movement with more force at the end of the load arm of the lever, where the work is done. Alternatively, if the muscle is located relatively far from the joint, the muscle contraction leads to relatively rapid but weak movement at the end of the load arm. Strength is increased if many muscle fibers work side by side in a bulky muscle. Slow contraction of a muscle often allows the muscle to generate a stronger contraction, whereas rapid contractions lend speed but little strength.

3. a. The walking elephant would experience less pressure drag and less friction drag than the swimming whale. Water is both denser and more viscous than air, greatly increasing both the pressure and friction drag experienced when moving at a given speed. In addition, whales swim more rapidly than elephants walk, which will also cause them a further increase in the pressure and friction drag that whales face.

b. The flying falcons have to deal with a lot of pressure drag, but relatively little friction drag, whereas the swimming penguins have to deal with a lot of both types of drag. Although air is less dense than water, falcons fly fast enough so that pressure drag becomes an issue, much as it does for the aquatic penguins. For this reason, both have streamlined shapes. The greater viscosity of water ensures that friction drag remains a much bigger problem for penguins than for falcons.

4. Fungi lack the muscles, cilia, and flagella that allow other organisms to move. In addition, they have rigid cell walls that restrict movement. To some extent, fungi spread as their rootlike hyphae grow through their food. Much more important, however, are the tiny spores which float long distances in even gentle air currents. Fungi spread primarily because their spores can exploit the movement of the air around them.

Chapter 28

1. There are several possible reasons why pitcher plants supplements the mineral nutrients that they obtain from the soil with nutrients absorbed from the insects that they catch, including the following: 1) The soil in which pitcher plants grow may not provide enough nutrients for them to survive. 2) The roots with which pitcher plants absorb minerals from the soil may not provide enough surface area with which to absorb an adequate supply of mineral nutrients. 3) Pitcher plants may have evolved carnivory because this allows them access to richer nutrient sources than are available to most plants. The natural history of pitcher plants shows that Reason 1 is most likely to explain carnivory in pitcher plants. Pitcher plants, like almost all other carnivorous plants, live in extremely nutrient-poor soils.

2. If humans fed exclusively on plant leaves, our digestive systems might have evolved differently from the ones that we have in several ways: 1) Our teeth would have been better suited to grinding up the tough cell walls of plant tissues. Most herbivores have well-developed grinding teeth with broad, flat surfaces. 2) Our small intestine would have evolved to provide a greater surface area for absorption to compensate for the low nutrient content in most plant tissues. Most herbivorous mammals have very long small intestines that allow them to absorb most of the nutrients from plant tissues. 3) If we fed on very nutrient-poor plant tissues, we may have evolved a mutualistic relationship with microbes to help break down the cellulose in the cell walls of plant tissues. The microbes often live in specialized structures in the guts of animals, as is the case in the modified stomachs of sheep.

3. To survive and grow, we need to obtain essential nutrients such as vitamins, minerals, and amino acids from our food.

The balance of these nutrients in our own bodies differs from those in the foods that we eat, forcing us to mix and match foods so that our intake corresponds to what we need. Vitamins are essential for our nutritional well-being, but we must obtain them from our food. Most foods contain only a few of the vitamins that we need. Similarly, not all foods contain all of the amino acids essential for our survival

Chapter 29

1. Terrestrial producers and terrestrial consumers are similar in that both lose valuable water while exchanging gases. As a result, both have gas exchange mechanisms that allow them to limit water loss. The carbon dioxide used by producers occurs in air at much lower concentrations than the oxygen used by consumers. As a result, plants must keep carbon dioxide levels very low inside their cells to maintain a concentration gradient of carbon dioxide into their cells.

2. When we begin to exercise, our muscles start to contract more vigorously. As a result, the demand for oxygen, particularly in the muscles, increases. We supply this increased oxygen demand by increasing the rate at which we breathe in and out. This keeps the oxygen concentration in our alveoli high, allowing for the rapid diffusion of oxygen into our blood. An increase in our heart rate once we start exercising increases the speed with which the oxygen in the blood, bound to hemoglobin, moves to the active muscles. The rapid movement of oxygenated blood past the metabolizing muscle cells ensures that the oxygen can diffuse rapidly into the muscle cells.

3. The major advantages of obtaining oxygen from air instead of water are, first, that air contains a greater proportion of oxygen than does water, and, second, that air offers much less resistance to ventilation or to moving over the gas exchange surface. An air-breathing aquatic mammal gets the oxygen that it needs from relatively few liters of air, as compared to the number of liters of water that a hypothetical, gill-bearing mammal would have to process to get an equivalent amount of oxygen. The animals that rely on gills, moreover, must have a large gill surface area in contact with relatively viscous water (Chapter 27), which means that gills represent a potentially great source of friction drag. The air-breathing mammal, with internal lungs, has no such problem.

Chapter 30

1. The highest blood pressure (approximately 120 mm Hg) occurs when the left ventricle contracts, pumping blood into arteries to be carried throughout the body. When this ventricle relaxes and refills, the pressure drops to about 80 mm Hg. Blood pumped from the right ventricle goes to the lungs. Since it has a shorter distance to travel, less pressure needs to be generated: 25 mm Hg drops to 8 mm during relaxation. As the blood passes into capillaries, friction drag lowers the pressure to 35 mm Hg. Pressure of blood leaving the smaller-diameter capillaries drops to 10 mm Hg. Resistance to blood flow is increased when the diameter of a vessel is decreased, slowing down the blood flow. Blood that must travel uphill to return to the heart depends on skeletal muscles to push it upward against gravity. Valves in the veins keep it from flowing backward. The consistent decrease in blood pressure throughout the human circulatory system is necessary to allow the blood to complete a circuit of the circulatory system by always flowing from a region of higher blood pressure to a region of lower blood pressure.

2. A lower demand for oxygen from your running muscles would lead to a reduced rate at which the circulatory system supplies the oxygen. After sitting down following your run, your heart rate would decrease, reducing the blood pressure generated by the ventricles. Because the steepness of the pressure gradient in the circulatory system determines how fast the blood flows, the drop in ventricular pressure leads to a decreased blood flow. In addition, local blood, and therefore oxygen, supply is adjusted by the opening and closing of the blood vessels that supply tissues. By reducing the diameter of the blood vessels feeding your running muscles, these muscles would receive less blood once you stopped running.

3. Advantages of a closed circulatory system: It allows more control over where the exchange of gases and nutrients occurs. Organs can be large because blood goes through the organs instead of flowing into spaces around them. Disadvantages of a closed circulatory system: A strong pump is needed to supply enough pressure to push blood through vessels, overcome friction drag, and lift blood against the flow of gravity. This requires energy.

4. The plant vascular system must lift sap under high tensions and pressures, and it must accomplish this without the muscle tissue that propels blood through animal circulatory systems. Natural selection can only act on available variation. Because only animals have muscle tissue, and the chances of a mutation in plants leading to a similar contractile tissue are slim, the evolution of internal transport systems in plants has followed a very different path. Plants rely on different means of moving fluid: osmotic pressure moves the phloem sap, and tension generated by evaporation of water from the leaves' surfaces moves the xylem sap. Although the movement of phloem sap requires active transport, and therefore the expenditure of energy on the part of the plant, xylem transport is accomplished using the sun's energy. Even if a plant were to evolve a muscle-like tissue capable of pumping fluid, it is not clear that natural selection would favor the spread of this feature, given the pressures that plants would need to generate and that their current xylem transport requires little energy expenditure on the part of the plant.

Chapter 31

1. With organisms, such as endotherms, that produce large amounts of metabolic heat, this form of heat can figure prominently in their heat budgets. Thus, endotherms such as humans rely heavily on metabolic heat as a heat input. Because organisms generate metabolic heat by spending energy, they can conserve energy by using other heat sources, including conductive and radiant heat inputs, to reduce the amount of metabolic heat they must generate to maintain body temperatures. Heat loss in organisms that depend on metabolic heat can be costly. Consequently, endotherms tend to minimize the conductive, convective, and radiative heat

loss to the environment with various insulating body covers, including fat, fur, and feathers.

2. Organisms do not regulate internal temperature or water when they cannot or when they obtain little benefit from doing so. Aquatic organisms generally do not regulate their body temperature. The high rate of conductive heat loss in water makes it virtually impossible to maintain a body temperature that differs much from the surrounding water. Organisms that live in a very constant environment often stand to gain little by having the machinery to regulate body temperature. Bodies of water often change in temperature little, if at all, from day to night, or even over the course of a year. Therefore, even though aquatic organisms cannot regulate body temperature, they suffer little, because their body temperature—like that of their environment—remains constant. Marine animals have solute concentrations in body water that closely match the concentration of salts in their environment. Thus, marine organisms have little need to regulate their body water.

3. Unlike ectotherms, humans do not rely on environmental heat sources to maintain their body temperature. This comes at a great energetic cost that results from the energy used to generate the metabolic heat. Thus, humans must eat much more, relative to their body weight, then ectotherms. Endothermy, however, allows us to maintain constant body temperatures and levels of activity during hot days and cool nights and during hot summers and cold winters.

Chapter 32

1. Recognizing their own tissues as different from the cells of invaders makes it possible for organisms to protect themselves against diseases that invade their body. In the absence of an ability to distinguish self from non-self, organisms could only kill invading cells by killing many of their own cells as well.

2. A dead or weakened disease organism in the vaccine has identifying proteins, called antigens, on its surface. A few of our white blood cells have surface proteins that can recognize these antigens. B lymphocytes are stimulated to reproduce themselves rapidly and produce antibodies, which are proteins that can circulate in the blood and neutralize a specific antigen. These antibodies remain in the bloodstream and can attack quickly if the virulent form of the organism enters the body. T lymphocytes are present in two forms: helper T cells, which allow other white blood cells (neutrophils and macrophages) to attach to and engulf the antigen, and killer T cells, which consume any cells damaged by the invader. Once stimulated by the vaccine, both B and T lymphocytes will "remember" and respond faster to a second exposure to the antigen.

3. We defend ourselves against disease organisms using our immune system. Our immune system mounts a defense specifically geared to a particular disease organism and it mobilizes rapidly when an attack by that disease organism occurs. In contrast, plants defend themselves against herbivores by mounting a nonspecific defense that works against many different kinds of herbivores or, quite often, by relying on a defense that is always present in the plant, whether it is under attack or not.

Chapter 33

1. Auxins suppress the growth of side branches, while cytokinins promote growth of side branches. Auxins, cytokinins, and gibberellins stimulate growth of fruit, but they slow ripening. ABA and ethylene promote ripening, but can also slow plant growth under stressful conditions, thus conserving energy.

2. A menstrual cycle begins as an egg develops in response to estrogen levels that increase over a two-week period. The high estrogen level triggers the pituitary gland to release follicle-stimulating hormone and luteinizing hormone. These two hormones act together to promote the release of an egg from the ovary. The high luteinizing hormone levels stimulate the release of another hormone, progesterone, that prepares the uterus for embryo implantation following the release of the egg. The high progesterone levels cause the pituitary gland to stop releasing follicle-stimulating hormone and luteinizing hormone. If the egg is not fertilized by a sperm, the progesterone level decreases after about 12 days. Without progesterone in the system, the uterine lining sloughs off and estrogen levels once again begin to increase.

3. In the medical use of hormones, the hormones taken by a patient replace missing or defective hormones. Thus the hormones are used to return the body to its normal way of functioning. In their use to enhance athletic performance, hormones are added to a normally functioning system. As a result, the hormones taken to enhance performance take the body outside its normal range of functioning, which may place a dangerous stress on the body.

Chapter 34

1. The advantage of a reflex arc response is its immediacy in an emergency situation. Signals filtered through the brain require longer response time but are needed for more complex responses. Example: A reflex action would let you pull away from a pinprick, but further processing by the cerebral cortex would allow you to inject a needed vaccine or medication, such as insulin.

2. In response to an action potential traveling down the axon, the dendrites of a neuron release chemical neurotransmitters from their tips. The neurotransmitters are released into the gap, or synapse, separating the neuron from adjacent neurons. Because of the narrow width of the synapse, the neurotransmitter diffuses rapidly. Receptors on the dendrites of adjacent neurons bind the neurotransmitter. Binding of the neurotransmitter to the receptors elicits a response in the second neuron. In some cases, this response is a new action potential that then travels the length of the second neuron's axon. In other cases, the binding of the neurotransmitter inhibits the formation of an action potential.

3. In image-forming eyes, a lens gathers and focuses light that reflects off of an object on to light-sensitive cells, which are gathered together into a retina. The interaction between light energy and light-sensitive chemicals inside the light-sensitive cells triggers an action potential that travels down nerve cells to the brain. The brain then interprets the nerve signals coming from various parts of the retina as an image.

Chapter 35

1. Behavior in animals depends on the interaction of the nervous, muscular, and endocrine systems. Of these systems, the nervous and muscular systems are unique to animals. The nervous system allows animals to quickly integrate sensory information, and to coordinate a movement in response by stimulating contraction of the appropriate muscles.

2. Evidence indicates that the ability to learn depends both on opportunities to learn from others or from experience (environmental effects), and from a genetic predisposition to learn (genetic effects). Organisms that live in groups or spend some time growing up with their parents can learn by imitating the other individuals of their own species. Organisms that live alone from birth or hatching can learn by remembering and selectively repeating only those behaviors that allow them to better find food or shelter or mates. For organisms to learn, however, they must have a nervous system and brain that is sufficiently well developed to allow them to interpret and remember what they have experienced. The development of these structures is under genetic control. In addition, species capable of learning often tend to excel in learning the types of behaviors most important to their success. Thus, organisms that hide away food throughout their territory are extremely good at remembering where things are.

3. Fixed behaviors that might be important to group members might involve responses to stimuli about which an organism might never have a chance to learn. Thus, a group-living animal might have a fixed behavior that led it to avoid snake-like objects or to hide at the sight of a silhouette that resembles that of a predatory bird. There might not be a chance to learn the appropriate behaviors from others, because one incorrect response could easily be fatal. Behaviors associated with surviving during the first few days following birth or hatching, before the creature has a chance to learn, are fixed behaviors.

4. Communication can make group life possible, it can help it find and evaluate a mate, it can help it learn from others. Most group-living organisms spend much time remaining in contact with each other through calls of various sorts. The calls can also inform group members of danger or of the discovery of food. Courtship behavior in humans, birds, frogs, and insects often involves elaborate sequences of actions that communicate the quality of one individual to the other. Learning in humans and other primates depends strongly on the ability of parents and other family members to communicate with the young.

Chapter 36

1. Humans produce eggs and sperm in distinct tissues (the gonads) within each individual. Plants have complex life cycles that involve the production of eggs and sperm by separate individuals called gametophytes. In flowering plants, these gametophyte individuals live within the tissues of the sporophyte that produced them, which makes it look on the surface as though the sporophyte produces the eggs and sperm within tissues.

2. Flowers attract insects, birds, and mammals. These animals may accumulate pollen on their bodies when they come into contact with the flower. This pollen can be carried long distances and eventually dropped off onto a flower of the same species. Wind, however, will not carry the pollen far, so without the diverse methods of pollination the flowering plant uses, wind-pollinated plants must live in close proximity to others of their species.

3. The structure of the female parts of a flower can be modified in many ways to increase competition among male gametes. One prominent mechanism is to increase the length of the style that separates the stigma (where the pollen lands) from the ovaries (where the embryo sacs reside). Lengthening the style lengthens the race between individual pollen tubes that grow down through the style tissue, making it easier for small differences in pollen quality to translate into earlier arrival of the winning sperm. A second approach is to increase the number of pollen grains landing on the stigma at any time by encouraging animal pollination. Having many pollen grains on the stigma at once is like having a large field of runners in the race. The best of a large field of runners is likely to be better than the best in a small field of runners. All of this male–male competition benefits the female by increasing the likelihood that the sperm that fertilizes the embryo sac is of high quality.

4. Aquatic habitats allow for external fertilization of eggs. Sperm and eggs can survive in seawater for some time because seawater has a solute concentration that resembles that inside the sperm and egg. As environments become more challenging (e.g., freshwater with much lower solute concentrations than those inside the sperm and eggs, and the dry terrestrial environment), the life span of sperm and unfertilized eggs outside the organism is much reduced. This means that sperm must either be deposited outside of the organism directly onto the eggs so that fertilization can take place quickly, or fertilization must take place internally. Most terrestrial organisms rely on internal fertilization.

Chapter 37

1. The increasingly narrowly defined fates of cells as development progresses help to ensure that fewer developmental mistakes will be made. A cell that retains a completely unlimited fate could give rise to descendant cells of any of a number of functional types. Such cells, if present late in development, could lead to organs that are supposed to fulfill a particular function containing a variety of functionally unrelated cells.

2. During development, the fates of cells depend on the identity of neighboring cells, their position within the developing organism as a whole, and the stage of development. They gain this information from a variety of chemicals released by other cells in the developing organism, including hormones and morphogens. In some cases, the presence or absence of the chemicals provides the necessary information; in others it is their concentration.

3. Small genetic changes during development can lead to large changes in form or function. Differences in human and chimpanzee head shapes may result in large part from timing of events during development of the head. The loss of webs in chicken feet results from the action of a single gene during development.

Chapter 38

1. As men and especially women age, the chances that the sperm or eggs that they produce will lead to a viable zygote decrease. As couples put off having children, therefore, they decrease the chances that they will manage to successfully conceive.

2. If the hipbones of human females reflected only the challenges posed by childbirth, we would expect that the ability of females to walk upright in an efficient way would be reduced. Because walking upright has also been very important to human success, it is unlikely that natural selection would ever have favored a hipbone design determined solely by the demands of childbirth.

3. The chance of getting pregnant during any given menstrual cycle is low because many things have to happen in just the right way for a fertilized egg to implant successfully in the uterus: The brief period in which the single egg is released during each cycle in the right portion of the oviduct must coincide with the even briefer period that sperm remain viable inside the female reproductive system. If a sperm manages to fertilize an egg, the resulting zygote must be genetically viable, which it often is not. Even if viable, many developing embryos never make it to the uterus, and a quarter of the embryos that do never manage to implant in the uterus wall.

Chapter 39

1. The potential or natural location of a biome is where conditions are such that the biome could, in principle, be found, while the actual location is where the biome currently is found. The potential locations of terrestrial biomes are most strongly influenced by climate, particularly by temperature and precipitation, while the potential locations of aquatic biomes are strongly influenced by their terrestrial biomes and by climate. The actual locations of both terrestrial and aquatic biomes are strongly influenced by human actions.

2. Atmospheric wind convection cells and oceanic water currents carry the results of local events (a volcanic eruption or an oil spill) to distant areas around Earth. For example, oil spilled into an ocean current next to one continent's shore may be carried by that current and end up coating the shores of other continents. This distant event can alter local ecological interactions when shorebirds are killed when they become coated with oil from that spill. In such a case, the birds may no longer keep certain organisms (their food) under control, and they are no longer available as a food source for their predators.

3. Climate can directly exclude species from a region, as when the lack of rainfall excludes many plant and animal species from desert biomes. Climate also can exclude species from a region indirectly, as when a species that can tolerate the climate is excluded by other organisms that are better adapted to the region's climate.

Chapter 40

1. An organism needs food and water, not too many predators, shelter, and acceptable climate conditions. Both the physical and biotic environments, and the interaction between them, will have an influence on the suitability of a given habitat for a given organism. Disturbances such as fire, flood, earthquakes, windstorms, and droughts will also determine whether a given habitat is suitable for a given organism.

2. Species distribution relies heavily on habitat distribution, and habitats are not evenly distributed throughout Earth's environment. In addition, plant distribution is often clumped due to the fact that plants cannot move to disperse themselves, and their offspring can only get so far away. This factor may cause animals that feed on these plants to have a distribution matching that of the plants.

3. If a species evolved before the continents began drifting apart, it is likely to have a wider distribution since it may have existed on each (current) landmass before they separated. A species that evolved later may be more isolated, since it might not have the ability to disperse the great distances between continents. In addition, the differences in climate between the continents is greater now than when they were closer together, and while organisms that existed on the supercontinent had a long amount of geologic time to adapt to these changes, organisms that evolved later might have had to be able to tolerate a more sudden change in climate as they attempted to colonize a new continent.

4. Studying the reproductive and social habits of the organism would help determine whether the distribution was, for instance, naturally clumped (as plants or family groups), or whether the offspring travel widely but cannot survive in other habitats. The fossil record for a species would help determine whether it evolved before or after the breakup of the continents. Finally, experiments could be performed to examine the impact of other species. For example, a species could be moved to a region where it does not occur and competitor species could either be removed or left alone to see if the competitors or local environmental conditions prevented the species from living there.

5. Thirty-five million years ago, there was no tundra in North America and tropical forests extended much further north than they do now. By 60,000 years ago, ice and tundra covered much of North America and the tropical forests were gone. Since then, the climate has warmed, causing tundra to retreat north and tropical forest to reclaim portions of North America.

Chapter 41

1. Coordinates for graph, in the notation (*x* coordinate, *y* coordinate):

(1, 150) (2, 225) (3, 337.5) (4, 506.3) (5, 759.4)

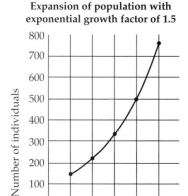

Expansion of population with exponential growth factor of 1.5

2. a. Some factors that limit population growth include available habitat, available food and water, disease, weather, natural disturbances, and predators.

b. Species new to an area often do not have an established predator. In addition, they have not yet reached the carrying capacity of their habitat.

3. A density-dependent factor is one whose intensity increases as the density of the population increases. An example of a density-dependent factor is the fact that disease spreads more rapidly in densely populated areas. A density-independent factor is not affected by the density of the population. Temperature is a density-independent factor. If the temperature drops below what a certain kind of plant can tolerate, it does not matter how dense that population of plants is; they will still die. Fire is also a density-independent factor.

4. If the pattern of population growth is understood, managers may be able to manipulate or protect the factors that most directly affect the population growth rate. If a population of organisms is more successful because, for example, it has adequate access to water, a manager would be sure that nearby rivers are not drained off for agricultural needs.

5. Limit reproduction to no more than one child per parent; reduce the consumption of unnecessary goods; reuse and recycle items to promote sustainable use of resources; work to develop and follow environmentally friendly policies and activities (for instance, using energy-efficient cars and light bulbs); purchase goods that have a lower impact on the environment, including organically grown clothing and food items.

Chapter 42

1. Mutualisms are common because their costs to organisms are outweighed by the benefits they provide organisms. Yucca plants may lose a few seeds to the offspring of their moth pol-

linators, but they still end up with more seeds than if the moth had not mediated pollination to begin with.

2. The inferior competitor is still using resources that the superior competitor needs, thus possibly limiting its distribution and/or abundance.

3. The plant community would most likely change to have fewer species. When the rabbits are removed, the grass that is no longer being eaten can assert its dominance as the superior competitor and will probably drive some of the other species in that area to extinction.

Chapter 43

1. a. Interactions among species, such as competition, mutualism, and consumer–victim interactions, have a large effect on the distribution and abundance of species found in a community. Some species, called keystone species, have a particularly strong effect in that, relative to their abundance, they have a disproportionately large impact on the type and abundance of other species in the community.

b. Disturbances such as fire occur so often in many ecological communities that the communities are constantly changing in response to a previous disturbance event. Depending on the type and severity of disturbance, a given community may or may not be able to recover from a particular disturbance event.

c. The climate is a main factor in determining what organisms can live in a given area. If the climate changes, the habitat changes, leading to changes in the success of various organisms in that area.

d. The effect of continental drift on ecological communities is largely based on the climate changes that occur as the continents move to different longitudes (or areas in a different wind or water current pattern).

2. Primary succession occurs in a newly created habitat. Secondary succession is the process by which communities recover from disturbance. The newly created habitat on which primary succession occurs can result from a "disturbance," such as the retreat of a glacier, but such disturbances are of a more catastrophic nature than those found in secondary succession.

3. The addition of beard grass to Hawaii has increased the rate and area of fires on the island. This is due to the large amount of dry matter the grass produces, which burns more easily and hotter than the native plants. In this way, one species in the community has altered the frequency of fire.

4. Change is a part of all ecological communities. However, human-caused change is unique in that we can consider the impact of our actions and we can decide whether or not to take the actions that cause community change. Whether or not a particular change is viewed as ethically acceptable will depend on the type of change, the reason for it, and the perspective of the person evaluating the change. For example, a person might find it ethically acceptable to alter a region so as to produce a long-term source of food for the growing human population, yet not ethically acceptable to take actions that result in short-term economic change but cause long-term economic loss and ecological damage.

Chapter 44

1. Energy captured by producers from an external source such as the sun is stored in the bodies of producers in chemical forms, like carbohydrates. At each step in a food chain, a portion of the energy captured by producers is lost from the ecosystem as metabolic heat. This steady loss prevents energy from being recycled.

2. Decomposers are essential because they recycle nutrients from dead organisms to the physical environment, where the nutrients may again be used by another organism.

3. Nutrients can cycle on a global level. When sulfur dioxide pollution is put into the atmosphere in one area of the world, wind convection cells move that pollution around the world where it can affect other ecosystems.

4. Internal agreements on inputs to nutrient cycles can in principle provide protection to the economic interests of nations (and they have in practice). Such agreements may also help nations realize that their actions can have far-reaching effects.

5. Human economic activity is interwoven with several key ecosystem services. Pollination is essential for the production of crop (and other) plants. Floodplains act as safety valves for major floods if we do not build on them or separate them from the body of water they help control, and wetlands and estuaries act as cost-free waste filtration systems. We rely on nutrient cycling to keep us alive. When ecosystem services such as these are damaged, human economic interests are damaged as well.

Chapter 45

1. Major types of global change caused by humans include land and water transformation, changes to the chemistry of Earth (for example, changes to nutrient cycles), increases in biological invasions, and increases in the extinction rate of species.

2. Human-caused global changes often happen at a much more rapid rate than changes due to natural causes. The speed of continental drift or natural climate change is much slower than the measurable changes humans have caused to carbon dioxide levels in the atmosphere or to the amount of nitrogen fixed. Also, humans have a choice about the global changes we cause.

3. Some examples of actions that are good for the global environment: Reduce the quantity of nonfood items purchased; reuse items until they are no longer usable; buy used items rather than getting everything new. Recycle paper, plastic, glass, and metal. Bring reusable cloth bags when shopping. Rarely use paper cups, plates, or towels. Plant trees and other native plants, especially those that help feed the native wildlife. Reduce water use by not leaving water running when brushing teeth, by adjusting the water level of washing machines to match the size of the load, and by using water-saving fixtures. Reduce fossil fuel use by choosing a gas-efficient car and by using household heating and air conditioning only as needed. Use compact fluorescent light bulbs, and turn off lights that are not in use. Support organic farmers by purchasing organically grown food. Some examples of actions that are harmful for the global environment: Use nonrecyclable products and products that are not from recycled sources. Drive alone rather than carpooling or taking public transportation. Use a gas- or electric-powered lawn mower. Make excessive purchases of nonessential items (gadgets, many sets of clothes, extra cars, vacation homes, etc.). Drive a low-mileage vehicle, such as a sport utility vehicle or truck, out of choice, not necessity.

4. For people to have a sustainable impact on Earth, we must reduce the rate of growth of the human population and we must reduce the (per person) rate of resource use. To achieve these goals, many aspects of human society would have to change. For example, our views of nature would have to change from one of a limitless source of goods and materials that can be exploited for short-term economic gain to one that accepts limits and seeks always to take only those actions that can be sustained for long periods of time. Many specific actions would follow from such a change in our view of the world, such as: an increase in recycling; the development and use of renewable sources of energy; a decrease in urban sprawl; an increase in the use of technologies with low environmental impact (such as organic farming); and a concerted effort to halt the ongoing extinction of species.

Photography Credits

Table of Contents Unit 1: © Fred Bavendam/Minden Pictures. Unit 2: © Jim Frazier/Mantis Wildlife Films/Oxford Scientific Films. Unit 3: © Dr. Gopal Murti/SPL/Science Source/Photo Researchers, Inc. Unit 4: © Tui De Roy/Minden Pictures. Unit 5: © Stephen Dalton/Photo Researchers, Inc. Unit 6: © Art Wolfe.

Chapter 1 *nanobes*: courtesy Dr. Philippa Uwins, University of Queensland. 1.2: Don Gettinger, courtesy K.C. Larson. 1.3: Ash Kaushesh, courtesy K.C. Larson. 1.4: © Stan Osolinsky/Dembinsky Photo Assoc. 1.4 *inset*: © Gail Nachel/Dembinsky Photo Assoc. 1.5: © Inga Spence/Tom Stack & Assoc. 1.7: © F. Stuart Westmorland/Photo Researchers, Inc. 1.8: © Jeff Lepore/Photo Researchers, Inc. Box: © Roger Ressmeyer/CORBIS.

Chapter 2 *ice man*: © Paul Hanny/Gamma Liaison. 2.1a *Prince William and Prince Harry*: © Ken Goff/TimePix. 2.1a *adults*: © Tim Graham/CORBIS . 2.1b: © James S. Miller. 2.5a: © Mick Ellison. 2.5b, c: © Edward Heck.

Chapter 3 *foraminiferan*: © Dennis Kunkel Microscopy, Inc. 3.2 *Chlamydia and archaeans*: © Dr. Kari Lounatmaa/SPL/Science Source/Photo Researchers, Inc. 3.2 *Borrelia*: © David M. Phillips/Science Source/Photo Researchers, Inc. 3.2 *Streptomyces*: © Dr. Jeremy Burgess/SPL/Science Source/Photo Researchers, Inc. 3.2 *E. coli*: © Biology Media/Science Source/Photo Researchers, Inc. 3.4: © Breck P. Kent/Earth Scenes. 3.5: © Krafft/Hoa-qui/Photo Researchers, Inc. 3.6: © Claudia Adams/Dembinsky Photo Assoc. 3.7 *seaweed*: © Science VU/Visuals Unlimited. 3.7 *Paramecium*: © Dennis Kunkel Microscopy, Inc. 3.7 *Plasmodium*: © Omikron/Science Source/Photo Researchers, Inc. 3.7 *dinoflagellate*: © Dennis Kunkel Microscopy, Inc. 3.7 *diatoms*: © Biophoto Assoc./Science Source/Photo Researchers, Inc. 3.9 *ferns*: © Greg Vaughn/Tom Stack & Assoc. 3.9 *sequoia*: © Tom Stack/Tom Stack & Assoc. 3.9 *mosses*: © Michael P. Gadomski/National Audubon Society Collection/Photo Researchers, Inc. 3.9 *Rafflesia*: © Compost/Visage/Peter Arnold, Inc. 3.9 *orchids*: © Kjell B. Sandred/Visuals Unlimited. 3.12a: © Inga Spence/Tom Stack & Assoc. 3.12b: © Ron Goulet/Dembinsky Photo Assoc. 3.13 *Pilobolus*: © Carolina Biological Supply Co./Visuals Unlimited. 3.13 *stinkhorns*: © Sharon Cummings/Dembinsky Photo Assoc. 3.13 *Penicillium*: © Dennis Kunkel Microscopy, Inc. 3.15: © Noble Proctor/Science Source/Photo Researchers, Inc. 3.16: © Dr. Edward S. Ross. 3.17: © John Shaw/Tom Stack & Assoc. 3.18 *jellyfish*: © Brian Parker/Tom Stack & Assoc. 3.18 *sponges*: © Tom Zuraw/Animals Animals. 3.18 *clam*: © Dave Fleetham/Tom Stack & Assoc. 3.18 *flatworm*: © Newman & Flowers/National Audubon Society Collection/Photo Researchers, Inc. 3.18 *fire worm*: © Susan Blanchet/Dembinsky Photo Assoc. 3.18 *sea star*: © F. Stuart Westmorland/National Audubon Society Collection/Photo Researchers, Inc. 3.18 *Morpho*: © G. W. Willis/Animals Animals. 3.18 *frog*: © Michael Fogden/Oxford Scientific Films. 3.18 *fish*: © Andrew J. Martinez/National Audubon Society Collection/Photo Researchers, Inc. 3.18 *kangaroos*: © Gerald and Buff Corsi/Visuals Unlimited. 3.18 *chimpanzees*: © Zig Leszczynski/Animals Animals. 3.23a: © Gerry Ellis/Minden Pictures. 3.23b: © Michael Neveux. Box (*A*): © L. Pyot/OCV-PN/Eurelios. (*B*): © David Robinson, courtesy Nalini Nadkarni.

Chapter 4 *toads*: © William M. Partington/Photo Researchers, Inc. 4.1: © Mark Moffett/Minden Pictures. 4.5: © Greg Vaughn/Tom Stack & Assoc. 4.6a: David McIntyre. 4.6b: © Bernd Witich/Visuals Unlimited. 4.7a: © Jack Jeffrey. 4.7b: © Jim Battles/Dembinsky Photo Assoc. 4.8: © Harald Pauli. 4.9: courtesy J. H. Lawton/Centre for Population Biology.

Chapter 5 *Mars rock*: NASA. 5.5b: © Biophoto Associates/Science Source/Photo Researchers, Inc. 5.5c: © CNRI/SPL/Science Source/Photo Researchers, Inc.

Chapter 6 *Listeria*: courtesy Justin Skoble and Daniel A. Portnoy. 6.3 *nucleus*: © Dr. Gopal Murti/SPL/Science Source/Photo Researchers, Inc. 6.3 *envelope*: © Dennis Kunkel Microscopy, Inc. 6.4 *smooth ER*: © David M. Phillips/Visuals Unlimited. 6.4 *rough ER*: © K. G. Murti/Visuals Unlimited. 6.6: © Dennis Kunkel Microscopy, Inc. 6.7 *upper*: © David M. Phillips/Visuals Unlimited. 6.7 *lower*: © Dr. Gopal Murti/SPL/Science Source/Photo Researchers, Inc. 6.8: © Biophoto Associates/Science Source/Photo Researchers, Inc. 6.9: © Bill Longcore/Science Source/Photo Researchers, Inc. 6.10: © Dr. Kari Lounatmaa/SPL/Science Source/Photo Researchers, Inc. 6.11a: © K. G. Murti/Visuals Unlimited. 6.11b: courtesy Dr. Mark McNiven. 6.12: courtesy Dr. Louise Cramer. Box *left*: © CNRI/Photo Researchers, Inc. *right*: © Richard Rodewald/Biological Photo Service.

Chapter 7 *kit*: Photo by Frank Simonetti, courtesy Dr. Michael Echols, Medical Antique Collectors (http://www.braceface.com/medical/medical1.htm). Box *runner*: © M. Gratton/Vision/Science Source/Photo Researchers, Inc. *mouse*: © Stephen Dalton/Animals Animals. *C. elegans*: © Sinclair Stammers/SPL/Science Source/Photo Researchers, Inc.

Chapter 8 *brain*: © Alexander Tsiaras/Science Source/Photo Researchers. 8.3: © Dr. Jeremy Burgess/SPL/Science Source/Photo Researchers, Inc. 8.8 *brewery*: © E. Brenckle/Explorer/Science Source/Photo Researchers, Inc. 8.8 *leg*: © Denton Taylor. Box *bacteria*: © Microfield Scientific Ltd./SPL/Science Source/Photo Researchers, Inc. *lab*: © Jerry Mason/SPL/Science Source/Photo Researchers, Inc.

Chapter 9 *Bacillus*: © Dennis Kunkel Microscopy, Inc. 9.1 *alveoli*: © Oliver Meckes/Science Source/Photo Researchers. 9.1 *blood vessel*: © Science Photo Library/Science Source/Photo Researchers, Inc. Box: photo by Martin Schweig, courtesy the National Library of Medicine.

Chapter 10 *blood cells*: © Andrew Syred/SPL/Science Source/Photo Researchers, Inc. 10.1: © Peter Skinner/Science Source/Photo Researchers, Inc. 10.4a: © Leonard Lessin/Peter Arnold, Inc. 10.4b: © Alfred Pasieka/Peter Arnold, Inc. 10.5: Andrew S. Bajer, Univ. of Oregon. Box: © Andrew Paul Leonard/Science Source/Photo Researchers, Inc.

Chapter 11 *adenovirus*: © P. Webster/Custom Medical Stock Photo. 11.2: © K. G. Murti/Visuals Unlimited. 11.5: © Richard S. LaRocco.

Chapter 12 *Anastasia*: © Rykoff Collection/CORBIS. 12.1: © The Mendelianum. 12.7 *chestnut*: © Daniel Johnson/Norvia Behling. 12.7 *palomino*: © Norvia Behling. 12.7 *cremello*: courtesy Jennifer Weske-Monroe. 12.8: © NCI/Photo Researchers, Inc. 12.9: © Walter Chandoha. *Punnett*: The Master and Fellows of Gonville and Caius College, Cambridge.

Chapter 13 *chromosomes*: courtesy Dr. Hans Dauwerse. 13.9 *child*: courtesy Dr. George Herman Valentine. 13.9 *chromosomes*: courtesy Dr. Irene Uchida. Box: courtesy Dr. Pragna Patel. *Thomas Mogan* (page 226): © Bettman/CORBIS.

Chapter 14 *DNA model*: © Ken Eward/Biografx/Science Source/Photo Researchers, Inc. 14.8: © Kenneth Greer/Visuals Unlimited. 14.9 *carcinoma*: © Paul Parker/SPL/Science Source/Photo Researchers, Inc. 14.9 *melanoma*: © Ken Greer/Visuals Unlimited. 14.9 *sunbathers*: Royalty-Free/CORBIS. Box: courtesy the Principal and Fellows of Newnham College, Cambridge.

Chapter 15 *tRNA model*: © Ken Eward/Science Source/Photo Researchers, Inc. 15.7: © Ken Eward/Science Source/Photo Researchers, Inc. 15.10 *left*: © Dr. Tony Brain/SPL/Science Source/Photo Researchers, Inc. 15.10 *right*: © Meckes/Ottawa/Science Source/Photo Researchers, Inc. Box: © Wolfgang Kaehler/CORBIS.

Chapter 16 *cyclops*: Johann Tischbein, *Polyphemus*; photo © Carlos Parada, Maicar Förlag—GML. 16.5: courtesy Dr. F. Rudolf Turner. 16.1: Joseph L. DeRisi, Vishwanath R. Iyer, and Patrick O. Brown. Box: courtesy the Barbara McClintock Papers, American Philosophical Society.

Chapter 17 *GFP Bunny, Alba*, Eduardo Kac, 2000. Photo © Chrystelle Fontaine.

17.5: courtesy Applied Biosystems. 17.6: courtesy Huntington Potter, University of South Florida and David Dressler, University of Oxford. 17.9: courtesy of Cellmark Diagnostics, Inc., Germantown, Maryland. 17.11: Purdue University Entomology Department. 17.12: © Ted Thai/*Time* Magazine. Box: © Terry Whittaker/Photo Researchers, Inc.

Chapter 18 *autoradiograph*: © Geoff Tompkinson/SPL/Science Source/Photo Researchers, Inc. 18.1 *Aristotle and sperm drawing*: courtesy the National Library of Medicine. 18.1 *microscope*: © Volker Steger/Peter Arnold, Inc. 18.1 *Mendel*: the Mendelianum. 18.1 *Morgan*: © Bettman/CORBIS. 18.1 *Avery*: courtesy the National Library of Medicine. 18.1 *Watson and Crick*: © A. Barrington Brown/Photo Researchers, Inc. 18.1 *Berg, Sanger*: courtesy the National Library of Medicine. 18.1 *DNA sequence*: © James King-Holmes/ICRF/SPL/Science Source/Photo Researchers, Inc. 18.1 *DeLisi*: courtesy Boston University. 18.1 *sequencer*: courtesy Applied Biosystems. 18.1 *Collins*: courtesy the National Human Genome Research Institute. 18.1 *H. influenzae*: © NIBSC/SPL/Science Source/Photo Researchers, Inc. 18.1 *yeast*: © Andrew Syred/SPL/Science Source/Photo Researchers, Inc. 18.1 *C. elegans*: courtesy Paola Dal Santo and Erik M. Jorgensen, University of Utah. 18.1 *Venter*: © Volker Steger/SPL/Science Source/Photo Researchers, Inc. 18.1 *chromosomes*: © Scott Camazine/Oxford Scientific Films. 18.1 *Venter and Collins*: © Reuters NewMedia Inc./CORBIS. 18.1 *Science*: Reprinted with permission from *Science*, Vol. 291, no. 5507. © 2001, American Association for the Advancement of Science. 18.1 *Nature*: Reprinted by permission from *Nature*, Vol. 409, no. 6822. © 2001 Macmillan Magazines Ltd.

Chapter 19 *Galapagos*: © Breck P. Kent/Earth Scenes. 19.1: © Mark J. Thomas/Dembinsky Photo Assoc. 19.2: © M. W. F. Tweedie/The National Audubon Society Collection/Photo Researchers, Inc. 19.3: © Dr. Blanche C. Haning/The Lamplighter. 19.8: courtesy Peter Boag.

Chapter 20 *HIV model*: © Mehau Kulyk/SPL/Photo Researchers, Inc. 20.1: © Darrell Gulin/Dembinsky Photo Assoc. 20.3 *inset*: © Rodger Jackman/Oxford Scientific Films. 20.6: © Dominique Braud/Tom Stack & Assoc. 20.9: ©

James H. Karales/Peter Arnold, Inc. 20.10: courtesy Thomas Bates Smith.

Chapter 21 *cichlid*: © Mark Smith/Photo Researchers, Inc. 21.1: © Erick Greene. 21.3a: © Bill Beatty/Visuals Unlimited. 21.3b: © Robert A. Lubeck/Animals Animals. 21.4: © Merlin D. Tuttle/Bat Conservation International/Photo Researchers, Inc. 21.5a: © Rod Planck/Photo Researchers, Inc. 21.5b: © Jeff Lepore/Photo Researchers, Inc. 21.6: © Dr. Rich Spellenberg.

Chapter 22 *landscape*: © Rod Planck/Dembinsky Photo Assoc. *plants/lichens*: © David G. Campbell. 22.1a: © Stanley M. Awramik/Biological Photo Service. 22.1b: © Ken Lucas/Visuals Unlimited. 22.1c: courtesy Niles Eldredge. 22.1d: © B. Miller/Biological Photo Service. 22.1e: © Jacky Beckett, courtesy David Grimaldi, American Museum of Natural History. 22.4: © D. W. Miller. Box: © Mary Graziose.

Chapter 23 *Neandertal*: Field Museum, Chicago, photo © Tom McHugh/Science Source/Photo Researchers, Inc. 23.4a: © John Reader/SPL/Science Source/Photo Researchers, Inc. 23.4b: © Anup Shah/Dembinsky Photo Assoc. 23.6: © John Reader/SPL/Science Source/Photo Researchers, Inc. 23.9b: © Tom McHugh/Science Source/Photo Researchers, Inc. 23.10: © David Perrett & Duncan Rowland, U. of St. Andrews/SPL/Science Source/Photo Researchers, Inc.

Chapter 24 *lichens*: © Dr. Jeremy Burgess/SPL/Science Source/Photo Researchers, Inc. 24.1: © Jean-Gérard Sidaner M./Photo Researchers, Inc. 24.2: © Don Fawcett/Science Source/Photo Researchers, Inc. 24.7: © Thomas Kitchin/Tom Stack & Assoc.

Chapter 25 *Kong*: © Bettmann/CORBIS. 25.1a: © Dennis Kunkel Microscopy, Inc. 25.1b: © Marilyn Kazmers/Dembinsky Photo Assoc. 25.1c: © Roland Birke/Peter Arnold, Inc. 25.1d: © Adam Jones/Dembinsky Photo Assoc. 25.4: © Robert Winslow/Animals Animals. 25.5: © Jeffrey L. Rotman/Peter Arnold, Inc. 25.6a: © David Phillips/Visuals Unlimited. 25.6b: © Manfred Kage/Peter Arnold, Inc. 25.6c: © Alex Rakosy/Dembinsky Photo Assoc. 25.6d: © Stan Osolinski/Dembinsky Photo Assoc. Box *left*: © Dennis Kunkel Microscopy, Inc. *right*: Royalty-Free/CORBIS.

Chapter 26 *cat*: © Gerard Lacz/Peter Arnold, Inc. 26.1: © Jeffrey L. Rotman/Peter Arnold, Inc. 26.2a: © Christopher Small. 26.2b: © Martyn F. Chillmaid National Audubon Society Collection/Photo Researchers, Inc. 26.3a: © Lothar Lenz/OKAPIA/National Audubon Society Collection/Photo Researchers, Inc. 26.3b: © William J. Weber/Visuals Unlimited. 26.3b inset: © Bill Beatty/Visuals Unlimited. 26.4: © Biophoto Associates/Science Source/Photo Researchers, Inc. 26.5b: © Carl Purcell/Science Source/Photo Researchers, Inc. 26.5c: © Ted Levin/Earth Scenes. 26.6a: © Flip Nicklin/Minden Pictures. 26.6b: © François Gohier/National Audubon Society Collection/Photo Researchers, Inc. 26.6c: © Ralph Reinhold/Animals Animals. 26.6d: © Carr Clifton/Minden Pictures. 26.8b *Marasmius*: © Richard L. Carlton/National Audubon Society Collection/Photo Researchers, Inc. 26.8b *Amanita*: © Martin H. Chester/National Audubon Society Collection/Photo Researchers, Inc. 26.10: © Reuters/Pierre duCharme/Archive Photos. Box: © AFP/CORBIS.

Chapter 27 *sperm*: © Dennis Kunkel Microscopy, Inc. *sprinter*: © L. S. Stepanowicz/Visuals Unlimited. 27.1a: © Gerard Lacz/Peter Arnold, Inc. 27.1b: © Dennis Kunkel Microscopy, Inc. 27.1c: © Stephen P. Parker/National Audubon Society Collection/Photo Researchers, Inc. 27.2: © Quest/SPL/Science Source/Photo Researchers, Inc. 27.4: © Larime Photographic/Dembinsky Photo Assoc. 27.7a: © Ray Wolfe/National Audubon Society Collection/Photo Researchers, Inc. 27.7b: © Eric Haucke/Photo Researchers, Inc. 27.7c: © Hans Pfletschinger/Peter Arnold, Inc. 27.8b: © John Cancalosi/Peter Arnold, Inc. 27.9a: © Dennis Kunkel Microscopy, Inc. 27.9c: © Dr. Gopal Murti/SPL/Science Source/Photo Researchers, Inc. 27.10a: © Dennis Kunkel Microscopy, Inc. 27.10b: © Julius Adler/Visuals Unlimited. 27.11a: © R. Calentine/Visuals Unlimited. 27.11b: © Hans Pfletschinger/Peter Arnold, Inc. 27.11c: © Barrie E. Watts/Oxford Scientific Films. 27.11d: © David Cavagnaro/Peter Arnold, Inc.

Chapter 28 *burial mounds*: © Matt Meadows/Peter Arnold, Inc. *skeletons*: © Brian Seed. 28.5: © Dennis Kunkel Microscopy, Inc. 28.6a: © David M. Dennis/Tom Stack & Assoc. 28.7: © C. G. Gardener/Oxford Scientific Films. 28.8a: © William J. Weber/Visuals Unlimited. 28.8b: © Tom J. Ulrich/

Visuals Unlimited. 28.8c: © Carroll W. Perkins/Animals Animals. 28.11a: © M. I. Walker/Science Source/Photo Researchers, Inc. 28.11b: © Wally Eberhart/Visuals Unlimited. 28.11c: © Bill Lea/Dembinsky Photo Assoc. 28.13: © Ken Greer/Visuals Unlimited. Box *cow*: © Mitsuaki Iwago/Minden Pictures. *ants*: © Dr. Edward S. Ross.

Chapter 29 *Andeans*: © Art Wolfe/Photo Researchers, Inc. *vulture*: © Joe McDonald/Animals Animals. 29.4: © Manfred Kage/Peter Arnold, Inc. 29.5 *left*: © Dr. John D. Cunningham/Visuals Unlimited. 29.5 *right*: © Gerald Van Dyke/Visuals Unlimited. 29.7: © David Sieren/Visuals Unlimited. 29.10: © Tom McHugh/National Audubon Society Collection/Photo Researchers, Inc. 29.11a: © David R. Frazier/Photo Researchers, Inc. 29.11b, c: © Ken Wagner/Visuals Unlimited. 29.11d: © Renee Lynn/Photo Researchers, Inc. Box *left*: © David M. Dennis/Tom Stack & Assoc. *right*: © John Innes Institute/SPL/Science Source/Photo Researchers, Inc.

Chapter 30 *giraffes*: © Inga Spence/Tom Stack & Assoc. 30.2: © Dennis Kunkel Microscopy, Inc. 30.11: © Biodisc/Visuals Unlimited. Box *left*: © Claudia Adams/Dembinsky Photo Assoc. *right*: © Fred Bavendam/Peter Arnold, Inc.

Chapter 31 *cactus*: © John Lemker/Earth Scenes. 31.1: © Tom Uhlman/Visuals Unlimited. 31.2 *cells*: courtesy Dr. Edward N. Ashworth. 31.2 *tree*: © Walter H. Hodge/Peter Arnold, Inc. 31.3 *humans*: © Laura Zito/Photo Researchers, Inc. 31.3 *mole rats*: © Raymond A. Mendez. 31.4 *left*: © Stan Flegler/Visuals Unlimited. 31.4 *center and right*: © David M. Phillips/Visuals Unlimited. 31.5: © Dr. John D. Cunningham/Visuals Unlimited. Box: Image # 5799(4), photo © D. Finnin/American Museum of Natural History.

Chapter 32 *Poster*: courtesy Nan Sinauer. 32.2a: © London School of Hygiene/SPL/Science Source/Photo Researchers, Inc. 32.2b: © Andrew Syred/SPL/Science Source/Photo Researchers, Inc. 32.2c: © Cherie Pittillo/The Lamplighter. 32.2d: © Milton Rand/Tom Stack & Assoc. 32.4a: © Inga Spence/Tom Stack & Assoc. 32.4b: © Christian Grzinek/Okapia/Science Source/Photo Researchers, Inc. 32.5: © Ed Reschke/Peter Arnold, Inc. 32.9: © Donald Specker/Earth Scenes. 32.10a: © Willard Clay Photography,

Inc. 32.10b: © Art Wolfe. 32.10c: © Stephen J. Krasemann/Nature Conservancy/National Audubon Society Collection/Photo Researchers, Inc. 32.11 *turtle*: © Jeff Lepore/National Audubon Society Collection/Photo Researchers, Inc. 32.11 *cacti*: © Dr. Edward S. Ross. 32.12a: © S. J. Krasemann/Peter Arnold, Inc. 32.12b: © Nigel Dennis/National Audubon Society Collection/Photo Researchers, Inc. 32.13 *left*: © Dr. Thomas Eisner. 32.13 *right*: © Doug Wechsler. Box *child*: © Eamonn McNulty/SPL/Science Source/Photo Researchers, Inc. *grasshopper*: © Bill Beatty/Visuals Unlimited.

Chapter 33 *caterpillar*: © Gary Meszaros/Dembinsky Photo Assoc. *chrysallis*: © Sharon Cummings/Dembinsky Photo Assoc. *butterfly*: © Claudia Adams/Dembinsky Photo Assoc. 33.4: courtesy Michael Clayton. 33.5a: © M. L. Dembinsky/Dembinsky Photo Assoc. 33.5b: © Brad Mogen/Visuals Unlimited. 33.6: © Laurence Pringle/National Audubon Society Collection/Photo Researchers, Inc. 33.8a: © SIU/Visuals Unlimited. 33.8b: © L. S. Stepanowicz/Visuals Unlimited. 33.9: © Petit Format/Nestle/Science Source/Photo Researchers, Inc. 33.11 *left*: © Gary Meszaros/Dembinsky Photo Assoc. 33.11 *center*: © Sharon Cummings/Dembinsky Photo Assoc. 33.11 *right*: © Claudia Adams/Dembinsky Photo Assoc. Box: © G. Vandystadt/Photo Researchers, Inc.

Chapter 34 *amputee*: © Art Stein/Photo Researchers, Inc. 34.3 *taste bud*: © P. Motta/Photo Researchers, Inc. 34.3 *catfish*: © Doug Wechsler. 34.6: © Dennis Kunkel Microscopy, Inc. Box: © Carolina Biological Supply Co./Visuals Unlimited.

Chapter 35 *swallowtail*: © S. McKeever/National Audubon Society Collection/Photo Researchers, Inc. 35.4a: © Ben Osborne/Oxford Scientific Films. 35.4b: © Ray Richardson/Animals Animals. 35.8a: © Cristopher Crowley/Tom Stack & Assoc. 35.8b: © Stan Osolinski/Dembinsky Photo Assoc. 35.8c: © Michael J. Minardi/Peter Arnold, Inc. 35.9a: © Thomas Kitchin/Tom Stack & Assoc. 35.9b: © David Hall/Photo Researchers, Inc. 35.9c: © Glen Woolfenden. 35.10: © Gunter Ziesler/Peter Arnold, Inc. Box: © François Gohier/National Audubon Society Collection/Photo Researchers, Inc.

Illustration and Text Credits

The following illustrations are, in full or in part, *from* Purves, W. K., Sadava, D., Orians, G. H., and Heller, H. C. 2001. *Life: The Science of Biology*, Sixth Edition. Sinauer Associates, Inc., Sunderland, MA, and W. H. Freeman and Co., New York: 2.6, 3.3, 19.7, 21.7, 27.2, 28.4, 29.1, 30.8, 31.6, 32.5, 33.1, 34.1, 34.5, 34.9.

The following illustrations are, in full or in part, *after* Purves, W. K., Sadava, D., Orians, G. H., and Heller, H. C. 2001. *Life: The Science of Biology*, Sixth Edition. Sinauer Associates, Inc., Sunderland, MA, and W. H. Freeman and Co., New York: 1.11, 2.3, 5.9, 10.5, 14.1, 16.2, 16.6, 17.8, 19.9, 20.7, 20.10, 23.8, 27.9, 28.7, 30.2, 30.3, 33.3, 33.10, 34.3, 34.4, 35.7, 42.7.

Chapter 2 2.5*b,c*: © Edward Heck, American Museum of Natural History.

Chapter 4 4.2: after data from E. O. Wilson, 1992. *The Diversity of Life*. The Belknap Press of Harvard University Press, Cambridge, MA.

Chapter 13 13.6: from *Time*; data from Dr. Victor A. McKusick, Johns Hopkins University.

Chapter 18 18.2: after Gibson, G. and Muse, S. V. 2002. *A Primer of Genome Science*, Sinauer Associates, Sunderland, MA.

Chapter 20 20.8: after Curtis, C. F., Cook, L. M., and Wood, R. J. 1978. "Selection for and against insecticide resistance and possible methods of inhibiting the evolution of resistance in mosquitoes." *Ecological Entomology* 3: 273–287. 20.9: after Karn, M.N. and Penrose, L.S., 1951. "Birth weight and gestation time in relation to maternal age, parity, and infant survival." *Annals of Eugenics* 16: 147–164. 20.11: after Gould, J. H. and Keeton, W. T. 1996. *Biological Science*, Sixth Edition. W. W. Norton and Co., New York.

Chapter 21 21.2: after Carroll, S. P. and Boyd, C. 1992. Host radiation in the soapberry bug—natural history with the history. *Evolution* 46(4) 1052–1069, Figure 1. 21.8: After Purves, W. K., Orians, G. H. , and Heller, H. C. 1992. *Life: The Science of Biology*, Third Edition. Sinauer Associates, Sunderland, MA.

Chapter 22 22.3: after Holland, H. D. 1999. When did the Earth's atmosphere become toxic? A reply. *The Geochemical News* 100: 20–22. 22.4: © D. W. Miller. 22.9: after Hickman, C. P., Jr. and Roberts, L. S. 1994. *Biology of Animals*. William C. Brown, Dubuque, IA. 22.10: after Benton, M. J. 1995. "Diversification and extinction in the history of life." *Science* 268: 52–58.

Chapter 23 23.9: after McBrearty, S. and Brooks, A. S. 2000. The revolution that wasn't: a new interpretation of the origin of modern human behavior. *Journal of Human Evolution* 39: 453–563. Table 23.1: after Ruff, C. B., Trinkaus, E., and Holliday, T. W. 1997. "Body mass and encephalization in *Pleistocene Homo*." *Nature* 387: 173–176; Strickberger, M. W. 1996. *Evolution*. Jones and Bartlett, Boston; Jurmain, R., Nelson, H., and Trevathan, W. 1997. *Introduction to Physical Anthropology*. West/Wadsworth, Belmont, MA.

Chapter 24 24.7: after Hawksworth, D. L. and Rose, F. 1973. *Lichens as Pollution Monitors*. Edward Arnold, London.

Chapter 25 25.3: after Niklas, K. J. 1994. *Plant Allometry: The Scaling of Form and Process*. University of Chicago Press, Chicago. 25.5: after Martin, R. D. 1981. "Relative brain size and basal metabolic rate in terrestrial vertebrates." *Nature* 293: 57–60.

Chapter 26 26.7: after data from Calder, W. A. 1984. *Size, Function, and Life History*. Harvard University Press, Cambridge, MA; Schmidt-Nielsen, K. 1984. *Scaling*. Cambridge University Press, Cambridge, England. 26.8: after Vogel, S. *Life's Devices*. Princeton University Press, Princeton, NJ. 26.11: after Diamond, J. M. 1988. "Why cats have nine lives." *Nature* 332: 586–587.

Chapter 27 Box Figure: after Tucker, V. A. 1975. "The energetic cost of moving about." *American Scientist* 63: 413–419.

Chapter 28 28.1: after Raven, J. 1986. *Biology*. McGraw-Hill, New York; Begon, M., Harper, J.L., and Townshend, C.R. 1990. *Ecology: Individuals, Populations, and Communities*, Second Edition. Blackwell Scientific Publishing, Boston; and Vander, A. J., Sherman, J. H., and Luciano, D. S. 1991. *Human Physiology: The Mechanisms of Body Function*, Sixth Edition. McGraw-Hill, New York. 28.5: After Larcher, W. 1995. *Physiological Plant Ecology: Ecophysiology and Stress Physiology of Functional Groups*, Third Edition. Springer Verlag, New York; Raven, J. 1986. *Biology*. McGraw-Hill, New York; and Salisbury, F. B. and Ross, C. W. 1985. *Plant Physiology*, Third Edition. Wadsworth Publishing Co., Belmont, CA. 28.10: map data after Faltz, G. and Rotthauwe, H. W. 1977. "The human lactose polymorphism: physiology and genetics of lactose absorption and malabsorption." In Steinberg, A. G., Bearn, A. G., Motolsky, A. G., and Childs, B. (Eds.) *Progress in Medical Genetics, NS, II*, pp. 205–249. Saunders, Philadelphia.

Chapter 29 29.2: after Graham, J. B. 1990. "Ecological, evolutionary, and physical factors influencing aquatic animal respiration." *American Zoologist* 30: 137–146. 29.9: after data in Welch, W. R. 1984. "Temperature and humidity of expired air: Interspecific comparisons and significance for loss of respiratory heat and water from endotherms." *Physiological Zoology* 57: 366–375. 29.11:After Raven, J. 1986. *Biology*. McGraw-Hill, New York; Salisbury, F. B. and Ross, C. W. 1985. *Plant Physiology*, Third Edition. Wadsworth Publishing Co., Belmont, CA.

Chapter 31 31.3: after data in Guyton, A. C. and Hall, J. E. 1996. *Textbook of Medical Physiology*, Ninth Edition. W. B. Saunders Co., Toronto; Louw, G. 1992. *Physiological Animal Ecology*. Longman Scientific and Technical, Harlow, England.

Chapter 32 32.6: after Guyton, A. C. and Hall, J. E. 1996. *Textbook of Medical Physiology*, Ninth Edition. W. B. Saunders Co., Toronto.

Chapter 37 37.1: after Sze, L. C. 1953. "Changes in the amount of deoxyribonucleic acid in the development of *Rana pipiens*." *Journal of Experimental Zoology* 122: 577–601.

Chapter 38 38.2: after Lovejoy, C. O. 1981. The origin of man. *Science* 211: 341–350. 38.4: After Hagerman, R. J. 1997. "Growth and development." In Hay, W. W., Groothius, J. R., Hayward, A. R., and Levin, M. J. (Eds.). *Current Pediatric Diagnosis and Treatment*. Appleton & Lange, Stamford, Connecticut.

Chapter 39 39.3: after Miller, G. T., Jr. 1996. *Living in the Environment*. Wadsworth Publishing Co., Belmont CA. 39.13: after *Our Changing Planet: The FY 1997 U.S. Global Change Research Program*. Supplement to the President's Fiscal Year 1997 Budget; NOAA Office of Global Programs. 39.7: After Raven, P. H., Evert, R. F., and Eichhorn, S. E. 1999. *Biology of Plants*, Sixth Edition. W. H. Freeman and Co./Worth Publishers, New York; based on A. W. Kiichler.

Chapter 40 40.4: top map from Pielou, E. C. 1979. *Biogeography*. John Wiley and Sons, Inc., New York. Reprinted by permission. 40.8: after Dorf, E. 1960. "Climatic changes of the past and present." *American Scientist* 48: 341–364.

Chapter 41 41.5: after Gause, G. F. 1934. *The Struggle for Existence*. Williams and Wilkins, Baltimore, MD. 41.6: after data from Alliende, M. C. and Harper, J. L. 1989. "Demographic studies of a dioecious tree. I. Colonization, sex, and age structure of a population of *Salix cinerea*." *Journal of Ecology* 77: 1029–1047. Box: © Frans Lanting/Minden Pictures. 41.8: after data from Scheffer, V. B. 1951. "The rise and fall of a reindeer herd." *Scientific Monthly* 73: 356–362. 41.10: After Goudie, A. S. and Viles, H. 1997. *Earth Transformed: An Introduction to the Human Impact on the Environment*. Blackwell Publishers, Malden, MA.

Chapter 42 42.3: after Myers, J. H. and Bazely, D. 1991. "Thorns, spines, prickles, and hairs: Are they stimulated by herbivory and do they deter herbivores?" In Tallamy, T. W. and Raupp, M. J. (Eds.) *Phytochemical Induction by Herbivores*, pp. 325–344. Wiley, New York. 42.6: after Kenward, R. E. 1978. "Hawks and doves: Factors affecting success and selection in goshawk attacks on wood-pigeons." *Journal of Animal Ecology* 47: 449–460. 42.8: after Ricklefs, R. E. and Miller, G. L. 2000. *Ecology*, Fourth Edition. W. H. Freeman and Co., New York; based on DeBach, P. and Sundby, R. A. 1963. "Competitive displacement between ecological homologues." *Hilgardia* 34: 105–106. 42.9: after Lack, D. 1947. *Darwin's Finches*. Cambridge University Press, Cambridge, England.

Chapter 43 Display quotation from Ernest Hemingway from *Green Hills of Africa*, Charles Scribner's Sons, New York, p. 284. 43.3: after data from Paine, R. T. 1974. "Intertidal community structure: Experimental studies on the relationship between a dominant competitor and its principal predator." *Oecologia* 15: 93–120. 43.4: after Krebs, C. J. 1994. *Ecology*, Fourth Edition. HarperCollins College Publishers, New York; based on Olson, J. S. 1958. "Rates of succession and soil changes on southern Lake Michigan sand dunes." *Botanical Gazette* 119: 125–170.

Chapter 44 44.3: from Campbell, N. A. (Ed.) © 1987, 1990, 1993, 1996. *Biology*, Fourth Edition. The Benjamin/Cummings Publishing Co., Inc. Reprinted by permission. 44.8b: after Krebs, C. J. 1994. *Ecology*, Fourth Edition. HarperCollins College Publishers, New York; based on Likens, G. E., Bormann, F. H., Johnson, N. M., Fisher, D. W., and Pierce, R. S. 1970. "Effects of forest cutting and herbicide treatment on nutrient budgets in the Hubbard Brook watershed-ecosystem." *Ecological Monographs* 40: 23–47. Table 44.1: after Whittaker, R. H. 1975. *Communities and Ecosystems*. MacMillan, London.

Chapter 45 45.2: from Dahl, T. E. 1990. *Wetland losses in the United States, 1780s to 1980s*. U.S. Department of Interior, Fish and Wildlife Service, Washington, D. C. 45.4: from Vitousek, P. M. 1994. "Beyond global warming: Ecology and global change." *Ecology* 75: 1861–1876. Reprinted by permission of Ecological Society of America. 45.7: after Post, W. M., Peng, T.-H., Emanuel, W. R., King, A. W., Dale, V. H., and DeAngelis, D. L. "The global carbon cycle." *American Scientist* 78: 310–326. 45.9: data from the National Climatic Data Center, 1997. 45.10: from *IPCC Second Assessment Report: Climate Change*. 1995. A report of the Intergovernment Panel on Climate Change. IPCC, Geneva, Switzerland. 45.11: from Noss, R. F. 1992. "Issues of scale in conservation biology." In Fiedler, P. L. and Jain, S. K. (Eds.) *Conservation Biology: The Theory and Practice of Nature Conservation, Preservation, and Management*, pp. 239–250. Chapman and Hall, New York. Reprinted by permission.

Fine Art Credits

Frontispiece Tom Uttech, *Bimawanadiwag Awessiiag*, 1996. Oil on canvas and wood, 57 × 61½ in. Courtesy of Schmidt Bingham Gallery, New York City.

Unit 1 Alexandre-Isidore Leroy de Barde, *Still Life with Exotic Birds*, 1804. Gouache and watercolor, Musée du Louvre, Paris.

Unit 2 Ross Bleckner, *Untitled*, 2000. Oil on linen, 48 × 36 in. Courtesy of the artist and Lehmann Maupin, New York.

Unit 3 Paul Giovanopoulos, *Man 2 & Man 1*, © 1987.

Unit 4 Philip Taaffe, *Passage II*, 1998. Mixed media on canvas, 26.5 × 36.5 in. Private collection, New York.

Unit 5 Robert Kushner, *Chrysanthemum Brocade*, 1994. Oil, glitter, and acrylic on canvas, 72 × 72 in. Private collection, courtesy DC Moore Gallery, New York City.

Unit 6 Kit Williams, *The Death of Spring*, 1980. 51 × 76 cm.

Chapter 1 Joan Miró, *Carnival of Harlequin*, 1924–25. Oil on canvas, 26 × 36.63 in. Albright-Knox Art Gallery, Buffalo, New York, © 2002 Successió Miró/Artists Rights Society (ARS), New York/ADAGP, Paris.

Chapter 2 Alexis Rockman, *The Bounty*, 1991. Photograph: Gorney Bravin and Lee.

Chapter 3 Michael Lucero, *Termite (Soldier)*, 69 × 22 × 13 in.; *Hercules Beetle*, 46 × 26 × 12 in.; *Unidentified Fish*, 52 × 24 × 9 in., clay with glazes; 1986. Photograph: Ellen Page Wilson.

Chapter 4 Martin Johnson Heade, *Study of an Orchid*, 1872. Oil on canvas, 18 × 23 in. Collection of the New York Historical Society.

Chapter 5 Tom Friedman, *Untitled*, 1999. Sugar cubes, 48 × 17 × 10 in. Photograph: Oren Slor, courtesy of Feature Inc., New York.

Chapter 6 Dale Chihuly, *Sky Blue Basket Set with Cobalt Lip Wraps*, 1992, 17 × 15 × 16 in. Photograph: Terry Rishel.

Chapter 7 Ben Shahn, *Helix and Crystal*, 1957. 53 × 30 in., tempera, San Diego Museum of Art. © Estate of Ben Shahn/ Licensed by VAGA, New York City.

Chapter 8 Lois Guarino, *The Prayer* (from The Fertility Series), 1992. Copyright: Lois Guarino/Courtesy: Robert Mann Gallery.

Chapter 9 Ross Bleckner, *Healthy Spot*, 1996. Oil/linen, 84 × 72 in. Photograph: Mary Boone Gallery, New York.

Chapter 10 Hilma af Klint, *Group 4. The Ten Greatest no. 7 Manhood (Lnr 108)*, 1907. Tempera on paper. The Hilma af Klint Foundation, Stockholm, Sweden.

Chapter 11 Eugene Von Bruenchenhein, *EVB 390 Untitled, no. 659 (GS261)*, 1957, Oil on Masonite, 24 × 24 in. Photograph: Ricco/Maresca Gallery.

Chapter 12 Jean Dubuffet, *Le Strabique*, 1953. Butterfly wing collage, 9¾ × 7 in. Collection of Arne and Milly Glimcher. © 2002 Artists Rights Society (ARS), New York/ADAGP, Paris. Photograph: Ellen Page Wilson, courtesy of PaceWildenstein.

Chapter 13 Rosamond Purcell, *It's How You Play The Game*, 1992.

Chapter 14 Hilma af Klint, *Group 9 Series UW. Dove no. 25*, 1915. Tempera on paper. The Hilma af Klint Foundation, Stockholm, Sweden.

Chapter 15 Dannielle Tegeder, *Chromosome Constellation*, 2000. Courtesy of the artist.

Chapter 16 Maria Nazor, *Little Misunderstandings of No Importance*, 1986. 76 × 53 in. Private collection.

Chapter 17 Alexis Rockman, *The Farm*, 2000. Acrylic and oil on wood, 72 × 84 in. Private collection, courtesy of Gorney Bravin and Lee, New York. Photograph: Oren Slor, New York City.

Chapter 18 Lawrence Berzon, *The Collector*, detail, 2000. Courtesy Caelum Gallery.

Chapter 19 Mary Frank, *Natural History*, 1985. Monoprint. DC Moore Gallery, New York City. Photograph: Ralph Gabriner.

Chapter 20 Ismael Vargas, *Tiempos Idos Que Felices Llamo*, 1986. Oil on cloth. Courtesy of Arte Actual Mexicano Galería, Monterrey, Mexico.

Chapter 21 Suzanne Stryk, *Tree of Life*, 1997. Gouache, 25 × 20 in.

Chapter 22 William Morris, *Artifact Series #11 (Man and Beast)*, 1988. Glass, 18 × 108 × 60 in.

Chapter 23 Matt Wainwright, *Head Count*, 1998. Watercolor and India ink on paper, 12 × 15 in. Courtesy of the artist.

Chapter 24 James Davis, *Museum of Natural History I*, 1996. Oil on canvas, 68 × 84 in. Courtesy of the artist.

Chapter 25 Roberto Marquez, *Theory of King David*, 1999. Oil on canvas, 72 × 60 in. Courtesy Riva Yares Gallery.

Chapter 26 Mansur ibn Muhammad Fagih Ilyas, Skeletal system from *Five Anatomical Figures*. Original, late 14th century illustration from Tashrih Munsori. Opaque watercolor, 16 × 9½ in. The British Library.

Chapter 27 Jess, *The Mouse's Tale*, 1951. Collage and gouache on paper, 47 × 32 in. Collection, San Francisco Museum of Modern Art, gift of Frederic P. Snowden.

Chapter 28 René Magritte, *Treasure Island (L'Ile aux Tresors)*, 1942. 60 × 80 cm. Musées Rouaux des Beaux-arts, Brussels. Photograph: Herscovici/Art Resource, New York. © 2002 C. Herscovici, Brussels/Artists Rights Society (ARS), New York.

Chapter 29 Vivian Torrence, *Greek Air*, 1989. http://home.t-online.de/home/vivian.t

Chapter 30 *Circulatory System,* late thirteenth-century manuscript. Pen and wash on parchment. 10.5 × 7.5 in. The Bodleian Library, Oxford.

Chapter 31 Roy de Forest, *The Importance of Being Earnest,* 1964. Courtesy George Adams Gallery.

Chapter 32 Eva Lee, *Virus Study,* 1999. Watercolor on paper, 14 × 11 in. Photograph: Karen Ostrom.

Chapter 33 Roberto Juarez, *Kether Place,* 2001. Mixed media on canvas, 58 × 60 in. ©Roberto Juarez. Courtesy Robert Miller Gallery, New York.

Chapter 34 Sharon Ellis, *Afternoon,* 1998. Courtesy of the artist and Christopher Grimes Gallery, Santa Monica, California.

Chapter 35 Anique Taylor, *Bliss,* 1989. Pencil, 4 × 6 in.

Chapter 36 Alex Grey, *Pregnancy,* 1988–89. From *Sacred Mirrors: Visionary Art of Alex Grey.* Oil on linen, 50 × 56 in. © Alex Grey. www.alexgrey.com

Chapter 37 Dale Chihuly, *Shell Pink Basket Set with Oxblood Lip Wraps,* 1995. 9 × 22 × 22 in. Photograph: Claire Garoutte.

Chapter 38 Helen Lundeberg, *Double Portrait of the Artist in Time,* 1935. Oil on fiberboard, 47¾ × 40 in. Copyright Smithsonian American Art Museum, Washington, D.C./Art Resource, New York. Smithsonian American Art Museum, Washington, D.C., USA.

Chapter 39 Dale Chihuly, *Sky Blue Soft Cylinder with Golden Yellow Lip Wrap,* 1993. Glass, 17 × 21 in. Photograph: Claire Garoutte.

Chapter 40 Erik La Gattuta, *Encouraging the Coelacanth,* 2000. Oil on canvas, 70 × 60 in. Photograph: Tom Van Eynde.

Chapter 41 Rosamond Purcell, *European Moles,* 1992.

Chapter 42 Melissa Miller, *Zebras and Hyenas,* 1985. Oil on linen, 72 × 84 in. Courtesy Jack S. Blanton Museum of Art, The University of Texas at Austin. Purchased through the generosity of Mari and James A. Michener, 1985. Photograph: Bill Kennedy.

Chapter 43 Jim Waid, *Daybreak,* 2000. Acrylic on canvas. Courtesy of Riva Yares Gallery.

Chapter 44 David Ligare, *Landscape with Rainbow,* 1999. Oil on canvas, 20 × 29 in. Courtesy of Hackett-Freedman Gallery, San Francisco.

Chapter 45 Oliver Burston, *Hypervoxel Embolism,* 1999. © 1999 Oliver Burston/debut art.

Glossary

acid   A chemical compound that can give up a hydrogen ion. Compare *base* and *buffer*.

acid rain   Rainfall with a low pH. Acid rain is a consequence of the release of sulfur dioxide and other pollutants into the atmosphere, where they are converted to acids that then fall back to Earth in rain or snow.

actin   A protein found in muscle tissue and in bacterial flagella.

actin filament   A protein fiber composed of actin monomers. Actin filaments are part of the cell's cytoskeleton.

action potential   An electrical signal generated by the flow of ions across the plasma membrane of a neuron. Action potentials are self-strengthening and can travel down a neuron in only one direction.

activation energy   The small input of energy required for a chemical reaction to proceed.

active carrier protein   A protein in the plasma membrane of a cell that, using energy from an energy storage molecule such as ATP, changes its shape to transfer a molecule across the plasma membrane. Compare *passive carrier protein*.

active forager   A forager that can move toward its food.

active site   The specific region on the surface of an enzyme where substrate molecules bind.

active transport   Movement of molecules that requires an input of energy. Compare *passive transport*.

adaptation   A characteristic of an organism that improves the organism's ability to survive and reproduce in its environment.

adaptive evolution   The process by which natural selection causes the quality of an adaptation to improve over time.

adenosine triphosphate   See *ATP*.

adrenaline   A hormone produced by the adrenal gland in response to stress, which activates a G protein signal cascade in cells.

aerobic   Of or referring to a metabolic process or organism that requires oxygen gas. Compare *anaerobic*.

aerobic respiration   A general term used to describe a series of oxidation reactions that use oxygen to produce ATP.

allele   An alternative version of a gene. Each allele has a DNA sequence different from that of all other alleles of the same gene.

allele frequency   The proportion (percentage) of a particular allele in a population.

allometric relationship   A relationship between body size and another functional or structural property of an organism.

allopatric speciation   The formation of new species from populations that are geographically isolated from one another. Compare *sympatric speciation* and *geographic isolation*.

altruism   A behavior that aids another individual while harming the individual performing the behavior.

alveolus (pl. alveoli)   Any of the small pockets in the mammalian lung in which most gas exchange takes place.

amino acid   A chemical compound that has an amino group, a carboxyl group, and a variable side chain attached to a single carbon atom. Proteins are polymers of amino acids.

ammonia   A chemical produced by the metabolism of proteins. If found in high concentrations in a cell, it can poison the cell.

anaerobic   Of or referring to a metabolic process or organism that does not require oxygen gas. Compare *aerobic*.

anaphase   The stage of mitosis during which sister chromatids separate and move to opposite poles of the cell.

angiosperms   The flowering plants, a group that includes most plants on Earth; named for the protective tissues covering the plant's embryo in the seed. Compare *gymnosperms*.

Animalia   The kingdom that is made up of animals, multicellular organisms that have evolved specialized tissues, organs and organ systems, body plans, and behaviors.

antenna complex   An arrangement of chlorophyll molecules in the thylakoid membrane of a chloroplast that harvests energy from sunlight.

antibody   A protein that is produced by a B lymphocyte in the human immune system and binds specifically to a particular antigen.

anticodon   A sequence of three nucleotide bases on a transfer RNA molecule that can bind to a particular codon on a messenger RNA molecule. Compare *codon*.

antigen   A characteristic membrane protein or chemical produced by an invading organism that is recognized by particular lymphocytes and the antibodies produced by those lymphocytes.

apical meristem   The region of rapidly dividing cells in the tips of plant branches and roots that gives rise to new stem and root tissues.

Archaea   A kingdom of microscopic, single-celled organisms that arose after the Bacteria.

artery   A blood vessel that carries blood from the heart. Compare *vein*.

arthropods   A group of animals characterized by a hard outer skeleton; includes millipedes, crustaceans, insects, and spiders.

artificial selection   A process in which only individuals that possess certain characters are allowed to breed; used to guide the evolution of crop plants and domestic animals in ways that are advantageous for people.

asexual reproduction   The production of genetically identical offspring without the exchange of genetic material with another individual. Compare *sexual reproduction*.

atherosclerosis   A narrowing of the blood vessels in humans that results from the formation of fatty plaques.

atmospheric nutrient cycle   A type of nutrient cycle in which the nutrient enters the atmosphere easily. Compare *sedimentary cycle*.

atom   The smallest unit of a chemical element that still has the properties of that element.

ATP   Adenosine triphosphate, a molecule that is commonly used by cells to store energy and to transfer energy from one chemical reaction to another.

atrium (pl. atria)   The chamber in a heart that receives blood from the body and pumps it into the ventricle.

autosome   Any chromosome that is not a sex chromosome. Compare *sex chromosome*.

axon   The portion of a neuron that acts as a sort of biological wire to carry action potentials through the organism.

B lymphocyte   A lymphocyte that produces antibodies.

Bacteria   A kingdom of microscopic, single-celled organisms that was the first of the kingdoms to arise.

bacterial flagellum (pl. flagella)   A rotary motor unique to bacteria that, in response to the flow of hydrogen ions across the bacterial plasma membrane, propels a bacterium. Compare *eukaryotic flagellum*.

base   (1) A chemical compound that can accept a hydrogen ion. Compare *acid* and *buffer*. (2) A nitrogen-containing molecule that is part of a nucleotide. See *nucleotide base*.

base pairing   A process in which complementary bases form hydrogen bonds with each other. In DNA, A pairs with T, and C pairs with G; in RNA, U replaces T.

behavior   A response to a stimulus; particularly, a response that involves movement.

behavioral mutualism   A mutualism in which each of two interacting species has altered its behavior to benefit the other.

benign   Of or referring to a relatively harmless cancerous growth that is confined to a single tumor and does not spread to other tissues in the body. Compare *malignant*.

bile   A substance produced by the liver that helps break up large fat globules in the small intestine.

biodiversity   The variety of organisms on Earth or in a particular location, ranging from the genetic variation and behavioral diversity of individual organisms or species through the diversity of ecosystems.

biological evolution   See *evolution*.

biomass   The mass of organisms per unit of area.

biome   A major terrestrial or aquatic life zone, defined either by its vegetation (terrestrial biomes) or by the physical characteristics of the environment (aquatic biomes). There are seven terrestrial biomes (tundra, boreal forest, temperate forest, chaparral, grassland, desert, and tropical forest) and eight aquatic biomes (lake, river, wetland, estuary, intertidal zone, coral reef, ocean, and benthic zone).

biosphere   All living organisms on Earth, together with the environments in which they live.

biosynthetic   Of or referring to chemical reactions that manufacture complex molecules in living cells. Compare *catabolic*.

birth control   Methods that allow individuals to go through the motions of reproducing without actually producing offspring.

bivalent   A pair of homologous chromosomes. Bivalents form during prophase I of meiosis.

blastula   The relatively undifferentiated stage early in animal development just prior to the formation of the germ line.

blood clot   A mesh formed of protein that captures blood cells and seals a break in a blood vessel.

bone   A support structure found in most vertebrate animals, formed by connective cells that secrete a mineral matrix rich in calcium. The core of a bone serves as the site of red and white blood cell production. Compare *cartilage*.

brain   A complex structure made up of neurons that serves as the major clearinghouse for information in the nervous system.

buffer   A chemical compound that can both give up and accept hydrogen ions. Buffers can maintain the pH of water within specific limits. Compare *acid* and *base* (1).

bundle sheath cells   Specialized cells in which fixed carbon is stored for C_4 photosynthesis.

C_3 photosynthesis   The most common photosynthetic pathway. See *photosynthesis*.

C_4 photosynthesis   A photosynthetic pathway in which the dark reactions of photosynthesis are separated into two steps, which are carried out at separate locations in the leaf.

CAM photosynthesis   A photosynthetic pathway in which the dark reactions of photosynthesis are separated into two steps, which are carried out at separate times of day.

Cambrian explosion   A major increase in the diversity of life on Earth that occurred about 530 million years ago, during the Cambrian period. The Cambrian explosion lasted 5 to 10 million years; during this time large and complex forms of most living animal phyla appeared suddenly in the fossil record.

cancer   A group of diseases caused by rapid and inappropriate cell division.

canopy   The habitat in the branches of forest trees.

capillary   A tiny blood vessel in a closed circulatory system, across the walls of which all exchange with the surrounding tissues takes place.

carbohydrate   Any of a class of organic compounds that includes sugars and their polymers, in which each carbon atom is linked to two hydrogen atoms and an oxygen atom. See also *sugar*.

carbon fixation   The process by which carbon atoms from CO_2 gas are incorporated into sugars; occurs in the chloroplasts of plants.

carcinogen   A physical, chemical, or biological agent that causes cancer.

carnivore   A consumer that relies on living animal tissues for nutrients. Compare *herbivore*.

carrier   An individual that carries a disease-causing allele but does not get the disease.

carrying capacity   The maximum population size that can be supported indefinitely by the environment in which the population is found.

cartilage   A support structure found in animals that consists of connective cells that secrete a matrix rich in the protein collagen. Compare *bone*.

catabolic   Of or referring to a chemical reaction that breaks down complex molecules to release energy for use by the cell. Compare *biosynthetic*.

catabolism   A general term for the series of chemical reactions that break down complex molecules to release energy for use by the organism.

catalyst   A molecule that speeds up a specific chemical reaction without being consumed in the process. Enzymes are protein catalysts.

cell   The smallest self-contained unit of life, enclosed by a membrane.

cell communication   The process by which one cell can affect the activities of another via signaling molecules such as hormones and growth factors.

cell division cycle   A series of distinct stages in the life cycle of a cell that culminate in cell division.

cell specialization   An organizational principle stating that different cells in a multicellular organism differ in their structure and function.

cell wall   A support layer that lies outside the plasma membrane of the cells of many prokaryotes, fungi, and plants.

cellulose   A carbohydrate produced by plants and some other organisms that makes up much of their cell walls.

central nervous system   The component of the nervous system devoted to the exchange of information among neurons.

centromere   A physical constriction that holds sister chromatids together.

centrosome   A protein structure in the cytosol that helps organize the mitotic spindle and defines the two poles of a dividing cell.

cerebral cortex The highly folded outer portion of the human brain that is specialized for coordinating information from sensory cells with the activity of muscles.

cerebrum The largest part of the human brain; includes the cerebral cortex.

channel protein A protein in the plasma membrane of a cell that forms an opening through which certain molecules can pass.

character A feature of an organism, such as height, flower color, or the chemical structure of a protein.

character displacement A process by which intense competition between species causes the forms of the competing species to evolve to become more different over time.

chemical compound An association of atoms of different chemical elements linked by covalent bonds.

chemical pathway A series of enzyme-catalyzed chemical reactions in which the products of one reaction serve as the substrates for the next.

chemical reaction A process that rearranges atoms in chemical compounds.

chitin A carbohydrate that serves as an important support material in the cell walls of fungi and in the skeletons of animals.

chlorophyll The green pigment that is used to capture energy from light in photosynthesis.

chloroplast An organelle found in plants and algae that is the primary site of photosynthesis.

chromatid Either of two identical, side-by-side copies of a chromosome that are linked at the centromere.

chromatin The combination of DNA and proteins that makes up chromosomes.

chromosome Any of several elongated structures found in the nucleus of a cell, each composed of DNA packaged with proteins. Chromosomes become visible under the microscope during mitosis and meiosis.

chromosome theory of inheritance A theory, supported by much experimental evidence, stating that genes are located on chromosomes.

cilium (pl. cilia) A hairlike structure found in some eukaryotes that uses a swimming motion to propel the organism or to move fluid past the organism. Compare *eukaryotic flagellum*.

citric acid cycle A series of oxidation reactions that produce high-energy electrons stored in NADH, and CO_2 as waste. In eukaryotic cells, the citric acid cycle takes place in mitochondria.

cladistics A dominant school of systematics, in which shared derived features are used to determine evolutionary relationships. Compare *evolutionary taxonomy*.

climate The prevailing weather conditions experienced in an area over relatively long periods of time (30 years or more).

climax community A community, typical of a given climate and soil type, whose species are not replaced by other species. A climax community is the end point of succession for a particular location; in many cases, however, ongoing disturbances such as fire or windstorms prevent the formation of a stable climax community.

clone (of a gene) A copy of a gene or other DNA sequence.

closed circulatory system An internal transport system found in animals in which blood vessels carry blood through all the organs and back to the heart. Compare *open circulatory system*.

clumped distribution A distribution pattern in which organisms tend to be clustered together throughout the environment. Compare *random distribution* and *regular distribution*.

codon A group of three nucleotide bases in a messenger RNA molecule. Each codon specifies either a particular amino acid or a signal to start or stop the translation of a protein. Compare *anticodon*.

collagen A protein produced by the connective cells of animals that resists tension.

colon The portion of the animal gut in which undigested food is prepared for release from the body. Also called *large intestine*.

communication A type of behavior that allows one individual to exchange information with another.

community An association of populations of different species that live in the same area.

companion cell A cell that uses energy to pump sugars into phloem cells of the plant vascular system against a concentration gradient.

competition An interaction between two species in which each has a negative effect on the other.

complement A group of blood proteins that concentrate in damaged tissues, stimulating macrophage and neutrophil production and activity.

compression A squeezing force that tends to push molecules together. Compare *tension*.

concentration gradient A change in the concentration of molecules from one location to another.

conduction The transfer of heat transferred by direct contact between two materials.

connective cell A type of animal cell that connects and supports other cells in the body. Many connective cells secrete a matrix of nonliving material.

consumer An organism that obtains its energy by eating other organisms or their remains. Consumers include herbivores, carnivores, and decomposers. Compare *producer*.

consumer–victim interaction An interaction between two species in which one species benefits (the consumer) and the other species is harmed (the victim). Consumer–victim interactions include the killing of prey by predators, the eating of plants by herbivores, and the harming or killing of a host by a parasite or pathogen.

continental drift The movement of Earth's continents over time.

convection The physical movement of heat in air or water.

convergent feature A feature shared by two groups of organisms not because it was inherited from a common ancestor, but because it arose independently in the two groups.

copulation A behavior in which a male animal places his sperm directly into the body of a female; the typical mode of sperm transfer in land animals.

covalent bond A strong chemical linkage between two atoms based on the sharing of electrons. Compare *hydrogen bond* and *ionic bond*.

crossing-over The physical exchange of genes between homologous chromosomes, in which part of the genetic material inherited from one parent is replaced with the corresponding genetic material inherited from the other parent.

cross-link A connection between two molecules that generally increases resistance to compression or tension, and often increases stiffness.

cytokinesis The stage following mitosis, during which the cell physically divides into two daughter cells.

cytoplasm The contents of the cell enclosed by the plasma membrane, but, in eukaryotes, excluding the nucleus. Compare *cytosol*.

cytoskeleton A complex network of protein filaments found in the cytosol of eukaryotic cells. The cytoskeleton maintains cell shape and is necessary for the physical processes of cell division and movement.

cytosol The contents of the cell enclosed by the plasma membrane, but, in eukaryotes, excluding all organelles. Compare *cytoplasm*.

dark reactions A series of chemical reactions that directly use CO_2 to synthesize sugars. The dark reactions do not require light and take place in the stroma of chloroplasts. Compare *light reactions*.

decomposer An organism that breaks down dead tissues into simple chemical components, thereby returning nutrients to the physical environment.

deletion mutation A mutation in which one or more bases are removed from the DNA sequence of a gene. Compare *insertion mutation* and *substitution mutation*.

dendrite A portion of a neuron that receives signals from adjacent cells.

density-dependent Of or referring to a factor, such as food shortage, that limits the growth of a population more strongly as the density of the population increases. Compare *density-independent*.

density-independent Of or referring to a factor, such as weather, that can limit the size of a population but does not act more strongly as the density of the population increases. Compare *density-dependent*.

deoxyribonucleic acid See *DNA*.

depolarization A decrease in the charge difference across a portion of the plasma membrane of a neuron; initiates an action potential.

dermal tissue system The tissue system in plants that forms the outer covering of the body.

deuterostome Any of a group of animals, including sea stars and vertebrates, in which the second opening to develop in the early embryo becomes the mouth. Compare *protostome*.

development The process by which an organism grows from a single cell to its adult form.

differentiation The process by which a cell develops physical characteristics that make it suited for a specific function.

diffusion The passive movement of a molecule from areas of high concentration of that molecule to areas of low concentration of that molecule.

digestion The chemical breakdown of food.

diploid Of or referring to a cell or organism that has two complete sets of homologous chromosomes ($2n$). Compare *haploid*.

directional selection A type of natural selection in which individuals with one extreme of a heritable phenotypic character have an advantage over other individuals in the population, as when large individuals produce more offspring than small and medium-sized individuals. Compare *disruptive selection* and *stabilizing selection*.

dispersal An act in which individuals travel relatively long distances away from other members of their species. Dispersal may be active, as when individuals walk, swim, or fly, or it may be passive, as when organisms are transported by wind or water.

disruptive selection A type of natural selection in which individuals with either extreme of a heritable phenotypic character have an advantage over individuals with an intermediate phenotype, as when both small and large individuals survive at a higher rate than medium-sized individuals. Compare *directional selection* and *stabilizing selection*.

distribution The geographic area over which a species is found. Also called *range*.

disturbance An event, such as a fire or windstorm, that kills or damages some organisms in a community, thereby creating an opportunity for other organisms to become established.

DNA Deoxyribonucleic acid, a polymer of nucleotides that stores the information needed to synthesize proteins in living organisms.

DNA fingerprinting The use of DNA analyses to identify individuals and determine the relatedness of individuals.

DNA hybridization Base pairing of DNA from two different sources.

DNA library A collection of an organism's DNA fragments that is stored in a host organism, such as a bacterium.

DNA polymerase The key enzyme that cells use to copy their DNA; used in DNA technology to make many copies of a gene or other DNA sequence.

DNA probe A short sequence of DNA (usually tens to hundreds of nucleotide bases long) that can pair with a particular gene or other specific region of DNA.

DNA repair A three-step process in which damage to DNA is repaired. Damaged DNA is first recognized, then removed, and then replaced with newly synthesized DNA.

DNA replication The duplication, or copying, of a DNA molecule. DNA replication begins when the hydrogen bonds connecting the two strands of DNA are broken, causing the strands to unwind and separate. Each strand is then used as a template for the construction of a new strand of DNA.

DNA segregation The process by which the DNA of a dividing cell is divided equally between two daughter cells.

DNA sequence The sequence or order in which the nucleotide bases adenine (A), cytosine (C), guanine (G), and thymine (T) are arranged throughout all or part of an organism's DNA.

DNA technology The set of techniques that scientists use to manipulate DNA.

domain A level of biological classification above kingdoms. The three domains are the Bacteria, the Archaea, and the Eucarya.

dominant Of or referring to an allele that determines the phenotype of an organism when paired with a different (recessive) allele. Compare *recessive*.

double helix The structure of DNA, in which two long strands of covalently bonded nucleotides are held together by hydrogen bonds and twisted into a spiral coil.

doubling time The time it takes a population to double in size. Doubling time can be used as a measure of how fast a population is growing.

drag A force that resists the motion of any moving object.

ecology The scientific study of interactions between organisms and their environment.

ecosystem A community of organisms, together with the physical environment in which the organisms live. Global patterns of air and water circulation link all the world's organisms into one giant ecosystem, the biosphere.

ectoderm The germ layer in animals that forms the exterior of the gastrula and gives rise to the epidermis and nerve tissue. Compare *endoderm* and *mesoderm*.

ectotherm An organism that relies on environmental heat for most of its heat input. Compare *endotherm*.

egg The large, sedentary, haploid gamete that is produced by sexually reproducing female eukaryotes. Compare *sperm*.

El Niño event A series of worldwide changes in weather triggered by a seasonal warm-water current, known as El Niño, that flows south along the coast of Ecuador. Usually, El Niño has a temporary and local effect, but once every 2 to 10 years it triggers floods, fires, disease outbreaks, crop failures, and more throughout much of Earth.

electron A negatively charged particle found in atoms. Each atom contains a characteristic number of electrons. Compare *proton*.

electron transport chain (ETC) A group of membrane-associated proteins that can both accept and donate electrons. The transfer of electrons from one ETC protein to another releases energy that is used to manufacture ATP in both chloroplasts and mitochondria.

element A substance made up of only one type of atom. The physical world is made up of 92 different elements.

embryo sac The female gametophyte produced by flowering plants. The embryo sac remains within the pistil of a flower produced by a spore-producing individual. Compare *pollen grain*.

endocrine system The collection of glands and other tissues in animals that produce and release signaling molecules called hormones into the circulatory system.

endoderm The germ layer in animals that forms the interior of the gastrula and gives rise to the gut and associated organs. Compare *ectoderm* and *mesoderm*.

endoplasmic reticulum (ER) An organelle composed of many interconnected membrane sacs and tubes; the major site of protein and lipid synthesis in eukaryotic cells.

endosperm The nutritive tissue produced within a plant seed.

endotherm An organism that relies on metabolic heat for most of its heat input. Compare *ectotherm*.

energy carrier A molecule that can store energy and donate it to another molecule or a chemical reaction. ATP is the most commonly used energy carrier in living organisms.

enzyme A protein that acts as a catalyst, speeding the progress of chemical reactions. All chemical reactions in living organisms are catalyzed by enzymes.

epidermis The waterproof outer layer of a multicellular organism.

epithelial cell A type of animal cell that forms the outer surface of the skin and lines the lungs and guts.

ER See *endoplasmic reticulum*.

essential amino acid An amino acid that a consumer cannot synthesize and must therefore obtain from its food.

estrogen Any of a group of steroid hormones that maintain female sexual and behavioral characteristics. Compare *testosterone*.

ETC See *electron transport chain*.

Eucarya The domain that encompasses the eukaryotes.

eukaryote A single-celled or multicellular organism in which each cell has a distinct nucleus and cytoplasm. All organisms other than the Bacteria and the Archaea are eukaryotes. Compare *prokaryote*.

eukaryotic flagellum (pl. flagella) A hairlike structure found in eukaryotes that propels organisms by means of waves passing from its base to its tip. Compare *bacterial flagellum* and *cilium*.

evolution Change over time in a lineage of organisms. More specifically, evolution can be defined as: (1) change in the genetic characteristics of populations over time;or (2) the history of the formation and extinction of species over time.

evolutionary innovation A key adaptation of a group that originated in that group.

evolutionary radiation Rapid expansion of a group of organisms to form new species and higher taxonomic groups.

evolutionary taxonomy The most traditional school of systematics, in which scientists use expert knowledge of a group to determine evolutionary relationships. Compare *cladistics*.

evolutionary tree A diagrammatic representation showing the order in which different lineages arose, with the lowest branches having arisen first.

experiment A controlled manipulation of nature designed to test a hypothesis.

exploitation competition A type of competition in which species compete indirectly for shared resources, with each reducing the amount of a resource available to the other. Compare *interference competition*.

exponential growth A type of rapid population growth in which a population increases by a constant proportion from one generation to the next.

external skeleton A skeleton that surrounds the soft tissues of the animal it supports. Compare *internal skeleton*.

F_1 generation The first generation of offspring in a genetic cross. Compare *P generation*.

F_2 generation The second generation of offspring in a genetic cross.

fast muscle fiber A type of muscle fiber that contracts quickly, but can sustain a contraction for only a short time. Compare *slow muscle fiber*.

fat A chemical compound that consists of glycerol linked to three fatty acids. Fats are solid at room temperature and can be used by living organisms to store energy.

fatty acid A chemical compound composed primarily of a long hydrocarbon chain; found in lipids and fats.

female An individual of a sexually reproducing species that produces eggs. Compare *male*.

fermentation A series of catabolic reactions that produce small amounts of ATP without the use of oxygen. Fermentation is similar to glycolysis, except that pyruvate is converted to other products, such as ethanol or lactic acid.

fertilization The fusion of haploid gametes (egg and sperm) to produce a diploid zygote (the fertilized egg).

first law of thermodynamics The law stating that energy can be neither created nor destroyed, only transformed or transferred from one molecule to another.

fixation The removal from a population of all alleles at a genetic locus except one; the allele that remains has a frequecy of 100 percent.

fixed behavior A predictable response to a particular, often simple, stimulus that does not involve learning. Compare *learned behavior*.

flower A specialized reproductive structure that is characteristic of the plant group known as the angiosperms, or flowering plants.

fluid mosaic model A model that describes the cell membrane as a mobile phospholipid bilayer with imbedded proteins that can move laterally in the plane of the membrane.

food chain A single sequence of feeding relationships describing who eats whom in a community. Compare *food web*.

food web A summary of the movement of energy through a community. A food web is formed by connecting all of the food chains in the community to one another. Compare *food chain*.

fossil Preserved remains of or an impression of a formerly living organism. Fossils document the history of life on Earth, showing that past organisms were unlike living forms, that many organisms have gone extinct, and that life has evolved through time.

frameshift A change in how the information in a DNA sequence is translated by the cell that results when a deletion or insertion mutation is not a multiple of three base pairs (a codon).

friction drag Drag that results from the interaction of a fluid with the surface of a moving object. Compare *pressure drag*.

fulcrum The point around which a lever pivots.

functional group A specific arrangement of atoms that helps define the properties of a chemical compound.

Fungi The kingdom of mushroom-producing species, yeasts, and molds, which usually live as decomposers.

G protein Any of a class of proteins that participate in signal cascades, including those activated by hormones. G protein activity is regulated by the guanine nucleotides GDP (guanosine diphosphate) and GTP (guanosine triphosphate).

G_0 phase The stage of the cell division cycle during which the cell pauses between mitosis and S phase. No preparations for S phase are made during this period.

G_1 phase The stage of the cell division cycle following mitosis and before S phase. The cell makes preparations for DNA synthesis during the G_1 phase.

G_2 phase The stage of the cell division cycle following S phase and before mitosis.

gamete   A haploid sex cell, which fuses with another sex cell during fertilization. Eggs and sperm are gametes.

gametophyte   The egg-producing or sperm-producing individual in the life cycle of a plant.

ganglion (pl. ganglia)   A structure consisting of densely packed neurons that allows the exchange of signals between many different neurons.

gastrulation   The movement of the germ layers (endoderm, mesoderm, and ectoderm) during animal development to the positions appropriate for the tissues to which these layers give rise.

gel electrophoresis   A process in which DNA fragments are placed in a gelatin-like substance (a gel) and subjected to an electrical charge, which causes the fragments to move through the gel. Small DNA fragments move farther than large DNA fragments, thus causing the fragments to separate by size.

gene   A DNA sequence that contains information for the synthesis of a protein or an RNA molecule. Genes are located on chromosomes.

gene cascade   A process in which the protein products of different genes interact with one another and with signals from the environment, thereby turning on other sets of genes in some cells, but not in other cells. Organisms use gene cascades to control how genes are expressed during development.

gene expression   The synthesis of a gene's protein or RNA product. Gene expression is the means by which a gene influences the cell or organism in which it is found.

gene flow   The exchange of alleles between populations.

gene therapy   A treatment approach that seeks to correct genetic disorders by fixing the genes that cause them.

genetic bottleneck   A drop in the size of a population that results in low genetic variation or causes harmful alleles to reach a frequency of 100 percent in the population.

genetic code   A code in which each set of three nucleotide bases in messenger RNA specifies either an amino acid or a signal to start or stop the construction of a protein. The genetic code allows the cell to use the information in a gene to build the protein called for by that gene.

genetic cross   A controlled mating experiment, usually performed to examine the inheritance of a particular character.

genetic drift   A process in which alleles are sampled at random over time, as when chance events cause certain alleles to increase or decrease in the population. The genetic makeup of a population undergoing genetic drift changes at random over time

rather than being shaped in a nonrandom way by natural selection.

genetic engineering   A three-step process in which a DNA sequence (often a gene) is isolated, modified, and inserted back into an individual of the same or a different species. Genetic engineering is commonly used to change the performance of the genetically modified organism, as when a crop plant is engineered to resist attack from an insect pest.

genetic linkage   The situation in which different genes that are located close to one another on the same chromosome do not follow Mendel's law of independent assortment.

genetic screening   The examination of an individual's genes to assess current or future health risks and status.

genetic variation   The genetic differences among the individuals of a population.

genetics   The scientific study of genes.

genome   All the DNA of an organism, including its genes; in eukaryotes, the term "genome" refers to a haploid set of chromosomes, such as that found in a sperm or egg.

genomics   The study of the structure and expression of entire genomes and how they change during evolution.

genotype   The genetic makeup of an organism. Compare *phenotype*.

genotype frequency   The proportion (percentage) of a particular genotype in a population.

geographic isolation   The physical separation of populations from one another by a barrier such as a mountain chain or a river. Geographic isolation often causes the formation of new species, as when populations of a single species become physically separated from one another and then accumulate so many genetic differences that they become reproductively isolated from one another. Compare *allopatric speciation* and *sympatric speciation*.

germ layer   Any of three layers of cells that differentiate early in animal development and give rise to predictable tissues and organs during subsequent development. The three germ layers are endoderm, mesoderm, and ectoderm.

germ line   The cells in animals that remain undifferentiated during early development and give rise to eggs or sperm at maturity.

giant convection cell   A large and consistent air circulation pattern in which warm, moist air rises in one area and cool, dry air sinks in another area. Earth has four of these giant convection cells, which play a major role in shaping its climate.

gill   The gas exchange surface of aquatic animals, typically consisting of finely folded sheets of thin epithelial tissue. Compare *lung*.

gizzard   A portion of the animal gut specialized for grinding food against rocks or sand collected from the environment.

gland   An organ specialized for producing and releasing hormones.

global change   Worldwide change to the environment. There are many causes of global change, including climate change caused by the movement of continents and changes in land and water use caused by humans.

global warming   A worldwide increase in temperature. Earth appears to be entering a period of global warming caused by human activities; specifically, by the release of large quantities of greenhouse gases such as carbon dioxide into the atmosphere.

glucose   A monosaccharide sugar that is the primary metabolic fuel in most cells.

glycolysis   A series of catabolic reactions that split glucose to produce pyruvate, which is used in either oxidative phosphorylation or fermentation.

Golgi apparatus   An organelle composed of flattened membrane sacs, which routes proteins and lipids to various parts of the eukaryotic cell.

gonadotropin   Any of a group of hormones produced by the pituitary gland that regulate the development and function of the reproductive organs.

greenhouse gas   Any of several gases in the atmosphere that function like the windows on a car: They let in sunlight but trap heat.

ground meristem   The tissue in the flowering plant embryo that gives rise to the cells of the ground tissue system. Compare *procambium* and *protoderm*.

ground tissue system   The tissue system in plants that forms most of the photosynthetic tissue in leaves and surrounds the vascular tissue system.

growth factor   Any of a class of signaling molecules that bind to cell surface receptors and promote cell division.

gut   The digestive system of animals.

gut inhabitant mutualism   A mutualism involving organisms that live in the digestive tract of a host, receiving food from the host and digesting foods that the host otherwise could not use.

gymnosperms   A group of plants that includes pine trees and other conifers, ginkgos, and cycads. Gymnosperms were the first plants to evolve seeds. Compare *angiosperms*.

habitat   A characteristic place or type of environment in which an organism lives.

haploid Of or referring to a cell or organism that has only one complete set of homologous chromosomes (*n*). Compare *diploid*.

Hardy–Weinberg equation An equation ($p^2 + 2pq + q^2 = 1$) that predicts the genotype frequencies in a population that is not evolving.

heart attack The death of muscle tissue in the heart that results from the blockage of arteries supplying blood to the heart.

helper T lymphocyte A lymphocyte that stimulates killer T lymphocytes and B lymphocytes.

hemoglobin An oxygen-binding pigment that is used to carry oxygen from the gas exchange surfaces to the tissues.

herbivore A consumer that relies on living plant tissues for nutrients. Compare *carnivore*.

hermaphrodite An individual that can produce both eggs and sperm.

heterozygote An individual that carries one copy of each of two different alleles (for example, an *Aa* individual). Compare *homozygote*.

heterozygote advantage An evolutionary advantage that heterozygotes (*Aa* individuals) enjoy because they leave more offspring than homozygotes (*AA* or *aa* individuals).

homeostasis The process of maintaining appropriate and constant conditions inside cells.

homeotic gene A master-switch gene that plays a key role in the control of gene expression during development. Each homeotic gene controls the expression of a series of other genes whose protein products direct the development of an organism.

hominid Any of a group of primates that encompasses humans and our now extinct humanlike ancestors. Compare *hominoid*.

hominoid Any of a group of primates whose living relatives include gibbons and the great apes (orangutans, gorillas, chimpanzees, and humans). Compare *hominid*.

homologous chromosomes The two members of a specific chromosome pair found in diploid cells, one of which was received from the mother and the other from the father.

homozygote An individual that carries two copies of the same allele (for example, an *AA* or an *aa* individual). Compare *heterozygote*.

hormone A signaling molecule released into the circulatory system of an animal or the vascular system of a plant that, in small amounts, affects the functioning of other, target tissues.

host An organism in which a parasite or pathogen lives.

housekeeping gene A gene that has an essential role in the maintenance of cellular activities and that is expressed by most cells in the body.

Human Genome Project (HGP) A publicly funded effort on the part of an international consortium created by the U.S. National Institutes for Health and the U.S. Department of Energy to determine the sequence of the human genome.

hybrid An offspring that results when two different species mate.

hybridize To cause hybrid offspring to be produced.

hydrocarbon A chemical compound that contains only hydrogen and carbon atoms.

hydrogen bond A chemical linkage between a hydrogen atom, which has a slight positive charge, and another atom with a slight negative charge. Compare *covalent bond* and *ionic bond*.

hydrophilic Of or referring to molecules or parts of molecules that interact freely with water. Hydrophilic molecules dissolve easily in water, but not in fats or oils. Compare *hydrophobic*.

hydrophobic Of or referring to molecules or parts of molecules that do not interact freely with water. Hydrophobic molecules dissolve easily in fats and oils, but not in water. Compare *hydrophilic*.

hydrostatic skeleton A support structure that depends on the interaction between a fluid under pressure and an elastic membrane for stiffness.

hypersensitive response A response by a plant to invasion by parasites in which infected cells die in order to isolate healthy tissues from the invading organisms.

hypha (pl. hyphae) In fungi, a threadlike absorptive structure that grows through a food source. Mats of hyphae form mycelia, the main bodies of fungi.

hypothalamus (pl. hypothalami) A structure at the base of the vertebrate brain that controls the release of hormones by the pituitary gland. Along with the pituitary gland, the hypothalamus helps regulate interactions between the nervous and endocrine systems.

hypothesis (pl. hypotheses) A possible explanation of how a natural phenomenon works. A hypothesis must have logical consequences that can be proved true or false.

immune system An assortment of defensive proteins and white blood cells in the circulatory system of vertebrate animals that help destroy invading parasites.

in vitro fertilization The fertilization of an egg outside the body; used as a treatment for infertility.

incomplete dominance The situation in which heterozygotes (*Aa* individuals) are intermediate in phenotype between the two homozygotes (*AA* and *aa* individuals) for a particular allele.

independent assortment of chromosomes The random distribution of maternal and paternal chromosomes into gametes during meiosis.

individual A single organism, usually physically separate and genetically distinct from other individuals.

inflammation The characteristic swelling and reddening of an area of tissue damage. Inflammation is followed by several responses, most of which isolate any invading parasites and recruit white blood cells and complement to the damaged area.

insect Any of a group of six-legged arthropods that includes grasshoppers, beetles, ants, and butterflies; the most species-rich group of animals on Earth.

insertion mutation A mutation in which one or more bases are inserted into the DNA sequence of a gene. Compare *deletion mutation* and *substitution mutation*.

intercellular air space The spaces between the spongy parenchyma cells in a leaf, through which carbon dioxide enters and oxygen and water leave the leaf.

interference competition A type of competition in which one organism directly excludes another from the use of resources. Compare *exploitation competition*.

intermembrane space The space between the inner and outer membranes of the chloroplast or the mitochondrion.

internal skeleton A skeleton that lies within the soft tissues of the animal it supports. Compare *external skeleton*.

interphase The period of time between two successive mitotic divisions, during which most of the preparations for cell division occur.

introduced species A species that does not naturally live in an area but has been brought there either accidentally or on purpose by humans. Also known as nonnative species.

intron A sequence of bases within a gene that does not specify part of the gene's final protein or RNA product. Enzymes in the nucleus must remove introns from mRNA, tRNA, and rRNA molecules for these molecules to function properly.

ion An atom or group of atoms that has either gained or lost electrons, and therefore has a negative or positive charge.

ionic bond A chemical linkage between two atoms based on the electrical attraction between positive and negative charges. Compare *covalent bond* and *hydrogen bond*.

joint   A flexible connection between the rigid elements that make up an animal skeleton.

karyotype   The specific number and shapes of chromosomes found in the cells of a particular species.

keystone species   A species that, relative to its own abundance, has a large effect on the presence and abundance of other species in a community.

kidney   An organ that regulates the body's supply of water and solutes and helps dispose of metabolic wastes.

killer T lymphocyte   A lymphocyte that binds to cells bearing a particular antigen and helps destroy them.

kinase   An enzyme that catalyzes the addition of phosphate groups to proteins.

kinetochore   A plaque of protein on the centromere of a chromosome where spindle microtubules attach during mitosis and meiosis.

kingdom   The largest taxonomic category in the Linnaean hierarchy. Generally six kingdoms are recognized: Bacteria, Archaea, Protista, Plantae, Fungi, and Animalia.

land transformation   Changes made by humans to the land surface of Earth that alter the physical or biological characteristics of the affected regions. Compare *water transformation*.

language   A means of communicating complex and often abstract ideas to other individuals. Language may be unique to humans.

large intestine   See *colon*.

law of equal segregation   Mendel's first law, which states that the two copies of a gene separate during meiosis and end up in different gametes.

law of independent assortment   Mendel's second law, which states that when gametes form, the separation of alleles for one gene is independent of the separation of alleles for other genes. We now know that this law does not apply to genes that are linked.

learned behavior   A behavior acquired by trial and error or by watching others. Compare *fixed behavior*.

lens   A structure that concentrates light on light-detecting sensory cells.

lever   A rigid structure that pivots about a fulcrum; it can either increase the force generated at the expense of speed or increase the speed generated at the expense of force.

ligament   A collagen-rich connective structure that attaches bone to bone in vertebrate skeletons. Compare *tendon*.

ligase   An enzyme that can connect two DNA fragments to each other; used in DNA technology when a gene from one species is inserted into the DNA of another species.

light reactions   A series of chemical reactions that harvest energy from sunlight and use it to produce energy-rich compounds such as ATP and NADPH. The light reactions occur at the thylakoid membranes of chloroplasts and produce oxygen gas as a waste product. Compare *dark reactions*.

lignin   A complex molecule largely responsible for the rigidity of the cell walls of woody plants.

lineage   A group of closely related individuals, species, genera, or the like, depicted as a branch on an evolutionary tree.

linked genes   See *genetic linkage*.

Linnaean hierarchy   The classification scheme used by biologists today to organize and name the organisms in the great tree of life.

lipid   A hydrophobic molecule that contains fatty acids; a key component of cell membranes. See also *phospholipid bilayer*.

load arm   The section of a lever that extends from the fulcrum to the point at which work is done. Compare *power arm*.

locus (pl. loci)   The physical location of a gene on a chromosome.

lumen   The space enclosed by the membrane of an organelle.

lung   The gas exchange structure of terrestrial animals, formed by an infolding of epidermal tissue. Compare *gill*.

lymph   The fluid that filters out of the capillaries and flows between the cells in vertebrates.

lymphatic system   A network of vessels that carries lymph from the body to the blood vessels.

lymphocyte   Any of several types of white blood cells that bind to specific antigens and then contribute in various ways to the destruction of the cells that bear those antigens.

lysosome   A specialized vesicle with an acidic lumen, containing enzymes that break down macromolecules.

macroevolution   The rise and fall of major taxonomic groups due to evolutionary radiations that bring new groups to prominence and mass extinctions in which groups are lost; the history of large-scale evolutionary changes over time. Compare *microevolution*.

macromolecule   A large organic molecule formed by the bonding together of small organic molecules.

macronutrient   A chemical element required by an organism in relatively large amounts. Compare *micronutrient*.

macrophage   A type of white blood cell that phagocytoses invading parasites marked by complement or antibodies.

male   An individual of a sexually reproducing species that produces sperm. Compare *female*.

malignant   Of or referring to a cancerous growth that begins as a single tumor and then spreads to other tissues in the body with life-threatening consequences. Compare *benign*.

map-based sequencing   A genome sequencing technique that begins by identifying the locations of known DNA polymorphisms on each chromosome, then uses the resulting genetic map as a guide to reassemble sequenced DNA fragments. Compare *whole-genome shotgun sequencing*.

mass extinction   A period during which large numbers of species become extinct throughout most of Earth.

matrix (pl. matrices)   A coating of nonliving material, released by the cells of multicellular animals, that often holds the cells together.

meiosis   A specialized process of cell division in eukaryotes during which diploid cells divide to produce haploid cells. Meiosis has two division cycles and occurs exclusively in cells that produce gametes. Compare *mitosis*.

meiosis I   The first cycle of cell division in meiosis. Meiosis I produces haploid daughter cells, each with half the chromosome number of the diploid parent cell.

meiosis II   The second cycle of cell division in meiosis. Meiosis II is essentially mitosis, but in a haploid cell.

membrane   A phospholipid bilayer with associated proteins that encloses both cells and organelles.

menopause   The end of a human female's reproductive life.

menstrual cycle   A series of hormonally controlled, cyclical changes that take place in the reproductive system of human females, which include the shedding of the lining of the uterus (menstruation) about every 28 days.

mesoderm   The germ layer present in most animals that lies between the endoderm and ectoderm in the gastrula and gives rise to muscle tissue, connective tissue, and the kidney. Compare *ectoderm* and *endoderm*.

messenger RNA (mRNA)   A type of RNA that specifies the order of amino acids in a protein.

metabolic heat   Heat generated as a product of respiration.

metabolic pathway   A series of enzyme-controlled chemical reactions in a cell in which the product of one reaction becomes the substrate for the next.

metabolism   All the chemical reactions that occur in a cell.

metamorphosis (pl. metamorphoses)   A dramatic transformation during animal development from a reproductively immature to a reproductively mature form, involving great change in the form and function of the animal.

metaphase   The stage of mitosis during which chromosomes become aligned at the equator of the cell.

microevolution   Changes in allele or genotype frequencies in a population over time; the smallest scale at which evolution occurs. Compare *macroevolution*.

micronutrient   A chemical element required by an organism in tiny amounts. Compare *macronutrient*.

microtubule   A protein fiber composed of tubulin monomers. Microtubules are part of the cell's cytoskeleton.

milk-letdown reflex   The reflexive release of milk by a human mother in response to the suckling behavior of an infant.

mismatch error   The insertion of an incorrect base during the DNA replication process that is not detected and corrected.

mitochondrion (pl. mitochondria)   An organelle with a double membrane that is the site of oxidative phosphorylation. Mitochondria break down simple sugars to produce most of the ATP needed by eukaryotic cells.

mitosis   The process of cell division in eukaryotes that produces two daughter nuclei, each with the same chromosome number as the parent nucleus. Compare *meiosis*.

mitotic spindle   An arrangement of microtubules that guides the movement of chromosomes during mitosis.

module   A repeated unit of a modular organism, such as a plant.

molecule   An arrangement of two or more atoms linked by chemical bonds.

monomer   A molecule that can be linked with other related molecules to form a larger polymer.

monosaccharide   A simple sugar that can be linked to other sugars, forming a polysaccharide. Glucose is the most common monosaccharide in living organisms.

morphogen   A chemical signal that influences the developmental fate of a cell.

morphology   The form and structure of an organism.

most recent common ancestor   The ancestral organism from which a group of descendants arose.

motor protein   A protein that uses the energy of ATP to move organelles or proteins along cytoskeletal filaments.

mRNA   See *messenger RNA*.

multicellular   Made up of more than one cell.

multiregional hypothesis   A hypothesis stating that anatomically modern humans evolved from *Homo erectus* populations scattered throughout the world. According to this idea, worldwide gene flow caused different human populations to evolve modern characteristics simultaneously and to remain a single species. Compare *out-of-Africa hypothesis*.

muscle cell   A type of animal cell that can contract and plays a central role in the ability of animals to move.

muscle fiber   The basic unit of muscle tissue, consisting of a collection of myofibrils.

mutation   A change in the sequence of an organism's DNA. New alleles arise only by mutation, so mutations are the original source of all genetic variation.

mutualism   An interaction between two species in which both species benefit.

mutualist   An organism that interacts with another organism to the mutual benefit of both.

mycelium (pl. mycelia)   The main body of a fungus, composed of hyphae.

mycorrhiza (pl. mycorrhizae)   A mutualism between a fungus and a plant, in which the fungus provides the plant with mineral nutrients while receiving organic nutrients from the plant.

myelin   A fatty material that forms an insulating sheath around the axon of a neuron and greatly speeds the rate at which action potentials move along the axon.

myofibril   A unit of a muscle fiber consisting of many sarcomeres attached end-to-end.

myoglobin   An oxygen-binding pigment that helps store oxygen in muscle tissue.

myosin   A protein found in muscle tissue. In the sarcomere, the heads of myosin molecules "walk" along actin filaments to cause contraction.

natural selection   An evolutionary mechanism in which those individuals in a population that possess particular heritable characters survive and reproduce at a higher rate than other individuals in the population because of those characters. Natural selection is the only evolutionary mechanism that consistently improves the survival and reproduction of the organism in its environment.

nerve   A major communication pathway in the animal body made up of the axons of many individual neurons bundled together.

nerve cell   See *neuron*.

nervous system   A high-speed internal communication system in animals that is made up of specialized cells called neurons.

net primary productivity (NPP)   The amount of energy that producers fix by photosynthesis, minus the amount lost as heat in respiration. NPP is usually measured as the amount of new biomass produced by photosynthetic organisms per unit of area during a specified period of time. Compare *secondary productivity*.

neuron   A type of animal cell that is highly specialized for transmitting action potentials from one part of the body to another. Also called *nerve cell*.

neurotransmitter   Any of various signaling molecules that transmit signals across the gaps, or synapses, that separate neurons from their target cells.

neutrophil   A type of white blood cell that destroys invading bacterial cells by phagocytosis.

nitric oxide   A gas that functions as a signaling molecule.

nitrogen fixation   The process by which nitrogen gas (N_2), which is readily available in the atmosphere but cannot be used by plants, is converted to ammonium (NH_4^+), a form of nitrogen that can be used by plants. Nitrogen fixation is accomplished naturally by bacteria and by lightning, and by humans in industrial processes such as the production of fertilizer.

noncoding DNA   A segment of DNA that does not code for protein or RNA. *Introns* and *spacer DNA* are two common types of noncoding DNA.

noncovalent bond   Any chemical linkage between two atoms that does not involve the sharing of electrons. Hydrogen bonds and ionic bonds are examples of noncovalent bonds.

nonpolar   Of or referring to a molecule or a portion of a molecule that has an equal distribution of electrical charge across all its constituent atoms. Nonpolar molecules do not form hydrogen bonds and therefore tend not to dissolve in water. Compare *polar*.

nonrandom mating   A situation in which individuals within a subset of a population are more likely to mate with each other than they are to mate with a randomly selected member of the population.

nonspecific response   A defensive response found in most animals that leads to the destruction of cells not recognized as belonging to the organism. Compare *specific response*.

NPP   See *net primary productivity*.

nuclear envelope   The double membrane that encloses the nucleus of a eukaryotic cell.

nuclear pore   A channel in the nuclear envelope that allows selected molecules to move into and out of the nucleus.

nucleotide Any of the organic compounds that serve as the chemical building blocks of nucleic acids such as DNA and RNA. A nucleotide has a sugar–phosphate backbone and one of four nitrogen-containing molecules called bases (see *nucleotide base*). Nucleotides are linked together to form a single strand of DNA or RNA.

nucleotide base Any of the five nitrogen-rich compounds found in nucleotides. The four nucleotide bases found in DNA are adenine (A), cytosine (C), guanine (G), and thymine (T); in RNA, uracil (U) replaces thymine.

nucleus (pl. nuclei) The organelle in a eukaryotic cell that contains the genetic blueprint in the form of DNA.

nutrient In an ecosystem context, an essential element required by a producer. Such essential elements include carbon, hydrogen, oxygen, nitrogen, phosphorus, and many others. See also *macronutrient*, *micronutrient*.

nutrient cycle The cyclical movement of a nutrient between organisms and the physical environment. There are two main types of nutrient cycles: atmospheric and sedimentary. See also *atmospheric nutrient cycle* and *sedimentary nutrient cycle*.

oncogene A mutated gene that promotes excessive cell division, leading to cancer.

open circulatory system An internal transport system in which blood flows through the body cavity that surrounds the organs. Compare *closed circulatory system*.

operator In prokaryotes, a regulatory DNA sequence that controls the transcription of a gene or group of genes.

organ A self-contained collection of tissues, usually of a characteristic size and shape, that is organized for a particular function.

organ system A group of organs that work together to carry out a particular function.

organelle A distinct, membrane-enclosed structure in a eukaryotic cell that has a specific function.

organic Of or referring to a carbon-containing compound of biological origin.

orientation behavior A sequence of behaviors that moves an organism toward or away from the source of a stimulus.

osmotic pressure The pressure generated when water diffuses down a concentration gradient.

out-of-Africa hypothesis A hypothesis stating that anatomically modern humans evolved in Africa within the past 200,000 years, then spread throughout the rest of the world. According to this idea, as they spread from Africa, modern humans completely replaced older forms of *Homo*

sapiens, including advanced forms such as the Neandertals. Compare *multiregional hypothesis*.

ovary The reproductive structure in female animals that produces eggs.

oviduct The tube down which eggs produced by female animals pass after leaving the ovary.

oxidation The process by which a chemical compound loses electrons. Compare *reduction*.

oxidative phosphorylation The shuttling of electrons down an electron transport chain in mitochondria that results in the production of ATP.

oxygen-binding pigment Any of various complex molecules used by animals to increase the oxygen capacity of their body fluids.

P generation The parent generation of a genetic cross. Compare F_1 *generation* and F_2 *generation*.

p53 A tumor suppressor gene that is inactive in many different types of cancer cells.

palisade parenchyma The tissue in plant leaves that is specialized for photosynthesis.

parasite An organism that lives in or on another organism (its host) and obtains nutrients from that organism. Parasites harm and may eventually kill their hosts, but do not kill them immediately.

passive carrier protein A protein in the plasma membrane of a cell that, without the input of energy, changes its shape to transport a molecule across the membrane from the side of higher concentration to the side of lower concentration. Compare *active carrier protein*.

passive transport Movement of molecules from areas of lower concentration to areas of higher concentration without the expenditure of energy. Compare *active transport*.

pathogen An organism or virus that infects a host and causes disease, harming and in some cases killing the host.

PCR See *polymerase chain reaction*.

PDGF (platelet-derived growth factor) A growth factor produced by specialized blood cells that induces cell division at sites of tissue damage; the first growth factor discovered.

pedigree A chart that shows genetic relationships among family members over two or more generations of a family's history.

penis A reproductive structure used by male animals to introduce sperm directly into the female's reproductive tract.

PEP carboxylase An enzyme involved in C_4 photosynthesis that binds low concentrations of carbon dioxide.

pH A scale from 1 to 14 that indicates the concentration of hydrogen ions in a solution. A pH of 7 is neutral; values below 7 indicate acids, and values above 7 indicate bases.

phagocytosis A process by which one cell engulfs and digests another.

phenotype The observable physical characteristics of an organism. Compare *genotype*.

pheromone A chemical signal produced by one individual to communicate its identity and location to another individual.

phloem A tissue composed of living cells through which a plant transports the products of photosynthesis. Compare *xylem*.

phospholipid A lipid molecule with an attached phosphate group. Phospholipids are the major components of all biological membranes.

phospholipid bilayer A double layer of phospholipid molecules arranged so that their hydrophobic "tails" lie sandwiched between their hydrophilic "heads." A phospholipid bilayer forms the basic structure of all biological membranes.

phosphorylation The addition of a phosphate group to an organic molecule.

photorespiration The binding of the enzyme rubisco to oxygen, rather than carbon dioxide, when O_2 concentrations are high relative to CO_2 concentrations.

photosynthesis The process by which the chloroplasts of plants capture energy from sunlight and use it to synthesize sugars from carbon dioxide and water.

photosystem A large complex of proteins and chlorophyll that captures energy from sunlight. Two distinct photosystems (I and II) are present in the thylakoid membranes of chloroplasts.

pistil The reproductive structure in spore-producing plant individuals in which the female gametophyte is produced.

pituitary gland A gland associated with the vertebrate brain that releases hormones that control the release of hormones by other glands. Along with the hypothalamus, the pituitary helps regulate interactions between the nervous and endocrine systems.

placenta A structure found in most mammals that transfers nutrients and gases from the blood of the mother to the blood of the developing infant in her uterus.

Plantae The kingdom that encompasses plants.

plaque A fatty growth that develops on the walls of arteries as a result of excess cholesterol in the human diet.

plasma The fluid portion of animal blood.

plasma membrane   The phospholipid bilayer that surrounds the cell.

plasmid   A small circular segment of DNA found naturally in bacteria. Plasmids can be used as vectors in the formation of a DNA library and are involved in natural gene transfers among bacteria.

polar   Of or referring to a molecule or a portion of a molecule that has an uneven distribution of electrical charge. Polar molecules can easily interact with water molecules and are therefore soluble. Compare *nonpolar*.

pollen grain   The mobile male gametophyte produced by plants. Compare *embryo sac*.

pollen tube   The structure produced by a pollen grain to enable the two sperm cells to reach an unfertilized embryo sac.

pollinator mutualism   A mutualism in which an animal transfers pollen grains from one plant to the female reproductive organs of another plant of the same species and receives food as a reward for this service.

polymer   A large organic molecule composed of many monomers linked together.

polymerase chain reaction (PCR)   A method that uses the DNA polymerase enzyme to make multiple copies of a targeted sequence of DNA in a test tube.

polyploidy   A condition in which an organism has three or more complete sets of chromosomes (rather than the usual two complete sets). Polyploidy can cause new species to form rapidly without geographic isolation.

polysaccharide   A polymer composed of many linked monosaccharides. Starch and cellulose are examples of polysaccharides.

population   A group of interacting individuals of a single species located within a particular area.

population density   The number of individuals in a population, divided by the area covered by the population.

population size   The total number of individuals in a population.

power arm   The section of a lever that extends from the fulcrum to the point at which force is applied. Compare *load arm*.

predator   An organism that kills other organisms for food.

pregnancy   In female mammals, the state of having an embryo implanted in the uterine wall.

pressure drag   Drag that results from a difference in the pressures in front of and behind a moving object. Compare *friction drag*.

primary consumer   An organism that eats a producer. Compare *secondary consumer*.

primary succession   Ecological succession that occurs in newly created habitat, as when an island rises from the sea or a glacier retreats, exposing newly available bare ground. Compare *secondary succession*.

primate   An order of mammals whose living members include lemurs, tarsiers, monkeys, humans, and other apes. Primates share characteristics such as flexible shoulder and elbow joints, opposable thumbs or big toes, forward-facing eyes, and brains that are large relative to body size.

procambium   The tissue in the developing plant embryo that gives rise to the cells of the vascular tissue system. Compare *ground meristem* and *protoderm*.

producer   An organism that uses energy from an external source such as the sun to produce its own food without having to eat other organisms or their remains. Compare *consumer*.

productivity   The mass of plant matter that a plot can produce from the available nutrients and sunlight.

progesterone   A steroid hormone that helps maintain pregnancy in female mammals.

prokaryote   A single-celled organism that does not have a nucleus. All prokaryotes are members of the kingdoms Bacteria or Archaea. Compare *eukaryote*.

prometaphase   The stage of mitosis during which chromosomes become attached to the mitotic spindle.

promoter   The DNA sequence in a gene to which RNA polymerase binds to begin transcription.

prophase   The stage of mitosis during which chromosomes first become visible under the microscope.

protein   A linear polymer of amino acids linked together in a specific sequence. Most proteins are folded into complex three-dimensional shapes.

Protista   The oldest eukaryotic kingdom, consisting of a diverse collection of mostly single-celled but some multicellular organisms.

protoderm   The tissue in the developing plant embryo that gives rise to the cells of the dermal tissue system. Compare *ground meristem* and *procambium*.

proton   A positively charged particle found in atoms. Each atom contains a characteristic number of protons. Compare *electron*.

proton gradient   An imbalance in the concentration of protons across a membrane.

proto-oncogene   A gene that promotes cell division in response to normal growth signals.

protostome   Any of a group of animals, including insects, worms, and snails, in which the first opening to develop in the early embryo becomes the mouth. Compare *deuterostome*.

pseudopodium (pl. pseudopodia)   A dynamic protrusion of the plasma membrane that enables some cells to move. The extension of pseudopodia depends on actin filaments inside the cell.

puberty   In humans, the transition from childhood to reproductive maturity.

Punnett square   A diagram in which the possible types of male and female gametes are listed on two sides of a square, providing a graphical way to predict the genotypes of the offspring produced in a genetic cross.

radiant heat   Heat transferred in the form of light energy.

rain shadow   An area on the side of a mountain facing away from moist prevailing winds where little rain or snow falls.

rainforest   A forest that receives high rainfall.

random distribution   A distribution pattern in which organisms tend to be positioned in an unpredictable manner throughout the environment. Compare *clumped distribution* and *regular distribution*.

range   See distribution.

real group   A group consisting of all the descendants of a single common ancestor.

receptor   A protein that facilitates the transmission of a signal after binding to a specific signaling molecule. Receptors may be found inside the cell or embedded in the plasma membrane.

recessive   Of or referring to an allele that does not have a phenotypic effect when paired with a dominant allele. Compare *dominant*.

recombination   A collective term for the processes of fertilization, crossing-over, and independent assortment of chromosomes, all of which result in new combinations of alleles.

reduction   The process by which a chemical compound gains electrons. Compare *oxidation*.

reflex arc   A neuronal connection in the spinal cord between a sensory cell and a muscle cell that allows a simple stimulus to be translated rapidly into a motion.

regular distribution   A distribution pattern in which organisms tend to be positioned in an even or uniform manner throughout the environment. Compare *clumped distribution* and *random distribution*.

regulatory DNA sequence   A DNA sequence that can turn the expression of a particular gene or group of genes on or off. Regulatory DNA sequences interact with regulatory proteins to control of gene expression.

regulatory protein   A protein that signals whether or not a particular gene or group of genes should be expressed. Regulatory proteins interact with regulatory DNA sequences to control gene expression.

repressor protein   A protein that prevents the expression of a particular gene or group of genes.

reproduction   The act of producing more individuals.

reproductive isolation   A condition in which barriers prevent or strongly limit two or more populations from reproducing with one another. Reproductive isolation can occur in many different ways, but it always has the same effect: No or few genes are exchanged between reproductively isolated populations or species.

restriction enzyme   Any of a number of enzymes that cut DNA molecules at a specific target sequence; a key tool of DNA technology.

restriction fragment length polymorphism (RFLP) analysis   A method of DNA technology in which restriction enzymes are used to cut an organism's genome into small pieces, which are sorted by size using gel electrophoresis. Next, a DNA probe is used to form a profile, whose pattern depends on the number and size of the fragments that can bind to the probe.

retina   A field of light-sensing cells in the eye that allows organisms to form an image.

ribonucleic acid   See *RNA*.

ribosomal RNA (rRNA)   A type of RNA that is an important component of ribosomes.

ribosome   A particle composed of proteins and RNA at which new proteins are synthesized. Ribosomes can be either attached to the endoplasmic reticulum or free in the cytosol.

rivet hypothesis   A hypothesis that likens an ecosystem slowly losing species to an airplane slowly losing its rivets, and suggests that once enough species are lost, the ecosystem can suddenly collapse.

RNA   Ribonucleic acid; a polymer of nucleotides that is necessary for the synthesis of proteins in living organisms.

RNA polymerase   The key enzyme in DNA transcription, which links together the nucleotides of the RNA molecule specified by a gene.

root hair   An absorptive structure formed by epidermal cells produced near the root tips of a plant.

rooting reflex   The reflexive search for a nipple by a human infant when its cheek or lips are touched lightly.

rough ER   A region of the endoplasmic reticulum that has attached ribosomes. Compare *smooth ER*.

rRNA   See *ribosomal RNA*.

rubisco   The enzyme that catalyzes the first reaction of carbon fixation in photosynthesis.

S phase   The stage of the cell division cycle during which the cell's DNA is replicated.

sap   The fluid that carries nutrients through the vascular system of plants.

sarcomere   A unit of a myofibril consisting of an arrangement of actin and myosin filaments that extends between two Z discs.

saturated   Of or referring to a fatty acid that has no double bonds between its carbon atoms. Compare *unsaturated*.

science   A method of inquiry that provides a rational way to discover truths about the natural world.

scientific method   A series of steps in which a scientist develops a hypothesis, tests its predictions by performing experiments, and then changes or discards the hypothesis if its predictions are not supported by the results of the experiments.

second law of thermodynamics   The law stating that all systems, such as a cell or the universe, tend to become more disordered, and that the creation and maintenance of order in a system requires the transfer of disorder to the environment.

secondary chemical   Any of various chemicals that are used for defense by plants and animals, but are not essential to their basic metabolism.

secondary consumer   An organism that eats a primary consumer. Compare *primary consumer*.

secondary productivity   The rate of new biomass production by consumers per unit of area. Compare *net primary productivity*.

secondary succession   Ecological succession that occurs as communities recover from disturbance, as when a forest grows back when a field ceases to be used for agriculture. Compare *primary succession*.

sedimentary nutrient cycle   A type of nutrient cycle in which the nutrient does not enter the atmosphere easily. Compare *atmospheric cycle*.

seed   A structure produced by a plant in which a plant embryo is encased in a protective covering.

seed dispersal mutualism   A mutualism in which an animal eats a fruit provided by a plant, then later deposits the seeds contained within the fruit far from the parent plant.

seed leaf   One of a pair of leaves that develops in a flowering plant embryo while it is still within the seed.

sex chromosome   Either of a pair of chromosomes that determines the gender of an individual. Compare *autosome*.

sex-linked   Of or referring to genes located on a sex chromosome. Genes located on the X chromosome are called *X-linked*; genes located on the Y chromosome are called *Y-linked*.

sexual reproduction   Reproduction in which genes from two individuals are recombined. Compare *asexual reproduction*.

shared ancestral feature   A feature shared by members of a group of organisms that was not unique to their most recent common ancestor but is shared by many other organisms as well. Compare *shared derived feature*.

shared derived feature   A feature unique to a common ancestor that is passed down to all of its descendants, clearly defining them as a group. Compare *shared ancestral feature*.

signal cascade   A series of stepwise protein activations triggered by a signaling molecule. Signal cascades occur inside the cell and greatly amplify the effect of the original signal.

signal integration   The process by which several different signaling molecules can affect the components of one signal cascade.

signal transduction   The triggering of a specific chemical reaction inside the cell in response to an external signal.

signaling molecule   A molecule produced and released by one cell that affects the activities of another cell (referred to as a target cell). Signaling molecules enable the cells of a multicellular organism to communicate with one another and coordinate their activities.

single nucleotide polymorphisms (SNPs)   Single base-pair differences among the genomes of individuals.

sit-and-wait forager   A forager that waits in one place for food to come to it.

slow muscle fiber   A type of muscle fiber that is specialized for slow, sustained contraction. Compare *fast muscle fiber*.

small intestine   The portion of the animal gut where digestion and most nutrient absorption takes place.

smooth ER   A region of the endoplasmic reticulum that does not have attached ribosomes. Compare *rough ER*.

soluble   Of or referring to a chemical compound that will dissolve in water.

solutes   Materials dissolved in water.

spacer DNA A region of noncoding DNA that separates two genes. Spacer DNA is common in eukaryotes but not in prokaryotes.

speciation The process in which one species splits to form two or more species that are reproductively isolated from one another.

species A group of interbreeding natural populations that is reproductively isolated from other such groups.

specific response An immune response in which an organism recognizes and responds much more rapidly to a parasite or infectious disease organism to which it has had previous exposure. Compare *nonspecific response*.

sperm The small, mobile, haploid gamete that is produced by sexually reproducing male eukaryotes. Compare *egg*.

spinal cord A dense collection of neuron cell bodies and axons that carries information between the brain and the hindmost part of a vertebrate. The spinal cord of vertebrates forms an intermediate site of information processing that acts as a buffer between the brain and the muscles and sensory neurons.

spiracle An opening to the system of tracheoles that supplies oxygen to insect tissues.

spongy parenchyma The tissue in plant leaves that is specialized for gas exchange and the evaporation of water that drives the movement of xylem sap.

spore (1) The reproductive cell of a fungus, which is typically encased in a protective coating that shields it from drying or rotting. (2) A mobile structure produced by plants such as ferns and mosses from which a gametophyte develops.

stabilizing selection A type of natural selection in which individuals with intermediate values of a heritable phenotypic character have an advantage over other individuals in the population, as when medium-sized individuals survive at a higher rate than small or large individuals. Compare *directional selection* and *disruptive selection*.

stamen The reproductive structure in spore-producing individuals of flowering plants that produces pollen grains.

stem cell An undifferentiated cell that can divide to produce one or more specialized cell types.

steroid Any of a class of hydrophobic signaling molecules that can pass through the plasma membrane of the target cell.

stigma The reproductive structure in spore-producing flowering plant individuals on which pollen grains land and develop pollen tubes.

stoma (pl. stomata) An opening to the intercellular air spaces in plant leaves that opens and closes to regulate gas exchange and evaporative water loss.

stomach The portion of the animal gut that is specialized for mixing and storing food.

stroma The space enclosed by the inner membrane of the chloroplast, in which the thylakoid membranes are situated.

substitution mutation A mutation in which one base is replaced by another at a single position in the DNA sequence of a gene. Compare *deletion mutation* and *insertion mutation*.

substrate The specific molecule on which an enzyme acts. Only the substrate will bind to the active site of the enzyme.

succession A process in which species in a community are replaced over time. For a given location, the order in which species are replaced over time is fairly predictable.

sugar An organic compound that has the general chemical formula $(CH_2O)_n$.

surface area–to–volume ratio The ratio between the surface area of an organism and the volume of cells that it must supply.

sustainable In ecology, of or referring to an action or process that can continue indefinitely without using up resources or causing serious damage to ecosystems.

symbiosis A relationship in which two or more organisms of different species live together in close association.

sympatric speciation The formation of new species from populations that are not geographically isolated from one another. Compare *allopatric speciation* and *geographic isolation*.

synapse The narrow gap that separates a neuron from its target cell.

synovial fluid A fluid that forms a cushion in vertebrate joints.

systematist A scientist who studies evolutionary relationships among organisms and builds evolutionary trees.

T cell Any of a specialized class of white blood cells that form part of the vertebrate immune system. See *helper T lymphocyte*, *killer T lymphocyte*.

target cell A cell that receives and responds to a signaling molecule.

telophase The stage of mitosis during which chromosomes arrive at the opposite poles of the cell and new nuclear envelopes begin to form around each set of chromosomes.

tendon A collagen-rich connective structure that attaches muscle to bone in vertebrate skeletons. Compare *ligament*.

tension A stretching force that tends to pull molecules apart. Compare *compression*.

testis (pl. testes) The reproductive structure in male animals that produces sperm.

testosterone One of a group of steroid hormones that maintains male sexual and behavioral characteristics. Compare *estrogen*.

thalamus (pl. thalami) The portion of the vertebrate brain that relays sensory information to the cerebral cortex.

thrust The force generated by an organism that propels it forward.

thylakoid membrane The membrane that encloses the thylakoid space inside a chloroplast. The thylakoid membrane houses both photosystems and their associated electron transport chains.

thylakoid space The space enclosed by the thylakoid membrane inside a chloroplast; the innermost compartment of the chloroplast.

tissue A collection of coordinated and specialized cells that together fulfill a particular function for the body. Animals have four basic tissue types, corresponding to the four basic cell types: connective tissue, epithelial tissue, muscle tissue, and nerve tissue.

tissue system Any of three basic collections of different plant cell types that, in various arrangements, make up all parts of a plant. The three tissues systems are the dermal, ground, and vascular tissue systems.

tracheole A tube that carries gases to and from insect cells.

transcription Synthesis of an RNA molecule from a DNA template. Transcription is the first of two major steps in the process by which genes specify proteins; it produces mRNA, tRNA, and rRNA molecules, all of which are essential in the production of proteins. Compare *translation*.

transfer RNA (tRNA) A type of RNA that transfers the correct amino acid to the ribosome for each amino acid specified by a messenger RNA during protein synthesis.

transformation A change in the genotype of a cell as a result of the incorporation of external DNA by the cell.

translation The conversion of a sequence of bases in a messenger RNA molecule to a sequence of amino acids in a protein. Translation occurs at the ribosomes and is the second of two major steps in the process by which genes specify proteins. Compare *transcription*.

transposon A DNA sequence that can move from one position on a chromosome to another, or from one chromosome to another; known informally as a "jumping gene."

tRNA   See *transfer RNA*.

trophic level   A level or step in a food chain. Trophic levels begin with producers and end with predators such as lions that eat other organisms but are not fed upon by other predators.

true-breeding variety   A subset of individuals within a species that are homozygous for a particular trait. When individuals of a true-breeding variety mate with other individuals of the same variety, all of the offspring resemble the parents for the trait that defines the variety.

tubulin   The protein monomer that makes up microtubules.

tumor suppressor   A gene that inhibits cell division under normal conditions.

unsaturated   Of or referring to a fatty acid that has one or more double bonds between its carbon atoms. Compare *saturated*.

uterus (pl. uteri)   The organ in female mammals in which the zygote implants and develops.

vaccine   A preparation of killed or weakened organisms (or viruses) that is used to stimulate the vertebrate immune system as a protection against future attack by that organism (or virus).

vacuole   A large water-filled vesicle found in plant cells. Vacuoles help maintain the shape of plant cells and can also be used to store food molecules.

vagina   The reproductive structure in female animals into which the penis of the male deposits sperm.

vascular tissue system   The tissue system in plants that is devoted to internal transport.

vasodilation   The expansion of blood vessels, leading to increased blood flow.

vector   In genetics, a piece of DNA that is used to transfer a gene or other DNA fragment from one organism to another.

vein   A blood vessel that carries blood to the heart. Compare *artery*.

ventricle   The chamber of a heart that receives blood from the atrium and contracts to propel blood away from the heart.

vertebrates   A group of animals that has backbones. Vertebrates include fish, amphibians, mammals, birds, and reptiles.

vesicle   A small, membrane-enclosed sac found in the cytosol of eukaryotic cells.

virus   An infectious particle consisting of nucleic acids and proteins. A virus cannot reproduce on its own, and must instead use the cellular machinery of its host to reproduce.

viscous   Of or referring to a fluid in which adjacent molecules stick to one another, leading to a syrupy consistency.

vitamin   An organic compound required by consumers in small amounts.

water transformation   Changes made by humans to the waters of Earth that alter their physical or biological characteristics. Compare *land transformation*.

weather   Temperature, precipitation, wind speed, humidity, cloud cover, and other physical conditions of the lower atmosphere at a specific place over a short period of time.

white blood cell   Any of several types of cells that are part of the vertebrate immune system and provide protection against invading parasites.

whole-genome shotgun sequencing   A genome sequencing technique in which the entire genome is broken up into many small fragments, which are then sequenced and reassembled by computer based strictly on their overlapping ends. Compare *map-based sequencing*.

X-linked   See *sex-linked*.

xylem   A tissue composed of dead cells through which a plant transports mineral nutrients and water from the soil to the leaves. Compare *phloem*.

Z disc   The disc that separates the sarcomeres in a myofibril.

zygote   The diploid cell formed by the fusion of two haploid gametes; a fertilized egg.

Index